AF817467

Causes and Consequences of Globally Warm Climates in the Early Paleogene

Edited by

Scott L. Wing
Department of Paleobiology
Smithsonian Institution
Washington, D.C. 20560
USA

Philip D. Gingerich
Department of Geological Sciences and Museum of Paleontology
University of Michigan
Ann Arbor, Michigan 48109-1079
USA

Birger Schmitz
Department of Earth Sciences, Marine Geology
Göteborg University, Box 460, SE 405 30
Göteborg, Sweden

and

Ellen Thomas
Department of Earth and Environmental Sciences
Wesleyan University
Middletown, Connecticut 06459-0139
USA

THE
GEOLOGICAL
SOCIETY
OF AMERICA

Special Paper 369

3300 Penrose Place, P.O. Box 9140 ▪ Boulder, Colorado 80301-9140, USA

2003

Published by The Geological Society of America, Inc.
3300 Penrose Place, P.O. Box 9140, Boulder, Colorado 80301
www.geosociety.org

Printed in U.S.A.

GSA Books Science Editor: Abhijit Basu

Library of Congress Cataloging-in-Publication Data

Causes and consequences of globally warm climates in the early Paleogene
/ edited by Scott L. Wing . . . [et al.].
p. cm.—(Special Paper ; 369)
Includes bibliographical references and index.
ISBN 0-8137-2369-8 (softcover)
1. Paleoclimatology—Paleogene. 2. Paleoecology—Paleogene. 3. Climatic changes—Environmental aspects. 4. Global warming—Environmental aspects. I. Wing, Scott L. II. Special papers (Geological Society of America) ; 369.

QC884.C38 2003
551.6′09′01—dc21

2003044359

Cover, top: Paleocene-Eocene boundary, Sand Creek Divide, Washakie County, Wyoming. The bright red badlands in the foreground expose highly mature paleosols that formed in the earliest Eocene during the short period of great warmth thought to have been caused by a sudden release of methane clathrates from oceanic sediments. The Bighorn and Pryor Mountains are on the horizon. **Bottom:** Evening in the Eocene. The reconstruction shows a caterpillar feeding on a leaf of the sycamore relative, *Macginitiea gracilis.* In the middle ground, two cow-size *Coryphodon* cross the point bar of a small stream, and crocodilians rest on the bank. Dense forests cover the floodplain. The scene represents the lower Eocene Willwood Formation and is based on sedimentological and paleontological studies conducted in the Bighorn Basin, Wyoming. Art by Rebecca Horwitt.

10 9 8 7 6 5 4 3 2 1

Contents

Preface

Evidence for warmer climates during the early Cenozoic has existed for over a century, and the Paleogene has long been known to be the most recent period of extreme global warmth. The difference from present temperatures is most noticeable at high latitudes. However, the implications of such warm conditions only began to be appreciated in the mid-1980s, and their causes are still actively debated. Over the past 10–15 years, marine and terrestrial paleoclimate estimates have become more precise, proxy data have been obtained from new regions, and paleoclimatic modeling has become more sophisticated. Furthermore, improvements in correlation and chronostratigraphy (including tuning Paleogene records to Milankovitch frequencies) have refined our understanding of the timing and pace of climatic fluctuations and biotic change in the early Cenozoic.

In spite of these advances, geoscientists still find it difficult to explain deep ocean temperatures of 10–12 °C, the low gradients of temperature across latitude, and the relatively low seasonality of continental interiors during the early Cenozoic. The effects of high ocean temperatures and low latitudinal temperature gradients on organisms and ecosystems are only beginning to be understood. In the past, warm climates were thought to be relatively invariant on Milankovitch and shorter time scales because of the lack of rapid feedback mechanisms such as polar ice caps and sea ice, but recent work has revealed episodes of geologically rapid climate change even during the warm Paleogene, which has initiated new investigations into the mechanisms and biological consequences of rapid global warming, and on possible climate-biota feedback.

The overall goal of this volume is to illustrate the current state of knowledge of the globally warm early Paleogene, and to forge links among research conducted in different regions from different perspectives. To accomplish this, the volume is organized into four sections: (1) physical and geochemical causes of climate change; (2) the effect of climatic change on marine biota; (3) the effect of climate change on terrestrial biota; and (4) the stratigraphic framework of the Paleogene.

The first section, "Climate Mechanisms and Models," includes six papers that discuss physical and geochemical forcing of globally warm climates, and the possible interaction between marine and terrestrial systems in producing rapid climate change. Two modeling studies address the potential for Milankovitch orbital cycles to cause fluctuations in continental climate, and the interaction of atmospheric and oceanic circulation in producing globally warm climates during the Eocene. Three papers are devoted to understanding how biogeochemical cycles may have been involved in creating and modulating the rapid global warming that occurred near the Paleocene-Eocene boundary. These papers examine marine and terrestrial systems, as well as the linkages between them. The final paper in this section discusses how stomata of fossil leaves can be used to estimate atmospheric pCO_2 levels in the Late Cretaceous and Paleogene, and the implications of such estimates for the role of CO_2 in generating and modifying warm climates.

The second section, "Marine Biotas, Environments, and Climate Change," contains 13 papers that focus on various aspects of marine environments, faunas, and floras. There are five regional studies describing Paleogene biotas and paleoenvironments of the peri-Tethyan region of Russia, some focused on events near the Paleocene-Eocene boundary. These data were until now largely unknown in the international community, but because they characterize a vast region, these studies may strongly influence global geochemical models. A group of five papers examines biotic and isotopic records through the Paleocene-Eocene interval and interprets them in terms of productivity, ocean circulation, temperature and seasonality. These records come from paleoenvironments ranging from lagoonal to deep sea, and from locations around the world, but with a focus on western Europe. This section also includes papers describing the ecological and evolutionary response of planktic foraminifera to changes in Paleocene surface water characteristics, evidence for changing sea level and faunal change in the Paleocene of Egypt, and isotopic reconstructions of seasonality in the North Sea region during the late Paleocene.

The third and longest section of the volume, "Continental Biotas, Environments, and Climate Change," has 15 papers concerning the reconstruction of Paleogene terrestrial environments and the effects of globally warm climates on the terrestrial biota. Five papers are narrowly focused on the biotic effects of rapid global warming near the Paleocene-Eocene boundary in locations including western Europe, China, New Zealand, and North America. Three of these papers concern plants, and two focus on mammals. Four papers address Paleogene floral change in a more general context, and cover regional patterns in India, Antarctica, Australia, and North America. Three more papers look at regional change in Paleogene vertebrate faunas from Europe and Asia, and discuss the implications for migration between these regions. Two papers deal primarily with the biostratigraphic use of Paleocene and Eocene vertebrate faunas,

Participants in the conference "Climate and Biota of the Early Paleogene" on an outcrop of the early Eocene Willwood Formation. The conference, from which this volume emerged, was held in Powell, Wyoming, in July 2001. Participants: Arvid Aase, Abdel-Kireem Mohamed Rashad, Laia Alegret, Andrew Anderson, Robert L. Anemone, Estíbaliz Apellaniz, Bruce Archibald, Amanda Ash, Humberto Astibia, Marie-Pierre Aubry, Marc Augé, Evelyn Augé, Juan Ignacio Baceta, Ainara Badiola Kortabitarte, Richard S. Barclay, Kenneth Christopher Beard, Vladimir Beniamovski, Bill Berggren, Gilen Bernaola Bilbao, Karen L. Bice, Marley Bice, David Blackburn, Jonathan I. Bloch, Gabriel Bowen, Douglas M. Boyer, Henk Brinkhuis, John H. Burris, Laurel M. Bybell, Fernando Caballero Santamaria, Amy Chew, Mark Clementz, William C. Clyde, Margaret Collinson, Jeff P. Crabaugh, Benjamin S. Cramer, Katherine Siân Davies-Vollum, David Dilcher, Shelley Dowson, Ilse Draxler, Regan E. Dunn, Christian Dupuis, Andrea Dutton, Hans Egger, Martin B. Farley, Kristina Faul, John Finarelli, Herbert Frankenhaeuser, Jens Lorenz Franzen, Elisabeth Franzen, Henry Fricke, Yuri Gavrilov, Thomas Gibson, Philip D. Gingerich, Yury Gladenkov, Marc Godinot, Margaret C. Gorringe, Anjali Goswami, David R. Greenwood, Gregg F. Gunnell, Elaine Hale, Haidi Jean Louise Hancock, Guy J. Harrington, Jason Hicks, Christopher J. Hollis, Patricia A. Holroyd, Jerry Hooker, Rebecca Horwitt, Brian Huber, Matthew Huber, Richard J. Hunt, Gretchen L. Hurley, J. Howard Hutchison, Carlos Jaramillo, Kirk Johnson, Rachel Keefe, Elizabeth Kennedy, Paul Kester, Intizar Khan, Paul L. Koch, Elizabeth Kowalski, Valentin Krassilov, Mary J. Kraus, Zlatko Kvacek, David J. Large, Kira Lawrence, Michael B. Leite, Jason A. Lillegraven, Nieves Lopez-Martinez, Herbert Lutz, Mandela A. Lyon, Roberto Magioncalda, Carine Magioncalda, Steven R. Manchester, Ernest A. Mancini, Finnegan Marsh, Natalia Maslova, Casey McKinney, R.C. Mehrotra, Kathleen M. Muldoon, Jason Mussell, Douglas J. Nichols, Richard D. Norris, Michael D. Nowak, Tom Oesleby, Khaled Ouda, Tatiana Oreshkina, Xabier Orue-Etxebarria, Tonya Penkrot, Pablo Pelaez-Campomanes, Vera Mikhailovna Podobina, Victoriano Pujalte, Frédéric Quillévéré, Eleonora Petrovna Radionova, Robert G. Raynolds, David K. Rea, Tom Rea, Susan G. Riggins, Ursula Röhl, Kenneth D. Rose, Dana Royer, Ali Sadek, Ashraf Mohammed Samir, Juila Sankey, Anindya Sarkar, Stephen Schellenberg, David Schankler, Judith A. Schiebout, Jessica Scott, Ross Secord, Jacob O. Sewall, Ekaterina A. Shcherbinina, Lucinda J. Shellito, Mary T. Silcox, Lisa C. Sloan, John Richard Smith, Krister Smith, Thierry Smith, Robert P. Speijer, Hassan Soliman, Baruch F. Spiro, Rainer Springhorn, Etienne Steurbaut, Thomas A. Stidham, Suzanne Strait, Liz Fet, Lou Taylor, Deborah J. Thomas, Ellen Thomas, Bruce H. Tiffney, Theo Tiffney, Robin M. G. Tiffney, Ting Suyin, Aradhna Tripati, Bridget S. Wade, Wei-Ming Wang, Yuanqing Wang, Paul D. White, Michael C. Wiemann, Volker Wilde, Peter Wilf, Scott L. Wing, Jack A. Wolfe, Thomas E. Yancey, James Zachos, Shawn Zack, and Holley M. Zupon. (Photo by S. Wing.)

and a final paper examines changes in paleosols during the Paleocene-Eocene transition in southern Texas.

The fourth and final section, "Integrated Stratigraphic Framework," consists of two papers, one that treats the history of study and formal chronostratigraphic terminology of the Paleocene-Eocene boundary interval, and a second that refines the timing of events near the Paleocene-Eocene boundary by applying cyclostratigraphy to cores taken from ODP Site 1051 on Blake Nose off the Florida coast.

This volume is notable in that it contains a diversity of diversities; it is broad geographically, disciplinarily, taxonomically, and temporally. The geographic diversity is impressive.

"Climate and Biota of the Early Paleogene" conference participants descending the end of Polecat Bench into a sequence of uppermost Paleocene and lowermost Eocene paleosols. (Photo by S. Wing.)

A "Climate and Biota of the Early Paleogene" conference bus on a dirt road after a sudden rainstorm in the badlands. (Photo by F. Marsh.)

There are papers describing data from every continental area and ocean basin. The variety of approaches is equally broad. There are studies that model the atmosphere, oceans, and biogeochemical systems, and those that use geochemical, paleontological, and/or lithological data to reconstruct changes in environments ranging from the deep sea floor to terrestrial uplands. The taxonomic range of organisms surveyed for their response to climate change is also impressive, including land plants, mammals and other vertebrates, marine mollusks, and marine microfossil groups such as diatoms, dinoflagellates, calcareous nannoplankton, and foraminifera. A final aspect of the diversity of the volume is that, although all the papers focus on the Paleogene, they provide information on many intervals within that span. Some papers are geographically broad but temporally coarse, while others focus on fine-scale temporal resolution of particular events in a few local records.

The diversity of approaches, regions and subjects of study, and degree of temporal resolution displayed in this volume reflects a conscious effort by those who work on Paleogene climates and biota to integrate their research. The study of past global change in earth systems requires all researchers to be aware of results that emerge from studies outside their areas of expertise. The conference from which this volume emerged ("Climate and Biota of the Early Paleogene" held in Powell, Wyoming, in July 2001) was the most recent in a series of conferences devoted to the Paleogene, all of which have had the goal of encouraging the integration of research across the usual lines of terrestrial versus marine, geochemical versus paleontological, and modeling versus observational. We, as editors, feel strongly that the insights developed in this volume justify the effort by the authors to interpret their data in the light of scientific knowledge being developed on many fronts.

Over the past 10 years, one particular climatic event during the Paleogene has attracted much attention: the rapid warming followed by more gradual cooling that occurred during a period of less than 200,000 years near the Paleocene-Eocene boundary. This warming event and its associated negative excursion in carbon isotopes was first recognized in a deep-sea section (Kennett and Stott, 1991), and referred to as Late Paleocene Thermal Maximum (LPTM) by Zachos et al. (1993) because it was in the Paleocene according to both calcareous nannoplankton and planktic foraminiferal zonations. At the time, however, there was no agreement on the definition and recognition of the Paleocene-Eocene boundary (Aubry et al., this volume).

When the carbon isotope excursion (CIE) was subsequently discovered on land (Koch et al., 1992), its onset coincided with the Paleocene-Eocene boundary long recognized on the basis of continental vertebrate biostratigraphy. Now the onset of the CIE, virtually synchronous with the initial rapid increase in temperature, is the principal criterion used by the International Subcommission on Paleogene Stratigraphy to recognize the Paleocene-Eocene boundary (Luterbacher et al., 2000; Aubry et al., this volume). Much of the 110–220 k.y. long temperature "spike" previously known as the LPTM falls, more or less by definition, in the earliest part of the Eocene. Some authors in this volume, following Schmitz (2000), now use the phrase Initial Eocene Thermal Maximum (IETM) to refer to the event. Others, emphasizing the coincidence of the event with the Paleocene-Eocene boundary, refer to it as the Paleocene-Eocene Thermal Maximum (PETM). In the absence of agreement among the editors and authors, we have allowed both IETM and PETM. It is important for readers to understand that both refer to the same event.

The symposium from which this volume was produced was supported by the National Science Foundation (EAR 0119092); by the Office of the Director, National Museum of Natural History, Smithsonian Institution; by the Evolution of Terrestrial Ecosystems Program of the Smithsonian Institution; by the Museum of Paleontology, University of Michigan; and by the generous hospitality of Northwest College, the Churchill family, and other residents of Powell, Wyoming. We thank all of these, and the editorial staff of the Geological Society of America books division for their help. We also thank the many authors who made contributions to publications costs. This is Evolution of Terrestrial Ecosystems publication number 120.

REFERENCES CITED

Kennett, J.P., and Stott, L.D., 1991, Abrupt deep sea warming, paleoceanographic changes and benthic extinctions at the end of the Paleocene: Nature, v. 353, p. 225–229.

Koch, P.L., Zachos, J.C., and Gingerich, P.D., 1992, Correlation between isotope records in marine and continental carbon reservoirs near the Paleocene/Eocene boundary: Nature, v. 358, p. 319–322.

Luterbacher, H. P., Hardenbol, J., and Schmitz, B., 2000, Decision of the Voting Members of the International Subcommission on Paleogene Stratigraphy on the Criterion for the Recognition of the Paleocene/Eocene Boundary: Newsletter of the International Subcommission on Paleogene Stratigraphy, v. 9, November 2000, Tübingen, p. 13.

Schmitz, B., 2000, Is the CIE in the late Paleocene or early Eocene? Newsletter of the International Subcommission on Paleogene Stratigraphy, v. 9, November 2000, Tübingen, p. 19.

Zachos, J.C., Lohmann, K.C., Walker, J.C.G., and Wise, S.W., 1993, Abrupt climate change and transient climates during the Paleogene: A marine perspective: Journal of Geology, v. 101, p. 191–213.

Scott L. Wing
Philip D. Gingerich
Birger Schmitz
Ellen Thomas

Geological Society of America
Special Paper 369
2003

Marine-terrestrial linkages at the Paleocene-Eocene boundary

Santo Bains
Department of Earth Sciences, University of Oxford, Oxford, OX1 3PR, UK
Richard D. Norris
Department of Geology and Geophysics, Woods Hole Oceanographic Institution, Woods Hole, Massachusetts 02543, USA
Richard M. Corfield
Department of Earth Sciences, University of Oxford, Oxford, OX1 3PR, UK
Gabriel J. Bowen
Department of Earth Sciences, University of California, Santa Cruz, California 95062, USA
Philip D. Gingerich
Department of Geological Sciences and Museum of Paleontology, University of Michigan, Ann Arbor, Michigan 48109, USA
Paul L. Koch
Department of Earth Sciences, University of California, Santa Cruz, California 95062, USA

ABSTRACT

A fossil-bearing continental sequence that spans the Paleocene-Eocene boundary (ca. 55 Ma) can now be accurately correlated to expanded deep-sea oceanic sediments at an extremely high resolution (~10 k.y.), thus facilitating detailed investigations into abrupt global climate change and its influence on the evolution of terrestrial organisms. Here we show that the onset of this extremely warm interval is associated with a stepped terrestrial carbon isotope ($\delta^{13}C$) excursion. This suggests that a pulsed sublimation of submarine gas hydrate accumulations at this time may have caused a rapid venting of significant quantities of light carbon through the ocean/atmosphere interface. Major mammalian turnover occurred near the onset of the ensuing greenhouse event, and this also appears to have occurred in a sequential fashion, although the changes we see in population composition and morphology lag the major features of the global $\delta^{13}C$ record by some ~10–20 k.y., which could represent the duration required for evolutionary mechanisms to occur due to greenhouse-associated stresses. Additionally, we have evidence that increased soil respiration rates occurred in response to the core episode of global warmth. Paleocene-Eocene boundary carbon cycle perturbations were apparently as remarkable in the atmospheric and terrestrial reservoirs as they were in the oceans, and these changes had a dramatic effect on terrestrial biota.

INTRODUCTION

At the Paleocene-Eocene (P-E) epoch boundary, ocean water $\delta^{13}C$ became more negative by ~2.5‰ in a stepped manner over <52 k.y. (Norris and Röhl, 1999; Röhl et al., 2000). The rapidity, amplitude, and structure of this major carbon isotope event (CIE) have been explained by the pulsed introduction into the exchangeable carbon reservoir of large volumes of isotopically light methane (CH_4 with $\delta^{13}C$ ~ –60‰; Kvenvolden, 1988)—a potent greenhouse gas—from subliming gas hydrate

Bains, S., Norris, R.D., Corfield, R.M., Bowen, G.J., Gingerich, P.D., and Koch, P.L., 2003, Marine-terrestrial linkages at the Paleocene-Eocene boundary, *in* Wing, S.L., Gingerich, P.D., Schmitz, B., and Thomas, E., eds., Causes and Consequences of Globally Warm Climates in the Early Paleogene: Boulder, Colorado, Geological Society of America Special Paper 369, p. 1–9.

accumulations (Dickens et al., 1995; Bains et al., 1999; Dickens, 2001). As this occurred, the Earth's climate warmed by as much as 5–7 °C via greenhouse mechanisms (e.g., Kennett and Stott, 1991; Eldholm and Thomas, 1993; Rea et al., 1995; Fricke et al., 1998; Sloan and Thomas, 1998; Bains et al., 1999), which may have been a major cause of continental faunal migration corridors opening (Gingerich, 1989; Clyde and Gingerich, 1998; Beard and Dawson, 1999; Peters and Sloan, 2000) and marine and terrestrial taxonomic turnover increases (e.g., Gingerich, 1989; Kelly et al., 1996, 1998; Clyde and Gingerich, 1998).

In response to the rapidly changing environment, oceanic phytoplankton bloomed and an elevated marine carbon pump probably helped the Earth to recover quickly to near pre-event conditions (Bains et al., 2000; Zachos and Dickens, 2000). According to recent models, a similar greenhouse-induced biological feedback system may also have occurred on land (Beerling, 2000). Given the reality of current anthropogenic global warming, and the uncertainties over future ecological change, the terrestrial floral and faunal response to the transient P-E boundary greenhouse world warrants thorough investigation. We have thus conducted, and here present, a high-resolution continental geochemical and paleontological study of this exceptional interval in the Earth's history.

HIGH-RESOLUTION TERRESTRIAL STABLE ISOTOPE AND FOSSIL MAMMAL RECORDS

The $\delta^{13}C$ and $\delta^{18}O$ values of ancient soil nodule carbonates can provide valuable information about terrestrial environmental change from the geological record. The $\delta^{13}C$ of soil nodule carbonate can be influenced by several factors (Cerling, 1984, 1991, 1999; Cerling et al., 1991; Cerling and Quade, 1993). However, since a major component of soil gas (the source of carbon in soil nodule carbonate) is supplied by root respiration and the bacterial decay of litter, it can reflect the $\delta^{13}C$ of overlying plants if other factors are carefully accounted for (see discussion in Koch et al., 1995). The $\delta^{13}C$ of plants reflects that of atmospheric CO_2, which in turn is dominated by gas exchange with the oceans. Koch et al. (1992, 1995) have previously demonstrated that at the P-E boundary, shifts in soil nodule carbonate $\delta^{13}C$ can indeed be correlated to shifts in the $\delta^{13}C$ of the surface ocean (with a regular fractionation). Building on this knowledge, we have now constructed an extremely detailed continental $\delta^{13}C$ record across the P-E boundary, which is of a similar temporal resolution to those that have recently been created from deep-sea carbonates (Bains et al., 1999) and an independent data set also derived from the Bighorn Basin sequence (Bowen et al., 2001). The new high-resolution terrestrial CIE records show virtually identical trends, and are used here for the first time to assess atmospheric and terrestrial carbon reservoir consequences of gas hydrate decomposition, changes in soil respiration/production and terrestrial climate feedbacks, as well as to further address the timing and nature of mammalian turnover across this interval. The $\delta^{18}O$ of soil nodule carbonate is used here to show that diagenetic effects due to deep burial do not affect our isotopic record.

We measured the $\delta^{13}C$ and $\delta^{18}O$ composition of 485 carbonate soil nodules (see Table DR1[1]) collected from an ~80 m thick stratigraphic section at the southern end of Polecat Bench (Bighorn Basin, Wyoming, at, and in, the vicinity of University of Michigan locality SC-67; Gingerich, 1989). By moving west around the southern tip of Polecat Bench, we were able to map and sample a continuous composite sequence of fine-grained mudstones and well-developed paleosols deposited across the late Paleocene through to the early Eocene, successfully avoiding the Clarkforkian–Wasatchian sandstone unit, which is not laterally continuous here. We collected fossil soil nodules from the narrow spines of steep ridges, digging out ones that were just emerging out of the rock surface as well as those that were deeply buried. The average within-soil standard deviation is 0.8‰ and 0.5‰ for $\delta^{13}C$ and $\delta^{18}O$ respectively, in accordance with typical within-nodule and within-soil isotopic differences reported in other studies of Bighorn Basin paleosol carbonates (Koch et al., 1992, 1995). In soils, carbonate precipitated <30 cm from the soil-atmosphere boundary can have a $\delta^{13}C$ value that is directly influenced by atmospheric CO_2, rather than purely by soil gas derived from plants. This mechanism can cause shifts of ≥10‰ from the base to the top of a soil unit, an effect most pronounced in desert environments (Quade et al., 1989; Cerling and Quade, 1993). However, in the data presented here, no obvious association was found between soil thickness and $\delta^{13}C$ value, and carbonate nodules collected from <30 cm thick soils do not have unusual $\delta^{13}C$ values. Nevertheless, as a precaution, we calculate and plot average $\delta^{13}C$ values for only carbonate nodules collected at ≥30 cm below surfaces of sufficiently thick paleosols. These data points follow the same trend as that of the entire data set (Fig. 1). Some soil nodules contained diagenetic sparry calcite. Co-occurring spar and micrite do not differ significantly in terms of $\delta^{13}C$, but spar $\delta^{18}O$ values are significantly more negative than those of micrite (mean difference –6.7‰). For our record, sparry calcite was carefully avoided when drilling samples. For isotopic analysis, several hundred micrograms of sample were drilled from micrite in soil nodules and cleaned of possible organic contaminants using 10% H_2O_2 and acetone. Samples were reacted at 90 °C in a common phosphoric acid bath carbonate preparation device interfaced with a mass spectrometer in the Department of Earth Sciences, University of Oxford. The average time between reaction of samples was 30 minutes, which is sufficient to completely react all carbonate phases. Analytical precision better than 0.1‰

[1]GSA Data Repository item 2003047, Table DR1, High-resolution stable carbon and oxygen isotope data from the Bighorn Basin, is available on request from Documents Secretary, GSA, P.O. Box 9140, Boulder, CO 80301-9140, USA, editing@geosociety.org, or at www.geosociety.org/pubs/ft2003.htm.

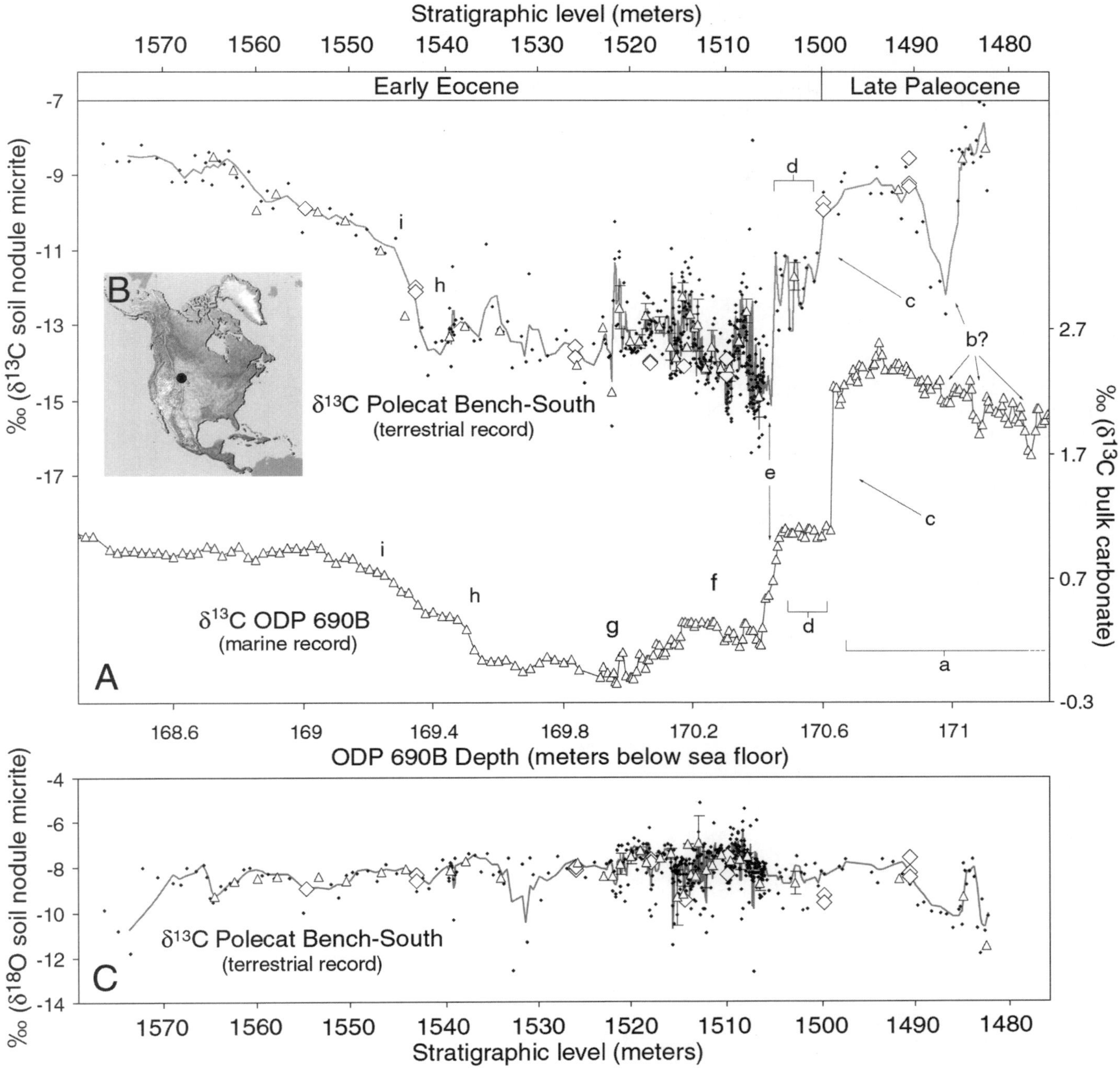

Figure 1. Detailed terrestrial $\delta^{13}C$ and $\delta^{18}O$ and marine $\delta^{13}C$ records across the P-E boundary. Terrestrial data are derived from carbonate soil nodules from an expanded section at the southern end of Polecat Bench, northern Wyoming, and marine data are from ODP 690B bulk carbonates (Bains et al., 1999). A: Labels a through i were assigned to the $\delta^{13}C$ record of ODP 690B in Bains et al. (1999), and here labels c, d, e, h and i have also been used to denote the corresponding $\delta^{13}C$ stable isotope events found in the terrestrial realm, and are referred to in the text. For Polecat Bench data, stratigraphic level has been plotted as depth in meters above the K-T boundary. The record decreases in age from right to left and includes previously published data from this locality (large open diamonds; Koch et al., 1995). ODP 690B data has been plotted against depth as meters below the seafloor. Open triangles in the Polecat Bench–South record represent averaged data from nodules collected at depths ≥30 cm from the surface of ancient paleosols. The previous data of Koch et al. (1995) has been excluded from the trendline (which is a 3 point moving average). B: Map of modern North America showing the approximate location of the Polecat Bench section. C: $\delta^{18}O$ data from soil nodule micrite from Polecat Bench. Because our data are consistent with the low-resolution trend of Koch et al. (1995) we are confident that our samples have not been affected by diagenesis or reworking.

for both carbon and oxygen isotopes was maintained throughout all analysis. Carbon and oxygen data are reported in δ values: $\delta^{13}C$ or $\delta^{18}O = [(R_{sample}/R_{standard}) - 1] \times 1000$, where R_{sample} and $R_{standard}$ are the ratios of $^{13}C/^{12}C$ or $^{18}O/^{16}O$ for the sample and the standard. Results were calibrated to Pee Dee belemnite (V-PDB) by reference to in-house Carrara Marble standard and are reported in per mil (‰).

We have also collected, and below interpret, a new highly detailed fossil record from the Polecat Bench outcrop. The P-E boundary is generally thought to coincide with an abrupt change from a mammal fauna dominated by archaic lineages to a fauna dominated by modern orders including the first Primates (Gingerich, 1989). In North America, this faunal change happens at the Clarkforkian–Wasatchian land-mammal age boundary (Rose, 1981; Gingerich, 1989), which occurs near the CIE (Koch et al., 1992, 1995). We find that this faunal transition occurred in a sequential nature, with changes in earliest Wasatchian mammals lagging features of the new high-resolution $\delta^{13}C$ record.

SIMILARITIES BETWEEN THE TERRESTRIAL AND MARINE CIE

Comparisons between the detailed marine P-E boundary $\delta^{13}C$ records of Ocean Drilling Program (ODP) Southern Ocean Site 690B and North Atlantic Site 1051B (Bains et al., 1999) and our high-resolution terrestrial $\delta^{13}C$ record from the Bighorn Basin, Wyoming (Fig. 1) will allow us to investigate the mechanisms of "whole-Earth" climate change at this time. Apparent similarities in the early stages of the boundary excursion include a marked initial negative drop in $\delta^{13}C$ (feature "c"), a plateau (feature "d"), followed by a further drop (feature "e"). Coming out of the event, a rapid increase in $\delta^{13}C$ (feature "h") precedes an inflection toward a more gradual positive trend (feature "i"). Based on correlation of features "c" and "h", assuming constant sedimentation rates, and using an orbital age model for the ODP 690B record (Röhl et al., 2000), the $\delta^{13}C$ excursion and recovery from the Polecat Bench record (from ~ –10 to +30 m) spans about four precession cycles, or ~84 k.y. This approximation yields a sedimentation rate of ~475 m/m.y., which is very comparable to a long-term rate of ~470 m/m.y. calculated for ~1200 m of sediments spanning the P-E boundary from the Fort Union and Willwood formations of this region (Gingerich, 2000; Wing et al., 2000), suggesting that our isotopic correlation is probably quite precise.

DIFFERENCES BETWEEN THE TERRESTRIAL AND MARINE CIE

The similarities between the terrestrial and marine records indicate that both of these reservoirs were significantly perturbed by the proposed sublimation of gas hydrate accumulations at the P-E boundary; however, intriguing differences do exist that raise questions about various aspects of carbon cycle dynamics. Of particular interest is the larger amplitude of the CIE in the terrestrial record (Fig. 2). Excursions "c" and "e" are similar in both records in terms of their almost instantaneous nature, however, their magnitude in the Polecat Bench record (~ –2‰ and –3.5‰, respectively) are much greater than those

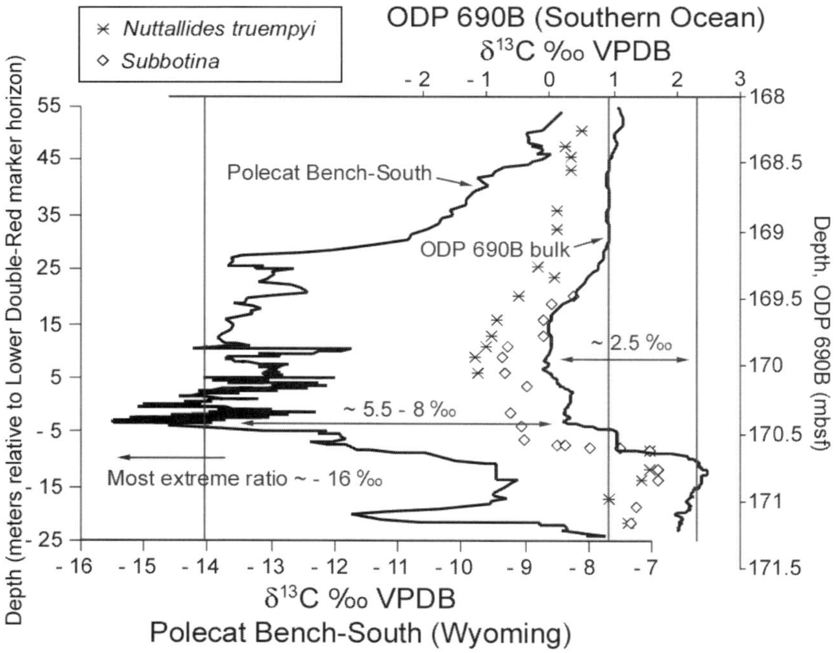

Figure 2. The differing magnitudes of the P-E boundary carbon isotope event (CIE) recorded in terrestrial and marine sediments. Trends have been plotted to illustrate the large magnitude of the CIE recorded from carbonate soil nodules collected from Wyoming (~5.5‰–8‰), compared with the CIE recorded in marine bulk carbonate sediments (Bains et al., 1999) and foraminiferal records of benthic dwelling *Nuttallides truempyi* and planktonic *Subbotina patagonica* and *Subbotina varianta* (Kennett and Stott, 1991) from Southern Ocean Site ODP 690B (~2.5‰).

observed in the bulk carbonate record from ODP 690B (~ −1.2‰ and −1‰, respectively). Also, the total magnitude of the $\delta^{13}C$ excursion on land (~5.5‰–8‰) remains larger than the marine record $\delta^{13}C$ excursion (~2.5‰) for tens of thousands of years throughout the core of the CIE (features "d" through "i").

It is possible that a small fraction of the liberated CH_4 from subliming gas hydrate reservoirs bubbled through the water column and oxidized to CO_2 in the atmosphere. Because the atmosphere contains much less carbon than the oceans, such a transfer of very light carbon to this reservoir would have a great effect on its $\delta^{13}C$ composition. In fact, according to calculations that we performed, which were based on previously published equations (Dickens et al., 1995; Dickens, 2001), the introduction of ~50 Gt and 970 Gt of ~ −60‰ $\delta^{13}C$ carbon into the atmospheric and oceanic reservoirs respectively could have caused the initial isotope event "c." A second injection of ~100 Gt to the atmosphere and ~760 Gt to the oceans would explain excursion "e." Under contemporary oceanographic conditions, however, equilibration of carbon isotopes between the atmosphere and marine bicarbonate occurs rapidly (~1 k.y.). The persistently larger magnitude of the $\delta^{13}C$ excursion on land suggests either that oceanic and atmospheric circulation were dramatically different across the P-E boundary, or that some other mechanism may have been operating to maintain the greater magnitude of the excursion in terrestrial systems.

INCREASED SOIL RESPIRATION RATES AT THE P-E BOUNDARY

We suggest that increased atmospheric CO_2 concentrations and surface temperatures following the initial sublimation of gas hydrates may have led to elevated rates of soil respiration, thus causing the magnified $\delta^{13}C$ excursion on land (as discussed above). Cerling (1999) has shown that an increase in soil respiration rates, which are often associated with increased primary production, lead to a decrease in the $\delta^{13}C$ of soil CO_2 and soil carbonates. A conservative estimate (using the method of Cerling, 1999, and assuming that the atmospheric CIE magnitude was ~ −4.25‰ larger than the CIE recorded in the oceans—as suggested by our data) suggests at least a doubling of soil respiration rates in the Bighorn Basin following the P-E boundary.

However, there are other variables besides soil respiration rates and the $\delta^{13}C$ of overlying plants and atmosphere that can significantly influence the $\delta^{13}C$ of soil carbonates (Cerling, 1984, 1991, 1999; Cerling et al., 1991; Cerling and Quade, 1993). These include (i) differences in carbon isotope fractionation between C3 and C4 plants; (ii) atmospheric pCO_2; (iii) soil porosity; and (iv) plant rooting depth. A shift in the overlying biomass of C4 versus C3 plants can produce large shifts in the $\delta^{13}C$ of soil carbonates, but C4 plants are not believed to have become abundant until the Neogene (Thomasson et al., 1986; Quade et al., 1989; Quade and Cerling, 1995). Changes in the concentration of CO_2 in the atmosphere also affects the penetration of this gas into soils and the $\delta^{13}C$ of soil CO_2 (Cerling, 1991). However, atmospheric pCO_2 was probably 1000 ppmv or less during the P-E transition (Koch et al., 1992; Beerling, 2000; Royer et al., 2001), which is too low to have had a major effect on the $\delta^{13}C$ of soil CO_2 and pedogenic carbonates (Cerling, 1999). One estimate of P-E transition pCO_2 based on boron isotopes does suggest higher levels (Pearson and Palmer, 2000), but an increase in atmospheric pCO_2 would decrease the magnitude of the $\delta^{13}C$ excursion in soils, which is the opposite of our observations. Although changes in rooting depth and soil porosity have not been thoroughly studied in P-E boundary soils, exploration of a soil carbon isotope model suggests minimal sensitivity to these parameters unless they undergo dramatic changes in value, such as those seen between tropical forest and desert soils (Cerling, 1984, 1991; Cerling et al., 1991; Cerling and Quade, 1993). Given the similarity in depositional environments and sedimentary rates across the boundary, we conclude that such drastic changes in these soil parameters are unlikely, and that the Bighorn Basin record is consistent with an increase in soil respiration rates during the $\delta^{13}C$ excursion. The exact reasons for such a large increase in soil productivity are not fully understood, but it seems likely that it may have been linked to the increase in Bighorn Basin temperatures of 5–7 °C that has been recorded at the P-E boundary (Fricke et al., 1998) and an increase in atmospheric pCO_2 following CH_4 input and oxidation. However, there is some evidence that soil decomposition rates can remain constant with temperature (Giardina and Ryan, 2000), and so other factors such as changes in precipitation and vegetation patterns may also have been important.

Beerling (2000) used vegetation-biogeochemistry and carbon isotope mass balance models to study the response of global terrestrial ecosystems to different scenarios for increased atmospheric pCO_2 and temperature during the P-E boundary event. He inferred increases in terrestrial carbon storage ranging from 7% to 200% of the pre-excursion carbon in vegetation biomass and soils. In his vegetation-biogeochemistry models, the increase in vegetation biomass was always much greater than that in soils, and it was accomplished primarily by increased plant productivity, rather than by an expansion of plant cover on the land surface. Increased plant productivity, with attendant increases in root respiration and bacterial decay of litter, should be expressed as a decrease in soil carbonate $\delta^{13}C$, which is what we observed here.

NEGATIVE CLIMATE FEEDBACK INDUCED BY INCREASED SOIL DEVELOPMENT

Of further interest is that amplified respiration should increase soil pCO_2, leading to an increase in the rate of soil development and silicate weathering. More productive soils have higher pCO_2 due to root respiration and bacterial oxidation of organic matter; this should lead to lower soil pH and therefore higher chemical weathering rates (e.g., Berner, 1993). This effect would be consistent with the marked increase in soil development as observed in the Bighorn Basin at the P-E boundary (Gingerich, 1989). Silicate weathering reactions would have consumed CO_2, serving as a direct negative feedback that decreased global CO_2 levels at this time (Zachos and Dickens, 2000). Increased silicate weathering would also have served as an indirect negative feedback by augmenting the quantity of nutrients supplied to oceanic continental margins (Bains et al., 2000; Zachos and Dickens, 2000), thus supporting a vigorous biological "pump" caused by increased oceanic productivity (Bains et al., 2000). Our new data, along with evidence for a synchronized enlargement of marine biomass (Bains et al., 2000), support the proposition that elevated temperatures and atmospheric CO_2 concentrations induced intricate negative feedback processes that increased carbon storage and subsequent burial in the oceans and on land, thereby withdrawing CO_2 from the atmosphere and cooling the planet within ~84 k.y.

MAMMALIAN RESPONSES TO A SEQUENTIAL PERTURBATION OF THE CARBON CYCLE

The P-E boundary in North America has long been known to coincide with abrupt change from archaic Clarkforkian to more modern Wasatchian mammals (Granger, 1914; Rose, 1981). The general patterns of dwarfing and faunal turnover at this boundary are described by Gingerich (this volume). We sampled 14 fossil localities that span seven successive stratigraphic intervals in the same composite section as our new $\delta^{13}C$ record (stippled in Figure 3) to assess how this paleontological transition compares to the climate and environmental changes that we have discussed above. The first three intervals yield distinctively Clarkforkian mammals (*Probathyopsis praecursor, Apheliscus nitidus, Aletodon gunnelli,* and *Haplomylus simpsoni,* in addition to a high proportion of *Copecion brachypternus* and *Ectocion osbornianus*), which, taken together, are characteristic of late Clarkforkian age (Cf-3). The fourth stratigraphic interval is represented by a single locality, SC-404 (although additional localities representing this interval are known nearby; Gingerich, 1989), and yields characteristically Wasatchian endocarps of the dicot *Celtis phenacodorum* and a dentary of *Meniscotherium priscum,* but no later Wasatchian indicators; hence it is probably Wasatchian in age (here referred to as "Wa-0?") and could represent a transient interval of first arrivals and/or appearances of a new dominant community. The fifth stratigraphic interval, from locality SC-67 and others, yields the classic Wa-0 earliest Wasatchian fauna, with many species found to be dwarfed relative to congeners found earlier or later in the same stratigraphic sequence (Gingerich, 1989; Clyde and Gingerich, 1998). The sixth interval is indeterminately Wa-0 or Wa-1 in age (referred to as "Wa-0 or Wa-1" in Figure 3). Finally, the seventh stratigraphic interval, from locality SC-68 and others, yields a large fauna with typical early Wasatchian (Wa-1) indicators like *Haplomylus speirianus, Cantius ralstoni,* and *Diacodexis metsiacus.* Higher strata nearby yield large Wa-2 through Wa-5 faunas (Gingerich, 2000).

Comparison of our new high-resolution $\delta^{13}C$ record to biotic change across the P-E boundary from the Polecat Bench shows that the initial stepping in the negative $\delta^{13}C$ excursion (features "c", "d", and "e") coincides with the interval of localities SC-71 and others, and is clearly of archaic Clarkforkian Cf-3 mammal age. The appearance of the new Wa-0? *Celtis-Meniscotherium* association (in locality SC-404) lags behind the onset of the initial $\delta^{13}C$ excursion (feature "c") by some 15–20 k.y., while the appearance of the classic (dwarfed) Wa-0 mammals (at locality SC-67 and others), including the first representatives of Artiodactyla, Perissodactyla, and Primates, follows the maximum $\delta^{13}C$ excursion by some 10 k.y. The definitive appearance of "normal" Wa-1 Wasatchian mammals follows the $\delta^{13}C$ recovery (features "h" and "i") by some 10–20 k.y. The sequential nature of these events strongly suggests that continental biota were changing in response to abrupt climate change caused by severe carbon cycle perturbations, and the time-lags offer insight into the duration necessary for such evolutionary mechanisms to occur.

The distinctive dwarfing of the classic Wa-0 fauna (which apparently persisted only through extreme P-E boundary climate conditions) strongly suggests an evolutionary response to the global warming and vegetation changes described above. However, the synchronized and abrupt appearance of major groups like Artiodactyla, Perissodactyla, Primates, and hyaenodontid Creodonta, at the base of Wa-0 time suggests immigration of species from elsewhere. The source area for the new North American Wa-0 mammals may be Africa by way of Europe (Estravís and Russell, 1989), Europe itself (Godinot, 1996), Africa by way of Asia (Gingerich, 1986, 1989), Asia by way of Europe (Hooker and Dashzeveg, this volume), South Asia (Krause and Maas, 1990), or Central Asia (Bowen et al., 2002). Dispersal into North America would have required the opening of high-latitude connections between northern continents (Gingerich, 1989; Clyde and Gingerich, 1998; Beard and Dawson, 1999; Peters and Sloan, 2000). Since this happened just after the time of maximum $\delta^{13}C$ excursion values, and probably just after the peak of greenhouse induced global warming (Bains et al., 1999), the ~20 k.y. lag before Wa-0 mammals appeared in North America (see also Gingerich, this volume) may represent the time required for progressive colonization of intervening areas. A ~20 k.y. lag for mammalian dispersal between continents is approximately two orders of magnitude less than mammalian dispersal lags in the range of 14 to 0.7 m.y. reviewed by Woodburne and Swisher (1995).

CONCLUSIONS

Our high-resolution terrestrial P-E boundary stable isotope record contains small-scale structures that may correspond to multiple gas hydrate dissociation events recently described from

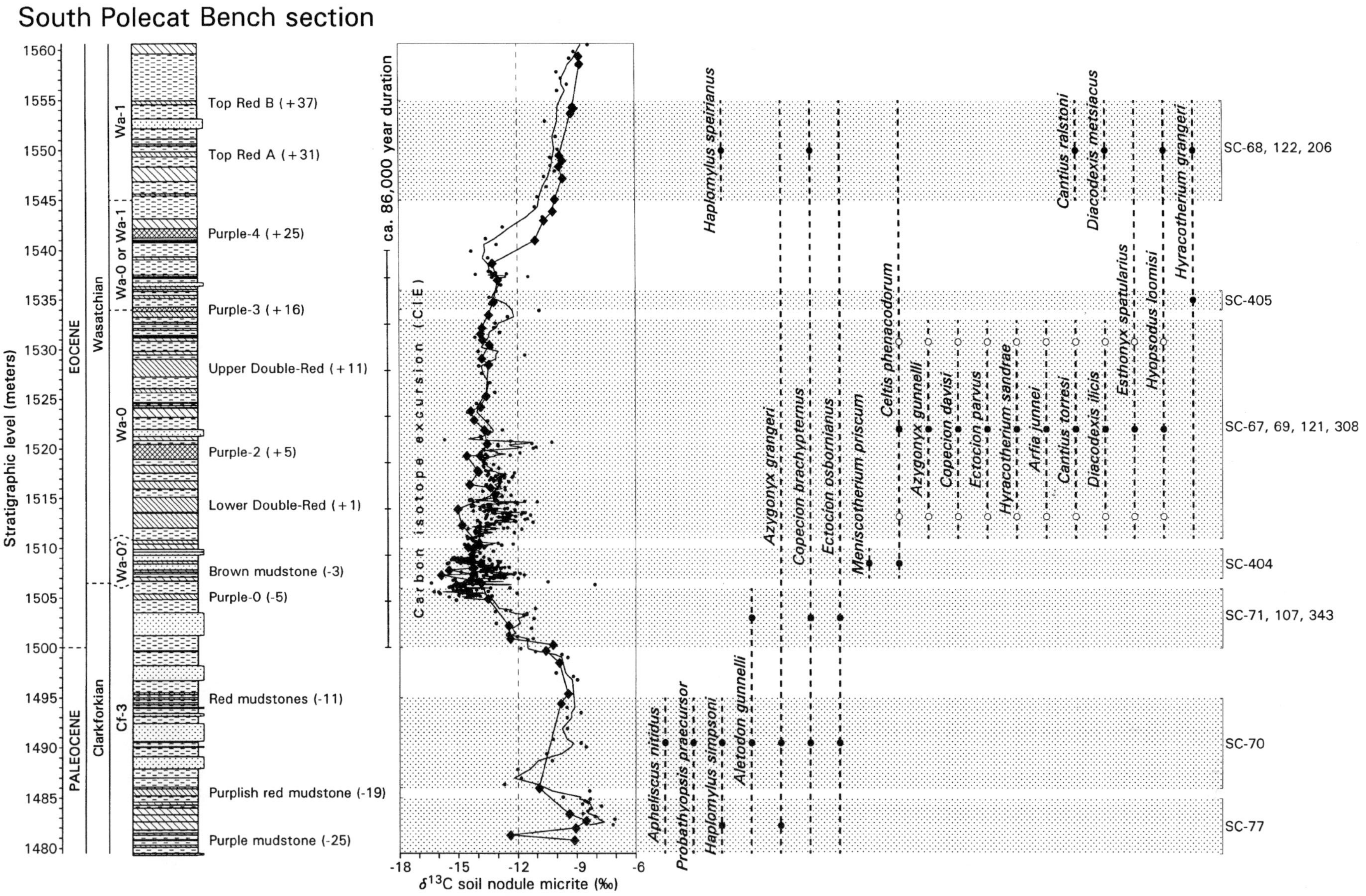

Figure 3. Comparison of high-resolution $\delta^{13}C$ isotope stratigraphy and biotic change at the southern end of Polecat Bench. Data from Bowen et al. (2001) are also included in this figure for comparative purposes (large filled diamonds). Fourteen mammal-bearing fossil localities are known (SC-77, etc.), representing seven distinct stratigraphic intervals (stippled). These can be grouped into four time-successive associations based on their biota (Cf-3, Wa-0?, Wa-0, and Wa-1 on the ordinate). The interval labeled "Wa-0 or Wa-1" on the ordinate is currently undersampled, and could be either Wa-0 or Wa-1 in age. The 86 k.y. time interval shown was derived from interpolation within chron 24R in the Polecat Bench–Sand Coulee stratigraphic section. This matches extremely closely with the 84 k.y. four-precession-cycle duration (4 × 21 k.y.) of the core interval of the CIE (Norris and Röhl, 1999; Röhl et al., 2000) as determined by comparison with the marine CIE. Note that the beginning of the CIE is in the late Clarkforkian, overlapping the stratigraphic interval of localities SC-71, etc. The most negative $\delta^{13}C$ values are in the earliest zone of the Wasatchian (Wa-0?), represented by SC-404 with the first *Celtis* and *Meniscotherium.* The remainder of the CIE spans the interval of the classic dwarfed Wa-0 fauna with Perissodactyla, Artiodactyla, and Primates from localities SC-67, etc. Wasatchian mammals of standard size are known from localities SC-68, etc., in Wa-1 after carbon isotope values returned to normal (here ~9‰).

detailed marine records. Comparisons of these data suggest that the continental sequence is probably continuous, and that the terrestrial carbon cycle was also drastically thrown into disequilibrium for a core interval of ~84 k.y. The persistently larger magnitude of the terrestrial CIE through this 84 k.y. interval suggests that soil respiration rates in this region may have increased due to the changing environment. Amplified silicate weathering would have occurred due to elevated respiration rates, and would have contributed to any negative climate feedbacks that may have been occurring at this time, by withdrawing CO_2 from the atmosphere. New mammal fossil collections, from the same continental P-E boundary sequence from which we have derived our isotope record, show that the transition from Clarkforkian to Wasatchian faunal assemblages occurred in several stages. Given the stepped perturbation of the carbon cycle that we document, and that the mammal stages lag the main features of the $\delta^{13}C$ excursion by ~10–20 k.y., evolutionary innovations and immigration events at this time were almost certainly forced by environmental changes.

ACKNOWLEDGMENTS

Thanks especially to J. Cartlidge and S. Wyatt for their expert help in the operation of the mass spectrometers in the Stable Isotope Laboratories at the Department of Earth Sciences, University of Oxford. We also thank O. Green, and S. Baker for other technical assistance and E. Linsey for maps of paleomagnetic collecting localities on Polecat Bench. This contribution benefited greatly from the extremely constructive comments of three anonymous reviewers. This work was supported by a Natural Environment Research Council grant to Bains, a Jesus College, Oxford, JRF to Bains, and a NSF-OCE grant to Norris.

REFERENCES CITED

Bains, S., Corfield, R.M., and Norris, R.D., 1999, Mechanisms of climate warming at the end of the Paleocene: Science, v. 285, p. 724–727.

Bains, S., Norris, R.D., Corfield, R.M., and Faul, K.L., 2000, Termination of global warmth at the Palaeocene/Eocene boundary through productivity feedback: Nature, v. 407, p. 171–174.

Beard, K.C., and Dawson, M.R., 1999, Intercontinental dispersal of Holarctic land mammals near the Paleocene/Eocene boundary: Paleogeographic, paleoclimatic and biostratigraphic implications: Bulletin de la Société Géologique de France, v. 170, p. 697–706.

Beerling, D.J., 2000, Increased terrestrial carbon storage across the Palaeocene–Eocene boundary: Palaeogeography, Palaeoclimatology, Palaeoecology, v. 161, p. 395–405.

Berner, R.A., 1993, Paleozoic atmospheric CO_2: Importance of solar radiation and plant evolution: Science, v. 261, p. 68–70.

Bowen, G.J., Koch, P.L., Gingerich, P.D., Norris, R.D., Bains, S., and Corfield, R.M., 2001, Refined isotope stratigraphy across the continental Paleocene–Eocene boundary on Polecat Bench in the northern Bighorn Basin, *in* Gingerich, P.D., ed., Paleocene–Eocene stratigraphy and biotic change in the Bighorn and Clarks Fork basins, Wyoming: University of Michigan Papers on Paleontology, v. 33, p. 73–88.

Bowen, G.J., Clyde, W.C., Koch, P.L., Ting, S., Alroy, J., Tsubamoto, T., Wang, Y., and Wang, Y., 2002, Mammalian dispersal at the Paleocene/Eocene boundary: Science, v. 295, p. 2062–2065.

Cerling, T.E., 1984, The stable isotopic composition of modern soil carbonate and its relationship to climate: Earth and Planetary Science Letters, v. 71, p. 229–240.

Cerling, T.E., 1991, Carbon dioxide in the atmosphere: evidence from Mesozoic and Cenozoic paleosols: American Journal of Science, v. 291, p. 377–400.

Cerling, T.E., 1999, Stable carbon isotopes in palaeosol carbonates: Special Publication of the International Association of Sedimentologists, v. 27, p. 43–60.

Cerling, T.E., and Quade, J., 1993, Stable carbon and oxygen isotopes in soil carbonates, *in* Swart, P., et al., eds., Continental isotopic indicators of climate: Washington, D.C., American Geophysical Union, p. 217–231.

Cerling, T.E., Solomon, D.K., Quade, J.A., and Bowman, J.R., 1991, On the isotopic composition of carbon in soil carbon dioxide: Geochimica et Cosmochimica Acta, v. 55, p. 3403–3406.

Clyde, W.C., and Gingerich, P.D., 1998, Mammalian community response to the latest Paleocene thermal maximum: An isotaphonomic study in the northern Bighorn Basin, Wyoming: Geology, v. 26, p. 1011–1014.

Dickens, G.R., 2001, Carbon addition and removal during the Late Palaeocene Thermal Maximum: basic theory with a preliminary treatment of the isotope record at ODP Site 1051, Blake Nose, *in* Kroon, D., et al., eds., Western North Atlantic Palaeogene and Cretaceous Palaeoceanography: Geological Society [London] Special Publication 183, p. 293–306.

Dickens, G.R., O'Neil, J.R., Rea, D.K., and Owen, R.M., 1995, Dissociation of oceanic methane hydrate as a cause of the carbon isotope excursion at the end of the Paleocene: Paleoceanography, v. 10, p. 965–971.

Eldholm, O., and Thomas, E., 1993, Environmental impact of volcanic margin formation: Earth and Planetary Science Letters, v. 117, p. 319–329.

Estravís, C., and Russell, D.E., 1989, Découverte d'un nouveau Diacodexis (Artiodactyla, Mammalia) dans l'Eocène inférieur de Silveirinha, Portugal: Palaeovertebrata, v. 19, p. 29–44.

Fricke, H.C., Clyde, W.C., O'Neil, J.R., and Gingerich, P.D., 1998, Evidence for rapid climate change in North America during the latest Paleocene thermal maximum: oxygen isotope compositions of biogenic phosphate from the Bighorn Basin (Wyoming): Earth and Planetary Science Letters, v. 160, p. 193–208.

Giardina, C.P., and Ryan, M.G., 2000, Evidence that decomposition rates of organic carbon in mineral soil do not vary with temperature: Nature, v. 404, p. 858–861.

Gingerich, P.D., 1986, Early Eocene *Cantius torresi*—oldest primate of modern aspect from North America: Nature, v. 320, p. 319–321.

Gingerich, P.D., 1989, New earliest Wasatchian mammalian fauna from the Eocene of northwestern Wyoming: Composition and diversity in a rarely sampled high–flood plain assemblage: University of Michigan Papers on Paleontology, v. 28, p. 1–97.

Gingerich, P.D., 2000, Paleocene–Eocene boundary and continental vertebrate faunas of Europe and North America: GFF, v. 122, p. 57–59.

Godinot, M., 1996, Le renouvellement des faunas de mammifères en Europe et en Amérique du Nord autour de la limite Paléocène-Eocène: Strata, v. 8, p. 18–20.

Granger, W., 1914, On the names of lower Eocene faunal horizons of Wyoming and New Mexico: Bulletin of the American Museum of Natural History, v. 33, p. 201–207.

Kelly, D.C., Bralower, T.J., Zachos, J.C., Premoli Silva, I., and Thomas, E., 1996, Rapid diversification of planktonic foraminifera in the tropical Pacific (ODP Site 865) during the late Paleocene thermal maximum: Geology, v. 24, p. 423–426.

Kelly, D.C., Bralower, T.J., and Zachos, J.C., 1998, Evolutionary consequences of the latest Paleocene thermal maximum for tropical planktonic foraminifera: Palaeogeography, Palaeoclimatology, Palaeoecology, v. 141, p. 139–161.

Kennett, J.P., and Stott, L.D., 1991, Abrupt deep sea warming, palaeoceanographic changes and benthic extinctions at the end of the Palaeocene: Nature, v. 353, p. 225–229.

Koch, P.L., Zachos, J.C., and Gingerich, P.D., 1992, Correlation between isotope records in marine and continental carbon reservoirs near the Palaeocene/Eocene boundary: Nature, v. 358, p. 319–322.

Koch, P.L., Zachos, J.C., and Dettman, D.L., 1995, Stable isotope stratigraphy and paleoclimatology of the Paleogene Bighorn Basin (Wyoming, USA): Palaeogeography, Palaeoclimatology, Palaeoecology, v. 115, p. 61–90.

Krause, D.W., and Maas, M.C., 1990, The biogeographic origins of late Paleocene–early Eocene mammalian immigrants to the western interior of North America, *in* Bown, T.M., and Rose, K.D., eds., Dawn of the age of mammals in the northern part of the Rocky Mountain interior, North America: Boulder, Colorado, Geological Society of America Special Paper 243, p. 71–105.

Kvenvolden, K.A., 1988, Methane hydrate: a major reservoir of carbon in the shallow geosphere?: Chemical Geology, v. 71, p. 41–51.

Norris, R.D., and Röhl, U., 1999, Carbon cycling and chronology of climate warming during the Palaeocene/Eocene transition: Nature, v. 401, p. 775–778.

Pearson, P.N., and Palmer, M.R., 2000, Atmospheric carbon dioxide concentrations over the past 60 million years: Nature, v. 406, p. 695–699.

Peters, R.B., and Sloan, L.C., 2000, High greenhouse gas concentrations and polar stratospheric clouds: A possible solution to high-latitude migration at the Latest Paleocene Thermal Maximum: Geology, v. 28, p. 979–982.

Quade, J., and Cerling, T.E., 1995, Expansion of C4 grasses in the late Miocene of Northern Pakistan: evidence from stable isotopes in paleosols: Palaeogeography, Palaeoclimatology, Palaeoecology, v. 115, p. 91–116.

Quade, J., Cerling, T.E., and Bowman, J.R., 1989, Systematic variations in the carbon and oxygen isotopic composition of pedogenic carbonate along elevation transects in the southern Great Basin, United States: Geological Society of America Bulletin, v. 101, p. 464–475.

Rea, D.K., Zachos, J.C., Owen, R.M., and Gingerich, P.D., 1995, Global change at the Paleocene–Eocene boundary: climatic and evolutionary consequences of tectonic events: Palaeogeography, Palaeoclimatology, Palaeoecology, v. 79, p. 117–128.

Röhl, U., Bralower, T.J., Norris, R.D., and Wefer, G., 2000, New chronology for the late Paleocene thermal maximum and its environmental implications: Geology, v. 28, p. 927–930.

Rose, K.D., 1981, The Clarkforkian land-mammal age and mammalian faunal composition across the Paleocene–Eocene boundary: University of Michigan Papers on Paleontology, v. 26, p. 1–197.

Royer, D.L., Wing, S.L., Beerling, D.J., Jolley, D.W., Koch, P.L., Hickey, L.J., and Berner, R.A., 2001, Paleobotanical evidence for near present-day levels of atmospheric CO_2 during part of the tertiary: Science, v. 292, p. 2310–2313.

Sloan, L.C., and Thomas, E., 1998, Global climate of the late Paleocene epoch: Modeling the circumstances associated with a climatic "event," *in* Aubry, M.-P., et al., eds., Late Paleocene–early Eocene climatic and biotic events in the marine and terrestrial records: New York, Columbia University Press, p. 138–157.

Thomasson, J.R., Nelson, M.E., and Zakrzewski, R.J., 1986, A fossil grass (Gramineae: Chloridoideae) from the Miocene with Kranz anatomy: Science, v. 233, p. 876–878.

Wing, S.L., Bao, H., and Koch, P.L., 2000, An early Eocene cool period? Evidence for continental cooling during the warmest part of the Cenozoic, *in* Huber, B.T., et al., eds., Warm climates in Earth history: Cambridge, UK, Cambridge University Press, p. 197–237.

Woodburne, M.O., and Swisher, C.C., 1995, Land mammal high-resolution geochronology, intercontinental overland dispersals, sea level, climate, and vicariance, *in* Berggren, W.A., et al., eds., Geochronology, time scales, and global stratigraphic correlations: A unified framework for an historical geology: SEPM (Society for Sedimentary Geology) Special Publication 54, p. 335–364.

Zachos, J.C., and Dickens, G.R., 2000, An assessment of the biogeochemical feedback response to the climatic and chemical perturbations of the LPTM: GFF, v. 122, p. 188–189.

Manuscript Accepted by the Society August 13, 2002

Printed in the USA

Geological Society of America
Special Paper 369
2003

Excess barite accumulation during the Paleocene-Eocene Thermal Maximum: Massive input of dissolved barium from seafloor gas hydrate reservoirs

Gerald R. Dickens
Thomas Fewless
Department of Earth Science, Rice University, Houston, Texas 77005, USA
Ellen Thomas
Department of Earth and Environmental Sciences, Wesleyan University, Middletown, Connecticut 06459-0139, USA
Timothy J. Bralower*
Department of Geological Sciences, University of North Carolina, Chapel Hill, North Carolina 27599-3315, USA

ABSTRACT

A –2‰ to –3‰ excursion characterizes carbon isotope records across the Paleocene-Eocene Thermal Maximum (PETM), probably signifying transfer of ~2000 gigatons of CH_4 carbon from seafloor gas hydrate reservoirs to the ocean and atmosphere. During at least 30 k.y. of this carbon cycle perturbation, euhedral barite apparently accumulated in deep-sea sediment faster than before or after. This excess burial of "biogenic barite" has been argued to reflect elevated global primary productivity in surface waters, an interpretation in disagreement with microfossil assemblages. In this paper, we develop a numerical model for the marine Ba cycle to explore possible mechanisms for widespread barite accumulation. This model immediately highlights a fundamental problem with the productivity explanation: Ba has a short residence time in the ocean (~8000 yr) so that excess Ba output on a global scale over periods >1000 yr signifies excess Ba input. However, a greater Ba input from conventional sources, hydrothermal waters and rivers, seems untenable. Assuming that available Ba records represent increased global output, we suggest an unconventional explanation for widespread barite accumulation. Prior to the PETM, gas hydrate reservoirs stored enormous masses of CH_4 and dissolved Ba^{2+}. During the PETM, substantial quantities of gas hydrate converted to free CH_4 gas, increasing pore pressures, and releasing significant amounts of CH_4 and dissolved Ba^{2+} to intermediate waters of the ocean. Consequently, dissolved Ba^{2+} concentrations in the deep ocean rose, a smaller fraction of sinking barite particles dissolved, and "biogenic barite" accumulation increased. The model satisfactorily explains available records but forces us to link components of the marine CH_4 and Ba cycles, an endeavor with few constraints.

*Also at: Department of Geosciences, Pennsylvania State University, University Park, Pennsylvania 16802, USA.

Dickens, G.R., Fewless, T., Thomas, E., and Bralower, T.J., 2003, Excess barite accumulation during the Paleocene-Eocene Thermal Maximum: Massive input of dissolved barium from seafloor gas hydrate reservoirs, *in* Wing, S.L., Gingerich, P.D., Schmitz, B., and Thomas, E., eds., Causes and Consequences of Globally Warm Climates in the Early Paleogene: Boulder, Colorado, Geological Society of America Special Paper 369, p. 11–23.

INTRODUCTION

Prodigious amounts of ^{12}C-rich carbon suddenly entered the ocean and atmosphere during the Paleocene-Eocene Thermal Maximum (PETM) as evidenced by an abrupt –2‰ to –3‰ $\delta^{13}C$ excursion in global carbon isotope records (Fig. 1A) and intense carbonate dissolution in deep-sea sediments (e.g., Kennett and Stott, 1991; Bralower et al., 1995, 1997; Thomas and Shackleton, 1996; Katz et al., 1999). Thermal dissociation of oceanic gas hydrates followed by release and oxidation of ~2000 gigatons (Gt = 10^{15} g) of CH_4 carbon provides the most satisfactory explanation for this carbon cycle perturbation (Dickens et al., 1995; Kaiho et al., 1996; Katz et al., 1999; Dickens, 2001a). An especially important issue is how the excess carbon exited the ocean and atmosphere after initial injection, however it was released (Dickens, 1999). High-resolution carbon isotope records display a relatively fast logarithmic increase in $\delta^{13}C$ following the prominent negative excursion (Fig. 1A), suggesting a period of greater net carbon output to the lithosphere (Dickens, 1999, 2001b; Norris and Röhl, 1999). This carbon removal phase lasted between 150 and 220 k.y. (Katz et al., 1999; Norris and Röhl, 1999; Röhl et al., 2000) but its cause remains uncertain. Several plausible processes exist, including expansion of the terrestrial biosphere, enhanced burial of marine organic carbon in open ocean or marginal basins, and recharge of seafloor gas reservoirs (cf. Kaiho et al., 1996; Bains et al., 2000; Beerling, 2000; Zachos and Dickens, 2000; Dickens, 2000, 2001a, b; Speijer and Wagner, 2002).

One potential mechanism to enhance carbon output after onset of the PETM is through greater marine biological productivity, with an associated increase in export production. A large carbon input to the ocean or atmosphere, through CH_4 or otherwise, should elevate atmospheric pCO_2 (Walker and Kasting, 1992; Dickens, 2000) and continental weathering (Walker et al., 1981; Berner et al., 1983). Greater nutrient delivery to the oceans, a boon of weathering, might then increase primary production and organic carbon fluxes from surface waters to the seafloor (Zachos and Dickens, 2000).

As a test for a weathering induced "PETM marine productivity feedback," Bains et al. (2000) have presented an intriguing set of geochemical records. These authors have shown a strong antithetical relationship between the $\delta^{13}C$ of marine carbonate and the accumulation of euhedral barite across the first 30–40 k.y. of the PETM at two widely separated, open-ocean locations, Ocean Drilling Program (ODP) Site 690 in the Southern Ocean (Fig. 1A) and ODP Site 1051 in the North Atlantic. Bains et al. (2000) state that additional deep-water sections have this unusual geochemical relationship across the PETM, and Schmitz et al. (1997) demonstrated such a link at the Ben Gurion deep-water section in Egypt. Euhedral barite in pelagic sediment relates to sinking marine organic matter (e.g., Paytan et al., 1996). Thus, one interpretation for the observed relationship is that enhanced weathering increased primary productivity and the burial flux of organic carbon for ~30–40 k.y. after massive carbon input (Bains et al., 2000). Some records appear to support this overall explanation. Specifically, across the PETM, kaolinite accumulation increased around many continents (Robert and Kennett, 1994; Kaiho et al., 1996; Gawenda et al., 1999; Harrington and Kemp, 2001; Schmitz et al., 2001), osmium isotope ratios ($^{187}Os/^{188}Os$) rose in seawater (Ravizza et al., 2001), and heterotrophic dinoflagellate cysts abounded in shallow marine waters (Crouch et al., 2001). However, pelagic

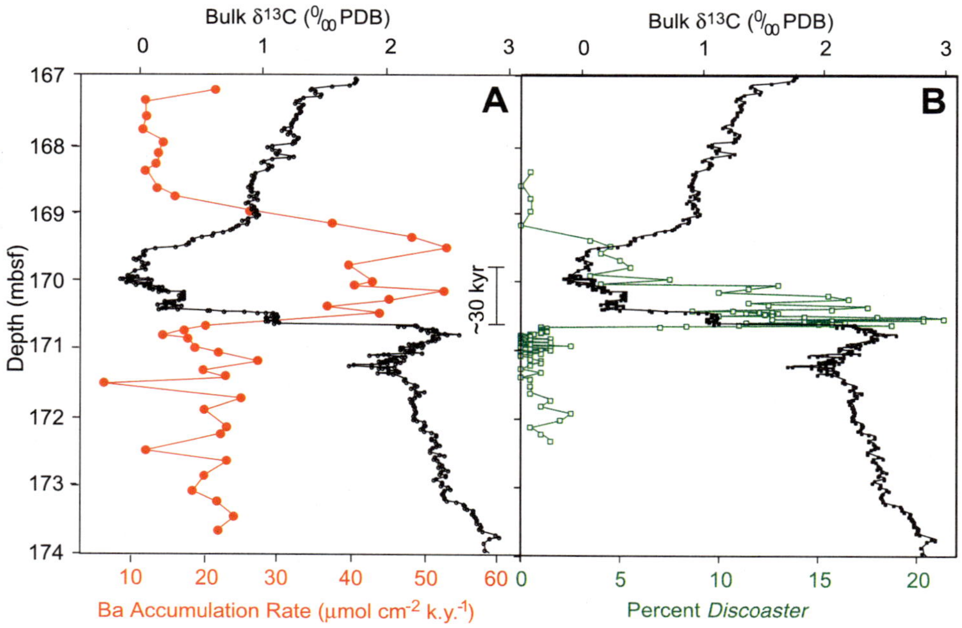

Figure 1. Three environmental records across the Paleocene-Eocene Thermal Maximum at Ocean Drilling Program (ODP) Site 690, Maud Rise, South Atlantic Ocean. A: Carbon isotopes of bulk carbonate (black) and estimated sedimentary Ba accumulation rate (Bains et al., 2000). B: Carbon isotopes of bulk carbonate (black) and discoaster abundance (Bralower et al., 2002).

microfossil assemblages, including those at sites with elevated barite accumulation, often indicate decreased surface water productivity during the PETM (e.g., Kelly et al., 1996; Thomas, 1998, this volume; Thomas et al., 1999; Bolle et al., 2000; Bralower, 2002). For example, at ODP Site 690, the abundance of discoasters, acknowledged oligotrophic taxa (e.g., Haq and Lohmann, 1976; Aubry, 1998; Boersma et al., 1998), increases markedly over the interval of elevated barite accumulation at the start of the PETM (Fig. 1B).

In this paper, we develop a numerical model for the marine barium cycle to elucidate potential causes for widespread increases in euhedral barite accumulation. Our primary goal is to present at least one plausible simulation for the PETM whereby biogenic barite accumulation increases globally without a rise in marine productivity. We quickly conclude that the short residence time of Ba in the ocean precludes elevated productivity as a primary explanation for prolonged increases in barite accumulation. However, a sudden increase in the barite saturation state of the deep ocean, caused by massive release of dissolved Ba^{2+} from marine gas hydrate deposits, provides a satisfactory explanation.

THE BASIC MARINE BARIUM CYCLE

Any explanation for a significant widespread increase in Ba accumulation during the PETM necessitates an understanding of the marine barium cycle. At a basic level, this cycle (Fig. 2)

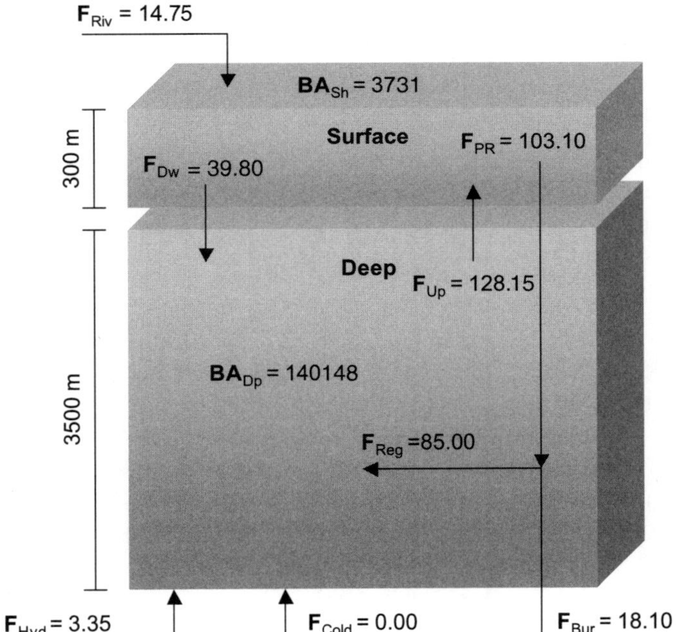

Figure 2. Basic steady-state box model for the present-day marine barium cycle (adapted from Wolgemuth and Broecker, 1970; Paytan and Kastner, 1996). A flux of Ba from cold seeps (F_{Cold}) has been included in the model but set at zero for consistency with previous literature. Masses are in Gmol; fluxes are in Gmol/yr.

contains two ocean reservoirs—surface water and deep water—and a series of external and internal mass fluxes (Wolgemuth and Broecker, 1970; Paytan and Kastner, 1996). Rivers add relatively large quantities of dissolved Ba^{2+} to surface water. Barite then precipitates in microenvironments within sinking and decomposing organic matter, leading to a particulate rain of euhedral barite crystals from surface water to deep water (Chow and Goldberg, 1960; Dehairs et al., 1980, 1992; Bishop, 1988; Stroobants et al., 1991). Although some of this "biogenic barite" accumulates into the sediment (e.g., Schmitz, 1987; Dymond et al., 1992; Gingele and Dahmke, 1994; Paytan et al., 1996), a large fraction dissolves on the seafloor or within shallow sediment (e.g., Chow and Goldberg, 1960; Francois et al., 1995; McManus et al., 1998, 1999; Klump et al., 2001). Deep water thus receives substantial amounts of Ba^{2+} through regeneration of barite (Wolgemuth and Broecker, 1970; Payton and Kastner, 1996). Submarine hydrothermal systems (e.g., Elderfield and Schultz, 1996; Monnin et al., 2001) and cold seeps (Torres et al., 1996a) introduce additional dissolved Ba^{2+} to deep water. As a consequence of this overall cycling, Ba^{2+} concentrations in deep water greatly exceed those in surface water (e.g., Paytan and Kastner, 1996; Monnin et al., 1999). Advection of water by thermohaline circulation returns Ba^{2+} from deep water to surface water, closing the mass balance loop (Wolgemuth and Broecker, 1970; Paytan and Kastner, 1996). This basic view of the marine Ba cycle excludes metal oxyhydroxide accumulation, an additional and potentially important Ba sink in deep-sea sediments (Schroeder et al., 1997).

Temporal variations in one or more of the Ba fluxes, external or internal, can change Ba concentrations in surface and deep reservoirs, as well as other Ba fluxes, including sedimentary barite accumulation. These changes can be described through a series of flux equations and assessed by numerical methods as for other geochemical cycles (e.g., Berner et al., 1983; Kump and Arthur, 1999; Dickens, 2001a, 2000b). Paytan and Kastner (1996) presented the most recent steady-state box model for the modern marine Ba cycle. After adding a Ba^{2+} flux from cold seeps, their model (Fig. 2) provides a framework for developing an appropriate set of equations to evaluate Ba fluxes in the time domain.

A NUMERICAL MODEL FOR THE MARINE BARIUM CYCLE

Given the basic barium cycle (Fig. 2), the equation to describe changes in surface water Ba abundance (BA_{sh}) is:

$$dBA_{Sh}/dt = F_{Riv} + F_{Up} - F_{Dw} - F_{PR} \quad (1)$$

where F_{Riv}, F_{Up}, F_{Dw}, and F_{PR}, are Ba fluxes associated with rivers, upwelling, downwelling and particulate rain, respectively, and t is time. Likewise, the equation to describe changes in deep water Ba abundance (BA_{Dp}) is:

$$dBA_{Dp}/dt = F_{Reg} + F_{Hyd} + F_{Cold} + F_{Dw} - F_{Up} \quad (2)$$

where F_{Reg}, F_{Hyd} and F_{Cold} are Ba fluxes associated with benthic regeneration, seafloor hydrothermal vents and cold seeps, respectively, and other terms are the same as above. As in many box models for global biogeochemical cycles, all of the flux terms have been normalized for area, in this case the ocean or 2.40×10^{14} m^2 (Appendix 1).

Processes outside the surface and deep waters dictate the magnitude of the three external inputs, F_{Riv}, F_{Hyd}, and F_{Cold}. By contrast, processes within the ocean control the magnitude of the internal fluxes, F_{Dw}, F_{Up}, F_{PR}, $F_{Reg,}$ and the single external output, F_{Bur}.

Downwelling and upwelling fluxes depend on dissolved Ba^{2+} concentrations in surface ($[Ba]_{Sh}$) and deep water ($[Ba]_{Dp}$), and the advective flow between these water masses (Fig. 2). The appropriate expressions for these fluxes are (Paytan and Kastner, 1996):

$$F_{Dw} = [Ba]_{Sh}\,C \tag{3}$$

and

$$F_{Up} = [Ba]_{Dp}\,C \tag{4}$$

where C is the mass of water that exchanges between surface and deep reservoirs over time.

The sinking of organic matter incorporates Ba from the surface water column (Dehairs et al., 1980, 1992; Bishop, 1988; Stroobants et al., 1991; McManus et al., 1999). Consequently, the particulate rain should be related to the downward supply of solid organic carbon (P_{Org}) and dissolved Ba^{2+} concentrations in shallow to intermediate water (Dymond et al., 1992). A straightforward expression for this process is:

$$F_{PR} = [Ba]_{Sh}\,D\,P_{Org} \tag{5}$$

where D is a coefficient linking Ba concentrations in surface water to Ba contents in organic matter before it sinks into deep water and decomposes. For simplicity, we have set D as a constant (Appendix 1). Clearly, this assumption masks the true complexity of Ba uptake because the amount of Ba in sinking organic carbon varies (e.g., Francois et al., 1995; Dehairs et al., 2000). Although we could incorporate more complicated expressions for D into our model (e.g., Dymond et al., 1992), as we shall see, the actual relationship between dissolved Ba^{2+} in shallow water and solid Ba in organic matter is somewhat irrelevant when considering global Ba accumulation over significant time.

The key parameter of interest to this paper—the burial flux of Ba (F_{Bur})—is the fraction of the particulate rain that escapes dissolution at or near the seafloor. This flux can be coupled to aforementioned fluxes as follows:

$$F_{Bur} = F_{PR} - F_{Reg} \tag{6}$$

or

$$F_{Bur} = f\,F_{PR} \tag{7}$$

where f represents the fraction (0–1) of the particulate barite rain incorporated into the sediment. A critical observation is that this fraction depends on the barite saturation state of bottom water (Monnin et al., 1999; Klump et al., 2001) so that "bottom water Ba concentration has a significant influence on barite preservation" (Francois et al., 1995, p. 301). A simple expression to describe this effect is:

$$f = k\,[Ba]_{Dp}/[Ba]_{Sat} \tag{8}$$

where $[Ba]_{Sat}$ is the dissolved Ba^{2+} concentration at barite saturation, and k is a "time factor" to increase dissolution because barite dissolves on and in the upper few centimeters beneath the seafloor (e.g., Paytan and Kastner, 1996; McManus et al., 1998). As such, at low $[Ba]_{Dp}$, a greater fraction of barite dissolves, decreasing F_{Bur}, and increasing $F_{Re.g.}$. This expression, especially with constant k, oversimplifies barite preservation and dissolution because it omits other processes that can affect seafloor barite accumulation, such as changes in sedimentation rate (Dymond et al., 1992; Francois et al., 1995; McManus et al., 1999). However, the expression satisfactorily explains observations in modern deep waters, such as the greater barite accumulation beneath Ba-rich deep waters (Francois et al., 1995; Monnin et al., 1999).

All the internal Ba fluxes depend on dissolved Ba^{2+} concentrations. These, of course, are dictated by the total abundances of Ba in surface and deep waters:

$$[Ba]_{Sh} = BA_{Sh}/M_{Sh} \tag{9}$$

and

$$[Ba]_{Dp} = BA_{Dp}/M_{Dp} \tag{10}$$

where M_{Sh} and M_{Dp} are the masses of water in the surface and deep ocean reservoirs, respectively. For simplicity, we assume that the surface and deep water masses are rectangular prisms with similar horizontal areas (Wolgemuth and Broecker, 1970). Thus:

$$[Ba]_{Sh} = M_{Sh}/Az_{Sh}\rho_{sw} \tag{11}$$

and

$$[Ba]_{Dp} = M_{Dp}/Az_{Dp}\rho_{sw} \tag{12}$$

where z_{Sh} and z_{Dp} are the depths of the surface and deep water reservoirs, respectively, ρ_{sw} is the density of seawater.

MODEL SIMULATIONS AND THE PRODUCTIVITY PROBLEM

Initial conditions

In their steady-state box model of the modern marine Ba cycle, Paytan and Kastner (1996) have quantified all Ba masses and fluxes except discharge from cold seeps (F_{Cold}). Their fluxes (Appendix 1) are similar to those in previous models of the marine Ba cycle (Wolgemuth and Broecker, 1970; Dehairs et al., 1980), and we assume they are reasonably accurate. Implicit in this assumption is that modern Ba fluxes from cold seeps are trivial, a potentially incorrect view (McManus et al., 1999). Three of the parameters introduced above (D, *f,* $[Ba]_{Sat}$) are not expressly included in the box model by Paytan and Kastner (1996) but are necessary for coupling the Ba fluxes in numerical simulations. Assuming steady-state conditions, these parameters can be derived easily from given masses and fluxes (Appendix 1). The downward supply of solid organic carbon (P_{Org}) is initially set at 7500 Gmol/yr. This is an average value taken from the range of P_{Org} in current models of the Holocene carbon cycle (see discussion by Dickens, 2001b).

Two time "constants" can be calculated with the given steady-state fluxes and concentrations (Appendix 1). The total mass of water ($M_{Sh} + M_{Dp} = 1.31 \times 10^{21}$ kg) exchanges between the shallow and deep ocean in ~1200 yr. The residence time of Ba in the ocean is ~8000 yr.

Increased productivity

The raising of global primary productivity and the downward flux of organic matter by 15%, 50% and 100% results in predictable, although perhaps nonintuitive, changes in the global accumulation of biogenic Ba (Fig. 3). At first, in all simulations, F_{Bur} increases. However, F_{Bur} begins to decrease immediately afterwards. This result stems from the rapid cycling of water and the limited amount of Ba in the ocean. Removal of Ba via an enhanced particulate rain of barite quickly depletes the supply of Ba in the small, surface water reservoir (BA_{Sh}) and reduces the downward transport of dissolved Ba^{2+} (F_{Dw}). Although upwelling (F_{Up}) replenishes the surface water Ba reservoir, this comes at the expense of Ba in deep water (BA_{Dp}), which then decreases barite preservation.

Our model of the Ba cycle (Fig. 2), and any ensuing numerical simulations, contains numerous potential errors, particularly regarding variations in the internal cycling of Ba or adjustments for the PETM. For example, if the particulate rain of barite (F_{PR}) is made largely independent of surface Ba concentrations ($[Ba]_{Sh}$), as suggested by Francois et al. (1995), the duration and shape of significantly elevated Ba accumulation will change. Interestingly, however, appropriate changes in key parameters generally decrease F_{Bur}. For example, the 4–6 °C increase in deep ocean temperature and the pronounced dissolution of deep-sea carbonate during the PETM (e.g., Thomas and Shackleton, 1996; Katz et al., 1999) would raise barite solubility, either directly or by lowering sedimentation rates (Francois et al., 1995; Monnin, 1999). Likewise, weakened ocean-atmosphere circulation (i.e., a lower C, equations 3, 4), which may have occurred during the PETM (e.g., Kennett and Stott, 1991; Hovan and Rea, 1992; Thomas and Shackleton, 1996), would reduce F_{Up}, the critical short-term supply of Ba to surface water. This, in turn, would decrease the amount of dissolved Ba^{2+} available to organic matter.

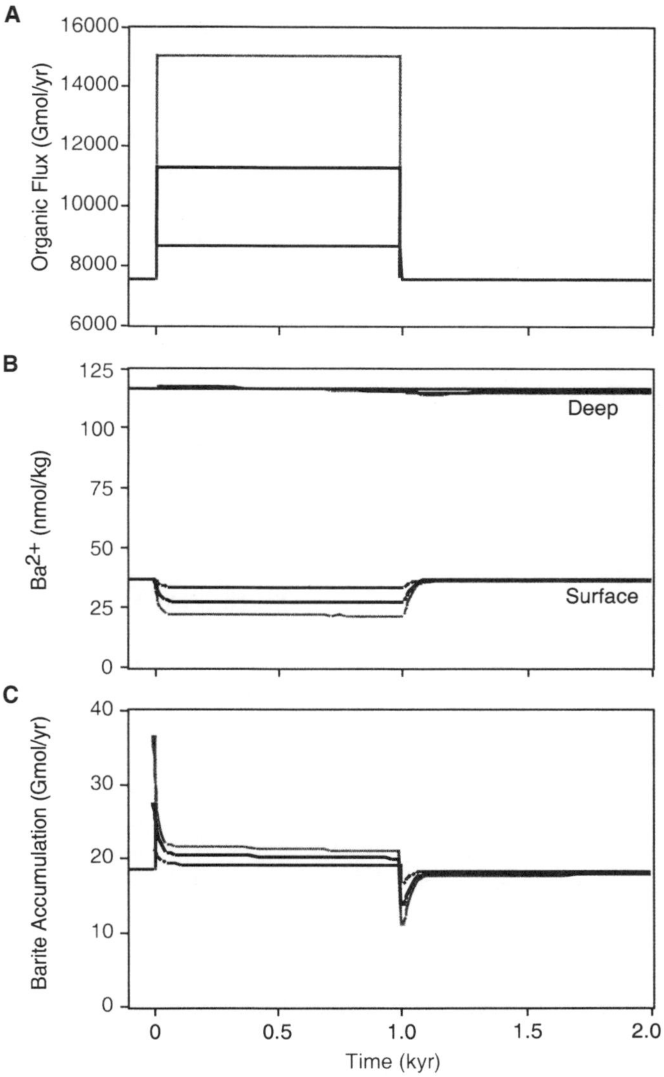

Figure 3. Numerical simulations of global deep-sea barium accumulation given 15%, 50% and 100% increases in primary productivity. A: Productivity input. B: Dissolved Ba^{2+} concentrations in surface and deep ocean reservoirs. C: Ba accumulation.

Importantly, and as alluded to by McManus et al. (1999, p. 60), irrespective of how internal Ba fluxes are expressed in detail (equations 3–8), a significant global increase in Ba outputs (>10%) over appreciable time (>1000 yr) necessitates elevated Ba inputs because Ba has a relatively short residence time. Consider an extreme simulation where all parameters internal to the marine Ba cycle are adjusted to maximize Ba outputs. In this case, a doubling of global Ba outputs will deplete all Ba in the ocean within 8000 yr unless Ba inputs increase ($[BA_{Sh} + BA_{Dp}]/F_{Bur} = 7950$ yr). A similarly short residence time likely existed in the Paleocene, because barite is close to saturation in the modern ocean (Monnin et al., 1999); that is, the ocean cannot store significantly more Ba than at present without invoking wholesale changes in SO_4^{2-} caused by other, unidentified processes. If truly global, a significant increase in biogenic barite accumulation on time scales greater than several thousand years should not be viewed as representing elevated primary productivity but rather as an increase in Ba inputs. This crucial concept has been stressed previously for other "biolimited" species with short marine residence times (e.g., phosphorous and silica; Delaney and Filippelli, 1994; Dickens and Owen, 1999).

Conventional external fluxes

Given conventional models of the marine Ba cycle (Fig. 2), a significant increase in global biogenic barite accumulation implies an increase in Ba from hydrothermal effluents or riverine discharge. But neither of these inputs is a satisfactory source of Ba for 30–40 k.y. of the Paleocene.

Seafloor volcanism peaked in the late Paleocene and early Eocene with emplacement of flood basalts in the North Atlantic Igneous Province (e.g., Eldholm and Thomas, 1993; Heister et al., 2001). This intensified volcanism, although partly subaerial, potentially raised global hydrothermal fluxes to the ocean (Olivarez and Owen, 1989). However, the volcanism, and by inference any excess hydrothermal activity, lasted for several millions of years, preceding and postdating the PETM (Heister et al., 2001). There is no evidence for a special, short-term burst of seafloor hydrothermal activity precisely coinciding with the onset of the PETM.

Some shallow marine environments received a greater discharge of siliciclastic material, especially kaolinite, across the PETM (Robert and Kennett, 1994; Kaiho et al., 1996; Gawenda et al., 1999; Harrington and Kemp, 2001; Schmitz et al., 2001). Several authors (e.g., Robert and Kennett, 1994; Gawenda et al., 1999) have suggested this high kaolinite flux reflects higher humidity and intensified chemical weathering, which might signify globally increased riverine dissolved loads during the PETM. However, as emphasized by Thiry (2000), the formation of sufficient kaolinite by chemical weathering would take several million years. Consequently, the short-term kaolinite spike is much better explained by mechanical erosion (Thiry, 2000; Schmitz et al., 2001). A similar argument may apply to rapid changes in riverine Ba fluxes. Barite and common minerals hosting significant Ba (e.g., feldspar) have very slow dissolution rates (e.g., Dove and Czank, 1995; White et al., 2001), which may preclude a rapid (30–40 k.y.) weathering-induced discharge of dissolved Ba^{2+} from continents. However, we cannot rigorously exclude the possibility of a short-term fluvial input of Ba-rich clays that released dissolved Ba^{2+} upon contact with seawater.

A significant global increase in deep-sea barite accumulation >30–40 k.y. probably implies a sudden increase in a Ba flux outside of current models for the marine Ba cycle. Interestingly, such a flux would have existed precisely at the onset of the PETM if release of seafloor CH_4 caused the dramatic perturbations in the global carbon cycle.

METHANE AND BARIUM CYCLING IN MARINE GAS HYDRATE DEPOSITS

Methane cycling

Gas hydrate reservoirs are dynamic systems (Fig. 4) where carbon cycles between dissolved gas, gas hydrate and free gas over time (e.g., Xu and Ruppel, 1999; Davie and Buffett, 2001; Dickens, 2001a; Kastner, 2001). Within SO_4^{2-}-depleted pore waters, several microbially mediated reactions generate new CH_4 from sedimentary organic matter. This dissolved gas migrates laterally and vertically. Eventually, at appropriate pressure and temperature conditions, and sufficient gas concentrations, gas hydrate precipitates in pore space. Sedimentation over time slowly buries these solid clathrates of CH_4 and water to higher temperatures because of the geothermal gradient. At some pressure and temperature, gas hydrate is no longer stable and dissociates to gas-saturated water and free CH_4 gas bubbles. Although much of the CH_4 released at depth recycles through the above processes, some CH_4 escapes to shallow sediments. In a few places, such as where faults intersect the seafloor, high advection rates lead to CH_4 venting and aerobic oxidation of methane by microbes in the water column (Paull et al., 1995; Suess et al., 1999; Valentine et al., 2001). In many places, however, the upward flux of CH_4 encounters SO_4^{2-} diffusing down from the seafloor. This drives anaerobic oxidation of methane by Archaea-Bacteria consortia (Hinrichs et al., 1999; Boetius et al., 2000) at a sulfate/methane interface (SMI) in the sediment column (Borowski et al., 1996, 1999; Fossing et al., 2000). As exemplified by studies on the Blake Ridge (Fig. 4), the SMI often occurs between 10 and 40 mbsf above gas hydrate reservoirs (e.g., Borowski et al., 1999).

Barium cycling

Most Ba in pelagic sediment occurs in aluminosilicate phases, Fe-Mn oxyhydroxides, or discrete microscopic euhedral crystals of barite (e.g., Dehairs et al., 1980; Dymond et al., 1992; Gingele and Dahmke, 1994; McManus et al., 1998). Although the Ba in aluminosilicates is essentially immobile during early diagenesis, the Ba in metal oxyhydroxides and barite is susceptible to fluctuations in pore water chemistry. In particular, barite will be affected by the SO_4^{2-} content of surrounding pore water (Von Breymann et al., 1990, 1992; Torres et al., 1996b). The solubility of barite is very low in deep marine waters (K_{sp} ~1.0 and 2.5 × 10 ^{10}; Church and Wolgemuth, 1972; Monnin, 1999). Thus, in sediment/pore water systems with even minor concentrations of dissolved SO_4^{2-}, most labile Ba exists in barite with only small amounts as dissolved Ba^{2+} (Blount, 1977; McManus et al., 1998). However, when pore waters become depleted in SO_4^{2-}, the solubility of barite in-

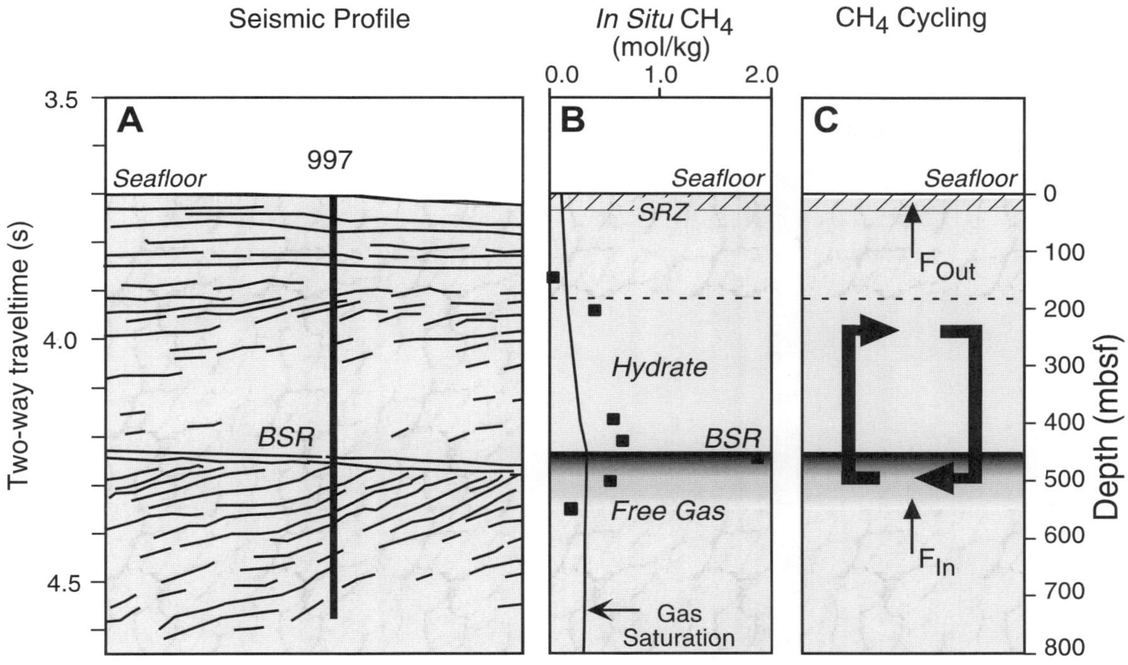

Figure 4. Methane cycling in a modern gas hydrate reservoir, Ocean Drilling Program (ODP) Site 997, Blake Ridge, southeast United States margin (adapted from Paull et al., 1996; Dickens et al., 1997; Dickens, 2001a). A: Schematic of seismic reflection profile showing site location and the bottom-simulating reflector (BSR), an acoustic interface between overlying gas hydrate and underlying free gas bubbles. B: Methane solubility curves and in situ gas concentrations at Site 997. C: Inferred CH_4 cycling at Site 997.

creases greatly and dissolved Ba^{2+} concentrations can increase by four orders of magnitude (Brumsack and Gieskes, 1983; Von Breymann et al., 1990, 1992).

The coupled cycling of carbon and sulfur in and above marine gas hydrate deposits profoundly affects Ba cycling in surrounding sediment and water (Figs. 5, 6). During sediment burial, particulate barite moves downward from SO_4^{2-} rich pore water into SO_4^{2-} depleted pore water where it slowly dissolves. In turn, dissolved Ba^{2+} slowly diffuses upward from SO_4^{2-}-depleted pore water into SO_4^{2-} rich pore water where it precipitates as barite. In most gas hydrate systems (e.g., the Blake Ridge, Figures 4, 5; Peruvian Margin, Figure 6), microbial processes consume SO_4^{2-} and CH_4 beneath the seafloor, and upward Ba^{2+} diffusion exceeds downward barite burial. The continual cycling of Ba therefore traps enormous amounts of dissolved Ba^{2+} beneath shallow sedimentary "barite fronts" (Von Breymann et al., 1990, 1992; Gingele and Dahmke, 1994; Torres et al., 1996b; Dickens, 2001c). When CH_4 charged fluids of gas hydrate reservoirs escape to the ocean, however, they carry significant amounts of dissolved Ba^{2+}, sometimes leading to precipitation of barite mounds and chimneys on the seafloor (e.g., Torres et al., 1996a; Fu and Aharon, 1998; Naehr et al., 2000).

A LOW-PRODUCTIVITY MODEL FOR EXCESS BARITE ACCUMULATION

A straightforward model for excess Ba accumulation across the PETM can be constructed by considering gas hydrate reservoirs as repositories of CH_4 *and* dissolved Ba^{2+} (Fig. 7). Prior to the PETM and similar to the present-day situation, the upper few hundred meters of sediments on continental slopes at appropriate pressures and temperatures contained enormous quantities of CH_4 as solid gas hydrate (e.g., Kvenvolden, 1988, 1999; Dickens, 2001d). These hydrates were greatly enriched in ^{12}C (e.g., Kvenvolden, 1995) and surrounded by pore waters with extremely high dissolved Ba^{2+} concentrations (e.g., Figure 6). For reasons that remain unclear, but which may have involved the passage of a critical ocean-atmosphere threshold condition and sinking of waters at a warmer source region, temperatures at intermediate water depths rose by 4–6 °C (Kennett and Stott, 1991; Bralower et al., 1995; Thomas and Shackleton, 1996; Katz et al., 1999). This warming propagated into the sediments, converting once solid gas hydrates into free gas bubbles (Dickens et al., 1995). Free gas formation increased pore pressures at depth, leading to sediment failure and the release of massive quantities of CH_4 (e.g., Kayen and Lee, 1991; Paull et al., 1996;

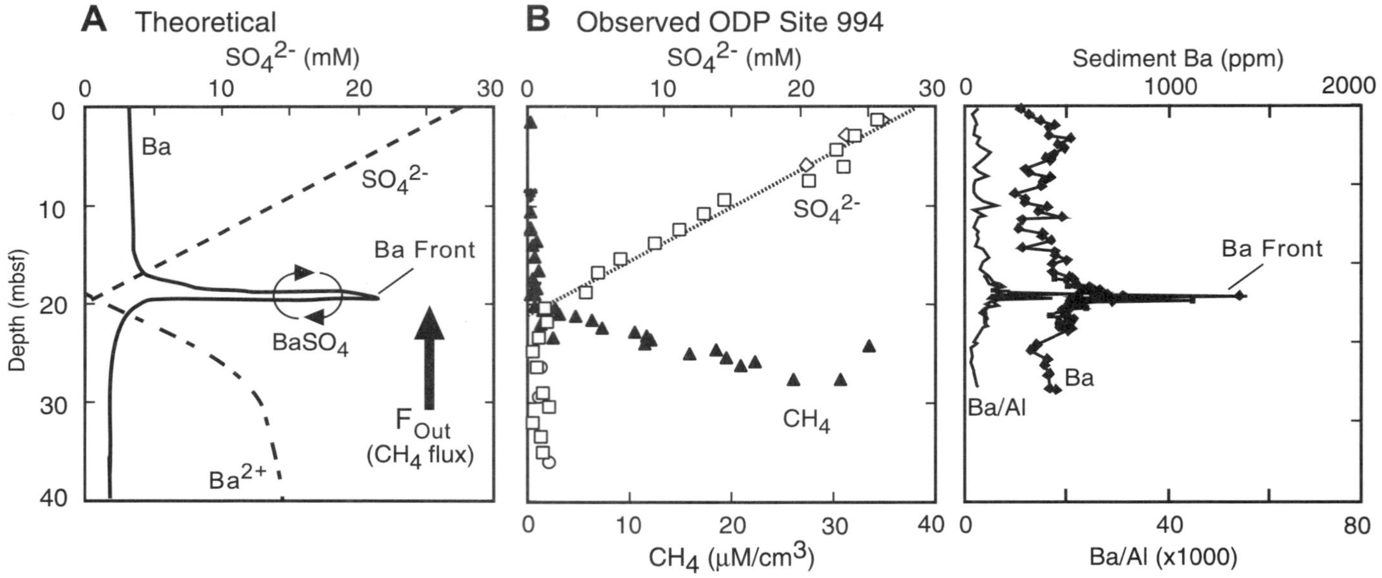

Figure 5. Theoretical and observed cycling of Ba above a gas hydrate reservoir. A: Schematic of expected Ba cycling (adapted from Torres et al., 1996b). B: Measured concentrations of headspace CH_4, dissolved SO_4^{2-} and sedimentary Ba at ODP Site 994, Blake Ridge (Dickens, 2001c).

Katz et al., 1999) *and* dissolved Ba^{2+} into the ocean. In the Paleocene, CH_4 and Ba release would have occurred on slopes between 900 and 2000 m water depth (Fig. 7) because all gas hydrate in sediment at these depths would dissociate with a 4–6 °C rise in bottom water temperature (Dickens et al., 1995; Dickens, 2001d). The gaseous CH_4 reacted to produce ^{12}C-enriched CO_2, either via OH radicals in the atmosphere or via aerobic oxidation in the water, and the Ba^{2+} would increase deep ocean Ba^{2+} concentrations provided that it did not precipitate locally upon contact with SO_4^{2-} rich waters. Consequently, discrete barite particles raining from shallow water would experience lower dissolution, and "biogenic" barite would accumulate faster, independent of productivity changes.

Barium release from gas hydrate reservoirs is an appealing mechanism to explain excess biogenic barite accumulation during the PETM because it is a direct and necessary expectation given release of CH_4 as an explanation for the prominent negative carbon isotope excursion. At issue is whether this mechanism is plausible; that is, can massive discharge of methane-laden pore waters supply the requisite flux of Ba to significantly change global dissolved Ba^{2+} concentrations?

Limited information exists to quantitatively couple the seafloor cycles of CH_4 and Ba. Sites 995 and 997 on the Blake Ridge (Fig. 4) remain the only scientific drill holes where CH_4 abundances have been determined in a gas hydrate reservoir at in situ conditions by multiple and direct methods (Dickens et al., 1997, 2000). At these sites, pore space CH_4 concentrations average ~0.55 mol/kg in sediment horizons with gas hydrate and free gas (Dickens et al., 1997, 2000). Sites 685 and 688 on the Peruvian Margin (Fig. 7) remain the only locations with gas hydrate where dissolved Ba^{2+} concentrations have been determined (cf. Kvenvolden and Kastner, 1990; Von Breymann et al., 1990). Here, dissolved Ba^{2+} concentrations average ~0.45 mmol/kg in intervals with gas hydrate and free gas (Von Breymann et al., 1990). Thus, with available data, we estimate a ~1200:1 molar ratio of CH_4 to Ba in marine gas hydrate deposits.

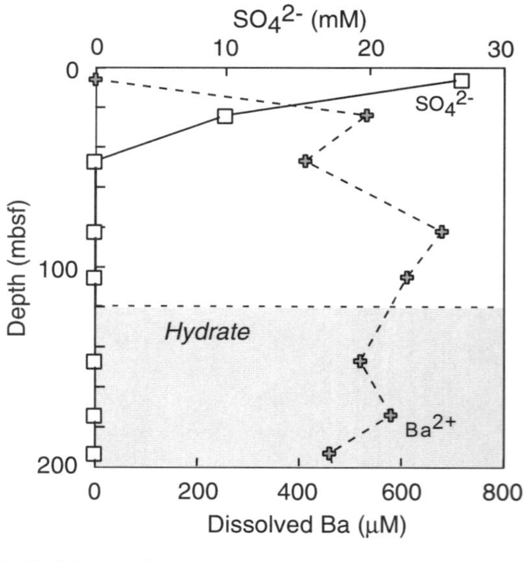

Figure 6. Solid and dissolved Ba concentrations at ODP Site 688, Peruvian Margin (Von Breymann et al., 1990). Gas hydrate, predominantly composed of CH_4, occurs beneath ~100 m (Kvenvolden and Kastner, 1990).

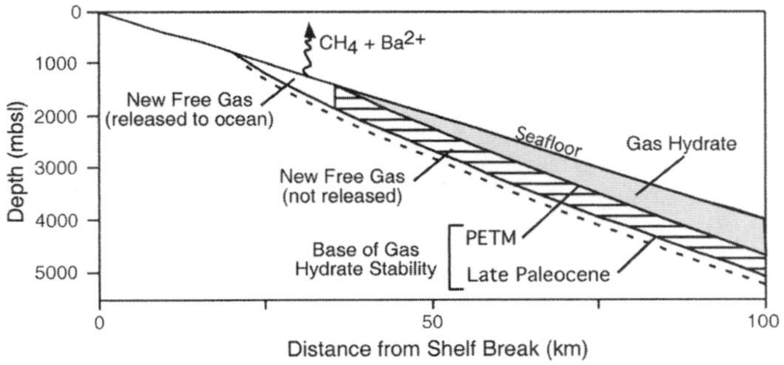

Figure 7. Schematic showing hypothesized release of CH_4 and Ba from the seafloor during the PETM.

Given release of ~2000 Gt (1.62×10^5 Tmol) of CH_4 carbon from the seafloor during the PETM (Dickens, 2001a, 2001b), ~19 Gt (135 Tmol) of Ba could have entered the deep ocean. Over 10–30 k.y., the apparent duration of massive carbon injection (Kennett and Stott, 1991; Katz et al., 1999; Norris and Röhl, 1999; Röhl et al., 2000), the new "cold seep" flux would raise global Ba fluxes from 14 Gmol/yr by an additional 4–14 Gmol/yr if all of the Ba stayed dissolved in seawater and did not precipitate locally as barite. Numerical simulations of global biogenic Ba accumulation (F_{Bur}) after injecting 135 Tmol of Ba into the deep ocean >10–30 k.y. (Fig. 8) differ significantly from those driven by elevated productivity (Fig. 3). As with enhanced productivity, F_{Bur} increases rapidly in all simulations. However, this elevated output of Ba continues over considerable time, mimicking the duration of CH_4 and Ba inputs. Importantly, because expulsion of ^{12}C-rich fluid drives the Ba input, the records of seawater $\delta^{13}C$ and sedimentary Ba accumulation (1) vary antithetically at the onset of the PETM (nominally 30–40 k.y.) even if productivity dropped significantly, and (2) become decoupled afterwards when $\delta^{13}C$ slowly increases and barite accumulation plummets. The second result occurs because the residence time of carbon in the ocean (~100 k.y.) is much longer than that of barium.

Anomalous barite accumulation across the PETM can be linked to massive CH_4 input through less direct scenarios (e.g., Bains et al., 2000) if the existing Ba records are not globally representative. However, changes in the loci of sedimentary barite deposition do not necessarily imply variations in surface water productivity. In particular, benthic foraminifera assemblages and other information strongly suggest that deep ocean dissolved O_2 contents dropped across the PETM (Kaiho et al., 1996; Bralower et al., 1997; Thomas, 1998; Katz et al., 1999). This O_2 deficiency, perhaps driven by CH_4 oxidation (Dickens, 2000), could have dissolved metal oxyhydroxides in the water column and on the seafloor (e.g., Johnson et al., 1992; Dickens and Owen, 1994). Thus, a potentially important Ba output could have diminished, a change that should have raised deep ocean dissolved Ba^{2+} concentrations and enhanced the preservation of biogenic barite in certain locations.

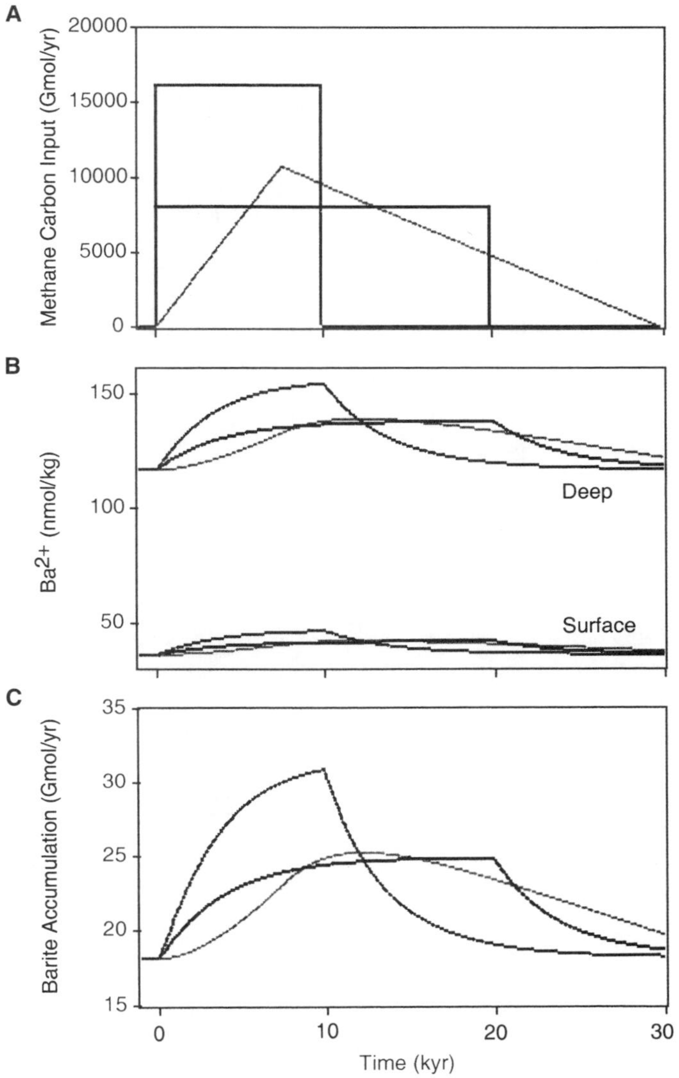

Figure 8. Numerical simulations of global deep-sea barium accumulation assuming three different injections of ~2000 Gt of CH_4 carbon and a 1200:1 molar CH_4:Ba ratio in pore space of gas hydrate reservoirs. A: Methane input. B: Dissolved Ba^{2+} concentrations in surface and deep ocean reservoirs. C: Ba accumulation.

CONCLUSIONS AND FUTURE RESEARCH

Schmitz et al. (1997) and Bains et al. (2000) have documented coeval, antithetical changes in carbonate $\delta^{13}C$ and Ba accumulation across the first 30–40 k.y. of the PETM in several sections deposited in deep water (Fig. 1). The prominent negative $\delta^{13}C$ excursion probably signifies massive release of ^{12}C-rich CH_4 from seafloor gas hydrate deposits to the ocean and its subsequent removal to the lithosphere (e.g., Dickens et al., 1995; Dickens, 2001a, 2001b). The excess accumulation of Ba, caused by an increase in the flux of discrete euhedral barite crystals, suggests a greater burial of biogenic Ba at these sites (Schmitz et al., 1997; Bains et al., 2000). Following previous work (e.g., Schmitz, 1987), Bains et al. (2000) have argued that this observation reflects a time of globally elevated marine primary productivity, presumably driven by higher atmospheric pCO_2 (via CH_4 oxidation), enhanced terrestrial weathering, and additional nutrient delivery to the ocean. However, this interpretation immediately raises two problems. First, it is inconsistent with other proxies for productivity, such as microfossil assemblages, which suggest oligotrophic conditions in the open ocean during the PETM. Second, and as emphasized in this paper, there is a serious mass balance problem. The ocean contains a relatively small amount of Ba with a short residence time. Consequently, a widespread increase in biogenic Ba burial over an extended time interval (>1000 yrs) requires an increase in the external supply of Ba to the ocean, even if primary productivity rose.

We assume that current models for the marine Ba cycle, although primarily developed for the Holocene and somewhat unconstrained, apply to Paleogene conditions. Thus, there appear to be two general explanations for the available barite records across the PETM: (1) decreased Ba accumulation at unidentified locations counterbalanced the excess Ba accumulation observed in pelagic records, or (2) some external flux of Ba to the ocean increased significantly. Additional records of Ba accumulation are needed to address the first possibility. In particular, Ba accumulation records are needed from locations where suboxic conditions could have remobilized metal oxyhydroxides and associated Ba. The second possibility may require a basic reconstruction of the marine Ba cycle whereby massive release of CH_4 from gas hydrates somehow contributes considerable dissolved Ba^{2+} to the ocean. Trace element analyses of sediment and waters in and above gas hydrate reservoirs, albeit limited (Von Breymann et al., 1990; Torres et al., 1996a, b; Fu and Aharon, 1998; Naehr et al., 2000; Dickens, 2001c), suggest that they store abundant dissolved Ba^{2+} (Figs. 5, 6). We thus speculate that massive quantities of dissolved Ba^{2+} (nominally 135 Tmol) vented into the ocean during the start of the PETM (Fig. 8), and this increased biogenic barite accumulation in deep-sea sediment through enhanced preservation. This scenario explains the unusual coupling between $\delta^{13}C$ and Ba records and the presence of oligotrophic microfossil assemblages across intervals with excess biogenic barite (Fig. 1).

Although our explanation for excess biogenic barite is plausible in the general sense, we note at least one significant problem in the details. Given a marine Ba cycle with Holocene masses and fluxes (Fig. 3), an inferred 10–30 k.y. injection of ~2000 Gt of CH_4 carbon (cf. Katz et al., 1999; Röhl et al., 2000; Dickens, 2001a), and the postulated ~1200:1 molar ratio of CH_4 to Ba in marine gas hydrate deposits, global Ba fluxes and accumulation can at most double. However, Bains et al. (2000) indicate a three-fold increase of biogenic barite burial at ODP Site 690 and a six-fold increase at Site 1051. Although faster CH_4 injection would increase Ba fluxes and accumulation, excessively rapid Ba release would surpass local barite saturation, leading to direct precipitation of authigenic barite on the seafloor (e.g., Torres et al., 1996a; Fu and Aharon, 1998; Naehr et al., 2000) instead of enhanced preservation of biogenic barite. One solution to this problem is to consider barite dissolution analogous to that of carbonate whereby a "barite lysocline" moves vertically in response to changes in saturation (Monnin et al., 1999). In this case, barite accumulation during massive CH_4 injection at the PETM might increase significantly more at sites farthest from saturation (young, deep waters) than at sites closest to saturation (old, intermediate waters). This possibility cannot be rigorously assessed with available data, but could be evaluated with future Ba accumulation records across the PETM and along depth transects.

Isotopic signatures consistent with thermal dissociation of gas hydrate and release of CH_4 have been documented across other brief intervals of the geological record, notably events in the Aptian and Toarcian (Hesselbo et al., 2000; Jahren et al., 2001; Padden et al., 2001). If the mechanism proposed here is correct, seafloor Ba fluxes also may have increased significantly during these times. We note, however, that barite preservation would be complicated in O_2 deficient deep ocean waters with SO_4^{2-} reduction, conditions that did not occur in the open ocean during the PETM but may have during earlier CH_4 release events (Hesselbo et al., 2000).

ACKNOWLEDGMENTS

The inspiration behind this paper came through disparate, offhand comments to G. Dickens by good colleagues: Steve D'Hondt who pointed out the short residence time of Ba in the oceans, Rick Murray who mentioned that barite preservation depends on bottom water Ba concentrations, and Jim McManus and Adina Paytan who introduced the basics of the Ba cycle. We thank three reviewers and an editor for very constructive comments, especially given the simple and highly speculative nature of the paper. Captain Rick and his tremendous blessings expanded our lexicon. Funding for this work comes from NSF grants EAR 0120727 (G. Dickens, E. Thomas and T. Bralower), OCE 0117950 (G. Dickens) and EAR 9814604 (T. Bralower).

APPENDIX 1. NOMENCLATURE AND PARAMETERS FOR THE BASIC MODEL OF THE MARINE BARIUM CYCLE.

Primary parameters		*Units*	*Standard conditions*†
BA_{Sh}	Abundance of Ba in surface water	Gmol (= 10^9 mol)	3731
BA_{Dp}	Abundance of Ba in deep water	Gmol	140148
F_{Riv}	River Ba flux	Gmol/yr *	14.75
F_{Up}	Upwelling Ba flux	Gmol/yr *	128.15
F_{Dw}	Downwelling Ba flux	Gmol/yr *	39.80
F_{PR}	Particulate rain Ba flux	Gmol/yr *	103.10
F_{Hyd}	Hydrothermal Ba flux	Gmol/yr *	3.35
F_{Cold}	Cold Seep Ba flux	Gmol/yr *	0.00 ¶

Primary parameters		*Units*	*Standard conditions*[†]
F_{Reg}	Benthic regeneration Ba flux	Gmol/yr *	85.00
F_{Bur}	Burial Ba flux	Gmol/yr *	18.10
$[Ba]_{Sh}$	Dissolved Ba^{2+} in surface water	nmol/kg_{sw}	36
$[Ba]_{Dp}$	Dissolved Ba^{2+} in deep water	nmol/kg_{sw}	116
$[Ba]_{Sat}$	Dissolved Ba^{2+} at saturation	mol/kg_{sw}	330
C	Exchange of surface and deep water	kg_{sw}/yr	1.105×10^{18} §
D	Coefficient between Ba^{2+} in shallow water and solid Ba in organic matter	kg_{sw}/mol (carbon)	3.8157×10^{-4}
P_{Org}	Sinking organic carbon flux	Gmol/yr *	7500 [F]
f	Fraction of F_{PR} buried	(0–1)	0.17556
k	"Time factor" for barite preservation	(0–1)	0.4994
M_{Sh}	Mass of surface water	kg	1.04×10^{20}
M_{Dp}	Mass of deep water	kg	1.21×10^{21}
A	Area of surface and deep water	m^2	2.40×10^{14}
z_{Sh}	Depth of surface water	m	300
z_{Dp}	Depth of deep water	m	3500
ρ_{sw}	Density of seawater	kg/m^3	1030

Notes:
*Fluxes have been normalized by area (A).
†Except where noted below, standard conditions refer to the Holocene steady-state marine barium cycle as presented by Paytan and Kastner (1996).
¶Cold seep flux has been set at zero for consistency with previous models.
§Exchange flux has been modified from Paytan and Kastner (1996) to normalize for area (A) and include density of seawater.
[F]Organic carbon burial flux taken from average for Holocene carbon cycle (see Dickens, 2001b). Digits in values have been retained for purposes of numerical calculations and do not imply significance.

REFERENCES CITED

Aubry, M.-P., 1998, Early Paleogene calcareous nannoplankton evolution: A tale of climatic amelioration, *in* Aubry, M.-P., et al., eds., Late Paleocene and early Eocene climatic and biotic evolution: New York, Columbia University Press, p. 158–203.

Bains, S., Norris, R.D., Corfield, R.M., and Faul, K.L., 2000, Termination of global warmth at the Paleocene–Eocene boundary through productivity feedback: Nature, v. 407, p. 171–174.

Beerling, D.J., 2000, Increased terrestrial carbon storage across the Paleocene–Eocene boundary: Palaeogeography, Palaeoclimatology, Palaeoecology, v. 161, p. 395–405.

Berner, R.A., Lasaga, A.C., and Garrels, R.M., 1983, The carbonate-silicate geochemical cycle and its effect on atmospheric carbon dioxide over the past 100 million years: American Journal of Science, v. 283, p. 641–683.

Bishop, J.K.B., 1988, The barite–opal–organic carbon association in oceanic particulate matter: Nature, v. 332, p. 341–343.

Blount, C.W., 1977, Barite solubilities and thermodynamic quantities up to 300°C and 1400 bars: American Mineralogist, v. 62, p. 942–957.

Boersma, A., Premoli Silva, I., and Hallock, P., 1998, Trophic models for the well-mixed and poorly mixed warm oceans across the Paleocene–Eocene epoch boundary, *in* Aubry, M.-P., et al., eds., Late Paleocene–early Eocene biotic and climatic events in the marine and terrestrial records: New York, Columbia University Press, p. 204–213.

Boetius, A., Ravenschlag, K., Schubert, C.J., Rickert, D., Widdel, F., Gieseke, A., Amann, R., Jørgensen, B.B., Witte, U., and Pfannkuche, O., 2000, A marine microbial consortium apparently mediating anaerobic oxidation of methane: Nature, v. 407, p. 623–626.

Bolle, M.P., Pardo, A., Adatte, T., Von Salis, K., and Burns, S., 2000, Climatic evolution on the southeastern margin of the Tethys (Negev, Israel) from the Paleocene to the early Eocene: Focus on the late Paleocene thermal maximum: Journal of the Geological Society of London, v. 157, p. 929–941.

Borowski, W.S., Paull, C.K., and Ussler, W., III, 1996, Marine pore water sulfate profiles indicate in situ methane flux from underlying gas hydrate: Geology, v. 24, p. 655–658.

Borowski, W.S., Paull, C.K., and Ussler, W., III, 1999, Global and local variations of interstitial sulfate gradients in deep-water, continental margin sediments: Sensitivity to underlying methane and gas hydrates: Marine Geology, v. 159, p. 131–154.

Bralower, T.J., 2002, Evidence for surface water oligotrophy during the late Paleocene Thermal Maximum: Nannofossil assemblage data from Ocean Drilling Program Site 690, Maud Rise, Weddell Sea: Paleoceanography, v. 17, paper no. 10.1029/2001PA000662.

Bralower, T.J., Zachos, J.C., Thomas, E., Parrow, M., Paull, C.K., Kelly, D.C., Premoli Silva, I., Sliter, W.V., and Lohmann, K.C., 1995, Late Paleocene to Eocene paleoceanography of the equatorial Pacific Ocean: Stable isotopes recorded at Ocean Drilling Program Site 865, Allison Guyot: Paleoceanography, v. 10, p. 841–865.

Bralower, T.J., Thomas, D.J., Zachos, J.C., Hirschmann, M.M., Röhl, U., Sigurdsson, H., Thomas, E., and Whitney, D.L., 1997, High-resolution records of the late Paleocene thermal maximum and circum-Caribbean volcanism: Is there a causal link?: Geology, v. 25, p. 963–966.

Brumsack, H.J., and Gieskes, J.M., 1983, Interstitial water trace metal chemistry of laminated sediments from the Gulf of California, Mexico: Marine Chemistry, v. 14, p. 89–106.

Chow, T.J., and Goldberg, E.D., 1960, On the marine geochemistry of barium: Geochimica et Cosmochimica Acta, v. 20, p. 192–198.

Church, T.M., and Wolgemuth, K., 1972, Marine barite saturation: Earth and Planetary Science Letters, v. 15, p. 35–44.

Crouch, E.M., Heilmann-Clausen, C., Brinkhuis, H., Morgans, H.E.G., Rogers, K.M., Egger, H., and Schmitz, B., 2001, Global dinoflagellate event associated with the late Paleocene thermal maximum: Geology, v. 29, p. 315–318.

Davie, M.K., and Buffett, B.A., 2001, A numerical model for the formation of gas hydrate below the seafloor: Journal of Geophysical Research, v. 106, p. 497–514.

Dehairs, F., Chesselet, R., and Jedwab, J., 1980, Discrete suspended particles of barite and the barium cycle in the ocean: Earth and Planetary Science Letters, v. 49, p. 528–550.

Dehairs, F., Baeyens, W., and Goeyens, L., 1992, Accumulation of suspended barite at mesopelagic depths and export production in the Southern Ocean: Science, v. 258, p. 1332–1335.

Dehairs, F., Fagel, N., Antia, A.N., Peinert, R., Elskens, M., and Goeyens, L., 2000, Export production in the Bay of Biscay as estimated from barium-barite in settling material: A comparison with new production: Deep-Sea Research I, v. 47, p. 583–601.

Delaney, M.L., and Filippelli, G.M., 1994, An apparent contradiction in the role of phosphorus in Cenozoic chemical mass balances for the world ocean: Paleoceanography, v. 9, p. 513–527.

Dickens, G.R., 1999, The blast in the past: Nature, v. 401, p. 752–755.

Dickens, G.R., 2000, Methane oxidation during the late Paleocene thermal maximum: Bulletin de la Société Géologique de France, v. 171, p. 37–49.

Dickens, G.R., 2001a, Modeling the global carbon cycle with a gas hydrate capacitor: Significance for the latest Paleocene thermal maximum, *in* Paull C.K., and Dillon, W.P., eds., Natural gas hydrates, occurrence, distribution and detection: Washington, D.C., American Geophysical Union Geophysical Monograph Series, v. 124, p. 19–38.

Dickens, G.R., 2001b, Carbon addition and removal during the late Paleocene thermal maximum: Basic theory with a preliminary treatment of the isotope record at ODP Site 1051, Blake Nose, *in* Kroon, D., et al., eds., Western North Atlantic Palaeogene and Cretaceous palaeoceanography: Geological Society [London] Special Publication 183, p. 293–305.

Dickens, G.R., 2001c, Sulfate profiles and barium fronts in sediment on the Blake Ridge: Present and past methane fluxes through a large gas hydrate reservoir: Geochimica et Cosmochimica Acta, v. 65, p. 529–543.

Dickens, G.R., 2001d, The potential volume of oceanic methane hydrates with variable external conditions: Organic Geochemistry, v. 32, p. 1132–1193.

Dickens, G.R., and Owen, R.M., 1994, Late Miocene–early Pliocene manganese redirection in the central Indian Ocean: Expansion of the intermediate water oxygen minimum zone: Paleoceanography, v. 9, p. 169–181.

Dickens, G.R., and Owen, R.M., 1999, The latest Miocene–early Pliocene biogenic bloom: A revised Indian Ocean perspective: Marine Geology, v. 161, p. 75–91.

Dickens, G.R., O'Neil, J.R., Rea, D.K., and Owen, R.M., 1995, Dissociation of oceanic methane hydrate as a cause of the carbon isotope excursion at the end of the Paleocene: Paleoceanography, v. 10, p. 965–971.

Dickens, G.R., Paull, C.K., Wallace, P., and the ODP Leg 164 Scientific Party, 1997, Direct measurement of in situ methane quantities in a large gas hydrate reservoir: Nature, v. 385, p. 426–428.

Dickens, G.R., Wallace, P., Paull, C.K., and Borowski, W.S., 2000, Detection of methane gas hydrate in the pressure core sampler (PCS): Volume-pressure-time relations during controlled degassing experiments, *in* Paull, C.K., et al., eds., Proceedings of the Ocean Drilling Program, Scientific Results: Texas A & M University, College Station, Texas, USA, v. 164, p. 113–128.

Dove, P.M., and Czank, C.A., 1995, Crystal chemical controls on the dissolution kinetics of the isostructural sulfates celestite, anglesite, and barite: Geochimica et Cosmochimica Acta, v. 59, p. 1907–1915.

Dymond, J., Suess, E., and Lyle, M., 1992, Barium in deep-sea sediments: A geochemcial proxy for paleoproductivity: Paleoceanography, v. 7, p. 163–181.

Elderfield, H., and Schultz, A., 1996, Midocean ridge hydrothermal fluxes and the chemical composition of the ocean: Annual Review of Earth and Planetary Sciences, v. 24, p. 191–224.

Eldholm, O., and Thomas, E., 1993, Environmental impact of volcanic margin formation: Earth and Planetary Science Letters, v. 117, p. 319–329.

Fossing, H., Ferdelman, T.G., and Berg, P., 2000, Sulfate reduction and methane oxidation in continental margin sediments influenced by irrigation (southeast Atlantic off Namibia): Geochimica Cosmochimica Acta, v. 64, p. 897–910.

Francois, R., Honjo, S., Manganini, S.J., and Ravizza, G.E., 1995, Biogenic barium fluxes to the deep sea: Implications for paleoproductivity reconstruction: Global Biogeochemical Cycles, v. 9, p. 289–303.

Fu, B., and Aharon, P., 1998, Sources of hydrocarbon-rich fluids advecting on the seafloor in the northern Gulf of Mexico: Transactions Gulf Coast Association of Geological Societies, v. 48, p. 73–81.

Gawenda, P., Winkler, W., Schmitz, B., and Adatte, T., 1999, Climate and bioproductivity control on carbonate turbidite sedimentation (Paleocene to earliest Eocene Gulf of Biscay, Zumaia): Journal of Sedimentary Research, v. 69, p. 1253–1261.

Gingele, F., and Dahmke, A., 1994, Discrete barite particles and barium as tracers of paleoproductivity in South Atlantic sediments: Paleoceanography, v. 9, p. 151–168.

Haq, B.U., and Lohmann, G.P., 1976, Early Cenozoic calcareous nannoplankton biogeography of the Atlantic Ocean: Marine Micropaleontology, v. 1, p. 119–194.

Harrington, G.J., and Kemp, S.J., 2001, U.S. Gulf Coast vegetation dynamics during the latest Paleocene: Palaeogeography, Palaeoclimatology, Palaeoecology, v. 167, p. 1–21.

Heister, L.E., O'Day, P.A., Brooks, C.K., Neuhoff, P.S., and Bird, D.K., 2001, Pyroclastic deposits within the East Greenland Tertiary flood basalts: Journal of the Geological Society of London, v. 158, p. 269–284.

Hesselbo, S.P., Gröcke, D.R., Jenkyns, H.C., Bjerrum, C.J., Farrimond, P., Bell, H.S.M., and Green, O.R., 2000, Massive dissociation of gas hydrate during a Jurassic oceanic anoxic event: Nature, v. 406, p. 392–395.

Hinrichs, K.-U., Hayes, J.M., Sylva, S.P., Brewer, P.G., and DeLong, E.F., 1999, Methane-consuming archaebacteria in marine sediments: Nature, v. 398, p. 802–805.

Hovan, S.A., and Rea, D.K., 1992, Paleocene Eocene boundary changes in atmospheric and oceanic circulation: A Southern Hemisphere record: Geology, v. 20, p. 15–18.

Jahren, A.H., Arens, N.C., Sarmiento, G., Guerrero, J., and Amundson, R., 2001, Terrestrial record of methane hydrate dissociation in the Early Cretaceous: Geology, v. 29, p. 159–162.

Johnson, K.S., Berelson, W.M., Coale, K.H., Coley, T.L., Elrod, V.A., Fairey, W.R., Iams, H.D., Kilgore, T.E., and Nowicki, J.L., 1992, Manganese flux from continental margin sediments in a transect through the oxygen minimum: Science, v. 257, p. 1242–1245.

Kaiho, K., Arinobu, T., Ishiwatari, R., Morgans, H.E.G., Okada, H., Takeda, N., Tazaki, K., Zhou, G., Kajiwara, Y., Matsumoto, R., Hirai, A., Niitsuma, N., and Wada, H., 1996, Latest Paleocene benthic foraminiferal extinction and environmental changes at Tawanui, New Zealand: Paleoceanography, v. 11, p. 447–465.

Kastner, M., 2001, Gas hydrates in convergent margins: Formation, occurrence, geochemistry, and global significance, *in* Paull, C.K., et al., eds., Natural gas hydrates, occurrence, distribution, and detection: Washington, D.C., American Geophysical Union Geophysical Monograph Series, v. 124, p. 67–86.

Katz, M.E., Pak, D.K., Dickens, G.R., and Miller, K.G., 1999, The source and fate of massive carbon input during the latest Paleocene Thermal Maximum: Science, v. 286, p. 1531–1533.

Kayen, R.E., and Lee, H., 1991, Pleistocene slope instability of gas hydrate-laden sediment on the Beaufort Sea Margin: Marine Geotechnology, v. 10, p. 125–141.

Kelly, D.C., Bralower, T.J., Zachos, J.C., Premoli Silva, I., and Thomas, E., 1996, Rapid diversification of planktonic foraminifera in the tropical Pacific (ODP Site 865) during the late Paleocene thermal maximum: Geology, v. 24, p. 423–426.

Kennett, J.P., and Stott, L.D., 1991, Abrupt deep sea warming, paleoceanographic changes, and benthic extinctions at the end of the Paleocene: Nature, v. 353, p. 319–322.

Klump, J., Hebbeln, D., and Wefer, G., 2001, High concentrations of biogenic barium in Pacific sediments after Termination, I: A signal of changes in productivity and deep water chemistry: Marine Geology, v. 177, p. 1–11.

Kump, L.R., and Arthur, M.A., 1999, Interpreting carbon isotope excursions: Carbonates and organic matter: Chemical Geology, v. 161, p. 181–198.

Kvenvolden, K.A., 1988, Methane hydrate: A major reservoir of carbon in the shallow geosphere: Chemical Geology, v. 71, p. 41–51.

Kvenvolden, K.A., 1995, A review of the organic geochemistry of methane in natural gas hydrate: Organic Geochemistry, v. 23, p. 997–1008.

Kvenvolden, K.A., 1999, Potential effects of gas hydrate on human welfare: Proceedings of the National Academy of Sciences, v. 96, p. 3420–3426.

Kvenvolden, K.A., and Kastner, M., 1990, Gas hydrates of the Peruvian outer continental margin: *in* Seuss, E., et al., eds., Proceedings of the Ocean Drilling Program, Scientific Results: Texas A & M University, College Station, Texas, USA, v. 112, p. 517–526.

McManus, J., Berelson, W.M., Klinkhammer, G.P., Johnson, K.S., Coale, K.H., Anderson, R.F., Kumar, N., Burdige, D.J., Hammond, D.E., Brumsack, H.J., McCorkle, D.C., and Rushdi, A., 1998, Geochemistry of barium in marine sediments: Implications for its use as a paleoproxy: Geochimica et Cosmochimica Acta, v. 62, p. 3453–3473.

McManus, J., Berelson, W.M., Hammond, D.E., and Klinkhammer, G.P., 1999, Barium cycling in the North Pacific: Implications for the utility of Ba as a paleoproductivity and paleoalkalinity proxy: Paleoceanography, v. 14, p. 53–61.

Monnin, C., 1999, A thermodynamic model for the solubility of barite and celestite in electrolyte solutions and seawater to 200 °C and to 1 kbar: Chemical Geology, v. 153, p. 187–209.

Monnin, C., Jeandel, C., Cattaldo, T., and Dehairs, F., 1999, The marine barite saturation state of the world's oceans: Marine Chemistry, v. 65, p. 253–261.

Monnin, C., Wheat, C.G., Dupre, B., Elderfield, H., and Mottl, M.M., 2001, Barium geochemistry in sediment pore water and formation waters of the oceanic crust on the eastern flank of the Juan de Fuca Ridge (ODP Leg 168): Geochemistry, Geophysics, Geosystems, v. 2, paper no. 2000 GC000073.

Naehr, T.H., Stakes, D.S., and Moore, W.S., 2000, Mass wasting, ephemeral fluid flow, and barite deposition on the California continental margin: Geology, v. 28, p. 315–318.

Norris, R.D., and Röhl, U., 1999, Carbon cycling and chronology of climate warming during the Paleocene–Eocene transition: Nature, v. 401, p. 775–778.

Olivarez, A.M., and Owen, R.M., 1989, Plate tectonic reorganizations: Implications regarding the formation of hydrothermal ore deposits: Marine Mining, v. 14, p. 123–138.

Padden, M., Weissert, H., and de Rafelis, M., 2001, Evidence for Late Jurassic release of methane from gas hydrate: Geology, v. 29, p. 223–226.

Paull, C.K., Ussler, W., III, Borowski, W.S., and Spiess, F.N., 1995, Methane-rich plumes on the Carolina continental rise: Associations with gas hydrates: Geology, v. 23, p. 89–92.

Paull, C.K., Buelow, W.J., Ussler, W., III, and Borowski, W.S., 1996, Increased continental-margin slumping frequency during lowstands above gas hydrate-bearing sediments: Geology, v. 24, p. 143–146.

Paytan, A., and Kastner, M., 1996, Benthic barium fluxes in the central equatorial Pacific: Implications for the oceanic Ba cycle: Earth and Planetary Science Letters, v. 142, p. 439–450.

Paytan, A., Kastner, M., and Chavez, F.P., 1996, Glacial to interglacial fluctuations in productivity in the equatorial Pacific as indicated by marine barite: Science, v. 274, p. 1355–1357.

Ravizza, G., Norris, R.N., Blusztajn, J., and Aubry, M.-P., 2001, An osmium isotope excursion associated with the late Paleocene thermal maximum: Evidence of intensified chemical weathering: Paleoceanography, v. 16, p. 155–163.

Robert, C., and Kennett, J.P., 1994, Antarctic subtropical humid episode at the Paleocene–Eocene boundary: Clay mineral evidence: Geology, v. 22, p. 211–214.

Röhl, U., Bralower, T.J., Norris, R.D., and Wefer, G., 2000, New chronology for the late Paleocene thermal maximum and its environmental implications: Geology, v. 28, p. 927–930.

Schmitz, B., 1987, Barium, equatorial high productivity, and the wandering of the Indian continent: Paleoceanography, v. 2, p. 63–77.

Schmitz, B., Charisi, S.D., Thompson, E.I., and Speijer, R.P., 1997, Barium, SiO_2 (excess), and P_2O_5 as proxies of biological productivity in the Middle East during the Paleocene and the latest Paleocene benthic extinction event: Terra Nova, v. 9, p. 95–99.

Schmitz, B., Pujalte, V., and Núñez-Betelu, K., 2001, Climate and sea-level perturbations during the incipient Eocene Thermal Maximum: Evidence from siliciclastic units in the Basque Basin (Ermua, Zumaia and Trabakua Pass), northern Spain: Palaeogeography, Palaeoclimatology, Palaeoecology, v. 167, p. 299–320.

Schroeder, J.O., Murray, R.W., Leinen, M., Pflaum, R.C., and Janecek, T.R., 1997, Barium in equatorial Pacific carbonate sediment: Terrigenous, oxide, and biogenic associations: Paleoceanography, v. 12, p. 125–146.

Speijer, R.P., and Wagner, T., 2002, Transgressive black shales associated with the late Paleocene Thermal Maximum (LPTM): Organic geochemical and micropaleontologic evidence from the southern Tethyan Margin, *in* Koeberl, C., and MacLeod, K., eds., Catastrophic events and mass extinctions: Impacts and beyond: Boulder, Colorado, Geological Society of America Special Paper, v. 356, p. 533–550.

Stroobants, N., Dehairs, F., Goeyens, L., Vanderheijden, N., and Van Grieken, R., 1991, Barite formation in the Southern Ocean water column: Marine Chemistry, v. 35, p. 411–421.

Suess, E., Torres, M.E., Bohrmann, G., Collier, R.W., Greinerft, J., Linke, P., Rehder, G., Tréhu, A., Wallman, K., Winckler, G., and Zuleger, E., 1999, Gas hydrate destabilization: Enhanced dewatering, benthic material turnover, and large methane plumes at the Cascadia Margin: Earth and Planetary Science Letters, v. 170, p. 1–15.

Thiry, M., 2000, Palaeoclimatic interpretation of clay minerals in marine deposits: An outlook from the continental origin: Earth Science Reviews, v. 49, p. 201–221.

Thomas, D.J., Bralower, T.J., and Zachos, J.C., 1999, New evidence for subtropical warming during the late Paleocene thermal maximum: Stable isotopes from Deep Sea Drilling Project Site 527, Walvis Ridge: Paleoceanography, v. 14, p. 561–570.

Thomas, E., 1998, The Paleocene–Eocene benthic foraminiferal extinction and stable isotope anomalies, *in* Aubry, M.-P., et al., eds., Late Paleocene–early Eocene biotic and climatic events in the marine and terrestrial records: New York, Columbia University Press, p. 214–243.

Thomas, E., and Shackleton, N.J., 1996, The Paleocene–Eocene benthic foraminiferal extinction and stable isotope anomalies, *in* Knox, R.W.O'B., et al., eds., Correlations of the early Palaeogene in northwest Europe: Geological Society [London] Special Publication 101, p. 401–411.

Torres, M.E., Bohrmann, G., and Suess, E., 1996a, Authigenic barites and fluxes of barium associated with fluid seeps in the Peru subduction zone: Earth and Planetary Science Letters, v. 144, p. 469–481.

Torres, M.E., Brumsack, H.J., Bohrmann, G., and Emeis, K.-C., 1996b, Barite fronts in continental margin sediments: A new look at barium remobilization in the zone of sulfate reduction and formation of heavy barites in diagenetic fronts: Chemical Geology, v. 127, p. 125–139.

Valentine, D.L., Blanton, D.C., Reeburgh, W.S., and Kastner, M., 2001, Water column methane oxidation adjacent to an area of active hydrate dissociation, Eel River Basin: Geochimica et Cosmochimica Acta, v. 65, p. 2633–2640.

Von Breymann, M.T., Emeis, K.-C., and Camerlenghi, A., 1990, Geochemistry of sediments from the Peru upwelling area: Results from Sites 680, 682, 685, and 688, *in* Suess, E., et al., eds., Proceedings of the Ocean Drilling Program, Scientific Results: Texas A & M University, College Station, Texas, USA, v. 112, p. 491–503.

Von Breymann, M.T., Brumsack, H., and Emeis, K.-C., 1992, Deposition and diagenetic behavior of barium in the Japan Sea, *in* Sciotto, K.A., et al., eds., Proceedings of the Ocean Drilling Program, Scientific Results: Texas A & M University, College Station, Texas, USA, v. 127/128, p. 651–665.

Walker, J.C.G., and Kasting, J.F., 1992, Effects of fuel and forest conservation on future levels of atmospheric carbon dioxide: Palaeogeography, Palaeoclimatology, Palaeoecology, v. 97, p. 151–189.

Walker, J.C.G., Hays, P.B., and Kasting, J.F., 1981, A negative feedback mechanism for the long-term stabilization of Earth's surface temperature: Journal of Geophysical Research, v. 86, p. 9776–9782.

White, A.F., Bullen, T.D., Schulz, M.S., Blum, A.E., Huntington, T.G., and Peters, N.E., 2001, Differential rates of feldspar weathering in granitic regoliths: Geochimica et Cosmochimica Acta, v. 65, p. 847–869.

Wolgemuth, K., and Broecker, W.S., 1970, Barium in seawater: Earth and Planetary Science Letters, v. 8, p. 372–378.

Xu, W., and Ruppel, C., 1999, Predicting the occurrence, distribution, and evolution of methane gas hydrate in porous marine sediments: Journal of Geophysical Research, v. 104, p. 5081–5095.

Zachos, J.C., and Dickens, G.R., 2000, An assessment of the biogeochemical feedback response to the climatic and chemical perturbations of the LPTM: GFF, v. 122, p. 188–189.

Manuscript Accepted by the Society August 13, 2002

Printed in the USA

Geological Society of America
Special Paper 369
2003

Early Paleogene oceans and climate: A fully coupled modeling approach using the NCAR CCSM

Matthew Huber
Danish Center for Earth System Science, Niels Bohr Institute, Juliane Maries Vej 30, 2100 Copenhagen O, Denmark
Lisa C. Sloan
Cindy Shellito
Earth Sciences Department, University of California Santa Cruz, Santa Cruz, California 95060, USA

ABSTRACT

Attempts to understand early Paleogene climate using models have provided insights, but have consistently had difficulty explaining essential climatic features. A main limitation of those studies has been the lack of interaction between dynamical ocean and atmospheric models. After reviewing previous model studies and quantitatively defining the "low gradient problem," we present results from the first fully coupled general circulation model simulations in equilibrium with reconstructed early Paleogene topography, bathymetry, vegetation, and pCO_2 (560 ppm). Although our results should be understood as preliminary, predictions of both Paleogene sea surface temperatures and salinities are, for the first time, made without assuming an arbitrary value of the oceanic or atmospheric heat transport. Model-predicted bottom water temperatures are 7 °C warmer than modern. Deep convection occurs in the North Atlantic and Tethys, forming warm (12–15 °C), salty water masses (35.5 ppt). Through most of the Southern Ocean, vigorous wind-driven upwelling and a stable water column caused by low salinities (~32 ppt) produce a temperature inversion that may explain proxy interpretations of "warm salty deep water" formation in the tropics. Estimates of seawater oxygen isotopic ratios reveal substantially different patterns than usually predicted. Ocean heat transport is little-changed from modern values and warming of tropical temperatures (by ~3 °C) is about half that occurring in high latitudes. The mean annual temperature and seasonality patterns produced by the simulation are biased toward colder and more seasonal values than those reconstructed from temperature proxies.

INTRODUCTION

Climate proxies depict the early Paleogene as having warm (6–12 °C) deep ocean and polar temperatures, year-round above-freezing temperatures in continental interiors, near-modern (23–27 °C) tropical sea surface temperatures (SSTs) (Zachos et al., 2001; Zachos et al., 1994; Greenwood and Wing, 1995; Markwick, 1998; Wilf, 2000; Crowley and Zachos, 2000; Bralower et al., 1995), and modern-to-higher greenhouse gas concentrations (e.g., pCO_2 of 400–4000 ppm) (Retallack, 2001; Royer et al., 2001; Pearson and Palmer, 2000). This combination of small temperature gradients (the "low gradient problem"), reduced continental seasonality (the "equable climate problem"), and modern or higher greenhouse gas concentrations

Huber, M., Sloan, L.C., and Shellito, C., 2003, Early Paleogene oceans and climate: A fully coupled modeling approach using the NCAR CCSM, *in* Wing, S.L., Gingerich, P.D., Schmitz, B., and Thomas, E., eds., Causes and Consequences of Globally Warm Climates in the Early Paleogene: Boulder, Colorado, Geological Society of America Special Paper 369, p. 25–47.

poses some of the most puzzling and important problems in the study of climate (Barron, 1987; Valdes, 2000a).

It is possible that existing early Paleogene tropical SST reconstructions are biased toward cool values, and that an upward revision of those values might help resolve this debate (Huber and Sloan, 2000; Pearson et al., 2001). But, barring that, the one commonly advocated theory that might appear to solve these problems relies on oceanic heat transport and greenhouse gas concentrations much greater than present values (Rind and Chandler, 1991; Barron et al., 1993). Consequently, attention has focused on the challenging problem of modeling the paleo-ocean circulations and, especially, on placing bounds on ocean heat transport for past greenhouse climates (Schmidt and Mysak, 1996; Lyle, 1997; Brady et al., 1998; Bice et al., 2000a).

The advancement of our understanding of climate requires theories and models that are independent of the data, in order to avoid circularity, and to make predictions testable against those proxy data. Previous attempts to address the question of early Paleogene ocean heat transport and temperature gradients have been limited in their ability to properly represent interaction between ocean circulations and the atmosphere (e.g., Sloan and Rea, 1996; Bice et al., 2000a). In this study, we couple dynamical atmospheric, ocean, and sea-ice component models to achieve a greater level of realism (and a decreased degree of circularity) than is possible in a framework in which climate components can not interact with each other. The coupled simulations are performed with higher-than-modern pCO_2, 560 ppm (at the lower-to-mid range of proxy estimates), and boundary conditions appropriate for the early Paleogene. This study is a continuation and expansion of a study described in Huber and Sloan (2001).

In the following section, we summarize the major modeling methods used previously and the relevant results of early Paleogene studies performed with those models. In the third section we outline a method, predicated on the results of previous uncoupled studies, for performing fully coupled simulations of past climates. The fourth section addresses important questions about early Paleogene climates that have not been explored previously in a fully coupled and self-consistent framework. Those questions include: (1) Were ocean currents weaker or stronger in an early Paleogene greenhouse climate? (2) Is the ocean salinity distribution different than usually assumed? (3) What were the locations and watermass properties of deep water formation? (4) Was meridional oceanic heat transport greater than modern in an early Paleogene simulation that is both fully coupled and includes an equilibrated deep ocean circulation? (5) Can the small temperature gradients evidenced in the climate proxy record be maintained in a fully coupled model? We discuss the implications and limitations of this study in the fifth section. This is followed by a summary of the results.

A SUMMARY OF PREVIOUS MODELING STUDIES

Three types of dynamical numerical model have been used to address the questions posed by early Paleogene climate: (1) atmospheric general circulation models (AGCMs) driven by fixed SST distributions; (2) AGCMs coupled to mixed-layer, "slab" oceans, with specified mixed-layer thickness and meridional heat fluxes or diffusivities; and (3) ocean (O) GCMs usually driven by SST, precipitation minus evaporation (*P–E*), and wind fields taken from the former types of experiments (cf. DeConto et al., 2000). We review results from these studies because they provide the motivation for the use of a comprehensive general circulation model simulations (CGCMs) in this study.

Type 1 models

In studies of type 1, the SST distribution is specified (Fig. 1), and the object is to explore what atmospheric features (e.g., rainfall, wind, outgoing longwave radiation, and terrestrial temperature patterns), exist in equilibrium with such a distribution, and the sensitivity of these features to other factors (e.g., topography). The key physical assumption in these experiments is that the thermal inertia of the ocean is so much larger than that of the atmosphere that the ocean temperatures may be considered constant. By changing the SST distribution that drives the model, assumptions about the role of ocean heat transport also can be tested because the ocean heat transport *implied* by a fixed SST distribution is calculable as a residual from an AGCM simulation (Fig. 1).

In other work by the authors, we have found that high-latitude SST changes lead to small (relative to the data-model bias) responses over high-latitude landmasses, mostly within 1–2 model cells of the ocean, excepting western Europe and northeast North America where larger responses are produced (Huber and Sloan, 1999a; Sloan et al., 2001). Increasing tropical SSTs by 5 °C from 26 to 31 °C leads to extratropical winter warming (up to 5 °C), but the model-predicted sensitivity is not substantial enough to explain year-round above freezing temperatures (Huber, 2001). When tropical SSTs are decreased and high-latitude SSTs are increased at the same time, the total effect is a nearly linear combination of the separate forcings (Huber and Sloan, 1999a, 1999b; Huber, 2001). These low SST gradients, consistent with the traditional interpretation of the geologic record, have little net impact on winter temperatures in extratropical continental interiors. Even when extremely warm SSTs are imposed in high latitudes, below-freezing continental interior winter temperatures occur over most of North America, Asia, and Antarctica (Huber and Sloan, 1999a, 1999b; Sloan et al., 2001). The SST sensitivity results suggest that extratropical oceans with specified warm SSTs do not solve the equable climate problem, unless tropical SSTs were significantly warmer than 31 °C.

When SSTs are specified, pCO_2 may be changed as a sensitivity parameter although the results must be interpreted cautiously in light of the fact that the top-of-the-atmosphere radiative balance may not be consistent between experiments. In practice, changes in pCO_2 from 560 to 1120 ppm have small impacts on the radiative balance and a small effect on land surface

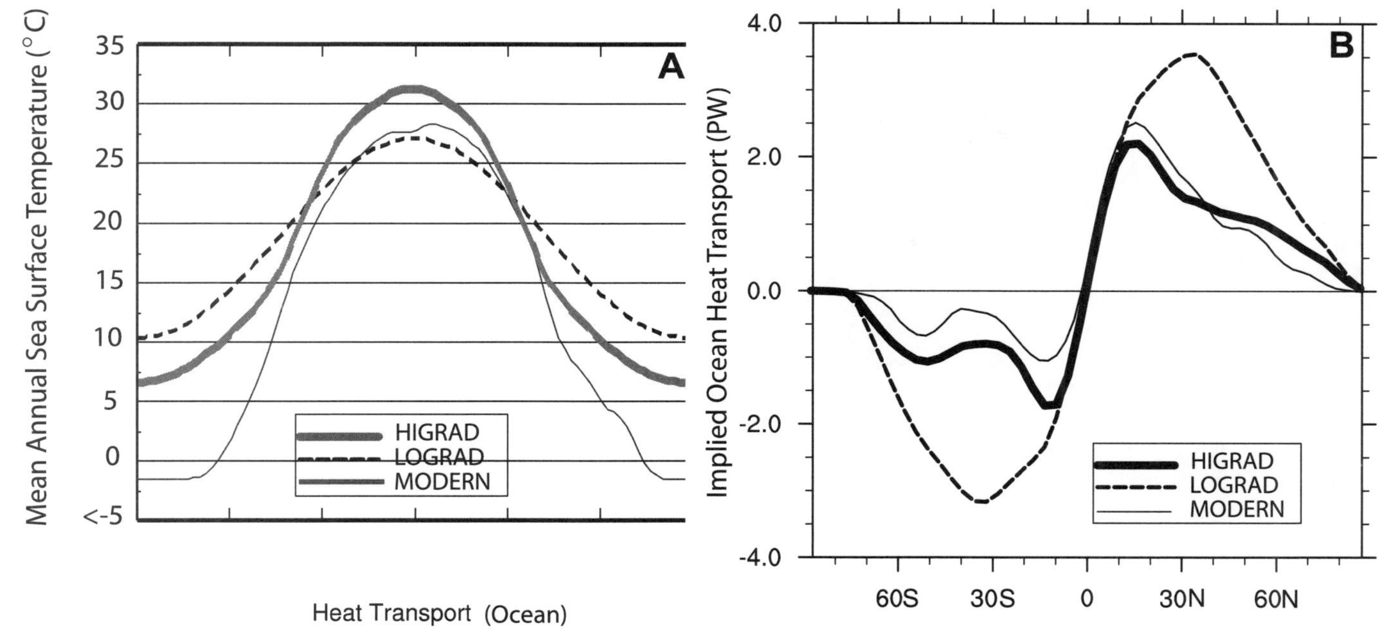

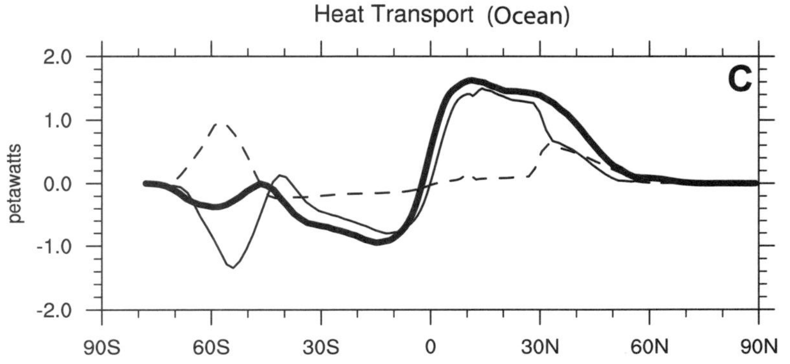

Figure 1. A: Zonal annual mean sea surface temperature (SST). The approximate range of early-to-mid Eocene SSTs from proxies including possible biases are spanned by the HIGRAD and LOGRAD distributions (estimates from Zachos et al., 1994; Crowley and Zachos, 2000). The degraded modern model predicted SSTs are shown with a thin line. B: Implied residual ocean heat transport calculated from the SST distributions (as described in the text). C: The global (thick line), Pacific and Indian Ocean (thin line), and Atlantic (dashed line) values of poleward ocean heat transport (positive northwards) are shown.

temperatures. Temperatures of some regions increase by ~2 °C for each doubling of pCO_2 within the range of 280 to1120 ppm (Huber and Sloan, 1999b). The relatively small response of extratropical continental interior winter temperatures has not been sensitive to AGCM formulation. For example, the mismatch between model result and proxies was only marginally improved when CCM3.6 was used instead of the older models GENESIS versions 1 and 2 (cf. Sloan et al., 2001). This same mismatch is further confirmed by Valdes (2000b), using both the UKMO and UGAMP AGCMs.

Type 1 experiments are fundamentally limited because they do not incorporate two-way interaction with the ocean. But, we can conclude from them that specification of warm extratropical SSTs—essentially assuming greater ocean heat transport and infinite ocean heat capacity—does little by itself to warm winter temperatures in continental interiors. Therefore, *if the ocean has a direct role in maintaining warm winter temperatures, it is only through coupled (two-way) interaction with the atmosphere.*

If tropical SST proxies are not biased to too cool values, then either (1) the ocean is not the key to the equable climate problem, or (2) some aspect of coupling a realistic ocean to the atmosphere is the key to solving the equable climate problem.

An initial approach to the second option is to couple the atmosphere to a simplified ocean.

Type 2 models

In studies of type 2, SSTs are predicted, so it is possible to test SST sensitivity to pCO_2 (or other parameter) variations. It is explicitly assumed the ocean can be represented as a fixed-depth mixed layer, i.e., that the important ocean thermal inertia is in the upper 50 m, and ocean poleward heat transport is at specified levels (usually modern or higher). With early Paleogene boundary conditions, and near-modern pCO_2 and ocean heat transport values, these studies produce zonally averaged SSTs and meridional SST gradients nearly identical to modern. With higher pCO_2, tropical SSTs increase (to ~32 °C for 1000 ppm) without changing meridional gradients substantially. Sea ice forms seasonally in the Northern and Southern Hemispheres even at 1000 ppm (Shellito et al., personal commun., 2001). Studies that demonstrate substantial high-latitude amplification of temperature response to increases in pCO_2, or other forcings, generally do so because of the nonlinearity introduced by crossing a threshold from extensive sea-ice cover to little or no sea-ice cover (e.g., Sloan and Rea, 1996; Otto-Bliesner and Upchurch, 1997; Peters and Sloan, 2000). Proxy data imply (but do not rule out) that it was unlikely that sea-ice was present at all during past, warm "greenhouse" climates (e.g., Markwick, 1998), therefore this sensitivity may not be representative of actual warm Paleogene climate systems.

In the discussion of the importance of ocean heat transport, a distinction must be made between the potential warming effect of an individual current on a specific coastal region (e.g., <0.1 PW forcing) and the level of ocean heat transport change necessary to maintain relatively cool tropics (e.g., >1.0 PW). The former requires the penetration of a current to high latitudes in order to explain the polar warmth indicated by each isolated, coastal paleoclimate proxy locality. The latter has a major impact on tropical temperatures, which we have argued are the key factor for understanding winter warmth within continental interiors. The important effect of specified ocean heat transport increases in type 2 experiments is that it allows for the existence of higher pCO_2 (raising temperatures globally) without the production of overly warm tropical SSTs (Barron et al., 1993; Sloan et al., 1995).

Type 3 models

Early Paleogene uncoupled OGCM experiments (type 3) have successfully shown that the ocean circulation is sensitive to freshwater distributions (Bice et al., 1997) and land-sea distributions (Bice et al., 2000a). However, while an uncoupled OGCM modeling framework is useful for testing the sensitivity of ocean flow patterns to forcing, it may not be well suited for investigating some other important questions, such as poleward heat transport.

When the SST distribution is fixed, the implicit assumption is that the atmosphere has an infinite heat capacity relative to the ocean, opposite to the assumption in type 1 experiments. In a steady state and zonal average sense, the surface temperature gradient must map into the vertical temperature gradient, thus bottom water temperatures in such experiments cannot be cooler than the coldest surface value. In type 1 experiments, neither the coldest temperatures (upper atmospheric), nor terrestrial temperatures are fixed, so winds, heat transport, and surface land temperatures can be independently verified. In a fixed SST OGCM experiment, however, it is not clear how to explore the factors maintaining temperature distributions and heat fluxes. This highlights one fundamental difference between the ocean and the atmosphere—the atmosphere has two "free" boundaries, one at the top and one at the bottom, whereas the ocean has the top as its only "free" boundary. If this "free" boundary is specified as in type 3 studies, the problem becomes overdetermined. As this "fixed" assumption is relaxed, by using long restoring times, so-called "bulk forcing" (Brady et al., 1998; Doney et al., 1998), or other methods (Nong et al., 2000), the top boundary specification better approximates a real atmosphere, and, we argue, better estimates ocean heat transport.

An additional difficulty is presented by the definition of poleward heat transport itself. In the real atmosphere-ocean system, there are two ways to calculate poleward ocean heat transport, and if the measurements (or model) are perfect, the methods should yield identical results. The direct method involves integrating the product of poleward and equatorward ocean velocities and temperatures throughout the ocean to calculate the integral of the ocean heat transport divergence. Schematically, $F \propto V\Delta T$, where F is the poleward heat flux, V is the meridional overturning rate, and ΔT, is the vertical temperature gradient. The other, indirect, method for calculating ocean heat transport involves calculating the net energy flux through the *surface* of the ocean and integrating meridionally to obtain the so-called implied ocean heat transport (Peixoto and Oort, 1992; Huber and Sloan, 1999a; Trenberth and Caron, 2001). In steady state, the heat lost or gained through the sea surface (the net sensible, latent heat fluxes and incoming and outgoing radiation) at the ocean surface must be balanced exactly by heat transported by ocean currents. The direct and indirect methods should give identical results in any system, because they are mathematically identical (via Gauss's Divergence Theorem), and in coupled models without "flux corrections" this requirement is nearly exactly met (Hack, 1998). In uncoupled OGCM sensitivity studies, there is no mechanism to guarantee that the implied and direct ocean heat transports agree—frequently, they do not. This inconsistency is important for ocean gateway change studies of ocean heat transport.

In studies of the effects of ocean gateway changes on ocean heat transport it has been common to use the same (or nearly the same) surface forcing (i.e. SSTs, winds, surface moisture flux) throughout the experiments, as land-sea distributions and gateways are changed, and then to compare differences in ocean heat transport (e.g., Mikolajewicz et al., 1993; Bice et al., 2000a). Using the direct method, substantial changes in ocean heat transport may occur as ocean currents are altered by differences in deep water formation patterns and bathymetry. The implied ocean heat transport, which is just as valid a measure of ocean heat transport, can not change since the fields on which it depends have not been allowed to change (or the change is very small).

This is a logical and physical inconsistency that obscures the interpretation of these experiments, because the surface climate state used to drive the ocean model is incompatible with the surface climate that the simulated ocean would create if the boundary conditions were free to evolve (as preliminarily explored by Mikolajewicz et al., 1993, and reexamined in Mikolajewicz and Crowley, 1997). A minimum requirement for realism in any ocean model simulation is that it be a possible solution in a coupled system, because the real atmosphere and ocean are coupled. Thus, the ocean heat transport produced by the circulation should not be substantially out of equilibrium with the heat transport implied by the surface forcing. The results of Bice et al. (2000a) for the early Paleogene are out of equilibrium by >2 PW, which is important for understanding the comparison of our results to that study below.

Another problem with uncoupled OGCM experiments is that, in the absence of salinity proxies, the surface freshwater balance must be arbitrarily specified, usually from AGCM results. A consequence is that one of the two major variables for determining ocean density gradients is unconstrained, either by proxies, or by the physics of ocean-atmosphere exchanges in which interactions are two-way. A further possible pitfall in such uncoupled studies, is an inconsistency between the ocean model circulation and the noninteractive moisture balance forcing that drives it.

As an example, consider a steady state, zonally averaged overturning circulation in equilibrium with a fixed SST and surface salinity distribution in which sinking occurs at high latitudes in cooler regions and fills the deep ocean with cold water. If evaporation is increased in a warmer (e.g., subtropical) region so that deep water formation occurs in this region, then the deep ocean will fill with warmer water. The volume average ocean temperature will therefore increase, although there has been no change in the heat flux at any boundaries. As an artifact of the assumed infinite atmospheric thermal inertia in the top boundary condition, heat has been added to the ocean without removing heat from any real source. In this way, conservation of energy (of the climate system) can be violated if proper account is not taken of the true fluxes into and out of the system.

Studies of modern climate using uncoupled OGCMs avoid these difficulties because the forcing at the top boundary is well known and the goal is to understand the ocean flows in equilibrium with this distribution—these flows are also largely known and thus the models are testable. However, in the Paleogene, constraints on ocean overturning rates, zonal temperature gradients, heat transport, current strengths or directions are weak because of poor coverage, small gradients in carbon and oxygen isotopes, and the absence of a proxy for salinity. Consequently, there are few means to provide an independent check on the predictions of uncoupled OGCMs, and no guarantee of energetic self-consistency in the simulations. An OGCM-produced circulation, which would collapse if coupled to an atmosphere, cannot reveal much about any real past climate. Therefore, regardless of whether a fully coupled model is used, a minimum requirement of OGCM experiments should be careful attention to the compatibility of fluxes passed through the ocean surface between the ocean and atmosphere.

A definition of the low gradient problem

Attention to the fluxes between the atmosphere and ocean allow us to make a quantitative definition of the low gradient problem (cool tropics paradox). The core problem is that atmospheric models generate weak poleward heat transport when driven by small meridional temperature gradients. To maintain the small gradients, the ocean is required to make up the remaining heat transport. In other words, the ocean heat transport implied by the indirect method increases to levels much greater than modern. However, as the imposed SST gradient decreases, the ocean heat flux (F) is expected to decrease quickly with decreasing temperature gradients, because $F \propto V\Delta T$. Ocean heat transports calculated from the direct and indirect methods therefore diverge strongly as the imposed SST gradient decreases, and ocean heat transport greater than modern appears to be both required and ruled out by small temperature gradients.

There is a range, however, of smaller-than-modern temperature gradients that do not imply ocean heat transport increases over modern values. The HIGRAD SST distribution (described above and Figure 1) with an equator-to-pole SST difference, ΔT, of 24 °C requires near-modern values of F, whereas the LOGRAD distribution with a ΔT of 15 °C requires an F ~2–3 times as large (Fig. 1). In other words, warm polar temperatures of 10 °C and tropical temperatures of 34 °C can exist in equilibrium with near-modern ocean transport values, but if the tropical temperatures are 25 °C, the climate is solidly in the "paradox" regime (D'Hondt and Arthur, 1996).

THIS MODELING STUDY

General issues

The most obvious missing physics of the experimental frameworks reviewed above is the lack of two-way interactive coupling of complete ocean and atmosphere models—and the best current explanation for the maintenance of these key early Paleogene climatic features is atmosphere-ocean interaction. In a fully coupled approach, assumptions based on modern day are not explicitly incorporated into the simulation (usually in type 2 experiments) and circularity is avoided because temperature proxy interpretations are not used to drive the experiment (usually in type 1 and 3 experiments). Furthermore, energetic consistency is built into the modeling framework. CGCM paleoclimate modeling (type 4) is a new and rapidly evolving field, so the simulations presented here should be considered as first attempts. Further refinement and validation will be required in future work. We describe here the three main challenges. The first lies in the ability of the CGCM to achieve and maintain a steady state climate for distant past conditions. The second is a product of uncertainty in initial and boundary conditions. The third difficulty is that the deep ocean has a characteristic time scale of two thousand years, a length of time over which it is computationally infeasible to integrate a CGCM. In this section, we describe the difficulties in performing CGCM experiments of the distant past, summarize a method that succeeds in resolving these difficulties, and outline the experiments performed.

The CCSM and climate drift

Due to deficiencies in model formulation, most CGCMs require so-called "flux corrections" to maintain a stable steady state climate for modern conditions. Without these corrections, the modeled climate drifts toward unrealistic values of important climate parameters. These models are not convergent on a realistic climate state even for perfectly known initial and boundary conditions and consequently cannot be used for paleoclimate modeling in the deep past. It is therefore critical to use a CGCM that does not require "flux corrections" for modern day when attempting to model the early Paleogene.

We simulate early Paleogene climate with the Community Climate System Model (CCSM) developed at the National Center for Atmospheric Research (NCAR). CCSM (formerly CSM) version 1 (v.1) is described and validated for modern conditions in Boville and Gent (1998) and references therein. The results presented here are produced with an updated version 1.4, described for modern studies in Boville et al. (2001) and Blackmon et al. (2001). The atmospheric and land models are the Community Climate Model (CCM) 3.6 and the Land Surface Model (LSM) 1.2 at T31 (~3.75° by 3.75°) resolution. These are coupled to NCAR's Ocean Model (NCOM) 1.5 and the Community Sea Ice Model (CSIM) 2.2. The latter models are on a stretched grid with 0.9° equatorial and 1.8° high-latitude meridional grid spacing; zonal resolution is 3.6°. Components are connected through the Flux Coupler v. 4.0.5. As described in the references above, the major differences from CCSM v.1 are (1) better resolved tropical ocean circulations and thermocline structure (2) improvements in conservation properties of Flux Coupler interpolation between component models, and (3) an error in the sea-ice in v.1 has been fixed, leading to an order of magnitude decrease in the already weak model "drift."

The inclusion of state of the art treatments of boundary layer (KPP) and isopycnal eddy mixing (GM) in the ocean are notable features of this model (Boville and Gent, 1998). The model also predicts accurately modern El Niño–Southern Oscillation phenomenon (Otto-Bliesner and Brady, 2001), the Australian–East Asian monsoon, the North Atlantic Oscillation (B. Otto-Bliesner, 2001, personal commun.) and their remote climate responses (Meehl and Arblaster, 1998). CCSM also produces a response to future global warming well in the range produced by other models, with equilibrium sensitivity of ~2.1 °C for a doubling of pCO_2 (Dai et al., 2001; Meehl et al., 2000). Thus, the ocean-atmosphere interactions and their terrestrial impacts that are important in modern climate are well reproduced in this model, indicating that it is realistically sensitive to external forcing and internal variability. CCSM has also been used with paleoclimate settings in a series of Last Glacial Maximum (Liu et al., 2002; Shin et al., 2002) and Cretaceous simulations (Otto-Bliesner et al., 2002).

Initial and boundary conditions and the effects of uncertainty

Process and importance of degradation. Early Paleogene climate proxies that might be used as initial conditions for a modeling experiment are sparsely distributed in time and space, have inherent uncertainties associated with them, and may not constrain quantities that are important for climate modeling (e.g., salinity). There is also a good deal of uncertainty in boundary conditions, such as topographic or bathymetric reconstructions. Our knowledge of these boundary and initial conditions can never be as great for the early Paleogene as they are for today. Therefore we restrict this investigation to questions that can be answered with the level of detail available for the early Paleogene. We assess which questions can be answered with a process we call "degradation."

Approximation connotes that the essential elements of a system have been captured despite simplification. Given that the boundary and initial conditions are only roughly known and without universal agreement on the essential elements of the climate problem, our approximations are unavoidably wrong to a degree. We prefer the term degradation, because it better describes the process applied to validate this method for coupled model spin-up. Degradation provides one (minimum) estimate of the errors introduced into model predictions by misrepresenting multiple aspects of the system. Although the CCSM is currently one of the best described and validated CGCMs (in terms of number of recent publications), a "perfect" model for modern day might not be perfect for the correct physical reasons (instead, tuning might be responsible), nor nearly as accurate with less accurate boundary conditions. The process of degradation includes changing both the boundary conditions in the modern control case and some "detuning" of the model (used in both modern and paleo-cases). The modern-day bathymetry was also degraded as described below.

In the land surface model (LSM), modern soil sand/silt/clay fractions were set to the same global average values used in Sloan and Rea (1996) and three elements of tuning were turned off in modern and Eocene cases. These tunings make Arabian desert sand darker than average sand, and make snow that falls on land-ice south of 60°S, and north of 40°N between 45 and 165° east longitude (Siberia) more reflective. If the exact color of sand in a desert or the precise albedo of snow or the presence of small islands (like Iceland) actually play a large role in climate, then: (1) this detuning and degrading procedure should show it, and (2) it would raise serious questions about our ability to model past climates.

This degradation has been applied to a modern-day case in a manner analogous to that in which the paleoclimate cases are likely to be degraded. In so doing, many possibly important parameters have been changed at once. A weakness of this decision is that comparing the degraded modern case to standard modern case results is not a single parameter sensitivity study. Single-parameter sensitivity studies have been performed that indicate the effects are small enough to be well within observational error for the early Paleogene. As described below, and in Huber (2001), the degraded modern simulation quickly reaches a near-modern state that is close to modern observations. The strength of degradation is that by changing so many things at once, we more closely match the kind of errors that are likely to be introduced in paleoclimate work—since it is unlikely that only one feature is incorrectly represented. Despite its limitations, degradation addresses the question of whether even the "perfect" model can converge on a realistic modern climate state from poor knowledge of initial and boundary conditions.

Creation of early Paleogene and "degraded" modern boundary conditions. The vegetation and topographic distributions for early Paleogene conditions are described in Sewall et al. (2000). The creation of a bathymetric data set for the early Paleogene began with land-sea distributions created by Sewall et al. (2000). Those land sea distributions were then remapped from a 2° × 2° distribution to the stretched grid of the ocean model. Slight modifications were made to ensure proper resolution of ocean gateways and narrow embayments. Isochron maps created by Royer et al. (1992) were then used to identify the paleolocations of Chron 25 and other late Paleocene–early Eocene isochrons, in order to constrain the location of midocean ridges during the early Paleogene. This information was supplemented by other tectonic reconstruction data in the Pacific (Lonsdale, 1998), Tethys (Dercourt et al., 1986; Zacher and Lupu, 1999) and the Indian Ocean (Zachos et al., 1992; Wilson and Moss, 1999). The locations of passive and active margins consistent with our tectonic reconstruction were mapped. Then, using a modern analogue approach, 6 depth classes appropriate for these tectonic regimes were derived from the modern ETOPO5 data set (NOAA, 1988).

These depth estimates were then spot checked with paleodepth estimates of specific ODP and DSDP cores based on faunal assemblage estimates (Zachos et al., 1992, 1993; Corfield and Norris, 1996; Miller et al., 1987), and found to be in good agreement. The depth of the Panamanian sill was picked somewhat arbitrarily (at 2500–3000 m) given the considerations that there is no evidence of a constriction of flow during the Paleogene, but in the late Paleocene, there is evidence of land faunal migration from S. America into N. America (Marshall et al., 1997). The bathymetric data set was smoothed using a 5 point smoother and then rechecked against the corederived paleodepth estimates mentioned above.

To evaluate the potentially large error introduced in the method of deriving paleobathymetry, analogous filters were applied to modern-day bathymetry. Modern bathymetry was discretized into the same 6 classes; seamounts and islands, including Iceland and Japan were removed, and the bathymetry was passed through a 5 point smoother, in addition to the detuning described above. These Eocene and degraded modern bathymetries are shown in Figure 2. In the simulations, the viscosity/diffusivity values are set at extremely low values (similar to those in Otto-Bliesner and Brady, 2001), which should allow the simulation to provide a more accurate representation of tropical current systems and gateways.

Initial conditions. Early Paleogene simulations were performed with two different initial conditions with identical boundary conditions. The LOGRAD distribution was created from the SST record of Zachos et al. (1994) for the early Eocene. The Zachos et al. (1994) record was used as a basis for the zonally constant meridional SST values in LOGRAD. The coldest temperature in that data set was used as a first estimate of the deep ocean temperature, and SSTs were matched to this deep ocean value with a smooth fit. Salinity was initially set to a global average value of 34.7 ppt. The ocean heat transport implied by this distribution is several times modern (Fig. 1). The HIGRAD distribution was chosen assuming that the interpretation of tropical SSTs were biased toward too cold values in the Zachos et al. (1994) reconstruction and so tropical SSTs were made warmer and polar SSTs were made cooler to ensure the same global average temperature. An alternative interpretation of the HIGRAD distribution is that it is a better match to late middle Eocene or Paleocene values. The ocean heat transport implied by the HIGRAD distribution is near modern values. Deep ocean temperatures were specified in the same manner as in the LOGRAD distribution. Initial ocean conditions for the degraded present case are zonally averaged temperature and salinity taken from Levitus and Boyer (1994) and Levitus et al. (1994) and a state of rest.

The boundary conditions for the early Paleogene cases are described above and include pCO_2 of 560 ppm. The different Paleogene cases produce nearly identical surface properties, so we describe the case that began with HIGRAD initial conditions in this study. We discuss the results of only one in depth, because the two case results are similar, though some discussion of the second experiment is included in the Discussion and Summary sections. We have also performed several degraded modern control cases to explore the sensitivity of boundary and initial conditions, and spin-up procedure as described above and in Huber and Sloan (2001) and Huber (2001). The modern control case includes degraded modern boundary condition described above and a preindustrial value of pCO_2, 280 ppm.

Deep ocean spin-up procedure

A fundamental difficulty with performing paleoclimate CGCM experiments is integrating the entire system, including the deep ocean, to quasi-steady state. It is not currently feasible with existing computational resources to attempt the ideal solution, which is to integrate several 3000 year simulations with a CGCM. Therefore, an iterative accelerated spin-up technique was developed to allow several thousand years of deep ocean simulation to be carried out in only several hundred years of surface ocean computation.

The procedure draws on two existing procedures: iteration of coupled and uncoupled modeling steps, and deep ocean acceleration. The iteration technique has been used and validated in a somewhat different form than that described here by Kutzbach and Liu (1997), Liu et al. (1999), Vavrus et al. (2000), Shin et al. (2002), and Otto-Bliesner et al. (2002), but their techniques are somewhat different, and the results of different techniques have yet to be quantitatively compared. The combination of uncoupled and accelerated ocean model integrations and fully

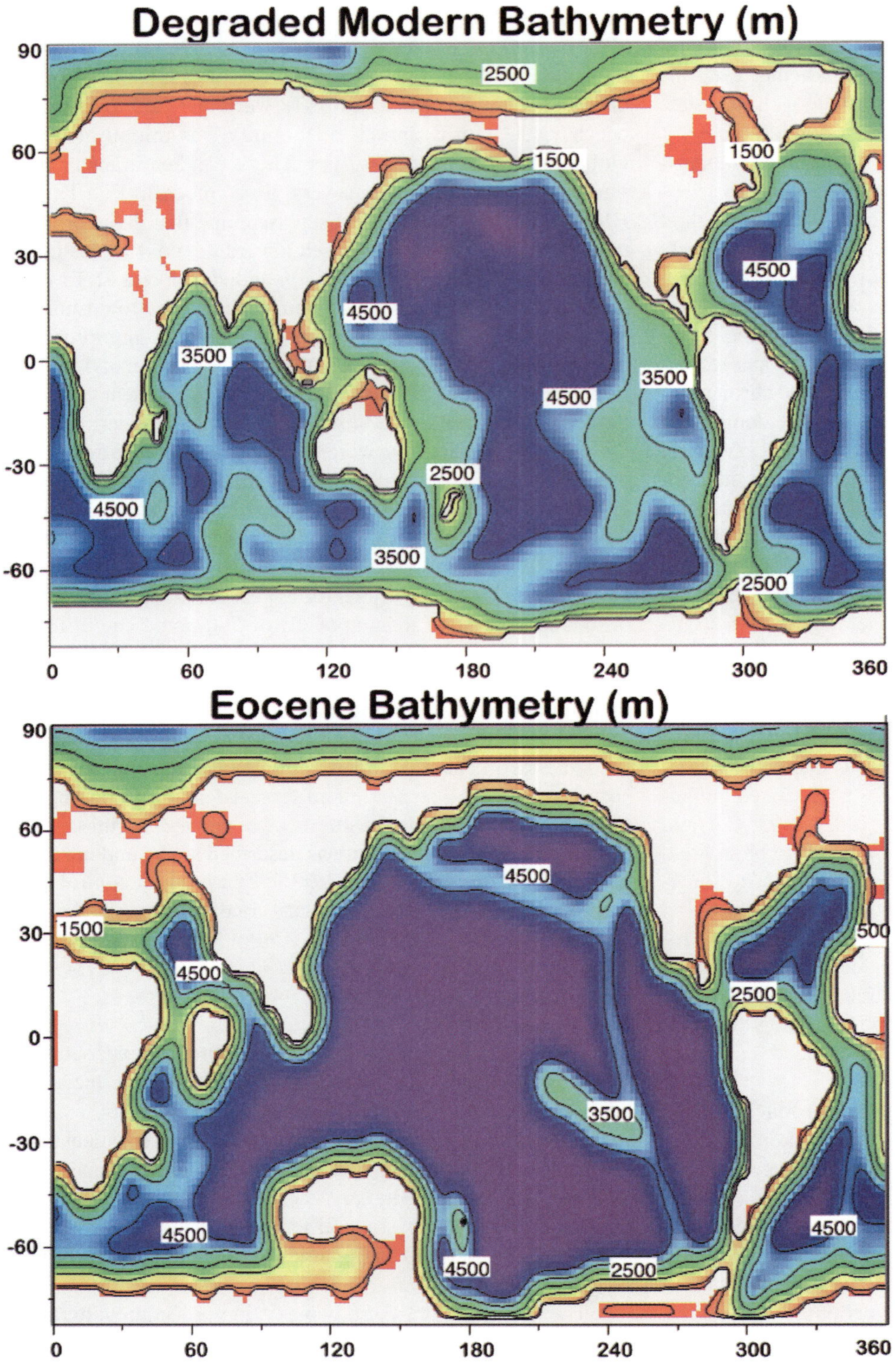

Figure 2. Degraded modern (top) and Eocene (bottom) bathymetry at 2° × 2° resolution. Contour intervals are 1000 m below 1500 m and 500 m above.

coupled unaccelerated integrations is quite common and is the usual way that CCSM integrations are spun up for modern day (see Boville and Gent, 1998; Bryan, 1998), but the process has not been systematically iterated, as it is in this study. The exact spin-up procedure is described in Huber and Sloan (2001).

SST biases in the modern (degraded) control case are <2 °C over most of the ocean, and ocean heat transport values are within the error for existing modern-day estimates (Ganachaud and Wunsch, 2000; Trenberth and Caron, 2001). The major temperature errors are in the stratus cloud regions at the eastern edges of the tropics (~+2 °C) and in the Ross Sea (~+8 °C). The control simulation's major weakness is too-weak formation of Antarctic Bottom water and a sluggish Antarctic Circumpolar current (ACC). CCSM v.1, when integrated for long periods, tends toward overly strong Antarctic Bottom water formation and ACC transport as described in Bryan (1998). Sensitivity studies (not shown) demonstrate that the Ross Sea temperature bias, weak ACC, and weak Antarctic Bottom Water formation are due to sea-ice and albedo feedbacks and spin-up procedures. Sea-ice was not likely to have been present in the warm polar early Paleogene climate and does not play an important role in our results described below, so this sensitivity minimally affects our early Paleogene results. The changes in bathymetry associated with degradation are also important for changes in the ACC transport, appearing to account for about a 25 Sv reduction (~1/5 of the total transport). This is in keeping with current understanding of the unique dynamics of the ACC, and is probably not likely to be a dominant source of uncertainty in our Paleogene simulations. The results of those studies show, in agreement with many previous studies, that in a suitably high resolution and inviscid ocean model the details of bathymetry have small effects on most large-scale ocean currents.

The total length of the early Paleogene integration in the case described here is 2200 deep water years (DWYs), and there was little significant difference between successive iterations, thus indicating convergence (not shown). As described in Huber and Sloan (2001), the iterative method proves to be robust and leads to an equilibrated coupled simulation that reproduces present climate well in the degraded modern case, and a stable and convergent solution in the early Paleogene cases. Extensions of the early Paleogene experiments to a total length of 6000 DWYs, completed during the revisions of the manuscript, produce substantially the same results as those described here, further demonstrating the results' robustness (not shown). In order to cover the results in sufficient depth we focus only on results in Atlantic and Indian Ocean regions; results from the Pacific are covered in other work (e.g., Huber, 2002). Means of the last 20 yr are discussed below.

RESULTS

The global average surface temperature difference between the degraded and the early Paleogene case is 5.1 °C. This indicates a strong sensitivity of mean temperatures to the changes in boundary conditions (e.g., vegetation, topography), given that the effect of higher pCO_2 alone should be ~2.1 °C. Global and volume average ocean temperature in the early Paleogene simulation is 8.1 °C, and the global average deep ocean temperatures are ≥ 6 °C, demonstrating that realistic boundary conditions and 560 ppm pCO_2 are sufficient to reproduce generally warm deep ocean temperatures, albeit at the lower range of proxy estimates (e.g., late-middle Eocene, Zachos et al., 2001). There is very little "drift" in the simulation: The temperature trend for the last 90 yr is <.024 °C/100 yr and salinity 0.0003 ppt/100 yr. These values represent quasi-steady state values and the trends are extremely small even relative to modern CGCM integrations (e.g., Blackmon et al., 2001) (the salinity trend is statistically indistinguishable from zero).

Meridional overturning rates in the North Atlantic are steady at ~10 Sv, half that of the control simulation, but there is active deep convection at 55°N latitude with near-surface water penetrating to ~3000 m (Fig. 3). In the upper ocean (500 m) this water mass has a temperature of 13 °C and a salinity of 35.56 ppt (Fig. 4). This "young" North Atlantic Deep Water does not dominate the abyssal Atlantic because significant quantities of this water mass flow into the Pacific at 1–2 km depth through the Panamanian Seaway. The temperature of this NADW at depth is 9.5 °C and the salinity is 35.4 ppt (Fig. 4). In the South Atlantic, vigorous upwelling in the subpolar gyres leads to watermass transformation at the surface in high latitudes, where heat loss and freshening create a watermass analogous to Antarctic Intermediate Water (T = 3–6 °C, S = 33.0–33.75 ppt) (Figs. 3 and 4). Some of this water is exported into the North Atlantic at ~1000 m, but most of it mixes with warmer and saltier water from the North Atlantic to form a deep Antarctic water mass (T = 6 °C, S = 33.75 ppt) (Fig. 4).

The processes operating in Tethys and the Proto-Indian Ocean are not substantially different from those in the Atlantic. Deep convection occurs in the northern portion of Tethys (45°–50°N) (Fig. 3). Seasonal cooling and latent heat fluxes from offshore winds lead to an upper ocean Tethyan watermass with a temperature of ~15 °C and a salinity of 35.5 ppt (Fig. 4). This water flows southwards at depths of 1000–2000 m with a temperature of 10 °C and salinity of 35.5 ppt (Figs. 3 and 4). In the Southern Hemisphere the process and pattern of deep and intermediate water formation is similar to that described for the Atlantic.

Meridional ocean heat transport is similar to modern values (Fig. 1C). The majority of poleward heat transport by the oceans occurs in the Pacific. Atlantic heat transport is neglible in the Northern Hemisphere and directed equatorward in the Southern Hemisphere (as it is today). Peak heat transport values are higher in the Northern Hemisphere (1.6 PW) than in the Southern (0.8), but the distribution is more hemispherically balanced than in the present-day control run. Figures 4A–E present map views of annual mean near-surface or "mixed layer" properties. We present these vertically averaged values (upper two levels), rather than model surface temperatures both as a crude representation of the likely depth habits of foraminifera (e.g., Schmidt, 1999a; Poulsen et al., 1999; Bice et al., 2000b) and to increase the robustness of the model result interpretation.

Mixed-layer temperature, salinity, and velocities

Far North Atlantic temperature values are ~15 °C and Southern Ocean temperatures are predominantly 1–3 °C (Fig. 4A). Tropical mixed-layer temperatures are above 29 °C with peak values of 31 °C. Salinity reaches very low values (31.5 ppt)

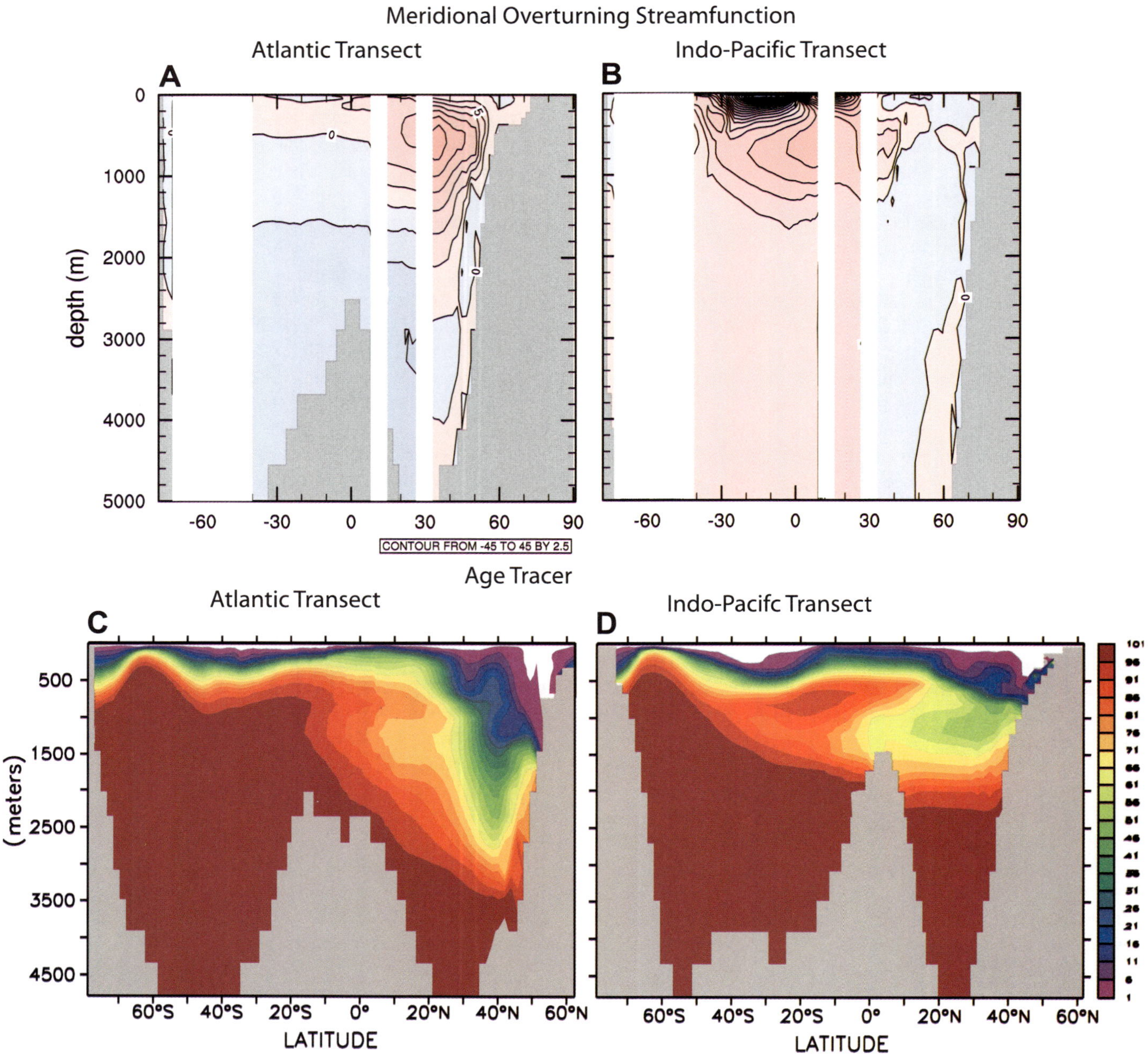

Figure 3. A: Zonal, annual average meridional overturning streamfunction (in Sv) in the Atlantic (A) and Pacific (B). Red indicates clockwise overturning, blue indicates counterclockwise. On the bottom an "ideal age" tracer used as a CFC-like qualitative indicator of the ventilation and transport of water masses in the Atlantic (C) and Pacific (D). The oldest value is 100 and represents values that have not been in contact with the surface within 100 yr, whereas younger values indicate that some or all of the water has been in contact with the surface within the period indicated by the color bar.

in the Southern Ocean and off the west coast of India (Fig. 4B). A low-salinity tongue is advected by currents northwestwards from India, into Tethys and thence into the Atlantic. The rest of Tethys is dominated by an anticyclonic gyre circulation that generates an eastward-directed current along its northern margin. The Benguela Current system advects cool, fresh water (17 °C, 32.5 ppt) northwestward to Brazil. Subtropical gyre evaporation zones have high salinities (>36.5), but are not significantly different from modern values. Strong northward cross-equatorial transport of warm, salty water is generated in this upper layer dominated by Ekman dynamics. A vigorous Gulf Stream-analogue transports warm (18 °C), salty (35.5 ppt) water north

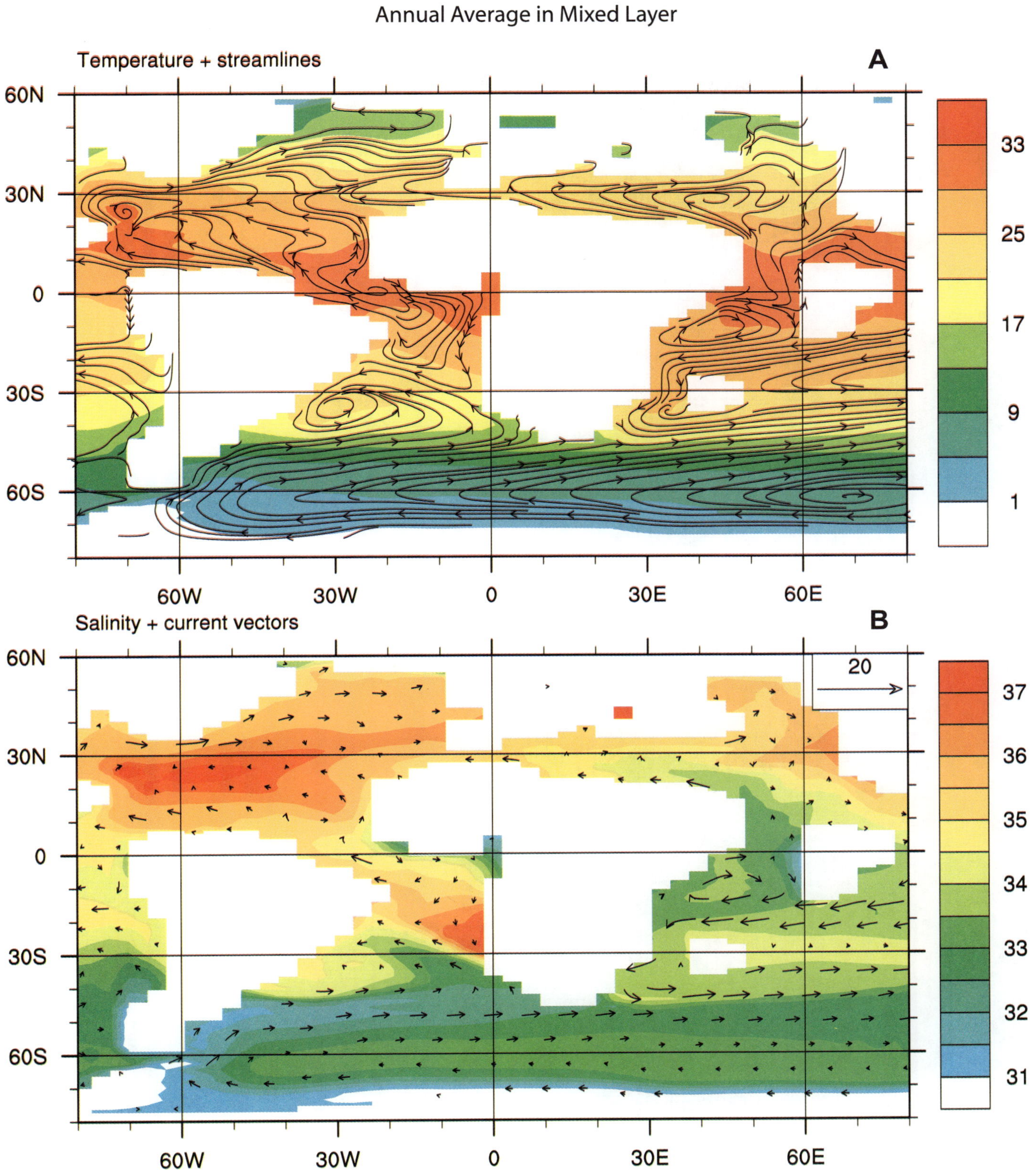

Figure 4 *(on this and following three pages).* Tracer properties and velocities at different levels. In the isotopic calculations we use a correction of –0.98‰ for ice-free conditions and an additional offset of –0.22‰ to convert from VSMOW to VPDB. Values are indicated by the color map and vectors located on the figures.

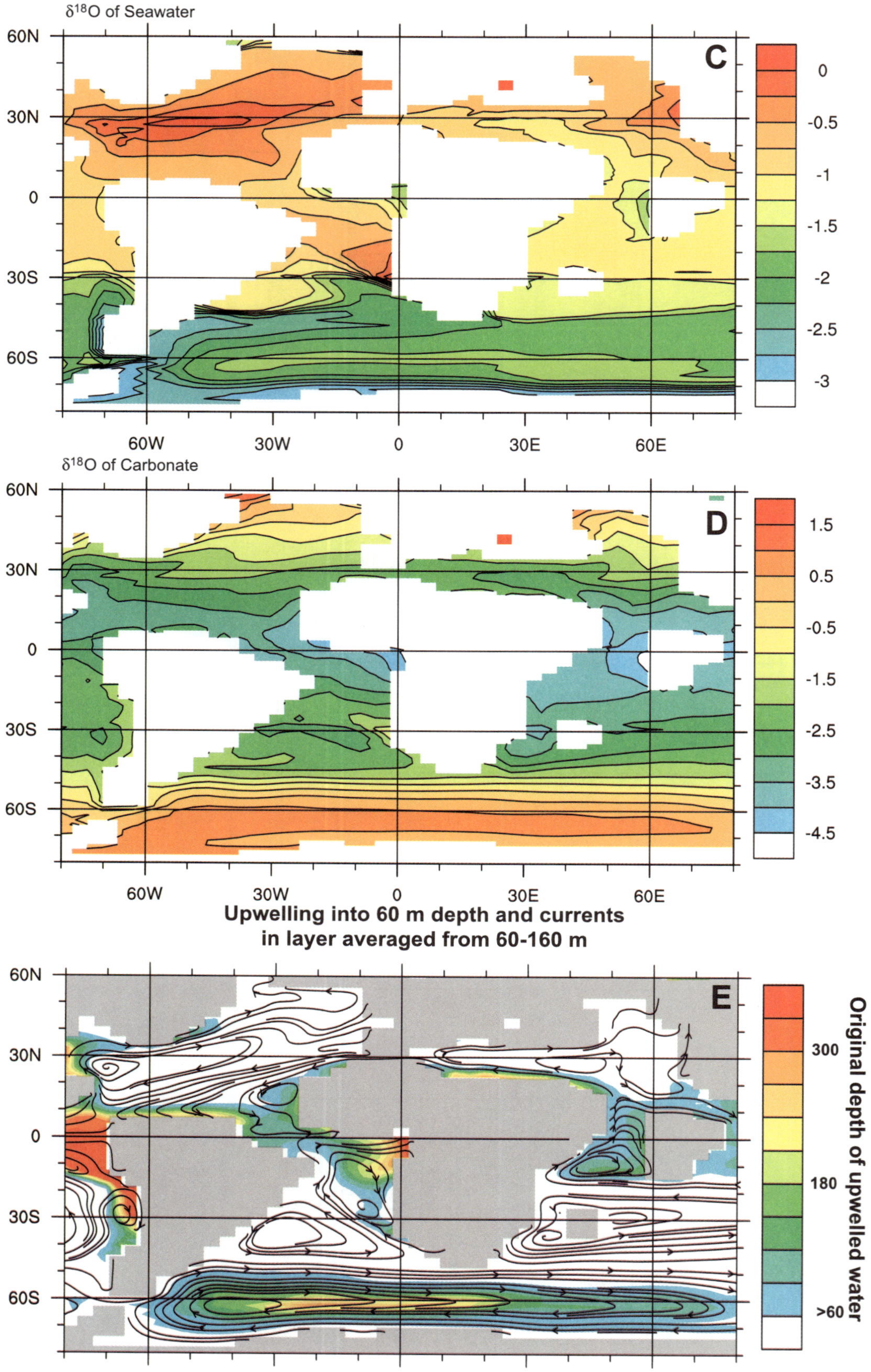

Figure 4. (*continued*)

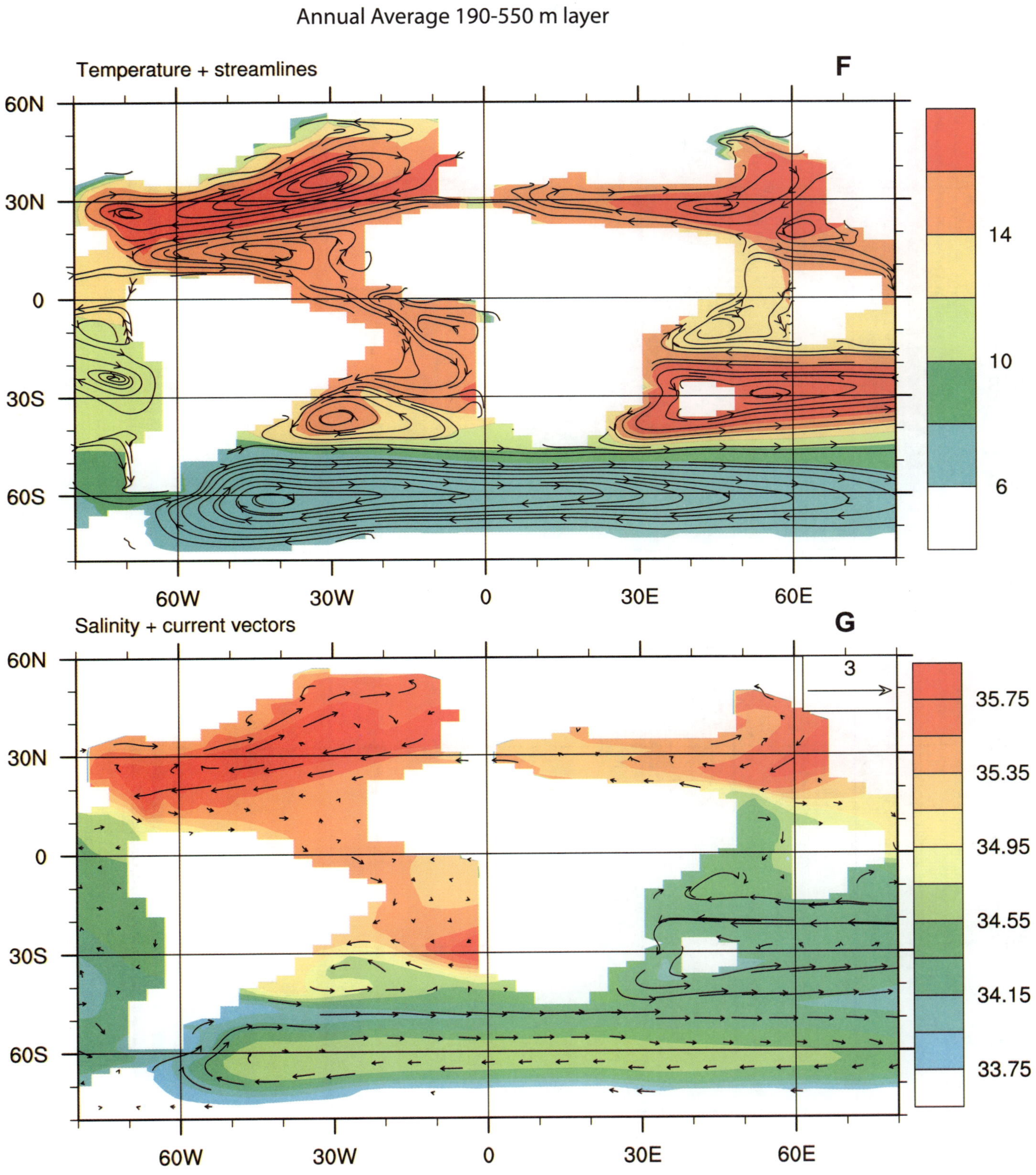

Figure 4. (*continued*)

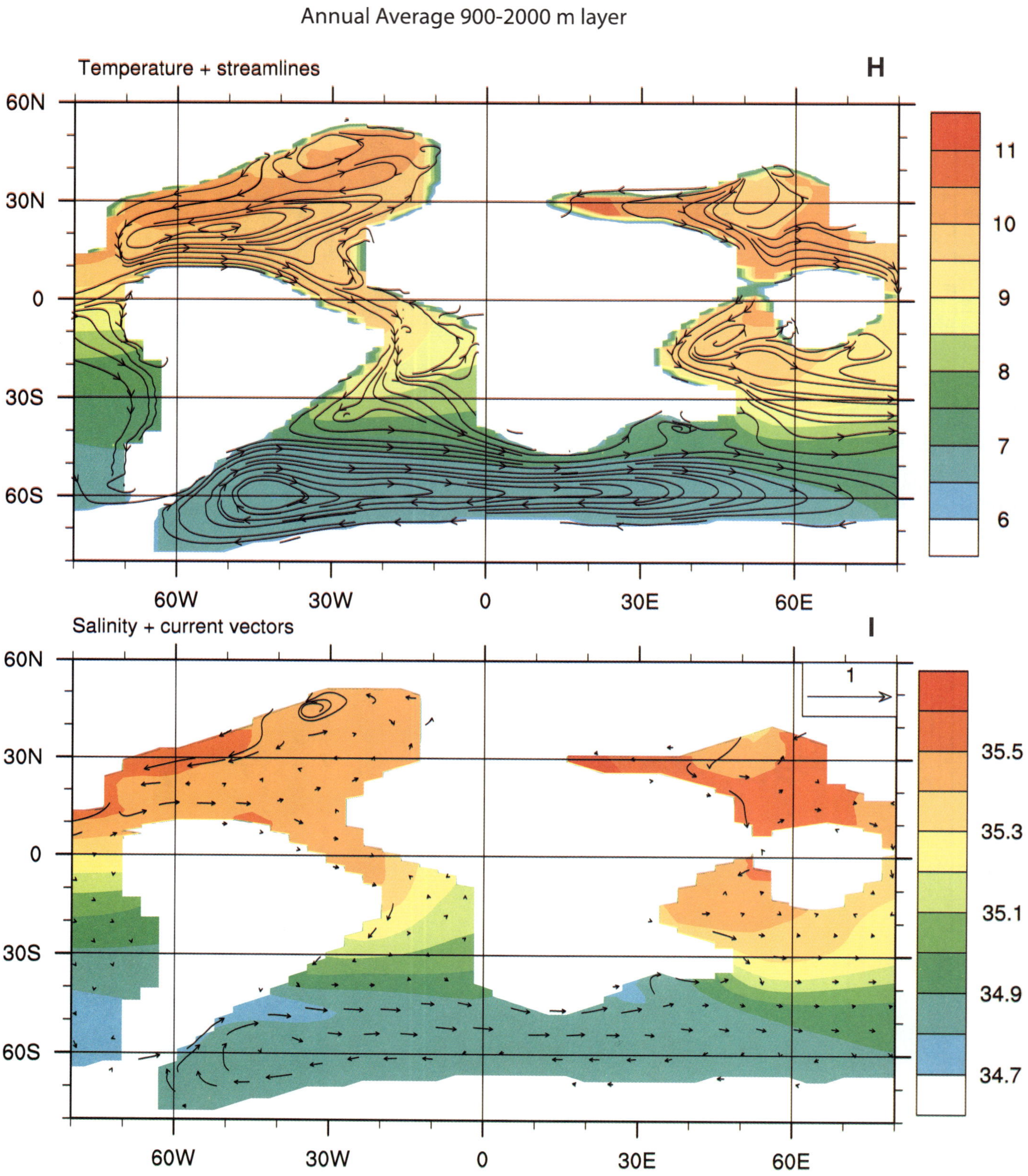

Figure 4. (*continued*)

and eastwards. A subpolar gyre, a Labrador Current-analogue, transports cooler, fresher water southwards along the east coast of North America (this feature is seen more clearly at other depths [below]).

Oxygen isotopes

In Figure 4C, an estimate of seawater $\delta^{18}O$ (δ_w hereafter) is plotted. The δ_w distribution is calculated from the empirical relation between seawater salinity, and δ_w gradients proposed by Fairbanks et al. (1997) for the equatorial region (within 30° of the equator), and the relationship of Broecker (1989) in the extratropics. The strengths and weaknesses of this general technique are discussed in Schmidt (1999b) and Bice et al. (2000b). One improvement of our experimental framework over others is the use of a CGCM, so salinity is predicted from two-way ocean-atmosphere interactions, although we can not avoid the uncertainties introduced by features not well represented by the model, e.g., freshwater continental runoff (Bice et al., 1997). Values in the Southern Ocean are notably more negative than previous estimates (Zachos et al., 1994; Bice et al., 2000b), and Northern Hemisphere values more positive. A point-wise comparison is done below.

To facilitate direct comparison of our results with foraminiferal calcium carbonate $\delta^{18}O$ values (δ_c hereafter), δ_c is explicitly calculated from model output in the following way. The calculated δ_w and model predicted temperatures (averaged through the mixed layer) and the commonly used Erez and Luz (1983) relationship were used to produce our estimate for δ_c (similar to Schmidt, 1999a). In general the results are more negative (~2‰) than most proxies in the tropics and more positive (1‰) in the high latitudes. Depth habit of the planktonic foraminifera is a key area of uncertainty in any comparison because predicted values of δ_c are >1‰ more positive at 50 m depth in the tropics than at the surface (similar results were obtained by Poulsen et al., 1999). Some of the tropical results are in agreement with the new estimates of Pearson et al. (2001).

Upwelling

Another important property that can be predicted by the model is the "age" of water upwelled into the mixed layer, which may be compared with previous independent model-derived estimates of upwelling and with upwelling and productivity proxies. In Figure 4E, we show the depth from which water is upwelled into 60 m depth. Greater depths are indicative of more vigorous upwelling and greater nutrient input into the mixed layer, assuming the deep ocean was a net supplier of nutrients. The coastal Gulf of Mexico, Blake Nose, the North Coast of Africa and South America, India, the Benguela Current system and Niger experience significant upwelling on an annually averaged basis. Most of the Southern Ocean ~60°S also is vigorously upwelled. Subtropical gyres including the Brazil Current system, the North Atlantic gyre, and the Agulhas-South Equatorial Current complex experience downwelling as expected. These results are in agreement with the completely independent upwelling estimates of Huber and Sloan (2000), and Handoh et al. (1999) and generally in agreement with proxies, summarized in Huber and Sloan (2000) and Sloan and Huber (2001).

Upper ocean

In the above-thermocline upper ocean (an average from 190 m to 550 m), Sverdrup dynamics dominate the velocity distribution with wind-driven geostrophic gyres in evidence (Figs. 4F and 4G). Sharp fronts are evident between the Southern Ocean subpolar gyre (6 °C, 34.5 ppt) and the subtropical gyres (Brazil Current 13 °C, 34.75 ppt and Agulhas 16 °C, 34.35 ppt). It is noteworthy that there is a sharp gradient in temperature and salinity in the South Atlantic at the subpolar front, but there is no sharp gradient in salinity at the subpolar front in the southern Indian Ocean. Transport along the eastern margin of South America is southward across the equator, bringing a signal of warm (14 °C, 35.4 ppt) and "young" water (some of the water has been at the surface within 100 yr) into the South Atlantic. In the North Atlantic, a strong Gulf Stream flows north and east, but it is noteworthy that the front between this current and the Labrador Current analogue extends down to the location of ODP cores from Blake Nose. This suggests that substantial variability (>4 °C or 1‰ $\delta^{18}O$) may occur at this site due to slight changes in the position of the Gulf Stream. This may explain some of the variability noted by Wade et al. (2000) at Blake Nose.

Deep ocean

The majority of "deep" ocean proxy data are from paleodepths <3000 m and ocean properties become more homogenous with depth, so we average vertically between 900 and 2000 m (Figs. 4H and 4I). The hemispheric asymmetry in temperature and salinity, with cooler, fresher values in the Southern Hemisphere, is apparent at this depth, although the contrasts are weaker. There is southward transport of warm, salty water into the Southern Hemisphere to balance the northward transport in the upper layers. Temperature decreases with depth in the North Atlantic, but in the South Atlantic, deeper values are significantly warmer than upper ocean and surface values. This demonstrates that the sort of "inversion" in temperature seen in Southern Ocean sites (e.g., 689 and 690 during the PETM) can be explained by warm deep water formation at high latitudes rather than at low latitudes in contradiction to the Proteus Ocean hypothesis by Kennett and Stott (1990).

Mean annual temperature range

In the near surface ocean, the seasonal cycle of temperature is an important factor for understanding climate and the interpretation of climate proxies (e.g., Andreasson and Schmitz,

1998). The seasonal temperature cycle also provides a connection between ocean processes and the equable climate problem, so the mean annual temperature range (MATR) over both land and sea is shown in Figure 5. MATR over land and ocean is similar in our early Paleogene and modern control case with some exceptions. Large decreases in the magnitude of MATR occur over land in North Africa, Siberia, Antarctica, and the Arctic Ocean. Increases in MATR occur in Australia and central Canada. MATR over the oceans is virtually unchanged from near-modern values.

Pointwise comparison with data

Mean annual temperatures (MAT) predicted by the model are compared to data in Figure 6. Spatially collocated values for mean annual SST predicted by the model and those produced from the Zachos et al. (1994) δ_c compilation have similar values except in regions poleward of 40° latitude (Fig. 6A). The model underpredicts SSTs in high latitudes by 5–10 °C. When the model-predicted δ_w distribution is used instead of the empirical function of latitude proposed by Zachos et al. (1994), estimated SST values decrease by 2–3 °C in the Southern Ocean and by 5–10 °C in the Southern Hemisphere midlatitude oceans, whereas there is little difference in the Northern Hemisphere (Fig. 6B). Although inclusion of the model-predicted δ_w distribution improves (by 20%) the model-data mismatch at low temperatures, it weakens the correlation at temperatures between 20 and 25 °C (Fig. 6C).

A similar bias is seen in the comparison with terrestrial MAT values (Fig. 6D). The model accurately captures the MAT in one compilation of values from Wyoming and Baja California. Other compilations of MAT in Wyoming, including the high-resolution records of Wing et al. (2000) and Wilf (2000), show substantial variability, from near the model-predicted value to significantly higher values. These warmer values fall along a line that is characteristic of other (low-resolution) terrestrial MAT estimates from the early Eocene. In general, compared to these warm estimates, the model is offset uniformly by 10 °C.

DISCUSSION AND IMPLICATIONS

Comparison to results of previous coupled models

The basic patterns and processes produced in this study are not qualitatively different from those that appear in the modern day, and the early Paleogene results are generally in agreement with previous coupled model studies. Since this is the first fully coupled model study of the early Paleogene, there are some meaningful comparisons we can make with other coupled model studies.

The study by Saravanan and McWilliams (1995) included a simplified 2-layer atmosphere with rudimentary treatments of moist processes, clouds, and radiation coupled to a low-resolution "sector" ocean model, i.e., a zonally averaged ocean that includes an ad hoc treatment of important features such as western boundary currents, and no treatment of sea-ice. Although not focused on past warm climates, with modern pCO_2 concentrations, this study produced three possible ocean circulation modes, and indicated a near-constant total poleward heat transport (atmosphere + ocean) and small changes in temperature gradients (ΔT = 28–34 °C), with higher pCO_2. Using a different kind of simplified model, Schmidt and Mysak (1996) conducted a study of past warm climates within a framework in which simple "sector" oceans were connected in an idealized ocean basin configuration. This ocean model was coupled to an atmospheric model in which atmospheric heat fluxes were constrained to maintain a steady climate in equilibrium with the initial (specified) SST distribution. Because this framework assumes the existence of whatever atmospheric fluxes are necessary for steady state, this technique cannot discern whether a climate state with small temperature gradients is possible. Nor can greenhouse gas concentrations be meaningfully changed, because as Schmidt and Mysak acknowledge, "the atmospheric model can be considered to take account of whatever CO_2 level is necessary to maintain the specified initial climate." Instead this technique allows the estimation of the ocean and atmospheric heat fluxes necessary to maintain different temperature profiles, and the deep ocean temperature and overturning circulations in equilibrium with those values. The existence of Indian Ocean sourced salty deep water produced in the Schmidt and Mysak (1996) study is consistent with the Tethyan watermass formed in our study.

Bush and Philander (1997) performed a fully coupled simulation of the past warm climate of the Cretaceous that produced surface temperature gradients similar to those of our study, but that simulation was only integrated for 35 yr as opposed to 2200 yr in this study. Otto-Bliesner et al. (2002) have performed fully coupled simulations of Cretaceous climate also with NCAR's CCSM, but the spin-up technique and boundary conditions, including pCO_2, are different than in our study (their simulation is for 100 Ma). Some key features of our results and those of Otto-Bliesner et al. (2002) are similar: Their deep ocean temperatures of 8–10 °C (for 1120 ppm pCO_2) compare favorably with our results (given our lower pCO_2); warm deep water forms in the extratropics in both experiments, the meridional temperature gradients produced are also similar (compare with Huber and Sloan, 2001). Importantly, their simulation shows extremely vigorous Southern Ocean deep water formation (>20Sv), and with it, somewhat increased heat transport in the Southern Hemisphere (relative to the values in our simulation), although in general their heat transport values and ours agree.

To our knowledge no simulation with coupled dynamical models has produced a low-gradient climate, i.e. a ΔT as low as 15 °C. The ΔT and ocean heat transport produced in this study is well within the range expected from past coupled work.

Overturning, deep water formation and multiple equilibria

Averaged over all basins, the meridional overturning circulation is weaker than modern (see Huber and Sloan, 2001), but

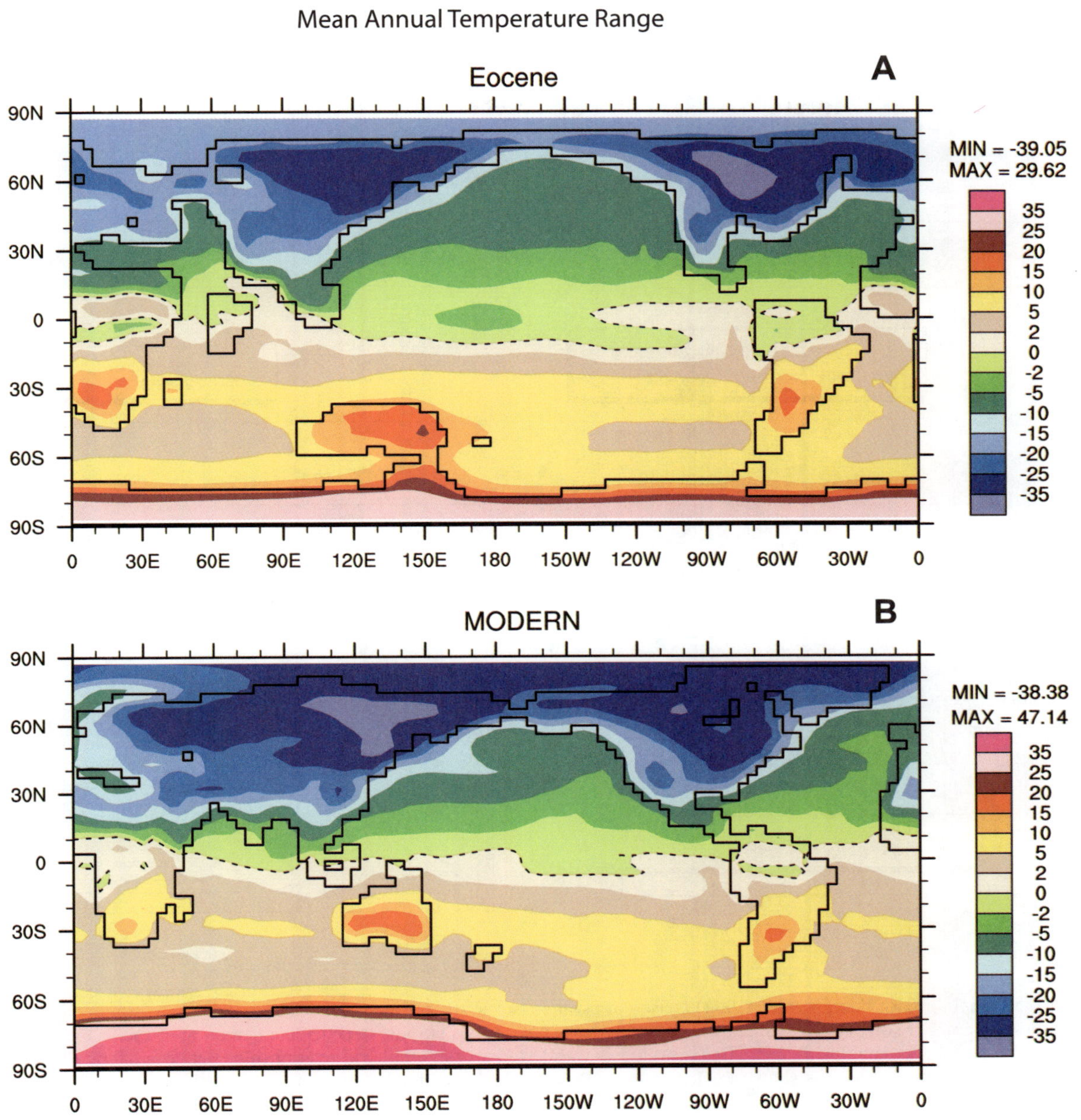

Figure 5. Mean annual temperature range for the early Paleogene (A) and the modern "control" case (B). Dashed line is the zero line. The absolute value of these fields is the relevant climate parameter.

within the individual basins considered here, the ocean velocities and overturning circulations are not dramatically reduced compared to modern (compare with Gent et al., 1998). Deep convection occurs in this simulation in the North Atlantic and Tethys (similar to Barron and Petersen, 1991), and these warm, salty water masses penetrate and mix into the Southern Ocean water masses that are cooler and fresher. In the simulation focused on here, the Southern Ocean is extremely fresh and deep ocean convection is suppressed (although see below for the second simulation).

The net effect is a noticeable hemispheric asymmetry in tracer properties and flow patterns. This bears some resemblance to the "bipolar" ocean circulation proposed by Haupt and Seidov (2001), with a cool overturning cell in one hemisphere and a warm cell in the other. The actual flow is more complex, and one impact of the complex flow patterns is the import of warm, salty NADW into the Southern Ocean, which leads to warmer temperatures at depth in the South Atlantic. In general, the distribution provides an explanation for the isotopic pattern noted in the North Atlantic by Charisi and Schmitz (1996) and

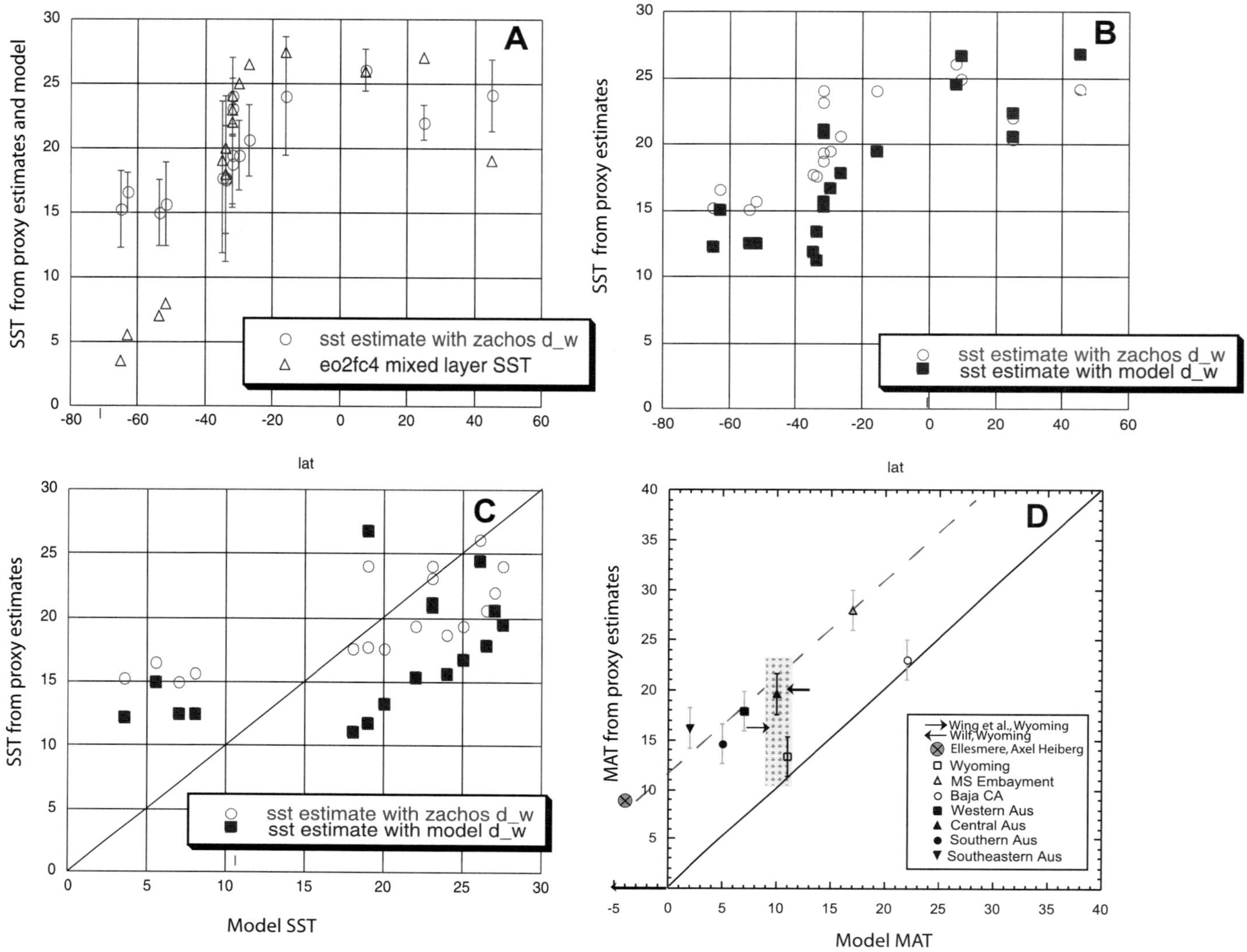

Figure 6. In A the compilation of Zachos et al. (1994) is used to estimate sea surface temperature (SST; circles) and compared with model predictions (triangles). A comparison of SST estimates with the Zachos et al. (1994) latitude dependent calculation of the isotopic composition of seawater and with those recalculated using the salinity produced in this study, from empirical relationships between salinity and isotopic composition of seawater (described in text) is shown in B. The magnitude of the difference between these two estimates is used to create the error bars in A, thus the error indicated is only that introduced by different assumptions about seawater oxygen isotopic gradients. C: Crossplot of model predicted temperatures against proxy estimates of temperature incorporating the Zachos et al. (1994) isotopic composition of seawater (circles) and those predicted in this study. A perfect model would produce values along the black diagonal. In D, mean annual temperature (MAT) from terrestrial proxies are plotted against MAT from the model. The data are from compilations by Sloan and Barron (1992), Greenwood and Wing (1995), Wilf (2000), Wing et al. (2000). The grey stippled zone represents the range of data in the latter two high-resolution studies and the arrows represent the means of those studies.

Corfield and Norris (1996) and in the South Atlantic by Kennett and Stott (1990).

There is some evidence from "aging" gradients of carbon isotopes that deep water tended to form in the Southern Ocean (e.g., Miller et al., 1987; Corfield and Norris, 1996). Because of the potential for multiple equilibria our results do not rule out a switch in modes whereby with the same initial conditions the simulation might evolve to a state with sinking in the Southern Ocean. In fact, the case beginning with LOGRAD initial conditions (not discussed here in detail) produces deep convection in the Southern Ocean, in addition to the NADW and Tethyan sources. The climatic differences, including ocean heat transport and surface temperature, between the two simulations is relatively small with <1 °C temperature change over most of the Earth and <0.1 PW over most of the oceans. The changes in deep water formation may, however, be important in explaining isotopic records. Existing carbon isotope reconstructions are very sparsely distributed, tending to be underrepresented in the North Atlantic; the model-predicted flow patterns are complex, indicating that the simple water mass trajectories assumed in these reconstructions may be questioned. For example, the model results indicate that, at some times in the early Paleogene, the deep central Pacific may have been bathed in "young" water that recently flowed in from the North Atlantic, thus aging gradients between the two regions might be erased by short circuits in the ocean circulation.

Ocean heat transport

Neither the details of deep water formation properties, nor any of the changes to boundary conditions, prove to be critical to understanding how much heat the ocean transports, given the remarkable consistency of the ocean heat transport produced in these simulations. Heat transport produced by the early Paleogene coupled model is not very different than the heat transport implied by the initial HIGRAD SST distribution, but it is somewhat reduced in high latitudes, which is consistent with the cooling there. The other simulation, which we did not describe, began with the LOGRAD SST distribution and rapidly converged on nearly the same SST pattern described in this study, with a nearly identical pattern of ocean heat transport, despite a different meridional overturning structure. Thus, the very vigorous ocean heat transport implied by the LOGRAD distribution is not maintainable in a fully coupled model with early Paleogene conditions—instead, near-modern values are robustly predicted. This implies that previous studies that used type 2 models, which assume near-modern ocean heat transport, are valid in that assumption. This means that results from those previous simulations remain valid and the type 2 experimental framework may still be applied to good advantage in future attempts to understand early Paleogene climate.

It also raises serious questions about studies of type 3 models that produce vastly different ocean heat transports than implied by the atmospheric model output with which they are driven. Our Southern Hemisphere ocean heat transport results are significantly different in magnitude, although not in direction, from those produced for the early Paleogene by Bice et al. (2000a). In fact, none of the fully coupled dynamical models described above produce ocean heat transports of the magnitude found in the Bice et al. (2000a) study. As argued earlier, we believe that the results of their study are most likely due to the SST boundary condition, because it assumes that the atmosphere is an infinite source of heat and it ignores negative feedbacks (i.e. up to 40 $W/m^2/°K$) imposed by the atmosphere (Rahmstorf and Willebrand, 1995; Nong et al., 2000). This is confirmed in an OGCM study by Najjar et al. (2002), which, using bathymetry and other conditions nearly identical to those in the Bice et al. (2000a) study, predicts values of Southern Ocean heat transport in agreement with those here. The main improvement of the Najjar et al. (2002) study is the implementation of a less artificial way of imposing the SST boundary condition. The agreement between our study and that of Najjar et al. (2002) is further indication that the results presented here are robust, which suggests that caution should be used in the interpretation of links between gateway changes and large ocean heat transport changes.

The low gradient and equable climate problems

In equilibrium with early Paleogene boundary conditions and with greenhouse gas concentrations at the low end of proxy estimates, deep ocean temperatures are 7 °C warmer than in the degraded modern control case. Tropical SSTs are ~3 °C warmer. Thus, ΔT is 3–4 °C lower in the early Paleogene simulation, but temperatures in high latitudes are still too cold (by 5–10 °C) and tropical temperatures (by 0–5 °C) too warm relative to existing proxies. The salinity distribution produced by the simulation suggests that existing proxy interpretations may overestimate high-latitude SSTs by 3–5 °C in the Southern Ocean, but underestimate them in the North Atlantic so the net effect is probably small.

Greenhouse gas concentrations in the upper range predicted from some proxies, e.g., Pearson and Palmer (2001), may alleviate the cold bias at high latitudes, but only at the expense of further raising tropical SSTs. The results of Pearson et al. (2001) are encouraging in that regard, but it is not clear whether the upward revision of tropical SSTs suggested from their results is sufficient to allow a resolution of the low gradient and equable climate problems. Our results disprove the speculation by Bice et al. (2000a) that a 2 °C warming above present tropical-subtropical SST values might solve the problem of polar warmth, via latent heat transport feedbacks. Our simulation, which includes latent heat transport feedbacks, produces a 3 °C tropical warming, but does not solve the equable climate problem or the problem of polar warmth.

Other variables need to be constrained to more precisely define the data-model mismatch in the tropics, because possible sampling biases introduce a large degree of uncertainty in tropical SST estimates. Our simulations indicate that if planktonic foraminifera record tropical temperatures from >40 m depth, rather than 6 m, then the low gradient problem disappears (in keeping with the results of Poulsen et al., 1999). This depth habit may be difficult to reconcile with evidence for the presence of photosymbionts associated with many foraminifera in existing reconstructions. Alternatively, our results indicate that a sampling bias toward warm season values in high latitudes and cold (upwelling) season temperatures in low latitudes could also partially resolve the low gradient problem. Warm terrestrial extratropical winter temperatures tend to rule out the existence of a large high-latitude bias in planktonic foraminiferal SST estimates, but the cool bias may still exist in the tropics.

Caveats

Because we have not constrained certain climate properties, e.g., SSTs, to be near values close to those reconstructed from proxies or made explicit assumptions based on modern relationships (with the exception of the isotope calculations), the simulations should be useful as an independent prediction for comparison with proxy interpretations. This also means that some of the predictions may not be as close to proxy interpretations as those from simulations that were more constrained. This is evident when comparing the model results of Bice et al. (2000b) to those here.

An important source of uncertainty in model interpretation is the treatment of runoff, as pointed out by Bice et al. (1997). In this version of CCSM, runoff is collected over all land points and distributed in a weighted fashion over the ocean, with the weighting dependent on the precipitation amount. In other words, CCSM puts most runoff where rainfall is greatest, tending to increase spatial gradients in salinity, but otherwise tending to mask the geographically local effects of runoff. Sensitivity to this will be tested in future work, but we note that the degraded modern case, which begins with zonally constant salinity distributions, produced reasonable salinity distributions and deep water formation properties (i.e., in the North Atlantic not the North Pacific) with the same runoff treatment.

Another source of uncertainty is sensitivity to initial conditions and the technique for spin-up. Two early Paleogene simulations have been performed, with very different initial conditions, that both reached very similar upper ocean properties, current structures, temperature gradients, and poleward flux characteristics, so we think this is not an important source of uncertainty. In work in progress we are experimenting with other techniques for accelerating the coupled model to an equilibrated deep ocean, with good preliminary agreement with the results here. The success of the degraded modern case (Huber and Sloan, 2001) gives us further confidence in the methods.

Vertical diffusion in the ocean is currently a controversial and important issue in understanding heat transport and the meridional overturning. Changes in oceanic vertical diffusion have been proposed by Lyle (1997) to explain past warm climates. The treatment of vertical diffusion in CCSM includes a formulation that depends on vertical density gradients (Richardson number dependence) capturing the essence of the Lyle mechanism, but we find no evidence of important consequences from this dependency. Nevertheless, it is too early to rule out changes in diffusion due to changes in tidal dissipation or bathymetric roughness (e.g., Egbert and Ray, 2001) as an important contributor to past warm climates.

Implications

These results strongly suggest that the ocean circulation is not the solution to the low gradient or equable climate problems. Nor are atmospheric feedbacks, at least as currently exist in this comprehensive model. Our coupled simulations lend credence to the oft-noted compensation of atmospheric and oceanic fluxes, and the near constancy of meridional temperature gradients suggested by climate theory (cf. Saravanan and McWilliams, 1995). The fact that the model is in keeping with these latter characterizations—but at odds with some early Paleogene climate proxies—lends support to the notion that there is an important component of climate that we still do not properly understand, or that the proxy data are being interpreted incorrectly. If the model is essentially correct, then high greenhouse concentrations (e.g., pCO_2 >1000 ppm) may solve the equable climate problem, but only if tropical SSTs are warmer than most current reconstructions. It remains to be seen how large the discrepancy between SSTs predicted by the coupled model, and those from proxy reconstructions, will be under higher greenhouse gas concentrations.

SUMMARY

Results from previous work described here demonstrate that (1) changes to land surface characteristics have important consequences at the local and regional scales, but not at large scales; (2) specification of warm polar SSTs does little, in and of itself, to warm continental interiors during the winter; (3) midlatitude continental interior temperatures are at least as sensitive—if not more sensitive—to changes in tropical SSTs as to extratropical SSTs.

Results from the simulations in this study demonstrate that:

1. In this model, a transition occurs between an equator-to-pole SST gradient of 24 and 15 °C, at which point the temperature gradient becomes a major climate problem.

2. This early Paleogene CGCM simulation produces continental interior temperatures that are unrealistically cold, showing that a complete and interactive treatment of the ocean is not, by itself, a resolution to the equable climate problem. Model-predicted mean tropical temperatures are closer to those of Pearson et al. (2001) than to those of Zachos et al. (1994).

3. Although zonally averaged overturning is reduced in the early Paleogene simulation relative to modern, the overturning and current systems within individual basins can be as vigorous as modern.

4. Deep waters can be relatively warm (7 °C) and these form at high latitudes. Deep convection of warm (12–15 °C) and saline (35.5 ppt) water occurs in the Northern Hemisphere. The Southern Ocean is relatively fresh (~32 ppt) and surface temperatures are cooler (1–4 °C). In the two simulations described here, one undergoes deep convection in the Southern Hemisphere and the other does not, indicating there may be two quasi-stable states of the early Paleogene system.

5. In all cases, the simulation produces temperature gradients and ocean heat transport within 30% of modern, which suggests that it is unlikely that increased meridional heat transport was responsible for small temperature gradients in the early Paleogene.

ACKNOWLEDGMENTS

This chapter emerged as an outgrowth of Matthew Huber's Ph.D. dissertation, and he gratefully acknowledges the support and encouragement of his supervisor, L.C. Sloan, and that of the rest of his committee: P. Duffy, G. Glatzmaier, P. Koch, R.T. Pierrehumbert, and J. Zachos. This work was supported by grants to L.C. Sloan by the National Science Foundation (NSF) and the D. and L. Packard Foundation. Gary Shaffer, through the Danish Center for Earth System, provided support to Matthew Huber during the writing of the chapter. The National Center for Atmospheric Research (NCAR), through the NSF, provided the NCAR CCSM, computer time, and graphics tools. Special thanks are owed to B. Otto-Bliesner, E. Brady, and C. Shields at NCAR.

REFERENCES CITED

Andreasson, F.P., and Schmitz, B., 1998, Tropical Atlantic seasonal dynamics in the early middle Eocene from stable oxygen and carbon isotope profiles of mollusk shells: Paleoceanography, v. 13, p. 183–192.

Barron, E.J., 1987, Eocene equator-to-pole surface ocean temperatures: A significant climate problem?: Paleoceanography, v. 2, p. 729–732.

Barron, E.J., and Peterson, W.H., 1991, The Cenozoic ocean circulation based on ocean general circulation model results: Palaeogeography, Palaeoclimatology, Palaeoecology, v. 83, p. 1–28.

Barron, E., Peterson, W., Pollard, D., and Thompson, S., 1993, Past climate and the role of ocean heat transport: Model simulations for the Cretaceous: Paleoceanography, v. 8, p. 785–798.

Bice, K.L., Barron, E.J., and Peterson, W.H., 1997, Continental runoff and early Cenozoic bottom-water sources: Geology, v. 25, p. 951–954.

Bice, K.L., Scotese, C.R., Seidov, D., and Barron, E.J., 2000a, Quantifying the role of geographic change in Cenozoic ocean heat transport using uncoupled atmosphere and ocean models: Palaeogeography, Palaeoclimatology, Palaeoecology, v. 161, p. 295–310.

Bice, K.L., Sloan, L.C., and Barron, E.J., 2000b, Comparison of early Eocene isotopic paleotemperatures and the three-dimensional OGCM temperature field: The potential for use of model-derived surface water $\delta^{18}O$, *in* Huber, B.T., et al., eds., Warm climates in Earth history: Cambridge, UK, Cambridge University Press, p. 79–131.

Blackmon, M., Boville, B., Bryan, F., Dickinson, R., Gent, P., Kiehl, J., Moritz, R., Randall, D., Shukla, J., Solomon, S., Bonan, G., Donley, S., Fung, I., Hack, J., Hunkle, E., Hurrell, J., Kutzbach, J., Meehl, J., Otto-Bliesner, B., Saravanan, R., Schneider, E.K., Sloan, L., Spall, M., Taylor, K., Tribbin, J., and Washington, W., 2001, The Community Climate System Model: Bulletin of the American Meteorological Society, v. 82, p. 2357–2376.

Boville, B.A., and Gent, P.R., 1998, The NCAR Climate System Model Version One: Journal of Climate, v. 11, p. 1115–1130.

Boville, B.A., Kiehl, J.T., Rasch, P.J., and Bryan, F.O., 2001, Improvements to the NCAR CSM-1 for transient climate simulations: Journal of Climate, v. 14, p. 164–179.

Brady, E.C., DeConto, R.M., and Thompson, S.L., 1998, Deep water formation and poleward ocean heat transport in the warm climate extreme of the Cretaceous (80 Ma): Geophysical Research Letters, v. 25, p. 4205–4208.

Bralower, T.J., Zachos, J.C., Thomas, E., Parrow, M., Paull, C.K., Kelly, D.C., Premoli Silva, I., Sliter, W.V., and Lohmann, K.C., 1995, Late Paleocene to Eocene paleoceanography of the equatorial Pacific Ocean: Stable isotopes recorded at Ocean Drilling Program Site 865, Allison-Guyot: Paleoceanography, v. 10, p. 841–865.

Broecker, W.S., 1989, The salinity contrast between the Atlantic and Pacific Oceans during glacial time: Paleoceanography, v. 4, p. 207–212.

Bryan, F.O., 1998, Climate drift in a multicentury integration of the NCAR Climate System Model: Journal of Climate, v. 11, p. 1455–1471.

Bush, A.B.G., and Philander, S.G.H., 1997, The Late Cretaceous: Simulation with a coupled atmosphere-ocean general circulation model: Paleoceanography, v. 12, p. 495–516.

Charisi, S., and Schmitz, B., 1996, Early Eocene palaeoceanography and palaeoclimatology of the eastern North Atlantic: Stable isotope results for DSDP Hole 550, *in* Knox, R.W.O'B., et al., eds., Correlation of the early Paleogene in northwest Europe: Geological Society [London] Special Publication 101, p. 420–442.

Corfield, R., and Norris, D.R., 1996, Deep water circulation in the Paleocene ocean, *in* Knox, R.W.O'B., et al., eds., Correlation of the early Paleogene in northwest Europe: Geological Society [London] Special Publication 101, p. 443–456.

Crowley, T.J., and Zachos, J.C., 2000, Comparison of zonal temperature profiles for past warm time periods, *in* Huber, B.T., et al., eds., Warm climates in Earth history: Cambridge, UK, Cambridge University Press, p. 50–76.

Dai, A., Wigley, T.M.L., Boville, B.A., Kiehl, J.T., and Buja, L.E., 2001, Climates of the twentieth and twenty-first centuries simulated by the NCAR Climate System Model: Journal of Climate, v. 14, p. 485–519.

DeConto, R.M., Thompson, S.L., and Pollard, D., 2000, Recent advances in paleoclimate modeling: Toward better simulations of warm paleoclimates, *in* Huber, B.T., et al., eds., Warm climates in Earth history: Cambridge, UK, Cambridge University Press, p. 21–49.

Dercourt, J., Zonenshain, L.R., Ricou, L.E., Kazmin, V.G., Lepichon, X., Knipper, A.L., Grandjaquet, C., Bortshikov, I.M., Geyssant, J., LepVrier, C., Pechersky, D.H., Boulin, J., Sibuet, J.L., Savostin, J.A., Sorokhtin, O., Westphal, M., Bazhenov, M.L., Lauer, J.P., and Bijudural, B., 1986, Geologic evolution of the Tethys belt from the Atlantic to the Pamirs since the Lias: Tectonophysics, v. 123, p. 241–315.

D'Hondt, S., and Arthur, M.A., 1996, Late Cretaceous oceans and the cool tropic paradox: Science, v. 271, p. 1838–1841.

Doney, S.C., Large, W.G., and Bryan, F.O., 1998, Surface ocean fluxes and water-mass transformation rates in the coupled Climate System Model: Journal of Climate, v. 9, p. 1420–1441.

Egbert, G.D., and Ray, R.D., 2000, Significant dissipation of tidal energy in the deep ocean inferred from satellite altimeter data: Nature, v. 405, p. 775–778.

Erez, J., and Luz, B., 1983, Experimental paleotemperature equation for planktonic foraminifera: Geochimica et Cosmochimica Acta, v. 47, p. 1025–1031.

Fairbanks, R.G., Evans, M.A., Rubenstone, J.L., Mortlock, R.A., Broad, K., Moore, M.D., and Charles, C.D., 1997, Evaluating climate indices and their geochemical proxies measured in corals: Coral Reefs, v. 16, p. 93–100.

Ganachaud, A., and Wunsch, C., 2000, Improved estimates of global ocean circulation, heat transport and mixing from hydrological data: Nature, v. 408, p. 453–457.

Gent, P.R., Bryan, F.O., Danabasoglo, G., Doney, S.C., Holland, W.R., Karge, W.G., and McWilliams, J.C., 1998, The NCAR Climate System Model global ocean component: Journal of Climate, v. 11, p. 1287–1306.

Greenwood, D.R., and Wing, S.L., 1995, Eocene continental climates and latitudinal temperature gradients: Geology, v. 23, p. 1044–1048.

Hack, J.J., 1998, Analysis of the improvement in implied meridional ocean energy transport as simulated by the NCAR CCM3: Journal of Climate, v. 11, p. 1237–1244.

Handoh, I.C., Bigg, G.R., Jones, E.J.W., and Inoue, M., 1999, An ocean modeling study of the Cenomanian Atlantic: Equatorial paleo-upwelling, organic-rich sediments, and the consequences for a connection between the proto-North and South Atlantic: Geophysical Research Letters, v. 26, p. 223–226.

Haupt, B.J., and Seidov, D., 2001, Warm deep-water ocean conveyor during Cretaceous time: Geology, v. 29, p. 295–298.

Huber, M., 2001, Modeling early Paleogene climate: From the top of the atmosphere to the bottom of the ocean [Ph.D. thesis]: Santa Cruz, University of California, p. 220.

Huber, M., 2002, Straw man 1: A preliminary view of the tropical Pacific from a global coupled climate model simulation of early Paleogene climate: Proceedings of the Ocean Drilling Program, Initial Reports, Leg 199, Texas A & M University, College Station, Texas, USA.

Huber, M., and Sloan, L.C., 1999a, Warm climate transitions: A general circulation modeling study of the late Paleocene thermal maximum (ca. 56 Ma): Journal of Geophysical Research, v. 104, p. 16633–16655.

Huber, M., and Sloan, L.C., 1999b, The challenges of modeling extreme warm climates: Old problems, new models: Eos (Transactions, American Geophysical Union), Spring Meeting Supplement, v. 80, p. F488.

Huber, M., and Sloan, L.C., 2000, Climatic responses to tropical sea surface temperature changes on a "greenhouse" Earth: Paleoceanography, v. 15, p. 443–450.

Huber, M., and Sloan, L.C., 2001, Heat transport, deep waters, and thermal gradients: Coupled simulation of an Eocene "greenhouse" climate: Geophysical Research Letters, v. 28, p. 3481–3484.

Kennett, J.P., and Stott, L.D., 1990, Proteus and Proto-Oceanus: Paleogene oceans as revealed from Antarctic stable isotopic results, *in* Barker, P.F., et al., eds., Proceedings of the Ocean Drilling Program, Scientific Results: Texas A & M University, College Station, Texas, USA, v. 113, p. 865–880.

Kutzbach, J.E., and Liu, Z., 1997, Response of the African monsoon to orbital forcing and ocean feedbacks in the middle Holocene: Science, v. 278, p. 440–443.

Levitus, S., and Boyer, T.P., 1994, World ocean atlas: Temperature: Washington, D.C., National Oceanic and Atmospheric Administration, v. 4, 117 p.

Levitus, S., Burgett, R., and Boyer, T.P., 1994, World ocean atlas: Salinity: Washington, D.C., National Oceanic and Atmospheric Administration, v. 3, 99 p.

Liu, Z., Gallimore, R.G., Kutzbach, J.E., Xu, W., Golubev, Y., Behlilng, P., and Selin, R., 1999, Coupled modeling of long-term climate change with the equilibrium asynchronous coupling scheme: Climate Dynamics, v. 15, p. 325–240.

Liu, Z., Shin, S., Otto-Bliesner, B., Kutzbach, J.E., Brady, E.C., and Lee, D., 2002, Tropical cooling at the Last Glacial Maximum and extratropical ocean ventilation: Geophysical Research Letters, v. 29, paper no. 1409.

Lonsdale, P., Paleogene history of the Kula plate, 1998, Offshore evidence and onshore implications: Geological Society of America Bulletin, v. 110, p. 733–754.

Lyle, M., 1997, Could early Cenozoic thermohaline circulation have warmed the poles?: Paleoceanography, v. 12, p. 161–167.

Markwick, P.J., 1998, Fossil crocodilians as indicators of Late Cretaceous and Cenozoic climates: Implications for using palaeontological data in reconstructing palaeoclimate: Palaeogeography, Palaeoclimatology, Palaeoecology, v. 137, p. 205–271.

Marshall, L.G., Sempere, T., and Butler, R.F., 1997, Chronostratigraphy of the mammal-bearing Paleocene of South America: Journal of South American Earth Sciences, v. 10, p. 49–70.

Meehl, G.A., and Arblaster, J.M., 1998, The Asian-Australian monsoon and El Niño Southern Oscillation in the NCAR Climate System Model: Journal of Climate, v. 11, p. 1356–1385.

Meehl, G.A., Collins, W.D., Boville, B.A., Kiehl, J.T., Wigley, T.M.L., and Arblaster, J.M., 2000, Response of the NCAR Climate System Model to increased CO_2 and the role of physical processes: Journal of Climate, v. 13, p. 1879–1898.

Mikolajewicz, U., and Crowley, T.J., 1997, Response of a coupled ocean/energy balance model to restricted flow through the central American isthmus: Paleoceanography, v. 12, p. 429–441.

Mikolajewicz, U., Maierreimer, E., Crowley, T.J., and Kim, K.Y., 1993, Effect of Drake and Panamanian Gateways on the circulation of an ocean model: Paleoceanography, v. 8, p. 409–426.

Miller, K.G., Janacek, T.R., Katz, M.E., and Keil, D.J., 1987, Abyssal circulation and benthic foraminiferal changes near the Paleocene–Eocene boundary: Paleoceanography, v. 2, p. 741–761.

Najjar, R.G., Nong, G.T., Seidov, D., and Peterson, W.H., 2002, Modeling geographic impacts on early Eocene ocean temperature: Geophysical Research Letters, v. 29, paper no. 1750.

NOAA Data Announcement 88-MGG-02, Digital relief of the Surface of the Earth, National Geophysical Data Center, Boulder, Colorado, 1988.

Nong, G.T., Najjar, R.G., Seidov, D., and Peterson, W.H., 2000, Simulation of ocean temperature change due to the opening of the Drake Passage: Geophysical Research Letters, v. 27, p. 2689–2692.

Otto-Bliesner, B.L., and Brady, E.C., 2001, Tropical Pacific variability in the NCAR Climate System Model: Journal of Climate, v. 14, p. 3587–3607.

Otto-Bliesner, B.L., and Upchurch, G.R., 1997, Vegetation-induced warming of high-latitude regions during the Late Cretaceous period: Nature, v. 385, p. 804–807.

Otto-Bliesner, B.L., Brady, E.C., and Shields, C., 2002, Late Cretaceous ocean: Coupled simulations with the NCAR CSM: Journal of Geophysical Research, v. 107, 10.1029/2001JD000821.

Pearson, P.N., and Palmer, M.R., 2000, Atmospheric carbon dioxide concentrations over the past 60 million years: Nature, v. 406, p. 695–699.

Pearson, P.N., Ditchfield, P.W., Singano, J., Harcourt-Brown, K.G., Nicholas, C.J., Olsson, R.K., Shackleton, N.J., Hall, M.A., 2001, Warm tropical sea surface temperatures in the Late Cretaceous and Eocene epochs: Nature, v. 413, p. 481–487.

Peixoto, J.P., and Oort, A.H., 1992, Physics of climate: American Institute of Physics, New York, p. 520.

Peters, R.B., and Sloan, L.C., 2000, High concentrations of greenhouse gases and polar stratospheric clouds: A possible solution to high-latitude faunal migration at the latest Paleocene thermal maximum: Geology, v. 28, p. 979–982.

Poulsen, C.J., Barren, E.J., Peterson, W.H., and Wilson, P.A., 1999, A reinterpretation of mid-Cretaceous shallow marine temperatures through model-data comparison: Paleoceanography, v. 14, p. 679–697.

Rahmstorf, S., and Willebrand, J., 1995, The role of temperature feedback in stabilizing the thermohaline circulation: Journal of Physical Oceanography, v. 25, p. 787–805.

Retallack, G.J., 2001, A 300 million year record of atmospheric carbon dioxide from fossil plant cuticles: Nature, v. 411, p. 287–290.

Rind, D., and Chandler, M., 1991, Increased ocean heat transports and warmer climate: Journal of Geophysical Research, v. 96, p. 7437–7461.

Royer, D.L., Wing, S.L., Beerling, D.J., Jolley, D.W., Koch, P.L., Hickley, L.J., and Berner, R.A., 2001, Paleobotanical evidence for near present-day levels of atmospheric CO_2 during part of the Tertiary: Science, v. 292, p. 2310–2313.

Royer, J.Y., Müller, R.D., Gahagan, L.M., Lawrer, L.A., Mayes, C.L., Nünberg, D., and Sclüter, J.G., 1992, A global isochron chart: University of Texas Institute for Geophysics, Technical Report No. 117, 38 p.

Saravanan, R., and McWilliams, J.C., 1995, Multiple equilibria, natural variability, and climate transitions in an idealized ocean-atmosphere model: Journal of Climate, v. 10, p. 2296–2323.

Schmidt, G.A., 1999a, Forward modeling of carbonate proxy data from planktonic foraminifera using oxygen isotope tracers in a global ocean model: Paleoceanography, v. 14, p. 482–497.

Schmidt, G.A., 1999b, Error analysis of paleosalinity calculations: Paleoceanography, v. 14, p. 422–429.

Schmidt, G.A., and Mysak, L.A., 1996, Can increased poleward oceanic heat flux explain the warm Cretaceous climate?: Paleoceanography, v. 11, p. 579–593.

Sewall, J.O., Sloan, L.C., Huber, M., and Wing, S., 2000, Climate sensitivity to changes in land surface characteristics: Global and Planetary Change, v. 26, p. 445–465.

Shin, S.I., Liu, Z.Y., Otto-Bliesner, B., Brady, E.C., Kutzbach, J.E., and Harrison, S.P., 2003, A simulation of the Last Glacial Maximum climate using the NCAR-CCSM: Climate Dynamics, v. 20, p. 127–151, DOI: 10.1007/500382-002-0260-X.

Sloan, L.C., and Barron, E.J., 1992, A comparison of Eocene climate model results to quantified paleoclimatic interpretations: Palaeogeography, Palaeoclimatology, Palaeoecology, v. 93, p. 183–202.

Sloan, L.C., and Huber, M., 2001, Eocene oceanic responses to orbital forcing on precessional time scale: Paleoceanography, v. 16, p. 101–111.

Sloan, L.C., and Rea, D.K., 1996, Atmospheric CO_2 of the early Eocene: A general circulation modeling sensitivity study: Palaeogeography, Palaeoclimatology, Palaeoecology, v. 119, p. 275–292.

Sloan, L.C., Walker, J.C.G., and Moore, T.C., 1995, Possible role of oceanic heat transport in early Eocene climate: Paleoceanography, v. 10, p. 347–356.

Sloan, L.C., Huber, M., Crowley, T.J., Sewall, J.O., and Baum, S.K., 2001, Effect of sea surface temperature configuration on model simulations of "equable" climate in the early Eocene: Palaeogeography, Palaeoclimatology, Palaeoecology, v. 167, p. 321–335.

Trenberth, K.E., and Caron, J.M., 2001, Estimates of meridional atmosphere and ocean heat transports: Journal of Climate, v. 14, p. 3433–3443.

Valdes, P.J., 2000a, Warm climate forcing mechanisms, *in* Huber, B.T., et al., eds., Warm climates in Earth history: Cambridge, UK, Cambridge University Press, p. 3–20.

Valdes, P.J., 2000b, Paleoclimate modeling, *in* Mote, P., and O'Neill, A., eds., Numerical modeling of the global atmosphere in the climate system: Dordrecht, Netherlands, Kluwer Academic Publishers, p. 118–134.

Vavrus, S., Gallimore, R., and Liu, Z., 2000, A mixed-flux equilibrium asynchronous coupling scheme for accelerating convergence in ocean-atmosphere models: Climate Dynamics, v. 16, p. 821–831.

Wade, B.S., Kroon, D., and Norris, R.D., 2000, Upwelling in the late middle Eocene at Blake Nose?: GFF, v. 122, p. 174–175.

Wilf, P., 2000, Late Paleocene–early Eocene climate changes in southwestern Wyoming: Paleobotanical analysis: Geological Society of America Bulletin, v. 112, p. 292–307.

Wilson, M.E.J., and Moss, S.J., 1999, Cenozoic palaeogeographic evolution of Sulawesi and Borneo: Palaeogeography, Palaeoclimatology, Palaeoecology, v. 145, p. 303–337.

Wing, S.L., Bao, H., and Koch, K.L., 2000, An early Eocene cool period? Evidence for continental cooling during the warmest part of the Cenozoic, *in* Huber, B.T., et al., eds., Warm climates in Earth history: Cambridge, UK, Cambridge University Press, p. 197–237.

Zacher, W., and Lupu, M., 1999, Pitfalls on the race for an ultimate Tethys model: International Journal of the Earth Sciences, v. 88, p. 111–115.

Zachos, J.C., Rea, D.K., Seto, K., Nomura, R., and Niitsuma, N., 1992, Paleogene and early Neogene deep water paleoceanography of the Indian Ocean as determined from benthic foraminifer stable carbon and oxygen isotope records, *in* Synthesis of results from scientific drilling in the Indian Ocean: American Geophysical Union Geophysical Monograph Series, v. 70, p. 351–385.

Zachos, J.C., Lohmann, K.C, Walker, J.C.G., and Wise, S.W., 1993, Abrupt climate change and transient climates during the Paleogene: A marine perspective: Journal of Geology, v. 101, p. 191–213.

Zachos, J.C., Stott, L.D., and Lohmann, K.C, 1994, Evolution of early Cenozoic marine temperatures: Paleoceanography, v. 9, p. 353–387.

Zachos, J., Pagani, M., Sloan, L., Thomas, E., and Billups, K., 2001, Trends, rhythms, and aberrations in global climate 65 Ma to present: Science, v. 292, p. 686–694.

Manuscript Accepted by the Society August 13, 2002

Printed in the USA

Geological Society of America
Special Paper 369
2003

Carbon and oxygen isotope records from paleosols spanning the Paleocene-Eocene boundary, Bighorn Basin, Wyoming

Paul L. Koch*
Department of Earth Sciences, University of California, Santa Cruz, California, 95064, USA
William C. Clyde
Department of Earth Sciences, University of New Hampshire, Durham, New Hampshire 03824, USA
Robert P. Hepple
2380 Hilgard Avenue, Berkeley, California 94709, USA
Marilyn L. Fogel
Geophysical Laboratory, Carnegie Institute of Washington, Washington, D.C. 20015, USA
Scott L. Wing
Department of Paleobiology, Smithsonian Institution, Washington, D.C. 20560, USA
James C. Zachos
Department of Earth Sciences, University of California, Santa Cruz, California, 95064, USA

ABSTRACT

The isotopic composition of paleosol carbonate and organic matter were investigated in the Bighorn Basin, Wyoming, to explore changes in the carbon cycle and climate across the Paleocene-Eocene boundary. In three different measured sections, soil carbonate $\delta^{13}C$ values change in phase with marine surface water carbonates on both long (~7 m.y.) and short (~100 k.y.) time scales. The carbon cycle perturbations at the Paleocene-Eocene Boundary Thermal Maximum (PETM) and the Eocene Warm Interval (EWI) are recorded in multiple sections, providing unambiguous links between marine and continental deposits. The PETM and EWI $\delta^{13}C$ excursions in the Bighorn Basin are larger than those in the surface ocean, but the reasons for this amplification are unclear. Organic matter samples from the Bighorn Basin yield noisy $\delta^{13}C$ records that do not mirror global changes, perhaps due to diagenetic alteration or postformational contamination. The $\delta^{18}O$ values of soil carbonate are subject to multiple climatic influences that are often antagonistic. Although the $\delta^{18}O$ shifts at the PETM and EWI are small, the shift at the PETM is statistically significant in two of the measured sections. Assuming a plausible range of values for the meteoric water $\delta^{18}O$/mean annual temperature relationship, the perturbation in soil carbonate $\delta^{18}O$ at the PETM is consistent with an increase in meteoric water $\delta^{18}O$ of ~2‰ and changes in local temperature of 3–7 °C.

* pkoch@es.ucsc.edu

Koch, P.L., Clyde, W.C., Hepple, R.P., Fogel, M.L., Wing, S.L., and Zachos, J.C., 2003, Carbon and oxygen isotope records from paleosols spanning the Paleocene-Eocene boundary, Bighorn Basin, Wyoming, *in* Wing, S.L., Gingerich, P.D., Schmitz, B., and Thomas, E., eds., Causes and Consequences of Globally Warm Climates in the Early Paleogene: Boulder, Colorado, Geological Society of America Special Paper 369, p. 49–64.

INTRODUCTION

The Paleocene-Eocene (P-E) boundary is marked by biotic, climatic and environmental events that had a profound influence on Cenozoic Earth history. Oxygen isotope records from marine microfossils reveal a long-term warming of bottom waters and mid- to high-latitude surface waters spanning from ca. 59 to 50 million years ago (Ma) (reviewed in Zachos et al., 2001) (Fig. 1A). By the end of this long-term trend, global climates had reached a peak warming (referred to here as the Eocene Warm Interval [EWI]) that would not be surpassed for the rest of the Cenozoic. The $\delta^{13}C$ values of carbonates from ocean surface and deep waters dropped ~3.0‰ from the late Paleocene through the early Eocene, reflecting a long-term shift in the cycling of carbon among Earth surface carbon reservoirs (Fig. 1B). A brief (~150,000 year) pulse of extreme warming and a dramatic drop in $\delta^{13}C$ values were superimposed on these long-term trends near the P-E boundary (ca. 55 Ma) (Kennett and Stott, 1991; Röhl et al., 2000; Zachos et al., 2001). We will refer to this event as the Paleocene-Eocene Boundary Thermal Maximum (PETM). The short-lived drop in the $\delta^{13}C$ of marine carbonates at the PETM is best explained by the release and oxidation of at least 1500 gigatons of ^{13}C-depleted carbon from seafloor methane hydrate deposits, followed by sequestration of this excess carbon during carbon cycle reequilibration (Dickens et al., 1997; Dickens, 2000). The short-term warming at the PETM may reflect (in part) a greenhouse response to addition of this new carbon.

Events on land were no less profound. The most complete and best-studied terrestrial records spanning the P-E boundary occur in Wyoming. Physiognomic analyses of Wyoming floras indicate a long-term increase in mean annual temperature (MAT) of ~10 °C from the late Paleocene (Tiffanian [Ti] North American Land Mammal Age [NALMA]) to the late early Eocene (late Wasatchian NALMA), reaching temperatures of 20–25 °C (Wing, 1998; Wilf, 2000). Temperature increase was not monotonic, however. Isotopic data from fossil mammals and paleosol carbonates are consistent with short-term warming at the boundary between the Clarkforkian (Cf) and Wasatchian (Wa) NALMAs at 55 Ma (Koch et al., 1995; Fricke et al., 1998). Oxygen isotope records and floral analysis suggest that conditions subsequently cooled for perhaps 700,000 yr before warming renewed, reaching maximum values during the EWI in the late Wasatchian (Bao et al., 1999; Wing et al., 2000). There were significant faunal turnovers at the Ti-Cf and Cf-Wa boundaries at ca. 56.2 and 55 Ma, respectively (Gingerich, 1989; Clyde and Gingerich, 1998; Alroy et al., 2000), and a major floral turnover was associated with the EWI (Wing et al., 2000). Lastly, a sudden drop in the $\delta^{13}C$ of soil carbonates and mammal teeth in the northern Bighorn Basin (the Clarks Fork region) (Fig. 2) coincides precisely with the Cf-Wa boundary. This indicates that the earliest Wasatchian faunas (zone Wa 0), which contain the first primates, artiodactyls, and perissodactyls, were synchronous with the short-term warming and carbon cycle perturbation at the PETM (Koch et al., 1992; Bowen et al., 2001).

Here, we expand on prior isotopic work on Bighorn Basin paleosols (Koch et al., 1992, 1995), extending our records temporally to encompass middle-to-late Wasatchian paleosols, and geographically to incorporate data from sections in the McCullough Peaks region in the north-central Bighorn Basin and several regions in the southern Bighorn Basin. Carbon isotope analysis of paleosol carbonates and organic carbon sheds light on the timing of carbon cycle events on land and the response of continental carbon reservoirs to these events. Oxygen isotope analysis of paleosol carbonates serves as a monitor of climatically driven changes in the isotopic composition of meteoric precipitation and soil waters, as well as soil temperature. We end by synthesizing our new results with previous work on P-E climates in the Bighorn Basin.

CONTROLS ON CARBON AND OXYGEN ISOTOPES IN PALEOSOLS

The isotope systematics of soils and paleosols have been reviewed elsewhere (Cerling and Quade, 1993; Koch et al., 1995; Koch, 1998); these papers are the basis for our discussion. Soil carbonate carbon is derived from soil CO_2, which is supplied by decay of organic matter, soil respiration, and atmospheric CO_2. At depths >30 cm in even moderately productive soils, atmospheric inputs are negligible. The $\delta^{13}C$ of soil carbonate mirrors the isotopic composition of the overlying flora, offset by ~+15‰ due to fractionations associated with carbonate equilibria, diffusion of CO_2 from the soil, and decomposition/respiration.

The carbon isotope composition of plants is dependent on photosynthetic pathway (C3, C4 or CAM), environmental factors that influence photosynthetic rate (e.g., light intensity and water stress), and the $\delta^{13}C$ of atmospheric CO_2, the ultimate source of fixed plant carbon (Ehleringer and Monson, 1993). There is no compelling evidence that C4 or CAM plants were common in Paleogene floras (Jacobs et al., 1999). A recent meta-analysis of the carbon isotope fractionation between atmospheric CO_2 and C3 plants suggests that a difference of ~19‰ is to be expected under a range of conditions (Arens et al., 2000). Thus, the $\delta^{13}C$ of soil carbonates should roughly track changes in the $\delta^{13}C$ of atmospheric CO_2.

Finally, CO_2 in the atmosphere undergoes rapid exchange (tens of years) with the surface ocean, which in turn exchanges carbon with the massive deep ocean reservoir on longer time scales (hundreds to thousands of years) (Sundquist, 1993). At equilibrium, the difference in $\delta^{13}C$ between carbonate in the surface ocean and atmospheric CO_2 is ~9‰ (Mook, 1986), though the magnitude of this fractionation is affected by temperature and the $\delta^{13}C$ of local DIC can be affected by productivity. On geologic time scales, carbon exchange between the oceans and atmosphere is essentially instantaneous. Thus, shifts in the $\delta^{13}C$ of marine systems are transmitted via the atmosphere to terrestrial plants, soils, and mammals (Koch et al., 1992). Isotopic coupling of marine, atmospheric, and terrestrial carbon reservoirs is increasingly being used to correlate stratigraphic records

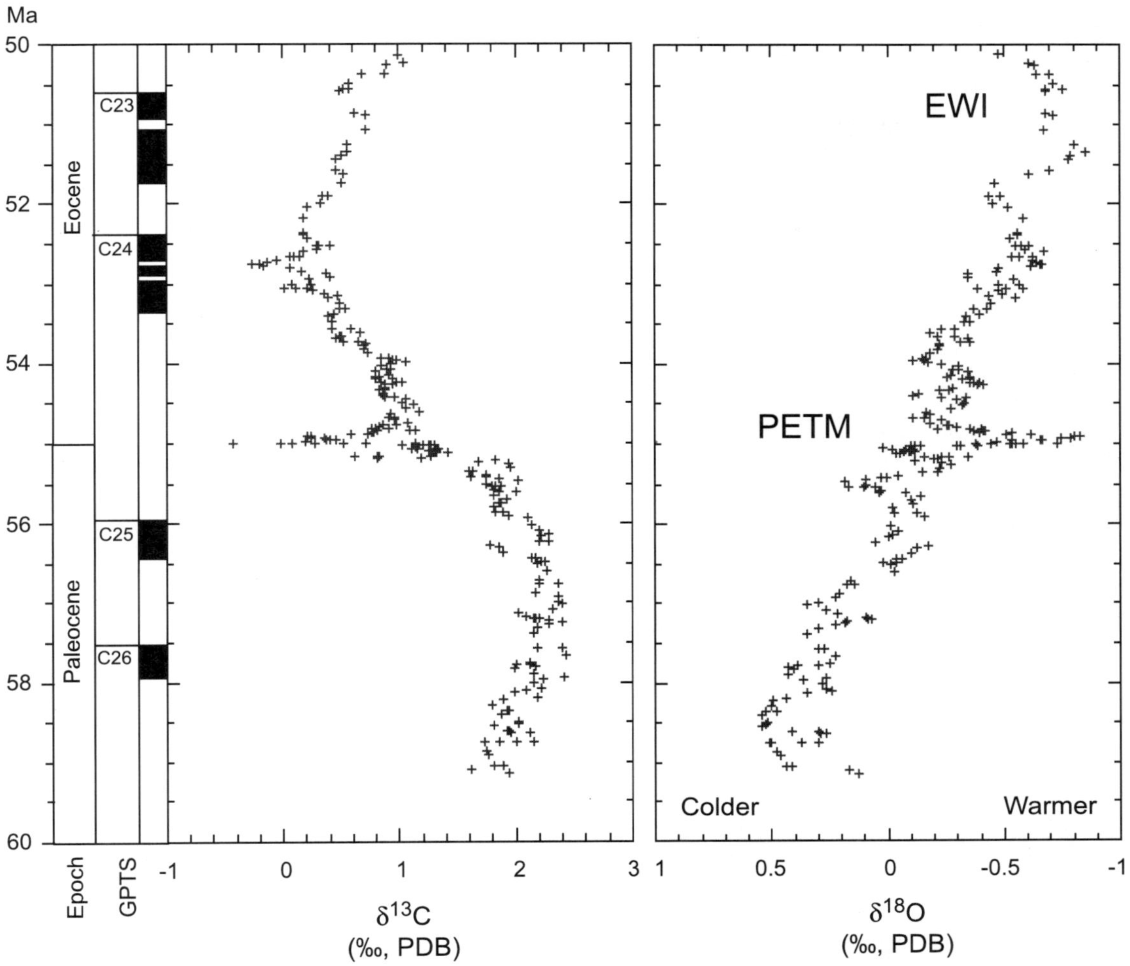

Figure 1. Carbon and oxygen isotope data for benthic foraminifera spanning the Paleocene-Eocene boundary. The figure is based on data compiled in Zachos et al. (2001). EWI—Eocene Warm Interval; GPTS—geomagnetic polarity time scale; PDB—Peedee belemnite; PETM—Paleocene-Eocene Thermal Maximum.

at times of global change, which are often marked by carbon cycle perturbations (MacLeod et al., 2000; Hesselbo et al., 2000; Jahren et al., 2001).

The oxygen isotope composition of soil carbonate is a function of the $\delta^{18}O$ of soil water and the temperature dependent fractionation of oxygen isotopes during carbonate formation. The $\delta^{18}O$ of modern, deeply formed soil carbonate is correlated with the $\delta^{18}O$ of meteoric water, though typically offset by 2‰–10‰ relative to equilibrium with meteoric water owing to evaporative enrichment of soil water. Shifts in the $\delta^{18}O$ of meteoric water, in turn, may be correlated to changes in atmospheric temperature (with lower $\delta^{18}O$ values at lower temperatures), or to changes in the source of water reaching a region. Overall, temporal changes in the $\delta^{18}O$ of soil carbonate may reflect the combined effects of several factors, precluding rigorous estimation of meteoric or climatic parameters. However, consideration of the potential magnitudes of these effects can provide some quantitative constraints on the magnitude of climatic perturbations.

MATERIALS AND METHODS

Authigenic carbonate nodules were sampled near the base of paleosol B-horizons from freshly trenched surfaces that had no evidence of modern soil development. Nodules were collected from at least 30 cm below the preserved top of the B horizon to minimize the influence of atmospheric CO_2 on $\delta^{13}C$ values and evaporation on $\delta^{18}O$ values. Plant fragment and cuticle samples were collected from fossil plant quarries or from organic-rich layers encountered during paleosol sampling. Blocks of organic-rich sediment were wrapped in aluminum foil to minimize contamination during transport.

Paleosol carbonate samples from the McCullough Peaks region were collected by W.C. Clyde during the course of paleo-

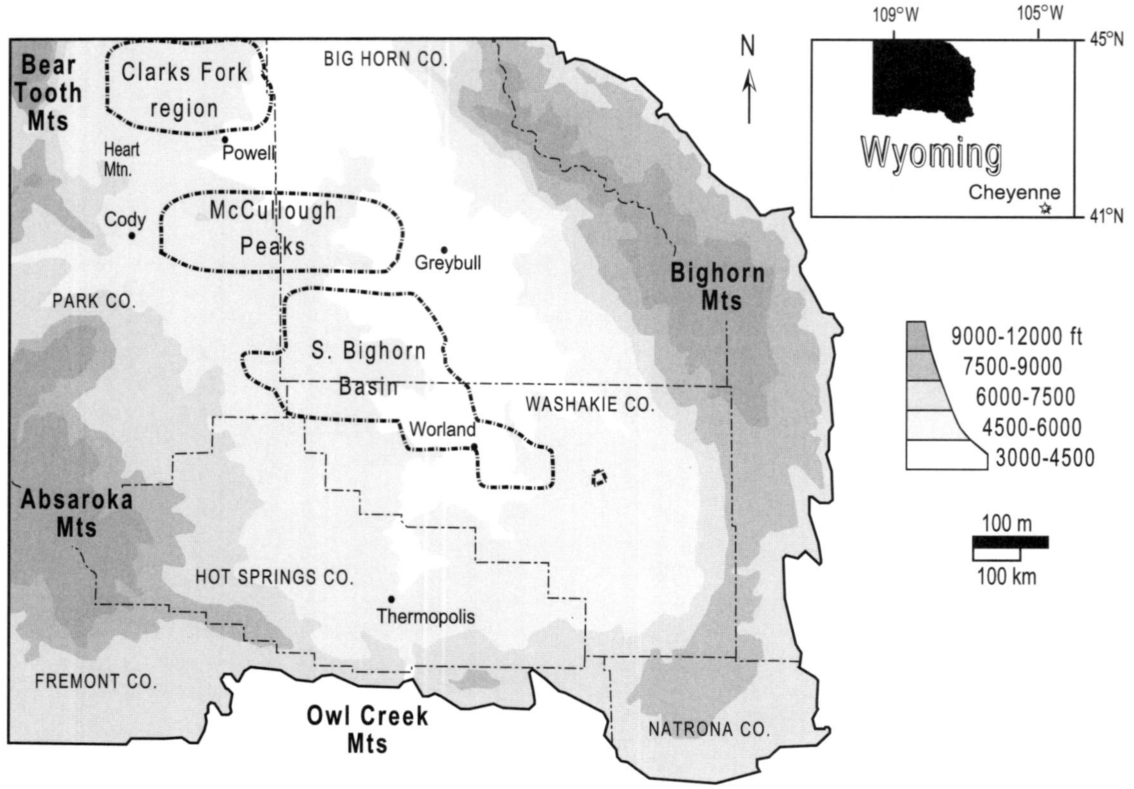

Figure 2. Map of the Bighorn Basin, Wyoming, showing the three regions of the basin where samples were collected from measured sections.

magnetic sampling. Samples are referenced by meter level relative to the Cretaceous-Tertiary boundary in the Southeast McCullough Peaks composite section (Clyde, 1997, 2001). Sediments below the Cf-Wa contact (~1195 m), which coincides with the Fort Union–Willwood Formation contact, do not contain paleosol carbonates.

Paleosol carbonate and organic matter samples in the southern Bighorn Basin were collected by P.L. Koch from fossil mammal or plant localities established by workers from Yale (YPM localities), Johns Hopkins/USGS-Denver (D localities), and the Smithsonian Institution (SLW and MBF localities), or from new measured sections (PK localities). Sampling localities were distributed among different subregions in the southern Bighorn Basin (Elk Creek/Antelope Creek, Fifteen Mile Creek, Worland East, North Butte) (Bown et al., 1994; Wing et al., 2000). Localities are referenced by meter level relative to the Cf-Wa boundary in the southern Bighorn Basin composite section (Bown et al., 1994). As in the McCullough Peaks region, sediments below the Cf-Wa boundary, which again coincides with the Fort Union–Willwood Formation contact, do not contain paleosol carbonates.

Organic matter samples from the Clarks Fork region were collected at localities established by workers from the Smithsonian (SLW and DCF localities), Yale (LJH localities) and the University of Michigan (SC localities). Localities in the Clarks Fork region are referenced by meter level relative to the Cretaceous-Tertiary boundary in the Clarks Fork Basin composite section (Butler et al., 1981). Paleosol carbonate data for the Clarks Fork region were reported in Koch et al. (1995). We have replotted those data here, substituting a subsample of data from a new, high-resolution record at the Cf-Wa boundary for the interval from 1480 to 1560 m (Bowen et al., 2001). Unlike the situation in the McCulloughs Peak and southern Bighorn Basin sections, the Fort Union–Willwood Formation occurs well below the Cf-Wa boundary (at 1070 m), and paleosol carbonates occur in the Fort Union formation into the Tiffanian NALMA.

Carbonate nodules were polished flat on a lapidary wheel, washed, and then dried in a low-temperature oven. Samples (~100 μg) were drilled from the polished surface under a binocular microscope using a mounted dental drill. Primary micritic carbonate and secondary diagenetic spar were sampled. When available, at least two micrite samples were drilled from each of two nodules from every soil that was sampled in the McCullough Peaks and southern Bighorn Basin sections. Samples were roasted in vacuo at 400 °C for 1 hr to remove organic contaminants and analyzed with a gas source mass spectrometer following reaction with 100% phosphoric acid at 90 °C using an automated Isocarb device. McCullough Peaks samples were analyzed using the Optima mass spectrometer in the Department of Geological Sciences, Princeton University. Southern Bighorn Basin samples were analyzed using either the Optima or the Prism mass spectrometers in the Departments of Earth and Ocean Sciences, University of California, Santa Cruz. Carbonate isotope values are reported in delta notation relative to the V-PDB standard: $\delta = [(R_{SAMPLE}/R_{V\text{-}PDB}) - 1] \times 1000$, where R_{SAMPLE} and $R_{V\text{-}PDB}$ are the isotope ratios of the sample and standard, respectively. Analytical precision, based on repeated analysis of an in-house standard, was better than 0.1‰ for $\delta^{13}C$ and $\delta^{18}O$.

Two types of organic matter were analyzed. We analyzed fossil plant fragments and cuticles isolated from organic-rich rocks via digestion with HF. Samples were collected from fresh surfaces inside large blocks of lithified sediment. Plant fragments were stored in precombusted, teflon-lidded glass vials, whereas cuticle was suspended in alcohol. Cuticles were dried to volatilize alcohol prior to analysis. We also analyzed dispersed organic matter occluded within soil carbonate nodules. To extract this material, powdered paleosol carbonate was reacted with 1N HCl, then the insoluble residue was collected by centrifugation and rinsed repeatedly in distilled, deionized water. The residue was freeze-dried and stored in precombusted, teflon-lidded glass vials.

To obtain CO_2 for $\delta^{13}C$ analysis, organic samples were sealed in precombusted, evacuated quartz tubes with CuO and Cu metal, combusted at 900 °C for 1 hour, then cooled at a controlled rate. CO_2 was isolated cryogenically. CO_2 was analyzed on a Finnegan 252 at the Geophysical Lab or on the Optima at University of California, Santa Cruz. Repeated analyses of in-house gelatin standards prepared and analyzed in this manner had a standard deviation <0.15‰.

AGE MODEL

The Tiffanian and Clarkforkian NALMAs have been divided into numbered zones (e.g., Ti 1–6 and Cf 1–3) based on generic first occurrences and within lineage transformation events (Gingerich, 1983). Gingerich (1991) developed a numbered zonation of the Wasatchian based on first occurrences of perissodactyl species (supported by generic first occurrences in other orders [Gingerich, 2000]) that captured long-recognized concepts of the zones of the Wasatchian (Sandcouleean = Wa 0–2, Graybullian = Wa 3–5, Lysitean = Wa 6, Lostcabinian = Wa 7). This zonation was extended to the McCullough Peaks region by Clyde (1997, 2001). Unfortunately, zonation of the Wasatchian-aged sediments in the southern Bighorn Basin was last investigated in detail by Schankler (1980), and the perissodactyls from this region have not received systematic revision. Wa 0, Wa 5, 6, and 7 can be recognized in this region, but zones Wa 1–4 are grouped into Schankler's *Haplomylus-Ectocion* range zone.

Model ages for zone boundaries and tie points used to construct the age model are shown in Table 1. Calculation of an age model for the three sections follows Age Model 1 in Wing et al. (2000), incorporating revisions to the McCullough Peaks section presented in Clyde (1997, 2001). Briefly, reliable paleomagnetic data are available for the Clarks Fork and McCullough Peaks regions, but not for the southern Bighorn Basin. Yet, in the Clarks Fork region, the top of the section is unconstrained by magnetic reversals, whereas in the McCullough Peaks region, the base of the section lacks paleomagnetic constraints. We developed a hybrid model that incorporates estimates for the ages of two key events: The base of Wa 5 (*Bunophorus* Interval Zone) and the base of Wa 0, which coincides with the onset of the $\delta^{13}C$ excursion.

The logic of the model is as follows. Age estimates for magnetic reversals at the base and top of C24N.3n are used to calculate a sedimentation rate in the McCullough Peaks section that is then used to estimate an age for the base of Wa 5 by extrapolation. The age estimate for the base of Wa 5 is then used as a calibration point in the Clarks Fork Basin section. Assuming linear sedimentation between the top of C25N and the base of Wa 5, which are both present in the Clarks Fork Basin section, an age of 54.97 Ma is estimated for the base of Wa 0, which coincides with the $\delta^{13}C$ excursion. This age is then used as a calibration point at the base of the McCullough Peaks and the southern Bighorn Basin sections. Other calibration points are provided by age estimates for magnetic reversals in the Clarks Fork Basin and McCullough Peaks sections, and by an ash in the southern Bighorn Basin section (Table 1).

Although the age model has several potential sources of error, three lines of evidence suggest it is robust. First, the age estimate for the $\delta^{13}C$ excursion (54.97 Ma) is within the range (54.93–54.98 Ma) for the excursion in marine records, which was estimated via astronomical calibration (Norris and Röhl, 1999). Likewise, the age for the base of C24N.2n was not used to calibrate the age model, yet its estimated age in the McCullough Peaks section differs from the age reported in Cande and Kent (1995) by just 40,000 yr. Second, variations in sedimentation rate in the portions of the McCullough Peaks and Clarks Fork Basin sections that provide the calibration points are small (Table 1). Finally, age estimates for the bases of zones in different regions of the basin are similar. Differences between the Clarks Fork and McCullough Peaks regions are 0 and 170,000 yr for Wa 3 and Wa 4, respectively. Differences between McCullough Peaks and southern Bighorn Basin for the bases of Wa 6 and Wa 7 are larger (190,000 yr).

RESULTS

Data reliability was established in four ways. First, we collected multiple samples (from 2 to 4) from most nodules and calculated maximum within-nodule isotopic difference. The average maximum difference for McCullough Peaks nodules (n = 88) was 0.2‰ for both isotopes, whereas for southern Bighorn Basin nodules (n = 85), the average maximum difference for both isotopes was 0.4‰. Second, for most paleosols, we ana-

TABLE 1. AGE CALIBRATION

Event	Clarks Fork Basin Level (m)	Clarks Fork Basin Age (Ma)	McCullough Peaks Level (m)	McCullough Peaks Age (Ma)	Southern Bighorn Basin Level (m)	Southern Bighorn Basin Age (Ma)
Ash fall	N.A.	N.A.	N.A.	N.A.	**634**	**52.80**
Base of Wa 7 (*Lambdotherium* FAD)	N.A.	N.A.	2540	52.72	591	52.91
Top of C24N.2n	N.A.	N.A.	**2515**	**52.76**	N.A.	N.A.
Base of C24N.2n	N.A.	N.A.	2460	52.84	N.A.	N.A.
Top of C24N.3n	N.A.	N.A.	**2420**	**52.90**	N.A.	N.A.
Base of Wa 6 (*Heptodon* FAD)	N.A.	N.A.	2257	53.13	430	53.32
Base of C24N.3n	N.A.	N.A.	**2100**	**53.35**	N.A.	N.A.
Base of Wa 5 (*Bunophorus* FAD)	**2240**	**53.49**	2000	**<u>53.49</u>**	**365**	**53.49**
Base of Wa 4	2020	53.94	1665	54.11	N.A.	N.A.
Base of Wa 3 (*Homogalax* FAD)	1750	54.50	1450	54.50	N.A.	N.A.
Base of Wa 2	1645	54.72	1315	54.75	N.A.	N.A.
Base of Wa 1	1540	54.93	1205	54.96	N.A.	N.A.
Base of Wa 0	1520	**<u>54.97</u>**	**1195**	**54.97**	**0**	**54.97**
Base of Cf 3	1335	55.36	N.A.	N.A.	N.A.	N.A.
Base of Cf 2	1140	55.76	N.A.	N.A.	N.A.	N.A.
Top of C25N	**1070**	**55.90**	N.A.	N.A.	N.A.	N.A.
Base of Cf 1	935	56.17	N.A.	N.A.	N.A.	N.A.
Base of C25N	**820**	**56.39**	N.A.	N.A.	N.A.	N.A.
Base of Ti 6	815	56.41	N.A.	N.A.	N.A.	N.A.
Base of Ti 5	510	57.52	N.A.	N.A.	N.A.	N.A.
Top of C26N	**500**	**57.55**	N.A.	N.A.	N.A.	N.A.
	Level (m)	Sed. rate (m/1000 yr)	Level (m)	Sed. rate (m/1000 yr)	Level (m)	Sed. rate (m/1000 yr)
	2240–1070	0.48	2515–2420	0.65	634–365	0.39
	1070–820	0.51	2420–2000	0.72	365–0	0.25
	820–500	0.28	1195–2000	0.54		

Note: Level/age estimates in bold are the calibration points used for each composite section. Ages in bold that are underlined were determined in one section, then used as calibration points in the other two sections. Ages for all chron boundaries except the base of C24n.2n are from Cande and Kent (1995). N.A.—not applicable.

lyzed data from more than one nodule (up to 6). Using mean values for nodules (the average of all subsamples from the same nodule), we calculated the maximum within-soil difference for both isotopes. Within-soil differences in the McCullough Peaks soils (n = 41) were 0.5‰ and 0.4‰ for $\delta^{13}C$ and $\delta^{18}O$, respectively, whereas within-soil differences for southern Bighorn Basin soils (n = 40) were slightly higher (0.6‰ for both isotopes).

Third, in the southern Bighorn Basin, we analyzed sparry calcite in fractures to estimate isotopic values for the diagenetic end-member (GSA Data Repository Appendix 1[1]). In most cases, spar had much lower $\delta^{18}O$ values than micrite, indicating a higher temperature of formation. Spar $\delta^{13}C$ varied from values similar to those for micritic cements to very ^{13}C-depleted values. Thus, as in the study of the Clarks Fork region (Koch et al., 1995), contamination by diagenetic spar can be diagnosed by unusually negative $\delta^{18}O$ values, but the effect on $\delta^{13}C$ values is less pronounced.

[1]GSA Data Repository item 2003048, Appendices 1–3, is available on request from Documents Secretary, GSA, P.O. Box 9140, Boulder, CO 80301-9140, USA, editing@geosociety.org, or at www.geosociety.org/pubs/ft2003.htm.

Finally, soil development differs conspicuously among Bighorn Basin paleosols, and these variations might have an impact on isotope values from soil carbonates. In Data Repository Appendix 1, we report paleosol maturity values determined by Tom Bown for a subset of the southern Bighorn Basin paleosols. The maturity scale varies from stage 0 (least mature) to stage 5 (most mature) and can be modified to indicate whether soils are hydromorphic (Bown and Kraus, 1987). Increasing maturity, as indicated by increasing horizonation of the soil and increasing B horizon thickness, is thought to indicate increasing time of formation, though climate change (temperature, moisture) may affect the rate of soil maturation. The low density of maturity data prohibits quantitative analysis, but a cursory examination is informative. Samples from soils of similar maturity at adjacent stratigraphic levels (e.g., YPM 389 and YPM389 + 2 m, both stage 1 paleosols at 170 and 172 m, respectively) exhibit isotopic differences of the same magnitude as samples from soils of different maturity at adjacent levels (e.g., YPM 212 [stage 2H], YPM 290N [stage 1], and YPM 363 [stage 2] at 230, 210 and 190 m, respectively). Although the influence of soil maturity and hydromorphy certainly warrant further study, they impart no clear signal on the isotopic data reported here.

Mean $\delta^{13}C$ and $\delta^{18}O$ values for nodules are plotted versus stratigraphic level in Figures 3–5 along with 3-point running averages that illustrate trends. As reported in Koch et al. (1995), paleosol carbonate $\delta^{13}C$ values from the Clarks Fork region drop by ~3‰ from the late Paleocene (Ti 5/C25R) to the early Eocene (base of Wa 5) (Fig. 3). Superimposed on the long-term trend is a short-term drop of almost 8‰ to values as low as −16‰. Details of this excursion are explored in Bowen et al. (2001); we note here that the initiation of the excursion is nearly coincident with the first appearance of the three orders of modern mammals and the unusual Wa 0 fauna that marks the beginning of the Wasatchian NALMA. Oxygen isotope data show no long-term trends and typically range between −8‰ to −9‰. There is a suggestion that Wa 0 beds have less negative $\delta^{18}O$ values (Fig. 3). Nevertheless these higher values are balanced by adjacent low $\delta^{18}O$ values, so the running average does not show a significant deflection. ANOVA demonstrates that mean values for nodules are not significantly different between Wa 0 and beds below or above this zone.

The paleosol carbonate isotope record from the McCullough Peaks region begins with the Wa 0 beds and extends to the late Wasatchian (Wa 7) (Fig. 4; Data Repository Appendix

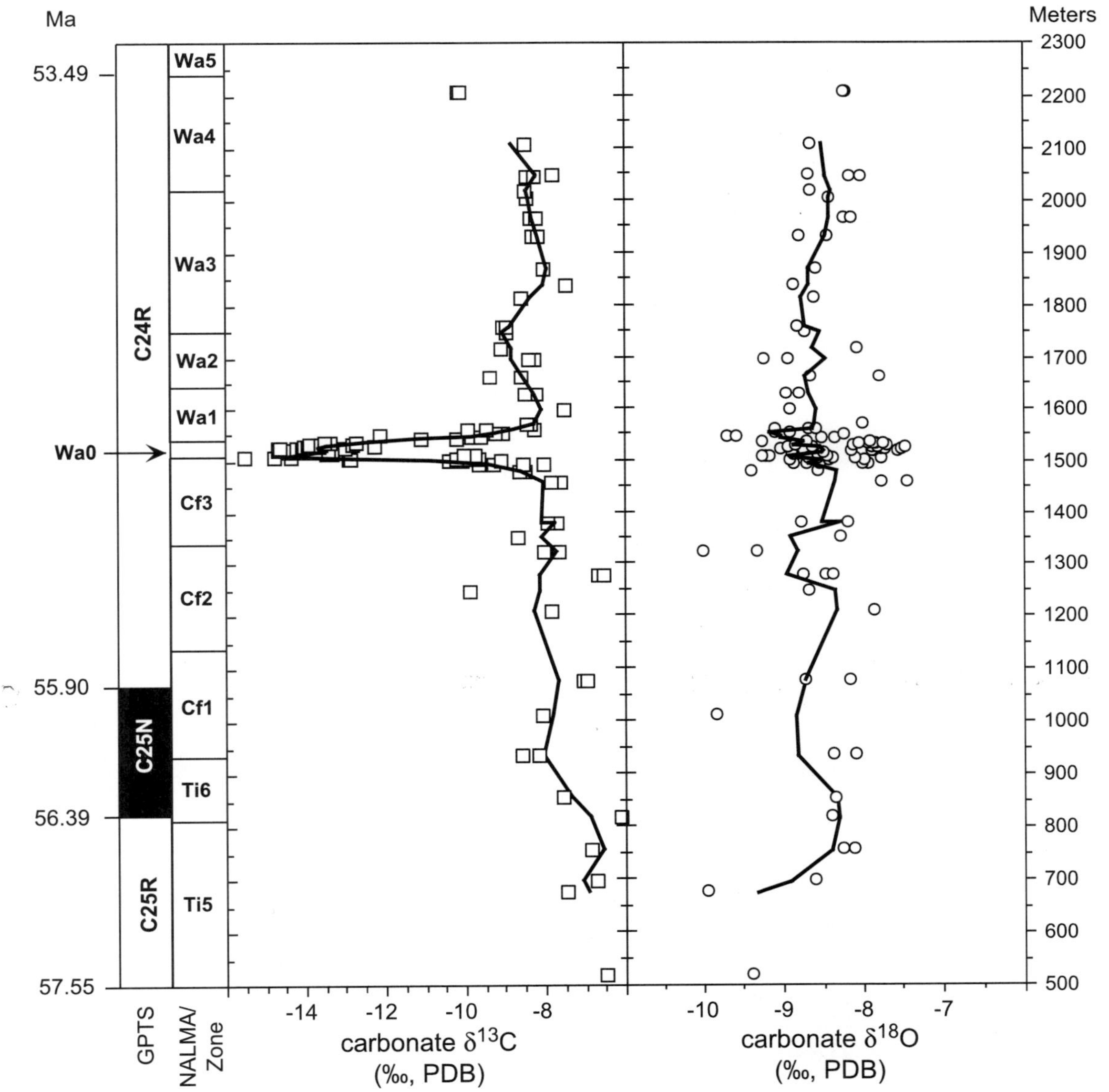

Figure 3. Carbon and oxygen isotope data for paleosol carbonates from the Clarks Fork region plotted versus meter level. The lines are 3-point running averages for meter-level mean values. Most data are from Koch et al. (1995), except for data near the Cf-Wa boundary, which are from Bowen et al. (2001). GPTS—geomagnetic polarity time scale; NALMA—North American Land Mammal Age; PDB—Peedee belemnite.

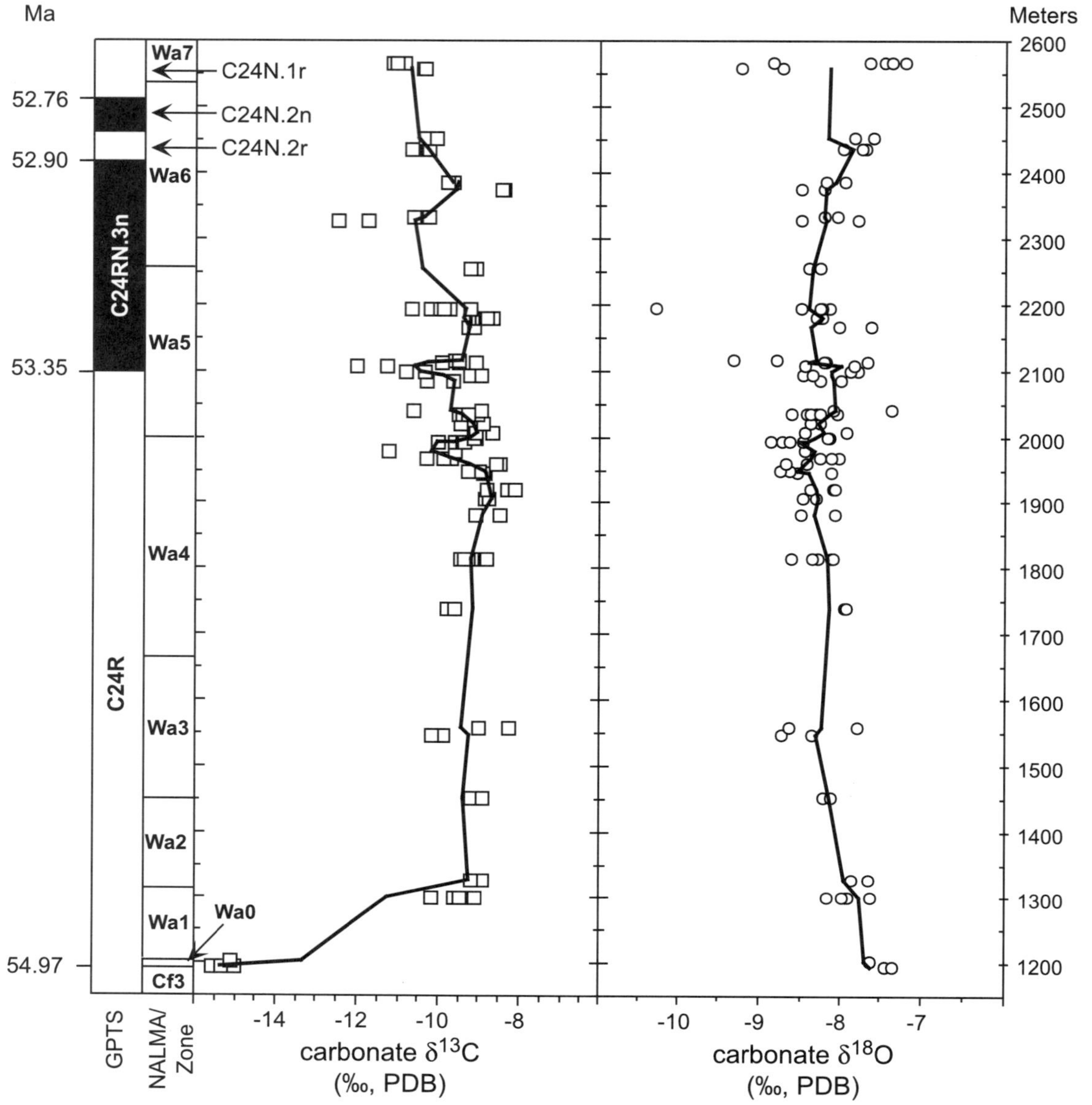

Figure 4. Carbon and oxygen isotope data for paleosol carbonates from the McCullough Peaks region plotted against meter level. Each data point represents the average of at least two analyses from each paleosol carbonate nodule. The lines are 3-point running averages for meter-level mean values. GPTS—geomagnetic polarity time scale; NALMA—North American Land Mammal Age; PDB—Peedee belemnite.

2; see footnote 1). The $\delta^{13}C$ record at the base of the section is similar to that in the Clarks Fork region. Wa 0 values are extremely low (~ –16‰), but values rise rapidly to –8‰ to –10‰ in Wa 1–4. $\delta^{13}C$ values show a second, more gradual decrease beginning in the middle of Wa 4, oscillating to a series of lows in Wa 5–6. The $\delta^{18}O$ record shows greater structure than in the Clarks Fork region. Wa 0 carbonates are consistently ^{18}O-enriched (mean nodule $\delta^{18}O$ value ±1σ: –7.6 ± 0.2‰, n = 4) relative to those from Wa 2–5 (–8.3 ± 0.4‰, n = 79). Values rise slightly in Wa 6–7 but show greater variability (–8.1 ± 0.5‰, n = 23). Differences in mean among these groups are significant (ANOVA, $p < 0.001$). Posthoc tests reveal that the Wa 0 to Wa 2–5 and Wa 2–5 to Wa 6–7 differences are significant; the difference between Wa 0 and Wa 6–7 is not (Scheffé's test).

The paleosol carbonate record from the southern Bighorn Basin also begins at Wa 0 and extends to Wa 7 (Fig. 5). $\delta^{13}C$ values for the Wa 0 beds are strongly negative, but not as ^{13}C-depleted as those from the other two sections. Following the rapid rise above the Wa 0 beds, values rise more gradually to the middle of the Upper *Haplomylus-Ectocion* zone (middle Wa

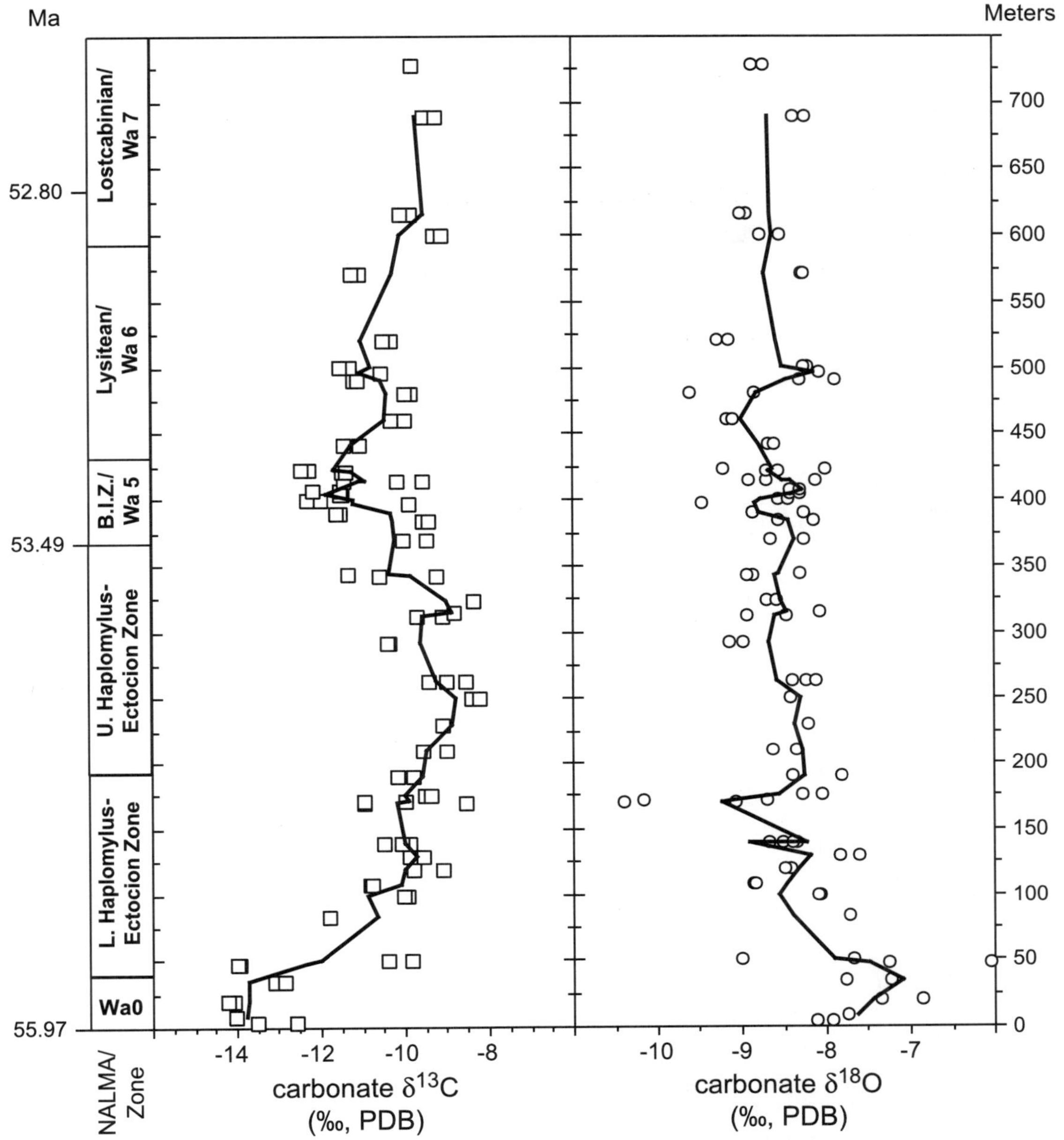

Figure 5. Carbon and oxygen isotope data for paleosol carbonates from the southern Bighorn Basin plotted against meter level. Each data point represents the average of at least two analyses from each paleosol carbonate nodule. The lines are 3-point running averages for meter-level mean values. NALMA—North American Land Mammal Age; PDB—Peedee belemnite.

4?), then drop gradually to a low in the *Bunophorus* Interval Zone (Wa 5). Values for the remainder of the section stay somewhat lower than the peak values in the *Haplomylus-Ectocion* zone. As in the McCullough Peaks section, carbonates from Wa 0 are consistently ^{18}O-enriched (−7.4 ± 0.6‰, n = 9) relative to samples from higher in the section (−8.6 ± 0.5‰, n = 84). This difference in mean is significant (t-test, $p < 0.0001$). Unlike the McCullough Peaks record, there is no return to higher $\delta^{18}O$ values in Wa 6–7.

Organic carbon records are plotted versus model age in Figure 6. Organic carbon within paleosol carbonates from the Clarks Fork region shows no change from the base of Ti 5 to the base of Wa 5. We sampled organics in carbonates from the Wa 0 beds intensively (SC 67 and SC 351, DR Appendix 3 [see footnote 1]). They show among the lowest values for the sequence (−25.3 ± 0.2‰ for Wa 0 soils vs. −24.8 ± 0.6‰ for the non-Wa 0 soils). This difference in mean is statistically significant (t-test, $p < 0.001$).

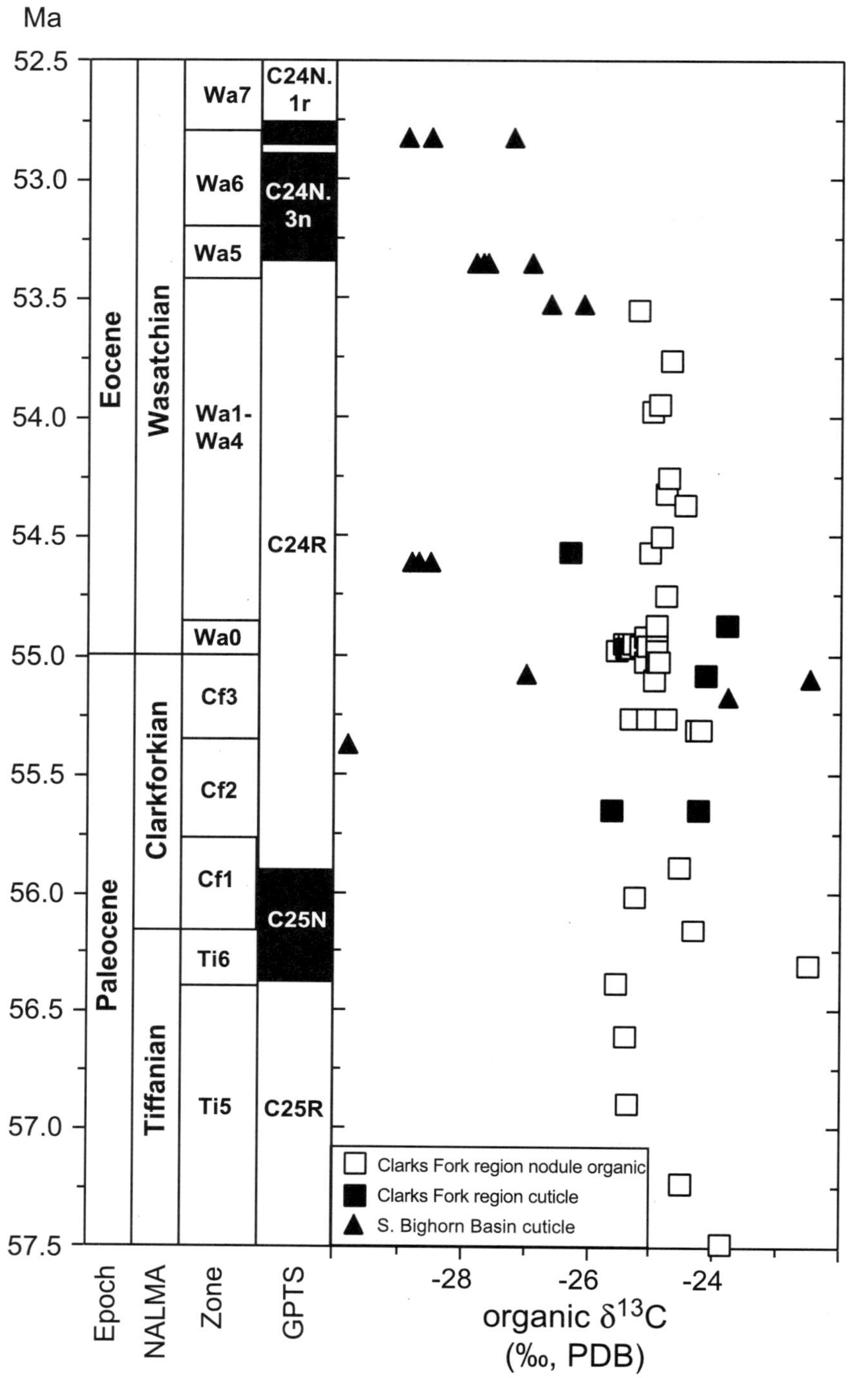

Figure 6. Carbon isotope data for cuticle and organic matter occluded in paleosol carbonates from the Bighorn Basin plotted against model age. GPTS—geomagnetic polarity time scale; NALMA—North American Land Mammal Age; PDB—Peedee belemnite.

Fossil cuticles and plant fragments from organic-rich beds are typically ^{13}C-depleted relative to nodule-occluded organic matter from adjacent horizons. In addition, cuticle and plant fragments from adjacent layers show a greater spread of values. We suspect that some of this variance is related to preservation. Well-preserved cuticle (pale brown with good cellular/morphologic preservation) typically has lower $\delta^{13}C$ values than poorly preserved samples (carbonized/black samples with poor morphologic preservation). This phenomenon merits further study, but at present, it limits the utility of plant fragments as a data source in our system.

DISCUSSION

Carbon isotopes: Correlation and the carbon cycle

If our conceptual model of the controls on soil carbonate $\delta^{13}C$ values is correct, then coeval carbonates from different parts of the Bighorn Basin should have similar values, and secular trends on land should be of the same direction and roughly the same magnitude as those in marine surface waters. Both these issues are best viewed by combining soil carbonate $\delta^{13}C$ values from the different sections and plotting them relative to model age (Fig. 7). Where the three sections overlap in time (Wa 0–4), they show similar values, with two exceptions. First, in the southern Bighorn Basin, $\delta^{13}C$ values rise to postexcursion values over ~500,000 yr, whereas in the other two sections, the rise occurs in ~100,000 yr. This may reflect a problem with the assumption of constant sedimentation rate in the lower 365 m of the southern Bighorn Basin section, which underpins the age model. However the variations in sedimentation rate needed to bring the southern Bighorn Basin into accord with the other two records are large. Alternatively, we may be detecting differences in soil properties that affect isotope values (Cerling and Quade, 1993) or in the type of C3 vegetation overlying this region. The second difference is that Clarks Fork soil carbonates are consistently ^{13}C-enriched by 1‰–2‰ relative to those from the other two sections in the interval from 54.25 to 53.75 Ma. Again, this difference could relate to within-basin differences in soil properties or vegetation. For example, we might suspect that the

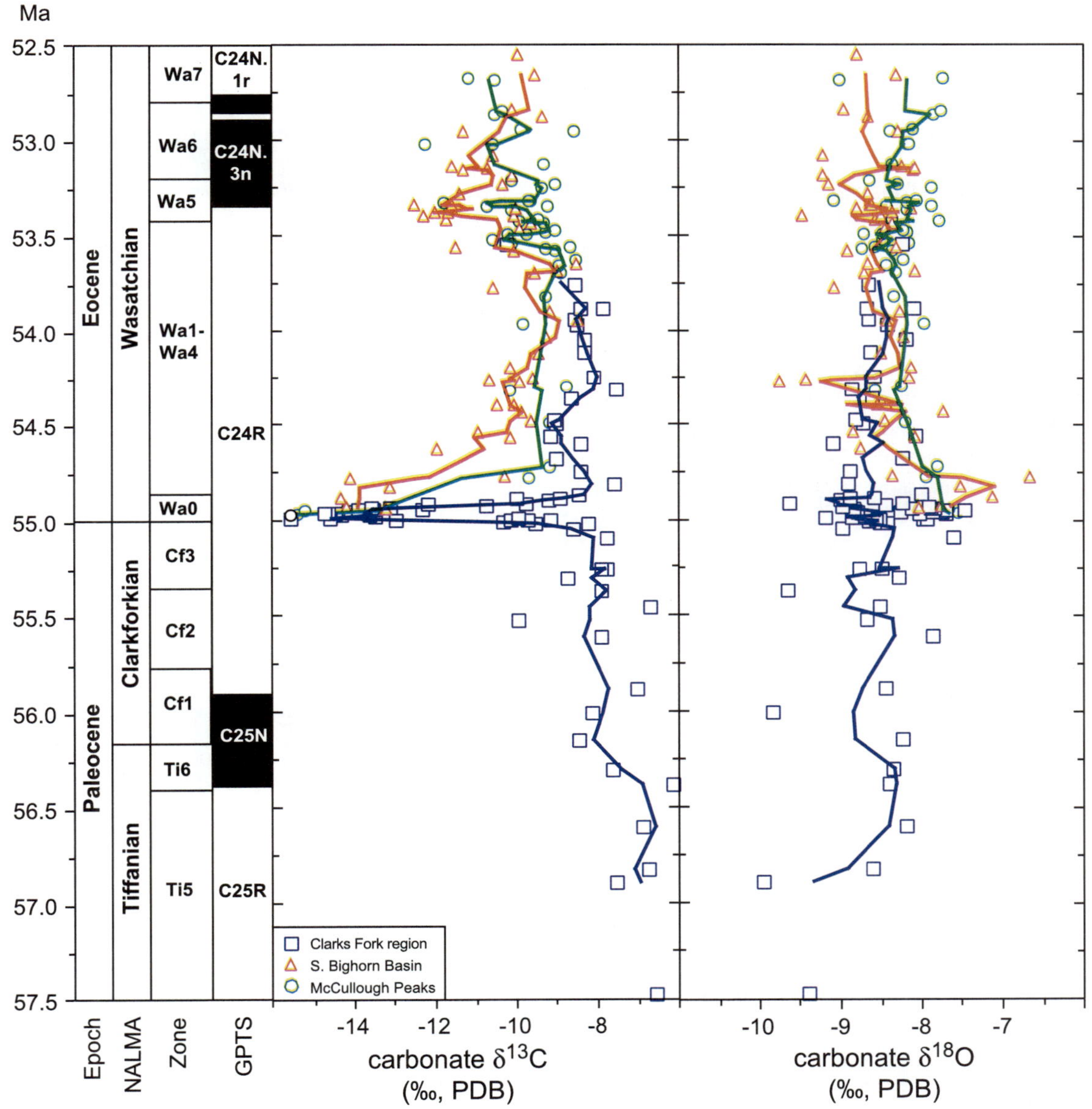

Figure 7. Carbon and oxygen isotope data for paleosol carbonates from three different regions within the Bighorn Basin plotted against model age. Each data point represents the average values of a particular meter level. In the southern Bighorn Basin and McCullough Peaks region, data from at least two different paleosol carbonate nodules were used to generate the level mean. The lines are 3-point running averages for each measured section. GPTS—geomagnetic polarity time scale; NALMA—North American Land Mammal Age; PDB—Peedee belemnite.

Clarks Fork region was drier than the rest of the Bighorn Basin during this interval, which would contribute to higher $\delta^{13}C$ values through lower soil productivity and water stress effects on plant isotope compositions (Cerling and Quade, 1993; Ehleringer and Monson, 1993). The higher incidence of hydromorphic soils and organic-rich shales in this interval in the southern Bighorn Basin is intriguing, but the issue clearly needs further, careful study.

Despite these small within-basin deviations, the long-term drop in $\delta^{13}C$ values on land matches the long-term marine shift well. Values on land drop by 4‰, from ~ –7‰ in C25R to values of ~ –11‰ in C24N.3n. Over this same temporal interval, values of marine surface and deep waters drop by ~2.5‰ (Fig. 1). Thus, the shift in continental records is of the same direction, but slightly magnified. There is a suggestion that $\delta^{13}C$ values begin to rise in C24N.2n, as they do in the marine record, but Bighorn Basin terrestrial sections are truncated by the cessation of paleosol formation.

Another intriguing feature of the Bighorn Basin $\delta^{13}C$ record relates to variability. The range in $\delta^{13}C$ values in Wa 5–7 (~4‰) is higher than the range in Wa 2–4 or in the Clarkforkian and Tiffanian (~2‰). Inspection reveals that increased variability is not solely a result of combining data from independent sections. The reason for increased variability is unclear, but it may relate to the rise in temperatures toward the EWI. Clyde et al. (2001) have recently reported highly variable paleosol carbon and oxygen isotope records from early Bridgerian-aged rocks (ca. 50 Ma) in the Green River Basin, Wyoming. These soils were formed during the EWI. Isotopic cycles in these soils seem to follow 100,000 year eccentricity cycles, suggesting that climatic forcing of vegetation change or the hydrologic cycle during this time of extreme global warming may be affecting soil formation processes. High-resolution (soil-by-soil) analysis of records from the top of the McCullough Peaks section, which has excellent paleomagnetic control, could reveal whether variability in this section is following Milankovich cycles, and it might point to the proximate link between isotopic variability and climate change.

The duration, direction, and time of onset of the short-term $\delta^{13}C$ excursion at the Cf-Wa boundary all demonstrate unambiguously that it is associated with the short-term carbon cycle perturbations at the PETM in marine systems. Still, the magnitude of the event on land (~8‰) is much greater than that in marine surface waters (2.5‰–4‰) (Kennett and Stott, 1991; Zachos et al., 2001). Potential causes for the amplified magnitude of the excursion on land are discussed in Bowen et al. (2001), but at present, the issue is unresolved. A second striking feature of the excursion on land is that its magnitude is reduced by ~2‰ in the southern Bighorn Basin relative to the other two sections. One possible explanation is that the Wa 0 beds in the southern Bighorn Basin do not record the onset of the excursion, where $\delta^{13}C$ values are lowest (Figs. 3–5). It is also possible that the base of Wa 0 is preserved in underlying Fort Union Formation lithologies where no pedogenic carbonate is preserved. The Fort Union–Willwood Formation contact is time transgressive in the northern part of the basin (e.g., the contact lies in the Clarkforkian NALMA in the Clarks Fork region but at the base of Wa 0 in the McCullough Peaks). This pattern may extend into the southern basin as well. Another possible explanation for the reduction in magnitude in the southern Bighorn Basin is time averaging. The Wa 0 paleosols in this area are among the most mature found anywhere in the basin. Maturity in part reflects the amount of time the original soil was exposed and active (Kraus and Bown, 1993). It is possible that Wa 0 paleosols in the southern Bighorn Basin have time averaged the $\delta^{13}C$ excursion more than the Wa 0 paleosols in the northern parts of the basin, where short-term accumulation rates are higher (and thus exposure times are shorter). Tests of these alternative explanations await development of better constraints on short-term sediment accumulation rates in these areas.

Plants are a key link in the chain connecting isotopic records from marine to atmospheric to continental carbon reservoirs. If sedimentary organic carbon records are derived from decayed ancient vegetation, they should show both the long-term and short-term shifts in $\delta^{13}C$ values. This prediction is not borne out by data from organic matter occluded in paleosol carbonates in the Bighorn Basin (Fig. 6). Nodule organic $\delta^{13}C$ values are invariant up-section, with a slight deflection toward lower values in Wa 0 soils. How can paleosol carbonates show a large decrease during this period of carbon cycle perturbation if soil organic matter does not? We can envision scenarios whereby soils receive a pulse of ^{12}C-enriched CO_2 directly from the atmosphere, perhaps via microbial oxidation of atmospheric methane. Yet this (and other) scenarios that explain the $\delta^{13}C$ excursion on land as a phenomenon restricted to soils are falsified by isotopic data from Bighorn Basin fossil mammals (Koch et al., 1995). Mammals obtain carbon only from plants, and they show an excursion as large as that seen in paleosol carbonates. We conclude that Paleocene-Eocene plants in the Bighorn Basin did show large $\delta^{13}C$ shifts in response to changes in the $\delta^{13}C$ of the atmosphere, but that these shifts are not recorded in the organic matter occluded in paleosol carbonates.

Paleosol nodules contain only trace amounts of organic matter. The most likely explanation for the invariant nodule organic data is that nearly all the original organic carbon has been removed by postburial oxidation. We are probably measuring traces of either modern Earth surface or basinal fluid organic carbon that is carried into nodules as a dissolved phase, and then sorbed to clay surfaces. Bird et al. (1994) demonstrated through stable isotope and ^{14}C analysis that even densely cemented pedogenic nodules are not closed with respect to organic carbon exchange, so the addition of trace amounts of organic carbon is not surprising. Values of –25‰ are within the range expected for degraded carbon from the contemporary C3 ecosystem that covers the land surface in Wyoming.

A clear implication is that great care must be taken when interpreting organic carbon records from organic-poor nodules. In the Paris Basin, Sinha et al. (1995) and Magioncalda et al.

(2001) studied bulk sediment organic matter, fragments of wood, and organic matter occluded within soil carbonate nodules. They detected a 6‰ drop in the $\delta^{13}C$ in "Sparnacian"-aged sediments that they correlated to the PETM. This result supports our contention that land plants exhibited a conspicuous $\delta^{13}C$ drop at the PETM. Clearly, the safest approach to constraining $\delta^{13}C$ shifts in plants is to use organic biomarkers characteristic of particular taxa, or at least to use clearly nontransported, ancient carbon sources such as well-preserved leaf cuticles or wood (e.g., Arens and Jahren, 2000; Hesselbo et al., 2000). Whenever possible, these records should be compared to associated carbon phases like soil carbonate or biogenic phosphate to evaluate potential diagenesis (as was done in the study by Sinha et al. [1995]).

Oxygen isotopes: Proxies for temperature and moisture

The clearest signal in the $\delta^{18}O$ records is at the PETM. In two of the three records (southern Bighorn Basin and McCullough Peaks), values are ^{18}O-enriched by at least 1‰ during the PETM. This enrichment in carbonate $\delta^{18}O$ values is consistent with several different scenarios.

The increase could reflect an increase in evaporative water loss from soils due to a decrease in regional precipitation. This scenario is not consistent with climate modeling results, which suggest an enhanced hydrological cycle during the PETM (Huber and Sloan, 1999). In addition, Wa 0 soils are extremely mature relative to soils higher or lower in the section. Increased maturity, which may reflect an increased intensity of chemical weathering, is more consistent with increased precipitation, higher temperatures, and greater soil productivity, not drier, less productive soils. Finally, this mechanism, which explains the shift in soil carbonate $\delta^{18}O$ values solely as a response to change in soil moisture ignores independent evidence that the $\delta^{18}O$ of meteoric water changed during the PETM. Oxygen isotope analysis of mammalian tooth enamel and gar scale ganoine suggest that meteoric water in the Bighorn Basin was ^{18}O-enriched by 2‰–3‰ during Wa 0 relative other intervals in the section (Fricke et al., 1998). In summary, we consider increased soil evaporation an unlikely source for the ^{18}O-enrichment in Wa 0/PETM soil carbonates.

An alternate explanation, which focuses on the shift in the $\delta^{18}O$ of Bighorn Basin meteoric water, is that the source of vapor masses supplying the region altered in response to the global climatic perturbations at the PETM. Today, precipitation reaching midcontinental North America that is derived from the Gulf of Mexico is ^{18}O-enriched relative to water arriving from the Pacific Ocean. Temporal shifts in vapor source region have been detected in $\delta^{18}O$ time series from Holocene and Pleistocene deposits (Amundson et al., 1996; Chamberlain and Poage, 2000). This difference is a result of the cooler temperatures of evaporation in the north-central Pacific (which cause vapor masses to be ^{18}O-depleted), and the strong rainout effects associated with transport over the mountains of the western United States. Although the altitude of the western United States in the Paleogene is a subject of controversy, it is very likely that altitudes to the west of Wyoming were greater than altitudes between Wyoming and Gulf of Mexico. We expect the same general pattern of difference in $\delta^{18}O$ values between Pacific and Gulf precipitation at the P-E boundary, though perhaps of lower magnitude. Thus, enhanced moisture supply from the Gulf of Mexico and Mississippi Embayment during the PETM could lead to ^{18}O-enrichment of meteoric waters and the soil minerals that form from them.

Another mechanism to change the $\delta^{18}O$ of meteoric water is to shift the seasonality of precipitation. Today, in regions where MAT is <20 °C, cool season precipitation is ^{18}O-depleted relative to warm season precipitation. In tropical regions, wet season precipitation is ^{18}O-depleted relative dry season precipitation. Thus, shifts in either the seasonality of rainfall or soil carbonate formation could contribute to ^{18}O-enrichment in the PETM.

None of the preceding scenarios entail a regional increase in temperature, which might be expected given the sharp global rise in ocean surface temperatures at the PETM (Huber and Sloan, 1999). Soil carbonate $\delta^{18}O$ values can serve as a very rough proxy for temperature, assuming that factors mentioned above are invariant. Temperature has antagonistic effects on the $\delta^{18}O$ of soil carbonates. In modern temperate and high-latitude regions, there is a strong positive correlation between the $\delta^{18}O$ of meteoric water and MAT, reflecting chiefly increased rainout of ^{18}O from vapor masses at lower temperatures (Dansgaard, 1964). In regions with a MAT between 0 and 20 °C, the slope for the spatial relationship between $\delta^{18}O$ of meteoric water and MAT is ~+0.6‰ per °C ($r > 0.8$) (Rozanski et al., 1993). The effect of temperature on the oxygen isotope fractionation between water and calcite is weaker and of the opposite sign; the slope for the relationship between the $\delta^{18}O$ of calcite and the temperature of formation is ~ –0.23‰ per °C (Friedman and O'Neil, 1977). We can add these effects to solve for the changes in temperature and meteoric water $\delta^{18}O$ value that are mutually consistent with the observed change in soil carbonate $\delta^{18}O$ values. Using this approach, the observed increase in soil carbonate $\delta^{18}O$ value of 1.25‰ is consistent with an increase in meteoric water $\delta^{18}O$ value of 2.0‰, and both are consistent with an increase in MAT of 3.4 °C (Table 2).

There has been considerable debate about the applicability of the spatially derived meteoric water $\delta^{18}O$/MAT relationship for reconstructing temporal shifts in MAT (Boyle, 1997; Jouzel et al., 1997). The meteoric water $\delta^{18}O$/MAT slope is sensitive to the pole-to-equator temperature gradient, as well as the temperature of evaporation. Although independent changes in these factors can produce complex results, we might expect that in a world with a shallower pole-to-equator temperature gradient, the slope of the relationship would be lower as well (Fricke and O'Neil, 1999). Table 2 evaluates the effects of assuming shallower slopes on the estimated changes in MAT and meteoric water $\delta^{18}O$ values. Using slopes ranging from +0.6‰ to +0.4‰ per °C, we estimate PETM changes in the meteoric water $\delta^{18}O$ values ranging from +2‰ to +3‰, consistent with estimates for

TABLE 2. ESTIMATING CHANGE IN MEAN ANNUAL TEMPERATURE AND METEORIC WATER $\delta^{18}O$ AT THE PETM FROM CHANGE IN SOIL CARBONATE $\delta^{18}O$

Meteoric water $\delta^{18}O$/MAT slope (‰, SMOW/°C)	Estimated soil calcite $\delta^{18}O$/MAT slope* (‰, SMOW/°C)	Estimated ΔMAT# (°C)	Estimated $\delta^{18}O_{meteoric\ water}$§ (‰, SMOW)
+0.6	+0.37	+3.4	+2.0
+0.5	+0.27	+4.6	+2.3
+0.4	+0.17	+7.4	+2.9

Notes: Δ—change in; MAT—mean annual temperature.

*Calculated by adding the temperature-dependent fractionation of oxygen isotopes between calcite and water (–0.23‰/°C) to the meteoric water/MAT slope.

#Calculated by dividing the shift in soil carbonate $\delta^{18}O$ (+1.25) by the value in column 2.

§Calculated by multiplying columns 1 and 3.

Bighorn Basin meteoric water from Fricke et al. (1998). The estimated rise in temperature ranges from 3.4 to 7.4 °C.

The lack of a conspicuous response in soil carbonate $\delta^{18}O$ values to the onset of the EWI is puzzling. Studies of leaf physiognomy suggest that MAT in the Bighorn Basin was ~4 °C higher at 52.5 Ma than it was at 54.5 Ma (Wing et al., 2000) and a similar rise in MAT has been detected in the nearby Green River Basin (Wilf, 2000). Oxygen isotope compositions of tooth enamel, gar scales, and iron oxides also seem to track the EWI warming (Fricke et al., 1998; Bao et al., 1999). Using the relationships presented in Table 2, we would expect soil carbonate $\delta^{18}O$ values to increase by +1.5‰ to +0.7‰. There is a suggestion of a greater number of ^{18}O-enriched values near the top of the McCullough Peaks section (Fig. 4) but the record is noisy, so there is no pronounced affect on the running average of $\delta^{18}O$ values.

The fact that the EWI warming seems to be clearly recorded in the $\delta^{18}O$ of mammalian enamel and fish ganoine but not in coeval soil carbonates is intriguing. Enamel and ganoine are characterized by small, tightly packed crystals of relatively insoluble carbonate hydroxylapatite, whereas soil carbonates are composed of more soluble micritic calcium carbonate. It is possible that the oxygen in enamel and ganoine is more resilient to diagenetic recrystallization than the oxygen in soil carbonates. Alternatively the lack of change in soil carbonates may be a real signal, indicating that the multiple controls on $\delta^{18}O$ have somehow acted in a compensatory fashion.

CONCLUSION

The simple conceptual model of carbon isotope coupling among oceanic, atmospheric, and terrestrial carbon reservoirs is supported by this study. Soil carbonate $\delta^{13}C$ values change in phase with marine surface water carbonates on both the long (~7 m.y.) and short (~100 k.y.) time scales. These shifts are much greater than average within-nodule and within-soil isotopic differences. The PETM isotopic excursion is present in long measured sections from three regions in the Bighorn Basin separated by ~100 km. In addition, the two new sections (McCullough Peaks, southern Bighorn Basin) provide the first evidence in the Bighorn Basin for the long-term $\delta^{13}C$ minimum in C24N. The long-term $\delta^{13}C$ trend in the Bighorn Basin closely matches a paleosol carbonate record from the Aix-en-Provence basin in France (Cojan et al., 2000), demonstrating the global nature of these carbon cycle perturbations. Interestingly, the PETM $\delta^{13}C$ excursion in the Bighorn Basin is much larger than that in surface ocean carbonates or in Aix-en-Provence, but the reason for this amplification is unclear and under investigation at present. Finally, organic substrates from the Bighorn Basin yielded noisy $\delta^{13}C$ records that did not mirror these global changes, perhaps due to variable alteration of cuticle fragments or postformational overprinting in organic-poor soil carbonates.

The $\delta^{18}O$ values of meteoric water and soil carbonate are influenced by multiple, antagonistic factors. The observed responses in $\delta^{18}O$ values at the PETM and EWI are small, scarcely greater than average within-soil $\delta^{18}O$ differences. Still, the $\delta^{18}O$ perturbation at the PETM is statistically significant in two of the measured sections. Assuming a range of plausible values for the relationship between meteoric water $\delta^{18}O$ and temperature, this perturbation is consistent with independent evidence for changes in meteoric water and changes in local temperature of 3–7 °C. Still, more robust measures of change in temperature and precipitation, such as those provided by leaf physiognomic analysis, will be required before the terrestrial climatic response at the PETM can be evaluated with confidence.

ACKNOWLEDGMENTS

We thank H. Bao, J. Gorman, M. Johnston, G. Kardon, A. Marr, and K. Rose for assistance in the field, and M. Hogan, B. Johnson, M. Toperoff, and K. Young for assistance in the laboratory. We thank D. Dettman, S. Hesselbo, and L. Stott for thoughtful reviews of the manuscript. Dana Royer supplied cuticle samples for analysis. This work would not have been possible without the decades of geological and paleontological research on the Bighorn Basin led by Tom Bown, Ken Rose and Phillip Gingerich, and their assistance throughout the course of this study is greatly appreciated. This research was supported by NSF grants EAR 9105160 (to PLK, JCZ, and MLF) and EAR 9627953 (to PLK), and Princeton University.

REFERENCES CITED

Alroy, J., Koch, P.L., and Zachos, J.C., 2000, Global climate change and North American mammalian evolution: Paleobiology, v. 26 (supplement), p. 259–288.

Amundson, R., Chadwick, O., Kendall, C., Wang, Y., and DeNiro, M., 1996, Isotopic evidence for shifts in atmospheric circulation patterns during the late Quaternary in mid–North America: Geology, v. 24, p. 23–26.

Arens, N.C., and Jahren, A.H., 2000, Carbon isotope excursion in atmospheric CO_2 at the Cretaceous–Tertiary boundary: Evidence from terrestrial sediments: Palaios, v. 15, p. 314–322.

Arens, N.C., Jahren, A.H., and Amundson, R., 2000, Can C3 plants faithfully record the carbon isotopic composition of atmospheric carbon dioxide?: Paleobiology, v. 26, p. 137–164.

Bao, H., Koch, P.L., and Rumble, D., 1999, Paleocene-Eocene climatic variation in western North America: Evidence from the $\delta^{18}O$ of pedogenic hematite: Geological Society of America Bulletin, v. 111, p. 1405–1415.

Bird, M.I., Quade, J., Chivas, A.R., Fifield, L.K., Allan, G.L., and Head, M.J., 1994, The carbon isotope composition of organic matter occluded in iron nodules: Chemical Geology, v. 114, p. 269–279.

Bowen, G.J., Koch, P.L., Gingerich, P.D., Norris, R.D., Bains, S., and Corfield, R.M., 2001, A high-resolution isotope stratigraphy across the Paleocene–Eocene boundary at Polecat Bench, Wyoming: University of Michigan Papers on Paleontology, v. 33, p. 73–88.

Bown, T.M., and Kraus, M.J., 1987, Intergration of channel and flood-plain suites, I: Developmental sequence and lateral relations of alluvial paleosols: Journal of Sedimentary Petrology, v. 57, p. 587–601.

Bown, T.M., Rose, K.D., Simons, E.L., and Wing, S.L., 1994, Distribution and stratigraphic correlation of upper Paleocene and lower Eocene fossil mammal and plant localities of the Fort Union, Willwood, and Tatman formations, southern Bighorn Basin, Wyoming: U.S. Geological Survey Professional Paper 1540, p. 1–103.

Boyle, E.A., 1997, Cool tropical temperatures shift the global $\delta^{18}O$-T relationship: An explanation for the ice core $\delta^{18}O$-borehole thermometry conflict?: Geophyscial Research Letters, v. 24, p. 273–276.

Butler, R.F., Gingerich, P.D., and Lindsay, E.H., 1981, Magnetic polarity stratigraphy and biostratigraphy of Paleocene and lower Eocene continental deposits, Clarks Fork Basin, Wyoming: Journal of Geology, v. 89, p. 299–316.

Cande, S.C., and Kent, D.V., 1995, Revised calibration of the geomagnetic polarity timescale for the Late Cretaceous and Cenozoic: Journal of Geophysical Research ser. B, v. 100, p. 6093–6095.

Cerling, T.E., and Quade, J., 1993, Stable carbon and oxygen isotopes in soil carbonates, *in* Swart, P.K., et al., eds., Climate change in continental isotope records: American Geophysical Union Geophysical Monograph Series, v. 78, p. 344–345.

Chamberlain, C.P., and Poage, M.A., 2000, Reconstructing the paleotopography of mountain belts from the isotopic composition of authigenic minerals: Geology, v. 28, p. 115–118.

Clyde, W.C., 1997, Stratigraphy and mammalian paleontology of the McCullough Peaks, northern Bighorn Basin, Wyoming: Implications for biochronology, basin development, and community reorganization across the Paleocene–Eocene boundary [Ph.D. thesis]: Ann Arbor, University of Michigan, 271 p.

Clyde, W.C., 2001, Mammalian biostratigraphy of the McCullough Peaks area in the northern Bighorn Basin, Wyoming: University of Michigan Papers on Paleontology, v. 33, p. 109–126.

Clyde, W.C., and Gingerich, P.D., 1998, Mammalian community response to the latest Paleocene Thermal Maximum: An isotaphonomic study in the northern Bighorn Basin, Wyoming: Geology, v. 26, p. 1011–1014.

Clyde, W.C., Sheldon, N.D., Koch, P.L., Gunnell, G.P., and Bartels, W.S., 2001, Linking the Wasatchian–Bridgerian boundary to the Cenozoic Global Climate Optimum: New magnetostratigraphic and isotopic results from South Pass, Wyoming: Palaeogeography, Palaeoclimatology, Palaeoecology, v. 167, p. 175–199.

Cojan, I., Moreau, M.-G., and Stott, L.E., 2000, Stable carbon isotope stratigraphy of the Paleogene pedogenic series of southern France as a basis for continental-marine correlation: Geology, v. 28, p. 259–262.

Dansgaard, W., 1964, Stable isotopes in precipitation: Tellus, v. 16, p. 436–468.

Dickens, G.R., 2000, Methane oxidation during the late Paleocene Thermal Maximum: Bulletin de la Société Géologique de France, v. 171, p. 37–49.

Dickens, G.R., Castillo, M.M., and Walker, J.C.G., 1997, A blast of gas in the latest Paleocene: Simulating first-order effects of massive dissociation of oceanic methane hydrate: Geology, v. 25, p. 259–262.

Ehleringer, J.R., and Monson, R.K., 1993, Evolutionary and ecological aspects of photosynthetic pathway variation: Annual Review of Ecology and Systematics, v. 24, p. 411–439.

Fricke, H.C., and O'Neil, J.R., 1999, The correlation between $^{18}O/^{16}O$ ratios of meteoric water and surface temperature: Its use in investigating terrestrial climate change over geologic time: Earth and Planetary Science Letters, v. 170, p. 181–196.

Fricke, H.C., Clyde, W.C., O'Neil, J.R., and Gingerich, P.D., 1998, Evidence for rapid climate change in North America during the latest Paleocene thermal maximum: Oxygen isotope compositions of biogenic phosphate form the Bighorn Basin (Wyoming): Earth and Planetary Science Letters, v. 160, p. 193–208.

Friedman, I., and O'Neil, J.R., 1977, Compilation of stable isotope fractionation factors of geochemical interest: Data of Geochemistry. M. Fleischer. 440-KK: 1–12.

Gingerich, P.D., 1983, Paleocene–Eocene faunal zones and a preliminary analysis of Laramide structural deformation in the Clarks Fork Basin, Wyoming: Casper, Wyoming Geological Association, 34th Annual Field Conference, Guidebook, p. 185–195.

Gingerich, P.D., 1989, New earliest Wasatchian mammalian fauna from the Eocene of northwestern Wyoming: Composition and diversity in a rarely sampled, high–flood plain assemblage: University of Michigan Papers on Paleontology, v. 28, p. 1–97.

Gingerich, P.D., 1991, Systematics and evolution of early Eocene Perissodactyla (Mammalia) in the Clarks Fork Basin, Wyoming: Contributions from the Museum of Paleontology, University of Michigan, v. 28, p. 181–213.

Gingerich, P.D., 2000, Paleocene–Eocene boundary and continental vertebrate faunas of Europe and North America: GFF, v. 122, p. 57–59.

Hesselbo, S.P., Gröcke, D.R., Jenkyns, H.C., Bjerrum, C.J., Farrimond, P., Morgans-Bell, H.S., and Green, O.R., 2000, Massive dissociation of gas hydrate during a Jurassic oceanic anoxic event: Nature, v. 406, p. 392–395.

Huber, M., and Sloan, L.C., 1999, Warm climate transitions: A general circulation modeling study of the late Paleocene thermal maximum (similar to 56 Ma): Journal of Geophysical Research—Atmospheres, v. 104, p. 16633–16655.

Jacobs, B.F., Kingston, J.D., and Jacobs, L.L., 1999, The origin of grass-dominated ecosystems: Annals of the Missouri Botanical Garden, v. 86, p. 590–643.

Jahren, A.H., Arens, N.C., Sarmiento, G., Guerrero, J., and Amundson, R., 2001, Terrestrial record of methane hydrate dissociation in the Early Cretaceous: Geology, v. 29, p. 159–162.

Jouzel, J., Alley, R.B., Cuffey, K.M., Dansgaard, W., Grootes, P., Hoffmann, G., Johnsen, S.J., Koster, R.D., Peel, D., Shuman, C.A., Stievenard, M., Stuiver, M., and White, J., 1997, Validity of the temperature reconstruction from water isotopes in ice cores: Journal of Geophysical Research—Oceans, v. 102, p. 26471–26487.

Kennett, J.P., and Stott, L.D., 1991, Abrupt deep-sea warming, paleoceanographic changes, and benthic extinctions at the end of the Paleocene: Nature, v. 353, p. 225–229.

Koch, P.L., 1998, Isotopic reconstruction of past continental environments: Annual Review of Earth and Planetary Sciences, v. 26, p. 573–613.

Koch, P.L., Zachos, J.C., and Gingerich, P.D., 1992, Correlation between isotope records in marine and continental carbon reservoirs near the Paleocene–Eocene boundary: Nature, v. 358, p. 319–322.

Koch, P.L., Zachos, J.C., and Dettman, D.L., 1995, Stable isotope stratigraphy and paleoclimatology of the Paleogene Bighorn Basin (Wyoming, USA): Palaeogeography, Palaeoclimatology, Palaeoecology, v. 115, p. 61–89.

Kraus, M.J., and Bown, T.M., 1993, Short term sediment accumulation rates determined from Eocene alluvial paleosols: Geology, v. 21, p. 743–746.

MacLeod, K.G., Smith, R.M.H., Koch, P.L., and Ward, P.D., 2000, Timing of mammal-like reptile extinctions across the Permian–Triassic boundary in South Africa: Geology, v. 28, p. 227–230.

Magioncalda, R., Dupuis, C., Blamant, D., Fairon-Demaret, M., Perreau, M., Renard, M., Riveline, J., Roche, M., and Keppens, E., 2001, L'excursion isotopique du carbone organique ($\delta^{13}C_{org}$) dans les paléoenvironments continentaux de l'intervalle Paleocène–Eocène de Varangeville (Haute-Normandie): Bulletin de la Société Géologique de France, v. 172, p. 349–358.

Mook, W.G., 1986, ^{13}C in atmospheric CO_2: Netherlands Journal of Sea Research, v. 20, p. 211–223.

Norris, R.D., and Röhl, U., 1999, Carbon cycling and chronology of climate warming during the Paleocene–Eocene transition: Nature, v. 401, p. 775–778.

Röhl, U., Bralower, T.J., Norris, R.D., and Wefer, G., 2000, New chronology for the late Paleocene thermal maximum and its environmental implications: Geology, v. 28, p. 927–930.

Rozanski, K., Araguás-Araguás, L., and Gonfiantini, R., 1993, Isotopic patterns in modern global precipitation, *in* Swart, P.K., et al., eds., Climate change in continental isotope records: American Geophysical Union Geophysical Monograph Series, v. 78, p. 1–35.

Schankler, D.M., 1980, Early Cenozoic paleontology and stratigraphy in the Bighorn Basin, Wyoming: University of Michigan Papers on Paleontology, v. 24, p. 99–114.

Sinha, A., Aubry, M.-P., Stott, L.D., Thiry, M., and Berggren, W.A., 1995, Chemostratigraphy of the "lower" Sparnacian deposits (Argiles plastiques bariolées) of the Paris Basin: Israel Journal of Earth Sciences, v. 44, p. 223–237.

Sundquist, E.T., 1993, The global carbon dioxide budget: Science, v. 259, p. 934–941.

Wilf, P., 2000, Late Paleocene–early Eocene climate changes in southwestern Wyoming: Paleobotanical analysis: Geological Society of America Bulletin, v. 112, p. 292–307.

Wing, S.L., 1998, Late Paleocene–early Eocene floral and climatic change in the Bighorn Basin, Wyoming, *in* Aubry, M.-P., et al., eds., Late Paleocene–early Eocene climatic and biotic events in the marine and terrestrial records: New York, Columbia University Press, p. 380–400.

Wing, S.L., Bao, H., and Koch, P.L., 2000, An early Eocene cool period? Evidence for continental cooling during the warmest part of the Cenozoic, *in* Huber, B.T., et al., eds., Warm climates in Earth history: New York, Cambridge University Press, p. 197–237.

Zachos, J., Pagani, M., Sloan, L., Thomas, E., and Billups, K., 2001, Trends, rhythms, and aberrations in global climate 65 Ma to present: Science, v. 292, p. 686–693.

Manuscript Accepted by the Society August 13, 2002

Printed in the USA

Geological Society of America
Special Paper 369
2003

Terrestrial climatic response to precessional orbital forcing in the Eocene

K.T. Lawrence
L.C. Sloan*
J.O. Sewall
Department of Earth Sciences, University of California, Santa Cruz, California 95064, USA

ABSTRACT

The early Eocene, the most recent ice-free period in Earth's history, has been studied extensively with the goal of understanding the dynamics of such an extremely warm climate state. In this study, we incorporate more realistic orbital forcing, topography, vegetation, and sea surface temperatures than have previously been included in early Eocene modeling experiments. By considering two cases that represent the end members of a realistic precessional cycle, we attempt to constrain the magnitude of climatic variability due to this orbital parameter. Model results for the western interior of North America indicate very small (~1 °C) differences in mean annual temperature (MAT) and cold month mean (CMM) temperature in response to the imposed precessional cycle. Seasonally, the greatest temperature response to the orbital forcing occurs in summer (~6 °C). Mean annual precipitation (MAP) in the region shows little variability between cases either annually or seasonally (<10%) with the exception of autumn (~50%). Our MAT, CMM, MAP, and mean annual temperature range (MATR) results are generally consistent with proxy data interpretations. These results have implications for the interpretation of cyclic desiccation recorded in the lithologies of the Wilkins Peak member of the Eocene Green River Formation (GRF). The low modeled sensitivity of MAT and MAP to precessional forcing implies that the lithologic cycles found in the GRF may not be a consequence of dramatic swings in climate across a precessional cycle, but rather are a consequence of the extremely high climatic sensitivity of the sedimentary processes that formed the Wilkins Peak member. These results also have implications for the interpretation of recent high temporal resolution studies of Paleogene climate that indicate significant shifts in MAT (~1–12 °C) over relatively short (~1 m.y.) time scales. Our modeling cases indicate that precession could account for some portion of this short-term climate variability. However, variability due to precession is of higher frequency and lower amplitude than the MAT fluctuations inferred from the paleontological record, implying that the dominant cause(s) of these fluctuations remain(s) to be identified.

*Corresponding author: Lcsloan@es.ucsc.edu.

Lawrence, K.T., Sloan, L.C., and Sewall, J.O., 2003, Terrestrial climatic response to precessional orbital forcing in the Eocene, *in* Wing, S.L., Gingerich, P.D., Schmitz, B., and Thomas, E., eds., Causes and Consequences of Globally Warm Climates in the Early Paleogene: Boulder, Colorado, Geological Society of America Special Paper 369, p. 65–77.

INTRODUCTION

The early Eocene (ca. 50 Ma), the warmest period in the Cenozoic and the most recent ice-free interval in Earth's history (Zachos et al., 1994), offers an excellent opportunity to study the dynamics and processes that govern Earth's warm climate states. Much of our knowledge of early Eocene terrestrial climate is derived from proxy data recovered from fluvial and lacustrine deposits in western North America (Roehler, 1993; Wing and Greenwood, 1993; Greenwood and Wing, 1995; Koch et al., 1995; Wilf, 2000; Wing et al., 2000). Proxy data represent past climate at a given locality. However, due to limitations imposed by incomplete or discontinuous deposition and preservation of sediments, proxy-based characterization of climate over large geographic areas is challenging. Ideally, proxy data would be corroborated by paleoclimate model results, thus yielding a more geographically extensive description of paleoclimate characteristics. Modeling paleoclimates further instructs our comprehension of Earth's geologic history because it provides a means of investigating and understanding the nature of the physical and dynamical processes of past climate states. In this study, we attempt to make the linkage between climatic processes and climatic evidence in the geologic record by producing model estimates of early Eocene temperature and precipitation, by considering the implications of these results for understanding early Eocene climate, and by using these results to interpret high-frequency geologic proxy data sets from western North America.

To date, comparisons of model output and proxy data have focused on temperature characteristics of the early Eocene. These comparisons have found general agreement in mean annual temperature (Sloan and Barron, 1992; Sloan, 1994; Sloan et al., 1995; Sloan and Morrill, 1998). However, there are large discrepancies between data and model output for CMM temperature and MATR. Proxy data indicate monthly mean temperatures that are above-freezing, reduced annual temperature seasonality, and low meridional surface temperature gradient (equable) climate (Greenwood and Wing, 1995; Wilf, 2000), whereas models produce winter temperatures that are below freezing and great continental annual temperature seasonality (Sloan et al., 1995; Sloan and Morrill, 1998).

The most recent of these modeling studies (Sloan and Morrill, 1998), demonstrated that orbital forcing could help ameliorate these model/proxy data discrepancies by raising modeled CMM temperatures and significantly reducing MATR. However, the orbital parameters utilized in Sloan and Morrill (1998) were set to maximize seasonality, by specifying eccentricity at a maximum value and varying obliquity and precession. Although imposing these values of solar insolation helped to constrain the absolute extremes of climate variability due to orbital forcing, the occurrence of such celestial configurations in Earth's history is extremely rare. In addition, Sloan and Morrill (1998) used prescribed sea surface temperatures (SSTs) which influenced the modeled terrestrial responses to the imposed forcing. A more complete discussion of the history of model/proxy comparisons can be found in Sloan et al. (2001). Investigation of the early Eocene climatic response to realistic orbital forcing with calculated, rather than prescribed, SSTs is required to obtain a more realistic characterization of early Eocene climate.

Recent high-resolution paleobotanical studies of fossil plants derived from Paleogene strata in Wyoming support the historical proxy data interpretation of warm, equable early Eocene climate. However, these studies also identify previously undetected, high-frequency (<1 m.y.) variability (~1–12 °C) of MAT during the Paleogene (Wilf, 2000; Wing et al., 2000). The mechanism for these high-frequency climatic shifts is presently unknown. Modeling work to investigate potential causal processes has not previously been carried out.

Because precipitation is important in many biotic and abiotic systems, accurate characterization of paleoprecipitation is critical to describing and understanding past climates. Comparisons between model output and paleoprecipitation estimates have been limited, primarily because accurate interpretations from proxy data have proven to be quite challenging. Numerous attempts have been made to estimate the amount and type of paleoprecipitation, often generated using the Climate Leaf-Analysis Multivariate Program (CLAMP) or multiple-regression analysis of that data set (Wing and Greenwood, 1993; Wolfe, 1994). However, these studies have stressed the low accuracy of their paleoprecipitation estimates. Only recently have Wilf et al. (1998) elucidated a more reliable methodology for determining paleoprecipitation. This methodology identifies a relatively robust correlation ($r^2 = 0.760$) between the mean of the natural logarithms of the species' leaf areas and MAP. Given these improvements, we are now able to make more meaningful comparisons between proxy data and modeled precipitation estimates.

In this study, paleoprecipitation results are used in conjunction with paleotemperature calculations to interpret cyclic sedimentary variation in the Wilkins Peak Member of the Green River Formation (GRF) in Wyoming. The Wilkins Peak Member is a suite of Eocene age lacustrine rocks that contains 77 repeating cycles of three lithologies: Oil shale, evaporate trona or halite, and dolomitic mudstone (Roehler, 1993). These cycles are believed to reflect significant periodic climate change during the Eocene and have been attributed to variations in precessional forcing (Fischer and Roberts, 1991; Roehler, 1993). By modeling the end members of a high-amplitude precessional cycle, we gain insight into the climatic variations that potentially could have generated these lithologic oscillations.

Here, we attempt to generate the most realistic representation of early Eocene climate to date by imposing new topographic, orographic, vegetative, and orbital parameters as boundary conditions. We test the sensitivity of early Eocene climate to variations in solar insolation across a realistic precessional cycle. We consider discrepancies in estimated seasonality and cold season temperature between model output and proxy data. Additionally, we compare paleoprecipitation estimates derived from proxy data to those derived from model output, a comparison rarely made in paleoclimate modeling studies. Finally, we consider the implications of this work for the interpretation of the cyclically deposited lithologies of the Wilkins Peak member of the GRF and for the high-frequency variability documented by recent proxy data investigations. This paper is the terrestrially focused companion to Sloan and Huber (2001) which examined the oceanic response to orbital forcing across a precessional cycle.

MODEL DESCRIPTION

In this study, we used the Genesis v2 general circulation model (GCM), which was developed by Thompson and Pollard (1997) for the purpose of performing greenhouse and paleoclimate experiments. The atmospheric component of the model has a resolution of 3.75° latitude × 3.75° longitude (spectral resolution T31) with 18 vertical levels, which are unevenly distributed vertically, with the greatest density of layers near the surface. The atmospheric component is coupled to land surface and ocean submodels. The land surface model has a 2° latitude × 2° longitude resolution and includes vegetation, soil, and snow (Thompson and Pollard, 1997). The ocean model is a 50-m deep mixed layer slab ocean, in which meridional heat transport is calculated using a diffusion coefficient, which is governed by latitude and the distribution of land versus sea at each latitude. Heat flow under sea ice is set to 2 Wm^{-2} in the Northern Hemisphere and 10 Wm^{-2} in the Southern Hemisphere. The Norwegian Sea oceanic heat flux used by Thompson and Pollard (1997) was not used in our experiments since it was designed for the present orographic distribution. Thus, SSTs and sea ice are calculated by the model. Full diurnal and seasonal cycles are included in the model and the model allows for the prescription of greenhouse gases CO_2, CH_4, and N_2O. Genesis v2 performs well in modern climate simulations (Thompson and Pollard, 1997) and has been widely used previously by the paleoclimate modeling community (Kutzbach and Liu, 1997; Otto-Bliesner and Upchurch, 1997; Sloan and Huber, 2001).

EXPERIMENT DESCRIPTION

We devised a pair of model experiments to test the sensitivity of early Eocene climate to solar insolation variation across a realistic, high-amplitude precessional cycle. Precession defines the relationship between the solstices and aphelion and perihelion, the positions of the largest and smallest distances, respectively, between the Earth and Sun (see Berger, 1978, for more details). This aspect of orbital motion essentially controls the degree of seasonality experienced in each hemisphere through time.

Because the amount and distribution of solar radiation reaching the Earth is based on complex gravitational interactions of celestial bodies that have changed through time, precise calculations of natural early Cenozoic orbital variations have not been made. Therefore, we selected a high-amplitude precessional cycle from the time series of values calculated by Berger (1978) for the past 5 million years as representative of the maximum solar insolation differences across realistic precessional cycles (Fig. 1). Orbital parameters at one end member of our chosen precessional cycle (MINSEAS case)—eccentricity (0.0531), obliquity (22.78°), precession (90.00°)—were selected to produce a dampened seasonal insolation cycle in the Northern Hemisphere and an amplified seasonal cycle in the Southern Hemisphere. In this case, perihelion occurs in Northern Hemisphere winter. Conversely, orbital parameters for the MAXSEAS case—eccentricity (0.0524), obliquity (23.83°), and precession (270.00°)—represent the other end member of our chosen precessional cycle (~11,500 yr later) and yield an amplified seasonal cycle in the Northern Hemisphere and a dampened seasonal cycle in the Southern Hemisphere (see also explanation in Sloan and Huber [2001]). In this case, perihelion occurs in Northern Hemisphere summer.

Unlike most climate modeling sensitivity studies that alter only one climate parameter between modeling experiments, we vary eccentricity, obliquity, and precession across our chosen precessional cycle in keeping with values calculated by Berger (1978). Although this does result in small changes of obliquity (1.05°) and eccentricity (0.0007) between experiments, realistic

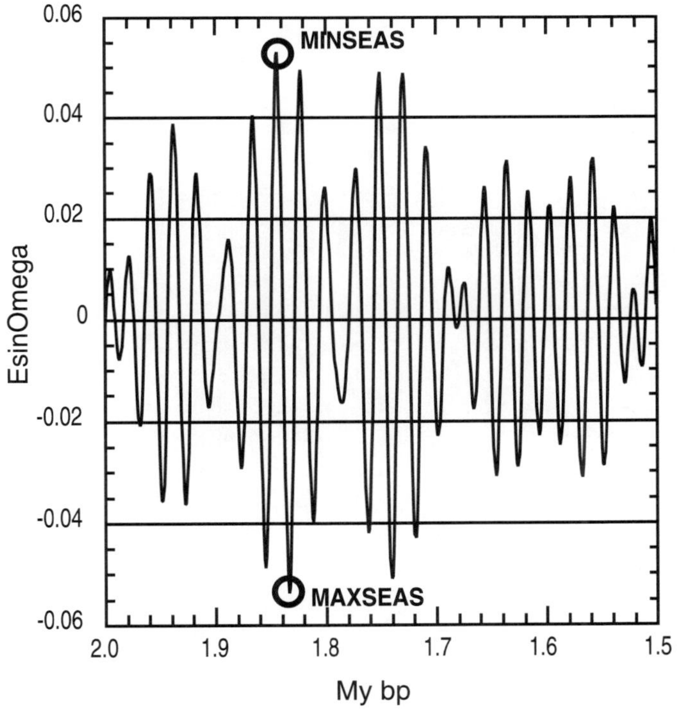

Figure 1. EsinOmega (precession as modulated by eccentricity) for the time interval 1.5–2 million years before present. E is the eccentricity of the Earth's orbit and Omega is the longitude of perihelion measured from the moving vernal equinox. Black circles denote the end members of a high-amplitude precessional cycle representative of the cycle used in this study.

responses to orbital parameter changes can only be obtained by changing all three orbital parameters as they would vary as a consequence of natural celestial motion. The obliquity and eccentricity changes noted above cause <5 W/m^2 difference in solar insolation (concentrated in the midlatitudes) between model cases on a mean annual basis. As in the companion paper (Sloan and Huber, 2001), we will hereafter refer to the sum of changes in orbital parameters (eccentricity, obliquity, and precession) as "precessional-driven forcing" and "prescribed orbital forcing."

Experiments MINSEAS and MAXSEAS used the paleogeography, topography, and vegetation reported in Sewall et al. (2000). For both experiments, SSTs were calculated. Therefore, ocean temperatures both respond to and influence atmospheric conditions. Atmospheric CO_2 concentration was set to doubled preindustrial levels (560 ppm), based on modeling work that investigated proposed Eocene greenhouse gas concentrations (Sloan and Rea, 1995). Greenhouse gases CH_4 and N_2O were set to 0.700 ppm and 0.285 ppm, respectively (preindustrial levels). All runs were started from the end of an equilibrated model case with modern orbital parameters and all other boundary conditions set to Eocene values. Each model experiment was run for 14 yr, and the final 4 yr of each experiment were averaged and utilized in analyses.

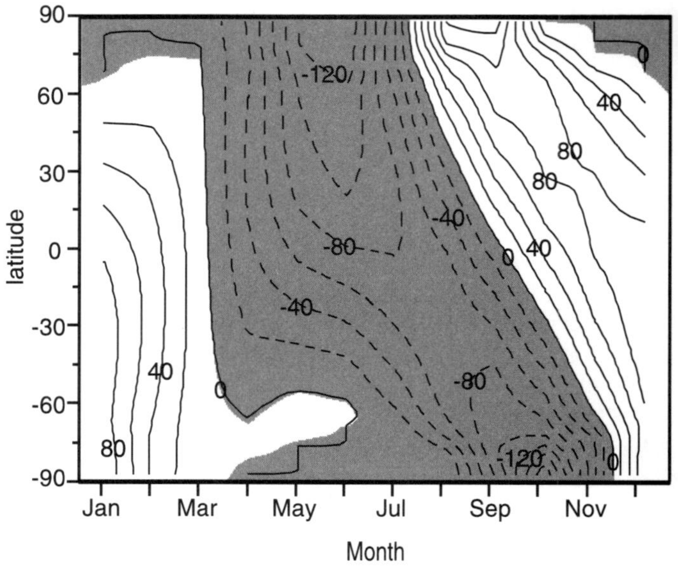

Figure 2. Average solar insolation difference at the top of the atmosphere, MINSEAS minus MAXSEAS. Shading and dashed lines indicate negative values. The contour interval is 20 W/m^2.

RESULTS

Solar insolation

In this sensitivity test, we investigate the climatic response to a change in the distribution of radiation reaching the top of the Earth's atmosphere (Fig. 2). Because solar radiation is the engine that drives the Earth's climate system, this alteration of solar forcing is the root cause of the climatic variations observed between our two model cases. The greatest range in insolation in the Northern Hemisphere is found in the MAXSEAS cases where the average incoming solar radiation is 214 W/m^2 for the three winter months, December-January-February (DJF), and 479 W/m^2 for the three summer months, June-July-August (JJA). Conversely, the MINSEAS case experiences a Northern Hemisphere seasonal range of solar insolation of 255 W/m^2 (DJF) to 417 W/m^2 (JJA). In the Southern Hemisphere, the greatest range in solar forcing is found in the MINSEAS case with a DJF three-month average of 492 W/m^2 versus a three month JJA average of 207 W/m^2. In contrast, the MAXSEAS case experiences a Southern Hemisphere range in seasonal insolation of 416 W/m^2 (DJF) to 260 W/m^2 (JJA). These contrasts in solar forcing are reflected in the enhanced seasonality in the Northern Hemisphere in the MAXSEAS case and in the Southern Hemisphere in the MINSEAS case (not shown).

Mean annual and seasonal temperature

Global MAT and seasonal temperatures exhibit low sensitivity to the imposed orbital forcing (Table 1). With the exception of seasonal temperatures over land only (JJA ~4 °C; DJF ~2 °C), relatively small differences in temperature (~1 °C or less) between model cases are found for all types of surfaces in both MAT and all seasons. Higher sensitivity of land surface temperatures during JJA and DJF is consistent with seasonal differences in the imposed insolation forcing (Fig. 2). Regions of highest insolation difference between the two cases occur dur-

TABLE 1. GLOBAL MEAN ANNUAL AND SEASONAL AVERAGE SURFACE TEMPERATURE FOR EACH EXPERIMENT AND THE DIFFERENCE IN TEMPERATURE BETWEEN EXPERIMENTS

(a) Experiment	MAT	DJF	MAM	JJA	SON
MINSEAS	18.5	18.0	19.0	19.1	18.2
MAXSEAS	18.9	17.3	18.2	20.5	19.4
Difference	–0.4	0.7	0.8	–1.4	–1.2
(b) Experiment	**MAT**	**DJF**	**MAM**	**JJA**	**SON**
MINSEAS	12.8	10.6	12.7	14.8	13.0
MAXSEAS	13.4	8.7	12.4	18.5	13.9
Difference	–0.6	1.9	0.3	–3.7	–0.9
(c) Experiment	**MAT**	**DJF**	**MAM**	**JJA**	**SON**
MINSEAS	20.6	20.5	21.2	20.6	20.0
MAXSEAS	20.8	20.4	20.2	21.1	21.4
Difference	–0.2	0.1	1.0	–0.5	–1.4

Note: All surfaces (land and oceans) (a), land surfaces only (b), and oceans only (c). MAT—mean annual temperature; DJF—December-January-February; MAM—March-April-May; JJA—June-July-August; SON—September-October-November.

ing JJA and DJF. For example, at 30°N JJA insolation is up to 120 W/m^2 greater in the MAXSEAS case and DJF insolation is 40–80 W/m^2 greater in the MINSEAS case.

Over much of the globe (~65°N to 90°S) MAT also shows low spatial sensitivity to the prescribed orbital forcing, varying on the order of 2 °C or less (Fig. 3). However, the greatest temperature variations in MAT are observed in the high latitudes (above 65°N) of the Northern Hemisphere (Fig. 3). The MAXSEAS case is up to 6 °C warmer than the MINSEAS case. This temperature difference between cases is largely a consequence of sea ice differences (not shown) and differences in shortwave radiation received at the surface between the cases (see explanation and figure in Sloan and Huber [2001]). In the MINSEAS case, year-round sea ice in the Northern Hemisphere prevents the warming of northern ocean waters and keeps the region cool relative to the MAXSEAS case. In contrast, summer heating in the MAXSEAS case melts almost all Northern Hemisphere sea ice resulting in warming of the ocean and consequently warmer mean annual temperatures at high latitudes (Fig. 3). The result of sea ice forming in our cases is not consistent with some paleoclimate data, although the evidence against sea ice is indirect and inconclusive. With higher GHG concentrations, sea ice would be entirely eradicated. Without sea ice, Arctic temperature differences between the cases would be less, due to the absence of ice-albedo feedbacks.

Cold month and warm month mean temperature

Mean January surface temperature, which represents the coldest month in the Northern Hemisphere and the warmest month in the Southern Hemisphere, is up to 10 °C different between the two cases, with the MINSEAS case being higher at almost all latitudes (75°N to 90°S) (Fig. 4A). Again, the warmth of the MAXSEAS case relative to the MINSEAS case at high latitudes (>75°N) is related to the extent and distribution of sea ice between the two cases, with the MINSEAS case having perennial ice in the Arctic while the MAXSEAS case has little to no ice during the summer months (Sloan and Huber, 2001). Higher January temperatures in the MINSEAS case (Fig. 4A) at almost all latitudes are consistent with the imposed orbital forcing. The MINSEAS case has greater seasonality in the Southern Hemisphere and therefore higher summer (January) temperatures than the MAXSEAS case. Conversely, the MINSEAS case has lower seasonality in the Northern Hemisphere and thus milder (warmer) winter (January) temperatures than the MAXSEAS case.

Mean July surface temperature shows a similar magnitude in temperature change between cases (up to 10 °C), however, the MAXSEAS has higher temperatures at almost all latitudes with the exception of two regions adjacent to the Antarctic continent (Fig. 4B). These regions are nucleation points for sea ice in the

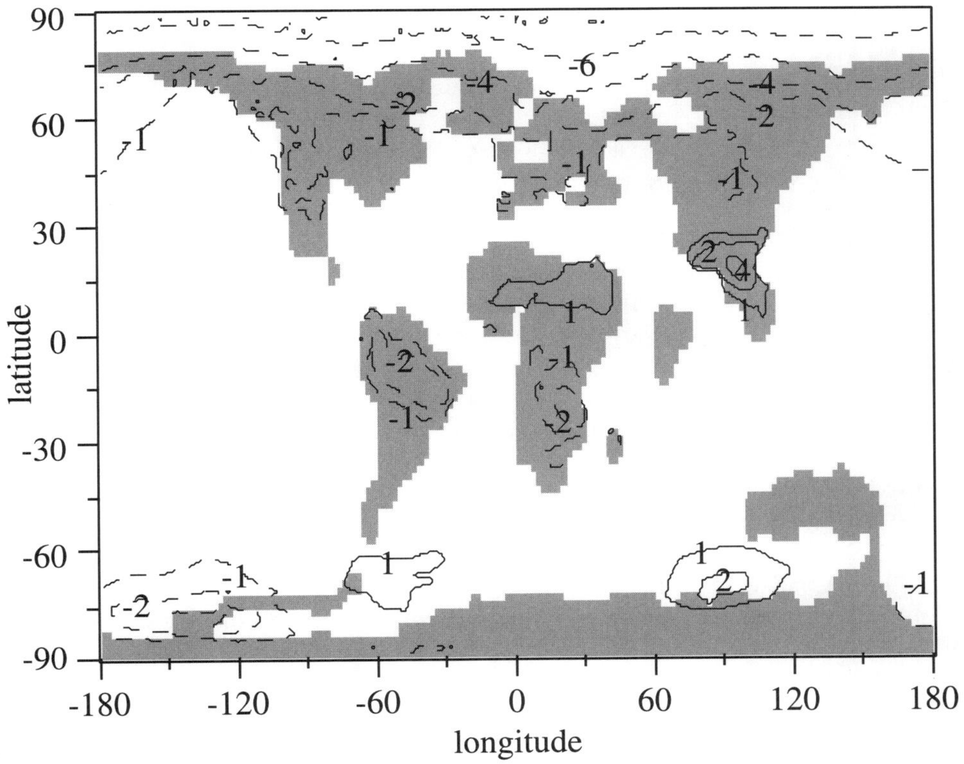

Figure 3. Mean annual surface temperature difference between our two end member experiments (MINSEAS minus MAXSEAS) in degrees C. Dashed lines indicate negative values. Contours are 1, 2, 4, and 6 degrees.

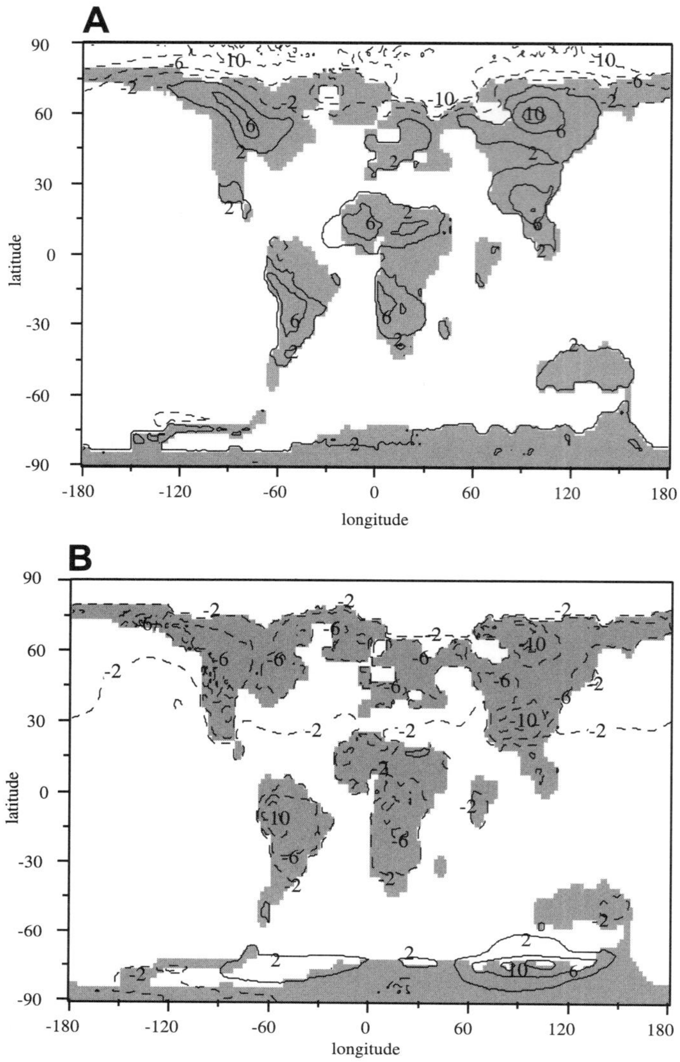

Figure 4. Mean January (A) and July (B) surface temperature difference (MINSEAS minus MAXSEAS) in degrees C. Dashed lines indicate negative values. Contours are from –10 to 10 at 4 degree intervals.

Southern Hemisphere. We attribute these temperature differences (where MINSEAS is warmer than MAXSEAS in Southern Hemisphere winter) to the difference in sea ice distribution and extent between the two cases, as discussed in Sloan and Huber (2001). Again, these temperature results are consistent with the imposed insolation differences. The MAXSEAS case has higher seasonality in the Northern Hemisphere and thus warmer summer (July) temperatures relative to the MINSEAS case (temperatures up to 22 °C in central North America, and up to 17 °C in Northern Asia). Conversely, the MAXSEAS case has low seasonality in the Southern Hemisphere and therefore a milder (warmer) winter (July) temperatures than the MINSEAS case (Fig. 4B). Relatively high-magnitude January and July temperature changes between cases indicate that, on a global scale, cold month and warm month mean temperatures are sensitive to precession-driven forcing.

Mean annual temperature range

Mean annual temperature range, calculated as the mean January surface temperature minus the mean July surface temperature, is an indicator of seasonality. MATR responds as expected to the prescribed orbital forcing in our cases. The MINSEAS case exhibits greater seasonality in the Southern Hemisphere relative to the MAXSEAS case, with a temperature range of ~20 °C (up to 10 °C greater than the MAXSEAS case) in parts of Africa and South America (Figs. 5A and B). In the Northern Hemisphere, the MAXSEAS case exhibits greater seasonality relative to the MINSEAS case, with temperature variations up to 60 °C (up to 10 °C greater than MINSEAS) in North America and up to 50 °C (up to 20 °C greater than MINSEAS) in Northern Asia (Figs. 5A and B). The MINSEAS case exhibits greater seasonality north of ~70° latitude (Fig. 5C) because summer radiation in the MAXSEAS case is used to melt sea-ice into water rather than to increase local air temperature; this lowers Northern Hemisphere high-latitude seasonality in that case relative to the MINSEAS case. In both experiments, MATR is greatest in continental interiors (up to 60 °C in the MAXSEAS case and up to 50 °C in the MINSEAS case), which are far from the moderating affect of oceanic water masses (Figs. 5A and B). Considerable differences in MATR (up to 20 °C) between cases (Fig. 5C) indicate that MATR is sensitive to the imposed forcing. Importantly, sensitivity of MATR to precessional orbital forcing occurs at all latitudes.

Precipitation

Global precipitation at the scale considered here shows low to moderate sensitivity to the imposed orbital forcing (Fig. 6). Based on the topographic boundary conditions, we have divided the continents into drainage basins (further subdivisions in North America are explained below). Most basins have <15% difference in annual precipitation between our modeling cases and only 2 basins (northern North America and northern Greenland) have annual precipitation differences of >20% between cases. Global precipitation patterns are generally consistent with prescribed orbital forcing. The case with greater seasonality in temperature for a particular hemisphere experiences higher precipitation for nearly all basins (in a sense, increased temperature drives increased precipitation in these areas). The MAXSEAS case has more precipitation for all drainage basins in the Northern Hemisphere except southeast Central America (4% less) (Basin 3) and western central North America (10% less) (Basin 4) (Fig. 6). Conversely, the MINSEAS case has more precipitation for all drainage basins in the Southern Hemisphere except western South America (3% less) and northern Australia–eastern Antarctica (2% less) (Fig. 6).

In western North America, drainage basins have been subdivided beyond the extent dictated by topographic slope (Fig. 6)

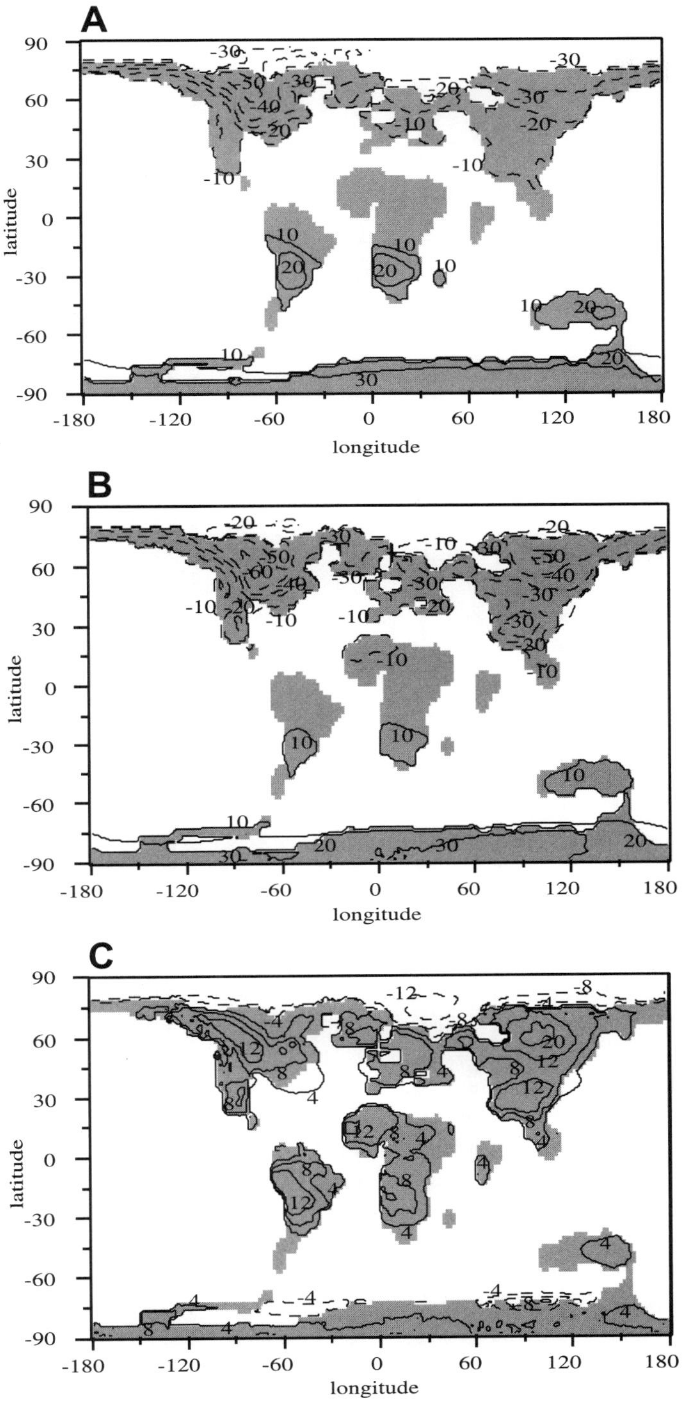

Figure 5. Mean annual temperature range in degrees C calculated as January minus July surface temperature. For the MINSEAS case (A), the MAXSEAS case (B), and the difference between cases (MINSEAS minus MAXSEAS) (C). Dashed lines indicate negative values. For A and B, contours are from –60 to 60 at intervals of 10 degrees, excluding 0. In C, contours are 4, 8, 12, and 20.

to allow for comparisons with early Eocene proxy data, which are densely concentrated in the region. Although the model's surface resolution precludes capturing the full orographic influence of precipitation (e.g., Sewall et al., 2000), we feel that these subdivisions are useful to investigate differences in precipitation character in the area.

Overall, seasonal sensitivity of precipitation to changing insolation distributions over time and across the basins is driven by a combination of feedbacks and influences of topography, latitude, longitude, surface pressure, and surface wind directions. For all regions, winds are intersecting the continent from sources located more (or less) over the oceans and bringing moister (or drier) air over the continent as a result. Also affecting the degree of moisture of the air is the direction that the air parcel takes: A more southerly source means warmer and moister conditions than a more northerly source. Although wind directions are largely similar (see Figure 5 of Sloan and Huber [2001]), there are small, important differences between the cases in each month that cause the precipitation differences observed in the subdivided basins. (These small differences in wind strength and orientation also drive substantially different wind-driven upwelling patterns, as demonstrated in Sloan and Huber [2001].)

In both experiments, the western coastal basins (Basins 1 and 2) experience a precipitation pattern much like that of the modern day, with low precipitation in the summer months and higher precipitation in the fall and winter months (Fig. 7). In contrast, in both experiments, basins adjacent to the Mississippi Embayment (Basins 3–6) have their highest precipitation in the summer and early fall. These precipitation results are consistent with seasonal wind patterns in both experiments (not shown; however, they are reported in Sloan and Huber [2001]). Strong onshore winds from the Pacific dominate during winter months. Conversely, during the summer, strong onshore monsoonal winds flow in off of the Mississippi Embayment and west to east airflow from the Pacific is significantly reduced.

Comparison of western North America hydrographs (Fig. 7) and the solar insolation difference plot for latitudes corresponding to those basins (Fig. 2) demonstrates that precipitation peaks are positively correlated with solar insolation. In the winter, when solar insolation for both experiments is similar, precipitation differences between the cases are small. For summer months, when MAXSEAS has significantly greater solar insolation, MAXSEAS has greater precipitation for most basins. In the fall, when MINSEAS has greater solar insolation, precipitation in that experiment is greater than in the MAXSEAS case. Comparisons of the absolute quantity of rainfall for each basin between the two cases indicate low sensitivity of MAP (only 5%–10% change) in response to precessional-driven forcing (Fig. 7). Additionally, similarity in the general character of the two hydrographs for most basins (Fig. 7) illustrates overall low to moderate sensitivity of seasonal precipitation to the imposed forcing. However, for each basin this general agreement is punctuated by at least one month in which precipitation between the two cases is dramatically different (~40% or greater) (Fig. 7),

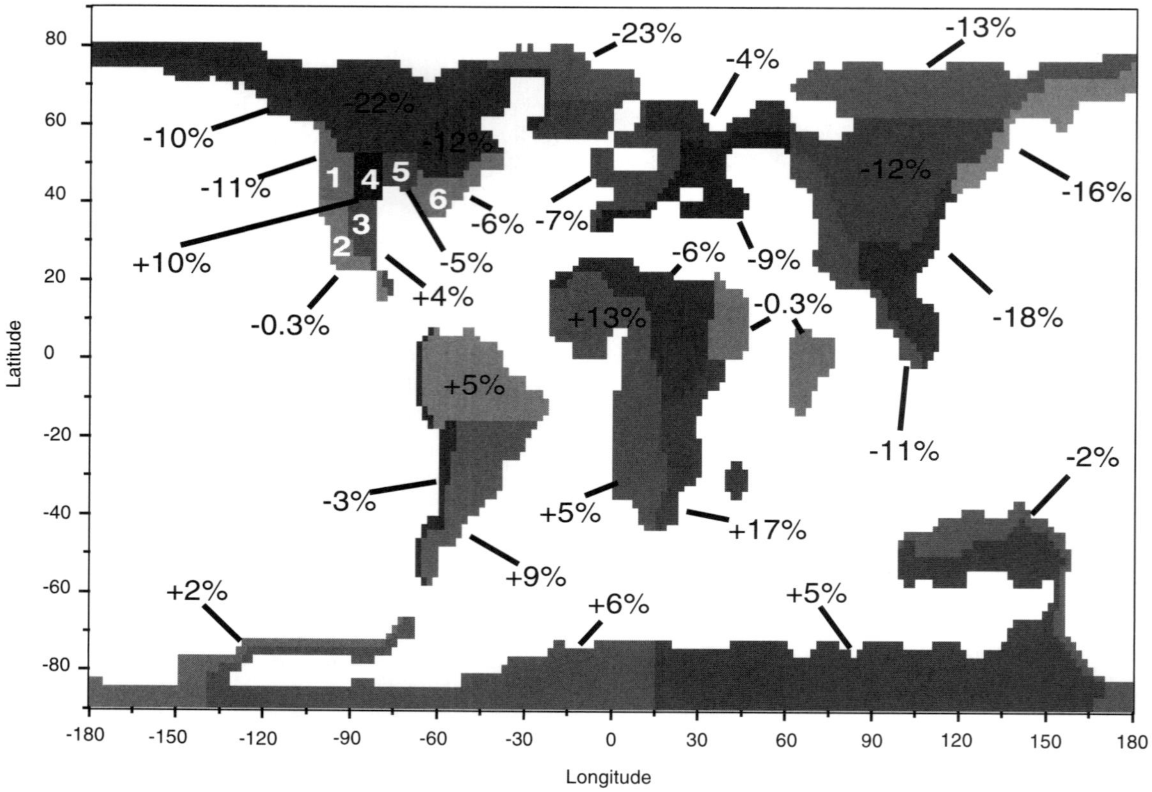

Figure 6. Eocene drainage basins and percent difference in total annual precipitation for each basin calculated as ((MINSEAS–MAXSEAS)/MINSEAS)*100. White numbers indicate North American drainage basins whose hydrographs are shown in Figure 7.

demonstrating that seasonal sensitivity to the imposed forcing is highly variable.

DISCUSSION AND INTERPRETATION

Comparisons to proxy data

Direct comparison of model results and proxy data is challenging because there are dramatic differences in scale between a GCM grid cell and a proxy data locality. This discordance in scale raises several questions. First, is the climate estimate derived from a proxy data source representative of climate in the general region in which that indicator was found or is it only representing the climate of a very localized environment? Furthermore, how reliably does a model grid cell portray the climate of the regions it represents? In this study, we approach this discordance in scale by selecting a sample box of grid cells (blue box shown in Figure 8, also Basin 4 in Figure 6) that corresponds to the paleo-location (approximately Eocene Wyoming and Colorado) with the highest concentration of early Eocene proxy data. However, comparisons are further complicated because proxy data are often derived from localities in intermontane basins whose estimated elevations range from ~200 m to 1 km (Wing and Greenwood, 1993) to 1–2.5 km (Chase et al., 1998) above sea level. The large size of model grid cells (typically ~90,000 km^2 each) averages out these intermontane elevations, producing average grid cell elevation values that are greater than those of the basins (Sewall et al., 2000). This scaling problem introduces a cold bias when model-produced temperatures are compared to proxy data. To address this resolution-introduced cold bias, we have corrected model MAT and CMM values to an elevation of 1 km using a paleolapse rate of 6 °C/km (Meyer, 1992). MAT and CMM values reported in Table 2 reflect this lapse rate correction. Using this lapse rate correction method, we obtain MAT and CMM results for Basin 4 that are comparable to proxy data (Table 2). Unfortunately, no comparable mode of correcting precipitation result for gridcell size elevational discrepancies exists. Thus, we have compared uncorrected model results to proxy data estimates of paleoprecipitation. Future regional modeling work, which will allow for more realistic topographic representation, should allow for more accurate comparisons of modeled and proxy data derived temperature and precipitation results.

For western North America (Basin 4), the prescribed orbital forcing results in an average difference of ~1 °C in MAT. Seasonal results are of a similar magnitude with the exception of

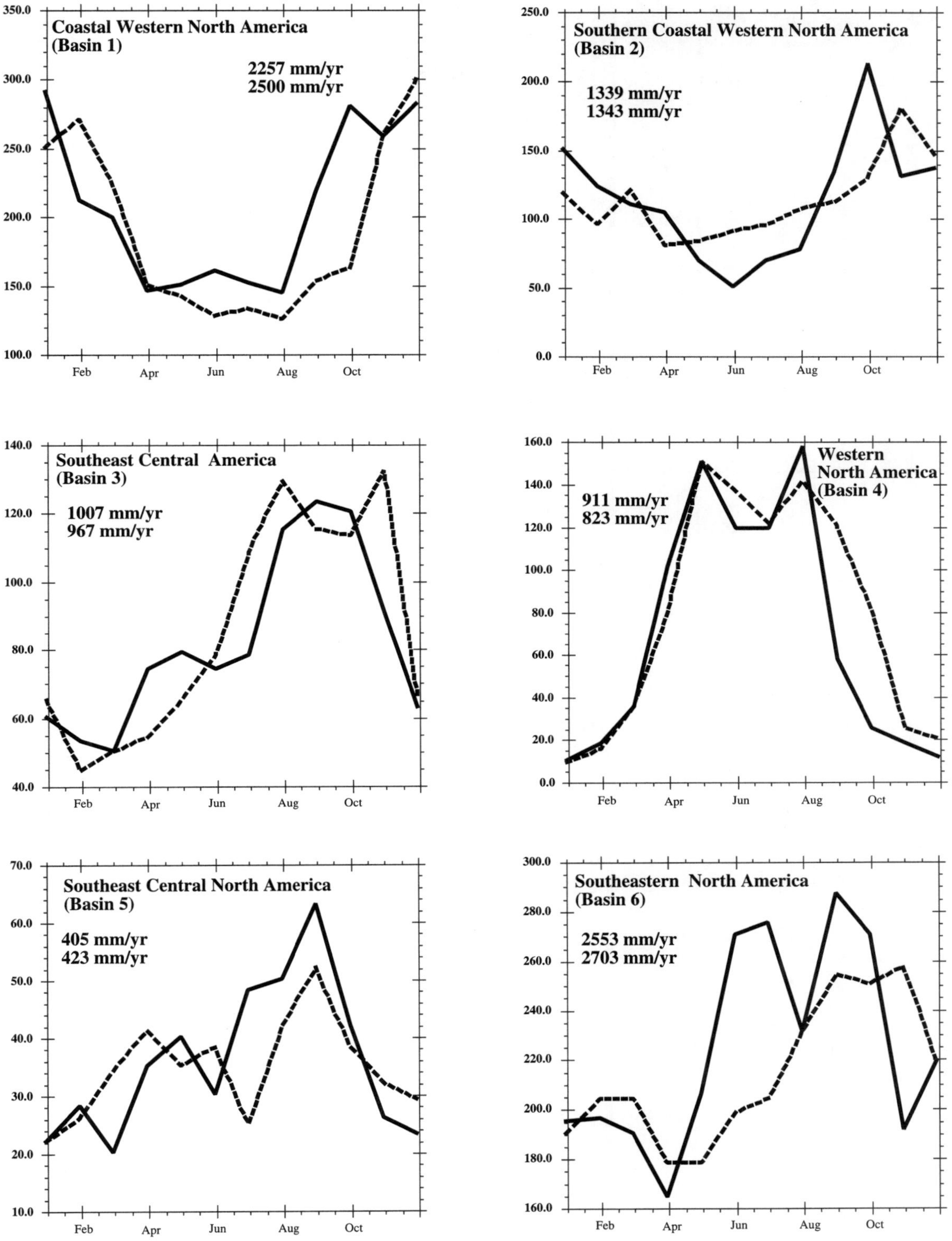

Figure 7. Average monthly precipitation (mm/month) for MINSEAS case (dashed line) and MAXSEAS case (black line) for North American drainage basins. The locations of basins are demarcated with white numbers in Figure 6. Note change in scale of y-axis between the plots. Numbers in box of each plot are annual precipitation totals (MINSEAS top number, MAXSEAS bottom number).

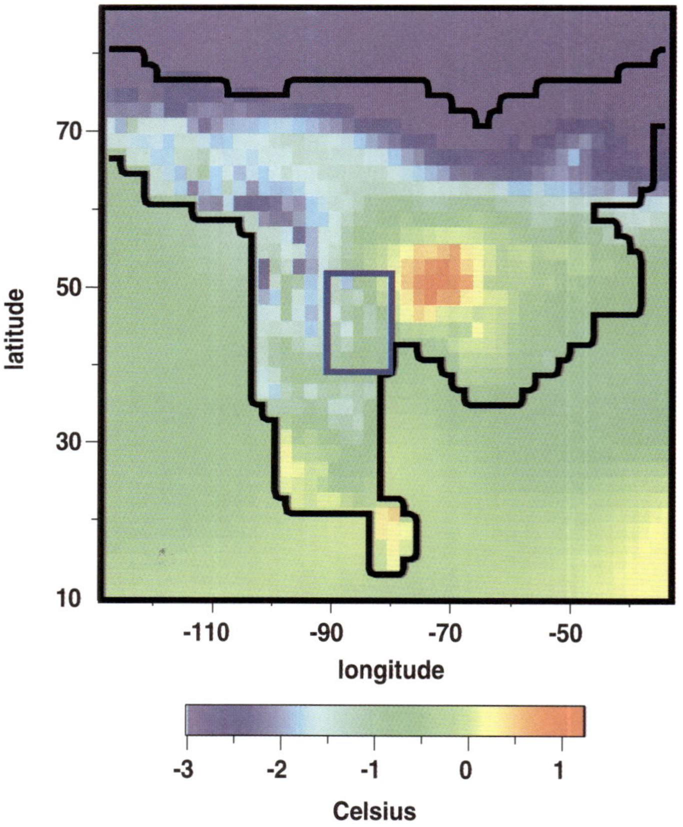

Figure 8. MAT difference between our experiments (MINSEAS minus MAXSEAS) for western North America in degrees C. The dark blue rectangle over a portion of western North America delineates the area used in generating the results reported in Table 2 and corresponds to drainage Basin 4 in Figures 6 and 7.

JJA, where the difference is ~6 °C. Spatially, MAT for North America varies by up to 4 °C; however, within our sample box, variability of MAT is ~2 °C (Fig. 8). In western North America, MATR is highly sensitive to precession-driven forcing (~8 °C difference between cases) whereas CMM is not very sensitive to precessional forcing (~2 °C difference between cases) (Table 2). Mean annual paleoprecipitation (MAP) results of ~85–95 cm/year are consistent with the most recent estimates of paleoprecipitation derived from proxy data (Wilf, 2000) for the late early Eocene (Table 2). For central western North America, the sensitivity of MAP and seasonal precipitation to precessional forcing is small (<10%), except for fall (September-October-November [SON] precipitation difference ~50%) (Fig. 7, Basin 4). This large precipitation difference corresponds to the autumn transition in the greatest quantity of solar radiation from the MAXSEAS case to the MINSEAS case (Fig. 2).

Implications for the Green River Formation

The Wilkins Peak Member of the Eocene age GRF in Wyoming contains three lacustrine lithologies that vary cyclically from oil shale to trona/halite to dolomitic mudstone (Surdam and Wolfbauer, 1975; Fischer and Roberts, 1991; Roehler, 1993). These lithologic units and are believed to have been deposited in a relatively flat, laterally extensive, closed basin (Eugster and Hardie, 1975). The cyclic lithologic variations have been interpreted as the periodic desiccation of paleolake Gosiute over a time period during which the lacustrine environment changed from a stagnant, shallow organic-rich lake, to a hypersaline evaporitic environment, followed by an intermittently desiccated/flooded playa (Eugster and Hardie, 1975; Surdam and Wolfbauer, 1975; Fischer and Roberts, 1991; Roehler, 1993). When the lithologic variations in this stratigraphic section are correlated with time, the frequency of desiccation is notably similar to that of precession (Fischer and Roberts, 1991; Roehler, 1993). Thus, these lithologic cycles are believed to reflect considerable periodic climate change across the precessional orbital cycle in an area well represented by proxy climate data.

To explore possible climatic causes from these cyclic variations, we have examined our model results of precipitation, evaporation, and net moisture availability (defined as net precipitation minus evaporation). In this region (Basin 4), annual precipitation and evaporation vary by ~10% between model cases (Fig. 7 and Figure 9A). However, the direction of change

TABLE 2. COMPARISON OF RESULTS FROM THIS STUDY WITH PROXY DATA FROM SITES IN WESTERN NORTH AMERICA

	MAT (°C)	MATR (°C)	CMM (°C)	MAP (cm/yr)
MINSEAS case	11.5$^{\gamma}$	18.9	2.0$^{\gamma}$	95
MAXSEAS case	12.4$^{\gamma}$	27.2	0.4$^{\gamma}$	85
Wing and Greenwood (1993)	9.1–18.6	8.2–17.5	0.6–8.5	116–277
Greenwood and Wing (1995)	9.1–23.9	8.2–18.7	0.5–8.5	
Wing et al. (2000)	10.8*–22.2*			
Wilf (2000)	11.5*–23.0*			80*–150*

Note: MATR—mean annual temperature range; MAP—mean annual precipitation.
$^{\gamma}$denotes mean annual temperature (MAT) and cold month mean (CMM) calculations that have been lapse rate corrected.
*denotes data derived from Wing et al. (2000) and from a time series of paleoprecipitation estimates in Wilf (2000) (not shown).

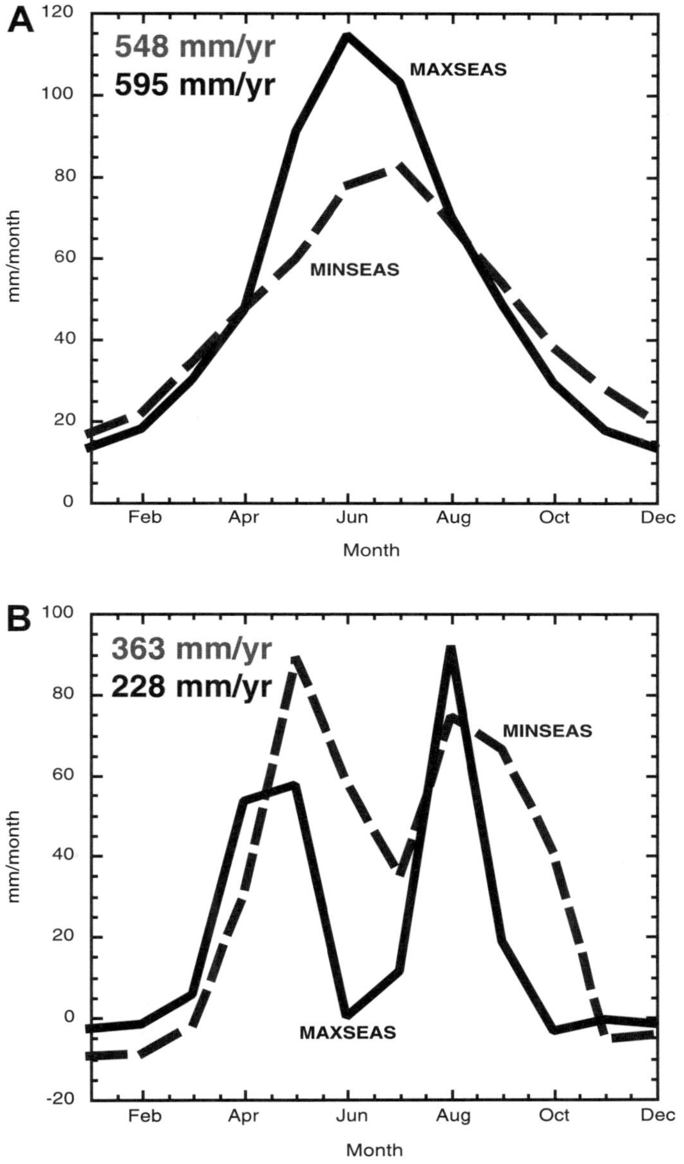

Figure 9. Western North America (Basin 4) evaporation (A) and net moisture availability (P-E) (B) in mm/month for MINSEAS case (dashed line) and MAXSEAS case (black line). Numbers in the upper left-hand corner of each plot are annual totals (MINSEAS top number, MAXSEAS bottom number).

between cases is opposite for precipitation and evaporation leading to a more limited annual moisture availability in the MAXSEAS case (~40% less than MINSEAS case) (Fig. 9B).

Although the Wilkins Peak member of the GRF varies cyclically between oil shale (humid), trona (arid), and mudstone (arid with episodic wet phases), our modeling strategy only describes two climate intervals across a half precession cycle (~11,500 yrs). Thus, we have examined the modeled climates and considered if they correspond to any of the climate states indicated by the observed lithologies. The model results exhibit modest changes in net moisture during the summer between cases (~15% change) (Fig. 9B). Neither case shows a net annual moisture deficit (Fig. 9B), which would be required to produce evaporitic trona or intermittently desicated mudstone deposits. However, the resolution differences between model grid cells and proxy data localities again become a confounding issue. Since these model results are the average of a number of grid cells that cover a significantly larger area than the GRF, the more localized climate corresponding to the deposition of the GRF may be averaged out. As a rough first comparison of model results and Wilkins Peak member lithologies, the more pronounced summer aridity of the MAXSEAS case might correspond to the trona or mudstone formations of the sedimentary record. The MINSEAS case, which has more net moisture and less net evaporation, could perhaps correspond to the formation of the oil shale lithology. However, more detailed consideration of climate states corresponding to the deposition of the lithologies of the GRF could be better addressed through regional paleoclimate modeling studies; in those studies the resolution of model grid cells would more closely correspond with the areal extent of proxy data localities.

If our model results are indeed representing the full range of variability induced by precession-driven insolation changes, the results of this study indicate that, for a region corresponding to Wyoming and Colorado in the Eocene, there is very little variation in MAT (Table 2 and Figure 8) and slightly more variation in MAP across a realistic precessional cycle (Table 2 and Figure 7, Basin 4). However, there is a modest change in net moisture (Fig. 9B) and the annual precipitation cycle does exhibit some pronounced variability in the autumn season (Fig. 7, Basin 4). These results imply that, if indeed the cyclic variations observed in the GRF are a consequence of precession, they are not a result of dramatic swings in climate across a precessional cycle. Rather, they are due to the high sensitivity of the Green River paleolake system and the sedimentary processes that formed the GRF to seasonal changes in precipitation and evaporation and to the relatively small shifts in mean climate that occur across a precessional cycle. Thus, more accurate characterization of mean climate for the Eocene is not likely to aid in determining the role of climate in producing lithologic variation in the GRF.

Implications for high-resolution Paleogene climate records

Recent paleobotanical work has generated high temporal climate records for the Paleogene (Wilf, 2000; Wing et al., 2000). The increased temporal sampling frequency of these studies has revealed a high-frequency (~1 m.y.) temperature fluctuation (~1–12 °C) of the mean of the equable state. Presently, little is known about the cause of these fluctuations. Milankovitch cycles, which oscillate on slightly shorter time scales (20–400 k.y.) (Berger, 1978) than the current temporal resolution of paleobotanical records (0.5–1 Ma), have been linked to major climate changes in Earth's history (e.g., the glacial/interglacial cycles of the Pleistocene) (Hays et al., 1976). Here, we have documented the climatic variability due to one of the

Milankovitch orbital parameters, precession, and have found low sensitivity of the mean climatic state to this imposed forcing. If our modeling results are capturing the full range of variability due to precession, this orbital parameter can account for no more than 2 °C of the variability observed in the high-resolution geologic records produced by Wing et al. (2000) and Wilf (2000). Thus, the dominant source(s) of this high-frequency variability remain(s) to be discovered.

CONCLUSIONS

The goal of this study was to test the sensitivity of terrestrial Eocene climate to precessional orbital forcing. Although we have investigated climatic sensitivity to the end members of only one set of realistic orbital parameters, these parameters bound the potential magnitude of Eocene climate variability across a high-amplitude, realistic precessional cycle. Thus, they provide a reasonable estimate of the potential distribution and magnitude of Eocene climatic change due to precessional oscillations. Our results show that, on a global scale, changes in insolation over a realistic precessional cycle have a relatively small effect on MAT and a larger impact on MAP, seasonal precipitation, MATR, and CMM. Model generated estimates for MAT, CMM, MATR, and MAP are generally consistent with the proxy data interpretations from the geologic record.

Based on modeled sensitivity to precession-driven forcing of MAT and MAP for western North America, we have inferred that the environments/processes that deposited the Wilkins Peak Member of the GRF may have been highly sensitive to seasonal precipitation changes and to small changes in mean climate. Further interpretation of cyclic desiccation in the GRF using modeling results would be best undertaken through regional modeling studies. Additionally, we have considered precessional orbital forcing as a potential partial explanation for the high-frequency variability reported in recent paleobotanical studies of the Paleogene. We have found that precession-driven climate variability can only potentially account for a small portion of the variability observed in paleontological record, if we have fully resolved the amount of temperature variation due to precession. Higher-resolution sampling of rock layers for paleobotanical data and more climate modeling experiments investigating potential forcing mechanisms including Milankovitch, tectonic, and possible biogeochemical forcing are needed to determine the dominant source(s) of the observed, high-frequency variability in the geologic record.

ACKNOWLEDGMENTS

We thank M. Huber, P. Koch, and S. Wing for their insightful comments and suggestions and C. Morrill for assistance in an earlier phase of this work. This research was supported by a grant from the National Science Foundation (EAR 9814883), a fellowship from the D. and L. Packard Foundation to LCS, a graduate student fellowship from the National Science Foundation to KTL, and a predoctoral NDSEG fellowship to JOS. Computing was done at the National Center for Atmospheric Research, which is supported by the National Science Foundation.

REFERENCES CITED

Berger, A., 1978, Long-term variations of daily insolation and Quaternary climatic changes: Journal of Atmospheric Science, v. 35, p. 2362–2367.

Chase, C.G., Gregory-Wodzicki, K.M., Parrish, J.T., and DeCelles, P.G., 1998, Topographic history of the Western Cordillera of North America and controls on climate, *in* Crowley, T.J., and Burke, K.C., eds., Tectonic boundary conditions for climate reconstructions: New York, Oxford University Press, p. 73–99.

Eugster, H.P., and Hardie, L.A., 1975, Sedimentation in an ancient playa-lake complex: The Wilkins Peak Member of the Green River Formation of Wyoming: Geological Society of America Bulletin, v. 86, p. 319–334.

Fischer, A.G., and Roberts, L.T., 1991, Cyclicity in the Green River Formation (Lacustrine Eocene) of Wyoming: Journal of Sedimentary Petrology, v. 61, p. 1146–1154.

Greenwood, D.R., and Wing, S.L., 1995, Eocene continental climates and latitudinal temperature gradients: Geology, v. 23, p. 1044–1048.

Hays, J.D., Imbrie, J., and Shackleton, N.J., 1976, Variations in the Earth's Orbit: Pacemaker of the ice ages: Science, v. 194, p. 1121–1132.

Koch, P.L., Zachos, J.C., and Dettman, D.L., 1995, Stable-isotope stratigraphy and paleoclimatology of the Paleogene Bighorn Basin (Wyoming, USA): Palaeogeography, Palaeoclimatology, Palaeoecology, v. 115, p. 61–89.

Kutzbach, J.E., and Liu, Z., 1997, Response of the African monsoon to orbital forcing and ocean feedbacks in the middle Holocene: Science, v. 278, p. 440–443.

Meyer, H.W., 1992, Lapse rate and other variables applied to estimating paleolatitudes from fossil floras: Palaeogeography, Palaeoclimatology, Palaeoecology, v. 99, p. 71–99.

Otto-Bliesner, B.L., and Upchurch, G.R., 1997, Vegetation-induced warming of high-latitude regions during the Late Cretaceous period: Nature, v. 385, p. 804–807.

Roehler, H.W., 1993, Eocene climate, depositional environments, and geography, greater Green River Basin, Wyoming, Utah, and Colorado: U.S. Geological Survey Professional Paper 1506-F, 76 p.

Sewall, J.O., Sloan, L.C., Huber, M., and Wing, S.L., 2000, Climate sensitivity to changes in land surface charateristics: Climatic Change, v. 26, p. 445–465.

Sloan, L.C., 1994, Equable climates during the early Eocene: Significance of regional paleogeography for North American climate: Geology, v. 22, p. 881–884.

Sloan, L.C., and Barron, E.J., 1992, A comparison of Eocene climate model results to quantified paleoclimatic interpretations: Palaeogeography, Palaeoclimatology, Palaeoecology, v. 93, p. 183–202.

Sloan, L.C., and Huber, M., 2001, Eocene oceanic responses to orbital forcing on precessional time scales: Paleoceanography, v. 16, p. 101–111.

Sloan, L.C., and Morrill, C., 1998, Orbital forcing and Eocene continental temperatures: Palaeogeography, Palaeoclimatology, Palaeoecology, v. 144, p. 21–35.

Sloan, L.C., and Rea, D.K., 1995, Atmospheric CO_2 of the early Eocene: A general circulation modeling sensitivity study: Palaeogeography, Palaeoclimatology, Palaeoceanography, v. 119, p. 275–292.

Sloan, L.C., Walker, J.C.G., and Moore, T.C., 1995, Possible role of oceanic heat transport in early Eocene climate: Paleoceanography, v. 10, p. 347–356.

Sloan, L.C., Huber, M., Crowley, T.J., Sewall, J.O., and Baum, S.K., 2001, Effect of sea surface temperature configuration on model simulations of "equable" climate in the early Eocene: Palaeogeography, Palaeoclimatology, Palaeoecology, v. 167, p. 321–335.

Surdam, R.C., and Wolfbauer, C.A., 1975, Green River Formation, Wyoming: A playa-lake complex: Geological Society of America Bulletin, v. 86, p. 335–345.

Thompson, S.L., and Pollard, D., 1997, Greenland and Antarctic mass balances for present and doubled atmospheric CO_2 from the GENESIS version-2 global climate model: Journal of Climate, v. 10, p. 871–900.

Wilf, P., 2000, Late Paleocene–early Eocene climate changes in southwestern Wyoming: Paleobotanical analysis: Geological Society of America Bulletin, v. 112, p. 292–307.

Wilf, P., Wing, S.L., Greenwood, D.R., and Greenwood, C.L., 1998, Using fossil leaves as paleoprecipitation indicators: An Eocene example: Geology, v. 26, p. 203–206.

Wing, S.L., and Greenwood, D.R., 1993, Fossils and fossil climate: The case for equable continental interiors in the Eocene: Philosophical Transactions of the Royal Society of London, series B, v. 341, p. 243–252.

Wing, S.L., Bao, H.M., and Koch, P.L., 2000, An early Eocene cool period? Evidence for continental cooling during the warmest part of the Cenozoic, *in* Huber, B.T., et al., eds., Warm climates in Earth history: Cambridge, UK, Cambridge University Press, p. 197–237.

Wolfe, J.A., 1994, Alaskan Palaeogene climate as inferred from the CLAMP database, *in* Boulter, M.C., and Fisher, H.C., eds., Cenozoic plants and climates of the Arctic: NATO ASI Series I, p. 223–237.

Zachos, J.C., Stott, L.D., and Lohmann, K.C, 1994, Evolution of early Cenozoic marine temperatures: Paleoceanography, v. 9, p. 353–387.

Manuscript Accepted by the Society August 13, 2002

Printed in the USA

Geological Society of America
Special Paper 369
2003

Estimating latest Cretaceous and Tertiary atmospheric CO_2 from stomatal indices

Dana L. Royer*
Department of Geology and Geophysics, Yale University, P.O. Box 208109, New Haven, Connecticut 06520-8109, USA

ABSTRACT

A quantitative understanding of the levels of atmospheric CO_2 in the geologic past sheds light on the operation of the carbon cycle and the biosphere, and aids in the prediction of future climate change. Here I present a CO_2 reconstruction for the very latest Cretaceous to early Eocene and middle Miocene based on the stomatal distributions in fossil and extant *Ginkgo* and *Metasequoia* cuticles. Although both of these intervals are representative of globally warm climates, my CO_2 reconstruction indicates near present-day values (300–450 ppmV) for both times. Although these data do not cast doubt on the theory of the greenhouse effect, they do suggest that other thermal forcings were more important during these intervals than they are today.

INTRODUCTION

Globally averaged surface temperatures have risen sharply in the last century (~1 °C in the Northern Hemisphere) (Mann et al., 1998, 1999), concomitant with a 30% increase in the concentration of the greenhouse gas CO_2 (Friedli et al., 1986). It is now clear that CO_2 is largely responsible for this warming (Mann et al., 1998; Crowley, 2000; Barnett et al., 2001; Harries et al., 2001; Levitus et al., 2001; Mitchell et al., 2001). On the longer time scale of 10^5 yr, Antarctic air temperatures and atmospheric CO_2 have been tightly coupled for at least the last four glacial-interglacial cycles (Petit et al., 1999), and it appears that CO_2 has also been driving these temperature variations (Shackleton, 2000).

On multimillion year time scales, proxy- and modeling-based estimates of CO_2 over the Phanerozoic largely correlate with geologic indicators of continent-scale glaciations, which are the most reliable index for the presence of "greenhouse" or "icehouse" climates (Crowley and Berner, 2001). This correlation is not present, however, during certain intervals. CO_2 estimates for the mid-Miocene thermal maximum (17–14.5 Ma), a brief interval of global warmth relative to today, suggest concentrations largely lower than the preindustrial value of 280 ppmV (Pagani et al., 1999a; Pearson and Palmer, 2000). In addition, CO_2 estimates derived from multiple proxies for other time periods are not consistent. For example, early Tertiary (65–50 Ma) CO_2 estimates range from <300 ppmV to >3000 ppmV (Ekart et al., 1999; Pearson and Palmer, 2000).

Here I apply a CO_2-calibrated set of stomatal indices from *Ginkgo biloba* and *Metasequoia glyptostroboides* to reconstruct paleo-CO_2 levels for the very latest Cretaceous to early Eocene (66–53 Ma) and the middle Miocene (~15.5 Ma). This study expands on the data presented by Royer et al. (2001b) and Beerling et al. (2002).

THE STOMATAL INDEX METHOD

Most modern vascular C_3 plants show an inverse relationship between the partial pressure of atmospheric CO_2 and stomatal index (Woodward, 1987; Beerling, 1999; Lake et al., 2001;

*E-mail: droyer@psu.edu. Present address: Department of Geosciences, Pennsylvania State University, University Park, Pennsylvania 16802, USA.

Royer, D.L., 2003, Estimating latest Cretaceous and Tertiary atmospheric CO_2 from stomatal indices, *in* Wing, S.L., Gingerich, P.D., Schmitz, B., and Thomas, E., eds., Causes and Consequences of Globally Warm Climates in the Early Paleogene: Boulder, Colorado, Geological Society of America Special Paper 369, p. 79–93.

Royer, 2001). Stomatal index (SI) is the percentage of epidermal cells that are stomatal complexes (guard cells + stomatal pore), and is defined as the following (Salisbury, 1927):

$$SI = \frac{SD}{SD + ED} \times 100$$

where SD = stomatal density (mm^{-2}) and ED = epidermal cell density (mm^{-2}). The mechanism(s) linking CO_2 to SI is not fully understood, but probably involves the strong selective pressure in plants to maximize carbon fixation per unit of water transpired (water-use efficiency) (e.g., Woodward, 1987). For example, in a rising CO_2 regime plants can increase their water-use efficiency by reducing their stomatal conductance, thereby reducing water loss. Stomatal conductance is largely a function of stomatal pore area, which in turn is controlled by the size of individual stomatal pores and the density of those pores. On the time scale of several growing seasons or longer, stomatal density is typically more sensitive to changes in atmospheric CO_2 than individual pore size (e.g., Royer, 2001). Furthermore, on multimillion year time scales stomatal density and SI across all plant taxa inversely respond to CO_2 (Beerling and Woodward, 1997; McElwain, 1998; Royer, 2001).

In contrast to stomatal density, SI is area independent, which normalizes for the effects of cell expansion. Stomatal index is essentially a measure of the number of stomatal complexes that develop per unit number of epidermal cells. Plant water-potential strongly influences cell expansion (e.g., water stressed plants generally reduce the size of their epidermal cells [Royer, 2001]), but not the rate of stomatal initialization. Thus, changes in a plant's water budget can affect stomatal density, but not stomatal index. Both experimental and field studies show that SI is largely independent of water potential, irradiance, and temperature, and is primarily a function of CO_2 (e.g., Kürschner et al., 1998; Beerling, 1999; Royer, 2001).

Stomatal indices (and stomatal densities) have been experimentally shown to respond to the partial pressure of CO_2, not concentration (as expressed as mole fraction, e.g., parts per million) (Woodward and Bazzaz, 1988). This finding is corroborated by a positive correlation in many plants between elevation and stomatal density (Körner and Cochrane, 1985; Woodward, 1986; Woodward and Bazzaz, 1988; Beerling et al., 1992) and SI (Rundgren and Beerling, 1999). It is important, therefore, to control for elevation when estimating CO_2 from stomatal properties. The relationship between elevation and partial pressure in the lower atmosphere is roughly as follows (linear fit to equation 11.8 in Jones, 1992, p. 308):

$$P = -10.6E + 100$$

where E = elevation in kilometers and P = the percentage of partial pressure relative to sea level. If one does not control for elevation (i.e., assumes concentration = partial pressure), CO_2 concentration estimates will be underestimated by 3% for plants growing at 250 m elevation, for example, and 11% for plants at 1000 m.

Choice of species

Stomatal responses to atmospheric CO_2 are usually species-specific (Woodward and Kelly, 1995; Beerling and Kelly, 1997; Royer, 2001). For example, based upon genera represented by >1 species in the compilation of Royer (2001) (n = 7), only one shows no significant intrageneric variation in SI ($\alpha = 0.05$; chi-squared test). In order to quantitatively reconstruct paleo-CO_2, then, the SI-CO_2 relationship in an extant species must be applied in the fossil record to the same species. This requirement has previously hindered the applicability of this method for fossil leaves predating the late Miocene (10 Ma) (van der Burgh et al., 1993; Kürschner et al., 1996). Here I use the lineages represented today by *Ginkgo biloba* and *Metasequoia glyptostroboides,* which greatly extends the temporal applicability of the method.

Morphologically indistinguishable forms of *G. biloba* extend back to the Early Cretaceous, leading many authors to consider the fossil and modern forms conspecific (Tralau, 1968; see Royer et al., 2003). This fossil form is most commonly identified as *G. adiantoides,* however many other designations are considered equivalent, e.g., *G. spitsbergensis* (Tralau, 1968). Similarity in the environment of growth of modern and Tertiary fossil *Ginkgo* also supports their conspecifity (Royer et al., 2003). *G. gardneri* is the only *Ginkgo* species of Tertiary age in the Northern Hemisphere sufficiently different from *G. adiantoides-biloba* to warrant a separate species (Manum, 1966; Tralau, 1968). *G. gardneri* occurs at one early Paleogene site on the Isle of Mull, Scotland (e.g., Boulter and Kvaček, 1989).

Chen et al. (2001) report that the SI in modern *G. biloba* is statistically independent of leaf size, long versus short shoots, and male versus female trees, and only minor differences (<5%) due to canopy position and the timing of leaf development are present. These results further suggest that stomatal indices in *Ginkgo* are reliable indicators of atmospheric CO_2 concentration.

Ginkgo foliage is distinctive and readily identified. A *Ginkgo* leaf consists of a petiole and fan-shaped dichotomously veined lamina (Fig. 1). The veins only very rarely anastomose (Arnott, 1959). Vein spacing is typically ~1 mm. Stomata are found only on the abaxial side of the leaf (hypostomatous), and are distributed randomly in the intercostal regions. Subsidiary cells are moderately papillose and partially cover the stomatal pore. The leaf is usually divided into two lobes with wavy margins. The petiole contains two vascular strands, each of which vascularizes one of the lobes (Chamberlain, 1935; Gifford and Foster, 1989). Mucilage cavities are common and conspicuous, even in fossils. They are 1–5 mm in length, and are oriented parallel to venation (Chamberlain, 1935). The *G. adiantoides-biloba* form is differentiated from other ginkgo-forms (which largely occur earlier in the rock record) by its less dissected lamina (Tralau, 1968), however deeply lobed leaves do occur occa-

Figure 1. A leaf of *Ginkgo biloba.* Collected from New Haven, Connecticut (USA). Scale bar = 1 cm.

sionally in *G. biloba* (Critchfield, 1970). The only other plants with open dichotomizing venation and bilobed lamina are ferns such as *Adiantum* (e.g., Gifford and Foster, 1989). These can be distinguished from *Ginkgo* by the single vascular strand in their petioles, the presence of sori abaxially near the tips of their leaves, and their more closely spaced veins.

Metasequoia also has a long fossil record of morphologically similar forms back to the Late Cretaceous (Liu et al., 1999). Many workers consider this fossil form, *M. occidentalis,* closely related to, or conspecific with the modern *M. glyptostroboides* (Chaney, 1951; Christophel, 1976; Liu et al., 1999). *Metasequoia* needles are flat, ~15–20 mm long and ~2 mm wide. The leaf morphology of *Metasequoia* is very similar to *Sequoia* and *Taxodium,* however *Metasequoia* needles are opposite decussate while the needles in the other two genera are spirally alternate. *Metasequoia* also tends to have more rounded leaf tips, a higher angle of divergence between the leaf and shoot (~90°), and a more highly twisted base than *Sequoia* and *Taxodium* (Chaney, 1951).

The fossil records of *Ginkgo* and *Metasequoia* indicate that both were common components of mid- to high-latitude temperate forests (Tralau, 1967; Mai, 1994). In contrast to multistory tropical forests, intracanopy CO_2 gradients are small in temperate forests and should not bias stomatal-based CO_2 estimates (Royer, 2001).

Construction of training set

Leaves from dated herbarium sheets of *Ginkgo biloba* and *Metasequoia glyptostroboides* were used to assess the response of their SIs to the anthropogenic rise in atmospheric CO_2 concentration over the last 145 yr. Because both *Ginkgo* and *Metasequoia* have deciduous leaf habits, all leaves are assumed to have grown only during the year recorded on the sheets. The CO_2 concentrations that the leaves developed in were taken from ice core derived (Friedli et al., 1986) and direct atmospheric (Keeling et al., 1995) measurements. The herbarium specimens came mostly from cultivated individuals in the United States, China, and Japan. All trees grew at <250 m in elevation, with 84% growing at elevations <75 m. Therefore, no correction is needed to convert CO_2 partial pressure to concentration. For each herbarium sheet, five mature leaves were randomly selected. Stomatal index was calculated for three fields of view (0.1795 mm^2 each) on every leaf. Measurements were made in the intercostal regions near the centers of the leaves, which in other species show the least variation in SI (Salisbury, 1927; Royer, 2001). Stomatal observations were made from acetate peels and cleared leaves (light microscopy), and unaltered leaf tissue (epifluorescence microscopy). No systematic variation in SI among the three preparation techniques was noted. During the growing season of 1999 (and 2000 for *Ginkgo*), SI was measured from trees growing in New Haven (Connecticut, USA), Ann Arbor (Michigan, USA), and London (UK). Sampling details are given in Table 1.

Stomatal indices were also measured on saplings growing in greenhouses at four discrete CO_2 concentrations. For each CO_2 treatment, three potted saplings were placed in each of four independent greenhouses (see Beerling et al., 1998, and Beerling and Osborne, 2002, for details). For the 430/445 ppmV and 790/800 ppmV treatments, SI was measured after CO_2 exposure times of both one and two growing seasons. The *Metasequoia* saplings were one year old after their first season of CO_2 treatments. Two sets of *Ginkgo* saplings were used, aged one and six years old after their first season of CO_2 treatments. For the 350 and 560 ppmV treatments, SI was determined from three-year-old *Ginkgo* saplings after a CO_2 exposure time of two growing seasons (Beerling et al., 1998). Sampling details are given in Table 1.

Application to fossil cuticles

Stomatal indices were calculated from 28 very latest Cretaceous to middle Miocene–aged *Ginkgo* cuticle-bearing sites and three middle Miocene–aged *Metasequoia* cuticle-bearing sites. As with the training set material, SI was determined for three fields of view on each fossil leaf. At least five leaves were measured per site (see Table 2). This ensures that at least as many leaves per site were used to reconstruct paleo-CO_2 as were used per herbarium sheet to establish the training sets. This contrasts with other studies, where <5 fossil leaves are commonly measured per fossil site (e.g., Retallack, 2001). For *Ginkgo,* epifluorescence microscopy on unaltered cuticle was employed. Fossil *Metasequoia* cuticle failed to fluoresce, and so leaves were first cleared using 1% sodium hypochlorite (bleach) and then analyzed using light microscopy. For voucher specimen information for both the training and fossil data sets, consult Royer (2002).

The paleoelevations of all regions sampled in this study are most likely <1000 m. For example, the Late Cretaceous Hell

TABLE 1. DETAILS OF SAMPLING FOR TRAINING SETS

A. Historical Collections

Species	Year(s) collected	CO_2 (ppmV)	Mode of preparation	Trees†	Leaves†	Field of views†
G. biloba	1856–1996	288–361	E	1	5	15
	1999	367	A	6	13	46
	2000	369	C	3	10	30
M. glyptostroboides	1947–1980	310–338	A, C	1	5	15
	1999	367	A	4	21	46

B. Greenhouse Experiments

Species	Exposure to CO_2 (years)	CO_2 (ppmV)	Mode of preparation	Trees†	Leaves†	Field of views†
G. biloba	2	350§	A	8	16	160
	1	430#	C	8	16	48
	1	430**	C	8	8	24
	2‡	445#	C	8	8	24
	2‡	445**	C	8	8	24
	2	550§	A	8	16	160
	1	790#	C	8	16	48
	1	790**	C	7	7	21
	2‡	800#	C	8	8	24
	2‡	800**	C	8	8	24
M. glyptostroboides	1	430**	C	8	16	48
	2‡	445**	C	7	7	21
	1	790**	C	8	16	48
	2‡	800**	C	8	8	24

Note: A—acetate peels (light microscopy); C—cleared leaves (light microscopy); E—unaltered leaves (epifluorescence microscopy).

†Total number per CO_2 value or herbarium sheet.

§From Beerling et al. (1998). Three-year-old saplings.

#Six-year-old (for 430 and 790 ppmV treatments) and seven-year-old (for 445 and 800 ppmV treatments) saplings.

**One-year-old (for 430 and 790 ppmV treatments) and two-year-old (for 445 and 800 ppmV treatments) saplings.

‡Includes the exposure time from the previous growing season (either 430 or 790 ppmV).

Creek Formation in the Williston Basin contains marine and brackish facies (Frye, 1969; Murphy et al., 1999), indicating paleoelevations near sea level. In the Bighorn Basin, which produced the largest number of fossil cuticle sites, there are few constraints on paleoelevation (see discussion in Chase et al., 1998). The highest estimates for the basin during this time range from 1500 to 2500 m (Chase et al., 1998), however other workers consider these estimates to be too high because they think it unlikely that climate at these higher elevations would be warm enough to support the large number of subtropical and frost intolerant plants that are known from basinal deposits (S.L. Wing, 2001, personal commun.). Also, estimates of mean annual temperature for the low-lying Williston Basin during this time are similar to estimates for the Bighorn Basin (15–17 °C) (Hickey, 1977; Wing et al., 2000). Since both basins are at similar paleolatitudes, the most parsimonious solution is a similar low paleoelevation for the Bighorn Basin. Due to the lack of reliable estimates for the Bighorn Basin and the inferred low paleoelevations for the other sites used in this study, no corresponding corrections were applied to the CO_2 estimates.

Most (88%) *Ginkgo*-bearing sites used in this study have been interpreted as representing streamside and crevasse splay environments (Royer et al., 2003). These highly disturbed environments usually support well-watered, open-canopy forests. Mesic conditions should improve the fidelity of the SI-CO_2 signal, while open canopies should remove any potential intra-canopy CO_2 gradients and provide a closer analog to the open grown trees used to construct the training set.

The following sources were used to date the fossil sites: Williston Basin (Hell Creek Formation)—Hicks et al. (2002); Bighorn Basin (Fort Union and Willwood Formations)—Wing et al. (1995) and Age Model 2 in Wing et al. (2000); Denver Basin (Dawson Formation)—Raynolds et al. (2001, p. 25); Spitsbergen (Firkanten Formation)—Manum (1963) and Kvaček et al. (1994); Hanna Basin (Hanna Formation)—Higgins (2000); Alberta (Paskapoo Formation)—Fox (1990); Isle of

TABLE 2. SAMPLING DETAILS AND CO_2 RECONSTRUCTION FOR FOSSIL DATA

Site	Location†	Depository§	Age (Ma)	*n*#	SI (%)	CO_2 (ppmV)
*A. GINKGO ADIANTOIDES**						
DMNH 566	WB	DMNS	65.5	31	8.32	385
LJH 7423	BHB	USNM	64.5	5	9.48	339
LJH 7659	BHB	YPM	64.5	15	9.32	344
DMNH 2360	DB	DMNS	64.1	5	9.90	329
Basilika**	S	PMO	64.0	8	9.42	341
DMNH 2644	HB	DMNS	61.0	5	10.93	313
Burbank	A	UA	58.5	7	7.55	451
Joffre Bridge	A	UA	58.5	5	7.96	409
SLW 0025	BHB	USNM	57.3	7	9.01	353
LJH 7132	BHB	YPM	56.4	5	8.75	363
SLW 991	BHB	USNM	55.9	5	10.97	312
SLW 992	BHB	USNM	55.9	8	10.80	314
SLW 993	BHB	USNM	55.9	8	11.43	307
LJH 72141–1	BHB	USNM,YPM	55.8	12	10.63	317
SLW 9155	BHB	USNM	55.7	10	11.21	309
SLW 9411	BHB	USNM	55.6	8	11.50	306
SLW 9434	BHB	USNM	55.4	7	12.23	299
SLW 9715	BHB	USNM	55.3	12	8.23	390
SLW 9050	BHB	USNM	55.3	5	12.18	300
SLW 9936	BHB	USNM	55.3	15	11.77	303
SLW 8612	BHB	USNM	55.3	7	12.41	298
Ardtun Head††	M	BNHM, PMO,YPM	55.2	13	6.54	826
SLW 9812	BHB	USNM	55.1	22	8.53	373
SLW 9915	BHB	USNM	54.8	8	8.83	360
SLW LB	BHB	USNM	53.9	5	9.29	345
SLW H	BHB	USNM	53.5	9	10.22	323
LJH 9915	BHB	YPM	53.4	15	9.38	342
Juliaetta (P6)	I	YPM	16.5	14	8.14	396
B. METASEQUOIA OCCIDENTALIS						
Clarkia (P33a)	I	YPM	15.3	6	11.59	310
Clarkia (P33b)	I	YPM	15.3	10	10.94	316
Emerald Creek (P37a)	I	YPM	15.2	10	10.95	316

*Unless noted otherwise.

†WB—Williston Basin (North Dakota, USA); BHB—Bighorn Basin (Wyoming and Montana, USA); DB—Denver Basin (Colorado, USA); S—south-central Spitsbergen (Norway); HB—Hanna Basin (Wyoming, USA); A—south-central Alberta (Canada); M—Isle of Mull (Scotland, UK); I—north-central Idaho (USA).

§Depository for cuticle preparates: DMNS—Denver Museum of Nature & Science; USNM—National Museum of Natural History, Smithsonian Institution; YPM—Yale Peabody Museum; PMO—Oslo Paleontological Museum; UA—University of Alberta; BNHM—British Natural History Museum.

#Number of leaves measured for calculation of SI.

***Ginkgo spitsbergensis* used for stomatal counts, which is considered conspecific with *G. adiantoides* by Tralau (1968).

††*G. gardneri* used for stomatal counts.

Mull (Staffa Group)—Royer et al. (2001b); Idaho (Latah Formation)—Reidel and Fecht (1986).

THE STOMATAL RATIO METHOD

An alternative, semiquantitative, stomatal-based CO_2 proxy compares the SI in a fossil species to its nearest living equivalent (NLE) (McElwain and Chaloner, 1995, 1996; McElwain, 1998; McElwain et al., 1999). NLE's are defined ecologically and structurally, not taxonomically. The stomatal ratios of the extant:fossil species are directly translated into CO_2 estimates in a linear fashion, such that a stomatal ratio of three equates to a CO_2 estimate of 900 ppmV if the leaves of the extant trees grew in a 300 ppmV CO_2 atmosphere. This method is not species-specific, and so can be applied to sediments dating back to the Early Devonian (McElwain and Chaloner, 1995), however it assumes that the SIs in the related sets of species respond to CO_2 in the following hyperbolic fashion: $CO_2 \alpha (SI)^{-1}$.

To provide an independent check on the early Paleogene *Ginkgo*-derived CO_2 estimates, the stomatal indices in *Platanus guillelmae* from the *Ginkgo*-bearing site SLW H were compared to its NLE's, *P. occidentalis* and *P. orientalis* (and their hybrid *P.* × *acerfolia*), using the stomatal ratio method. *P. occidentalis* and *P. orientalis* are morphologically similar to the broadly trilobed *P. guillelmae,* and both the modern and fossil species are most commonly found in disturbed streamside environments (Royer et al., 2003).

RESULTS AND DISCUSSION

The stomatal indices in both *Ginkgo* and *Metasequoia* show a strong linear response to the anthropogenic increases in atmospheric CO_2 over the last 145 yr ($r^2 = 0.98$, $P < 0.001$ for *Ginkgo;* $r^2 = 0.68$, $P < 0.001$ for *Metasequoia*). For CO_2 concentrations above present-day levels, the responses are nonlinear (Fig. 2; Table 3). This nonlinear response is predicted by plant physiological principles (Kürschner et al., 2001) and supported by recent genetic work (Gray et al., 2000). Overall, both nonlinear regressions are highly significant (Fig. 2).

A discontinuity exists in the *Ginkgo* training set between the experimentally derived SI measurements at elevated CO_2 (>370 ppmV) and the rest of the training set (which includes one experimentally derived measurement at 350 ppmV) (Fig. 2). Plants often require >1 growing season for their SIs to respond to CO_2 concentrations above present-day levels (Royer, 2001), and indeed, the stomatal indices in *Ginkgo* exposed to CO_2 treatments for two growing seasons are lower than the same plants from the first growing season (Fig. 2; Table 3). In the case of the 440 ppmV CO_2 treatment, the difference is significant [$F(1,118) = 10.61$; $P < 0.001$; one-way ANOVA]. This suggests that the discontinuity is an artifact of insufficient CO_2 exposure time.

The CO_2 reconstruction based upon 31 *Ginkgo*- and *Metasequoia*-bearing fossil sites is shown in Figure 3 and Table 2. Except for one estimate near the Paleocene-Eocene boundary, all reconstructed CO_2 concentrations (300–450 ppmV) are close to present-day values. Estimates based upon <5 cuticle fragments are less reliable (Beerling and Royer, 2002a), however in general they conform with their more reliable coeval estimates (Table 4).

There is some stratigraphic evidence that the site associated with the one high CO_2 estimate (Ardtun Head, Isle of Mull, Scotland) dates to the Initial Eocene Thermal Maximum (IETM; previously known as LPTM) (Royer et al., 2001b; cf. Jolley et al., 2002). The IETM was brief (~10^5 yr; Kennett and Stott, 1991; Bains et al., 1999; Norris and Röhl, 1999) but significant (global mean surface temperatures rose ~2 °C [Huber and Sloan, 1999]). Concomitant with warming, there was a worldwide negative excursion in $\delta^{13}C$, ranging in magnitude from –2.5‰ in the benthic marine record (Kennett and Stott, 1991; Bains et al., 1999) to –6‰ in the terrestrial record (Koch et al., 1992; Sinha

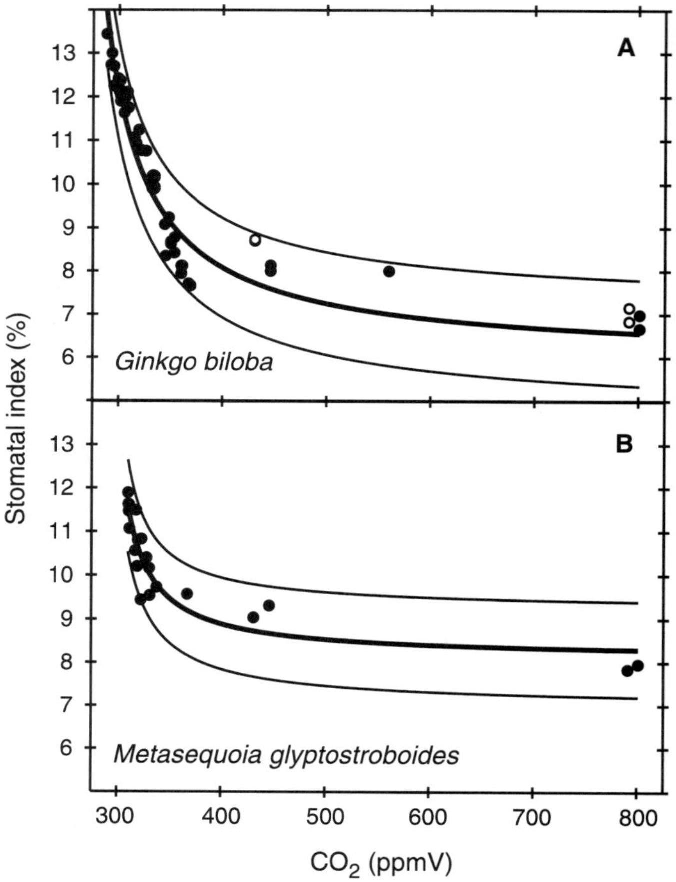

Figure 2. Training sets for (A) *Ginkgo biloba* ($n = 44$) and (B) *Metasequoia glyptostroboides* ($n = 20$). Thick lines represent regressions {*Ginkgo:* $r^2 = 0.93$, $F(1, 37) = 243$, $P < 0.0001$, $SI = 2000 \times [CO_2 - 196.1]/[333.7 \times CO_2 - 83000]$; *Metasequoia:* $r^2 = 0.86$, $F(1, 17) = 54$, $P < 0.0001$, $SI = 10000 \times [CO_2 - 273.7]/[1230.1 \times CO_2 - 350000]$}. Thin lines represent ±95% prediction intervals. For *Ginkgo,* experimental results based upon a CO_2 exposure time of one growing season (○) were not used in regression (see text).

and Stott, 1994; Bowen et al., 2001). The leading hypothesis to explain this event involves the dissociation of isotopically light (~ –60‰) methane hydrates along continental margins (Dickens et al., 1995; Bains et al., 1999; Katz et al., 1999) and their subsequent oxidation to carbon dioxide. However, all previous attempts to discern this hypothesized atmospheric CO_2 spike using other CO_2 proxies have failed (Koch et al., 1992; Stott, 1992; Sinha and Stott, 1994). Although it is tempting to correlate Ardtun Head and its corresponding 500 ppmV spike in atmospheric CO_2 to the IETM, further stratigraphic work is required to resolve this issue. If it is not of IETM age, then it is recording another high-CO_2 event. Alternatively, Ardtun Head is the one site where *G. gardneri* was used (see Table 2), and this species may respond to CO_2 differently than *G. adiantoides.*

TABLE 3. RAW DATA FOR TRAINING SETS

A. GINKGO

Date	CO_2 (ppmV)	SI (%)	Herbarium*	Source
1856	288	13.44	YPM	N.D.
1880	292	12.73	YPM	Hartford, Connecticut
1883	293	13.00	MBG	Botanic Garden, Cambridge, Massachusetts
1900	295	12.25	NA	Washington, D.C.
1900	295	12.70	YPM	Yale Forestry School, New Haven, Connecticut
1910	299	12.42	BBG	Orange, New Jersey
1915	300	12.13	MBG	Hannibal, Missouri
1918	301	12.37	BBG	Arnold Arboretum, Boston, Massachusetts
1920	302	11.90	MGB	Nanking, Kiangsu Province, China
1921	302	12.02	MGB	Missouri Botanical Garden, St. Louis, Missouri
1922	302	12.13	BBG	Purdys, New York
1928	304	11.93	NA	Maymont Park, Richmond, Virginia
1932	305	11.63	MA	Arnold Arboretum?, Boston, Massachusetts
1937	306	11.98	MBG	Japan
1942	308	12.11	NA	Berkeley, California
1943	309	11.75	MBG	Japan
1958	315	11.07	MBG	Tower Grove Park, St. Louis, Missouri
1960	317	10.93	MBG	Peking, China
1965	319	11.24	MA	Arnold Arboretum, Boston, Massachusetts
1968	322	10.77	NA	Arnold Arboretum, Boston, Massachusetts
1971	326	10.76	MA	Knox College, Galesburg, Illinois
1977	333	9.92	NA	U.S. Capitol, Washington, D.C.
1977	333	10.17	NA	Blandy Experimental Farm, Boyce, Virginia
1984	344	9.08	MBG	China (Yellow Plateau Team)
1985	345	8.35	BBG	Brooklyn Botanic Gardens, New York City, New York
1987	348	9.24	NA	Morton Arboretum, Lisle, Illinois
1988	350	8.68	MBG	Clemson University, Clemson, South Carolina
N.A.	350	8.63	Exp	3-year-old saplings in greenhouses† (2§)
1990	354	8.42	MBG	Missouri Botanical Garden, St. Louis, Missouri
1990	354	8.78	MBG	Oomoto Kameyama Botanical Garden; Kyoto, Japan
1995	360	7.94	YPM	Yale Peabody Museum, New Haven, Connecticut
1995	360	8.12	MBG	Missouri Botanical Garden, St. Louis, Missouri
1996	361	8.12	MBG	Oomoto Kameyama Botanical Garden; Kyoto, Japan
1999	367	7.71	Modern	New Haven, Connecticut
2000	369	7.66	Modern	New Haven, Connecticut
N.A.	431	8.70	Exp	6-year-old saplings in greenhouses (1§)
N.A.	431	8.73	Exp	1-year-old saplings in greenhouses (1§)
N.A.	446	8.13	Exp	7-year-old saplings in greenhouses (2§)
N.A.	446	8.00	Exp	2-year-old saplings in greenhouses (2§)
N.A.	560	8.00	Exp	3-year-old saplings in greenhouses† (2§)
N.A.	791	7.16	Exp	6-year-old saplings in greenhouses (1§)
N.A.	791	6.84	Exp	1-year-old saplings in greenhouses (1§)
N.A.	801	6.99	Exp	7-year-old saplings in greenhouses (2§)
N.A.	801	6.67	Exp	2-year-old saplings in greenhouses (2§)

B. METASEQUOIA

Date	CO_2 (ppmV)	SI (%)	Herbarium*	Source
1947	310	11.89	UC	E Szechuan / NW Hubei, China
1948	310	11.63	UC	E Szechuan / NW Hubei, China
1949	311	11.47	UC	Berkeley, California
1950	311	11.62	UC	Bar Harbor, Maine
1953	312	11.07	UC	San Rafael, California
1961	317	10.56	UC	Tallahassee, Florida
1962	318	11.49	YPM	Bethany, Connecticut
1964	319	10.20	MA	Pennsylvania
1965	320	10.81	UC	Washington, D.C.
1968	323	9.44	UC	Fukuchiyama-city, Japan
1969	323	10.83	BBG	Brooklyn Botanic Gardens, New York

continued

TABLE 3. *continued*

B. METASEQUOIA				
Date	CO_2 (ppmV)	SI (%)	Herbarium*	Source
1973	328	10.40	NA	Turlock, California
1975	331	10.16	UC	University of Oregon, Eugene, Oregon
1976	331	9.54	UC	University of Oregon, Eugene, Oregon
1980	338	9.73	UC	Western Hubei, China
1999	367	9.57	modern	New Haven, Connecticut; London, UK
N.A.	431	9.04	exp	1-year-old saplings in greenhouses (1[§])
N.A.	446	9.31	exp	2-year-old saplings in greenhouses (2[§])
N.A.	791	7.84	exp	1-year-old saplings in greenhouses (1[§])
N.A.	801	7.96	exp	2-year-old saplings in greenhouses (2[§])

Note: N.D.—no data; N.A.—not applicable.

*YPM—Peabody Museum of Natural History (Yale University); MGB—Missouri Botanical Gardens; NA—U.S. National Arboretum; BBG—Brooklyn Botanic Gardens; MA—Morris Arboretum (University of Pennsylvania); UC—University Herbarium (University of California, Berkeley); modern—collections of fresh leaves; exp—collections from greenhouses. All cuticle preparates deposited at YPM.

†From Beerling et al. (1998).

§Number of growing seasons saplings were exposed to CO_2 treatments.

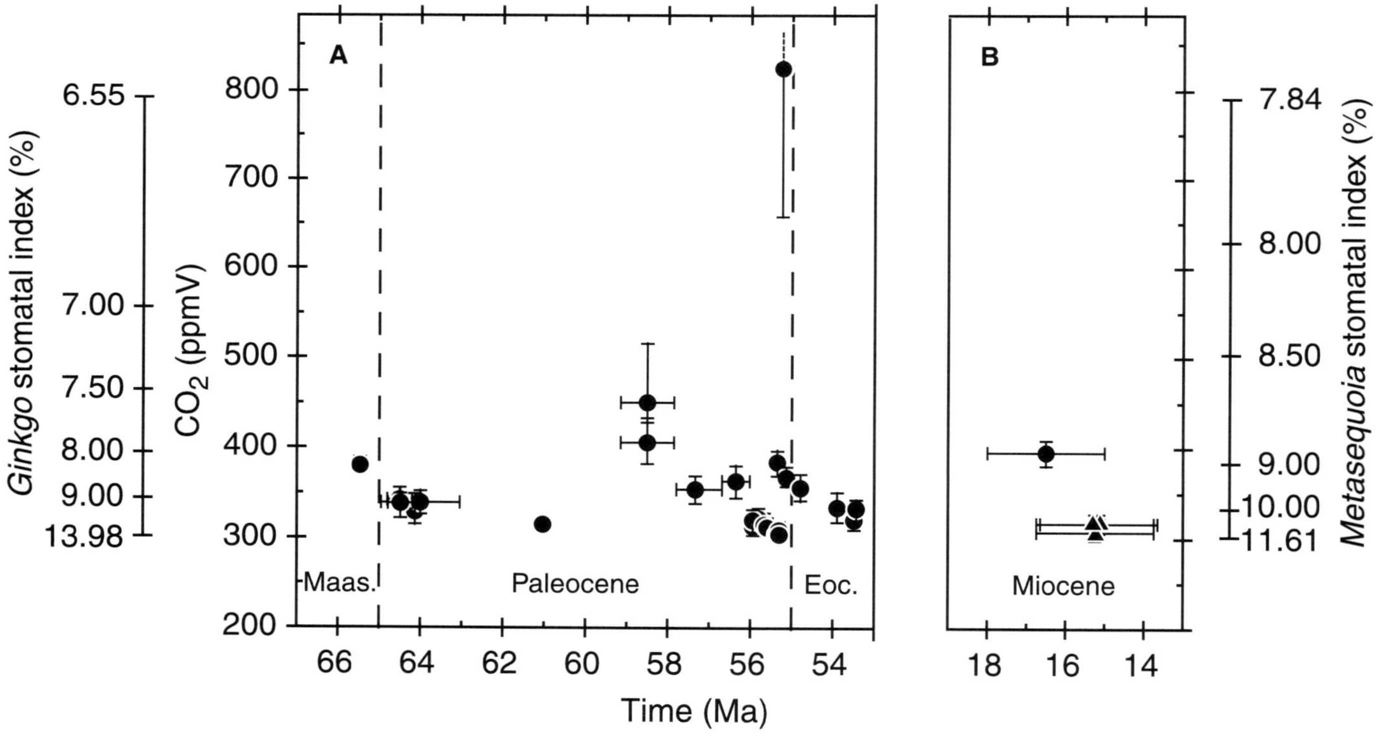

Figure 3. Reconstruction of paleo-CO_2 for the (A) very latest Cretaceous to early Eocene and (B) middle Miocene based upon SI measurements of *Ginkgo* (●) and *Metasequoia* (▲) fossil cuticles. Errors represent ±95% confidence intervals. Regressions used to solve for CO_2 were rewritten from the training set-derived regressions (*Ginkgo:* $CO_2 = 2000 \times [415 \times SI - 1961]/[3337 \times SI - 20000]$; *Metasequoia:* $CO_2 = 70000 \times [50 \times SI - 391]/[12301 \times SI - 100000]$).

TABLE 4. SAMPLING DETAILS AND CO_2 RECONSTRUCTION FOR FOSSIL *GINKGO* DATA BASED UPON SITES WITH <5 CUTICLE FRAGMENTS

Site	Location*	Depository†	Age (Ma)	*n*§	SI (%)	CO_2 (ppmV)
DMNH 571	WB	DMNS	65.9	1	7.03	554
KJ 87134	WB	YPM	65.8	3	7.09	536
DMNH 1489	WB	DMNS	65.4	2	8.42	379
N.D.	BHB	YPM	~63	3	8.60	370
UF 18255	PRB	FMNH	~58	3	11.24	309
DMNH 907	WB	DMNS	56	2	8.89	358
LJH 8411	E	YPM	55	2	8.29	386
Klärbecken	G	BNHM	~2	3	12.29	299

Note: N.D.—no data.

*WB—Williston Basin (North Dakota, USA); BHB—Bighorn Basin (Wyoming and Montana, USA); PRB—Powder River Basin (Wyoming, USA); E—Ellesmere Island (Canada); G—Germany.

†Depository for cuticle preparates: DMNS—Denver Museum of Nature & Science; YPM—Yale Peabody Museum; FMNH—Florida Museum of Natural History; BNHM—British Natural History Museum.

§Number of leaves measured for calculation of SI.

Ginkgo *and* Metasequoia *cuticle as reliable recorders of atmospheric CO_2*

There are numerous factors indicating that the CO_2 reconstruction presented here is accurate. As discussed above, both genera are restricted in the fossil record to mid- and high-latitude temperate forests, and, in addition, *Ginkgo* is preferentially found in disturbed riparian settings. These characteristics should strengthen the applicability of the fossil cuticles to their respective training sets. Also, both *Ginkgo* and *Metasequoia* are highly conservative phylogenetically, and, in the case of *Ginkgo,* the slope of the SI-CO_2 relationship derived from modern *Ginkgo* (Fig. 2) is similar to the SI-CO_2 relationship derived from a multimillion year sequence of fossil SI measurements and coeval pedogenic carbonate-derived CO_2 estimates (Beerling and Royer, 2002b). This suggests that (1) the SI-CO_2 relationships in *Ginkgo* and *Metasequoia* have remained unchanged throughout the Cenozoic; and (2) the short-term (largely phenotypic) responses captured in the training sets are similar to the long-term (largely genotypic) responses reflected in the fossils.

As a cross-check on the early Paleogene *Ginkgo*-derived estimates, I compared the SIs of *Platanus guillelmae* from site SLW H with the SIs of its nearest living equivalents *P. occidentalis* and *P. orientalis* using the stomatal ratio method (see above). Although this method can only generate semiquantitative estimates of CO_2, all of the estimates for this site fall between 350 and 390 ppmV (Table 5). This range compares favorably with the *Ginkgo*-derived CO_2 estimate from the same site (323 ppmV), indicating, again, that *Ginkgo* faithfully records CO_2.

The SIs of nearly all the measured fossil material fall within the range of high CO_2 sensitivity in the training sets (Fig. 2; Table 2). This contrasts with the study of Retallack (2001), where the stomatal indices in 82% of the fossil sites fall outside the total range captured in his training set, and so these CO_2 estimates are less reliable. In addition, I have intensively sampled several intervals, for example 12 sites from the last 1 m.y. of the Paleocene. These characteristics allow for more precise determinations of paleo-CO_2 relative to other CO_2 proxies, and, because SIs respond to CO_2 on the time scale of 10^0–10^2 yr, the ability to discern rapid shifts in atmospheric CO_2 (Royer et al., 2001a).

A few of the fossil SIs fall outside the densely sampled regions in the training sets. This suggests that the SIs in *G. biloba* and *G. adiantoides* are not constrained to a limited range (8%–12%) irrespective of atmospheric CO_2 levels. Additional evidence that the SIs in *Ginkgo* are highly adaptable comes from Mesozoic studies where SIs in *Ginkgo* are as low as 2.6% (McElwain and Chaloner, 1996; McElwain et al., 1999; Chen et al., 2001; Retallack, 2001).

The only possible confounding factor that I have discerned in my fossil data set is latitude (Fig. 4). Except for the Spitsbergen site, there is a negative correlation between latitude and the SI in *Ginkgo.* Even within the densely sampled Bighorn Basin, there is a small (slope = –2.2 using SI as the dependent variable) but moderately significant (r^2 = 0.43; P = 0.002) correlation. Differences in irradiance (Lake et al., 2001) or temperature (Wagner, 1998) may be driving this response, however more data are required before any firm conclusions can be drawn.

Paleoclimatic implications

Fig. 5 is a compilation of most published proxy- and modeling-based CO_2 estimates for the last 66 m.y. There is very good agreement among the methods for the Neogene, strongly indicating that CO_2 levels were not much different from the

TABLE 5. CO_2 RECONSTRUCTION FROM STOMATAL RATIOS IN *PLATANUS*

A. FOSSILS

Species	Site	Age (Ma)	*n**	SI (%)
P. guillelmae	SLW H	53.5	7	12.89

B. NEAREST LIVING EQUIVALENTS

Species	Site	Date	*n**	SI (%)	Stomatal ratio†	Paleo-CO_2 (ppmV)§
P. occidentalis	North Carolina, USA	1891	6	15.35	1.19	350
P. orientalis	Iraq	1975	5	15.22	1.18	390
P. × acerfolia	Connecticut, USA	2001	5	13.28	1.03	381
P. occidentalis	Connecticut, USA	2001	6	13.44	1.04	386

*Number of leaves measured for calculation of SI.

†Stomatal ratio—$SI_{modern} / SI_{fossil}$.

§CO_2 reconstructions for site SLW H. These estimates should be considered semi-quantitative only. Values calculated by multiplying the stomatal ratio by the corresponding CO_2 concentration in which the leaves developed (see text for details).

present day. CO_2 reconstructions for the Paleogene, however, are inconsistent, ranging from <300 ppmV to >3000 ppmV. These inconsistencies are largely stratified by method, with the highest CO_2 estimates derived from the boron proxy and the lowest from this stomatal study (Fig. 5).

What do these CO_2 reconstructions mean in terms of paleoclimate? Assuming that we understand the role of CO_2 as a greenhouse gas in regulating global temperature, it is possible to convert a given concentration of atmospheric CO_2 into a prediction of global mean surface temperature (GMST). I have done this for the Late Cretaceous to early Paleogene (69–50 Ma) and middle Miocene (19–13 Ma) CO_2 reconstructions, the two intervals for which I have stomatal data, using the CO_2-GMST sensitivity study of Kothavala et al. (1999). This general circulation model study was calibrated to present-day conditions. Although it may be instructive to use sensitivity studies calibrated to early Paleogene and middle Miocene conditions, the CO_2-GMST relationship is most accurately known for present-day conditions, and applying a sensitivity study calibrated to the present day allows testing the sole effects of CO_2 on GMST. In other words, if the CO_2-derived GMST predictions do not match the geologic indicators of GMST for a given period, then factors in addition to CO_2 (e.g., paleogeography, vegetative feedbacks) are required to explain the GMST for that period.

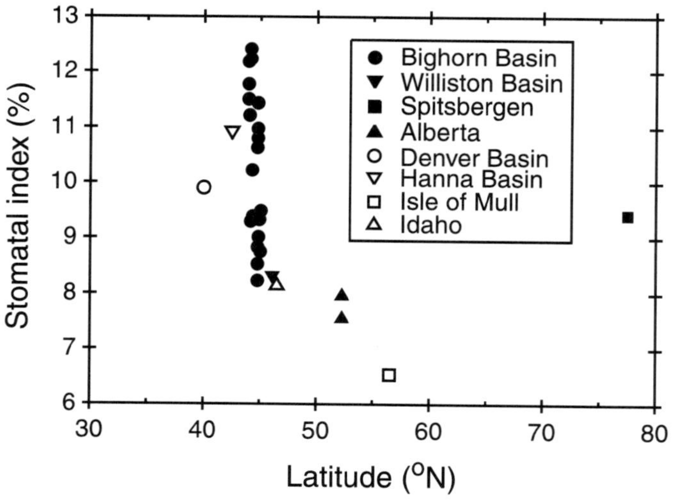

Figure 4. Correlation between latitude and stomatal index for fossil sites. Present-day latitudes are plotted, but should reflect their corresponding paleolatitudes to within 5° (Schettino and Scotese, 2001).

The CO_2-derived GMST predictions are shown in Figure 6. As with the raw CO_2 data, there is general agreement for the middle Miocene, with most predictions within 1 °C of the present-day GMST. Predictions for the early Paleogene are not consistent, ranging from near present-day values to >10 °C warmer than the present-day GMST.

Both the early Paleogene and middle Miocene are intervals of global warmth relative to today. The physiognomic characters of early Paleogene fossil floras indicate temperatures much higher than today at mid and high latitudes (e.g., Hickey, 1980; Spicer and Parrish, 1990; Wing and Greenwood, 1993; Wolfe, 1994; Greenwood and Wing, 1995; Wilf, 2000). The comparison of these fossil plants with their nearest living relatives (NLRs) also indicates warmer temperatures (e.g., Hickey, 1977; Spicer and Parrish, 1990). For example, frost intolerant palms occur at paleolatitudes up to 20° north of their current northern limit in North America (Greenwood and Wing, 1995). The early Paleogene distributions of large tortoises (Hutchison, 1982) and crocodilians (Markwick, 1994, 1998) in the United States also indicate temperatures warmer than the present day.

Although the above data constrain continental temperatures, $\delta^{18}O$ data from calcitic marine shells provide temperature information for the oceans. Oceans are the largest reservoirs of

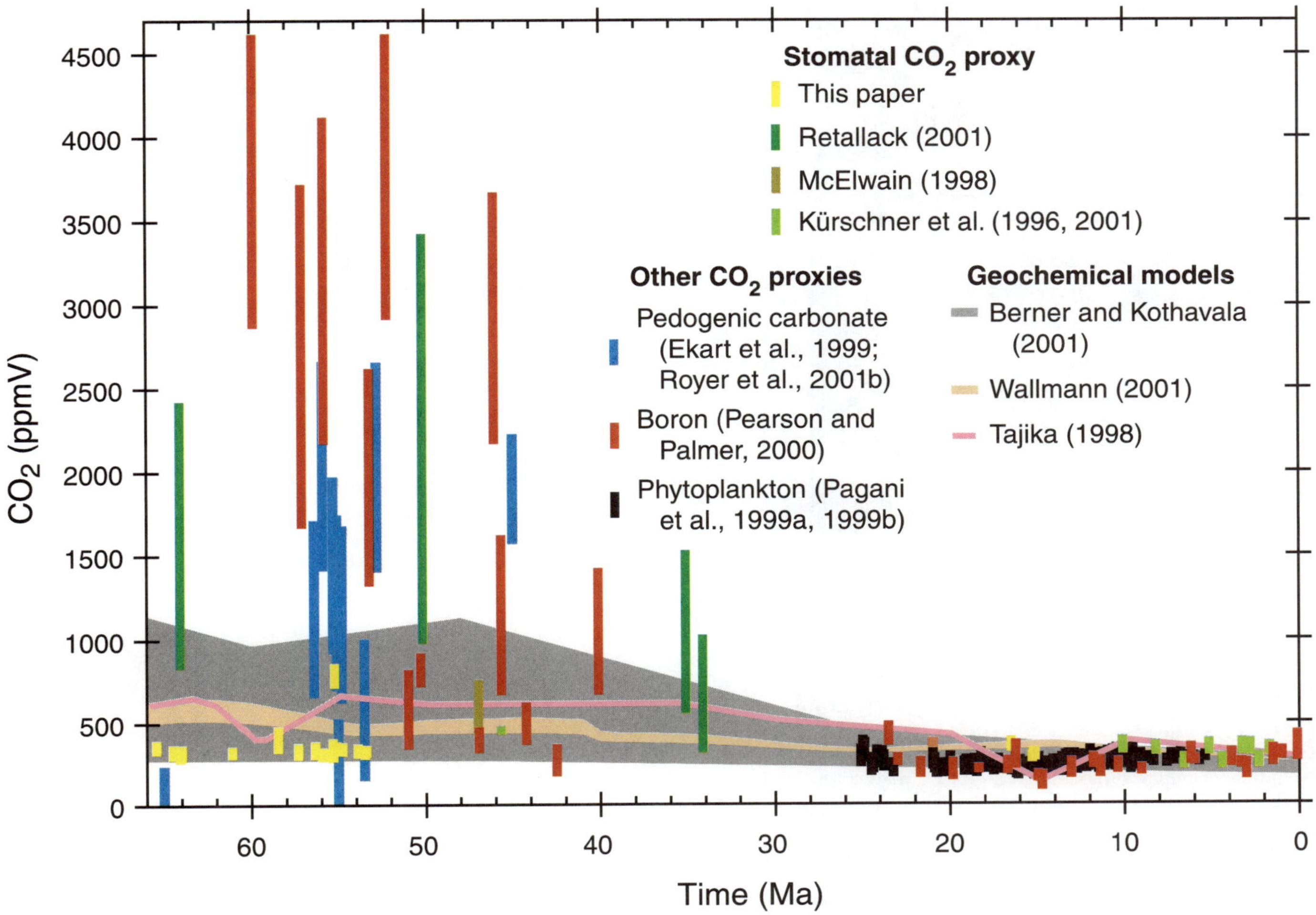

Figure 5. A compilation of CO_2 reconstructions for the Tertiary. In the case of Retallack (2001), only those sites for which ≥5 cuticle fragments were counted are plotted. No pedogenic carbonate-derived estimates are plotted for the Neogene because their margins of error are much larger relative to the other methods (cf. Ekart et al., 1999; Royer et al., 2001a).

heat in the biosphere, and therefore provide the best proxy for globally averaged temperatures. Based upon the $\delta^{18}O$ compilations of Zachos et al. (1994) and Zachos et al. (2001), benthic temperatures were 8–12 °C warmer than the present day during the early Paleogene, and high-latitude near-surface temperatures were >15 °C warmer. Using late Paleocene near surface paleotemperature data and present-day levels of atmospheric CO_2 as input into a GCM, O'Connell et al. (1996) calculated a GMST 3 °C warmer than the present day.

During the mid-Miocene thermal maximum (~17–14.5 Ma), benthic waters warmed 1–2 °C to temperatures 6 °C warmer than the present day (Lear et al., 2000; Zachos et al., 2001). Plant distributions (White et al., 1997; Utescher et al., 2000) and leaf physiognomic characters (Wolfe, 1994) also indicate substantial warming at midlatitudes.

If my stomatal-based CO_2 reconstruction is correct, then, factors in addition to carbon dioxide are required to explain these two intervals of global warmth, and these intervals may not serve as good analogs for understanding future climate change. It is crucial to emphasize that low concentrations of atmospheric CO_2 during intervals of global warmth do not counter the theory of the greenhouse effect (for this, high CO_2 levels are required during globally cool intervals); instead, they indicate that other thermal forcings were more important at those times than they are today.

Traditionally, modelers interested in warm climates have struggled to match their output with the geologic data (Barron, 1987; Sloan and Barron, 1990, 1992), particularly when using low CO_2 levels (Sloan and Rea, 1995). Recent work, however, has begun to resolve these discrepancies (Otto-Bliesner and Upchurch, 1997; Sloan and Morrill, 1998; Sloan and Pollard, 1998; DeConto et al., 1999; Upchurch et al., 1999; Sewall et al., 2000; Sewall and Sloan, 2001; Sloan et al., 2001). It will be crucial to understand the various roles CO_2 plays during globally warm pe-

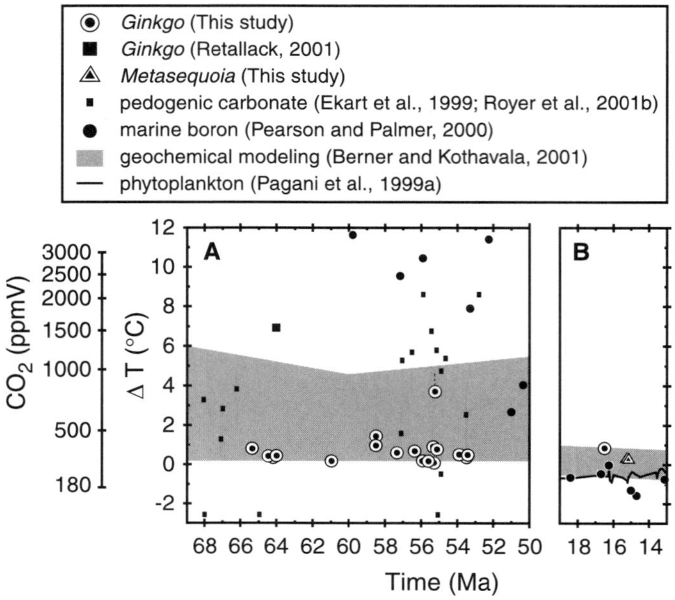

Figure 6. Estimates of CO_2 from Figure 5 and their corresponding model-determined temperature departures (ΔT) of global mean surface temperature (GMST) from present day for the (A) very latest Cretaceous to early Eocene and (B) middle Miocene. Paleo-GMST calculated using the CO_2-temperature sensitivity study of Kothavala et al. (1999). Present-day reference GMST calculated using the preindustrial CO_2 value of 280 ppmV (14.7 °C). The error range of GMST predicted from the geochemical modeling-based CO_2 predictions of Berner and Kothavala (2001) corresponds to the model's sensitivity analysis. Most error ranges derived from my stomatal data set are smaller than the plotted symbols. Error ranges derived from the phytoplankton methods are not shown.

riods, particularly as atmospheric CO_2 rises to levels in the very near future that are perhaps unprecedented for the past 65 m.y.

ACKNOWLEDGMENTS

I thank R. Berner, L. Hickey, D. Beerling, S. Wing, W. Kürschner, and J. McElwain for helpful conversations and comments. I also thank R. Dunn for providing material from site DMNH 2644. This work was supported by a National Science Foundation Graduate Research Fellowship and Department of Energy grant DE-FGO2-95ER14522.

REFERENCES CITED

Arnott, H.J., 1959, Anastomoses in the venation of *Ginkgo biloba:* American Journal of Botany, v. 46, p. 405–411.

Bains, S., Corfield, R.M., and Norris, R.D., 1999, Mechanisms of climate warming at the end of the Paleocene: Science, v. 285, p. 724–727.

Barnett, T.P., Pierce, D.W., and Schnur, R., 2001, Detection of anthropogenic climate change in the world's oceans: Science, v. 292, p. 270–274.

Barron, E.J., 1987, Eocene equator-to-pole surface ocean temperatures: A significant climate problem?: Paleoceanography, v. 2, p. 729–739.

Beerling, D.J., 1999, Stomatal density and index: Theory and application, *in* Jones, T.P., and Rowe, N.P., eds., Fossil plants and spores: Modern techniques: London, The Geological Society, p. 251–256.

Beerling, D.J., and Kelly, C.K., 1997, Stomatal density responses of temperate woodland plants over the past seven decades of CO_2 increase: A comparison of Salisbury (1927) with contemporary data: American Journal of Botany, v. 84, p. 1572–1583.

Beerling, D.J., and Osborne, C.P., 2002, Physiological ecology of Mesozoic polar forests in a high CO_2 environment: Annals of Botany, v. 89, p. 329–339.

Beerling, D.J., and Royer, D.L., 2002a, Fossil plants as indicators of the Phanerozoic global carbon cycle: Annual Review of Earth and Planetary Sciences, v. 30, p. 527–556.

Beerling, D.J., and Royer, D.L., 2002b, Reading a CO_2 signal from fossil stomata: New Phytologist, v. 153, p. 387–397.

Beerling, D.J., and Woodward, F.I., 1997, Changes in land plant function over the Phanerozoic: Reconstructions based on the fossil record: Botanical Journal of the Linnean Society, v. 124, p. 137–153.

Beerling, D.J., Chaloner, W.G., Huntley, B., Pearson, J.A., Tooley, M.J., and Woodward, F.I., 1992, Variations in the stomatal density of *Salix herbacea* L. under the changing atmospheric CO_2 concentrations of late- and post-glacial time: Philosophical Transactions of the Royal Society of London, v. B336, p. 215–224.

Beerling, D.J., McElwain, J.C., and Osborne, C.P., 1998, Stomatal responses of the "living fossil" *Ginkgo biloba* L. to changes in atmospheric CO_2 concentrations: Journal of Experimental Botany, v. 49, p. 1603–1607.

Beerling, D.J., Lomax, B.H., Royer, D.L., Upchurch, G.R., and Kump, L.R., 2002, An atmospheric pCO_2 reconstruction across the Cretaceous–Tertiary boundary from leaf megafossils: Proceedings of the National Academy of Sciences USA, v. 99, p. 7836–7840.

Berner, R.A., and Kothavala, Z., 2001, GEOCARB III: A revised model of atmospheric CO_2 over Phanerozoic time: American Journal of Science, v. 301, p. 182–204.

Boulter, M.C., and Kvaček, Z., 1989, The Paleocene flora of the Isle of Mull: Special Papers in Palaeontology, v. 42, p. 1–149.

Bowen, G.J., Koch, P.L., Gingerich, P.D., Norris, R.D., Bains, S., and Corfield, R.M., 2001, Refined isotope stratigraphy across the continental Paleocene–Eocene boundary on Polecat Bench in the northern Bighorn Basin, *in* Gingerich, P.D., ed., Paleocene–Eocene stratigraphy and biotic change in the Bighorn and Clarks Fork Basins, Wyoming: University of Michigan Papers on Paleontology, v. 33, p. 73–88.

Chamberlain, C.J., 1935, Gymnosperms: Structure and evolution: New York, Dover, 484 p.

Chaney, R.W., 1951, A revision of fossil *Sequoia* and *Taxodium* in western North America based on the recent discovery of *Metasequoia:* Transactions of the American Philosophical Society, v. 40, p. 171–263.

Chase, C.G., Gregory-Wodzicki, K.M., Parrish, J.T., and DeCelles, P.G., 1998, Topographic history of the Western Cordillera of North America and controls on climate, *in* Crowley, T.J., and Burke, K.C., eds., Tectonic boundary conditions for climate reconstructions: Oxford University Press, Oxford Monographs on Geology and Geophysics, v. 39, p. 73–99.

Chen, L.-Q., Cheng-Sen, L., Chaloner, W.G., Beerling, D.J., Sun, Q.-G., Collinson, M.E., and Mitchell, P.L., 2001, Assessing the potential for the stomatal characters of extant and fossil *Ginkgo* leaves to signal atmospheric CO_2 change: American Journal of Botany, v. 88, p. 1309–1315.

Christophel, D.C., 1976, Fossil floras of the Smoky Tower locality, Alberta, Canada: Palaeontographica, v. B157, p. 1–43.

Critchfield, W.B., 1970, Shoot growth and heterophylly in *Ginkgo biloba:* Botanical Gazette, v. 131, p. 150–162.

Crowley, T.J., 2000, Causes of climate change over the past 1000 yr: Science, v. 289, p. 270–277.

Crowley, T.J., and Berner, R.A., 2001, CO_2 and climate change: Science, v. 292, p. 870–872.

DeConto, R.M., Hay, W.W., Thomson, S.L., and Bergengren, J., 1999, Late Cretaceous climate and vegetation interactions: Cold continental interior paradox, *in* Barrera, E., and Johnson, C.C., eds., Evolution of the Cretaceous ocean-climate system: Boulder, Colorado, Geological Society of America Special Paper 332, p. 391–406.

Dickens, G.R., O'Neil, J.R., Rea, D.K., and Owen, R.M., 1995, Dissociation of oceanic methane hydrate as a cause of the carbon isotope excursion at the end of the Paleocene: Paleoceanography, v. 10, p. 965–971.

Ekart, D.D., Cerling, T.E., Montañez, I.P., and Tabor, N.J., 1999, A 400 million year carbon isotope record of pedogenic carbonate: Implications for paleoatmospheric carbon dioxide: American Journal of Science, v. 299, p. 805–827.

Fox, R.C., 1990, The succession of Paleocene mammals in western Canada, *in* Bown, T.M., and Rose, K.D., eds., Dawn of the Age of Mammals in the northern part of the Rocky Mountain interior, North America: Boulder, Colorado, Geological Society of America Special Paper 243, p. 51–70.

Friedli, H., Lötscher, H., Oeschger, H., Siegenthaler, U., and Stauffer, B., 1986, Ice core record of the $^{13}C/^{12}C$ ratio of atmospheric CO_2 in the past two centuries: Nature, v. 324, p. 237–238.

Frye, C.I., 1969, Stratigraphy of the Hell Creek Formation in North Dakota: North Dakota Geological Survey Bulletin, v. 54, p. 1–65.

Gifford, E.M., and Foster, A.S., 1989, Morphology and evolution of vascular plants (3rd edition): New York, W.H. Freeman, 626 p.

Gray, J.E., Holroyd, G.H., van der Lee, F.M., Bahrami, A.R., Sijmons, P.C., Woodward, F.I., Schuch, W., and Hetherington, A.M., 2000, The *HIC* signalling pathway links CO_2 perception to stomatal development: Nature, v. 408, p. 713–716.

Greenwood, D.R., and Wing, S.L., 1995, Eocene continental climates and latitudinal temperature gradients: Geology, v. 23, p. 1044–1048.

Harries, J.E., Brindley, H.E., Sagoo, P.J., and Bantges, R.J., 2001, Increases in greenhouse forcing inferred from the outgoing longwave radiation spectra of the Earth in 1970 and 1997: Nature, v. 410, p. 355–357.

Hickey, L.J., 1977, Stratigraphy and paleobotany of the Golden Valley Formation (early Tertiary) of western North Dakota: Geological Society of America Memoir 150, p. 1–181.

Hickey, L.J., 1980, Paleocene stratigraphy and flora of the Clarks Fork Basin, *in* Gingerich, P.D., ed., Early Cenozoic paleontology and stratigraphy of the Bighorn Basin, Wyoming: University of Michigan Papers on Paleontology, v. 24, p. 33–49.

Hicks, J.F., Johnson, K.R., Obradovich, J.D., Tauxe, L., and Clark, D., 2002, Magnetostratigraphy and geochronology of the Hell Creek and basal Fort Union Formations of southwestern North Dakota and a recalibration of the age of the Cretaceous-Tertiary boundary, *in* Hartman, J.H., et al., eds., The Hell Creek Formation and the Cretaceous-Tertiary boundary in the northern Great Plains: An integrated continental record of the end of the Cretaceous: Boulder, Colorado, Geological Society of America Special Paper 361, p. 35–55.

Higgins, P., 2000, Reevaluation of the boundary between the Torrejonian and the Tiffanian North American land mammal "ages," with description of new fauna from the Hanna Basin, south-central Wyoming [Ph.D. thesis]: Laramie, University of Wyoming, 222 p.

Huber, M., and Sloan, L.C., 1999, Warm climate transitions: A general circulation modeling study of the late Paleocene Thermal Maximum (ca. 56 Ma): Journal of Geophysical Research, v. 104, p. 16,633–16,655.

Hutchison, J.H., 1982, Turtle, crocodilian, and champsosaur diversity changes in the Cenozoic of the north-central region of western United States: Palaeogeography, Palaeoclimatology, Palaeoecology, v. 37, p. 149–164.

Jolley, D.W., Clarke, B., and Kelley, S., 2002, Paleogene time scale miscalibration: Evidence from the dating of the North Atlantic igneous province: Geology, v. 30, p. 7–10.

Jones, H.G., 1992, Plants and microclimate: Cambridge, UK, Cambridge University Press, 428 p.

Katz, M.E., Pak, D.K., Dickens, G.R., and Miller, K.G., 1999, The source and fate of massive carbon input during the latest Paleocene Thermal Maximum: Science, v. 286, p. 1531–1533.

Keeling, C.D., Whorf, T.P., Wahlen, M., and van der Plicht, J., 1995, Interannual extremes in the rate of atmospheric carbon dioxide since 1980: Nature, v. 375, p. 666–670 (see www.cmdl.noaa.gov/ccg/figures/co2mm_mlo.jpg for update).

Kennett, J.P., and Stott, L.D., 1991, Abrupt deep-sea warming, palaeoceanographic changes, and benthic extinctions at the end of the Paleocene: Nature, v. 353, p. 225–229.

Koch, P.L., Zachos, J.C., and Gingerich, P.D., 1992, Correlation between isotope records in marine and continental carbon reservoirs near the Paleocene–Eocene boundary: Nature, v. 358, p. 319–322.

Körner, Ch., and Cochrane, P.M., 1985, Stomatal responses and water relations of *Eucalyptus pauciflora* in summer along an elevational gradient: Oecologia, v. 66, p. 443–455.

Kothavala, Z., Oglesby, R J., and Saltzman, B., 1999, Sensitivity of equilibrium surface temperature of CCM3 to systematic changes in atmospheric CO_2: Geophysical Research Letters, v. 26, p. 209–212.

Kürschner, W.M., van der Burgh, J., Visscher, H., and Dilcher, D.L., 1996, Oak leaves as biosensors of late Neogene and early Pleistocene paleoatmospheric CO_2 concentrations: Marine Micropaleontology, v. 27, p. 299–312.

Kürschner, W.M., Stulen, I., Wagner, F., and Kuiper, P.J.C., 1998, Comparison of palaeobotanical observations with experimental data on the leaf anatomy of durmast oak [*Quercus petraea* (Fagaceae)] in response to environmental change: Annals of Botany, v. 81, p. 657–664.

Kürschner, W.M., Wagner, F., Dilcher, D.L., and Visscher, H., 2001, Using fossil leaves for the reconstruction of Cenozoic paleoatmospheric CO_2 concentrations, *in* Gerhard, L.C., et al., eds., Geological perspectives of global climate change: AAPG Studies in Geology 47, p. 169–189.

Kvaček, Z., Manum, S.B., and Boulter, M.C., 1994, Angiosperms from the Palaeogene of Spitsbergen, including an unfinished work by A.G. Nathorst: Palaeontographica, v. B232, p. 103–128.

Lake, J.A., Quick, W.P., Beerling, D.J., and Woodward, F.I., 2001, Signals from mature to new leaves: Nature, v. 411, p. 154.

Lear, C.H., Elderfield, H., and Wilson, P.A., 2000, Cenozoic deep-sea temperatures and global ice volumes from Mg/Ca in benthic foraminiferal calcite: Science, v. 287, p. 269–272.

Levitus, S., Antonov, J.I., Wang, J., Delworth, T.L., Dixon, K.W., and Broccoli, A.J., 2001, Anthropogenic warming of Earth's climate system: Science, v. 292, p. 267–270.

Liu, Y.-J., Li, C.-S., and Wang, Y.-F., 1999, Studies on fossil *Metasequoia* from northeast China and their taxonomic implications: Botanical Journal of the Linnean Society, v. 130, p. 267–297.

Mai, D.H., 1994, Two conifers: *Tetraclinis* Mast. (Cupressaceae) and *Metasequoia* Miki (Taxodiaceae): Relicts or palaeoclimatic indicators of the past, *in* Boulter, M.C., and Fisher, H.C., eds., Cenozoic plants and climates of the Arctic: Berlin, Springer Verlag, p. 199–214.

Mann, M.E., Bradley, R.S., and Hughes, M.K., 1998, Global-scale temperature patterns and climate forcing over the past six centuries: Nature, v. 392, p. 779–787.

Mann, M.E., Bradley, R.S., and Hughes, M.K., 1999, Northern hemisphere temperatures during the past millennium: Inferences, uncertainties, and limitations: Geophysical Research Letters, v. 26, p. 759–762.

Manum, S., 1963, Notes on the Cretaceous–Tertiary boundary in Basilikaen, Vestspitsbergen, and a new record of *Ginkgo* from the Spitsbergen Tertiary: Norsk Polarinstitutt, Årbok 1962, p. 149–152.

Manum, S., 1966, *Ginkgo spitsbergensis* n. sp. from the Paleocene of Spitsbergen and a discussion of certain Tertiary species of *Ginkgo* Europe and North America: Norsk Polarinstitutt, Årbok 1965, p. 49–58.

Markwick, P.J., 1994, "Equability," continentality, and Tertiary "climate": The crocodilian perspective: Geology, v. 22, p. 613–616.

Markwick, P.J., 1998, Fossil crocodilians as indicators of Late Cretaceous and Cenozoic climates: Implications for using palaeontological data in reconstructing palaeoclimate: Palaeogeography, Palaeoclimatology, Palaeoecology, v. 137, p. 205–271.

McElwain, J.C., 1998, Do fossil plants signal palaeoatmospheric CO_2 concentration in the geological past?: Philosophical Transactions of the Royal Society of London, v. B353, p. 83–96.

McElwain, J.C., and Chaloner, W.G., 1995, Stomatal density and index of fossil plants track atmospheric carbon dioxide in the Paleozoic: Annals of Botany, v. 76, p. 389–395.

McElwain, J.C., and Chaloner, W.G., 1996, The fossil cuticle as a skeletal record of environmental change: Palaios, v. 11, p. 376–388.

McElwain, J.C., Beerling, D.J., and Woodward, F.I., 1999, Fossil plants and global warming at the Triassic–Jurassic boundary: Science, v. 285, p. 1386–1390.

Mitchell, J.F.B., Karoly, D.J., Hegerl, G.C., Zwiers, F.W., Allen, M.R., and Marengo, J., 2001, Detection of climate change and attribution of causes, *in* Houghton, J.T., et al., eds., Climate change 2001: The scientific basis: Contribution of Working Group I to the Third Assessment Report of the Intergovernmental Panel on Climate Change: Cambridge, UK, Cambridge University Press, 881 p.

Murphy, E.C., Hoganson, J.W., Johnson, K.R., and Hartman, J.H., 1999, Lithostratigraphy of the Hell Creek Formation in North Dakota: Geological Society of America Abstracts with Programs, v. 31, no. 7, p. A70–71.

Norris, R.D., and Röhl, U., 1999, Carbon cycling and chronology of climate warming during the Paleocene–Eocene transition: Nature, v. 401, p. 775–778.

O'Connell, S., Chandler, M.A., and Ruedy, R., 1996, Implications for the creation of warm saline deep water: Late Paleocene reconstructions and global climate model simulations: Geological Society of America Bulletin, v. 108, p. 270–284.

Otto-Bliesner, B.L., and Upchurch, G.R., 1997, Vegetation-induced warming of high-latitude regions during the Late Cretaceous period: Nature, v. 385, p. 804–807.

Pagani, M., Arthur, M.A., and Freeman, K.H., 1999a, Miocene evolution of atmospheric carbon dioxide: Paleoceanography, v. 14, p. 273–292.

Pagani, M., Arthur, M.A., and Freeman, K.H., 1999b, Late miocene atmospheric CO_2 concentrations and the expansion of C_4 grasses: Science, v. 285, p. 876–879.

Pearson, P.N., and Palmer, M.R., 2000, Atmospheric carbon dioxide concentrations over the past 60 million years: Nature, v. 406, p. 695–699.

Petit, J.R., Jouzel, J., Raynaud, D., Barkov, N.I., Barnola, J.-M., Basile, I., Bender, M., Chappellaz, J., Davis, M., Delaygue, G., Delmotte, M., Kotlyakov, V.M., Legrand, M., Lipenkov, V.Y., Lorius, C., Pépin, L., Ritz, C., Saltzman, E., and Stievenard, M., 1999, Climate and atmospheric history of the past 420,000 yr from the Vostok ice core, Antarctica: Nature, v. 399, p. 429–436.

Raynolds, R.G.H., Johnson, K.R., Arnold, L.R., Farnham, T.M., Fleming, R.F., Hicks, J.F., Kelley, S.A., Lapey, L.A., Nichols, D.J., Obradovich, J.D., and Wilson, M.D., 2001, The Kiowa core, a continuous drill core through the Denver Basin bedrock aquifers at Kiowa, Elbert County, Colorado: U.S. Geological Survey Open-File Report 01-185: http://geology.cr.usgs.gov/pub/open-file-reports/ofr-01-0185/ (October 2002).

Reidel, S.P., and Fecht, K.R., 1986, The Huntziger flow: Evidence of surface mixing of the Columbia River Basalt and its petrogenetic implications: Geological Society of America Bulletin, v. 98, p. 664–677.

Retallack, G.J., 2001, A 300-million-year record of atmospheric carbon dioxide from fossil plant cuticles: Nature, v. 411, p. 287–290.

Royer, D.L., 2001, Stomatal density and stomatal index as indicators of paleoatmospheric CO_2 concentration: Review of Palaeobotany and Palynology, v. 114, p. 1–28.

Royer, D.L., 2002, Estimating latest Cretaceous and Tertiary atmospheric CO_2 concentration from stomatal indices [Ph.D. thesis]: New Haven, Yale University, 164 p.

Royer, D.L., Berner, R.A., and Beerling, D.J., 2001a, Phanerozoic atmospheric CO_2 change: Evaluating geochemical and paleobiological approaches: Earth Science Reviews, v. 54, p. 349–392.

Royer, D.L., Wing, S.L., Beerling, D.J., Jolley, D.W., Koch, P.L., Hickey, L.J., and Berner, R.A., 2001b, Paleobotanical evidence for near present-day levels of atmospheric CO_2 during part of the Tertiary: Science, v. 292, p. 2310–2313.

Royer, D.L., Hickey, L.J., and Wing, S.L., 2003, Ecological conservatism in the "living fossil" *Ginkgo:* Paleobiology, v. 29, p. 84–104.

Rundgren, M., and Beerling, D., 1999, A Holocene CO_2 record from the stomatal index of subfossil *Salix herbacea* L. leaves from northern Sweden: The Holocene, v. 9, p. 509–513.

Salisbury, E.J., 1927, On the causes and ecological significance of stomatal frequency, with special reference to the woodland flora: Philosophical Transactions of the Royal Society of London, v. B216, p. 1–65.

Schettino, A., and Scotese, C.R., 2001, New internet software aids paleomagnetic analysis and plate tectonic reconstructions: Eos (Transactions, American Geophysical Union), v. 82, p. 530–536 (software at: www.itis-molinari.mi.it/plot.htm).

Sewall, J.O., and Sloan, L.C., 2001, Equable Paleogene climates: The result of a stable, positive Arctic oscillation?: Geophysical Research Letters, v. 28, p. 3693–3695.

Sewall, J.O., Sloan, L.C., Huber, M., and Wing, S., 2000, Climate sensitivity to changes in land surface characteristics: Global and Planetary Change, v. 26, p. 445–465.

Shackleton, N.J., 2000, The 100,000-year ice-age cycle identified and found to lag temperature, carbon dioxide, and orbital eccentricity: Science, v. 289, p. 1897–1902.

Sinha, A., and Stott, L.D., 1994, New atmospheric pCO_2 estimates from paleosols during the late Paleocene–early Eocene global warming interval: Global and Planetary Change, v. 9, p. 297–307.

Sloan, L.C., and Barron, E.J., 1990, "Equable" climates during Earth history?: Geology, v. 18, p. 489–492.

Sloan, L.C., and Barron, E.J., 1992, A comparison of Eocene climate model results to quantified paleoclimatic interpretations: Palaeogeography, Palaeoclimatology, Palaeoecology, v. 93, p. 183–202.

Sloan, L.C., and Morrill, C., 1998, Orbital forcing and Eocene continental temperatures: Palaeogeography, Palaeoclimatology, Palaeoecology, v. 144, p. 21–35.

Sloan, L.C., and Pollard, D., 1998, Polar stratospheric clouds: A high-latitude warming mechanism in an ancient greenhouse world: Geophysical Research Letters, v. 25, p. 3517–3520.

Sloan, L.C., and Rea, D.K., 1995, Atmospheric carbon dioxide and early Eocene climate: A general circulation modeling sensitivity study: Palaeogeography, Palaeoclimatology, Palaeoecology, v. 119, p. 275–292.

Sloan, L.C., Huber, M., Crowley, T.J., Sewall, J.O., and Baum, S., 2001, Effect of sea surface temperature configuration on model simulations of "equable" climate in the early Eocene: Palaeogeography, Palaeoclimatology, Palaeoecology, v. 167, p. 321–335.

Spicer, R.A., and Parrish, J.T., 1990, Late Cretaceous–early Tertiary palaeoclimates of northern high latitudes: A quantitative view: Journal of the Geological Society of London, v. 147, p. 329–341.

Stott, L.D., 1992, Higher temperatures and lower oceanic pCO_2: A climate enigma at the end of the Paleocene Epoch: Paleoceanography, v. 7, p. 395–404.

Tajika, E., 1998, Climate change during the last 150 million years: Reconstruction from a carbon cycle model: Earth and Planetary Science Letters, v. 160, p. 695–707.

Tralau, H., 1967, The phytogeographic evolution of the genus Ginkgo L.: Botaniska Notiser, v. 120, p. 409–422.

Tralau, H., 1968, Evolutionary trends in the genus Ginkgo: Lethaia, v. 1, p. 63–101.

Upchurch, G.R., Otto-Bliesner, B.L., and Scotese, C.R., 1999, Terrestrial vegetation and its effects on climate during the latest Cretaceous, *in* Barrera, E., and Johnson, C.C., eds., Evolution of the Cretaceous ocean-climate system: Boulder, Colorado, Geological Society of America Special Paper 332, p. 407–426.

Utescher, T., Mosbrugger, V., and Ashraf, A.R., 2000, Terrestrial climate evolution in northwest Germany over the last 25 million years: Palaios, v. 15, p. 430–449.

Van der Burgh, J., Visscher, H., Dilcher, D.L., and Kürschner, W.M., 1993, Paleoatmospheric signatures in Neogene fossil leaves: Science, v. 260, p. 1788–1790.

Wagner, F., 1998, The influence of environment on the stomatal frequency in *Betula:* Utrecht, Netherlands, Laboratory of Palaeobotany and Palynology Contributions Series, v. 9, p. 1–102.

Wallmann, K., 2001, Controls on the Cretaceous and Cenozoic evolution of seawater composition, atmospheric CO_2, and climate: Geochimica et Cosmochimica Acta, v. 65, p. 3005–3025.

White, J.M., Ager, T.A., Adam, D.P., Leopold, E.B., Liu, G., Jetté, H., and Schweger, C.E., 1997, An 18-million-year record of vegetation and climate change in northwestern Canada and Alaska: Tectonic and global climatic correlates: Palaeogeogrpahy, Palaeoclimatology, Palaeoecology, v. 130, p. 293–306.

Wilf, P., 2000, Late Paleocene–early Eocene climate changes in southwestern Wyoming: Paleobotanical analysis: Geological Society of America Bulletin, v. 112, p. 292–307.

Wing, S.L., and Greenwood, D.R., 1993, Fossils and fossil climate: The case for equable continental interiors in the Eocene: Philosophical Transactions of the Royal Society of London, v. B341, p. 243–252.

Wing, S.L., Alroy, J., and Hickey, L.J., 1995, Plant and mammal diversity in the Paleocene to early Eocene of the Bighorn Basin: Palaeogeography, Palaeoclimatology, Palaeoecology, v. 115, p. 117–155.

Wing, S.L., Bao, H., and Koch, P.L., 2000, An early Eocene cool period? Evidence for continental cooling during the warmest part of the Cenozoic, *in* Huber, B.T., et al., eds., Warm climates in Earth history: Cambridge, UK, Cambridge University Press, p. 197–237.

Wolfe, J.A., 1994, Tertiary climatic changes at middle latitudes of western North America: Palaeogeography, Palaeoclimatology, Palaeoecology, v. 108, p. 195–205.

Woodward, F.I., 1986, Ecophysiological studies on the shrub *Vaccinium myrtillus* L. taken from a wide altitudinal range: Oecologia, v. 70, p. 580–586.

Woodward, F.I., 1987, Stomatal numbers are sensitive to increases in CO_2 from preindustrial levels: Nature, v. 327, p. 617–618.

Woodward, F.I., and Bazzaz, F.A., 1988, The responses of stomatal density to CO_2 partial pressure: Journal of Experimental Botany, v. 39, p. 1771–1781.

Woodward, F.I., and Kelly, C.K., 1995, The influence of CO_2 concentration on stomatal density: New Phytologist, v. 131, p. 311–327.

Zachos, J.C., Stott, L.D., and Lohmann, K.C., 1994, Evolution of early Cenozoic marine temperatures: Paleoceanography, v. 9, p. 353–387.

Zachos, J., Pagani, M., Sloan, L., Thomas, E., and Billups, K., 2001, Trends, rhythms, and aberrations in global climate 65 Ma to present: Science, v. 292, p. 686–693.

Manuscript Accepted by the Society August 13, 2002

Printed in the USA

Geological Society of America
Special Paper 369
2003

Middle to upper Eocene dysoxic-anoxic Kuma Formation (northeast Peri-Tethys): Biostratigraphy and paleoenvironments

Vladimir N. Beniamovski*
Geological Institute of Russian Academy of Sciences, Pyzhevsky 7, 119017, Moscow, Russia
Alexander S. Alekseev
Department of Paleontology, Geological Faculty, Moscow State University, 119992, Moscow, Russia
Mariya N. Ovechkina
Paleontological Institute of Russian Academy of Sciences, Profsoyuznaya 123, 117868, Moscow, Russia
Hedi Oberhänsli*
GeoForschungZentrum Potsdam, Telegrafenberg, D-14473, Potsdam, Germany

ABSTRACT

In the late middle to earliest late Eocene, a vast dysoxic to anoxic basin extended from the Crimea in the West through the northern Caucasus to the Aral Sea in the East, and a huge amount of organic carbon was deposited in the organic-rich sediments of the Kuma Formation. We document the paleo-environmental conditions in the basin using benthic foraminiferal and organic carbon data. Oxygen levels were lowest during the deposition of the middle part of the Kuma Formation, a time of relatively cool climate. We estimate the time of deposition and its duration as ranging from planktonic foraminiferal zones upper P11 through P14, and calcareous nannofossil zones CP13c to lowermost CP15 (~44–36.5 Ma). We argue that this large-scale storage of organic carbon over ~7 m.y. resulted from a complex interaction of tectonic and climatic factors, and that it may have influenced the global carbon budget in the ocean-atmosphere system. It probably was at least a contributing factor to the global cooling in the middle Eocene that led to the establishment of ice sheets on the Antarctic continent by the end of the Eocene.

INTRODUCTION

By the late middle Eocene, a large epicontinental dysoxic basin, which at times became anoxic, was established on the southern margin of the Russian Platform and the Scythian Platform (Fig. 1). The sediments deposited in this basin form the sapropelitic, widespread Kuma Formation, described in the early twentieth century as "coffee-brown marls with *Lyrolepis caucasica* (fish scales)" and also referred to as the "fish formation" (Vassoevich, 1934). The Kuma Formation can be traced over 2000 km, from the northwest coast of the Black Sea in the west to the Aral Sea in the east; its surface area is ~500, 000 km^2, its thickness varies between 20 and 50 m.

Nannofossil studies in the Northern Caucasus (Kheu Section; Muzylöv, 1980; Bugrova et al., 1988) suggested that the Kuma Formation is time equivalent to the upper part of the *Discoaster bifax* Subzone [CP14a or NP16; Okada and Bukry (1980) and Martini (1971) respectively], the *D. saipanensis* Subzone (CP14b or NP17), and the lowermost part of the *Chiasmolithus oamaruensis* Subzone (lowermost CP15 or NP18). Sedimentation in this dysoxic-anoxic basin thus mainly occurred during the Bartonian.

* E-mails: Beniamovski: ben@geo.tv-sign.ru; Oberhänsli: oberh@gfz-potsdam.de.

Beniamovski, V.N., Alekseev, A.S., Ovechkina, M.N., and Oberhänsli, H., 2003, Middle to upper Eocene dysoxic-anoxic Kuma Formation (northeast Peri-Tethys): Biostratigraphy and paleoenvironments, *in* Wing, S.L., Gingerich, P.D., Schmitz, B., and Thomas, E., eds., Causes and Consequences of Globally Warm Climates in the Early Paleogene: Boulder, Colorado, Geological Society of America Special Paper 369, p. 95–112.

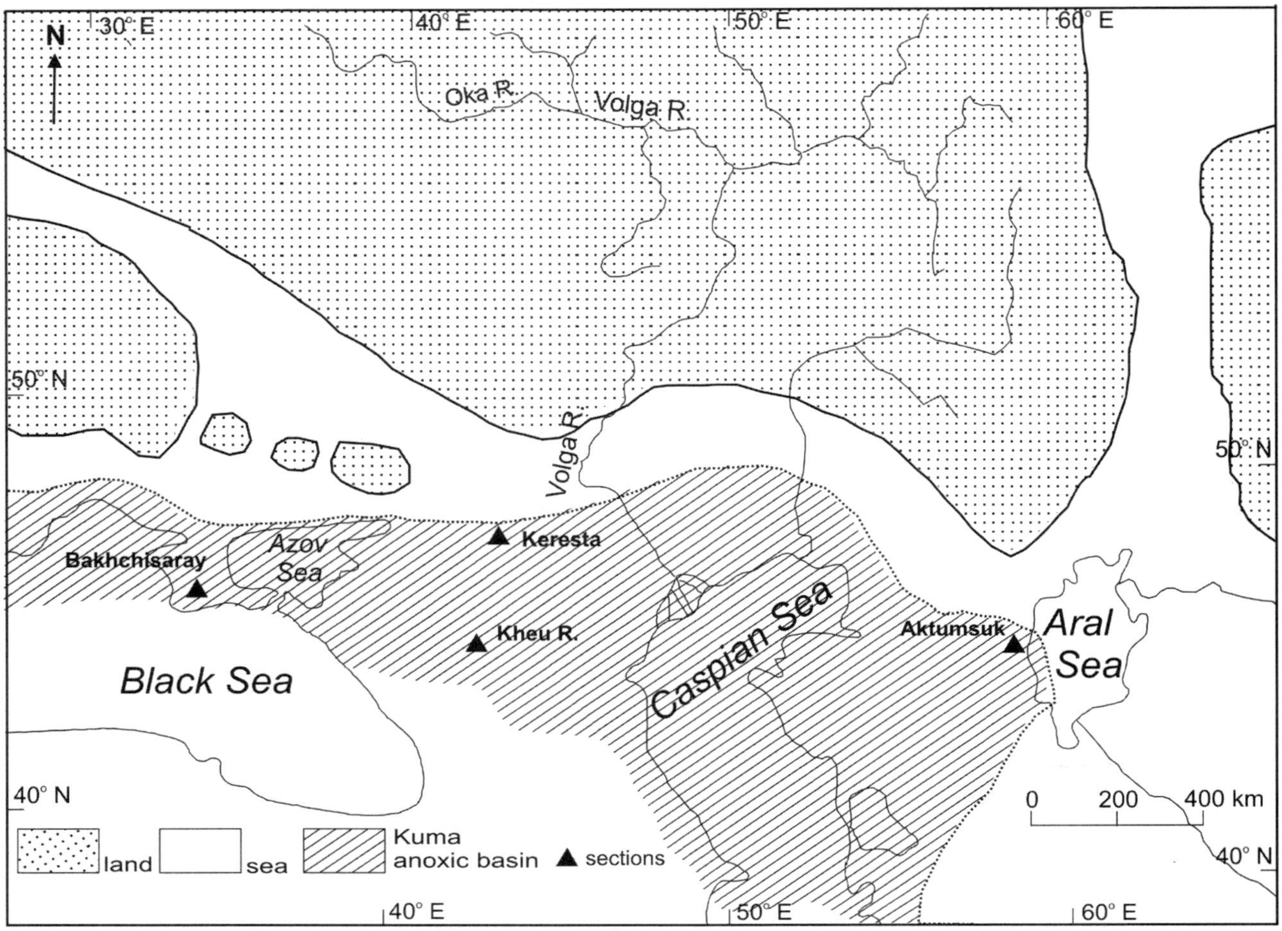

Figure 1. Outline of the late middle Eocene Kuma anoxic basin in the northeastern Peri-Tethys and locations of the four studied sections in the modern geographic setting.

The most prominent feature of the Kuma Formation is its high organic carbon content, with total organic carbon content (TOC) between 2 and 15 wt% (Zhabrev et al., 1966; Eisenstadt et al., 1967; Gavrilov, 2000). We suggest that burial of such a large mass of organic carbon must have affected the dynamics of the global carbon cycle. In this paper we discuss (i) the age of the onset of the dysoxia-anoxia (ii) the depositional environments in the Kuma Basin, and (iii) the duration of the period of deposition of the organic-rich facies.

LITHOSTRATIGRAPHY

We present information on the planktonic foraminiferal and calcareous nannofossil zonation in four sections: (i) the Bakhchisaray Section, north of Bakhchisaray in the Crimea (44°47′ N, 33°53′E) (ii) the Kheu Section, southeast of Nalchik, near Gerpegesh Village (43°24′ N, 43°38′ E) (iii) the Keresta Section in the Rostov Region, near Pervomayskoe Village (46°20′ N, 43°42′ E), and (iv) the Aktumsuk Section, on the southwest coast of the Aral Sea (44°37′ N, 58°20′ E) (Fig. 1).

The composite Bakhchisaray Section (Fig. 2), the most complete section of the Kuma Formation sapropelites in the Crimea, was first described by Dubois de Montpéreux (1837) and later by Menner (1971). The formation was sampled in 1998 in the quarry of a concrete factory (samples 1–46) and in a ravine cutting the southwest slopes of the Kazantash Mountain (samples 47–111). This composite section contains the Keresta Formation in its lower part (quarry outcrop), overlain by the Kuma Formation (quarry and ravine outcrops) and the Belaya Glina Formation (outcrop at the steep hill face). The lower part of the Keresta Formation (32 m exposed) consists of marls, its upper part of alternating marls and limestone. The Kuma Formation conformably overlays the Keresta Formation. Between the quarry and the ravine sections a few meters of the lowermost Kuma Formation are not exposed because of the dense vegetation (Fig. 2).

The Kuma Formation (57 m thick) mainly consists of grey and black marls (light coffee-colored when weathered) with grey-green intercalations in its uppermost part (Fig. 2). Fine (mm-scale) lamination and rhythmical alternation (~0.5 m) be-

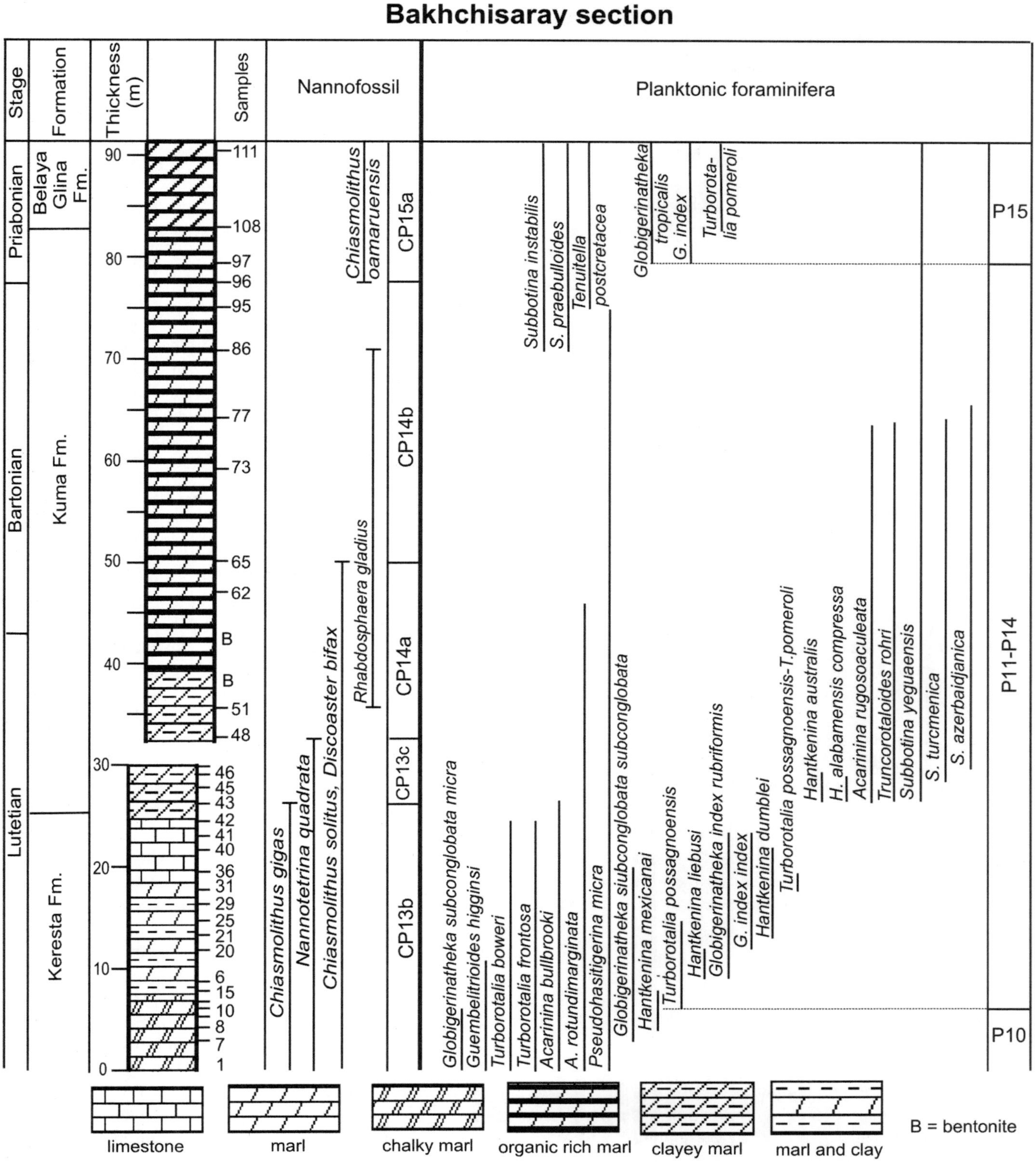

Figure 2. Distribution of planktonic foraminifera and most important datum levels of calcareous nannofossils in the upper middle through lower upper Eocene in the Bakhchisaray Section (Crimea).

tween relatively light marls and 3–5 cm thick beds of darker marls are typical for the middle part of the Kuma Formation, as is the occurrence of abundant fish remains (scales and fragmented bones). Several bentonite layers occur in the lower part of the Kuma Formation. In the upper part of the Kuma Formation plant debris becomes abundant. Only a few bivalves (*Lucina* sp.) were found. The overlying Belaya Glina Formation is up to 150 m thick and consists of green-grey marls, which are white when dry, and intensively bioturbated.

The most complete Northern Caucasus section of the Kuma Formation, the Kheu Section, is continuously exposed on the banks of the Kheu River (Gavrilov et al., 2000). We discuss the Keresta Formation, consisting of ~15 m of white limestones and marls, and the Kuma Formation, which consists of 55 m of coffee- to dark-brown marls (Fig. 3). The lithology is very similar to that in the Crimea. Twenty bentonite interlayers (several mm to 10 cm thick) occur in the lowermost part (CP14a).

The Keresta Section, the stratotype of the Keresta Formation, represents the northernmost Kuma facies (Figs. 1, 4). The sediments were recovered in a drill-hole on the bank of the Keresta Ravine in 1995; recovery was 100%. The core was described and sampled by Muzylöv et al. (1999). The Keresta Formation is 10 m thick and consists of white clayey marls and calcareous marls in the lower part, and thin limestone beds (1 m thick) at the top. The Solonka Formation is the time equivalent of the Kuma Formation. It consists of 6 m of brown-grey calcareous clays in its lower part, with green-grey clays in the topmost 5 m, and it unconformably overlies the Keresta Formation. There are no bentonites in this section.

We use the lithostratigraphic nomenclature of northern Turkmenistan (Saperson and Zheleznov, 1962; Muzylöv et al., 1990; Khodzhakhmedov, 2001) for the description of the Aktumsuk Section, the easternmost studied outcrop. The Ilyaly Formation (20 m thick; equivalent to the Keresta Formation) consists mainly of white marls and soft limestones, and contains several bentonite beds (Fig. 5). The Kurtysh Formation is 15 m thick and time equivalent to the organic-rich Kuma Formation. It consists of light-grey and brownish marls with numerous bentonites and is unconformably overlain by the white marls of the Akhchakaya Formation (>7 m thick; equivalent to the Belaya Glina Formation).

BIOSTRATIGRAPHY

In order to develop a more detailed time frame for the onset and duration of dysoxic-anoxic deposition, we studied planktonic foraminiferal and calcareous nannofossil assemblages. Some biostratigraphic markers of both groups present problems in correlation to global zonation schemes. These problems may be caused by the relatively high paleolatitude (41° to 44°N) of the sections (Fig. 6), or, more likely, by the existence of relatively shallow gateways to the open oceanic basins located to the south and west. The planktonic foraminiferal assemblages include few marker species and not all calcareous nannofossil markers have the stratigraphic ranges found in the pelagic Tethyan and subtropical open oceanic sites. Therefore independent regional zonations have been proposed by several authors (Subbotina, 1953; Shutskaya, 1970; Krasheninnikov and Muzylöv, 1975).

Foraminiferal samples were washed by standard methods, over a screen with a mesh width of 50 µm. The species identification was made in the size fraction of >150 µm. Calcareous nannofossils were studied using a standard method of decantation; smear slides were examined under a light microscope (Zeiss Axiolab, × 1500). Two hundred nannofossil specimens were counted per sample.

Calcareous nannofossils

The middle Eocene presents one of the most intricate problems of Paleogene nannofossil zonation and biostratigraphic correlation in the northeastern Peri-Tethys. The zonal concepts of Okada and Bukry (1980) are most suitable for this basin. We could recognize most of their subzones and correlate them to Martini's (1971) zonation. There is a problem, however, with the determination of the boundary between Zones NP15/NP16 or CP13/CP14 (Shcherbinina, 2000). The Last Occurrence (LO) of *Nannotetrina* is at the boundary between the Keresta and the Kuma Formations in most of the studied sections, in agreement with Perch-Nielsen (1985), who pointed out that the LO of *Nannotetrina* occurred close to the LO of *Rhabdosphaera gladius,* the marker of the top of Zone NP15. In many sections in the northeastern Peri-Tethys, however, the latter species ranges as high as Zone CP14 (NP16).

In the northeastern Peri-Tethys, the First Occurrence (FO) of large *Reticulofenestra umbilicus* (~14–16 µm) cannot be used as a marker for this boundary because it usually first occurs much earlier, in Subzone NP13b. In addition, all taxa in the nannofossil assemblage from the Aktumsuk Section are very small, and *R. umbilicus* never exceeds 10–12 µm. We conclude that coccolith size appears to be ecologically controlled and thus should not be used as a stratigraphical marker. The reticulofenestrid bloom, at the time when large specimens of *Reticulofenestra* become common worldwide (Shcherbinina, 2001), occurs close to the LO of *Nannotetrina.* The LO of the genus *Nannotetrina* thus can be used to define the NP15/NP16 zonal boundary as well as the CP13/CP14 zonal boundary in the northeastern Peri-Tethys, and provides the possibility to correlate both zonal schemes.

The base of the Keresta Formation corresponds to the bottom of Zone NP15. At this level, *Nannotetrina quadrata* occurs while the zonal markers of NP14 (*Discoaster sublodoensis* and *Rhabdosphaera inflata*) disappear near it in all sections. Subzone CP13b is clearly defined by the range of *Chiasmolithus gigas.* This species, however, is absent at the northern periphery of the basin, so that Subzone CP13b could not be identified in the bore hole at Keresta. Its range also seems reduced in the Aktumsuk Section, possibly because of its low abundance.

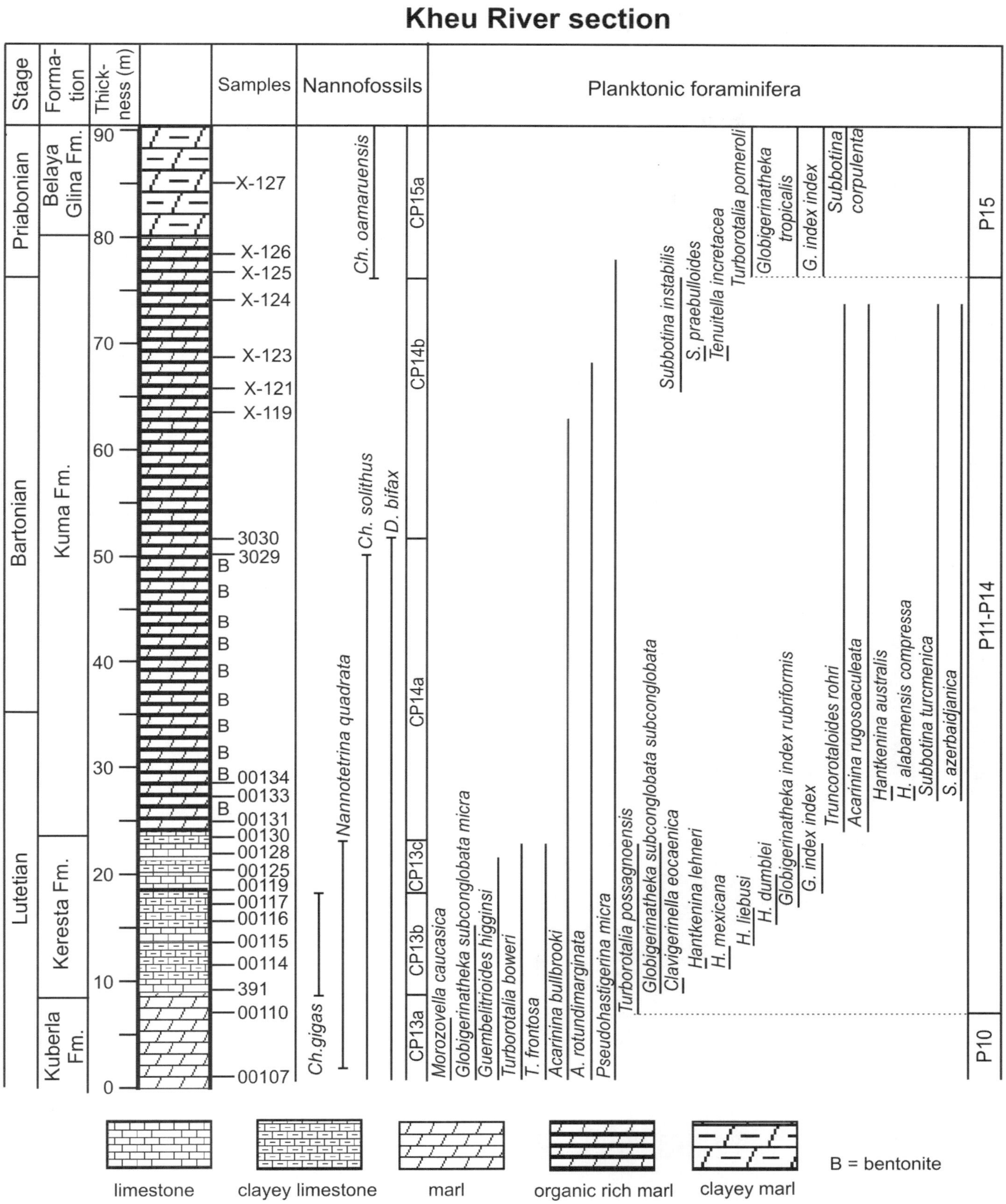

Figure 3. Distribution of planktonic foraminifera and most important datum levels of calcareous nannofossils in the upper middle through lower upper Eocene of the Kheu Section (Northern Caucasus).

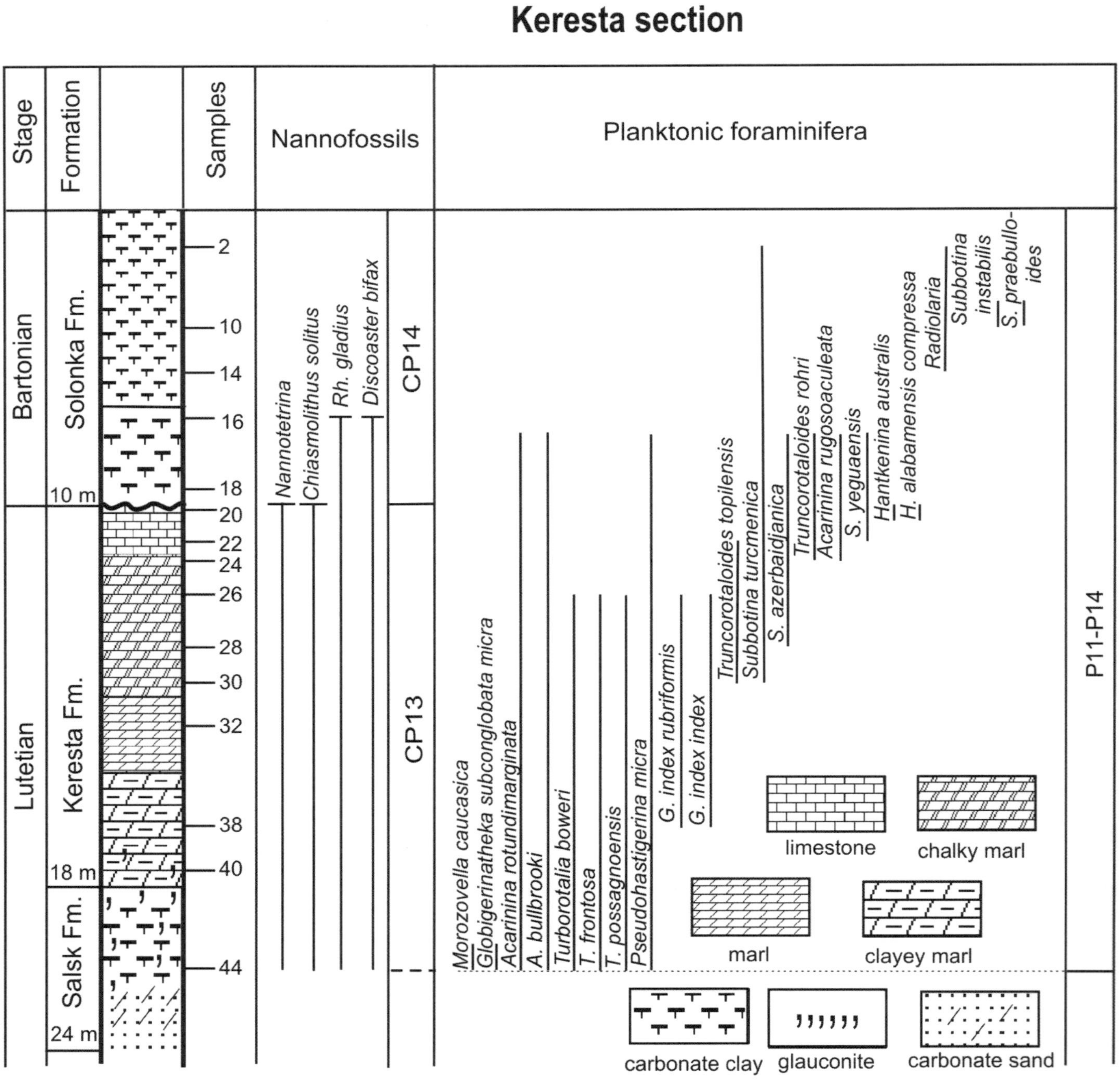

Figure 4. Distribution of planktonic foraminifera and most important datum levels of calcareous nannofossils in the upper middle Eocene of the Keresta Section (Pre-Caucasus).

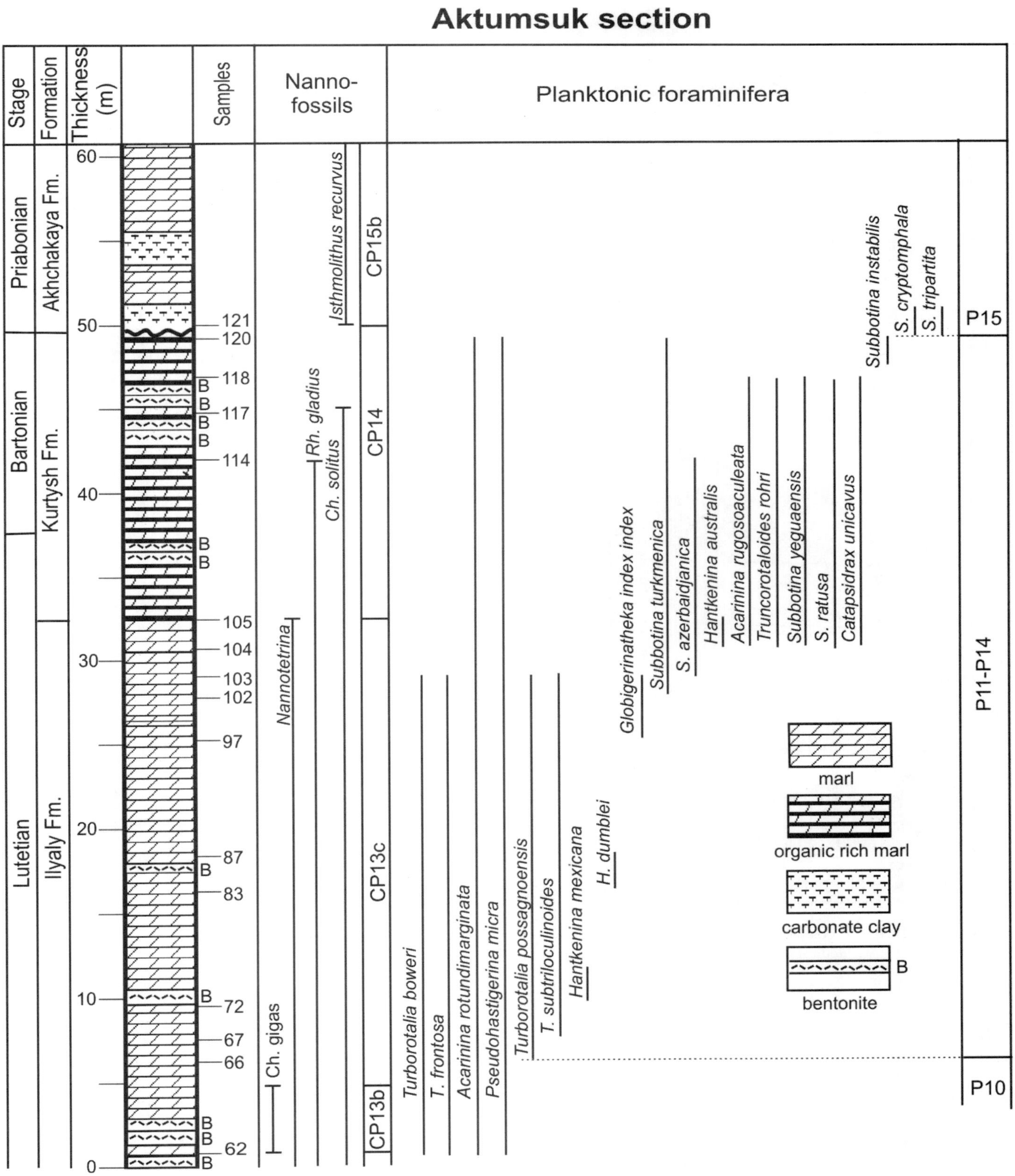

Figure 5. Distribution of planktonic foraminifera and most important datum levels of calcareous nannofossils in the upper middle and lower upper Eocene of the Aktumsuk Section (western shore of the Aral Sea).

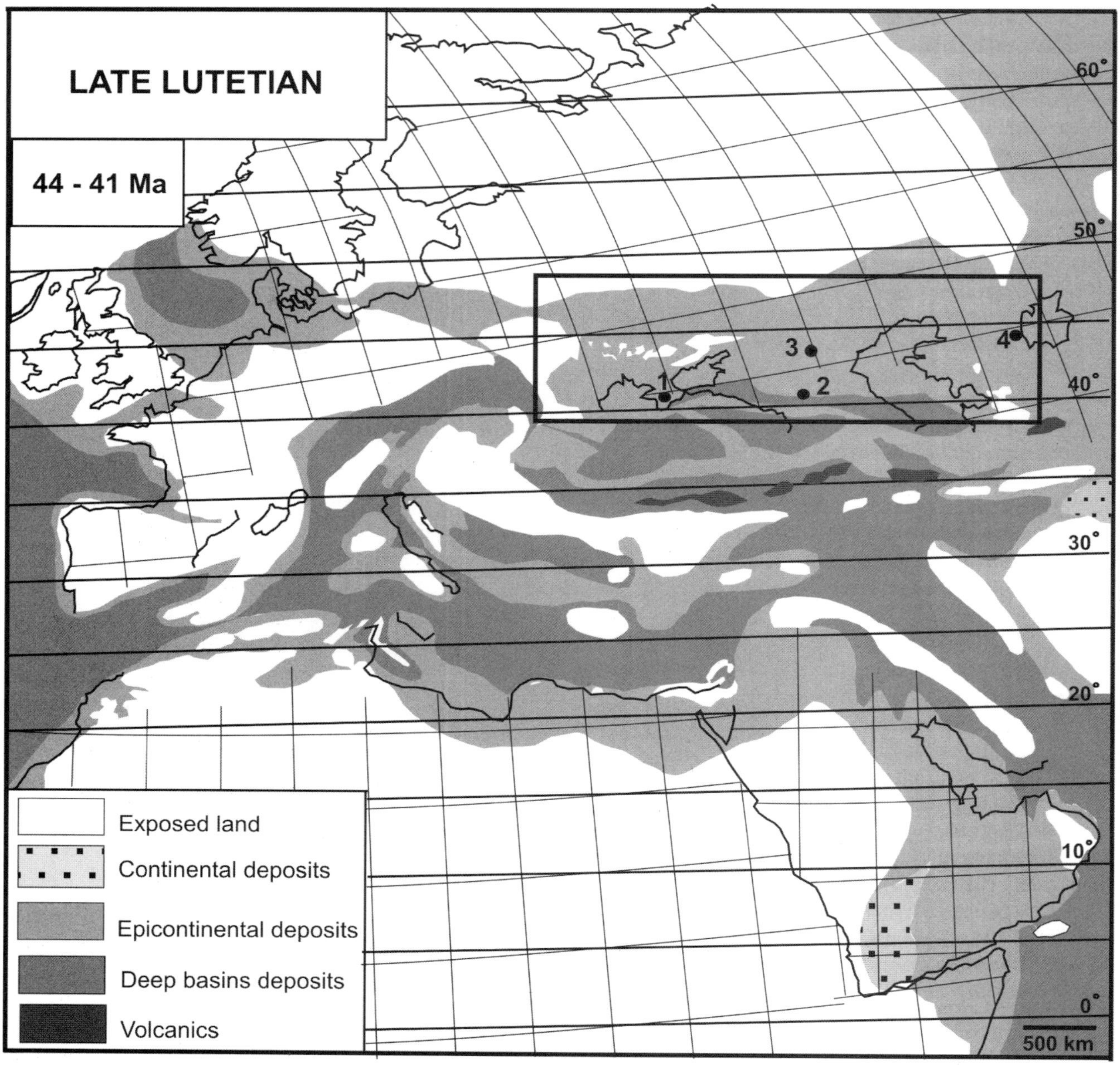

Figure 6. The paleo-position of the studied sections shown in a simplified paleogeographic map of the Peri-Tethyan area, based on the map for the late Lutetian in the Peri-Tethys Atlas (Meulenkamp et al., 2000).

Planktonic foraminifera

An independent planktonic foraminiferal zonation for the southern part of the former USSR (Crimea-Caucasus Province) was proposed by Subbotina (1953) and Shutskaya (1970), and later modified and correlated to the nannoplankton zonation by Krasheninnikov and Muzylöv (1975). Only a few biostratigraphic events can be used to correlate the planktonic foraminiferal zonation for the Crimea-Caucasus Province with standard oceanic zones. The zonal scheme (including subzones) was recently modified by Beniamovski (2001) and Beniamovski et al. (2001), based on a detailed study of planktonic foraminifera in the Bakhchisaray and Kheu River sections and other sections in the northeastern Peri-Tethys. These authors documented the succession of such taxa as *Turborotalia, Subbotina, Hantkenina,* and *Globigerinatheka* in the middle Eocene (Figs. 7 and 8).

In this paper we do not use this recently proposed zonation because it has only regional importance with few possibilities for global correlation. We use several faunal events in the plank-

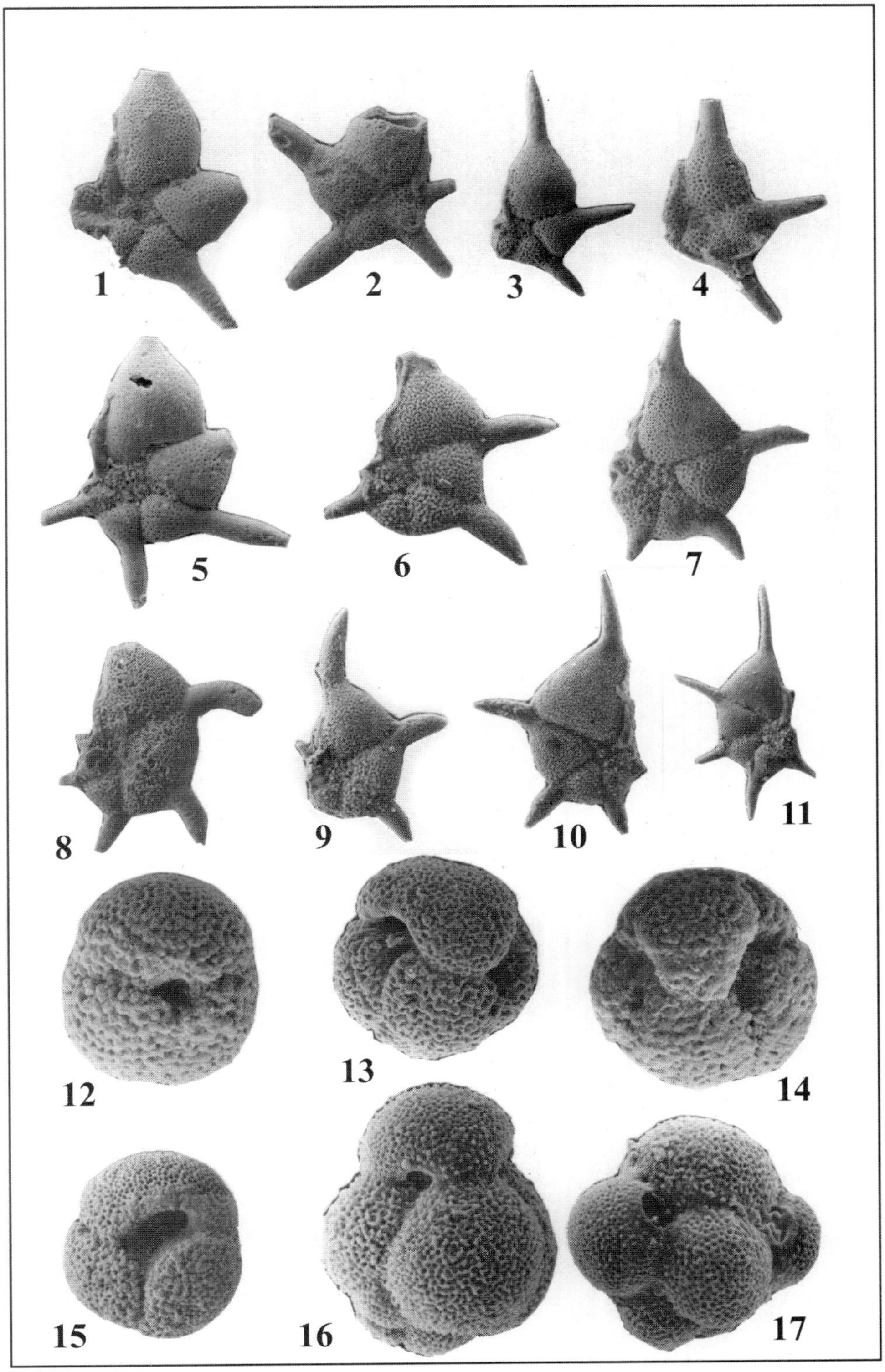

Figure 7. Planktonic foraminifera from the middle Eocene of the Crimea (Bakhchisaray Section), Pre-Caucasus (Keresta Section) and western coast of the Aral sea (Aktumsuk Section). 1–4—*Hantkenina mexicana* Cushman: Bakhchisaray: 1—sample 9 (× 70), 2—sample 34 (× 70), 3—sample 19 (× 70), 4—9 (× 70); 5—*Hantkenina liebusi* Shokhina, sample 19 (× 100); 6—*Hantkenina alabamensis compressa* Parr, sample 43 (× 70); 7—10 *Hantkenina australis* Finlay: 7—Bakhchisaray, sample 43 (× 70), 8—Bakhchisaray, sample 43 (× 70), 9—Bakhchisaray, sample 43 (× 70), 10—Aktumsuk, sample 104 (× 70); 11—*Hantkenina dumblei* Weinzierl and Applin, Bakhchisaray, sample 43 (× 80); 12—*Globigerinatheka micra* (Shutskaya), sample 1 (× 120); 13–14 *Globigerinatheka subconglobata* (Shutskaya) transition to *Globigerinatheka index* (Finlay), 13—Keresta, sample 38 (× 80), 14—Bakhchisaray, sample 19 (× 100), 15—*Globigerinatheka index* (Finlay), Bakhchisaray, sample 38 (× 100), 16—*Catapsydrax unicavus* Bolli, Loeblich and Tappan, Aktumsuk, sample 104(× 130), 17—*Catapsydrax dissimilis* (Cushman and Bermudes), sample 104 (× 100).

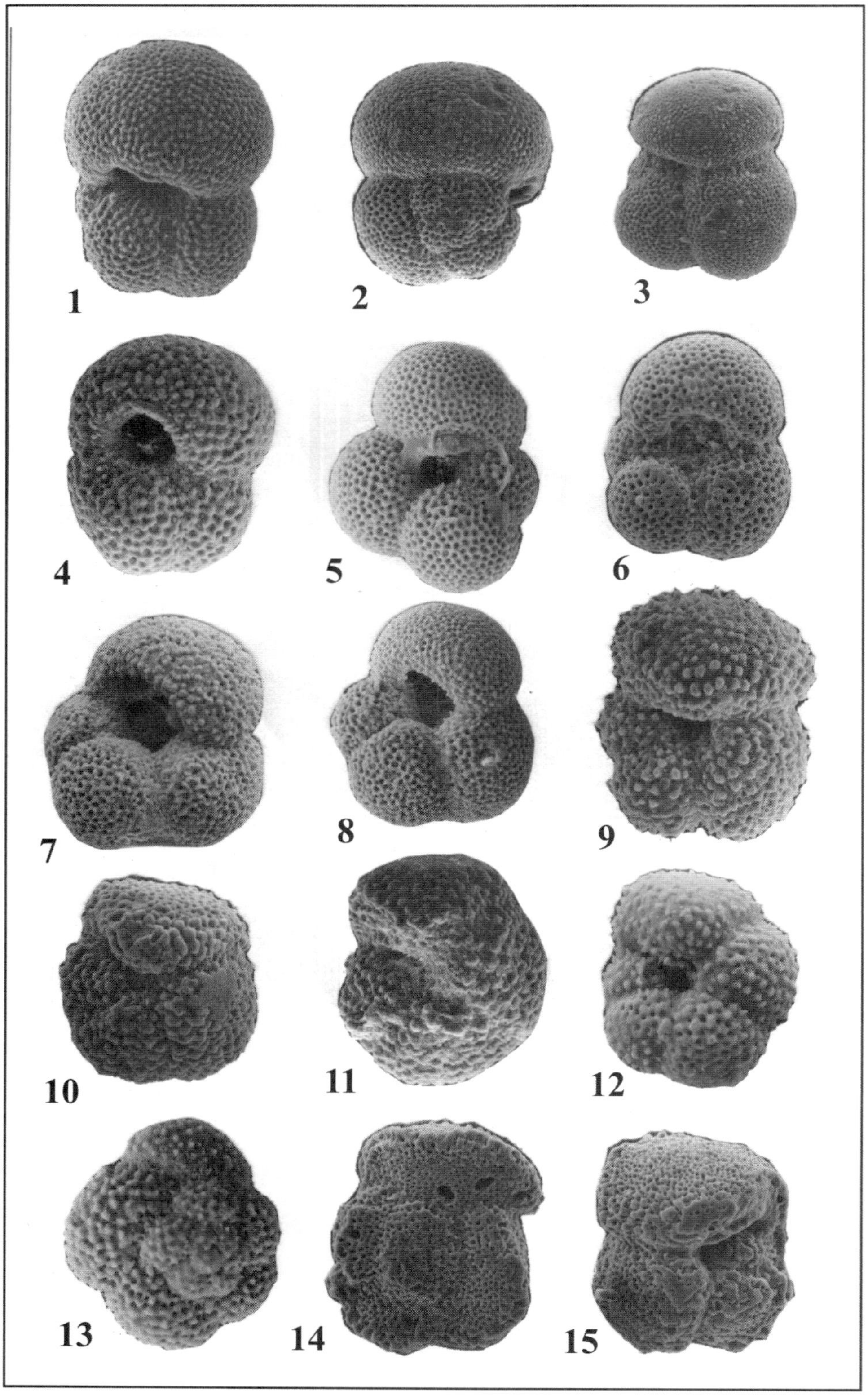

Figure 8. Planktonic foraminifera from the middle Eocene of the Crimea (Bakhchisaray Section) and Pre-Caucasus (Keresta Section). 1, 2—*Turborotalia boweri* (Bolli), Keresta, sample 36 (× 100); 3—*Turborotalia possagnoensis* (Toumarkine and Bolli), Keresta, sample 36 (× 80); 4—*Turborotalia frontosa* (Subbotina), Bakhchisaray, sample 1 (× 150); 5—*Subbotina turcmenica* (Khalilov), Bakhchisaray, sample 48 (× 130); 6—*Subbotina azerbaidjanica* (Khalilov), Keresta, sample 20 (× 120); 7—*Subbotina yeguaensis* (Weinzierl and Applin), Bakhchisaray, sample 48 (× 130); 8—*Subbotina turcmenica* (Khalilov), Keresta, sample 20 (× 120); 9—*Acarinina triplex* Subbotina, Keresta, sample 44 (× 130); 10—*Acarinina bullbrooki* Bolli, Keresta, sample 40 (× 130); 11—*Acarinina rotundimarginata* Subbotina, Bakhchisaray, sample 98/1 (× 130); 12, 13—*Acarinina rugosoaculeata* Subbotina, Bakhchisaray, sample 48 (× 180); 14, 15—*Truncorotaloides topilensis* (Cushman), Keresta, sample 30 (× 100).

tonic foraminiferal succession in order to correlate the formations in the northeastern Peri-Tethys to the zonation of Berggren et al. (1995) (Fig. 9).

In the Bakhchisaray section (Fig. 2), the P10/P11 boundary could be drawn at about the level of sample 10 in the Keresta Formation, at the first occurrence of *Turborotalia possagnoensis*. According to Berggren et al. (1995), this species has its earliest appearance just below Zone P11. *Globigerinatheka micra* has its last appearance in the earliest part of Zone P11 (Bolli, 1972). This species has its highest occurrence in sample 11 in the Bakhchisaray section.

There is a sharp faunal break at the top of the Keresta Formation: all species of *Globigerinatheka* disappear in samples 40 and 41. *Turborotalia frontosa* and *T. boweri* have their highest appearance in sample 42, and *Acarinina bullbrooki* in sample 43. Five species (*Hantkenina alabamensis compressa, H. australis, Truncorotaloides rohri, Acarinina rugosoaculeata* and *Subbotina yeguaensis*) have their lowest appearance in sample 43.

This sharp break may have been caused by change from one to another part of the composite section between samples 42 and 43. The two parts of the section are ~600 m apart, and the area between the sections is not exposed, so that we might have missed several meters of section because of uncertainty in the correlation. This hypothesis, however, is contradicted by data in an unpublished manuscript by Grossheim (1974), describing the planktonic foraminiferal succession in two Bakhchisaray reference boreholes. One of these was drilled in a field close to the quarry and the second ~15 km to the northeast, close to Kochergino Village. In both cores, the highest occurrence of *Globigerinatheka micra* is 20–21 m below the base of the Kuma Formation, which is very similar to its location in the Bakhchisaray composite section. We thus conclude that the sharp faunal break is probably not caused by the break in the section between samples 42 and 43.

The break could also be caused by the presence of an unconformity between the Keresta and Kuma formations. The contact between the light-colored Keresta marls and the brown or coffee-colored marls of the Kuma Formation is abrupt, but there are no signs of an unconformity. In addition, all the studied sections show the same rapid faunal change. We thus consider it most probable that the rapid faunal event was caused by a rapid environmental shift, i.e., a change from a subtropical Keresta basin to a relatively cold, anoxic Kuma basin (see below).

The LOs of *Turborotalia frontosa* and *T. boweri* coincide with the top of Zone P12 (e.g., Toumarkine and Luterbacher, 1985; Premoli Silva and Boersma, 1988). In Italy (Umbria), the species have a coeval LO with *Morozovella aragonensis* at the top of Zone P11 (Nocchi et al., 1988). In the Bakhchisaray section, however, the use of this criterion for the top of Zone P11 would place the base of nannofossil Subzone CP13c in the upper part of zone P11 (sample 43), in disagreement with Berggren et al. (1995). We think that this discrepancy could be caused by an unusually late LO of *Chiasmolithus gigas* in the Peri-Tethys. This species is a cold-water taxon, and thus might have an extended range in boreal regions. We cannot, however, exclude reworking as cause for the extension of nannofossil ranges, because redeposited Cretaceous taxa (*Micula decussata* and *Eiffellithus turiseiffelii*) occur in the lowermost (sample 55a) and upper (sample 95) parts of the Kuma Formation (Ovechkina, 1999).

The Bakhchisaray section shows the typical succession of *Hantkenina* species, from the more "primitive" *H. mexicana, H. liebusi* and *H. dumblei* to the more "progressive" *H. alabamensis compressa* and *H. australis*. Their ranges, however, are very short relative to their ranges in oceanic regions (e.g., Coxall et al., 2000). We thus cannot use their FOs and LOs for correlation, and we cannot place the P11/P12 Zonal boundary in this section. The Belaya Glina Formation is placed in Zone P15, suggesting that in the Bakhchisaray section the whole Kuma Formation could possibly be correlated to the uppermost (?) part of oceanic standard Zone P11, Zones P12–14, and lowermost P15.

In the Kheu River section (Fig. 3), we recognized the same sequence of foraminiferal events as in the Crimea. The FO of *Turborotalia possagnoensis* was used as marker of the P10/P11 zonal boundary. All *Turborotalia* species had their LO at the same level as that of *Globigerinatheka subconglobata* and *G. index,* at the top of the Keresta Formation. The planktonic fauna characterized by *Truncorotaloides rohri, Acarinina rugosoaculeata, Subbotina turcmenica,* and associated forms occur first in the lowermost Kuma Formation.

The Keresta section (Fig. 4) is more condensed than the two sections described above. Sediments from the Keresta Formation, with an erosional lower boundary, are placed in Zone P11: *Turborotalia possagnoensis* occurs from the bottom of the formation. The sharp faunal change, however, at the LO of all *Turborotalia* species, occurs within the uppermost Keresta Formation, ~1.5 m below its upper boundary. *Truncorotaloides rohri* and *Acarinina rugosoaculeata* also have their LO in this interval. The Solonka Formation, which is time equivalent to most of the Kuma Formation, contains only few planktonic foraminifera, and has a relatively rich assemblage (*Subbotina turcmenica, S. yeguaensis, Acarinina bullbrooki* and others) only in its lowermost part. Higher in the section radiolarians are the predominant component of the microfauna, because of the shallow setting and the low temperatures.

The Aktumsuk Section (Fig. 5), situated far to the northeast, is very similar to the Keresta Section. Most of the Ilyaly Formation (from sample 66 upward) is placed in Zone P11, and the most important faunal break is in the uppermost part of this formation (sample 103). The Kurtysh Formation, representing sedimentation in anoxic environments in the Aral Sea region, accumulated in Zones P12–P14.

PALEOENVIRONMENTS

We used quantitative analysis of calcareous nannoplankton (this paper; Shcherbinina, 2000) and planktonic and benthic foraminifera in the Bakhchisaray and Kheu sections to identify paleoenvironmental changes in the Kuma Basin.

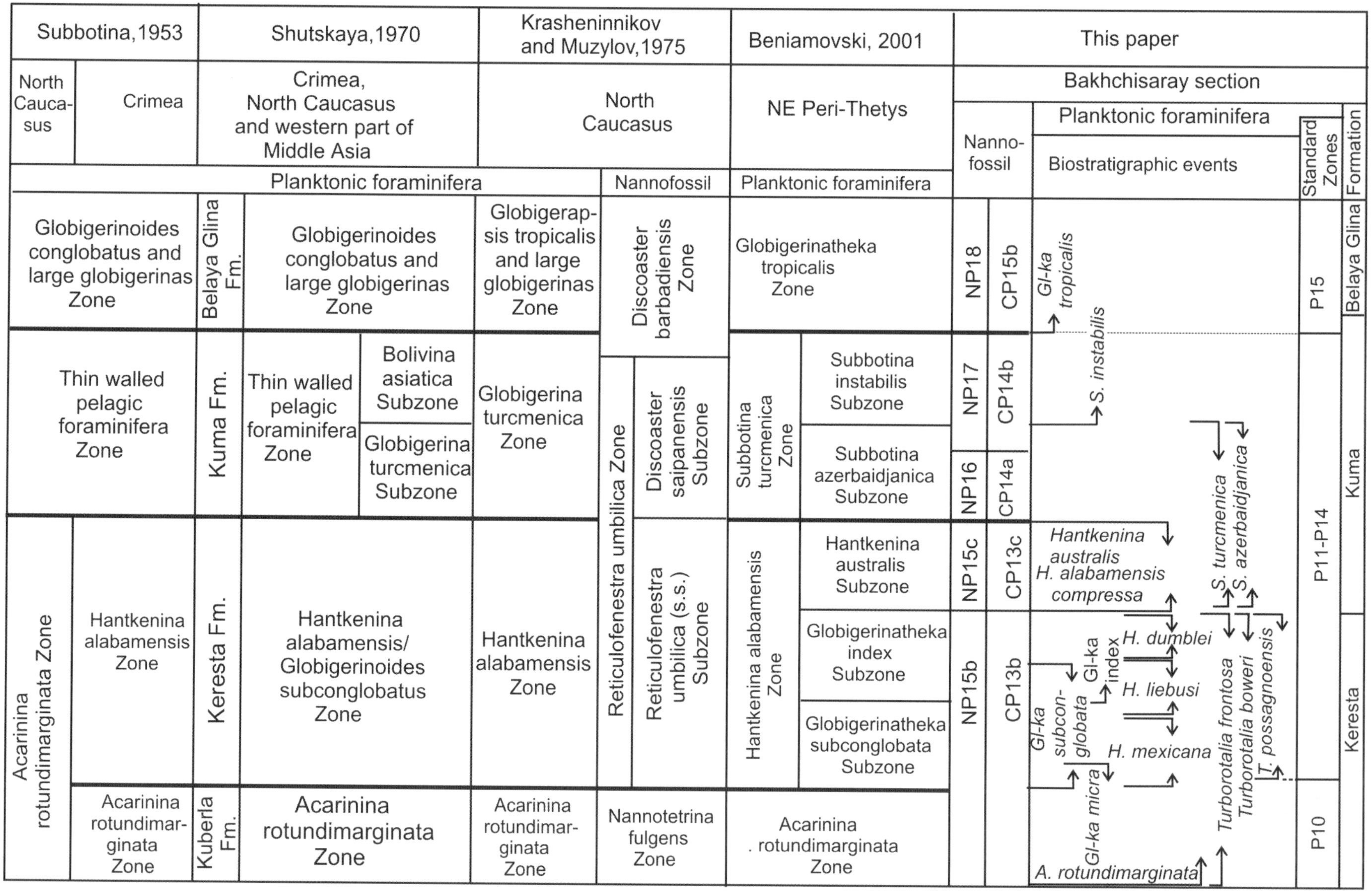

Figure 9. Correlation of planktonic foraminiferal and calcareous nannofossil zonations in the middle and upper Eocene of the northeastern Peri-Tethys.

Planktonic foraminifera

In the Kuma Formation, planktonic foraminifers are abundant but have a low species diversity. The few species mainly belong to the genus *Subbotina,* although *Pseudohastigerina* commonly form >30% of the assemblage, and biserial and triserial taxa are common at a few levels. *Hantkenina australis* and *H. alabamensis compressa* are abundant at the base of the Kuma Formation. Both taxa are cold-water surface dwellers, and characteristic for areas with high productivity and possibly relatively low salinity. Biserial taxa prefer a poorly oxygenated environment (Boersma et al., 1987; Premoli Silva and Boersma, 1988), and the hantkeninids inhabited intermediate water-depths within the oxygen minimum level (Boersma et al., 1987). In most samples the planktonic/benthic ratio is >90%, but this decreases drastically when the organic carbon content increases up to 8%.

Benthic foraminifera

The species richness of benthic foraminifera in 75 samples from the Kuma Formation in the Bakhchisaray section varies between 13 and 20 species (Fig. 10), while the mean "coefficient of specialization" (defined as the relative abundance of the three most abundant species) is high, with values >80% in many samples (Nezlina, 1996). The most important taxa are bolivinids, which constitute >80% of the total fauna in most samples, except for the lowest and the uppermost part of the section. Common species are *Brizalina karakemerica, Br. jacksonensis, Br. budugensis, Br. tuberna* and *Bolivina crenulata.*

The low diversity, high dominance, and abundance of bolivinid species all indicate deposition in dysoxic conditions (e.g., Kaiho, 1994). Benthic foraminiferal species indicative of dysoxic conditions are abundant throughout the section, while

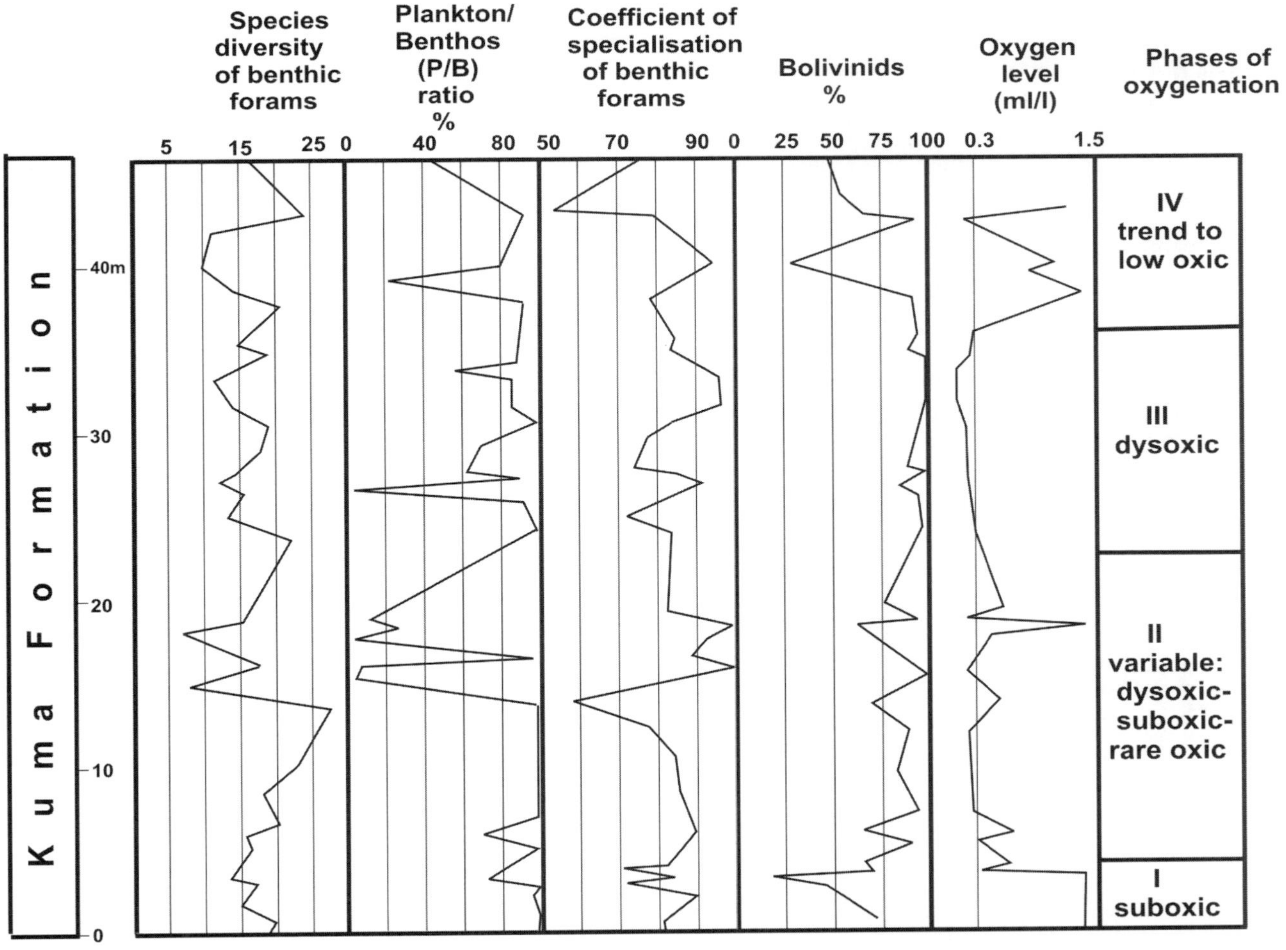

Figure 10. Main indices of benthic foraminiferal assemblages and phases of bottom water oxygenation in the Kuma Basin (Bakhchisaray Section).

suboxic species (Kaiho, 1994) are restricted to the lowest and uppermost samples. The oxygen index (OI, calculated from the benthic species assemblage according to Kaiho, 1994) varies from 0 to +2 in the lower part of the Kuma Formation, decreases sharply to –40 to –50 in the middle part, and increases again in the uppermost samples, where *Baggina iphigenia* is an important component. We can distinguish 4 phases in the deposition of the Kuma Formation (Fig. 10), using the oxygen index to estimate the oxygen content in bottom waters (Kaiho, 1994).

During deposition of the lowermost part of the Kuma Formation (Phase I, Figure 10) the bottom waters in the basin were suboxic to moderately oxygenated (1.5 mL/L). The oxygen content decreased to <0.5 ml/L (severely dysoxic) during the deposition of most of the Kuma Formation (Phases II and III, Figure 10). When the oxygen levels started to decline (Phase I/Phase II boundary), the foraminiferal number increased to a maximum, suggesting that productivity was very high. During Phase II, oxygenation fluctuated from suboxic to dysoxic or possibly even anoxic (~0.3–1.5 mL/L), as indicated by the occurrence of abundant *Brizalina* species with different morphology, and the absence of benthic foraminifera in a few samples. During Phase III, conditions were more uniformly dysoxic and the occurrence of *Cibicidoides* aff. *perlucidus* indicates a slight warming of bottom waters. During the last several hundred thousand years of Kuma Basin history (Phase IV, Figure 10), the oxygen content increased to suboxic levels (~1.5 mL/L), *Br. jacksonensis* disappeared and was replaced by the sculptured *Br. costifera* and *Br. antegressa,* both indicators of more oxygenated conditions.

In the Kheu section the species richness of benthic foraminifera is between 24 and 26 (coefficient of specialization 25%–50%) in the upper part of the Keresta Formation, where benthic assemblages are dominated by indicators of oxic conditions such as *Heterolepa eocaena* and *Brotzenella granosa,* so that the samples are characterized by a high oxygen index (66–97). In the Kuma Formation, the species diversity of benthic foraminifera drastically decreases to 13 and then down to 0, and the coefficient of specialization increases to 56%–100%. Only a few taxa, mainly nodosariids and bolivinids (*Nodosaria* ex gr. *elegans, N. ewaldi, Brizalina karakemerica, B. jacksonensis*), are abundant. These taxa are classified as suboxic and dysoxic indicators by Kaiho (1994). In the lowest part of the Kuma Formation the suboxic nodosariids dominated (66%–77%), and the oxygen index decreased first to about –4, then to –15. Above the lower part of the Kuma Formation the environment of deposition was very variable: in some samples suboxic nodosariids are dominant, in others dysoxic bolivinids (up to 100%), with fluctuations in the oxygen index from –36 to –44. In the middle and upper parts of the Kuma Formation benthic foraminifera are absent, suggesting anoxic conditions. In the overlying Belaya Glina Formation, consisting of light gray marls, benthic assemblages are rich, with a species richness of ~41 (coefficient of specialization 27%). In these samples benthic foraminiferal assemblages consist of a mixture of oxic, suboxic and dysoxic species, including *Baggina iphigenia, Melonis* cf. *affinis, Robulus diaphanus,* and *Latibolivina reticulosa,* so that the oxygen index is between 22 and 66. During deposition of the lower part of the Kuma Formation in the Kheu Section the bottom waters were suboxic (~1 mL/L of dissolved oxygen), but they later became dysoxic to anoxic (<0.5 mL/L). In the Belaya Glina Basin the bottom waters were well oxygenated (2.5 mL/L).

Calcareous nannofossils

Changes in the relative abundances of warm water taxa such as *Discoaster* and *Sphenolithus* on the one hand, and cool water taxa such as *Chiasmolithus* species and *Reticulofenestra* on the other hand, indicate changes in sea surface temperature (e.g., Muzylöv, 1980; Pospichal and Wise, 1990; Wei and Wise, 1990; Madile and Monechi, 1991). The environmental preferences of many taxa were discussed in detail by Dmitrenko (1993), using a biogeographic map of the distribution of the most important Eocene nannofossil species. According to this author, all species of discoaster and most species of *Sphenolithus* and *Helicosphaera* are warm-water taxa, whereas different species of *Chiasmolithus* differ in their environmental preferences, with *Ch. grandis* a tropical species, *Ch. bidens* subtropical, *Ch. altus* and *Ch. oamaruensis* temperate. *Reticulofenestra* is mainly subtropical to temperate.

In all sections there is a series of changes in the nannofossil assemblages in the lowermost Kuma Formation (Shcherbina, 2000). This change occurs not exactly at the base of the Kuma Formation in the Bakhchisaray section (Fig. 11), but close to the CP13c/CP14 Zonal boundary. At this level, *Discoaster* species sharply decrease in relative abundance to a few percent, and *Reticulofenestra* species increase in abundance to a maximum of 60%–70%. Species of *Chiasmolithus* become more abundant than those of *Discoaster.* These assemblages dominated by cool-water species are typical for most of the Kuma Formation. The transition from the Kuma into the overlying Belaya Glina formation coincides with slight warming in the Bakhchisaray Section.

Organic carbon accumulation

High organic carbon contents are typical for the Kuma Formation in all sections (Zhabrev et al., 1966; Eisenstadt et al., 1967; Gavrilov, 2000). In the Bakhchisaray section, the basal part of the Kuma Formation (coffee marls) has a low C_{org} content (~0.2%), but the organic carbon content sharply increases up to a maximum of 8% in the cyclic lower interval (Fig. 12). In the upper part of the Kuma Formation C_{org} values are somewhat lower again (2%–4%). The distribution of C_{org} is similar in the Kheu section (Gavrilov et al., 2000). The mean C_{org} content of the complete Kuma Formation is ~3%.

A rough estimate, using this value and the average extent (500,000 km^2) and a low estimate of average thickness (20 m) of the Kuma Formation, indicates that a huge amount of carbon (~6×10^{16} g of C_{org}) is stored in this formation. The average C_{org}

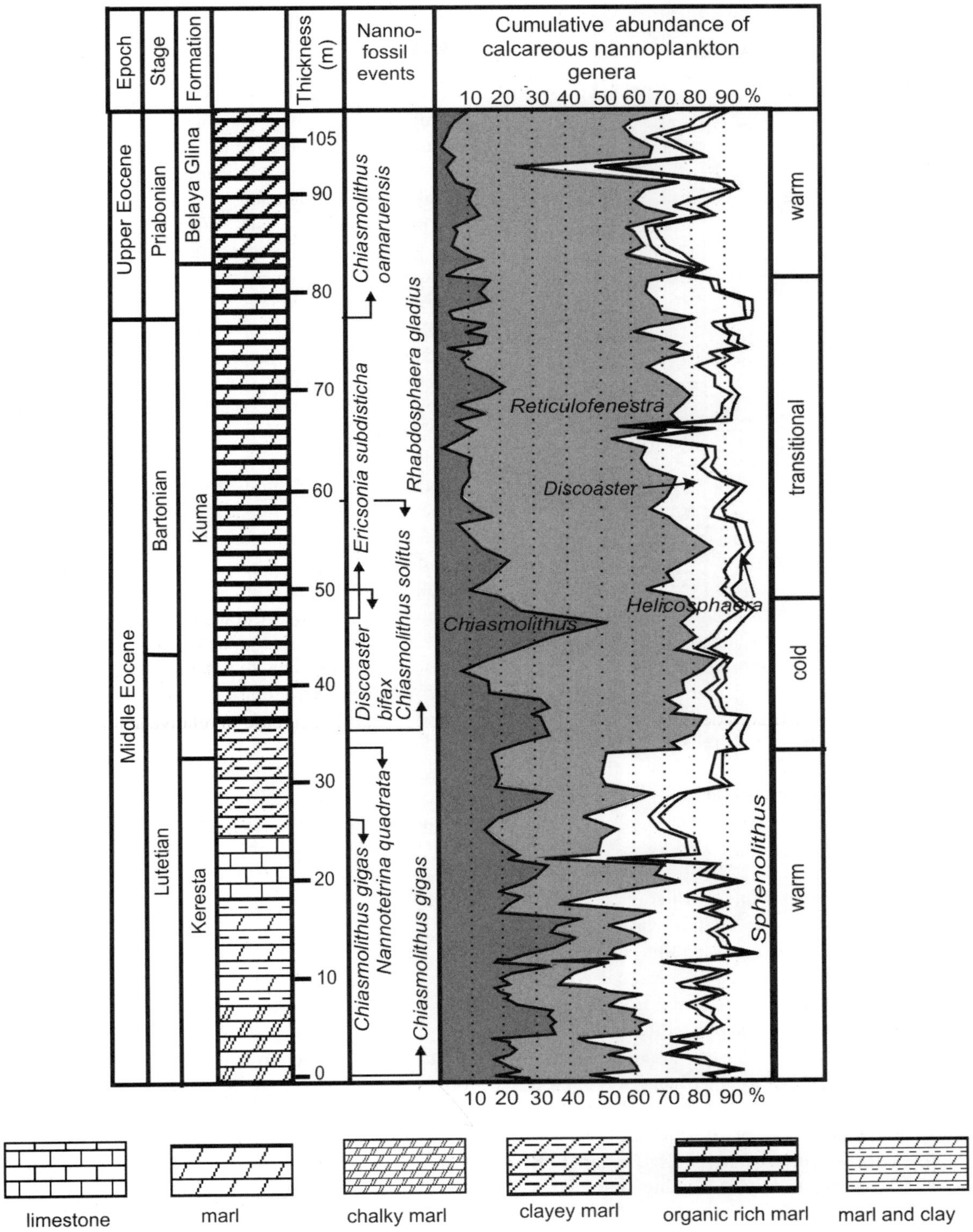

Figure 11. Quantitative analysis of calcareous nannofossils in the upper middle through lower upper Eocene of the Bakhchisaray Section.

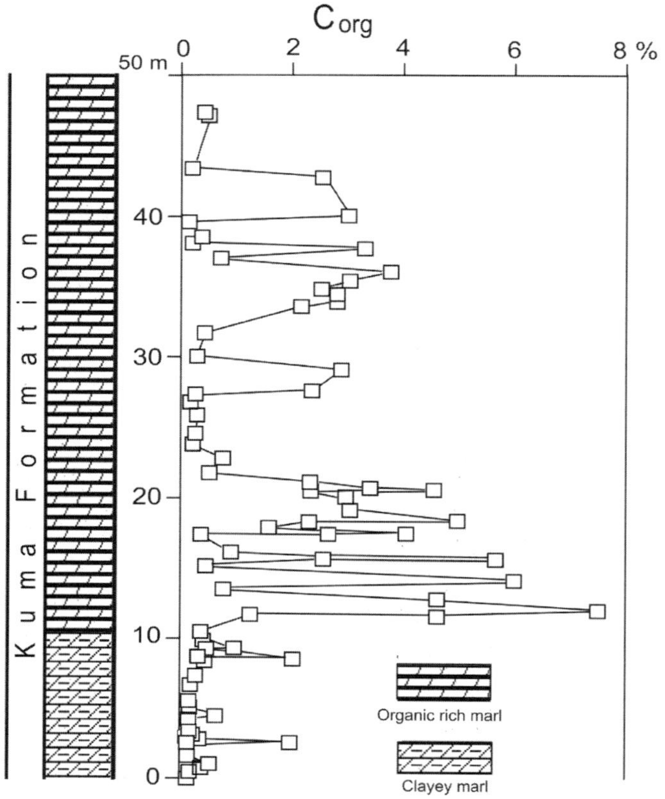

Figure 12. Stratigraphic distribution of organic carbon in the Kuma Formation of the Bakhchisaray section.

accumulation rate during the 7 m.y. of persistent dysoxic-anoxic conditions was ~1×10^{15} g per 100 k.y., but in the early Bartonian (CP14a) it was 2–3 times as high.

Volcanism

Numerous bentonite beds occur in all studied Kuma sections except the Keresta section. In the northeastern part of the Kuma Basin (Mangyshlak) there are also thin, crystal-bearing volcanic ash layers (A. Alekseev, 2002, personal commun.). This volcanic material was probably derived from the subduction- and rift-related Eocene volcanic belt, which stretched from the eastern Pontides in Turkey to eastern Iran (Kazmin et al., 1986; Koronovsky et al., 1997; Okay and Mahintürk, 1997; Nikishin et al., 2001). This volcanic activity reached its maximum intensity during the middle Eocene (Bartonian), and ash was transported to the north by monsoonal winds. The input of volcanic material could have contributed to the high productivity in the Kuma Basin (Muzylöv, 1996).

DISCUSSION

A vast dysoxic to anoxic basin (the Kuma Basin) became established in northeastern Peri-Tethys during the latest Lutetian and persisted through the Bartonian and into the earliest Priabonian. In the easternmost and northernmost parts of the basin highly dysoxic to anoxic conditions were limited to the early Bartonian. The anoxia in this basin probably had a complex origin. Tectonic events in the Caucasus fold belts narrowed the straits connecting the basin to western Siberian and northern European basins, thus making water circulation less vigorous, so that stratification could occur. The fast-growing middle Eocene volcanic belt (Nikishin et al., 2001) added volcanic material that may have contributed to increased surface productivity. In addition, the onset of dysoxic and anoxic conditions coincided with cooling (persisting into the Priabonian), as seen from the changes in planktonic foraminiferal and calcareous nannofossil assemblages. In the Crimea, cool and humid climates during the deposition of the Kuma Formation are also documented in palynological data (Panova, 1978).

In open ocean environments in the Southern Hemisphere (Kerguelen Plateau) the oldest ice-rafted material was reported from the middle Eocene, topmost CP13 and CP14a (Barron et al., 1991a, 1991b), and more intensive glaciation occurred from the end of the latest middle Eocene into the beginning of the late Eocene. The Kuma anoxic basin in the Crimea became established during latest nannofossil Zone CP13, approximately at the onset of glaciation in the Antarctica. At this same time, planktonic foraminiferal assemblage and isotope data indicate an increasing input of run off in the Kuma Basin (Beniamovski et al., 2001) with the continental run off probably enriched in nutrients. This interpretation is in agreement with the occurrence of low-diversity dinocyst assemblages dominated by the genera *Horologinella* and *Paucilobimorpha* (Zaporozhets, 2001), as also described from the Asse Clays in South Holland (de Coninck, 1986). The larger volume of runoff might have enhanced stratification of the water column as well as high productivity, which both would have contributed to the establishment of highly dysoxic to anoxic conditions.

CONCLUSIONS

Our observations document that the onset of dysoxic and anoxic conditions as indicated by the sedimentological and micropaleontological features of the Kuma Formation (and its lateral facies equivalents) are close to synchronous throughout the Kuma Basin. In the Crimea (Bakhchisaray Section), the oldest organic-rich deposits are uppermost Lutetian (ca. 44 Ma, Berggren et al., 1995). Persistently oxic conditions were reestablished in the early Priabonian (~36.5 Ma, Berggren et al., 1995). In the northern Caucasus, these events occurred at the same time as in the Crimea. In the Aktumsuk Section (close to the Aral Sea), dysoxia started at the same time as in the Crimea and the Caucasus, but the timing of the reestablishment of oxic conditions is not known because the top of the Kurtysh Formation (Kuma facies equivalent) is truncated. In the northern part of the Kuma Basin (Keresta Section), dysoxic-anoxic conditions were present only during the early Bartonian (~41.3–37.0 Ma, Berggren et al., 1995).

Eocene calcareous nannoplankton assemblages in the northeastern Peri-Tethys are characteristic for midlatitudes, and the relative abundance of warm water *Discoaster* species decreased rapidly during the transition of the Keresta into the Kuma Formation, at the same place in the section where the planktonic foraminiferal faunas also indicate a rapid cooling. Most of the Kuma Formation accumulated during a relatively cool period. Regional cooling in the Peri-Tethys region thus started rapidly in the later part of Zone P11, at the CP13c/CP14 zonal boundary (ca. 44 Ma, Berggren et al., 1995), synchronous with the global cooling trend (e.g., Zachos et al., 2001). During the deposition of the Kuma Formation (~44–36.5 Ma) an estimated 6×10^{16} g C_{org} was stored in the dysoxic-anoxic sediments. The carbon isotope curve in Zachos et al. (2001) shows an excursion of ~0.5‰ to heavier values at ca. 40 Ma, about the midpoint of the deposition of the Kuma Formation. We speculate that the carbon storage during the latest Lutetian through Bartonian may have been at least one of the causes of lowered atmospheric levels of CO_2 and global cooling that led to the establishment of the East Antarctic ice sheet in the earliest Oligocene.

ACKNOWLEDGMENTS

The authors are greatly indebted to V.A. Musatov and N.G. Muzylöv who provided the samples from the Keresta Section. We thank Ekaterina A. Shcherbinina for fruitful discussions on the nannofossil distribution in the studied sections, and E. Thomas for improving the manuscript. We also thank the reviewers K. von Salis, L. Bybell and T. Wagner for many corrections and suggestions. The studies were supported financially by grants of INTAS 93-2509, RFBR 00-05-64917 and 00-05-64593, Russian Federal Program "Integration."

REFERENCES CITED

Barron, J.A., Baldauf, J.G., Barrera, E., Caulet, J.-P., Huber, B.T., Keating, B.H., Lazarus, D., Sakai, H., Thierstein, H.R., and Wei, W., 1991a, Biochronological and magnetochronologic synthesis of Leg 119 sediments from Kerguelen Plateau and Prydz Bay, Antarctica, *in* Barron, J., et al., eds., Proceedings of the Ocean Drilling Program, Scientific Results: Texas A & M University, College Station, Texas, USA, v. 119, p. 813–847.

Barron, J.A., Larsen, B., and Baldauf, J.G., 1991b, Evidence for late Eocene to early Oligocene Antarctic glaciation and observations on late Neogene glacial history of Antarctica: Results from Leg 119, *in* Barron, J., et al., eds., Proceedings of the Ocean Drilling Program, Scientific Results: Texas A & M University, College Station, Texas, USA, v. 119, p. 869–891.

Beniamovski, V.N., 2001, Detailed stratigraphic scheme for the lower Paleogene of the Crimea-Caucasus region, *in* Gladenkov, Yu.B., and Kuznetsova, K.I., eds., Toward detailed stratigraphic schemes and paleogeographic reconstructions: Moscow, Publishing House GEOS, p. 210–223 (in Russian).

Beniamovski, V.N., Shcherbinina, E.A., Alekseev, A.S., Ovechkina, M.N., Nezlina, M.Yu., and Oberhänsli, H., 2001, Biostratigraphy and environmental trends in the middle–late Eocene deposits of Crimea, *in* Ash, A., and Wing, S.L., eds., Climate and biota of the early Paleogene: International Meeting July 3–8, 2001, Northwest College, Powell, Wyoming, USA, Abstract volume, p. 12.

Berggren, W.A., Kent, D.V., Swisher, C.C., III, and Aubry, M.-P., 1995, A revised Cenozoic geochronology and chronostratigraphy, *in* Berggren, W.A., et al., eds., Geochronology, time scales, and global stratigraphic correlations: A unified temporal framework for an historical geology: SEPM (Society for Sedimentary Geology) Special Publication 54, p. 129–212.

Boersma, A., Premoli Silva, I., and Shackleton, N.J., 1987, Atlantic Eocene planktonic foraminiferal paleohydrographic indicators and stable isotope paleogeography: Paleoceanography, v. 2, p. 287–331.

Bolli, H.M., 1972, The genus *Globigerinatheka* Brönnimann: Journal of Foraminiferal Research, v. 2, p. 109–136.

Bugrova, E.M., Nikolaeva, I.A., Panova, L.A., and Tabachnikova, I.P., 1988, On the zonal subdivision of Paleogene of southern regions of USSR: Sovetskaya Geologiya, no. 4, p. 96–107 (in Russian).

Coxall, H.K., Pearson, P.N., Shackleton, N.J., and Hall, M., 2000, Hantkeninid depth adaptation: An evolving life strategy in a changing ocean: Geology, v. 28, p. 87–90.

de Coninck, J., 1986, Organic walled phytoplankton from the Bartonian and Eo-Oligocene transitional deposits of the Woensdrecht borehole, southern Netherlands: Mededelingen Rijks Geologische Dienst, no. 40–2, 49 p.

Dmitrenko, O.B., 1993, Cenozoic biogeography of Atlantic and Indian oceans on nannoplankton: Moscow, Nakua, 175 p. (in Russian).

Dubois de Montpéreux, F., 1837, Lettre sur les principaux phénomènes géologiues du Caucase et de la Crimée, addressée à M. Élie de Beaumont: Bulletin de la Société Géologique de France, v. 8, p. 371–388.

Eisenstadt, G.E.-A., Koltypin, S.N., Razmyslova, S.S., Sokolova, E.I., Torgovanova, V.B., Pinchuk, I.A., Pomarnatsky, M.A., and Semina, I.V., 1967, The oil-bearing formations of Peri-Caspian depression: Trudy Vsesoyuznogo neftyanogo nauchno-issledovatelskogo geologorazvedochnogo instituta (VNIGRI), v. 253, 321 p. (in Russian).

Gavrilov, Yu.O., 2000, Lithology and geochemistry of oxic and anoxic sedimentary environments in ancient basins (Eocene of Cis-Caucasus), *in* Proceedings, First All-Russian Lithological Meeting, Moscow, December 19–21, 2000: Moscow, Publishing House GEOS, p. 168–170 (in Russian).

Gavrilov, Yu.O., Shcherbinina, E.A., and Muzylöv, N.G., 2000, A Paleogene sequence in central North Caucasus: A response to paleoenvironmental changes: GFF, v. 122, p. 51–53.

Grossheim, V.A., editor, 1974, Bakhchisaray Paleogene stratotype section (its fauna and flora): Interdepartmental Stratigraphic Committee of USSR, Leningrad, unpublished report, 287 p. (in Russian).

Kaiho, K., 1994, Benthic foraminiferal dissolved-oxygen index and dissolved-oxygen levels in the modern ocean: Geology, v. 22, p. 719–722.

Kazmin, V.G., Sbortshikov, I.M., Ricou, L.-E., Zonenshain, L.P., Boulin, J., and Knipper, A.L., 1986, Volcanic belts as markers of the Mesozoic–Cenozoic active margins of Eurasia: Tectonophysics, v. 123, p. 123–152.

Khodzhakhmedov, K.A., 2001, Paleocene and Eocene zonal stratigraphy of Western Turkmenia on calcareous nannoplankton: Moscow, International Academy of Sciences on Nature and Society, 201 p. (in Russian).

Koronovsky, N.V., Lomize, M.G., Gushchin, A.I., Zarshchikov, A.A., Panina, L.V., and Romanovskaya, M.A., 1997, Main events in tectonic evolution of Caucasus segment of Mediterranean folded belt: Vestnik Moskovskogo Universiteta, Seriya 4, Geologiya, no. 4, p. 5–12 (in Russian).

Krasheninnikov, V.A., and Muzylöv, N.G., 1975, Correlation of planktonic foraminifera and nannoplankton zonations in Paleogene sections of Northern Caucasus: Voprosy Micropaleontologii, v. 18, p. 212–224 (in Russian).

Madile, M., and Monechi, S., 1991, Late Eocene to early Oligocene calcareous nannofossil assemblages from sites 699 and 703, Subantarctic South Atlantic Ocean, *in* Ciesielski, P.F., Kristoffersen, Y., et al., eds., Proceedings of the Ocean Drilling Program, Scientific Results: Texas A & M University, College Station, Texas, USA, v. 114, p. 179–192.

Martini, E., 1971, Standard Tertiary and Quaternary calcareous nannoplankton zonation, *in* Farinacci, A., ed., Proceedings of the 2nd Planktonic Conference, Rome, Italy, 1970: Tecnoscienza, v. 2, p. 739–785.

Menner, V.V., editor, 1971, XII European Micropaleontological Colloqium, Field trip guidebook, Part 1: Crimea: Moscow, Geological Institute of Academy of Sciences of USSR, 148 p. (in Russian).

Meulenkamp, J.E., Sissingh, W., Calvo, J.P., Daams, R., Studencka, B., Londeux, L., Cahuzac, B., Kovac, M., Nagymarosy, A., Rusu, A., Badescu, D., Beniamovski, V.N., Scherba, I.G., Roger, J., Platel, J.-P., Hirsch, F., Sadek, A., Abdel-Gawad, G.I., Ben Ismail-Lattrache, K., Zaghbib-Turki, D., Bouaziz, S., Karoui-Yaakoub, N., and Yaich, C., 2000, Late Lutetian, *in* Crasquin, S., coordinator, Atlas Peri-Tethys. Palaeogeographical maps. Explanatory Notes, *CCGM/CGMW:* Paris, p. 163–170.

Muzylöv, N.G., 1980, Stratigraphy of Paleogene of the South of USSR by nannoplankton (the North Caucasus and the Crimea): Trudy Geologicheskogo Instituta Akademii Nauk SSSR, v. 348, 94 p. (in Russian).

Muzylöv, N.G., 1996, Early Paleogene episodes of eruptive activity in the East Peri-Tethys and their possible influence on the marine microbiota, *in* Kuznetsova, K.I., and Muzylöv, N.G., eds., Iskopaemye mikroorganizmy kak osnova stratigrafii, korrelyatsii i paleobiogeografii fanerozoya: Trudy Geologicheskogo Instituta Rossiskoi Akademii Nauk, no. 501, p. 128–132.

Muzylöv, N.G., Kurgalimova, G.G., Abrosimov, P.N., and Khodzhakhmedov, K.A., 1990, Calcareous plankton of late Eocene–Early Oligocene deposits of Western Priaralie and North Turkmenia: Byulleten Moskovskogo Obshchestva Ispytatelei Prirody, Otdel Geologicheskii, v. 65, no. 3, p. 23–29 (in Russian).

Muzylöv, N.G., Stupin, S.I., and Musatov, V.A., 1999, Problem of Keresta Formation (Lutetian–Bartonian boundary interval in south of European Platform): Nedra Povolzhiya i Prikaspiya, no. 19, p. 11–16 (in Russian).

Nezlina, E.Yu., 1996, Benthic foraminfera of Kuma Formation (middle Eocene, Globigerina turcmenica Zone) of Bakhchisaray area [master's thesis]: Moscow State University, Geological Faculty, Department of Paleontology, 80 p. (in Russian).

Nikishin, A.M., Ziegler, P.A., Panov, D.I., Nazarevich, B.P., Brunet, M.-F., Stephenson, R.A., Bolotov, S.N., Korotaev, M.V., and Tikhomirov, P.L., 2001, Mesozoic and Cainozoic evolution of the Scythian Platform–Black Sea–Caucasus domain, *in* Ziegler, P.A., et al., eds., Peri-Tethys Memoir 6: Peri-Tethyan rift/wrench basins and passive margins: Mémoires du Muséum National d'Histoire Naturelle, v. 186, p. 295–346.

Nocchi, M., Parisi, G., Monaco, P., Monechi, S., and Madile, M., 1988, Eocene and Early Oligocene micropaleontology and paleoenvironments in southeast Umbria, Italy: Palaeogeography, Palaeoclimatology, Palaeoecology, v. 67, p. 181–244.

Okada, H., and Bukry, D., 1980, Supplementary modification and introduction of code numbers to the low-latitude coccolith biostratigraphic zonation (Bukry, 1973, 1975): Marine Micropaleontology, v. 5, p. 321–325.

Okay, A.I., and Mahintürk, Ö., 1997, Geology of the Eastern Pontides, *in* Roobinson, A.G., ed., Regional and petroleum geology of the Black Sea and surrounding region: American Association of Petroleum Geologists Memoir 68, p. 291–311.

Ovechkina, M.N., 1999, Calcareous nannoplakton of upper middle and upper Eocene of Bakhchisaray area of southwestern Crimea [master's thesis]: Moscow State University, Geological Faculty, Department of Paleontology, 130 p. (in Russian).

Panova, L.A., 1978, Distribution of sporae and pollen in Paleogene of Bakhchisaray stratotype section, *in* Nosovsky, M.F., et al., eds., Stratigrafiya kainozoya severnogo Prichernomoriya i Kryma: Dnepropetrovsky Gosudarstvennyi Universitet, Sbornik Nauchnykh Trudov, no. 2, p. 69–80 (in Russian).

Perch-Nielsen, K., 1985, Cenozoic calcareous nannofossils, *in* Bolli, H.M., et al., eds., Plankton stratigraphy: Cambridge, UK, Cambridge University Press, p. 423–554.

Pospichal, J.J., and Wise, S.W., Jr., 1990, Paleocene to middle Eocene calcareous nannofossils of ODP sites 689 and 690, Maud Rise, Weddell Sea, *in* Barker, P.F., and Kennett, J.P., et al., eds., Proceedings of the Ocean Drilling Program, Scientific Results: Texas A & M University, College Station, Texas, USA, v. 113, p. 613–638.

Premoli Silva, I., and Boersma, A., 1988, Atlantic Eocene planktonic foraminiferal historical biogeography and paleohydrographic indices: Palaeogeography, Palaeoclimatology, Palaeoecology, v. 67, p. 315–356.

Saperson, E.I., and Zheleznov, A.A., 1962, The scale of division of Paleocene and Eocene of North Turkmenia: Doklady Akademii Nauk SSSR, v. 144, no. 4, p. 893–895 (in Russian).

Shcherbinina, E.A., 2000, Middle Eocene nannofossils and geological events of the northeastern Peri-Tethys: GFF, v. 122, p. 143–145.

Shcherbinina, E.A., 2001, What controlled *Chiasmolithus* distribution?, *in* Ash, A., and Wing, S.L., eds., Climate and biota of the early Paleogene: International Meeting July 3–8, 2001, Northwest College, Powell, Wyoming, USA, Abstract volume, p. 82.

Shutskaya, E.K., 1970, Stratigraphy, foraminifera, and palaeogeography of the lower Paleogene of Crimea, Precaucasus, and western part of the Middle Asia: Trudy Vsesoyuznogo nauchno-issledovatel'skogo geologorazvedochnogo neftyanogo instituta (VNIGNI), v. 70, 256 p. (in Russian).

Subbotina, N.N., 1953, Late Eocene lagenids and buliminids of South of USSR: Trudy Vsesoyuznogo neftyanogo nauchno-issledovatel'skogo geologorazvedochnogo instituta (VNIGRI), v. 69, p. 115–282 (in Russian).

Toumarkine, M., and Luterbacher, H., 1985, Paleocene and Eocene planktic foraminifera, *in* Bolli, H.M., et al., eds., Plankton stratigraphy: Cambridge, UK, Cambridge University Press, p. 87–154.

Vassoevich, N.B., 1934, On the horizon with Lyrolepis caucasica Roman: Trudy Neftyanogo geologo-razvedochnogo instituta, Series B, no. 47, p. 1–23 (in Russian).

Wei, W., and Wise, S.W., Jr., 1990, Middle Eocene to Pleistocene calcareous nannofossils recovered by Ocean Drilling Program Leg 133 in the Weddell Sea, *in* Barker, P.F., Kennett, J.P., et al., eds., Proceedings of the Ocean Drilling Program, Scientific Results: Texas A & M University, College Station, Texas, USA, v. 113, p. 639–666.

Zachos, J., Pagani, M., Sloan, L., Thomas, E., and Billups, K., 2001, Trends, rhythms, and aberrations in global climate change 65 Ma to present: Science, v. 292, p. 686–693.

Zaporozhets, N.I., 2001, Palynological assemblages of middle Eocene Kuma Horizon of Caucasus and its age analogues in adjacent regions: Stratigraphy and Geological Correlation, v. 9, no. 6, p. 83–103 (in Russian).

Zhabrev, D.V., Larskaya, E.S., Nosov, G.I., Vladarskaya, V.R., and Lazareva, V.M., 1966, Oil-bearing formations of Western Cis-Caucasia, main regularities of its origin and perspectives of oil and gas bearing: Trudy Vsesoyuznogo nauchno-issledovatel' skogo geologorazvedochnogo neftyanogo instituta (VNIGNI), v. 55, 296 p. (in Russian).

Manuscript Accepted by the Society August 13, 2002

Printed in the USA

Geological Society of America
Special Paper 369
2003

Late Paleocene–early Eocene dinoflagellate cyst records from the Tethys: Further observations on the global distribution of Apectodinium

Erica M. Crouch*
Henk Brinkhuis
Henk Visscher
Laboratory of Palaeobotany and Palynology, Department of Geobiology, Utrecht University, Budapestlaan 4, 3584CD Utrecht, The Netherlands
Thierry Adatte
Marie-Pierre Bolle
Institut de Géologie, 11 Emile Argand, Université de Neuchâtel, 2007 Neuchâtel, Switzerland

ABSTRACT

To further understand the distribution and abundance of *Apectodinium* through space and time, dinoflagellate cyst (dinocyst) records from two well-calibrated Paleogene sections in the Tethys (Tunisia and Uzbekistan) have been examined. *Apectodinium* was present in the southern Tethys (Tunisia) by the late Paleocene (planktonic foraminiferal Subzone P4a), confirming earlier findings that the genus evolved in lower latitudes. *Apectodinium* abundance was frequent to dominant from the upper Paleocene to lower Eocene in the Tethys. The initial Eocene thermal maximum (IETM) at lower latitudes appears to be characterized by high percentages of *Apectodinium*, although it may not always have dominated assemblages. Other dinocyst genera with thermophilic preferences may be abundant during the IETM. At mid to high latitudes, *Apectodinium* has been sporadically recognized during the upper Paleocene. Nevertheless, the IETM at mid to high latitudes was still characterized by *Apectodinium*-dominated assemblages. Additional *Apectodinium* acmes occurred in the lower Eocene, although as yet it is unclear if these are globally isochronous. Intervals of *Apectodinium*-dominated assemblages are indicative of environmental conditions that were apparently markedly different to "normal background settings" of the late Paleocene–early Eocene. Sea surface temperature (SST) was probably the main control in distribution since *Apectodinium* acmes at mid to high latitudes seem to have occurred during intervals of highest SSTs, such as the IETM. In addition, the probable heterotrophic *Apectodinium* "blooms" were intricately linked to enhanced runoff and increased delivery of nutrients in surface waters, probably as a result of intensification of the weathering cycle and other possible specific oceanic conditions.

*Present address: Institute of Geological and Nuclear Sciences, P.O. Box 30-368, Lower Hutt, New Zealand; E.Crouch @gns.cri.nz.

Crouch, E.M., Brinkhuis, H., Visscher, H., Adatte, T., and Bolle, M.-P., 2003, Late Paleocene–early Eocene dinoflagellate cyst records from the Tethys: Further observations on the global distribution of *Apectodinium, in* Wing, S.L., Gingerich, P.D., Schmitz, B., and Thomas, E., eds., Causes and Consequences of Globally Warm Climates in the Early Paleogene: Boulder, Colorado, Geological Society of America Special Paper 369, p. 113–131.

INTRODUCTION

The late Paleocene to early Eocene (~57.4–53.4 Ma) was characterized by globally warm climates, shallow latitudinal gradients and prevalent ice-free conditions (e.g., Stott et al., 1990; Zachos et al., 1994; Berggren et al., 1998). Superimposed on this equable climate was a short-lived, ~220 k.y. (Röhl et al., 2000), intense warming event currently referred to as the initial Eocene thermal maximum or IETM. This interval corresponds to a globally recognized negative 2‰–3‰ carbon isotope excursion (CIE) (Bralower et al., 1995; Koch et al., 1995) and a brief warming of deep ocean and high-latitude surface temperatures by 6–8 °C (Kennett and Stott, 1991; Thomas and Shackleton, 1996; Katz et al., 1999). The CIE is indicative of a massive injection of ^{12}C-rich carbon into the ocean-atmosphere system, most probably from methane released from seafloor gas hydrates (Dickens, 2000, 2001). Associated with the IETM are numerous changes in biota, including a prominent benthic foraminiferal extinction event (Thomas, 1998), turnovers in planktonic organisms (Kelly et al., 1996; Aubry, 1998), and a fundamental diversification of terrestrial mammal orders (Clyde and Gingerich, 1998). Although the IETM is best dated as beginning between 54.93 and 54.98 Ma (Norris and Röhl, 1999), this age is currently difficult to correlate with other stratigraphic events associated with the Paleocene-Eocene (P-E) transition. Hence, we use the age of 55.5 Ma for the onset of the IETM (Berggren and Aubry, 1998).

Marine dinoflagellates, surface dwelling plankton, also experienced significant changes during the upper Paleocene–lower Eocene. As part of their life cycle many dinoflagellates produce an organic-walled cyst (dinocyst) that can be preserved in the geological record. Ecologically, dinoflagellates appear to have either heterotrophic or autotrophic lifestyles (Fensome et al., 1993). The most profound change in dinocyst assemblages during the late Paleocene–early Eocene involved representatives of one particular genus, *Apectodinium.* Recently, an extraordinary event has been linked with the IETM: coeval with the beginning of the CIE, *Apectodinium* species suddenly dominate dinocyst assemblages at locations around the world (Bujak and Brinkhuis, 1998; Heilmann-Clausen and Egger, 2000; Bybell et al., 2001; Crouch et al., 2001).

Bujak and Brinkhuis (1998) compiled information regarding the distribution, abundance and ecological preference of *Apectodinium* through the upper Paleocene–lower Eocene from available global dinocyst data. Several important conclusions were established: (1) the oldest definite record of *Apectodinium* is from lower latitudes (Tethys) in the mid Paleocene, within calcareous nannoplankton Zone NP4 and planktonic foraminiferal Subzone P3a (2) *Apectodinium* may be abundant in lower latitudes during the upper Paleocene and lower Eocene (3) the motile stage of representatives of *Apectodinium* were probably thermophilic and heterotrophic, implying requirements of warm sea surface temperatures and a sufficient food source and (4) there is a lack of marine sections studied for dinocysts that contain well-calibrated ages and precise recognition of the IETM, particularly in equatorial regions and the Southern Hemisphere. Since this overview, dinocyst records have been documented from a lower-latitude sequence in the Northern Hemisphere (Austria; Fig. 1) and a mid to high latitude sequence in the Southern Hemisphere (New Zealand; Fig. 1). Both records are correlated with calcareous nannoplankton zones and the IETM (Egger et al., 2000; Heilmann-Clausen and Egger, 2000; Crouch et al., 2001).

Although these recent studies provide more information regarding *Apectodinium,* additional well-correlated dinocyst studies are needed to further document the spatial and temporal distribution of this genus, notably from equatorial regions. In this paper, dinocyst assemblages are examined from the lower-latitude setting of the Tethys. One section comes from the southern part of the basin (Elles section, Tunisia) and the other is from the northeast (Aktumsuk section, Uzbekistan) (Figs. 1 and 2). Both successions span the IETM and are correlated with carbon isotope records and calcareous nannoplankton and planktonic foraminiferal zonations.

The dinocyst record at Elles and Aktumsuk offer the possibility to increase our understanding of the distribution and relative abundance of *Apectodinium,* along with other significant dinocyst events, in the lower-latitude Tethys during the late Paleocene–early Eocene. Furthermore, by combining these results with other well-correlated dinocyst records from the Northern and Southern Hemispheres, a more up-to-date overview of the global pattern of *Apectodinium* during the upper Paleocene and lower Eocene can be provided.

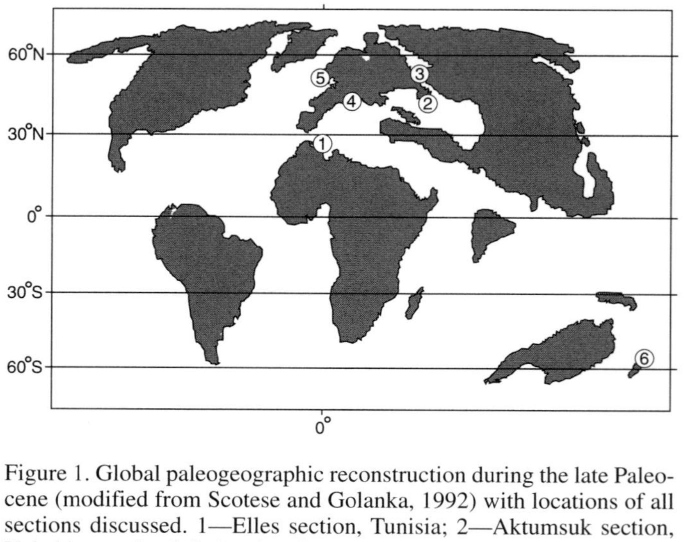

Figure 1. Global paleogeographic reconstruction during the late Paleocene (modified from Scotese and Golanka, 1992) with locations of all sections discussed. 1—Elles section, Tunisia; 2—Aktumsuk section, Uzbekistan; 3—Sokolovsky Quarry, northern Kazakhstan (Turgay Strait); 4—Anthering section, Austria; 5—North Sea Basin; 6—Tawanui section, New Zealand.

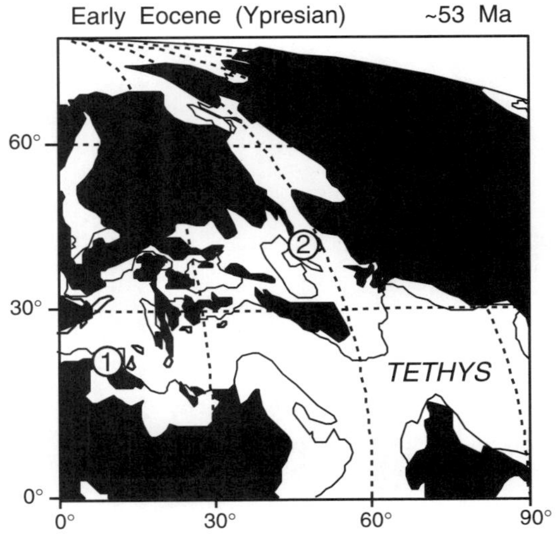

Figure 2. Paleogeographic reconstruction of the Tethys region during the early Eocene (ca. 53 Ma) with location of the Elles section, Tunisia (1) and Aktumsuk section, Uzbekistan (2). Map modified from Smith et al. (1994).

SETTING OF LOCATIONS

Elles section, Tunisia

The Elles section is located in the El Kef Basin, Tunisia (Figs. 1 and 2). During the Paleocene-Eocene, the El Kef Basin was situated in the northern tropical zone (~ paleolatitude 20°N) and in an open marine environment connected to the Tethys (Bolle et al., 1999, 2000b). The studied section at Elles is ~45 m thick and encompasses calcareous nannoplankton Zones NP7 to NP10 and planktonic foraminiferal Subzones P4a to P6a (Figs. 3 and 4). The IETM is thought to be associated with a ~2.7 m clay layer, within Subzone NP9b and Zone P5 (Bolle et al., 1999). The clay layer is characterized by a δ^{13}C shift of –1.3‰, decline in carbonate sedimentation, increased detrital input and high kaolinite/smectite ratio (Bolle et al., 1999). The negative shift in bulk rock carbon isotopic analyses, recognized in the clay layer, is less pronounced than the "typical" 2‰–3‰ excursion recorded at other marine locations during the IETM (e.g., Kennett and Stott, 1991). The most prominent negative shift in

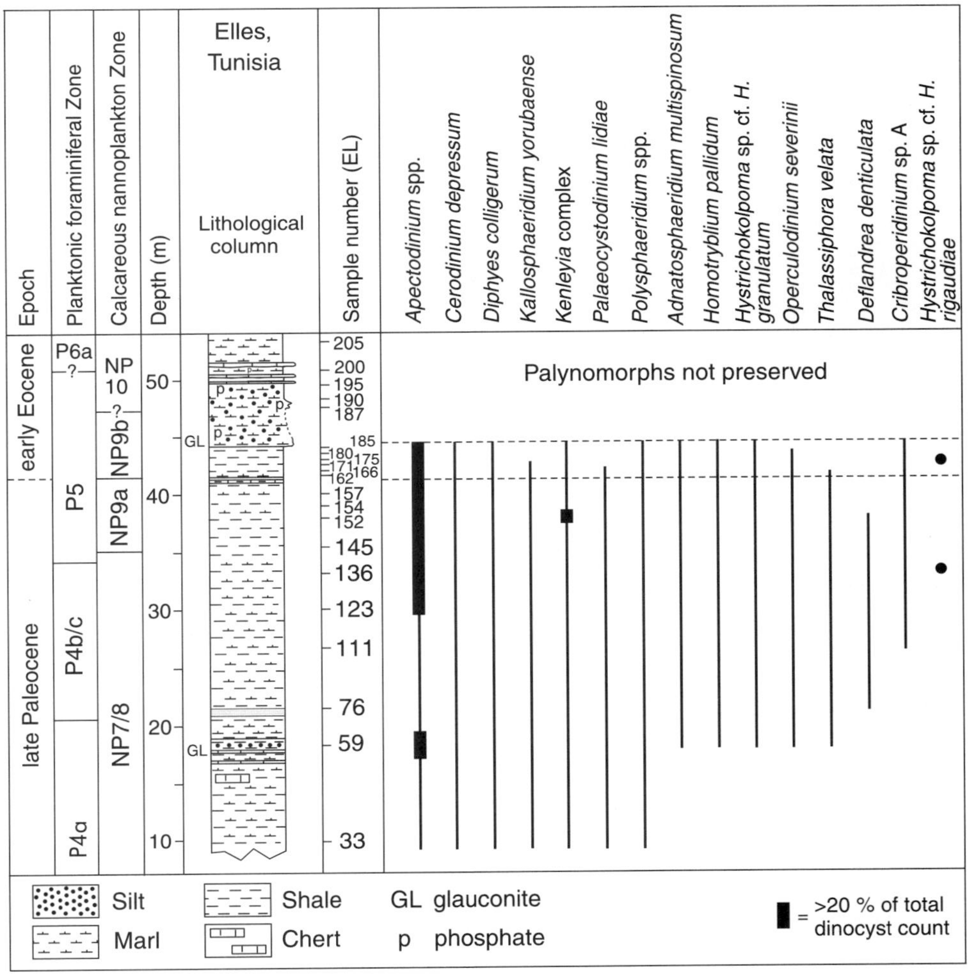

Figure 3. Stratigraphical distribution of selected dinocysts at Elles section, Tunisia. A summary lithological column and sample points are shown. Dashed line shows the proposed position of the initial Eocene thermal maximum (after Bolle et al., 1999). Calcareous nannoplankton and planktonic foraminiferal zonation scheme from Bolle et al. (1999).

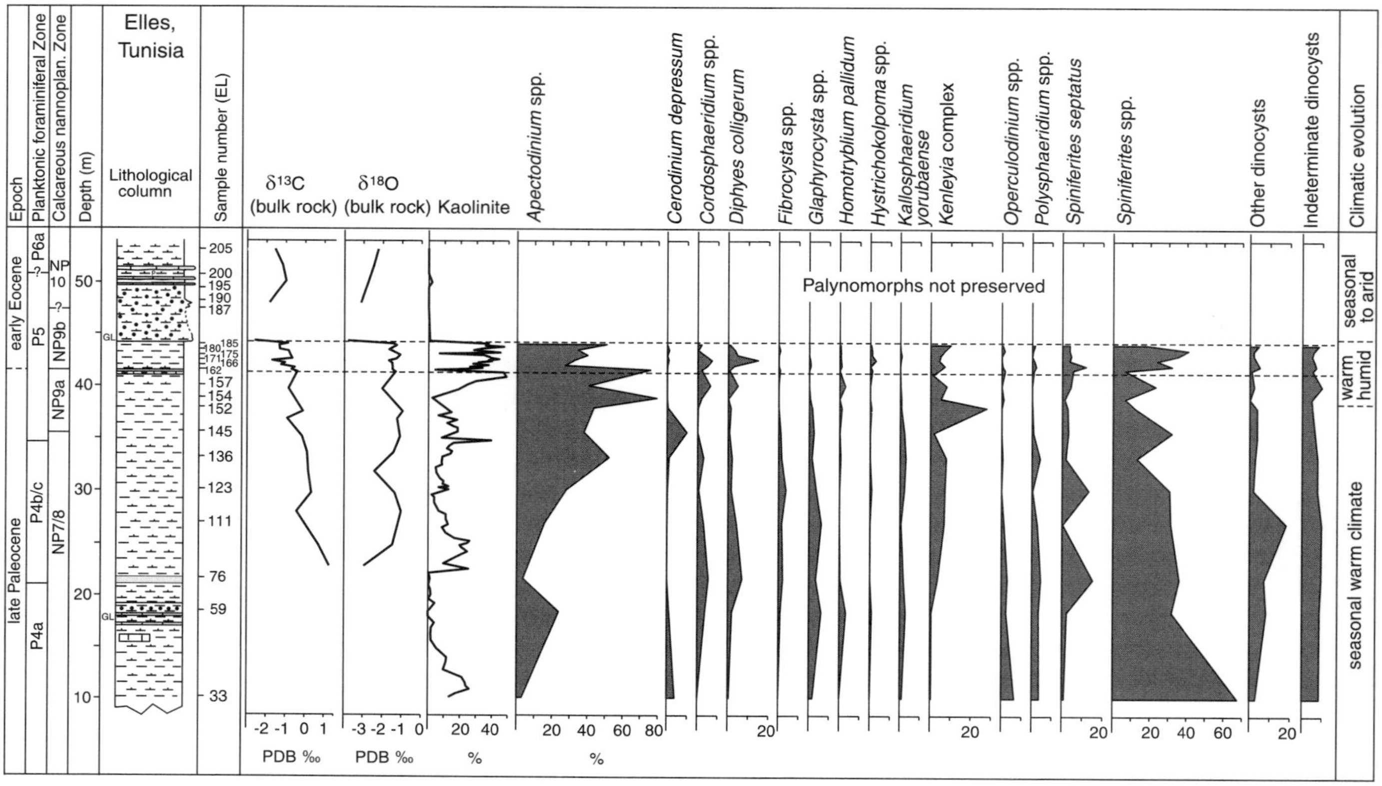

Figure 4. Relative abundance of dinocysts from the upper Paleocene to lowermost Eocene at Elles section, Tunisia. A summary lithological column and sample points are shown. Dashed line shows the proposed position of the initial Eocene thermal maximum (after Bolle et al., 1999). Calcareous nannoplankton and planktonic foraminiferal zonation scheme, and climatic evolution information from Bolle et al. (1999).

$\delta^{13}C$ (~1.7‰) is noted at the base of a ~6 m unit that overlies the clay layer (Fig. 4). This unit consists primarily of a glauconitic siltstone with phosphatic particles and francolite, which may have altered or overprinted the original marine isotopic signal (Bolle et al., 1999).

The section consists mainly of marls, shales, shaley marls and thin calcareous beds (Figs. 3 and 4). Two glauconitic siltstone layers are present that are considered to represent sequence boundaries associated with relative sea-level fall and possible increased tectonic activity (Bolle et al., 1999). The major components of the bulk rock sediment are phyllosilicates, with calcite becoming more abundant in the upper 9 m of section (Bolle et al., 1999). Although smectite is the major clay mineral component, kaolinite becomes increasingly more abundant in Zone NP9 (Fig. 4). Variations in clay minerals of the sediments indicate changing climatic conditions during the upper Paleocene–lower Eocene (Bolle et al., 1999). A low kaolinite content and abundant smectite suggest a seasonal warm climate during the upper Paleocene, while a very high kaolinite/smectite ratio across the P-E transition (Fig. 4) implies a warm and humid climate with enhanced terrestrial runoff. The disappearance of kaolinite and an increase in smectite and mica suggests development of arid to seasonal climatic conditions during the lowermost Eocene (Bolle et al., 1999).

Aktumsuk section, Uzbekistan

The Aktumsuk section is located southwest of the Aral Sea, Uzbekistan (Figs. 1 and 2), with a succession of ~32 m that spans calcareous nannoplankton Zone CC19 to NP14 (Bolle et al., 2000a). In this paper we focus on a ~3 m interval spanning the IETM, from calcareous nannoplankton Zones NP9 to NP11 (Figs. 5 and 6). The base of the succession is marked by a hiatus, with latest Paleocene sediments unconformably overlying Campanian white chalks (Bolle et al., 2000a). During the late Paleocene, Aktumsuk was located close to an emerged continental area that was subsequently flooded by a marine transgression at the end of the Paleocene (Gavrilov et al., 1997; Bolle et al., 2000a). During the lower Eocene, the marginal northeastern Tethys was marked by a period of high sea level and increasing water circulation, resulting in enhanced levels of oxygenation (Haq et al., 1988; Bolle et al., 2000a).

The IETM at Aktumsuk occurs within the lower part of a ~1 m organic-rich interval of grey to dark grey chalk, with cross

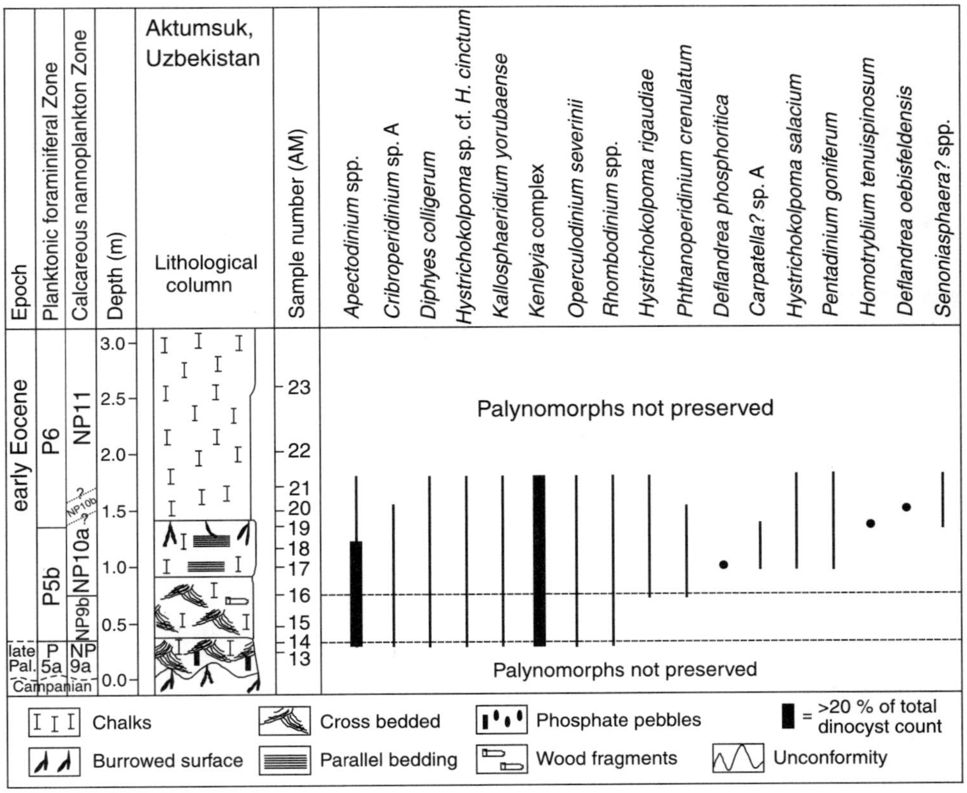

Figure 5. Stratigraphical distribution of selected dinocysts at Aktumsuk section, Uzbekistan. A summary lithological column and sample points are shown. Dashed line shows the proposed position of the initial Eocene thermal maximum (after Bolle et al., 2000a). Calcareous nannoplankton and planktonic foraminiferal zonation scheme from Bolle et al. (2000a).

bedding and wood fragments in the lower 0.5 m and parallel bedding in the upper 0.5 m (Figs. 5 and 6). The IETM is marked by a decrease of between ~1.8‰ and 5.9‰ in $\delta^{13}C$ isotopes of the bulk fine fraction (<63 µm), mixed planktonic foraminifera assemblages and total organic carbon (Fig. 6; Bolle et al., 2000a, 2000b). In addition, the IETM corresponds to minimum $\delta^{18}O$ values, high organic carbon content (maximum 8.44%) and an incursion of a transient subtropical planktonic foraminiferal fauna (Bolle et al., 2000a).

Distribution of clay minerals suggest that a temperate warm climate, with wet and arid seasons, was dominant during the upper Paleocene, and that a warm and humid climate was present during the IETM and lower Eocene (Bolle et al., 2000a). The presence of organic-rich sediments in the uppermost Paleocene can be correlated over a significant area of central and south Asia, the Caspian region and Caucasus (e.g., Gavrilov and Muzylöv, 1991; Gavrilov et al., 1997).

MATERIAL AND METHODS

Twenty-one rock samples were collected from across ~45 m of section at Elles (Figs. 3 and 4). The lowest sample was collected at ~10.2 m and a further nine samples were taken in the overlying ~30 m upper Paleocene interval. Six samples were taken from the ~2.7 m clay layer that is thought to represent the IETM and five samples from the overlying ~10 m of section. At Aktumsuk, 18 rock samples were collected from across ~5 m of section (Figs. 5 and 6). One sample was collected from below the organic-rich layer, six samples from within the ~1 m organic-rich layer and 11 samples from the overlying ~4 m of chalk.

All samples were prepared for palynological examination at the Laboratory of Palaeobotany and Palynology (LPP), Utrecht University, using standard techniques (e.g., Van Steenbergen, 1997). Between 1.5 and 10 g of sediment was crushed, and the carbonate and siliceous components removed by adding 30% HCl and 38% HF, respectively. Samples were then sieved over a 10 µm–mesh sieve and well-mixed representative fractions of the >10 µm residue were mounted on glass slides using a glycerine jelly medium.

Several samples from both sections contain no palynomorphs and these sediments have most likely experienced oxidation, either syn- or postdepositional. At Elles, the five samples (EL187 to EL205) collected above the ~2.7 m clay layer contain no palynomorphs. At Aktumsuk, one sample (AM13) below the organic-rich layer and the uppermost nine samples (AM22 to AM30) from the chalk are either barren or contain very sparse palynomorphs. When samples do contain palynomorphs they are generally well preserved and abundant. Quantitative analysis of dinocyst assemblages was completed on 16 samples from Elles and eight samples from Aktumsuk.

For all samples quantitatively examined, a minimum of 200 dinocysts were counted and the entire slide was scanned for the

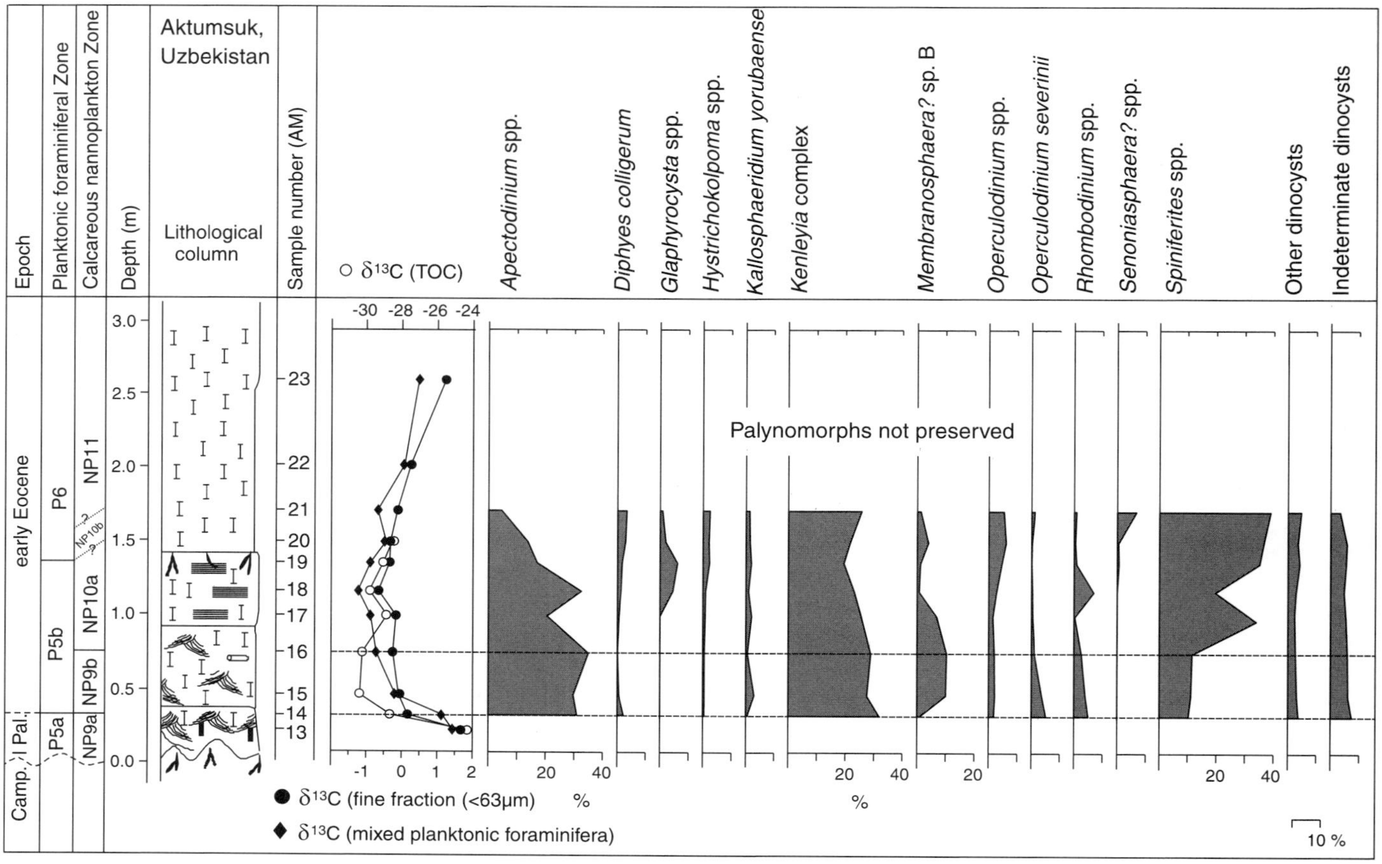

Figure 6. Relative abundance of dinocysts from the initial Eocene thermal maximum (IETM) and lower Eocene at Aktumsuk section, Uzbekistan. A summary lithological column and sample points are shown. Dashed line shows the proposed position of the initial Eocene thermal maximum (after Bolle et al., 2000a). Calcareous nannoplankton and planktonic foraminiferal zonation scheme from Bolle et al. (2000a).

presence of additional taxa. In this paper, most attention is given to stratigraphically significant dinocyst taxa, whose approximate ranges have been ascertained from previous research in lower-latitude regions (e.g., Jan du Chêne and Adediran, 1984; Brinkhuis and Zachariasse, 1988; Köthe, 1988; Brinkhuis et al., 1994; Iakovleva et al., 2001). A list of recorded species and brief taxonomic notes are provided in the Appendix. Light-photomicrographs were taken from permanent strew mounts. For photographed specimens England Finder coordinates are provided. All material is filed in the LPP collection of Utrecht University.

RESULTS

Elles section, Tunisia

The dinocyst assemblages at Elles are primarily dominated by either *Apectodinium* or *Spiniferites* species (Figs. 3 and 4), and their overall composition is consistent with an open marine environmental setting, as suggested by Bolle et al. (1999). Dinocysts thought to be indicative of oceanic water masses, such as *Impagidinium,* are not recorded while many of the dinocysts, including *Operculodinium* and *Spiniferites,* are typically characteristic of open marine, neritic water masses (e.g., Wall et al., 1977; Brinkhuis, 1994; Powell et al., 1996; Stover et al., 1996; van Mourik et al., 2001).

Apectodinium is present in low percentages at the base of the succession (Fig. 3). The morphology of representatives of *Apectodinium* becomes more diverse up-section, although *Apectodinium augustum* (an extreme morphotype with well-developed lateral and antapical horns) is not recorded. Interestingly, several specimens, otherwise assignable to *Apectodinium,* show an archeopyle that involves two plates (2a and 4″ plates; Plate 1, Figures 1 and 2). This contrasts to the standard 2a quadra archeopyle that characterizes *Apectodinium* (Costa and Downie, 1976; Williams et al., 1998).

The first occurrence of *Homotryblium pallidum* (Plate 2, Figure 3) and *Hystrichokolpoma* sp. cf. *H. granulatum* (Plate 2, Figures 5 and 6), within calcareous nannoplankton Zone NP7/8 and the upper part of planktonic foraminiferal Subzone P4a, are noteworthy (Fig. 3). Another specimen of *Hystrichokolpoma,* tentatively assigned to *H.* sp. cf. *H. rigaudiae* (Plate 2, Figure 4), occurs in samples EL136 and EL171 (Fig. 3). At present, the oldest occurrence of *Homotryblium* is in sediments tentatively assigned to calcareous nannoplankton Zone NP6/7 from a section in Kazakhstan (Iakovleva et al., 2001), and our results confirm that *Homotryblium* was present in the southern Tethys in sediments assigned to calcareous nannoplankton Zone NP7/8. *Hystrichokolpoma* species are most commonly recorded from sediments of an early Eocene age or younger, although some species (e.g., *H. granulatum, H. salacium, H. unispinum*) have previously been recognized from lower-latitude upper Paleocene successions (Jan du Chêne and Adediran, 1984; Iakovleva et al., 2001).

The relative abundance of *Apectodinium* increases upsection and is a dominant component from the upper part of calcareous nannoplankton Zone NP7/8 and planktonic foraminifera Subzone P4b/c (Fig. 4). Prior to the ~2.7 m clay layer, three peaks in *Apectodinium* abundance are recorded, with values of ~24%, ~52% and ~79%, respectively. *Apectodinium*-dominated assemblages (~75%) mark the base of the ~2.7 m clay layer, although the relative abundance drops to between ~27% and 51% in the remaining interval. Notably, the pattern of *Apectodinium* abundance shows some covariance with the record of kaolinite content (Fig. 4).

The *Kenleyia* complex is a common element of the dinocyst assemblage, with the highest abundance (~28%) recorded in the lower part of calcareous nannoplankton Subzone NP9a (Fig. 4). Specimens assigned to this complex show a large variation in morphology (Plate 3, Figures 4–9), and are generally characterized by the presence of apical and antapical protrusions, and variable development of the periphragm (see Appendix).

Aktumsuk section, Uzbekistan

Similar to Elles, the dinocyst assemblages at Aktumsuk are dominated by either *Apectodinium* or *Spiniferites* species, and specimens assigned to the *Kenleyia* complex (Figs. 5 and 6). The overall dinocyst assemblage recorded at Aktumsuk is consistent with a neritic marine setting, as suggested by Bolle et al. (2000a).

Apectodinium is present throughout the succession and several morphological variations are recognized, although the extreme morphotype *Apectodinium augustum* is absent. As recorded at Elles, several *Apectodinium* specimens show an archeopyle that involves two plates (2a and 4″ plates). Specimens assigned to *Rhombodinium,* another member of the Wetzelielloideae subfamily, are common throughout (Fig. 5). Possible reworked dinocysts include specimens tentatively assigned to *Senoniasphaera?,* a genus primarily restricted to the Cretaceous. Reworked calcareous nannoplankton, mainly of Late Cretaceous age, have previously been noted in the same interval at Aktumsuk (Bolle et al., 2000a).

The highest *Apectodinium* percentages (up to ~35%) are recorded in calcareous nannoplankton Subzones NP9b and NP10a (Fig. 6). Although *Apectodinium* is relatively abundant during the IETM, the percentages associated with the onset of the CIE are not as high as those recorded from the presumed equivalent horizon (i.e., base of the ~2.7 m clay layer) at Elles (Figs. 4 and 6). Specimens assigned to the *Kenleyia* complex (Plate 3, Figures 10–12) constitute ~20%–32% of the assemblages and are consistently more abundant than at Elles, which may be related to the shallower depositional environment at Aktumsuk.

DISCUSSION

The combined dinocyst results at Elles and Aktumsuk provide a reasonably detailed record from the Tethys during the late Paleocene (calcareous nannoplankton Zone NP7/8 and planktonic foraminiferal Subzone P4a) to early Eocene (calcareous nannoplankton Zone NP10 and planktonic foraminiferal Zone P6).

Apectodinium

Apectodinium was present in the southern Tethys by the late Paleocene, becoming a notable component of dinocyst assemblages in upper calcareous nannoplankton Zone NP7/8, and continued to be a frequent to superabundant component (~20%–80%) during the remaining part of the late Paleocene and into the earliest Eocene (Figs. 4 and 6). A decrease in *Apectodinium* abundance is recorded within calcareous nannoplankton Zone NP10a (Fig. 6). At the Jebel Boudabous section in central Tunisia (Bujak and Brinkhuis, 1998), the relative abundance of *Apectodinium* also remained high during the early Eocene, although there it extends into calcareous nannoplankton Zone NP11 before the abundance declines.

At Elles, the interval of frequent to abundant *Apectodinium* corresponds to a phase of elevated kaolinite percentages (Fig. 4). In particular, kaolinite is a more significant component of clay mineral composition during planktonic foraminiferal Subzone P4b/c and Zone P5, with the highest percentages (up to ~49%) recorded just prior to the ~2.7 m clay layer and within the clay layer. Kaolinite is a mineral that typically develops on well-drained tropical soils characterized by a warm and humid climate with high precipitation (Robert and Kennett, 1994). The apparent link between elevated kaolinite and *Apectodinium* percentages is, therefore, consistent with the hypothesis that the motile representatives of *Apectodinium* required both warm temperatures and an availability of nutrients (in this case primarily supplied by terrestrial runoff) and/or tolerance of such conditions.

The timing of the two peaks in *Apectodinium* abundance (~24% and ~52%) within calcareous nannoplankton Zone NP7/8 at Elles (Fig. 4), appear to be approximately coeval with other late Paleocene occurrences of *Apectodinium* at higher latitudes in the Northern Hemisphere. In the North Sea Basin (Fig. 1), two short-lived excursions of *Apectodinium* have been recorded in the upper Paleocene; one within the lower part of calcareous nannoplankton Zone NP8 (sequence Tht-3 of Powell et al., 1996) and the other close to the NP8/NP9 zonal boundary (sequence Tht-4) (Bujak and Brinkhuis, 1998; Fig. 7). These ex-

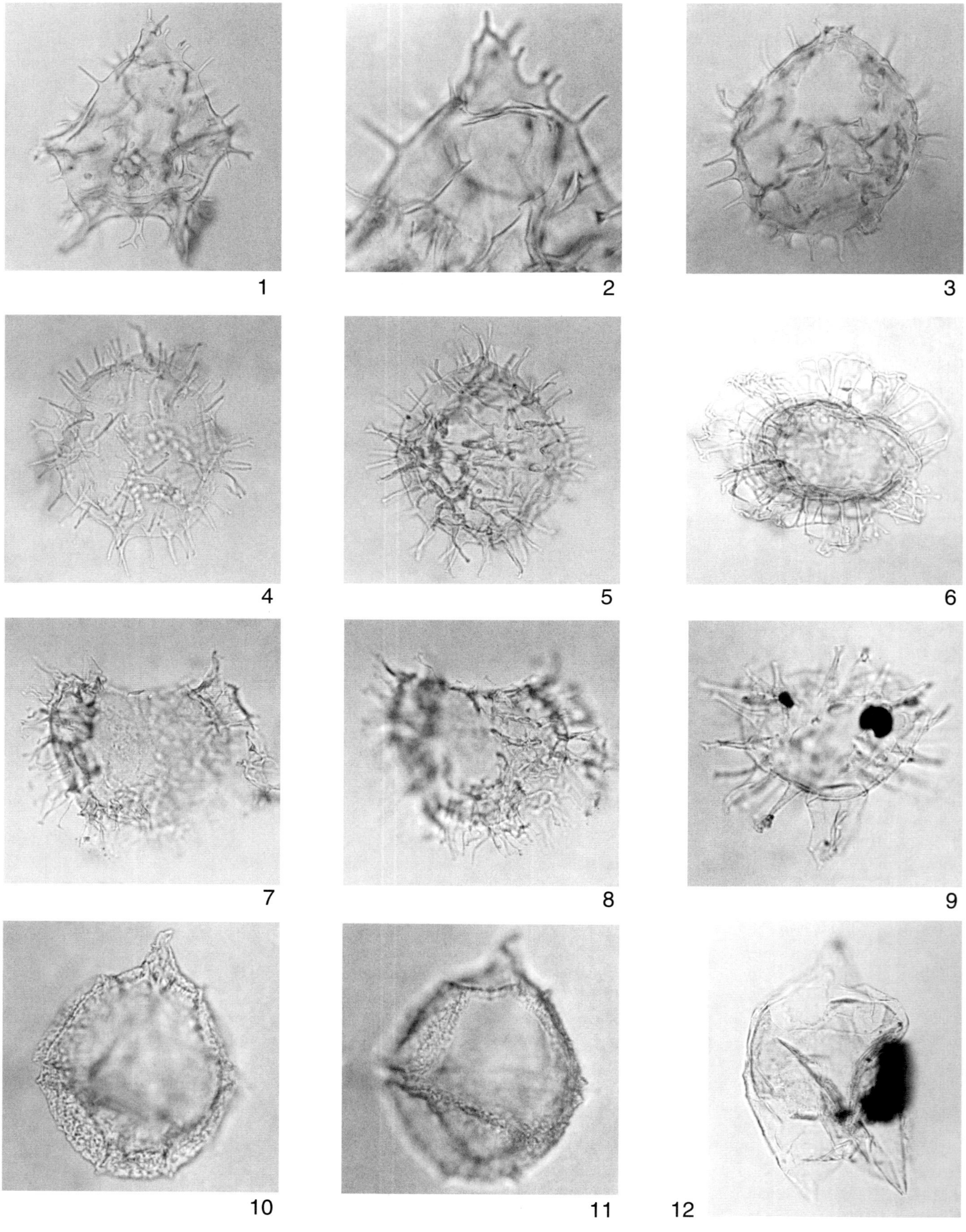
1
2
3
4
5
6
7
8
9
10
11
12

cursions in the North Sea Basin are thought to represent periods of warmer surface waters during the late Paleocene (Bujak and Brinkhuis, 1998). From the Turgay Strait, Northern Kazakhstan (Fig. 1), a peak in the relative abundance of *Apectodinium* (~28%) has been recorded in upper Paleocene sediments thought to be equivalent to sequence Tht-4 in the North Sea Basin (Iakovleva et al., 2001; Fig. 7). This coincidence in the timing of late Paleocene *Apectodinium* bioevents suggests there may have been a link between high numbers of *Apectodinium* in lower latitudes, short-lived incursions of *Apectodinium* into mid to high latitudes, and the possibility of brief intervals of climatic warming or "hyperthermals" (Thomas et al., 2000; Thomas and Zachos, 2000). At present, the occurrence of short-lived migrations of *Apectodinium* into the Southern Hemisphere during the late Paleocene is not well established (Crouch, 2001).

At Elles, the base of the ~2.7 m clay layer (level recognized as the onset of the CIE) is characterized by *Apectodinium*-dominated assemblages (Fig. 4). At Aktumsuk it is difficult to obtain an overview of assemblage changes from the late Paleocene to the IETM, although *Apectodinium* percentages do not reach comparably high values during the interval of lowest $\delta^{13}C$ values (Fig. 6). Although it is difficult to establish a definitive record of *Apectodinium* abundance at lower latitudes during the IETM from these two sections, the differences in relative abundance suggest that *Apectodinium* may not always have dominated the IETM in lower-latitude regions. This is in contrast to current evidence from higher latitudes, where the IETM is consistently characterized by *Apectodinium*-dominated assemblages (Bujak and Brinkhuis, 1998; Heilmann-Clausen and Schmitz, 2000; Bybell et al., 2001; Crouch et al., 2001).

The morphology of representatives of *Apectodinium* varies considerably at both sections, although the extreme morphotype *Apectodinium augustum* was not recorded. At present, records of this species are restricted to the North Sea Basin, northern Kazakhstan and northern Tethys (Austria), and confined to a narrow time interval around the IETM (Powell et al., 1996; Bujak and Brinkhuis, 1998; Egger et al., 2000; Iakovleva et al., 2001).

Other dinocysts

The presence of specimens assignable to *Rhombodinium* (Plate 4, Figures 6–9), another member of the Wetzelielloideae subfamily, during the IETM at Aktumsuk is of note. It is generally accepted that the evolutionary lineage of Wetzelielloids begins in the mid Paleocene with the appearance of *Apectodinium* and that *Wetzeliella* and allied genera, including *Rhombodinium* and *Wilsonidium,* are considered to characterize the early Eocene (e.g., Bujak and Brinkhuis, 1998). Recently, taxa assignable to *Wilsonidium* have been recorded from uppermost Paleocene sediments in northern Kazakhstan (Iakovleva et al., 2001) and this is currently the oldest stratigraphic occurrence of representatives of the *Wetzeliella* group. Our data provide further evidence of taxa assignable to the *Wetzeliella* group prior to the characteristic taxa of the lower Eocene.

Deflandrea phosphoritica (Plate 1, Figure 12) was recorded at Aktumsuk within calcareous nannoplankton Subzone NP10a (Fig. 5). In the lower Eocene this species was common in dinocyst assemblages worldwide (e.g., Stover et al., 1996), although the origin of the species is less well documented. At present, the earliest occurrence of *D. phosphoritica* is from the IETM at an upper Paleocene–lower Eocene section in New Zealand (Crouch, 2001), suggesting this species may have evolved in a Southern Hemisphere region. The earliest records of *D. phosphoritica* in the Northern Hemisphere are similar to its first occurrence at Aktumsuk, in the lower part of calcareous nannoplankton Zone NP10 (Heilmann-Clausen, 1985; Bujak and Brinkhuis, 1998).

The common to abundant occurrence of the *Kenleyia* complex at both sections confirms evidence from earlier studies (e.g., Fechner and Mohr, 1986; Brinkhuis and Leereveld, 1988; Brinkhuis and Zachariasse, 1988; Brinkhuis et al., 1994) that taxa assignable to this complex had a preference for warm water conditions. The abundant presence of the *Kenleyia* complex at Aktumsuk, a succession thought to have been deposited in a marginal marine setting (Bolle et al., 2000a, 2000b), supports indications that taxa assignable to this complex may have also had a preference for shallow to brackish waters (H. Brinkhuis, 2002, personal commun.).

LATE PALEOCENE–EARLY EOCENE GLOBAL DISTRIBUTION OF *APECTODINIUM*

In order to develop a detailed temporal and spatial account of *Apectodinium* distribution, the dinocyst record at Elles and Aktumsuk is compared with several well-calibrated upper

Plate 1. **1:** *Apectodinium* sp. Sample AM16, Slide 1 (E24/3). Aktumsuk section. Overall dimensions ~87 μm by 103 μm. Archeopyle that involves two plates; 2a and 4″ plates. **2:** *Apectodinium* sp. Sample AM16, Slide 1 (E24/3). Aktumsuk section. Vertical dimension ~60 μm; close up of archeopyle (2a and 4″ plates). **3:** *Apectodinium* sp. Sample EL145 (~35.75 m), Slide 1 (H24/0). Elles section. Overall dimensions ~80 μm by 85 μm. Archeopyle that involves two plates; 2a and 4″ plates. **4:** *Apectodinium* sp. Sample EL136 (~33.43 m), Slide 1 (R22/3). Elles section. Overall dimensions ~79 μm by 85 μm. **5:** *Apectodinium* sp. Sample EL162 (~41.8 m), Slide 1 (E20/0). Elles section. Overall dimensions ~90 μm by 100 μm. **6:** *Adnatosphaeridium multispinosum.* Sample EL59 (~18.4 m), Slide 1 (K20/3). Elles section. Overall dimensions ~78 μm by 98 μm. **7** and **8:** *Areoligera senonensis* group. Sample EL111 (~26.92 m), Slide 1 (U33/3). Elles section. Overall dimensions ~65 μm by 95 μm. **9:** *Diphyes colligerum.* Sample AM17, Slide 1 (P30/0). Aktumsuk section. Overall dimensions ~58 μm by 50 μm. **10** and **11:** *Cribroperidinium* sp. A. Sample AM19, Slide 1 (P35/0). Aktumsuk section. Overall dimensions ~50 μm by 61 μm. **12:** *Deflandrea phosphoritica.* Sample AM17, Slide 1 (R22/0). Aktumsuk section. Overall dimensions ~70 μm by 120 μm.

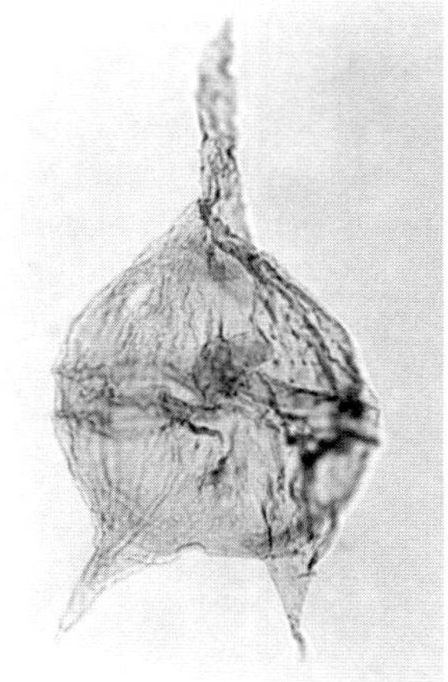
1

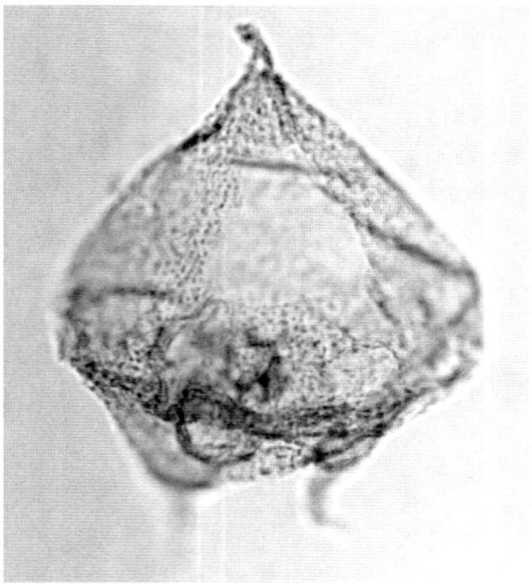
2

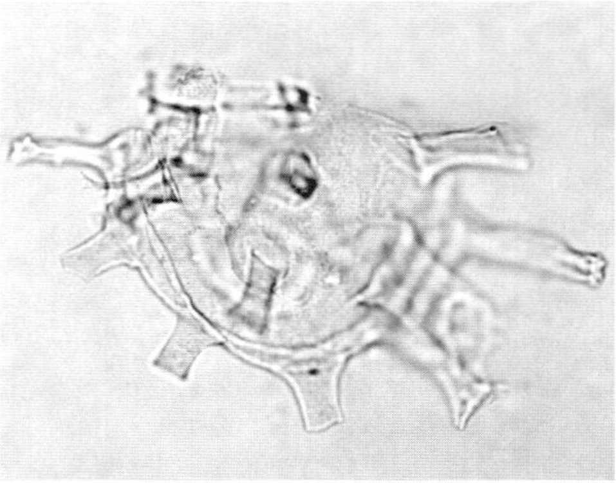
3

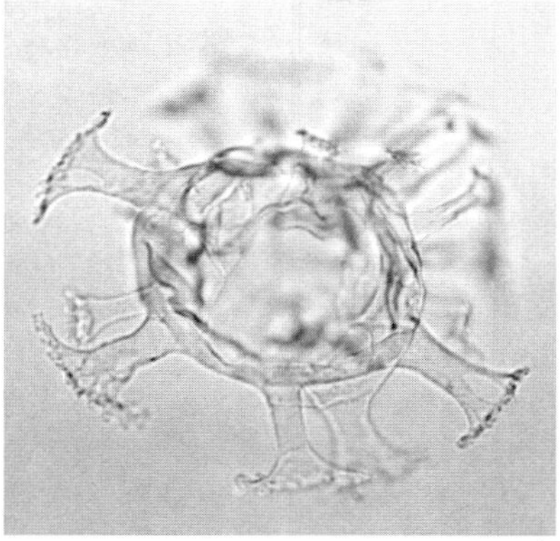
4

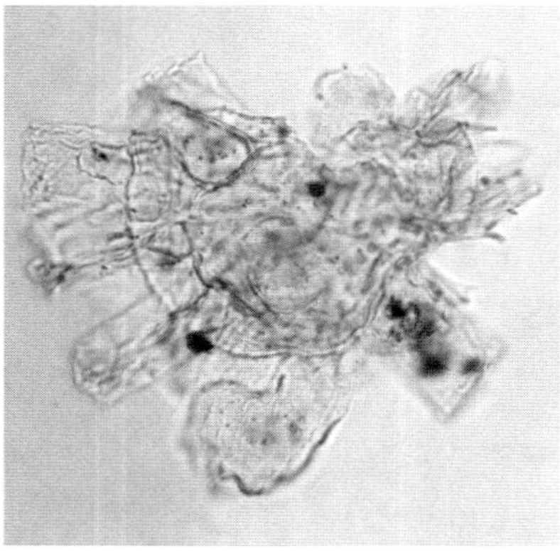
5

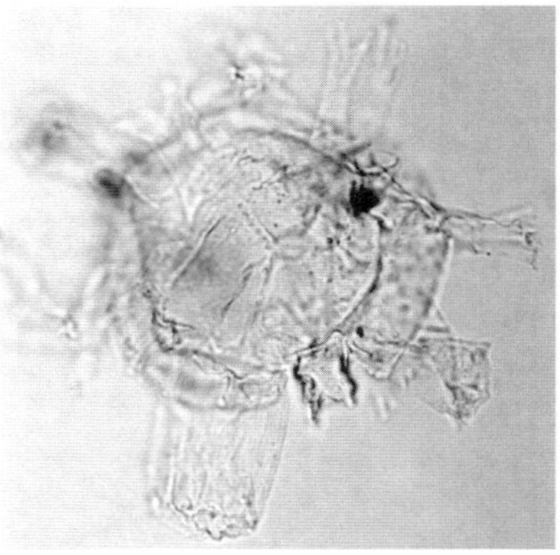
6

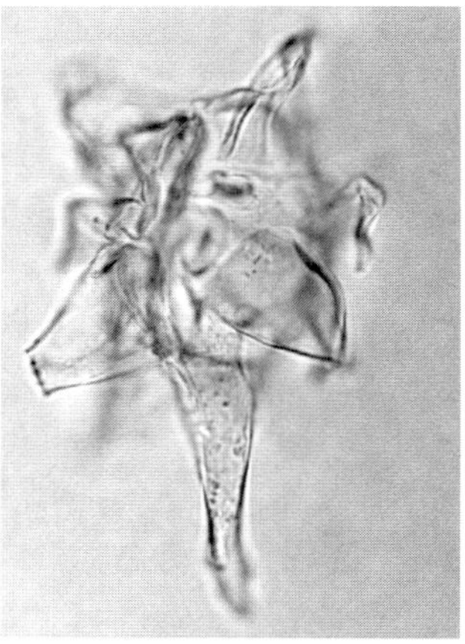
7

8

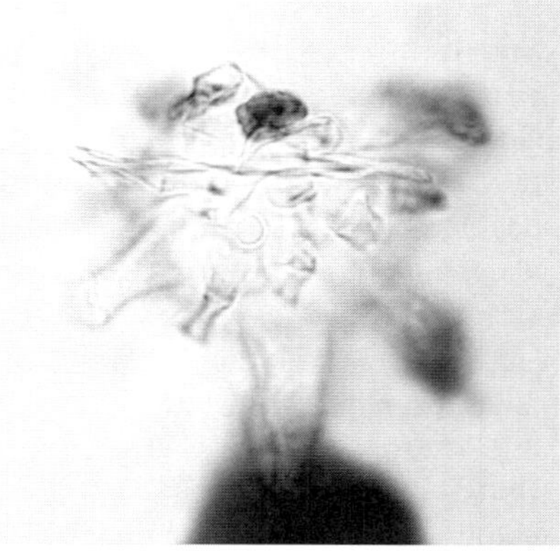
9

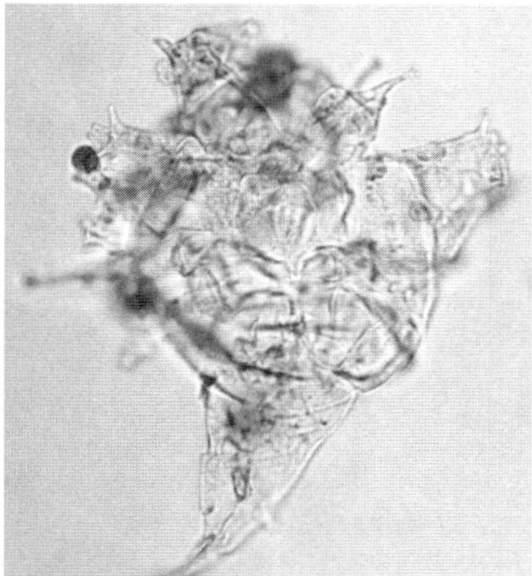
10

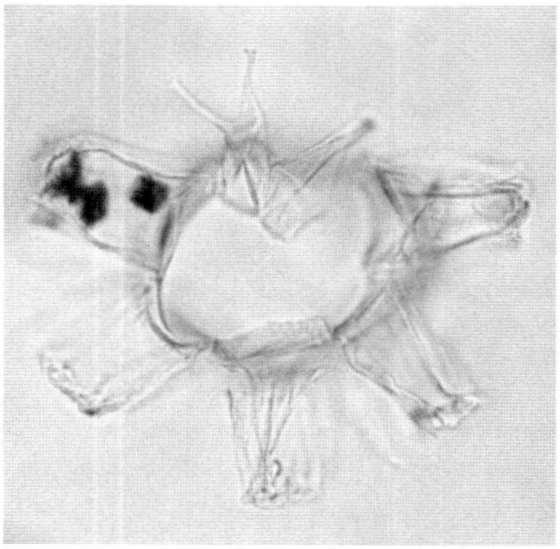
11

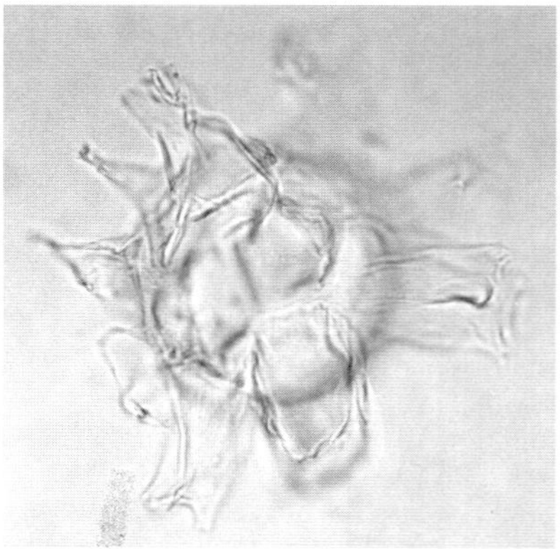
12

Paleocene–lower Eocene records from other regions. A schematic overview of the *Apectodinium* record, along with other dinocyst bioevents of possible global significance, from selected locations in the Northern and Southern Hemispheres is shown in Figure 7.

In the Northern Hemisphere, the best calibrated dinocyst records currently come from the North Sea Basin and the Tethys. From the North Sea Basin, information of dinocyst assemblages is taken from several upper Paleocene–lower Eocene studies (e.g., Heilmann-Clausen, 1985; Powell, 1992; Bujak and Mudge, 1994; Mudge and Bujak, 1996; Powell et al., 1996; Bujak and Brinkhuis, 1998). For the Tethys, the results from Elles and Aktumsuk are presented, along with information from an upper Paleocene–lower Eocene succession in Austria (Egger et al., 2000; Heilmann-Clausen and Egger, 2000; Crouch et al., 2001). In addition, dinocyst information is taken from a section in northern Kazakhstan (Iakovleva et al., 2001) that is located in the Turgay Strait (Fig. 1). During the Paleogene this seaway intermittently connected the Tethyan Basin and Boreal Arctic Sea-North Sea Basin (e.g., Benyamovsky et al., 1991). For the Southern Hemisphere, selected dinocyst information is taken from an upper Paleocene–lower Eocene succession in New Zealand (Crouch, 2001).

Lower latitudes

As established by Brinkhuis et al. (1994), and further documented in this paper and other recent studies (Bujak and Brinkhuis, 1998; Egger et al., 2000; Iakovleva et al., 2001), the earliest appearance of *Apectodinium* was in lower latitudes during the mid Paleocene. Well-correlated dinocyst records from these latitudes are currently restricted to the Tethys, although high percentages of *Apectodinium* have been recorded in upper Paleocene sections from other equatorial regions, including northern Africa (Jan du Chêne and Adediran, 1984), Pakistan (Köthe, 1988) and U.S. Gulf Coast (Harrington and Kemp, 2001).

At lower latitudes, the relative abundance of *Apectodinium* can be frequent to superabundant prior to and after the intense warming of the IETM (Fig. 7). At present it appears that the IETM at lower latitudes was characterized by high percentages of *Apectodinium,* although it may not always have dominated dinocyst assemblages. Moreover, other thermophilic genera, most likely dinoflagellates with autotrophic lifestyles (e.g., the *Kenleyia* complex), may have been a common component of assemblages during the IETM. Recent studies indicate that SSTs did not change markedly during the late Paleocene–early Eocene in equatorial regions (Lu et al., 1998; Thomas et al., 1999), which suggests that fluctuations in *Apectodinium* abundance in the Tethys may have been more intricately related to surface water conditions, including availability of nutrients, enhanced runoff and/or other specific oceanic conditions, rather than changes in SST.

Mid to high latitudes

Apectodinium has been recorded from upper Paleocene sediments at mid to high latitudes (North Sea Basin), although these occurrences appear to have been relatively short-lived and *Apectodinium* did not dominate dinocyst assemblages (Powell et al., 1996; Bujak and Brinkhuis, 1998). Further research is needed to ascertain whether the timing of *Apectodinium* incursions in the North Sea Basin during the late Paleocene (lower part of calcareous nannoplankton Zone NP8 and close to the NP8-NP9 zonal boundary; Fig. 7) can be recognized in other mid to high latitude regions. The relatively high percentage of *Apectodinium* (~28%) during the upper Paleocene at northern Kazakhstan (Fig. 7) suggests that the Turgay Strait was connected to the Tethyan Basin at this time (Iakovleva et al., 2001).

At present, the IETM is characterized at mid to high latitudes by the sudden appearance of *Apectodinium*-dominated assemblages. Subsequent to the IETM, the relative abundance of *Apectodinium* decreased at mid to high latitudes. However, additional *Apectodinium* acmes have been recorded during the lower Eocene in both Hemispheres (Fig. 7) and further research is needed to establish if these acmes were also synchronous on a global scale.

Correlation between lower latitude and mid to high latitudes

The frequent to superabundant presence of *Apectodinium* throughout the upper Paleocene and lower Eocene in equatorial regions makes it difficult to correlate *Apectodinium* acmes between lower latitudes and mid to high latitudes. However, cor-

Plate 2. **1:** *Deflandrea denticulaa.* Sample EL76 (~21.8 m), Slide 2 (W28/0). Elles section. Overall dimensions ~52 μm by 104 μm. **2:** *Cerodinium depressum.* Sample EL171 (~42.74 m), Slide 1 (O33/3). Elles section. Overall dimensions ~58 μm by 75 μm. **3:** *Homotryblium pallidum* Sample EL59 (~18.4 m), Slide 1 (P13/1). Elles section. Overall dimensions ~87 μm by 50 μm. **4:** *Hystrichokolpoma* sp. cf. *H. rigaudiae.* Sample EL136 (~33.43 m), Slide 1 (S19/0). Elles section. Overall dimensions ~100 μm by 88 μm. **5:** *Hystrichokolpoma* sp. cf. *H. granulatum.* Sample EL175 (~43.29 m), Slide 1 (P16/0). Elles section. Overall dimensions ~87 μm by 77 μm. **6:** *Hystrichokolpoma* sp. cf. *H. granulatum.* Sample EL152 (~37.38 m), Slide 1 (Q24/0). Elles section. Overall dimensions ~71 μm by 75 μm. **7:** *Hystrichokolpoma* sp. cf. *H. cinctum.* Sample AM20, Slide 1 (G34/0). Aktumsuk section. Overall dimensions ~50 μm by 80 μm. **8** and **9:** *Hystrichokolpoma* sp. cf. *H. cinctum.* Sample AM14, Slide 1a (T24/0). Aktumsuk section. Overall dimensions ~58 μm by 55 μm. **10:** *Hystrichokolpoma rigaudiae.* Sample AM16, Slide 1 (J30/2). Aktumsuk section. Overall dimensions ~75 μm by 92 μm. **11:** *Hystrichokolpoma* sp. cf. *H. salacium.* Sample AM17, Slide 2 (V26/3). Aktumsuk section. Overall dimensions ~73 μm by 65 μm. **12:** *Hystrichokolpoma* sp. cf. *H. salacium.* Sample AM19, Slide 1 (S30/2). Aktumsuk section. Overall dimensions ~60 μm by 60 μm.

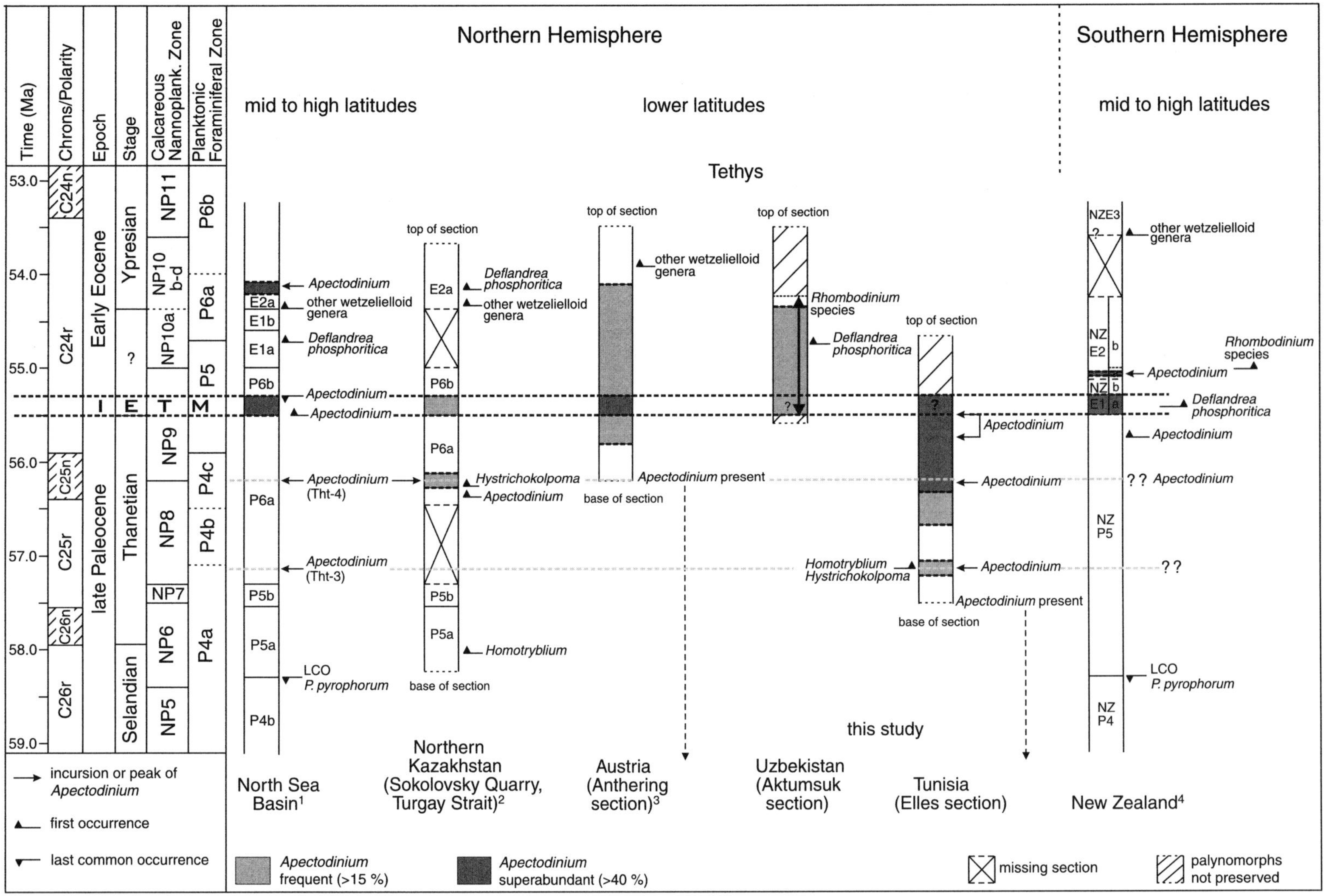

Figure 7. Global overview of the late Paleocene–early Eocene distribution and abundance of *Apectodinium,* and other selected dinocyst bioevents, from five Northern Hemisphere sites and one Southern Hemisphere site. Dashed line shows the position and duration of the initial Eocene thermal maximum (IETM). International chronology comes from Berggren et al. (1995), Berggren and Aubry (1998) and Aubry (2000). Dark grey color represents *Apectodinium* abundance >40% of dinocyst assemblage; light grey color represents *Apectodinium* abundance generally >15% of dinocyst assemblage. References: 1—Powell, 1992; Bujak and Mudge, 1994; Mudge and Bujak, 1996; Powell et al., 1996; Bujak and Brinkhuis, 1998. 2—Iakovleva et al., 2001. 3—Egger et al., 2000; Heilmann-Clausen and Egger, 2000; Crouch et al., 2001. 4—Crouch, 2000.

relation may be possible by further examination of intervals at lower latitudes where *Apectodinium* abundance increases relatively, and comparing this timing with incursions of *Apectodinium* into mid to high latitudes. One such example is the possible correlation between peaks in *Apectodinium* at the Elles section and incursions of *Apectodinium* into the North Sea Basin during the lower part of calcareous nannoplankton Zone NP8 and in close proximity to the NP8-NP9 zonal boundary (Fig. 7).

Correlation between mid to high latitudes in the Northern and Southern Hemispheres is potentially less complicated given that *Apectodinium*-dominated assemblages occur at specific horizons and are separated by intervals where *Apectodinium* may be absent or rare. The best example of this is the *Apectodinium* acme associated with the IETM, which at present is the most significant and globally synchronous dinocyst bioevent that occurred during the late Paleocene–early Eocene at mid to high latitudes. The presence of additional dinocyst taxa, which are confined to particular intervals within *Apectodinium*-dominated assemblages, offer the potential for further correlation between mid to high latitudes in both Hemispheres.

Environmental conditions

A special set of environmental conditions, notably different to the "normal background setting" of the upper Paleocene–lower Eocene, must have occurred at various times and locations to allow *Apectodinium* to dominate the dinocyst assemblages. Indeed, the character of the *Apectodinium* acme associated with the IETM suggests it acted as an opportunistic genus and was able to take advantage of the prevailing ecosystem better than other dinoflagellates. It is worth noting that the typical environment for *Apectodinium* was in lower latitudes, where it comprised a regular component of dinocyst assemblages. The genus was frequent to superabundant in these regions throughout the upper Paleocene–lower Eocene, implying that these "special" environmental conditions occurred more frequently in warmer water settings.

We suggest that *Apectodinium* was able to migrate into higher latitudes when SSTs were suitably warm, and that a marked increase in SST (such as at the IETM) was a crucial environmental factor controlling the occurrence of *Apectodinium* "blooms" in these regions. Moreover, it appears that intensification of biogeochemical feedbacks, such as chemical weathering, precipitation and runoff, during periods of intense warming may have resulted in enhanced delivery of nutrients to (marginal) ocean surface waters (Zachos and Dickens, 2000; Crouch et al., 2003; Ravizza et al., 2001). Intervals of abundant *Apectodinium* recognized at Tawanui, New Zealand (Crouch et al., 2003) and Elles, Tunisia (Fig. 4) appear to have occurred during phases of enhanced terrigenous input, increased runoff and elevated kaolinite percentages. This delivery of nutrients may have allowed primary productivity to increase and altered surface water conditions (e.g., freshwater lenses and stratified water masses).

Recent studies in different Hemispheres across the P-E transition suggest that nutrient availability in surface waters may have remained elevated subsequent to the IETM. In New Zealand, material derived from terrigenous sources remained elevated after the IETM (Crouch, 2001). Similarly, high concentrations of diatoms continued in the North Sea Basin above the IETM (Bujak and Brinkhuis, 1998). However, in both regions *Apectodinium*-dominated assemblages were replaced, near the end of the IETM, by a high abundance of other probable heterotrophic dinocysts, primarily taxa assignable to *Deflandrea* and *Cerodinium.* These dinocyst genera may be indicative of slightly less warm SSTs (Bujak and Brinkhuis, 1998; Crouch, 2001).

In summary, it appears that the principal environmental control on the presence and distribution of *Apectodinium* was warm SSTs. However, during intervals of *Apectodinium* "blooms" the crucial environmental parameters probably involved even higher SSTs and the interaction of this with the availability of nutrients, enhanced runoff and/or other "special" oceanic conditions (e.g., stratified water masses).

CONCLUSIONS

To further understand the distribution and abundance of *Apectodinium* through space and time, we have studied dinocyst records from two well-calibrated Paleogene successions in the Tethys. The combined records at Elles, Tunisia, and Aktumsuk, Uzbekistan, span the late Paleocene (calcareous nannoplankton Zone NP7/8 and planktonic foraminiferal Subzone P4a) to early Eocene (calcareous nannoplankton Zone NP10 and planktonic foraminiferal Zone P6). The main findings are:

1. *Apectodinium* was present in the southern Tethys by planktonic foraminiferal Subzone P4a, confirming earlier findings that this genus first evolved in lower latitudes.

2. *Apectodinium* abundance increased within planktonic foraminiferal Subzone P4b/c and remained a frequent to superabundant component of assemblages until the early Eocene (lower part of planktonic foraminiferal Zone P6).

3. At Elles, *Apectodinium*-dominated assemblages were present at the base of a ~2.7 m clay layer, which is thought to indicate the onset of the IETM. *Apectodinium* was abundant during the IETM at Aktumsuk, although percentages are generally lower than at Elles.

4. For the first time we recognize specimens assignable to *Apectodinium* that contain an archeopyle formation involving two plates (2a and 4″), rather than the standard 2a quadra archeopyle.

The Elles and Aktumsuk records have been compared to several well-correlated upper Paleocene–lower Eocene dinocyst records in other regions, to begin to develop a more detailed spatial and temporal overview of the distribution of *Apectodinium.* Our main observations are:

1. *Apectodinium* first appeared in lower latitudes and was frequent to superabundant during the late Paleocene and early Eocene. The IETM at lower latitudes was characterized by high percentages of *Apectodinium,* although it may not always have

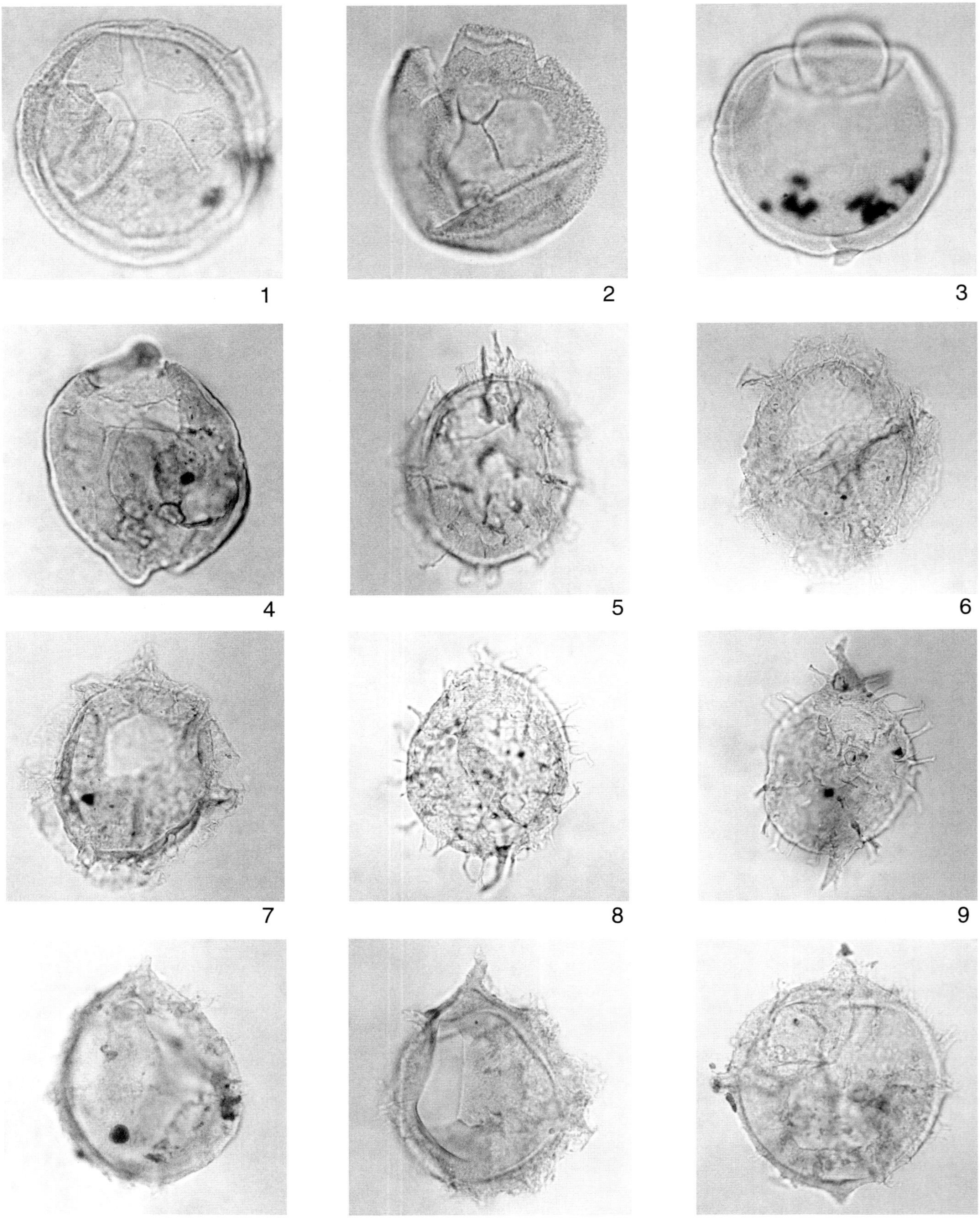
1
2
3
4
5
6
7
8
9
10
11
12

dominated assemblages. Other dinocyst genera with thermophilic preferences may have also been abundant during the IETM.

2. *Apectodinium* sporadically occurred at mid to high latitudes during the late Paleocene, although incursions are geologically brief and *Apectodinium* was not dominant. The IETM at mid to high latitudes was characterized by *Apectodinium*-dominated assemblages. Additional *Apectodinium* acmes occurred during the early Eocene, however as yet it is unclear if these are globally isochronous.

3. Intervals of *Apectodinium*-dominated assemblages are indicative of environmental conditions that were markedly different to "normal background settings" of the late Paleocene–early Eocene. We suggest SST was the main control on their distribution since *Apectodinium* acmes at mid to high latitudes seem to have occurred during intervals of highest SSTs, such as the IETM. *Apectodinium* "blooms" were probably also intricately linked to enhanced runoff and increased delivery of nutrients to surface waters. This may have been a result of intensification of the weathering cycle, and other possible specific oceanic conditions such as stratified water masses.

ACKNOWLEDGMENTS

Natasja Welters is thanked for palynological processing of samples from the Elles and Aktumsuk sections. We thank Jamie Powell and Henrik Nøhr-Hansen for constructive and helpful comments on the manuscript. This is NSG publication number 20020404.

APPENDIX. DINOFLAGELLATE CYST SPECIES LIST AND TAXONOMIC NOTES

The generic allocation of taxa follows that cited in Williams et al. (1998) unless otherwise stated. Illustrated taxa are followed by plate and figure references in brackets. Some new taxa have been recorded and they are discussed briefly. These new taxa are not formally described. Notes on certain genera are also provided.

Adnatosphaeridium multispinosum (Plate 1, Figure 6)
Apectodinium spp. (Plate 1, Figures 1–5)

Remarks: A variety of *Apectodinium* morphotypes have been recognized and several typical specimens are illustrated. Species identified included *A. homomorphum, A. hypercanthum, A. parvum* and *A. quinquelatum. Apectodinium augustum* (an extreme morphotype with well-developed lateral and antapical horns) was not recorded. Several specimens, otherwise assignable to *Apectodinium,* show an archeopyle that involves two plates, the 2a and 4″ plates (Plate 1, Figures 1–3).

Areoligera senonensis group (Plate 1, Figures 7 and 8)
Carpatella? sp. A (Plate 3, Figure 3)

Remarks: The specimens encountered in this study broadly resemble *Carpatella cornuta.* The development of an apical horn is absent in some forms and, hence, they are tentatively assigned to *Carpatella. Carpatella cornuta* is typically known from lower Danian sediments (Damassa, 1988) and may have also had a preference for generally warm water settings (Brinkhuis and Zachariasse, 1988; Brinkhuis et al., 1998).

Cerodinium depressum (Plate 2, Figure 2)
Cordosphaeridium spp.
Cribroperidinium sp. A (Plate 1, Figures 10 and 11)

Remarks: Cysts assignable to *Cribroperidinium* sp. A are small, proximate and subspherical in shape. A small apical horn is present. The archeopyle is precingular and formed by the loss of 3″. The periphragm is strongly granulate, forming an almost microreticulum, and paratabulation is discernible by perforated septa.

Deflandrea denticulata (Plate 2, Figure 1)
Deflandrea phosphoritica (Plate 1, Figure 12)
Deflandrea oebisfeldensis
Diphyes colligerum (Plate 1, Figure 9)
Fibrocysta spp.
Glaphyrocysta spp.
Homotryblium pallidum (Plate 2, Figure 3)
Homotryblium tenuispinosum
Hystrichokolpoma sp. cf. *H. cinctum* (Plate 2, Figures 7–9)

Remarks: Cysts assignable to *Hystrichokolpoma* sp. cf. *H. cinctum* differ from *Hystrichokolpoma cinctum* by being smaller and by having granules developed on the processes.

Hystrichokolpoma sp. cf. *H. granulatum* (Plate 2, Figures 5 and 6)

Remarks: Cysts assignable to *Hystrichokolpoma* sp. cf. *H. granulatum* closely resemble *Hystrichokolpoma granulatum,*

Plate 3. **1:** *Kallosphaeridium yorubaense.* Sample EL123 (~30.04 m), Slide 1 (R18/0). Elles section. Overall dimensions ~54 μm by 53 μm. **2:** *Kallosphaeridium yorubaense.* Sample AM19, Slide 1 (U34/1). Aktumsuk section. Overall dimensions ~60 μm by 65 μm. **3.** *Carpatella?* sp. A. Sample AM17, Slide 1 (F23/2). Aktumsuk section. Overall dimensions ~80 μm by 80 μm. **4:** *Kenleyia* complex. Sample EL111 (~26.92 m), Slide 1 (N32/0). Elles section. Overall dimensions ~55 μm by 70 μm. **5:** *Kenleyia* complex. Sample EL166 (~42.21 m), Slide 1 (W19/2). Elles section. Overall dimensions ~75 μm by 93 μm. **6:** *Kenleyia* complex. Sample EL145 (~35.75 m), Slide 1 (X26/0). Elles section. Overall dimensions ~80 μm by 95 μm. **7:** *Kenleyia* complex. Sample EL145 (~35.75 m), Slide 1 (F28/0). Elles section. Overall dimensions ~82 μm by 95 μm. **8:** *Kenleyia* complex. Sample EL123 (~30.04 m), Slide 1 (J16/4). Elles section. Overall dimensions ~65 μm by 85 μm. **9:** *Kenleyia* complex. Sample EL162 (~41.8 m), Slide 1 (R21/0). Elles section. Overall dimensions ~72 μm by 112 μm. **10:** *Kenleyia* complex. Sample AM15, Slide 1 (O28/4). Aktumsuk section. Overall dimensions ~75 μm by 90 μm. **11:** *Kenleyia* complex. Sample AM15, Slide 1 (Q22/3). Aktumsuk section. Overall dimensions ~85 μm by 100 μm. **12:** *Kenleyia* complex. Sample AM16, Slide 1a (T18/0). Aktumsuk section. Overall dimensions ~86 μm by 92 μm.

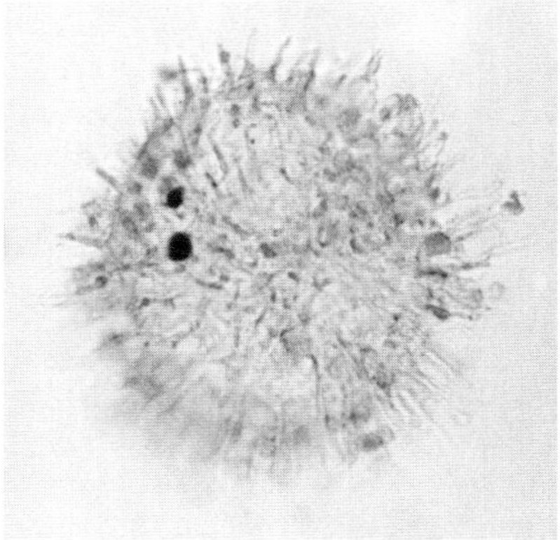

1

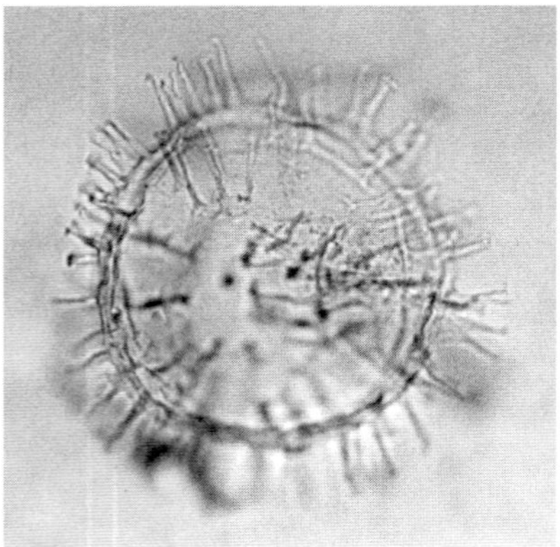

2

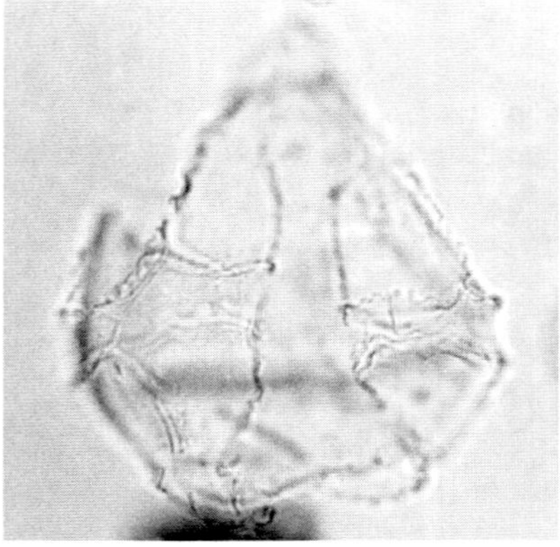

3

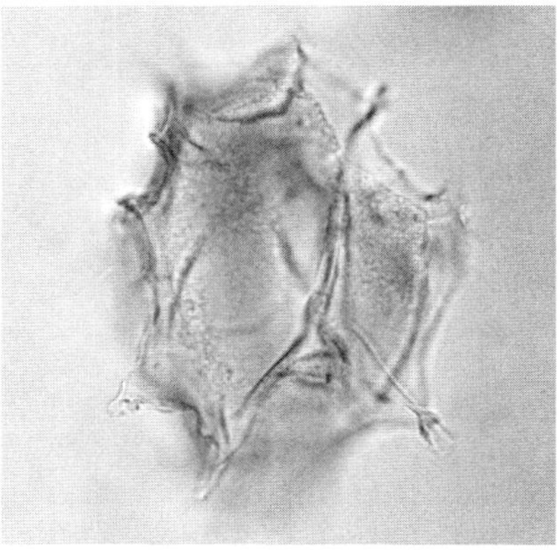

4

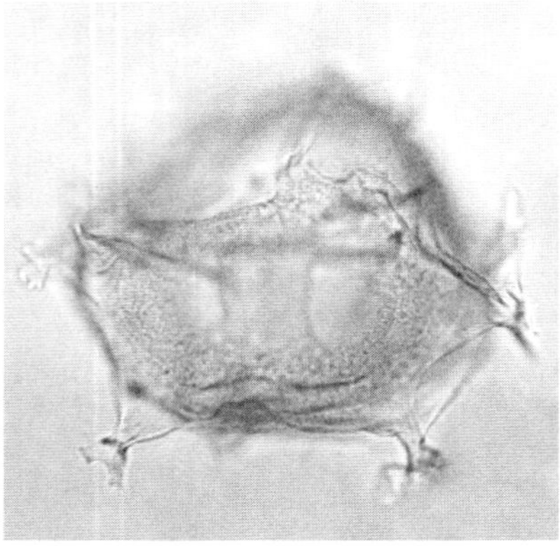

5

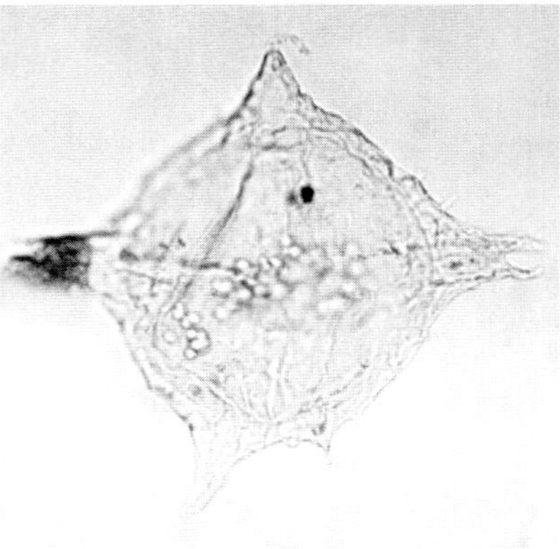

6

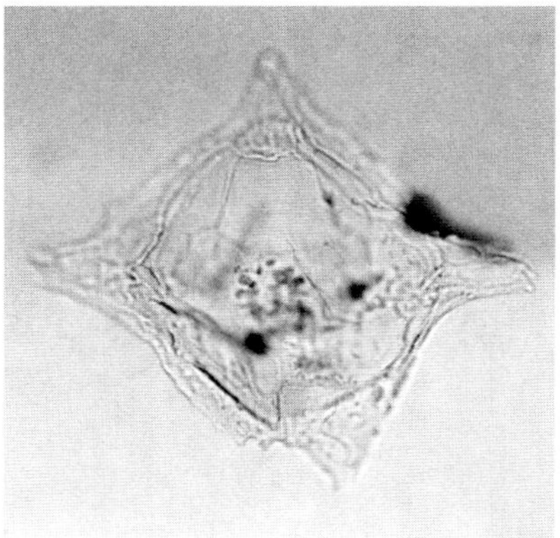

7

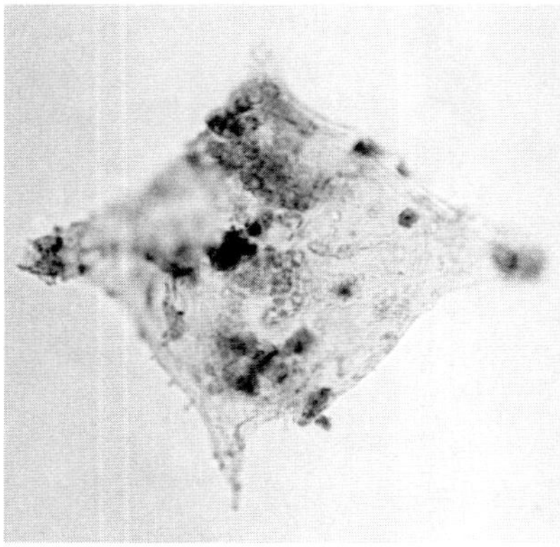

8

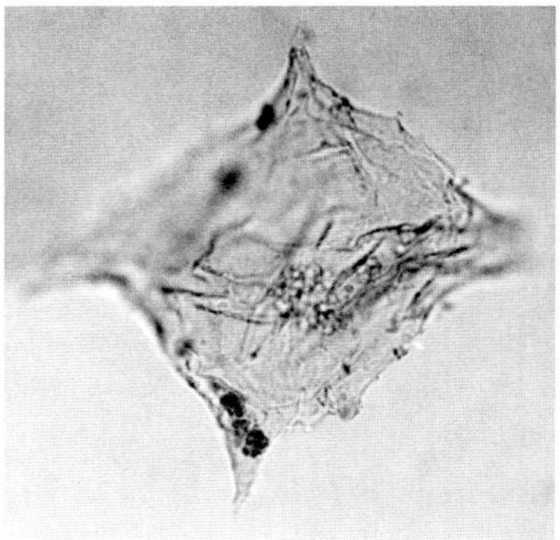

9

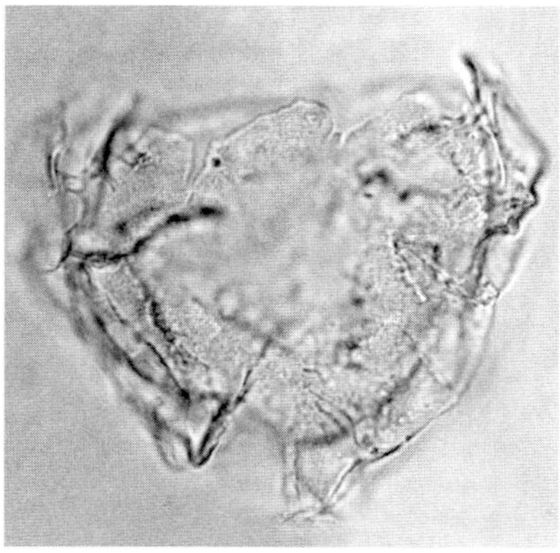

10

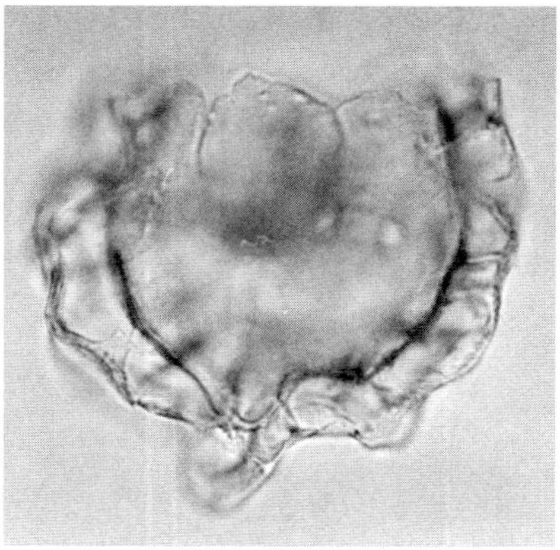

11

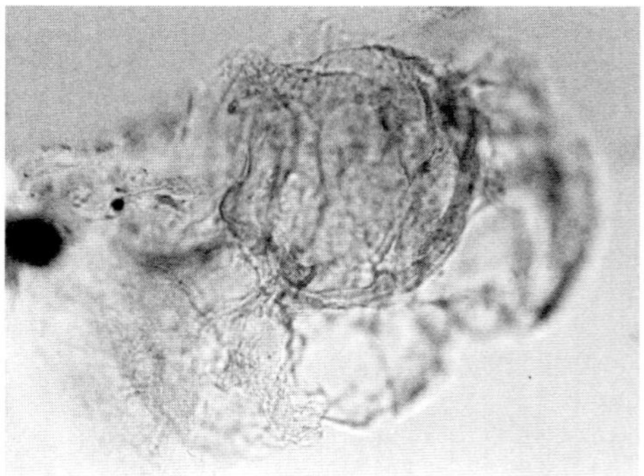

12

although they appear to be thinner walled and are generally poorly preserved making confident assignment to *Hystrichokolpoma granulatum* difficult.

Hystrichokolpoma rigaudiae (Plate 2, Figure 10)
Hystrichokolpoma sp. cf. *H. rigaudiae* (Plate 2, Figure 4)

Remarks: Cysts assignable to *Hystrichokolpoma* sp. cf. *H. rigaudiae* broadly resemble *Hystrichokolpoma rigaudiae,* but differ in having distinctly different process terminations. In addition, the antapical process is not as well developed.

Hystrichokolpoma salacium (Plate 2, Figures 11 and 12)
Hystrichokolpoma spp. (pars.)

Remarks: This includes all other taxa assignable to *Hystrichokolpoma.*

Kallosphaeridium yorubaense (Plate 3, Figures 1 and 2)
Kenleyia complex (Plate 3, Figures 4–12)

Remarks: Cysts assignable to the *Kenleyia* complex exhibit a wide morphological variation and some of the main forms are illustrated. The cysts are characterized by having apical and antapical protrusions and a precingular (3″) archeopyle. There is substantial variation in the development of the periphragm, ranging from nearly smooth to the development of processes, spines and membranes. Some of the more extreme forms appear to cross generic boundaries, such as *Cordosphaeridium, Fibrocysta,* and *Muratodinium.* For the purpose of this study all of these morphological variations are placed in the *Kenleyia* complex.

Membranosphaera? sp. B De Coninck, 1975
Operculodinium severinii (Plate 4, Figure 1)
Operculodinium spp. (pars.)

Remarks: This includes all other taxa assignable to *Operculodinium.*

Palaeocystodinium lidiae
Pentadinium sp. cf. *P. goniferum* (Plate 4, Figures 4 and 5)
Phelodinium spp.
Phthanoperidinium crenulatum (Plate 4, Figure 3)
Polysphaeridium spp. (Plate 4, Figure 2)
Rhombodinium spp. (Plate 4, Figures 6–9)
Senegalinium spp.
Senoniasphaera? spp. (Plate 4, Figures 10 and 11)

Remarks: Specimens assigned here to *Senoniasphaera?* spp. closely compare with Upper Cretaceous morphotypes of *Senoniasphaera rotundata* and are possibly reworked in to Eocene sediments at Aktumsuk. Alternatively, they may represent a previously undescribed Eocene species of *Senoniasphaera.*

Spinidinium spp.
Spiniferites cornutus
Spiniferites septatus
Spiniferites spp. (pars.)

Remarks: This includes all other taxa assignable to *Spiniferites.*

Thalassiphora velata (Plate 4, Figure 12)

REFERENCES CITED

Aubry, M.-P., 1998, Early Paleogene calcareous nannoplankton evolution: A tale of climatic amelioration, *in* Aubry, M.-P., et al., eds., Late Paleocene–early Eocene climatic and biotic events in the marine and terrestrial records: New York, Columbia University Press, p. 158–203.

Aubry, M.-P., 2000, Where should the Global Stratotype Section and Point (GSSP) for the Paleocene-Eocene boundary be located?: Bulletin de la Société Géologique de France, v. 171, p. 461–476.

Benyamovsky, V.N., Levina, A.P., and Pronin, V.G., 1991, Lower Eocene sediments of the Turgay Passage: Izvestiya Vysshikh Uchebnykh Zavedeniy, Geologia i Razvedka, v. 7, p. 3–15.

Berggren, W.A., and Aubry, M.-P., 1998, The Paleocene–Eocene Epoch/Series boundary: Chronostratigraphic framework and estimated geochronology, *in* Aubry, M.-P., et al., eds., Late Paleocene–early Eocene climatic and biotic events in the marine and terrestrial records: New York, Columbia University Press, p. 18–36.

Berggren, W.A., Kent, D.V., Swisher, C.C., and Aubry, M.-P., 1995, A revised Cenozoic chronology and chronostratigraphy, *in* Berggren, W.A., et al., eds., Geochronology, time scales, and global stratigraphic correlation: SEPM (Society for Sedimentary Geology) Special Publication 54, p. 129–212.

Berggren, W.A., Lucas, S., and Aubry, M.-P., 1998, Late Paleocene–early Eocene climatic and biotic evolution: An overview, *in* Aubry, M.-P., et al., eds., Late Paleocene–early Eocene climatic and biotic events in the marine and terrestrial records: New York, Columbia University Press, p. 1–17.

Bolle, M.-P., Adatte, T., Keller, G., Von Salis, K., and Burns, S., 1999, The Paleocene–Eocene transition in the southern Tethys (Tunisia): Climatic and environmental fluctuations: Bulletin de la Société Géologique de France, v. 170, p. 661–680.

Plate 4. **1:** *Operculodinium severinii.* Sample AM15, Slide 1 (P23/1). Aktumsuk section. Overall dimensions ~80 μm by 85 μm. **2:** *Polysphaeridium* sp. Sample EL59 (~18.4 m), Slide 1 (J24/0). Elles section. Overall dimensions ~70 μm by 72 μm. **3:** *Phthanoperidinium crenulatum.* Sample AM16, Slide 1 (R27/2). Aktumsuk section. Overall dimensions ~54 μm by 63 μm. **4:** *Pentadinium* sp. cf. *P. goniferum.* Sample AM17, Slide 1 (R28/4). Aktumsuk section. Overall dimensions ~65 μm by 83 μm. **5:** *Pentadinium* sp. cf. *P. goniferum.* Sample AM17, Slide 2 (L28/0). Aktumsuk section. Overall dimensions ~100 μm by 90 μm. **6:** *Rhombodinium* sp. Sample AM17, Slide 2 (R37/0). Aktumsuk section. Overall dimensions ~93 μm by 88 μm. **7:** *Rhombodinium* sp. Sample AM18, Slide 1 (W18/2). Aktumsuk section. Overall dimensions ~95 μm by 86 μm. **8:** *Rhombodinium* sp. Sample AM15, Slide 1 (M26/0). Aktumsuk section. Overall dimensions ~98 μm by 92 μm. **9:** *Rhombodinium* sp. Sample AM16, Slide 1 (K26/2). Aktumsuk section. Overall dimensions ~96 μm by 95 μm. **10:** *Senoniasphaera?* sp. Sample AM19, Slide 1 (X20/2). Aktumsuk section. Overall dimensions ~72 μm by 66 μm. **11:** *Senoniasphaera?* sp. Sample AM21, Slide 1 (P20/0). Aktumsuk section. Overall dimensions ~65 μm by 60 μm. **12:** *Thalassiphora velata.* Sample EL111 (~26.92 m), Slide 1 (U34/4). Elles section. Overall dimensions ~125 μm by 100 μm.

Bolle, M.-P., Pardo, A., Hinrichs, K.-U., Adatte, T., Von Salis, K., Burns, S., Keller, G., and Muzylöv, N., 2000a, The Paleocene–Eocene transition in the marginal northeastern Tethys (Kazakhstan and Uzbekistan): International Journal of Earth Sciences, v. 89, p. 390–414.

Bolle, M.-P., Pardo, A., Adatte, T., Tantawy, A.A., Hinrichs, K., Von Salis, K., and Burns, S., 2000b, Climatic evolution on the southern and northern margins of the Tethys from the Paleocene to early Eocene: GFF, v. 122, p. 31–32.

Bralower, T.J., Zachos, J.C., Thomas, E., Parrow, M., Paull, C.K., Kelly, D.C., Premoli Silva, I., Sliter, W.V., and Lohmann, K.C., 1995, Late Paleocene to Eocene paleoceanography of the equatorial Pacific Ocean: Stable isotopes recorded at Ocean Drilling Program Site 865, Allison Guyot: Paleoceanography, v. 10, p. 841–865.

Brinkhuis, H., 1994, Late Eocene to early Oligocene dinoflagellate cysts from the Priabonian type-area (Northeast Italy): Biostratigraphy and paleoenvironmental interpretation: Palaeogeography, Palaeoclimatology, Palaeoecology, v. 107, p. 121–163.

Brinkhuis, H., and Leereveld, H., 1988, Dinoflagellate cysts from the Cretaceous–Tertiary boundary sequence of El Kef, northwest Tunisia: Review of Palaeobotany and Palynology, v. 56, p. 5–19.

Brinkhuis, H., and Zachariasse, W.J., 1988, Dinoflagellate cysts, sea-level changes, and planktonic foraminifera across the Cretaceous–Tertiary boundary at El Haria, northwest Tunisia: Marine Micropaleontology, v. 13, p. 153–191.

Brinkhuis, H., Romein, A.J.T., Smit, J., and Zachariasse, W.J., 1994, Danian–Selandian dinoflagellate cysts from lower latitudes with special reference to the El Kef section, northwest Tunisia: GFF, v. 116, p. 46–48.

Brinkhuis, H., Bujak, J.P., Smit, J., Versteegh, G.J.M., and Visscher, H., 1998, Dinoflagellate-based sea surface temperature reconstructions across the Cretaceous–Tertiary boundary: Palaeogeography, Palaeoclimatology, Palaeoecology, v. 141, p. 67–83.

Bujak, J., and Mudge, D., 1994, A high-resolution North Sea dinocyst zonation: Journal of the Geological Society of London, v. 151, p. 449–462.

Bujak, J., and Brinkhuis, H., 1998, Global warming and dinocyst changes across the Paleocene–Eocene epoch boundary, *in* Aubry, M.-P., et al., eds., Late Paleocene–early Eocene climatic and biotic events in the marine and terrestrial records: New York, Columbia University Press, p. 277–295.

Bybell, L.M., Gibson, T.G., Heilmann-Clausen, C., and Zachos, J.C., 2001, Events associated with the late Paleocene thermal maximum in the northeastern U.S., *in* Ash, A.W., and Wing, S.L., eds., Climate and biota of the early Paleogene: International Meeting July 3–8, 2001, Northwest College, Powell, Wyoming, USA, Abstract volume, p. 18.

Clyde, W.C., and Gingerich, P.D., 1998, Mammalian community response to the latest Paleocene thermal maximum: An isotaphonomic study in the northern Bighorn basin, Wyoming: Geology, v. 26, p. 1011–1014.

Costa, L.I., and Downie, C., 1976, The distribution of the dinoflagellate *Wetzeliella* in the Paleogene of northwestern Europe: Palaeontology, v. 19, p. 591–614.

Crouch, E.M., 2001, Environmental Change at the time of the Paleocene–Eocene biotic turnover [Ph.D. thesis]: Utrecht, Netherlands, Utrecht University, Laboratory of Palaeobotany and Palynology Contributions Series, v. 14, 216 p.

Crouch, E.M., Dickens, G.R., Brinkhuis, H., Aubry, M.-P., Hollis, C.J., Rogers, K.M., and Visscher, H., The *Apectodinium* acme and terrestrial discharge during the Paleocene-Eocene thermal maximum: New palynological, geochemical and calcareous nannoplankton observations at Tawanui, New Zealand: Palaeogeography, Palaeoclimatalogy, Palaeoecology, in press.

Crouch, E.M., Heilmann-Clausen, C., Brinkhuis, H., Morgans, H.E.G., Rogers, K.M., Egger, H., and Schmitz, B., 2001, Global dinoflagellate event associated with the late Paleocene thermal maximum: Geology, v. 29, p. 315–318.

Damassa, S.P., 1988, *Carpatella cornuta* Grigornovich 1969 (Dinophyceae): A member of the Aptiana-Vetriosum complex: Palynology, v. 12, p. 167–177.

De Coninck, J., 1975, Microfossiles à paroi organique de l'Ypresian de Bassin Belge: Professional Paper, Service Géologique de Belgique, v. 12, 165 p.

Dickens, G.R., 2000, Methane oxidation during the late Paleocene thermal maximum: Bulletin de la Société Géologique de France, v. 171, p. 37–49.

Dickens, G.R., 2001, Carbon addition and removal during the late Paleocene thermal maximum: Basic theory with a preliminary treatment of the isotope record at Ocean Drilling Program Site 1051, Blake Nose, *in* Kroon, D., et al., eds., Western North Atlantic Paleogene and Cretaceous palaeoceanography: Geological Society [London] Special Publication 183, p. 293–305.

Egger, H., Heilmann-Clausen, C., and Schmitz, B., 2000, The Paleocene–Eocene boundary interval of a Tethyan deep-sea section and its correlation with the North Sea basin: Bulletin de la Société Géologique de France, v. 171, p. 207–216.

Fechner, G.G., and Mohr, B., 1986, Zur tertiören Dinoflagellaten-Gattung *Carpatella* Grigorovich 1969 (Gonyaulacystaceae): Paläontologische Zeitschrift, v. 60, p. 181–188.

Fensome, R.A., Taylor, F.J.R., Norris, G., Sarjeant, W.A.S., Wharton, D.I., and Williams, G.L., 1993, A classification of fossil and living dinoflagellates: Micropaleontology, Special Publication, v. 7, p. 1–351.

Gavrilov, Y.O., and Muzylöv, N.G., 1991, The geochemistry of sapropelic interbeds in the Paleogene of the Central Caucasus: Lithology and Mineral Resources, v. 26, p. 485–590.

Gavrilov, Y.O., Kodina, L.A., Lubchenko, I.Y., and Muzylöv, N.G., 1997, The late Paleocene anoxic event in epicontinental seas of Peri-Tethys and formation of the Sapropelite Unit: Sedimentology and geochemistry: Lithology and Mineral Resources, v. 32, p. 427–450.

Haq, B.U., Hardenbol, J., and Vail, P.R., 1988, Mesozoic and Cenozoic chronostratigraphy and cycles of relative sea-level change, *in* Wilgus, C.K., et al., eds., Sea-level changes: An integrated approach: Society of Economic Paleontologists and Mineralogists Special Publication 42, p. 71–108.

Harrington, G.J., and Kemp, S.J., 2001, U.S. Gulf Coast vegetation dynamics during the latest Paleocene: Palaeogeography, Palaeoclimatology, Palaeoecology, v. 167, p. 1–21.

Heilmann-Clausen, C., 1985, Dinoflagellate stratigraphy of the uppermost Danian to Ypresian in the Viborg 1 borehole, central Jylland, Denmark: Danmarks Geologiske Undersøgelse, series A, v. 7, p. 1–69.

Heilmann-Clausen, C., and Egger, H., 2000, The Anthering outcrop (Austria): A key-section for correlation between Tethys and northwestern Europe near the Paleocene–Eocene boundary: GFF, v. 122, p. 69.

Heilmann-Clausen, C., and Schmitz, B., 2000, The late Paleocene thermal maximum $\delta^{13}C$ in Denmark?: GFF, v. 122, p. 70.

Iakovleva, A.I., Brinkhuis, H., and Cavagnetto, C., 2001, Late Paleocene–early Eocene dinoflagellate cysts from the Turgay Strait, Kazakhstan: Correlations across ancient seaways: Palaeogeography, Palaeoclimatology, Palaeoecology, v. 172, p. 243–268.

Jan du Chêne, R.E., and Adediran, S.A., 1984, Late Paleocene to early Eocene dinoflagellates from Nigeria: Cahiers de Micropaléontologie, Centre National de la Recherche Scientifique, v. 3, p. 5–38.

Katz, M.E., Pak, D.K., Dickens, G.R., and Miller, K.G., 1999, The source and fate of massive carbon input during the latest Paleocene thermal maximum: Science, v. 286, p. 1531–1533.

Kelly, D.C., Bralower, T.J., Zachos, J.C., Premoli Silva, I., and Thomas, E., 1996, Rapid diversification of planktonic foraminifera in the tropical Pacific (ODP Site 865) during the late Paleocene thermal maximum: Geology, v. 24, p. 423–426.

Kennett, J.P., and Stott, L.D., 1991, Abrupt deep sea warming, paleoceanographic changes and benthic extinctions at the end of the Paleocene: Nature, v. 353, p. 225–229.

Koch, P.L., Zachos, J.C., and Dettmann, D.L., 1995, Stable isotope stratigraphy and paleoclimatology of the Paleogene Bighorn basin (Wyoming, USA): Palaeogeography, Palaeoclimatology, Palaeoecology, v. 115, p. 61–89.

Köthe, A., 1988, Biostratigraphy of the Surghar Range, Salt Range, Sulaiman Range and the Kohat area, Pakistan, according to Jurassic through Paleogene calcareous nannofossils and Paleogene dinoflagellates: Geologisches Jahrbuch, v. B71, p. 3–87.

Lu, G., Adatte, T., Keller, G., and Ortiz, N., 1998, Abrupt climatic, oceanographic, and ecologic changes near the Paleocene–Eocene transition in the deep Tethys basin: The Alademilla section, southern Spain: Eclogae Geologicae Helvetiae, v. 91, p. 293–306.

Mudge, D.C., and Bujak, J.P., 1996, Paleocene biostratigraphy and sequence stratigraphy of the UK central North Sea: Marine and Petroleum Geology, v. 13, p. 295–312.

Norris, R.D., and Röhl, U., 1999, Carbon cycling and chronology of climate warming during the Paleocene–Eocene transition: Nature, v. 401, p. 775–778.

Powell, A.J., 1992, Dinoflagellate cysts of the Tertiary System, *in* Powell, A.J., ed., A stratigraphic index of dinoflagellate cysts: British Micropaleontological Society series, London, Chapman and Hall, p. 155–251.

Powell, A.J., Brinkhuis, H., and Bujak, J.P., 1996, Upper Paleocene–lower Eocene dinoflagellate cyst sequence biostratigraphy of southeast England, *in* Knox, R.W.O'B., et al., eds., Correlation of the early Paleogene in northwest Europe: Geological Society [London] Special Publication 101, p. 145–183.

Ravizza, G., Norris, R.N., and Blusztajn, J., 2001, An osmium isotope excursion associated with the late Paleocene thermal maximum: Evidence of intensified chemical weathering: Paleoceanography, v. 16, p. 155–163.

Robert, C., and Kennett, J.P., 1994, Antarctic subtropical humid episode at the Paleocene–Eocene boundary: Clay-mineral evidence: Geology, v. 22, p. 211–214.

Röhl, U., Bralower, T.J., Norris, R.D., and Wefer, G., 2000, New chronology for the late Paleocene thermal maximum and its environmental implications: Geology, v. 28, p. 927–930.

Scotese, C.P., and Golanka, J., 1992, Paleogeographic Atlas: PALEOMAP progress report 20–0692, Arlington, University of Texas, 34 p.

Smith, A., Smith, D., and Funnel, B., 1994, Atlas of Mesozoic and Cenozoic coastlines: Cambridge, UK, Cambridge University Press, 99 p.

Stott, L.D., Kennett, J.P., Shackleton, N.J., and Corfield, R.M., 1990, The evolution of Antarctic surface waters during the Paleogene: Inferences from stable isotopic composition of planktonic foraminifera, ODP Leg 113, *in* Barker, P.F., et al., eds., Proceedings of the Ocean Drilling Program, Scientific Results: Texas A & M University, College Station, Texas, USA, v. 113, p. 849–864.

Stover, L.E., Brinkhuis, H., Damassa, S.P., de Verteuil, L., Helby, R.J., Monteil, E., Partridge, A.D., Powell, A.J., Riding, J.B., Smelror, M., and Williams, G.L., 1996, Mesozoic–Tertiary dinoflagellates, acritarchs, and prasinophytes, *in* Jansonius, J., and McGregor, D.C., eds., Palynology: Principles and applications: American Association of Stratigraphic Palynologists Foundation, p. 641–750.

Thomas, D.J., Bralower, T.J., and Zachos, J.C., 1999, New evidence for subtropical warming during the late Paleocene thermal maximum: Stable isotope evidence from DSDP Site 527, Walvis Ridge: Paleoceanography, v. 14, p. 561–570.

Thomas, E., 1998, Biogeography of the late Paleocene benthic foraminiferal extinction, *in* Aubry, M.-P., et al., eds., Late Paleocene–early Eocene climatic and biotic events in the marine and terrestrial records: New York, Columbia University Press, p. 214–243.

Thomas, E., and Shackleton, N.J., 1996, The Paleocene–Eocene benthic foraminiferal extinction and stable isotope anomalies, *in* Knox, R.W.O'B., et al., eds., Correlation of the early Paleogene in northwest Europe: Geological Society [London] Special Publication 101, p. 401–444.

Thomas, E., and Zachos, J.C., 2000, Was the late Paleocene thermal maximum a unique event?: GFF, v. 122, p. 169–170.

Thomas, E., Zachos, J.C., and Bralower, T.J., 2000, Deep-sea environments on a warm Earth: Latest Paleocene–early Eocene, *in* Huber, B.T., et al., eds., Warm climates in Earth history: Cambridge, UK, Cambridge University Press, p. 132–160.

van Mourik, C.A., Brinkhuis, H., and Williams, G.L., 2001, Mid to late Eocene organic-walled dinoflagellate cysts from ODP Leg 171B, offshore Florida, *in* Kroon, D., et al., eds., Western North Atlantic Paleogene and Cretaceous palaeoceanography: Geological Society [London] Special Publication 183, p. 225–251.

Van Steenbergen, A., 1997, Palynological techniques for rock sampling: Utrecht, Netherlands, Laboratory of Palaeobotany and Palynology internal report, 25 p.

Wall, D., Dale, B., Lohmann, G.P., and Smith, W.K., 1977, The environmental and climatic distribution of dinoflagellate cysts in modern marine sediments from regions in the North and South Atlantic Oceans and adjacent seas: Marine Micropaleontology, v. 2, p. 121–200.

Williams, G.L., Lentin, J.K., and Fensome, R.A., 1998, The Lentin and Williams index of fossil dinoflagellates: American Association of Stratigraphic Palynologists Foundation, Dallas, 817 p.

Zachos, J.C., and Dickens, G.R., 2000, An assessment of the biogeochemical feedback response to the climatic and chemical perturbations of the LPTM: GFF, v. 122, p. 188–189.

Zachos, J.C., Stott, L.D., and Lohmann, K.C., 1994, Evolution of early Cenozoic marine temperatures: Paleoceanography, v. 9, p. 353–387.

Manuscript Accepted by the Society August 13, 2002

Printed in the USA

Geological Society of America
Special Paper 369
2003

Paleoproductivity of the northwestern Tethyan margin (Anthering Section, Austria) across the Paleocene-Eocene transition

Hans Egger*
Geologische Bundesanstalt, Rasumofskygasse 23, 1031 Wien, Austria
Juliane Fenner*
Bundesanstalt für Geowissenschaften und Rohstoffe, Postfach 510153, 30631 Hannover, Germany
Claus Heilmann-Clausen*
Geologisk Institut, Aarhus Universitet, 8000 Aarhus C, Denmark
Fred Rögl*
Naturhistorisches Museum, Burgring 7, 1010 Wien, Austria
Reinhard F. Sachsenhofer*
Universität Leoben, Institut für Geowissenschaften, Franz-Josef-Strasse 18, 8700 Leoben, Austria
Birger Schmitz*
Department of Earth Sciences, Marine Geology, Göteborg University, SE 405 30 Göteborg, Sweden

ABSTRACT

Sediments in the expanded Anthering Section at the northwestern Tethyan margin were deposited in an abyssal environment, at the continental rise to the south of the European plate. The section contains deposits from calcareous nannoplankton Zones NP9 and NP10 and displays the global negative carbon isotope excursion (CIE) in the upper part of Zone NP9. Associated with the CIE is a strong three-fold increase in the rate of hemipelagic sedimentation. This suggests an increased input of siliciclastic suspended material into the basin, which is indicative of enhanced continental run-off at that stratigraphic level. Concurrent acmes of siliceous plankton and dinoflagellate cysts indicate that a high input of dissolved nutrients affected even open marine settings and caused eutrophication of surface waters. The associated high flux of organic carbon to the seafloor resulted in oxygen-depleted conditions and caused the total extincton of benthic faunas. However, impoverished foraminifera faunas indicate that a change of ecological conditions started shortly before the CIE.

INTRODUCTION

A prominent decrease in the $^{13}C/^{12}C$ ratio of global carbon reservoirs (negative carbon isotope excursion—CIE) took place ca. 55 Ma ago, and is explained as the result of massive methane release due to the dissociation of gas hydrates (see Katz et al., 1999, for a review). This process may have been triggered by a rise in ocean bottom water temperature of between 4 and 8 °C which can be inferred from oxygen isotope values of foraminifera (e.g., Kennett and Stott, 1991). Coeval intense deep-sea carbonate dissolution indicates a shoaling of the calcite compensation depth as a result of the oxidation of methane to carbon dioxide (Dickens et al., 1995). The CIE is associated with a global extinction event of deep-sea benthic foraminifera (see

*E-mails: Egger—eggjoh@cc.geolba.ac.at; Fenner—J.Fenner@bgr.de; Heilmann-Clausen—claus.heilmann@geo.au.dk; Rögl—fred.roegl@nhm-wien.ac.at; Sachsenhofer—sachsenh@unileoben.ac.at; Schmitz—birger@gvc.gu.se.

Egger, H., Fenner, J., Heilmann-Clausen, C., Rögl, F., Sachsenhofer, R.F., and Schmitz. B., 2003, Paleoproductivity of the northwestern Tethyan margin (Anthering Section, Austria) across the Paleocene-Eocene transition, *in* Wing, S.L., Gingerich, P.D., Schmitz, B., and Thomas, E., eds., Causes and Consequences of Globally Warm Climates in the Early Paleogene: Boulder, Colorado, Geological Society of America Special Paper 369, p. 133–146.

Thomas, 1998, for a review), a rapid diversification of planktonic foraminifera (Lu and Keller, 1993), a global bloom of the dinoflagellate genus *Apectodinium* (Crouch et al., 2001), a turnover in calcareous nannoplankton (Bybell and Self-Trail, 1994), and a major turnover in land mammals (Wing et al., 1991). In the South Atlantic and Antarctica and in some Tethyan sections a high influx of kaolinite has been interpreted as an indicator for a change to more humid climatic conditions or, alternatively, as an effect of enhanced erosion under more arid conditions (see Thiry, 2000, for a review).

Over extensive areas in the southern and northern Tethys and Peri-Tethys, the CIE is coeval with sapropelic sedimentation (Speijer et al., 1997; Gavrilov et al., 2000; Schmitz et al., 1997a). Biogenic barium, which can be used as a proxy for paleoproductivity, indicates a dramatic increase in surface water fertility in the Eastern Tethys (Schmitz et al., 1997b) and in the North and South Atlantic (Bains et al., 2000). However, studies of microfossil assemblages, including these at sites where elevated barite accumulation has been documented, suggest that surface productivity actually decreased during the $\delta^{13}C$ excursion (Dickens et al., 2001). From the expanded and continuous sedimentary record of the Anthering Section (Fig. 1) an acme of siliceous plankton (Egger et al., 2000b) can be interpreted either as an indicator of high surface-water productivity or as a result of better preservation conditions for siliceous plankton.

The main objective of this paper is to explore the significance of the faunal and floral changes at the level of the CIE at Anthering. Since the influx of suspended terrestrial material into the basin is essential for the preservation of organic matter as well as for the level of dissolved nutrients in the surface water, we calculated hemipelagic sedimentation rates that can be used as a proxy for the amount of fluvial discharge. We studied temporal relationships between high sedimentation rates, acmes of planktic assemblages and the benthic foraminifera extinction event (BFEE). In order to discriminate between effects of preservation and signals of changed primary productivity, we focus on the ecological significance of *Apectodinium* and large diatom species (*Trinacria* spp. and *Craspedodiscus* spp.).

ENVIRONMENTAL SETTING

The 250 m thick upper Paleocene to lower Eocene deposits of the Austrian Anthering Section, spanning calcareous nannoplankton Zones NP9 and NP10, represent the most expanded marine sedimentary record of this interval known to date. These sediments comprise the youngest part of the Rhenodanubian Flysch that was deposited on the continental rise to the south of the European plate (Fig. 2), which was the main source for the siliciclastic detritus entering the basin. The section is composed of carbonaceous mud-turbidites with intervening hemipelagic shales, which indicate deposition below the calcite compensation depth. Paleo-water depth estimations by Butt (1981), using foraminifera assemblages, range between 3000 and 5000 m.

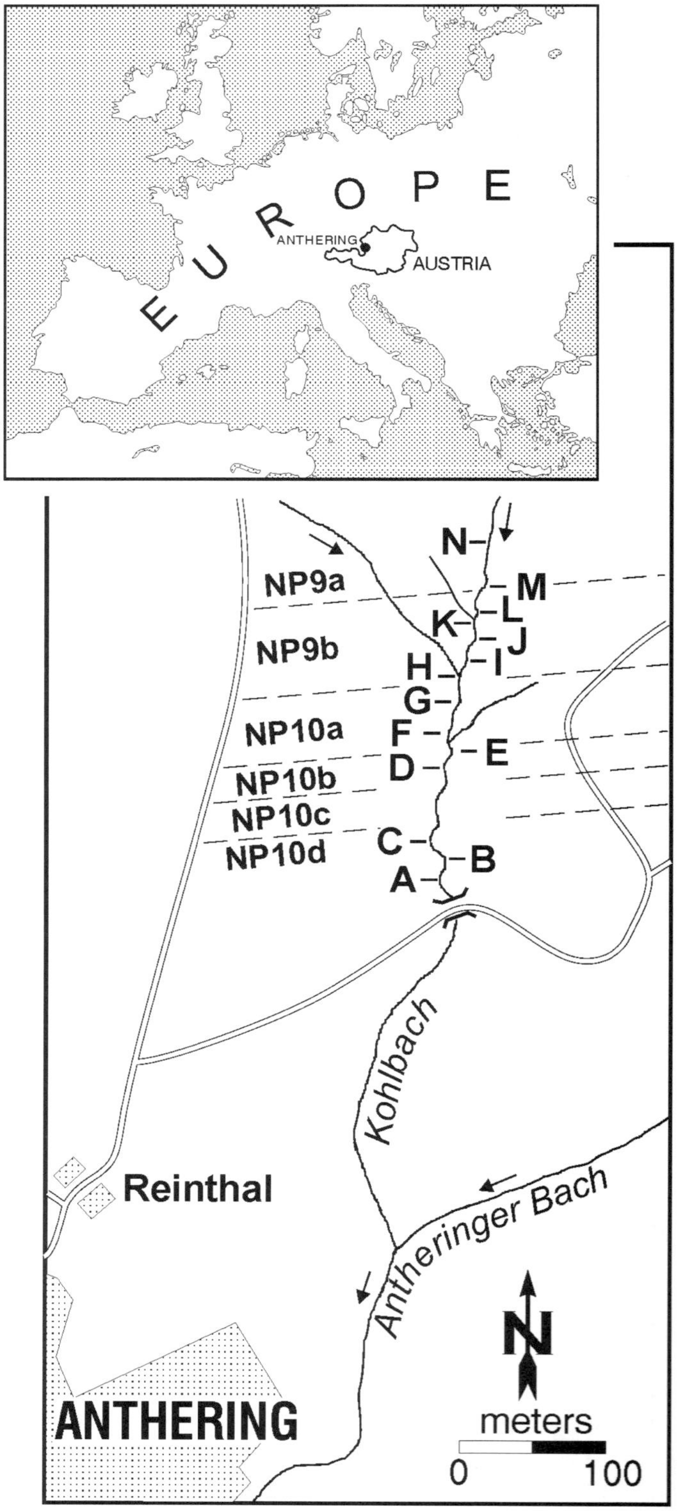

Figure 1. Position of the Anthering section, location of outcrops (A–N) and biostratigraphy of outcrops in terms of the calcareous nannoplankton standard zonation (NP Zones) of Martini (1971) and their further subdivision by Aubry (1996).

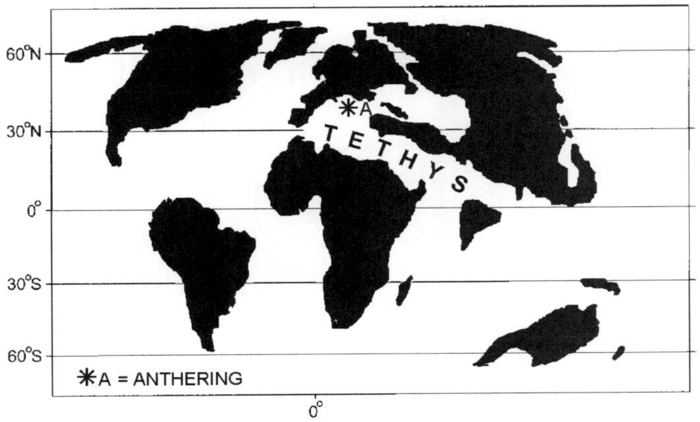

Figure 2. Paleogeographic reconstruction for the late Paleocene (modified from Scotese and Golanka, 1992) showing the location of Anthering.

For the Paleogene, a warm seasonal climate with alternating wet and dry conditions is indicated by the strong dominance of smectite in the clay mineral assemblages of hemipelagic shale. Close to the CIE, slightly increased percentages of kaolinite suggest a change to a more humid climate with high precipitation rates and associated enhanced continental run-off (Egger et al., 2002). This interpretation is consistent with terrestrial palynomorph assemblages in samples taken immediately above the onset of the CIE. These contain elements of the subtropical/tropical vegetation similar to the recent flora of Southeast Asia and southeast North America (Draxler, 2000).

SAMPLES AND METHODS

The stratigraphical framework of our study is based on the standard nannofossil zonation of Martini (1971), the subdivision of Zone NP10 by Aubry (1996) and the position of the CIE in the upper part of zone NP9 (Egger et al., 2000a). To show the temporal variations in turbidity current frequency, we counted the number of turbidite beds per meter of hemipelagic interturbidite sediment (Gawenda et al., 1999). These sediments are devoid of carbonate and, therefore, can be easily distinguished from carbonate-bearing turbidites (see Hesse, 1975, for a review). Additionally, we measured logs at a scale of 1:10 to assess the number of turbidites and the percentages of hemipelagic shale. Since only ~40% of the entire Anthering Section is exposed, we extrapolated these results to the complete thicknesses of the nannoplankton subzones to calculate the compacted hemipelagic and turbiditic sedimentation rates.

For micropaleontological, clay mineralogical and organic geochemical studies we made splits of the same 33 hemipelagic samples; these were taken at least 10 cm below the outcrop surface. Total organic carbon (TOC) contents were determined on decarbonated samples using a LECO 300 CS™ analyzer. Additionally, Rock Eval analyses (Espitalié et al., 1977) of 10 whole rock samples with increased percentages of organic matter were performed in duplicate using a Rock-Eval II+ instrument. Standard notations are used: S1 and S2 in milligrams of hydrocarbons (HC) per gram of rock; Tmax expressed in °C and the total organic carbon in wt%. The hydrogen index (HI = 100*S2/TOC) and oxygen index (OI = 100*S3/TOC) are expressed in mgHC/gTOC and mgCO_2/gTOC. Semiquantitative maceral evaluation of samples with a TOC content of >1% was performed using whole rock samples orientated perpendicular to bedding and a point counter. Between 500 and 1000 points per polished block were analysed using reflected white light and blue light excitation. Vitrinite reflectance was determined following established procedures and using a Leitz MPV-SP microscope.

Absolute dinoflagellate cyst abundances (cysts per gram sediment) were calculated following the method of Stockmarr (1971) which involves the addition of tablets with a known number of Lycopodium spores to the sediment samples during palynological processing. Semiquantitative abundances of agglutinating foraminifera and siliceous plankton from sieving residues (>125 μm) were determined from 32 shale samples. 200 g of each sample were processed for these analyses. In addition, the smaller size fractions were checked for remains of diatoms by treating the bulk sediment sample with 10% flouric acid and by further concentration using heavy liquid separation. Beside the taxonomic determination of the diatom valves, the relative abundance of the remains of the different siliceous microfossil groups and their preservation were examined. The samples were analysed with a lightmicroscope at a magnification of 1000x and with a scanning electron microscope.

For the paleoecological interpretation the deep water agglutinating foraminifera (DWAF) were grouped into assemblages ("morphogroups") according to modern analogues of life habitats and feeding strategies (e.g., Jones and Charnock, 1985; Kaminski et al., 1989; Kaminski et al., 1999). The diversity of DWAF has been calculated by using the Shannon-Wiener index (Lipps et al., 1979).

RESULTS

Sedimentation rates

The onset of the CIE is characterized by the presence of the thickest hemipelagic layers of the entire Anthering Section. About 45% of the rocks in the upper part of NP9 are shales, whereas the average percentage of shales in the overlying NP10 is only 14%, and even less in the lower part of NP9. The threefold increase in shale thickness can be explained either as a result of a decrease in the frequency of turbidity currents entering the basin or as a result of enhanced hemipelagic sedimentation. Sedimentation rates for the entire Anthering Section, incorporating both the turbidites and the hemipelagites, have been estimated as 9.1 cm/k.y.$^{-1}$ (Egger et al., 2000a). The sedimentation rate for the hemipelagic shale has been estimated as 1.8 cm/k.y.$^{-1}$. As this rate can be used as a proxy for the amount of fluvial discharge, we tried to get a better resolution by assessing this rate for the individual nannoplankton subzones of the section.

A four-fold subdivision of calcareous nannoplankton zone NP10 has been proposed by Aubry (1996) who estimated the duration of the subzones 10a, 10b, 10c, and 10d as 630 k.y., 200 k.y., 240 k.y., and 320 k.y. For zone NP9 we used estimations by Aubry et al. (1996) who calculated the duration of this zone as 1 Ma and the position of the CIE as 530 k.y. above the base of NP9.

Using these estimations we have calculated the rate of hemipelagic sedimentation (Fig. 3) as 1.3 cm/k.y. for the lower part of zone NP9 (120 m thick below the CIE), as 4.3 cm/k.y. for the upper part of NP9 (45 m thick), as 1.3 cm/k.y. for NP10a (50 m thick), as 0.9 cm/k.y. for NP10b (12 m thick) and as 0.7 cm/k.y. for NP10c (16 m thick). The sediments of subzone NP10d attain a thickness of 25 m, however, in contrast to previously published results (Egger et al., 2000a), we found *Tribrachiatus contortus* in our new samples from the very top of the section, suggesting

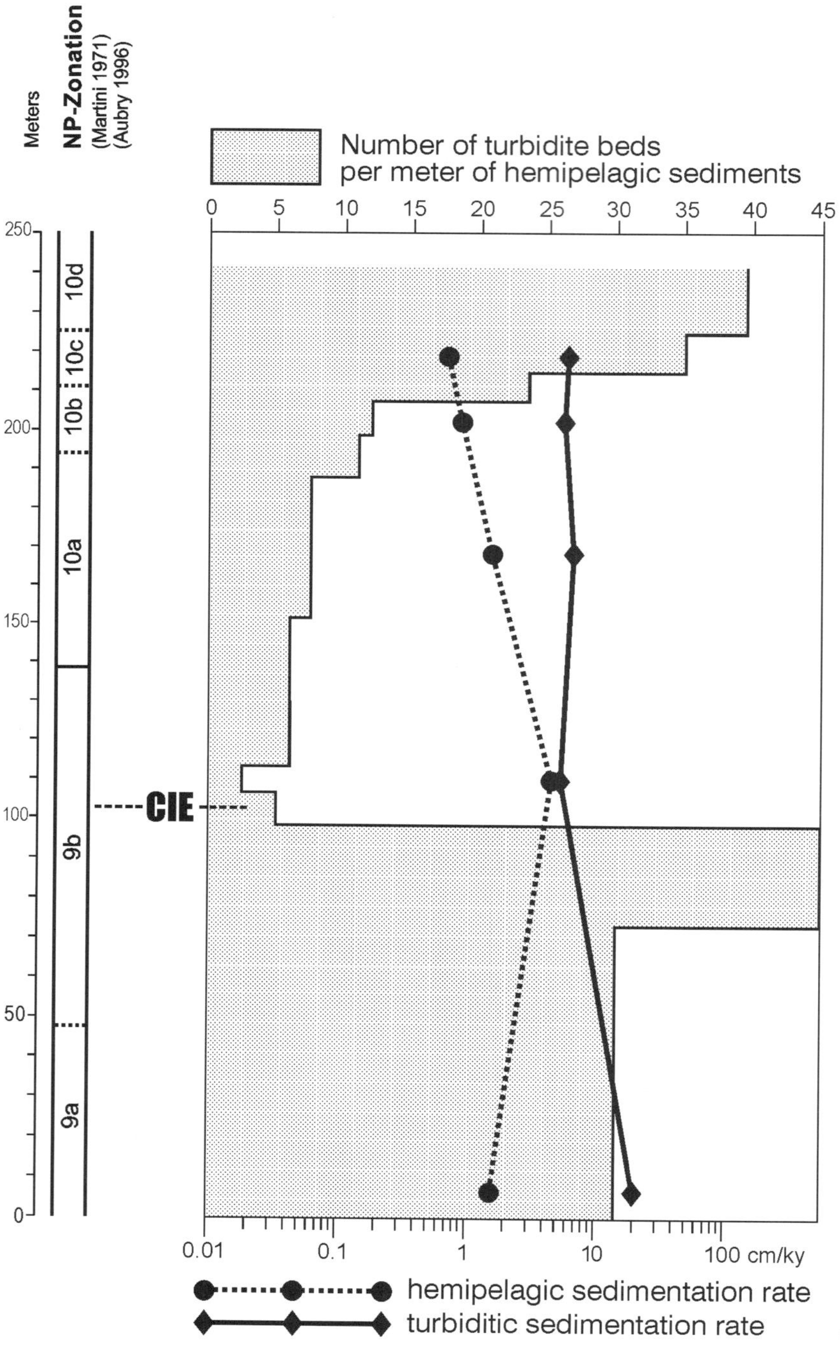

Figure 3. Relative frequency of turbidity currents and estimated turbidite and hemipelagic rates of sedimentation in nannofossil zones in the Anthering section. CIE—carbon isotope excursion.

that subzone NP10d is not complete. Therefore we did not estimate the rate of sedimentation for this subzone.

The strong increase in the rate of shale sedimentation at the CIE, by at least a factor of 3, might be partly an effect of shale deposition by bottom currents, as we found a few thin shale layers that show a transition into siltstones at their base. However, since the vast majority of shales do not display indications of redeposition, it seems unlikely that redeposition is a substantial source of the shale material. Obviously, the dramatically increased rate of hemipelagic sedimentation at the CIE suggests a high input of siliciclastic suspension into the basin. At the level of the CIE clay mineral assemblages of hemipelagic shale display a distinct increase of smectite and kaolinite at the expense of illite and chlorite (Egger et al., 2002). This indicates a decrease of bedrock erosion in the adjoining land areas. Well-developed smectitic soils with a mixture of kaolinite are mostly restricted to subtropical climates with a well-marked dry season (see Thiry, 2000, for a review). During the rainy season continental erosion of such areas is very pronounced (see van der Zwaan, 2002, for a review) and will result in a strong increase in hemipelagic sedimentation rates (Schmitz et al., 2001).

Turbiditic sedimentation rates display a decrease from 20 cm/k.y. in the lower part of zone NP9 to ~5 cm/k.y. in the upper part (Fig. 3). This is probably the result of a strong reduction of the siliciclastic sand-fraction in the turbidites: This fraction forms ~30% of the rocks in the lower part of NP9, but only 5% in the upper part of zone NP9 and in NP10. These different percentages of sandstone are used to distinguish between the older Altlengbach Formation and the younger Anthering Formation (Egger, 1995). The decrease in the input of the sand-fraction together with the coeval increase in the input of the clay-fraction suggests reduced availability of coarse material. This is consistent with the change from a temperate climate, with pronounced mechanical erosion, to a more humid climate with increased rates of soil formation.

Just before the CIE the rate of turbidite sedimentation dramatically increased and led to the deposition of the thickest turbidites of the entire Anthering Section. The calcareous nannoplankton assemblages of these thick turbidites consist almost exclusively of allochthonous Cretaceous species (Egger et al., 2000a). This suggests that during the latest Paleocene nonlithified Cretaceous sediments were mobilized as gravity flows by a very effective trigger mechanism. We suspect that the melting of gas hydrates out of the pore spaces of the Cretaceous sediments is responsible because methane can only escape from buried gas hydrate reservoirs through sediment failures (e.g., Katz et al., 1999).

Within zone NP10 no significant change in the rate of turbidite sedimentation occurred indicating that sediment accumulation in the source area of the turbidites remained stable. As the turbidites consist predominantly of penecontemporaneous calcareous plankton, it is likely that no significant change in the productivity of calcareous plankton took place within zone NP10. The concurrent increase in the relative frequency of turbiditic events was obviously a result of a decrease in the rate of hemipelagic sedimentation (Fig. 3).

Organic petrology and Rock-Eval pyrolysis

Vitrinite reflectance (~0.40% Rr) and Rock-Eval parameter Tmax (~418 °C) indicate a low maturity of the organic matter at Anthering (Table 1). As revealed by optical microscopy, the organic matter consists of marine material as well as terrestrially derived material. Samples ranging from 1.0 to 1.2 wt% TOC contain 1.8–2.5 vol% optically resolvable particles (macerals), and thus most of the organic matter is visible by optical microscopy. Marine-derived liptinite macerals, including liptodetrinite and lamalginite, are more common than sporinite, vitrinite, recycled vitrinite, and inertinite macerals of terrigenous origin. The ratio between marine and terrigenous components ranges from 1.1 to 1.5. Many liptinite macerals are characterized by orange fluorescence, which is unusual considering its low maturity, and might be an effect of oxidation. Some vitrinite particles show oxidation cracks. However, since the accompanying

TABLE 1. ROCK-EVAL AND VITRINITE REFLECTANCE DATA

Sample	TOC (wt.%)	S1 (mgHC/g rock)	S2 (mgHC/g rock)	S3 ($mgCO_2$/ g rock)	HI (mgHC/g rock)	OI ($mgCO_2$/g TOC)	T_{max} (°C)	Vitrinite Reflectance (%Rr/s)
DA66/1	0.45	0.05	0.22	0.65	48	144	–	–
EF2/98	1.22	0.19	0.82	0.27	67	23	412	0.38 / 0.05
E2/99	1.06	0.08	0.62	0.30	58	28	418	–
E1/99	1.03	0.08	0.72	0.39	69	38	418	0.36 / 0.05
JA3/99	1.01	0.08	0.58	0.46	58	46	421	0.44 / 0.06
J15/99	0.73	0.05	0.44	0.36	60	49	420	–
JF2/98	0.67	0.08	0.18	0.15	27	22	419	–
Jf3/98	0.65	0.08	0.37	0.27	57	42	418	–
J85/97	0.53	0.09	0.13	0.18	25	35	420	–
LF2/98	0.38	0.10	0.04	0.22	10	58	–	–

Note: TOC—total organic carbon; HI—hydrogen index; OI—oxygen index; Rr—mean random reflectivity; s—standard deviation.

pyrite framboids and pyritized radiolaria, which are particularly abundant in samples of outcrop E (E1/99, E2/99 and EF1/98), are unweathered. Oxidation must have occurred prior to the deposition of the organic debris. Reworking of vitrinite is also indicated by the common occurrence of particles with high reflectance, which does not equate with the low maturity of the samples. In particular, sample JA3/99 contains high amounts of inertinite.

Rock-Eval parameters are summarized in Table 1. HI values of the studied samples are below 70 mgHC/gTOC. In order to eliminate a possible mineral-matrix effect, S2 is plotted versus TOC (Langford and Blanc-Valleron, 1990) in Figure 4. According to the regression line, the true average hydrogen index is ~90 mgHC/gTOC. This value is typical for type III kerogen, which is essentially derived from terrestrial plants. This classification is inconsistent with the optical results, which revealed liptinite as the predominant maceral. Therefore we suspect that the low HI values are a result of the predepositional oxidation of both vitrinite and liptinite within the water column.

On the basis of the organic carbon contents of the hemipelagic shales, the Anthering section displays a threefold subdivision. Below the CIE the mean TOC value is calculated as 0.27%, from the CIE to the lower part of Zone NP10 this value increases to 0.58% while further up-section a distinct decrease to 0.16% can be identified. This varying content of organic carbon may reflect changes in primary productivity, changes in the preservation conditions of organic matter or varying oxygen levels of bottom waters.

Agglutinating foraminifera

The rich and diverse agglutinated faunas of the Anthering section offered a superb opportunity for a palecological study across the Paleocene–Eocene boundary interval in a well-dated abyssal setting. Individual samples contain up to 65 species and >700 specimens (Table 2 and Figure 5). More than 90 species were identified and grouped into four morphogroup assemblages (tubular genera, infaunal passive deposit feeders, active

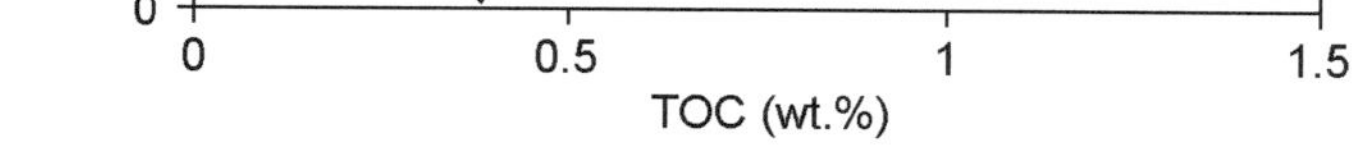

Figure 4. S2 vs. total organic carbon (TOC) diagram (Langford and Blanc-Valleron, 1990). The true average hydrogen index of samples from the Anthering Formation is 88 mgHC/gTOC.

deposit feeders, epifaunal active herbivores and omnivores). Distributional patterns of morphogroups of agglutinating foraminifera are related, more or less directly, to food supply and food utilization processes (Jones and Charnock, 1985).

At Anthering, tubular forms comprise the genera *Nothia, Rhabdammina, Rhizammina, Psammosiphonella* and *Bathysiphon.* These typical "flysch-type" elements have been interpreted as sessile suspension feeders (morphogroup A of Jones and Charnock). However, the ecological interpretation of some of these deep-sea genera is still under discussion (Gooday et al., 1997), e.g., the life habitat of *Nothia* has been reinterpreted as epibenthic detrivore (Geroch and Kaminski, 1992). Epi- and infaunal passive deposit feeders (morphogroup B1) comprise *Sascammina, Psammosphaera, Hormosina, Hormosinella, Trochamminoides, Paratrochamminoides, Lituotuba, Hyperammina* and *Kalamopsis.* Another epifaunal and shallow infaunal group of active deposit feeders (morphogroup B2) corresponds to the *Ammodiscus-Glomospira* assemblage of "Biofacies B" (Kuhnt et al., 1989). It consists of the genera *Ammodiscus, Glomospira* and *Rzehakina.* The B3 assemblage of epifaunal active herbivores and omnivores (*Haplophragmoides, Trochammina* s.l.) may be restricted to omnivores in this deep-sea environment. The C-morphogroup of infaunal forms (*Gerochammina, Karrerulina, Reophax, Subreophax, Spiroplectammina*) are negligible in the abyssal setting of the Anthering section. The genera *Recurvoides* and *Thalmannammina* were summarized as *Recurvoides*-assemblage. The microhabitat preferences of this assemblage are questionable. In the Cretaceous "Hatteras Fauna" of the Fardes Formation in southern Spain it co-occurs with *Glomospira* and *Ammodiscus,* and might, therefore, be indicative of oxygen deficient conditions (Kaminski et al., 1999). In our samples we did not find this correlation because the highest percentages of the *Recurvoides*-assemblage occur in high-diversity faunas without any indication of oxygen depletion. It is noteworthy that the *Recurvoides*-assemblage usually forms >10% of the agglutinated faunas within nannoplankton zone NP9, whereas in zone NP10 this percentage is much lower.

The highest diversity and the highest abundance of agglutinated specimens occur in the lower part of the section (samples NF2 to LF1). These assemblages display balanced proportions of infaunal, epifaunal and suspension feeding species. The high diversity of these agglutinated faunas is seen as typical for oligotrophic, food-limited environments where the various microhabitats are fully occupied. Several taxa have their last occurrences in this part of the section: *Ammodiscus cretaceus, Aschemocella* cf. *carpathica, A. grandis, Haplophragmoides horridus, H. suborbicularis, Hormosina trinitatensis, Karrerulina* cf. *coniformis, Paratrochamminoides heteromorphus, P. multilobus, Recurvoides walteri, Remesella varians, Rzehakina complanata, R. epigona, R. fissistomata, Spiroplectammina* cf. *dentata, Spiroplectammina spectabilis, Thalmannammina* nov. sp., *Thurammina papillata.*

Further up-section (samples J85 to JaF1) impoverished faunas with a predominance of the genus *Glomospira* appear. This

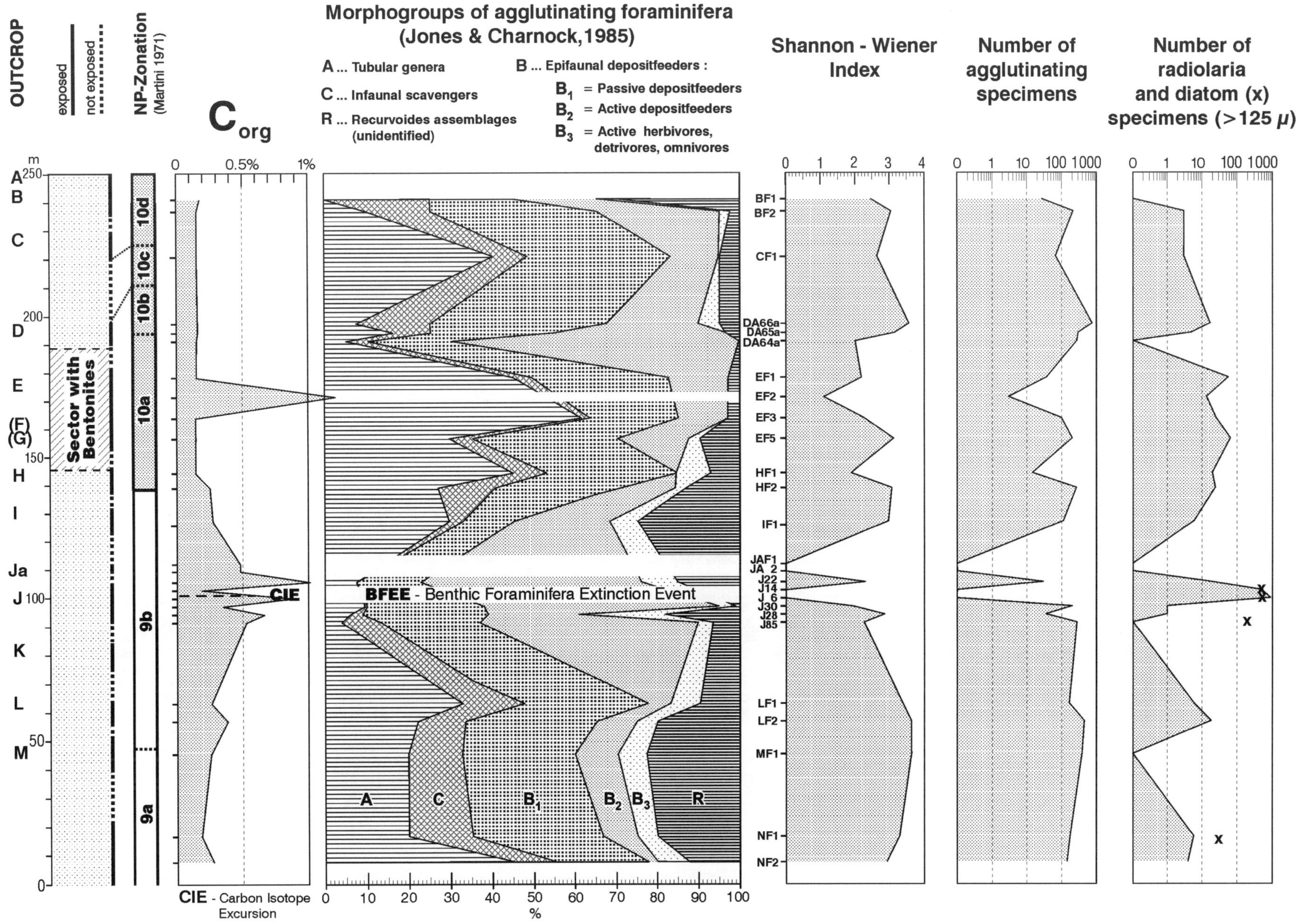

Figure 5. Distribution of organic carbon, agglutinating foraminifera and siliceous plankton within the Anthering section.

TABLE 2. DISTRIBUTION OF DEEP-WATER AGGLUTINATING FORAMINIFERA (NUMBER OF SPECIMENS) IN THE ANTHERING SECTION

sample	*Ammobaculites? sp.*	*Ammodiscus cretaceus* (REUSS)	*Ammodiscus latus* GRZYBOWSKI	*Ammodiscus planus* LOEBLICH	*Ammodiscus tenuissimus* GRZYBOWSKI	*Ammodiscus glabratus* CUSHMAN & RENZ	*Ammosphaeroidina pseudopauciloculata* (MYATLYUK)	*Annectina grzybowskii* (JURKIEWICZ)	*Aschemocella* cf. *carpathica* (NEAGU)	*Aschemocella grandis* (GRZYBOWSKI)	*Bathysiphon sp.*	*Bathysiphon? annulatus* (ANDREAE)	*Buzasina pacifica* (KRASHENINNIKOV)	*Caudammina? sp.*	*Dendrophrya sp.*	*Gerochammina conversa* (GRZYBOWSKI)	*Glomospira charoides* (JONES & PARKER)	*Glomospira diffundens* CUSHMAN & RENZ	*Glomospira glomerata* (GRZYBOWSKI)	*Glomospira gordialis* (JONES & PARKER)	*Glomospira serpens* (GRZYBOWSKI)	*Glomospira? irregularis* (GRZYBOWSKI)	*Haplophragmoides horridus* (GRZYBOWSKI)	*Haplophragmoides suborbicularis* (GRZY.)	*Haplophragmoides walteri* (GRZYBOWSKI)	*Haplophragmoides spp.*	*Hormosina* cf. *pilulifera* BRADY	*Hormosina trinitatensis* (CUSHMAN & RENZ)	*Hyperammina dilatata* GRZYBOWSKI	*Hyperammina nuda* SUBBOTINA	*Hyperammina sp.*	*Kalamopsis grzybowskii* (DYLAZANKA)	*Karrerulina* cf. *coniformis* (GRZYBOWSKI)	*Karrerulina horrida* (GRZYBOWSKI)	*Lituotuba? vermetiformis* (GRZYBOWSKI)	*Nothia communis* (PFLAUMANN)	*Nothia excelsa* (GRZYBOWSKI)	*Nothia grilli* (NOTH)	*Nothia latissima* (GRZYBOWSKI)	*Nothia robusta* (GRZYBOWSKI)	*Nothia subalpina* (PFLAUMANN)	*Paratrochamminoides deformis* (GRZYBOWSKI)	*Paratrochamminoides draco* (GRZYBOWSKI)	*Paratrochamminoides gorayskii* (GRZYBOWSKI)	*Paratrochamminoides heteromorphus* (GRZYBOWSKI)	*Paratrochamminoides multilobus* (DYLAZANKA)	*Paratrochamminoides olszewskii* (GRZYBOWSKI)	*Paratrochamminoides spp.*
BF1/98	–	–	–	–	–	–	–	1	–	–	–	–	–	–	–	4	5	–	–	1	–	–	–	–	–	–	–	–	–	–	–	–	–	–	–	–	–	–	–	–	–	–	1	–	–	–	–	–
BF2/98	–	–	1	20	–	–	–	–	–	–	7	–	1	–	–	7	54	3	10	6	–	5	–	–	–	–	–	–	–	13	–	–	–	8	–	–	–	–	–	–	–	–	6	–	–	–	7	9
CF1/98	–	–	–	–	–	–	–	–	–	–	–	–	–	–	–	1	6	–	1	1	–	3	–	–	–	–	2	–	–	4	–	–	–	2	–	1	–	1	–	–	18	–	1	–	–	–	–	–
DA66A	–	–	–	14	–	–	–	7	–	–	–	4	1	–	–	–	81	20	37	13	4	20	–	–	–	10	4	–	–	46	–	–	–	–	9	14	9	17	–	16	29	12	30	–	–	–	36	25
DA65A	–	–	–	3	–	–	–	15	–	–	6	2	–	–	–	–	58	17	20	4	1	6	–	–	–	–	–	–	–	13	–	1	–	–	–	–	–	4	–	10	9	2	12	–	–	–	19	9
DA64a	–	–	–	2	–	–	–	3	–	–	–	–	–	–	–	–	135	15	23	7	–	13	–	–	–	–	–	–	–	–	–	–	–	–	2	–	–	–	–	–	–	–	5	–	–	5	8	–
EF1/98	–	–	–	–	–	–	–	–	–	–	–	–	–	–	–	–	5	–	–	–	–	–	–	–	–	–	–	–	–	3	–	–	–	–	–	6	–	–	–	6	3	–	–	–	–	–	–	–
EF2/98	–	–	–	–	–	–	–	–	–	–	–	–	–	–	–	–	–	–	–	–	–	–	–	–	–	–	–	–	–	1	–	–	–	–	–	–	1	–	–	–	–	–	–	–	–	–	–	–
EF3/98	–	–	2	1	–	–	–	–	–	–	–	–	–	–	–	–	6	–	1	1	–	–	–	–	–	–	–	–	–	–	–	–	–	–	–	9	–	1	–	15	30	–	1	–	–	–	–	4
EF5/98	–	–	–	2	–	–	–	1	–	–	–	1	–	–	–	–	22	7	8	3	–	8	–	–	–	–	–	–	–	–	–	–	–	–	4	1	–	6	–	22	19	–	12	–	–	–	13	8
HF1/98	–	–	–	–	–	–	–	–	–	–	–	–	–	–	–	–	1	–	–	–	–	1	–	–	–	–	–	–	–	–	–	–	–	–	–	–	–	–	–	2	5	–	1	–	–	–	–	2
HF2/98	–	–	–	3	–	–	1	–	–	–	9	–	2	–	–	–	32	–	12	1	–	3	–	–	–	3	1	–	–	4	–	–	–	–	–	–	8	5	–	7	28	–	13	–	–	–	13	–
IF1/98	–	–	–	1	–	–	–	–	–	–	–	–	3	–	–	–	14	2	2	4	1	1	–	–	–	1	–	–	–	–	–	–	–	–	–	4	–	7	–	7	11	–	1	2	–	–	3	–
JaF1/98	–	–	–	–	–	–	–	–	–	–	–	–	–	–	–	–	–	–	–	–	–	–	–	–	–	–	–	–	–	–	–	–	–	–	–	–	–	–	–	–	2	–	–	–	–	–	–	–
J22/98	–	–	–	–	–	–	–	–	–	–	–	–	3	–	–	–	2	1	–	2	–	–	–	–	–	2	–	–	–	–	–	–	–	–	–	–	–	–	–	–	–	–	–	–	–	–	–	–
J17/98	–	–	–	–	–	–	1	–	–	–	–	–	–	–	–	–	3	–	–	3	–	2	–	–	–	1	–	–	–	–	–	–	–	–	–	–	–	–	–	–	1	–	–	–	–	–	–	–
J9/97	–	–	–	–	–	–	–	–	–	–	–	–	–	–	–	–	1	1	–	1	–	–	–	–	–	–	–	–	–	–	–	–	–	–	–	–	–	–	–	–	–	–	–	–	–	–	–	–
J30/97	–	–	–	2	–	–	2	–	–	–	–	–	2	–	–	–	85	6	–	22	–	1	–	–	–	7	–	–	–	12	2	–	–	–	–	–	2	–	–	–	–	–	–	–	–	–	–	–
J28/98	–	–	–	–	–	–	–	–	–	–	–	–	–	–	–	–	3	–	–	2	2	1	–	–	2	2	2	–	–	–	–	–	–	–	–	–	–	–	–	–	–	–	–	–	–	–	–	–
J85/97	–	–	–	2	2	–	4	–	–	–	1	–	6	–	–	–	68	37	–	18	–	1	–	–	–	3	–	–	–	7	–	–	1	1	–	–	–	4	–	–	–	–	–	–	–	–	–	–
LF1/98	–	–	–	1	–	–	1	–	9	8	6	–	5	–	–	–	–	–	–	1	2	4	–	–	3	–	4	3	1	1	–	–	–	–	–	–	7	12	–	1	–	–	–	–	–	1	3	3
LF2/98	–	14	4	–	6	1	15	5	3	21	–	–	12	2	2	–	–	1	–	–	9	21	–	1	10	8	18	–	–	2	1	2	–	–	–	28	5	–	–	3	12	–	2	–	9	1	8	1
MF1/98	–	5	2	–	4	4	1	6	–	6	–	–	9	4	8	20	3	–	–	1	13	10	–	1	18	3	17	1	–	1	9	–	–	–	1	11	4	–	–	6	11	–	7	–	–	–	2	9
NF1/98	–	–	–	2	–	–	–	–	–	6	–	–	3	–	–	2	5	1	–	–	4	6	–	–	–	7	12	–	–	5	–	3	1	–	1	5	–	3	–	–	6	–	–	–	–	–	–	3
NF2/98	3	–	–	–	–	–	–	1	–	3	3	–	2	–	–	–	–	–	–	2	–	5	3	–	–	–	7	–	–	–	–	–	–	–	–	1	–	27	7	1	4	–	–	–	–	–	–	4

TABLE 2. ***Continued***

sample	*Plectorecurvoides parvus* KRASHENINNIKOV	*Psammosiphonella cylindrica* (GLAESSNER)	*Psammosiphonella* sp.	*Psammosphaera irregularis* (GRZYBOWSKI)	*Recurvoidella lamella* (GRZYBOWSKI)	*Recurvoides anormis* MYATLYUK	*Recurvoides immane* (GRZYBOWSKI)	*Recurvoides nucleolus* (GRZYBOWSKI)	*Recurvoides pseudoregularis* MYATLYUK	*Recurvoides* cf. *recurvoidiformis* NEAGU & TOCOR.	*Recurvoides retroseptus* (GRZYBOWSKI)	*Recuvoides smugarensis* MYATLYUK	*Recurvoides varius* MYATLYUK	*Recurvoides walteri* (GRZYBOWSKI)	*Recurvoides spp.*	*Remesella varians* (GLAESSNER)	*Reophax elongatus* (GRZYBOWSKI)	*Reophax nodulosus* (GRZYBOWSKI)	*Rhabdammina spp.*	*Rhizammina* cf. *algaeformis* BRADY	*Rhizammina spp.*	*Rzehakina complanata* (GRZYBOWSKI)	*Rzehakina epigona* (RZEHAK)	*Rzehakina fissistomata* (GRZYBOWSKI)	*Saccammina grzybowskii* (SCHUBERT)	*Saccammina placenta* (GRZYBOWSKI)	*Silicobathysiphon spp.*	*Spiroplectammina* cf. *dentata* (ALTH)	*Spiroplectammina spectabilis* (GRZYBOWSKI)	*Subreophax pseudoscalaria* SAMUEL	*Subreophax scalaria* (GRZYBOWSKI)	*Subreophax splendidus* (GRZYBOWSKI)	*Subreophax spp.*	*Thalmannammina subturbinata* (GRZYBOWSKI)	*Thalmannammina n.sp.* (large)	*Thuramina papillata* BRADY	*Trochammina? spp.*	*Trochamminoides dubius* (GRZYBOWSKI)	*Trochamminoides folius* (GRZYBOWSKI)	*Trochamminoides grzybowskii* KAMINSKI & GEROCH	*Trochamminoides proteus* (KARRER)	*Trochamminoides septatus* (GRZYBOWSKI)	*Trochamminoides subcoronatus* (GRZYBOWSKI)	*Trochamminoides variolarius* (GRZYBOWSKI)	*Trochamminoides spp.*
BF1/98	–	–	–	–	4	–	–	3	–	–	–	–	–	–	4	–	2	–	–	–	–	–	–	–	1	–	–	–	–	–	1	–	–	–	–	–	–	2	1	–	–	–	–	–	2
BF2/98	–	4	–	–	–	–	–	–	1	–	–	–	–	–	1	–	–	–	–	17	4	–	–	–	2	3	1	–	–	–	23	3	8	5	–	–	1	12	14	–	6	–	15	–	2
CF1/98	–	–	–	–	–	–	–	–	–	–	–	–	–	–	2	–	–	–	–	–	6	–	–	–	2	4	–	–	–	–	1	–	1	1	–	–	–	2	–	–	–	–	2	2	–
DA66A	–	–	–	4	19	–	–	–	–	–	1	9	–	–	9	–	11	29	6	13	5	–	–	–	15	23	14	–	–	7	30	20	33	18	–	–	2	15	34	–	1	–	28	9	2
DA65A	–	4	–	–	1	–	–	–	–	–	–	1	–	–	4	–	4	6	–	4	5	–	–	–	–	4	1	–	–	–	5	13	–	4	–	–	–	5	12	–	–	–	4	13	–
DA64a	–	1	–	–	–	–	–	–	–	–	–	–	–	–	–	–	–	–	1	–	6	–	–	–	–	1	7	–	–	2	–	–	6	–	–	–	–	13	–	–	–	–	13	–	–
EF1/98	–	–	–	5	–	–	–	–	–	–	–	–	–	–	–	–	–	–	–	–	–	–	–	–	–	1	–	–	–	–	–	1	–	1	–	–	–	1	–	–	–	–	–	2	–
EF2/98	–	–	–	–	–	–	–	–	–	–	–	–	–	–	1	–	–	–	–	–	–	–	–	–	–	–	–	–	–	–	–	–	–	–	–	–	–	–	–	–	–	–	–	–	–
EF3/98	–	–	–	7	–	–	–	–	–	–	–	2	–	–	–	–	–	–	–	7	–	–	–	–	–	7	–	–	–	–	–	–	–	1	–	–	–	2	–	–	–	–	–	–	–
EF5/98	–	–	3	4	–	–	3	–	–	–	2	4	–	–	4	–	–	–	–	4	–	–	–	–	3	5	–	–	–	–	4	4	1	7	–	–	–	2	–	–	1	–	–	8	–
HF1/98	–	–	–	–	–	–	–	–	–	–	–	–	–	–	1	–	–	–	2	–	–	–	–	–	–	–	–	–	–	–	–	–	–	–	–	–	–	–	–	–	–	–	–	–	–
HF2/98	8	–	8	–	–	–	7	–	–	–	–	5	–	–	5	–	–	–	–	–	2	–	–	–	9	8	–	–	–	4	–	26	1	4	–	–	–	–	11	–	1	–	1	–	–
IF1/98	10	–	–	1	1	–	–	6	–	–	–	4	–	–	10	–	–	–	–	–	3	–	–	–	4	–	–	–	–	–	–	1	1	3	–	–	–	1	–	–	–	–	–	–	–
JaF1/98	–	–	–	–	–	–	–	–	–	–	–	–	–	–	–	–	–	–	–	–	–	–	–	–	–	–	–	–	–	–	–	–	–	–	–	–	–	–	–	–	–	–	–	–	–
J22/98	–	–	–	–	1	–	–	–	–	–	–	–	–	–	2	–	–	–	–	–	4	–	–	–	5	2	–	–	–	–	–	–	–	–	–	–	3	–	–	–	–	–	–	–	–
J17/98	–	–	–	–	–	–	–	–	–	–	–	–	–	–	2	–	–	–	–	–	–	–	–	–	2	–	–	–	–	–	–	–	–	–	–	–	–	–	–	–	–	–	–	–	–
J9/97	–	–	–	–	–	–	–	–	–	–	–	–	–	–	2	–	–	–	–	–	–	–	–	–	2	–	–	–	–	–	–	–	–	–	–	–	–	–	–	–	–	–	–	–	–
J30/97	–	–	6	–	–	–	–	–	–	–	–	–	–	–	2	–	–	–	–	11	–	–	–	–	41	1	–	–	–	–	–	1	–	–	–	–	1	–	–	–	–	–	–	–	–
J28/98	–	–	2	1	–	–	1	–	–	–	1	–	–	–	1	–	–	–	–	–	–	–	–	–	1	1	–	–	–	–	–	–	1	1	–	–	–	1	–	–	2	–	–	1	–
J85/97	–	–	–	–	1	–	–	–	–	–	–	–	–	–	11	–	–	–	3	–	–	–	–	–	49	1	2	–	–	–	15	6	–	–	–	–	1	–	–	–	–	–	–	–	–
LF1/98	–	5	6	11	–	–	–	–	1	–	–	–	–	–	5	–	1	–	14	–	–	–	1	–	5	6	–	–	–	–	1	1	3	3	6	–	3	2	2	–	1	–	–	1	–
LF2/98	–	6	23	13	–	7	7	–	–	5	–	–	1	–	21	–	–	–	14	–	–	1	1	–	7	24	2	–	1	2	10	6	6	–	25	1	1	4	7	–	4	–	4	–	–
MF1/98	–	–	10	–	–	3	7	–	–	–	–	–	2	1	9	–	–	2	18	–	17	–	1	1	4	23	3	–	–	–	8	3	10	–	37	–	6	2	5	–	1	–	13	1	1
NF1/98	–	2	4	–	–	–	–	–	–	–	3	5	1	3	20	–	–	1	13	–	2	–	–	–	6	13	–	1	11	–	1	1	1	–	4	–	–	–	3	–	–	1	–	–	–
NF2/98	–	–	5	–	–	–	–	–	3	–	5	–	–	–	11	1	–	–	17	–	–	–	–	–	–	6	–	–	2	–	–	1	2	–	–	1	–	–	1	–	–	–	2	2	1

"*Glomospira* event" has been observed at numerous localities in the Tethys and northern North Atlantic (see Kaminski et al., 1996, for a review). Kaminski et al. (1989) speculated that the predominance of *Glomospira* indicates areas of high surface productivity that caused low oxygen levels at the seafloor. However, this assemblage occurs also in well-oxidized sediments and, therefore, it may be opportunistic rather than a reliable indicator for high productivity (Galeotti et al., 2000; Kaminski et al., 1996). With the onset of the CIE, even this opportunistic assemblage disappeared and over a period of at least 180,000 yr the benthic communities suffered severely from unfavorable habitat conditions.

Between samples HF2 to EF1 the majority of the hemipelagic layers have an organic carbon content between 0.14% and 0.17% (0.15% on average), but several black shale layers (up to 1.22% TOC) occur. This suggests periodic eutrophication of the sea water probably by volcanic ashfall as closely spaced bentonites were found in that part of the section (Egger et al., 2000a). The black shales are usually devoid of benthic foraminifera and contain common framboidal pyrite indicating anoxic conditions (Egger et al., 1997). The agglutinating faunas of these layers are not as rich and diverse as those from further down the section. *Glomospira glomerata* has its first appearance in this part of the section. The faunal assemblage changed to a predominance of passive deposit feeders (B1-assemblage) and tubular genera (A-assemblage). These assemblages are dominant along the continental rises where bottom currents or distal turbidity currents occur (Kaminski et al., 1996).

In the uppermost part of the Anthering section (samples DA64a to BF1) a strong increase in the number of species and specimens of the DWAF, with relatively balanced assemblages, occurs indicating the return of ecological conditions similar to those at the base of the section.

Siliceous plankton

Throughout the Anthering section the fossil remains of siliceous plankton (radiolaria, diatoms as well as rare ebridians and silicoflagellates) have been replaced by pyrite. Silica dissolution prior to this replacement, and damage caused by the pressure of pyrite crystals growing inside the shells, can make identification difficult. In particular, radiolarians are very poorly preserved and are all taxonomically indeterminate spheroidal or lenticular spumellarians (Christopher Hollis, 2001, personal commun.). If pyrite fillings only are preserved, the outline and shape of diatom frustules can be recognized, but a specific and often generic determination is impossible. However, in the more robust frustules, even relatively fine pores and cribra covering the areolae are preserved, and thus allow species determination.

Most samples have diatom floras dominated by the taxa *Paralia sulcata* var. *biseriata, Paralia sulcata* var. *crenulata, Coscinodiscus antiquus,* and by species of the genera *Auloplicata* and *Stephanopyxis.* The recent relatives of the latter two genera occur in coastal-neritic as well as in oceanic environments. This may also be the case for the less common species of the genera *Hemiaulus* (e.g., *H. peripterus), Actinoptychus* and *Sceptroneis.* Species of the genus *Trochosira,* which are also rather rare, are considered to have been fully planktonic, whereas specimens of *Craspedodiscus, Trinacria, Sheshukovia* and *Aulacodiscus* probably indicate a coastal-neritic environment. Other genera can be considered to have been fully benthic, e.g., species of the genera *Auliscus* and *Arachnodiscus.* In neritic assemblages, resting spores should be abundant, but in the studied samples only single specimens of resting spores were found. These belong to the form groups *Xanthiopyxis, Pterotheca* and *Bicornis.* As resting spores are most resistant to dissolution, their scarcity indicates that the encountered diatoms represent an oceanic assemblage (Fenner, 1994). The minor admixture of coastal and neritic specimens may have been caused by storm events that whirled up freshly deposited sediment in shallow regions, which thereafter settled out from suspension beyond the shelf edge.

Dinoflagellates

Information on the distribution of organic-walled dinoflagellate cysts in the Anthering section has previously been briefly published by Egger et al. (1997, 2000a), Heilmann-Clausen and Egger (2000) and Crouch et al. (2001). With results obtained during the present study the data can be summarized as follows. Common genera and species occurring throughout the section are *Apectodinium* spp., *Areoligera* spp., *Glaphyrocysta* spp., *Spiniferites* spp., *Polyspaeridium zoharyi, Homotryblium tenuispinosum, Operculodinium centrocarpum,* and *Phthanoperidinium crenulatum. Lingulodinium machaeorophorum* occurs sporadically and is usually rare.

The overall composition of the dinoflagellate assemblages allows a simple subdivision of the section into three parts: The lower and upper interval are characterized by generally low dominance and relatively high species richness. These two intervals are separated by a middle interval coinciding with the CIE (outcrops J and JA). There the genus *Apectodinium,* which usually accounts for 5%–20% of the dinoflagellate assemblages, is dominant and reaches abundances up to 69% in hemipelagic samples. The genus *Apectodinium* includes several intergrading species, which form a closely related group (Harland, 1979). In spite of the strong dominance, the species richness remains relatively high within the CIE interval.

Quantitative dinoflagellate cyst data from hemipelagic layers at outcrop J (Fig. 6) reveal there is a 10-fold to 40-fold increase in the total number of cysts within the CIE interval (where *Apectodinium* dominates). Above the CIE, counts reveal fluctuations in cyst numbers with a general trend toward reduction, which parallel the declining percentages of *Apectodinium.* The total number of cysts is low in sample J4 from the CIE-interval, with dominant *Apectodinium,* but this sample is from a layer with a minimum in the TOC (Fig. 6) which suggests increased oxygenation at this level. The *Apectodinium* percentages are in-

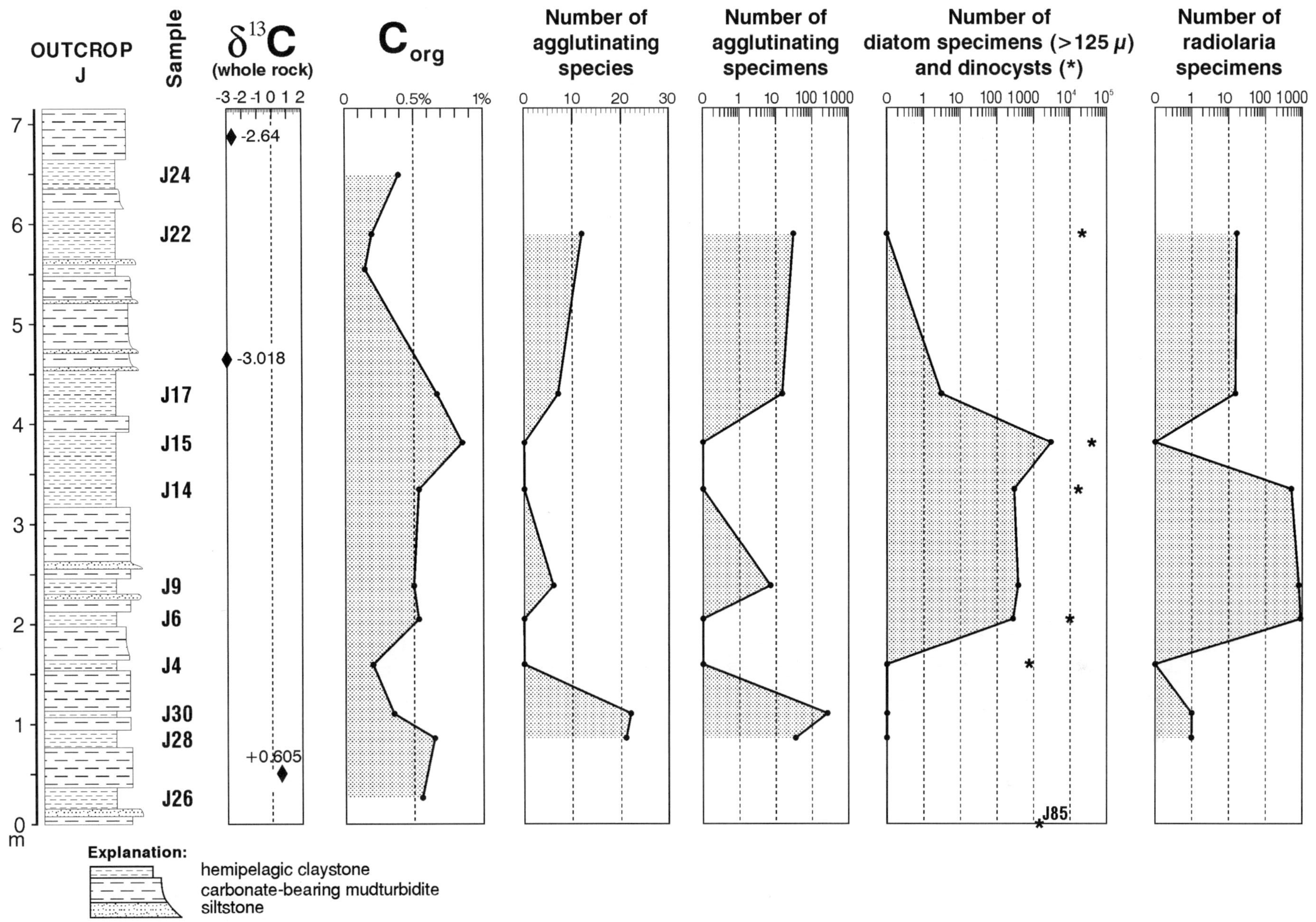

Figure 6. Detailed section of the onset of the negative carbon isotope event (CIE) displaying abundances of agglutinating foraminifera and siliceous plankton in 200 g sediment and abundances of dinoflagellates in 1 g sediment. Sample J85 was taken ~4 m below the base of the measured section.

creased in sample J26 (33%) and J28 (62%), but no data of absolute dinoflagellate abundances are available for these two samples.

Relying on information from modern cyst production (e.g., Dale, 1996), the Anthering Section must have been deposited below neritic waters, or waters that originated in the neritic zone. The genus *Impagidinium,* which is today purely oceanic, is present in several samples (especially in outcrop N), but usually rarer than 1%–2%. Such low occurrences indicate the neritic/oceanic boundary interval (Dale, 1996). It should be noted that the boundary between neritic and oceanic waters today does not strictly follow bathymetry and presence of neritic waters need not conflict with the interpretation of the depositional environment at Anthering being an abyssal basin floor (Egger et al., 2000a). It is also well known that neritic cysts are today transported over long distances with currents, and are deposited in various basinal parts of the Atlantic Ocean (e.g., Dale, 1996).

The continuous presence of *Polysphaeridium zoharyi* and *Homotryblium tenuispinosum* is evidence of a rather constant, and significant, mixing of the water masses at Anthering. *Polysphaeridium zoharyi* today mainly characterizes equatorial lagoons (Dale, 1996), and the extinct *Homotryblium* is a dominant form in several well-documented inner neritic, probably lagoonal settings of various ages (e.g., Köthe, 1990; Brinkhuis, 1994; Dybkjær and Rasmussen, 2000; and personal observation of Heilmann-Clausen in the basal Oligocene Heide Sand, Belgium).

DISCUSSION

The main factors controlling the abundance of dinoflagellate cysts and other organic matter in sediments are oxygenation of bottom waters, primary productivity, and rate of sedimentation. The latter two factors are also essential for the preservation of siliceous plankton, which is strongly influenced by dissolution, because ocean waters are highly undersaturated with respect to silica. At Anthering, our calculations of sedimentation rate provide evidence for a pronounced increase in the rate of hemipelagic sedimentation with the onset of the CIE and the coeval benthic foraminifera extinction event indicates dysoxic to anoxic conditions at the seafloor. The increase in the number of dinoflagellate cysts and siliceous plankton, and the increase in the total amount of organic carbon in the hemipelagic sediment at the level of the CIE, might be a result of these factors, at least in part. However, an increase in the rate of sedimentation, and a decrease in the oxygenation of bottom waters, cannot explain the changes in the composition of microfossil assemblages within the CIE-interval.

In 4 out of 32 samples abundant specimens of pyritized large diatoms (>125 μm) of the genera *Craspedodiscus* and *Trinacria* were found. They are restricted to samples J9, J14, J15 and J17 immediately above the CIE. These samples represent 1.30 m of shale, equivalent to an episode of 30 k.y. according to our estimated sedimentation rate. In this part of the section, the preservation of siliceous plankton is much better than in other parts at similar sedimentation rates. However, the occurrence of large diatoms cannot just be a matter of preservation because due to their size, these specimens should be less prone to dissolution than small diatoms (Mitlehner, 1996). Therefore, we assume that high abundance of siliceous plankton caused the better preservation as the result of an increased input of biosiliceous particles. This is consistent with the large quantity of small, roundish diatoms (not identifyable with the light microscope) and common specimens of the genus *Hemiaulus,* which suggest high surface water productivity (Fenner, 1991).

The occurrence of *Craspedodiscus* spp. and *Trinacria* spp. in deep-water deposits at Anthering is highly remarkable as these genera are usually restricted to neritic environments. We can rule out redeposition of these specimens because in that case, resting spores and benthic species would have been redeposited in considerable amounts. Their scarcity indicates an oceanic assemblage (Fenner, 1994). This suggests that water-depth was not the limiting factor for the occurrence of *Craspedodiscus* spp. and *Trinacria* spp. Probably, the preference of these genera for neritic settings was due to the higher level of dissolved nutrients in these areas.

The high abundance of large diatom specimens correlates with the onset of the *Apectodinium* acme. This *Apectodinium* event may reflect exceptionally high sea-surface temperatures and/or a significant increase in surface-water productivity (Brinkhuis et al., 1994; Bujak and Brinkhuis, 1998; Crouch et al., 2001). Data from the southern, marginal part of the North Sea Basin suggest that the distribution of *Apectodinium* was also strongly influenced by distance from the coast. This is evident from a comparison of abundances in post-CIE deposits within the NP10 chronozone in three boreholes in the Hamburg region: *Apectodinium* is more abundant in the relatively coastal Gartow and Lingen-1 boreholes (Köthe, 1990) but much rarer in the more offshore Wursterheide borehole (Heilmann-Clausen and Costa, 1989). It thus appears that *Apectodinium* flourished in certain inner neritic environments, as formerly suggested by Stover et al. (1996). However, during the CIE extreme abundances of *Apectodinium* were found also in the central parts of the North Sea basin. There, organic-rich, anoxic layers in part of the Sele Formation, and equivalent Stolleklint Clay (Heilmann-Clausen and Schmitz, 2000), indicate that waters in the entire basin were nutrient-rich with a high plankton productivity.

It can be summarized that the concurrent acmes of predominantly neritic organisms (*Craspedodiscus* spp., *Trinacria* spp. and *Apectodinium* spp.) in the sedimentary record of the open marine setting at Anthering are most likely an effect of a strong increase in nutrient supply. Biogene carbonate production primarily controls the volume of sediment available for incorporation in turbidity currents. At Anthering, the rate of turbidite sedimentation remained stable during zone NP10, suggesting that the productivity of calcareous plankton did not change, apparently in disagreement with the assumed high productivity of dinoflagellates and siliceous plankton. This is, however, consistent with modern oceans (Ziveri et al., 1995), where in the presence of high nutrient levels, diatoms and dinoflagellates are the most abundant phytoplankton and the productivity and flux of coccolithophorides decreases.

The benthic foraminifera extinction event (BFEE) in the upper part of Zone NP9 and its potential causes, have been discussed extensively (see Thomas, 1998, for a review). Around the CIE at Anthering, the morphogroup B assemblage (*Glomospira*-fauna) dominates. The ecological significance of this assemblage is still unclear. *Glomospira* may be an opportunistic genus in the sense that it is a good recolonizer of azoic sediments (Kaminski et al., 1996).

CONCLUSIONS

The Anthering Section provides good evidence for an increase in the rate of hemipelagic sedimentation and primary surface productivity at the level of the CIE. This suggests that eutrophication of surface waters was triggered by enhanced continental run-off as a result of the establishment of subtropical climate conditions. The associated increase in the input of nutrients to the ocean basins was enough to have affected even open marine settings. This increased nutrient supply caused the migration of planktonic taxa, like *Apectodinium* spp., *Trinacria* spp. and *Craspedodiscus* spp., from neritic environments to settings further away from the coast. As a result of the associated episodically increased flux of organic carbon to the seafloor, a total extinction of deep-sea benthic foraminifera faunas took place.

ACKNOWLEDGMENTS

For their helpful reviews and comments we are grateful to Wolfgang Kuhnt, Simonetta Monechi, James Powell and Ellen Thomas. We are indebted to Christopher Hollis for examining the radiolarian faunas, Miroslav Bubik for discussion and comparative material of DWAF, and Peter Klein for the determination of total organic carbon contents. We thank Silvia Spezziferri and Johann Hohenegger for their help in the statistical interpretation of the DWAF and Jacek Ruthner for his technical assistance in preparing the figures.

REFERENCES CITED

Aubry, M.-P., 1996, Toward an upper Paleocene–lower Eocene high-resolution stratigraphy based on calcareous nannofossil stratigraphy: Israel Journal of Earth Sciences, v. 44, p. 239–253.

Aubry, M.-P., Berggren, W.A., Stott, L., and Sinha, A., 1996, The upper Paleocene–lower Eocene stratigraphic record and the Paleocene-Eocene boundary carbon isotope excursion: Implications for geochronology, *in* Knox, R.W.O'B., et al., eds., Correlation in the early Paleogene in northwest Europe: Geological Society [London] Special Publication 101, p. 353–380.

Bains, S., Norris, R.D., Corfield, R.M., and Faul, K.L., 2000, Termination of global warmth at the Paleocene–Eocene boundary through productivity feedback: Nature, v. 407, p. 171–174.

Brinkhuis, H., 1994, Late Eocene to early Oligocene dinoflagellate cysts from the Priabonian type-area (Northeast Italy): Biostratigraphy and paleoenvironmental interpretation: Palaeogeography, Palaeoclimatology, Palaeoecology, v. 107, p. 121–163.

Brinkhuis, H., Romein, A.J.T., Smit, J., and Zachariasse, J.-W., 1994, Danian–Selandian dinoflagellate cysts from lower latitudes with special reference to the El Kef section, northwest Tunisia: GFF, v. 116, p. 46–48.

Bujak, J.P., and Brinkhuis, H., 1998, Global warming and dinocyst changes across the Paleocene–Eocene epoch boundary, *in* Aubry, M.-P., et al., eds., Late Paleocene–early Eocene: New York, Columbia University Press, p. 277–295.

Butt, A., 1981, Depositional environments of the Upper Cretaceous rocks in the northern part of the Eastern Alps: Cushman Foundation for Foraminiferal Research, Special Publication, v. 20, 121 p.

Bybell, L.M., and Self-Trail, J.M., 1994, Evolutionary, biostratigraphic, and taxonomic study of calcareous nannofossils from a continuous Paleocene–Eocene boundary section in New Jersey: U.S. Geological Survey Professional Paper 1554, 73 p.

Crouch, E.M., Heilmann-Clausen, C., Brinkhuis, H., Morgans, H.E.G., Rogers, K.M., Egger, H., and Schmitz, B., 2001, Global dinoflagellate event associated with the late Paleocene Thermal Maximum: Geology, v. 29, p. 315–318.

Dale, B., 1996, Dinoflagellate cyst ecology: Modeling and geological applications, *in* Jansonius, J., and McGregor, D.C., eds., Palynology: Principles and applications: American Association of Stratigraphic Palynologists Foundation, v. 3, p. 1249–1275.

Dickens, G.R., O'Neil, J.R., Rea, D.K., and Owen, R.M., 1995, Dissociation of oceanic methane hydrate as a cause of the carbon isotope excursion at the end of the Paleocene: Paleoceanography, v. 10, p. 965–971.

Dickens, G.R., Thomas, E., and Bralower, T.J., 2001, A low productivity model for excess barite accumulation during the late Paleocene thermal maximum, *in* Ash, A.W., and Wing, S.L., eds., Climate and biota of the early Paleogene: International Meeting July 3–8, 2001, Northwest College, Powell, Wyoming, USA, Abstract volume, p. 27.

Draxler, I., 2000, First results on pollen and spores from the Paleocene–Eocene boundary interval of the Anthering section, Austria: GFF, v. 122, p. 42.

Dybkjær, K., and Rasmussen, E.S., 2000, Palynological dating of the Oligocene–Miocene successions in the Lillebælt area, Denmark: Bulletin of the Geological Society of Denmark, v. 47, p. 87–103.

Egger, H., 1995, Die Lithostratigraphie der Altlengbach-Formation und der Anthering-Formation im Rhenodanubischen Flysch (Ostalpen, Penninikum): Neues Jahrbuch für Geologie und Paläontologie Abhandlungen, v. 196, p. 69–91.

Egger, H., Bichler, M., Draxler, I., Homayoun, M., Huber, H.-J., Kirchner, E.Ch., Klein, P., and Surenian, R., 1997, Mud turbidites, black shales, and bentonites from the Paleocene–Eocene boundary: The Anthering Formation of the Rhenodanubian flysch (Austria): Jahrbuch Geologische Bundesanstalt, v. 140, p. 29–45.

Egger, H., Heilmann-Clausen, C., and Schmitz, B., 2000a, The Paleocene–Eocene boundary interval of a Tethyan deep-sea section and its correlation with the North Sea Basin: Société Géologique de France Bulletin, v. 171, p. 207–216.

Egger, H., Homayoun, M., and Rögl, F., 2000b, Deep-sea sedimentation and paleoproductivity in the Paleogene of a Tethyan basin: GFF, v. 122, p. 44–45.

Egger, H., Homayoun, M., and Schnabel, W., 2002, Tectonic and climatic control of Paleogene sedimentation in the Rhenodanubian Flysch basin (Eastern Alps, Austria): Sedimentary Geology, v. 152, p. 247–262.

Espitalié, J., Laporte, J.L., Madec, M., Marquis, F., Leplat, P., Paulet, J., and Boutefeu, A., 1977, Méthode rapide de caractérisation des roches mères de leur potentiel pétrolier et de leur degré d'évolution: Revue Institut France Pétrologique, v. 32, p. 23–43.

Fenner, J., 1991, Taxonomy, stratigraphy, and palaeoceanographic implications of Paleocene diatoms, *in* Ciesielski, P., and Kristoffersen, Y., eds., Proceedings of the Ocean Drilling Program, Scientific Results: Texas A & M University, College Station, Texas, USA, v. 114, p. 123–154.

Fenner, J., 1994, Diatoms of the Fur Formation, their taxonomy and biostratigraphic interpretation: Results from the Harre borehole, Denmark: Aarhus Geoscience, v. 1, p. 99–163.

Galeotti, S., Angori, E., Coccioni, R., Ferrari, G., Galbrun, B., Monechi, S., Premoli-Silva, I., Speijer, R., and Turi, B., 2000, Integrated stratigraphy across the Paleocene–Eocene boundary in the Contessa Road section, Gubbio (central Italy): Société Géologique de France Bulletin, v. 171, p. 355–366.

Gavrilov, Y.O., Shcherbinina, E.A., and Muzylöv, N.G., 2000, A Paleogene sequence in central North Caucasus: A response to paleoenvironmental changes: GFF, v. 122, p. 51–53.

Gawenda, P., Winkler, W., Schmitz, B., and Adatte, T., 1999, Climate and bioproductivity control on carbonate turbidite sedimentation (paleocene to earliest Eocene, Gulf of Biscay, Zumaia, Spain): Journal of Sedimentary Petrology, v. 69, p. 1253–1261.

Geroch, S., and Kaminski, M.A., 1992, The morphology, palecology and systematics of Nothia excelsa (Grzybowski), a deep-water agglutinated foraminifera: Annales Societas Geologorum Poloniae, v. 62, p. 255–265.

Gooday, A.J., Shires, R., and Jones, A.R., 1997, Large, deep-sea agglutinated foraminifera: Two different kinds of organization and their possible ecological significance: Journal of Foraminiferal Research, v. 27, p. 278–291.

Harland, R., 1979, The *Wetzeliella (Apectodinium) homomorpha* plexus from the Paleogene/earliest Eocene of northwest Europe: IV International Palynological Conference, Lucknow (1976–1977), v. 2, p. 59–70.

Heilmann-Clausen, C., and Costa, L.I., 1989, Dinoflagellate Zonation of the Uppermost Paleocene? to lower Miocene in the Wursterheide Research Well, northwest Germany: Geologisches Jahrbuch, Ser. A, v. 111, p. 431–521.

Heilmann-Clausen, C., and Egger, H., 2000, The Anthering outcrop (Austria): A key section for correlation between Tethys and northwestern Europe near the Paleocene–Eocene boundary: GFF, v. 122, p. 69.

Heilmann-Clausen, C., and Schmitz, B., 2000, The late Paleocene thermal maximum $\delta^{13}C$ excursion in Denmark?: GFF, v. 122, p. 70.

Hesse, R., 1975, Turbiditic and nonturbiditic mudstone of Cretaceous flysch sections of the East Alps and other basins: Sedimentology, v. 22, p. 387–416.

Jones, R.W., and Charnock, M.A., 1985, "Morphogroups" of agglutinating foraminifera: Their life positions and feeding habits and potential applicability in (paleo)ecological studies: Revue Paléobiologique, v. 4, p. 311–320.

Kaminski, M.A., Gradstein, F.M., and Berggren, W.A., 1989, Palaeogene benthic foraminifera biostratigraphy and Palaeoecology at Site 647, southern Labrador Sea, *in* Srivstava, S.P., et al., eds., Proceedings of the Ocean Drilling Program, Scientific Results: Texas A & M University, College Station, Texas, USA, v. 105, p. 705–730.

Kaminski, M.A., Kuhnt, W., and Radley, J.D., 1996, Paleocene–Eocene deep water agglutinated foraminifera from the Numidian Flysch (Rif, Northern Morocco): Their significance for the palaeoceanography of the Gibraltar gateway: Journal of Micropalaeontology, v. 15, p. 1–19.

Kaminski, M.A., Kuhnt, W., and Moullade, M., 1999, The evolution and paleobiogeography of abyssal foraminifera since the Early Cretaceous: A tale of four faunas: Neues Jahrbuch für Geologie und Paläontologie Abhandlungen, v. 212, p. 401–439.

Katz, M.E., Pak, D.K., Dickens, G.R., and Miller, K.G., 1999, The source and fate of massive carbon input during the latest Paleocene thermal maximum: Science, v. 286, p. 1531–1533.

Kennett, J.P., and Stott, L.D., 1991, Abrupt deep-sea warming, palaeoceanographic changes, and benthic extinctions at the end of the Paleocene: Nature, v. 353, p. 225–229.

Köthe, A., 1990, Paleogene Dinoflagellates from Northwest Germany: Geologisches Jahrbuch, Serie A, v. 118, p. 3–111.

Kuhnt, W., Kaminski, M.A., and Moullade, M., 1989, Late Cretaceous deepwater agglutinated foraminiferal assemblages from the North Atlantic and its marginal seas: Geologische Rundschau, v. 78, p. 1121–1140.

Langford, F.F., and Blanc-Valleron, M.-M., 1990, Interpreting Rock-Eval pyrolysis data using graphs of pyrolyzable hydrocarbons versus total organic carbon: AAPG (American Association of Petroleum Geologists) Bulletin, v. 74, p. 799–804.

Lipps, J., Berger, W.H., Buzas, M.A., Douglas, R.G., and Ross, C.A., 1979, Foraminiferal ecology and palecology: Society of Economic Paleontologists and Mineralogists, Short Course 6, Houston, Texas, 198 p.

Lu, G., and Keller, G., 1993, The Paleocene–Eocene transition in the Antarctic Indian Ocean: Inference from planktic foraminifera: Marine Micropaleontology, v. 21, p. 101–142.

Martini, E., 1971, Standard Tertiary and Quaternary calcareous nannoplankton zonation, *in* Farinacci, A., ed., Proceedings of the 2nd Planktonic Conference, Rome, Italy, 1970: Tecnoscienza, v. 2, p. 739–785.

Mitlehner, A.G., 1996, Palaeoenvironments in the North Sea Basin around the Paleocene–Eocene boundary: Evidence from diatoms and other siliceous microfossils, *in* Knox, R.W.O'B., et al., eds., Correlation of the early Paleogene in northwestern Europe: Geological Society [London] Special Publication 101, p. 255–273.

Schmitz, B., Asaro, F., Molina, E., Monechi, S., von Salis, K., and Speijer, R., 1997a, High-resolution iridium, $\delta^{13}C$, $\delta^{18}O$, foraminifera, and nannofossil profiles across the latest Paleocene benthic extinction event at Zumaya, Spain: Palaeogeography, Palaeoclimatology, Palaeoecology, v. 133, p. 49–68.

Schmitz, B., Charisi, S.D., Thompson, E.I., and Speijer, R.P., 1997b, Barium, SiO_2 (excess), and P_2O_5 as proxies of biological productivity in the Middle East during the Paleocene and the latest Paleocene benthic extinction event: Terra Nova, v. 9, p. 95–99.

Schmitz, B., Pujalte, V., and Nunez-Betelu, K., 2001, Climate and sea-level perturbations during the Initial Eocene Thermal Maximum: Evidence from siliciclastic units in the Basque Basin (Ermua, Zumaia and Trabakua Pass), northern Spain: Palaeogeography, Palaeoclimatology, Palaeoecology, v. 165, p. 299–320.

Scotese, C.P., and Golanka, J., 1992, Paleogeographic atlas: Paleomap progress report: Arlington, University of Texas, 34 p.

Speijer, R.P., Schmitz, B., and van der Zwaan, G.J., 1997, Benthic foraminiferal extinction and repopulation in response to latest Paleocene Tethyan anoxia: Geology, v. 25, p. 683–686.

Stockmarr, J., 1971, Tablets with spores used in absolute pollen analysis: Pollen et Spores, v. 13, p. 615–631.

Stover, L.E., Brinkhuis, H., Damassa, S.P., De Vertuil, L., Helby, R.J., Monteil, E., Partridge, A.D., Powell, A.J., Riding, J.B., Smelror, M., and Williams, G.L., 1996, Mesozoic–Tertiary dinoflagellates, acritarchs and prasinophytes, *in* Jansonius, J., and McGregor, D.C., eds., Palynology: Principles and applications: American Association of Stratigraphic Palynologists Foundation, v. 2, p. 641–750.

Thiry, M., 2000, Palaeoclimatic interpretation of clay minerals in marine deposits: An outlook from the continental origin: Earth Science Reviews, v. 49, p. 201–221.

Thomas, E., 1998, Biogeography of the late Paleocene benthic foraminifera extinction, *in* Aubry, M.-P., et al., eds., Late Paleocene–early Eocene: New York, Columbia University Press, p. 214–235.

van der Zwan, C.J., 2002, The impact of Milankovitch-scale climatic forcing on sediment supply: Sedimentary Geology, v. 147, p. 271–294.

Wing, S.L., Bown, T.M., and Obradovic, J.D., 1991, Early Eocene biotic and climatic change in interior western North America: Geology, v. 19, p. 1189–1192.

Ziveri, P., Thunell, R.C., and Rio D., 1995, Export production of coccolithophores in an upwelling region: Results from San Pedro Basin, Southern California Borderlands: Marine Micropaleontology, v. 24, p. 335–358.

MANUSCRIPT ACCEPTED BY THE SOCIETY AUGUST 13, 2002

Printed in the USA

Geological Society of America
Special Paper 369
2003

Paleocene-Eocene boundary events in the northeastern Peri-Tethys

Yuri O. Gavrilov
Ekaterina A. Shcherbinina
Geological Institute of Russian Academy of Sciences, Pyzhevsky 7, Moscow 119017, Russia
Hedi Oberhänsli
GeoForschungZentrum Potsdam, Telegrafenberg, D-14473, Potsdam, Germany

ABSTRACT

In the northeastern Peri-Tethys epicontinental basin stretching from the Black Sea to Tadjikistan, a sapropelitic bed with total organic carbon (TOC) contents up to 20% and enriched in redox-sensitive elements was deposited during the Initial Eocene Thermal Maximum (IETM), recognized by the occurrence of a negative carbon isotope excursion in organic carbon and carbonate. In the sapropelitic bed, calcareous plankton decreased in abundance, a benthic foraminiferal extinction occurred, dinocysts were abundant, and there were abnormally thick-walled dinocysts and asymmetric calcareous nannoplankton species. We argue that the sapropelitic bed accumulated as a result of high productivity, triggered by an influx of nutrients that reached the basin when organic- and P-rich sediments in low-lying coastal areas were flooded during a rapid transgression. High productivity led to oxygen depletion, decimating benthic foraminiferal assemblages. Organic biomarkers suggest that at least in some places even waters in the photic zone became anoxic. The dysoxia/anoxia might have been exacerbated by stratification, resulting from high runoff in the humid western region, high evaporation in the arid eastern area. Anoxia/dysoxia ended gradually. We suggest that deposition of large amounts of organic matter in the northeastern Peri-Tethys may have been a factor in terminating the IETM.

INTRODUCTION

During the last decade many scientists have investigated events that occurred during the latest Paleocene and earliest Eocene (e.g., Kennett and Stott, 1991; Zachos et al., 1993; Dickens et al., 1995; Thomas and Shackleton, 1996; Schmitz et al., 1996; Aubry et al., 1998; Koch et al., 1992; Wing et al., 1995; Katz et al., 1999). The most prominent event during this period is the Initial Eocene Thermal Maximum (IETM), which has been documented in sections deposited in such different environments as the open ocean, marginal basins, and on the continents. During this extreme warm period marine and continental biota show rapid change, coeval with distinct negative excursions in oxygen and carbon isotopic values in carbonate and organic matter.

In some areas, sediments deposited during the IETM are enriched in organic matter (e.g., Gavrilov et al., 1997; Speijer et al., 1997). One of these areas is the northeastern Peri-Tethys, where organic-rich sediments can be traced over >2500 km, from the Caucasus to Central Asia (Muzylöv et al., 1989; Muzylöv and Shcherbinina, 1995). The properties of the organic-rich sediments, called the sapropelitic bed, are relatively constant over this large area, although they accumulated in different environments in different parts of the region.

The origin of the sapropelitic bed has been discussed extensively, after Muzylöv et al. (1989) first documented that its

E-mails: gavrilov@geo.tv-sign.ru; shcherbinina@geo.tv-sign.ru; oberh@gfz-potsdam.de.

Gavrilov, Yu.O., Shcherbinina, E.A., and Oberhänsli, H., 2003, Paleocene-Eocene boundary events in the northeastern Peri-Tethys, *in* Wing, S.L., Gingerich, P.D., Schmitz, B., and Thomas, E., eds., Causes and Consequences of Globally Warm Climates in the Early Paleogene: Boulder, Colorado, Geological Society of America Special Paper 369, p. 147–168.

deposition was synchronous over a large area (e.g., Gavrilov and Muzylöv, 1991; Muzylöv, 1994; Muzylöv et al., 1996; Gavrilov et al., 1997; Stupin and Muzylöv, 2001), but no comprehensive discussion of its deposition has been presented. We describe the sedimentological, geochemical, and paleoecological changes during the IETM in the northeastern Peri-Tethys, and propose a scenario for the accumulation of the sapropelitic bed.

METHODS

The mineralogy and geochemistry of the sapropelitic bed and its surrounding sediments were studied in many sections (Fig. 1), six of which are described in this paper, while two more are mentioned. Clay minerals were analyzed using X-ray diffraction. A range of elements and compounds were analyzed in the Laboratory of Analytical Chemistry at the Geological Institute RAS, Moscow: total organic carbon (TOC), total carbonate, S, P, Fe, Ti (wet chemistry), V, Cr, Ni, Co, Pb, Cu, Ga, Ge, Mo, Ag, Zn, Sn (quantitative spectral analysis using an atomic emission specrometer), Sc, Se, Br, Sb, Cs, Hf, Au, Th, Y, La, Ce, Nd, Sm, Eu, Tb, Yb, Lu (instrumental neutron activation analysis, INAA), and Nb, Zr, Rb, Sr, and Ba (X-ray fluorescence analysis). In two sections (Kheu in the northern Caucasus and Kurpai in Tadjikistan), the organic matter was examined by express pyrolithic analysis using Rock-Eval II, chromatography, molecular mass-spectrometry, and IR-spectrometry.

Nannofossils were studied in the bulk sediment using light microscopy. We counted the relative abundance of major nannofossil taxa. In most samples, 300 specimens were counted, but in a few samples with a low abundance of nannofossils 100 specimens only could be counted. In some intervals, there were so few specimens of nannofossils (<100 specimens per sample) that no statistically significant counts could be made.

RESULTS

We present results from six most important sections in the northeastern Peri-Tethys, covering a large area in the central and eastern Caucasus, the western shores of the Aral Sea, the western parts of the Kopetdag Mountains, and the Tadjik depression (Fig. 1). Generally, the sapropelitic bed is dark in color and fissile on weathering (Figs. 2, 3). The lower boundary is always sharp, and in some sections erosional, whereas the upper boundary is more gradual. In most sections the sapropelitic bed can be subdivided into three units, with a more calcareous middle unit that is less rich in TOC between lower and upper units with high TOC. Toward the northern margin of the basin, TOC content decreases significantly in the sapropelitic bed. Locally, sediments coeval to the bed contain abundant fish detritus, and in some localities in Central Asia (e.g., the Kyzyl Kum Desert, Tadjik depression) the bed is either associated with or replaced by phosphorite-bearing deposits.

Lithology and paleontology

In the Kheu River section (Central Caucasus) the sapropelitic bed is ~0.5 m thick, and exposed along the riverbank near the village of Gerpegezh (20 km southeast of the city of Nalchik)

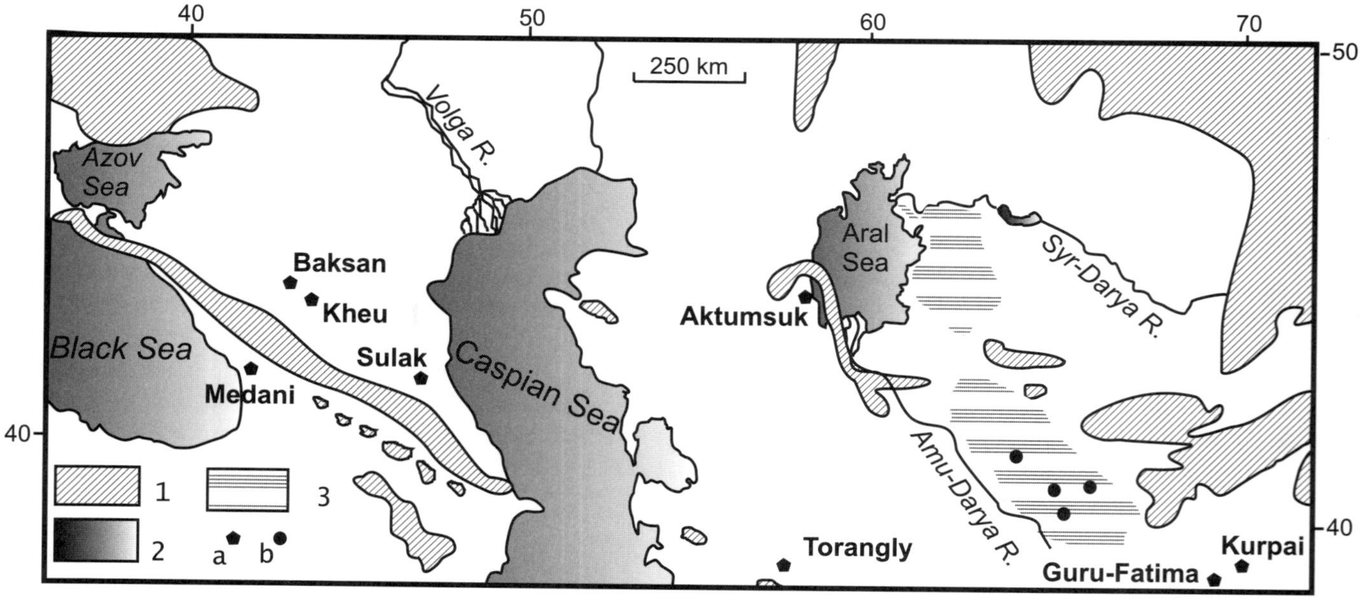

Figure 1. Studied localities in the northeastern Peri-Tethys epicontinental basin. 1—land, 2—sea, 3—area of oil shale distribution, a—localities of the studied sections, b—localities of studied boreholes.

Figure 2. Sapropelitic bed in the Paleocene-Eocene boundary sediments of the western Aral Sea region (Aktumsuk Cape).

within bluish-gray calcareous clays. The bed is dark brown (Plate 1, Figures 1–6) and contains abundant fish remains, ostracods, sulfide nodules, and rare pyritized wood fragments. It is underlain by a noncalcareous light brown clay (0.1 m) with scarce nannofossils and no foraminifera (Gavrilov et al., 2000). Within this layer are the first occurrences (FOs) of the calcareous nannofossils *Transversopontis pulcher* and *Campylosphaera eodela* (Fig. 4A, Appendix A). The FO of the latter species marks the bottom of Subzone CP8b in pelagic sections (Okada and Bukry, 1980).

Within the sapropelitic bed there is a poorly preserved and low-diversity nannoflora, dominated by various species of *Toweius*. Compared to the underlying clays, the planktic/benthic foraminiferal ratio in the bed is high (Stupin and Muzylöv, 2001). Nannofossils are rare and foraminifera are absent in the lowermost 5 cm of the sapropelitic bed where the TOC is highest. Slightly higher in the bed, *Pontosphaera* and the calcareous dinocyst *Thoracosphaera* become significant components of the assemblage. In the more calcareous middle part of the bed, the nannofossil content and diversity increase, and the first, rare rhomboasters (*R. bramlettei* s.l.), occur. In the upper, TOC-rich part of the bed, rhomboasters and discoasters are abundant (including the FO of *Discoaster araneus*). The abundance of benthic foraminifera, mainly calcareous species, is low (Stupin and Muzylöv, 2001).

In the Sulak River section (Dagestan, eastern Caucasus, near the Chirkey Hydroelectric Power Station), the sapropelitic bed is charcoal-black and occurs intercalated in a series of alternating dark red and green marls (Fig. 3). It is 0.3 m thick, with the lowermost 0.1 m slightly disturbed, possibly as the result of minor slumping. The nannofossil assemblage has a high relative abundance of *Fasciculithus* and *Discoaster* species (Fig. 5A, Appendix B). In some intervals, we observed abundant calcareous detritus and reworked Cretaceous nannofossils.

The underlying homogeneous green sediments are calcareous in their lowermost part, with abundant nannofossils. Toward the top of these sediments the carbonate content decreases and nannofossils disappear. In the lowermost part of the sapropelitic bed (0.20 m), there are almost no nannofossils. In the uppermost part, various species of *Toweius* dominate, and rhomboasters, pontosphaerids, *Discoaster araneus* and asymmetric *D. anartios* first appear. The total number of *Thoracosphaera* specimens

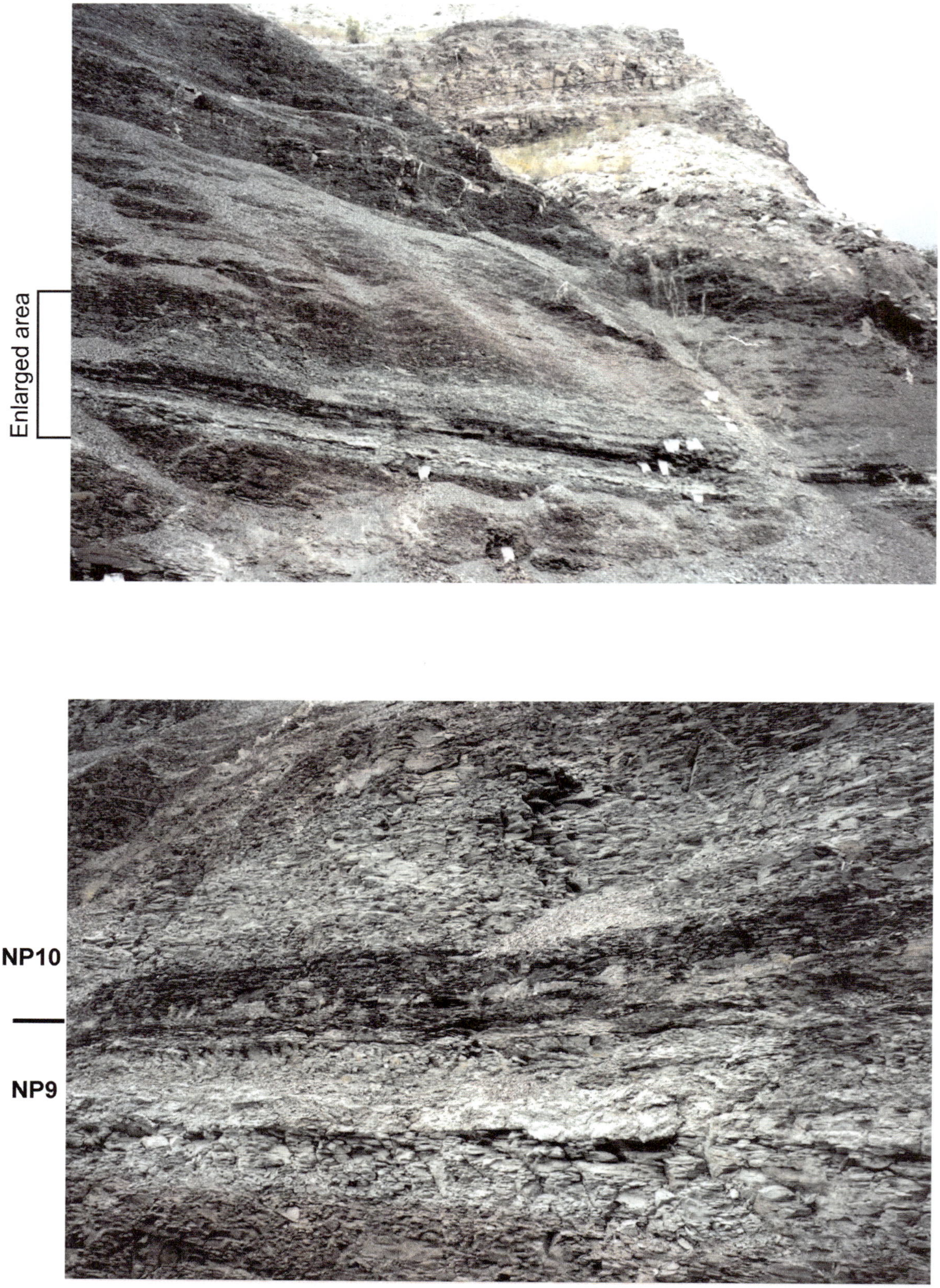

Figure 3. Sapropelitic bed in the Paleocene-Eocene boundary sediments of Dagestan (Eastern Caucasus, Sulak River).

increases up-section within the sapropelitic bed, and *Campylosphaera eodela* first occurs above the bed. A sharp increase in the abundance of calcareous detritus at the base of the overlaying sediments, the reduced thickness of the bed, and the lack of a threefold division suggest that erosion removed the uppermost part of the bed.

In the Aktumsuk Cape section (western Aral Sea, Uzbekistan, at the base of the cliff of the Usturt Plateau), the sapropelitic bed is 1 m thick, dark gray, and intercalated in white limestones (Fig. 2). It is finely laminated and shows significant fluctuations in TOC. Sharp negative excursions in both $\delta^{13}C$ and $\delta^{18}O$ are marked in the bed (Bolle et al., 2000). The underlying thin limestone bed, which has been assigned to calcareous nannofossil zone NP9 (Martini, 1971), overlies upper Campanian sediments with a sharp erosional contact (Fig. 4B, Appendix C). The high carbonate concentration is the result of a high content of calcareous detritus (Plate 1, Figures 7, 8), and nannofossils are less common than in the sapropelitic bed. Various pontosphaerids and the first species of rhabdosphaerids (*Rhabdosphaera* cf. *sola*) occur. As in most other sections, TOC is most enriched in the lowermost part (0.40 m) of the sapropelitic bed where the beginning of carbon isotope excursion and FO of *R. bramlettei* were established. Unique for the studied region, species of *Coccolithus* become more abundant in this section, while species of *Toweius* are rare, as was observed in other regions near the Paleocene-Eocene transition (Tantawy et al., 2000; Monechi et al., 2000).

TOC is lower and the nannofossil abundance increases in the calcareous middle part of the sapropelitic bed, but the nannofossil assemblage does not change significantly. Nannofossils and planktonic foraminifera become more abundant in the organic-rich uppermost part of the sapropelitic bed. The abundance of *Thoracosphaera* and calcareous debris increases sharply in the overlaying sediments. The sapropelitic bed has common benthic foraminifera, equally divided between agglutinated and calcareous taxa (Stupin and Muzylöv, 2001), although the latter are more common in its middle part.

In the Torangly section (western Kopetdag Mountains, Turkmenistan, exposed in the northern slope of the Kurendag anticline), the SB is 1.2 m thick and consists of brown clays, rich in fish scales, intercalated in gray, slightly calcareous clays (Fig. 5B, Appendix D). TOC is relatively low, possibly because the sampled sediment was rather weathered. In this section, the first rhomboasters occur in sediments just below the sapropelitic bed, which are not enriched in TOC but have increased concentrations of some elements (e.g., Mo, Zn) characteristic of the sapropelitic bed in other sections (Fig. 7). We suggest that there was an erosional event at the time of deposition of the lowermost part of the sapropelitic bed, so that only the middle and upper parts of the bed are present, as is also the case in the Baksan River section (Muzylöv et al., 1996). The abundance of *Fasciculithus,* which is believed to be an oligotrophic taxon, sharply increases in the sapropelitic bed, while planktonic foraminifera are absent. *C. eodela* first occurs in the overlying sediments. In the upper part of the bed, nannofossils are very rare, and calcareous species dominate the benthic foraminiferal assemblage (Stupin and Muzylöv, 2001).

In the Kurpai section (Tadjik depression, Tadjikistan; Figure 6A), the sapropelitic bed is 1.3 m thick and consists of black, laminated clays with rather poorly preserved and low-diversity nannofossil assemblages dominated by *Fasciculithus* and *Discoaster* species. It is intercalated in slightly calcareous sediments with a significant fraction of terrigenous silty material. In the underlying noncalcareous layers nannofossils are absent and tiny calcareous benthic foraminifera are present. In the lowermost part of the sapropelitic bed, the $CaCO_3$ content increases, mainly due to the presence of calcareous detritus and, to a minor extent, calcareous benthic foraminifera, but nannofossils are absent. Within the light-colored intercalation nannofossils are present, and *Fasciculithus* is abundant (>50%) together with *Coccolithus.* The uppermost part of the sapropelitic bed is an oil shale with very high TOC, and nannoplankton and foraminifera are absent. In the sediments overlying the bed, nannofossils including pontosphaerids are rare and calcareous detritus is common.

In the Guru-Fatima section (Tadjik depression) the sapropelitic bed is a 1.3 m thick, black oil shale with TOC >15%, intercalated in low-carbonate gray clays (Fig. 6B). There is a prominent erosional surface at the NP8/NP9 boundary below the bed. In the lowermost part of the bed, where TOC is highest, nannofossils are rare and their diversity is low. *Braarudosphaera bigelowii* (57%), a long-ranging nannoplankton species (Lower Cretaceous–Quaternary) which is abundant in sporadic blooms, is dominant. Calcareous detritus is abundant. Nannoplankton are not common but diverse (though species of *Toweius* dominate the assemblage) higher in the sapropelitic bed, despite the low $CaCO_3$ content, and oxic benthic foraminifera are common (Stupin and Muzylöv, 2001). Up-section, TOC decreases gradually, and calcareous nannofossils become rare or are absent. In the uppermost part of the bed, rare nannofossils occur, and the abundance of *Toweius* sharply decreases while various *Fasciculithus* and *Discoaster* species become more abundant. The overlying sediments contain rare various nannofossils and planktic foraminifera only.

In the Kheu river (N.I. Zaporozhets, 2002, personal commun.), Medani (Transcaucasia, Georgia) and Torangly (Akhmetiev and Zaporozhets, 1996) sections, organic-walled dinocysts show good preservation without reworking within the sapropelitic bed and surrounding sediments. Scarce *Apectodinium* spp. are found in the sediments surrounding the bed. Within the bed, however, *Apectodinium homomorphum, A. augustum, A. parvum, A. quinquelatum, A. hyperacanthum, A. paniculatum,* and *A. sumissum,* are abundant, and co-occur with *Kenleyia, Rhombodinium, Hystrichokolpoma,* and *Fibrocysta.* This assemblage was assigned to the dinocyst *Apectodinium hyperacanthum* zone (Powell, 1992). The sapropelitic bed thus corresponds to the *Apectodinium* acme observed in

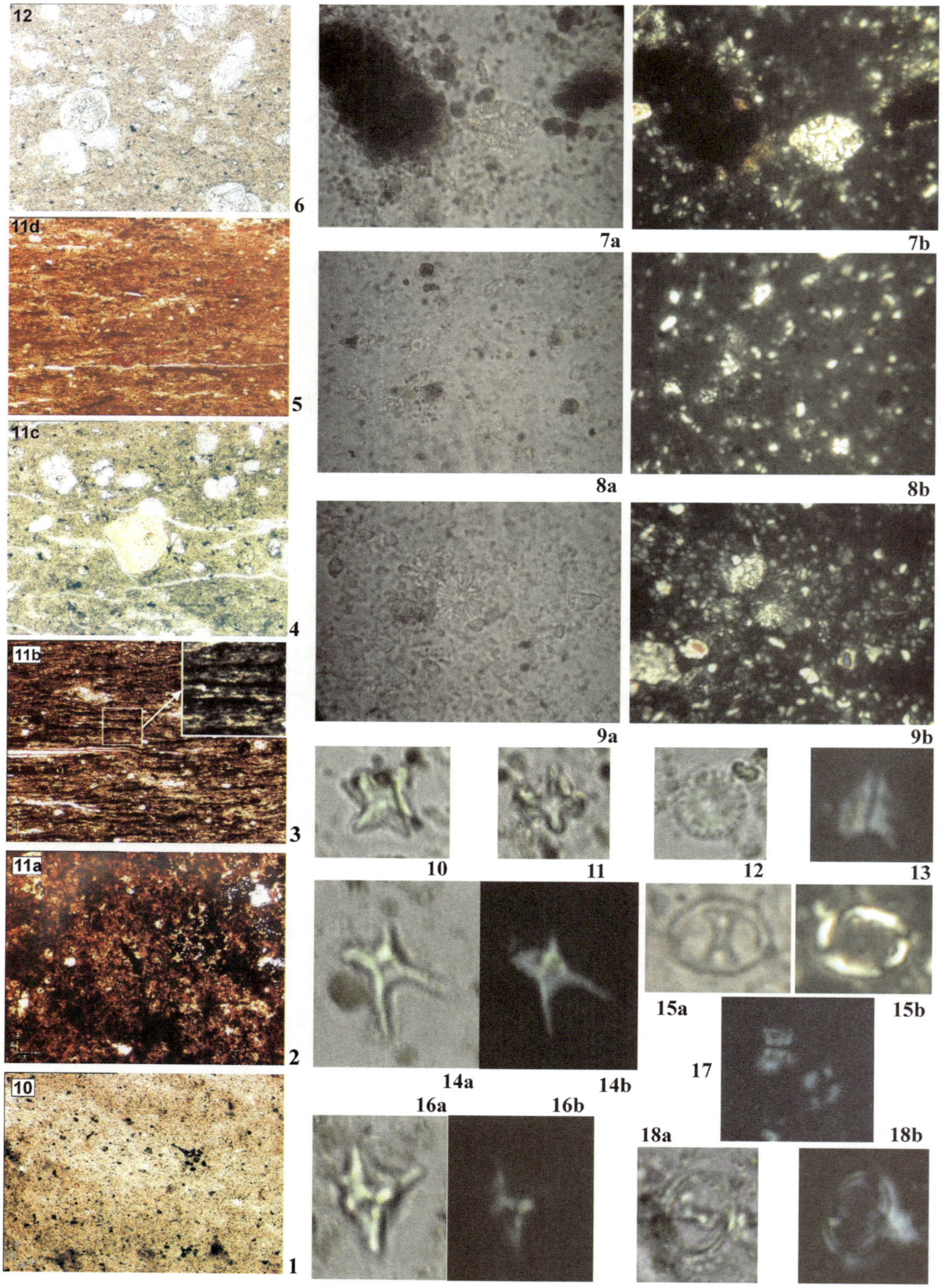
12
6
11d
5
11c
4
11b
3
11a
2
10
1
7a
7b
8a
8b
9a
9b
10
11
12
13
14a
14b
15a
15b
16a
16b
17
18a
18b

many parts of the world during the IETM (e.g., Bujak and Brinkhuis, 1998; Crouch et al., 2000; Steurbaut et al., 2000; Crouch et al., 2001).

The sapropelitic bed contains unusual forms of the genus *Apectodinium* with extremely thick-walled peri-cysts, which constitute up to 60% of the dinocyst assemblage. These forms may be endemic to the northeastern Peri-Tethys and indicate eutrophic conditions (Akhmetiev and Zaporozhets, 1996). Organic-walled plankton remains are abundant in the bed, but their total amount is only a part (25% in Kheu river section; 65% in Medani section) of the bulk amorphous matrix, which consists of solid organic matter and other algal remains.

Mineralogical and geochemical characteristics of the sapropelitic bed

We analyzed the clay fraction in samples from all sections. Clay mineral assemblages consist of varying proportions of smectite, illite, mixed-layered illite-smectite, chlorite, and rarely kaolinite. In most cases there is no significant difference between the clay mineral assemblages in the sapropelitic bed and those in the surrounding sediments. Kaolinite is absent in most sections, and found only in the Aktumsuk section (northwestern Uzbekistan) and, to a lesser extent, in the Medani section (Western Georgia). In these sections, the sapropelitic bed and surrounding sediments contain kaolinite, but it is slightly more abundant in the bed.

The oil shale collected near the Amu-Daria River (Fig. 1) contains an admixture of palygorskite, which is a common mineral in the Paleocene–lower Eocene sediments along the eastern margin of the basin (Fergana embayment, eastern Uzbekistan), indicating that the climate was arid in these regions. In the Torangly (Turkmenistan) and Sulak River (Dagestan) sections Paleocene-Eocene sediments contain illite with a low percentage of expanding interlayer clays (5–10%) and chlorite, as a result of deep burial.

The TOC in the sapropelitic bed is highly variable (Fig. 7). In the Amu Darya region (Suzak oil shale zone; Figure 1), the TOC content is higher than 20%, but in the Lower Volga region it does not exceed 1%. In the areas of predominantly carbonate sedimentation (Caucasus and Aral Sea region), where the sapropelitic bed is organic-rich, the TOC is usually negatively correlated with the $CaCO_3$ content, but this correlation does not hold in the regions of essentially terrigenous sedimentation (Central Asia).

Fe, P, S are enriched in the sapropelitic bed whereas Mn is slightly depleted. The minor elements can be divided into two groups. V, Cu, Mo, Ni, Ag, Au, Se, Co, Zn, Cr, Sn, Ga, Ge, Br, Re, Te, Tl, Bi, Sb, Cd are generally enriched to a variable extent in the bed. Some of these elements show enrichment throughout the area, others in some sections only. Ti, Zr, Nb, Rb, Sr, Sc, Cs, Hf, and the REE have a very low concentration in the sapropelitic bed (Gavrilov et al., 1997). The pattern of elemental enrichment in the bed is similar to that in the Quaternary sapropels of the Black Sea and Mediterranean (e.g., Fomina and Volkov, 1970; Volkov, 1973; Calvert, 1990; Pruysers et al., 1991; Sutherland et al., 1984).

In the Kheu section, the organic matter in the sapropelitic bed is enriched in ^{12}C ($\delta^{13}C$–30.3‰ to –30.8‰) relative to the organic matter in the surrounding sediment (–27‰ to –28‰; Figure 8). The carbonates show the same isotopic excursion (Kodina et al., 1995; Gavrilov et al., 1997), as also reported for the Aktumsuk section (Bolle et al., 2000).

Organic geochemical analyses show a strong increase in the abundance of marine, planktonic organic matter in the sapropelitic bed relative to the surrounding sediments, but the sources of the organic matter vary by location (Gavrilov et al., 1997). In the Kheu River section, organic matter is mainly derived from marine plankton, but there is a minor amount of terrestrial organic material. Terrestrial organic matter is absent in the Kurpai section. This variability in the composition of organic matter may reflect the regional variability in setting, run off, and climate. Climate was more humid in the western region, and a terrestrial organic matter was derived from the vegetated coastal areas, whereas the eastern regions were much more arid, and there was minimal input from vegetation.

Biomarkers derived from the anaerobic photosynthetic green sulphur bacteria *Chlorobium* (Summons and Powell,

Plate 1. Figures 1–6. Thin sections of different layers of sapropelitic bed and surrounding sediments in the Kheu river section. Numbers in the top left corner correspond to numbers of samples in Figure 7: (1) "presapropelitic" clay (10 cm); (2) lowermost sapropelitic layer (5 cm) rich in organic matter in the major extent; (3) thinly laminated organic-rich layer (20 cm); (4) calcareous interlayer (10 cm); (5) uppermost organic-rich layer (15 cm); (6) overlying marl. Figures 7–9. Smear slides of sapropelitic bed and underlying sediments of Aktumsuk section: (7–8) underlying sediments; (9) uppermost part of sapropelitic bed, sample 58 (similar states of discoaster preservation and a high concentration of calcareous detritus can be seen in all these samples). 7: *Thoracosphaera operculata,* ×400; a—transmitted light, b—cross-polarized light. 8: *Discoaster multiradiatus,* ×400; a—transmitted light, b—cross-polarized light. 9: *Discoaster multiradiatus,* ×400; a—transmitted light, b—cross-polarized light. Figures 10–18. Selected nannofossil species in the sapropelitic bed of the Kheu River section: (10) *Rhomboaster bramlettei,* ×2000, sample 11a, transmitted light; (11) *Rhomboaster bramlettei,* ×2000, sample 11b, transmitted light; (12) *Discoaster lenticularis,* ×2000, sample 11b, transmitted light; (13) *Fasciculithus thomasii,* ×2000, sample 11b, cross-polarized light; (14) *Rhomboaster bramlettei,* ×2000, sample 11a, a—transmitted light, b—cross-polarized light; (15) *Neochiastozygus concinnus,* ×2000, sample 11b, a—transmitted light, b—cross-polarized light; (16) *Rhomboaster bramlettei,* ×2000, sample 11b, a—transmitted light, b—cross-polarized light; (17) *Fasciculithus involutus, Toweius callosus,* ×2000, sample 11b, cross-polarized light, (18) *Placozygus sigmoides,* ×2000, sample 11b, a—transmitted light, b—cross-polarized light.

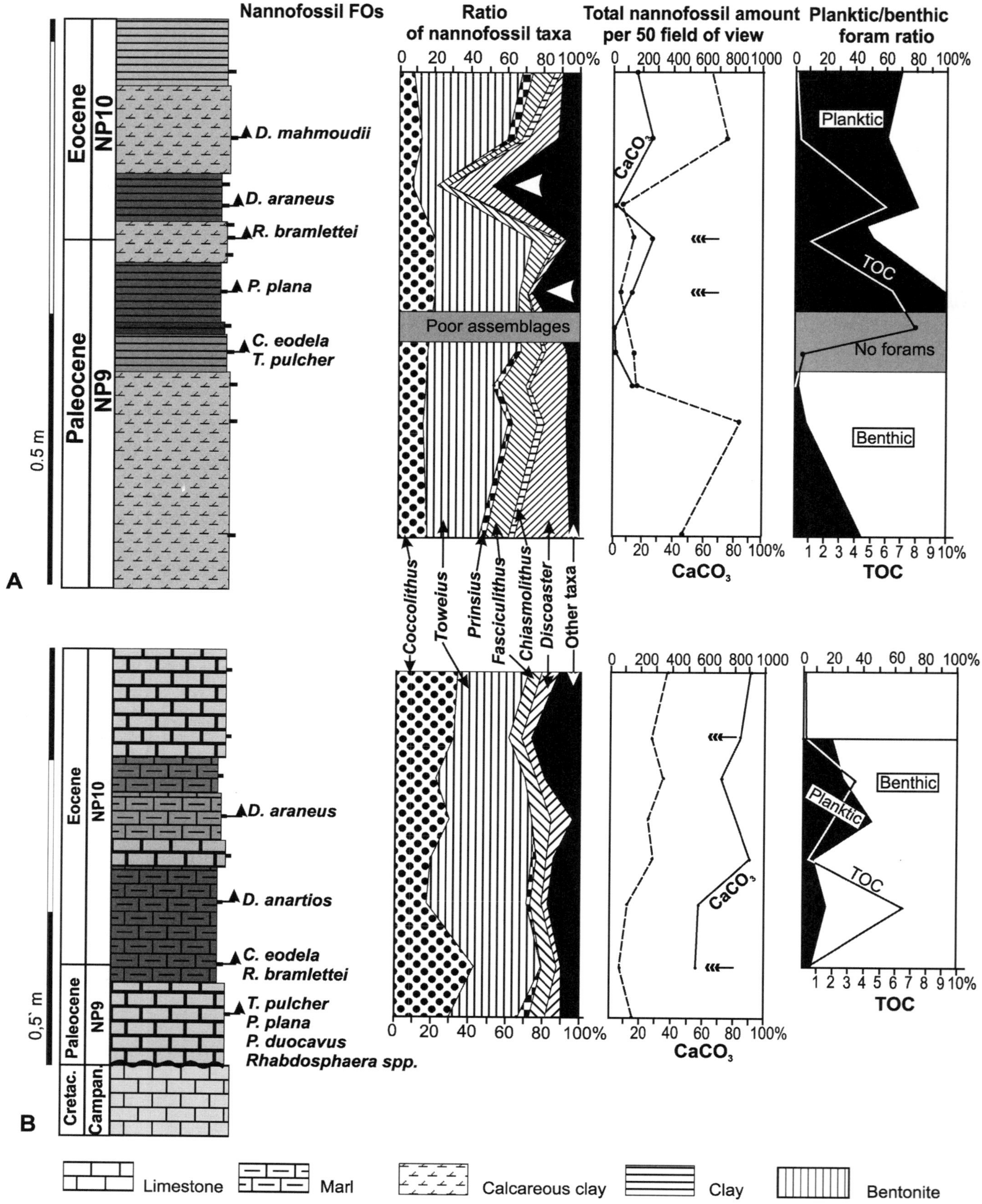

Nannofossil FOs
Ratio of nannofossil taxa
Total nannofossil amount per 50 field of view
Planktic/benthic foram ratio
Eocene
NP10
Paleocene
NP9
0.5 m
D. mahmoudii
D. araneus
R. bramlettei
P. plana
C. eodela
T. pulcher
Poor assemblages
Planktic
TOC
No forams
Benthic
CaCO3
A
Coccolithus
Toweius
Prinsius
Fasciculithus
Chiasmolithus
Discoaster
Other taxa
D. anartios
C. eodela
R. bramlettei
T. pulcher
P. plana
P. duocavus
Rhabdosphaera spp.
0,5 m
Cretac.
Campan.
B
Limestone
Marl
Calcareous clay
Clay
Bentonite

1986, 1987) were detected in the organic matter of the sapropelitic bed in the Kurpai section (Kodina et al., 1995). These biomarkers indicate anaerobic conditions and the presence of H_2S in the photic zone of the water column, as in the present Black Sea (Sinninghe Damste et al., 1993; Repeta et al., 1989; Repeta and Simpson, 1991). Organic geochemical data thus show that relative to the surrounding sediments the organic matter in the sapropelitic bed is more dominated by marine planktonic organic matter, and at least in some parts of the basin waters were truly anoxic in the photic zone, and H_2S was present.

Biostratigraphy and environments (planktonic and benthic microfossils)

Recent studies (Muzylöv, 1994; Muzylöv et al., 1989, 1996; Muzylöv and Shcherbinina, 1995) documented that the sapropelitic bed corresponds in age to the boundary between late Paleocene Subzones CP8a/CP8b of Okada and Bukry (1980), based on the FO of *Rhomboaster cuspis* within the sapropelitic bed. According to Martini (1971), the FO of *R. bramlettei* (base of Zone NP10) marks the Paleocene-Eocene boundary. Recently, however, the taxonomy of rhomboaster has been revised, with the proposition to establish two subspecies of *Rhomboaster bramlettei, R. bramlettei bramlettei* and *R. bramlettei cuspis* (Bybell and Self-Trail, 1994, 1997; von Salis et al., 2000). In this revision, the base of Zone NP10 is close to that boundary between Subzones CP8a/CP8b. In addition, the base of the global Carbon Isotope Excursion has now been proposed as the Paleocene-Eocene (P-E) boundary by the International Subcommission on Paleogene Stratigraphy (Luterbacher et al., 2000). If we follow this revision, the P-E boundary falls at the base of the sapropelitic bed. We do not have isotope data in all studied sections, and in the figures for this paper have still placed the P-E boundary at the NP9/NP10 zonal boundary.

Consistent biotic changes in our sections include a major change from mainly calcareous-walled phytoplankton in sediments below the sapropelitic bed to mainly organic-walled plankton remains within the bed, separated in some sections (Kheu, Sulak, Aktumsuk, and Kurpai) by a thin layer of sediment just below the bed which lacks both planktic and benthic calcareous microfossils. Some nannofossil taxa (pontosphaerids and rhabdosphaerids) which are common in epicontinental sediments in subsequent epochs, evolved at this time as seen in some sections (Figs. 4, 5, and Muzylöv, 1994).

The sapropelitic bed itself contains both nannofossils and planktic foraminifers, but the nannofossil assemblages within the bed differ from those in the surrounding sediments. Generally, within the sapropelitic bed the nannofossil diversity is low, and one or two taxa dominate. Sections with carbonate sediments (Kheu, Aktumsuk) show a strong dominance of *Toweius,* and those with mostly terrigenous sediments (Torangly, Kurpai, and Guru-Fatima sections) contain assemblages dominated by *Fasciculithus,* probably reflecting shallow water depth and oligotrophic environments.

The nannofloral changes in the northeastern Peri-Tethys sections thus differ from those in sections in Spain (Trabakua Pass section), where the IETM occurs in a dissolution horizon (Orue-Etxebarria et al., 1996), and is marked by a low abundance of the cool water species of the genera *Chiasmolithus* and *Prinsius* and a high abundance of *Fasciculithus.* An abrupt decline in abundance of *Fasciculithus* and the appearance of abundant *Zygrhablithus* above the base of the IETM have been observed in epicontinental and oceanic sections (Aubry et al., 1996; Orue-Etxebarria et al., 1996; Monechi et al., 2000; Bralower and Premoli-Silva, 2002). In our sections, *Zygrhablithus* is generally absent and *Fasciculithus* are common, showing decreased abundance above the sapropelitic bed in the Guru Fatima section only.

In most sections, the lowermost part of the sapropelitic bed which contains the highest TOC content is characterized by a poor nannoflora, and an increased abundance of *Thoracosphaera.* A bloom of *Braarudosphaera* is found only in the Guru-Fatima section, situated in the arid eastern part of the basin. These two taxa are abundant after the Cretaceous-Paleogene boundary, when other nannofossil taxa were rare. Occasional *Braarudosphaera* blooms have been recognized in mid-Oligocene sediments from the South Atlantic, possibly due to voluminous freshwater influx resulting from the melting of Antarctic ice. In Miocene and Pliocene sediments in the Mediterranean and Black Sea, *Braarudosphaera*-rich sediments are associated with sapropelitic layers (L.A. Golovina, 2001, personal commun.). Bukry (1974) concluded that *B. bigelowii* may thrive at relatively low salinities, but Perch-Nielsen (1977) speculated that the taxon is euryhaline, and thus more common where salinity fluctuations limit the occurrence of other species. Experimental data on modern *Thoracosphaera heimii* in culture show a maximum growth rate and productivity at 27 °C in a nutrient-rich medium (Karwath et al., 2000), suggesting that the sapropelitic bed accumulated at relatively high temperatures, in agreement with the occurrence of abundant warm-water discoasters within the bed. Besides, this species has its maximal productivity during periods of stratification of the upper water column and a well-developed thermocline (Vink et al., 2002).

The microfossil assemblages thus indicate that conditions were warm during the deposition of the sapropelitic bed, and locally salinity variations occurred. In addition, the occurrence of some nannofossil forms such as rhomboasters, asymmetric discoasters, morphotypes of *Fasciculithus* with a very high curved cone, and extremely thick-walled *Neochiastozygus* sp. testify to the occurrence of unusual environmental circumstances.

Figure 4. Lithological and microfossil changes near the Paleocene-Eocene boundary. A: Kheu River section (Central Caucasus). B: Aktumsuk Cape section (western Aral Sea). Arrows indicate levels of enrichment in calcitic detritus. Data on foraminifera after Stupin and Muzylöv, 2001. Note that the Paleocene-Eocene boundary was drawn at the NP9/NP10 zonal boundary.

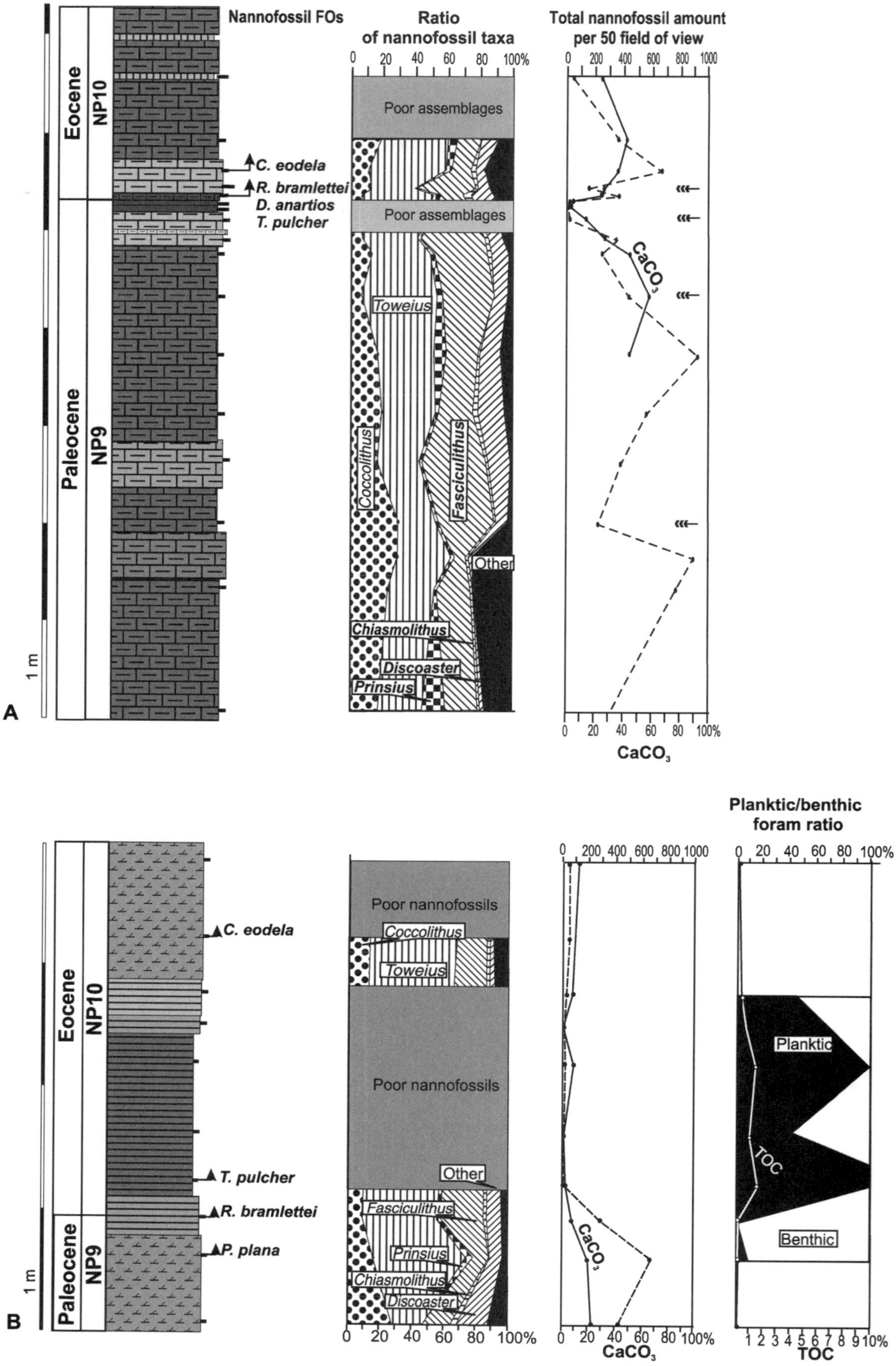
A
Nannofossil FOs
Eocene
NP10
Paleocene
NP9
1 m
C. eodela
R. bramlettei
D. anartios
T. pulcher
Ratio
of nannofossil taxa
0 20 40 60 80 100%
Poor assemblages
Toweius
Coccolithus
Fasciculithus
Other
Chiasmolithus
Discoaster
Prinsius
Total nannofossil amount
per 50 field of view
0 200 400 600 800 1000
CaCO3
0 20 40 60 80 100%
CaCO3
B
Eocene
NP10
Paleocene
NP9
1 m
C. eodela
T. pulcher
R. bramlettei
P. plana
Poor nannofossils
Coccolithus
Toweius
Poor nannofossils
Other
Fasciculithus
Prinsius
Chiasmolithus
Discoaster
0 20 40 60 80 100%
0 200 400 600 800 1000
CaCO3
0 20 40 60 80 100%
CaCO3
Planktic/benthic
foram ratio
0 20 40 60 80 100%
Planktic
TOC
Benthic
1 2 3 4 5 6 7 8 9 10%
TOC

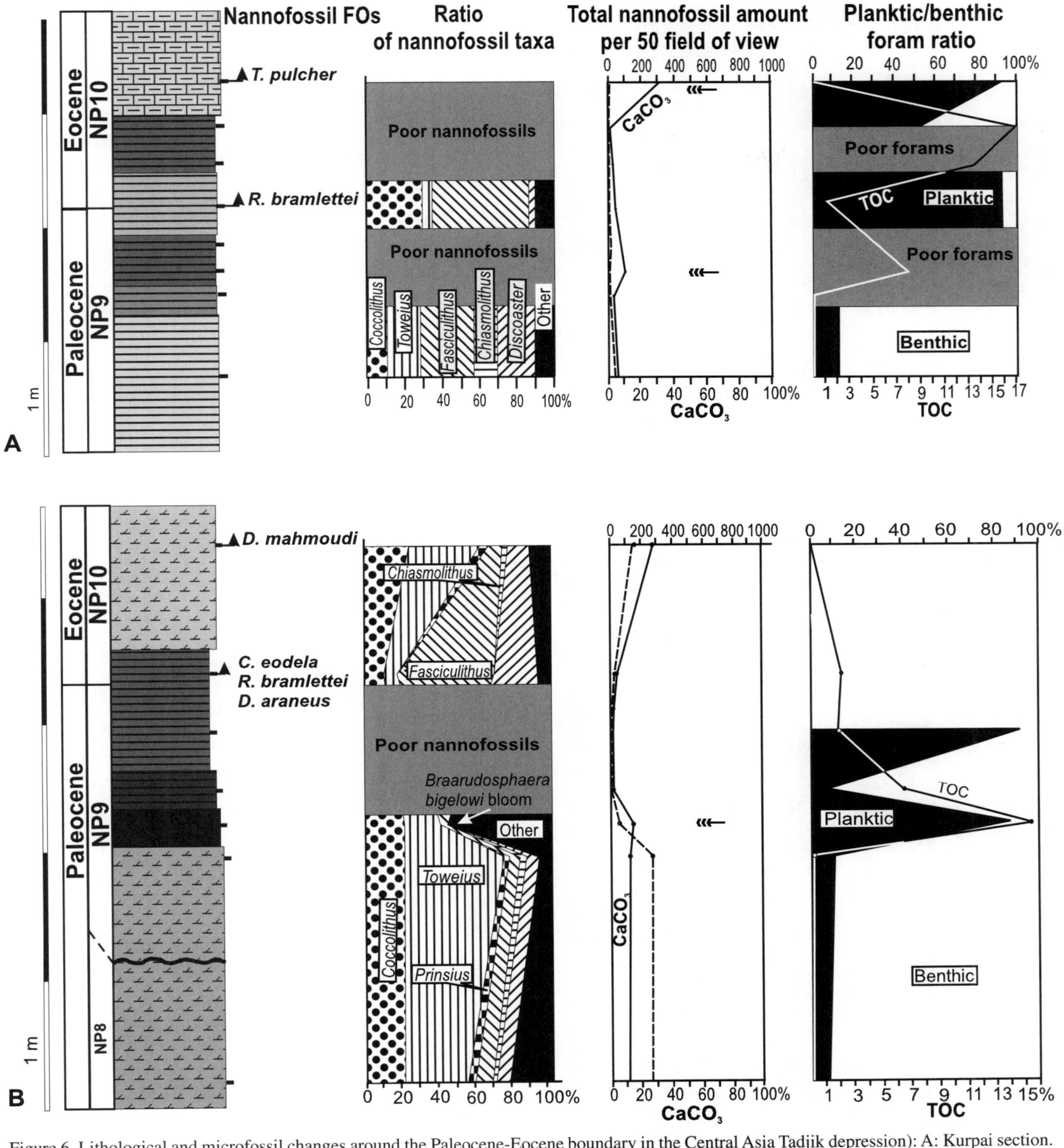

Figure 6. Lithological and microfossil changes around the Paleocene-Eocene boundary in the Central Asia Tadjik depression): A: Kurpai section. B: Guru-Fatima section. Arrows indicate levels of enrichment in calcitic detritus. Data on foraminifera after Stupin and Muzylöv, 2001. Note that the Paleocene-Eocene boundary was drawn at the NP9/NP10 zonal boundary. For legend see Figure 4.

Figure 5. Lithological and microfossil changes around the Paleocene-Eocene boundary: A: Sulak River section (Eastern Caucasus). B: Torangly section (Turkmenistan). Arrows indicate levels of enrichment in calcitic detritus. Data on foraminifera after Stupin and Muzylöv, 2001. Note that the Paleocene-Eocene boundary was drawn at the NP9/NP10 zonal boundary. For legend see Figure 4.

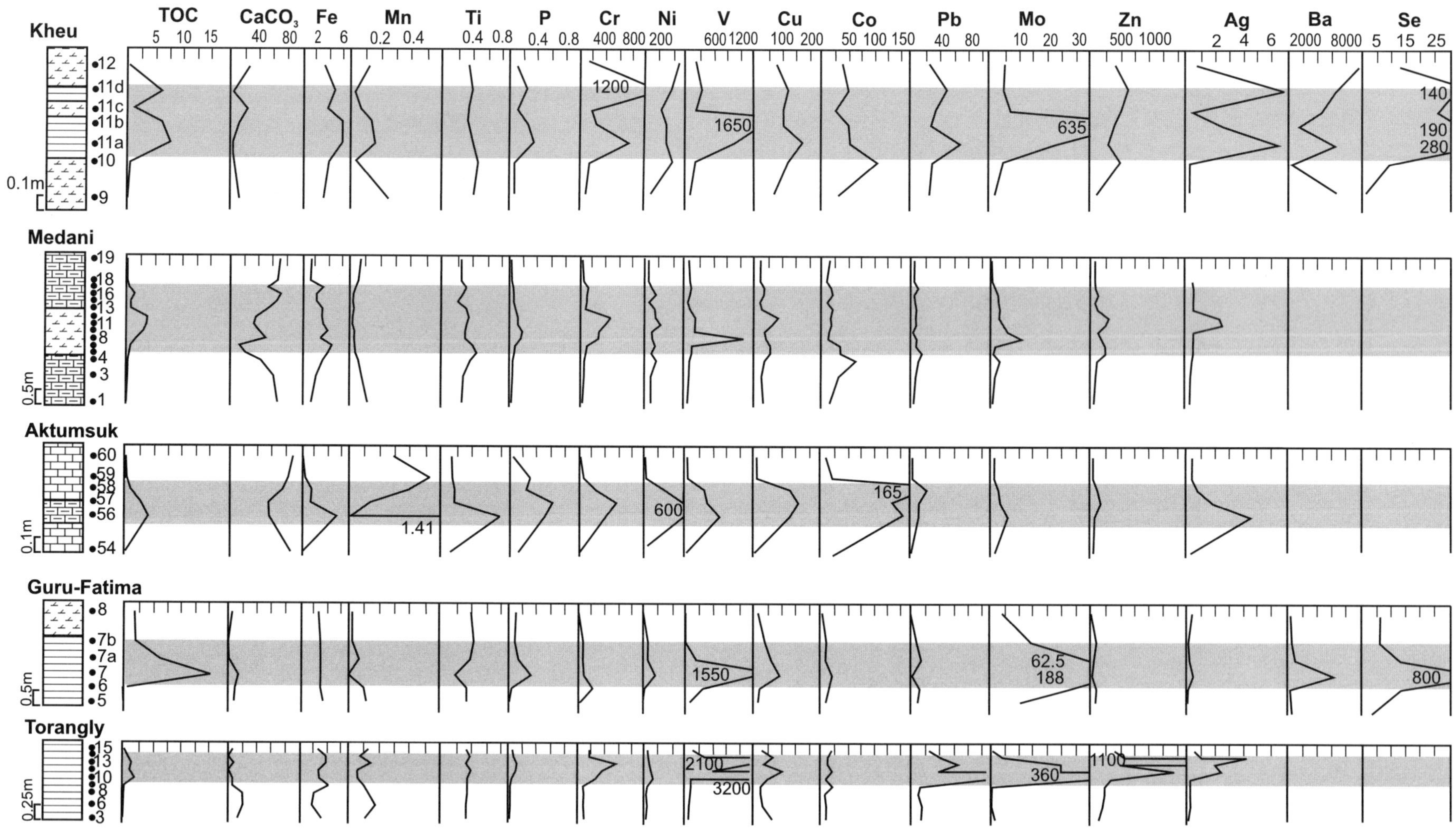

Figure 7. Concentrations of various compounds and elements in the sapropelitic bed in selected sections. The concentrations of C_{org}, Ca CO_3, P, Ti, Fe, Mn, are given in percent, those of the other elements in ppm.

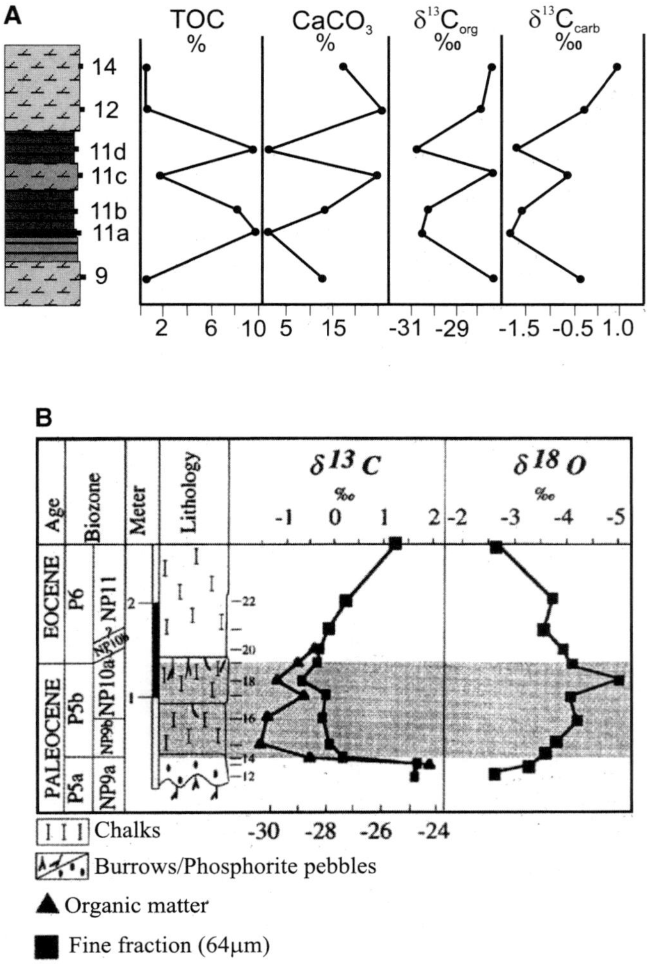

Figure 8. Oxygen and carbon isotopes in the sapropelitic bed: A: Kheu River section. B: Aktumsuk section (Bolle et al., 2000).

The planktic/benthic foraminifera ratio increased sharply in the sapropelitic bed, with benthic biota represented mostly by dysoxic species of benthic foraminifera, indicating varying degrees of dysoxia/anoxia within the basin (Stupin and Muzylöv, 2001). The benthic fauna recovered partially in the middle part of the sapropelitic bed, concurrent with the recovery of calcareous plankton and a decrease in TOC, suggesting that dysoxia/anoxia weakened temporarily. The changes in benthic foraminiferal assemblages in the sapropelitic bed of the northeastern Peri-Tethys thus resemble those documented in coeval sediments in Egypt (Speijer et al., 1997; Speijer, 2003). In the Torangly section, buliminids, generally thought to indicate a high food supply, were abundant (Speijer et al., 1997).

The calcareous microplankton apparently recovered during the deposition of the sapropelitic bed, after the "presapropelitic" crisis (thin interval without microplankton). We do not consider that carbonate dissolution caused the absence of calcareous plankton in this layer; in fact, dissolution appears to have been somewhat more important in the sapropelitic bed. Remains of organic-walled plankton dominated during the accumulation of the bed. The acme in dinoflagellate *Apectodinium* cysts in sediments deposited during the IETM has usually been explained as the result of high temperatures as well as high nutrient levels (e.g., Crouch et al., 2001). Dinoflagellates form a major part of primary planktonic production in oceans, and their blooms usually occur in warm, eutrophic waters. Dinoflagellates are also known for producing toxins in the so-called "red tide" blooms, causing fish and plankton kills (e.g., Taylor and Seliger, 1979; Steidinger and Baden, 1984).

When the sapropelitic bed accumulation gradually came to an end, the nannofloras recovered slowly in abundance and diversity. In the Torangly section the benthic foraminifera also show a gradual recovery (Speijer et al., 1997), as they did at ODP Site 1051 (Katz et al., 1999) and at several other sites globally (Thomas, 1998). Low-diversity nannofloras persisted for a long time after the accumulation of the sapropelitic bed, although the relative abundances of the most important taxa recovered to pre–sapropelitic bed levels. *Discoaster mahmoudi* appeared just after the deposition of the sapropelitic bed, with a FO apparently synchronous over the Tethys, as well as in a number of sections in Spain and Egypt (Monechi et al., 2000).

Depositional scenario for the sapropelitic bed

The deposition of the sapropelitic bed in the northeastern Peri-Tethys was the regional expression of the global IETM event in the unique setting of a large epicontinental basin. The regional features developed in response to the global events (climatic warming, carbon isotope excursion [CIE], benthic extinction event) documented to have occurred on land, in open ocean, and shelf areas. The sapropelitic bed is characterized by a light carbon isotopic composition of organic matter and carbonates (Kodina et al., 1995), corresponding to globally light values of carbon isotopes in the ocean-atmosphere system during the IETM (e.g., Kennett and Stott, 1991; Thomas and Shackleton, 1996), which probably resulted from the large-scale dissociation of gas hydrates (e.g., Dickens et al., 1995). The CIE in the northern Peri-Tethys section is similar in magnitude to that in open ocean sections, and we thus assume that values were probably not influenced by regional factors to a major degree.

The duration of the accumulation of the sapropelitic bed can be only approximately estimated by comparison with rates of sediment deposition elsewhere. Sapropelites may accumulate rapidly; for example, a sapropelitic bed of similar thickness as the sapropelitic bed accumulated in the Black Sea over a period of ~4–5 k.y. (Vinogradov et al., 1962). Mediterranean sapropels of similar thickness are also thought to have been accumulated over several thousands of years, and the accumulation rate of sapropels is 3–5 times greater than that of surrounding sediments (Sutherland et al., 1984). A lower Eocene succession of sapropelitic bed (10–15 cm) accumulated within 2 k.y. maximum (Oberhänsli and Beniamovskii, 2000). An approximate estimate obtained by assuming that the lamination of the sapropelitic bed represents annual laminae (Plate 1, Figure 3) indicates that the whole of the bed accumulated in ~10–12 k.y. However, correlation of the carbon isotope record of the sapropelitic bed (Fig. 8) with records in oceanic sections suggests that the main part of the bed formed during the peak of the CIE, which lasted ~60 k.y. (3 precession cycles; Röhl et al., 2000). We can presently not explain this disagreement.

The formation of organic-rich sediments is usually thought to result from increased biological productivity (e.g., Pedersen and Calvert, 1991). We agree with these authors that high productivity was the most important factor in the accumulation of the sapropelitic bed, while stratification of the water column may have played a role, but was not the main cause of low-oxygen conditions. Models proposed for the genesis of the Pliocene-Pleistocene eastern Mediterranean sapropels (e.g., Kidd el al., 1978; Calvert, 1983; Sutherland et al., 1984) and the Black Sea (Strakhov, 1971) are probably not relevant to the deposition of the sapropelitic bed, because the northeastern Peri-Tethys was a very large, epicontinental, relative shallow-water basin, separated from the open ocean by a system of archipelagoes, and thus had an oceanographic setting very different from that of the eastern Mediterranean and the Black Sea.

The sapropelitic bed accumulated during a rapid transgression, as documented by its sharp, sometimes erosional lower boundary. This transgression was preceded by a short regression (Muzylöv et al., 1996; Gavrilov et al., 1997), very similar to interpretations of the setting of coeval TOC-rich sediments at the southern Peri-Tethyan margin (Speijer, 2003; Speijer and Morsi, 2001). These authors argue that both increased productivity and increased preservation of organic matter were important factors in the deposition of the organic-rich sediments, and that the high productivity was driven by increased upwelling of nutrient-rich waters through offshore Ekman transport.

In contrast to these authors, we think that upwelling cannot have been efficient throughout the large, shallow epicontinental basin of the northern Peri-Tethys, and argue that high productivity throughout this vast region was mainly caused by an increased supply of nutrients into the basin from coastal regions during a rapid transgression (Fig. 9) (Gavrilov, 1994; Gavrilov and Kopaevich, 1996; Gavrilov et al., 1997).

This increased nutrient flux resulted from a rather complex series of events. During the rapid regression just before the IETM, large lowland areas with little relief formed along the periphery of the basin and around archipelagoes in a relatively shallow epicontinental sea. Non- or weakly lithified sediments in the newly exposed, widespread coastal plains underwent soil formation, and the soils became enriched in nutrients. In humid areas, such as the Caucasus and adjacent areas, lakes and water-logged bogs developed rapidly in the extensive coastal plains. In present-day peat bogs, P is very soluble and becomes enriched in surface waters (Kovalev, 1985). Water-logged basins contain abundant dissolved and particulate organic matter from the decomposition of vegetation.

We argue that a rapid transgression flooded the coastal plains at the beginning of the formation of the sapropelitic bed (Fig. 9). Because most of the northeastern Peri-Tethys was a peneplain formed in the Late Cretaceous and earliest Paleogene, even minor sea-level fluctuations resulted in a major change in basin outline. The sea-level rise was at least several tens of meters (Haq et al., 1987; Gavrilov et al., 1997) in agreement with estimates by Speijer and Morsi (2001) and Speijer (2003) for the southern margin of Peri-Tethys. During the sea-level rise, particulate and dissolved organic matter was transported into the sea from the drowned peat bogs and soils. The particulate organic matter was partially redeposited, enhancing enrichment in TOC in some regions (e.g., the Kheu river section). The dissolved organic matter and nutrients (including P which is enriched in most sapropelitic bed sections) were transported into the basin, triggering a strong increase in productivity of organic-walled and possibly prokaryote phytoplankton. The blooms of organic-walled phytoplankton may have adversely affected other plankton groups and fish, as observed in present-day "red tides" in eutrophic coastal regions.

The hypothesis that the transgression played a major role in triggering increased productivity was also invoked to explain the origin of Pennsylvanian black shales (Wenger and Baker, 1986). In addition, data on events during the filling of reservoirs behind dams, which simulates a transgression, support this hypothesis: During the initial filling of reservoirs, phytoplankton productivity is extremely high because large amounts of nutrients are washed from the flooded soils into the basin (Petrova, 1990).

In some sections in the eastern arid part of the basin (e.g., the Kurpai section) only a negligible amount of terrestrial organic matter accumulated in the sapropelitic bed, but the bed is closely associated with phosphate-bearing rocks and phosphorites. The lowermost part of the Suzak Formation (underlying the oil shale unit) contains abundant phosphate concretions. We argue that a significant amount of P, concentrated in soils and produced by weathering in pre–sapropelitic bed time, was supplied to the basin during the transgression, triggering phytoplankton blooms and causing high organic matter concentrations.

The high productivity of organic-walled plankton supplied most of the organic material present as TOC in the sapropelitic bed, while productivity of calcareous plankton was low, although we may underestimate calcareous plankton productivity as a result of slight dissolution. This high productivity was the major cause of the anoxia/dysoxia in the water column, as indicated by the impoverished benthic foraminiferal faunas, sediment lamination, and organic biochemical tracers. In addition, it has been argued that open-ocean waters were overall low in oxygen during the IETM (e.g., Thomas, 1998), and these global low oxygen levels could have influenced oxygen levels in marginal basins (e.g., Speijer, 2003).

We cannot, however, completely exclude the possibility that the anoxia/dysoxia was exacerbated by stratification and resulting stagnation. Increased runoff of fresh water could have caused stratification in western parts of the basin. The only ev-

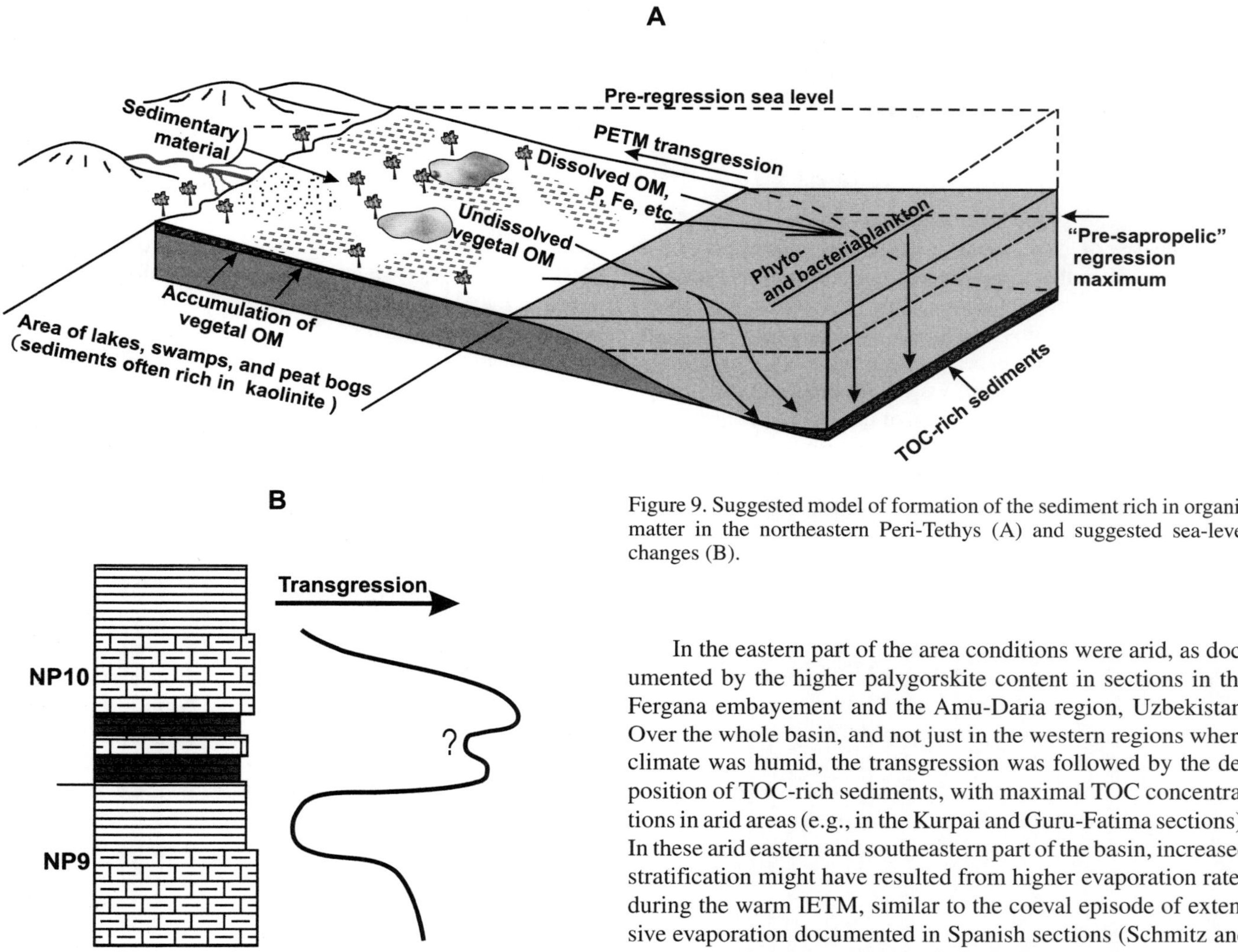

Figure 9. Suggested model of formation of the sediment rich in organic matter in the northeastern Peri-Tethys (A) and suggested sea-level changes (B).

idence for such an increase in runoff can be seen in the low values of oxygen isotopes in bulk carbonates in the sapropelitic bed in the Aktumsuk section (Fig. 8). But the thick-walled dinocyst abundance, the blooms of the calcareous dinocyst *Thoracosphaera,* and diverse nannofossil assemblages indicate eutrophic environments and normal salinity, and the biota thus do not indicate lowered salinity. At other times, the biota very clearly documented increased runoff. For instance, well-documented episodes of high freshwater influx in the northeastern Peri-Tethys (e.g., in the middle Oligocene part of the Maykop Formation) are characterized by the occurrence of abundant monospecific nannofossil assemblages associated with small thin-walled dinocyst remains (M. Akhmetiev, 2002, personal commun.). We thus have no solid evidence for increased runoff resulting in lowered salinity, and conclude that deposition of the sapropelitic bed was probably not caused by stagnation resulting from stable stratification as a result of increased runoff, even in the more humid western parts of the basin.

In the eastern part of the area conditions were arid, as documented by the higher palygorskite content in sections in the Fergana embayement and the Amu-Daria region, Uzbekistan. Over the whole basin, and not just in the western regions where climate was humid, the transgression was followed by the deposition of TOC-rich sediments, with maximal TOC concentrations in arid areas (e.g., in the Kurpai and Guru-Fatima sections). In these arid eastern and southeastern part of the basin, increased stratification might have resulted from higher evaporation rates during the warm IETM, similar to the coeval episode of extensive evaporation documented in Spanish sections (Schmitz and Pujalte, 2001). Hypersaline lagoons and lakes may have formed during pre–sapropelitic bed regression, and supplied saline water to the marine basin during the following transgression, possibly increasing local salinity and causing density stratification. Stable density stratification thus may have been a contributing factor to the high concentrations of organic matter in sediments (e.g., TOC >20% in the Suzak oil shales).

The accumulation of organic matter in the sediments caused sulfate-reduction and the generation of H_2S that diffused into the water, accumulated close to the bottom, and locally reached the photic zone, as shown by the biomarkers. The development of anoxic and dysoxic conditions caused extinction of benthic fauna, the appearance of anomalous planktonic taxa and possibly impoverished planktonic assemblages where these conditions reached the photic zone. The intensity of dysoxia/anoxia varied in different parts of the basin, as indicated by differences in benthic foraminiferal assemblages (Stupin and Muzylöv, 2001).

The presence of H_2S affected the geochemical signature of the sediments, too. In all sections, the sapropelitic bed is strongly enriched in Mo with respect to the surrounding sediments (Fig. 7), and enrichment patterns are similar for Se and

Re (Gavrilov et al., 1997). These elements (Mo, Se, Re) are redox-sensitive and form insoluble sulfides in the presence of H_2S in the water column. This process caused the observed enrichment in the sapropelitic bed, in combination with other mechanisms for the concentration of minor elements, such as sorption by clay particles and organic matter, and possibly diagenetic enrichment.

In some sections, we observed that biotic changes started below the base of the sapropelitic bed, where a thin layer of sediment contains neither plankton nor benthos. We have no lithological or geochemical evidence of major taphonomic changes during diagenesis in this layer (e.g., calcite redistribution). We suggest that a dramatic decrease in productivity may have occurred during the regression, which occurred before the rapid transgression that led to deposition of the sapropelitic bed. During this regression, the waters had a high turbidity in some parts of the basin, as a result of erosion of weakly lithified sediments, as documented by the occurrence of detrital carbonate and reworked Cretaceous nannofossils in some sections. The increased turbidity would have decreased light penetration in the water column, thus reducing primary productivity, and influencing floral and faunal composition.

After the peak of the transgression, the rich supply of nutrients gradually decreased with the deposition of the organic matter in the sapropelitic bed, and the intense blooms of various forms of plankton ended, ending the dysoxic to anoxic conditions. Besides, global ocean waters became more oxygenated (e.g., Thomas, 1998), which may have assisted in helping to alleviate the environmental conditions in the marginal basins (e.g., Speijer, 2003). In addition, the world cooled gradually (e.g., Zachos et al., 2001), which made it possible to dissolve more oxygen in the waters.

There has been much speculation regarding the cause of the end of the IETM, and the prevention of a runaway greenhouse effect. It has been argued that increased productivity on land (Beerling, 2000) or in open ocean (Bains et al., 2000) was responsible for the removal of carbon dioxide from the atmosphere, thus ending the IETM. We suggest that the deposition of huge amounts of organic material in the sapropelitic bed in the vast, northeastern Peri-Tethys may very well have contributed to the ending of the IETM and restoration of more normal environmental conditions.

ACKNOWLEDGMENTS

We thank N. Muzylöv who provided the samples of the Western Kopetdag and from the Tadjik depression sections, N.I. Zaporozhets who has kindly given her results from the dinocyst study from the Kheu River section, and M.A. Akhmetiev for a very fruitful discussion on dinocyst assemblages. We are grateful to T. Bralower, K. Bice, P. Luger, H. Brinkhuis, and M. Kruge for their thorough review of the paper and helpful comments. We also thank E. Thomas for considerable help in improving the manuscript.

This research was supported by the Russian Foundation for Basic Research, projects no. 00-05-64593 and 01-05-64805 and INTAS grant no. 93-2509.

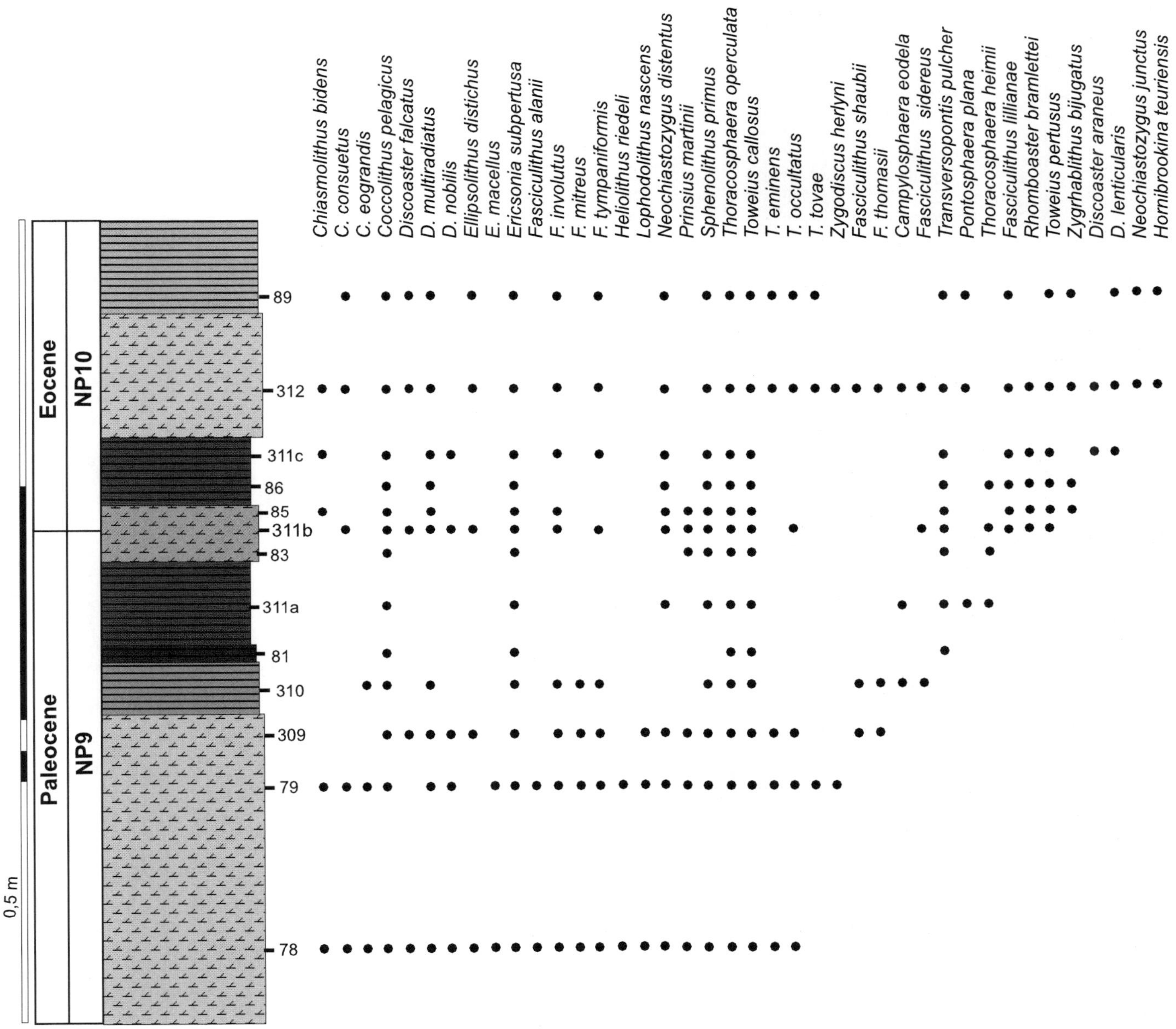

Appendix A. Nannofossil range chart of the sapropelitic bed of Kheu River section. Note that the Paleocene-Eocene boundary was drawn at the NP9/NP10 zonal boundary.

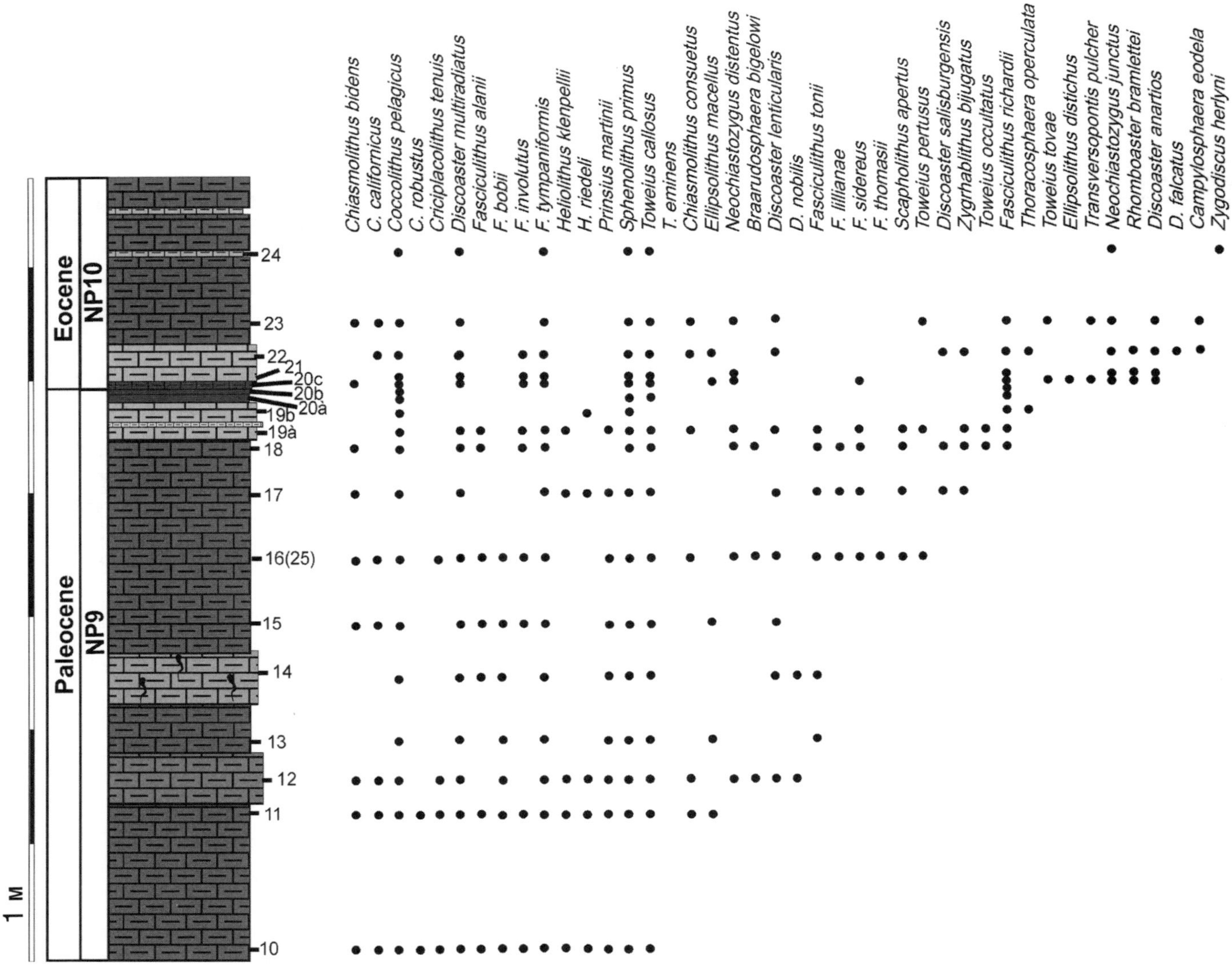

Appendix B. Nannofossil range chart of the sapropelitic bed of Sulak River section. Note that the Paleocene-Eocene boundary was drawn at the NP9/NP10 zonal boundary.

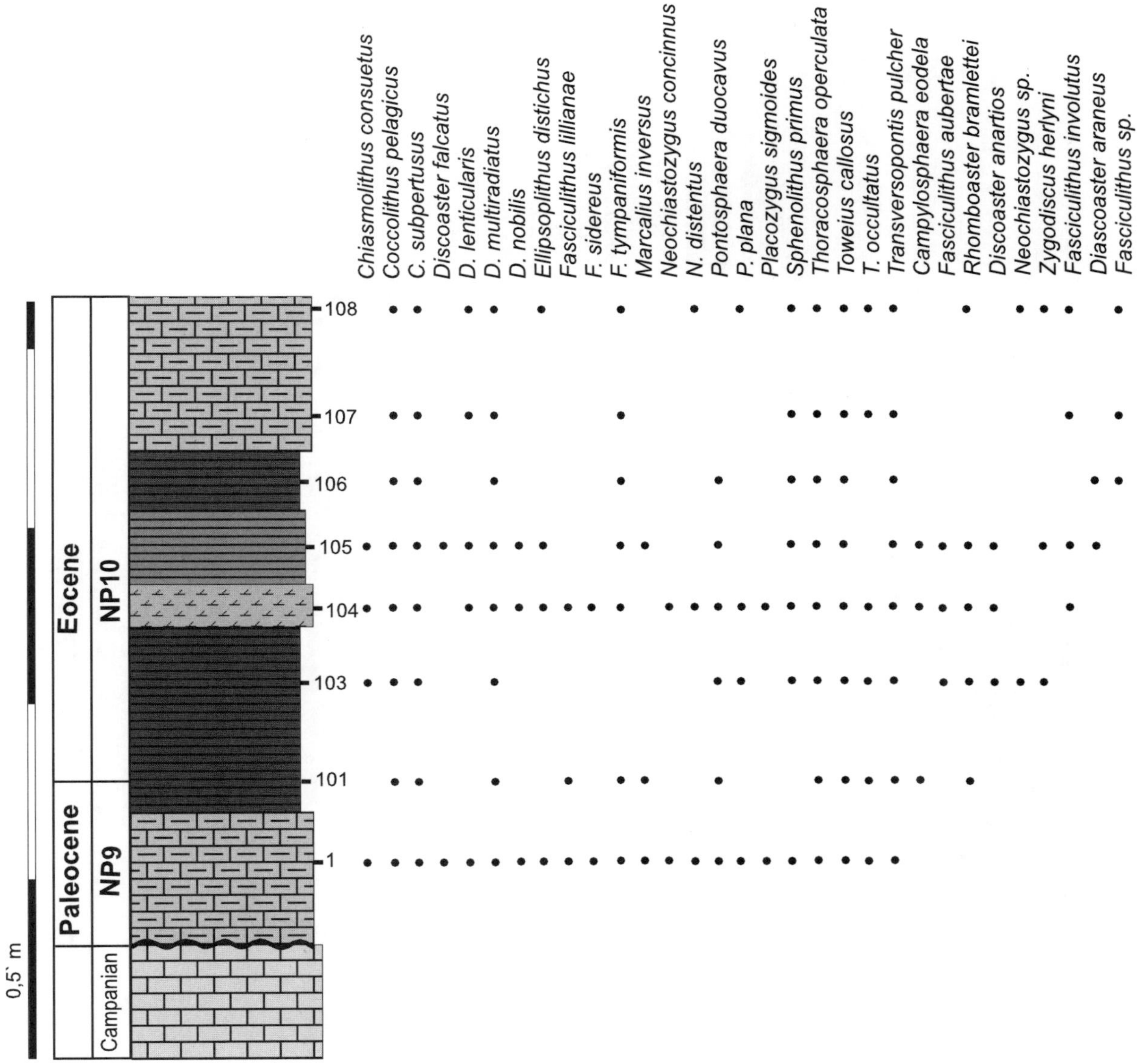

Appendix C. Nannofossil range chart of the sapropelitic bed of Aktumsuk section. Note that the Paleocene-Eocene boundary was drawn at the NP9/NP10 zonal boundary.

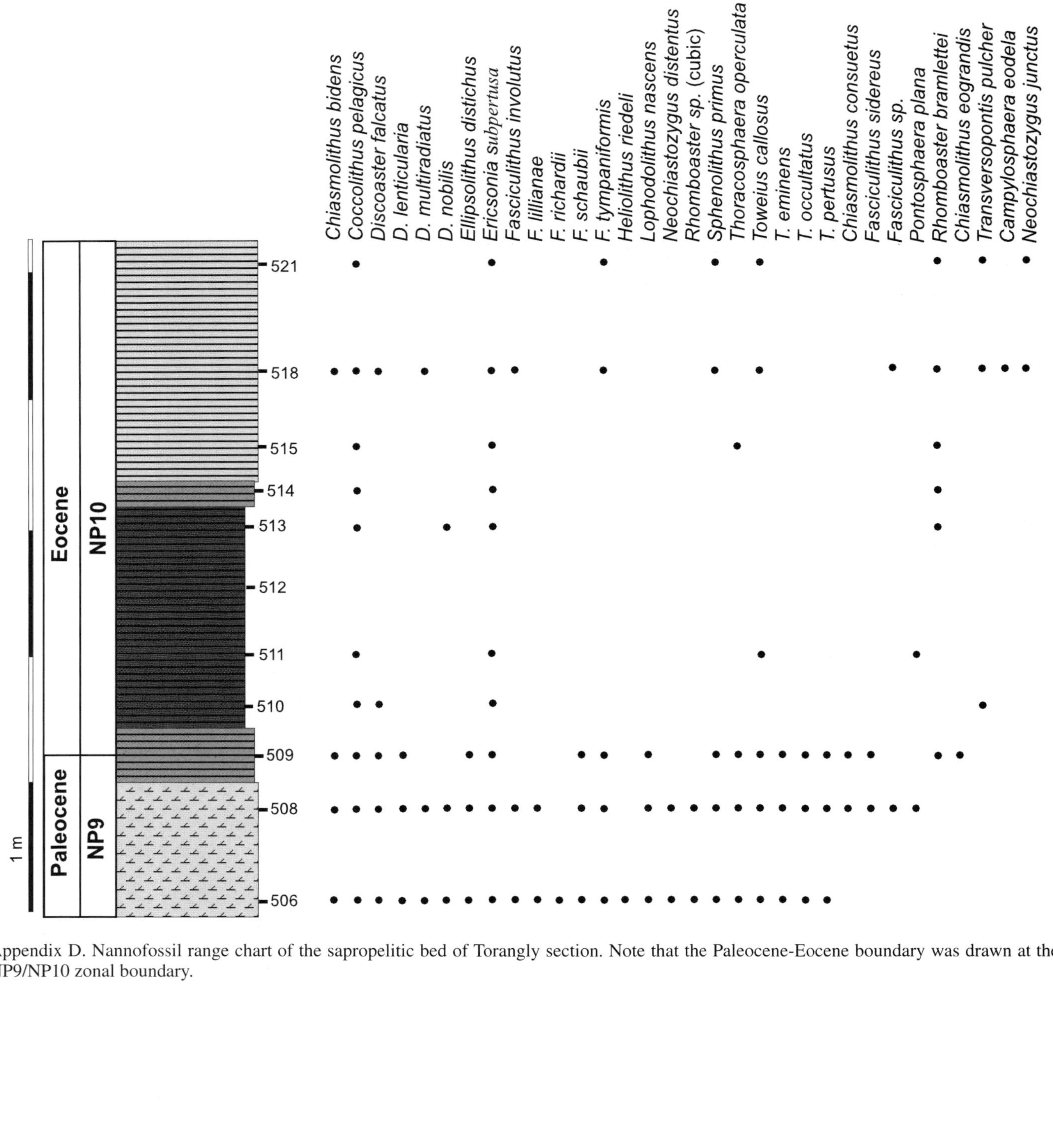

Appendix D. Nannofossil range chart of the sapropelitic bed of Torangly section. Note that the Paleocene-Eocene boundary was drawn at the NP9/NP10 zonal boundary.

REFERENCES CITED

Akhmetiev, M.A., and Zaporozhets, N.I., 1996, Changes in dinocysts assemblages as reflecting in ecosystem transformations in the Russian Platform, Crimea, Caucasus, and Turanian plate during Paleogene and early Miocene, *in* Kuznetsova, K.I., and Muzylöv, N.G., eds., Fossil microorganisms as the basis of the Phanerozoic stratigraphy, correlation, and paleobiogeography: Moscow, GEOS, p. 55–69 (in Russian).

Aubry, M.-P., Berggren, W.A., Stott, L., and Sinha, A., 1996, The upper Paleocene–lower Eocene stratigraphic record and the Paleocene-Eocene boundary carbon isotope excursion: Implications for geochronology, *in* Knox, R.W.O'B., et al., eds., Correlation of the early Paleogene in northwestern Europe: Geological Society [London] Special Publication 101, p. 353–380.

Aubry, M.-P., Lucas, S.G., and Berggren, W.A., editors, 1998, Late Paleocene–early Eocene climatic and biotic events in the marine and terrestrial records: New York, Columbia University Press, 513 p.

Bains, S., Norris, R.D., Corfield, R.M., and Faul, K., 2000, Termination of global warmth at the Paleocene/Eocene boundary through productivity feedback: Nature, v. 407, p. 171–174.

Beerling, D., 2000, Increased terrestrial carbon storage across the Paleocene-Eocene boundary: Palaeogeography, Palaeoclimatology, Palaeoecology, v. 161, p. 395–405.

Bolle, M.-P., Pardo, A., Adatte, T., Tantawy, A.A., Hinrichs, K.-N., von Salis, K., and Burns, S., 2000, Climatic evolution on the southern and northern margins of the Tethys from the Paleocene to the early Eocene: GFF, v. 122, p. 31–32.

Bralower, T.J., and Premoli-Silva, I., 2002, Leg 198 Summary: Bralower, T.J., et al., eds., Ocean Drilling Program, Preliminary Report, Leg 198, http://www.dp.tamu.edu/publications/198_IR/VOLUME/CHAPTERS/IR 198_01.PDF.

Bujak, J.P., and Brinkhuis, H., 1998, Global warming and dinocyst changes across the Paleocene/Eocene epoch boundary, *in* Aubry, M.-P., et al., eds., Late Paleocene–early Eocene biotic and climatic events in the marine and terrestrial records: New York, Columbia University Press, p. 277–295.

Bukry, D., 1974, Coccoliths as paleosalinity indicators: Evidence from Black Sea, *in* Degens, E.T., and Ross, D.A., eds., Black Sea: Geology, chemistry, and biology: Tulsa, Oklahoma, American Association of Petroleum Geologists Publishing, p. 353–363.

Bybell, L.M., and Self-Trail, J.M., 1994, Evolutionary, biostratigraphic, and taxonomic study of calcareous nannofossil from a continuous Paleocene/Eocene boundary section in New Jersey: U.S. Geological Survey Professional Paper 1554, 36 p.

Bybell, L.M., and Self-Trail, J.M., 1997, Late Paleocene and early Eocene calcareous nannofossils from three boreholes in an onshore-offshore transect from New Jersey to the Atlantic continental rise, *in* Miller, K.J., and Snyder, S.W., eds., Proceedings of the Ocean Drilling Program, Scientific Results: Texas A & M University, College Station, Texas, USA, v. 150X, p. 91–110.

Calvert, S.E., 1983, Geochemistry of Pleistocene sapropels and associated sediments from the Eastern Mediterranean: Oceanologica acta, v. 6, no. 3, p. 255–267.

Calvert, S.E., 1990, Geochemistry and origin of the Holocene sapropel in the Black Sea, *in* Ittekkot, V., et al., eds., Facets of modern biogeochemistry: Berlin, Springer-Verlag, p. 326–352.

Crouch, E., Bujak, J.P., and Brinkhuis, H., 2000, Southern and Northern Hemisphere dinoflagellate cyst assemblage changes in association with the late Paleocene thermal maximum: GFF, v. 122, p. 40–41.

Crouch, E.M., Heilmann-Clausen, C., Brinkhuis, H., Morgans, H.E.G., Rogers, K.M., Egger, H., and Schmitz, B., 2001, Global dinoflagellate event associated with the late Paleocene thermal maximum: Geology, v. 29, p. 315–318.

Dickens, G.R., O'Neil, J.R., Rea, D.K., and Owen, R.M., 1995, Dissociation of oceanic methane hydrate as a cause of the carbon isotope excursion at the end of Paleocene: Geology, v. 25, p. 259–262.

Fomina, L.S., and Volkov, I.I., 1970, Rare earth elements in the Black Sea sediments: Litologia i poleznye iskopaemye, no. 2, p. 148–160 (in Russian).

Gavrilov, Yu.O., 1994, The possible causes of sediment rich in organic matter accumulation in relation with eustatic sealevel changes, *in* Semikhatov, M.A., and Rozanov, A.Yu., eds., Ecosystem restructures and the evolution of biosphere: Moscow, Nedra, p. 305–311 (in Russian).

Gavrilov, Yu.O., and Kopaevich, L.F., 1996, Geochemical, biochemical, and biotic consequences of eustatic fluctuations: Stratigraphy and Geological Correlation, v. 4, no. 4, p. 3–14.

Gavrilov, Yu.O., and Muzylöv, N.G., 1991, The geochemistry of sapropelite interbeds in the Paleogene of Central Caucasus: Lithology and Mineral Resources, v. 6, p. 84–98.

Gavrilov, Yu.O., Kodina, L.A., Lubchenko, I.Yu., and Muzylöv, N.G., 1997, The Late Paleocene anoxic event in epicontinental seas of Peri-Tethys and formation of the sapropelite unite: Sedimentology and geochemistry: Lithology and Mineral Resources, v. 5, p. 492–517.

Gavrilov, Yu.O., Shcherbinina, E.A., and Muzylöv, N.G., 2000, A Paleogene sequence in central North Caucasus: A response to paleoenvironmental changes: GFF, v. 122, p. 51–53.

Haq, B.U., Hardenbol, J., and Vail, P.R., 1987, Chronology of fluctuating sea levels since Triassic: Science, v. 235, p. 1156–1167.

Karwath, B., Janofske, D., Tetjen, F., and Willems, H., 2000, Temperature effects on growth and cell size in the marine calcareous dinoflagellate Thoracosphaera heimii: Marine Micropaleontology, v. 39, no. 1–4, p. 43–51.

Katz, M.E., Pak, D.K., Dickens, G.R., and Miller, K.G., 1999, The source and fate of massive carbon input during latest Paleocene thermal maximum: Science, v. 286, p. 1531–1533.

Kennett, J.P., and Stott, L.D., 1991, Abrupt deep-sea warming, paleoceanographic changes, and benthic extinction at the end of Paleocene: Nature, v. 353, p. 225–229.

Kidd, R.B., Cita, M.B., and Ryan, W.B.F., 1978, Stratigraphy of Eastern Mediterranean sapropel sequences recovered during DSDP Leg 42A and their paleoenvironmental significance, Hsü, K.J., et al., eds., Initial reports of the Deep Sea Drilling Project: Washington, D.C., U.S. Government Printing Office, v. 42, p. 421–443.

Koch, P.L., Zachos, J.C., and Gingerich, P.D., 1992, Coupled isotopic changes in marine and continental carbon reservoirs at the Paleocene-Eocene boundary: Nature, v. 358, p. 319–322.

Kodina, L.A., Huang, Y., Gavrilov, Yu.O., Jones, M., and Eglinton, G., 1995, Environment of upper Paleocene black shale deposition in southern Russia and adjacent regions as revealed by isotope and biomarker study, *in* Grimalt, J.O., and Dorronsoro, C., eds., Organic geochemistry: Developments and applications to energy, climate, environment and human history: Selected papers from the 17th International Meeting on Organic Geochemistry, Donostia-San Sebastian, The Basque Country, 4th–8th September 1995, p. 192–194.

Kovalev, V.A., 1985, Bolotnye mineralogo-geokhimicheskie systemy (Mineralogical and geochemical systems of bogs): Minsk, Nauka i Tekhnika, 327 p. (in Russian).

Luterbacher, H.P., Hardenbol, J., and Schmitz, B., 2000, Decision of the voting members of the International Subcommission on Paleogene Stratigraphy on the criterion for the recognition of the Paleocene-Eocene boundary: Newsletter of the International Subcommission on Paleogene Stratigraphy, Tübingen, November 2000, v. 9, p. 13.

Martini, E., 1971, Standard Tertiary and Quaternary calcareous nannoplankton zonation, *in* Farinacci, A., ed., Proceedings of the 2nd Planktonic Conference, Rome, Italy, 1970: Tecnoscienza, v. 2, p. 739–785.

Monechi, S., Angori, E., and Speijer, R., 2000, Upper Paleocene stratigraphy in the Mediterranean region: Zonal markers, diachronism, and preservational problems: GFF, v. 122, p. 108–110.

Muzylöv, N.G., 1994, Paleocene–middle Eocene anoxic events, *in* Semikhatov, M.A., and Rozanov, A.Yu., eds., Ecosystem restructures and the evolution of biosphere: Moscow, Nedra, p. 151–159 (in Russian).

Muzylöv, N.G., and Shcherbinina, E.A., 1995, The main geological events of Thanetian–lower Ypresian of the South of the Former USSR: IGCP (International Geological Correlation Program) Project 308 meeting Paleocene-Eocene boundary events, Beer Sheva, Israel, 8–15 January 1995, p. 6–7.

Muzylöv, N.G., Benyamovskii, V.N., and Tabachnikova, I.P., 1989, Sapropel interlayers in lower Paleocene deposits of the southern USSR: Izvestia Akademii Nauk SSSR, Seria Geologicheskaya, v. 11, p. 117–119 (in Russian).
Muzylöv, N.G., Benjamovsky, V.N., Gavrilov, Yu.O., Shcherbinina, E.A., and Stupin, S.I., 1996, Paleontological and geochemical features of the Central Caucasian upper Paleocene sapropelite, *in* Kuznetsova, K.I., and Muzylöv, N.G., eds., Fossil microorganisms as the basis of the Phanerozoic stratigraphy, correlation, and paleobiogeography: Moscow, GEOS, p. 117–127.
Oberhänsli, H., and Beniamovskii, V.N., 2000, Dysoxic bottom water events in the Peri-Tethys during the late Ypresian: A result of changes in the evaporation/precipitation balance in adjacent continental regions: GFF, v. 122, p. 21–123.
Okada, H., and Bukry, D., 1980, Supplementary modification and introduction of code numbers to the low-latitude coccolith biostratigraphic zonation (Bukry, 1973, 1975): Micropaleontology, v. 5, no. 3, p. 321–325.
Orue-Etxebarria, X., Apellaniz, E., Baceta, J.I., Coccioni, R., Di Leo, R., Dinares-Turell, S., Galeotti, S., Monechi, S., Nunez-Betelu, K., Pares, J.M., Payros, A., Pujalte, V., Samso, J.M., Serra-Kiel, J., Schmitz, B., and Tosquella, J., 1996, Physical and biostratigraphic analysis of two prospective Paleocene-Eocene boundary stratotypes in the intermediate-deep water Basque Basin, western Pyrenees: The Trabakua Pass and Ermua sections: Neues Jahrbuch für Geologie und Paläontologie, Abhandlungen, v. 200, p. 1–64.
Pedersen, T.F., and Calvert, S.E., 1991, Anoxia vs. productivity: What controls the formation of organic-carbon-rich sediments and sedimentary rocks?: Discussion: AAPG (American Association of Petroleum Geologists) Bulletin, v. 75, p. 500–501.
Perch-Nielsen, K., 1977, Albian to Pleistocene calcareous nannofossils from the western South Atlantic, *in* Stupko, P.R., ed., Initial reports of the Deep Sea Drilling Project: Washington, D.C., U.S. Government Printing Office, v. 39, p. 699–824.
Petrova, N.A., 1990, Phytoplankton successions related to the antropogenic eutrophic processes in large lakes: Leningrad, Nauka, 199 p. (in Russian).
Powell, A.J., 1992, Dinoflagellate cysts of the Tertiary System, *in* Powell, A.J., ed., A stratigraphic index of dinoflagellate cysts: London, Chapman and Hall, p. 155–251.
Repeta, D.J., and Simpson, D.J., 1991, The distribution and recycling of chlorophyll, bacteriochlorophyll and carotenoids in the Black Sea: Deep-Sea Research, v. 38, Supplementary 2, p. 969–984.
Repeta, D.J., Simpson D.J., Jorgensen, B.B., and Jannasch, H.W., 1989, Evidence of anoxigenic photosynthesis from the distribution of bacteriochlorophylls in the Black Sea: Nature, v. 342, p. 69–72.
Röhl, U., Bralower, T.J., Norris, R.D., and Wefer, G., 2000, New chronology for the late Paleocene thermal maximum and its environmental implications: Geology, v. 28, p. 927–930.
Schmitz, B., and Pujalte, V., 2001, Correlating the Paleocene-Eocene boundary in the Pyrenees, Spain, by carbon isotopes: A transect from deep marine to continental environments, *in* Ash, A.W., and Wing, S.L., eds., Climate and biota of the early Paleogene: International Meeting July 3–8, 2001, Northwest College, Powell, Wyoming, USA, Abstract volume, p. 81.
Schmitz, B., Heilmann-Clausen, C., King, C., Steurbaut, E., Andreasson, F.P., Corfield, R.M., and Cartlidge, J.E., 1996, Stable isotope and biotic evolution in the North Sea during the early Eocene: The Albaek Hoved section, Denmark, *in* Knox, R.W.O'B., et al., eds., Correlation of the early Paleogene in northwestern Europe: Geological Society [London] Special Publication 101, p. 275–306.
Sinninghe Damste, J.S., Wakeham, S.G., Kohnen, M.E.L., Hayes, J.M., and de Leeuw, J.W., 1993, A 6000-year sedimentary molecular record of chemocline excursions in the Black Sea: Nature, v. 362, p. 827–829.
Speijer, R.P., and Morsi, A.M., 2001, First detailed ostracod records indicating sea-level change associated with the LPTM (Eastern Desert, Egypt), *in* Ash, A.W., and Wing, S.L., eds., Climate and biota of the early Paleogene: International Meeting July 3–8, 2001, Northwest College, Powell, Wyoming, USA, Abstract volume, p. 88.
Speijer, R.P., and Wagner, T., 2001, Black shale deposition and sea-level change in the Tethys during the LPTM, *in* Ash, A.W., and Wing, S.L., eds., Climate and biota of the early Paleogene: International Meeting July 3–8, 2001, Northwest College, Powell, Wyoming, USA, Abstract volume, p. 89.
Speijer, R.P., Schmitz, B., and van der Zwaan, G.J., 1997, Benthic foraminifera extinction and repopulation in response to latest Paleocene anoxia: Geology, v. 25, p. 683–686.
Steidinger, K.A., and Baden, D.G., 1984, Toxic marine dinoflagellates, *in* Spector, D.L., ed., Dinoflagellates: New York, Academic, p. 201–261.
Steurbaut, E., De Conink, J., Dupui, C., and King, C., 2000, Dinoflagellate cyst events and depositional history of the Paleocene-Eocene boundary interval in the southern North Sea basin: GFF, v. 122, p. 154–155.
Strakhov, N.M., 1971, Geochemical evolution of the Black Sea in Holocene: Litologia i Poleznie iskopaemye, no. 3, p. 3–17 (in Russian).
Stupin, S.I., and Muzylöv, N.G., 2001, The late Paleocene Ecologic crisis in epicontinental basins of the eastern Peri-Tethys: Microbiota and accumulation conditions of sapropelitic bed: Stratigraphy and Geological Correlation, v. 9, no. 5, p. 501–507.
Summons, R.E., and Powell, T., 1986, Chlorobiaceae in Paleozoic seas revealed by biological markers, isotopes, and geology: Nature, v. 319, p. 763–765.
Summons, R.E., and Powell, T., 1987, Identification of aryl isoprenoids in source rocks and crude oils: Biological markers for the green sulphur bacteria: Geochimica and Cosmochimica Acta, v. 51, p. 557–566.
Sutherland, H.E., Calvert, S.E., and Morris, R.J., 1984, Geochemical studies of the recent sapropel and associated sediment from the Hellenic Outer Ridge, eastern Mediterranean Sea: Mineralogy and chemical composition: Marine Geology, v. 56, no. 1–4, p. 79–92.
Tantawy, A.A., Ouda, Kh., von Salis, K., and Saad El-Din, M., 2000, Biostratigraphy of Paleocene sections in Egypt: GFF, v. 122, p. 163–164.
Taylor, D.L., and Seliger, H.H., editors, 1979, Toxic dinoflagellate blooms: New York, Elsevier, 528 p.
Thomas, E., 1998, The biogeography of the late Paleocene benthic foraminiferal extinction, *in* Aubry, M.-P., et al., eds., Late Paleocene–early Eocene biotic and climatic events in the marine and terrestrial records: New York, Columbia University Press, p. 214–243.
Thomas, E., and Shackleton, N., 1996, The Paleocene-Eocene benthic foraminiferal extinction and stable isotope anomalies, *in* Knox, R.W.O'B., et al., eds., Correlation of the early Paleogene in northwest Europe: Geological Society [London] Special Publication 101, p. 401–441.
Vink, A., Brune, A., Häll, C., Zonnerveld, K.A.F., and Willems, H., 2002, On the response of calcareous dinoflagellates to oligotrophy and stratification of the upper water column in the equatorial Atlantic Ocean: Palaeogeography, Palaeoclimatology, Palaeogeography, v. 178, no. 1–2, p. 53–66.
Vinogradov, A.P., Grinenko, V.A., and Ustinov, V.I., 1962, Isotopic composition of sulfur in the Black Sea: Geokhimiya, no. 10, p. 851–873 (in Russian).
Volkov, I.I., 1973, Main regularities in the distribution of chemical elements in deepwater sediments of the Black Sea: Litologia i poleznye iskopaemye, no. 2, p. 3–22.
von Salis, K., Monechi, S., Bybell, L.M., Self-Trail, J.M., and Young, J., 2000, Remarks on calcareous nannofossil markers Rhomboaster and Tribrachiatus around the Paleocene-Eocene boundary: GFF, v. 122, p. 138–140.
Wenger, L.M., and Baker, D.R., 1986, Variation in organic geochemistry of anoxic-oxic black shale-carbonate sequences in the Pennsylvanian of the Midcontinent, USA: Organic Geochemistry, v. 10, no. 1–3, p. 85–92.
Wing, S.L., Alroy, J., and Hickey, L.J., 1995, Plant and mammal diversity in the Paleocene to early Eocene of the Bighorn Basin: Palaeogeography, Palaeoclimatology, Palaeoecology, v. 115, p. 117–156.
Zachos, J.C., Lochman, K.C., Walker, J.C.G., and Wise, S.W., Jr., 1993, Abrupt climate change and transient climates during Paleogene: A marine perspective: Journal of Geology, v. 101, p. 191–213.
Zachos, J., Pagani, M., Sloan, L., Thomas, E., and Billups, K., 2001, Trends, rhythms, and aberrations in global climate 65 Ma to present: Science, v. 292, p. 686–693.

Manuscript Accepted by the Society August 13, 2002

Printed in the USA

Geological Society of America
Special Paper 369
2003

Diatom turnover in the early Paleogene diatomite of the Sengiley section, Middle Povolzhie, Russia: A response to the Initial Eocene Thermal Maximum?

Tatiana V. Oreshkina*
Geological Institute, Russian Academy of Sciences, 7 Pyzhevsky per., Moscow 119017, Russia
Hedi Oberhänsli*
GeoForschungsZentrum, Projektbereich 3.3, Telegraphenberg, D-14473 Potsdam, Germany

ABSTRACT

The lower Paleogene sequence in the Middle Povolzhie (Ulyanovsk-Saratov syncline) of Russia is dominated by a terrigenous–rich siliceous facies with abundant and well-preserved siliceous plankton. Calcareous microfossils are virtually absent. The diatom successions of the Sengiley diatomite are correlated to diatom assemblages of the Fur Formation (Denmark) and the Polosataya Formation (Kazakhstan), and are calibrated to calcareous nannofossil and dinocyst zonations. The *Trinacria ventriculosa* Zone, which is found in the lower part of the Sengiley diatomite (Kamyshin Formation), is age equivalent to the upper part of calcareous nannofossil zones NP8 through NP9a. The upper part of the Sengiley diatomite (Kamyshin Formation; *Hemiaulus proteus* Zone) can be correlated to the uppermost part of the *Apectodinium augustum* dinocyst Zone, which corresponds to the NP9b calcareous nannofossil Subzone, in which the globally observed negative carbon isotope excursion occurs.

A major floral turnover in the upper third of the Sengiley diatomite is characterized by a succession of bioevents, including first appearances at generic and specific levels. Quantitative diatom analyses show pronounced compositional changes: (1) the decline of the meroplanktonic *Pyxidicula* group, typical of neritic conditions (2) the increase of the *Paralia* group, a marker for the proximity of the shoreline, and (3) a relatively high abundance of pelagic cosmopolitan species. Grain size analyses show an increase in terrigenous input in the upper part of Sengiley diatomite. The floristic and lithologic shifts indicate changes from stable, highly productive neritic conditions during a sea-level lowstand (Phase 1) to the onset of a transgression (Phase 2), which is followed by a highstand (Phase 3). The highstand phase coincides with enhanced exchange between the Tethys and the adjacent epicontinental seas. The diatom bioevents in Phases 2 and 3 in the Sengiley section are most probably coeval to the well-documented biotic changes of the Initial Eocene Thermal Maximum.

*oreshkina@geo.tv-sign.ru; oberh@gfz-potsdam.de.

Oreshkina, T. V., and Oberhänsli, H., 2003, Diatom turnover in the early Paleogene diatomite of the Sengiley section, Middle Povolzhie, Russia: A response to the Initial Eocene Thermal Maximum?, *in* Wing, S.L., Gingerich, P.D., Schmitz, B., and Thomas, E., eds., Causes and Consequences of Globally Warm Climates in the Early Paleogene: Boulder, Colorado, Geological Society of America Special Paper 369, p. 169–179.

INTRODUCTION

During the Thanetian, the Ulyanovsk-Saratov syncline was the proximal part of a gulf from the northern Peri-Tethys (Radionova et al., 1999). Sediments accumulating in the gulf consisted of diatomites and cherts, partly enriched in terrigenous components, with abundant diatoms, radiolaria, and silicoflagellates (Fig. 1).

One of the most important upper Paleocene–lower Eocene reference sections is a ~26-m-thick sequence near the village of Sengiley, consisting of pure, homogeneous diatomites. The stratigraphic study of Sengiley diatoms was pioneered by Jousé (1949, 1979, 1982). The Sengiley assemblages were investigated by many other authors (Gleser et al., 1977; Gleser, 1979; Strelnikova, 1992; Gleser, 1994, 1995; Khokhlova and Oreshkina, 1999). Due to the lack of calcareous plankton and paleomagnetic stratigraphy, the common approach to dating the Sengiley diatomite has been to correlate the observed diatom assemblages with diatom zones from areas where calibration with calcareous nannoplankton zones is possible.

The Sengiley diatom flora is similar to assemblages in sediments deposited at high latitudes of the Southern Hemisphere, such as the Falkland Plateau (Deep Sea Drilling Project [DSDP] Site 327A; Gombos, 1976), the South Atlantic (DSDP Site 700; Fenner, 1991), the Cape Basin (DSDP Site 524; Gombos, 1984), the East Indian Range (DSDP Site 214; Jousé, 1982; Mukhina, 1976), and the eastern Indian Ocean (Ocean Drilling Project [ODP] Site 752; Fourtanier, 1991). Such nearly global correlations help to date Sengiley diatom successions, and place them in the late Paleocene (Selandian–Thanetian) to early Eocene (NP5–NP10).

Radiolarian data (Kozlova, 1994; Khokhlova and Oreshkina, 1999), also based on the correlation of regional zonal divisions with pelagic zones, constrain the age of Sengiley diatomite to the equivalent of calcareous nannofossil zones CP3–CP8 (NP4–NP9). Our reexamination of the diatom successions will attempt to confine the stratigraphic position of the diatom assemblages in this area more rigorously, using new data from key sections in the neighboring regions of the northern Peri-Tethys. In addition, we intend to reconstruct the paleoenvironments using the relative abundance of dominant diatom groups and grain size distribution.

GEOLOGICAL SETTING AND STRATIGRAPHY

The Paleocene-Eocene sedimentary section (~300 m thick) in the Ulyanovsk–Saratov syncline in the middle Volga area (Fig. 1), includes five lithological units: (i) the lower Syzran Subgroup (100 m thick, Danian); (ii) the upper Syzran Subgroup (30 m, Selandian); (iii) the Saratov Formation (70 m, Thanet-

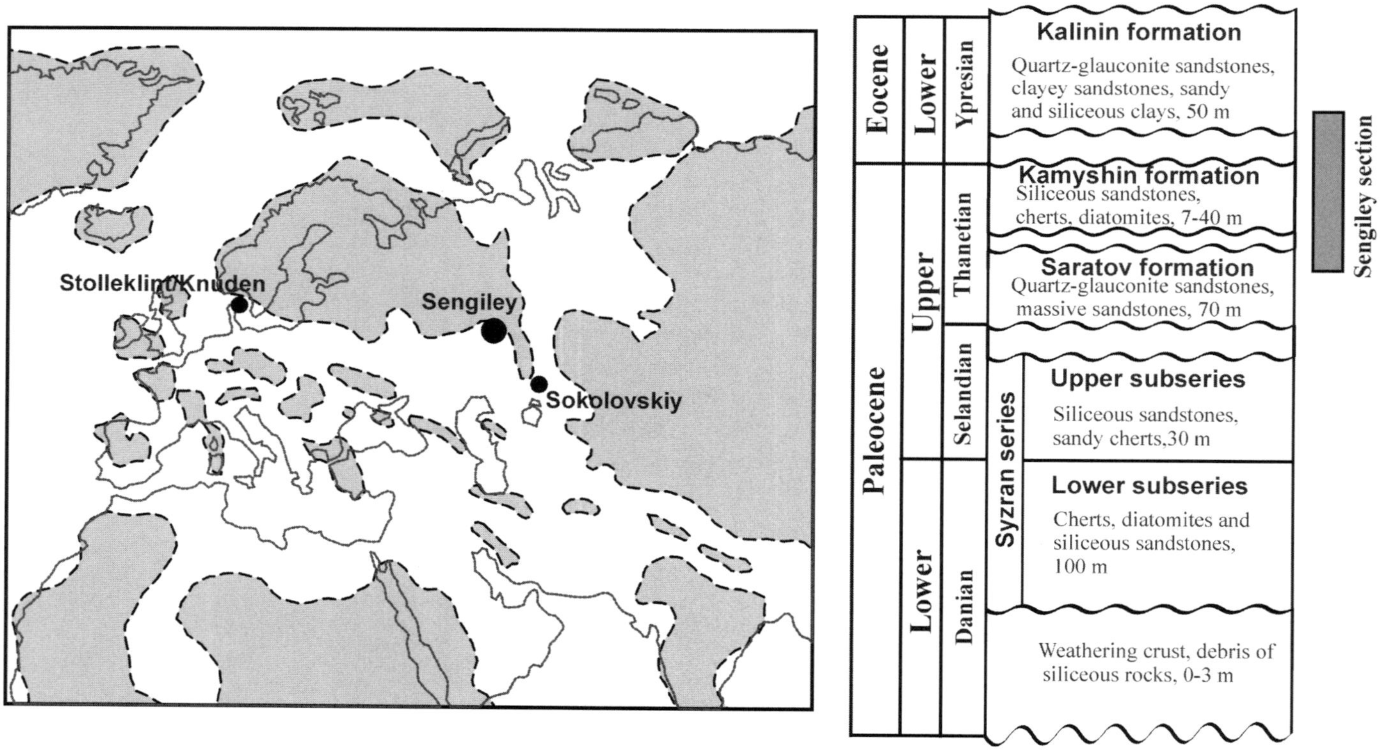

Figure 1. Location of the Sengiley section (Russia), and the correlative sections of the Stolleklint-Knuden (Denmark) and Sokolovskiy (Kazakhstan), shown in the paleogeographic setting of the area in the latest Paleocene. The figure on the right shows the Tertiary succession in the Ulyanovsk-Saratov syncline together with the stratigraphic position of the Sengiley section (shaded area).

ian); (iv) the Kamyshin Formation (7–40 m, Thanetian); and (v) the Kalinin Formation (50 m, Ypresian).

The Sengiley section (Figs. 1, 2) is located 7 km northwest of the village of Sengiley (Fig. 1) and is presently located ~400 m above sea level. Three depositional cycles are recognized in a well-exposed quarry at the hill Granoye Ukho. In this outcrop, Cretaceous deposits (Maastrichtian) are overlain by a rhythmically layered terrigenous-siliceous sequence (30–40 m), attributed to the Saratov Formation. This unit is represented by alternating dark-gray, silicified, thin-layered glauconitic sandstones and siliceous clays with chert nodules. Preliminary sampling of diatoms and dinocysts from this series has shown that diatoms are poorly preserved and cannot be identified. The samples yield dinocysts typical of the *Alisocysta margarita* Zone, a time equivalent to calcareous nannofossil zone NP8 (Aleksandrova, 2001, personal commun.).

The overlying Sengiley diatomite is ascribed to the Kamyshin Formation and is ~26 m thick. It is divided into two parts. The lower part (some 19 m thick), is made up of white, massive diatomite with many grayish intercalations, in places with glauconite. Samples 109–72 from the lower part display well-preserved diatoms, silicoflagellates, and radiolarians. The upper 6–7 m interval is composed of light-gray, massive, clayey diatomite, resting conformably on the underlying sequence.

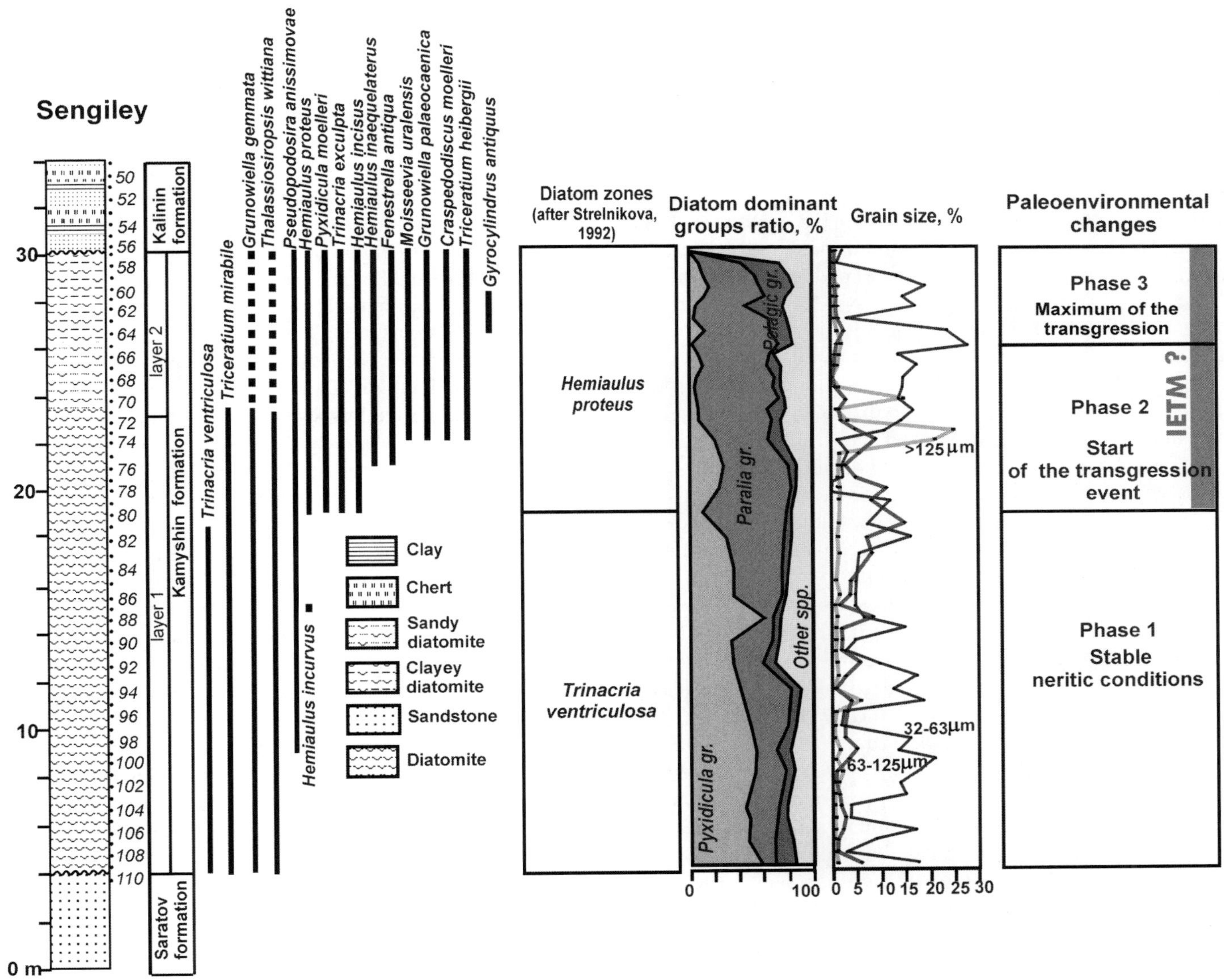

Figure 2. Stratigraphic ranges of selected diatom species in the studied Sengiley section, including the first appearances of the genera *Fenestrella, Moisseevia, Craspedodiscus,* and *Gyrocylindrus.* Also included are the biozonation, the relative abundances of dominant diatom taxa, grain size changes, and a summary of the paleoenvironmental interpretation. Stratigraphic interval in m is indicated on the left of the lithologic column, sample number on the right.

Samples 71–58 from this interval contain well-preserved diatoms, silicoflagellates, and radiolarians.

The diatomites are unconformably overlain by the Kalinin Formation (11 m in thickness) which is represented by alternating brownish green clays, grayish-green siliceous sandstones, brownish sandy cherts, and dark gray cherts (Samples 56–42). No microfossils have been found in this unit.

MATERIAL AND METHODS

In 1994, the Sengiley section was sampled for microfossils studies during the field trip of the Geological Institute, Russian Academy of Sciences and the Simbirsk Geological Survey. Sample locations with a sampling interval of ~0.5 m are shown in Figure 2.

We used a standardized procedure of laboratory treatment and diatom slide preparation (Jousé et al., 1974). Species identifications were checked at 1250× magnification. Of 67 samples processed, 51 yielded siliceous microfossils, while 17 were barren. In order to calculate the relative abundance of dominant diatom species, 300–350 diatom valves were counted in each sample. For grain size analyses, 5 g of sediment was disaggregated in distilled water mixed with 3 ml of 10% H_2O_2 and placed on a shaker for several days. Then, the samples were wet sieved. Subsequently the different size fractions were dried and weighed.

DIATOM BIOEVENTS

Almost all the samples from the Sengiley diatomite contain abundant and excellently preserved diatom and silicoflagellate assemblages (>65 diatom species and ~10 silicoflagellate species, Fig. 4). The diatom and silicoflagellate floral lists with basionyms, synonyms and stratigraphic occurrences are given in Khokhlova and Oreshkina (1999). The stratigraphic framework of the Sengiley diatomites is based on the diatom zonation for Northern latitudes after Strelnikova (1992).

The *Trinacria ventriculosa* Zone, corresponding to calcareous nannofossil Zones NP8-NP6 after Strelnikova (1992), is recognized in the lower part of the diatom sequence (Samples 109–80). The zonal assemblage consistently includes *Triceratium mirabile* Jousé, *Grunowiella gemmata* (Grunow) Van Heurck, *Thalassiosiropsis wittiana* (Pantocsek) Hasle, *Trinacria ventriculosa* (Schmidt) Gleser, and *Pyxidicula ferox* (Greville) Strelnikova et Nikolaev. In the upper part of the zone, sporadic *Hemiaulus incurvus* Shibkova are found and *Pseudopodosira anissimovae* (Gleser et Rubina) Strelnikova first appears.

The *Hemiaulus proteus* Zone, corresponding to calcareous nannofossil zone NP 9 after Strelnikova (1992) is recognized in the upper part of the diatomite sequence (Sample 80–56). Its base coincides with the first appearance of *Hemiaulus proteus* Heiberg, but upsection a number of new taxa gradually appear (Fig. 2). Simultaneously, the abundance of *Thalassiosiropsis wittiana* and *Triceratium mirabile* declines to a few specimens. The genus *Hemiaulus* becomes more diverse, as *Hemiaulus proteus* Heiberg, *H. inaequilaterus* Gombos, and *H. incisus* Hajos successively have their first appearances. In the genus *Grunowiella*, the species *G. palaeocaenica* Jousé has its lowest occurrence. A turnover in the genus *Pyxidicula* occurred, as evidenced by the upsection decline of this genus and the first appearance of *Pyxidicula moelleri* (Schmidt) Strelnikova et Nikolaev.

A remarkable feature of this assemblage is the appearance of a taxon identified as *Triceratium heibergii* Grunow sensu Gombos (1984). Its relict morphologic features resemble those of the early Paleocene species *Trinacria heibergiana* (Grunow) Gleser. Characteristically, its stratigraphic range is limited to short interval within calcareous nannofossil zone NP9 (Gombos, 1984).

There is also a turnover in taxonomic composition at the generic level. Several new genera appear, such as *Craspedodiscus, Fenestrella, Moisseevia,* and *Gyrocylindrus.* The last two genera are monotaxic and short-lived. The genus *Moisseevia* represented by *M. uralensis* (Jousé) Strelnikova (1997) was described based on a revision of the species *Coscinodiscus uralensis* Jousé from Paleogene localities in the Russian territory, including the Sengiley diatomite. This taxon has features in common with both the planktonic genus *Coscinodiscus* and the benthic genus *Aulacodiscus.* The identity of this taxon as *Coscinodiscus josefinus* Grunow (1884) is suggested by Gleser and Stepanova (1994) and Strelnikova (1997) in marine deposits of the Barentz sea near Frantz-Josef Land. Homann (1991), in her monograph on diatoms of the Fur Formation, Denmark, placed *Coscinodiscus josefinus* Grunow in synonymy of *Aulacodiscus suspectus* Schmidt (1876). These taxons look identical, but no formal comparison of holotypes has been carried out.

The genus *Gyrocylindrus* Strelnikova et Nikolaev, 1995 (= *Cylindrospira* Mitlehner, 1995) can be distinguished by its particular morphology, which renders it similar to genera that belong to different orders. It shows similarity to *Cerataulina* Peragallo, the freshwater genus *Aulacoseira* Thwaites, and the fossil tubular genera *Gladius* and *Pyxilla* (Strelnikova and Nikolaev, 1995; Mitlehner, 1995). Only the presence of the linking, flattened pads on top of the valve make it possible to tentatively attribute it to the order *Biddulphiales* (Strelnikova and Nikolaev, 1995; Mitlehner, 1995). *Gyrocylindrus antiquus* has an extremely short stratigraphic range. It was endemic to the European part of the northern Peri-Tethys, as it has been reported only from two localities, the Fur Formation in Denmark and the Sengiley diatomite. In conclusion, the Sengiley section exhibits a significant turnover in the diatom taxonomy in the upper part of the diatomite sequence.

BIOSTRATIGRAPHIC CORRELATION

The diatom successions from the Sengiley section reveal affinities to assemblages reported from sediments in the Southern Ocean, and resemble diatom associations from neighboring regions of the northern Peri-Tethys, especially the Fur Formation in northern Denmark (Homann, 1991; Fenner, 1994;

Mitlehner, 1996), the Sambiiskaya Formation in the Kaliningrad Region, Russia (Strelnikova et al., 1978), the Veshenskaya and Buzinovskaya Formations, the Voronezh depression (Strelnikova, 1992), the Polosataya Formation in the Sokolovsky quarry, the Turgay basin (Radionova et al., 2001), and the Serovskaya/Irbitskaya Formations on the eastern slope of the Sub-Polar Urals (Kozlova and Strelnikova, 1984). Only from the Fur Formation, Denmark and the Polosataya Formation in the Sokolovsky quarry (Kazakhstan), however, are data from other plankton groups such as dinocysts (in the Fur Formation) and calcareous nannoplankton and dinocysts (in the Sokolovsky quarry) available to construct a biostratigraphic age model, making it possible to correlate these regional zonations to pelagic zonations.

A correlation of the Sengiley diatom successions with diatom assemblages from the StolleKlint-Knuden sections, Fur Formation, and Polosataya Formation in the Sokolovsky quarry is presented in Figure 3.

A virtually identical set of species in samples from the Fur Formation and the Sengiley diatomite and, in particular, the presence of stratigraphic key species such as the short-lived *Gyrocylindrus antiquus* (= *Cylindrospira simsii* Mitlehner, 1995) along with *Craspedodiscus moelleri, Hemiaulus proteus,* and *Moisseevia uralensis,* make it possible to correlate the upper part of the Sengiley sequence, in which the *Hemiaulus proteus* Zone is recognized, reliably with the lower part of the Stolle-Klint-Knuden section. This interval corresponds to the dinocyst *Apectodinium augustum* Zone and Viborg Zone 6 (Heilmann-Clausen, 1985, 1995).

The intervals of biogenic silica accumulation in the Sengiley section and the StolleKlint-Knuden section do not completely coincide. In the StolleKlint-Knuden section, the lower part of the *Apectodinium augustum* Zone, which is characterized by the *Apectodinium* spp. acme, there are no diatoms. The upper part of the StolleKlint-Knuden section, representing Viborg Zone 7 (= *Glaphirocysta ordinata*), is not represented in the Sengiley section.

Correlation with the Sokolovsky quarry section (Radionova et al., 2001) is less reliable due to differences in the diatom floras, with the diatoms from the Sokolovsky section indicating

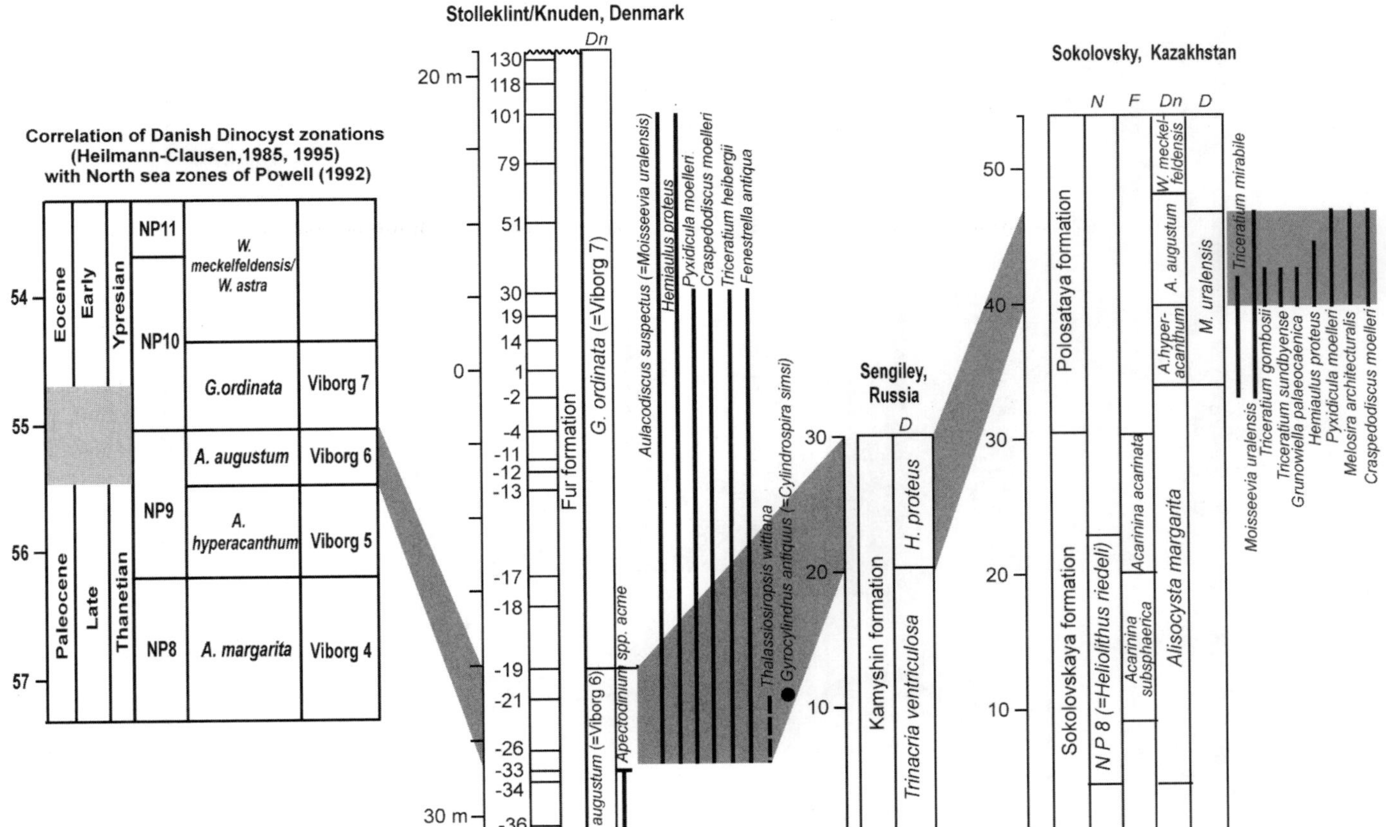

Figure 3. Correlation of the Sengiley section with the StolleKlint-Knuden and Sokolovsky sections. For the StolleKlint-Knuden section, the dinocyst zonation (Dn) is after Heilmann-Clausen (1985, 1995). Stratigraphic ranges of selected diatoms are after Homann (1991). For the Sengiley section, diatom (D) zonation is after Strelnikova (1992). For the Sokolovsky section, nannofossil (N), foraminifera (F), dinocystes (Dn), and diatom (D) zonations are after Radionova et al. (2001). The grey-shaded intervals indicate the correlation by diatoms.

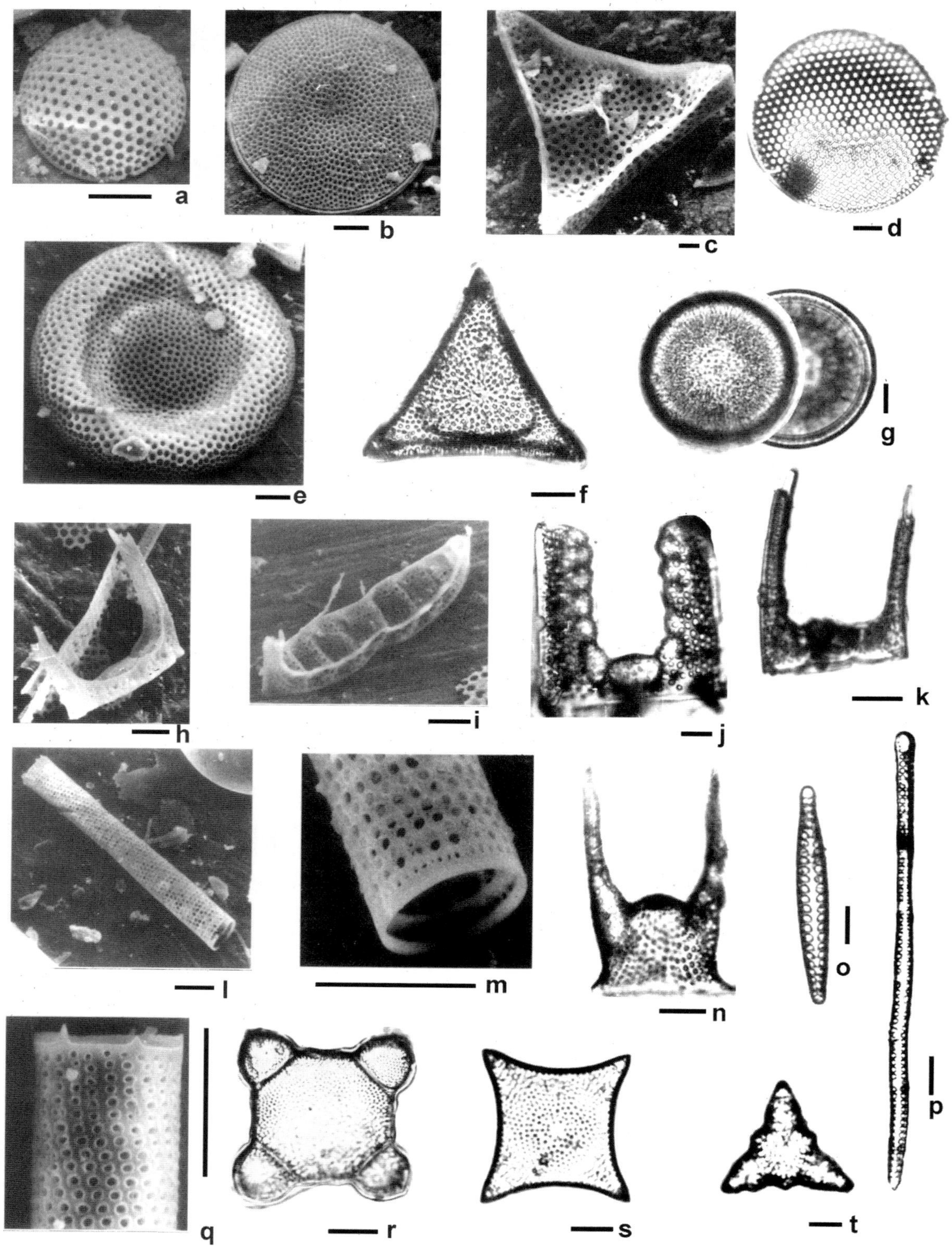
a
b
c
d
e
f
g
h
i
j
k
l
m
n
o
p
q
r
s
t

more offshore paleoenvironments. The data generally suggest that the diatom assemblages of the upper part of the *Moisseevia uralensis* Zone of Radionova et al. (2001) in the Sokolovskiy section can be correlated with the *Hemiaulus proteus* diatom Zone in the Sengiley section. Particularly, the occurrences of *Craspedodiscus moelleri, Hemiaulus proteus, Moisseevia uralensis,* and *Triceratium mirabile* suggest biostratigraphic equivalence. The presence of dinocysts corresponding to those in the *Apectodinium augustum* Zone in the Sokolovsky quarry section also suggests correlation with the lower part of the StolleKlint-Knuden section in Denmark (Fig. 3).

Radionova et al. (2001) noted the similarity of the Sokolovsky diatom assemblages to those from ODP Site 752 in the eastern Indian Ocean (Fourtanier, 1991). This deep-sea section, which recovered Paleocene-Eocene transitional strata, provides data to correlate diatom and nannoplankton assemblages directly with the negative carbon isotope excursion (CIE), and benthic extinction event (BEE) (Nomura, 1991). The Paleocene-Eocene boundary has been placed at the lower boundary of the carbon isotope excursion. These events thus coincide with boundaries of Subzones B and C of the *Hemiaulus incurvus* diatom Zone (upper part of CP8), as proposed by Fourtanier (1991). Characteristically, *Triceratium mirabile* from ODP Site 752, has a dashed range just above the CIE and BEE events. This provides indirect evidence that, in the Sengiley section, the Paleocene-Eocene biotic crisis correspond roughly to the *Trinacria ventriculosa–Hemiaulus proteus* zonal boundary, which would make this equivalent to the Paleocene-Eocene boundary.

Stratigraphic assignment of the lower part of the Sengiley section, corresponding to the *Trinacria ventriculosa* Zone, is more questionable. The Sengiley diatomite might correspond to the entire *Apectodinium augustum* dinocyst Zone, but possibly also covers the upper part of the *Apectodinium hyperacanthum* dinocyst Zone. These stratigraphic data constrain the phase of considerable taxonomic rearrangement to the top of the *Apectodinium* spp. acme and the upper limit of the *Apectodinium augustum* Zone. According to the data of Egger et al. (2000) on the Anthering section in Austria and Heilmann-Clausen and Schmitz (2000) data on Danish sections in Olst-Hinge in eastern Jylland and in the DGI 83 101 borehole in Osterrenden, Store Bolt, the level of the global negative CIE corresponds to the upper part of the *Apectodinium augustum* Zone (= NP9b).

PALEOENVIRONMENTAL CHANGES

In general, the taxonomically rich diatom associations of the Sengiley diatomite are typical of epicontinental seas, as they are dominated by neritic diatoms. This category of diatoms is represented by meroplanktonic, tychopelagic, and benthic species.

Meroplanktonic species are distinguished by their ability to produce resting spores, enabling them to survive environmental stress, such as extreme seasonal changes. Among this group of diatoms, representatives of the genus *Pyxidicula* [*P. ferox* (Greville) Strelnikova et Nikolaev, *P. turris* (Greville et Arnott) Strelnikova et Nikolaev, *P. corona* (Ehrenberg) Strelnikova et Nikolaev, *P. moelleri* (A. Schmidt) Strelnikova et Nikolaev] and the closely related genus *Costopyxis* [*C. antiqua* (Jousé) Gleser] are predominant. Representatives of other meroplanktonic diatoms are also common: *Briggera sibirica* (Grunow) Ross et Sims, *Stellarima microtrias* (Ehrenberg) Hasle et Sims, and *Fenestrella antiqua* (Grunow) Swatman. The group of so-called extinct spore-genera *Odontotropis* (*O. cristata* Grunow, *O. carinata* Grunow), *Pterotheca* (*P. kittoniana* Grunow, *P. major* Jousé), *Pseudopyxilla* [*P. rossica* (Pantocsek) Forti], *Kentrodiscus* (*K. fossilis* Pantocsek), and *Xanthiopyxis* spp., which are represented only by resting spores, is also taxonomically diverse.

In the group of tychopelagic and benthic species, *Paralia sulcata* (Ehrenberg) Cleve, is characteristic of present-day diatom assemblages of near-shore environments. Extinct species of this genus, such as *Paralia grunowii* Gleser and *P. selecta* (A. Schmidt) Gleser, and taxonomically related species of the fossil genus *Anuloplicata* [*A. concentrica* (Grunow) Gleser, *A. ornata* (Grunow) Gleser] are predominant in samples from the Sengiley section. *Hyalodiscus radiatus* (O'Meara) Grunow, *Aulacodiscus distinguendos* Hustedt, *A. probabilis* Schmidt, *A. schmidtii* Witt, *Pseudopodosira westii* W. Smith, *P. anissimovae* (Gleser et Rubina) Strelnikova, *Pseudostictodiscus angulatus* Grunow, *Rattrayella oamaruensis* (Grunow) De Toni, *R. rotundata* (Shibkova) Gleser, *Rhaphoneis morsiana* Grunow, *R. simbirskiana* Grunow, *Actinoptychus* spp., *Arachnoidiscus* spp., and *Rutilaria* sp. are also common.

Trinacria [*T. ventriculosa* (A. Schmidt) Gleser, *T. excavata* (Ehrenberg) Grunow, *Tr. regina* Heiberg], *Trinacria exculpta* Heiberg, *Eunotogramma weissii* Ehrenberg, and *Trochosira spinosa* Kitton, which are fully planktonic but gravitate toward the near-shore zone, are important components of the flora.

Figure 4. Diatom marker-species of the Sengiley section. a: *Pyxidicula moelleri* (Schmidt) Strelnikova et Nikolaev, sample 58. b: *Moisseevia uralensis* (Jousé) Strelnikova, sample 58. c: *Triceratium heibergii* Grunow sensu Gombos, (1977), sample 58. d: *Thalassiosiropsis wittiana* (Pantocsek) Hasle, sample 100. e: *Craspedodiscus moelleri* sample 58. f: *Trinacria ventriculosa* (Schmidt) Gleser, sample 103. g: *Pseudopodosira anissimovae* (Gleser et Rubina) Strelnikova, sample 100. h: *Hemiaulus inaequilaterus* Gombos, sample 64. i: *Hemiaulus proteus* Heiberg, sample 58. j: *Hemiaulus incurvus* Shibkova, sample 85. k: *Hemiaulus curvatulus* Strelnikova, sample 67. l, m, q: *Gyrocylindrus antiquus* Strelnikova et Nikolaev, sample 62. n: *Hemiaulus incisus* Hajos, sample 58. o: *Grunowiella gemmata* (Grunow) Van Heurck, sample 100. p: *Grunowiella palaeocaenica* Jousé, sample 58. r: *Trinacria exculpta* (Heiberg) Hustedt, sample 67. s: *Trinacria regina* Heiberg, sample 61. t: *Triceratium mirabile* Jousé, sample 98. Scale bar indicates 10 μm.

Pelagic species, which, as a rule, have a subglobal distribution constitute a relatively small group. These are the previously mentioned *Triceratium mirabile* Jousé, *Grunowiella gemmata* (Grunow) Van Heurck, *G. palaeocaenica* Jousé, *Thalassiosiropsis wittiana* (Pantocsek) Hasle, *Hemiaulus incurvus* Shibkova, *H. proteus* Heiberg, *H. inaequilaterus* Gombos, *H. incisus* Hajos, *Triceratium heibergii* Grunow sensu Gombos, and representatives of the genera *Coscinodiscus, Craspedodiscus, Moisseevia, Proboscia,* and *Rhizosolenia.*

Within the Sengiley sequence there are quantitative trends in the proportion of dominant diatom groups (Fig. 2, Table A1): The *Paralia-Anuloplicata* and *Pyxidicula* groups. The lower part of the diatom sequence is dominated by the meroplanktonic *Pyxidicula* assemblage (Fig. 2, Table A1) and their close environmental counterparts, such as *Trochosira.* This floral composition suggests that paleoenvironmental conditions were stable, in a neritic setting (Phase 1). Pelagic constituents are *Triceratium mirabile* and *Grunowiella gemmata, Thalassiosiropsis wittiana.*

The middle part of the sequence shows a gradual increase in the tychopelagic *Paralia-Anuloplicata* group and their environmental counterparts, such as the genus *Pseudopodosira,* with a decreasing abundance of the genus *Pyxidicula.* In the upper part of the sequence the relative abundance and species richness of the pelagic diatoms and silicoflagellates increased (Phases 2 and 3), and the short-lived, endemic species *Gyrocylindrus antiquus* with its unusual morphological features appeared. Sporadic freshwater forms similar to the genera *Navicula* and *Aulacoseira* are present, and the relative abundance of the benthic genera *Raphoneis, Rutilaria,* and *Pseudostictodiscus* increases toward the upper part of the Sengiley diatomite. Grain size analysis shows an increase in terrigenous input in the upper part of the Sengiley section (Fig. 2, Table A2).

The changes in the diatom assemblages in the Sengiley diatomites, i.e., a change in the ratio of the dominant *Pyxidicula-Paralia* groups with a concurrent increase in the relative abundance and species richness of pelagic taxa, reflect the global biotic crisis at the Paleocene-Eocene boundary (Fig. 2). The assemblage characteristic of relatively stable paleoenvironmental conditions in the near-shore zone of an epicontinental basin, dominated by the characteristic neritic genus *Pyxidicula* (Phase 1 in Figure 2), is replaced by an assemblage dominated by *Paralia sulcata* and a group of species closely related to this taxon in ecological characteristics.

The environmental affinity of *Paralia sulcata* is ambiguous. Although this species is characteristic of coastal near-bottom (benthic) environments, it can also be highly abundant in planktonic associations. According to McQuoid and Hobson (1998), this species is easily detached from the substrate when vigorous mixing occurs in near-shore waters. An increasing abundance of this species is usually related to a transgressive phases in basin evolution. According to Zong (1997), *Paralia sulcata* is tolerant of a wide range in salinity (5%–35%) and depths (>1 m), but prefers warm water enriched in organic matter. Such environmental conditions occur during a transgressive phase, because organic matter is reworked from a previously exposed shelf area. A coarsening of grain size (samples 85–73) is in agreement with a strengthening of bottom currents at the beginning of the transgressive phase (Fig. 2, Table A2). These floral and lithological changes thus may reflect the onset (Phase 2) and the highstand (Phase 3) of a transgressive cycle in basin evolution as well as a warmer climate.

We thus suggest that the turnover in diatom flora in the Sengiley section, as in many other marine and on-land sections, is the local reflection of a global event, the Initial Eocene Thermal Maximum (IETM). The significant changes in composition and structure of diatom assemblages from epicontinental basins of the northern Peri-Tethys may reflect a more vigorous exchange of surface waters with the Tethys, possibly as a result of sea-level rise, as well as warming.

CONCLUSIONS

In the Sengiley section, the IETM is reflected in: (i) a considerable turnover of the diatom flora at generic and species levels (ii) the appearance of short-lived taxa (iii) showing both new and relict morphologic features due to peculiar paleoenvironmental conditions, and an increasing communication with neighboring epicontinental basins and the Tethys. These floristic changes are accompanied by changes in current dynamics (possibly as a result of sea-level rise) as documented by changes in grain size distribution.

ACKNOWLEDGMENTS

We would like to thank Dr. John Barron and Dr. Juliane Fenner for their helpful reviews and suggestions. We also appreciate stylistic and organizational improvements suggested by Dr. Ellen Thomas. The financial assistance to attend the International meeting "Climate and biota of the early Paleogene" was provided by Smithsonian Institution. We are grateful to our colleagues from the Simbirsk Geological Survey for their help during the field trip. This work is a contribution to the Russian Foundation for Basic Research (project no. 05-64917) and the INTAS Foundation (project no. 93-2509).

APPENDICES

TABLE A1. RELATIVE ABUNDANCE OF DIATOM GENERA IN THE SENGILEY SECTION

Sample no. and depth (m)	Number of counted diatoms	% *Paralia+Anuloplicata*	% *Actinoptychus*	% *Rhaphones*	% *Pseudopodosira*	% *Hyalodiscus*	% *Trochosira*	% *Odontella*	% *Pyxidicula*	% *Pterotheca*	% *Pseudopyxilla*	% *Trinacria*	% *Stellarima*	% *Grunowiella*	% *Coscinodiscus*	% *Moisseevia*	% *Proboscia*	% *Hemiaulus*	% *Triceratium*	% *Thalassiosiropsis*
58 (29.5 m)	340	35.6	0.0	0.9	17.1	1.2	2.4	0.0	6.8	1.5	0.0	0.9	0.3	11.8	0.6	1.2	0.0	17.6	0.0	0.0
59 (29.0 m)	367	42.0	0.3	1.1	10.9	0.8	4.1	0.3	9.8	0.8	0.0	2.2	0.3	11.4	1.1	0.5	0.5	11.2	0.0	0.0
60 (28.5 m)	358	40.0	0.0	0.0	7.8	0.3	3.1	0.0	17.0	0.8	0.3	1.7	0.0	4.2	0.6	1.1	1.1	17.9	0.3	0.0
61 (28.0 m)	356	51.4	0.7	0.7	10.4	0.3	0.6	1.4	12.6	2.5	0.3	2.3	0.0	5.9	0.6	0.8	0.3	6.7	0.0	0.0
62 (27.5 m)	368	43.4	0.0	0.8	15.8	0.5	1.6	0.8	6.0	1.6	0.0	2.2	0.0	13.6	0.8	0.3	1.4	7.1	0.0	0.0
63 (27.0 m)	346	63.3	0.3	0.9	15.0	0.0	0.9	0.3	4.1	0.3	0.3	0.9	0.0	5.5	0.0	0.6	2.9	3.5	0.3	0.0
64 (26.5 m)	357	53.9	0.7	0.3	7.3	2.2	3.9	0.0	11.8	0.6	0.4	2.5	0.3	2.8	0.8	0.8	1.4	7.3	0.3	0.0
65 (26.0 m)	349	70.5	0.3	0.3	9.5	0.6	0.9	0.9	2.9	0.6	0.0	2.6	0.0	4.0	0.6	0.3	1.2	3.4	0.0	0.0
66 (25.5 m)	368	58.2	2.2	0.5	7.9	0.3	9.0	0.5	7.3	1.9	0.5	7.6	0.0	1.1	0.0	0.3	0.8	1.1	0.0	0.0
67 (25.0 m)	398	55.0	0.8	0.3	4.3	0.8	2.0	1.8	13.8	2.3	0.0	7.8	0.0	1.8	0.0	0.5	0.5	6.3	0.3	0.0
68 (24.5 m)	357	63.8	1.4	0.3	3.9	1.7	7.6	1.1	7.3	1.4	0.6	3.1	0.0	0.3	0.0	0.6	0.8	2.8	0.3	0.0
69 (24.0 m)	370	62.7	1.1	0.5	6.2	1.4	5.7	1.4	6.5	2.4	0.3	4.9	0.0	1.1	0.0	1.4	0.0	3.8	0.0	0.0
70 (23.5 m)	358	67.6	1.1	0.3	4.2	2.2	4.2	1.7	8.10	2.0	0.3	4.2	0.0	0.8	0.0	0.6	0.6	1.4	0.0	0.0
71 (23.0 m)	341	53.1	0.6	0.3	6.5	3.2	2.1	1.8	8.5	1.5	0.6	10.3	0.0	1.8	0.0	0.9	2.4	4.4	0.0	0.6
72 (22.5 m)	372	59.4	0.8	0.0	5.4	2.5	4.8	0.8	8.9	4.0	0.8	4.8	0.0	1.1	0.0	0.8	1.6	2.4	0.0	0.3
74 (21.5 m)	377	43.8	1.1	0.5	1.6	4.5	3.7	3.7	20.4	4.8	0.5	5.6	0.5	0.5	0.0	0.8	1.1	5.8	0.3	0.0
76 (20.5 m)	384	48.2	0.0	0.0	2.9	3.4	1.6	1.6	28.4	0.0	1.0	3.7	0.3	1.6	0.0	0.0	0.8	3.4	1.3	0.3
78 (19.5 m)	346	57.2	0.0	0.0	2.3	1.5	1.7	0.0	24.6	0.3	0.6	4.6	0.9	0.3	0.0	0.0	0.3	1.5	0.6	0.0
80 (18.5 m)	370	68.1	0.0	0.0	6.5	2.2	0.5	0.0	10.3	0.5	0.0	5.1	0.5	0.3	0.0	0.0	0.5	1.6	1.7	0.3
82 (17.5 m)	388	40.2	1.6	0.0	0.5	4.6	3.4	0.5	35.1	1.8	1.0	1.6	0.8	0.0	0.0	0.0	1.0	2.6	2.6	0.3
84 (16.5 m)	332	36.1	0.3	0.0	0.9	3.3	5.4	0.6	37.1	1.2	0.6	3.0	0.9	0.0	0.0	0.0	0.9	0.9	3.3	1.5
86 (15.5 m)	365	37.5	1.6	0.0	4.9	3.0	2.7	0.3	36.7	0.8	0.6	3.8	0.3	0.3	0.0	0.0	0.0	1.1	1.9	0.6
88 (14.5 m)	369	6.8	0.5	0.0	3.0	4.6	3.8	2.7	58.3	1.6	0.3	5.4	1.4	0.3	0.0	0.0	0.5	2.2	2.7	3.4
90 (13.5 m)	350	33.4	0.0	0.0	4.0	4.0	6.9	0.9	35.7	0.6	0.3	4.9	0.9	0.0	0.0	0.0	0.7	0.9	2.6	1.7
92 (12.5 m)	346	19.7	0.0	0.0	2.0	7.5	8.7	1.2	35.6	0.3	0.0	2.6	7.5	0.0	0.0	0.0	0.9	1.7	5.5	0.9
94 (11.5 m)	318	20.4	0.0	0.0	3.1	1.9	16.7	0.9	41.2	0.3	0.3	4.4	0.0	0.6	0.0	0.0	0.9	0.9	6.0	0.0
97 (10.0 m)	339	26.8	0.3	0.0	0.3	2.7	4.1	0.3	49.6	1.8	0.0	3.2	0.3	0.9	0.0	0.0	0.3	1.2	3.8	0.9
99 (9.0 m)	345	15.1	0.3	0.0	0.3	3.8	10.7	0.3	51.6	0.0	0.6	3.5	0.6	2.0	0.0	0.0	0.0	1.2	7.3	0.9
101 (8.0 m)	342	23.1	0.0	0.0	0.0	2.3	3.2	0.0	53.5	0.3	0.6	2.9	1.2	2.1	0.0	0.0	0.6	3.5	1.2	1.8
104 (6.5 m)	340	23.8	0.0	0.0	0.0	2.7	12.7	0.9	44.7	2.4	0.3	1.5	0.3	2.7	0.0	0.0	0.9	0.9	2.1	0.3
105 (6.0 m)	332	22.6	0.6	0.0	0.6	2.4	7.5	1.5	49.4	0.6	0.3	3.0	0.3	3.0	0.0	0.0	0.6	1.8	3.3	0.6
107 (5.0 m)	321	21.5	0.0	0.0	1.3	3.4	7.2	0.0	42.1	0.3	1.7	3.4	0.6	3.4	0.0	0.0	0.9	5.3	4.4	0.3
109 (4.0 m)	321	8.4	0.0	0.0	1.0	3.4	3.4	0.6	57.3	0.9	0.6	3.4	0.0	5.6	0.0	0.0	1.6	3.1	5.3	2.5

TABLE A2. GRAIN SIZE ANALYSIS OF THE DIATOMITE OF THE SENGILEY SECTION

Sample no. and depth (m)	Weight (g)	>125µm (%)	63–125µm (%)	32–63µm (%)
57 (30.0 m)	2.50	0.31	0.83	3.28
58 (29.5 m)	2.50	0.41	0.89	1.38
59 (29.0 m)	2.50	0.29	0.68	14.01
60 (28.5 m)	2.50	0.58	0.96	19.34
61 (28.0 m)	2.50	0.78	0.86	13.69
62 (27.5 m)	2.50	0.97	1.42	17.40
63 (27.0 m)	2.57	0.98	1.58	4.45
64 (26.5 m)	2.50	0.85	3.25	24.85
65 (26.0 m)	2.50	1.78	0.98	27.45
66 (25.5 m)	2.50	1.30	0.74	14.31
67 (25.0 m)	2.50	0.62	0.84	18.39
68 (24.5 m)	2.50	0.09	0.20	13.20
69 (24.0 m)	2.50	1.30	0.96	14.84
70 (23.5 m)	2.50	15.77	2.65	13.52
71 (23.0 m)	2.50	2.26	2.13	17.20
72 (22.5 m)	2.50	4.08	1.66	14.26
73 (22.0 m)	3.03	25.66	5.07	11.10
74 (21.5 m)	2.50	22.10	9.22	1.22
75 (21.0 m)	2.50	1.50	6.14	4.21
76 (20.5 m)	2.50	1.70	2.40	1.52
77 (20.0 m)	2.50	2.16	4.52	2.40
78 (19.5 m)	2.50	2.04	10.96	-5.49
79 (19.0 m)	2.50	7.65	7.65	11.68
81 (18.0 m)	2.50	1.87	30.12	7.21
82 (17.5 m)	2.50	1.21	6.68	16.81
83 (17.0 m)	2.50	1.25	7.61	5.88
85 (16.5 m)	2.50	1.10	8.83	4.61
86 (15.5 m)	2.62	1.85	3.51	4.90
87 (15.0 m)	2.50	0.46	1.73	5.28
88 (14.50 m)	2.50	0.80	8.44	6.64
89 (14.0 m)	3.06	1.01	2.30	15.79
90 (13.5 m)	2.44	1.04	2.38	5.06
91 (13.0 m)	2.98	0.10	2.42	3.40
92 (12.5 m)	2.64	1.23	5.98	9.88
93 (12.0 .m)	2.70	5.25	13.05	5.61
94 (11.5 m)	2.69	0.45	0.87	12.46
95 (11.0 m)	2.79	5.86	3.53	18.67
96 (10.5 m)	2.54	1.19	3.43	3.12
97 (10. 0 m)	3.50	0.70	1.62	2.06
98 (9.5 m)	3.40	0.72	3.29	16.40
99 (9.0 m)	2.72	2.02	4.65	14.11
100 (8.5 m)	2.73	0.55	3.66	20.81
101(8.0 m)	3.03	1.60	2.29	17.63
102 (7.5 m)	2.70	1.42	0.45	13.86
103 (7.0 m)	2.84	0.42	1.60	15.23
104 (6.5 m)	2.97	0.92	2.13	4.37
105 (6.0 m)	3.03	0.84	2.79	4.11
106 (5.5 m)	2.81	0.53	2.41	17.41
107 (5.0 m)	2.68	0.55	1.61	8.59
108 (4.5 m)	2.59	0.96	2.13	3.22
109 (4.0 m)	2.93	1.09	2.63	18.18

REFERENCES CITED

Egger, H., Heilmann-Clausen, C., and Schmitz, B., 2000, The Paleocene/Eocene boundary interval of a Tethyan deep-sea section (Austria) and its correlation with the North Sea basin: Bulletin de la Société Géologique de France, v. 171, no. 2, p. 207–216.

Fenner, J., 1991, Taxonomy, stratigraphy, and paleoceanographic implications of Paleocene diatoms, *in* Ciesielski, P.F., Kristoffersen, Y., et al., eds., Proceedings of the Ocean Drilling Program, Scientific Results: Texas A & M University, College Station, Texas, USA, v. 114, p. 123–154.

Fenner, J., 1994, Diatoms of the Fur Formation, their taxonomy and biostratigraphic interpretations: Results from the Harre borehole, Denmark: Aarhus Geoscience, v. 1, p. 99–163.

Fourtanier, E., 1991, Paleocene and Eocene diatom stratigraphy and taxonomy of eastern Indian Ocean site 752, *in* Weissel, J., Peirce, J., Taylor, E., Alt, J., et al., eds., Proceedings of the Ocean Drilling Program, Scientific Results: Texas A & M University, College Station, Texas, USA, v. 121, p. 171–187.

Gleser, Z.I., 1979, Zonal subdivision of Paleogene deposits by diatoms: Sovetskaya geologia, no. 11, p. 19–30.

Gleser, Z.I., 1994, The correlation of Paleogene diatom florae evolution from epicontinental and oceanic areas: Stratigraphy and Geological Correlation, v. 2, no. 1, p. 103–107.

Gleser, Z.I., 1995, Diatom and silicoflagellate assemblages as indicators of synchronous occurrence of diatomites of Middle Volga region, Denmark, and lower Paleogene oceanic bottom sediments, *in* Barash, M.S., ed., Recent and fossil microplankton of the World Ocean: Moscow, Nauka, p. 57–63.

Gleser, Z.I., and Stepanova, G.V., 1994, Stratigraphic subdivision and correlation of Paleogene deposits of Kara Sea by diatoms and silicoflagellates: Regionalnaya geologia i metallogenia, St. Petersburg, no. 2, p. 148–153.

Gleser, Z.I., Kuznetsova, T.A., and Afanas'eva, N.I., 1977, About age and facial position of the Middle Volga region's diatomites: Sovetskaya Geologia, no. 8, p. 113–119.

Gombos, A.M., 1976, Paleogene and Neogene diatoms from the Falkland plateau and Malvinas outer basin Leg 36, DSDP, *in* Barker, P.F., Dalziel, I.W.D., et al., eds., Initial reports of the Deep Sea Drilling Project: Washington, D.C., U.S. Government Printing Office, v. 36, p. 575–687.

Gombos, A.M., 1984, Late Paleocene diatoms in the Cape basin, *in* Hsü, K.L., La Breque, J.L., et al., eds., Initial reports of the Deep Sea Drilling Project: Washington, D.C., U.S. Government Printing Office, v. 73, p. 495–512.

Grunow, A., 1884, Die Diatomeen von Franz Josefs-Land: Denkschriften der mathematisch-naturwissenschaftlichen Classe der kaiserlichen Akademie der Wissenschaften, Wien v. 48, p. 53–112.

Heilmann-Clausen, C., 1985, Dinoflagellate stratigraphy of the uppermost Danian to Ypresian in the Viborg 1 borehole, central Jylland, Denmark: Danmarks Geologiske Undersøgelse, no. A7, p. 1–69.

Heilmann-Clausen, C., 1995, Palægene aflejringer over Danskekalken, *in* Nielsen, O.B., ed., Danmarks geologi fra Kridt til i dag: Aarhus Geokompendier 1, p. 69–114.

Heilmann-Clausen, C., and Schmitz, B., 2000, The late Paleocene thermal maximum $\delta^{13}C$ excursion in Denmark?: GFF, v. 122, p. 70.

Homann, M., 1991, Die Diatomeen der Fur-Formation (Alttertiär) aus dem Limfjord-Gebiet, Nordjütland/Dänemark: Geologisches Jahrbuch, Reihe A, Heft 123, p. 3–285.

Jousé, A.P., 1949, Diatoms of the Tertiary deposits, *in* Krishtofovich, A.N., ed., Diatomoviy analiz: Leningrad, Gosgeolizdat, v. 1, p. 114–152.

Jousé, A.P., 1979, Diatom biostratigraphic zones of the Eocene: Nova Hedwigia, beiheft 64, p. 427–440.

Jousé, A.P., 1982, Paleocene diatoms and silicoflagellates from the Pacific, Indian, and Atlantic Oceans, *in* Jousé, A.P., and Krasheninnikov, V.A., eds., Morskaya mikropaleontologia: Moscow, Nauka, p. 131–145.

Jousé, A.P., Proshkina-Lavrenko, A.I., and Sheshukova-Poretzkaya, V.S., 1974, Method of research, *in* Proshkina-Lavrenko, A.I., ed., The diatoms of the USSR (fossil and recent): Leningrad, Nauka, v. 1, p. 50–79.

Khokhlova, I.E., and Oreshkina, T.V., 1999, Early Palaeogene siliceous microfossils of the Middle Volga region: Stratigraphy and palaeogeography: Geodiversitas, v. 21, no. 3, p. 429–451.

Kozlova, G.E., 1994, Radiolarian zonal scale of the boreal Palaeogene: Micropaleontology (special publication), no. 6, p. 90–93.

Kozlova, G.E., and Strelnikova, N.I., 1984, Significance of the section Ust'-Man' insk hole 19-V for zonal division of the West Siberian Paleogene, *in* Shatzkiy, S.B., et al., eds., Environment and life at the boundaries of Cenozoic epochs in Siberia and Far East: Nauka, Russia, Novosibirsk, p. 70–77.

McQuoid, M.R., and Hobson, L.A., 1998, Assessment of palaeoenvironmental conditions on southern Vancouver Island, British Columbia, Canada, using the marine tychoplankter *Paralia sulcata:* Diatom research, v. 13, no. 2, p. 311–321.

Mitlehner, A.G., 1995, *Cylindrospira,* a new diatom genus from the Paleogene of Denmark with palaeoecological significance: Diatom Research, v. 10, no. 2, p. 321–331.

Mitlehner, A.G., 1996, Palaeoenvironments in the North Sea Basin around the Paleocene-Eocene boundary: Evidence from diatoms and other siliceous microfossils, *in* Knox, R.W.O'B., et al., eds., Correlation of the early Paleogene in northwest Europe: Geological Society [London] Special Publication 101, p. 255–273.

Mukhina, V.V., 1976, Species composition of the late Paleocene diatoms and silicoflagellates in the Indian Ocean: Micropaleontology, v. 22, no. 22, p. 151–158.

Nomura, R., 1991, Paleoceanography of upper Maestrichtian to Eocene benthic foraminiferal assemblages at sites 752, 753, and 754, eastern Indian Ocean, *in* Weissel, J., Peirce, J., Taylor, E., Alt, J., et al., eds., Proceedings of the Ocean Drilling Program, Scientific Results: Texas A & M University, College Station, Texas, USA, v. 121, p. 3–24.

Powell, A.J., 1992, Dinoflagellate cysts of the Tertiary system, *in* Powell, A.J., ed., A stratigraphic index of dinoflagellates cysts: British Micropaleontological Society Publication series, London, Chapman and Hall, p. 155–251.

Radionova, E.P., Khokhlova, I.E., and Oreshkina, T.V., 1999, Configuration of the Palaeogene deposits of southern Russia: Geodiversitas, v. 21, no. 3, p. 421–428.

Radionova, E.P., Khokhlova, I.E., Beniamovsky, V.N., Shcherbinina, E.A., Iakovleva, A.I., and Sadchikova, T.A., 2001, Paleocene/Eocene transition in the northeastern Peri-Tethys area: Sokolovskii key section of the Turgay Passage (Kazakhstan): Bulletin de la Société Géologique de France, v. 172, no. 2, p. 245–256.

Schmidt, A., 1874–1959, Atlas der Diatomaceen-Kunde, R. Reisland, Leipzig, Berlin, 480 plates.

Strelnikova, N.I., 1992, Paleogene diatoms: St. Petersburg, St. Petersburg University Press, 321 p.

Strelnikova, N.I., 1997, A new genus *Moisseevia* (Bacillariophyta) from the Paleogene deposits of Russia: Botanisheskiy journal, v. 82, no. 3, p. 119–121.

Strelnikova, N.I., and Nikolaev, V.A., 1995, A new genus *Gyrocylindrus* (Bacillariophyta) from Paleogene deposits of Russia and Denmark: Botanicheskiy journal, v. 80, no. 8, p. 93–96.

Strelnikova, N.I., Kaplan, A.A., and Travina, M.A., 1978, Paleogene diatoms, silicoflagellates, ebridian of Kaliningrad district, *in* Jousé, A.P., ed., Morskaya micropaleontologia: Moscow, Nauka, p. 57–66.

Zong, I., 1997, Implications of *Paralia sulcata* abundance in Scottish isolation basins: Diatom research, v. 12, no. 1, p. 125–150.

Manuscript Accepted by the Society August 13, 2002

Printed in the USA

Geological Society of America
Special Paper 369
2003

Paleocene biota of the West Siberian plain

Vera M. Podobina*
Tomsk State University, 36, Lenin Avenue, Tomsk, 634050 Russia

ABSTRACT

In the Paleocene, most of the present West Siberian plain was occupied by a vast epicontinental sea, which reached from the present Arctic and Atlantic oceans in the North to the Aral Basin in the South. Climate and ocean circulation fluctuated and affected marine and continental biota. Benthic foraminifera that occurred over a large part of the sea throughout the Paleocene were used as representatives of the marine biota, and spores and pollen of terrestrial plants as representatives of the continental biota, in order to study these environmental and biotic changes. Foraminifera and palynomorphs were studied in sedimentary sequences of Danian, Selandian, and Thanetian age. These sequences have regional names, and encompass the upper part of the Gankinskaya Stage, the whole Talitskaya Stage, and the lower part of the Lyulinvorskaya Stage.

In the southwest part of the basin (the Omsk depression), upper Gankinskaya sediments (Maastrichtian to lowermost Danian) contain calcareous benthic foraminifera. Benthic foraminiferal assemblages suggest that sea level fell during the early Paleocene, so that the West Siberian basin shallowed significantly and decreased in size. During the earliest Danian, the basin in the Ust-Tym depression also became shallower, and "primitive" agglutinated foraminifera were dominant (e.g., *Bathysiphon, Glomospira, Ammodiscus*). Sea level rose during the Selandian boreal transgression (Talitskaya Stage), and agglutinated foraminifera occurred over large areas. In the Thanetian (represented by the uppermost Talitskaya and the base of the Lyulinvorskaya Stage) the basin again shallowed, and foraminifera became rare.

The analysis of palynomorphs (spores and pollen) in allochthonous assemblages in marine sediments documents that an alluvial plain with lakes and bogs was situated on the southeast coast of the Paleocene West Siberian marine basin. The climate became more arid during the Danian and late Thanetian, with a pronounced phase of drying at the boundary between the Talitskaya and Lyulinvorskaya Stages (early Thanetian).

INTRODUCTION

Western Siberia is the part of Siberia between the Urals in the west and the Yenisey River in the east (Fig. 1A). It is largely occupied by the West Siberian plain, which can be subdivided into different biomes, from the tundra in the North to the steppe in the South. The West Siberian plain is one of the largest low-lying plains on Earth, with a north-south dimension of ~2500 km, an east-west dimension of ~1900 km, and an area of ~2.6×10^6 km^2. Its surface is flat, with few outstanding features, and only minor differences in elevation.

The West Siberian plain developed on an isolated part of the lithosphere, and remained so for an appreciable length of time. It is a part of the Epihercynian West Siberian plate, with base-

*podobina@ggf.tsu.ru

Podobina, V.M., 2003, Paleocene biota of the West Siberian plain, *in* Wing, S.L., Gingerich, P.D., Schmitz, B., and Thomas, E., eds., Causes and Consequences of Globally Warm Climates in the Early Paleogene: Boulder, Colorado, Geological Society of America Special Paper 369, p. 181–204.

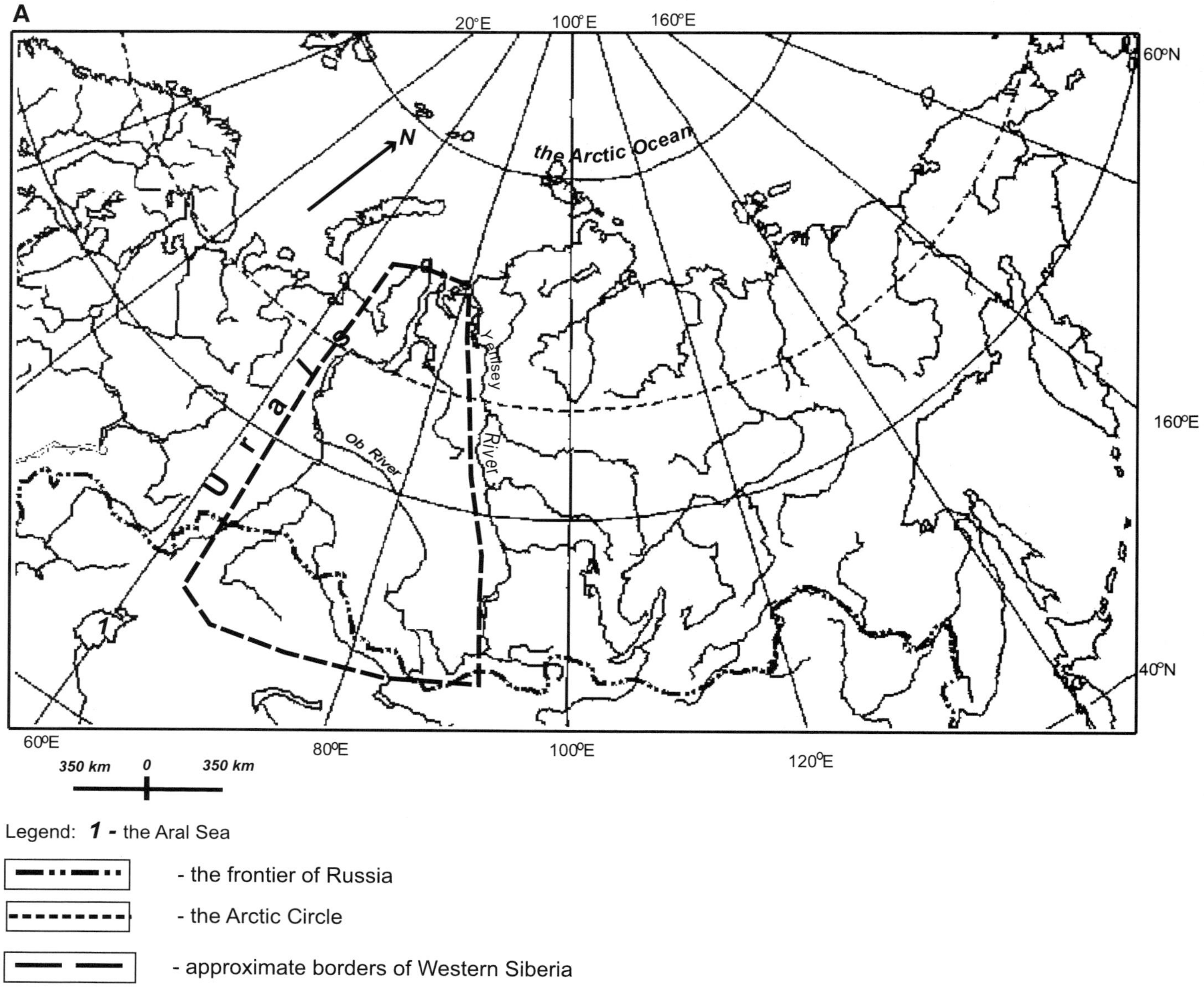

Figure 1. A: Location of the West Siberian Plain, with its most important rivers. B: Location of the sections and boreholes studied, and the extent of the Omsk and Ust-Tym depressions.

ment formed by intensely deformed Paleozoic rocks covered by superficial deposits of marine and continental, Mesozoic and Cenozoic sediments, including clay, sandstone, and marl. These superficial deposits reach >1000 m in thickness, in basement troughs even up to 3000–4000 m.

Throughout the first half of the Paleogene, epicontinental seas extended over the entire plain, but the sediments deposited in the basin vary in character with space and time, dependent upon climate as well as the landscape. Lithological and paleontological studies of the Paleogene deposits formed under the diverse conditions in this vast region have elucidated the overall paleo-environmental developments (Gurova and Kazarinov, 1962). Variations in sea level and climate resulted in varying width and depth of the connections to other basins, in turn resulting in changes in ocean currents, and migration of marine and continental biota.

Most of the Paleogene deposits in the West Siberian plain are buried under a thick cover of Neogene-Quaternary deposits, and can be studied only in boreholes (Shatskiy, 1973), many of which have been drilled in the southern part of the West Siberian plain. Their locations are spread evenly over this part of the plain, in Transuralia and, to a lesser extent, over the northern

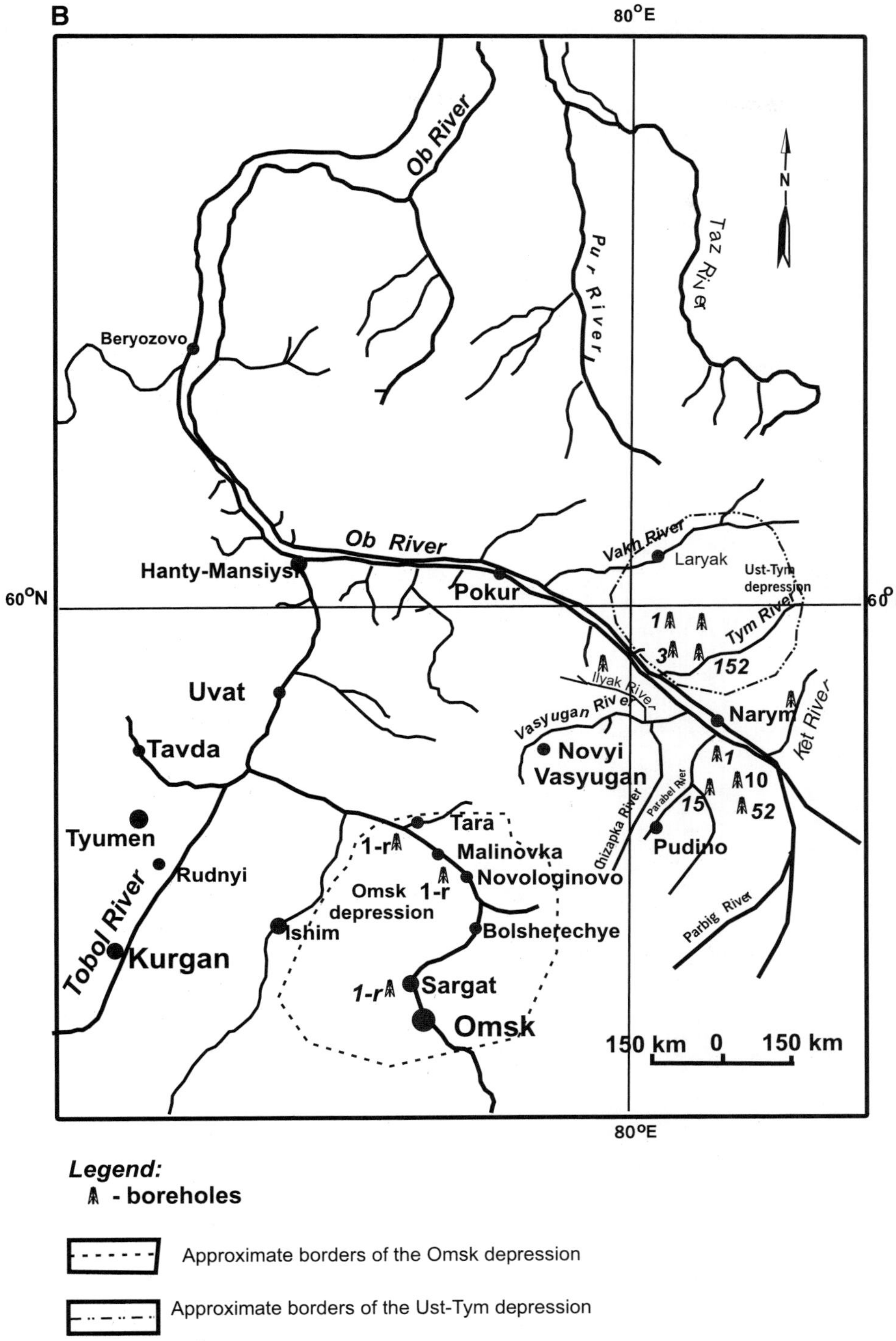

Figure 1. *(continued)*

plains (Figs. 1A, B). Cores from these boreholes have been studied over many years, so that the general stratigraphy of the Paleo-gene deposits is well known (e.g., Podobina, 1998).

In micropaleontological studies of the Paleogene of West Siberia, benthic foraminifera have been used most extensively to investigate paleogeographical changes, because they are widely distributed and evolved relatively rapidly (Podobina, 2000a). Mollusca are rare or absent. The palynological assemblages have been found only in sediments deposited in the shallow, marginal parts of the basin.

Foraminiferal and palynological assemblages have been studied in many different West Siberian sections from the Omsk and Ust-Tym depressions, but the two microfossil groups have usually been studied in separate sections and boreholes. The results of detailed investigations of the micropaleontology of the region, including data on several Paleogene sections, have been reported in numerous works published earlier (e.g., Podobina, 1998; Kabanova and Kostesha, 1988).

In this paper foraminiferal and palynological data from several borehole sections in the Omsk depression are combined for the first time (Fig. 2). The data in this paper and their paleoecological interpretation are preliminary, because this is probably the first attempt to combine and compare the results of foraminiferal and palynological studies of West Siberia, and correlate the early Paleogene foraminiferal and palynological assemblages from different areas of the West Siberian plain.

GEOLOGICAL SETTING; MATERIAL AND METHODS

Samples for the investigations of foraminiferal assemblages as well as palynomorphs were obtained from cores drilled in the Omsk and Ust-Tym depressions, in the Parbig and Ket river basins and other districts (Fig. 1), where these sediments have been preserved from erosion because of their location in basement depressions. The foraminiferal assemblages have been studied in many sections from this region (Podobina, 1998; Fig. 1B). In this paper, foraminiferal assemblages (Figs. 3, 4) are investigated from locations that also contain palynomorphs (Fig. 5) (Podobina et al., 1997).

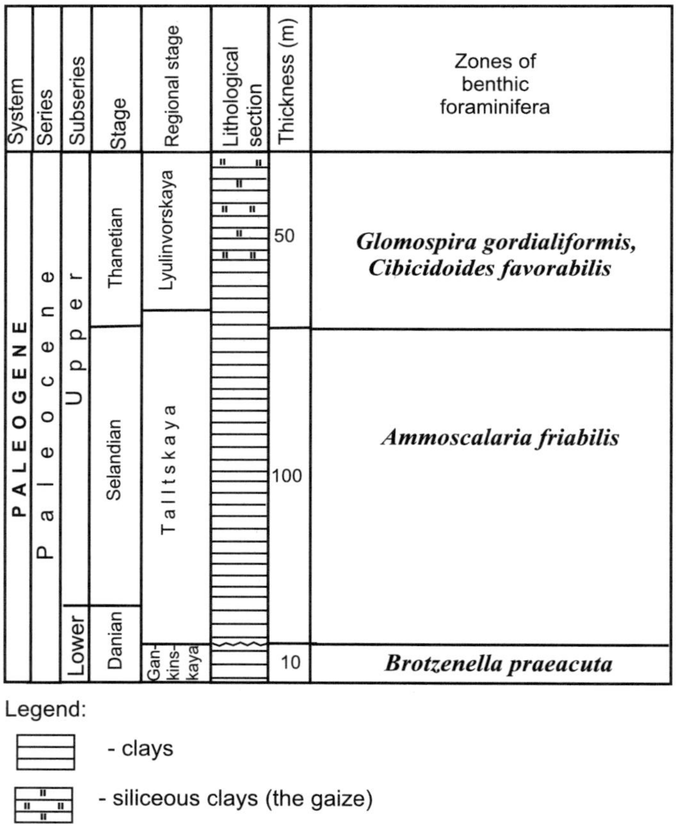

Figure 2. Representative section of the formations in the West Siberian plain, with benthic foraminiferal zonation.

Sediments deposited during the Gankinskaya, Talitskaya and Lyulinvorskaya regional stages vary considerably in lithology, with a typical Paleocene section with its foraminiferal zones from the Omsk depression illustrated in Figure 2. This paper is not intended to present detailed lithological descriptions, because these have been presented earlier in lithological columns and stratigraphy (e.g., Gurova and Kazarinov, 1962; Shatskiy, 1973). Within the sections in the Omsk depression, the Gankinskaya Stage consists of dark-grey and grey, slightly calcareous clays with silt, and with sandy, noncalcareous layers (~10 m thick) in its upper parts. The sediments were assigned to the lowermost Danian because of their planktonic foraminiferal faunas, which include *Subbotina varianta* (Subbotina), *S. trivialis* (Subbotina), *S. triculinoides* (Plummer) as well as a few other species (Subbotina and Kiselman, 1961; Podobina, 1998).

The Talitskaya Stage is represented by dark-grey, almost black, noncalcareous clays (10–100 m thick). The lowermost part of Lyulinvorskaya Stage consists of grey, siliceous rocks (35–110 m thick) (Podobina, 1975, 1998; Ilyenok, 1968; Ilyenok et al., 1989).

About 10,000 samples for the micropaleontological analysis were taken at 5 m intervals from each borehole in all three regional stages. Close to the stage boundaries samples were taken at 1–2 m intervals. All core samples were processed in the same way. The dry weight of the samples was ~100 g. Samples were soaked and boiled in a solution of baking soda for 3–5 hours, then washed and sieved into two size fractions: 1–3 mm, and between 1 mm and 0.63 mm. Both fractions were oven-dried (~300°C). Foraminifera were picked from the dried residues of both fractions for further investigations.

The dry weight of the samples for palynological analyses was ~50 g. They were crushed and treated with nitric acid (HNO_3 65% w/v, 3–7 min) and an alkaline solution (10% solution of NaOH). The fraction with the lowest density (macerate rich in organic matter) was separated by centrifuging in a heavy liquid (2.3–2.35 g/cm) for 40 min and then in distilled water for 5 min. All data on palynomorphs were provided by the palynological laboratory of Siberian Paleontological Scientific Center of Tomsk State University.

EARLY PALEOCENE

Danian

Marine Danian deposits occur in the region to the East of the Ural Mountains (Transuralia), and in various places in the central and southern regions of the West Siberian plain (Fig. 1A, B). The lower part of the Danian is represented by sediments of the Gankinskaya regional stage, which escaped erosion only in

System	Series	Stage	Regional stage	Zones and layers of benthic foraminifera	
				South-Western region (the Omsk depression)	Eastern region (the Ust-Tym depression)
Paleogene	Paleocene	Thanetian	Lyulinvorskaya	***Glomospira gordialiformis, Cibicidoides favorabilis***	Layers with ***Glomospira gordialiformis, Cyclammina coksuvorovae***
		Selandian	Talitskaya	***Ammoscalaria friabilis***	Layers with ***Cyclammina coksuvorovae***
				Layers with ***Cibicidoides incognitus***	
		Danian	Gankinskaya	***Brotzenella praeacuta***	Layers with ***Bathysiphon nodosarieformis, Glomospira charoides***

Figure 3. Paleocene foraminiferal zones and layers of the West Siberian plain.

the regions where Mesozoic and Cenozoic deposits occur in pronounced basement depressions (e.g., the Omsk depression). The deposits include sediments assigned to the *Brotzenella praeacuta* benthic foraminiferal zone (Subbotina and Kiselman, 1961; Podobina, 1998), characterized by the following species: *Gaudryina gigantica* Subbotina, *Clavulina parisiensis* Orb., *Lenticulina inusitata* (Kisselman), *Parrella lens* Brotzen, *Anomalina danica* (Brotzen), *Cibicides spiropunctatus* Galloway and Morrey, *Brotzenella praeacuta* (Vassilenko). Typical for this zone is the presence of species that survived from the late Maastrichtian, and that have a wide distribution in the southern part of West Siberia (Podobina, 2000b). The most characteristic of these Cretaceous-Paleocene species are *Heterostomella foveolata* (Marsson), *Valvulineria imitata* (Olsson), *Valvulinoides umovi* (Kyprianova), *Gyroidinoides turgidus* (Hagenow), *Eponides sibiricus* Neckaja, *Cibicides gankinoensis* Neckaja, *Anomalinoides pinguis* (Jennings), *A. neckajae* Vassilenko, *A. justus* Podobina, *Pullenia americana* Cushman, *Nonionellina pseudoinsecta* (Putrja), *Bulimina rozenkrantzi* Brotzen, *Reussella minuta* (Marsson), *Bolivina plaita* Carsey and other species (Figs. 3, 4). These faunas thus include both calcareous and agglutinated taxa.

In the eastern part of the basin (the Ust-Tym depression; Figure 1B), "simple" forms of the genera *Bathysiphon, Hyperammina, Ammodiscus, Glomospira* with relatively "primitive" construction (one or two chambers only) dominate (95%). The term "primitive" is used here to mean that these forms form relatively simple tests consisting of few chambers and without a complex wall structure. More "highly evolved" species co-occur (5%), such as *Heterostomella aff. foveolata* (Marsson) and *Spiroplectammina kasanzevi* Dain in the *Bathysiphon nodosarieformis* and *Glomospira charoides* assemblage (Plates 1–4, Figures 3, 4). The species composition in the western and eastern foraminiferal faunas suggest that the shallow basin had a relatively low salinity, as documented by the absence of planktonic forms and the common occurrence of such "primitive" genera as *Bathysiphon, Hyperammina, Ammodiscus* (Podobina, 1998).

Up-section in the Omsk depression (in the lowermost Talitskaya regional stage), the foraminiferal assemblage with *Cibicidoides incognitus* occurs (Freiman, 1960), which also contains planktonic foraminifera: *Subbotina varianta* (Subbotina), *S. trivialis* (Subbotina), *Morozovella* cf. *inconstens* (Subbotina) and other species (Podobina, 1998). These planktonic foraminifera indicate that these sediments can be dated as early Paleocene (Podobina, 1998). The *Cibicidoides incognitus* assemblage has been detected only in two sections in the Omsk depression.

The palynological assemblage was studied in the marine sediments containing the foraminifera and is thus allochthonous. In the uppermost part of the Gankinskaya Stage (the *Brotzenella praeacuta* Zone) the palynomorph assemblage is largely the same as that in Upper Cretaceous sediments, and many of these Cretaceous palynomorphs may be reworked. Sporophytes are dominated by bog mosses (30%): *Sphagnum regium* Drozh, and *Sph. australe* (Cook) Drozh. Occasionally

System	Series	Stage	Regional stage	Zones of benthic foraminifera	Characteristic foraminiferal assemblages
Paleogene	Paleocene	Thanetian	Lyulinvorskaya	***Glomospira gordialiformis, Cibicidoides favorabilis***	***Glomospira gordialiformis*** Podobina, ***Asanospira grzybowski*** (Mjatliuk), ***Adercotryma horrida*** (Grzybowski), ***Cyclammina coksuvorovae*** Uschakova, ***Trochammina intacta*** Podobina, ***Cibicidoides favorabilis*** (Vassilenko)
		Selandian	Talitskaya	***Ammoscalaria friabilis***	***Ammodiscus glabratus*** Cushman et Jarvis, ***Trochamminoides lamentabilis*** Podobina, ***Labrospira granulosa*** (Lipman), ***Haplophragmoides fastosus*** Podobina, ***Asanospira grzybowski*** (Mjatliuk), ***Cyclammina coksuvorovae*** Uschakova, ***Ammoscalaria friabilis*** (Ehremeeva), ***Ammomarginulina brevis*** (Lipman), ***Trochammina pentacamerata*** Lipman, ***T. intacta*** Podobina, ***Verneuilinoides paleogenicus*** (Lipman), ***Bolivinopsis scanica*** Brotzen, ***Donsissonia laxata*** Podobina, ***Eponides lunatus*** Brotzen, ***Gavelinella umbilicata*** (Brotzen), ***Cibicides simplex*** Brotzen, ***Cibicidoides proprius*** Brotzen, ***Nonionellina ovata*** (Brotzen)
		Danian	Gankinskaya	***Brotzenella praeacuta***	**Characteristic Danian species:** ***Bathysiphon nodosarieformis*** Subbotina, ***Glomospira charoides*** Parker & Jones, ***Trochamminoides labentabilis*** Podobina, ***Gaudryina gigantica*** (Subbotina), ***Clavulina parisiensis*** Orb., ***Parrella lens*** (Brotzen), ***Cibicides spiropunctatus*** Galloway & Morrey, ***Anomalina danica*** (Brotzen), ***Brotzenella praeacuta*** (Vassilenko), ***Subbotina varianta*** (Subbotina), ***S. trivialis*** (Subbotina), ***S. triloculinoides*** (Plummer) **Species inherited from Maastrichtian:** ***Heterostomella foveolata*** (Marsson), ***Valvulineria imitata*** (Olsson), ***Valvulinoides umovi*** (Kyprianova), ***Gyroidinoides turgidus*** (Hagenow), ***Eponides sibiricus*** Neckaja, ***Cibicides gankinoensis*** Neckaja, ***Anomalinoides pinguis*** (Jennings) ***neckajae*** Vassilenko, ***A. justus*** Podobina, ***Pullenia americana*** Cushman, ***Nonionellina pseudoinsecta*** (Putrja), ***Bulimina rozenkrantzi*** Brotzen, ***Reussella minuta*** (Marsson), ***Bolivina plaita*** Carsey

Figure 4. Paleocene foraminiferal assemblages of the West Siberian plain.

Gleichenia, Cyathea, Matonia, and *Osmunda* are present (5%–10%). Among the gymnosperm pollen, there are abundant *Taxodiaceae, Pinaceae* (25%), rare *Ephedra* (5%) and a few specimens (up to 3%) of *Cedrus, Podocarpus,* and *Caytonia.* Among the angiosperms, there are abundant *Orbiculapollis globosus* Chlonova (20%), and *Triprojectacites, Mancicorpus* and *Aquilapollenites* are rare (5%).

Up section, the relative abundance and diversity of pollen of the group of *Normapolles* Pfl. and *Ulmoideipites* And., which are typical for the Paleocene, increases to 20%. Rare *Myricaceae, Ulmaceae, Betulaceae, Triatriopollenites* spp., and *Triporopollenites robustus* Pfl. (up to 3%) occur consistently throughout the section (Fig. 5).

The analysis of these assemblages is rather difficult, because the assemblage was influenced by taphonomic processes. The assemblages, however, include palynomorphs characteristic for plant communities living on a coastal lacustrine-alluvial plain, locally occupied by bogs. The presence in the marine deposits of palynomorphs (miospores) that are not easily transported by winds (*Ephedra, Orbiculapollis, Sphagnum, Tax-*

System	Series	Stage	Regional stage	Palynological assemblages of eastern region
				The Parbig River basin
Paleogene	Paleocene	Thanetian	Lyulinvorskaya	Predominant: ***Triatripollenites exelsus*** (R.Rot.) Pfl., ***Myrica*** spp., Characteristic: ***Triporopollenites*** spp., ***Trudopollis pompeckji*** (R.Rot) Pfl., ***Castanea crenataeformis*** Samig., ***Quercus gracilis*** Boitz. Rare: ferns and gymnospermae
		Selandian	Talitskaya	Predominant:Taxodiaceae, Pinaceae, Normapolles Characteristic: ***Extratriporopollenites*** spp., ***Trudopollis menneri*** (Martyn) Zakl., ***T. connector*** Pfl., ***Nudopollis endangulatus*** Pfl., ***N. thirgartii*** Pfl., ***Oculopollis sibiricus*** Zakl., ***Basopollis*** sp., ***B. vestibulatus*** Pfl., ***Postnormapollis*** spp., Myricaceae, Ulmaceae, Juglandaceae, Betulaceae.
		Danian	Talitskaya	**Palynological assemblage 1***
			Gankinskaya	Predominant: Taxodiaceae, Pinaceae. Characteristic: ***Orbiculapollis globosus*** Chlonova, ***Wodehouseia*** spp., ***Tricolporites gracilis***, ***Aquilapollenites*** spp., ***Ulmoideipites*** spp., ***Ephedra*** sp. Rare: ***Normapolles*** Pfl. Abundant: ***Membranosphera maastrichtica*** Samoil.

***Palynological assemblage 1:**
Predominant: Taxodiaceae, Pinaceae
Characteristic: ***Normapolles*** spp., ***Orbiculapollis globosus*** Chlonova, ***Ulmoideipites tricostatus*** And., ***Trudopollis menneri*** (Martin) Zakl., ***T. nonperfectus*** Pfl., ***T. conrector*** Pfl., ***T. fossulotrudens***, ***Nudopollis enangulatus*** Pfl., ***Oculopollis giganteus*** Zakl., ***Oculopollis sibiricus*** Zakl., ***Anacolosidites*** sp., ***Triporopollenites robustus*** Pfl., ***Myrica*** spp., ***Triporopollenites*** spp.
Rare: ***Mancicorpus, Aquilapollenites***

Figure 5. Paleocene palynological assemblages of the West Siberian plain.

odium, etc.) indicates that these palynomorphs represent the vegetation of the nearest coast, especially because these taxa may be dominant in some samples (>30%).

Nagorskaya et al. (1978) studied the coeval continental deposits on the eastern bank of the river Ob in the Ket river basin, close to the Ust-Tym depression. These deposits are characterized by a palynological assemblage with *Triporopollenites robustus* pfl.–*Ulmoideipites tricostatus* And. The floral composition is very similar to that of the assemblage described above, but it represents a different environment, probably because these regions further away from the coast were less boggy.

There are few spores (5%) in this assemblage. Gymnosperm pollen is represented mainly by *Pinaceae* (16%) and to a lesser degree by *Taxodiaceae* (12%). Angiosperm pollen is highly diverse; *Ulmoideipites tricostatus* And., and *U. krempi* And. are the most abundant taxa (13%). There are few *Myrica intermedia* Glad. (5%), *Triporopollenites robustus* Pfl. (6%), *T.* spp., and single specimens of *Alnus* sp., *Platycarya* sp., *Juglans* sp., *Engelhardtia* sp.

The palynological assemblage in the lowermost Talitskaya stage of the northern West Siberian plain is characterized by the occurrence of *Trudopollis menneri* (Martyn) Zakl., *Nudopollis endangulatus* Pfl., and *Oculopollis giganteus* Zakl., as was first described by Ilyenok (1967) for the basin of the Taz river (Fig. 1B) in a continuous section of Cretaceous-Paleogene deposits (Podobina et al., 1997).

This palynological assemblage contains equal portions of spores and pollen of gymnosperm and angiosperm plants. Among the spores there is relatively much *Sphagnum* (20%), less *Lycopodium* (10%) and *Selaginella* (5%), but ferns are also present (3%). Gymnosperm pollen has common Taxodiaceae (25%), and lower percentages of Pinaceae (10%). Angiosperm

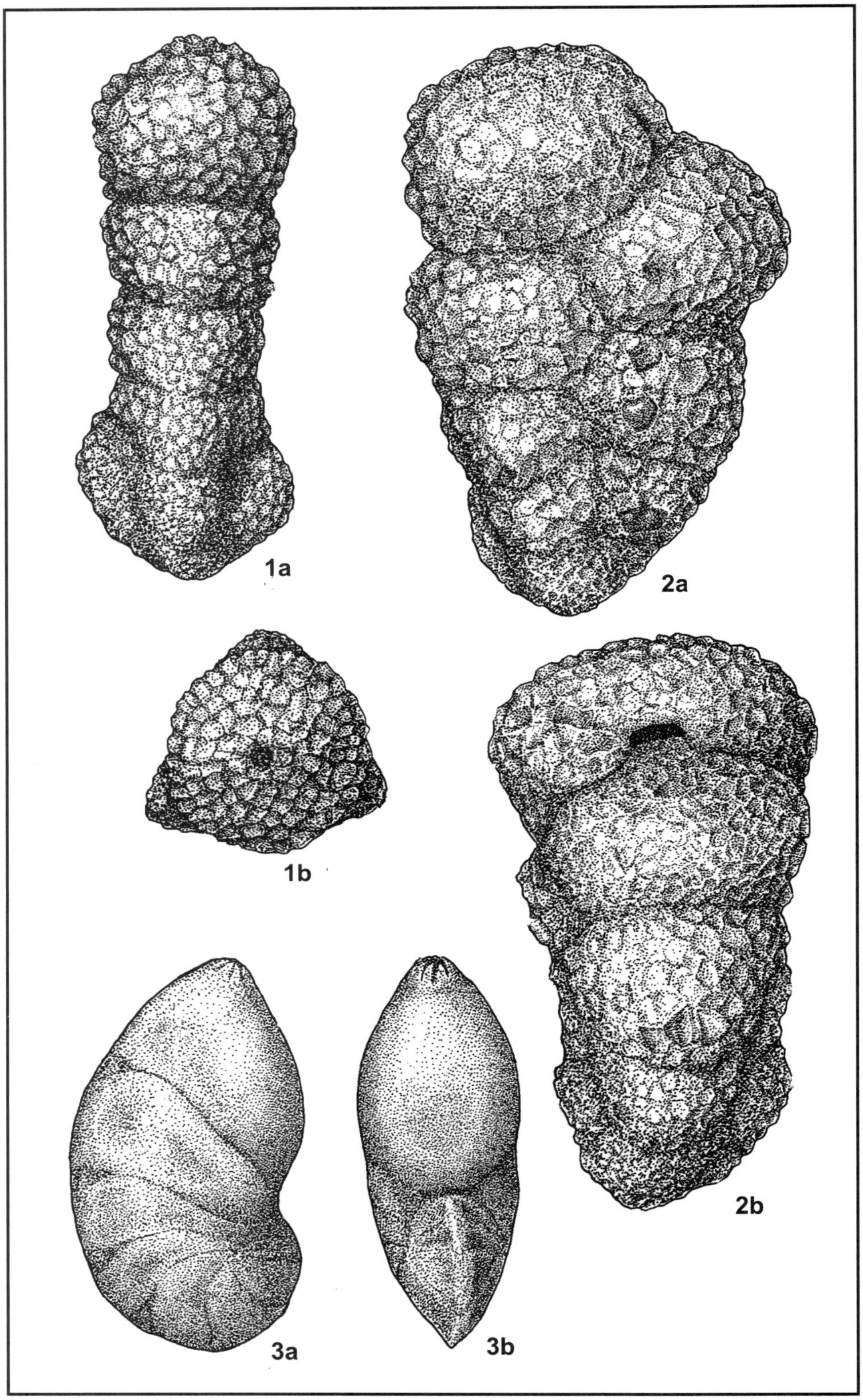

Plate 1. West Siberian Plain; early Paleocene; *Brotzenella praeacuta* zone, magnification ×80. (All specimens are deposited in the Micropaleontological Laboratory and Paleontological Museum of Tomsk State University.) Figure 1. *Clavulina parisiensis* Orbigny; a, lateral view; b, apertural view. Figure 2. *Gaudryina gigantica* (Subbotina); a, lateral view; b, apertural view. Figure 3. *Astacolus elatus* Podobina; a, lateral view; b, apertural view.

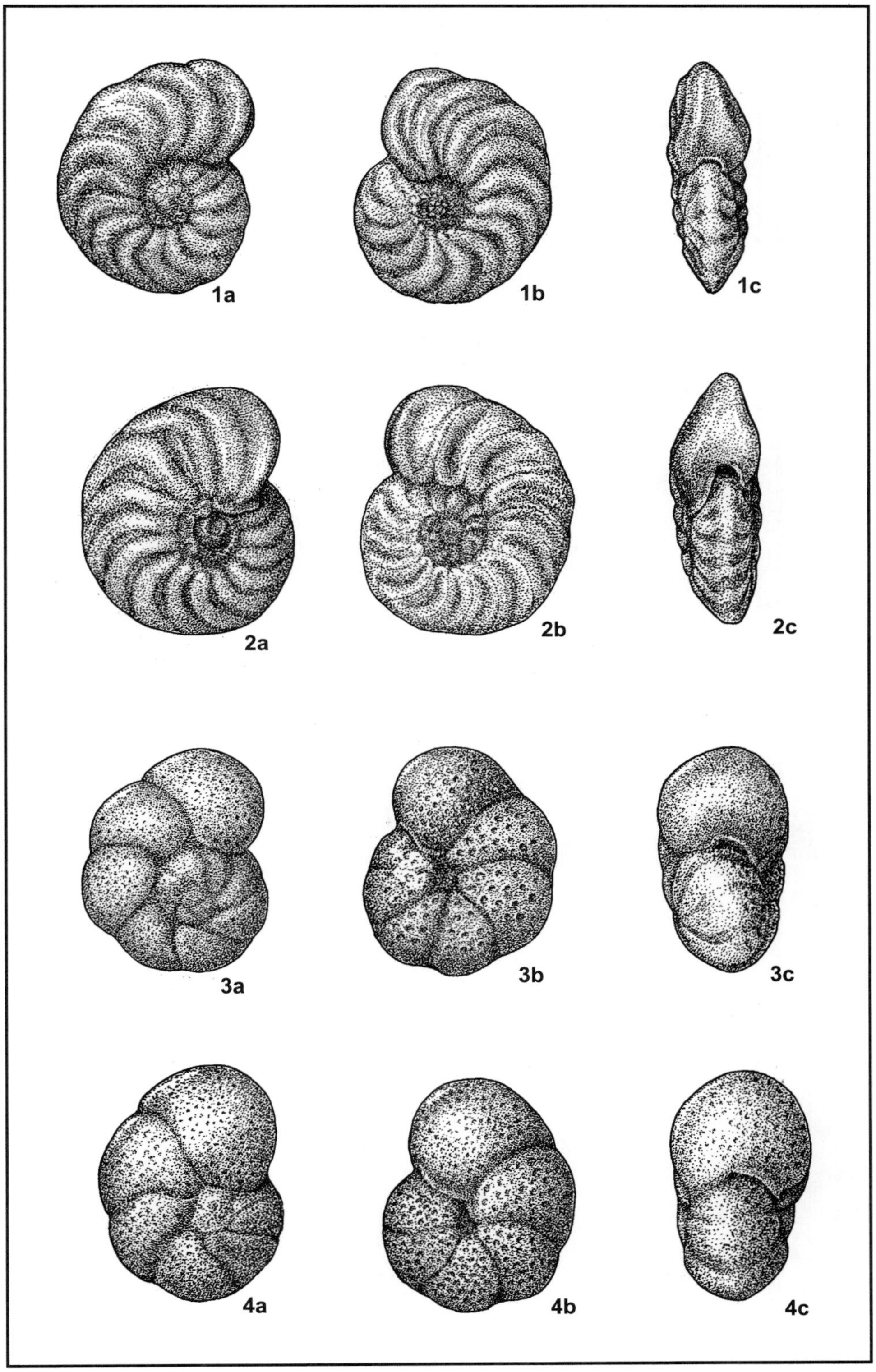

Plate 2. West Siberian Plain; early Paleocene; *Brotzenella praeacuta* zone, magnification ×80. (All specimens are deposited in the Micropaleontological Laboratory and Paleontological Museum of Tomsk State University.) Figures 1–2. *Brotzenella praeacuta* (Vassilenko); a, dorsal view; b, ventral view; c, apertural view. Figures 3–4. *Anomalina danica* (Brotzen); a, dorsal view; b, ventral view; c, apertural view.

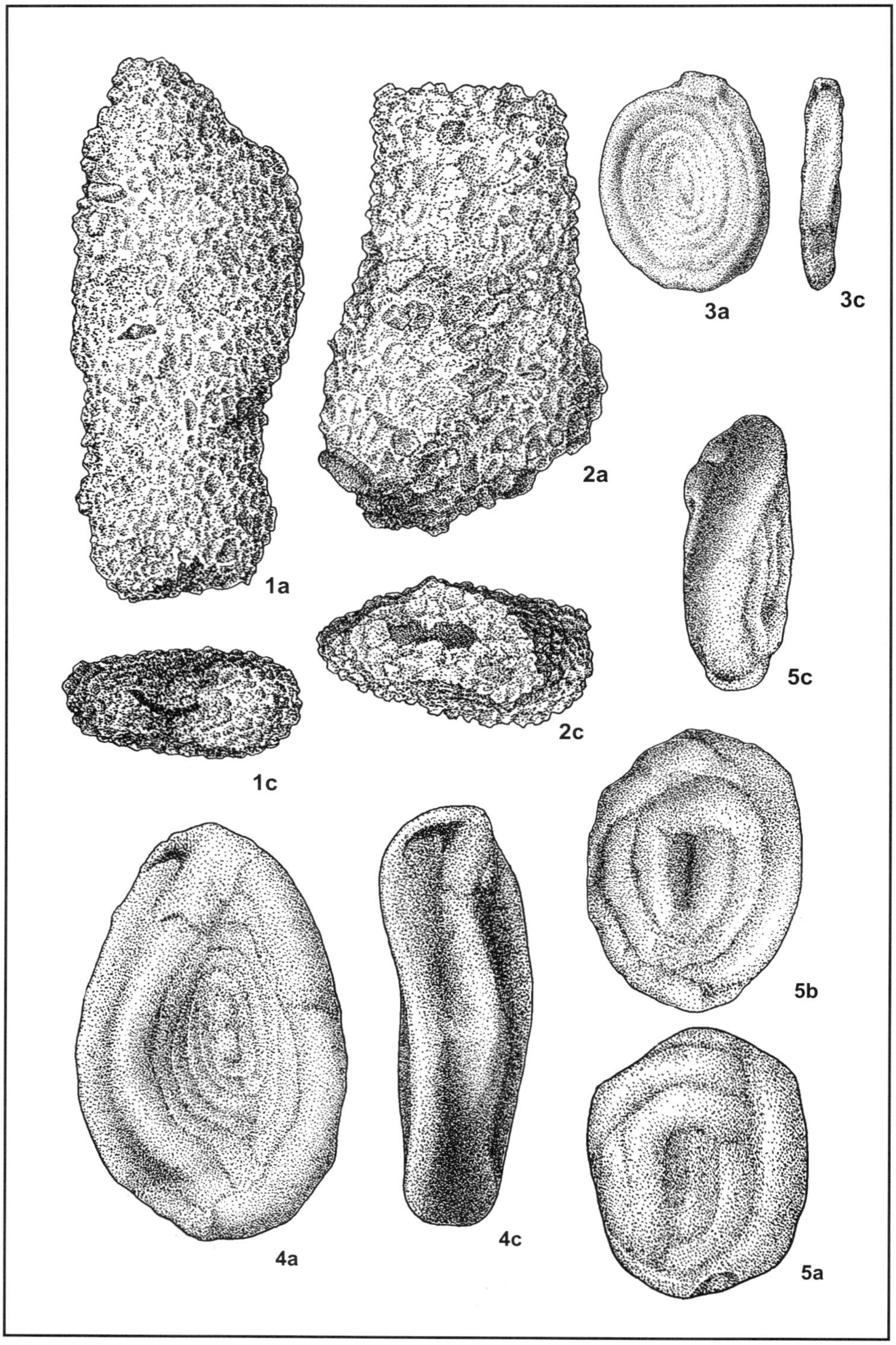

Plate 3. West Siberian Plain; early Paleocene; strata with *Bathysiphon nodosarieformis* and *Glomospira charoides,* magnification ×100. (All specimens are deposited in the Micropaleontological Laboratory and Paleontological Museum of Tomsk State University.) Figure 1. *Bathysiphon nodosarieformis* Subbotina; a, lateral view; c, apertural view. Figure 2. *Hyperammina elongata* Brady; a, lateral view; c, apertural view. Figure 3. *Ammodiscus glabratus* Cushman and Jarvis; a, lateral view; c, apertural view. Figure 4. *Ammodiscus incertus* (Orbigny); a, lateral view; c, apertural view. Figure 5. *Glomospira gordialiformis* Podobina and Kabanova; a, dorsal view; b, ventral view; c, apertural view.

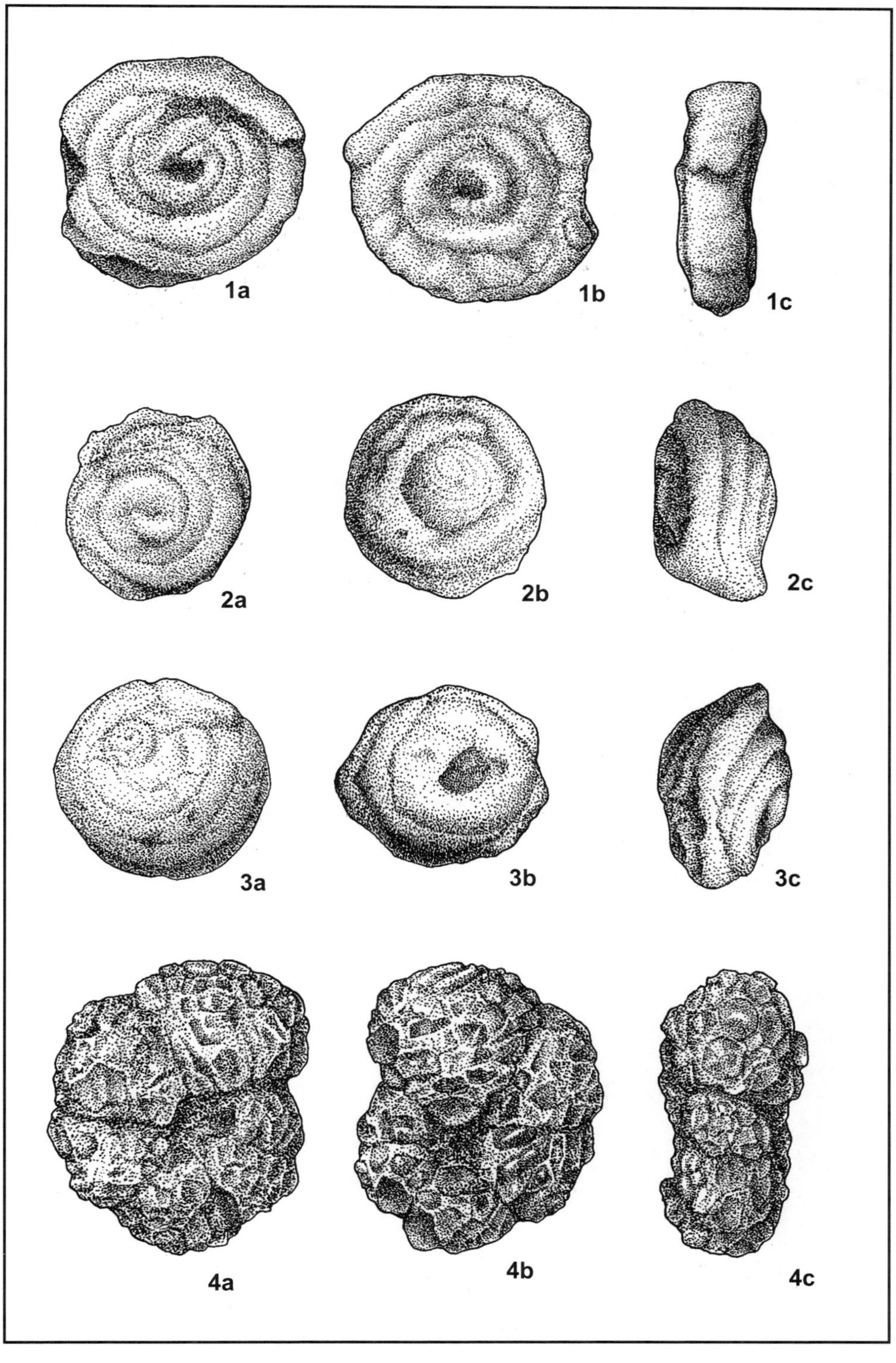

Plate 4. West Siberian Plain; early Paleocene; strata with *Bathysiphon nodosarieformis,* and *Glomospira charoides,* magnification ×80. (All specimens are deposited in the Micropaleontological Laboratory and Paleontological Museum of Tomsk State University.) Figure 1. *Glomospira gordialiformis* Podobina; a, dorsal view; b, ventral view; c, apertural view. Figures 2–3. *Glomospira charoides* (Parker and Jones); a, dorsal view; b, ventral view; c, apertural view. Figure 4. *Labrospira granulosa* (Lipman); a, dorsal view; b, ventral view; c, apertural view.

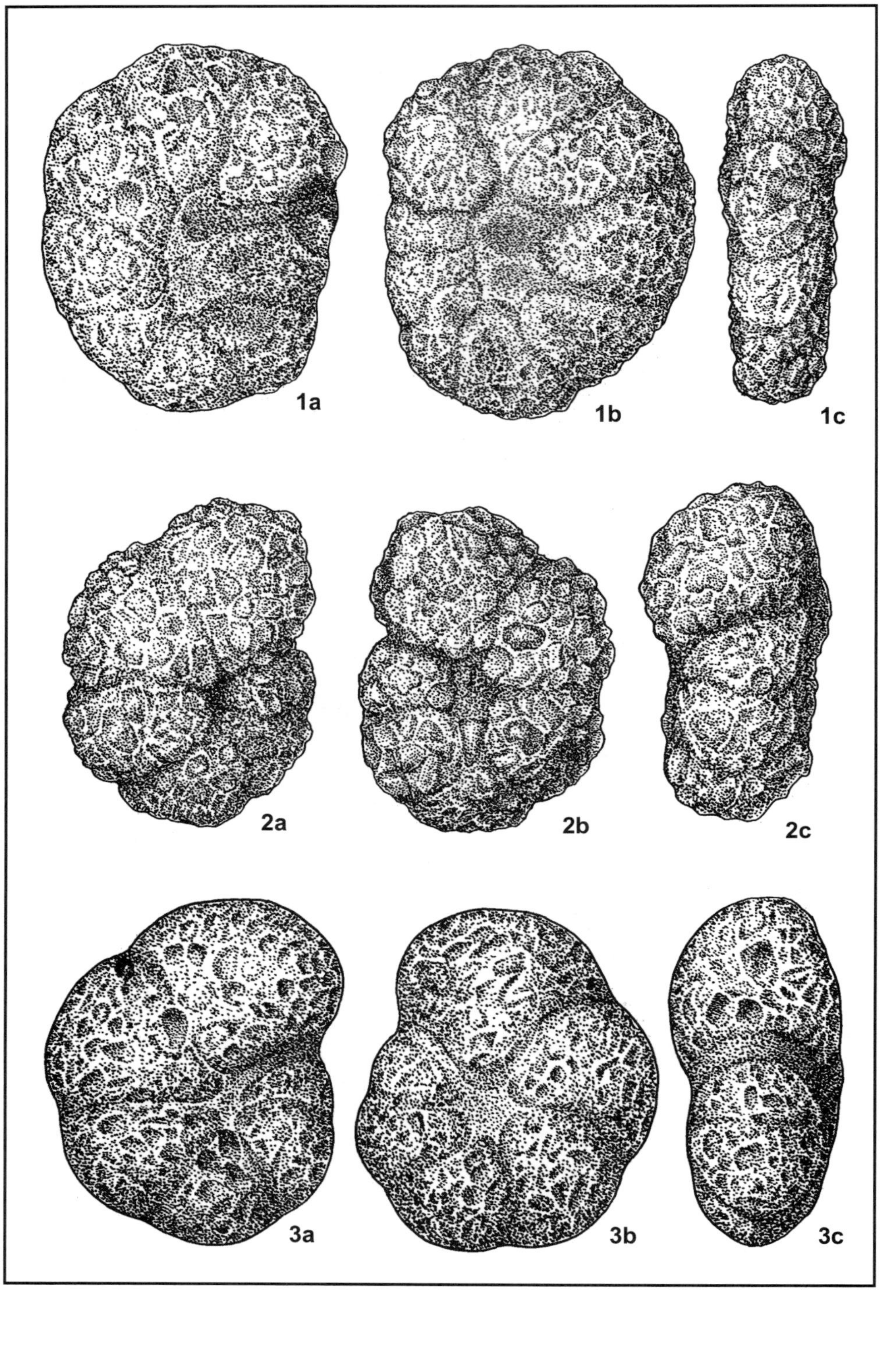

Plate 5. West Siberian Plain; late Paleocene; *Ammoscalaria friabilis* zone, magnification ×100. (All specimens are deposited in the Micropaleontological Laboratory and Paleontological Museum of Tomsk State University.) Figure 1. *Trochamminoides lamentabilis* Podobina; a, dorsal view; b, ventral view; c, apertural view. Figure 2. *Labrospira granulosa* (Lipman); a, dorsal view; b, ventral view; c, apertural view. Figure 3. *Haplophragmoides fastosus* Podobina; a, dorsal view; b, ventral view; c, apertural view.

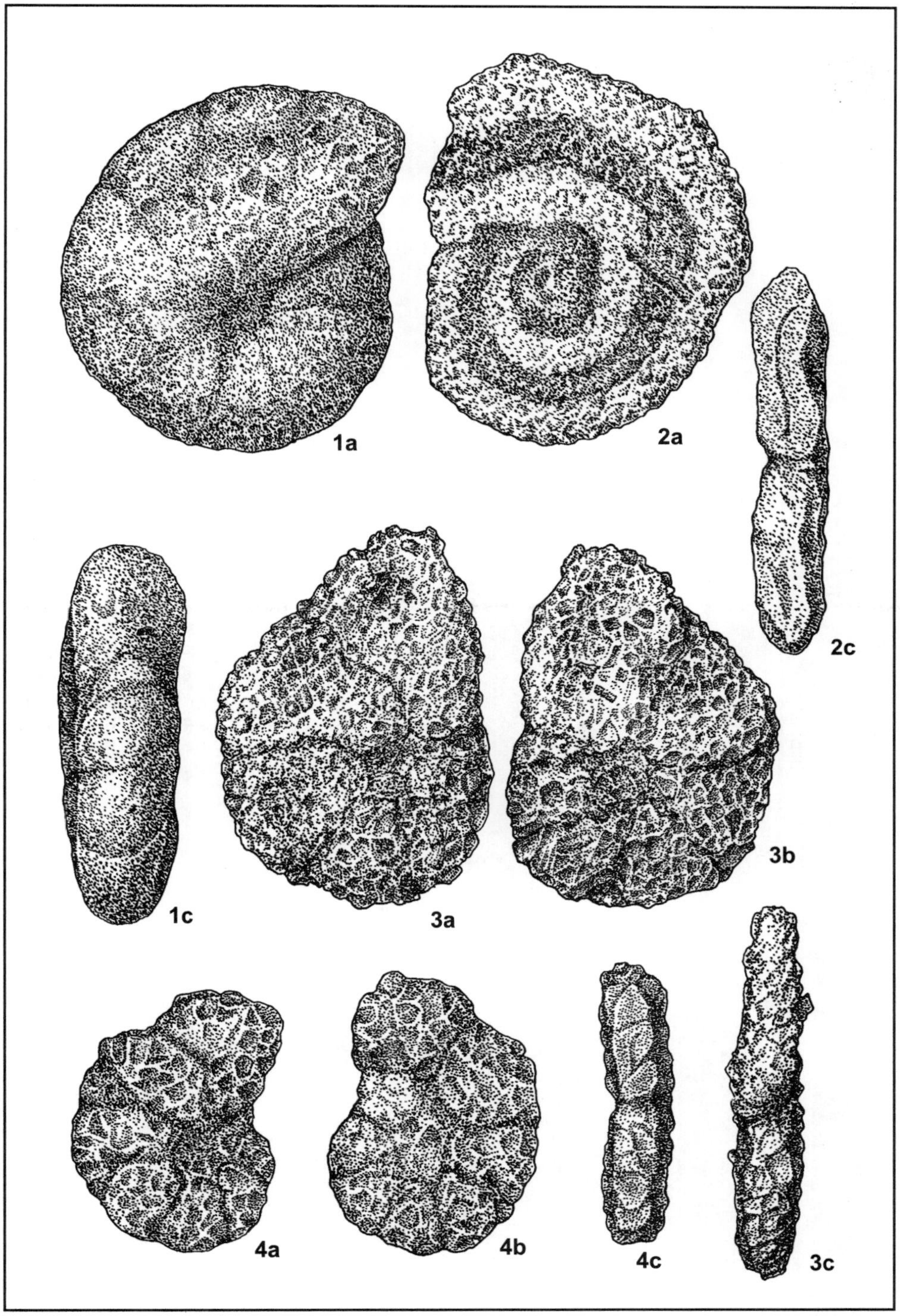

Plate 6. West Siberian Plain; late Paleocene; *Ammoscalaria friabilis* zone, magnification ×100. (All specimens are deposited in the Micropaleontological Laboratory and Paleontological Museum of Tomsk State University.) Figure 1. *Cyclammina coksuvorovae* Uschakova; a, dorsal view; c, apertural view. Figure 2. *Ammoscalaria friabilis* (Ehremeeva); a, dorsal view; c, apertural view. Figures 3–4. *Ammomarginulina brevis* (Lipman); a, dorsal view; b, ventral view; c, apertural view.

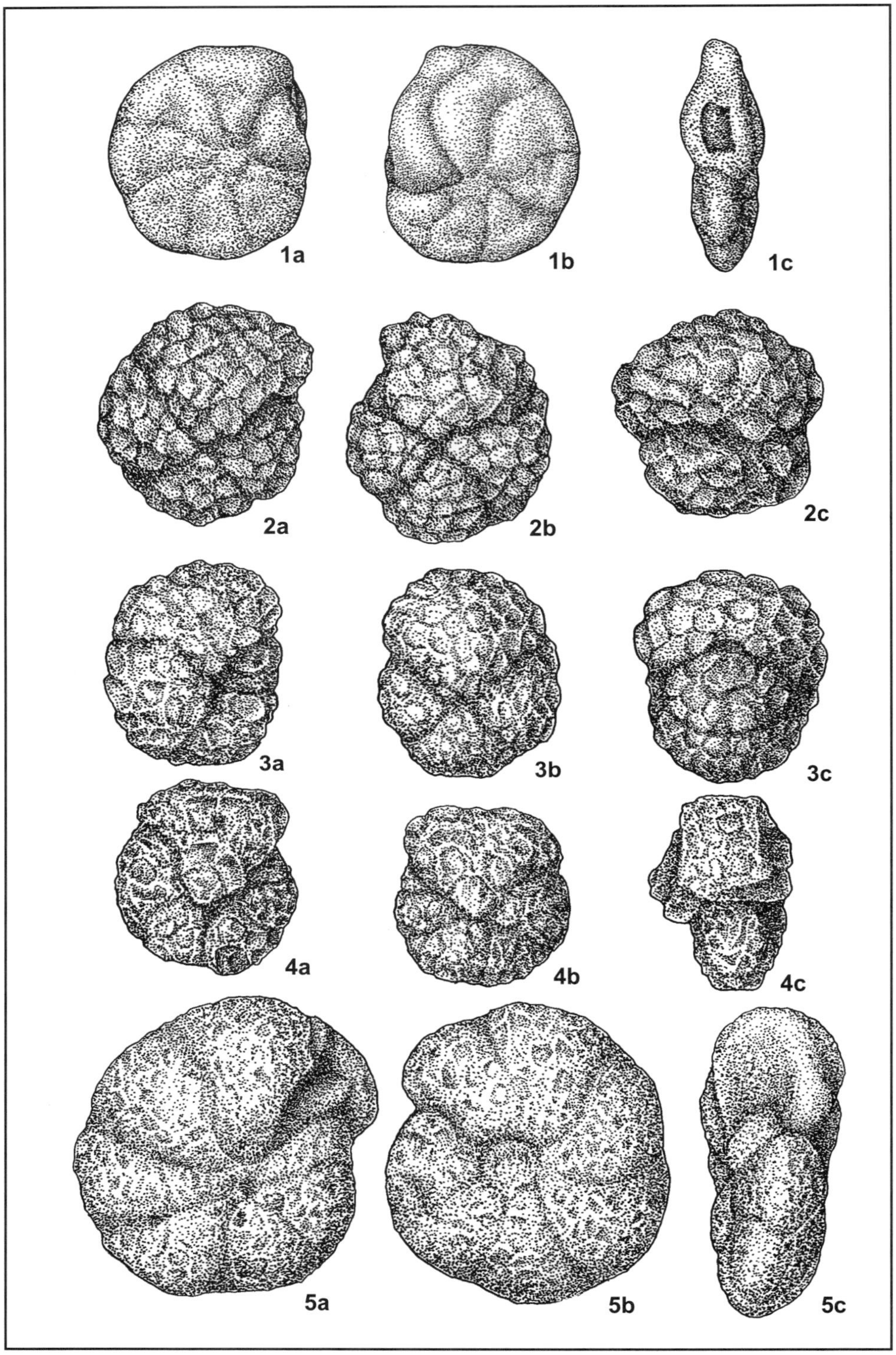

Plate 7. West Siberian Plain; late Paleocene; *Ammoscalaria friabilis* zone, magnification ×100. (All specimens are deposited in the Micropaleontological Laboratory and Paleontological Museum of Tomsk State University.) Figure 1. *Asanospira grzybowski* (Mjatliuk); a, dorsal view; b, ventral view; c, apertural view. Figures 2–4. *Adercotryma horrida* (Grzybowski); a, dorsal view; b, ventral view; c, apertural view. Figure 5. *Haplophragmoides stomatus* (Grzybowski); a, dorsal view; b, ventral view; c, apertural view.

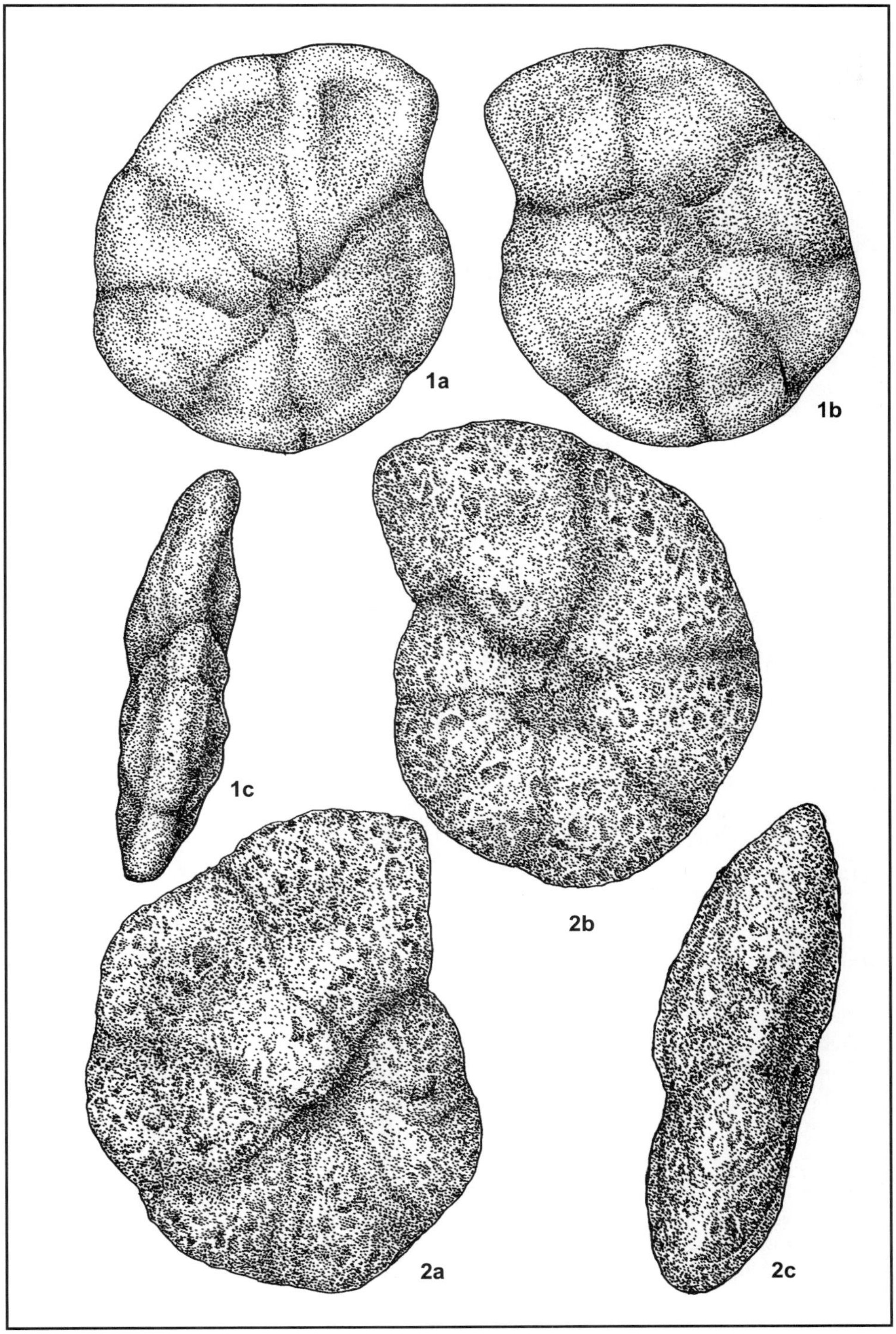

Plate 8. West Siberian Plain; late Paleocene; *Ammoscalaria friabilis* zone, magnification ×100. (All specimens are deposited in the Micropaleontological Laboratory and Paleontological Museum of Tomsk State University.) Figures 1–2. *Cyclammina coksuvorovae* Uschakova; a, dorsal view; b, ventral view; c, apertural view.

pollen is mainly represented by diverse *Normapolles,* some *Quercites sparsus* (Mart.) Samoil. (20%), and sparse *Salix, Myrica,* and *Betulaceae* (10%).

In the Parbig river basin in the southeastern reaches of West Siberia (Fig. 1B) the following coeval palynological assemblages occur: *Normapolles* spp., *Orbiculapollis globosus,* Chlonova, *Ulmoideipites tricostatus* And., *Trudopollis nonperfectus* Pfl., *T. conrector* Pfl., *T. fossulotrudens* Pfl., *Oculopollis sibiricus* Zakl., *Anacolosidites* sp., *Triporopollenites robustus* Pfl., *Myrica* spp., *Triporopollenites* spp.; *Mancicorpus* and *Aquilapollenites* are rare (up to 5%) (Fig. 5; Plates 9–11).

This palynological assemblage has been dated as early Paleocene (Danian), because it occurs in sediments containing *Cibicidoides incognitus* (Fig. 4). Planktonic foraminifera co-occurring with this species at other locations (see above) indicate that these samples have a Danian age. The assemblage represents the evolution of the flora throughout the early Paleocene, during which time the *Normapolles* vegetation became more common. During the whole time of deposition of the sediments with this assemblage, the southeastern coast of the basin remained a lacustrine-alluvial plain with bogs, from where spores and pollen were transported into the epicontinental marine basin.

LATE PALEOCENE

Selandian

Most of the lower part of the Talitskaya Stage is characterized by a foraminiferal assemblage with *Ammoscalaria friabilis* (Figs. 3, 4; Podobina, 1975, 1998). In the deeper, western parts of the basin (East of the Ural Mountains) and in the central region of the West Siberian plain, these sediments are thicker (80–100 m) than in the eastern regions (10–20 m). In northern Transuralia (close to the settlement of Beryozovo, Fig. 1B) the *Ammoscalaria friabilis* assemblage contains agglutinated taxa as well as abundant (30%) calcareous forms of the *Cibicidoides proprius* assemblage, many species of which are known from the Selandian stratotype of Denmark and Sweden (Brotzen, 1948; Frenzel, 2000).

The calcareous taxa (10%–15%) of the *Cibicidoides proprius* assemblage occur in the central West Siberian plain (the basin of the Vasyugan river) together with agglutinated ones. In these sediments *Ammoscalaria friabilis* is rare (5%), while the fine-grained tests of the agglutinated genera *Psammosphaera, Ammodiscus, Haplophragmoides* and *Trochammina* (Plates 5–8) are abundant (80%–85%; Podobina, 1998).

Toward the East, the thickness of the Talitskaya stage sediments decreases down to ~10–20 m, but the foraminiferal assemblage with *Ammoscalaria friabilis,* consisting mainly of agglutinated forms, occurs throughout the basins of the Ilyak, Chizhapka and other rivers (Fig. 1B). Each species is represented by several individuals (5–10 specimens in a sample), resulting in a diverse assemblage with fairly even distribution of specimens over species, suggesting that salinity and oxygen levels were normal marine. The predominance of medium-grained quartz in the agglutinated tests indicate that paleo-depths were about inner shelf.

Most of deposits of the Talitskaya stage with the *Ammoscalaria friabilis* assemblage, and the coeval deposits of the Central Priobye (the Ob riverside territory) are characterized by a palynological assemblage with *Trudopollis menneri* (Martyn) Zakl., *Anacolosidites insignis* Samoil., and *Triatriopollenites aroboratus* Pfl., taxa that survived from the early Paleocene. The assemblages are present in marine sediments, thus allochthonous, and may contain common reworked elements.

There are many *Sphagnum* spores (20%), as in the lower Paleocene, and Taxodiaceae and Pinaceae pollen are abundant and predominant (50%). Angiosperm pollen remained highly diverse, but the abundance of *Normapolles* pollen decreased to 20%. *Basopollis triangularis* Pfl. occurs as well as rare *Anacolosidites insignis* Samoil., *Regina exelsa* Samoil., *Projectoporites spinulosus* N. Mtchedlishvili. (5%–10%), which are typical late Paleocene taxa. The species richness of *Postnormapolles* species increased, and in marine and littoral deposits, microphytoplankton (microscopic algae) is present.

On the east bank of the river Ob the thickness of deposits with this palynological assemblage decreases to ~10 m. In the Ust-Tym depression, strata with palynological assemblages containing *Trudopollis menneri–Anacolosidites insignis–Triatriopollenites aroboratus* directly overlie Maastrichtian deposits. They are pinching out toward the East, where they remain only locally as separated lenses in depressions in the Cretaceous sediments.

Between the Ket and the Tym rivers (Fig. 1B), deposits with this assemblage occur in a lenticular body of dark-brown sands with thin lignite intercalations, overlain by dark-brown clay (Kabanova and Kostesha, 1988; Ilyenok et al., 1989). This body occurs within an eroded channel in Maastrichtian-Danian sediments, and is overlain by Oligocene channel deposits. In these eastern deposits there are few spores (10%–20%) and abundant pollen (80%–90%), with a floral composition very similar to that in the sediments described above, although the relative abundance of taxa differs. Gymnosperms are represented by Pinaceae (50%): Mostly *Pinus,* rarely *Picea, Abies, Keteleeria, Cedrus,* less common Taxodiaceae. Angiosperm pollen (30%) are more diverse.

Thanetian

The upper Paleocene Talitskaya sediments and the lower part of the Lyulinvorskaya sediments in the central region of Western Siberia contain a foraminiferal assemblage characterized by the index species *Glomospira gordialiformis* and *Cibicidoides favorabilis.* The agglutinated, sugary-white tests of the genera *Psammosphaera, Ammodiscus, Glomospira, Asanospira* and *Cyclammina* dominate (90%) the assemblage, and calcareous species are rare (10%), but the index species *Cibicidoides favorabilis* (Vassilenko) is present (Podobina, 1998). Calcareous

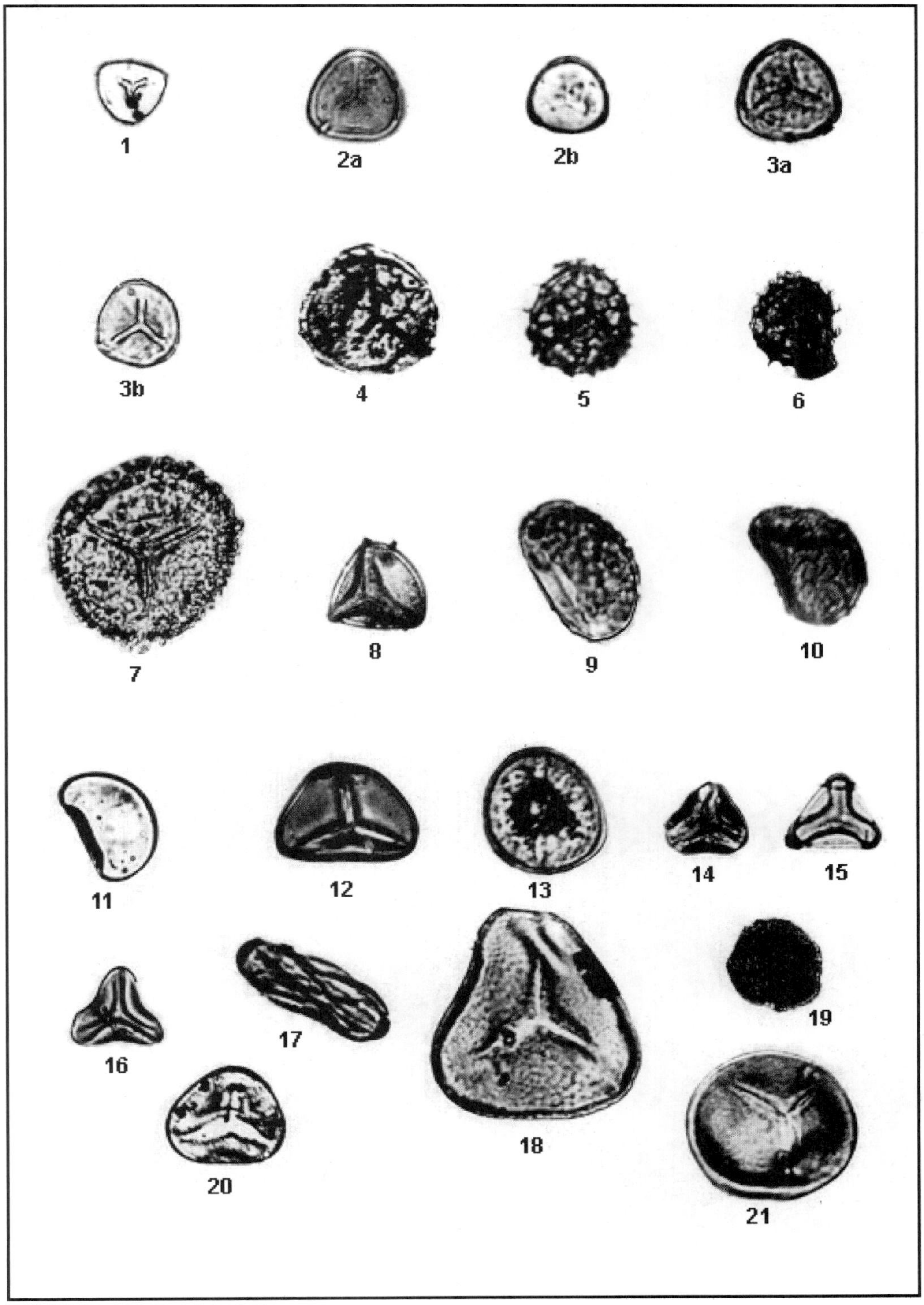

Plate 9. West Siberian Plain; early Paleocene; Sporae (magnification ×600). (All specimens are deposited in the Micropaleontological Laboratory and Paleontological Museum of Tomsk State University.) Figure 1. *Sphagnum australe* (Cook) Drozh; Figure 2a, b. *Sphagnum putillum* Drozh. et Purt. Figure 3a, b. *Sphagnum regium* Drozh. Figures 4–6. *Lycopodium* sp. Figure 7. *Selaginella* sp. Figure 8. *Cyathea* sp. Figures 9–10. Polypodidiaceae. Figures 11–12. *Adiantum* sp. Figure 13. *Cheiropleuria* sp. Figure 14. *Gleichenia umbonata* Bolch. Figure 15. *Gleichenia rara* Chlon. Figure 16. *Gleichenia* sp. Figure 17. *Schizaea dorogensis* (R. Pot.) Chlon. Figure 18. *Lygodium* sp. Figure 19. *Osmunda* sp. Figure 20. *Leiotriletes* Naum. Figure 21. *Stenozonotriletes* Naum.

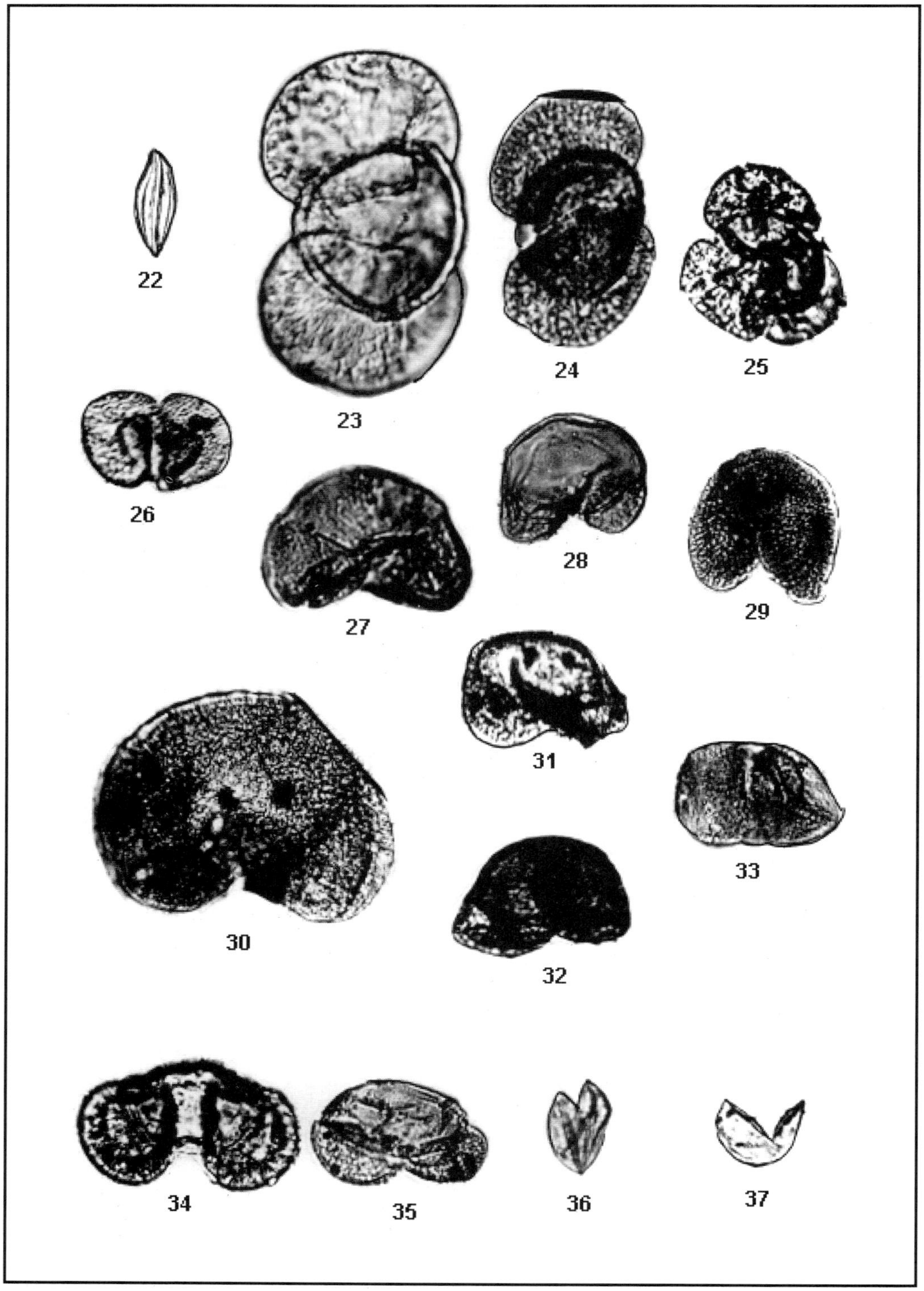

Plate 10. West Siberian Plain; early Paleocene; Gymnospermae Pollen (magnification ×600). (All specimens are deposited in the Micropaleontological Laboratory and Paleontological Museum of Tomsk State University.) Figure 22. *Ginkgo* sp. Figure 23. *Podocarpus unica* Bolch. Figures 24–26. *Podocarpus* sp. Figures 25–27. Pinaceae. Figure 28. *Cedrus parvisaccata* Sauer. Figures 29–32. *Cedrus* sp. Figure 33. *Pinus aequelis* (Naum.) Bolch. Figure 34. *Pinus vulgaris* (Naum.) Bolch. Figure 35. *Pinus* n./p. Dyploxylon. Figure 36. Taxodiaceae. Figure 37. *Taxodium* sp.

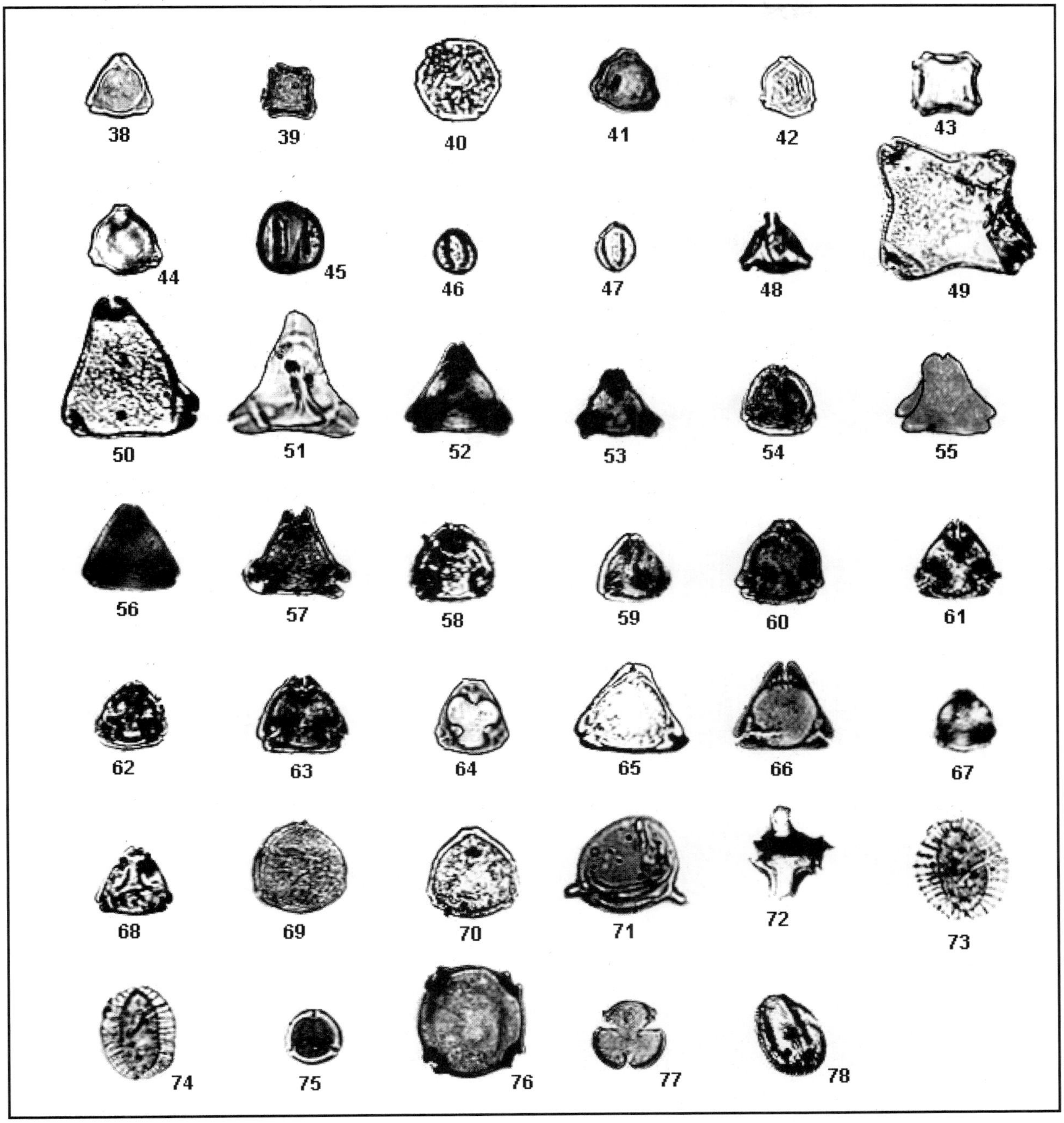

Plate 11. West Siberian Plain; early Paleocene; Angiospermae Pollen (magnification ×600). (All specimens are deposited in the Micropaleontological Laboratory and Paleontological Museum of Tomsk State University.) Figure 38. *Myrica* sp. Figure 39. *Myricacites* sp. Figure 40. *Juglans* sp. Figures 41–42. Betulaceae. Figure 43. *Alnus* sp. Figure 44. *Betula* sp. Figure 45. *Quercus* sp. Figures 46–47. *Quercites sparsus* (Martyn) Samoil. Figure 48. *Santalacites* sp. Figure 49. Proteacea. Figure 50. *Protea* sp. Figure 51. *Nudopollis endangulatus* Pfl. Figures 52–53. *Nudopollis* sp. Figure 54. *Basopollis vestibulatus* Zalkl. Figure 55. *Basopollis* sp. Figure 56. *Oculopollis praedicatus* Weil et Krieg. Figure 57. *Oculopollis torosus* Zakl. Figure 58. *Oculopollis* sp. Figure 59. *Vacuopollis* sp. Figure 60. *Trudopollis conrector* Pfl. Figures 61–64. *Trudopollis* sp. Figures 65–66. *Extratriporopollenites* sp. Figure 67. *Triatriopollenites aroboratus* Pfl. Figure 68. *Triatriopollenites.* Figure 69. *Trioropollenites robustum* Pfl. Figure 70. *Tricolpites* sp. Figure 71. *Triporina globosa* Chlon. Figure 72. *Aquilapolenites* sp. Figures 73–74. *Kryshtofoviana vera* Samoil. Figure 75. *Triporopollenites* sp. Figure 76. *Tetraporopollenites* sp. Figures 77–78. *Tricolpopollenites* sp.

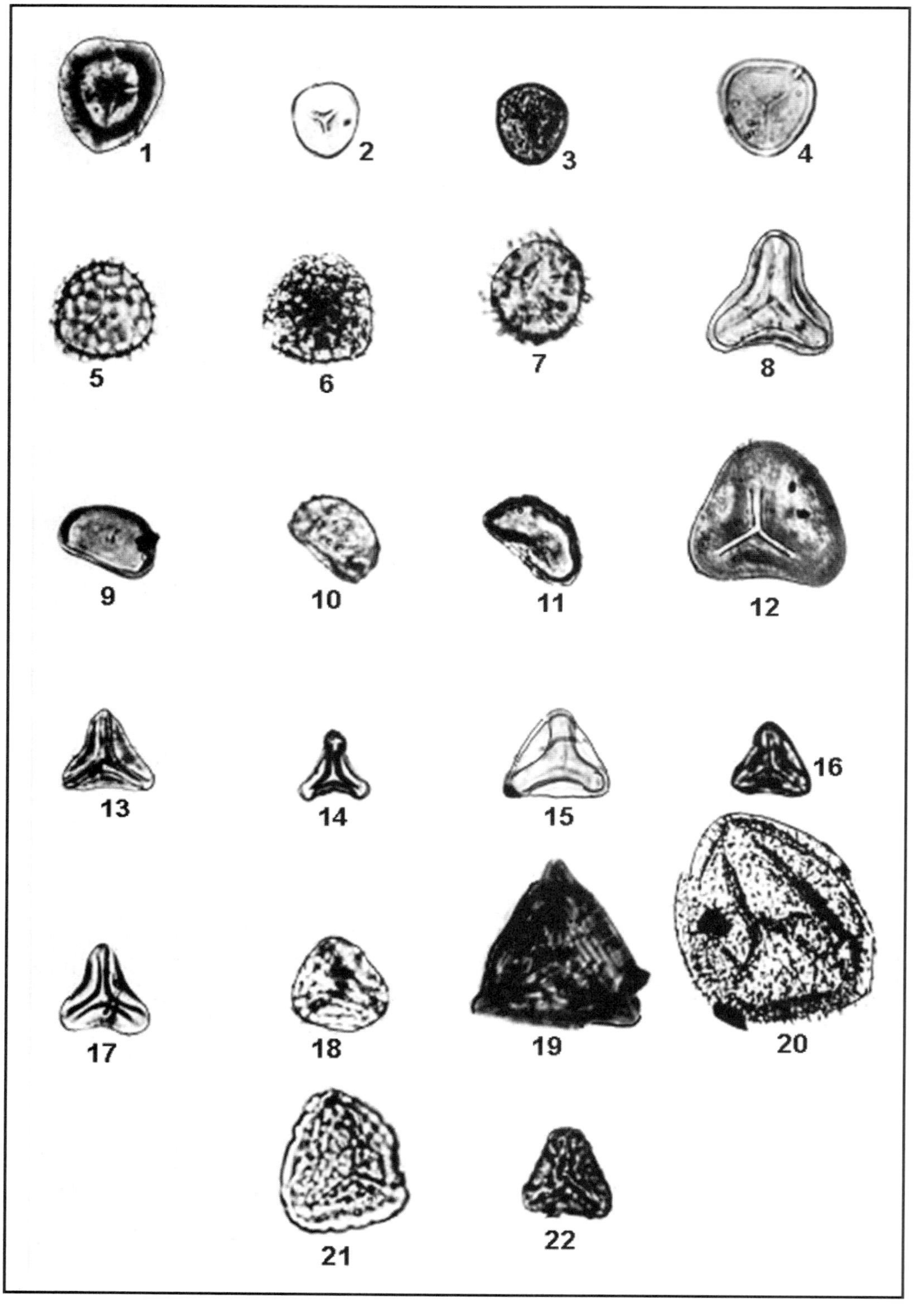

Plate 12. West Siberian Plain; late Paleocene; Sporae (magnification ×600). (All specimens are deposited in the Micropaleontological Laboratory and Paleontological Museum of Tomsk State University.) Figure 1. *Sphagnum australe* (Cook) Drozh. Figure 2. *Sphagnum putillum* Drozh. et Purt. Figure 3. *Sphagnum regium* Drozh. Figure 4. *Sphagnum putillum* Drozh. et Purt. Figures 5–6. *Lycopodium* sp. Figure 7. *Selaginella diuturna* Bolch. Figure 8. *Cyathea* sp. Figures 9–11. Polypodidiaceae. Figure 12. *Matonia* sp. Figure 13. *Gleichenia umbonata* Bolch. Figure 14. *Gleichenia laeta* Bolch. Figure 15. *Gleichenia rara* Chlon. Figures 16–17. *Gleichenia* sp. Figure 18. *Mohria* sp. Figure 19. *Aneimia* sp. Figure 20. *Osmunda* sp. Figures 21–22. *Lophotriletes* Naum (magnification ×600).

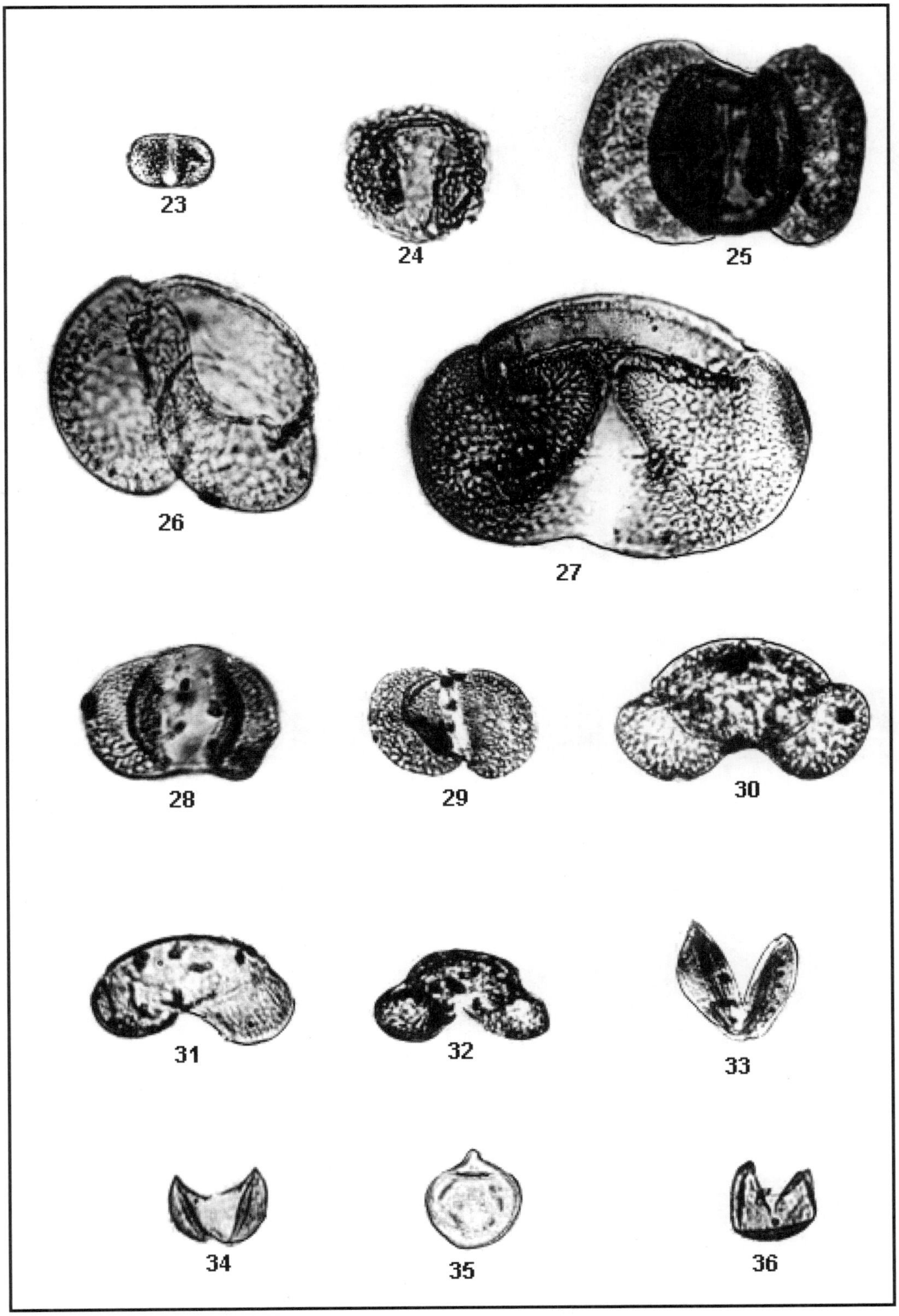

Plate 13. West Siberian Plain; late Paleocene; Gymnospermae Pollen (magnification ×600). (All specimens are deposited in the Micropaleontological Laboratory and Paleontological Museum of Tomsk State University.) Figure 23. *Caytonia* sp. Figure 24. *Dacrydiumites* sp. Figure 25. *Podocarpus* sp. Figure 26. Pinaceae. Figure 27. *Picea* sp. Figures 28–29. *Pinus* n./p. Haploxylon. Figure 30. *Pinus aralica* Bolch. Figures 31–32. *Pinus* n./p. *Diploxylon.* Figures 33–34. Taxodiaceae. Figures 35–36. *Taxodium* sp.

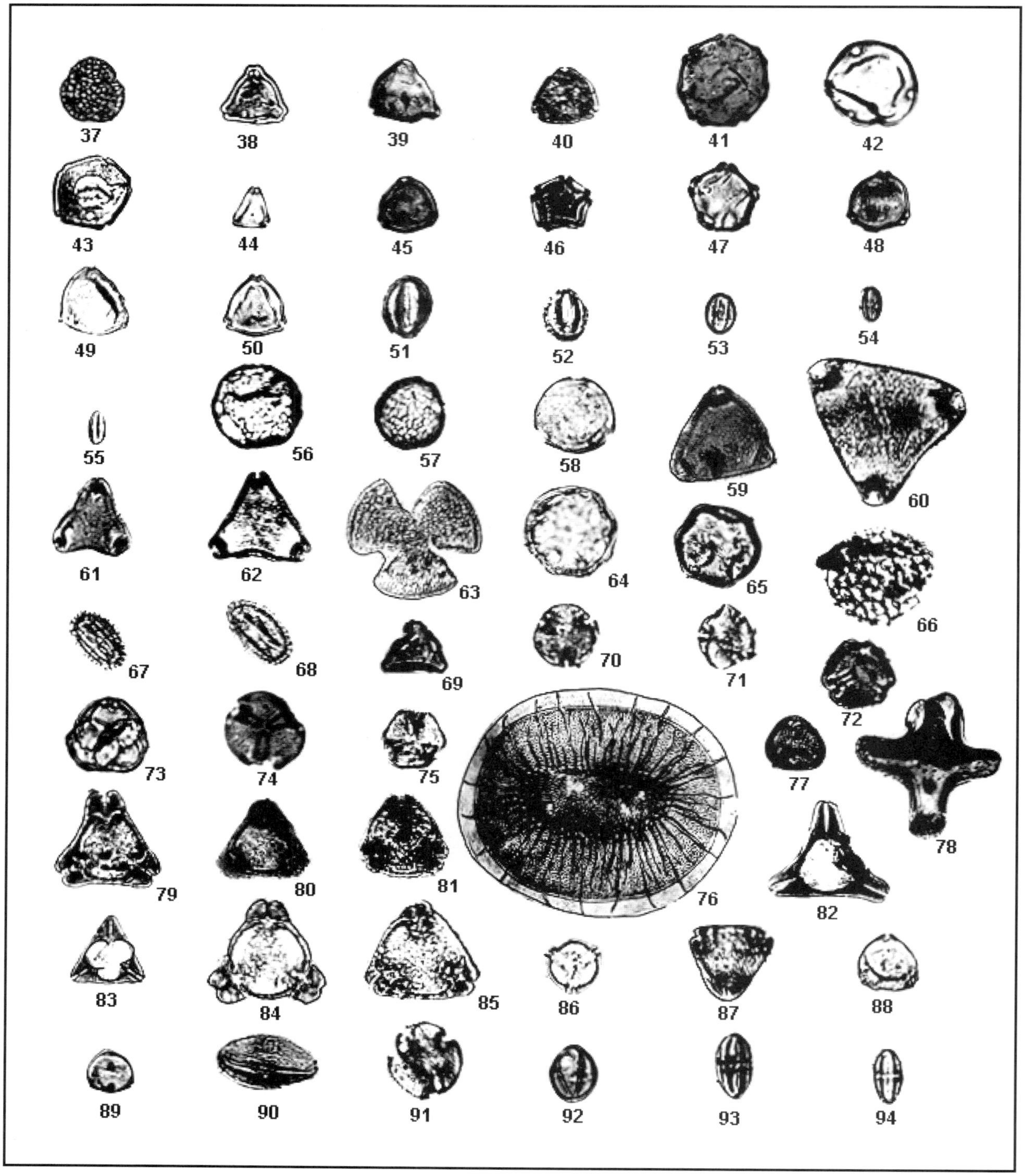

Plate 14. West Siberian Plain; late Paleocene; Angiospermae Pollen (magnification ×600). (All specimens are deposited in the Micropaleontological Laboratory and Paleontological Museum of Tomsk State University.) Figure 37. Solicaceae. Figures 38–40. *Myrica* sp. Figure 41. *Pterocarya* sp. Figure 42. *Carya* sp. Figure 43. *Juglans* sp. Figure 44. *Engelhardtia* sp. Figure 45. Betulaceae. Figures 46–47. *Alnus* sp. Figure 48. *Betula* sp. Figure 49. *Betulites* sp. Figure 50. *Corylus* sp. Figure 51. *Quercus* sp. Figures 52–53. *Quercites sparsus* (Martyn) Samoil. Figures 54–55. *Castanea* sp. Figures 56–57. Ulmaceae. Figure 58. Moraceae. Figures 59–62. Protaceae. Figure 63. Homamelidaceae. Figures 64–65. *Liquidambar* sp. Figure 66. Buxaceae. Figures 67–68. *Ilex* sp. Figure 69. *Santalacites* sp. Figures 70–71. *Tilia* sp. Figure 72. *Nyssa* sp. Figure 73. Ericaceae. Figures 74–75. Euphorbiaceae. Figure 76. *Regina exselsa* Samoil. Figure 77. *Anacolosidites insignis* Samoil. Figure 78. *Aquilapolenites* sp. Figure 79. *Trudopollis conrector* Pfl. Figure 80. *Trudopollis* sp. Figure 81. *Oculopollis* sp. Figures 82–83. *Nudopollis* sp. Figures 84–85. *Basopollis* sp. Figure 86. *Extratriporopollenites pompeckjic* Pfl. Figure 87. *Extratriporopollenites* sp. Figures 88–89. *Triorites* sp. Figures 90–94. *Tricolporopollenites* spp.

species including this index species dominate in the areas adjacent to the Urals (Ehremeyeva and Belousova, 1961; Subbotina, 1964). In that region, the representative species *Haplophragmoides excavatus* Cushman and Waters [=*Asanospira grzybowski* (Mjatliuk)], which is widely distributed, cooccurs with other agglutinated taxa.

In the eastern part of the plain, the upper Talitskaya sediments consist of lighter-colored clays, and the lower parts of the Lyulinvorskaya grey siliceous clays contain foraminiferal assemblages with *Glomospira gordialiformis* and *Cyclammina coksuvorovae,* dominated by the sugary-white, fine-grained tests of the genera *Bathysiphon, Glomospira, Ammodiscus, Asanospira* and *Cyclammina.* Calcareous forms are represented by rare specimens of *Eponides* and *Cibicidoides* only (Podobina, 1998). The foraminiferal species composition, as well as features of the test walls (fine-grained, sugary-white), point to a low salinity (10‰–15‰), at water depths corresponding to these on the outer shelf (Podobina, 1998).

In the palynological assemblage at the very top of the Talitskaya sediments, spores of sphagnum and gymnosperm pollen are predominant (~40% of the assemblage). The relative abundances of Taxodiaceae and Coniferae are about equal (20%) in most samples, although in some samples conifer pollen predominates (25%). *Classopollis* has its first appearance. In angiosperm pollen, the percentage of pollen of subtropical sclerophylls and *Postnormapollis* increases (up to 30%). This floral species composition indicates changes in ecology, probably a decrease in the extent of swamps and wetlands and an increase in more arid lands, as supported by Podobina (1995).

In the east and southeast of the West Siberian plain a palynological assemblage with *Triatriopollenites exelsus* (R.Rot) Pfl.–*Myrica* spp. occurs in the lower layers of the Lyulinvorskaya sediments. This assemblage differs qualitatively from the palynomorph assemblages described above, because it is dominated by angiosperm pollen (60%). There is much pollen of *Myrica* and of the artificial taxon *Triatriopollenites; Triporopollenites* are less abundant (20%) but diverse. *Trudopollis pompeckji* (R.Rot) Pfl., *Castanea crenataeformis* Samig., *C. stelmakae* Boitz., *Castanopsis pseudocingulum* (R.Rot.) Boitz., *Quercus gracilis* Boitz., which are common in early Eocene assemblages, have their first appearances. Ferns (10%–15%) and gymnosperms (25%) are relatively rare.

In contrast to the spectra of the relatively large (>40 m) pollen described above, the palynological assemblage with *Triatriopollenites exelsus* and *Myrica* spp. is rich in diverse species of small pollen (20–25 mm). The abundance of the artificial taxa restricts applying the principle of actualism in reconstructing the paleo-environments. Pollen of subtropical plants (Myricaceae, Moraceae, Hamamelidaceae, Cletraceae, Myrtaceae, Araliaceae, Aquifoliaceae, Nyssaceae, Loranthaceae, Oleaceae, Palmae, Vitaceae and others) account for about a third of the assemblage. In conclusion, a subtropical sclerophyllous plant association developed on the southeastern coast of the West Siberian basin during the late Paleocene (Plates 7–14).

CONCLUSIONS

In the early Paleocene, sediments of the Gankinskaya regional stage were deposited continuously in the southwest of the West Siberian plain, but sea level was falling so that the basin became shallower and during the Thanetian its size decreased slightly. Calcareous benthic foraminifera (occurring since the Maastrichtian) were preserved in the deepest southwestern part of the basin. The late Paleocene (Selandian) boreal transgression led to deposition of Talitskaya sediments with widespread agglutinated foraminifera.

In the western region of the West Siberian plain, close to the Ural Mountains, calcareous foraminiferal species occurred together with agglutinated taxa. Thanetian foraminifera with sugary-white, fine-grained agglutinated tests occur in the uppermost part of the Talitskaya and in the lowermost part of the Lyulinvorskaya regional stages. This assemblage is unusual because it contains a single calcareous species: *Cibicidoides favorabilis* (Vassilenko), especially in the western part of the West Siberian plain.

Data on spores and pollen suggest that in the early Paleocene, the southeastern coastal regions of the West Siberian epicontinental sea consisted of lacustrine-alluvial plains with local *Sphagnum* bogs. During the Danian and toward the end of the Thanetian the climate became drier. The largest changes in flora occurred at the boundary between the Talitskaya sediments and the lower Lyulinvorskaya sediments. During the deposition of the latter sediments, sclerophyllous subtropical plant communities were well developed, and the bogs were much less widespread.

ACKNOWLEDGMENTS

We gratefully acknowledge the scientific leaders of the conference "Climate and Biota of the early Paleogene," Drs. Scott Wing and Amanda Ash for the financial support and opportunity to participate in the work of the Conference and to publish our paper in this volume. Doctor V.M. Kabanova of Tomsk State University (Russia) made a substantial contribution to the work and allowed us to draw on her wide knowledge of the distribution and taxonomy of the plant species found in the Paleogene rocks of Western Siberia, providing us with the information on West Siberian palynological assemblages and assisting us in many ways with the palynological study. I am very much obliged to her as well as to the workers of the Palynological Laboratory of Siberian Scientific Paleontological Centre of Tomsk State University.

Reviews of the manuscript by Dr. M.A. Kaminski of University College, London (United Kingdom), Dr. G. Harrington of the University College of Cork (Ireland) and Dr. N. Frederiksen of the U.S. Geological Survey gave us much constructive criticism and suggestions for improvement, and the author appreciates and values their comments.

Special thanks go to Dr. Ellen Thomas of Wesleyan University (USA) for critical reading, valuable editorial assistance and help in improving the English text.

REFERENCES CITED

Brotzen, F., 1948, The Swedish Paleocene and its foraminiferal fauna: Sveriges Geologiska Undersökning Series C., No. 493, Årsbok, 42, No. 2, 140 p., 19 plates.

Ehremeyeva, A.I., and Belousova, N.A., 1961, Stratigraphy and foraminiferal fauna from Cretaceous and Paleogene deposits of the Urals eastern slope, of Transuralia and northern Kazakhstan: Materials on geology and mineral resources of the Urals, Volume 9: Moscow, Pervago, p. 3–189 (in Russian).

Freiman, E.V., 1960, On a new Danian–Paleocene foraminiferal assemblage of Western Siberia, *in* Halfin, L.L., ed., Proceedings of SNIIGGMS, Issue 8: Novosibirsk, Nauka, p. 212–215 (in Russian).

Frenzel, P., 2000, Die benthischen Foraminiferen der Rügener Schreibkreide (Unter-Maastricht, nordost-Deutschland): Neue Paläontologische Abhandlungen, v. 3, 350 p., 46 plates.

Gurova, T.I., and Kazarinov, V.P., 1962, Lithology and paleogeography of the West Siberian plain in connection with the oil and gas content: Moscow, Gostoptekhizdat, 296 p. (in Russian).

Ilyenok, L.L., 1967, Palynological characteristics of Paleocene in Western Siberia, *in* Samilovich, S.P., ed., Paleopalynological methods in stratigraphy: Leningrad, p. 188–198 (in Russian).

Ilyenok, L.L., Kabanova, V.M., and Kostesha, O.N., 1989, Correlation of Paleogene palynological assemblages from southeastern part of West Siberian plain, *in* Zykina, V.S., ed., Cenozoic of Siberia and the Far East: Novosibirsk, Nauka, p. 75–82 (in Russian).

Kabanova, V.M., and Kostesha, O.N., 1988, Palynostratigraphy of Cenozoic deposits of the Ket-Tym interfluve, *in* Podobina, V.M., ed., Materials on paleontology and stratigraphy of Western Siberia: Tomsk, University Publishing House, p. 126–135 (in Russian).

Nagorskaya, E.P., Aleksandrova, L.V., and Ilyenok, L.L., 1978, Paleocene and Eocene of the Ket Priobye, *in* Shatskiy, S.B., ed., Paleogene and Neogene of Siberia (paleontology and stratigraphy): Novosibirsk, Nauka, p. 93–100 (in Russian).

Podobina, V.M., 1975, The Upper Cretaceous and Paleogene foraminifera of the West Siberian plain and their significance for the stratigraphy: Tomsk, University Press, 220 p. (in Russian).

Podobina, V.M., 1995, Paleozoogeographic regionalization of Northern Hemisphere Late Cretaceous basin based on foraminifera, *in* Kaminski, M.A., et al., eds., Proceedings of the 4th International Workshop on Agglutinated Foraminifera (Krakow Poland, 1993): Grzybowski Foundation, Special Publication, v. 3, p. 239–247.

Podobina, V.M., 1998, Paleogene foraminifera and biostratigraphy of Western Siberia: Tomsk, NTL Press, 338 p. (in Russian).

Podobina, V.M., 2000a, Paleogene agglutinated foraminifera of the West Siberian biogeographical province, *in* Hart, M.V., et al., eds., Proceedings of the 5th International Workshop on Agglutinated Foraminifera (Plymouth, UK, September 6–16, 1997): Grzybowski Foundaton, Special Publication, v. 7, p. 387–396.

Podobina, V.M., 2000b, Upper Cretaceous foraminifera and biostratigraphy of Western Siberia: Tomsk, NTL Press, 388 p. (in Russian).

Podobina, V.M., Kabanova, V.M., and Kseneva, T.G., 1997, Change of micropaleontological assemblages at the Cretaceous-Paleogene boundary in Western Siberia: Palaeobotanist, v. 46(1/2), p. 156–160.

Shatskiy, S.B., 1973, Marine and continental Paleogene of Siberia: Novosibirsk, Nauka, 140 p. (in Russian).

Subbotina, N.N., editor, 1964, Foraminifera of Cretaceous and Paleogene deposits of West Siberian plain: Leningrad, Nedra, 456 p. (in Russian).

Subbotina, N.N., and Kiselman, E.N., 1961, The microfaunal zone with Anomalina praecuta in the West Siberian plain, *in* Rostovtsev, N.N., ed., Resolutions and proceedings of the Interdepartmental Session on the Revision and Refinement of Stratigraphical Charts of West Siberian plain: Leningrad, Gostoptekhizdat, p. 243–252 (in Russian).

Manuscript Accepted by the Society August 13, 2002

Printed in the USA

Geological Society of America
Special Paper 369
2003

Basal Ilerdian (earliest Eocene) turnover of larger foraminifera: Age constraints based on calcareous plankton and $\delta^{13}C$ isotopic profiles from new southern Pyrenean sections (Spain)

Victoriano Pujalte*
Xabier Orue-Etxebarria
Departamento de Estratigrafía y Paleontología, Facultad de Ciencias, Basque Country University, Ap. 644, E-48080 Bilbao, Spain
Birger Schmitz
Department of Earth Sciences, Marine Geology, Göteborg University, Box 460, SE-405 30 Göteborg, Sweden
Josep Tosquella
Departamento de Geodinámica y Paleontología, Facultad de Ciencias Experimentales, Universidad de Huelva, E-21071 Huelva, Spain
Juan I. Baceta
Aitor Payros
Gilen Bernaola
Fernando Caballero
Estíbaliz Apellaniz
Departamento de Estratigrafía y Paleontología, Facultad de Ciencias, Basque Country University, Ap. 644, E-48080 Bilbao, Spain

ABSTRACT

An important early Paleogene larger foraminiferal turnover (LFT) has long been recognized and used to draw the base of the Ilerdian stage. Yet, its chronostratigraphy is disputed: Most accounts placed it well before the Initial Eocene Thermal Maximum (IETM). More recently, it was suggested that the LFT and the IETM might be coeval; in turn, this possibility was questioned because it relied on correlation of inner neritic and deep-marine sections, and only in the latter the IETM had been recognized.

To resolve this controversy, two new sections have been studied in the Pyrenees. In the Urrobi section, composed of proximal outer-platform carbonates, the LFT occurs within a conspicuous 70-cm-thick layer of sandy marls, informally named the "*Nummulites* Band." The Mintxate section consists of distal outer-platform carbonates and thin marls, but includes a prominent 2-m-thick marly layer correlatable with the *Nummulites* Band. Planktic foraminifera demonstrate that this marly layer, and the *Nummulites* Band, pertains to the middle part of Zone P5, the interval during which the IETM took place. This dating is confirmed with calcareous nannofossils from Mintxate and with analysis of carbon isotopes of bulk samples from both sections.

*Corresponding author: gpppunav@lg.ehu.es.

Pujalte, V., Orue-Etxebarria, X., Schmitz, B., Tosquella, J., Baceta, J.I., Payros, A., Bernaola, G., Caballero, F., and Apellaniz, E., 2003, Basal Ilerdian (earliest Eocene) turnover of larger foraminifera: Age constraints based on calcareous plankton and $\delta^{13}C$ isotopic profiles from new southern Pyrenean sections (Spain), *in* Wing, S.L., Gingerich, P.D., Schmitz, B., and Thomas, E., eds., Causes and Consequences of Globally Warm Climates in the Early Paleogene: Boulder, Colorado, Geological Society of America Special Paper 369, p. 205–221.

The biotic impact of the IETM was thus greater than previously reported, having also led to migration and/or specialization of larger foraminifera. These organisms formed widespread accumulations on the early Paleogene carbonate platforms girding the Tethys Ocean. In these successions, the LFT is easily recognizable and constitutes an excellent marker for the base of the Eocene Series.

INTRODUCTION

The Initial Eocene Thermal Maximum (IETM) was a comparatively short episode (~100 k.y., Rölh et al., 2000) of very warm global temperature, superimposed on the warm late Paleocene–early Eocene global climate. The IETM, thought to have been forced by massive emissions of methane into the atmosphere (Dickens et al., 1997), is recorded in coeval deep-marine and terrestrial sequences by a prominent negative excursion of $\delta^{13}C$ values (Carbon Isotope Excursion, CIE), the base of which has been proposed as the Paleocene-Eocene (P-E) boundary by the International Subcommission on Paleogene Stratigraphy (Luterbacher et al., 2000). The IETM was also a period of important biotic turnovers, highlighted by an important immigration of mammals in northern continents (Clyde and Gingerich, 1998) and by a major global extinction of deep-marine benthic foraminifera (Thomas, 1990, 1998), as well as transient events in planktic foraminifera (Lu and Keller, 1993; Kelly et al., 1996), calcareous nannofossils (Aubry, 1998) and dinoflagellates (Bujak and Brinkhuis, 1998).

An important turnover of early Paleogene larger foraminifera (LFT) was noticed long ago (Hottinger, 1960), but was generally thought to be older than (and unrelated to) the IETM (e.g., Hottinger, 1998; Molina et al., 2000). More recently, Pujalte et al. (2000) and Orue-Etxebarria et al. (2001) suggested that the LFT and the IETM could be coeval. However, these two papers were based on the comparison of 3 sections (Campo, Ermua and Zumaia) that are far apart and composed of very different biofacies and lithofacies (Fig. 1). Campo, the parastratotype of the Ilerdian stage and one of the localities where the LFT was first identified, is an inner platform section rich in larger foraminifera (mainly alveolinids). Ermua and Zumaia are deep marine sections with abundant planktic foraminifera and calcareous nannofossils, but Ermua also contains larger foraminifera within carbonate turbidites derived from coeval carbonate platforms (Orue-Etxebarria et al., 1996). The position of the base of the CIE is known in Zumaia and Ermua (Schmitz et al., 1997, 2001), but not in Campo. Therefore, the synchrony between the LFT and the IETM proposed by Pujalte et al. (2000) and Orue-Etxebarria et al. (2001) relies on the correlation between these sections, a correlation questioned by others (e.g., Arenillas and Molina, 2000).

In this paper we report the results of biostratigraphic and isotopic analyses of two additional sections, Urrobi and Mintxate, situated in a (paleo)geographic position intermediate between Campo and Ermua-Zumaia (Fig. 1). The new data provide direct evidence of the time-equivalence between the LFT and the IETM, and help to illustrate the characteristics of the LFT in the nummulitid faunas in a proximal outer-platform setting.

GEOLOGICAL SETTING

Lower Paleogene deposits of a large range of facies, from terrestrial red beds to deep marine hemipelagites, outcrop widely in the Pyrenees. These rocks were deposited in an east-west elongated gulf, opening westwards into the Bay of Biscay, and surrounded by carbonate platforms to the south, east and west (Fig. 1A), and uplifted and exposed in the course of the middle Eocene–Oligocene Pyrenean Orogeny. Upper Thanetian–lower Ilerdian deposits of the Campo section mainly consist of carbonates of shallow-marine and transitional facies, characteristic of an inner platform setting (Payros et al., 2000; Orue-Etxebarria et al., 2001, and references therein). The coeval Ermua section is composed of carbonate breccias, carbonate turbidites and hemipelagic limestones, a typical base-of-slope association (Pujalte et al., 1998; Schmitz et al., 2001), whereas the Zumaia section consists of alternations of hemipelagic marls and limestones, accumulated on the basin floor (Canudo et al., 1995; Pujalte et al., 1998; Baceta et al., 2000).

The sections here studied, Urrobi and Mintxate, are both in the southwest Pyrenees, in the northern part of Navarra province. The former is in a recently abandoned quarry, between the villages of Nagore and Orbaiz, in the southern part of a domal structure dissected by the Urrobi river (Fig. 1B). The latter is along the local road that enters the Mintxate gorge, between the villages of Isaba and Uztarroz, in the southern flank of an unnamed anticline (Fig. 1B). The two sections are representative of, respectively, proximal and distal parts of the outer carbonate platform, as demonstrated by their facies (Fig. 2) and their paleogeographic position: Mintxate is situated just 1 km south of the platform margin (marked by a rapid facies transition to upper-slope marl-limestone alternations rich in calcareous plankton), Urrobi ~15 km to the south of the platform margin (Fig. 1A).

STRATIGRAPHY

A lithostratigraphic summary of the two sections is shown in Figure 2, along with a cross-section obtained by detailed mapping of the intervening outcrops. Three unconformity-bounded depositional sequences (DS) are recognized, coded DS Th-1, DS Th-2, and DS Il-1 (Pujalte et al., 2000). The bulk of DS Th-

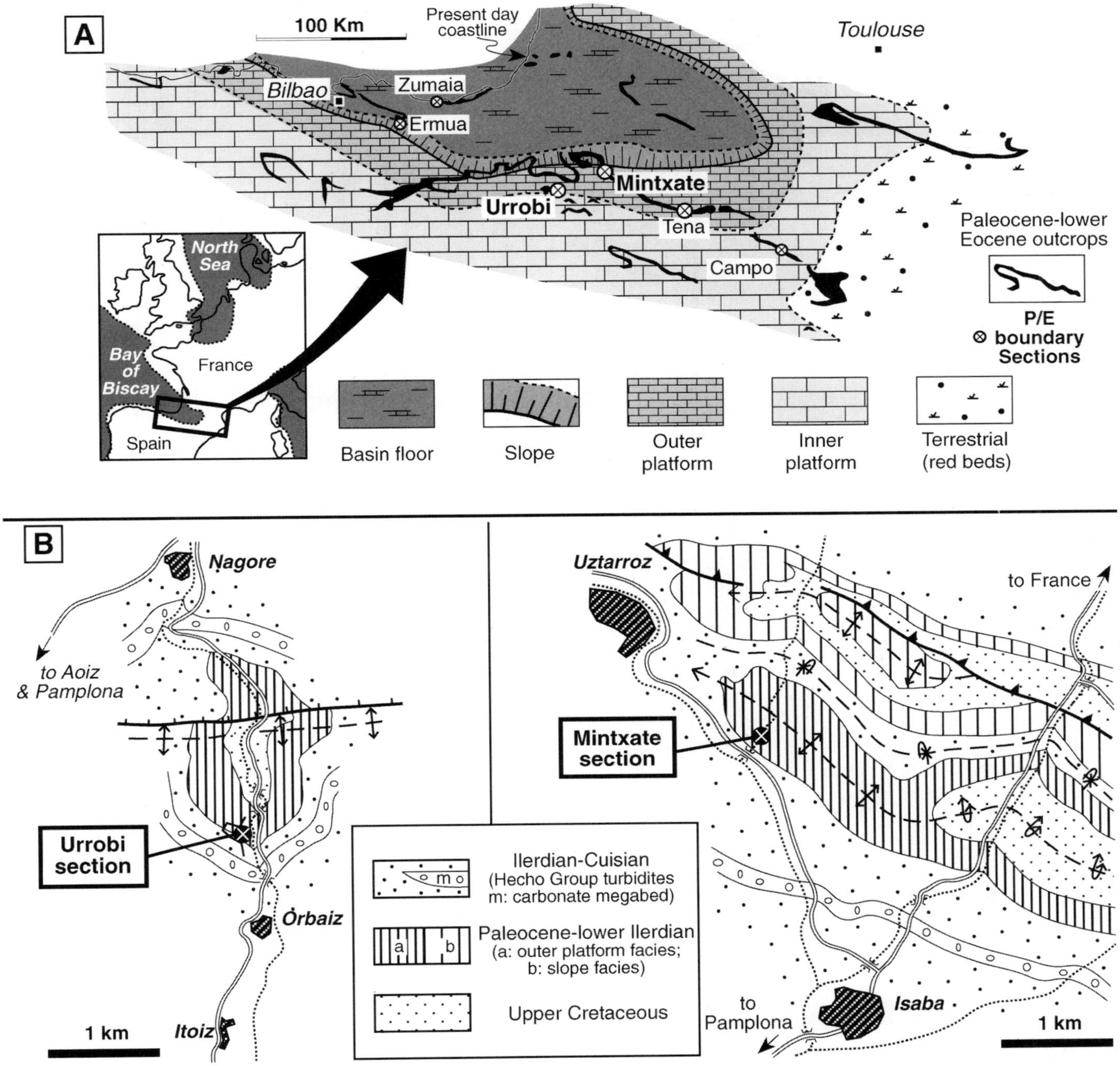

Figure 1. A: late Paleocene–early Eocene paleogeography of the Pyrenean area, outcrop map of Paleocene–lower Eocene strata and location of sections. B: Geological map and location of the Urrobi and Mintxate sections.

1 is made up of massive coralgal carbonates that grade northward (seaward) into well-bedded wackestone. Additionally, a few meters of sandstone occur at the base of the sequence in the Urrobi section. The boundary between DS Th-1 and DS Th-2 is abrupt and irregular in Urrobi, denoting erosion and perhaps emersion, whereas in Mintxate it is a flat surface with frequent borings, indicative of a period of nondeposition under marine conditions. In the Urrobi section, DS Th-2 is 11 m thick, and consists of calcareous sandstones and sandy grainstones rich in larger foraminifera (mainly nummulitids, see below). In the Mintxate section DS Th-2 is only 6.5 m thick, and composed of sandy wackestones alternating with sandy marls with planktic foraminifera and calcareous nannofossils.

In the two studied sections, the boundary between DS Th-2 and DS Il-1 is a sharp surface, locally lined by a weakly developed ferruginous crust. In Urrobi, this surface is overlain by

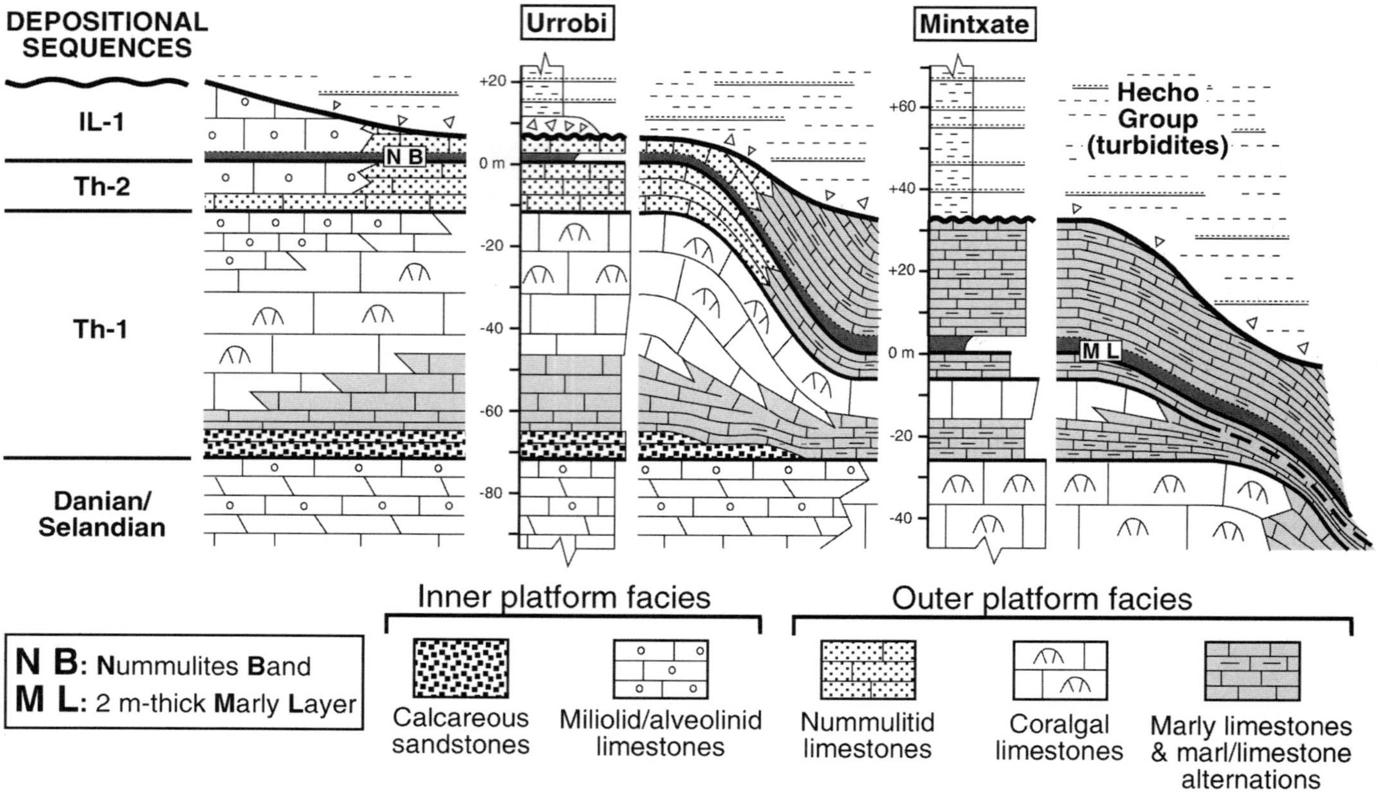

Figure 2. Reconstructed architecture of the Thanetian and early Ilerdian depositional sequences of the South Pyrenean area across the Urrobi and Mintxate sections.

a 0.7 m thick layer of dark grey sandy marls with abundant larger foraminifera. This layer, informally named the *Nummulites* Band, is a widespread feature in proximal outer platform successions of the southern Pyrenees. Being less resistant to erosion than the rocks above and below it, the *Nummulites* Band forms a conspicuous notch in the landscape that facilitates its identification. It has thus been traced northwards into a 2-m thick marly layer (partly covered) situated in the base of DS Il-1 in the Mintxate section (Fig. 2). The remainder of DS Il-1 in Urrobi is composed of bioclastic grainstones rich in nummulitids, whereas in Mintxate it is made up of a monotonous stack of wackestones and mudstones. In both sections, the upper part of the sequence is sharply truncated by an erosional unconformity, which records the onset of the foreland basin stage in the area, with the concurrent transformation of the former carbonate platform into a deep-water flysch trough (Labaume et al., 1985; Payros et al., 1999).

MATERIAL AND METHODS

Larger foraminifera were studied in the Urrobi section, where they are ubiquitous. Those within the *Nummulites* Band were picked from washed samples and studied following a two-step procedure: First, their outer test features (diameter and shape, morphology and arrangement of septal filaments and granules, etc.) were examined with a binocular microscope; then, they were split along the equatorial section to study their inner features such as number of whorls, rate of opening of the spire (whorl radius), number of chambers per whorl, septal and chamber shape, and the proloculus diameter of megalospheric forms. They could thus be confidently classified at the specific level. However, this procedure is not feasible in the well-indurated limestones that make up most of the succession. Therefore, larger foraminifera were searched directly in the field with a hand lens and then in thin sections cut from 52 selected samples. In many cases 2 thin sections were cut per sample, oriented at right angles to obtain axial and equatorial sections of the larger foraminifera. The vertical distribution of the samples was unequal, the sampling being especially dense in the interval comprising a few meters below and above the DS Th-2/DS Il-1 boundary (Fig. 3).

To search for planktic foraminifera, all of the few marly intercalations of the two studied sections were sampled. Only two samples could be obtained from Urrobi, respectively from the base and the top of the *Nummulites* Band (URR-1 and URR-2). From Mintxate, 11 samples were collected, 7 from DS Th-2, and 4 from the lower part of DS Il-1. At Mintxate the 2-m thick marly layer at the base of DS Il-1 is poorly exposed, so 3 additional

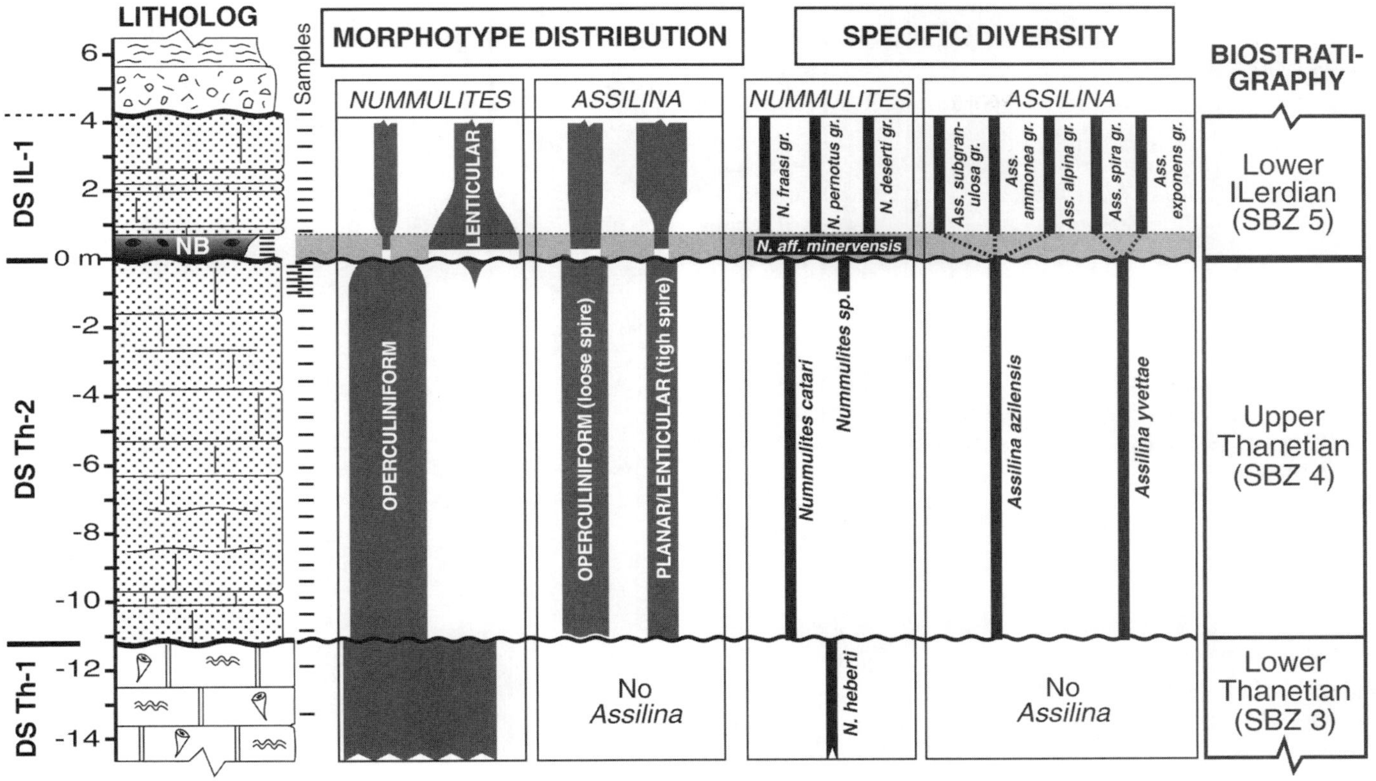

Figure 3. Morphotype distribution of specific diversity of nummulitids in the Urrobi section, derived mainly from thin section studies (location of samples to the right of the litholog; five samples from DS Th-1were studied, three of them collected below the part of the section shown in the figure).

samples from the lower and middle parts of this layer were collected in an auxiliary section situated to the north of Isaba (Fig. 1B). About 500 grams of each sample were washed and screened to obtain residues of a 100–500 µm size range, which were searched under a binocular microscope.

The 11 samples from Mintxate and the three from Isaba were also processed for calcareous nannofossils. Smear slides were prepared from solutions of raw materials following the pipette strew procedure. The slides were studied using a Leica DMPL light microscope, routinely at a 1500× magnification, although a 2000× magnification was occasionally used to recognize some small-sized species. Since autochthonous calcareous nannofossils were comparatively scarce in all samples, we searched systematically for them in at least six tracts per slide to ensure a representative composition of the assemblages.

Isotopic analyses were performed on whole-rock samples following procedures as in studies on the Zumaia and Ermua sections (Schmitz et al., 1997, 2001). A total of 44 samples across ~50 m of the Urrobi section were analyzed as well as 25 samples across ~35 m of the Mintxate section. The isotopic profiles extend much further down in the Thanetian than the foraminiferal profiles. About 3–4 g of sediment was ground in an agate mortar. A fraction of 600–1000 mg was separated and roasted in vacuum at 400 °C. The isotopic analyses were performed with a VG Prism Series II mass spectrometer attached to an Isocarb preparation system. For 12 NBS-19 standards in a carousel with 32 samples the mean values and standard deviations were typically 1.96 ±0.02‰ for $\delta^{13}C$ and –2.16 ±0.04‰ for $\delta^{18}O$. Calibration to PDB was via NBS-19 ($\delta^{13}C$ = 1.95‰ Vienna PDB and $\delta^{18}O$ = –2.20‰ VPDB).

RESULTS

Biozonation with larger foraminifera and the turnover of nummulitids

Almost all of the 52 studied thin-sections from Urrobi include nummulitids, and therefore our study has been centered on this particular group of larger foraminifera. In addition, orthophragminids occur in 38 of the thin-sections, while alveolinids, *Miscellanea* sp. and *Sphaerogypsina* sp. appear occasionally. As explained above, accurate species determinations could only be attained for the nummulitids of the *Nummulites* Band. However, the study of thin-sections has permitted the distinction of the four main types of nummulitids occurring in the section, namely lenticular *Nummulites* (= *Nummulites sensu stricto*), operculiniform *Nummulites,* operculiniform *Assilina* (loose spire, formerly *Operculina*) and tight spire *Assilina* (= *Assilina sensu stricto*). In a few cases, the study of thin-sections allowed species identifications. The results are summarized in two columns in Figure 3: The left-hand column (Morphotype Distribution) is a semiquantitative plot of the relative abundance of the four nummulitid types, while the right-hand column is a qualitative plot of those nummulitid species determined with a reasonable degree of confidence. The latter plot incorporates also earlier data of Tosquella (1995), from Urrobi and neighboring sections.

Samples from DS Th-1 contain only one type of operculiniform nummulitid, characterized by an evolute growth pattern, thin and smooth test walls and, in axial sections, parallel to near parallel outer walls, all features typical of *Nummulites heberti.* In addition, DS Th-1 samples include abundant orthophragminids, some of which can be classified as *Discocyclina seunesi,* and less often *Miscellanea* sp. The association can thus be assigned to Shallow Benthic Zone (SBZ)-3 of Serra-Kiel et al. (1998), attributed to the early Thanetian.

Samples from the bulk of DS Th-2 still comprise a large proportion of operculiniform *Nummulites* (probably, *Nummulites catari*), but also include two other morphotypes with evolute growth patterns as in *Assilina* (Fig. 4A). The most common, thick-walled and with loose spire, presents in axial sections a small but distinctive widening of the test in the lowermost part of each spire whorl, where the outer walls meet the next innermost whorl, features characteristic of the operculiniform *Assilina azilensis.* The other morphotype is attributable to *Assilina yvettae,* because it has a tight spire, lenticular test shape and a radial exterior pattern of discontinuous segments that reflects the shape of internal septa. This nummulitid association has a significant addition in the uppermost 80 cm of DS Th-2. This addition, which heralds the LFT, consists in the lowest appearance of lenticular *Nummulites* in the studied succession (Figs. 3 and 4B). Nevertherless, the cooccurrence of *Nummulites catari, Assilina azilensis* and *Assilina yvettae* places DS Th-2 within the late Thanetian SBZ-4 of Serra-Kiel et al. (1998).

The nummulitid association of the *Nummulites* Band is somewhat peculiar. First, it largely comprises a single species of lenticular *Nummulites,* which makes up between 60%–80% of the total nummulitid population (Fig. 3). This species has been classified as *Nummulites* aff. *minervensis* (*sensu* Schaub, 1981),

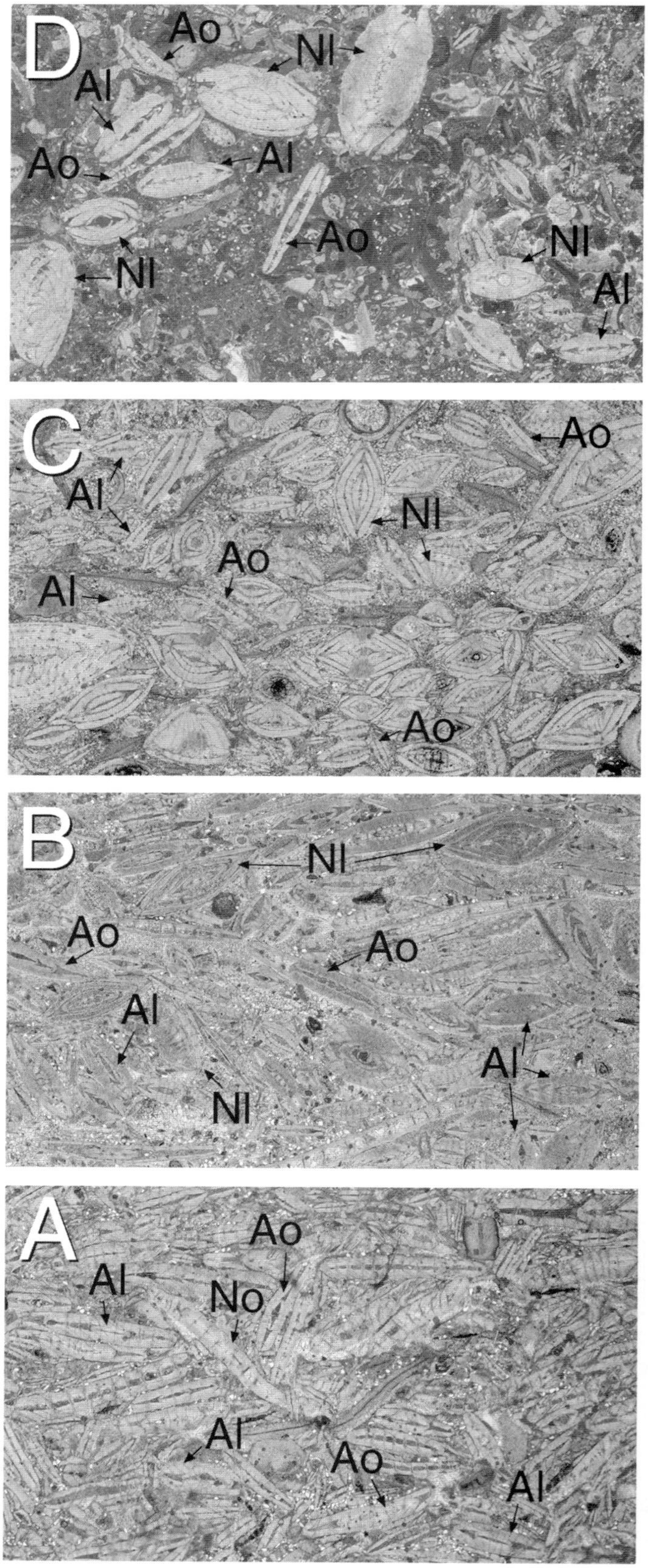

Figure 4. Representative larger foraminiferal assemblages in thin sections from the Urrobi section. The field of view in each photo is 15 × 9 mm. A: Middle part of DS Th-2: Assemblage composed of lenticular ("true") *Assilina* (Al), operculiniform *Assilina* (Ao), and operculiniform *Nummulites* (No). B: Uppermost 10 cm of DS Th-2: Similar to 4A, but including several specimems of lenticular *Nummulites* (Nl). C: Upper part of the *Nummulites* Band: Assemblage dominated by A-forms of lenticular *Nummulites* (Nl). D: From DS IL-1, 2 m above the *Nummulites* Band: Operculiniform *Assilina* (Ao), lenticular *Assilina* (Al) and lenticular *Nummulites* (Nl) in similar proportions.

its defining features being illustrated in the Appendix. Second, it is dominated by A-forms (Fig. 4C), which are the smaller morphotypes of each nummulitid species, characterized by a large proloculus and originated by asexual reproduction (schizogony) from either larger B-forms or other A-forms (Appendix). Third, adult larger foraminifera are scarce and minute in the lower 10–15 cm of the *Nummulites* Band, but increase rapidly in abundance and size upwards. Such features suggest a quick colonization of a favorable ecosystem by a new, well-adapted species (excursion fauna?), with a strategy of asexual reproduction to better exploit it rapidly. Interestingly, the search for planktic foraminifera revealed that the URR-1 residue, from the bottom of the *Nummulites* Band, contained few juvenile tests of larger foraminifera, whereas these juvenile tests numbered ~6500 per gram in the URR-2 residue from the upper part of the Band, a further indication of a blooming event. However, more data are necessary to clarify this particular point.

The preeminence of lenticular *Nummulites* persists in the 1.5 m above the *Nummulites* Band, but higher up in the sequence the relative abundance of *Nummulites* and *Assilina* become almost equal (Figs. 3 and 4D). The nummulitid association of DS Il-1 is very different from the one in the underlying sequences, and much more diverse, comprising three groups of lenticular *Nummulites (fraasi, pernotus* and *deserti),* three of operculiniform *Assilina (subgranulosa, ammonea* and *alpina*), and two of *Assilina s.str.* (*spira* and *exponens)* (Figs. 3 and 4). This association pertains to the early Ilerdian SBZ-5 of Serra-Kiel et al. (1998), and is a clear proof that the LFT, the important faunal turnover noticed long ago by Hottinger (1960), Schaub (1969, 1981) and Hottinger and Schaub (1960), occurred between these two sets of samples. As a result of the LFT, Thanetian and Ilerdian nummulitid associations can be readily differentiated in thin sections (Fig. 4), and even with a hand-lens. It is not surprising therefore that specialists of larger foraminifera have recommended to place the P-E boundary at the base of the Ilerdian stage (Hottinger, 1998).

Dating the **Nummulites** *Band with planktic foraminifera*

The 2 residues from Urrobi, and 6 out of the 11 from Mintxate, contained planktic foraminifera, albeit few or very few and with poor to very poor preservation, so that only qualitative analysis was possible. The residues from the three Isaba samples had no determinable planktic foraminifera. The sample from the bottom of the *Nummulites* Band (URR-1 in Figure 5) contained just a few identifiable planktic foraminifera, most with a wide stratigraphic range (e.g., *Morozovella subbotinae, Morozovella edgari, Subbotina velascoensis* and *Subbotina tri-*

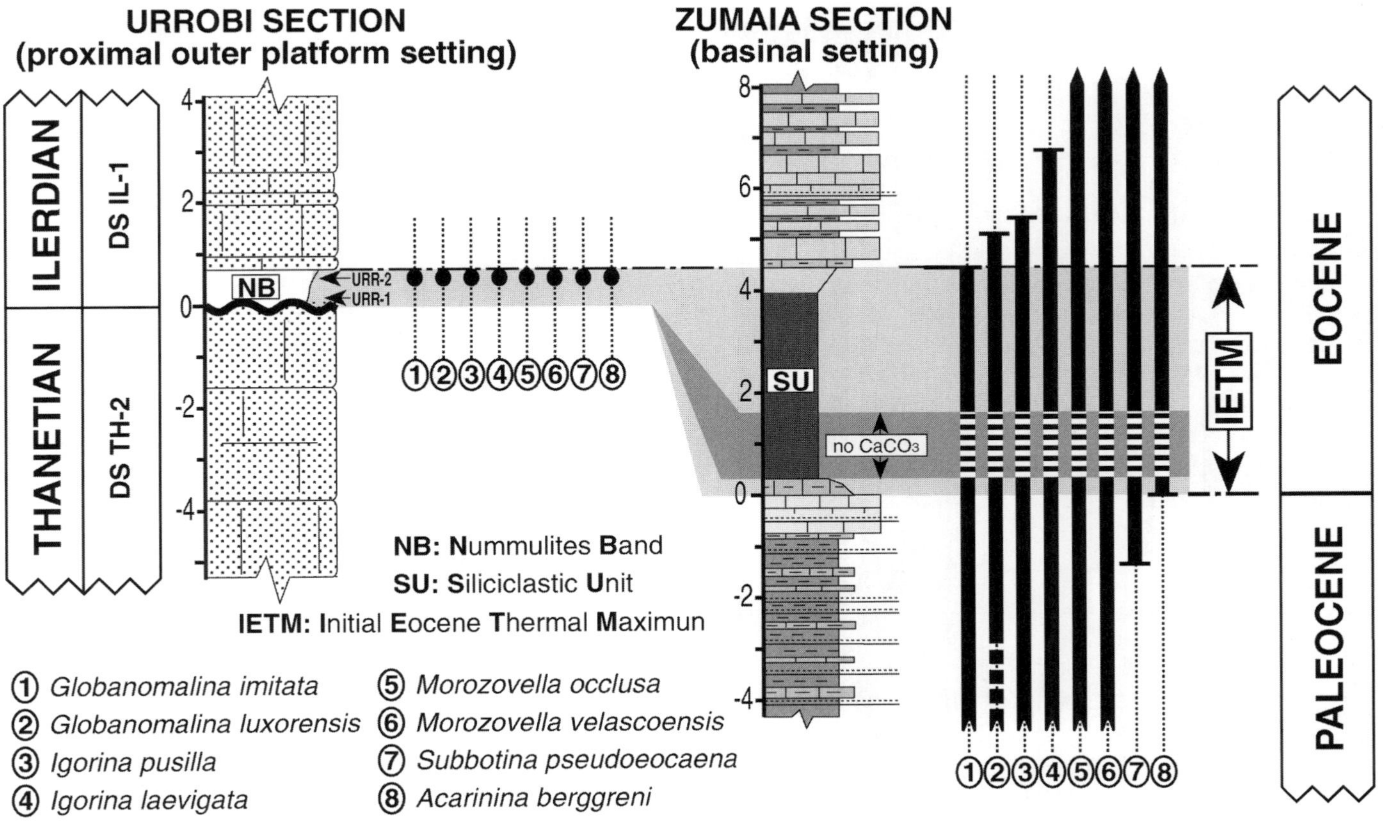

Figure 5. Correlation of the Urrobi and Zumaia sections based on planktic foraminifera.

angularis). However, the sample from the upper part of the Band (URR-2) yielded ~230 specimens per gram of dry residue, and allowed the determination of 20 species: The eight listed in Figure 5 (which are the most significant ones), and *Subbotina velascoensis, Subbotina linaperta, Subbotina triangularis, Morozovella subbotinae, Morozovella edgari, Morozovella acuta, Morozovella aequa, Acarinina nitida, Acarinina soldadoensis, Acarinina wilcoxensis, Acarinina pseudotopilensis* and *Acarinina subsphaerica.*

To determine its chronostratigraphic position, this association was compared with the vertical distribution of planktic foraminifera across the P-E boundary interval in the Zumaia section (Fig. 5) (Orue-Etxebarria et al., 2001). This section was deposited in a basinal setting (Fig. 1) and is considered to have a continuous or nearly continuous record (Canudo et al., 1995; Schmitz et al., 1997, 2001; Baceta et al., 2000). It mainly consists of hemipelagic marls and limestones, but includes a ~4 m thick intercalation of calcite-depleted claystones, referred to as the Siliciclastic Unit by Schmitz et al. (2001), which was deposited during the IETM (see below). In the URR-2 sample, species coexist that in Zumaia have their highest occurrences just above the Siliciclastic Unit (e.g., *Globanomalina imitata, Globanomalina luxorensis* and *Igorina laevigata*) and species that have lowest occurrences just below or at the bottom of it (e.g., *Subbotina pseudoeocaena* and *Acarinina berggreni*). Such cooccurrence implies that the upper part of the *Nummulites* Band, from where the URR-2 sample was collected, must have an age within the time span of the Siliciclastic Unit/IETM.

Preservation of planktic foraminifera in all Mintxate samples is very poor, and specimens are often difficult to separate from their surrounding rock matrix, particularly in samples from DS Il-1. Therefore, only from DS Th-2 samples (chiefly from MTX-2 and 6, stratigraphic position in Figure 6), a reasonable number of determinations could be obtained (14 different

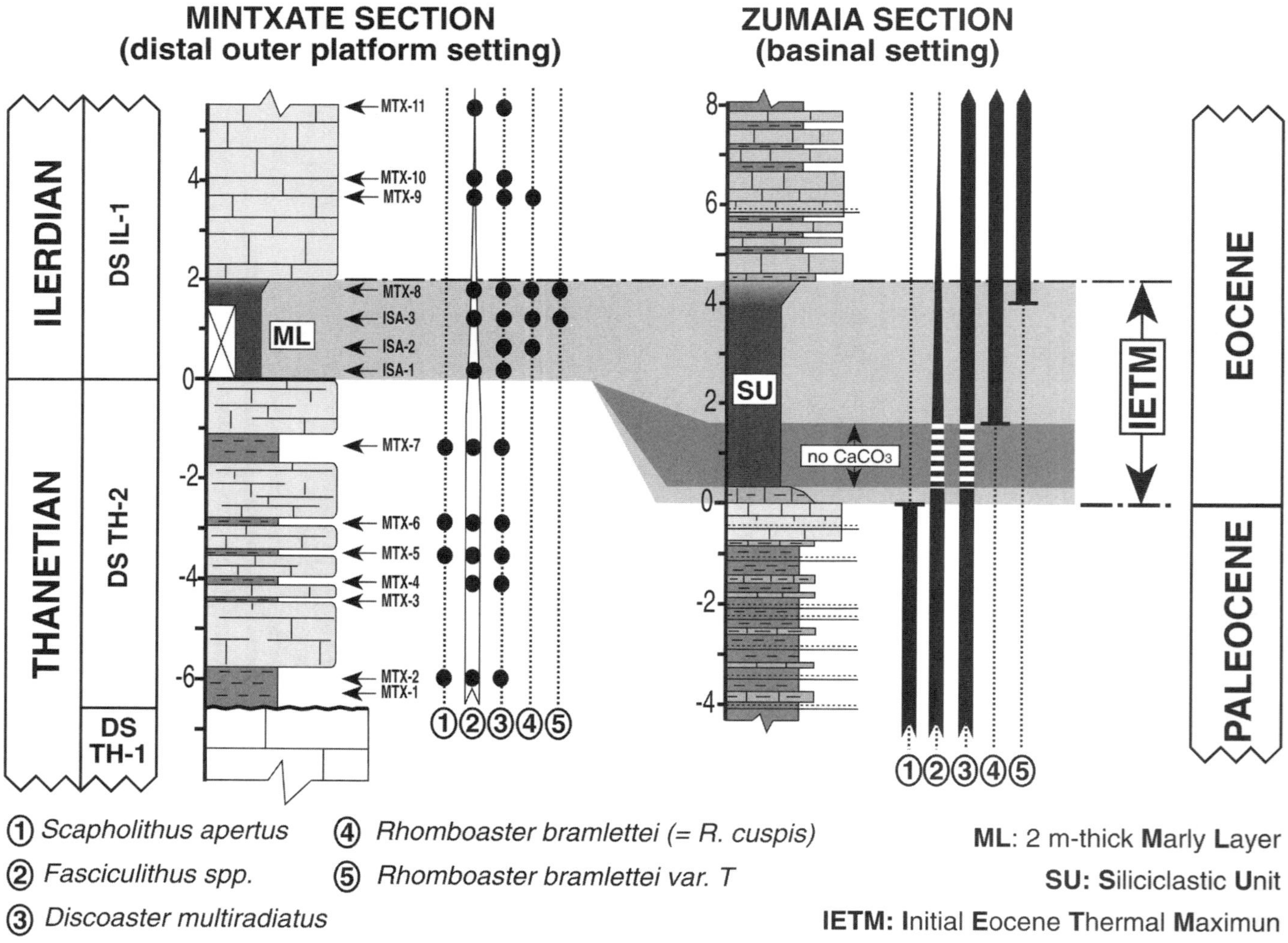

Figure 6. Correlation of the Mintxate and Zumaia sections based on calcareous nannofossils.

species). The most significant of these are *Morozovella subbotinae, Morozovella occlusa* and *Morozovella velascoensis,* which indicate the P5 Zone of Berggren and Norris (1997), and *Globanomalina pseudomenardii* that denotes a position below the IETM, since its highest occurrence in Zumaia occurs just below the Siliciclastic Unit. The remaining species from DS Th-2 samples are *Subbotina linaperta, Subbotina velascoensis, Globanomalina chapmani, Globanomalina imitata, Acarinina nitida, Acarinina subsphaerica, Acarinina soldadoensis, Morozovella aequa* and *Morozovella acuta.*

The results from DS Il-1 samples (MTX-8 to MTX-11 in Figure 6) are less conclusive, for only five species could be determined (*Subbotina linaperta, Subbotina velascoensis, Acarinina nitida, Acarinina soldadoensis* and *Morozovella aequa*). All of these span the P-E boundary, and therefore their occurrence does not contradict the correlation here proposed, but they have long stratigraphic ranges.

Dating the 2-m thick marly layer with calcareous nannofossils

Autochthonous Paleogene species of calcareous nannofossils occur in most of the studied samples from Mintxate and in the three samples from Isaba, but their preservation is poor to very poor in all samples, and there are very few, usually fewer than one species per 10 fields of view (for comparison, coeval samples from Zumaia contain between 15 and 20 spc/f.v.). Therefore, only qualitative analyses could be made. The number of taxons per slide ranges between 18 and 26 in DS TH-2 samples, and between 6 and 21 in DS Il-1 samples. Altogether, 43 different species were identified. As with the planktic foraminifera, the vertical distribution of the most significant assemblages was compared with that in Orue-Etxebarria et al. (2001) (Fig. 6).

Calcareous nannofossil assemblages from samples MTX-1 to MTX-7 are dominated by *Coccolithus pelagicus* and *Toweius pertusus,* but also contain abundant *Discoaster multiradiatus* and *Zygrhablithus bijugatus.* In addition, *Fasciculithus* are common, six different species having been identified. One of these (*Fasciculithus alanii*), which occurs in sample MTX-5, is particularly meaningful, for in Zumaia and elsewhere (Monechi et al., 2000) it has its LAD just below the NP9/NP10 zonal boundary of Martini (1971), as does *Scapholithus apertus,* which appears in samples MTX-2, -5, -6 and -7, but not higher. This calcareous nannofossil association therefore places DS Th-2 within Zone NP9, and below the IETM (Fig. 6).

Assemblages from the Isaba (ISA-1, -2, -3) and MTX-8 samples, all from the 2-m marly layer, are dominated by *Coccolithus pelagicus,* with subordinate amounts of *Toweius pertusus, Discoaster multiradiatus, Zygrhablithus bijugatus* and *Sphenolithus primus.* The abundance and number of *Fasciculithus* specimens decrease relatively to those in DS Th-2 (just two species have been observed), a decline that in the Zumaia section coincides with the IETM (Fig. 6). More significant, the samples from this interval contain specimens of the *Rhomboaster-Tribrachiatus* lineage: Rhomboedric forms appear in sample ISA-1; clear *Rhomboaster bramlettei* "short arms" of Angori and Monechi (1996) occur in ISA-2; and *Rhomboaster bramlettei* "var. T" of Angori and Monechi (1996) were first observed in ISA-3. Thus, the number of samples studied at Mintxate-Isaba is much smaller than at Zumaia, but the pattern of appearance of these species is identical, and suggests a time equivalence between the marly layer and the Siliciclastic Unit (Fig. 6). A further decrease in Fasciculiths occurs immediately above the marly layer, with just one species (*Fasciculithus tympaniformis*) in MTX-11, similar to observations in Zumaia (Fig. 6).

Stable isotopes

Reconstructions of past isotopic variations in sea water should ideally be based on analyses of well-preserved, monospecific calcite tests or shells. However, in Pyrenean P-E sections it is usually not possible to recover such material, and in its absence whole-rock analyses may give at least some information. Previous studies have shown that $\delta^{13}C$, particularly in lithified limestone, may accurately reflect paleoseawater isotopic trends (Schmitz et al., 1997, 2001).

The P-E marine limestones in the Pyrenean area were indurated during early diagenesis and have been closed systems with respect to carbon exchange. The limestones in the Zumaia and Ermua sections give $\delta^{13}C$ trends that mimic records measured on well-preserved marine planktic and deep-benthic foraminiferal tests (e.g., Corfield and Norris, 1998). Oxygen isotopes in the same limestones have been much more affected by diagenesis and rarely give any meaningful information about original conditions.

At Urrobi and Mintxate the samples from DS Th-1and Th-2 intervals dominated by hard limestone give $\delta^{13}C$ values similar to those in global $\delta^{13}C$ curves for the middle to late Paleocene (Figs. 7–8). In both sections, $\delta^{13}C$ values as high as 3‰ are measured in the lower and middle part of DS Th-1. Such high $\delta^{13}C$ values characterize the lower–middle Thanetian at Zumaia and other sections worldwide, and are unusual elsewhere in the stratigraphic record (e.g., Schmitz et al., 1997; Thompson and Schmitz, 1997; Charisi and Schmitz, 1998). The $\delta^{13}C$ values show a prominent decline by ~3‰ from the upper part of the DS Th-1 to the top of DS Th-2, in good agreement with the global $\delta^{13}C$ trend for the interval through the upper Thanetian to the base of the Eocene. The general $\delta^{13}C$ pattern throughout DS Th-1 and Th-2 at Urrobi and Mintxate together with the micropaleontological data in Figures 5 and 6 convincingly show that the P-E boundary must lie near or at the boundary between DS Th-2 and Il-1.

The origin of the negative $\delta^{13}C$ excursion of ~1‰–2‰ in the lower part of the *Nummulites* Band at Urrobi is ambiguous. The decline may reflect diagenesis or the global CIE at the P-E boundary (or both). The occurrence of this $\delta^{13}C$ decline in sandy

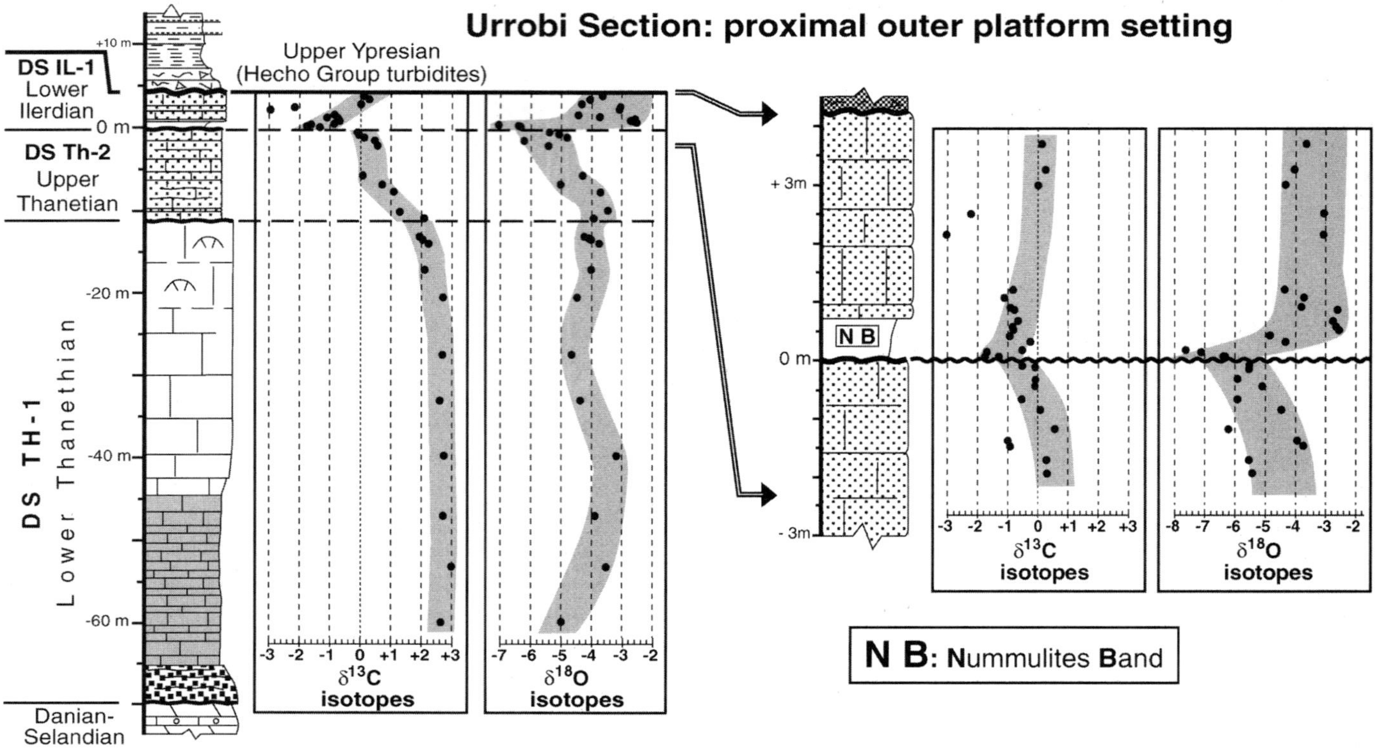

Figure 7. Carbon and oxygen isotope profiles from whole-rock samples across the Urrobi section: Left, general profile; right, detailed profile across the DS Th-2/Il-1 boundary.

marls and its coincidence with a prominent decline in $\delta^{18}O$ indicates that it may primarily be a diagenetic artifact. Spurious low $\delta^{13}C$ values at other levels within 2 m above and below the DS Th-2/Il-1 boundary support the hypothesis that diagenetic calcite (vein calcite?) plays a role in this section. However, in the upper part of the *Nummulites* Band and in the immediately overlying beds, $\delta^{13}C$ values as low as –1‰ occur in limestone, a lithology that is more favorable for preserving $\delta^{13}C$ signals. Here the low $\delta^{13}C$ values coincide also with more positive $\delta^{18}O$ values supporting that the later part of the CIE is registered. A similar trend is seen at Mintxate where one of the lowest $\delta^{13}C$ values in the section is measured in the limestone immediately overlying the 2-m marly layer at the base of DS Il-1.

DISCUSSION

Is the LFT a true evolutionary event?

In lower Paleogene successions of the southern Pyrenees, boundaries of larger foraminiferal biozones often coincide with sequence boundaries (e.g., Eichenseer and Luterbacher, 1992; Pujalte et al., 2000), as also observed in the Urrobi section (Fig. 3). This coincidence is due to the fact that most sequence boundaries in shallow-marine settings are also hiatal surfaces, at which evolutionary lineages of larger foraminifera become telescoped (e.g., Luterbacher, 1998, his Figure 8).

The important change in the nummulitid association that marks the LFT in the Urrobi section occurs at a sequence boundary (i.e., the DS Th-2/Il-1, or Thanetian–Ilerdian boundary, Figure 3), the hiatal character of which has been demonstrated in several South Pyrenean sections. In the Campo section, for instance, where the LFT is reflected by an abrupt change of alveolinid faunas, this boundary is a sharp surface with abundant *Microcodium* remains, proof of subaerial exposure (Molina et al., 1992; Payros et al., 2000; Orue-Etxebarria et al., 2001). If the Thanetian–Ilerdian hiatus would involve an extended time gap, the LFT might be an artifact. However, the combination of biostratigraphic and isotopic data from Urrobi and Mintxate indicates that the DS Th-2/Il-1 hiatus is comparatively small and the LFT represents a true faunal turnover.

Two lines of reasoning support this conclusion. First, the lowest occurrence of lenticular *Nummulites* in the uppermost 80 cm of DS Th-2 in Urrobi documents that the Thanetian part of this section is complete or near complete. According to Orue-Etxebarria et al. (1996), in the Ermua section (location in Figure 1) lenticular *Nummulites* first occur in turbidites intercalated in a comparatively thin unit of pure limestone (<1 m) of characteristic greenish coloring (glauconite). This "green limestone"

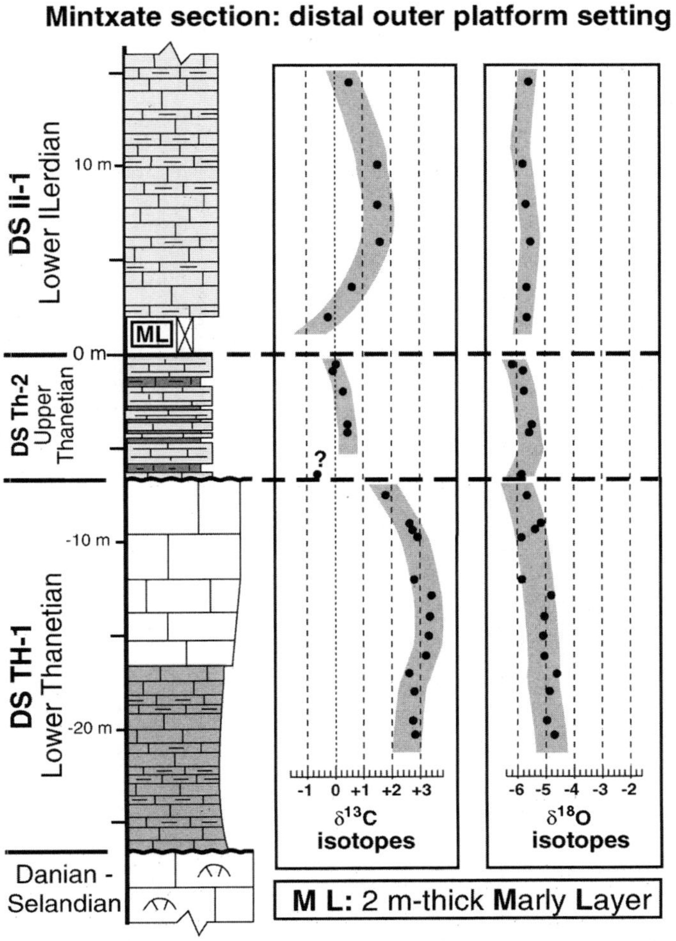

Figure 8. Carbon and oxygen isotope profiles from whole-rock samples across the Mintxate section.

unit is found in many other sections of the Basque basin, including Zumaia (Schmitz et al., 2001), always capping the Paleocene succession (Fig. 9). Second, the biostratigraphic information (Figs. 5 and 6), and the isotopic profiles (Figs. 7 and 8), place the Urrobi *Nummulites* Band and the Mintxate 2-m marly layer within the same time span as the Zumaia Siliciclastic Unit. Therefore, the DS Th-2/Il-1 boundary hiatus can only encompass a fraction of the latter unit.

The duration of the time gap of the DS Th-2/Il-1 hiatus can be further constrained by comparing isotopic information from the Zumaia section (Schmitz et al., 1997, 2001) with the isotopic profile in Röhl et al. (2000). At Zumaia, there is an ~1‰ gradual decline in $\delta^{13}C$ below the Siliciclastic Unit. In the 30–40 cm of marls between the "green limestone" and the Siliciclastic Unit, there is a rapid decrease in $\delta^{13}C$ from +1.2‰ to –0.6‰ that, according to Schmitz et al. (1997, 2001), marks the onset of the CIE and the IETM. Within the Siliciclastic Unit itself, $\delta^{13}C$ values are extremely negative (down to –4.5‰), although their reliability is questionable due to the low $CaCO_3$ content of this unit. In the uppermost part of the unit, where the $CaCO_3$ content reaches 25%, the isotopic value remains negative (~ –2‰), returning to positive values at the base of the overlying alternation of limestones and marls. Higher up, $\delta^{13}C$ values increase for ~1 m, then stabilize around +1‰ in the remaining part of the succession.

According to Röhl et al. (2000), $\delta^{13}C$ values decreased rapidly at the onset of the IETM at ODP Site 690 and remained low for ~84 k.y. (4 precessional cycles). After that, $\delta^{13}C$ values increased for nearly 150 k.y. (7 precessional cycles), rapidly at first, then more slowly. These two parts of the IETM have a distinct lithology, with the lower one characterized by low $CaCO_3$ contents and high sedimentation rates, whereas sediments of the upper part have high (>85%) $CaCO_3$ contents. We therefore speculate that the Siliciclastic Unit of Zumaia probably reflects the lower (low-calcareous) part of the IETM interval at Site 690, and might thus represent an 84 k.y. time span. If correct, our guess would imply that the extent of the Thanetian-Ilerdian hiatus at Urrobi and Mintxate would be <50–60 k.y. (Fig. 9), a comparatively short time by geological standards, and strong evidence to qualify the LTF as a true faunal turnover.

Chronostratigraphic implications of the LFT

Larger foraminifera were important rock builders in warm and temperate Paleogene shallow seas, sometimes contributing to up to 80% of the sediment volume (Hottinger, 1997). They abound therefore in the extensive carbonate sequences deposited on the ancient margins of the Paleogene Tethys Ocean, from the Pyrenees to Pakistan (Pignatti et al., 1998). Larger foraminifera are also of prime importance for biostratigraphic zonation of these sequences, the term "Nummulitique" having in fact been used as a synonym of Paleogene (e.g., Mangin, 1960). However, a reliable intercalibration between larger foraminifera and calcareous plankton time scales need still to be worked out, although the paper by Serra-Kiel et al. (1998) represents a valuable attempt in that direction. For this reason, when a zonation is made with larger foraminifera, stage names different from the standard ones are sometimes used. This is the case, for instance, of the Ilerdian and Cuisian stages. The Cuisian Stage was defined by Dollfus (1880), using the "Sables de Cuise," in the Paris Basin, as stratotype. The Ilerdian Stage was introduced by Hottinger and Schaub (1960), in the southern Pyrenees, to encompass the strata situated between deposits correlatable with the Thanet Sands below and the Sables de Cuise above. The stage name is derived from "Ilerda," the latin name of the Lleida province of northeast Spain, where the stage stratotype is situated (Tremp section).

Specialists in larger foraminifera have customarily used the base of the Ilerdian stage (defined by the LFT) as the P-E boundary in shallow carbonate sequences (e.g., Boukhary et al., 1995; Tari-Kovacic et al., 1998; Marjanac et al., 1998, and references

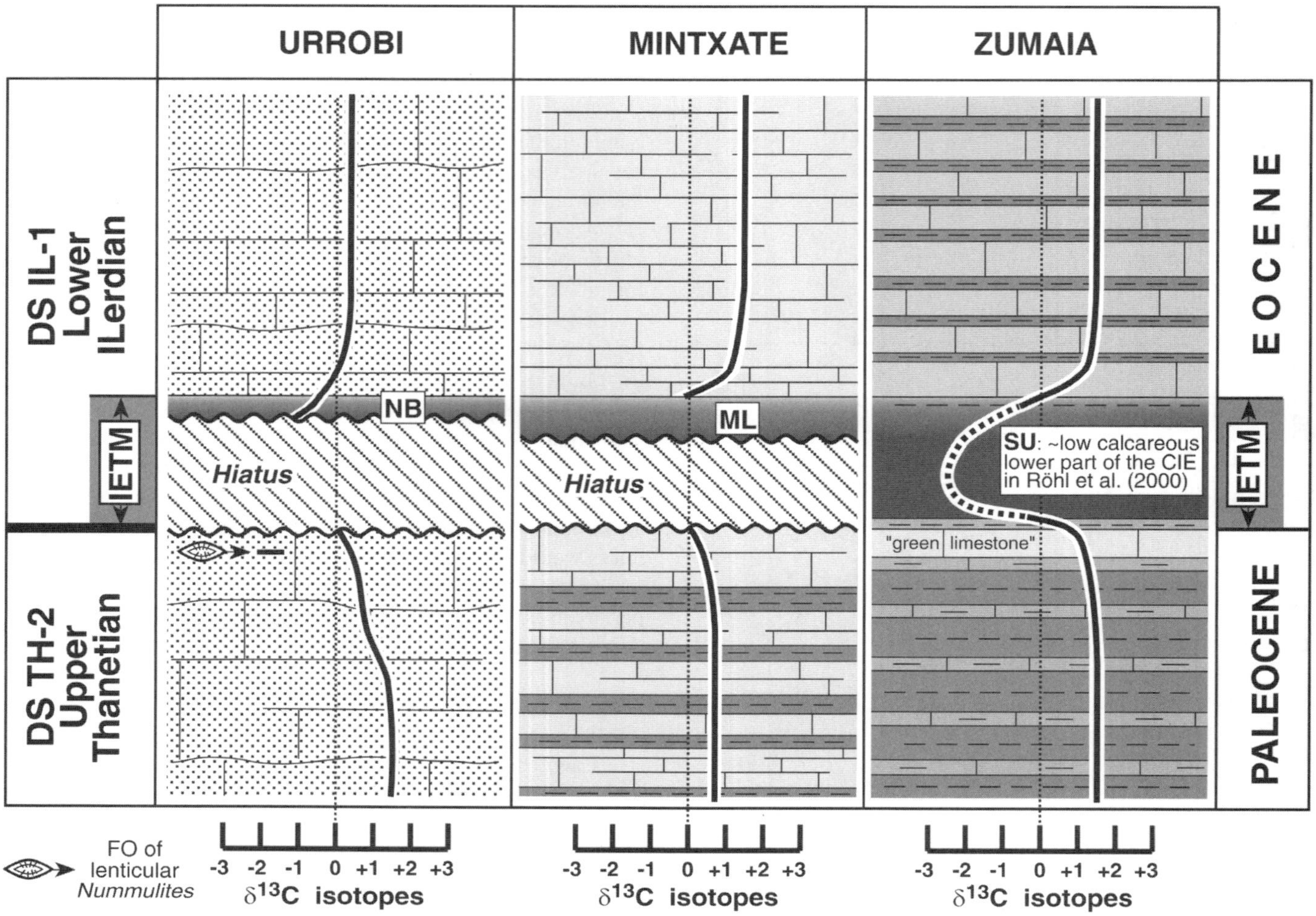

Figure 9. Tentative time correlation between the Urrobi, Mintxate and Zumaia sections.

therein). It was widely accepted, however, that the LFT had occurred near the P4-P5 planktic foraminiferal zonal boundary of Berggren et al. (1995), and was therefore much older than both the IETM and the lower limit of the Ypresian stratotype (e.g., Molina et al., 1992; Hottinger, 1998; Serra-Kiel et al., 1998) (Fig. 10A). However, as discussed in this paper, the base of the Ilerdian stage does coincide with the CIE, the base of which has been proposed as the new criterion to mark the P-E Epoch boundary (Luterbacher et al., 2000). Therefore, by lowering the base of the Ypresian Stage to the same level, as suggested by some members of the Paleocene-Eocene Boundary Working Group (in Aubry, 2000), the undesirable dual usage of the P-E boundary could be easily eliminated (Fig. 10B). In the resulting scheme, the Ilerdian and Cuisian could become substages, essentially synonyms of early and late Ypresian (Fig. 10B). However, if a new stage is introduced between the base of the CIE and the classical base of the Ypresian Stage, as advocated by Aubry et al. (1999), the two different subdivisions of the early Eocene, and the two nonequivalent nomenclatures, would persist.

LFT and IETM: A causal link?

Larger foraminifera are K-strategist, complex life forms that were practically eliminated by the Cretaceous-Paleogene boundary catastrophe. They reappeared in mid Paleocene times as a community of small-sized individuals that slowly diversified at generic level during Thanetian times. At the beginning of the Ilerdian, the LFT took place, characterized by a decrease in generic diversity and a rise of specific diversity, as a few genera (*Nummulites, Assilina, Alveolina*) became dominant and rapidly diversified at specific level. It was also characterized by a marked increase in test size and of adult dimorphism (Hottinger, 1997, 1998, and above). According to Hottinger (1997), this evolution of larger foraminifera can be explained in terms of a long range, global community maturation cycle. Community maturation operates at local, regional or global scales, and at short, intermediate or extended time scales. In accordance with the community maturation theory (Margalef, 1963), an ecological community matures during intervals of unchanged environ-

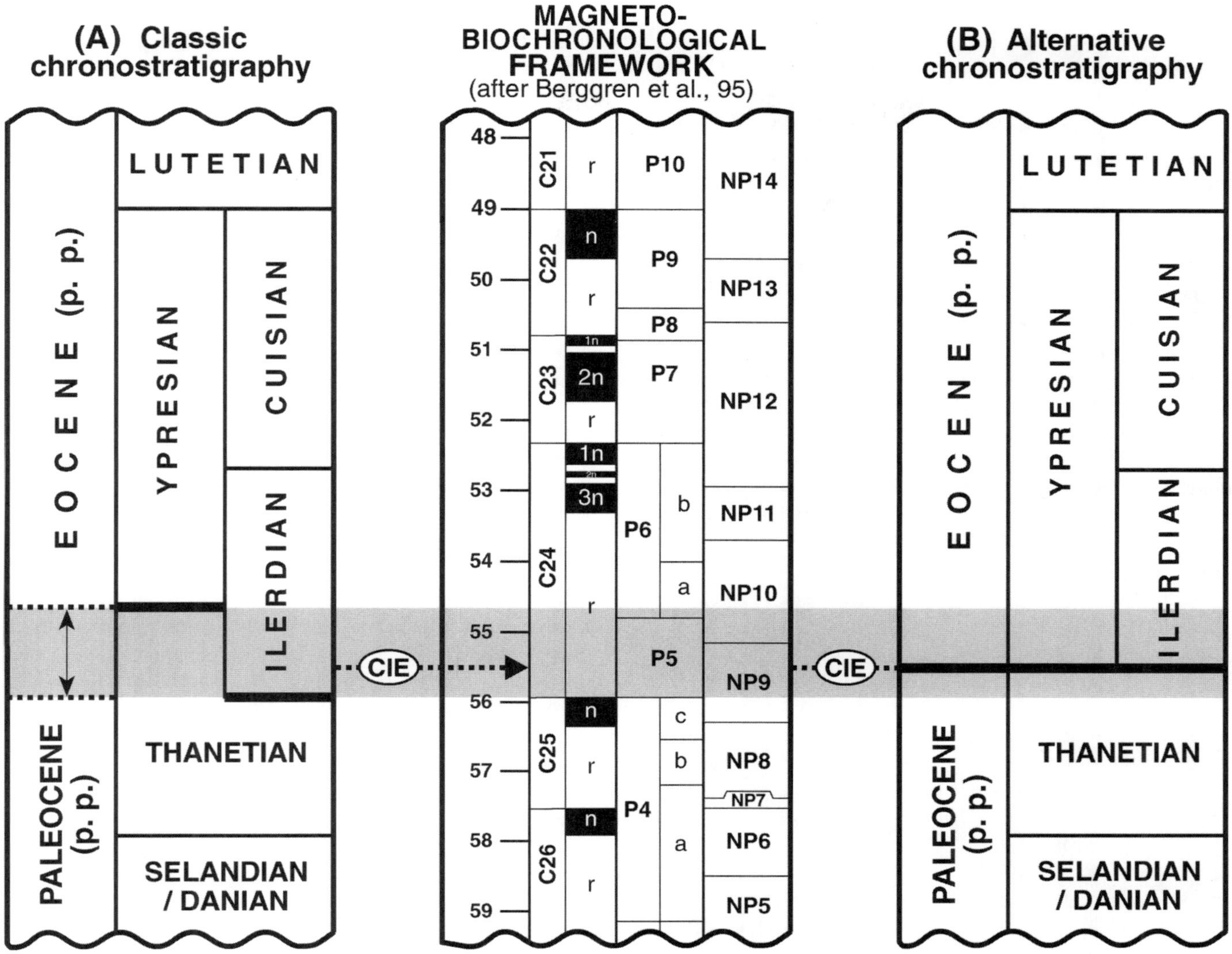

Figure 10. Chronostratigraphic schemes across the late Paleocene–early Eocene interval calibrated to the geomagnetic and calcareous plankton scales of Berggren et al. (1995): (A) as hitherto used (e.g., Hottinger, 1998); (B) after placing the base of the Ilerdian at its correct position, and lowering the classical base of the Ypresian to the CIE.

mental conditions, while environmental changes modulate or disrupt its development.

The synchrony of the LFT and the IETM raises the question of their possible genetic link. One hypothesis, inspired by the Lessepsian Migration into the Mediterranean Sea subsequent to the opening of the Suez Canal (Por, 1978), is that the LFT represents a Tethysian-wide expansion, prompted by the IETM, of a mature but formerly isolated late Thanetian community. Interestingly, Hottinger (1997, p. 501) pointed out that "already during late Paleocene some genera [of larger foraminifera] develop in a unique faunal province, as for instance Iran," a quotation suggesting that at least one of such communities might indeed have existed. Another possible explanation is that conditions during the IETM were stressful for some larger foraminiferal species, which would decline, and favorable for others, which would expand. More complex scenarios can be imagined, involving an evolution of symbiotic algae during the IETM, because many larger foraminifera are symbiont-bearing.

Clearly, a database much more extensive than the one so far available would be necessary to properly understand the origin of the LFT. We think, however, that any hypothesis on that line should take into account the following facts: (1) The LFT represents a rapid phase of maturation of the larger foraminiferal community; (2) The short time (in geological terms) during which the LFT was completed suggests external forcing; and (3) the occurrence of a possible "bloom species," represented in the proximal outer platform by *Nummulites* aff. *minervensis*.

CONCLUSIONS

We report the results of a biostratigraphic and isotopic analysis of two new southern Pyrenean P-E boundary sections, Urrobi and Mintxate. The new data illustrate the characteristics of the LFT in an outer platform setting, and provide direct proofs of the time-equivalence of this turnover and the IETM.

Confirmation of the LFT and IETM synchrony is important for several reasons. First, the LFT is easily recognizable in the field, and may be used to approximate the location of the IETM in the extensive shallow marine carbonate successions of the Tethys Ocean, even when isotopic data are lacking. Second, such confirmation may contribute to clarify the chronostratigraphic nomenclature across the P-E boundary, creating a common language for shallow marine and oceanic (standard) scales in the early Eocene. Finally, it implies that the IETM affected the evolution of larger foraminifera, forcing either large-scale migrations and/or specializations, and therefore that the biotic impact of the thermal event reached into shallow marine settings. Many details of how such a biotic impact came about, however, remain elusive.

ACKNOWLEDGMENTS

We express our gratitude to J. Serra-Kiel (University of Barcelona) and A. Robador (Instituto Geológico y Minero de España, Madrid) for calling our attention toward the Urrobi and Mintxate sections. Suggestions by reviewers H. Luterbacher and E. Molina, and careful editorial work by E. Thomas, are greatly appreciated. Funds for field and laboratory work were provided by Research Projects UPV 121-310-EB/191/98, UPV 121-310-G/12/99 and DGICYT-PB98-0243 to Spanish geologists, and by the Swedish Natural Science Research Council to B. Schmitz. Participation of J.I. Baceta was supported by a Ramón y Cajal Research Grant from the Spanish Science and Technology Ministry.

APPENDIX: SOME REMARKS ABOUT *NUMMULITES* AFF. *MINERVENSIS*

Systematic description

Species with small to medium test dimensions (Fig. A1). Megalospheric A-forms have lenticular to biconvex tests and fairly sharp peripheries; microspheric B-forms have lenticular tests with markedly sharp peripheries. The septal filaments, arranged in a radiating pattern, have nearly straight shapes in A-forms and slightly curved or moderately sinusoidal shapes in B-forms. The smallest specimens of both forms present a pillar with an irregular morphology in the polar zone of their tests (Fig. A1), which gradually disappear as they increase in size. Equatorial sections show a very regular coiling of the spire, which is fairly tight for the initial whorls, especially in the B-forms. Septa are thin, straight to slightly inclined, becoming gently curved toward the periphery, with a regular and tight spiral distribution (Fig. A1). Chambers are subromboidal to arcuate in morphology, slightly higher than long. The marginal cord is uniformly thin, being ~1/3–1/4 of chamber heights. The proloculus of the megalospheric forms are isolepidines.

Chrono- and biostratigraphical position

In his classic monograph on Tethysian nummulitids, Schaub (1981) quoted the occurrence of Ilerdian *Nummulites* in various localities of the Navarra province, including the Urrobi section. These fossils were supplied to him by colleagues working in the area, namely H.D. Kind, E. Mutti, H.P. Luterbacher, J. Ferrer, and J. Rosell, and were almost certainly collected from the *Nummulites* Band, because this is the only horizon of the upper Thanetian–lower Ilerdian interval where larger foraminifera can be liberated from the rock matrix. At least, this was the case with the Urrobi sample (H.P. Luterbacher, 2002, personal commun.).

Schaub (1981) determined that the Navarra specimens had features similar to *Nummulites minervensis,* although with larger test dimensions. He concluded therefore that they were a transitional form between *Nummulites minervensis* (an early Ilerdian-2 species) and *Nummulites carcasonensis* (a middle Ilerdian-1 species). Consequently, he named the Navarra specimens as *Nummulites* aff. *minervensis,* and assigned them to the early–middle Ilerdian.

However, later studies by Robador et al. (1991) found that, in the southern Pyrenean Tena section (Fig. 1), sandy marls with *Nummulites* aff. *minervensis* (i.e., the *Nummulites* Band) are directly overlain by limestones with *Alveolina avellana* and *Alveolina cucumiformis,* two characteristic species of the basal Ilerdian according to Hottinger (1960) and Serra-Kiel et al. (1998). This fact has two implications: (1) *Nummulites* aff. *minervensis* occurs in the basal Ilerdian, a conclusion reinforced by the present study; (2) *Nummulites* aff. *minervensis* can not be considered a more evolved species of *Nummulites minervensis,* and its taxonomic position should be revised.

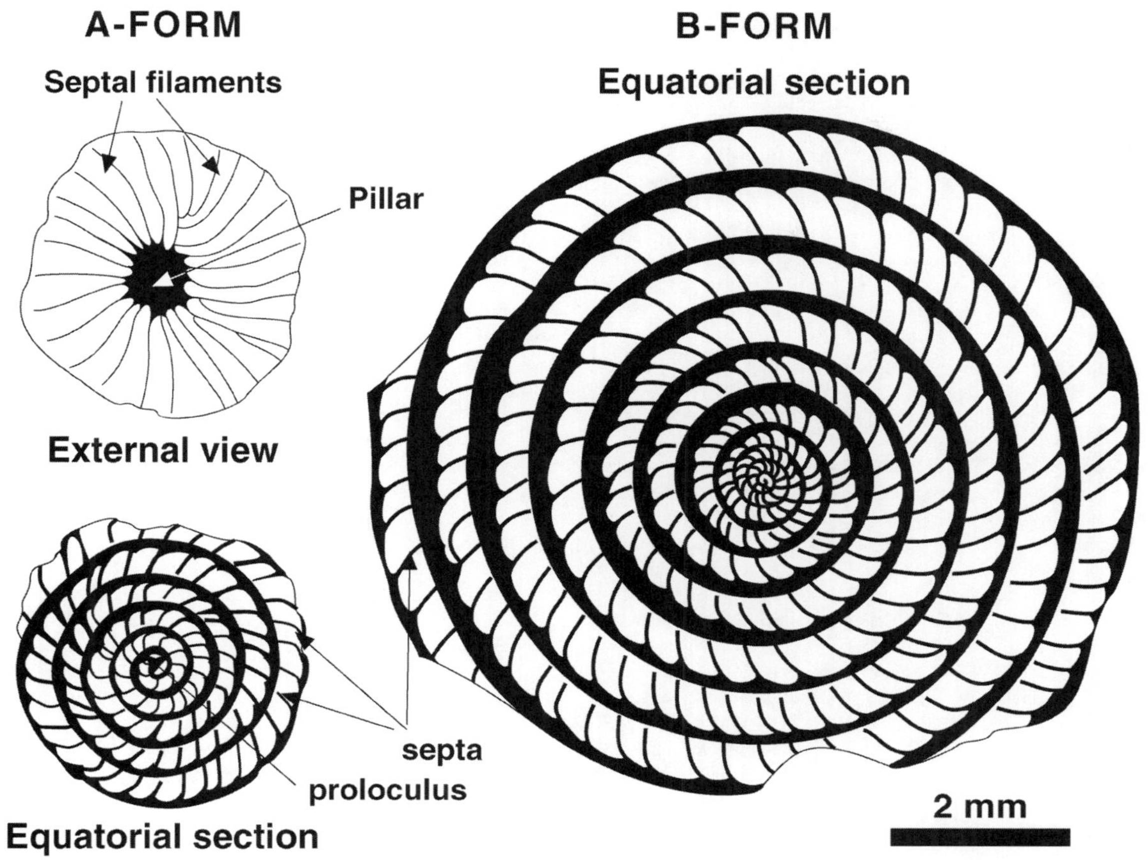

A-FORMS (values in mm)						B-FORMS (values in mm)			
Whole test dimensions Diametre: 2.5 - 4.0 Thickness: 1.0 - 1.5						Whole test dimensions Diametre: 7.0 - 9.5 Thickness: 2.5 - 3.5			
Proloculus diametre: 0.175-0.275									
Whorl n° in equatorial section						Whorl n° in equatorial section			
2	3	4	5	6		8	9	10	11
0.8-0.9	1.0-1.3	1.3-1.7	1.7-2.1	1.9-2.5	Radius	2.8-3.4	3.5-4.0	4.1-4.5	4.3-4.6
16	20-24	28	32		Septa x whorl	32-34	34-36	45-48	52-54

Figure A1. Camera lucida drawing and ranges of typical parameters of *Nummulites* aff. *minervensis.*

REFERENCES CITED

Angori, E., and Monechi, S., 1996, High-resolution nannofossil biostratigraphy across the Paleocene–Eocene boundary at Caravaca (southern Spain): Israel Journal of Earth Sciences, v. 44, p. 207–216.

Arenillas, I., and Molina, E., 2000, Reconstrucción paleoambiental con foraminíferos planctónicos y cronoestratigrafía del tránsito Paleoceno-Eoceno de Zumaya (Guipúzcoa): Revista Española de Micropaleontología, v. 32, no. 3, p. 283–300.

Aubry, M.P., 1998, Early Paleocene calcareous nannoplankton evolution: A tale of climatic amelioration, *in* Aubry, M.P., et al., eds., Late Paleocene–early Eocene biotic and climatic events in the marine and terrestrial record: New York, Columbia University Press, p. 158–203.

Aubry, M.P., 2000, Working Group on the Paleocene/Eocene boundary, Meeting of 16–17 September: Newsletter of the International Subcommission on Paleogene Stratigraphy, Tübingen, November 2000, v. 9, p. 20–36.

Aubry, M.P., Berggren, W.A., Van Couvering, J.A., and Steininger, F., 1999, Problems in chonostratigraphy: Stages, series, unit and boundary stratotypes, global stratotype section and point and tarnished golden spikes: Earth Science Reviews, v. 46, p. 99–148.

Baceta, J.I., Pujalte, V., Dinarès-Turell, J., Payros, A., Orue-Etxebarria, X., and Bernaola, G., 2000, The Paleocene–Eocene boundary interval in the Zumaia Section (Gipuzkoa, Basque Country): Magnetostratigraphy and high-resolution lithostratigraphy: Revista de la Sociedad Geológica de España, v. 13, no. 2, p. 375–391.

Berggren, W.A., and Norris, R.D., 1997, Biostratigraphy, phylogeny, and systematics of Paleocene trochospiral planktic foraminifera: Micropaleontology, v. 43, no. 1, p. 1–116.

Berggren, W.A., Kent, D.V., Swisher, C.C., III, and Aubry, M.P., 1995, A revised Cenozoic geochronology and chronostratigraphy, *in* Berggren, W.A., et al., eds., Geochronology, time scales, and global stratigraphic correlations: SEPM (Society for Sedimentary Geology) Special Publication 54, p. 129–212.

Boukhary, M., Bassiouni, M.E.A., and Hussein-Kamel, Y., 1995, *Nummulites Luterbacheri* n. sp., unexpected large *Nummulites* from basal Ilerdian (early Eocene) from El Quss Abu Said, Farafra oasis, Western Desert, Egypt: Revue de Micropaleontology, v. 38, p. 285–298.

Bujak, J.P., and Brinkhuis, H., 1998, Global warming and dinocyst changes across the Paleocene–Eocene epoch boundary, *in* Aubry, M.P., et al., eds., Late Paleocene–early Eocene biotic and climatic events in the marine and terrestrial record: New York, Columbia University Press, p. 214–243.

Canudo, J.I., Keller, G., Molina, E., and Ortiz, N., 1995, Planktic foraminiferal turnover and $\delta^{13}C$ isotopes across the Paleocene–Eocene transition at Caravaca and Zumaya, Spain: Palaeogeography, Palaeoclimatology, Palaeoecology, v. 114, p. 75–100.

Charisi, S.D., and Schmitz, B., 1998, Paleocene to early Eocene paleoceanography of the Middle East: The $\delta^{13}C$ and $\delta^{18}O$ isotopes from foraminiferal calcite: Paleoceanography, v. 13, p. 106–118.

Clyde, W.C., and Gingerich, P.D., 1998, Mammalian community response to the latest Paleocene thermal maximum: An isotaphonomic study in the northern Bighorn Basin, Wyoming: Geology, v. 26, p. 1011–1014.

Corfield, R.M., and Norris, R.D., 1998, The oxygen and carbon isotopic context of the Paleocene–Eocene Epoch boundary, *in* Aubry, M.P., et al., eds., Late Paleocene–early Eocene biotic and climatic events in the marine and terrestrial record: New York, Columbia University Press, p. 124–137.

Dickens, G.R., Castillo, M.M., and Walker, J.C.G., 1997, A blast of gas in the latest Paleocene: Simulating first-order effects of massive dissociation of oceanic methane hydrate: Geology, v. 25, p. 259–262.

Dollfus, G.F., 1880, Essai sur l'extension des terrains tertiaires dans le Bassin anglo-parisien: Bulletin Société Géologique Normandie, Le Havre, v. 6, p. 584–605.

Eichenseer, H., and Luterbacher, H.P., 1992, The marine Paleogene of the Tremp region (northeast Spain): Depositional sequences, facies history, biostratigraphy and controlling factors: Facies, v. 27, p. 119–152.

Hottinger, L., 1960, Recherche sur les Alvéolines du Paléocène et de l'Éocène: Mémoires Suisses de Paléontologie, v. 75/76, p. 1–243.

Hottinger, L., 1997, Shallow benthic foraminiferal assemblages as signals for depth of their deposition and their limitations: Bulletin de la Société Géologique de France, v. 168, no. 4, p. 491–505.

Hottinger, L., 1998, Shallow benthic foraminifera at the Paleocene–Eocene boundary: Strata, v. 9, p. 61–64.

Hottinger, L., and Schaub, H., 1960, Zur Stufeneinteilung des Paleocaens und Eocaens: Einführung der Stufen Ilerdien und Biarritzien: Ecologae Geologicae Helvetiae, v. 53, p. 454–479.

Kelly, D.C., Bralower, T.J., Zachos, J.C., Premoli Silva, I., and Thomas, E., 1996, Rapid diversification of planktonic foraminifera in the tropical Pacific (ODP Site 865) during the late Paleocene thermal maximum: Geology, v. 24, p. 423–426.

Labaume, P., Seguret, M., and Seyve, C., 1985, Evolution of turbiditic foreland basin and analogy with an accretionary prism: Example of the Eocene South Pyrenean Basin: Tectonics, v. 4, p. 661–685.

Lu, G., and Keller, G., 1993, The Paleocene–Eocene transition in the Antarctic Indian Ocean: Inference from planktic foraminifera: Marine Micropaleontology, v. 21, p. 101–142.

Luterbacher, H., 1998, Sequence stratigraphy and limitations of biostratigraphy in the marine Paleogene strata of the Tremp Basin (central part of the southern Pyrenean Foreland Basin, Spain), *in* de Graciansky, P.C., et al., eds., Mesozoic and Cenozoic sequence stratigraphy of European basins: SEPM (Society for Sedimentary Geology) Special Publication 60, p. 303–309.

Luterbacher, H.P., Hardenbold, J., and Schmitz, B., 2000, Decision of the voting members of the International Subcommission on Paleogene Stratigraphy on the criterion for the recognition of the Paleocene–Eocene boundary: Newsletter of the International Subcommission on Paleogene Stratigraphy, Tübingen, November 2000, v. 9, p. 13.

Mangin, J.Ph., 1960, Le Nummulitique sud-Pyrénéen a l'ouest de l'Aragon: Pirineos, v. 51–58, 656 p.

Margalef, R., 1963, On certain unifying principles in ecology: American Naturalist, v. 47, p. 357–374.

Marjanac, T., Babac, D., Benic, J., Cosovic, V., Drobne, K., Marjanac, L., Pavlovec, R., and Velimirovic, Z., 1998, Eocene carbonate sediments and sea-level changes on the northeast part of Adriatic carbonate platform (Island of Hvar and Peljesac Peninsula, Croatia), *in* Hottinger, L., and Drobne, K., eds., Paleogene shallow benthos of the Tethys: Dela-Opera SAZU, Ljubljana, v. 34, no. 2, p. 203–242.

Martini, E., 1971, Standard Tertiary and Quaternary calcareous nannoplankton zonation, *in* Farinacci, A., ed., Proceedings of the 2nd Planktonic Conference, Rome, Italy, 1970: Tecnoscienza, v. 2, p. 739–785.

Molina, E., Canudo, J.I., Guernet, C., McDougall, K., Ortiz, N., Pascual, J.O., Parés, J.M., Samso, J.M., and Tosquella, J., 1992, The stratotypic Ilerdian revisited: Integrated stratigraphy across the Paleocene–Eocene boundary: Revista Española de Micropaleontología, v. 35, no. 2, p. 143–156.

Molina, E., Angori, E., Arenillas, I., Monechi, S., and Schmitz, B., 2000, Integrated stratigraphy across the Paleocene–Eocene boundary at Campo, Spain: GFF, v. 122, p. 1060–1070.

Monechi, E., Angori, E., and Von Salis, K., 2000, Calcareous nannofossil turnover around the Paleocene–Eocene transition at Alamedilla (southern Spain): Bulletin de la Société Géologique de France, v. 171, no. 4, p. 477–489.

Orue-Etxebarria, X., Apellaniz, E., Baceta, J.I., Coccioni, R., Di Leo, R., Dinares-Turell, J., Galeotti, S., Monechi, S., Núñez-Betelu, K., Parés, J.M., Payros, A., Pujalte, V., Samso, J.M., Serra-Kiel, J., Schmitz, B., and Tosquella, J., 1996, Physical and biostratigraphic analysis of two prospective Paleocene–Eocene Boundary stratotypes in the intermediate-deep water Basque Basin, western Pyrenees: The Trabakua Pass and Ermua sections: Neues Jahrbuch für Geologie und Paläontologie Abhandlungen, v. 201, no. 2, p. 179–242.

Orue-Etxebarria, X., Pujalte, V., Bernaola, G., Apellaniz, E., Baceta, J.I., Payros, A., Nuñez-Betelu, K., Serra-Kiel, J., and Tosquella, J., 2001, Did the late Paleocene Thermal Maximum affect the evolution of larger foraminifers? Evidence from calcareous plankton of the Campo section (Pyrenees, Spain): Marine Micropaleontology, v. 41, no. 1/2, p. 45–71.

Payros, A., Pujalte, V., and Orue-Etxebarria, X., 1999, The South Pyrenean Eocene carbonate megabreccias revisited: New interpretation based on evidence from the Pamplona Basin: Sedimentary Geology, v. 125, p. 165–194.

Payros, A., Pujalte, V., Baceta, J.I., Bernaola, G., Orue-Etxebarria, X., Apellaniz, A., Caballero, F., and Ferrandez, C., 2000, Lithostratigraphy and sequence stratigraphy of the upper Thanetian to middle Ilerdian strata of the Campo section (southern Pyrenees, Spain): Revision and new data: Revista de la Sociedad Geológica de España, v. 13, no. 2, p. 51–64.

Pignatti, J.S., BartholoNdy, J., Cahuzac, B., Drobne, K., Ferrández, C., Hottinger, L., Jauhri, A.K., Matteucci, R., Papazzoni, C.A., Serra-Kiel, J., Sirel, E., and Tosquella, J., 1998, Paleogene larger foraminifera reference list: Slovenska Akademija Znanosti in Umetnosti, Dela-Opera SAZU, Ljubljana, v. 34, no. 1, 298 p.

Por, F.D., 1978, Lessepsian migration: The influx of Red Sea biota into the Mediterranean by way of the Suez Canal: Ecology studies: Berlin, Springer-Verlag, v. 23, p. 228.

Pujalte, V., Baceta, J.I., Orue-Etxebarria, X., and Payros, A., 1998, Paleocene strata of the Basque Country, western Pyrenees, northern Spain: Facies and sequence development in a deep-water starved basin, *in* de Graciansky, P.C., et al., eds., Mesozoic and Cenozoic sequence stratigraphy of European basins: SEPM (Society for Sedimentary Geology) Special Publication 60, p. 311–325.

Pujalte, V., Baceta, J.I., Payros, A., Orue-Etxebarria, X., and Schmitz, B., 2000, Upper Paleocene–lower Eocene strata of the western Pyrenees, Spain: A shelf-to-basin correlation: GFF, v. 122, p. 129–130.

Robador, A.J., Samsó, J.M., Serra-Kiel, J., and Tosquella, J., 1991, Field guide, *in* Barnolas, A., et al., eds., Introduction to the early Paleogene of the South Pyrenean Basin, International Geological Correlation Program (IGCP) 286 "Early Paleogene Benthos" First Meeting. Instituto Tecnológico Geominero de España, Madrid, p. 131–159.

Röhl, U., Bralower, T.J., Norris, R.D., and Wefer, G., 2000, New chronology for the late Paleocene thermal maximum and its enviromental implications: Geology, v. 28, p. 927–930.

Schaub, H., 1969, L'Ilerdien: état actuel du problème: Mémoires Bureau Recherches Géologiques et Minières, v. 69, p. 259–266.

Schaub, H., 1981, *Nummulites* et *Assilines* de la Tethys Paléogène: Taxinomie, phylogénesè et biostratigraphie: Mémoires Suisses de Paléontologie, v. 104–106, 236 p.

Schmitz, B., Asaro, F., Molina, E., Monechi, S., von Salis, K., and Speijer, R.P., 1997, High-resolution iridium $\delta^{13}C$, $\delta^{18}O$, foraminifera and nannofossil profiles across the latest Paleocene benthic extinction event at Zumaya, Spain: Palaeogeography, Palaeoclimatology, Palaeoecology, v. 133, p. 49–69.

Schmitz, B., Pujalte, V., and Nuñez-Betelu, K., 2001, Climate and sea-level perturbations during the Initial Eocene Thermal Maximum: Evidence from siliciclastic units in the Basque Basin (Ermua, Zumaia and Trabakua pass), northern Spain: Palaeogeography, Palaeoclimatology, Palaeoecology, v. 165, no. 3/4, p. 299–320.

Serra-Kiel, J., Hottinger, L., Caus, E., Drobne, K., Ferrandez, C., Jauhri, A.K., Less, G., Pavlovec, R., Pignatti, J., Samso, J.M., Schaub, H., Sirel, E., Strougo, A., Tambareau, Y., Tosquella, J., and Zakrevskaya, E., 1998, Larger foraminiferal biostratigraphy of the Tethyan Paleocene and Eocene: Bulletin de la Société Géologique de France, v. 169, no. 2, p. 281–299.

Tari-Kovacic, V., Kalac, K., Lucic, D., and Benic, J., 1998, Stratigraphy analysis of Paleogene beds in some off-shore wells (Central Adriatic area, Croatia), *in* Hottinger, L., and Drobne, K., eds., Paleogene shallow benthos of the Tethys: Dela-Opera SAZU, Ljubljana, v. 34, no. 2, p. 203–242.

Thomas, E., 1990, Late Cretaceous–early Eocene mass extinctions in the deep sea, *in* Sharpton, V.L., and Ward, P.D., eds., Global catastrophes in Earth history: Boulder, Colorado, Geological Society of America Special Paper 247, p. 481–495.

Thomas, E., 1998, The biogeography of the late Paleocene benthic foraminiferal extinction, *in* Aubry, M.P., et al., eds., Late Paleocene–early Eocene biotic and climatic events in the marine and terrestrial record: New York, Columbia University Press, p. 214–243.

Thompson, E.I., and Schmitz, B., 1997, Barium and the late Paleocene $\delta^{13}C$ maximum: Evidence of increased marine surface productivity: Paleoceanography, v. 12, p. 239–254.

Tosquella, J., 1995, Els nummulitinae del Paleocè-Eocé inferior de la Conca Sudpirenaica [Ph.D. thesis]: University of Barcelona, Barcelona, Spain, 582 p.

Manuscript Accepted by the Society August 13, 2002

Printed in the USA

Geological Society of America
Special Paper 369
2003

Ecological development of acarininids (planktonic foraminifera) and hydrographic evolution of Paleocene surface waters

Frédéric Quillévéré*
UFR des Sciences de la Terre, Université Claude Bernard–Lyon1,
27-43 Boulevard du 11 Novembre 1918, 69 622 Villeurbanne cedex, France
Richard D. Norris
308 Vaughn Hall, Scripps Institution of Oceanography,
University of California San Diego, La Jolla, California 92093-0244, USA

ABSTRACT

Paleogene planktonic foraminiferal communities diversified in surface ocean habitats by the end of the Danian (61.3 Ma). The specialized depth ecologies were related to the initiation of photosymbiosis as a trophic strategy within the last representatives of *Praemurica inconstans.* Our data confirm that *Acarinina* and *Morozovella* lived in the surface mixed layer, since they record lower $\delta^{18}O$ and higher $\delta^{13}C$ than contemporaneous *Globanomalina* and *Subbotina.* However, prior to 56.5 Ma, the data from the southeastern Indian Ocean (Ocean Drilling Program Hole 761B) show a clear separation in both oxygen and carbon isotopes between *Morozovella* and *Acarinina,* suggesting that the latter calcified slightly deeper in the photic zone than the former. The water column during the Paleocene appears to have been more stratified, in terms of plankton communities, than previously realized and the resulting creation of different depth habitats played an important role in plankton diversification. The surface waters over the subtropical Indian Ocean probably had a shallower mixed layer and a stronger thermal gradient than, for example, that of the mid-Pacific where the $\delta^{18}O$ offsets between different species of planktonic foraminifera are less pronounced. The different genera of planktonic foraminifera established distinct surface-ocean, shallow thermocline, and deep thermocline depth habitats that remained occupied by related species groups throughout the Paleocene. The invasion of planktonic foraminifera into the oligotrophic gyres during the middle Paleocene led to a significant diversification that was fueled by photosymbiosis in the planktonic foraminifera.

INTRODUCTION

The Paleocene (65–55.5 Ma) was a time of global warming, increasing oceanic productivity and increasing depth-related specialization of the planktonic foraminiferal communities (Shackleton et al., 1985; Corfield and Shackleton, 1988; Stott et al., 1990; Corfield and Cartlidge, 1992; Zachos et al., 1993, 1994, 2001; Corfield, 1994; Norris et al., 2001). These general patterns were associated with an explosive radiation of the trochospiral planktonic foraminifera following the Cretaceous-Paleogene (K-P) boundary mass extinction. Many of the major morphological and ecological patterns became established dur-

*frederic.guillevere@univ-lyon1.fr

Quillévéré, F., and Norris, R.D., 2003, Ecological development of acarininids (planktonic foraminifera) and hydrographic evolution of Paleocene surface waters, *in* Wing, S.L., Gingerich, P.D., Schmitz, B., and Thomas, E., eds., Causes and Consequences of Globally Warm Climates in the Early Paleogene: Boulder, Colorado, Geological Society of America Special Paper 369, p. 223–238.

ing the earliest Paleocene. The open ocean was originally recolonized by spinose (*Subbotina, Parasubbotina*), smooth (*Globanomalina*) and praemuricate (*Praemurica*) planktonic foraminifera that radiated within a hundred thousand years of the K-P boundary (Hemleben et al., 1991; Berggren and Norris, 1997; Olsson et al., 1992, 1999). These taxa are believed to have occupied deep environments (within or below the thermocline) of the low-latitude water column, throughout their evolution (Douglas and Savin, 1978; Boersma et al., 1979). A second radiation occurred ~3 m.y. later. In this radiation, at the end of the early Paleocene (late Danian–early Selandian), the last representatives of the genus *Praemurica* evolved toward the muricate genera (*Morozovella, Acarinina* and *Igorina*) (Norris, 1996; Kelly et al., 1996; Berggren and Norris, 1997; Olsson et al., 1999). These taxa are believed to have occupied the upper part of the water column (the surface mixed layer) throughout their evolution and their diversifications may have resulted from peculiar depth-related adaptations to the development of the oligotrophic provinces during the Paleocene (Shackleton et al., 1985; Norris, 1996; Berggren and Norris, 1997; Quillévéré et al., 2001).

One of the most distinctive and abundant groups of surface-dwelling planktonic foraminifera during the early Paleogene was the genus *Acarinina*. The acarininids were initially characterized by small and globular taxa with a muricate funnel-pore wall texture (e.g., Berggren and Norris, 1997), and subsequently, during the Paleocene, increased in size and acquired a distinctive morphology by elevating their coiling and developing robust muricae on the umbilical surface (Fig. 1). Recent morphometric and biogeographic studies by Quillévéré et al. (2000, 2001) have shown that the acarininids underwent two major diversification events during the Paleocene (Fig. 2). The genus originated in the high latitudes from the common ancestor *Acarinina strabocella,* and the initial diversification (appearance of *Acarinina nitida, A. subsphaerica* and *A. mckannai*), resulted from a migration event from high to low latitudes environments ca. 59.2 Ma. The later Paleocene diversification event (appearance of *A. soldadoensis* and *A. coalingensis*), occurred in the low-latitude oligotrophic and warm water masses ca. 56.5 Ma. Further diversification events occurred during the early and middle Eocene. By the Eocene, the acarininids became the most species-rich group of surface-dwelling planktonic foraminifera among the muricate genera (Keller et al., 1992).

During the late Paleocene expansion of the oligotrophic conditions in the global ocean, the acquisition of photosymbionts (most probably chrysophytes or dinoflagellates by analogy with modern foraminifera) may have facilitated the radiation of the acarininids, as it triggered the diversification of the morozovellids, their photosymbiotic sister group (Norris, 1996; Berggren and Norris, 1997). Indeed, photosymbiosis allowed the acarininids to thrive in low-nutrient water masses (Norris, 1996; Quillévéré et al., 2001). It is notable, however, that just as the taxonomic diversity within the morozovellids increased rapidly after the evolution of photosymbiosis (Norris, 1996), the initial diversification of the acarininids, was delayed. Additionally, the morozovellids remained restricted to the low-latitude oligotrophic provinces while the acarininids were diverse and abundant in high latitudes as well as in low latitudes. Quillévéré et al. (2001) suggested that the ecological deployment of the early acarininids within high-latitude mesotrophic settings may have accounted for their later radiation relative to the morozovellids. In this study, we explore stable isotopic evidence for depth-related specialization of *Acarinina* and compare it to ecological specializations of other Paleocene trochospiral planktonic foraminifera, including *Morozovella.* Because the stable isotopic composition of the acarininids is often used as a surface-water indicator in paleoceanographic studies, it is crucial to understand the level of interspecies isotopic differences and variability.

Figure 1. Scanning electron photomicrograph of one representative of the Paleocene acarininids. Umbilical view of *Acarinina soldadoensis* from Zone P4c in ODP Hole 761B. Scale bar = 100 µm.

Many workers have speculated that microevolutionary patterns in planktonic foraminifera were related to oceanographic and/or depth-related specializations (Norris et al., 1993; Schneider and Kennett, 1996; Pearson et al., 1997; Coxall et al., 2000; see also Pearson, 1998, for review). We document temporal isotopic profiles from several deep-sea sections, recorded in different oceanic provinces, in order to evaluate how evolutionary diversifications within the acarininids were related to the environmental changes of the surface waters during the Paleocene.

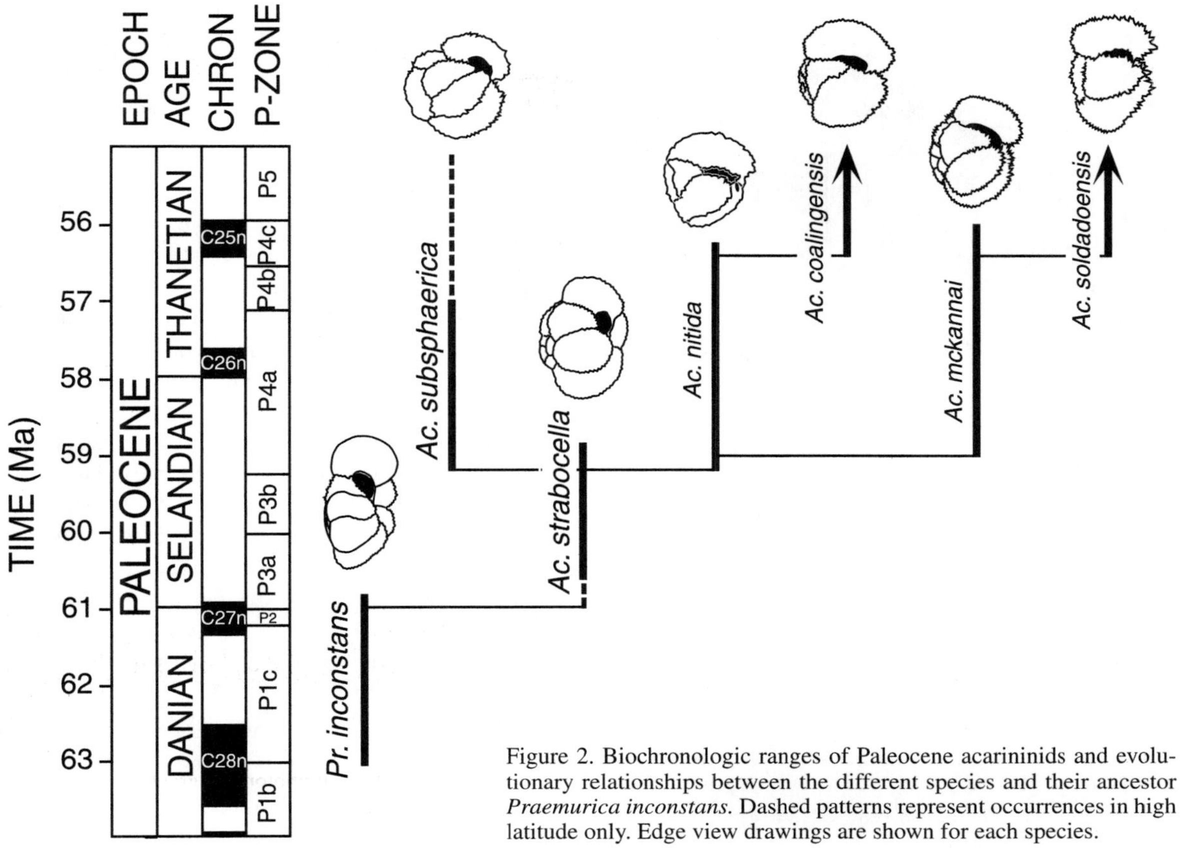

Figure 2. Biochronologic ranges of Paleocene acarininids and evolutionary relationships between the different species and their ancestor *Praemurica inconstans.* Dashed patterns represent occurrences in high latitude only. Edge view drawings are shown for each species.

MONITORING ECOLOGICAL ASSEMBLAGES USING STABLE ISOTOPE ANALYSIS

Many studies have shown that the ratios of both oxygen and carbon isotopes of planktonic foraminiferal tests (shells) denote gradients from surface to deeper waters (Emiliani, 1954; Douglas and Savin, 1978; Boersma et al., 1979; Corfield and Cartlidge, 1991). Oxygen isotope composition ($\delta^{18}O$) is a function of the calcification temperature among other variables including the $\delta^{18}O$ composition of the water (Epstein et al., 1953). As expected in a thermally stratified ocean, surface-dwelling planktonic foraminifera, which calcify in the warmest water masses, record lower $\delta^{18}O$ than coexisting deep-dwelling forms (Boersma et al., 1979; Boersma and Premoli Silva, 1983). The carbon isotope composition ($\delta^{13}C$) is controlled by the ratio of dissolved inorganic carbon in seawater and physiological processes such as respiration and symbiont photosynthesis (Spero and Williams, 1988; Spero et al., 1991; Spero, 1998), although biological isotopic fractionation effects may influence the carbon isotopic composition. Because surface waters are depleted in ^{12}C, as a result of photosynthesis, surface-dwelling planktonic foraminifera record higher $\delta^{13}C$ than coexisting deep-dwellers (Douglas and Savin, 1978; Boersma et al., 1979). The photosynthetic activity of symbiont-bearing planktonic foraminifera also has a major effect on $\delta^{13}C$ composition of their test. The natural, carbon-pump derived, surface-to-deep carbon gradient in the water column is amplified by the photosynthetic fractionation of algal symbionts in the surface-dwelling taxa while this amplification does not occur in deeper-dwelling asymbiotic taxa. In this context, stable isotope data provide a powerful means of understanding major aspects of the ecology of extinct planktonic foraminiferal species.

MATERIAL AND PROCEDURE

Samples were collected from three Paleocene deep-sea drill holes recovered from different paleoceanographic settings, spanning a wide range of latitude (~95°) (Fig. 3). Ocean Drilling Program (ODP) Hole 761B (16°44.23′S, 115°32.10′E; 2466 m water depth), located on the Wombat Plateau in the eastern Indian ocean, provides the most complete Paleocene section with excellently preserved foraminifera. One to three samples per core section (one every 0.4–1.5 m) were selected for isotopic analysis. Further samples (1 per core section; one every 1.5 m) were analyzed from (1) Deep Sea Drilling Project (DSDP) Holes 465 and 465A (33°49.23′N; 178°55.14′E; 2161 m water depth), located on the Hess Rise in the equatorial Pacific Ocean,

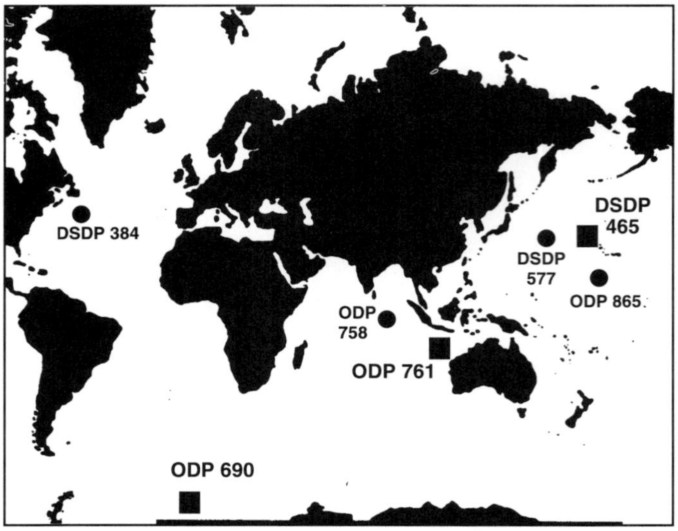

Figure 3. Location map of ODP and DSDP sites studied (black squares) and mentioned (black circles) in this paper.

and (2) ODP Hole 690C (65°9.629′S; 1°12.296′E; 2914 m water depth), located on Maud Rise in the southern part of the Atlantic Ocean. These two sites recovered, respectively, upper Paleocene and lower Paleocene sections, which also contain well-preserved and diverse planktonic foraminifera. Isotopic analyses were conducted on monospecific samples of all Paleocene species of *Acarinina* and a number of other reference species from the same sample, including deep-dwelling forms (*Subbotina trivialis, Subbotina triloculinoides, Subbotina triangularis, Globanomalina compressa, Globanomalina ehrenbergi, Globanomalina pseudomenardii, Praemurica pseudoinconstans, Praemurica inconstans*) and surface-dwelling forms (*Praemurica uncinata, Morozovella angulata, Morozovella velascoensis*). In order to preclude any size-dependent isotopic trends (see Shackleton et al., 1985; D'Hondt et al., 1994), all stable isotope measurements were performed on specimens picked from a restricted size range (212–250 μm). The number of specimens necessary to supply sufficient calcium carbonate and to minimize the effect of individual variation in our isotopic values ranged from 10 to 12. Individuals were ultrasonically cleaned in distilled water, weighed and analyzed with an OPTIMA AC117 Multiprep mass spectrometer at the University of Montpellier II (UM2), with the exception of samples from Cores 761B-18X and 761B-17X, which were analyzed with a Finnigan MAT 252 mass spectrometer and associated "Kiel" carbonate preparation system at the Woods Hole Oceanographic Institution (WHOI). Repeated analysis of in-house standards (Carrara Marble, Merk, B-1 Marine Carbonate, Atlantis-II) and NBS (National Bureau of Standards) 19 suggests that the standard machine error for oxygen and carbon isotopic measurements were respectively 0.1‰ and 0.05‰ at UM2 and 0.07‰ and 0.03‰ at WHOI. Replicate analyses of foraminiferal samples, performed in order to calibrate measurements from UM2 and WHOI, have yielded identical isotopic results within the reported standard errors.

Compilation of isotopic data such as attempted here are limited by the uncertainty of the chronologic framework for each deep-sea section. In this study, the age determinations for the levels that have provided isotopic analyses are based on the temporal interpretation of the Paleocene section recovered at each site following the methodology established by Aubry (1995). The temporal interpretation at Sites 761 and 690 (Quillévéré et al., 2002) were based on integrated bio- (calcareous nannoplankton and planktonic foraminifera) and magnetostratigraphy. At Site 465, the temporal interpretation (Quillévéré, 2000) was based on integrated biostratigraphy (calcareous nannoplankton and planktonic foraminifera). In each case, integrated stratigraphies have been calibrated to the Paleocene magnetobiochronology of Berggren et al. (2000). Although temporal interpretations have shown the occurrence of numerous stratigraphic gaps in the sections recovered at Sites 761, 465 and 690, all these sites considered together provide a nearly complete temporal record of the Paleocene.

Stable carbon and oxygen isotopic data for ODP Hole 761B, DSDP Hole 465/465A and ODP Hole 690C are provided in Appendix, as a function of the sample reference, age estimates and planktonic foraminiferal species we analyzed.

RESULTS AND DISCUSSION

Long-term isotopic trends

The planktonic foraminiferal long-term carbon isotope record from Hole 761B (Fig. 4A) shows a gradual increase (~2‰) and subsequent decrease in $\delta^{13}C$ during the Paleocene. The records from Holes 465/465A and 690C (Fig. 4B, C), although incomplete, show a similar pattern. At Hole 690C, $\delta^{13}C$ decreases by 0.5‰ during the early Paleocene (Biochrons P1a to P1c[*partim*]). These general trends are consistent with previously published records from DSDP Sites 384 (northwest Atlantic) and 577 (mid-Pacific), which have been used to define the paleoceanographic context of the Paleocene and in particular the Paleocene carbon isotope maximum (Corfield and Cartlidge, 1992; Corfield and Norris, 1996). Our isotopic compilation suggests that foraminiferal $\delta^{13}C$ values increased by the early Thanetian, where the heaviest values are recorded between 56.7 Ma and 57.0 Ma. Shackleton et al. (1985) and Corfield and Cartlidge (1992) inferred that the peak in $\delta^{13}C$ during the late Paleocene resulted from a global increase in surface water productivity, and this interpretation has been more recently supported by barium records (Thompson and Schmitz, 1997). Part of the $\delta^{13}C$ increase is probably due to increased fractionation in symbiont-bearing taxa such as *Morozovella* and *Acarinina* (see above). However, the $\delta^{13}C$ increase has also been seen in benthic foraminifera throughout the oceans (Miller et al., 1987; Corfield and Cartlidge, 1992; Corfield and Norris, 1996, 1998),

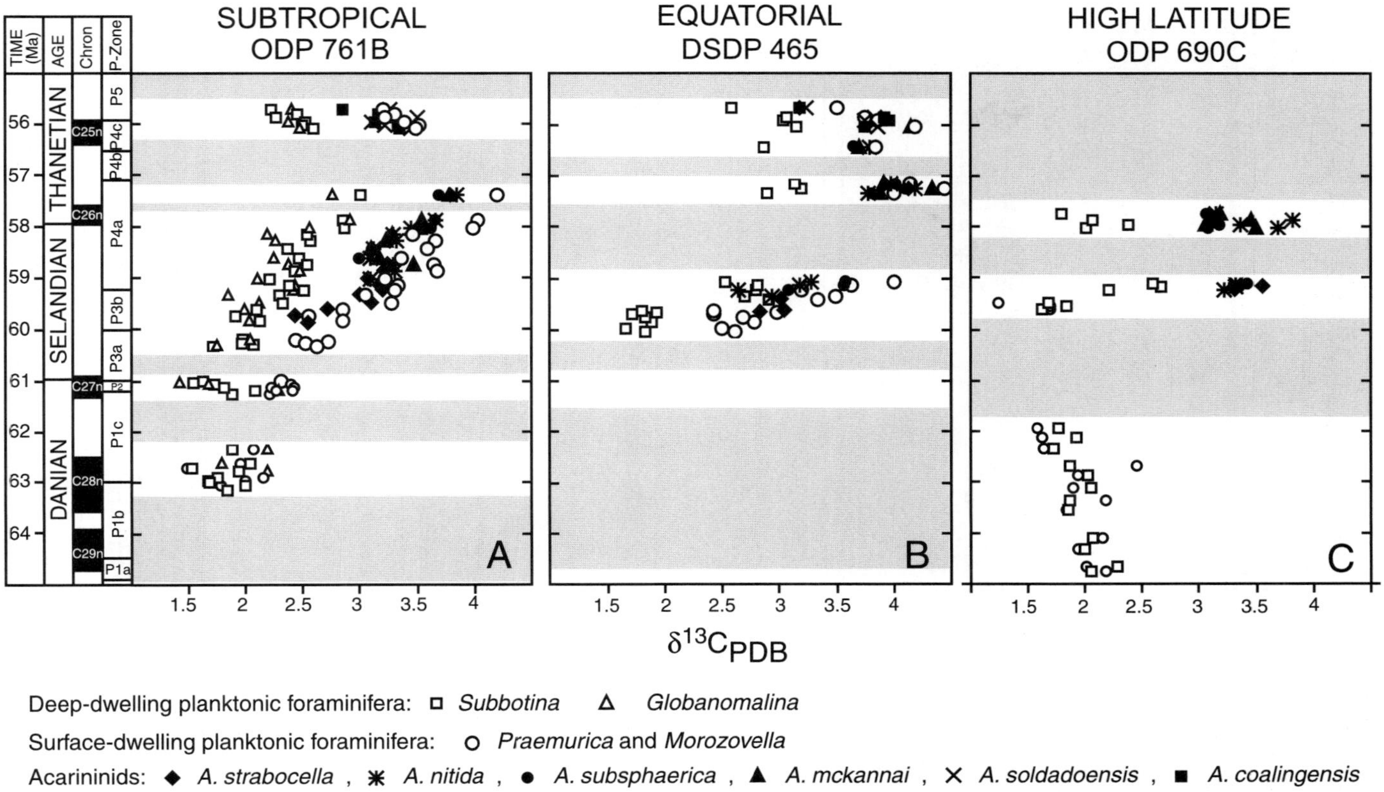

Figure 4. Monospecific planktonic foraminiferal carbon isotopes recorded at ODP Hole 761B (A), DSDP Holes 465/465A (B), and ODP Hole 690C (C) during the Paleocene. Hiatus in the sections are shown by grey patterns. Chronologic framework from Berggren et al. (2000).

suggesting that it results, at least partly, from a reservoir effect of large scale organic carbon burial.

Oxygen isotopes (Fig. 5) recorded by Paleocene planktonic foraminifera are relatively stable during the Danian and Selandian but generally decrease during the Thanetian. This decrease has been attributed to a long-term global warming that precedes the thermal maximum of the Paleocene-Eocene transition (Savin, 1977; Zachos et al., 1993). The $\delta^{18}O$ composition of surface-dwelling planktonic foraminifera clearly shows the effects of paleolatitude between Site 690 (~67°S during the Paleocene) in the Weddell Sea, Site 761 (~40°S) in the eastern Indian Ocean, and Site 465 (25°N) in the central Pacific. Prior to 57.0 Ma, surface water temperatures appear to have been cooler by ~4 °C in the subtropical eastern Indian Ocean than in the equatorial Pacific. The latitudinal thermal gradient dramatically decreased during the early Thanetian when surface waters at Site 690 became comparable to those of the subtropical Site 761.

The data presented here are consistent with a thermally stratified middle and late Paleocene ocean, as the oxygen isotopic values of the subbotinids and globanomalinids suggest cooler paleotemperatures than the shallower-dwelling morozovellids and acarininids. This confirms previous concepts of depth stratification for the early Paleogene planktonic foraminifera (Boersma et al., 1979; Berggren and Norris, 1997; Olsson et al., 1999). The pattern is consistent throughout the Paleocene at different oceanic locations, even at subantarctic ODP Hole 690C, where the surface ocean was strongly thermally stratified, as showed by a strong vertical gradient between acarininids and subbotinids.

Appearance of photosymbiosis

The sequence at Site 761 provides the opportunity to document the isotopic signature of the appearance of photosymbiosis in Paleogene trochospiral planktonic foraminifera (Fig. 4A and 5A). Photosymbioses involving various groups of planktonic algae have independently evolved several times in planktonic foraminifera and the acquisition of such a trophic strategy generally coincides with radiation among the host taxa (Norris, 1996, 1998). The evolution of photosymbiosis during the Paleogene apparently occurred in the common ancestor of the four major clades of muricate planktonic foraminifera, including two clades within *Morozovella* (the *M. velascoensis* and *M. subbotinae* lineages), as well as the genera *Acarinina* and *Igorina*. Previous studies (Kelly et al., 1996; Norris, 1996) suggested that Paleogene photosymbiosis first evolved either during the evolution of *Praemurica inconstans* to *Praemurica uncinata,* or during the range of *Pr. inconstans.* Our data are consistent with

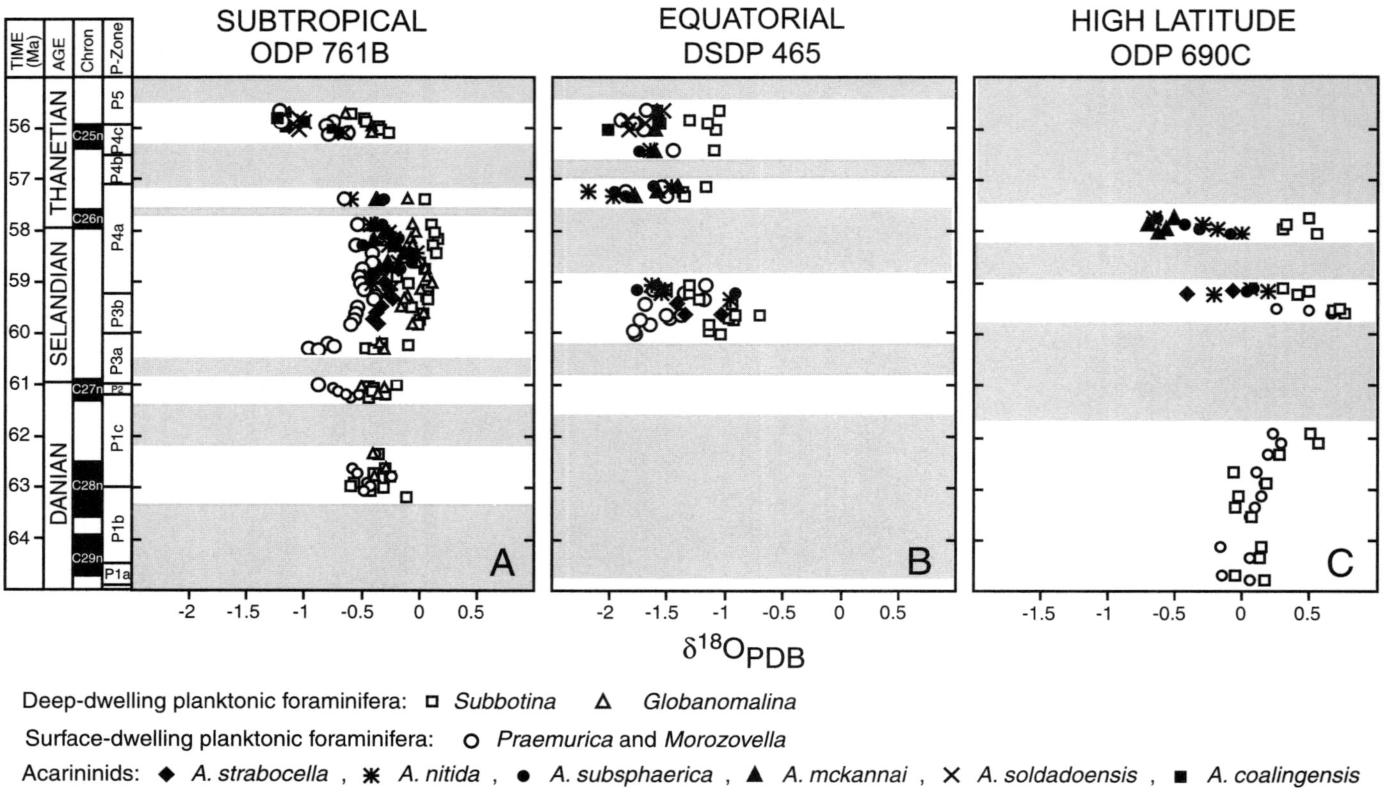

Figure 5. Monospecific planktonic foraminiferal oxygen isotopes recorded at ODP Hole 761B (A), DSDP Holes 465/465A (B), and ODP Hole 690C (C) during the Paleocene. Hiatus in the sections are shown by grey patterns. Chronologic framework from Berggren et al. (2000).

these previous results, showing a divergence in $\delta^{13}C$ of *Praemurica* and *Subbotina* at ~61.0 Ma and supporting a separation of the depth-habitat and trophic strategy of the planktonic foraminifera at this time (Fig. 6). At ~61.3 Ma, carbon and oxygen isotope compositions of *Pr. inconstans* began to deviate from those seen in *Subbotina.* This suggests that the planktonic foraminiferal communities started to stratify within the water column, with the last representatives of *Pr. inconstans* and *Pr. uncinata* occupying shallower habitats than contemporaneous *Subbotina.* Without ontogenetic information, distinguishing the occurrence of photosymbiosis in planktonic foraminifera is difficult. Norris (1996) showed that late representatives of *Pr. inconstans* from DSDP Hole 384 displayed moderate size-related covariation in $\delta^{13}C$ and light $\delta^{18}O$ similar to *Pr. uncinata.* We believe that the deviation of the carbon isotopic compositions recorded at Hole 761B, between ~61.3 and ~61.0 Ma, suggests that photosymbiosis in Paleogene planktonic foraminifera first originated in the younger representatives of the species *Pr. inconstans,* which may be considered as the best potential common ancestor for all the Paleogene symbiont-bearing planktonic foraminifera.

Depth habitat of the acarininids

Previous studies have demonstrated the existence of systematic differences between $\delta^{18}O$ and $\delta^{13}C$ compositions of surface- and deep-dwelling Paleocene planktonic foraminifera. Time series monospecific isotope data from Holes 761B, 465/465B and 690C (Fig. 4 and 5), agree with these established patterns. At low latitude as well as at high latitude, throughout the Paleocene, species of *Acarinina* consistently have heavier $\delta^{13}C$ and lighter $\delta^{18}O$ than coexisting species of *Subbotina* and *Globanomalina.* In addition, at low-latitude sites and during most of the late Paleocene, isotopic patterns in *Morozovella* broadly track trends in *Acarinina,* suggesting that the muricate taxa occupied the mixed layer and shallow thermocline. Previous comparisons of isotopic values between *Morozovella* and *Acarinina* have been inconsistent, although both genera are inferred to have been surface-dwellers.

Among surface-dwelling species, there appear to be subtle differences in ecology between acarininids and morozovellids. Interestingly, the acarininid species from Hole 761B, prior to 57.0 Ma, yield subtly lower $\delta^{13}C$ values (~0.2‰) than contem-

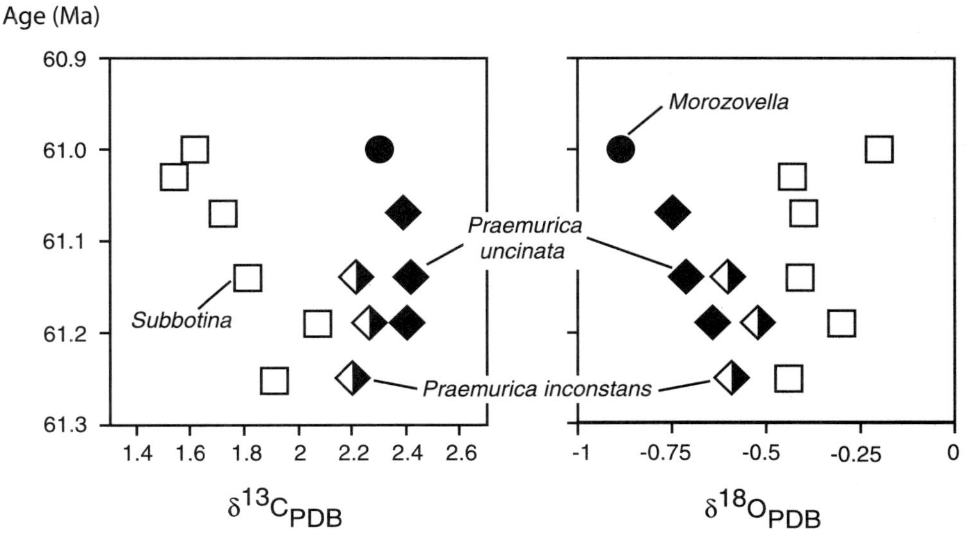

Figure 6. Oxygen and carbon stable isotopes for various planktonic foraminiferal species at the Danian-Selandian transition (between 61.3 and 60.9 Ma). The deviation in both $\delta^{13}C$ and $\delta^{18}O$ values between *Subbotina* and *Praemurica* documents the acquisition of photosymbionts in late representatives of *Praemurica inconstans,* and the increasing depth-related stratification of the planktonic foraminiferal communities in the water column. Time scale from Berggren et al. (2000).

poraneous morozovellids (Fig. 4A). At Hole 465, this offset is less pronounced but still present, at least before 56.5 Ma (Fig. 4B). Bralower et al. (1995) and Corfield and Norris (1998) also noted similar $\delta^{13}C$ differences between species of *Morozovella* and *Acarinina* in late Paleocene and early Eocene age sediments from the equatorial Pacific Ocean (respectively at DSDP Hole 577 and ODP Hole 865). Additionally, in studies of stable isotopic size-dependency of late Paleocene planktonic foraminifera from ODP Hole 758A (tropical Indian Ocean) and DSDP Hole 384 (northwest Atlantic), *Acarinina nitida* and *A. mckannai* were found to have recorded lighter $\delta^{13}C$ than *Morozovella velascoensis* (respectively D'Hondt et al., 1994; Norris, 1996). Small but consistent differences in $\delta^{13}C$ between planktonic foraminiferal species may result from differences in depth habitat, season of growth or vital effect. Corfield and Norris (1998) argued that these subtle differences cannot be ascribed to slight differences in depth habitat because the acarininids from Hole 577 did not record lighter $\delta^{18}O$ than coexisting morozovellids. They concluded that the offset was related to nonequilibrium fractionation, probably due to local metabolic enrichment of ^{13}C in morozovellids. As pointed out by Norris (1998), many organisms display a vital effect, in which food supply, reproductive activity, calcification rate, and association with symbionts alter the isotopic composition away from ratios present in the surrounding environment. Typically, these effects are more severe for $\delta^{13}C$ than for $\delta^{18}O$ since there is more oxygen (in sea water) than carbon (mostly dissolved bicarbonate) to interact with an organism's metabolism during shell growth (Wefer and Berger, 1991; Norris, 1998).

Our data from Hole 761B show that all the early species of *Acarinina,* prior to 57.0 Ma, record significantly heavier $\delta^{18}O$ than *Morozovella* (~0.2‰), suggesting that they occupied slightly cooler environments. At Site 384, located at a subtropical paleolatitude in the North Atlantic, Berggren and Norris (1997) found significantly lighter $\delta^{18}O$ and heavier $\delta^{13}C$ in *Morozovella* than in early species of *Acarinina.* Berggren and Norris (1997) even found in *A. subsphaerica* an apparent trend from $\delta^{18}O$ similar to *Globanomalina* to ratios like those of morozovellids from the Selandian to the early Thanetian. Our data from Site 761 (Fig. 5A) show that the middle Paleocene (Selandian) representatives of *A. strabocella, A. nitida* and *A. subsphaerica* have very positive $\delta^{18}O$ close to those of coexisting deep-dwellers. Therefore, we conclude that the earliest representatives of these species have grown in deeper waters or calcified at a colder season of the year than subsequent representatives.

When both $\delta^{13}C$ and $\delta^{18}O$ compositions are offset consistently, it is likely that depth habitat and/or seasonality may be a factor (Bralower et al., 1995). Although the isotopic offsets between acarininids and morozovellids are close to the uncertainty of the isotopic measurements, the pattern is consistent throughout the interval between 60.0 and 57.0 Ma, and may reflect a real ecologic pattern. Samples from Site 465, located at an equatorial paleolatitude, do not seem to have recorded such differences in oxygen compositions. Quillévéré et al. (2001) recently showed that the early forms of *Acarinina* may have favored relatively mesotrophic environments because the genus diversified in high latitudes before migrating into the oligotrophic low-latitude provinces of the Paleocene ocean. At least prior to their second Paleocene diversification event (appearance of *Acarinina soldadoensis* and *A. coalingensis*), which occurred in the (sub)tropical oligotrophic water masses at ~56.5 Ma, the acarininids may have been more restricted to mesotrophic and cooler habitats than the morozovellids. We believe that the isotopic offset between species of *Acarinina* and *Morozovella* may be attributable to minor differences in depth habitat, rather than the result of nonequilibrium fractionation or variation in season of growth. Hence, at least the Selandian and early Thanetian low-latitude species of *Acarinina* (*A. strabocella, A. nitida,* and *A. subsphaerica*) may have been adapted to slightly deeper subniches than those occupied by *Morozovella.* This pattern would explain why the acarininids were extremely abundant and diverse at high latitude during the Paleocene, while the morozovellids, their photosymbiotic sister group, were absent.

Paleogeographic settings and structure of the mixed layer

Why is the $\delta^{18}O$ variation, observed between *Acarinina* and *Morozovella* at Sites 761 and 384, not recorded at Site 465 and at the other equatorial sites previously studied? The answer may be related to differences in the physical structure of the surface ocean at these locations during the Paleocene. Today, maximum diversity of calcareous plankton communities (foraminifera and calcareous nannoplankton) is seen in the oligotrophic tropical oceanic gyres where the vertical distribution of species reflects a strong stratification of the surface waters. The thickness of the mixed layer and the strength of the thermocline vary with latitude and annual cycle of the surface wind. For example, in the Atlantic Ocean, the mixed layer is relatively shallow between 20°N and 10°S (Hastenrath and Merle, 1987). Additionally, throughout the year and for the basin as a whole, the mixed layer is deeper in the western Atlantic than in the eastern basins (Hastenrath and Merle, 1987; Ravelo and Fairbanks, 1992). The $\delta^{18}O$ offset measured between *Acarinina* and *Morozovella* at Sites 761 (this work) and 384 (Berggren and Norris, 1997) suggests that the structure of the mixed layer at these locations was different relative to that of Sites 465 (this work), 577 (Corfield and Norris, 1998), and 865 (Bralower et al., 1995). During the Paleocene, Sites 465, 577, and 865 were all located in the equatorial part of the Pacific Ocean, whereas Sites 761 and 384 were located in subtropical settings. We infer that the waters over Sites 465, 577 and 865 had a deeper mixed layer in which surface waters and thermocline waters were nearly the same temperature, whereas a higher thermal gradient existed in the upper ocean over Sites 384 and 761 (Fig. 7). Hence, the oxygen isotopic record in the equatorial Pacific did not record the differences in depth habitat as sensitively as those of the two subtropical sites.

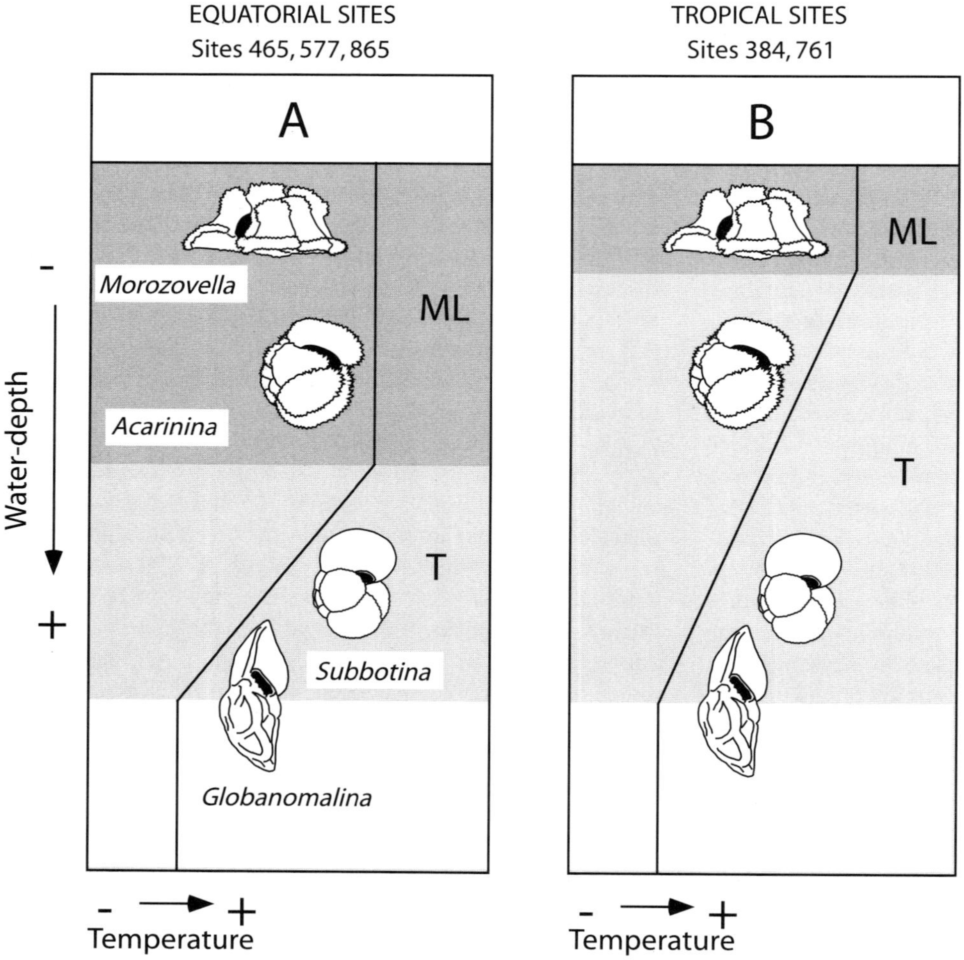

Figure 7. Proposed model for the depth stratification of Paleocene acarininids. Water depth and temperature in arbitrary units. (A) model for DSDP Sites 465, 577 and ODP Site 865, with a thick mixed layer and a low vertical thermal gradient. (B) model for DSDP Site 384 and ODP Site 761, with a shallow mixed layer and a steep vertical thermal gradient. ML—Mixed Layer, T—Thermocline.

It appears that much further research is needed on the stable isotopic characteristics of the early Paleogene surface-dwelling planktonic foraminifera (acarininids, morozovellids, and igorinids), at the species level, to establish whether $\delta^{13}C$ and $\delta^{18}O$ of their calcite reflect the isotopic composition of early Paleogene sea surface water masses as a whole. Corfield and Norris (1998) pointed out that significant carbon isotopic differences may have existed between different clades of *Morozovella.* These authors suggested that the differences were related to nonequilibrium fractionation effects. However, Berggren and Norris (1997), Kelly (1999), and Kelly et al. (2000) show significant differences in both carbon and oxygen isotopes between late Paleocene *Morozovella velascoensis* and *M. subbotinae,* suggesting that the latter species occupied a slightly deeper habitat. Interestingly, the species of the *M. subbotinae* lineage are the only morozovellids found occasionally (during the early Eocene climate optimum) in high-latitude sediments, at Site 690 (see Stott and Kennett, 1990). We conclude that the representatives of the *M. subbotinae* lineage were able to live in cooler and more mesotrophic environments that those of *M. velascoensis,* and we speculate that they may have favored slightly deeper environments at low latitudes. In terms of morphological features, representatives of the *M. subbotinae* lineage present several similarities with *Acarinina.* The species of the *M. subbotinae* lineage are more strongly muricate and more globular than those of the *M. velascoensis* lineage (Olsson et al., 1999). It is likely that as well as the acarininids, the morozovellid populations may have had different trophic strategies and depth-specializations closely related to their morphology. The low-latitude photic zone of the Paleocene ocean may have been more stratified in terms of plankton communities than previously realized.

CONCLUSIONS

Our results are consistent with previous isotopic studies that showed that acarininids and morozovellids were surface mixed-layer planktonic foraminifera, and that subbotinids and globanomalinids inhabited deeper parts of the water column, probably within or below the thermocline. Long-term Paleocene carbon and oxygen stable isotopic profiles at ODP Hole 761B, DSDP Holes 465 and 465A, and ODP Hole 690C are similar to those observed elsewhere. However, the waters over Site 761 probably had a shallower mixed layer and a stronger thermal gradient, based on the significant isotopic differences between morozovellids and early species of acarininids at this site.

Our data support an increasing depth stratification of the planktonic foraminiferal communities during the Paleocene, at least prior to 56.5 Ma. At the end of the Danian (~61.3 Ma), the carbon and oxygen isotope compositions of *Pr. inconstans* began to deviate from those of the other contemporaneous taxa. During the radiation in the Danian and Selandian, the planktonic foraminiferal communities developed specialized depth habitats that were maintained for the rest of the Paleocene. Other ecological innovations were also established near the end of the Danian, such as the acquisition of photosymbiosis within the last representatives of *Praemurica.* Although the acarininids, the morozovellids (and probably the igorinids) shared the same photosymbiotic strategy that they inherited from their common ancestor (the last representatives of *Praemurica inconstans*), they probably had minor differences in depth-ecology. We suggest that the subtropical water column at Site 761 may have had more distinct habitat separation than in the equatorial settings, and that the isotopic offset between acarininids and morozovellids may be more likely attributable to differences in depth habitat, rather than differences in season of growth or vital effect. In low latitude, the early acarininids (*Acarinina strabocella, A. nitida, A. subsphaerica* and *A. mckannai*) may have favored slightly deeper and more mesotrophic environments than contemporaneous morozovellids. We speculate that the water column during the middle and late Paleocene may have been more stratified in terms of plankton communities than previously expected. The global evolution toward oceanic oligotrophy during the Paleocene may have resulted in increasing diversification with extremely specialized planktonic foraminifera.

ACKNOWLEDGMENTS

We gratefully acknowledge C. Kelly, M. Leckie and R. Olsson for their review. We also thank I. Moussa (Institut des Sciences de l'Evolution de Montpellier) for assistance with the mass spectrometer. Stable isotope mass spectrometry at the Woods Hole Oceanographic Institution was supported by grants from the NSF-EAR and through a European Community Marie Curie Fellowship.

APPENDIX. SAMPLE HORIZONS, SPECIES ANALYZED, AND ISOTOPIC DATA

TABLE A1. ODP HOLE 761B

Sample reference (cm)	Depth (mbsf)	Age (Ma)	Species	$\delta^{13}C$	$\delta^{18}O$
21X-4: 50–52	175.2	63.16	*S. triloculinoides*	1.84	–0.11
21X-3: 105–107	174.25	63.06	*S. triloculinoides*	1.99	–0.42
			Pr. inconstans	1.79	–0.48
21X-3: 52–54	173.72	63.00	*S. triloculinoides*	1.68	–0.31
21X-3: 10–12	173.3	62.95	*S. triloculinoides*	1.67	–0.59
			Pr. inconstans	1.99	–0.42
21X-2: 105–107	172.75	62.89	*S. triloculinoides*	1.75	–0.56
			Pr. inconstans	2.15	–0.45
21X-2: 10–12	171.8	62.78	*S. triloculinoides*	1.94	–0.25
			Pr. inconstans	1.93	–0.24
			Gl. compressa	2.20	–0.38
21X-1: 97–99	171.25	62.72	*S. triloculinoides*	1.53	–0.40
			Pr. inconstans	1.48	–0.53
21X-1: 10–12	170.3	62.62	*S. triloculinoides*	2.04	–0.30
			Pr. inconstans	1.95	–0.58
			Gl. compressa	1.80	–0.28
20X-5: 102–104	167.8	62.34	*S. triloculinoides*	1.89	–0.35
			Pr. inconstans	2.07	–0.37
			Gl. compressa	2.20	–0.39
20X-4: 100–102	166.2	61.25	*S. triloculinoides*	1.89	–0.44
			Pr. inconstans	2.20	–0.59
20X-4: 10–12	165.3	61.19	*S. triloculinoides*	2.07	–0.30
			Pr. inconstans	2.27	–0.52
			Pr. uncinata	2.41	–0.64
			Gl. compressa	2.29	–0.35
20X-3: 100–102	164.7	61.14	*S. triloculinoides*	1.82	–0.41
			Pr. inconstans	2.22	–0.62
			Pr. uncinata	2.43	–0.71
20X-3: 10–12	163.8	61.07	*S. triloculinoides*	1.72	–0.40
			Pr. uncinata	2.39	–0.75
			Gl. compressa	1.68	–0.29
20X-2: 98–100	163.2	61.03	*S. triloculinoides*	1.54	–0.43
			Gl. ehrenbergi	1.43	–0.50
20X-2: 50–52	162.71	61.00	*S. triloculinoides*	1.62	–0.20
			M. angulata	2.31	–0.88
20X-2: 10–12	162.3	60.34	*S. triloculinoides*	1.73	–0.40
			M. angulata	2.62	–0.87
			Gl. ehrenbergi	1.76	–0.29
20X-1: 90–92	161.6	60.28	*S. triloculinoides*	1.97	–0.47
			M. angulata	2.52	–0.96
20X-1: 50–52	161.2	60.24	*S. triloculinoides*	2.05	–0.10
			M. angulata	2.72	–0.74
20X-1: 10–12	160.8	60.20	*S. triloculinoides*	2.01	–0.32
			M. angulata	2.44	–0.74
			Gl. ehrenbergi	2.04	–0.33
19X-4: 100–102	156.7	59.83	*S. triloculinoides*	2.13	–0.01
			M. angulata	2.85	–0.59
			A. strabocella	2.56	–0.35
			Gl. ehrenbergi	2.04	–0.06
19X-4: 10–12	155.8	59.75	*S. triloculinoides*	1.91	0.00
			M. angulata	2.55	–0.56
			A. strabocella	2.44	–0.40
19X-3: 10–12	154.3	59.61	*S. triloculinoides*	2.10	0.03
			M. angulata	2.86	–0.55
			A. strabocella	2.73	–0.37
			Gl. ehrenbergi	1.99	0.05

continued

TABLE A1. *Continued*

Sample reference (cm)	Depth (mbsf)	Age (Ma)	Species	$\delta^{13}C$	$\delta^{18}O$
19X-2: 10–12	152.8	59.48	*S. triloculinoides*	2.32	–0.07
			M. angulata	3.28	–0.54
			A. strabocella	3.11	–0.33
			Gl. ehrenbergi	2.12	–0.15
19X-1: 10–12	151.3	59.34	*S. triloculinoides*	2.30	0.07
			M. angulata	3.06	–0.39
			A. strabocella	3.00	–0.22
			Gl. ehrenbergi	1.85	–0.13
18X-6: 10–12	149.3	59.26	*S. triloculinoides*	2.51	0.06
			M. angulata	3.30	–0.38
			A. strabocella	3.21	–0.25
			Gl. pseudomenardii	2.43	–0.10
18X-5: 10–12	147.8	59.16	*S. triloculinoides*	2.38	0.07
			M. angulata	3.34	–0.48
			A. strabocella	3.17	–0.25
			A. nitida	3.28	–0.25
			Gl. pseudomenardii	2.44	0.02
18X-4: 10–12	146.3	59.02	*S. triloculinoides*	2.21	–0.10
			M. angulata	3.22	–0.50
			A. strabocella	3.07	–0.30
			A. nitida	3.08	–0.42
			Gl. pseudomenardii	2.11	0.12
18X-3: 10–12	144.8	58.89	*S. triloculinoides*	2.46	–0.20
			M. velascoensis	3.68	–0.53
			A. strabocella	3.22	–0.41
			A. nitida	3.31	–0.43
			Gl. pseudomenardii	2.48	0.08
18X-2: 10–12	143.3	58.76	*S. triloculinoides*	2.54	0.04
			M. velascoensis	3.65	–0.50
			A. strabocella	3.25	–0.28
			A. nitida	3.29	–0.24
			A. subsphaerica	3.25	–0.16
			A. mckannai	3.47	–0.33
			Gl. pseudomenardii	2.38	0.06
18X-1: 10–12	141.8	58.62	*S. triloculinoides*	2.47	0.00
			M. velascoensis	3.36	–0.41
			A. nitida	3.09	–0.18
			A. subsphaerica	3.00	–0.07
			A. mckannai	3.18	–0.26
			Gl. pseudomenardii	2.25	–0.01
17X-6: 10–12	139.8	58.44	*S. triloculinoides*	2.37	0.14
			M. velascoensis	3.59	–0.40
			A. nitida	3.12	–0.03
			A. subsphaerica	3.11	–0.07
			A. mckannai	3.10	–0.13
17X-5: 10–12	138.3	58.30	*S. triloculinoides*	2.57	0.12
			M. velascoensis	3.66	–0.55
			A. nitida	3.32	–0.26
			A. subsphaerica	3.25	–0.49
			A. mckannai	3.23	–0.49
			Gl. pseudomenardii	2.26	–0.05
17X-4: 10–12	136.8	58.17	*S. triloculinoides*	2.54	0.15
			M. velascoensis	3.47	–0.37
			A. nitida	3.30	–0.22
			A. subsphaerica	3.28	–0.19
			A. mckannai	3.29	–0.37
			Gl. pseudomenardii	2.19	–0.07

continued

TABLE A1. *Continued*

Sample reference (cm)	Depth (mbsf)	Age (Ma)	Species	$\delta^{13}C$	$\delta^{18}O$
17X-3: 10–12	135.3	58.03	*S. triloculinoides*	2.87	0.13
			M. velascoensis	3.99	−0.34
			A. nitida	3.44	−0.27
			A. subsphaerica	3.62	−0.26
			A. mckannai	3.54	−0.29
			Gl. pseudomenardii	2.57	−0.03
17X-2: 10–12	133.8	57.89	*S. triloculinoides*	2.86	0.10
			M. velascoensis	4.03	−0.54
			A. nitida	3.66	−0.42
			A. subsphaerica	3.66	−0.32
			A. mckannai	3.57	−0.38
			Gl. pseudomenardii	2.92	−0.06
17X-1: 10–12	132.3	57.39	*S. triloculinoides*	3.00	0.05
			M. velascoensis	4.20	−0.58
			A. nitida	3.83	−0.62
			A. subsphaerica	3.69	−0.31
			A. mckannai	3.80	−0.36
			Gl. pseudomenardii	2.77	−0.10
16X-4: 110–112	128.3	56.08	*S. triangularis*	2.59	−0.27
			M. velascoensis	3.50	−0.62
			A. soldadoensis	3.36	−0.65
			A. coalingensis	3.45	−0.70
			Gl. pseudomenardii	2.48	−0.41
16X-3: 110–112	127.8	56.04	*S. triangularis*	2.50	−0.34
			M. velascoensis	3.52	−0.77
			A. soldadoensis	3.34	−1.05
			A. coalingensis	3.22	−0.78
16X-3: 10–12	126.8	55.98	*S. triangularis*	2.52	−0.35
			M. velascoensis	3.40	−0.78
			A. soldadoensis	3.14	−1.07
			A. coalingensis	3.11	−1.16
			Gl. pseudomenardii	2.38	−0.41
16X-2: 103–105	125.2	55.88	*S. triangularis*	2.27	−0.47
			M. velascoensis	3.23	−0.75
			A. soldadoensis	3.26	−1.01
			A. coalingensis	3.50	−1.01
16X-2: 10–12	124.3	55.8	*S. triangularis*	2.44	−0.48
			M. velascoensis	3.30	−1.19
			A. soldadoensis	3.16	−1.04
			A. coalingensis	3.25	−1.23
16X-1: 8–10	122.8	55.73	*S. triangularis*	2.22	−0.59
			M. velascoensis	3.21	−1.16
			A. soldadoensis	2.85	−1.19
			A. coalingensis	3.26	−1.15
			Gl. pseudomenardii	2.41	−0.63

TABLE A2. ODP HOLES 465 AND 465A

Sample reference (cm)	Depth (mbsf)	Age (Ma)	Species	$\delta^{13}C$	$\delta^{18}O$
465-6X-5: 66–68	45.6	60.03	*S. triloculinoides*	1.84	−1.04
			M. angulata	2.61	−1.77
465-6X-4: 120–122	44.7	59.97	*S. triloculinoides*	1.66	−1.14
			M. angulata	2.51	−1.79
465-6X-3: 61–63	42.6	59.84	*S. triloculinoides*	1.88	−1.14
			M. angulata	2.78	−1.65
465-6X-2: 60–62	41.1	59.75	*S. triloculinoides*	1.83	−0.93
			M. angulata	2.68	−1.73
465A-1X-1: 141–146	40.41	59.70	*S. triloculinoides*	1.72	−0.98
			M. angulata	2.43	−1.48

continued

TABLE A2. *Continued*

Sample reference (cm)	Depth (mbsf)	Age (Ma)	Species	δ^{13}C	δ^{18}O
465-6X-1: 70–72	39.7	59.66	*S. triloculinoides*	1.93	–0.71
			M. velascoensis	2.99	–1.38
			A. strabocella	2.84	–1.34
465A-1X-1: 52–54	39.53	59.64	*S. triloculinoides*	1.80	–0.92
			M. angulata	2.43	–1.51
			A. strabocella	3.06	–1.03
465-5X-5: 60–62	36.1	59.43	*S. triloculinoides*	2.90	–0.93
			M. velascoensis	3.33	–1.69
			A. strabocella	3.03	–1.40
465-5X-4: 63–65	34.63	59.34	*S. triloculinoides*	2.70	–1.23
			M. velascoensis	3.50	–1.18
			A. nitida	2.94	–0.95
465-5X-3: 63–65	33.13	59.23	*S. triloculinoides*	2.80	–1.31
			M. velascoensis	3.19	–1.35
			A. nitida	2.65	–1.55
			A. subsphaerica	3.08	–0.91
465-5X-2: 62–64	31.64	59.15	*S. triloculinoides*	2.82	–1.50
			M. velascoensis	3.48	–1.62
			A. nitida	3.17	–1.52
			A. subsphaerica	3.57	–1.76
465-5X-1: 65–67	30.15	59.06	*S. triloculinoides*	2.53	–1.31
			M. velascoensis	4.00	–1.17
			A. nitida	3.28	–1.64
			A. subsphaerica	3.58	–1.59
465-4X-4: 62–64	25.12	57.34	*S. triloculinoides*	2.90	–1.35
			M. velascoensis	4.00	–1.50
			A. nitida	3.78	–1.98
			A. subsphaerica	3.81	–1.86
			A. mckannai	3.89	–1.77
465-4X-3: 60–62	23.64	57.25	*S. triloculinoides*	3.19	–1.46
			M. velascoensis	4.46	–1.85
			A. nitida	4.18	–2.18
			A. subsphaerica	4.14	–1.96
			A. mckannai	4.35	–1.58
465-4X-2: 60–62	22.18	57.15	*S. triloculinoides*	3.14	–1.16
			M. velascoensis	4.15	–1.56
			A. nitida	3.97	–1.47
			A. subsphaerica	4.02	–1.57
			A. mckannai	3.97	–1.44
465-4X-1: 59–61	20.57	57.15	*S. triloculinoides*	2.86	–1.10
			M. velascoensis	3.81	–1.45
			A. nitida	3.78	–1.64
			A. subsphaerica	3.67	–1.66
			A. mckannai	3.71	–1.66
465-3X-4: 60–62	15.6	56.04	*S. triangularis*	3.16	–1.08
			M. velascoensis	4.19	–1.70
			A. mckannai	4.16	–1.63
			A. soldadoensis	3.86	–1.83
			A. coalingensis	3.75	–2.02
465-3X-3: 60–62	14.1	55.92	*S. triangularis*	3.03	–1.16
			M. velascoensis	3.87	–1.79
			A. soldadoensis	3.75	–1.69
			A. coalingensis	3.96	–1.56
465-3X-2: 129–131	13.29	55.85	*S. triangularis*	3.06	–1.30
			M. velascoensis	3.75	–1.91
			A. soldadoensis	3.84	–1.85
			A. coalingensis	3.92	–1.58
465-3X-1: 59–61	11.09	55.67	*S. triangularis*	2.59	–1.05
			M. velascoensis	3.50	–1.64
			A. soldadoensis	3.24	–1.53
			A. coalingensis	3.18	–1.59

TABLE A3. ODP HOLE 690C

Sample reference (cm)	Depth (mbsf)	Age (Ma)	Species	$\delta^{13}C$	$\delta^{18}O$
15H-4: 37–39	247.77	64.76	*S. trivialis*	2.05	0.17
			Pr. pseudoinconstans	2.18	0.06
15H-3: 125–127	247.15	64.68	*S. trivialis*	2.29	–0.06
			Pr. pseudoinconstans	2.01	–0.15
15H-2: 34–36	244.74	64.32	*S. trivialis*	2.00	0.13
			Pr. pseudoinconstans	1.94	0.06
15H-1: 50–52	243.4	64.14	*S. trivialis*	2.06	0.15
			Pr. pseudoinconstans	2.15	–0.16
14H-5: 30–32	239.5	63.54	*S. trivialis*	1.86	0.06
			Pr. inconstans	1.84	0.05
14H-4: 50–52	238.2	63.35	*S. trivialis*	1.87	–0.06
			Pr. inconstans	2.18	0.09
14H-3: 50–52	236.7	63.12	*S. trivialis*	2.06	–0.02
			Pr. inconstans	1.90	0.15
14H-2: 32–34	235.02	62.87	*S. triloculinoides*	2.03	0.18
			Pr. inconstans	1.95	0.17
14H-1: 50–52	233.7	62.68	*S. triloculinoides*	1.88	–0.07
			Pr. inconstans	2.46	0.10
13H-6: 31–33	231.4	62.33	*S. triloculinoides*	1.72	0.27
			Pr. inconstans	1.64	0.19
13H-5: 31–33	229.9	62.11	*S. triloculinoides*	1.93	0.57
			Pr. inconstans	1.64	0.29
13H-4: 50–52	228.6	61.92	*S. triloculinoides*	1.77	0.51
			Pr. inconstans	1.58	0.22
13H-3: 50–52	226.9	59.60	*S. triloculinoides*	1.63	0.77
			Pr. inconstans	1.70	0.67
13H-2: 50–52	225.6	59.55	*S. triloculinoides*	1.84	0.67
			Pr. inconstans	1.70	0.50
13H-1: 50–52	224.1	59.50	*S. triloculinoides*	1.69	0.72
			Pr. inconstans	1.24	0.25
12H-3: 50–52	217.2	59.23	*S. triloculinoides*	2.21	0.41
			A. strabocella	3.32	–0.40
			A. nitida	3.23	–0.21
12H-2: 35–37	215.75	59.17	*S. triloculinoides*	2.67	0.49
			A. strabocella	3.56	–0.06
			A. nitida	3.32	0.19
			A. subsphaerica	3.35	0.04
12H-1: 50–52	214.4	59.12	*S. triloculinoides*	2.60	0.30
			A. nitida	3.32	0.05
			A. subsphaerica	3.41	0.08
11H-4: 55–57	209.25	58.04	*S. triloculinoides*	2.01	0.55
			A. nitida	3.69	–0.00
			A. subsphaerica	3.08	–0.09
			A. mckannai	3.51	–0.62
11H-3: 80–82	208.0	57.96	*S. triloculinoides*	2.39	0.30
			A. nitida	3.36	–0.19
			A. subsphaerica	3.18	–0.31
			A. mckannai	3.06	–0.57
11H-2: 80–82	206.5	57.86	*S. triloculinoides*	2.07	0.32
			A. nitida	3.82	–0.29
			A. subsphaerica	3.13	–0.42
			A. mckannai	3.46	–0.69
11H-2: 90–92	205.1	57.76	*S. triloculinoides*	1.80	0.49
			A. nitida	3.15	–0.66
			A. subsphaerica	3.07	–0.64
			A. mckannai	3.20	–0.50

REFERENCES CITED

Aubry, M.-P., 1995, From chronology to stratigraphy: Interpreting the lower and middle Eocene stratigraphic record in the Atlantic Ocean, *in* Berggren, W.A., et al., eds., Geochronology, time scales, and global stratigraphic correlations: SEPM (Society for Sedimentary Geology) Special Publication 54, p. 213–274.

Berggren, W.A., and Norris, R.D., 1997, Biostratigraphy, phylogeny, and systematics of Paleocene trochospiral planktic foraminifera: Micropaleontology Special Publication, v. 43, p. 1–116.

Berggren, W.A., Aubry, M.-P., van Fossen, M., Kent, D.V., Norris, R.D., and Quillévéré, F., 2000, Integrated Paleocene calcareous plankton magnetobiochronology and stable isotope stratigraphy: DSDP Site 384 (northwest Atlantic Ocean): Palaeogeography, Palaeoclimatology, Palaeoecology, v. 159, p. 1–51.

Boersma, A., and Premoli Silva, I., 1983, Paleocene planktonic foraminiferal biogeography and the paleoceanography of the Atlantic Ocean: Micropaleontology, v. 29, p. 355–381.

Boersma, A., Shackleton, N.J., Hall, M.A., and Given, Q.C., 1979, Carbon and oxygen isotope records at DSDP Site 384 (North Atlantic) and some Paleocene paleotemperatures and carbon isotope variations in the Atlantic Ocean, *in* Tucholke, B.E., Vogt, P.R., et al., eds., Initial reports of the Deep Sea Drilling Project: Washington, D.C., U.S. Government Printing Office, v. 43, p. 695–717.

Bralower, T.J., Zachos, J.C., Thomas, E., Parrow, M., Paull, C.K., Kelly, D.C., Premoli Silva, I., Sliter, W.V., and Lohmann, K.C., 1995, Late Paleocene to Eocene paleoceanography of the equatorial Pacific Ocean: Stable isotopes recorded at Ocean Drilling Program Site 865, Allison Guyot: Paleoceanography, v. 10, p. 841–865.

Corfield, R.M., 1994, Paleocene oceans and climate: An isotopic perspective: Earth Science Reviews, v. 37, p. 225–252.

Corfield, R.M., and Cartlidge, J.E., 1991, Isotopic evidence for the depth stratification of fossil and recent Globigerinina: A review: Historical Biology, v. 5, p. 37–63.

Corfield, R.M., and Cartlidge, J.E., 1992, Oceanographic and climatic implications of the Paleocene carbon isotope maximum: Terra Nova, v. 4, p. 443–455.

Corfield, R.M., and Norris, R.D., 1996, Deep water circulation in the Paleocene ocean, *in* Knox, R.W.O'B., et al., eds., Correlation of the early Paleogene in northwest Europe: Geological Society [London] Special Publication 101, p. 443–456.

Corfield, R.M., and Norris, R.D., 1998, The oxygen and carbon isotopic context of the Paleocene–Eocene Epoch boundary, *in* Aubry, M.-P., et al., eds., Late Paleocene–early Eocene climatic and biotic events in the marine and terrestrial records: New York, Columbia University Press, p. 124–137.

Corfield, R.M., and Shackleton, N.J., 1988, Productivity change as a control on planktonic foraminiferal evolution after the Cretaceous-Tertiary boundary: Historical Biology, v. 1, p. 323–343.

Coxall, H.C., Pearson, P.N., Shackleton, N.J., and Hall, M.A., 2000, Hantkeninid depth adaptation: An evolving life strategy in a changing ocean: Geology, v. 28, p. 87–90.

D'Hondt, S., Zachos, J.C., and Schultz, G., 1994, Stable isotopic signals and photosymbiosis in late Paleocene planktic foraminifera: Paleobiology, v. 20, p. 391–406.

Douglas, R.G., and Savin, S.M., 1978, Oxygen isotope evidence for the depth stratification of Tertiary and Cretaceous planktic foraminifera: Marine Micropaleontology, v. 3, p. 175–196.

Emiliani, C., 1954, Depth habitats of some species of pelagic foraminifera as indicated by oxygen isotope ratios: American Journal of Science, v. 252, p. 149–158.

Epstein, S., Buchsbaum, R., Lowenstam, H.A., and Urey, H.C., 1953, Revised carbonate-water isotopic temperature scale: Geological Society of America Bulletin, v. 64, p. 1315–1325.

Hastenrath, S., and Merle, J., 1987, Annual cycle of subsurface thermal structure in the tropical Atlantic Ocean: Journal of Physical Oceanography, v. 17, p. 1518–1538.

Hemleben, C., Muhlen, D., Olsson, R.K., and Berggren, W.A., 1991, Surface texture and the first occurrence of spines in planktonic foraminifera from the early Tertiary: Neues Jahrbuch für Geologie und Paläontologie, v. 128, p. 117–146.

Keller, G., MacLeod, N., and Barrera, E., 1992, Eocene–Oligocene faunal turnover in planktic foraminifera, and Antarctic glaciation, *in* Prothero, D.R., and Berggren, W.A., eds., Eocene and Oligocene climatic and biotic evolution: Princeton, New Jersey, Princeton University Press, p. 218–244.

Kelly, D.C., 1999, Ecospace partitioning during the late Paleocene radiation of the planktonic foraminiferal genus *Morozovella:* Inference from stable isotopes: Geological Society of America Abstracts with Programs, v. 31, no. 7, p. 122.

Kelly, D.C., Arnold, A.J., and Parker, W.C., 1996, Paedomorphosis and the origin of the Paleogene planktonic foraminiferal genus *Morozovella:* Paleobiology, v. 22, p. 266–281.

Kelly, D.C., Bralower, T.J., and Zachos, J.C., 2000, On the demise of the early Paleogene *Morozovella velascoensis* lineage: Terminal progenesis in the planktonic foraminifera: Geologiska Föreningens i Stockholm Förhandlingar, v. 122, p. 86–87.

Miller, K.G., Fairbanks, R.G., and Mountain, G.S., 1987, Tertiary oxygen isotope synthesis, sea-level history, and continental margin erosion: Paleoceanography, v. 2, p. 1–19.

Norris, R.D., 1996, Symbiosis as an evolutionary innovation in the radiation of Paleocene planktic foraminifera: Paleobiology, v. 22, p. 461–480.

Norris, R.D., 1998, Recognition and macroevolutionary significance of photosymbiosis in molluscs, corals, and foraminifera, *in* Norris, R.D., and Corfield, R.M., eds., Isotope paleobiology and paleoecology: The Paleontological Society Papers, v. 4, p. 68–100.

Norris, R.D., Corfield, R.M., and Cartlidge, J.E., 1993, What is gradualism? Cryptic speciation in globorotaliid foraminifera: Paleobiology, v. 22, p. 386–405.

Norris, R.D., Kroon, D., Huber, B.T., and Erbacher, J., 2001, Cretaceous–Palaeogene ocean and climate change in the subtropical North Atlantic, *in* Kroon, D., et al., eds., Western North Atlantic Palaeogene and Cretaceous palaeoceanography: Geological Society [London] Special Publication 183, p. 1–22.

Olsson, R.K., Hemleben, C., Berggren, W.A., and Liu, C., 1992, Wall texture classification of planktonic foraminifera genera in the lower Danian: Journal of Foraminiferal Research, v. 22, p. 195–213.

Olsson, R.K., Hemleben, C., Berggren, W.A., and Huber, B.T., 1999, Atlas of Paleocene planktonic foraminifera: Smithsonian Contributions to Paleobiology, v. 85., 255 p.

Pearson, P.N., 1998, Stable isotopes and the study of evolution in planktonic foraminifera, *in* Norris, R.D., and Corfield, R.M., eds., Isotope paleobiology and paleoecology: The Paleontological Society Papers, v. 4, p. 138–178.

Pearson, P.N., Shackleton, N.J., and Hall, M.A., 1997, Stable isotopic evidence for the sympatric divergence of *Globigerinoides trilobus* and *Orbulina universa* (planktonic foraminifera): Journal of the Geological Society of London, v. 154, p. 441–449.

Quillévéré, F., 2000, Etude morphométrique et isotopique ($\delta^{18}O$, $\delta^{13}C$) de la diversification Paléocène du genre *Acarinina:* Implications paléocéanographiques [Ph.D. thesis]: Académie de Montpellier, Université Montpellier II, Sciences et Techniques du Languedoc, Paléontologie, 199 p., 11 plates, appendices.

Quillévéré, F., Norris, R.D., Berggren, W.A., and Aubry, M.-P., 2000, 59.2 Ma and 56.5 Ma: Two significant moments in the evolution of acarininids (planktonic foraminifera): Geologiska Föreningens i Stockholm Förhandlingar, v. 122, p. 131–132.

Quillévéré, F., Norris, R.D., Moussa, I., and Berggren, W.A., 2001, Role of photosymbiosis and biogeography in the diversification of early Paleogene acarininids (planktonic foraminifera): Paleobiology, v. 27, p. 311–326.

Quillévéré, F., Aubry, M.-P., Norris, R.D., and Berggren, W.A., 2002, Paleocene oceanography of the eastern subtropical Indian Ocean: An integrated magnetobiostratigraphic and stable isotope study of ODP Hole 761B (Wombat Plateau): Palaeogeography, Palaeoclimatology, Palaeoecology, v. 184, p. 371–405.

Ravelo, A.C., and Fairbanks, R.G., 1992, Oxygen isotopic composition of multiple species of planktonic foraminifera: Recorders of the modern photic zone temperature gradient: Paleoceanography, v. 7, p. 815–831.

Savin, S.M., 1977, The history of the Earth's surface temperature during the past 100 million years: Annual Reviews of Earth Planetary Sciences, v. 5, p. 319–355.

Schneider, C.E., and Kennett, J.P., 1996, Isotopic evidence for interspecies habitat differences during evolution of the Neogene planktonic foraminiferal clade *Globoconella:* Paleobiology, v. 22, p. 282–303.

Shackleton, N.J., Corfield, R.M., and Hall, M.A., 1985, Stable isotope data and the ontogeny of Paleocene planktonic foraminifera: Journal of Foraminiferal Research, v. 15, p. 321–336.

Spero, H.J., 1998, Life history and stable isotope geochemistry of planktonic foraminifera, *in* Norris, R.D., and Corfield, R.M., eds., Isotope paleobiology and paleoecology: Paleontological Society Papers, v. 4, p. 7–36.

Spero, H.J., and Williams, D.F., 1988, Extracting environmental information from planktonic foraminiferal $\delta^{13}C$ data: Nature, v. 335, p. 717–719.

Spero, H.J., Lerche, I., and Williams, D.F., 1991, Opening the carbon isotope "vital effect" black box, 2: Quantitative model for interpreting foraminiferal carbon isotope data: Paleoceanography, v. 6, p. 639–655.

Stott, L.D., and Kennett, J.P., 1990, Antarctic Paleogene planktonic foraminifera biostratigraphy: ODP Leg 113, Sites 689 and 690, *in* Barker, P.F., Kennett, J.P., et al., eds., Proceedings of the Ocean Drilling Program, Scientific Results: Texas A & M University, College Station, Texas, USA, v. 113, p. 549–569.

Stott, L.D., Kennett, J.P., Shackleton, N.J., and Corfield, R.M., 1990, The evolution of Antarctic surface waters during the Paleogene inferences from the stable isotopic composition of planktonic foraminifers, ODP Leg 113, *in* Barker, P.F., Kennett, J.P., et al., eds., Proceedings of the Ocean Drilling Program, Scientific Results: Texas A & M University, College Station, Texas, USA, v. 113, p. 849–863.

Thompson, E.I., and Schmitz, B., 1997, Barium and late Paleocene $\delta^{13}C$ maximum: Evidence of increased marine surface productivity: Paleoceanography, v. 12, p. 239–254.

Wefer, G., and Berger, W.H., 1991, Isotope paleontology: Growth and composition of extant calcareous species: Marine Geology, v. 100, p. 207–248.

Zachos, J.C., Lohmann, K.C., Walker, J.C.G., and Wise, S.W., 1993, Abrupt climate change and transient climates during the Paleogene: A marine perspective: Journal of Geology, v. 101, p. 191–213.

Zachos, J.C., Stott, L.D., and Lohmann, K.C., 1994, Evolution of early Cenozoic marine temperatures: Paleoceanography, v. 9, p. 353–387.

Zachos, J.C., Pagani, M., Sloan, L., Thomas, E., and Billups, K., 2001, Trends, rythms, and aberrations in global climate 65 Ma to present: Science, v. 292, p. 686–693.

Manuscript Accepted by the Society August 13, 2002

Printed in the USA

Geological Society of America
Special Paper 369
2003

Early Paleogene transgressions: Stratigraphical and sedimentological evidence from the northern Peri-Tethys

E.P. Radionova*
V.N. Beniamovski*
Geological Institute of the Russian Academy of Sciences, Pyzhevsky, 7, Moscow 119017, Russia
A.I. Iakovleva*
Instituto de Geologia, UNAM, Ciudad Universitaria, Del. Coyacan Mexico D.T., 04510, Mexico
N.G. Muzylöv
T.V. Oreshkina*
E.A. Shcherbinina*
Geological Institute of the Russian Academy of Sciences, Pyzhevsky, 7, Moscow 119017, Russia
G.E. Kozlova*
All-Russian Petroleum Research Exploration Institute, Liteiny prospect, 39, St. Petersburg, Russia

ABSTRACT

Our study in lithologically diverse lower Paleogene deposits of the former southern USSR and West Siberia resolves three sedimentary provinces: (i) Southern (Crimea-Caucasus and Central Asia), dominated by biogenic calcareous sediments of the deepest basinal portions; (ii) Transitional (southern Russian craton and Turan plate), displaying calcareous and siliceous clayey deposits; and (iii) Northern (central Russian craton and West Siberian plate), showing biogenic siliceous and terrigenous-siliceous sediments of basinal margin. We used standard and regional scales based on seven microfossil groups (planktonic and benthic foraminifers, nannoplankton, radiolaria, diatoms, silicoflagellates, and dinocysts) to correlate various sediment facies among ~80 reference sections across the study area. Correlation was performed for the Transitional province, where the presence of both calcareous and siliceous facies affords the use of calcareous and siliceous planktonic and dinocyst scales with varying resolutions. Besides, in the Transitional province, we established regional sedimentary cycles corresponding to the late Thanetian, Ypresian, and late Lutetian–Bartonian. These cycles are traceable into the Northern province, enabling us to determine or refine ages of the sequences, and into the Southern province.

Each sedimentary supercycle falls into three units, often with hiatus at the base, correlatable to 3rd order eustatic cycles and featured by distinctive facies. The facies succession appears to reiterate through the upper Thanetian, Ypresian, and upper Lutetian–Bartonian cycles.

The lower units of each supercycle are composed of calcareous sediments rich in nannofossils and foraminifers and are traceable from the Southern province into the Transitional province, the late Lutetian unit extending as far as the Northern province.

*E-mail: Radionova—radionova@geo.tv-sign.ru; Beniamovski—ben@geo.tv-sign.ru; Iakovleva—iakovl@yahoo.com; Oreshkina—oreshkina@geo.tv-sign.ru; Shcherbinina—shcherbinina@geo.tv-sign.ru; Kozlova—ins@vnigri.spb.su.

Radionova, E.P., Beniamovski, V.N., Iakovleva, A.I., Muzylöv, N.G., Oreshkina, T.V., Shcherbinina, E.A., and Kozlova, G.E., 2003, Early Paleogene transgressions: Stratigraphical and sedimentological evidence from the northern Peri-Tethys, *in* Wing, S.L., Gingerich, P.D., Schmitz, B., and Thomas, E., eds., Causes and Consequences of Globally Warm Climates in the Early Paleogene: Boulder, Colorado, Geological Society of America Special Paper 369, p. 239–261.

The middle unit is high in organic matter in the Southern province and passes into terrigenous-siliceous sediments rich in radiolaria, diatoms, and dinocysts in the Transitional and Northern provinces. The upper unit is terrigenous-carbonate and, in places, terrigenous-siliceous in the Southern province, becoming terrigenous-siliceous and biogenic siliceous (spongolites, diatomites) in the Transitional and Northern provinces.

Calcareous plankton (lower unit) giving way to siliceous (upper unit) might record changes in basinal circulation due to Peri-Tethys communicating with the North Atlantic and/or Arctic. The sedimentary provinces shifted southward as the Peri-Tethys shrank and climate changed between the Thanetian and Bartonian.

INTRODUCTION

In the early Paleogene, the northeastern Peri-Tethys stretched from the present-day Caucasus in the south to the Baltic Sea in the European part of Russia, and its Asian part occupied the whole of West Siberia and reached as far south as present-day western Kazakhstan, Turkmenistan, and Uzbekistan. This vast region was a shelf area dominated by biogenic sedimentation of siliceous, calcareous, and mixed types. G. Leonov (1964) divided this region into two domains, southern and northern. The southern domain was characterized by deposition of nanno- and foraminiferal oozes, not infrequently with a considerable terrigenous component. A peculiar facies was represented by sediments enriched in organic matter. The northern domain of the basin displays marine diatom oozes, whose deposition has customarily been attributed to cold water incursions. Between the two domains, an area with a mixed sedimentation type was located. This is the reason, based on prevalent sediment type, we outline three provinces: (i) Northern, gravitating to the Arctic basin and dominated by siliceous sediments (ii) Southern, adjacent to the Tethys, with calcareous sediments, and (iii) Transitional, dominated by terrigenous sediments, in places with significant amounts of biogenic carbonate or silica.

By the beginning of the Paleogene, a large back-arc basin had been separated from the Tethys by a volcanic arc stretching from what is now the Balkans via the Pontus and further on through the Lesser Caucasus, Talysh, and Iran (Zonenshain and Le Pichon, 1987). Deep-water sags of this basin have been reconstructed on the site of the present-day Black Sea and South Caspian basins and of the Greater Caucasus Range. To the north,

→

Figure 1. Study region symbols: (a) boundary of the maximum development of Paleogene marine deposits. (b) Location of sections referred to in text and in figures, by region, follows. **Crimea:** 1—Bakhchisaray* (asterisks denote lithology and stratigraphic subdivision shown on the basis of our new data); 1k—Nasypkoy (Bugrova et al., 2002). **Cis-Caucasia:** 2—the Kuban R.; 3—the Maly Zelenchuk R. and 4—Medani village (Stupin and Muzylöv, 2001); 5—the Kheu R. (Krasheninnikov and Muzylöv, 1975; Gavrilov et al., 2000; Radionova and Khokhlova, 1994, and this study); 6—Keresta valley (*); 7—Well 619, Rostov Region (T.E. Ulanovskaya's material); 37—the Rubas-Chai R., Radionova and Khokhlova, 1994. **Baltic Region:** 8—Druskiniskajte Hole and 9—Induri Hole (Fursenko and Fursenko, 1961); 10—2 Pionerskaya Well (Strelnikova et al., 1978). **Pripyat trough:** 11—Teryukha village and 12—Velavoi village (Fursenko and Fursenko, 1961). **Dnieper-Donets basin:** 13—Hole 169 (Strelnikova, 1992); 14—Hole 371, northeast slope of the basin (V.K. Ryborak's material, stratigraphy according to T.V. Oreshkina); 15—Obukhov section, Kiev (T.E. Ulanovskaya's material, stratigraphy according to E.A.Shcherbinina). **Volga region:** 16—Sengilei section (Oreshkina and Khokhlova,1999; Kozlova et al., 1998b; Oreshkina, 2000); 17—Lomovka, and 18—Dyupa Valley (Kozlova, 1999). **North Peri-Caspian:** 21—Ozinki Mountain, and 22—the Solyanka R. (Kozlova, 1999); 23—Hole 14 and Hole 29 (Grachev et al., 1971; Pechonkina and Kholodilina, 1971); 24k—Hole 85 (A.S. Zastrozhny's material, stratigraphy according to V.N. Beniamovski). **Northeast Peri-Caspian:** 24—Hole 148, Kara-Chaganak (Kozlova et al., 1998); 25—Hole SP-1, Shaidinsky basin (Kozlova et al., 1998a; Bugrova et al., 1997 [*]); 26—Hole 221 Kamyskol (Kozlova, 1999). **Central and Eastern Peri-Caspian:** 27—7 Mialy Hole (*); 28—Hole 221, Shtuttuzinsky basin (*); 29—Well Hole, Kamsaktygol basin (*); 46—Lubenka Hole (Kozlova, 1999). **Turan plate:** 30—Hole 313, North Ustyurt (*); 31—Hole 206, Predustyurtsky Plain (*); 34—Usak section (Naidin et al., 1996); 35—Kaurtykapy section (H. Oberhänsli's material [*]); 36—Aktumsuk section (Gavrilov et al., 1997; Shcherbinina, 2000). **West Kopet Dagh:** 40—Kyzyl-Cheshme section (Muzylöv et al., 1987); 41—Maly Balkhan sections (Solun, 1975). **East Trans-Caspian:** 42—Hole 1228 and 43—Hole 1338, Buzachi Peninsula (*). **South Turgay:** 44—Holes 50 and 52 (Beniamovski et al., 1991, 1993; Vasilieva, 1994; Kozlova, 1999); 45—Holes 85 and 86 (Beniamovski et al., 1989, 1991), 47—Kundyzda Mtn. (*). **North Turgay:** 48—Belinsky pit (Beniamovski et al., 1989, 1991); 49—Sokolovsky pit (Radionova et al., 2001); 50—Emba river(*). **South Trans-Uralia:** 51—Korkino pit (*); 52—Hole IG-32, Kurgan (Vasilieva et al., 1994); 55—the Sary-Oba R. (Iakovleva, 1998). **North Trans-Uralia:** 57—Hole 19-U, Ust-Manya village (Kozlova and Strelnikova, 1984). **Polar Cis-Uralia:** 58—Hole 228, the Pechora R. (Yakovleva, 2000; Yakovleva et al., 2000c); 62—Hole 29, lower reaches of the Severnaya Sosva R. (Yakovleva et al., 2000b). **West Siberia Lowland:** 60—Hole 14k, upper reaches of the Konda R. (Kozlova, 1999); 61—Hole 4, the Vasyugan R. (Yakovleva et al., 2000c); 64—Kichigino village (Amon, 1994); 65—Hole 29, lower reaches of the Pur R. (Strelnikova, 1992; Iakovleva and Kulkova, 2001); 67—Hole 2, Vasyugan and 68—Hole 29, Chuzik–Kenga interfluve (Shatsky et al., 1973); 69—Hole 011-BP (*); 70—Hole 9, Ozerny village (*), 71—Hole 9 Petukhovo (Strelnikova, 1992); 72—Hole 1-OK (Ilyina et al., 1994; Kozlova, 1999); 73—Kamyshlov pit (*); 74—Irbit pit (*); 75—the Lavdinka R. (Kozlova, 1999); 77—the Pechal-Ky R. (Strelnikova, 1992); 78—Hole K-4, Kandinskoye (Ilyina et al., 1994). **Kara Sea:** 76—Hole 157 (Gleser and Stepanova, 1994). (c) Location of section profiles shown in Figures 3–7. (d) Facial provinces boundaries. (e) Direction of facial boundaries transition during Ypresian and Lutetian. Inset—Facial provinces.

there was a more shallow-water region, here referred to as the Southern province, which is represented by carbonate shelf facies—carbonate biogenic oozes (Ph. Allen and J. Allen, 1990). In Figures 1A and 1B, most sections of the Southern province correspond to this particular facies, only the southernmost sections (Kuban, Rubas-chay), in which the content of terrigenous material (including coarse clastics) increases appreciably, apparently representing slope deposits.

The Transitional province, apparently, can be viewed as inner shelf, in places separated from the more deep-water Southern province by minor carbonate shoals, composed of nummulitic limestone and red algae.

The Northern province is here interpreted as a seaway that opened both to the north and/or northwest toward the Arctic-Atlantic basin and to the south toward the Transitional province. This type of basin and the character of its link to the ocean remain largely unclear due to paucity of stratigraphic and sedimentologic data.

The initial objective of this study was purely stratigraphic, as we intended to correlate lower Paleogene siliceous sediments of the Northern province, whose subdivision had been based entirely on local siliceous plankton scales, with calcareous deposits of the Southern province, where standard nannoplankton scales are applicable in full measure, using stratigraphic data from the

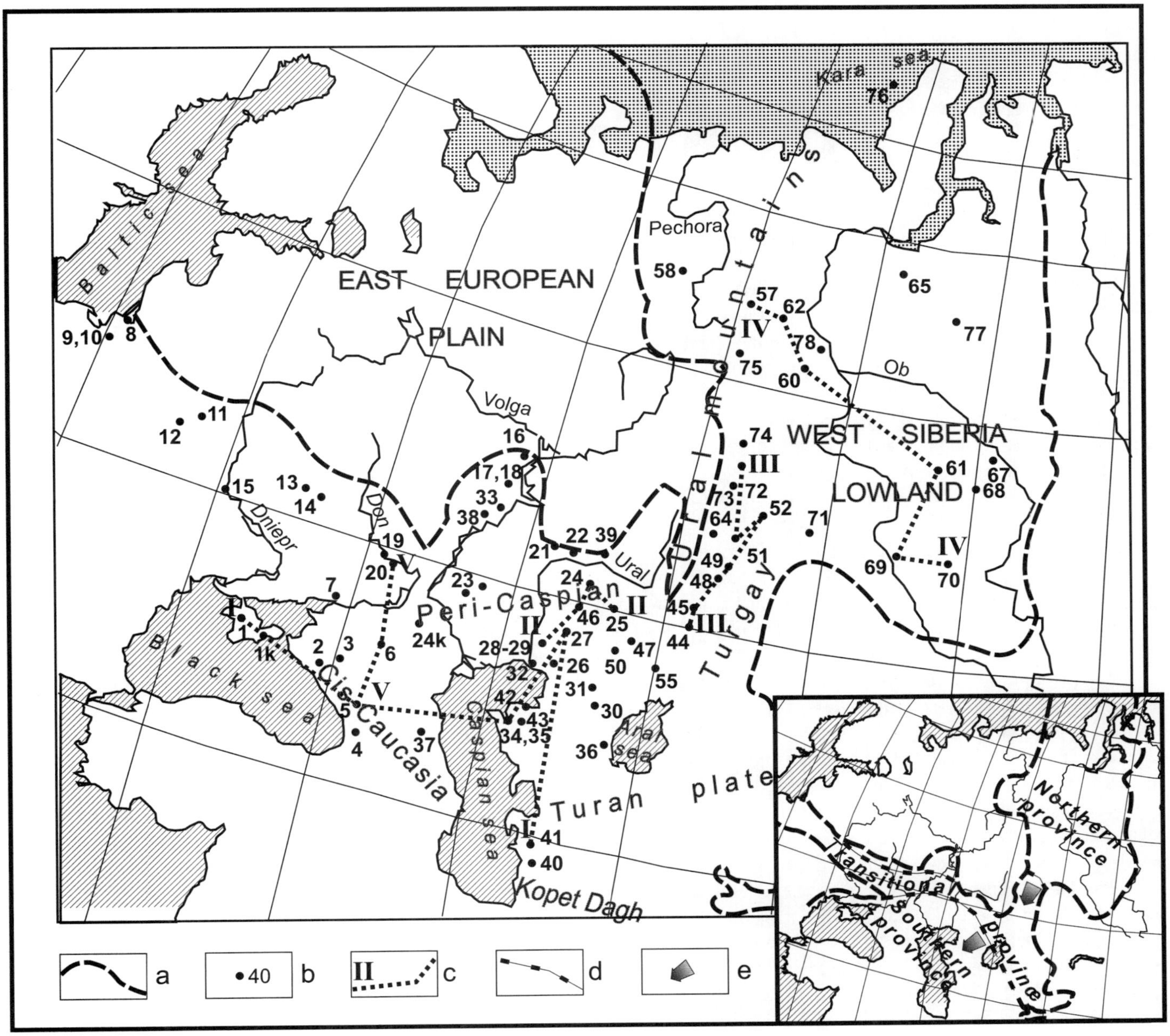

Transitional province, where all the plankton groups are represented to some extent or another. Customary stratigraphic techniques provided only a partial solution to this problem.

In addition, we made use of the cyclic and facies analysis methods. Sedimentary cycles were identified that are continuous in Thanetian-Bartonian strata throughout the region. Synchroneity of each cycle from section to section across the region was supported by stratigraphic data. It was found that distribution of each particular facies is also governed by a cyclicity that shows distinctive features in each particular province. One more objective of our integrated study was to unravel the scenario of changes in biogenic facies over the early Paleogene across the region.

DATA AND APPROACH USED

This study is based on >80 Paleogene sections across the study region (Fig. 1). The sections were correlated using an integrated zonal scale based on various microplankton groups. Principal key sections were revisited.

Stratigraphy

A comparative stratigraphic scheme for the Southern, Transitional, and Northern provinces is given in Figure 2.

Stratigraphically, the Southern province is well understood. Its deposits are subdivided reliably using the nannoplankton scales by Martini (1971) and Okada and Bukry (1980) and a regional foraminifer scale. The regional foraminiferal scale correlates well with the oceanic one (Krasheninnikov, 1982). We managed to identify beds with radiolarians or diatoms (Radionova and Khokhlova, 1994) within upper Paleocene, upper Ypresian, and Bartonian strata from a number of sections in the northern Cis-Caucasia, similar in their microfossil assemblages to correlative zones in the Transitional and Northern provinces.

Our integrated study of the Transitional province, in which both calcareous and siliceous planktonfossils are found at certain levels, permitted us to refine the stratigraphic ranges of some local zones based on siliceous plankton. The radiolarian *Petalospiris foveolata* Zone and the diatom *Coscinodiscus uralensis* Zone were found to correspond to the upper Thanetian, in the range of the nannoplankton Zone NP9b, and to the dinocyst *Apectodinium hyperocanthum–A. augustum* zones after (Radionova et al., 2001; Oreshkina and Khokhlova, 1999). The revision of the stratigraphic ranges of the Eocene radiolarian *Petalosyiris fiscella* to *Lychnocanium separatum* zones and diatom *Coscinodiscus payeri* to *Pyxilla oligocaenica* zones evoked considerable debate (Kozlova et al., 1998a; Radionova, 1998; Bugrova et al., 1997).

The Paleocene to Eocene section is incomplete. Regional hiatuses have been established at the base of the Thanetian and Ypresian, in the lower and middle Lutetian, and in the lower part of the Priabonian. To subdivide the deposits, we used five microplankton groups, which provided the mutual control. Stratigraphic correlation with the Southern province involved the use of nannoplankton zonation; in the southern part of the Transitional province both nannoplankton scales are applicable, while in the Northern, only Martini's scale can be used. In the northern part of the province, where terrigenous-calcareous deposits give way partly to terrigenous-siliceous strata, the upper Thanetian NP9, upper Ypresian NP13–NP14, or upper Bartonian NP17 Zones are not identified. Beds with planktonic foraminifers occur only in the Thanetian, lower Ypresian, upper Lutetian, and Bartonian strata.

The second group used in subdividing the deposits from the Transitional province and their correlation to the Northern one is dinocysts. The zonation shown in Figure 2 draws on Powell's (1992) scale for the Thanetian-Ypresian and on the scale by Andreeva-Grigorovich (1985) for the upper Ypresian to Bartonian strata.

In the Transitional province, on the Turan plate and in the Peri-Caspian depression, the lower Ypresian radiolarian *Petalospyris fiscella* or diatom *Coscinodiscus payeri* zones are not identified. Stratigraphic position of the radiolarian *Lychnocanium separatum,* dinocyst *Charlesdowniea coleothrypta rotundata,* and diatom *Pyxilla oligocaenica* zones remains open to discussion. The range of these zones is defined as either the upper Ypresian to lower Lutetian or lower Lutetian based on their correlation to nannoplankton Zone NP14 (Kozlova, 1999; Vasilieva, 1994) or even as the entire Lutetian (Gleser, 1996). The issue of the upper limit of all these zones is intertwined with that of the range of the hiatus within the Lutetian from the Transitional and Northern provinces, which remains debatable.

The Paleocene and Ypresian stratigraphy of the Northern province, developed from West Siberian sections, draws on siliceous microplankton groups, such as radiolarians, diatoms, silicoflagellates, and dinocysts, and stratigraphic ranges of most zonal divisions have been refined (Akhmetiev et al., 2001; Kozlova, 1999; Strelnikova, 1992). Hiatuses are recorded at the base of the Thanetian and at the Thanetian-Ypresian boundary from the lack of the dinocyst *Wetzeliella astra* Zone in the lower Lutetian.

The middle Eocene stratigraphy of the Northern province is based on sections in the Dnieper-Donets basin and north Peri-Caspian basin. Stratigraphic ranges of the radiolarian *Heliodiscus quadratus–Theocyrtis andriashevi* and the diatom *Paralia oamaruensis* regional zones in the Northern province were established as the middle Eocene based on radiolarian finds along with nannoplankton from sections in the Dnieper-Donets basin (Radionova et al., 1994; Khokhlova et al., 1999). Middle Eocene assemblages were found to be identical in the Dnieper-Donets basin and the northern Peri-Caspian region, Hole SP-1 (Radionova, 1998). The lack of reliable correlations with calcareous microplankton groups leaves it open to debate whether the upper limit of the occurrence of siliceous organisms is constrained to the middle Eocene or whether they are encountered in Upper Eocene deposits as well (Gleser, 1996; Kozlova, 1999).

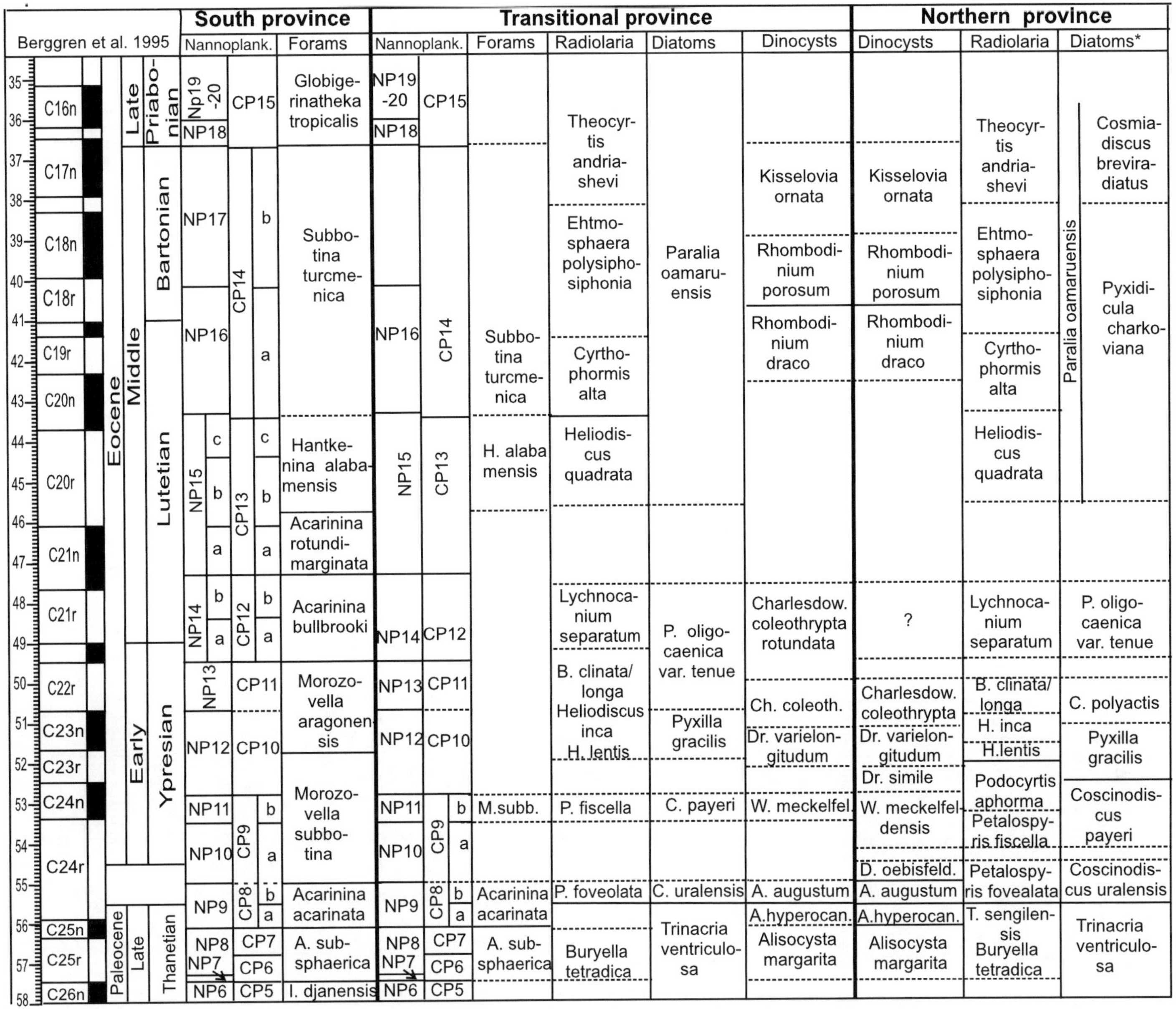

Figure 2. Zonal stratigraphic scheme correlating the Southern, Transitional, and Northern provinces. Zonal division of Paleogene deposits from the Southern province is based on standard nannoplankton scales (Martini, 1971; Okada and Bukry, 1980) and a regional foram scale (Krasheninnikov and Muzylöv, 1975; Iarkin, 1989, with 1999 amendments). To correlate the Southern and Transitional provinces, we used the standard nannoplankton scale and regional foraminiferal and radiolarian (Kozlova, 1999), diatom (Iarkin, 1989, with 1999 amendments; Kozlova et al., 1998), and dinocyst zones (Powell, 1992; Andreeva-Grigorovich, 1985). The stratigraphic range of zones whose boundaries were not documented is shown with dashed lines. To correlate the Transitional and Northern provinces, regional dinocyst and radiolarian scales were used. In subdividing Paleogene strata from the Northern province, zones are given after Akhmetiev et al. (2001) and Khokhlova et al. (1999) with minor modifications.

Cyclic analysis

In each province, the succession of lithostratigraphic units was documented by means of a series of stratigraphically correlated transects, and a certain cyclicity, reiterating in each transgressive-regressive cycle, was established. For this purpose, discontinuity surfaces, transgressive facies series, and regional stratigraphic hiatuses were established in each province.

The Southern province. To unravel cyclicity, we established the facies series from the most shallow-water nummulitic limestones, often deposited on erosion surfaces, to calcareous clays through to marls or clayey limestones of the interior basinal portions. Increase in carbonate content of the sediment in passing from calcareous clays to marls is due to increase in the calcareous plankton component of the sediment. This increase correlates to the periods of high bioproductivity of plankton, when plankton assemblages showed an increasing proportion of oceanic taxa; i.e., this marks eustatic rise of sea level in the basin.

Stratigraphic ranges of transgressive-regressive cycles in the Southern province were first proposed by Muzylöv (1996), and this study offers refined stratigraphic intervals for each cycle. The structure of the cycles from the Southern province is shown in Figure 3 through the Crimea to northern Cis-Caucasia to the southern Peri-Caspian to Mangyshlak to eastern Kopet Dagh (Transect I–I in Figure 1.)

The increase in the carbonate content upsection from the base of the transgressive Thanetian member persists across the Southern province. Thus, at the base of the lower cycle of Thanetian supercycle in the Kheu River section, one finds a clayey limestone member giving way upward to calcareous clays (Fig. 3). The same horizon displays a peak in carbonate content in the Nalchik beds, reaching 40% (Gavrilov et al., 2000). In Bakhshisaray section (Crimea), 7 Mialy Hole (Peri-Caspian), and Kyzyl-Cheshme section (western Kopet Dagh), its counterpart is marls.

A bituminous bed marks the middle cycle. We correlate the 65-cm-thick brown clay bed enclosing fish scales in the upper part of the Nalchik Formation, the Kheu section, with the pivotal event, referred to as the Initial Eocene Temperature Maximum, or IETM. The bituminous bed shows a reduced carbonate content and an increase in TOC (total organic carbon) up to 10%. Analysis of planktonic and benthic foraminifers suggests that the bituminous bed records anoxic to disoxic environments (Stupin and Muzylöv, 2001). This intercalation was first detected and reported by N.G. Muzylöv in the Southern province, from the Crimea to Central Asia, and it was addressed in rather numerous works (Muzylöv et al., 1989; Muzylöv et al., 1996; Gavrilov et al., 1997; Gavrilov et al., 2000; Gavrilov and Shcherbinina, this volume).

The upper cycle in the Kheu section is composed of alternating siliceous siltstones containing considerably recrystallized radiolarians and diatoms, and slightly calcareous argillite-like clays, with the carbonate content dropping to 10%. A considerable decrease in the carbonate content of the member overlying the bituminous layer is documented through most sections along Transect I–I

The lower cycle deposits correspond to nannoplankton Zones CP7–CP8a, the foraminifer *Acarinina subsphaerica* Zone, and the basal part of the *Acarinina acarinata* Zone. From the base of the marl member upward, planktonic foraminifers increase perceptibly in abundance and species diversity. In the eastern Kopet Dagh and Badkhyz, Thanetian deposits exhibit a facies transition, where shallow water nummulitic limestones with *Nummulites deserti* give way to the *Acarinina subsphaerica* Zone marls in deeper-water sections (Bugrova, 1991). A bed rich in organic matter occurs at the CP8a/CP8b boundary. Based on dinocysts, it was attributed to the *Apectodinium homomorphum* Zone (Akhmetiev and Zaporozhets, 1996). Alongside the index species, the bituminous bed contains other *Apectodinium* species, such as *A. augustum, A. parvum, A. quinquelatum, A. sumissum.* Currently, the appearance of *A. augustum* and *A. parvum* and the acme of the genus *Apectodinium* in the context of the Western European scale are used to identify the *Apectodinium augustum* Zone. The upper member matches nannoplankton Zone CP8b and the upper part of the foraminiferal *Acarinina acarinata* Zone. The fact that the top of the underlying Nalchik Group is lacking certain Zone CP8b nannoplankton species and the sharp lithologic boundary are interpreted to point to a minor erosional event at the base of the Ypresian deposits (Gavrilov et al., 2000)

In Mialy 7 Hole, a carbonate-free clay member, underlying the bituminous bed, yielded the *Buriella tetradica* Zone radiolarians, and further upsection, a member of alternating sand and clay yielded the *Petalospiris foveolata* Zone radiolarians (Kozlova, 1999). Correlating the span of this Zone to the foraminiferal *Acarinina acarinata* Zone in this section would help us to constrain stratigraphic position of this cycle in the Transitional and Northern provinces (Fig. 2).

All Ypresian cycles replicate the facies traits documented from the Thanetian supercycle.

In the Kheu section, Ypresian deposits are represented by the Cherkessk Formation, composed of soft, light greenish gray marls ~40 m thick.

The middle part of this sequence comprises green calcareous clays and contains a 15-m-thick interval (Zone CP10) with eight sapropel-like intercalations 10–25 cm thick each. Organic matter content of these intercalations is as high as 2%–3.5% (Gavrilov and Muzylöv, 1992). Foraminifers from these intercalations point to disoxic depositional environments (Oberhänsli and Beniamovski, 2000). Ypresian phase deposits terminate in a bluish marl member of Zone CP12, which again shows siliceous plankton, radiolaria and sparse diatoms (Radionova and Khokhlova, 1994).

The Cherkessk Formation displays a complete succession of nannoplankton zones CP9–CP12 and the foraminifer *Morozovella subbotinae* and *M. aragonensis* Zones.

In the Kheu section, the cyclic architecture of the Ypresian unit can be revealed only based on lithologic variations, although

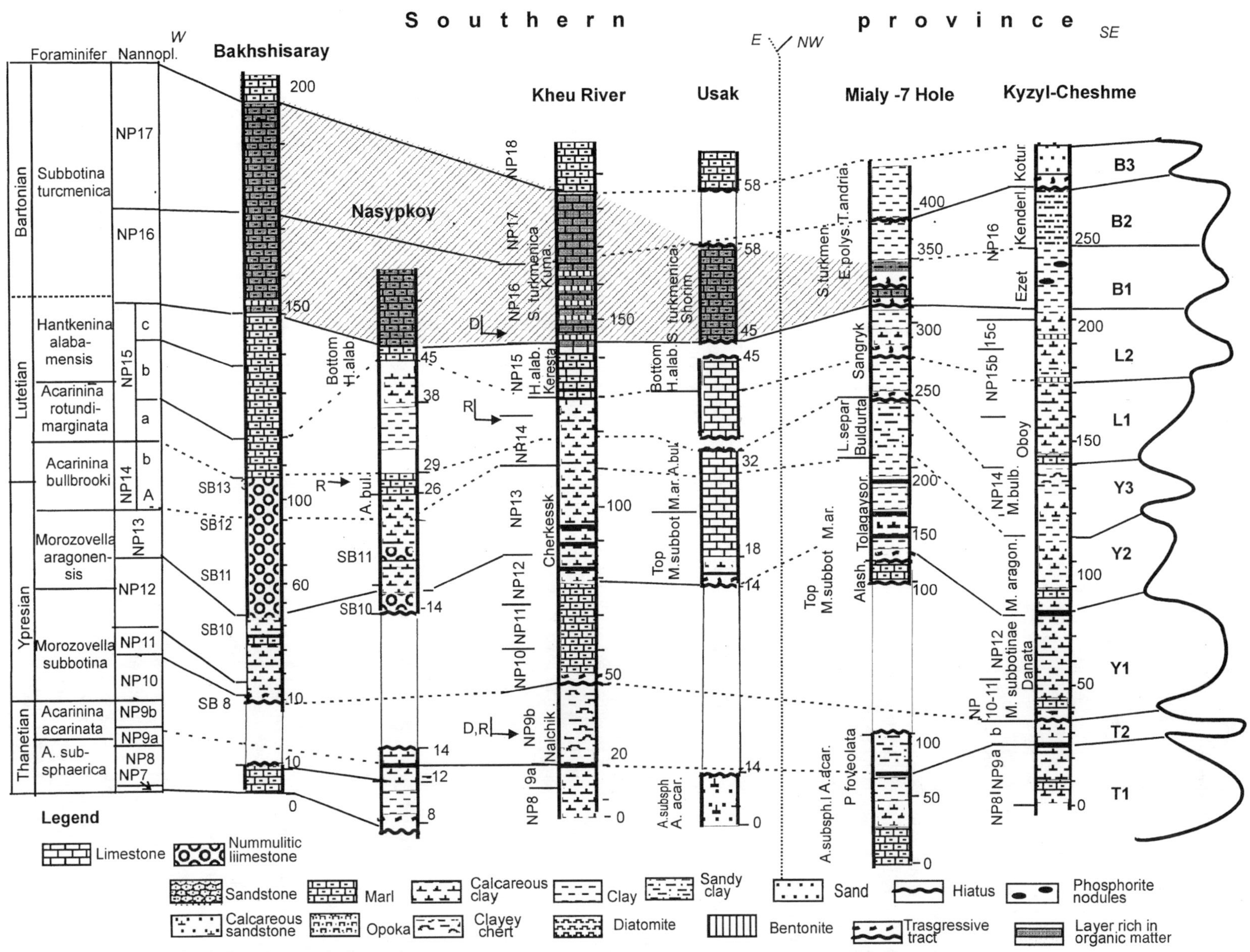

Figure 3. Correlation of key sections from the Southern province: Profile I–I in Figure 1. Zonal subdivision based on nannoplankton and foraminifers is shown in Figure 2 for the Southern province. Zonal subdivision from 7 Mialy Well based on foraminifers, carried out by V.N. Beniamovski, and based on radiolarians, by G.E. Kozlova, 1999. In lithologic columns, gray coloring depicts the "Kuma Facies" of middle Eocene age. Diagonal-striped shading shows "Kuma Facies" "distribution from section to section. T1, T2—Thanetian to initial Ypresian cycles; Y1, Y2, Y3—Ypresian to initial Lutetian (?) cycles; L, L1, L2—Lutetian cycles; B1, B2—Bartonian to initial Priabonian (?) cycles. R, D—radiolaria, diatoms in microplankton assemblage. SB8–SB13—Mediterranean Nummulitic assemblages in Bakhchisaray section according to E.Yu. Zakrevskaya (2002, personal commun. Legend is used for Fig. 3–7.

the transgressive character of the calcareous lower Ypresian sequence of the Southern province, corresponding to the *Morozovella subbotinae* Zone, was noted by many researchers (Shutskaya, 1970; Solun, 1975; Bugrova, 1991; Naidin and Beniamovski, 1994; and others). In Transect I–I, the full ranges of zones (as inferred from nannoplankton) across the Paleocene-Eocene transition are to be found only in the Kheu and Kyzyl-Cheshme sections, this cycle being in both cases represented by the most deep-water facies of multicolored marl. In the rest of the sections along this Transect, on the periphery of the Southern province, deposits of the lower cycle have an erosional base.

The same sections exhibit a distinct unconformity at the base of second cycles. In the Crimean sections, the second and third Ypresian cycles contain extensive nummulitic limestones. In the Nasypkoy section (Fig. 3), these have been encountered in a transgressive unit at the base of the middle cycle and are represented by an assemblage of Zone SB10 from the Mediterranean nummulitic scale (according to Bugrova et al., 2002). These same deposits in more deep-water Ypresian beds, represented by a sequence of calcareous clays with bituminous intercalations with fish scales at the level of Zone CP10, have been recorded from most sections along Transect I–I. A single bed of dark-colored clay with fish scales >0.5 m thick was documented at this stratigraphic level from western Kopet Dagh sections (see the Kyzyl-Cheshme section); this is the so-called "second fish bed" (Solun, 1975; Muzylöv et al., 1989). The upper part of the Ypresian strata and, apparently, the lower part of the Lutetian beds consist of terrigenous-calcareous and, occasionally, terrigenous-siliceous sediments with an abundant radiolarian assemblage. Up to 37% of radiolarians are recorded within the foraminiferal assemblage of the *Acarinina bulbrokii* Zone in this cycle in the Nasypkoy section. In 7 Mialy Hole, the upper cycle Buldurta Formation conformably overlies the Tolagaisor Formation and consists of noncalcareous gray and greenish gray clays with thin intercalations of siliceous silts. Here, the *Lichnocanium separatum* Zone radiolarians are abundant (Kozlova, 1999). This level is used for reference in correlations between provinces.

Cycles of the middle Eocene transgressive-regressive phase in the western areas of the Southern province largely mimic the structure of the Ypresian cycles. Along the Kheu River, the Keresta Formation conformably overlies the Cherkessk Formation. It is represented by rhythmically alternating compact limestone and soft greenish marl. The transgressive character for the Keresta Formation is suggested by its increased carbonate content and flysch-like rhythmicity, possibly pointing to the basin's deepening. In the Bakhchisaray section, the Keresta Formation replaces shallow-water nummulitic limestones of Ypresian–early Lutetian age. In western Cis-Caucasian sections, the Keresta Formation erosionally overlies the Cherkessk Formation (Grossheim, 1960). Unconformity at the base of the Keresta deposits is clearly traceable into the Peri-Caspian region and southern Turgay (Kurgalimova and Moksyakova, 1970; Naidin et al., 1994) and the Kyzyl-Cheshme section (Fig. 3), the western Kopet Dagh (Muzylöv et al., 1989)

Keresta Formation is correlatable to nannoplankton Zone CP13 and to the foraminifer *Acarinina rotundimarginata* and *Hantkenina alabamensis* zones and Kuma Formation corresponds to nannoplankton Zone CP14 and the foraminifer *Subbotina turkmenica* Zone, the lower and upper parts of the sequence yielding poorly preserved radiolarians and diatoms of the *Paralia oamaruensis* Zone (Radionova and Khokhlova, 1994).

In the Kheu River section, the Kuma Formation conformably overlies the Keresta Formation, but it is composed of a totally different type of deposits, thinly rhythmic soft clayey coffee-colored limestone. The rhythmicity of the formation is due to light colored clayey limestone intercalations alternating with dark colored ones, rich in organic matter and with fish scales. In the lower half, total organic carbon content is 2%–9% and in the upper half, it drops to 2.5%–4.5%. The $CaCO_3$ contents in light-colored intercalations are 20% greater than in dark-colored ones (Gavrilov et al., 2000). Kuma Formation is virtually barren of benthic foraminifers, which suggests a disoxic or anoxic environment (Beniamovski et al., 1999).

In both sequences, from the base of the Keresta Formation upwards, bentonite horizons are encountered.

In the western and central part of the Southern province, a somewhat distinctive structure is displayed by Bartonian deposits, represented along Transect I–I in the Bakhchisaray, Kheu, and Usak sections by the Kuma or Shorym Fms. Further east, in Mangyshlak and then in western and central Turkmenistan, considerable changes occur in the structure of the middle Eocene section (Korovina, 1970; Rodionova, 1963). In 7 Mialy Hole (Fig. 3), the Saurbay Formation has an erosional base and is composed of light brown marly clays with inclusions of light gray (ash?) material and contains a planktonic foraminifer assemblage of the *Subbotina turcmenica* Zone, which, nevertheless, exhibits some benthic forms, *Bifarina* ex gr. *adelae* and *Uvigerina castelata.* The upper half of the sequence can be envisaged as being an independent cycle. It is composed of greenish gray slightly sandy, low-calcareous to noncalcareous clays. This part of the sequence is dominated by agglutinated foraminifers and radiolarians of the *Ethmosphaera polysiphonia* Zone. Keguinkol Formation, overlying erosionally the Saurbay Formation and consisting of clays with radiolarians of the *Theocyrtis andriashevi* Zone, can be regarded as an independent third cycle.

In the western Kopet Dagh and on the Greater Balkhan (Solun, 1975), Bartonian deposits are represented by three lithologic units: The lower, Ezetsky Formation (Fig. 3, the Kyzyl-Cheshme section) is composed of dark greenish gray clays with phosphorite lenses and contains only planktic foraminifers. The Kenderlit Formation is transgressive and consists of calcareous silty clays with intercalations of Fe-dolomite concretions; the formation's lower horizons are barren of microfossils, and further upsection only benthic foraminifers occurr. The upper, Kotur Formation shows ample evidence of its regressive character. The formation is composed of brownish olive-colored clays intercalated by siltstone and sandstone. Clay intercalations contain abundant benthic foraminifers, including agglutinated ones, radiolarians, and diatoms. The siltstone intercalations often contain mollusks.

Therefore, in eastern region of Southern province, Bartonian deposits fall into two or three cycles. The "Kuma"-type disoxic facies are maintained only in the lower cycle of the Bartonian transgression. Deposits of both upper cycles have elevated silica contents due to abundant radiolarians and, to a lesser extent, diatoms. Although deposits of these cycles are also high in organic matter, the appearance of diversified benthos, such as foraminifers, mollusks, and nummulitic beds in the eastern Kopet Dagh and Badkhyz (Bugrova, 1991), suggests deposition in an oxic shallow-water basin. These facial types of middle Eocene deposits are close to those of the Transitional province.

The Transitional province. Although deciphering cyclicity, we established a transgressive series from the most shallow-water facies of calcareous sands with minor bioherms composed of *Cerripedia, Briozoa,* and large foraminifers, to calcareous clays. Regressive series are represented by transitions from silts to noncalcareous clays and clayey opokas.

Most cycles are marked by depositional breaks, established paleontologically. To study cyclicity patterns of Thanetian and Ypresian deposits, we constructed two roughly S–N trending Transects, II–II and III–III, respectively, which are shown in Figure 1.

Let us consider the stratigraphic ties and cyclic structure of this interval in, e.g., the Sokolovsky pit (Fig. 5), where Thanetian deposits, unconformably overlying the Mesozoic beds, are represented by two lithologic units: The Sokolovsky terrigenous-calcareous sequence and the Banded siliceous–clay sequence with a hiatus in-between. In the Transitional province, the Thanetian calcareous clay or silt-clay deposits correspond to nannoplankton Zone NP8 and often occur with a considerable stratigraphic unconformity. A hiatus at the base of the upper, terrigenous-siliceous sequence is also found across the region.

The Sokolovsky Formation was shown to be restricted to Zone NP8 of the standard nannoplankton scale and to the foraminifer *Acarinina subshpaerica* and lower *A. acarinata* zones. (Radionova et al., 2001). The radiolarian *Petalospiris foveolata,* dinocysts *Apectodinium hyperocanthum, A. augustum, Wetzeliella meckelfeldensis* zones were established in the Banded Formation. The Banded Formation contains a hiatus spanning two dinocyst zones, *Glaphyrocysta ordinata* and *Wetzeliella astra,* which correspond to nannoplankton Zone NP10.

Such architecture of the section is characteristic of the northern Turgay and the northern Peri-Caspian region alike, reflecting the transgressive-regressive cyclicity in the Thanetian to initial Ypresian, common to the entire region.

The architecture of Ypresian cycles from the Southern to Transitional province is portrayed by Peri-Caspian Transect II–II (Fig. 4). The lower cycle of Ypresian deposits is represented by marls in Kamyskol Hole 221, giving way to calcareous clays in Holes 148 and SP-1. Upsection, the Tolagaisor Formation in the Kamyskol Hole 221 has erosional base and is represented by calcareous clays with the bituminous intercalations. Further north, intercalations of beige colored slightly bituminous clays were recorded only in the Lubenka Hole 12. This interval in Hole 148 is composed of clays exhibiting erosion at the base, and in Hole SP 1, of opokas.

The upper part of Ypresian strata, the Buldurta Formation, composed of silty clay with erosional base, yields Zone NP14 nannoplankton at the bottom of some sections, giving way upsection to the *Lychnocanium separatum* Zone radiolarians. In northern sections, penetrated by Holes 148 and SP-1, the formation is represented by siliceous clays. The upper stratigraphic limit of this sequence is unclear.

The same structural peculiarities of the Ypresian cycles persist into the northern periphery of the Transitional province. As can be seen from Transect III–III through the north part of the Transitional province in the Turgay basin (Fig. 5), the lower, Ypresian, cycle—the Taup Formation—is represented by calcareous clays and yields nannoplankton Zone NP11 to the lower part of Zone NP12 from the Belinsky quarry and from Holes 48–50 (Beniamovski et al., 1989, 1991). The second cycle—the Tasaran Formation, erosionally overlying the Taup Formation—is represented by clays and contains Zone NP12 nannoplankton in its lower part and radiolarians of the *Heliodiscus inca* and *Buriella clinata–B. longa* Zone (Kozlova, 1999). The upper cycle of the Tasaran Formation has a clay-sand composition and bears the *Charlesdowniea coleothrypta–Ch. coleothrypta angulosa* Zone dinocysts (Vasilieva, 1994; Vasilieva et al., 1994).

As is seen from Figure 4, depicting Kamiskol Hole 221, Lutetian and Bartonian lithofacies become considerably more terrigenous than Ypresian facies. During the middle Eocene, geographic arrangement of all the provinces changed, facies shifting from north to south, and all the provinces becoming much more closely spaced.

Transect V–V (Fig. 7) shows how the middle Eocene lithology changes from the Southern to the Northern province. The Keresta Valley section provides an example of architecture of middle Eocene cycles in the Transitional province. Here, as in all other sections of the Peri-Caucasian region, the Keresta Formation rests on carbonate-free clays of late Ypresian to early Lutetian age (Shutskaya, 1970) and has the same structure, lithology, and stratigraphic range as in the Kheu section. The upper part of the section, the Solonskaya Formation, is bipartite. Its lower member is composed of Kuma-type coffee-colored clayey limestone, which yielded Zone CP14a nannoplankton and the *Subbotina turcmenica* Zone foraminifers. The upper member consists of carbonate-free clay with an abundant radiolarian assemblage of the *Ethmosphaera polysiphonia* Zone.

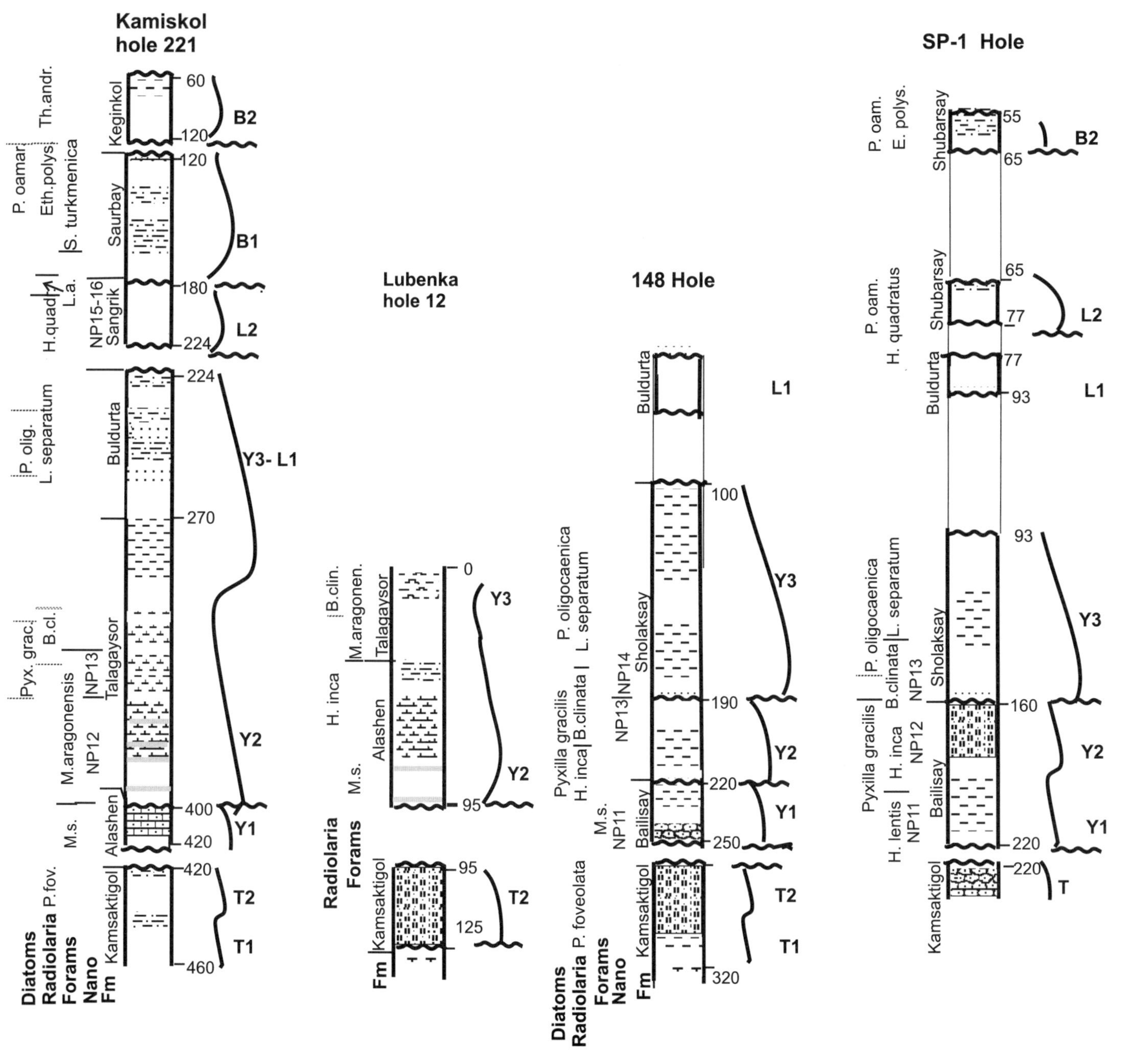

Figure 4. Correlation of Thanetian to Bartonian key sections in the Peri-Caspian basin, Transitional province. Transect II–II in Figure 1. Stratigraphic scheme, as in Figure 2 for Transitional province. In the 221 Kamyskol Hole, zonal subdivision is given after N.P. Tabachnikova (personal communication, 2001) based on nannoplankton, after V.N. Beniamovski based on foraminifers, after G.E. Kozlova based on radiolarians, and after Z.I. Gleser (personal communication, 2001) based on diatoms. Diagonal-striped shading shows facies boundary shift of the Transitional province to south in Lutetian.

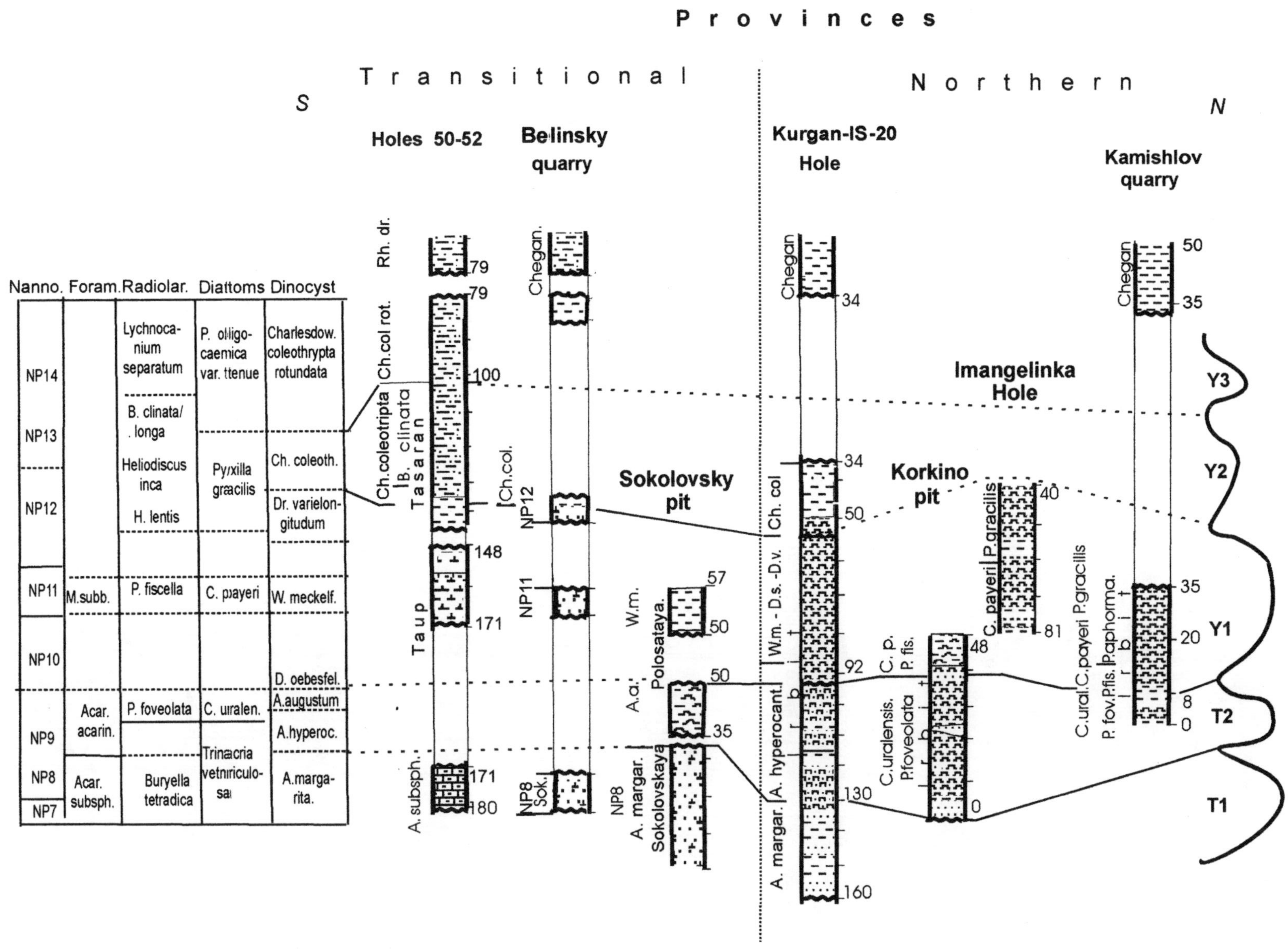

Figure 5. Correlation of key sections of Thanetian to Ypresian strata from the Transitional and Northern provinces along the S–N Transect through southern West Siberia and the north and central Turgay basin. Transect III–III in Figure 1. Stratigraphic scheme, as in Figure 2 for Transitional province. Transition from carbonate-terrigenous and siliceous-terrigenous strata of the Transitional province to the dominantly siliceous deposits of the Northern province. Breaks increase in magnitude where provinces adjoin each other, in the Turgay basin and the Sokolovsky pit and Belinsky quarry.

Comparison of the structures of the middle Eocene deposits from Mialy Hole 7, Kyzyl-Cheshme section (Fig. 3), 221 Kamiskol Hole (Fig. 4), and the Keresta Valley (Fig. 7) reveals identical regularities: (i) deposits of the lower, "Keresta," cycle in all three sections are represented by calcareous clays, (ii) Bartonian deposits are subdivided distinctly into at least two cycles, the lower being dominated by typical "Kuma" coffee-colored clays deposited in a disoxic basin, and the upper part of carbonate-free, siliceous-terrigenous strata. A zone with such structure stretches from North Cis-Caucasus Region via southern Peri-Caspian basin into the Kopet Dagh. Northwest and northeast of this zone, deposits of the northern facies type occur.

The Northern province. In this province, the transgressive depositional series is represented by transition from glauconitic sands to diatomites or diatomaceous clays, and the regressive succession by diatomaceous clays giving way to sandy siliceous strata (opokas), in turn grading into clayey sands, not infrequently with minor banks with *Bivalvia* and *Gastropoda.* Cyclicity of the Thanetian and Ypresian deposits can be studied along the transects through West Siberia, one trending N–S from central areas adjacent to the Urals to the Turgay passage (Fig. 5) and the other trending W–E from areas adjacent to the eastern Urals in the west to Omsk Region in the east (Fig. 6).

In the west of the West Siberian Lowland, Thanetian deposits are represented by Serov Formation composed of gray opoka-like clays and by diatomite customarily attributed to the Irbit Formation, exposed in a wide strip along the Urals' eastern slope, whereas further east, in the central West Siberian Lowland, these strata are essentially terrigenous, giving way to gaizes only in their upper part.

In Ust-Manya Hole 19 (Fig. 6), the Serov Formation, composed of opoka and opoka-like clays, unconformably overlies the clay-dominated Ivdel Formation. Within the formation, the radiolarian *Tripodiscinus sengilensis* Zone and the diatom *Trinacria ventriculosa* Zone are identified. We interpret this sequence as representing the lower Thanetian cycle. Transition to the Irbit Formation is identified by a considerable increase in biogenic silica content. The Irbit Formation exhibits the radiolarian *Petalospiris foveolata* Zone and the diatom *Coscinodiscus uralensis* Zone. Although the transition to the Irbit Formation diatomite is gradual throughout the zone east of the Urals, we documented the transgressive character of the Irbit strata in the Korkino section, where they rest on Triassic continental deposits (Fig. 5).

In Thanetian-Ypresian transitional strata, a change occurs in the lithology of the Irbit Formation in Ust-Manya Hole 19 and in some other holes. In the Kamyshlov section (Fig. 6), the radiolarian *foveolata-fiscella* transition and the diatom *uralensis-payeri* transition are restricted to a 0.5 m dark gray clay member with a considerable proportion of terrigenous material. Customarily, the changes in diatom, silicoflagellate, and radiolarian assemblages in these sections have been viewed as continuous. In this particular section, however, the upper part of the dinocyst *Deflandrea eobisfeldensis* Zone yields a mixed assemblage with reworked Cretaceous forms and two units of the dinocysts scale, *G. ordinata* and *W. astra,* lacking; upsection, the *Wetzelliela meckelfeldensis* Zone is identified. As appears from both transects, the absence of these dinocyst Zones is recorded from all the sections.

Ypresian deposits of the West Siberian Lowland, represented by the upper part of the Irbit Formation, Middle Lulinvor, Upper Lulinvor, and Nurolka formations, exhibit the maximum development of siliceous facies, although structural dissimilarities between the basin's western and eastern regions persist.

In the zone east of the Urals (Fig. 5), the middle and upper parts of the Irbit Formation are represented by diatomite, diatom clays, and gaize-like clays. Upsection, the proportion of terrigenous material increases, but both in the western and central parts of the basin siliceous lithologies with terrigenous admixture remain dominant. In its stratigraphic range, the lower Ypresian cycle corresponds to the radiolarian *Petalospyris fiscella–Heliodiscus lentis,* the diatom *Coscinodiscus payeri–Pyxilla gracilis* (bottom), and the dinocyst *Wetzelliela meckelfeldensis–Dracodinium varienlongitudum* (bottom) Zones, roughly correlative to nannoplankton Zones NP10 (top) to NP12 (bottom).

The second cycle is fully documented in sections near the central part of West Siberia, from Hole 14k in the Konda River basin (Fig. 6) and from Kurgan Hole IS-20 (Fig. 5). Deposits of this cycle are essentially clayey sequences of the Upper Lulinvor Formation and correspond to the radiolarian *Heliodiscus inca–Buriella clinata* (bottom) Zone, to beds with the *Coscinodiscus polyactis* Zones diatoms (top of the *Pyxilla gracilis* Zone to bottom of the *Pyxilla oligocaenica* Zone), and to the dinocyst *Dracodinium varielongitudum* (top) to *Charlesdowniea coleothrypta* (bottom) zones, which we correlate arbitrarily to nannoplankton Zones NP12 (top) to NP13 (bottom).

The transgressive character of this cycle is especially clear in northern (Chirva, 1971; Chirva and Lyubomirova, 1973) and eastern regions of West Siberia (Shatsky et al., 1973). In Hole 4, Vasyugan River section (Fig. 6), deposits of the Upper Lulinvor Formation erosionally overlie Paleocene strata and consist of light gray opoka-like clay and, less frequently, diatomite. They have an identical stratigraphic range in Holes 011-VA and 9 (Fig. 7) in southeastern West Siberia. In an easterly direction, they decrease in thickness gradually, giving way to glauconite sands, and pinch out on the east side of the Ob River in its middle reaches (Shatsky et al., 1973). Similar replacements are documented in a southern direction, along the Chizhapka River, a southern tributary to the Vasyugan River (Kriventsov, 1973).

The upper cycle is represented by Nurolka Formation deposits, composed of yellowish green clays with gaize-like intercalations, at times with silt accumulations (Shatsky et al., 1973). In northern sections of the zone adjacent to the Urals, in the lower reaches of the Ob River and along the Pur River, the Nurolka Formation is documented by the appearance of the *Pyxilla oligocaenica* Zone diatoms (Rubina, 1973; Chirva and Lyubomirova, 1973).

Deposits of the upper cycle are presented in Figure 6 as the

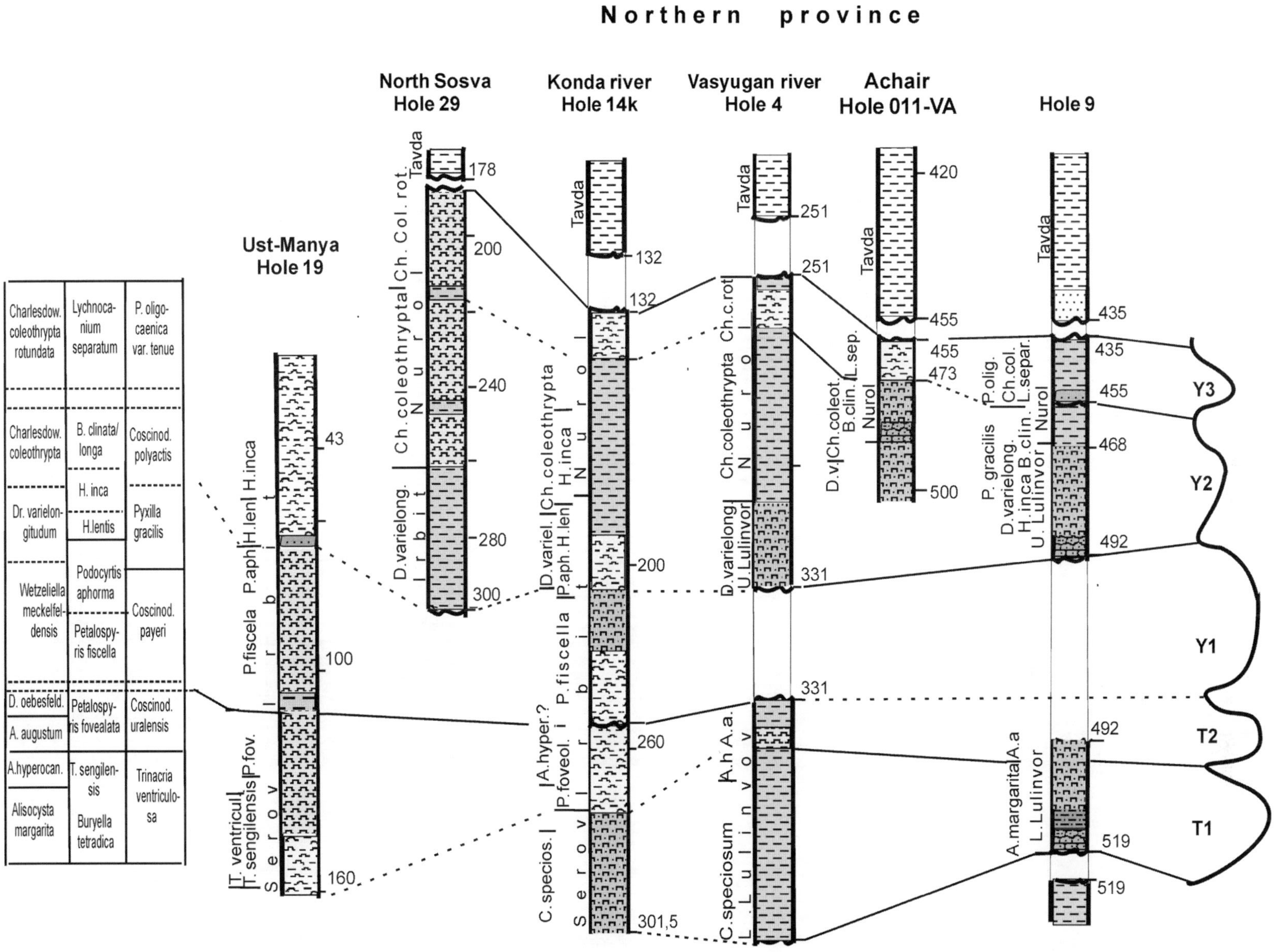

Figure 6. Correlation of Thanetian to Ypresian key sections from the Northern province along the E–W profile through the West Siberian Lowland. Transect IV–IV in Figure 1. Stratigraphic scheme, as in Figure 2 for Northern province. In western sections (Well 19, Ust-Manya) only lower and middle Ypresian strata are represented. All the eastern sections lack lower Ypresian strata, which implies gradual migration of the basin from west to east, in keeping with the cycles of Ypresian transgression (see also Figure 10).

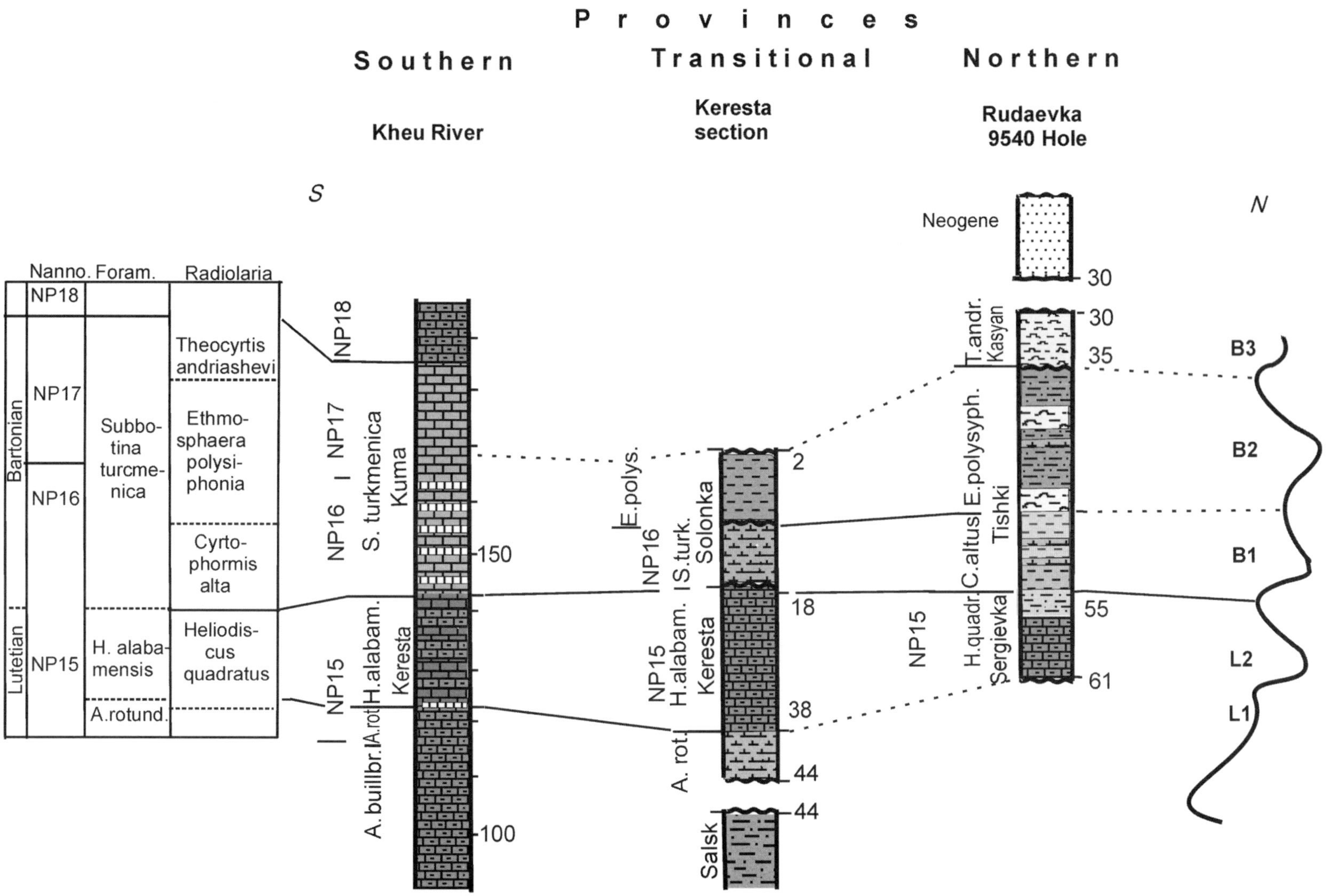

Figure 7. Correlation of Lutetian to Bartonian strata from the Southern, Transitional, and Northern provinces along the profile through the North Caucasus, Central Cis-Caucasia, Donets basin. Transect V–V in Figure 1. Stratigraphic scheme, as in Figure 2 for Transitional province.

upper strata of the Nurolka Formation, correlative to the dinocyst *Charlesdowniea coleothrypta rotundata* Zone, the radiolarian *Lychnocanium separatum* Zone, and the diatom *Pyxilla oligocaenica* Zone. In Hole 9, southeastern West Siberia, these deposits erosionally overlie those of the middle cycle.

In West Siberia, middle Eocene beds (the Tavda Formation) are represented by terrigenous deposits devoid of either calcareous or siliceous marine microplankton. These deposits belong to a basin of a different facies type and are not discussed here.

Middle Eocene deposits, represented by siliceous facies, are preserved only along the Eastern European periphery of the Peri-Tethyan basin and are addressed here in the context of Transect V–V through the north part of the Dnieper-Donets basin (Fig. 7).

DISCUSSION

Facies distribution through time

The cyclicity pattern established for each facies province persists across the study region. This pattern is illustrated on EW trending Transects I–I and IV–IV through the Southern and Northern provinces (Figs. 3, 6). N–S Transects II–II, III–III, and V–V (Figs. 4, 5, 7) show that the cycles remain laterally consistent in passing from the Southern to Transitional and from the Transitional to Northern provinces.

To summarize, we recognize in the study region three transgressive-regressive supercycles (phases): (i) Thanetian, covering nannoplankton Zones NP 8–9 (ii), Ypresian, covering Zones NP10 (top) to NP14, and (iii) upper Lutetian to Bartonian, covering Zones 15 (top) to NP17. Each supercycle had several cycles that we traced, while refining considerably the geographic pattern of various biogenic facies within the basin (Fig. 8).

In the Southern province, deposits of the lower cycle are represented in each phase by transgressive calcareous strata. More deep-water strata of this cycle are represented by marls, not infrequently multicolored, and more shallow-water strata, by limestones. Limestones become the most widespread during the late Lutetian, "Keresta" cycle. A characteristic feature of the middle part in all three phases is carbonate mud facies that are enriched in organic matter, as pointed out by Muzylöv (1994). In the Thanetian, this is a single intercalation, separating lithologically contrasting sequences; the Ypresian contains a series of bituminous layers; ultimately, the entire Bartonian sequence in the central part of the Southern province and the lower part of the Bartonian (the middle cycle) in the eastern regions become bituminous. The upper cycle of each phase exhibits a decrease in carbonate content, the appearance of clays with siliceous plankton, and, occasionally, purely siliceous facies.

In the Transitional province the deposits fall clearly into three sequences, separated by hiatuses or showing evidence of transgressive deposition: The lower, calcareous; the middle, composed of clay with intercalations rich in organic matter and replaced by a siliceous-calcareous sequence in the north of this province; and the upper, siliceous-clayey. The siliceous organisms appear in the middle cycle, the upper cycle being not infrequently represented by biogenic silica and siliciclastic deposits.

In the Northern province, during all three transgressive-regressive supercycles, the deposits maintained a similar facies aspect, and they are represented by diatomites, opoka, and clays containing siliceous plankton here. Geographic position of this province experienced considerable changes. In Thanetian time, it occupied West Siberia and the Middle Volga region; where as in Ypresian time, West Siberia and the western Arctic basin (Radionova and Khokhlova, 2000); and in middle Eocene time, it was restricted to the northern slopes of the Dnieper-Donets basin and Peri-Caspian depression.

Figure 8 shows the cycles of the Thanetian and Ypresian phases to be readily correlatable to oceanic transgressive-regressive cycles. Middle Eocene transgressions in the study region fit to oceanic ones less distinctly, biostratigraphic methods pinpointing confidently only the late Lutetian and early Bartonian transgressions. However, the boundaries of the supercycles here proposed distinctly match tectonic events at the Paleocene-Eocene transition (Kopp and Scherba, 1998), in the early and middle Lutetian (Kazmin et al., 1987), and at the Middle–upper Eocene transition (Meulenkamp and Sissingh, 2000), related to the southward jump of subduction in the Tethys and to mechanical coupling of the African-Apulian and Eurasian continental lithospheric plates since about the middle Eocene–late Eocene transition ~40–37 Ma ago.

Transgressive events through time

Time changes in facies distribution are especially spectacular if one compares the maps depicting all three transgressive-regressive phases (Figs. 9–11). In early Thanetian, Ypresian, and late Lutetian times, we restore clearly a continuous geographic distribution of carbonate shale, covering the Southern province, the distribution of this facies corresponding to the deepest part of the epeiric basin. Characteristically, the same areas were occupied by mud facies rich in organic matter. The Transitional province, composed of siliciclastic and carbonate mud, was the shallowest-water part of the basin—the inner shelf.

The greatest changes in geographic distribution are shown by siliceous facies. Note that the diatomite distribution area in Thanetian time was distinctly confined to the hypothesized basinal slope, likely suggesting either the existence of coastal upwellings or proximity of these basins to seaways. During the Ypresian, the siliceous deposition area migrated from west to east (Fig. 10). Although in the early Ypresian the entire sediment-occupied area was adjacent to the Urals, in the late Ypresian this area expanded into eastern West Siberia, i.e., this basin's role as a passage increased.

During the middle and late Ypresian transgressions, geographic links between the Arctic and Tethyan basins were

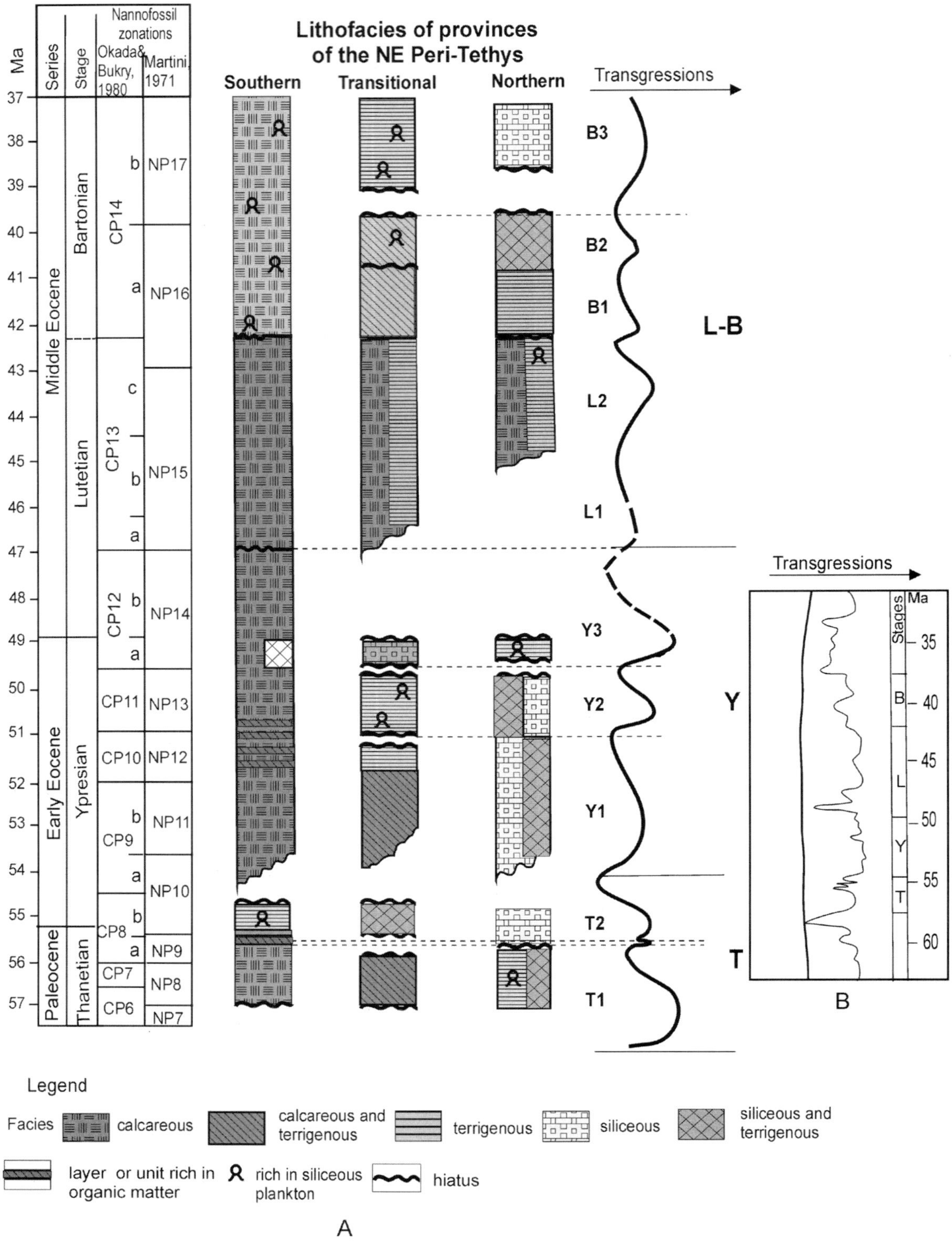

Figure 8. A: Dominant facies types of Thanetian to Bartonian strata from the Southern, Transitional, and Northern provinces and their cyclic interpretation. T, Y, L-B—Thanetian, Ypresian, Lutetian-Bartonian supercycles. In each supercycle, cycles are denoted by numbers. B: Sea-level curve for early Paleogene taken from Haq et al., 1987.

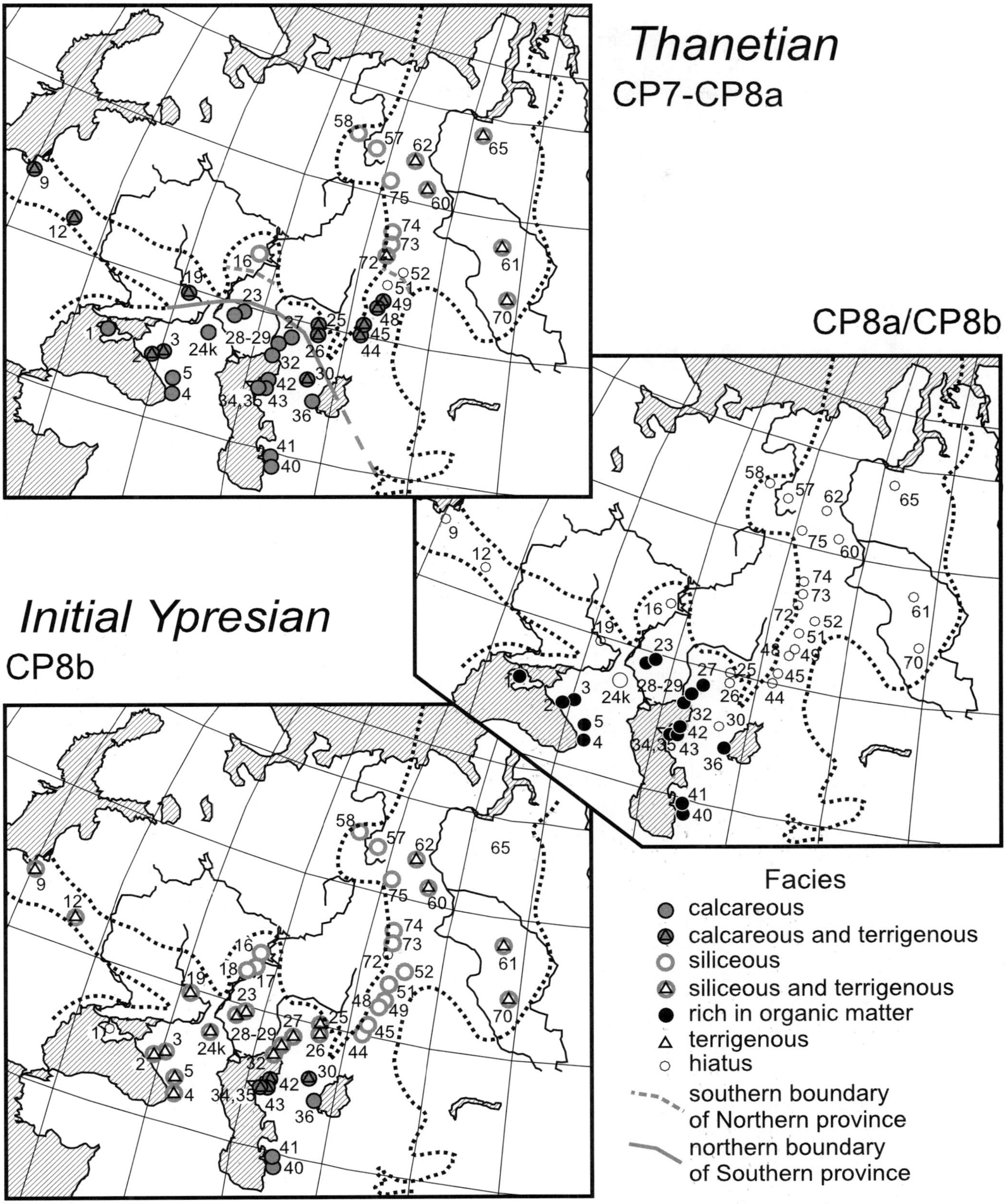

Figures 9–11 (on this and following pages). Maps showing distribution cycles of Thanetian, Ypresian, and Lutetian–Bartonian strata, constructed from the sections shown in Figure 1. Symbols are used for the dominant facies type of deposits and facial provinces boundaries.

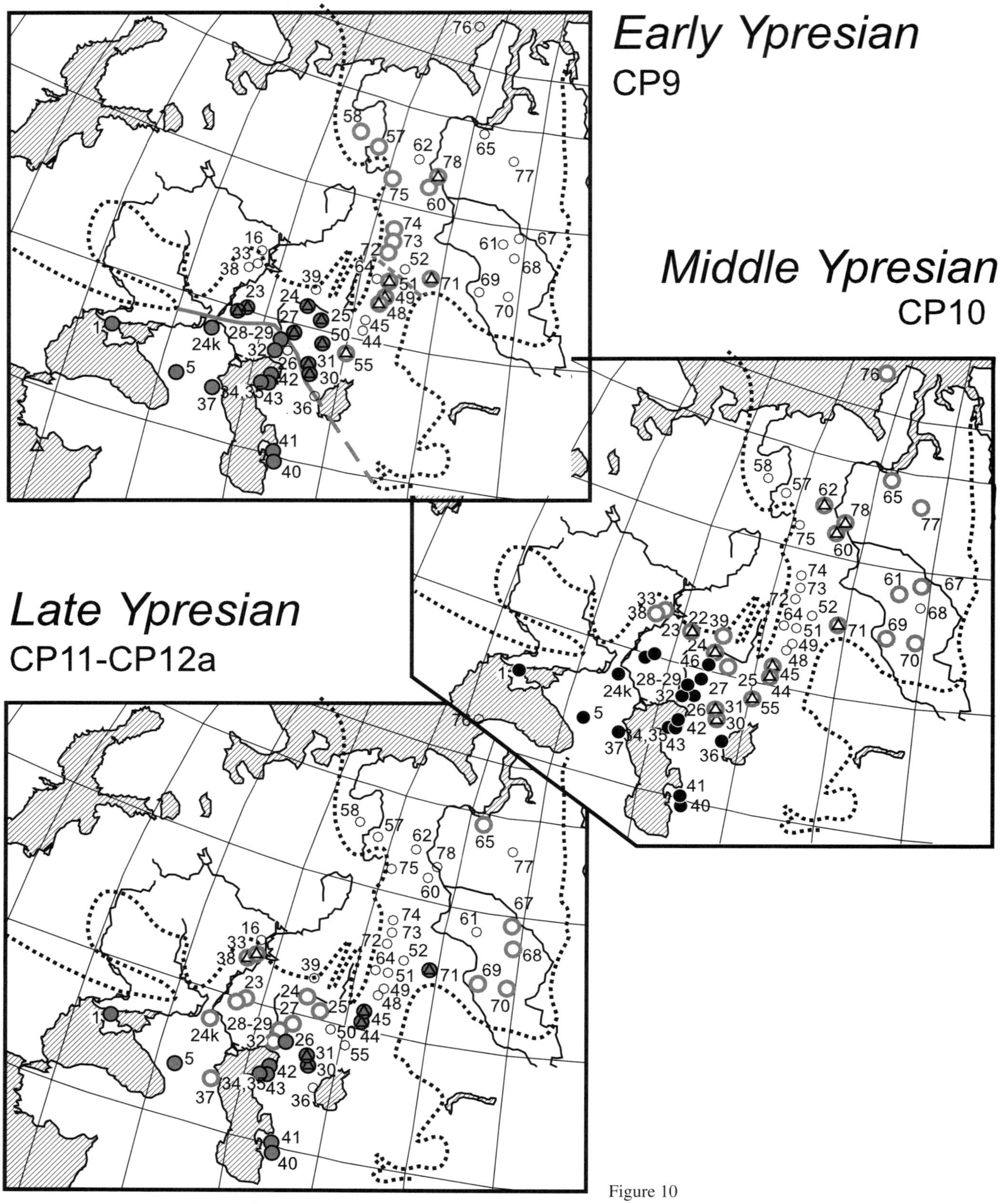
Early Ypresian
CP9
Middle Ypresian
CP10
Late Ypresian
CP11-CP12a

Figure 10

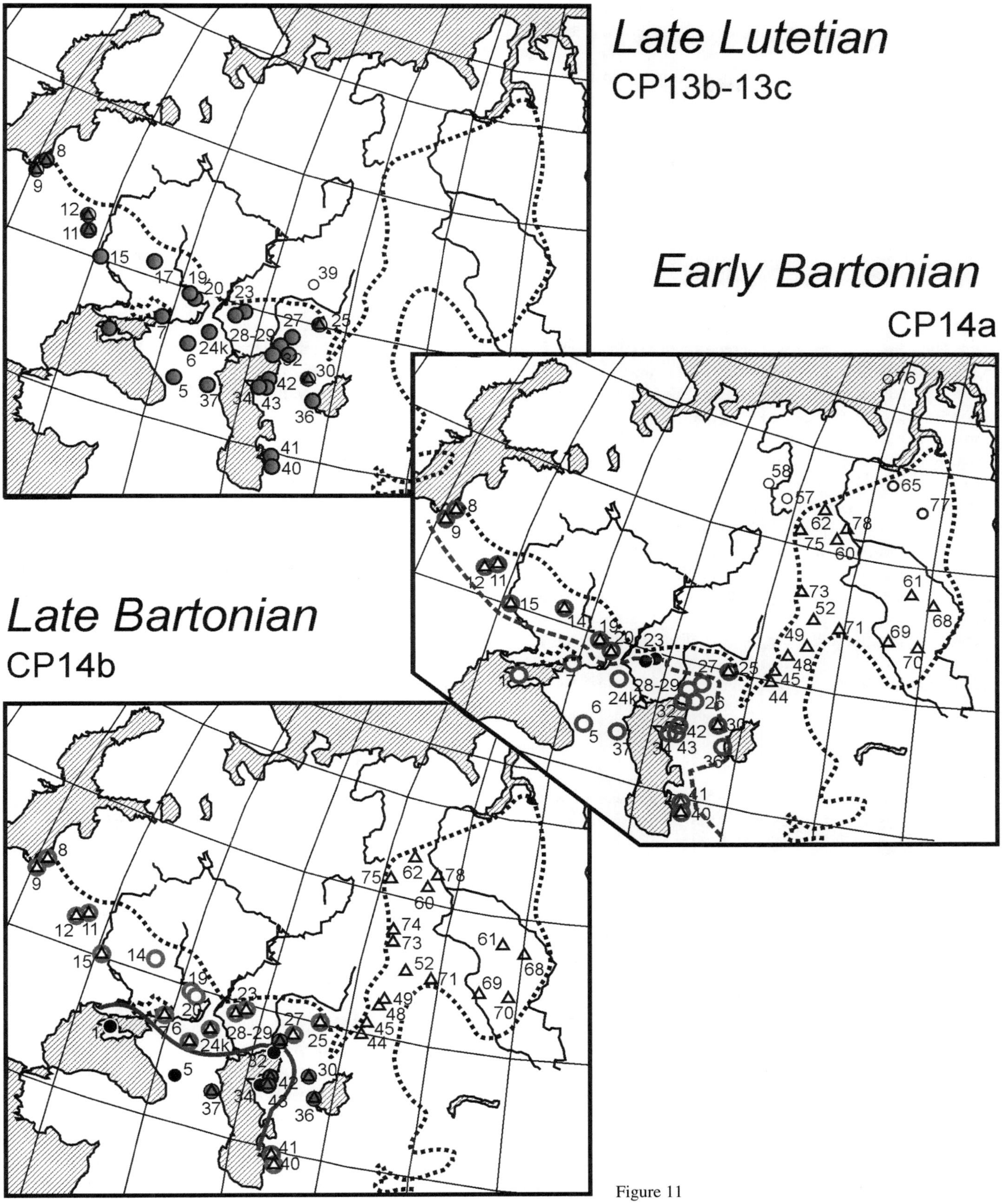
Late Lutetian
CP13b-13c
Early Bartonian
CP14a
Late Bartonian
CP14b

Figure 11

strongest, with the result that biogenic siliceous sediments then covered vast territories. Biogenic siliceous sediments penetrated into the Transitional and then into Southern provinces. Tectonic rearrangements of the early Lutetian led to a considerable regression. The link with the Arctic basin via West Siberia was cut and the geographic pattern of the late Lutetian to Bartonian transgression was essentially different.

The Southern province, as in the previous phase, shifted a significant distance southward. The Transitional province was restricted to the northern periphery of the Caucasus and the eastern Peri-Caspian. The Northern province was limited to within the Dneper-Donets basin, Lower Volga region, and eastern part of Turan plate (Fig. 11). Facies boundaries between the provinces were not sharp. Deposits of the lower upper Lutetian "Keresta" cycle in all three provinces have a similar facies aspect. They are characterized by green-colored calcareous facies with glauconite. At the Lutetian-Bartonian transition, the Southern province experienced a significant change in its depositional environment, and the disoxic "Kuma" basin, represented by coffee-colored bituminous marls, took shape there. Its formation is attributed to either a partial isolation from Tethys or to a considerable tectonic sagging of the basin, whose boundaries remained the same as previously.

The entire area of West Siberia turned into a gulf, and its links with normal marine basins were restricted (Shatsky, 1978). The distribution of siliceous facies and the siliceous biotic composition are suggestive of new northern links of the Peri-Tethyan basin with the Norwegian Sea via a system of seaways. The expansion of siliceous facies, in the early Bartonian, into northern Turkmenistan and, in the late Bartonian, into the Kopet Dagh and Central Asia, might point to new, southern, links with the Eastern Tethys.

Transgression scenario

Therefore, in the early Paleogene, the scenario of all three eustatic transgressions was the same in the entire northern Peri-Tethys and included several pulses. The transgression involved Tethyan water masses advancing from south to north, into the Northern province, and the development of a continuous circulation system including the Tethys, as evidenced by the appearance of numerous tropical microplankton species. During the same interval, the Northern province had a biota of its own, represented by siliceous microplankton. This phase gives no evidence of mutual penetration of the two biotas, which means that water masses of the Northern and Southern provinces did not interact. Then, a new phase in basinal evolution began, during which northern water masses transgressed to the south. This is evidenced by siliceous facies appearing in the Transitional province. Evidently, each transgression developed in several pulses, as evidenced by the hiatuses and lack of certain zonal assemblages of siliceous and organic walled organisms, observed in the uppermost Thanetian, Ypresian, and lower Bartonian strata in the Northern and Transitional provinces. Meanwhile, in the Southern province, Tethyan circulation was suffering destruction, with ensuing temporary anoxic environments.

During the terminal intervals of the Thanetian, Ypresian, and Bartonian, siliceous plankton penetrated into the Southern province, displacing considerably the calcareous plankton, with siliceous-terrigenous facies appearing in many sections of the Southern province. Therefore, during each transgression, the Arctic waters reached the Tethyan region proper, and water exchange took place between the two basins. Seaways between the Tethyan and Arctic basins changed throughout the early Paleogene. In the late Thanetian, the link was effectuated via the Baltic passage and the Dnieper-Donets basin on the one hand and via West Siberia and the Urals on the other. In the late Ypresian, the only link was via West Siberia, but it ensured an extensive exchange between the Arctic and Atlantic. In the Bartonian, the link was via the Norwegian Sea and the Dnieper-Donets basin. Evidently, the Tethyan basin communicated only with the North Atlantic.

CONCLUSIONS

The above correlation of Paleogene deposits between northern and southern areas of the eastern European craton, West Siberia, and Turan plate is as detailed as the resolving power of the methods used can afford. As discussed above, zonal stratigraphy of contrasting facies in continental sections has its limitations, due largely to transgressive-regressive cyclicity.

We are fully aware of the insufficient precision of the cyclic analysis techniques at hand. Needless to say, the use of seismostratigraphic techniques may refine considerably the cyclicity pattern for particular areas. However, across a study region as vast as ours, the use of zonal stratigraphy combined with the cyclic analysis, which draws on through-stratigraphy lithologic changes, supplemented in a number of sections by environmental analysis of microplankton assemblages, has enabled us (i) to introduce essential corrections to the synchronization of lower Paleogene deposits, (ii) to assess the ranges of hiatuses, and (iii) to propose a model for the Thanetian, Ypresian, and late Lutetian-to-Bartonian transgressions, which all followed the same scenario. This, still tentative, model requires a firm oceanologic justification, which is as yet to be elaborated.

ACKNOWLEDGMENTS

Our investigation of the links between northern and southern Paleogene domains in the northern Peri-Tethys was inspired by H. Oberhänsli, who formulated the study objective, *Climate Relevance of Sapropelite in a World Changing from a Nonglaciated to Glaciated Mode,* to our team in the context of the INTAS Project 93-2509. Our fieldwork, sample collection, and interpretation of regional geological data would hardly have been possible without the invaluable help from geoscientists and leaders of regional geological surveys, V.A. Martynov and V.D.

Dergachev (West Siberia), R.A. Segedin and T.R. Akopov (Western Kazakhstan), V.G. Pronin (Peri-Caspian Region), E.G. Sidorov, V.I. Musatov, and A.S. Zastrozhnov (Volga region), E.I. Kovalenko (Northern Caucasus), and Yu.I. Iosifova (the Dnieper-Donets basin), to all of whom we are deeply grateful. Thanks are due to E. Aleksandrova for her assistance with graphics and to I. Kravchenko-Berezhnoy for translating this paper into English. This work was supported by the Russian Foundation for Basic Research, project nos. 98-05-65060, 00-05-64917.

REFERENCES CITED

Akhmetiev, M.A., and Zaporozhets, N.I., 1996, Changes in dinocysts assemblages as reflecting in ecosystem transformations in the Russian Platform, Crimea, Caucasus, and Turan plate during Paleogene and early Miocene, *in* Kuznetsova, K.I., and Muzylöv, N.G., eds., Fossil microorganisms as the basis of the Phanerozoic stratigraphy, correlation, and paleobiogeography: Moscow, GEOS, p. 55–69 (in Russian).

Akhmetiev M.A., Alexandrova, G.N., Amon, E.O., Beniamovski, V.N., Bugrova, E.M., Vasilieva, O.N., Gleser, Z.I., Zhelezko, V.I., Zaporozhets, N.I., Kozlova G.E., Nikolaeva, I.A., Oreshkina, T.V., Panova, L.A., Radionova, E.P., Strelnikova, N.I., and Iakovleva, A.I., 2001, Biostratigraphy of the marine Paleogene of Western Siberia: Stratigraphy and Geological Correlation, v. 9, no. 2, p. 30–57.

Allen, Ph.A., and Allen, J.R., 1990, Basin analysis principles and applications: Oxford, Blackwell Scientific Publications, 451 p.

Amon, E.A., 1994, Radiolarians assocations from the Paleogene deposits of Kichigino section, South Cis-Urals, *in* Chuvashov, B.I., et al., eds., New data on Uralian upper Paleozoic and lower Cenozoic stratigraphy: Ekaterinburg, Uralian Branch of Academy of Science of Soviet Union, p. 151–162 (in Russian).

Andreeva-Grigorovich, A.S., 1985, Zonal subdivision of Paleocene–Eocene deposits of the Soviet Union based on dinocysts: Geological Journal, Kiev, v. 6, p. 112–119 (in Russian).

Beniamovski, V.N., Levina, A.P., Pronin, V.G., and Tabachnikova, I.P., 1989, The Paleogene deposits of Turgay depression: Izvestiya Vysshikh Uchebnykh Zavedeniy, Geologia i Razvedka, no. 5, p. 3–14 (in Russian).

Beniamovski, V.N., Levina, A.P., Pronin V.G., and Tabachnikova, I.P., 1991, The lower Eocene deposits of Turgay depression: Izvestiya Vysshikh Uchebnykh Zavedeniy, Geologia i Razvedka, no. 7, p. 3–16 (in Russian).

Beniamovski, V.N., Vasilieva, O.N., Levina, A.P., and Pronin, V.G., 1993, The Paleogene of south Trans-Urals, Paper 1: Paleogene deposits of the southern Turgay depression: Izvestiya Vysshikh Uchebnykh Zavedeniy, Geologia i Razvedka, no. 6, p. 22–36 (in Russian).

Beniamovski, V.N., Alekseev, A.S., Nezlina, E.Y., and Oberhänsli, H., 1999, The anoxic Kuma Formation (Bartonian) from the eastern Peri–Tethys: An environment study, *in* Schmitz, B., et al., eds., Early Paleogene warm climates and biosphere dynamics: GFF, v. 122 (abstracts).

Berggren, W.A., Kent, D.V., Swisher, C.C., III, and Aubry, M.P., 1995, A revised Cenozoic geochronology and chronostratigraphy, *in* Berggren, W.A., et al., eds., Geochronology, time scales, and global stratigraphic correlations: SEPM (Society for Sedimentary Geology) Special Publication 54, p. 129–212.

Bugrova, E.M., 1991, Paleogene Nummulites and Discocylines from eastern part of Central Asia: Izvestiya Akademii Nauk SSSR, Serya geologicheskaya, no. 1, p. 58–70 (in Russian).

Bugrova, E.M., Gleser, Z.I., Panova, L.A., Tabachnikova, I.P., Larionova, E.M., Gladkova, V.I., and Tkachenko, N.S., 1997, Biostratigraphy of the Paleogene key-section of the east Peri-Caspian region based on microfossils: All-Russian Institute of Geology (VSEGEI), St. Petersburg, Regional Geology and Metallogeny, no. 6, p. 64–76 (in Russian).

Bugrova, E.M., Zakrevskaya, E.Yu., and Tabachnikova, I.P., 2002, New Data on Paleogene biostratigraphy of the eastern Crimea: Stratigraphy and Geological Correlation, v. 10, no. 1, p. 83–93.

Chirva, S.A., 1971, Eocene sediments of the northern West Siberia and some problems of their correlation: Geology and Geophysics, no. 1, p. 44–55 (in Russian).

Chirva, S.A., and Lyubomirova, K.A., 1973, Marine and continental Paleogene correlation of the northern West Siberia, *in* Shatsky, S.B., ed. Marine and continental Paleogene of Siberia: Novosibirsk, Nauka, p. 69–77 (in Russian).

Fursenko, A.V., and Fursenko, K.B., 1961, The upper Eocene foraminifers in Belarus and their stratigraphic implications: Paleontology and Stratigraphy of the Belorusskaya Soviet Republic, sbornik 3, p. 246–361 (in Russian).

Gavrilov, Yu.O., and Muzylöv, N.G., 1992, The geochemistry of sapropelite interbeds in the Paleogene of the Central Caucasus: Lithology and Mineral Resources, v. 26, no. 6, p. 548–559.

Gavrilov, Y.O., Kodina, L.A., Lubchenko, I.Y., and Muzylöv, N.G., 1997, The Late Paleocene anoxic event in epicontinental seas of Peri-Tethys and formation of the sapropelite unit: Sedimentology and geochemistry: Lithology and Mineral Resources, v. 32, p. 427–450.

Gavrilov, Yu.O., Shcherbinina, E.A., and Muzylöv, N., 2000, A Paleogene sequence in the central North Caucasus: A response to paleoenvironmental changes, *in* Schmitz, B., et al., eds., Early Paleogene warm climates and biosphere dynamics: GFF, v. 122, p. 51–53.

Gleser, Z.I., 1996, Problems of Eocene siliceous phytoplankton zonation (exemplified by Caspian Eocene deposits): Stratigraphy and Geological Correlation, v. 4, no. 4, p. 83–92.

Gleser, Z.I., and Stepanova, G.V., 1994, Paleogene deposits of Kara sea subdivision and correlation based on diatoms and silicoflagellates: All-Russian Institute of Geology (VSEGEI), St. Petersburg, Regional Geology and Metallogeny, no. 2, p. 148–153 (in Russian).

Grachev, N.V., Zhizhchenko, B.P., Kolykhalova, L.A., and Kholodilina, T.S., 1971, The Paleogene deposits between Volga and Ural rivers, *in* Zhizhchenko, B.P., and Ivanova, V.A., eds., Cenozoic stratigraphy and paleogeogrphy of gas-bearing region of the southern Soviet Union, v. 31\39–32\40, p. 36–44 (in Russian).

Grossgeim, A.V., 1960, The Paleogene of the Northwest Caucasus: Transactions of the Krasnodar Branch of the All-Union Oil-and-Gas Research Institute, v. 4, p. 3–190.

Haq, B.U., Hardenbol, J., and Vail, P.R., 1987, Chronology of fluctuating sea levels since the triassic: Science, v. 235, p. 1156–1167.

Iakovleva, A.I., 1998, Eocene microphytofossils from the Sari-Oba section (north Aral region): Bulletin of the Moscow Naturalist Society, Geological section, v. 73, no. 3, p. 51–55 (in Russian).

Iakovleva, A.I., 2000, Biostratigraphical and paleogeographical role of the Paleocene–Eocene dinoflagellates in Western Siberia and adjacent regions (Petchora depression and Turgay Trough), *in* Schmitz, B., et al., eds., Early Paleogene warm climates and biosphere dynamics: GFF, v. 122, p. 82–83.

Iakovleva, A.I., and Kulkova, I.A., 2001, Paleocene–Eocene dinoflagellate cysts and continental palynomorphs of the Pur Downstream Basin (Northwestern Siberia): Biostratigraphical and paleoenvironmental implication: Revista Española de Micropaleontologica, v. 33 no. 1, p. 1–33.

Iakovleva, A.I., Cavagnetto, C., and Rousseau, D.D., 2000a, Paleocene–Eocene dinoflagellate stratigraphy of the borehole No. 4 (Central Western Siberia): AASP Palynology, v. 24, p. 187–200.

Iakovleva, A.I., Kulkova, I.A., and Cavagnetto, C., 2000b, Eocene microphytofossils (dinoflagellate cysts and continental palynomorphs) of Northwestern Siberia (Severnaya Sosva Basin): Newsletters on Stratigraphy, v. 38, no. 1, p. 13–38.

Iakovleva, A.I., Oreshkina, T.V., Alekseev, A.S., and Rousseau, D.D., 2000c, A new paleontological and paleogeographical data in the Pechora depression, northeastern European Russia: Comptes Rendus de l'Académie des Sciences de France, Série II, v. 330, p. 485–491.

Iarkin, V.I., editor, 1989, Regulations of the Interdepartmental Stratigraphic Committee (ISC) and its constant commissions (decisions of the XVI Plenum of the Paleogene system): All-Russian Institute of Geology (VSEGEI) Leningrad, v. 21, p. 51–54 (in Russian).

Ilyina, V.I., Kulkova, I.A., and Lebedeva, N.K., 1994, Dinoflagellate cysts of the marine Paleogene, *in* Chlonova, A.F., ed., Microphytofossils and detailed stpatigraphy of marine Mesozoic and Cenozoic of Siberia: Russian Academy of Sciences Siberian Branch, United Institute of Geology, Geophysics and Mineralogy Transaction, v. 818, p. 98–107 (in Russian).

Kazmin, V.G., Sborschikov, I.N., Ricou, L.A., Zonenshain, L.P., Poulen, J., and Knipper, A.L., 1987, Volcanic belt as indicator of the Mesozoic–Cenozoic activity of the Eurasia edge, *in* Monin, A.S., and Zonenshain, L.P., eds., Geological history of the Tethys Ocean: P.P. Shirshov Institut of Oceanology, p. 58–74.

Khokhlova, I.E., Radionova, E.P., Beniamovsi, V.N., and Shcherbinina, E.A., 1999, Eocene stratigraphy of key sections of the Dnieper-Donetz depression based on calcareous and siliceous microplankton: Geodiversitas, v. 21, no. 3, p. 453–476.

Kopp, M.L., and Scherba, I.G., 1998, Caucasian Basin in the Paleogene: Geotectonics, v. 2, p. 29–50.

Korovina, G.M., 1970, Zonal subdivision and facies of "Kuma" Horizon of upper Eocene "Bodrak" regional stage of West Turkmenistan, *in* Shutskaya, E.K., ed., Stratigraphy and paleontology Mezosoic and Paleogene deposits of Central Asia: Transactions of All-Russian Petroleum Exploration Institute (VNIGNI), v. 49, p. 122–128 (in Russian).

Kozlova, G.E., 1999, Paleogene boreal radiolarians from the Russia: St. Petersburg, All-Russian Petroleum Research Exploration Institute (VNIGRI), v. 9, 323 p. (in Russian).

Kozlova, G.E., and Strelnikova, N.I., 1984, Significance of the section of Ust-Maninsk hole 19-V for zonal division of the West Siberian Paleogene, *in* Shatsky, S.B., ed., Environment and life at the boundaries of Cenozoic epochs in Siberia and the Far East: Novosibirsk, Nauka, p. 70–78 (in Russian).

Kozlova, G.E., Strelnikova, N.I., and Khokhlova, I.E., 1998a, Marine Paleogene of the southwestern Ural region: The siliceous microplankton stratigraphy (Holes SP-1 and 148), *in* Knipper, A.L., et al., eds., The Urals: Fundamental problems of geodynamics and stratigraphy: Moscow, Nauka, Transactions of Geological Institute, v. 500, p. 193–205 (in Russian).

Kozlova, G.E., Radionova, E.P., Scherbinina, E.A., Khokhlova, I.E., Oreshkina, T.V., Beniamovski, V.N., and Iakovleva, A.I., 1998b, Late Paleocene changes in the northeastern Peri-Tethys: Comparative study of the Sokolovsky Quarry (Turgay Passage) and Sengiley (middle reaches of the Volga) sections: Strata, série 1, v. 9, p. 72–74.

Krasheninnikov, V.A., 1982, Paleogene Stratigraphy of the northwestern part of the Pacific Ocean: Transactions of the Geological Institute, v. 389, 142 p. (in Russian).

Krasheninnikov, V.A., and Muzylöv, N.G., 1975, Relationship of zonal schemes by planktonic foraminifera and nannofossils in the Paleogene sections of the North Caucasus: Moscow, Nauka, Voprosy micropaleontologii, v. 18, p. 212–224 (in Russian).

Kriventsov, A.V., 1973, Paleogene deposits of Chizhapka river: The Paleogene of central part of West Siberia, *in* Shatsky, S.B., ed., Marine and continental Paleogene of Siberia: Novosibirsk, Nauka, p. 40–46 (in Russian).

Kurgalimova, G.G., and Moksyakova, A.M., 1970, New data on the stratigraphic range of the lower part of the Tasaran Formation in North Cis-Aralian, *in* Shutskaya, E.K., ed., Mesozoic and Paleogene deposits of Central Asia stratigraphy and paleontology: Transactions of All-Russian Petroleum Exploration Institute (VNIGNI), v. 49, p. 129–134 (in Russian).

Leonov, G.P., 1964, The main problems of Paleogene stratigraphy of the Russian plate: Moscow State University Publication, Moscow, 552 p. (in Russian).

Martini, E., 1971, Standard Tertiary and Quaternary calcareous nannoplankton zonation, *in* Farinacci, A., ed., Proceedings of the 2nd Planktonic Conference, Rome, Italy, 1970: Tecnoscienza, v. 2, p. 739–785.

Meulenkamp, J.E., and Sissingh, W., 2000, Map 17–23 Tertiary, *in* Crasquin, S., coordinator, Explanatory notes, *in* Dercourt, J., et al., eds., Atlas Peri-Tethys paleogeographical maps: Paris, Commission for the Geologic Map of the World, p. 153.

Muzylöv, N.G., 1994, Anoxic events of Paleocene–Eocene, *in* Rozanov, A.Yu., and Semikhtov, M.A., eds., Ecosystem reorganizations and the evolution of biosphere: Moscow, Nedra, v. 1, p. 160–166 (in Russian).

Muzylöv, N.G., 1996, Early Paleogene episodes of eruptive activity in the east Peri-Tethys and their possible influence on the marine microbiota, *in* Kuznetsova, K.I., and Muzylöv, N.G., eds., Fossil microorganisms as the basis of the Phanerozoic stratigraphy, correlation, and paleobiogeography: Moscow, GEOS, Voprosy mikropaleotologii, v. 31: p. 128–132 (in Russian).

Muzylöv, N.G., Pyzhonkova, E.A., and Khodzhakhmedov, K.A., 1987, The lower Paleogene nannoplankton stratigraphy of South Turkmenistan (South Kopet Dagh, Lessor Balkhan, Gyaursky Kopet Dagh): Moscow, Nauka, Voprosy mikropaleotologii, v. 29, p. 114–128 (in Russian).

Muzylöv, N.G., Benyamovskii, V.N., and Tabachnikova, I.P., 1989, Sapropel interlayers in lower Paleocene deposits of the southern Soviet Union: Izvestia Akademii Nauk SSSR, Seria Geologicheskaya, v. 11, p. 117–119 (in Russian).

Muzylöv, N.G., Beniamovski, V.N., Gavrilov, Yu.O., Shcherbinina, E.A., and Stupin, S.I., 1996, Paleontological and geochemical features of the Central Caucasian upper Paleocene sapropelite, *in* Kuznetsova, K.I., and Muzylöv, N.G., eds., Fossil microorganisms as the basis of the Phanerozoic stratigraphy, correlation, and paleobiogeography: Moscow, GEOS, Voprosy mikropaleotologii, v. 31, p. 117–127 (in Russian).

Naidin, D.P., and Beniamovski, V.N., 1994, The Suvlukaya Paleogene section (Crimea): Stratigraphy and Geological Correlation, v. 2, no. 3, p. 75–86.

Naidin, D.P., Beniamovski, V.N., and Kopaevich L.F., 1994, A paleogeographic basis for stratigraphic construction: Moscow State University Publication, Moscow, 136 p. (in Russian).

Naidin, D.P., Beniamovski, V.N., and Kopaevich L.F., 1996, Paleocene reference sections of Mangyshlak: Stratigraphy and Geological Correlation, v. 4, no. 3, p. 46–61.

Oberhänsli, H., and Beniamovski, V.N., 2000, Planktonic and benthic assemblages from the late Ypresian–early Lutetian marly and TOC-rich sediments of the Central pre-Caucasus and west coast of the Aral Sea, *in* Schmitz, B., et al., eds., Early Paleogene warm climates and biosphere dynamics: GFF, v. 122, p. 121–123.

Okada, H., and Bukry, D., 1980, Supplementary modification and introduction of code numbers to the low-latitude coccolith biostratigraphic zonation (Bukry, 1973, 1975): Micropaleontology, v. 5, no. 3, p. 321–325.

Oreshkina, T.V., 2000, New data on early Eocene diatom succesion of the west Polar Urals margin: Paleoclimatic and paleobiogeographic implications, *in* Schmitz, B., et al., eds., Early Paleogene warm climates and biosphere dynamics: GFF, v. 122, p. 124–126.

Oreshkina, T.V., and Khokhlova, I.E., 1999, Early Paleogene siliceous microfossils of the Middle Volga region: Stratigraphy and paleogeography: Geodiversitas, v. 21, no. 3, p. 429–451.

Pechonkina, A.P., and Kholodilina, T.S., 1971, The data on Volga and Ural Paleogene deposits stratigraphy based on foraminifers, *in* Zhizhchenko, B.P., and Ivanova, V.A., eds., Cenozoic stratigraphy and paleogeogrphy of gas-bearing region of the southern Soviet Union, v. 31\39–32\40, p. 25–35 (in Russian).

Powell, A.J., 1992, Dinoflagellate cysts of the Tertiary System, *in* Powell, A.J., ed., A stratigraphic index of dinoflagellate cysts: London, Chapman and Hall, p. 155–251.

Radionova, E.P., 1998, Methods of construction of the Paleogene epicontinental zonal scales by diatoms (demonstrated with subdivision of Eocene sediments (Hole SP-1, northeastern pre-Caspian region), *in* Knipper, A.L., et al., eds., The Urals: Fundamental problems of geodynamics and stratigraphy: Moscow, Nauka, Transactions of Geological Institute, v. 500, p. 206–220 (in Russian).

Radionova, E.P., and Khokhlova, I.E., 1994, Paleogene episodes of biogenic silica accumulation in the northern Caucasus and adjacent Tethyan regions: Stratigraphy and Geological Correlation, v. 2, no. 5, p. 161–169.

Radionova, E.P., and Khokhlova, I.E., 2000, Was the North Atlantic connected via the Arctic with the Tethys in the early Eocene? Evidence from siliceous plankton, *in* Schmitz, B., et al., eds., Early Paleogene warm climates and biosphere dynamics: GFF, v. 122, p. 133–134.

Radionova, E.P., Oreshkina, T.V., Khokhlova, I.E., and Beniamovski, V.N., 1994, Eocene deposits on the northeastern slope of the Dnieper-Donets depression: Zonal stratigraphy and cyclic analysis: Stratigraphy and Geological Correlation, v. 2, no. 6, p. 85–102.

Radionova, E.P., Khokhlova, I.E., Beniamovski, V.N., Shcherbinina, E.A., Iakovleva, A.I., and Sadchikova, T.A., 2001, Paleocene–Eocene transition in the northeastern Peri-Tethys area: Sokolovskii key section of the Turgay Passage (Kazakhstan): Bulletin de la Société Géologique de France, v. 172, no. 2, p. 245–256.

Rodionova, M.K., 1963, Foraminifera from the upper Eocene deposits of West Turkmenistan, *in* Aliev, M.M., ed., Foraminifera from Cretaceous and Paleogene deposits of West Turkmenistan: Moscow, USSR Academy of Sciences, p. 58–92.

Rubina, N.V., 1973, Paleogene and Neogene stratigraphy of marine and continental deposits of West Siberia based on paleoalgology, *in* Shatsky, S.B., ed., Marine and continental Paleogene of Siberia: Novosibirsk, Nauka, p. 67–96.

Shatsky, S.B., 1978, Main problems of Paleogene stratigraphy and paleogeography of Siberia, *in* Shatsky, S.B., ed., Paleogene and Neogene of Siberia: Novosibirsk, Nauka, p. 3–21 (in Russian).

Shatsky, S.B., Aleksandrova, V.A., Bogdashev, V.A., Vasilev, I.P., Gorbovetz, A.N., Goriukhin, E.A., Kriventzov, A.V., Silvestrov, N.B., Stasov, V.I., Tolkachev, M.V., Fadeev, A.I., Freiman, E.V., and Shamakhov, A.V., 1973, The Paleogene of central part of West Siberia, *in* Shatsky, S.B., ed., Marine and continental Paleogene of Siberia: Novosibirsk, Nauka, p. 6–39 (in Russian).

Shcherbinina, E.A., 2000, Middle Eocene nannofossils and geological events of the northeastern Peri-Tethys, *in* Schmitz, B., et al., eds., Early Paleogene warm climates and biosphere dynamics: GFF, v. 122, p. 51–53.

Shutskaya, E.K., 1970, Stratigraphy, foraminifers, and paleogeography of the lower Paleogene in the Crimea, Ciscaucasus, and western part of Central Asia: Moscow, Transactions of All-Russian Petroleum Exploration Institute (VNIGNI), v. 70, 256 p. (in Russian).

Solun, V.M., editor, 1975, Paleogene of Turkmenistan (biostratigraphy and correlation): Ashkhabad, Alim, 195 p. (in Russian).

Strelnikova, N.I., 1992, Paleogene diatom algae: St. Petersburg State University Publication, St. Petersburg, 311 p. (in Russian).

Strelnikova, N.I., Kaplan, L.A., and Travina, M.A., 1978, Paleogene diatoms, silicoflagellates and ebrididea of Kaliningrad region, *in* Jouse, A.P., ed., Marine micropaleontology (diatoms, radiolarians, silicoflagellates, foraminifers, and calcareous nannoplankton): Moscow, Nauka, p. 57–65 (in Russian).

Stupin, S.I., and Muzylöv, N.G., 2001, The late Paleocene ecologic crisis in epicontinental basins of the eastern Peri-Tethys: Microbiota and accumulation conditions of sapropelitic bed: Stratigraphy and Geological Correlation, v. 9, no. 5, p. 87–93.

Vasilieva, O.N., 1994, Microphytoplankton association from the North Cis-Aralian and Usthurt regions (Kazakhstan), *in* Chuvashov, B.I., and Amon, E.A., eds., New data on Uralian late Paleozoic and lower Cenozoic stratigraphy: Ekaterinburg, Uralian Branch of Academy of Science of Soviet Union, p. 163–172 (in Russian).

Vasilieva, O.N., Zhelezko, V.I., and Lubcheko, G.N., 1994, Biostratigraphy of marine Paleogene deposits based on dinocysts and sharks in Kurgan town region, *in* Chuvashov, B.I., and Amon, E.A., eds., New data on Uralian late Paleozoic and lower Cenozoic stratigraphy: Ekaterinburg, Uralian Branch of Academy of Science of Soviet Union, p. 139–151 (in Russian).

Zonenshain, L.P., and Le Pichon, X., 1987, Deep-water basins of the Black and Caspian seas: Remnants of Mesozoic back-arc basins, *in* Monin, A.S., and Zonenshain, L.P., eds., Geological history of the Tethys Ocean: P.P. Shirshov Institute of Oceanology, p. 74–93 (in Russian).

Manuscript Accepted by the Society August 13, 2002

Geological Society of America
Special Paper 369
2003

A paleoenvironmental reconstruction of the early late Paleocene North Sea from intrashell δ18O and δ13C profiles of mollusks

Birger Schmitz
Elisabet I. Thompson
Lennart Bornmalm
Department of Earth Sciences, Marine Geology, Göteborg University, Box 460, SE-405 30 Göteborg, Sweden
Claus Heilmann-Clausen
Department of Earth Sciences, Aarhus University, DK-8000 Aarhus C, Denmark

ABSTRACT

Stable oxygen and carbon isotopic analyses of early Selandian (ca. 60 Ma) mollusk shells from Denmark (southeast margin of the North Sea Basin, paleolatitude ~47°N) were used to estimate intraannual temperature variations. We used shells from two water depths to constrain possible effects of freshwater influx. Assuming normal marine salinity we derived average temperatures for the winter of 16.9 °C and for the summer 22.5 °C at depths of 20–60 m, consistent with a subtropical climate. In contrast, *Turritella* spp., living at water depths of <20 m, record winter and summer temperatures of 23.4 °C and 28.5 °C, respectively, suggesting an unrealistically warm climate, especially in winter. The data indicate similar seasonal temperature ranges, 5–6 °C, at shallower as at greater depths. These estimates might reflect that *Turritella* did not calcify during the coldest months. Alternatively, they might indicate influx of local freshwater plumes. Positive $\delta^{13}C$ values (2.7‰–3.8‰) of *Turritella* shells, however, suggest that conditions were close to or fully marine. The deeper-dwelling mollusks were probably not affected by local freshwater influx, but we cannot rule out that the entire southern North Sea had slightly reduced (1–3 psu) salinity. In that case, the estimated deeper water temperatures are too high by 1–3.5 °C. Even with these uncertainties, the isotopic data suggest that summer surface water temperatures were between 22 °C and 28 °C, and winter temperatures not colder than 13 °C. Denmark thus had a warm subtropical climate, as had other locations at these paleolatitudes in western Europe during most of the early Paleogene.

INTRODUCTION

Past climates can be understood in detail by studying stable oxygen isotope ($\delta^{18}O$) profiles along the transect of growth in marine gastropod shells. These profiles may provide the climatically important mean annual range of temperature in addition to the mean annual temperature (Jones et al., 1983; Krantz et al., 1987; Krantz, 1990; Allmon et al., 1992, 1994; Jones and Allmon, 1995; Andreasson and Schmitz, 1996, 1998, 2000; Andreasson et al., 1999; Kobashi et al., 2001; Tripati et al., 2001).

The traditional way of obtaining ocean temperatures by $\delta^{18}O$ analyses of benthic and planktonic foraminifera provides at the very best an estimation of the ancient mean annual temperature. The isotopic record from foraminifera may be biased by factors such as vertical migration in the water column or sediment, preferential season of growth, or reliance on symbionts

Schmitz, B., Thompson, E.I., Bornmalm, L., and Heilmann-Clausen, C., 2003, A paleoenvironmental reconstruction of the early late Paleocene North Sea from intrashell $\delta^{18}O$ and $\delta^{13}C$ profiles of mollusks, *in* Wing, S.L., Gingerich, P.D., Schmitz, B., and Thomas, E., eds., Causes and Consequences of Globally Warm Climates in the Early Paleogene: Boulder, Colorado, Geological Society of America Special Paper 369, p. 263–274.

(see Grossman, 1987; Spero and Deniro, 1987; Berger and Wefer, 1988; Sautter and Thunell, 1991; Ravelo and Fairbanks, 1992; Spero and Lea, 1993; D'Hondt et al., 1994). Although gastropod isotopic records may provide seasonal resolution, interpretations are often complicated by uncertainties about the isotopic composition of nearshore water (Bice et al., 1996; Andreasson and Schmitz, 2000).

By establishing stable oxygen and carbon isotope ($\delta^{18}O$, $\delta^{13}C$) profiles from gastropod shells that formed at different water depths, we attempt to reconstruct the seasonal vertical distribution of water temperature and salinity at the southeast margin of the North Sea Basin during the early late Paleocene (early Selandian, ca. 60 Ma). Representing waters of ~20–60 m are the gastropod species *Natica detracta* and *Surcula hauniensis* as well as the scaphopod species *Dentalium rugiferum.* Representing water of ~0–20 m, is the gastropod genus *Turritella.* For comparison, we also studied the benthic foraminiferal genus *Lenticulina* spp. from two localities in the Danish Basin.

Previous $\delta^{18}O$ studies of high-latitude Paleocene mollusks (Bice et al., 1996; Tripati et al., 2001) have shown the importance of freshwater influx when interpreting the $\delta^{18}O$ record. By adding a vertical dimension to the intrashell $\delta^{18}O$ variations, we try to constrain any potential impact of continental runoff on the gastropod oxygen isotopic records. In addition, we use the stable carbon isotopic record from the mollusk shells to estimate freshwater admixture, since it has been shown that inorganic carbon in runoff generally is depleted in ^{13}C compared to marine carbon (Keith et al., 1964; Keith and Parker, 1965).

The North Sea Basin at this time is of particular interest as surface temperatures have never been reconstructed due to the absence or near absence of planktonic calcareous fossils. An earlier study of whole-shell mollusk $\delta^{18}O$ from the Paleogene North Sea Basin indicates an early Selandian temperature minimum (Buchardt, 1977, 1978). Evidence of a similar bottom-water cooling in the global oceans (Corfield and Cartlidge, 1992; Barrera and Huber, 1991; Seto et al., 1991; Thompson and Schmitz, 1997) was interpreted by Corfield and Cartlidge (1992) as a result of a climatic deterioration caused by a global increase in burial of marine organic carbon versus carbonate (Thompson and Schmitz, 1997). Fossil floral and faunal records on the North American continent also show a temperature decline at this time (Gingerich, 1980; Hickey, 1980; Wolfe, 1985; Frederiksen, 1994; Wing et al., 1995). Despite an early late Paleocene cooling, however, bottom water temperatures in the global ocean were higher than today, averaging ~7–10 °C (Zachos et al., 1994; Kaiho, 1998). At high northern and southern latitudes, respectively, shallow marine and surface water temperatures were higher than today by ~9–11 °C, whereas at lower latitudes surface water temperatures were higher by no more than a few degrees (Levitus and Boyer, 1994; Zachos et al., 1994; Bice et al., 1996), so that the pole-to-equator thermal gradient was lower than it is today (Boersma and Premoli Silva, 1983; Zachos et al., 1994; Markwick, 1998). Here we add information about the paleoclimate and paleoenvironment of the early late Paleocene northwestern Europe.

GEOLOGICAL AND STRATIGRAPHICAL SETTING

At the transition from the Danian to Selandian, in conjunction with the Laramide tectonic episode, chalk accumulation in the North Sea ended abruptly and was replaced by clastic-dominated deposition (King, 1989; Ziegler, 1990). At the southeastern margin of the North Sea Basin, at a paleolatitude of ~47°N (Scotese et al., 1988; Smith et al., 1994), two coeval lithofacies characterize the lower Selandian. Planktonic foraminifera and calcareous nannoplankton indicate a P3/NP4–5 age for these deposits (Hansen, 1968; Perch-Nielsen, 1979; Thomsen and Heilmann-Clausen, 1985; Thomsen, 1994).

The Lellinge Greensand facies is found predominantly in areas near the contemporary shoreline such as at the Vestre Gasværk locality (Larsen and Jørgensen, 1977) (Fig. 1). Sediments of the Kerteminde Marl facies, in Store Bælt and at Viborg, indicate deposition in deeper waters (Fig. 1). The lower Selandian

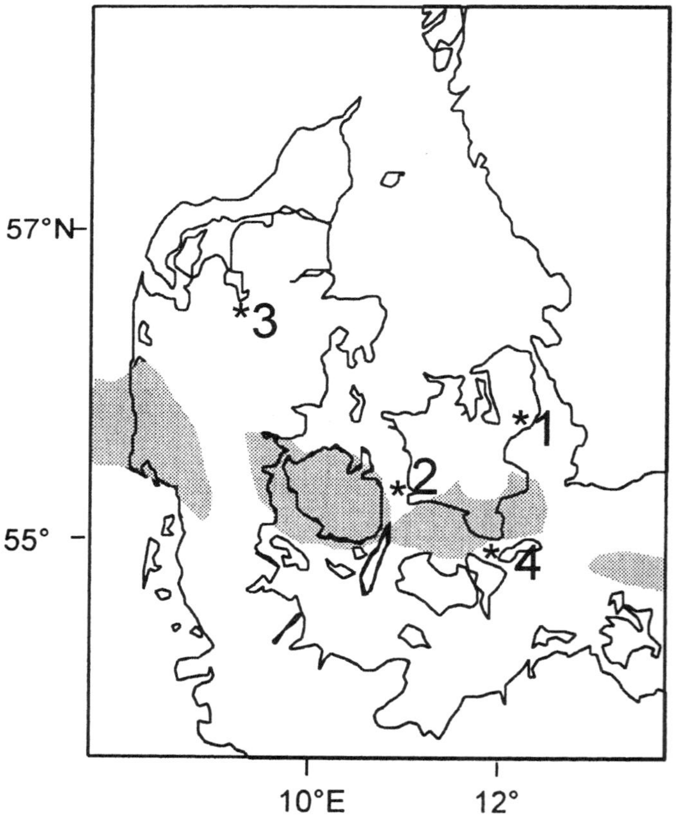

Figure 1. Map of the study area.*1 Vestre Gasværk, *2 Store Bælt Core No. 8604A, *3 Viborg Core No. 5 and *4 Bogø. Light gray areas represent the structural highs that existed in the region during the early Selandian. Longitude and latitude are for present-day Denmark.

deposits show fining upward (Heilmann-Clausen, 1989) concomitant with an early late Paleocene transgression that flooded central parts of the basin (Gramann, 1988). Coincident with this transgression is a wide, intrabasinal distribution of Midway type benthic foraminiferal fauna, indicative of a strong oceanic influence (Gramann, 1988). A connection between North Atlantic open ocean waters and North Sea basinal waters may have existed north of Scotland (Gramann and Kockel, 1988).

MATERIALS AND METHODS

Mollusk and foraminiferal fauna

We studied mollusks from Vestre Gasværk (Copenhagen), and from a large boulder of Selandian sandstone from the island Bogø and benthic foraminifera from Viborg Core No. 5 in Northern Jutland and Core 8604A in the Store Bælt Strait (Fig. 1). The mollusk fauna from the Vestre Gasværk section is diverse and excellently preserved (Ravn, 1939). The gastropods *Natica detracta* (Naticidae) and *Surcula hauniensis* (Turridae) and the scaphopod *Dentalium rugiferum* (Dentaliidae) from the upper part of the section (level 6.23–6.51 m according to Gry, 1935) were selected for $\delta^{18}O$ and $\delta^{13}C$ analyses. All three genera are today found from the Mediterranean to Norway. Naticidae and Turridae are commonly found at water depths from 10 m to 200 m (Hayward and Ryland, 1995). The majority of recent scaphopods occur at depths from 6 m to 1830 m (Barnes, 1980). In upper Eocene to lower Oligocene sediments from the northeastern Pacific, relatives of these three genera are codominant in the so-called Turrid Community, which may have lived at water depths shallower than 135 m (Hickman, 1984). Naticidae and Turridae are epifanual predators (Hickman, 1984; Gili and Martinell, 1994; Kelley and Hansen, 1996) and Dentaliidae are sedentary deposit feeders (Hayward and Ryland, 1995).

Large boulders of Selandian sandstone of the same age as the Lellinge Greensand occasionally wash up on the coasts of Denmark and Sweden (Andersen and Heilmann-Clausen, 1983). A specimen of the gastropod *Turritella* spp. (Turritellidae) was selected from one of these. Evidence of microalgae boring on shells found in the sandstone along with bedding planes with current-oriented *Turritella* shells suggest a habitat within the photic zone, close to the contemporary shoreline (Andersen and Heilmann-Clausen, 1983). Recent turritellids are epifaunal or shallow infaunal suspension-feeders and inhabit a wide range of environments, but appear to prefer depths of 10–100 m (Allmon, 1988). Living turritellids favor water temperatures of 15–20 °C, but may have preferred warmer waters in the past (Allmon, 1988).

According to Schnetler (2001), who summarized the regional paleontological and lithological information, the *Turritella* of the sandstone boulders lived in the littoral zone (above wavebase, tidal to subtidal), whereas the mollusks from the Vestre Gasværk locality lived below wave base, subtidal or deeper. The relative nearness to the Fyn structural high to the southwest and a possible shoreline at Ystad 80 km southeast (where *Turritella* sandstone boulders are common) suggests that depths at Vestre Gasværk were most likely not much greater than ~60 m (see Figure 1; Larsen and Jørgensen, 1977). We thus estimate a depth of 0–20 m for the *Turritella* and 20–60 m for the mollusks from Vestre Gasværk.

Danish mid-Paleocene marine deposits are almost barren of planktonic foraminifera, so that an isotope assessment of surface waters is difficult. The isotopic composition of bottom waters can, however, be provided by studying benthic foraminifera. For our study we analysed monogeneric assemblages of the benthic foraminifer *Lenticulina* spp. from both Viborg Core No.5 and Core 8604A. Judging from paleobathymetric reconstructions in Larsen and Jørgensen (1977) water depths at these localities were similar to that at Vestre Gasværk, however, we believe that waters at these localities were substantially deeper (middle to outer neritic depth), particularly for the Viborg core, which represents more central parts of the Danish Basin. This interpretation is supported by the isotopic results in this study, indicating colder conditions at the seafloor at Store Bælt and Viborg than at Vestre Gasværk. The Paleocene benthic foraminifera in Denmark show affinity to the so-called "Midway fauna" (Larsen and Jørgensen, 1977), which is described as: "a middle to outer shelf assemblage (50–200 m water depth) developed in a shale-marl environment" (Berggren and Aubert, 1975).

Sample preparation, oxygen and carbon isotope analyses

Prior to sampling, all gastropod shells were carefully examined under a light microscope to determine the degree of preservation. All specimens were excellently preserved as confirmed by X-ray diffraction analyses showing that the shells consisted of pure aragonite. Adhering sediment was gently removed. When necessary, the outer layer of the shell was polished to obtain a fresh surface to sample. The top layer of each shell was drilled with a 0.3 mm drill bit at ~1–2 mm intervals along a transect parallel to the direction of growth. Generally 60 µg of aragonite was used for analyses.

Some core samples were lithified, which made extraction of foraminifera difficult. Samples from Viborg Core No. 5 could be cleaned ultrasonically in deionized water. The Store Bælt samples were, however, soaked for approximately one hour in a detergent prior to thorough rinsing in deionized water over a 63 µm screen. The residue was left in 1000 ml distilled water for 24 h on an orbital shaker. Sieved residues from both cores were dried at 60 °C.

In almost all cases, the *Lenticulina* tests were filled in with secondary calcite, however, the test material appeared excellently preserved judging from examination under the light microscope. All tests were thus gently crushed, and the fragments were ultrasonically cleaned and dried. Only fragments completely free of infilling material were used. Approximately 20–30 μg of test fragments were used for each analysis.

Samples were roasted in vacuum at 400 °C for 30 minutes before analysis in a VG Prism II Series mass spectrometer with an Isocarb preparation system. Analyses were performed at the Department of Earth Sciences, Göteborg University. Isotope results are presented as δ per mil relative to the Vienna Pee Dee Belemnite (VPDB) standard. The mean values and standard deviations of 122 NBS-19 standards ($\delta^{18}O$ = –2.20‰, $\delta^{13}C$ = 1.95‰) were –2.161‰ ± 0.090‰ for $\delta^{18}O$ and 1.946‰ ± 0.035‰ for $\delta^{13}C$. The analytical precision of each NBS-19 standard after 12 replicate measurements was better than 0.013‰ for $\delta^{18}O$ and 0.007‰ for $\delta^{13}C$.

Isotopic considerations

The stable oxygen isotope composition of mollusk shells is usually in isotopic equilibrium with the ambient water (Wefer and Berger, 1991). Interpretations based on the stable carbon isotope record, however, should be made with caution because vital effects may influence $\delta^{13}C$ in the mollusk shell (Romanek and Grossman, 1989). Recent turritellids, however, precipitate their shell in equilibrium with the dissolved inorganic carbon (DIC) $\delta^{13}C$ of the ambient water (Rahimpour-Bonab et al., 1997).

The benthic foraminifera *Lenticulina* spp. precipitates its shell out of equilibrium with $\delta^{18}O$ in ambient water (Grossman, 1987). In the early Eocene North Sea, *Lenticulina* spp. records on average 0.48‰ more negative $\delta^{18}O$ values than *Oridorsalis umbonatus,* which calcifies in equilibrium with water (Shackleton et al., 1984; Schmitz et al., 1996). We have therefore adjusted our *Lenticulina* $\delta^{18}O$ data by +0.48‰ in order to compensate for disequilibrium effects.

Equations

To calculate the water temperatures for the mollusks we used the temperature equation for $\delta^{18}O$ in biogenic aragonite by Grossman and Ku (1986) (Eq. 1). For the foraminifera we applied the temperature equation for biogenic calcite by O'Neil et al. (1969) and Shackleton and Kennett (1974) (Eq. 2). In the equations $\delta^{18}O_{aragonite}$, $\delta^{18}O_{calcite}$, and $\delta^{18}O_{water}$ are relative to PDB.

$$T(°C) = 20.6 - 4.34\ (\delta^{18}O_{aragonite} - \delta^{18}O_{water}) \quad (1)$$

$$T(°C) = 16.9 - 4.38\ (\delta^{18}O_{calcite} - \delta^{18}O_{water}) + 0.10\ (\delta^{18}O_{calcite} - \delta^{18}O_{water})^2 \quad (2)$$

The linear relationship between $\delta^{18}O_{water}$ relative to SMOW and salinity (*S*) for the Holocene North Sea region is according to Frölich et al. (1988):

$$\delta^{18}O_{water} = S\ (0.272 \pm 0.009) - 8.9 \pm 0.11 \quad (3)$$

STABLE OXYGEN AND CARBON ISOTOPE RESULTS

Mollusks

The $\delta^{18}O$ profiles for all mollusks show cyclical variations (Figs. 2–4; Tables DR1–DR5[1]). The $\delta^{18}O$ records for all shells from Vestre Gasværk exhibit a striking similarity in amplitude (Fig. 2A–D). The carbon isotope records, however, show large differences between specimens (Fig. 3A–D).

The two *Natica* specimens (ND1 and ND2) show a similar, near-sinusoidal pattern in the $\delta^{18}O$ record (Fig. 2A–B). The highest values are on average 0‰ to 0.3‰, with the more positive values in the larger individual (ND2). The lowest values are around –1‰ to –1.2‰, but the smaller individual (ND1) records an extreme low of –1.5‰ close to the aperture. The $\delta^{13}C$ records for ND1 and ND2 are very different (Fig. 3A–B). Specimen ND1 exhibits an irregular cyclic pattern with minimum values averaging ~0‰ to –0.3‰ and maximum values between 0.7‰ to 1.4‰. The $\delta^{13}C$ record for ND2 is also cyclic. In the first half of the record, values begin at ~0‰, gradually decrease to ~ –1.0‰ and then steadily increase to a high of ~0.5‰. In the later half of the record, however, the lowest values average ~0‰ to –0.2‰ and the highest values ~0.5‰, and values increase overall by 0.6‰–0.7‰.

In the first part of the $\delta^{18}O$ record for the *Surcula* (SH) shell values range between 0.1‰ and –0.5‰, possibly because of rapid growth in the initial stage of the specimens life (Fig. 2C). Toward the aperture, however, the $\delta^{18}O$ record exhibits a cyclic pattern with the cycles growing shorter and the amplitude decreasing. Three distinct minima range between –1.2‰ and –1.5‰ and two maxima between 0‰ and –0.3‰. In the $\delta^{13}C$ record, a long semicycle with high amplitude is apparent (Fig. 3C). In the upper part of the shell values decrease gradually from –1‰ to ~ –3‰ in the middle part of the shell. Thereafter a gradual increase culminates in a high of ~1‰ close to the aperture.

The *Dentalium* (DR) $\delta^{18}O$ record shows an irregular cyclic pattern with relatively low amplitude (Fig. 2D). The amplitude is smaller in the initial as well as final stage of the scaphopods life cycle. The lowest $\delta^{18}O$ value of –1.2‰ is recorded at ~8 mm from the posterior end. Two maxima of ~0‰ are recorded at 4 mm and 12 mm, respectively. In the $\delta^{13}C$ record no cycle is distinguishable (Fig. 3D). Values in the first half of the record average slightly above 1.5‰, whereas in the later half of the record values lie just below 1.5‰.

For the *Turritella* (TURR) specimen the $\delta^{18}O$ record is significantly more negative than for the other shells, with an intra-annual variation from ~ –1.2‰ to ~ –2.6‰ (Fig. 4A). Despite high irregularity within the cycles there are two intervals of low $\delta^{18}O$ values, one at almost 30 mm and one at ~65 mm. Toward the aperture, the amplitude decreases. The $\delta^{13}C$ values fluctuate

[1]GSA Data Repository item 2003049, Tables DR1–DR6, is available on request from Documents Secretary, GSA, P.O. Box 9140, Boulder, CO 80301-9140, USA, editing@geosociety.org, or at www.geosociety.org/pubs/ft2003.htm.

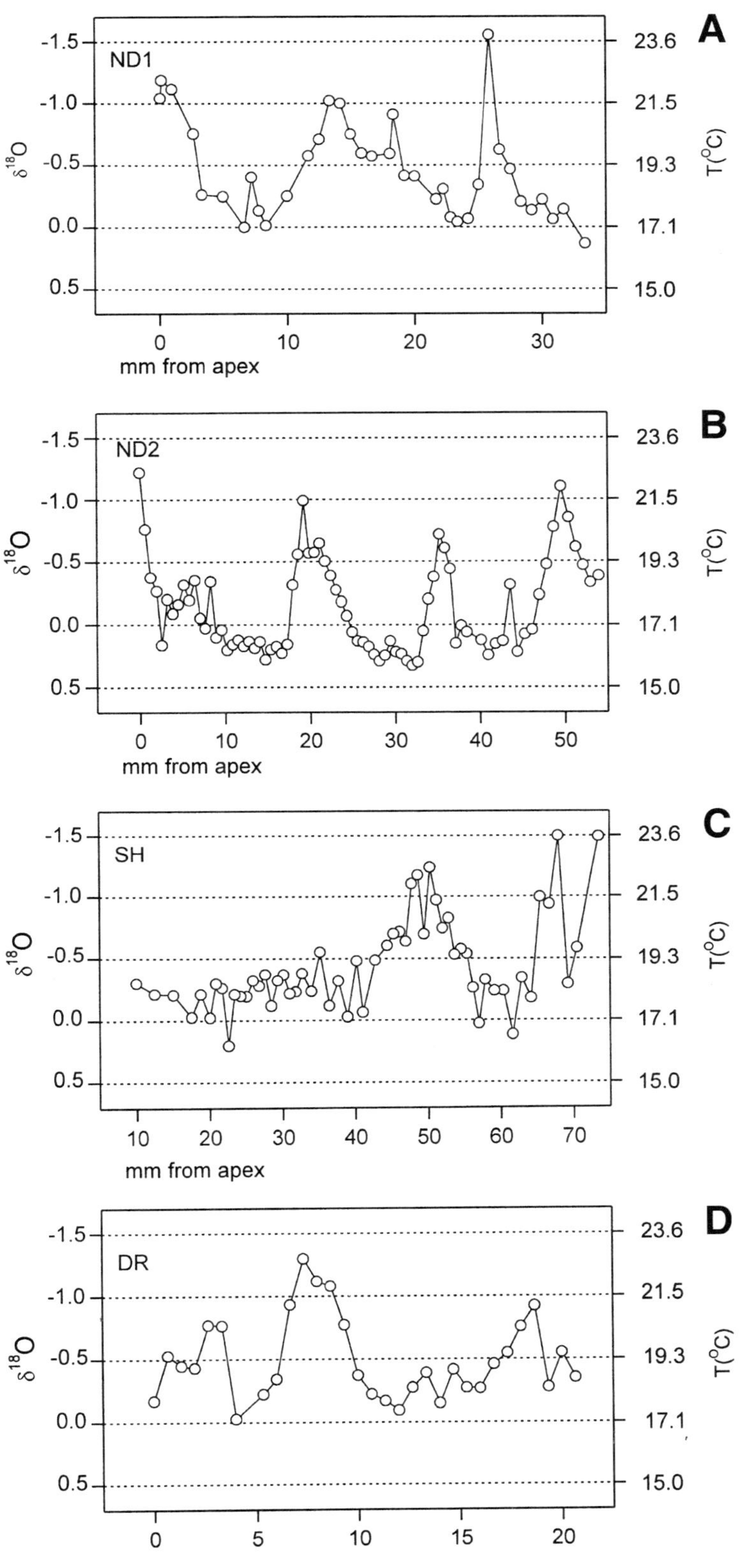

Figure 2. Stable oxygen isotope and temperature profiles for the deeper-dwelling gastropod and scaphopod species. A: ND1, *Natica detracta* 1. B: ND2, *Natica detracta* 2. C: SH, *Surcula hauniensis.* D: DR, *Dentalium rugiferum.* All stable isotope data in per mil relative to the VPDB standard. Temperature was calculated using the equation by Grossman and Ku (1986) and $\delta^{18}O_{water}$ of –0.8‰.

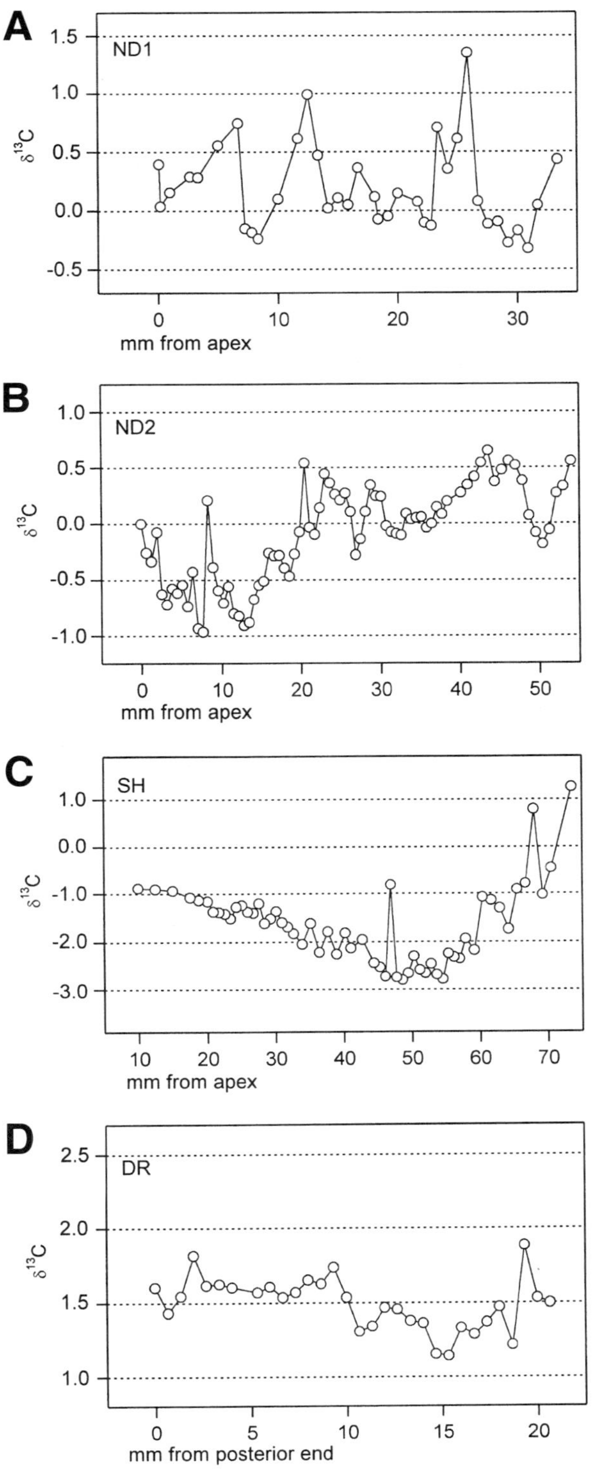

Figure 3. Stable carbon isotope profiles. A: ND1, *Natica detracta* 1. B: ND2, *Natica detracta* 2. C: SH, *Surcula hauniensis.* D: DR, *Dentalium rugiferum.* All stable isotope data in per mil relative to the VPDB standard. Note the differences in scale on $\delta^{13}C$ axis between the different shells.

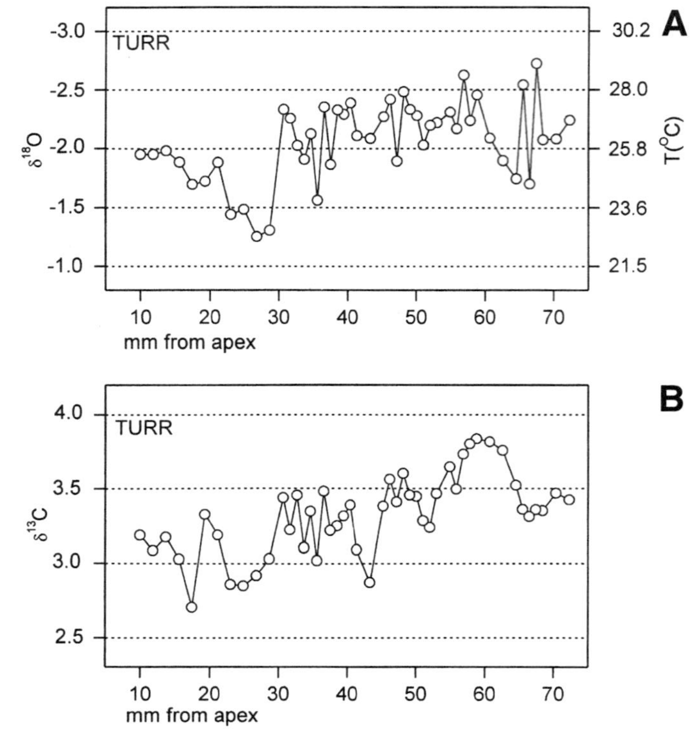

Figure 4. Stable isotope and temperature profiles for the surface-dwelling gastropod, TURR, *Turritella* spp. (A) $\delta^{18}O$ and temperature (B) $\delta^{13}C$. All stable isotope data in per mil relative to the VPDB standard. Temperature was calculated using the equation by Grossman and Ku (1986) and $\delta^{18}O_{water}$ of –0.8‰.

cyclically, with an overall increasing trend toward the aperture (Fig. 4B). The lowest and highest $\delta^{13}C$ values recorded for TURR are ~2.7‰ and ~3.8‰, respectively.

Foraminifera

In the Selandian section of the Store Bælt core, $\delta^{18}O$ varies between –0.5‰ and –1.5‰, whereas in the Viborg core, values range from –0.32‰ to –0.88‰ (Fig. 5; Table DR6, see footnote 1). In order to compare the benthic foraminifera data with that of the mollusks we averaged the $\delta^{18}O$ data for both cores, because the deposits from the three different locations in this study are basically of the same age. For the Store Bælt Core No. 8604A the mean $\delta^{18}O$ value was –0.38‰ ± 0.19‰ and in Viborg Core No. 5 it was –0.67‰ ± 0.28‰.

DISCUSSION

Sea water temperature and salinity

The influence of continental runoff on $\delta^{18}O$ records must be considered when interpreting records from marginally marine environments. In the *Turritella* sandstone, current-oriented shells and the nature of the dinoflagellate assemblage indicate near shore deposition, possibly in a lagoon (Andersen and Heilmann-Clausen, 1983). A freshwater component would bias the $\delta^{18}O$ record for the shallow-water dwelling *Turritella,* because an increasing freshwater component will have a similar effect on the shell $\delta^{18}O$ record as increasing temperatures. Such continental runoff, however, would not be dense enough to sink to any significant depth. For instance, during maximum discharge on the Amazon shelf dilution by fresh water is only observed in the top 20 m of the water column and bottom salinities of 35‰ closely follow the 20 m isobath (Geyer et al., 1996). The $\delta^{18}O$ records for the deeper-dwelling mollusks as well as the benthic foraminifera are thus probably essentially unaffected by local freshwater plumes.

When calculating paleotemperatures (Eq. 1 and 2), one must estimate a value for $\delta^{18}O_{water}$. Our estimate of –0.8‰ on the PDB scale for $\delta^{18}O_{water}$ in the early Selandian southeastern North Sea at a paleolatitude of ~47°N, is based on two considerations: (i) The Paleocene Earth was probably ice-free so a preglacial value of –1.2‰ relative to PDB is appropriate (Shackleton and Kennett, 1974); and (ii) $\delta^{18}O$ in ocean surface water is not uniform but deviates from $\delta^{18}O$ of surface mean ocean water (SMOW) in relation to latitude (Zachos et al., 1994). We assume that the meridional $\delta^{18}O_{water}$ gradient for open ocean waters during the Paleogene did not differ greatly from today (Zachos et al., 1994; Schmitz and Andreasson, 2001). Equation 3 implies that for the Holocene North Sea (latitude 54°N–59°N), at fully marine salinity [S = 35], $\delta^{18}O_{[SMOW]}$ is ~ +0.62‰. A similar value of +0.6‰ is estimated for the Bay of Biscay at latitude 45°N (Fairbanks et al., 1992). When using the present-day relationship between latitude and $\delta^{18}O$ in sea water (Equation 1 in Zachos et al., 1994), a lower $\delta^{18}O_{water}$ value of ~ +0.2‰ is obtained for latitude 45°N. We take the average of +0.6‰ and +0.2‰ and apply a value of +0.4‰ in order to compensate for the latitudinal effect.

Tables 1 and 2 show the estimated paleotemperatures based on all mollusks and benthic foraminifera. The maximum range in $\delta^{18}O$ in mollusk shells indicates the minimum seasonal water temperature range, because parts of the annual temperature range may not be registered due to seasonally reduced growth rates. For example, large gastropods such as *Strombus gigas* and *S. costatus* almost cease growing during the winter months (Wefer and Killingley, 1980), as does the bivalve *Spisula solidissima* (Dillwyn) (Jones et al., 1983). Biological factors such as sexual maturity may also influence the growth rate (Allmon et al., 1994). However, the similarity in absolute maximum and minimum $\delta^{18}O$ values for the different species of deeper-dwelling mollusks of Vestre Gasværk suggests that we see the maximum range of seasonal water temperature.

For DR and TURR, summer and winter temperatures were calculated by using the minimum and maximum $\delta^{18}O$ values in one single annual cycle. The $\delta^{18}O$ profiles for ND1, ND2, and SH exhibited several annual cycles, so an average of each individual's minima and maxima were used (Table 1). From Equa-

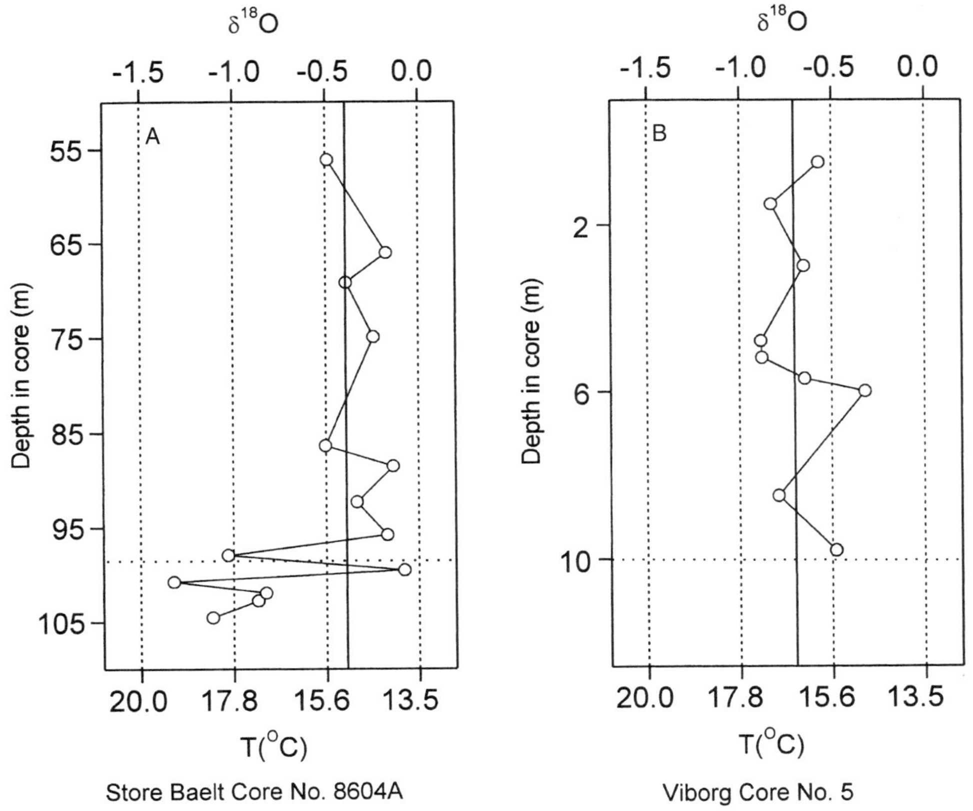

Figure 5. Stable oxygen isotope and temperature profile for the benthic foraminifera *Lenticulina* spp. in (A) Store Bælt Core No. 8604A and (B) Viborg Core No.5. The dotted horizontal line shows the position of the Danian-Selandian boundary in each core. The black vertical line shows the mean $\delta^{18}O$ value and mean temperature for the Selandian part of the two cores. A: mean $\delta^{18}O$ = –0.38‰, mean temperature = 15.1 °C. B: mean $\delta^{18}O$ = –0.67‰, mean temperature = 16.3 °C. All stable isotope data in per mil relative to the VPDB standard. Temperature was calculated using the equation by O'Neil et al. (1969) and $\delta^{18}O_{water}$ of –0.8‰.

TABLE 1. AVERAGE TEMPERATURES FOR EARLY SELANDIAN MOLLUSKS

Species	Minimum temperature* (°C)	Maximum temperature* (°C)	Δtemperature (°C)
Deeper water dwellers:			
Natica detracta 1	17.0; n = 3; σ ± 0.42	22.6; n = 3; σ ± 1.18	5.6
Natica detracta 2	15.9; n = 3; σ ± 0.20	21.5; n = 4; σ ± 0.94	5.6
Surcula hauniensis	17.5; n = 2; σ ± 1.27	23.2; n = 3; σ ± 0.64	5.7
Dentalium rugiferum	17.2; n = 1	22.8; n = 1	5.6
Average	16.8; n = 9; σ ± 0.84	22.4; n = 11; σ ± 0.84	5.6
Surface water dweller:			
Turritella spp.	23.4	28.5	5.1

*Based on the temperature equation for biogenic aragonite by Grossman and Ku (1986); $\delta^{18}O_{water}$ = –0.8‰.

TABLE 2. AVERAGE TEMPERATURES FOR EARLY SELANDIAN BENTHIC FORAMINIFERA *LENTICULINA* SPP

	Store Bælt Core No. 8604A	Viborg Core No.5
Average annual temperature* (°C)	15.1; n = 9; σ = ±1.2	16.9; n = 9; σ = ±0.84

*Based on the temperature equation for biogenic calcite by O'Neil et al. (1969); $\delta^{18}O_{water}$ = –0.8‰.

tion 1, we find that the summer temperatures recorded by the deeper-dwelling mollusks varied from 21.5 to 23.2 °C, with an average of 22.5 °C (Table 1). The corresponding winter temperatures are 15.9 to 17.2 °C, with an average of 16.9 °C (Table 1). Despite the differences in minimum and maximum temperatures, all individuals exhibit an almost identical seasonal temperature range, 5.6–5.7 °C (Δtemperature in Table 1). At the deeper sites of Viborg and Store Bælt annual temperature averages of ~16 °C and ~15 °C (Eq. 2) were recorded by the benthic foraminifera (Table 2). Coeval temperatures at Deep Sea

Drilling Project (DSDP) Hole 384 in the western North Atlantic, at a paleodepth of ~3400 m were estimated at ~11 °C (Thompson and Schmitz, 1997). Considering the great difference in depth between the Danish sites and DSDP Hole 384, the differences in bottom water temperature seam reasonable. These observations support the hypothesis that the deeper waters of the Danish Basin were not affected significantly by a freshwater component at this time.

For *Turritella,* Equation 1 yields a summer high of ~28.5 °C and winter minimum of ~23.4 °C, implying a seasonal temperature range of 5.1 °C (Table 1). The estimated surface paleotemperatures at the southeast margin of the North Sea appear unrealistically high. The fact that the seasonal temperature range does not decrease with water depth as observed in Holocene bivalves (Krantz et al., 1987), further indicates that estimated temperatures are erroneous. The *Turritella* based temperatures in Selandian coastal waters suggest a tropical climate, following the vegetational scheme by Wolfe (1978, 1980). However, paleoclimatic interpretations based on early late Paleocene botanical evidence and planktonic foraminiferal $\delta^{18}O$ data indicate subtropical conditions. Fossil floral evidence of the southern North American east coast shows that conditions were no more than paratropical with a mean annual temperature of ~20 °C (Wolfe, 1978, 1980) in the Mississippi embayment at a paleolatitude of ~30°N (Smith et al., 1994). At a paleolatitude of ~25°N to 10°N, sea surface temperatures of 21–25 °C were recorded based on planktonic foraminiferal $\delta^{18}O$ data (Boersma and Premoli Silva, 1983; Zachos et al., 1994). At 45°N in the western North Atlantic, a paleolatitude comparable to that of Selandian Copenhagen, the sea surface temperature was no more than ~22 °C (Boersma et al., 1979; Boersma and Premoli Silva, 1983). All other evidence thus points toward a subtropical climate at the southeastern margin of the North Sea in the early late Paleocene.

The bottom-water temperatures and their seasonal variations at Vestre Gasværk during the early Selandian are similar to conditions in subtropical regions of today. At localities in the eastern Mediterranean, for instance, at 34.5°N waters at 30–50 m depth show similar intra-annual temperature variations as estimated for Vestre Gasværk in the mid-Paleocene (Figs. 6–7) (Levitus and Boyer, 1994). At 30 m depth, temperatures are ~23 °C in summer and ~16 °C in winter and the seasonal temperature difference is 6–7 °C. For this comparison we have chosen a recent site at 34.5°N since this latitude can be considered representative for the Selandian world at 47°N, considering the markedly reduced latitudinal temperature gradients in the early Paleogene compared to the recent world. As in most oceanographic situations the intra-annual temperature differences in the recent eastern Mediterranean decrease with depth (Fig. 7).

In Figure 8 we compare Selandian North Sea and modern Mediterranean intra-annual temperature variations at the surface and at ~30 m depth. The offset between estimated Selandian and observed Mediterranean surface temperatures as compared to deeper water temperatures is larger during the winter than the summer.

Oxygen isotopic studies of fully marine early middle Eocene *Turritella* shells from the Paris Basin indicate surface water temperatures of 27–28 °C in the summer and 14–15 °C in the winter (Andreasson and Schmitz, 1996, 2000). A comparison with the Paris Basin data also shows that in particular the Selandian surface winter temperatures appear anomalously high. Although the early Selandian is generally considered to have had a somewhat cooler climate than the early middle Eocene (see previous section), it can not be ruled out that there were spatial and temporal climatic variations within both periods and that climatic conditions sometimes overlapped. The estimates of paleolatitudes of the southern North Sea in the early Selandian and the Paris Basin in the early middle Eocene also introduce uncertainty, but it is likely that paleolatitudes may only have differed by a few degrees.

The extremely high winter surface temperatures derived from the *Turritella* shells may reflect either a particularly strong freshwater influence on the isotopic signals during the winter, or incomplete registration of temperatures during this part of the

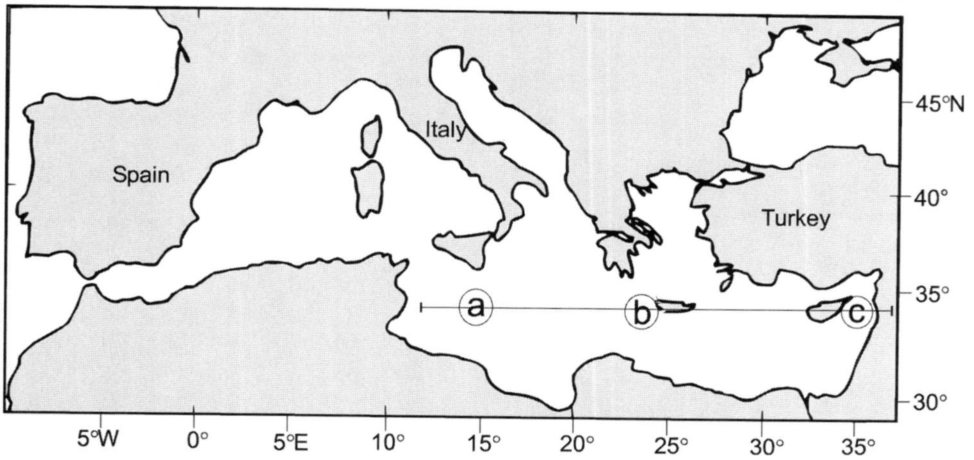

Figure 6. Map of the eastern Mediterranean, showing three marine sites along the 34.5°N transect for which temperature versus depth data are plotted in Figure 7.

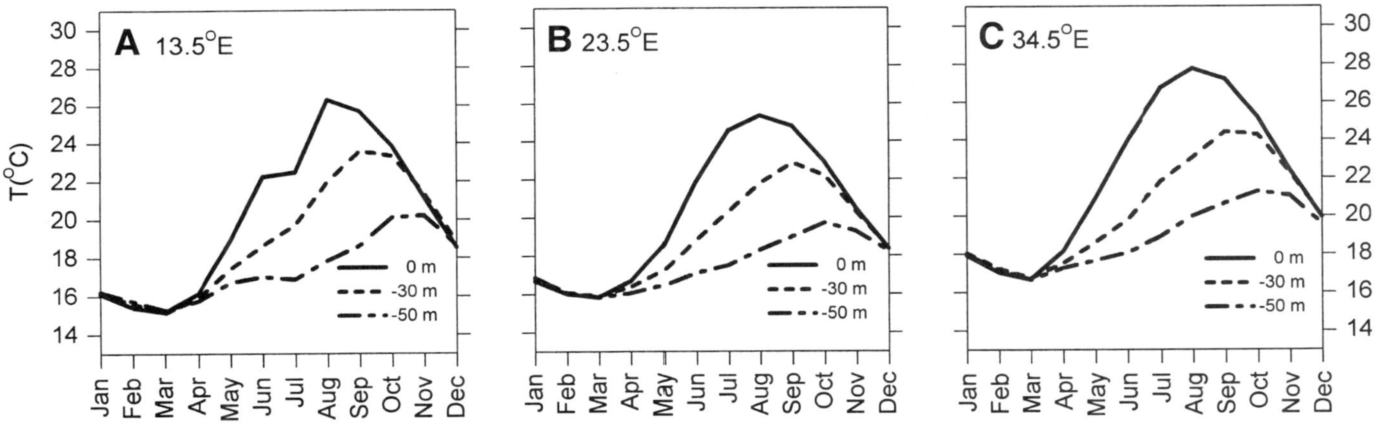

Figure 7. The intra-annual water temperature at three marine sites along the 34.5°N transect in the eastern Mediterranean (see Figure 6). Shown are the temperatures at the surface, –30 m and at –50 m. Data from Levitus and Boyer (1994).

year. If the difference in estimated surface temperatures in the Selandian North Sea compared to the recent eastern Mediterranean is entirely dependent on freshwater influence, salinity reductions of 1.7 psu in the summer and 5.6 psu in the winter could account for the temperature differences (Fig. 8). Larger amounts of continental runoff during the winter and evaporation of (lagoonal?) waters during the summer may have resulted in seasonal variations in the $\delta^{18}O$ of the water. Vertical temperature gradients were likely smaller in the early Paleogene ocean than in the recent Mediterranean. Since the exact depth of the Selandian deeper-dwelling mollusks is not known, however, we only attempt to illustrate the principles of seasonal temperature differences between surface and deeper waters.

Previous extensive studies of *Turritella* shells indicate that these gastropods usually register complete or almost complete annual cycles (Andreasson and Schmitz, 2000, and references therein), however, unusual paleoenvironmental conditions could have inhibited growth during winter months in the Selandian *Turritella* shells. Additional support for an incomplete annual cycle comes from the high carbon isotopic values, 2.7‰–3.8‰ measured in the *Turritella* shell (Fig. 2). Such high $\delta^{13}C$ values are characteristic for close to fully or fully marine conditions (Keith

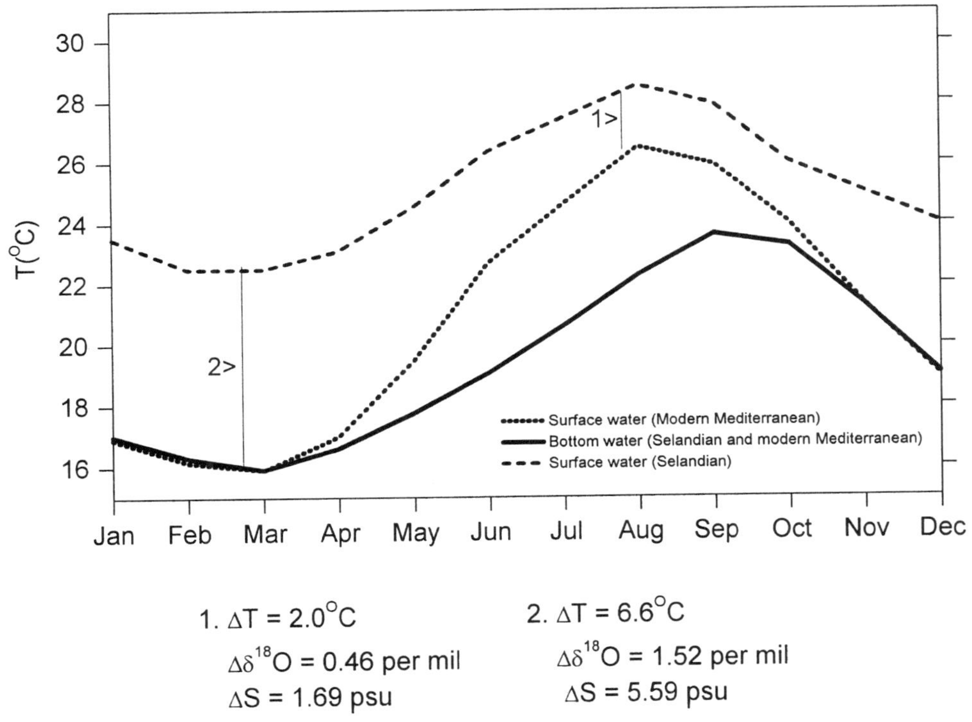

Figure 8. A comparison of the intra-annual temperature stratification for the early Selandian southeastern North Sea and the modern eastern Mediterranean. The Selandian temperature estimates are based on the oxygen isotopic results for the surface and deeper-dwelling mollusks in this study. Temperatures were calculated using the equation by Grossman and Ku (1986) and a $\delta^{18}O_{water}$ of –0.8‰. The Mediterranean data represent an average for the 30 m depth at the three sites in Figure 7 (Levitus and Boyer, 1994). See main text for further details.

et al., 1964; Keith and Parker, 1965; Andreasson and Schmitz, 2000; Schmitz and Andreasson, 2001). Studies of recent *Turritellas* show that they precipitate carbon isotopes in equilibrium with dissolved inorganic carbon in seawater (Rahimpour-Bonab et al., 1997).

Although it is likely that local freshwater plumes have not significantly affected the isotopes of the deeper-dwelling mollusks, it cannot be ruled out that the entire mid-Paleocene North Sea had slightly reduced salinities compared to the open ocean. The diverse marine biota in the North Sea at this time, paleogeographic reconstructions and the isotopic results from benthic foraminifera in this study suggest that such a possible salinity reduction can not have been large, i.e., probably <1–3 psu (see also Buchardt, 1977). In our isotopic temperature estimates a salinity reduction of 1–3 psu would give temperatures too high by 1.2–3.5 °C.

Although the present data can not constrain in detail the early Selandian coastal temperatures, we conclude on the basis of the evidence from both deeper-water and shallow-water mollusks that summer surface temperatures were not higher than 28 °C (as derived from *Turritella* shells) and not lower than ~22 °C. The minimum value is based on the summer temperature of 22.4 °C estimated from the deep-water mollusk isotopes and assuming that the reduction of temperature with depth and a possible small freshwater effect (<3.5 °C) essentially cancel each other out. Winter surface temperatures are likely to not have been lower than 13 °C, based on the deeper water minimum temperature of 16.8 °C, a possible maximum freshwater effect of 3.5 °C and a stable density-stratified water column (i.e., no colder temperatures at surface than at some depth). These winter temperatures are essentially similar to those estimated for the early middle Eocene Paris Basin (Andreasson and Schmitz, 2000) and do not concur with previous estimates of a generally cooler climate for the early Selandian compared to the rest of the early Paleogene period (e.g., Buchardt, 1977, 1978). Buchardt (1977, 1978) based his interpretation of a cooler early Selandian climate on whole-shell isotopic studies of the same mollusk fauna from the Vestre Gasværk, but did not consider the effect of depth on temperatures when comparing different faunas. These early studies also used temperature-isotope equations for calcite and not for aragonite, explaining major discrepancies in estimated absolute temperatures.

If winter temperatures were higher than 13 °C in the early Selandian, then summer surface temperatures in the 22–28 °C range appear reasonable. Previous studies of fully marine *Turritella* shells from the early middle Eocene Paris Basin show that, although mean annual temperatures were higher than today, the mean annual range in temperature was as large or only slightly smaller in the early Paleogene than it is today (Andreasson and Schmitz, 2000); the recent seasonal temperature contrast at Bordeaux in France at 45°N is ~15 °C (Rudloff, 1981).

CONCLUSIONS

The validity of temperature interpretations of intrashell (seasonal) stable isotopic variations in mollusk shells can be constrained by comparing results for shells from different water depths. By comparing the isotopic results for surface-dwelling (0–20) and deeper-dwelling mollusks (20–60 m) from early Selandian deposits in Denmark, and assessing uncertainties about salinity and depths, we show that summer surface water temperatures in the southeastern North Sea at a paleolatitude of 47°N lay in the range 22–28 °C and winter surface temperatures were not colder than 13 °C. This indicates a significantly warmer climate in this region than shown by previous isotopic studies of early Selandian mollusk whole-shell samples. Our data indicate a subtropical climate in this region essentially similar to the climate that prevailed at ~45°–47°N in western Europe during the major part of the early Paleogene.

ACKNOWLEDGMENTS

Supported by grants from the Swedish Natural Science Research Council. We thank the Geological Museum in Copenhagen and K.I. Schnetler for providing shell material, F.P. Andreasson for advice, and O. Gustafsson for help with the isotopic analyses. Valuable comments on an earlier version of the manuscript were provided by K. Bice, A. Dutton and E. Thomas.

REFERENCES CITED

Allmon, W.D., 1988, Ecology of Holocene turritelline gastropods (Prosobranchia, Turritellidae): Current knowledge and paleontological implications: Palaios, v. 3, p. 259–284.

Allmon, W.D., Jones, D.S., and Vaughan, N., 1992, Observations on the biology of *Turritella gonostoma* Valenciennes (Prosobranchia, Turritellidae) from the Gulf of California: Veliger, v. 35, p. 52–63.

Allmon, W.D., Jones, D.S., Aiello, R.L., Gowlett-Holmes, K., and Probert, K., 1994, Observations on the biology of *Maoricolpus roseus* (Quoy and Gaimard) (Prosobranchia Turritellidae) from New Zealand and Tasmania: Veliger, v. 37, p. 267–279.

Andersen, S.B., and Heilmann-Clausen, C., 1983, Petrografi og alder af den brune *Turritella*-sandsten, en Tertiær løsblock fra Østersøområdet: Dansk Geologisk Føreningen, Årsskrift for 1983, p. 17–24.

Andreasson, F.P., and Schmitz, B., 1996, Winter and summer temperatures of the early middle Eocene of France from *Turritella* $\delta^{18}O$ profiles: Geology, v. 24, p. 1067–1070.

Andreasson, F.P., and Schmitz, B., 1998, Tropical Atlantic seasonal dynamics in the early middle Eocene from stable oxygen and carbon isotope profiles of mollusk shells: Paleoceanography, v. 13, p. 183–192.

Andreasson, F.P., and Schmitz, B., 2000, Temperature seasonality in the early middle Eocene North Atlantic region: Evidence from stable isotope profiles of marine gastropod shells: Geological Society of America Bulletin, v. 112, p. 628–641.

Andreasson, F.P., Schmitz, B., and Jönsson, E., 1999, Surface-water seasonality from stable isotope profiles of *Littorina littorea* shells: Implications for paleoenvironmental reconstructions of coastal areas: Palaios, v. 14, p. 273–281.

Barnes, R.D., 1980, Invertebrate zoology (4th edition): Saunders College, Philadelphia, Pennsylvania, USA, 1089 p.

Barrera, E., and Huber, B.T., 1991, Paleogene and early Neogene oceanography of the southern Indian Ocean, Leg 119 foraminifer stable results *in* Barron, J., et al., eds., Proceedings of the Ocean Drilling Program, Scientific Results: Texas A & M University, College Station, Texas, USA, v. 119, p. 693–717.

Berger, W.H., and Wefer, G., 1988, Benthic deep-sea foraminifera: Possible consequences of infaunal habitat for paleoceanographic interpretation: Journal of Foraminiferal Research, v. 18, p. 147–150.

Berggren, W.A., and Aubert, J., 1975, Paleocene benthonic foraminiferal biostratigraphy, paleobiogeography, and paleoecology of Atlantic-Tethyan regions, Midway-type fauna: Palaeogeography, Palaeoclimatology, Palaeoecology, v. 18, p. 73–192.

Bice, K.L., Arthur, M.A., and Marincovich, L., Jr., 1996, Late Paleocene Arctic Ocean shallow-marine temperatures from mollusk stable isotopes: Paleoceanography, v. 11, p. 241–249.

Boersma, A., and Premoli Silva, I., 1983, Paleocene planktonic foraminiferal biogeography and the paleoceanography of the Atlantic ocean: Micropaleontology, v. 29, p. 355–381.

Boersma, A., Shackleton, N., Hall, M., and Given, Q., 1979, Carbon and oxygen isotope records at DSDP Site 384 (North Atlantic) and some Paleocene paleotemperatures and carbon isotope variations in the Atlantic Ocean, *in* Tucholke, B.E., et al., eds., Initial reports of the Deep Sea Drilling Project: Washington, D.C., U.S. Government Printing Office, v. 43, p. 695–717.

Buchardt, B., 1977, Oxygen isotope ratios from shell material from the Danish middle Paleocene (Selandian) deposits and their interpretations as paleotemperature indicators: Palaeogeography, Palaeoclimatology, Palaeoecology, v. 22, p. 209–230.

Buchardt, B., 1978, Oxygen isotope palaeotemperatures from the Tertiary period in the North Sea area: Nature, v. 275, p. 121–123.

Corfield, R.M., and Cartlidge, J.E., 1992, Oceanographic and climatic implications of the Paleocene carbon isotope maximum: Terra Nova, v. 4, p. 443–455.

D'Hondt, S., Zachos, J.C., and Schultz, G., 1994, Stable isotopic signals and photosymbiosis in late Paleocene planktic foraminifera: Paleobiology, v. 20, p. 391–406.

Fairbanks, R.G., Charles, C.D., and Wright, J.D., 1992, Origin of global meltwater pulses, *in* Taylor, R.E., et al., eds., Radiocarbon after four decades: An interdisciplinary perspective: New York, Springer-Verlag, p. 473–500.

Frederiksen, N.O., 1994, Paleocene floral diversities and turnover events in eastern North America and their relation to diversity models: Review of Palaeobotany and Palynology, v. 82, p. 225–238.

Frölich, K., Grabczak, J., and Rozanski, K., 1988, Deuterium and oxygen-18 in the Baltic Sea: Chemical Geology, v. 72, p. 77–83.

Geyer, W.R., Beardsley, R.C., Lentz, S.J., Candela, J., Limeburner, R., Johns, W.E., Castro, B.M., and Dias Soares, I., 1996, Physical oceanography of the Amazon shelf: Continental Shelf Research, v. 16, p. 575–616.

Gili, C., and Martinell, J., 1994, Paleobiogeography of turrid gastropods in the Pliocene of Catalonia: Acta Palaeontologica Polonica, v. 38, p. 349–358.

Gingerich, D., 1980, Evolutionary patterns in early Cenozoic mammals: Annual Review of Earth and Planetary Sciences, v. 8, p. 407–424.

Gramann, F., 1988, Major palaeontological events and biostratigraphical correlations: Geologisches Jahrbuch, Reihe A, v. 100, p. 410–427.

Gramann, F., and Kockel, F., 1988, Palaeogeographical, lithological, palaeoecological, and palaeoclimatic development of the northwest European Tertiary basin: Geologisches Jahrbuch, Reihe A, v. 100, p. 428–441.

Grossman, E.L., 1987, Stable isotopes in modern benthic foraminifera: A study of vital effect: Journal of Foraminiferal Research, v. 17, p. 48–61.

Grossman, E.L., and Ku, T.-L., 1986, Oxygen and carbon isotope fractionation in biogenic aragonite, temperature effects: Chemical Geology, v. 59, p. 59–74.

Gry, H., 1935, Petrology of the Paleocene sedimentary rocks of Denmark: Danmarks Geologiske Undersøgelse, Række 2, no. 61, 171 p.

Hansen, H.J., 1968, On the biostratigraphical age of the Lower Selandian of Denmark: Meddelser fra Dansk Geologisk Førening, v. 18, p. 277–284.

Hayward, J., and Ryland, J.S., 1995, Handbook of the marine fauna of northwest Europe: New York, Oxford University Press, 800 p.

Heilmann-Clausen, C., 1989, Paleogene aflejringer over danske kalken, *in* Nielsen, O.B., and Sandersen, P., eds., Danmarks geologi-fra øvre kridt til idag: Aarhus, Denmark, Aarhus Universitet, p. 43–87.

Hickey, L.J., 1980, Paleocene stratigraphy and flora of the Clarks Fork Basin, *in* Gingerich, P.D., ed., Early Cenozoic paleontology and stratigraphy of the Bighorn Basin, Wyoming: University of Michigan Papers on Paleontology, v. 24, p. 33–49.

Hickman, C.S., 1984, Composition, structure, ecology, and evolution of six Cenozoic deep-water mollusk communities: Journal of Paleontology, v. 58, p. 1215–1234.

Jones, D.S., and Allmon, W.D., 1995, Records of upwelling, seasonality, and growth in stable-isotope profiles of Pliocene mollusk shells from Florida: Lethaia, v. 28, p. 61–74.

Jones, D.S., Williams, D.F., and Arthur, M.A., 1983, Growth history and ecology of the Atlantic surf clam, *Spisula solidissima* (Dillwyn), as revealed by stable isotopes and annual shell increments: Journal of Experimental Marine Biology and Ecology, v. 73, p. 225–242.

Kaiho, K., 1998, Global climatic forcing of deep-sea benthic foraminiferal test size during the past 120 m.y.: Geology, v. 26, p. 491–494.

Keith, M.L., and Parker, R.H., 1965, Local variations of ^{13}C and ^{18}O content of mollusk shells from marine and freshwater environments: Marine Geology, v. 3, p. 115–129.

Keith, M.L., Anderson, G.M., and Eichler, R., 1964, Carbon and oxygen isotopic composition of mollusk shells from marine and freshwater environments: Geochimica et Cosmochimica Acta, v. 28, p. 1757–1786.

Kelley, H., and Hansen, T.A., 1996, Recovery of the naticid gastropod predator-prey system from the Cretaceous-Tertiary and Eocene-Oligocene, *in* Hart, M.B., ed., Biotic recovery from mass extinction events: Geological Society [London] Special Publication 102, p. 373–386.

King, C., 1989, Cenozoic of the North Sea, *in* Jenkins, D.G., and Murray, J.W., eds., Stratigraphical atlas of fossil foraminifera: Chichester, UK, Ellis Horwood Limited, p. 418–489.

Kobashi, T., Grossman, E.L., Yancey, T.E., and Dockery, D.T., 2001, Reevaluation of conflicting Eocene tropical temperature estimates: Molluskan oxygen isotope evidence for warm low latitudes: Geology, v. 29, p. 983–986.

Krantz, D.E., 1990, Mollusk-isotope records of Plio-Pleistocene marine paleoclimate, U.S. Middle Atlantic coastal plain: Palaois, v. 5, p. 317–335.

Krantz, D.E., Williams, D.F., and Jones, D.S., 1987, Ecological and paleoenvironmental information using the stable isotope profiles from living and fossil mollusks: Palaeogeography, Palaeoclimatology, Palaeoecology, v. 58, p. 249–266.

Larsen, A.R., and Jørgensen, N.O., 1977, Paleobathymetry of the lower Selandian of Denmark on the basis of foraminifera: Bulletin of the Geological Society of Denmark, v. 26, p. 175–184.

Levitus, S., and Boyer, T.P., 1994, World ocean atlas: Temperature: Washington, D.C., National Oceanic and Atmospheric Administration, v. 4, 117 p.

Markwick, J., 1998, Fossil crocodilians as indicators of Late Cretaceous and Cenozoic climates, implications for using palaeontological data in reconstructing palaeoclimate: Palaeogeography, Palaeoclimatology, Palaeoecology, v. 137, p. 205–271.

O'Neil, J.R., Clayton, R.N., and Mayeda, T.K., 1969, Oxygen isotope fractionation in divalent metal carbonates: Journal of Chemical Physics, v. 51, p. 5547–5558.

Perch-Nielsen, K., 1979, Calcareous nannofossil zonation at the Cretaceous–Tertiary boundary in Denmark, *in* Birkelund, T., and Bromley, R.G., eds., Cretaceous–Tertiary boundary events, I: The Maastrichtian and Danian of Denmark: University of Copenhagen, Copenhagen, Denmark, p. 115–135.

Rahimpour-Bonab, H., Bone, Y., and Moussavi-Harami, R., 1997, Stable isotope aspects of modern mollusks, brachiopods, and marine cements from cool-water carbonates, Lacepede shelf, South Australia: Geochimica et Cosmochimica Acta, v. 61, p. 207–218.

Ravelo, A.C., and Fairbanks, R.G., 1992, Oxygen isotopic composition of multiple species of planktonic foraminifera: Recorders of the modern photic zone temperature gradient: Paleoceanography, v. 7, p. 815–831.

Ravn, J.P.J., 1939, Études sur les mollusques du Paléocène de Copenhague, Det Kongelige Danske Videnskabernas Selskab, Biologiske Skrifter, Bind I, no. 1, 112 p.

Romanek, C.S., and Grossman, E.L., 1989, Stable isotope profiles of *Tridacna maxima* as environmental indicators: Palaios, v. 4, p. 402–413.

Rudloff, W., 1981, World-climates with tables of climatic data and practical suggestions: Stuttgart, Germany, Wissenschaftliche Verlagsgesellschaft, 632 p.

Sautter, L.R., and Thunell, R.C., 1991, Seasonal variability in the $\delta^{18}O$ and $\delta^{13}C$ of planktonic foraminifera from upwelling environment: Sediment trap results from the San Pedro Basin, southern California Bight: Paleoceanography, v. 6, p. 307–334.

Schmitz, B., and Andreasson, F.P., 2001, Air humidity and lake $\delta^{18}O$ during the latest Paleocene–earliest Eocene in France from recent and fossil freshwater and marine gastropod $\delta^{18}O$, $\delta^{13}C$, and $^{87}Sr/^{86}Sr$: Geological Society of America Bulletin, v. 113, p. 774–789.

Schmitz, B., Heilmann-Clausen, C., King, C., Steurbant, E., Andreasson, F.P., Corfield, R.M., and Cartlidge, J.M., 1996, Stable isotope and biotic evolution in the North Sea during the early Eocene, the Albæk Hoved section, Denmark, *in* Knox, R.W.O'B., and Corfield, R.M., eds., Correlation of the early Paleogene in northwest Europe: Geological Society [London] Special Publication 101, p. 275–306.

Schnetler, K.I., 2001, The Selandian (Paleocene) mollusk fauna from Copenhagen, Denmark: The Poul Harder 1920 collection: Geology of Denmark Survey Bulletin, v. 37, p. 1–85.

Scotese, C.R., Gahagan, L.M., and Larson, R.L., 1988, Plate tectonic reconstructions of the Cretaceous and Cenozoic ocean basins: Tectonophysics, v. 155, p. 27–48.

Seto, K., Nomura, R., and Niitsuma, N., 1991, Data report, oxygen and carbon isotope records of the upper Maestrichtian to lower Eocene benthic foraminifers at Site 752 in the eastern Indian Ocean, *in* Weissel, J., et al., eds., Proceedings of the Ocean Drilling Program, Scientific Results: Texas A & M University, College Station, Texas, USA, v. 121, p. 885–889.

Shackleton, N.J., and Kennett, J.P., 1974, Paleotemperature history of the Cenozoic and the initiation of Antarctic glaciation, oxygen and carbon isotope analyses in DSDP sites 277, 279 and 281, *in* Kennett, J.P., et al., eds., Initial reports of the Deep Sea Drilling Project: Washington, D.C., U.S. Government Printing Office, v. 29, p. 743–755.

Shackleton, N.J., Hall, M.A., and Boersma, A., 1984, Oxygen and carbon isotope data from Leg 74 foraminifers, *in* Moore, T.C., et al., eds. Initial reports of the Deep Sea Drilling Project: Washington, D.C., U.S. Government Printing Office, v. 74, p. 599–612.

Smith, A.G., Smith, D.G., and Funnell, B.M., 1994, Atlas of Mesozoic and Cenozoic coastlines: Cambridge, UK, Cambridge University Press, 99 p.

Spero, H.J., and Deniro, M.J., 1987, The influence of symbiont photosynthesis on the $\delta^{18}O$ and $\delta^{13}C$ values of planktonic foraminiferal shell calcite: Symbiosis, v. 4, p. 13–228.

Spero, H.J., and Lea, D.W., 1993, Intraspecific stable isotope variability in the planktic foraminifera *Globigerinoides sacculifer:* Results from laboratory experiments: Marine Micropaleontology, v. 22, p. 221–234.

Thomsen, E., 1994, Calcareous nannofossil stratigraphy across the Danian–Selandian boundary in Denmark: GFF, v. 116, p. 65–67.

Thomsen, E., and Heilmann-Clausen, C., 1985, The Danian–Selandian boundary at Svejstrup with remarks on the biostratigraphy of the boundary in the western Denmark: Bulletin of the Geological Society of Denmark, v. 33, p. 341–362.

Thompson, E.I., and Schmitz, B., 1997, Barium and the late Paleocene $\delta^{13}C$ maximum: Evidence of increased marine surface productivity: Paleoceanography, v. 12, p. 239–254.

Tripati, A., Zachos, J., Marincovich, L., and Bice, K., 2001, Late Paleocene Arctic coastal climate inferred from molluskan stable and radiogenic isotope ratios: Palaeogeography, Palaeoclimatology, Palaeoecology, v. 170, p. 101–113.

Wefer, G., and Berger, W.H., 1991, Isotope paleontology, growth, and composition of extant calcareous species: Marine Geology, v. 100, p. 207–248.

Wefer, G., and Killingley, J.S., 1980, Growth histories of Strombid snails from Bermuda recorded in their O-18 and C-13 profiles: Marine Biology, v. 60, p. 129–135.

Wing, S.L., Alroy, J., and Hickey, L.J., 1995, Plant and mammal diversity in the Paleocene to early Eocene of the Bighorn Basin: Palaeogeography, Palaeoclimatology, Palaeoecology, v. 115, p. 117–155.

Wolfe, J.A., 1978, A paleobotanical interpretation of Tertiary climates in the Northern Hemisphere: American Scientist, v. 66, p. 694–703.

Wolfe, J.A., 1980, Tertiary climates and florisitic relationships at high latitudes in the Northern Hemisphere: Palaeogeography, Palaeoclimatology, Palaeoecology, v. 30, p. 313–323.

Wolfe, J.A., 1985, Distribution of major vegetational types during the Tertiary, *in* Sundquist, E.T., and Broecker, W.S., eds., The carbon cycle and atmospheric CO_2: Natural variations Archean to present: Washington, D.C., American Geophysical Union Geophysical Monograph Series, v. 32, p. 357–375.

Zachos, J.C., Stott, L.D., and Lohmann, K.C., 1994, Evolution of early Cenozoic marine temperatures: Paleoceanography, v. 9, p. 353–387.

Ziegler, P.A., 1990, Geological atlas of western and central Europe (2nd edition): Den Haag, Netherlands, Shell Internationale Maatschappij B.V., 130 p.

Manuscript Accepted by the Society August 13, 2002

Printed in the USA

Geological Society of America
Special Paper 369
2003

Danian-Selandian sea-level change and biotic excursion on the southern Tethyan margin (Egypt)

Robert P. Speijer*
Department of Geosciences (FB5), Bremen University, P.O. Box 330440, 28334 Bremen, Germany

ABSTRACT

A study on Danian to Selandian foraminifera (planktic/benthic ratios, benthic assemblages) in 23 sections on a N-S transect in eastern Egypt documents a transient biotic excursion associated with sea-level fluctuation at ~60.5 Ma (Subbiochron P3a). In southern areas, the inner to middle neritic (30–70 m) *Neoeponides duwi* assemblage expanded into deeper parts of the shelf (~70–250 m), temporarily replacing the *Anomalinoides umboniferus* and *Angulogavelinella avnimelechi* assemblages. In the deeper parts of the basin (~400–600 m) in the northern part of the transect, the *Gavelinella beccariiformis* assemblage appears to have been persistent throughout the studied time interval. Biofacial and sedimentologic data suggest a relative sea-level fluctuation, possibly with an amplitude of 50–100 m, which may correlate with a eustatic sea-level cycle during the Danian-Selandian transition. During early sea-level rise, total organic carbon–enriched, partially laminated sediments, containing abundant fish-remains and planktic foraminifera (>99.5% planktics), were deposited, reflecting oxygen deficiency at the seafloor. The patterns of biotic and sea-level change at the Danian-Selandian transition strongly resemble those across the Paleocene-Eocene Thermal Maximum in the same basin, suggesting similar operative processes. Considering the close coincidence with a recently postulated brief period of oceanic warming at 60.5 Ma, the question arises whether the observed patterns in Egypt could in part be related to a global warming event, similar to the Paleocene-Eocene Thermal Maximum.

INTRODUCTION

The Danian and Selandian Stages are chronostratigraphically equivalent to the lower Paleocene and the lower upper ("middle") Paleocene, respectively (Berggren et al., 1995, 2000). In western Europe, an unconformity between the Danian and Selandian Stages marks a sequence boundary in the Cenozoic cycle charts (Se-1 of Hardenbol, 1994; Sel 1, Hardenbol et al., 1998). According to Hardenbol et al. (1998) this sequence boundary correlates with the sequence boundary between cycles TA-1.3 and TA-1.4, which may have resulted from a eustatic sea-level fall and subsequent rise (Haq et al., 1988). If this sequence boundary was indeed controlled by eustatic sea-level change, then the southern Tethyan (North African) margin must have recorded a similar sea-level history.

Tunisian and Libyan records do not provide support for a eustatic sea-level fluctuation during the Danian-Selandian transition (Kouwenhoven et al., 1997), but several records from Egypt seem to be in line with this part of the Haq et al. (1988) curve. Lüning et al. (1998) studied the stratigraphy of many Paleocene outcrops in eastern Sinai (Egypt), and correlated the local sequence boundary ThSin-1, with Sel 1 of Hardenbol et al. (1998) and with the TA1.3/TA1.4 sequence boundary of Haq et al. (1988). In the Aweina section (Eastern Desert, Egypt), Speijer and Schmitz (1998) documented a high-amplitude sea-level fluctuation from a succession of benthic foraminiferal assem-

*speijer@uni-bremen.de

Speijer, R.P., 2003, Danian-Selandian sea-level change and biotic excursion on the southern Tethyan margin (Egypt), *in* Wing, S.L., Gingerich, P.D., Schmitz, B., and Thomas, E., eds., Causes and Consequences of Globally Warm Climates in the Early Paleogene: Boulder, Colorado, Geological Society of America Special Paper 369, p. 275–290.

blages within the Danian-Selandian transition, and tentatively correlated it with Danian-Selandian sequence boundary Se-1 of Hardenbol (1994).

The foraminiferal succession in Aweina included the transient occurrence of the shallow water *Discorbis* (=*Neoeponides*) *duwi* assemblage between deeper shelf benthic assemblages (Speijer and Schmitz, 1998). New observations on Danian-Selandian benthic foraminiferal assemblages from 23 sections on a N-S transect in eastern Egypt reveal the incursion of the *Neoeponides* duwi assemblage in other sections in the southern part of the transect. In this paper, I present a summary of these observations, provide a new view on the relationship between biotic and sea-level patterns during the Danian-Selandian transition, compare these with those related to the Paleocene-Eocene Thermal Maximum in the same basin, and speculate on possible implications from this comparison.

Danian-Selandian transition in the type region, Denmark

In its type region, Denmark, Danian chalks and limestones are unconformably overlain by Selandian greensands and marls (e.g., Thomsen and Heilmann-Clausen, 1985). This transition to more terrestrially influenced deposition terminated a >35 m.y. period of hemipelagic carbonate deposition in the Danish basin

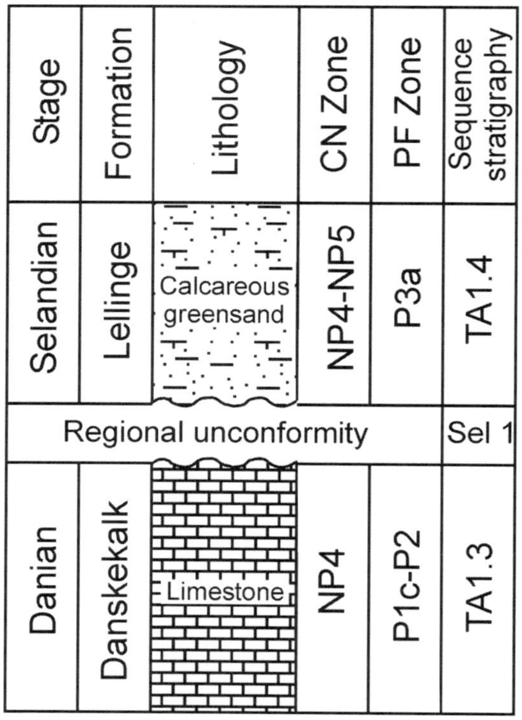

Figure 1. Lithostratigraphy, biostratigraphy, and sequence stratigraphy of the Danian-Selandian transition in the type region, Denmark. Biostratigraphic terminology as adopted in Berggren et al. (2000). Sequences TA1.3/1.4 after Haq et al. (1988) and sequence boundary Sel 1 after Hardenbol et al. (1998). PF—planktic foraminifera; CN—calcareous nannofossils.

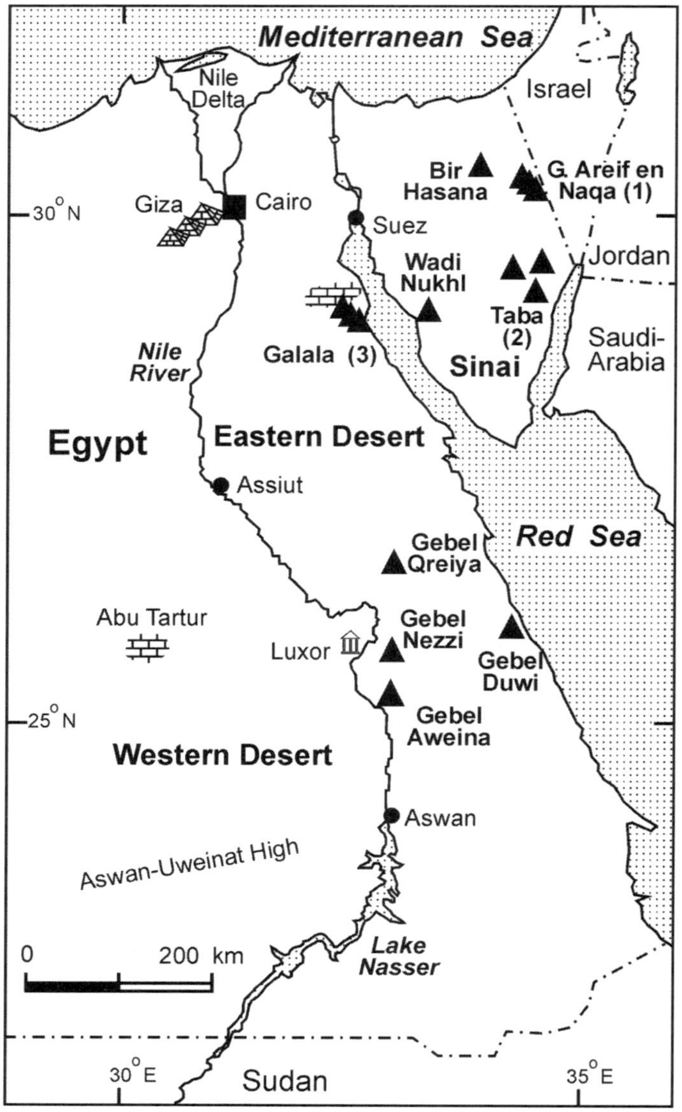

Figure 2. Location map with sections studied (triangles). Numbers refer to the following localities: (1) a transect composed of four sections across the Areif en Naqa anticline; (2) eight sections west of Taba; (3) a transect composed of five sections south of the Galala platform. See Table 1 for biostratigraphic information and references.

(Håkansson et al., 1974; Schmitz et al., 1998). This change in depositional regime between 60 and 61 Ma is related to the opening of the northeastern Atlantic and the uplift of west European landmasses (Berggren, 1994). Determining the exact biostratigraphic positions of the bounding surfaces of the unconformity marking the Danian-Selandian boundary in the type region, Denmark (Fig. 1), has proven difficult. Age and duration of the unconformity vary geographically, but there is general consensus that the top of the Danian correlates with a level within calcareous nannofossil Zone NP4 and planktic foraminiferal Subzone P1c or Zone P2 (e.g., Bignot et al., 1997). The base of the Selandian correlates with a level in the upper half of Zone NP4 or the lower part of Zone NP5 (Thomsen and Heilmann-Clausen, 1985; Thomsen, 1994; Stouge et al., 2000). Foraminifera indicative of Subzone P3a are associated with the base of the Selandian (Stouge et al., 2000), and this subzone ought to have no stratigraphic overlap with Zone NP5 (Berggren et al., 1995, 2000), so that it is plausible that the base of the Selandian correlates with the top part of Zone NP4 and not with Zone NP5. The stratigraphic gap at the Danian-Selandian boundary in the type region thus spans an interval within Zone NP4, correlative with at least the upper part of Zone P2 to the lower part of Subzone P3a. Berggren (1994) and Berggren et al. (1995, 2000) proposed to correlate the Danian-Selandian boundary with the P2/P3 zonal boundary, with an estimated age of 60.9 Ma, whereas Hardenbol et al. (1998) estimated the age for sequence boundary Sel 1 in western Europe at 60.7 Ma (lower Subzone P3a). Awaiting formal decision of the stratigraphic position of the Danian-Selandian boundary, I refer to the sea-level fluctuation recorded at Gebel Aweina within Subzone P3a (Speijer and Schmitz, 1998), correlative with Sel 1 of Hardenbol et al. (1998), as a Danian-Selandian transition event.

Geological setting

The sections studied are located in the Eastern Desert and in Sinai (Fig. 2, Table 1). This region was part of an extensive epicontinental basin—with various subbasins—covering Egypt during the Paleocene (Fig. 3). The area is traditionally subdivided into two main structural units, the so-called stable shelf also known as Nile Basin in the south, and the unstable shelf, also known as the Syrian Arc, in the north (Said, 1962; Shahar, 1994; Tawadros, 2001). The Nile Basin experienced little structural deformation during the early Paleogene, whereas the Syrian Arc is characterized by a fold belt that has been tectonically active since the Late Cretaceous (Shahar, 1994). During Danian-Selandian times, the southern margin of the Nile Basin was situated close to the present southern border of Egypt (Luger, 1988). Most of the area north of Aswan was characterized by hemipelagic marl and clay deposition in a generally northwest deepening basin. Uplifted areas of the Syrian Arc in northern Egypt (e.g., the Galala platform) and the local high of Abu Tartur in central Egypt interrupt this generally deepening trend. More terrestrially influenced deposition (deltaic, coastal) and phases of erosion characterized the basin's southern margin (Hendriks et al., 1985, 1987; Luger, 1985, 1988; Hewaidy, 1994). The microfossil material from hemipelagic parts of the basin studied here documents Danian-Selandian paleodepths ranging from ~70–150 m (Gebel Duwi) to ~400–600 m (Sinai).

MATERIAL AND METHODS

Samples from 23 sections in eastern Egypt (Table 1, Fig. 4) were collected during the last two decades by several teams from the Universities of Berlin, Bremen, Gothenburg and Utrecht in cooperation with partners in Egypt. In order to obtain fresh rock, the samples were generally collected from 10 to 30-cm-deep trenches or holes along the outcrop sections. Standard micro-

TABLE 1. STRATIGRAPHIC DATA AND PALEODEPTH OF STUDIED SECTIONS

Profile	Stratigraphy	Paleodepth*	Reference to stratigraphic data
Areif en Naqa, section A1b	NP2–NP3	400–600 m	Lüning (1997)
Areif en Naqa, section A5	NP2–NP3	400–600 m	Lüning (1997)
Areif en Naqa, section A8	NP2–NP5	400–600 m	Lüning (1997)
Areif en Naqa, section W	NP3–NP6	400–600 m	Lüning (1997)
Bir Hasana	NP3–NP5	400–600 m	Lüning (1997)
Sheikh Attiya	NP1–NP6	400–600 m	Lüning (1997); Lüning et al. (1998)
Gebel Misheiti	NP1–NP6	400–600 m	Lüning (1997); Lüning et al. (1998)
West Themed	NP3–NP4	400–600 m	Lüning (1997); Lüning et al. (1998)
Taba	NP3–NP4, NP6	400–600 m	Lüning (1997); Lüning et al. (1998)
Wadi Gureis	NP1–NP3, NP6	400–600 m	Lüning (1997); Lüning et al. (1998)
Egma Plateau	NP3–NP6	400–600 m	Lüning (1997); Lüning et al. (1998)
Gebel Umm Mafrud	NP3–NP6	400–600 m	Lüning (1997); Lüning et al. (1998)
North Nuweiba	NP3–NP5	400–600 m	Lüning (1997); Lüning et al. (1998)
Wadi Nukhl	P1a–P4	400–600 m	Anan (1992); this paper
Saint Paul 9	NP1–NP5	400–600 m	Scheibner (2001); Scheibner et al. (2001)
Bir Dakhl 2	NP3–NP5	400–600 m	Scheibner (2001); Scheibner et al. (2001)
Bir Dakhl 3	NP3–NP5	400–600 m	Scheibner (2001); Scheibner et al. (2001)
Wadi Tarfa 1	NP2–NP6	400–600 m	Scheibner (2001); Scheibner et al. (2001)
Wadi Tarfa 2	NP4–NP6	400–600 m	Scheibner (2001); Scheibner et al. (2001)
Gebel Qreiya	P1a–P4	150–250 m	Luger et al. (1998)); this paper
Gebel Nezzi	P1c–P4	150–250 m	This paper
Gebel Aweina	P1c–P4	150–250 m	Speijer and Schmitz (1998)
Gebel Duwi	P1b–P4	70–150 m	Faris (1984); this paper

Note: Most sections range higher into the upper Paleocene and lower Eocene.
*Paleodepth estimates in this study.

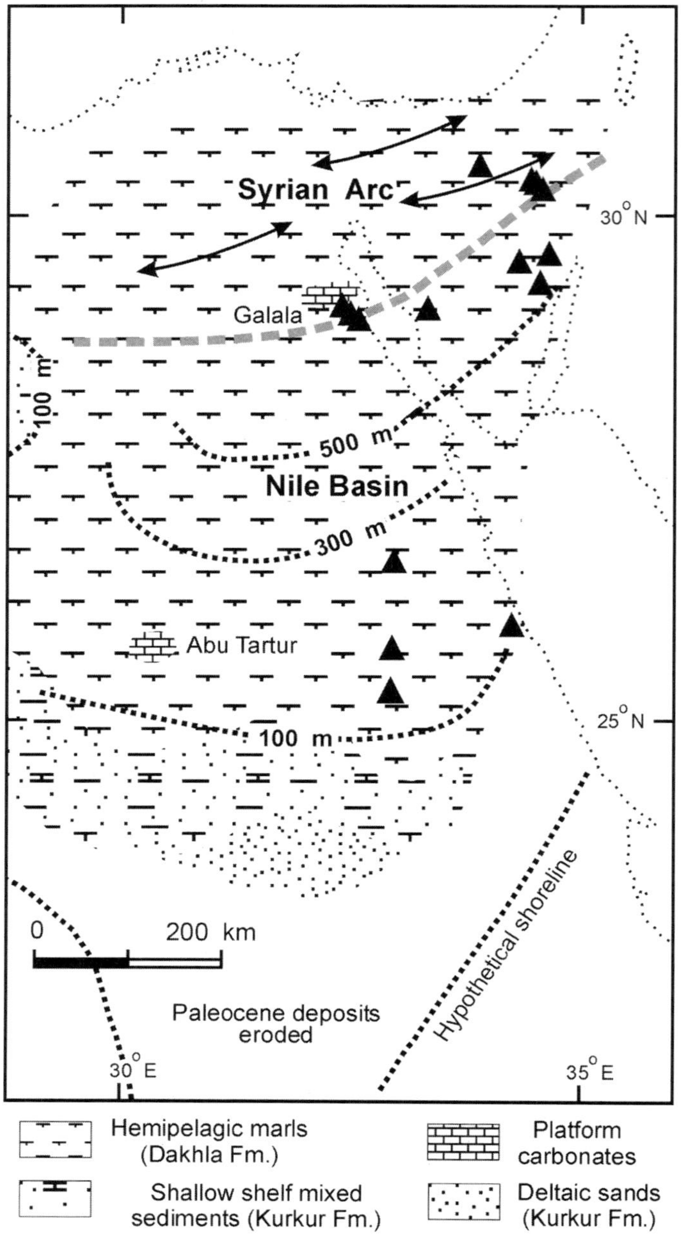

Figure 3. Reconstructed Paleocene paleogeography, lithofacies distribution and paleobathymetry (no correction for opening of Red-Sea rift). Triangles are sections studied (Fig. 2). Thick dashed grey line separates tectonically stable Nile Basin from tectonically active Syrian Arc. Arrows indicate general orientation of double-plunging anticlines (largely submarine swells; schematic) of the Syrian Arc. Isobaths are based on bio- and lithofacies distributions. Carbonate platforms (Abu Tartur and Galala) represent elevated structures on the deeper shelf. Lithofacies distribution modified from Luger (1988) with inclusion of data discussed in this study.

paleontologic sample-processing procedures were employed (e.g., Speijer et al., 1996) and the size fraction larger than 125 μm was used for foraminiferal analysis. Stratigraphic distribution and compositional data of the benthic foraminiferal assemblages were determined qualitatively. Only the main biotic patterns are discussed here.

Assemblages from beds directly overlying the Cretaceous-Paleogene boundary (e.g., Subzone P1a in Gebel Qreiya and Wadi Nukhl) containing large proportions of reworked Maastrichtian foraminifera were disregarded. Planktic/benthic (P/B) ratios, expressed as percentage planktics, were determined for the southern Eastern Desert sections (Tables 2, 3). The terms "neritic" and "bathyal" are used as representing paleodepths less than and in excess of ~200 m, respectively (Van Morkhoven et al., 1986). Note however, that the continental slope was situated north of the epicontinental basin studied (Mart, 1991).

Total organic carbon (TOC) and total carbon (C_{tot}) contents were measured on homogenized samples from Gebel Nezzi and Gebel Qreiya using Leco elemental analyzer facilities at Bremen University. TOC content (precision of measurement ±3%) was measured with a Leco CS-300 and C_{tot} with a Leco CS-200 (precision of measurement ±3%). $CaCO_3$ content (Table 2) was calculated from these measurements through: $CaCO_3 = 8.3333*(C_{tot}\text{-TOC})$.

LITHOSTRATIGRAPHY AND BIOSTRATIGRAPHY

In the study area, the Danian and Selandian Stages are represented by the upper part of the Maastrichtian-Paleocene Dakhla Formation (Said, 1962, 1990) (Fig. 5). This formation consists of rather monotonous brownish to grey marls and shales with occasional intercalated thin marly limestone beds. In Sinai, age-equivalent marls and shales are sometimes referred to as Esna Formation, overlying the Maastrichtian Sudr Formation (e.g., Said, 1990; Lüning et al., 1998). The Dakhla Formation is usually overlain by Thanetian limestones and chalks of the Tarawan Formation. In the Sinai sections, however, the Tarawan Formation is usually just a few meters thick or is absent. In the latter case, there is no clear lithologic boundary between the Dakhla Formation and the Esna Formation (*s.s.:* Thanetian–Ypresian marls and shales overlying the Tarawan Formation [Scheibner et al., 2001]). Danian and Selandian marginal marine successions in the Western Desert are represented by the Kurkur Formation. South of the Abu Tartur platform, this formation interfingers with marls of the basinal Dakhla Formation.

The studied interval spans planktic foraminiferal Zone P1 to the lower part of Zone P4, corresponding to calcareous nannofossil zones NP1-NP6 (Berggren et al., 1995, 2000). In this study, because of the rarity of the P4 zonal marker *Globanomalina pseudomenardii,* the base of Zone P4 is approximated by the lowest occurrence of *Morozovella velascoensis,* a species traditionally thought to appear approximately simultaneously with *G. pseudomenardii* (e.g., Blow, 1979; Toumarkine and Luterbacher, 1985). Recent studies, however, suggest that the

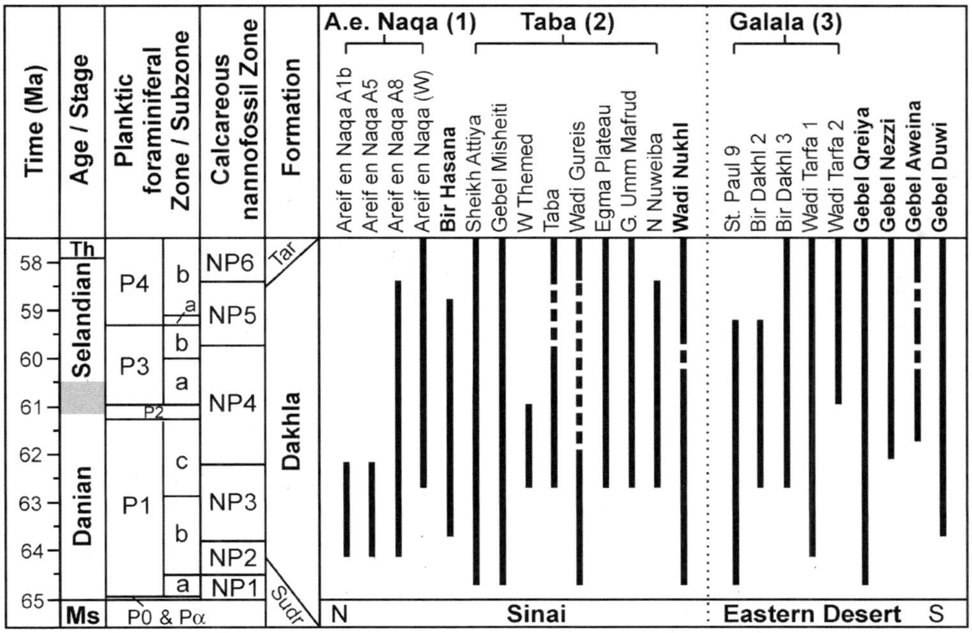

Figure 4. Biostratigraphic ranges of sections studied. Dashed lines indicate stratigraphic gaps. Ranges of the Galala transect (3) and all Sinai sections but Wadi Nukhl are based on calcareous nannofossil data (Table 1). Inconsistencies may exist with equivalent planktic foraminiferal zonal ranges (e.g., Wadi Gureis, Lüning et al., 1998). Biostratigraphic calibration of other sections is primarily based on planktic foraminifera. Bio- and chronostratigraphic scheme after Berggren et al. (2000). Grey band spanning the Danian-Selandian transition indicates the approximate extent of the stratigraphic gap between the two stages in Denmark (Fig. 1). Ms—Maastrichtian; Th—Thanetian; Tar—Tarawan; A.e. Naqa—Areif en Naqa.

lowest occurrence of *M. velascoensis* is within Subzone P3b (Olsson et al., 1999), and thus the stratigraphic range of Subzone P3b may be underestimated in the present study. Detailed biostratigraphic data of most of these sections can be found in previous reports (Table 1, Fig. 4). For Gebel Qreiya and Gebel Nezzi, new biostratigraphic data are presented in Table 2.

RESULTS

Sedimentology, $CaCO_3$, TOC

Sedimentary structures in the monotonous grey to grey-brown marl successions of the Dakhla Formation are rare, particularly in the successions in Sinai deposited at greater depths. In the Eastern Desert sections of Aweina, Nezzi, and Qreiya, the marls generally contain between 15 and 60% $CaCO_3$ (Table 2; Speijer and Schmitz, 1998). At 25–40 m above the Cretaceous-Paleogene boundary in these sections, an up to several-dm-thick dark-brown, foraminifera-rich marl bed is intercalated within Subzone P3a (Table 4). This partly laminated bed, containing abundant fish remains and planktic foraminifera, in part preserved as pyritic molds, seems to fill in an inconspicuous undulating channel-like surface in the marls. In Aweina, immediately overlying this bed common pyritic macrofossils are observed. In the Duwi section, 10 brownish more indurated beds are present in the Paleocene part of the Dakhla Formation. One of these, containing abundant secondary gypsum veins, was not sampled, but is situated in the same biostratigraphic position as the dark-brown marl beds in the other Eastern Desert sections. In the overlying interval, shark teeth, gastropods, bivalves, solitary corals and crinoid stems are common.

The TOC contents in "background marls" ranges generally between 0.1% and 0.2% in Nezzi and 0.1% and 0.3% in Qreiya (Table 2). The anomalous dark-brown bed in these sections, however, yields 0.75% and 2.0% TOC, respectively. This bed is sandwiched within shale beds with <10% $CaCO_3$, of which the overlying one also has slightly elevated TOC contents (0.95% in Qreiya and 0.27% in Nezzi).

Planktic/benthic (P/B) ratios and planktic foraminiferal assemblages

P/B ratios from Zone P3 (Table 3) represent the total range of P/B ratios observed in the Dakhla Formation very well. P/B ratios of the lower and upper parts of the Dakhla Formation in Nezzi and Qreiya are listed in Table 2 for comparison. Most P/B ratio values in Zone P3 fall within the range between 70% and 98% planktics. There appears to be no distinct gradient in P/B ratios from the shallower to the deeper sections. More significant are large variations within individual sections. Whereas in the Galala area P/B ratios <50% planktics are rare, these are more common in the Sinai sections and particularly in the southern Eastern Desert sections. In Qreiya and Nezzi, anomalous low P/B ratios below 20% planktics are generally associated with low-carbonate sediments. In combination with low absolute numbers of generally poorly preserved foraminifera, numerous fragments, large proportions of agglutinated and thick-shelled calcareous taxa, these low P/B ratios are an indication of postmortem dissolution. Loss of calcareous tests associated with reduced P/B ratios is observed in samples from all regions, but is particularly prominent within Zone P3 in the Aweina, Nezzi and Qreiya sections, especially in the shale beds

TABLE 2. FORAMINIFERAL AND GEOCHEMICAL DATA OF GEBEL QREIYA AND GEBEL NEZZI

Profile	Sample	Level* (m)	Biozone	% Planktics	% TOC	% $CaCO_3$
Gebel Qreiya	271185/30	54	P4	94.3	N.D.	N.D.
	271185/31	51	P4	98.0	0.21	81.6
	271185/32	48	P4	64.1	0.20	63.8
	271185/33	46.5	P4	48.7	0.15	42.9
	271185/34	44.5	P4	59.9	0.14	54.4
	271185/35	42	P4	63.1	0.18	50.4
	271185/36	40	P4 †	67.4	0.34	46.4
	271185/37	37.5	P3b	96.7	0.18	30.3
	271185/38	34.5	P3b	92.9	0.16	35.8
	271185/39	32	P3a	87.8	0.34	15.2
	271185/40	31.6	P3a	22.2	0.95 §	7.7
	271185/41	**31.4**	**P3a**	**99.5**	**2.02 §**	**19.5**
	271185/42	31	P3a	0 #	0.32 §	5.8
	271185/43	29	P3a	90.0	0.25	35.0
	271185/44	26.5	P3a	85.3	0.22	15.4
	271185/45	24	P2	93.5	0.24	24.1
	271185/46	22	P2	94.0	0.24	23.6
	281185/35	21	P2	90.9	0.12	27.4
	281185/36	19	P1c	91.2	0.07	29.7
	281185/37	17	P1c	86.9	0.07	35.2
	281185/38	14.5	P1c	92.0	0.14	55.8
	281185/26	13.5	P1c	75	0.10	58.5
	281185/25	12	P1c	86	0.08	53.2
	281185/24	10	P1b	69	0.08	52.8
	281185/23	9.5	P1b	81	0.09	54.7
	281185/22	8.0	P1b	47	N.D.	N.D.
	281185/21	5.5	P1b	65	0.02	57.2
	281185/20	3.5	P1b	10	0.10	12.3
Gebel Nezzi	51285/28	66	P4	80.3	0.06	81.9
	51285/27	61.5	N.D.	77.6	0.06	77.7
	51285/26	59.5	N.D.	63.0	0.06	72.1
	51285/25	57.5	N.D.	42.9	N.D.	N.D.
	51285/24	54	P4	92.5	0.06	78.7
	51285/23	50	P4 †	90.1	0.14	79.6
	51285/22	46	P3b	90.7	0.08	44.6
	51285/21	42	P3b	93.3	0.11 §	44.7
	51285/20	38.5	N.D.	1.9	0.27	1.9
	51285/19	**38**	**P3a**	**99.7**	**0.75 §**	**20.2**
	51285/18	37.5	N.D.	0 #	0.10	1.8
	51285/17	34	P3a	85.6	0.12 §	16.9
	51285/16	31.5	P2	86.5	0.08	25.5
	51285/15	29.5	P2	87.0	0.14	22.2
	51285/14	27	N.D.	92.3	0.11	30.2
	51285/13	25	N.D.	84.1	0.14	30.5
	51285/12	24.5	N.D.	90.1	0.15	44.4
	51285/11	24	N.D.	66.1	0.10	42.8
	51285/10	21.5	N.D.	85.5	0.10	49.6
	51285/9	19.5	N.D.	81.9	0.15	63.0
	51285/8	15	N.D.	25.9	0.14	46.8
	51285/7	11	N.D.	74.6	0.21	53.4
	51285/6	8.5	P1c	53.0	0.16	51.0
	51285/5	7	N.D.	12.3	0.30	37.2
	51285/4	5	N.D.	10.9	0.41	6.9
	51285/3	2.5	N.D.	7.9	N.D.	N.D.
	51285/2	1.5	N.D.	0.0	0.37	1.8

Note: Boldface data from anomalous dark-brown marl bed. N.D.—No data.

*Meters above Cretaceous-Paleogene boundary.

†The base of Zone P4 is approximated by the lowest occurrence of *M. velascoensis*.

§Average value from two sub-samples.

#Dissolution residues. Rare planktics only preserved as pyritic molds. Benthic assemblage consists of agglutinated taxa and fragments of thick-shelled calcareous taxa (Nodosarians).

TABLE 3. PERCENTAGE PLANKTICS (%P) WITHIN ZONE P3 IN STUDIED SECTIONS

Sinai *			Galala area*			Southern Eastern Desert		
Profile	Sample	%P	Profile	Sample	% P	Profile	Sample	%P
Areif en Naqa	A8-18	97	St Paul 9	S9-58	92.1	Gebel Qreiya	271185/37	96.7
	A8-17	94		S9-57	95.5		271185/38	92.9
	A8-16	93		S9-56	95.4		271185/39	87.8
				S9-55	96.2		271185/40	22.2
Areif en Naqa	W1-11	59		S9-54	96.3		271185/41	99.5
	W1-10	63		S9-53	94.2		271185/42	0†
	W1-9	89		S9-52	93.3		271185/43	90.0
	W1-8	72		S9-51	94.7		271185/44	85.3
	W1-7	61		S9-50	93.4			
	W1-6	98		S9-49	96.5	Gebel Nezzi	51285/23	90.1
	W1-5	88					51285/22	90.7
	W1-4	77	Bir Dakhl 2	D2-48	81.6		51285/21	93.3
	W1-3	90		D2-47	87.0		51285/20	1.9
	W1-2	87		D2-45	93.7		51285/19	99.7
				D2-46	93.0		51285/18	0†
Sheikh Attiya	C1-29	73		D2-44	92.2		51285/17	85.6
	C1-28	76		D2-43	91.0		51285/16	86.5
	C1-27	85		D2-42	91.4			
	C1-26	86		D2-41	96.6	Gebel Aweina	O95 + 750 cm	87.3
	C1-25	73		D2-40	86.1		O95 + 600 cm	92.7
	C1-24	85		D2-39	97.3		O95 + 450 cm	96.3
	C1-23	21		D2-38	97.9		O95 + 400 cm	94.6
	C1-22	67		D2-37	89.0		O95 + 275 cm	91.6
	C1-21	83		D2-36	85.8		O95 + 175 cm	81.4
	C1-20	97		D2-35	90.0		O95 + 70–80 cm	94.2
				D2-34	82.3		O95 + 65–70 cm	87.3
Gebel Misheiti	F1-43	82		D2-32	88.9		O95 + 55–60 cm	88.5
	F1-42	88		D2-33	47.2		O95 + 19–24 cm	55.4
	F1-41	78		D2-31	76.0		O95 + 0–2 cm	99.8
	F1-40	95		D2-30	84.8		O95 – 5–8 cm	70.4
							O95 – 9–15 cm	27.8
West Themed	K1-15	94	Bir Dakhl 3	D3-36	92.1		O95 – 18–25 cm	68.8
				D3-35	84.6		O95 – 30–35 cm	78.3
Taba	M1-36	97		D3-34	80.1		O95 – 90–100 cm	44.3
	M1-35	88		D3-33	94.7		O95 – 200 cm	72.4
	M1-34	96		D3-32	94.9		O95 – 300 cm	76.3
				D3-31	96.0		O95 – 400 cm	87.6
Wadi Gureis	P1-30	95		D3-30	93.8			
				D3-29	90.3	Gebel Duwi	DU95D/S+41	97.6
Egma Plateau	Q1-12	98		D3-28	88.3		DU95D/S+40.5	95.1
	Q1-11	53		D3-27	98.8		DU95D/S+40	70.0
				D3-26	84.3		DU95D/S+39A	67.5
Gebel Umm Mafrud	R1-8	52					DU95D/S+39	86.9
	R1-7	66	Wadi Tarfa 1	T1-42	98.8		DU95D/S+38	83.1
	R1-6	79		T1-41	97.3		DU95D/S+37	81.3
	R1-5	97		T1-40	95.8			
	R1-4	89		T1-39	94.0			
	R1-3	97		T1-38	94.6			
	R1-2	97		T1-37	94.8			
				T1-36	96.1			
North Nuweiba	T1-14	45		T1-35	90.7			
	T1-13	77		T1-34	68.3			
	T1-12	81		T1-33	90.3			
	T1-11	96		T1-32	92.4			
	T1-10	94		T1-31	86.9			
	T1-9	70						
	T1-8	93	Wadi Tarfa 2	T2-11	56.6			
	T1-7	89		T2-10	50.0			
				T2-9	89.4			
Wadi Nukhl	S718	98		T2-8	95.7			
	S716	95		T2-7	88.8			
				T2-6	95.8			
				T2-5	98.0			
				T2-4	94.5			
				T2-3	87.1			
				T2-2	82.7			

*Data from Sinai except for Wadi Nukhl from Lüning (1997) and data from Galala area from Scheibner (2001).

†Dissolution residues. Rare planktics only preserved as pyritic molds. Benthic assemblage consists of agglutinated taxa and fragments of thick-shelled calcareous taxa (Nodosarians).

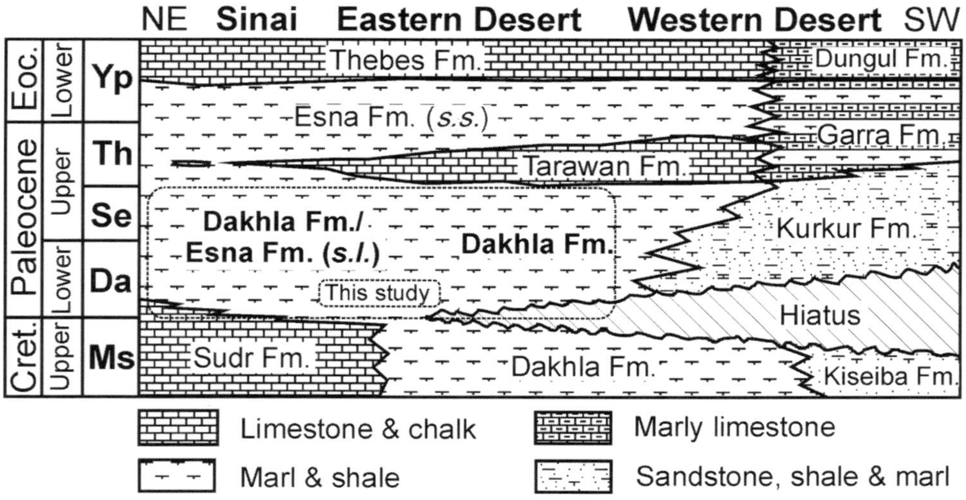

Figure 5. Schematic lithostratigraphic framework of the Paleocene in a northeast-southwest transect across Egypt. Only main lithostratigraphic units and lithologies are indicated and the lateral extent of marginal facies in southwest is exaggerated. Hiatuses other than the one spanning the Cretaceous-Paleogene boundary are not indicated. Cret.—Cretaceous; Eoc.—Eocene; Ms—Maastrichtian; Da—Danian; Se—Selandian; Th—Thanetian; Yp—Ypresian; Fm.—Formation. Drawing after Luger (1985), Hendriks et al. (1987), and Said (1990).

encompassing the anomalous dark-brown marl bed (0%–22% planktics). In contrast, the dark-brown bed itself, with >99.5% planktics, contains the highest P/B ratios, observed in all studied material.

Planktic foraminiferal assemblages in Zone P3 are diverse with most Paleocene genera like *Morozovella, Subbotina, Parasubbotina, Praemurica, Igorina,* and *Globanomalina* well represented. In contrast, samples from the dark-brown bed with P/B ratios of >99.5% planktics contain a poorly diverse planktic assemblage, largely constituted by nonkeeled genera with muricate and praemuricate tests, such as *Praemurica* and *Acarinina.* Particularly striking is the near absence of large morozovellids (*M. angulata, M. conicotruncata*), which are typical for planktic assemblages in Zone P3 in eastern Egypt. Samples from the interval overlying the dark-brown bed in Aweina also lack morozovellids and contain relatively large numbers of *Igorina tadjikistanensis.*

Danian-Selandian benthic foraminiferal assemblages

Three main Danian-Selandian benthic foraminiferal assemblages characterize three bathymetric regimes from ~70–150 to ~400–600 m in the studied part of the basin (Fig. 6). The bathymetric interval between ~250 and ~400 m (situated between Gebel Qreiya and the Galala transect) is not covered in this study. Shallower assemblages, characterizing deposition at up to ~70 m depth, were described from southern Egypt (Hewaidy, 1994; Schnack, 2000).

TABLE 4. DETAILED DATA ON SAMPLES CONTAINING THE *NEOEPONIDES DUWI* ASSEMBLAGE

Profile	Sample	Subzone	% Planktics	% TOC	% $CaCO_3$	Lithology	Macrofossils
Gebel Qreiya	271185/40	P3a	22.2	0.95*	7.7	Dark-grey shale	N.D.
	271185/41	P3a	99.5	2.02*	19.5	Dark-brown marl	N.D.
Gebel Nezzi	51285/20	P3a	1.9	0.27	1.9	Grey shale	N.D.
	51285/19	P3a	99.7	0.75*	20.2	Dark-brown marl	N.D.
Gebel Aweina†	O95 + 19–24 cm	P3a	55.4	N.D.	N.D.	Grey marl	Mollusks
	O95 + 0–2 cm	P3a	99.8	N.D.	N.D.	Dark-brown marl	No
	O95 – 5–8 cm	P3a	70.4	N.D.	N.D.	Grey marl	No
	O95 – 9–15 cm	P3a	27.8	N.D.	N.D.	Grey marl	No
	O92 + 26 m	P3a	10.9	N.D.	13	Grey shale	N.D.
	O92 + 25 m	P3a	2.9	N.D.	N.D.	Grey shale	N.D.
Gebel Duwi	DU95D/S+40	P3a	70.0	N.D.	N.D.	Grey shaley marl	No
	DU95D/S+39A	P3a	67.5	N.D.	N.D.	Grey marl	Shark teeth, mollusks, corals
	DU95D/S+18	P1b	82.5	N.D.	N.D.	Dark-grey shaley marl	No
	DU95D/S+12	P1b	92.5	N.D.	N.D.	Brown-grey marl	No

Note: N.D.—No data.

*Average value derived from two sub-samples.

†Data on O92 samples of Aweina from Charisi and Schmitz (1995) and Speijer and Schmitz (1998). The range of the two O92 samples corresponds to the interval from ~20 cm below to ~30 cm above the dark-brown marl bed.

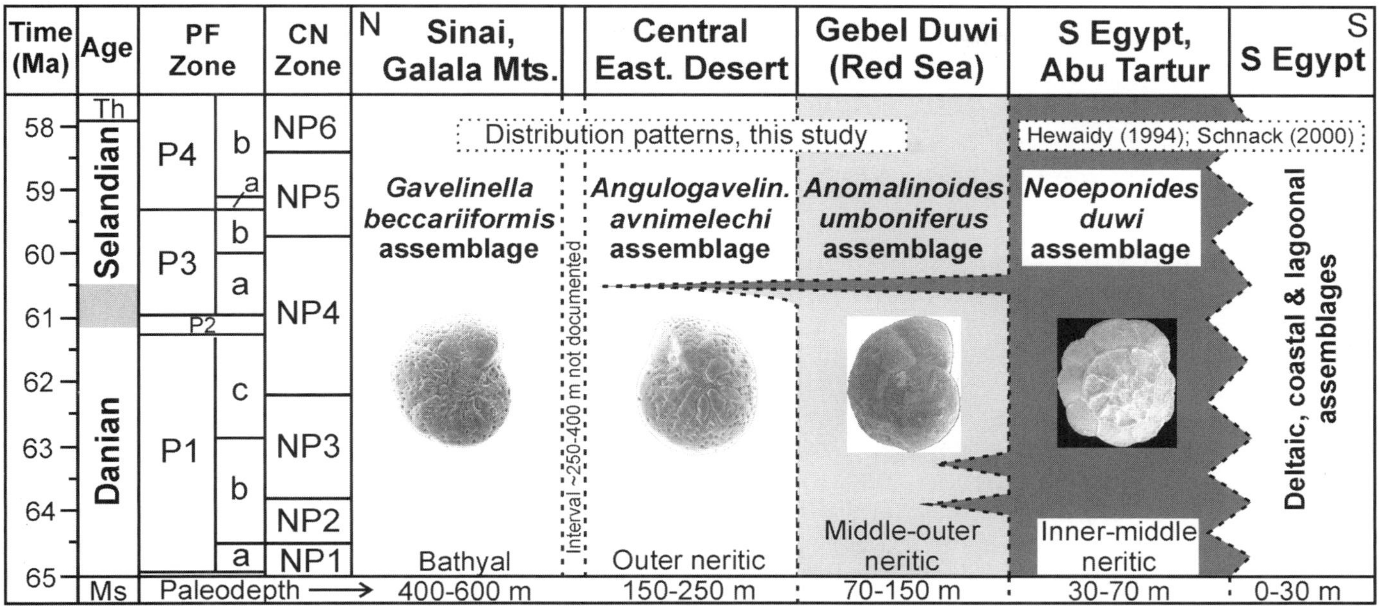

Figure 6. Schematic spatial and temporal distribution of main benthic foraminiferal assemblages in Danian to Selandian deposits of Egypt. Interfingering between the *Neoeponides duwi* and various shallower assemblages in southern Egypt (Hewaidy, 1994) is more complex than indicated. Bio- and chronostratigraphy as in Figure 4. PF—planktic foraminifera; CN—calcareous nannofossils; Ms—Maastrichtian; Th—Thanetian.

Anomalinoides umboniferus ***assemblage.*** The *Anomalinoides umboniferus* assemblage in Gebel Duwi consists, besides the common nominative species, of various *Anomalinoides* spp. (*A.* cf. *midwayensis, A. praeacutus*), *Cibicidoides* spp. (*C. pseudoacutus, C. rigidus*), *Lenticulina* spp., *Bulimina* spp. (*B. midwayensis, B. strobila*), *Siphogenerinoides eleganta, Valvalabamina depressa, Oridorsalis plummerae, Osangularia plummerae, Loxostomoides applinae,* and *Gyroidinoides* spp. Deep-sea taxa such as *Gavelinella beccariiformis, Nuttallides truempyi, Gyroidinoides globosus,* and *Bulimina trinitatensis* are absent and deep outer neritic taxa such as *Angulogavelinella avnimelechi, Anomalinoides susanaensis,* and *Anomalinoides affinis* rarely occur in this assemblage. The absence of deeper water species in combination with fairly high P/B ratios (average ~85% planktics) suggests water depths ranging from ~70 to 150 m (middle neritic to shallow outer neritic) for this assemblage. This assemblage is also found in a 1-m-thick interval immediately underneath the incursion of the *Neoeponides duwi* assemblage (see below) at Aweina (Speijer and Schmitz, 1998).

Angulogavelinella avnimelechi ***assemblage.*** The Aweina, Nezzi, and Qreiya sections are characterized by the *Angulogavelinella avnimelechi* assemblage. This assemblage is generally composed of most species of the *Anomalinoides umboniferus* assemblage listed above, but also contains larger numbers of trochamminids and deep outer neritic taxa like *A. avnimelechi, A. susanaensis,* and *A. affinis.* It also includes a small percentage of bathyal marker species such as *G. beccariiformis* and *N. truempyi,* indicating deep outer neritic to upper bathyal environments (~150–250 m; Speijer and Schmitz, 1998).

Gavelinella beccariiformis ***assemblage.*** The *Gavelinella beccariiformis* assemblage is found near the Galala Mountains and all over Sinai. The nominative species is the most common in almost all samples investigated. Otherwise this assemblage is composed of *N. truempyi, Osangularia plummerae, Gyroidinoides globosus, A. avnimelechi,* various *Anomalinoides* (*A. affinis, A. rubiginosus*), *Cibicidoides* (*C. pseudoacutus, C.* cf. *hyphalus*) and numerous infrequent species. Based on upper depth limits of ~500 m for *N. truempyi* and *B. trinitatensis* (Van Morkhoven et al., 1986), Speijer (1994) suggested a water depth of 500–700 m for this assemblage. Because these taxa have been found in low numbers in the Eastern Desert sections (see also Speijer and Schmitz, 1998), the upper depth limits for these taxa appear unrealistic for the Nile Basin. Consequently, the bathymetric range of the *G. beccariiformis* assemblage, occurring in the deepest parts of the basin, is estimated somewhat shallower (~400–600 m).

These depth-related assemblages are fairly uniform through time with respect to their general composition, although some species have local lowest or highest occurrences in the Danian-Selandian interval. Compositional variations are generally subtle and in some instances appear to relate to taphonomic processes, such as dissolution (see above) rather than to primary differences.

Neoeponides duwi ***assemblage.*** Superimposed on the general biotic stability in the investigated successions are a few stratigraphic levels with a strongly deviating benthic foraminiferal composition. In the Aweina section, the *Neoeponides duwi* assemblage occurred in an ~1-m thick interval within planktic foraminiferal Subzone P3a and calcareous nannofossil Zone NP4 (Speijer and Schmitz, 1998), and it includes the dark-brown marl bed. This same unusual assemblage occurs in short intervals in the Duwi, Nezzi and Qreiya sections. It generally consists of abundant *Neoeponides duwi,* composing up to ~60% of the benthic assemblage, *Haplophragmoides walteri,* trochamminids and small proportions of *Siphogenerinoides elegantus, Gyroidinoides girardanus, Anomalinoides praeacutus, A. umboniferus, Cibicidoides pseudoacutus, Bulimina strobila* and *Lenticulina* spp. This assemblage is not observed in any of the bathyal successions in the Galala area and in Sinai. In Nezzi and Qreiya, two localities similar to Aweina, the *N. duwi* assemblage is observed in the dark-brown marls and overlying shales within Subzone P3a. In the somewhat shallower Duwi succession, the assemblage occurs within Subzone P3a and further down in two Danian (Subzone P1b) horizons (Table 4). Within the present biostratigraphic constraints, the interval marked by the *N. duwi* assemblage in Subzone P3a seems to be a synchronous unit in the region.

The *N. duwi* assemblage is associated with highly variable P/B ratios. The nominative species is highly abundant (relatively and in absolute numbers) in foraminiferal assemblages with very low P/B ratios, e.g., Gebel Qreiya sample 271185/40, Gebel Nezzi sample 51285/20, and Aweina samples O92 + 25 and O92 + 26. *Neoeponides duwi* is much less abundant in absolute numbers, but similarly dominant among the benthic foraminifera in assemblages from the dark-brown beds (271185/41, 51285/19 and O95 + 0–2) characterized by >99.5% planktics and TOC enrichment. In the low-resolution successions of Nezzi and Qreiya, the appearance of the *N. duwi* assemblage seems to coincide with the extremely high P/B ratios. High-resolution data of Aweina, however, show that appearance of the *N. duwi* assemblage slightly predated the peak in the P/B ratios: A sample 9–15 cm below the dark-brown bed contains a typical *N. duwi* assemblage with abundant agglutinated taxa, suggesting that the appearance of the *N. duwi* assemblage is not directly linked to the conditions prevailing during deposition of the TOC-enriched dark-brown marl beds.

In order to understand the paleoenvironmental significance of the anomalous *N. duwi* assemblage, its stratigraphic and bathymetric distribution needs to be considered. This type of assemblage has previously been found in Danian–lower Selandian (Zones P1-P3) deposits in southern Egypt (Aswan-Uweinat high) and central Egypt, near the Abu Tartur plateau where it is associated with P/B ratios varying between 1 and 56% planktics (Hewaidy, 1994; Schnack, 2000). These areas are characterized by a carbonate-platform facies of the lower Paleocene Abu Tartur platform (Schnack, 2000) and by inner to middle neritic marls and shales (Hewaidy, 1994). Occasionally, *N. duwi* dominates the benthic foraminiferal assemblages in these shallow areas. *Haplophragmoides walteri,* another frequent component of the *N. duwi* assemblage, is a common constituent in inner neritic and restricted marine environments in the Paleocene of Tunisia and Texas (Kellough, 1965; Aubert and Berggren, 1976; Kouwenhoven et al., 1997), but also occurs in Paleocene to Eocene flysch-type deep-water deposits (Kaminski et al., 1988). Most other taxa in this assemblage appear to have had wide bathymetric distributions in Egypt, spanning from inner neritic to upper bathyal depths (Hewaidy, 1994; Speijer, 1994; Speijer et al., 1996; Speijer and Schmitz, 1998; Schnack, 2000). In Egypt, the *N. duwi* assemblage thus seems to typify inner to middle neritic (~30–70 m) settings and the invasions into deeper shelf domains, particularly those into the outer neritic realm should be considered as anomalies.

DISCUSSION

Sea-level change during the Danian–Selandian transition

Three different sea-level scenarios for the Danian-Selandian transition in the region are considered. In the first two scenarios, the interval with the *N. duwi* assemblage, indicative of inner-to-middle neritic deposition, constitutes a relative sea-level lowstand, as suggested by Speijer and Schmitz (1998). In the third scenario it constitutes the last phase of falling sea level and the subsequent rising sea level.

Scenario one. According to the first observations on the benthic foraminiferal record of the Aweina section, the presence of the *N. duwi* assemblage (two samples) coincided with a drop in the percentage of planktic foraminifera from ~90% to ~10% (Speijer and Schmitz, 1998). This suggested that a relative sea-level fall and a decrease in paleodepth from ~150–250 m to 30–70 m forced a biofacial replacement of the deeper *A. avnimelechi* assemblage by the shallower *N. duwi* assemblage. The return of the *A. avnimelechi* assemblage marked drowning of the deeper shelf and the return to former paleodepths. The new observations indicate that this overall pattern is valid on a more regional scale.

Scenario two. In a slightly modified interpretation, the *N. duwi* assemblage is thought to have been introduced from the shallower shelf into the deeper parts through increasing current activity with falling sea level. The presence of the *N. duwi* assemblage on top of the channel-like surfaces supports this view. This interpretation also differs from scenario one in that the paleodepth for the interval of the *N. duwi* assemblage cannot be accurately assessed, since the assemblage would be largely constituted by transported elements. As in scenario one, the return of the *A. avnimelechi* assemblage marks the return to former paleodepths.

There are several problems with considering the interval marked by the *N. duwi* assemblage as an exclusively lowstand signal as in these two scenarios. The dark-brown marl beds contain large numbers of planktic foraminifera in combination with extremely high P/B ratios (>99.5% planktics). Such occurrences are not compatible with a relative sea-level lowstand for the entire interval characterized by the presence of the *N. duwi* assemblage. These sediments are also enriched in organic carbon and fish remains, pointing to oxygen deficiency at the seafloor. This cannot be reconciled with deposition under high current activity.

Scenario three. The appearance of the *A. umboniferus* assemblage prior to the appearance of the *N. duwi* assemblage in the Aweina record indicates 50–100 m shallowing (Speijer and Schmitz, 1998). The subsequent replacement by the *N. duwi* assemblage, just below the dark-brown marl bed, reflects the shallowest phase at the end of a falling relative sea level. In sequence stratigraphic terminology (Posamentier et al., 1988; Van Wagoner et al., 1988), this could represent the late highstand systems tract (HST) of a relative sea-level cycle (Fig. 7). The channel-like surface may be interpreted as a type 2 sequence boundary, marked by minor submarine erosion (Emery and Myers, 1996). The overlying sediments, marked by rich foraminiferal assemblages, and at least in Duwi and Aweina common macrofossils, seem to represent a period of condensed sedimentation during the subsequent relative sea-level rise. This interval could represent the transgressive systems tract (TST) in the relative sea-level curve. Considering the absence of sand-sized terrigenous or obviously redeposited material, there are no good indications for the presence of a lowstand systems tract or shelf-margin systems tract between the HST and TST in the southern Eastern Desert sections, suggesting that there may be a hiatus of unknown duration associated with the bottom surface of the channel.

During the early part of the rising sea level, oxygen deficiency led to restricted bottom life reflected in extremely high P/B ratios, preservation of fish remains, dark coloration and TOC-enrichment of the sediments. Simultaneously, surface ecosystems were also perturbed, as indicated by the near disappearance of the morozovellids and the predominance of non-keeled praemuricate and muricate planktic genera. As sea level rose and seafloor oxygenation improved, the *N. duwi* assemblages rapidly disappeared and the normal bathymetric biofacial distribution (Fig. 6) was restored.

It is intriguing that *N. duwi* remained the dominant benthic taxon during the short oxygen-deficiency phase. Its persistence in this interval could suggest an opportunistic life mode, like has been suggested for other benthic foraminiferal taxa in black shales and sapropels (e.g., Sen Gupta and Machain-Castillo, 1993). However, most of these opportunistic taxa are characterized by small and thin-shelled tests, whereas *N. duwi* is one of the largest species in the benthic assemblages and has a thick and strongly decorated test (Fig. 8). Alternatively, despite their generally good preservation, it cannot be excluded that the specimens of *N. duwi* in the dark-brown bed represent a reworked fraction from the immediately underlying sediments. High-resolution data from Aweina show that the underlying 10–20 cm of sediment contain the *N. duwi* assemblage (Table 4), but sample density of the other sections does not allow to evaluate this alternative. Irrespective of the ecologic characteristics of *N. duwi* with respect to oxygen deficiency, with the available data scenario three provides the best explanation for the sequence of events in the Danian-Selandian transition in eastern Egypt.

Regional sea-level record

If the changes in water depth and relative sea level as recorded in the Eastern Desert sections are of regional or eustatic significance, they must also have left a record in other parts of the basin. In Sinai, Lüning et al. (1998) found micropaleontologic and sedimentologic evidence for a sequence boundary, correlative with the one observed in the Eastern Desert (Fig. 7),

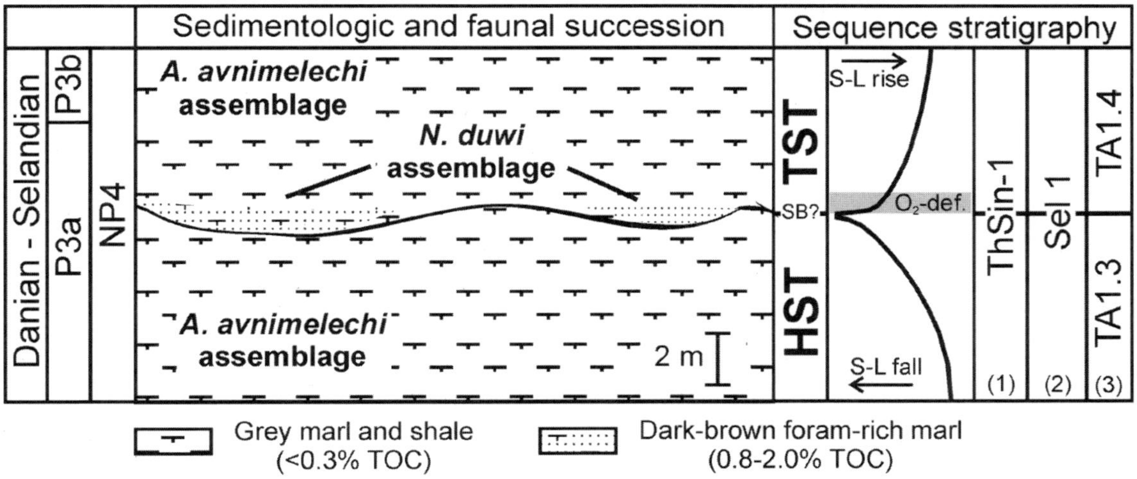

Figure 7. Schematic sedimentologic and benthic foraminiferal succession within outcrop across the Danian-Selandian transition in an outer neritic setting (e.g., Gebel Aweina). Note that in Gebel Aweina the lowest occurrence of the *N. duwi* assemblage is observed already 10–15 cm below the dark-brown foraminifera-rich marl bed. Interpreted relative sea-level history and systems tracts compared to cycle boundaries elsewhere: (1) Sinai (Lüning et al., 1998); (2) western Europe (Hardenbol et al., 1998); (3) "Global" (Haq et al., 1988). Grey band indicates oxygen deficiency at the base of the transgressive systems tract (TST). SB?—Type 2 sequence boundary?; HST—highstand systems tract; S-L—sea-level; TOC—total organic carbon.

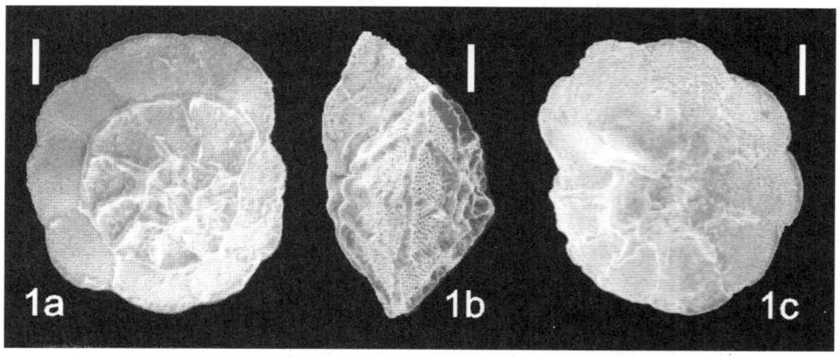

Figure 8. SEM photographs of *Neoeponides duwi* (Nakkady) from Subzone P3a in the Eastern Desert, Egypt. 1a—spiral side of specimen from Gebel Aweina (sample O95 + 19–24 cm); 1b—side view of specimen from Gebel Duwi (sample DU95DS+39A); 1c—umbilical side of another specimen from Gebel Duwi (sample DU95DS+39A). Scale bars are 100 µm.

reflected in a reduction in P/B ratios and biostratigraphic discontinuities. A sea-level fall within Subbiochron P3a and a subsequent unconformity was observed in the Western Desert (Luger, 1985; Schnack, 2000). In southern Egypt, the late Danian sea-level fall is manifested by a regressive sequence of deltaic deposits. An erosional contact marks the transition to the overlying transgressive sequence that correlates with the lower part of Zone P4 (Hendriks et al., 1985; Luger, 1985). Except for the Galala area, where there is yet no evidence for sea-level changes within Subzone P3a (Scheibner et al., 2000), the Danian-Selandian sea-level fall and subsequent rise can be monitored in various parts of the Nile Basin. Estimating the amount of relative sea-level change awaits a detailed quantitative analysis, but Speijer and Schmitz (1998) indicate that the amount of shallowing prior to the invasion of the *N. duwi* assemblage is probably on the order of 50–100 m. Water depths before and after this sea-level fluctuation were similar, so the subsequent deepening must have been in the same order of magnitude.

Eustatic sea-level change

In order to estimate the magnitude of eustatic changes, local subsidence must be assessed. Tectonic activity in the basin appears restricted to the late Paleocene, in particular to Zones NP5 to NP9 (Strougo, 1986; Guiraud and Bosworth, 1999; Kuss et al., 2000; Scheibner et al., 2000) and thus seems to play no role in the Danian-Selandian transition in Egypt. In Aweina, 45 m of compacted sediments accumulated during the interval between the base of the Paleocene (Biochron P1c) and the top of Biochron NP6, corresponding to ~4 m.y. (Speijer and Schmitz, 1998). Paleodepths over this time interval remained overall unchanged, and assuming no tectonic activity, the average subsidence (without correction for compaction) was 11 m/m.y. The total time involved in the Danian-Selandian sea-level event in Aweina (including hiatuses) is on the order of 1 m.y. or less. If the estimate of 50–100 m sea-level change and the assumption of no tectonic activity are generally correct, then the role of subsidence is negligible and the major component in the paleodepth and relative sea-level changes observed could be eustatic change.

The postulated type 2 sequence boundary in the Eastern Desert sections appears to correlate (Fig. 7) with the type 1 sequence boundary Sel 1 (Hardenbol et al., 1998) and the sequence boundary between third order cycles TA1.3 and TA1.4, for which the estimated amounts of sea-level fall and subsequent rise are 60 m and 30 m, respectively (Haq et al., 1988). The former value is within the estimated range for Aweina, but the latter value is only one half to one third of the estimate. This indicates that there is fair agreement in timing and magnitude of the Danian-Selandian sea-level fluctuation in the studied basin and globally. The magnitude of sea-level change (~50 m) suggests that glacio-eustatic processes may have been involved. Although the early Paleocene was a relatively cool period compared to the Late Cretaceous and late Paleocene (Oberhänsli and Hsü, 1986; Frakes et al., 1992; Zachos et al., 2001), independent evidence for the existence of ice sheets during the early to late Paleocene merely consists of a few records of early to "middle" Paleocene dropstones (Dalland, 1976; Leckie et al., 1995). Oxygen isotopic data, however, have been interpreted as indicating that the early Paleogene world was essentially free of ice-sheets (e.g., Zachos et al., 2001). A fairly good fit between eustatic cycles and the ice-volume record reconstructed from oxygen isotopic data currently only exists for middle Eocene and younger records (Miller et al., 1996; Abreu and Anderson, 1998). For now it remains enigmatic how this and other Paleocene large magnitude sea-level fluctuations (Haq et al., 1988) may have been eustatically controlled without good evidence for the waxing and waning of large ice-sheets.

Comparison with Paleocene-Eocene Thermal Maximum record in Egypt

In addition to indications for glacial periods within the early to early late Paleocene (Dalland, 1976; Leckie et al., 1995), it has been suggested that the first of a series of hyperthermals, short periods of extreme atmospheric and oceanic warmth, may have occurred at this time. Negative peaks in the deep-sea oxygen isotopic record suggest a total of perhaps six hyperthermals during the late Paleocene to early Eocene, of which the most prominent one was the Paleocene-Eocene Thermal Maximum (Thomas et al., 2000). Another hyperthermal may have occurred at 60.5 Ma (Thomas and Zachos, 2000), coinciding with a short-term polar-wards expansion of the distribution of the warm-water dinoflagellate genus *Apectodinium* (Bujak and Brinkhuis, 1998), at about the same time as the Danian-Selandian sea-level event. Whether or not these events truly correlate remains to be proven, but the Paleocene-Eocene Thermal Maximum appears to coincide with eustatic sea-level fluctuations (Crouch, 2001; Speijer and Morsi, 2002; Speijer and Wagner, 2002, and references therein) like the Danian-Selandian transition. The biotic and abiotic events at the Danian-Selandian transition thus warrant a comparison with those of the Paleocene-Eocene Thermal Maximum as observed in the Nile Basin.

The Paleocene-Eocene Thermal Maximum (~55.5 Ma) represents a brief period of global warming, particularly at high latitudes, associated with an ~3‰ negative carbon isotopic excursion (Zachos et al., 1993) that was likely related to massive methane release from continental margin sediments (e.g., Kennett and Stott, 1991; Dickens et al., 1995; Katz et al., 1999; Norris and Röhl, 1999). These climatic and oceanographic changes triggered major biotic extinctions and evolutionary innovations, particularly among deep-sea benthic foraminifera (e.g., Tjalsma and Lohmann, 1983; Thomas, 1998), marine micro- and nannoplankton (e.g., Kelly et al., 1996; Aubry, 1998; Crouch et al., 2001), and terrestrial mammals (e.g., Gingerich, 2000). In the Nile Basin, many of these were identified, including the carbon isotopic excursion, anomalous planktic foraminiferal assemblages characterized by numerous *Acarinina,* and a turnover in benthic foraminiferal assemblages (Speijer, 1994; Schmitz et al., 1996; Speijer et al., 1996, 2000; Charisi and Schmitz, 1998; Bolle et al., 2000; Speijer and Wagner, 2002). Some of the local paleoenvironmental changes were related to changes in sea level (Speijer and Morsi, 2002; Speijer and Wagner, 2002).

Several aspects of the Paleocene-Eocene Thermal Maximum in the Nile Basin are of particular interest in the comparison with the Danian-Selandian transition events (Table 5). Both events are marked by the onset of oxygen deficiency on the seafloor. During the Paleocene-Eocene Thermal Maximum, this occurred virtually basin-wide (middle neritic to bathyal), whereas during the Danian-Selandian transition apparently only the middle-outer neritic realms were affected. Both events are characterized by anomalous planktic foraminiferal assemblages. Although oxygen deficiency perturbed the seafloor biota, the otherwise common keeled *Morozovella* virtually disappeared from the surface waters and nonkeeled muricate and praemuricate taxa (*Praemurica, Acarinina*) became very abundant. This indicates transient environmental anomalies perturbing the ecosystems of the entire water column. Sedimentologic features of both events (minor submarine erosion, sediment starvation, omission surfaces) may have resulted from sudden sea-level rise—associated with circulation changes—after a sea-level fall. Speijer and Morsi (2002) correlated the Paleocene-Eocene Thermal Maximum with the transgressive systems tract (TST) of third order sea-level cycle TA2.3 of Haq et al. (1988). The Danian-Selandian transgressive phase appears to correlate with transgressive deposits of the Selandian transgression of cycle TA1.4.

Many details concerning the actual existence of a Danian-Selandian hyperthermal, correlations between the open ocean and the Nile Basin, the sequence-stratigraphic and oxygen-isotopic records, and ecologic changes still need to be clarified. This time interval has as yet received very little attention compared with the Paleocene-Eocene Thermal Maximum. Nevertheless, the many similarities in sedimentologic features and biotic patterns in the Nile Basin during these two time intervals

TABLE 5. BIOTIC AND SEDIMENTARY FEATURES OF DANIAN-SELANDIAN EVENT, PALEOCENE-EOCENE THERMAL MAXIMUM, AND BACKGROUND CHARACTERISTICS

Parameter	Paleocene background	Danian-Selandian transition*	PETM†
Predominant lithology	Grey to pale-brown marl	Dark-brown marl	Dark-brown marl
Discontinuities	Some paraconformities	Unconformity	Some distinct
Systems tract	Various	Transgressive	Transgressive
Fish remains	Rare	Abundant	Abundant
Peloids	Rare	Rare	Abundant
Macrofossils	Rare	Common	Rare
TOC	<0.3%	0.8–2.0%	1.5–2.7%
Foraminifera:			
% Planktics (>125 μm)	70–95% P§	2–99.8% P	~99% P
Morozovella #	Common to abundant	Rare	Rare
Praemurica and *Acarinina*	Common to abundant	Dominant	Dominant
Benthic assemblage	Diverse**	Oligotaxic	Oligotaxic
Dominant benthic taxon	Depth dependent	*Neoeponides duwi*	*Anomalinoides aegyptiacus*
Paleobathymetry	Middle neritic to bathyal	Not applicable††	Not applicable††
Oxygenation seafloor	Good to moderate	Poor	Poor (variable?)

Note: These features may not necessarily be present in every section studied. Paleocene-Eocene thermal maximum (PETM) characteristics from Speijer and Wagner (2002).

*Compilation of observations on sections Gebel Aweina, Gebel Duwi, Gebel Nezzi, Gebel Qreiya.

†Compilation of observations on the previous 4 sections plus Wadi Nukhl, and Ben Gurion (Israel).

§Lower percentages planktics (<50%) where dissolution has strongly affected the assemblages.

#*Morozovella* first appears in Danian Zone P2.

**Diversity increases with paleodepth.

††Paleobathymetric estimates based on benthic foraminifera are not reliable because of ecologic disturbance.

are significant, and suggest that at least in this basin similar processes have taken place. It is an intriguing consideration that the Danian-Selandian transition event may have been a precursor of the Paleocene-Eocene Thermal Maximum.

CONCLUSIONS

In Egypt, a marked relative sea-level fluctuation occurred during the Danian-Selandian transition at ~60.5 Ma. A local sequence boundary appears to correlate with sequence boundaries observed on other continental margins, suggesting eustatic control. In middle to outer neritic deposits in the Eastern Desert, the sea-level fluctuation is associated with an incursion of the anomalous *Neoeponides duwi* benthic foraminiferal assemblage, which is usually restricted to Danian-Selandian inner to middle shelf regions. During the early transgressive phase, dysoxic waters spread at middle to outer neritic depths. The patterns of environmental and biotic change associated with the Danian-Selandian transition in the Nile Basin show strong similarities with the ones associated with the Paleocene-Eocene Thermal Maximum in the same basin. The question arises whether the observed patterns may be linked to a hyperthermal during the Danian-Selandian as postulated on the basis of preliminary oxygen isotopic studies.

ACKNOWLEDGMENTS

Financial support was provided through the Deutsche Forschungsgemeinschaft (Sp-612/1 and 612/2). I am grateful to Christian Müller and Tom Wagner and for TOC measurements, to Peter Luger for providing samples and stratigraphic logs of the Qreiya and Nezzi sections and to Birger Schmitz for enabling to analyze samples from Aweina and Duwi. Thierry Adatte, Richard Fluegeman, Miriam Katz, Tanja Kouwenhoven, Christian Scheibner, Ellen Thomas, and an anonymous reviewer are thanked for constructive criticism and improvements of the text.

APPENDIX 1

Systematic part

Speijer (1994) provides taxonomic concepts and synonymies of most benthic foraminiferal taxa mentioned in the main text. A few notes on *Neoeponides duwi* are presented here. An extensive taxonomic discussion is presented elsewhere.

Neoeponides duwi (Nakkady)
(Fig. 8)

Synonymy

Discorbis pseudoscopos var. *duwi*—NAKKADY 1950, p. 689, Plate 90, Figures 5–7.
Discorbis pseudoscopos duwi Nakkady—ANAN and SHARABI 1988, Plate 2, Figure 15.
Discorbis duwi Nakkady—HEWAIDY 1994, Figure 13, nos. 7, 9 (not no. 8) (without description).
Discorbis duwiensis Nakkady—SCHNACK 2000, p. 49 of appendix, Plate 7, Figures 15, 16.

Neoeponides duwi lacks the characteristic apertural openings extending along the umbilical flaps in *Discorbis* and better fits with the generic concept of *Neoeponides* as defined by Reiss (1960). Note that Loeblich and Tappan (1987) provide a different view on the generic characters of *Neoeponides* (Hottinger et al., 1990).

Nakkady (1950) named *Discorbis pseudoscopos* var. *duwi* after the locality of Gebel Duwi. Following the general practice of naming species after localities by ending with "-ensis," Schnack (2000) proposed to change the name *Discorbis duwi* to *Discorbis duwiensis.* However, the International Code of Zoological Nomenclature (Chapter 7, Article 32) does not allow such a change (International Commission on Zoological Nomenclature, 1999).

REFERENCES CITED

Abreu, V.S., and Anderson, J.B., 1998, Glacial eustasy during the Cenozoic: Sequence stratigraphic implications: AAPG (American Association of Petroleum Geologists) Bulletin, v. 82, p. 1385–1400.

Anan, H.S., 1992, Maastrichtian to Ypresian stratigraphy of Abu Zenima section, west-central Sinai, Egypt: Middle East Research Center Ain Shams University, Earth Science Series, v. 6, p. 62–68.

Anan, H.S., and Sharabi, S.A., 1988, Benthonic foraminifera from the Upper Cretaceous–lower Tertiary rocks of the northwest Kharga Oasis, Egypt: Middle East Research Center Ain Shams University, Earth Science Series, v. 2, p. 191–218.

Aubert, J., and Berggren, W.A., 1976, Paleocene benthic foraminiferal biostratigraphy and paleoecology of Tunisia: Bulletin du Centre de Recherche Pau-SNPA, v. 10, p. 379–469.

Aubry, M.-P., 1998, Early Paleogene calcareous nannoplankton evolution: A tale of climatic amelioration, *in* Aubry, M.-P., et al., eds., Late Paleocene–early Eocene climatic and biotic events in the marine and terrestrial records: New York, Columbia University Press, p. 158–203.

Berggren, W.A., 1994, In defense of the Selandian: GFF, v. 116, p. 44–46.

Berggren, W.A., Kent, D.V., Swisher, C.C., III, and Aubry, M.-P., 1995, A revised Cenozoic geochronology and chronostratigraphy, *in* Berggren, W.A., et al., eds., Geochronology, time scales, and global stratigraphic correlation: SEPM (Society for Sedimentary Geology) Special Publication 54, p. 129–212.

Berggren, W.A., Aubry, M.-P., Van Fossen, M., Kent, D.V., Norris, R.D., and Quillévéré, F., 2000, Integrated Paleocene calcareous plankton magnetobiochronology and stable isotope stratigraphy, DSDP Site 384 (northwest Atlantic Ocean): Palaeogeography, Palaeoclimatology, Palaeoecology, v. 159, p. 1–51.

Bignot, G., Curry, D., and Pomerol, C., 1997, The resistible rise of the Selandian: Neues Jahrbuch für Geologie und Paläontologie, Monatshefte, v. 1997, p. 114–128.

Blow, W.H., 1979, The Cainozoic Globigerinida: A study of the morphology, taxonomy, evolutionary relationships, and the stratigraphical distribution of some Globigerinida (mainly Globigerinacea): Leiden, Netherlands, E.J. Brill, v. 1–3, 1413 p.

Bolle, M.P., Tantawy, A.A., Pardo, A., Adatte, T., Burns, S., and Kassab, A., 2000, Climatic and environmental changes documented in the upper Paleocene to lower Eocene of Egypt: Eclogae Geologicae Helvetiae, v. 93, p. 33–51.

Bujak, J.P., and Brinkhuis, H., 1998, Global warming and dinocyst changes across the Paleocene–Eocene Epoch boundary, *in* Aubry, M.-P., et al., eds., Late Paleocene–early Eocene climatic and biotic events in the marine and terrestrial records: New York, Columbia University Press, p. 277–295.

Charisi, S.D., and Schmitz, B., 1995, Stable ($\delta^{13}C$, $\delta^{18}O$) and strontium ($^{87}Sr/^{86}Sr$) isotopes through the Paleocene at Gebel Aweina, eastern Tethyan region: Palaeogeography, Palaeoclimatology, Palaeoecology, v. 116, p. 103–129.

Charisi, S.D., and Schmitz, B., 1998, Paleocene to early Eocene paleoceanography of the Middle East: The $\delta^{13}C$ and $\delta^{18}O$ isotopes from foraminiferal calcite: Paleoceanography, v. 13, p. 106–118.

Crouch, E.M., 2001, Environmental change at the time of the Paleocene-Eocene biotic turnover [Doctoral Thesis]: Utrecht, Netherlands, Utrecht University, Laboratory of Palaeobotany and Palynology Contributions Series, v. 14, 216 p.

Crouch, E.M., Heilmann-Clausen, C., Brinkhuis, H., Morgans, H.E.G., Rogers, K.M., Egger, H., and Schmitz, B., 2001, Global dinoflagellate event associated with the late Paleocene thermal maximum: Geology, v. 29, p. 315–318.

Dalland, A., 1976, Erratic clasts in the lower Tertiary deposits of Svalbard: Evidence of transport by winter ice: Årbok, Norsk Polarinstitutt, v. 1976, p. 151–165.

Dickens, G.R., O'Neil, J.R., Rea, D.K., and Owen, R.M., 1995, Dissociation of oceanic methane hydrate as a cause of the carbon isotope excursion at the end of the Paleocene: Paleoceanography, v. 10, p. 965–971.

Emery, D., and Myers, K., 1996, Sequence stratigraphy: London, Blackwell Science, 297 p.

Faris, M., 1984, Biostratigraphy of the Upper Cretaceous–lower Tertiary succession of Duwi Range, Quseir District, Egypt: Revue de Micropaléontologie, v. 27, p. 107–112.

Frakes, L.A., Francis, J.E., and Syktus, J.I., 1992, Climate modes of the Phanerozoic: The history of the Earth's climate over the past 600 million years: Cambridge, UK, Cambridge University Press, 274 p.

Gingerich, P.D., 2000, Paleocene/Eocene boundary and continental vertebrate faunas of Europe and North America: GFF, v. 122, p. 57–59.

Guiraud, R., and Bosworth, W., 1999, Phanerozoic geodynamic evolution of northeastern Africa and the northwestern Arabian Platform: Tectonophysics, v. 315, p. 73–108.

Håkansson, E., Bromley, R., and Perch-Nielsen, K., 1974, Maastrichtian chalk of northwest Europe: A pelagic shelf sediment, *in* Hsü, K.J., and Jenkyns, H.C., eds., Pelagic sediments: On land and under the sea: Special Publication of the International Association of Sedimentologists: Oxford, Blackwell, p. 211–233.

Haq, B.U., Hardenbol, J., and Vail, P., 1988, Mesozoic and Cenozoic chronostratigraphy and cycles of sea-level change, *in* Wilgus, C.K., et al., eds., Sea-level changes: An integrated approach: Society of Economic Paleontologists and Mineralogists Special Publication 42, p. 71–108.

Hardenbol, J., 1994, Sequence stratigraphic calibration of Paleocene and lower Eocene continental margin deposits in northwest Europe and the U.S. Gulf Coast with the oceanic chronostratigraphic record: GFF, v. 116, p. 49–51.

Hardenbol, J., Thierry, J., Farley, M.B., De Graciansky, P.-C., and Vail, P.R., 1998, Mesozoic and Cenozoic sequence chronostratigraphic framework of European basins, *in* De Graciansky, P.-C., et al., eds., Mesozoic and Cenozoic sequence stratigraphy of European basins: SEPM (Society for Sedimentary Geology) Special Publication 60, p. 3–13.

Hendriks, F., Luger, P., Kallenbach, H., and Schroeder, J.H., 1985, Faziesmuster und Stratigraphie der campanen bis paleozänen Schichtenfolgen im südlichen Obernil-Becken (Western Desert, Ägypten): Zeitschrift der Deutschen Geologischen Gesellschaft, v. 136, p. 207–233.

Hendriks, F., Luger, P., Bowitz, J., and Kallenbach, H., 1987, Evolution of depositional environments of southeast Egypt during the Cretaceous and lower Tertiary: Berliner Geowissenschaftliche Abhandlungen, Reihe A, Geologie und Paläontologie, v. 75, p. 49–82.

Hewaidy, A.G.A., 1994, Biostratigraphy and paleobathymetry of the Garra-Kurkur area, southwest Aswan, Egypt: Middle East Research Center Ain Shams University, Earth Science Series, v. 8, p. 48–73.

Hottinger, L., Reiss, Z., and Halicz, E., 1990, Comments on *Neoeponides* (foraminifera): Revue de Paléobiologie, v. 9, p. 335–340.

International Commission on Zoological Nomenclature, 1999, International code of zoological nomenclature (4th edition): London, The International Trust for Zoological Nomenclature, 306 p.

Kaminski, M.A., Gradstein, F.M., Berggren, W.A., Geroch, S., and Beckmann, J.P., 1988, Flysch-type agglutinated foraminiferal assemblages from Trinidad: Taxonomy, stratigraphy, and paleobathymetry: Abhandlungen der Geologischen Bundesanstalt, v. 41, p. 155–227.

Katz, M.E., Pak, D.K., Dickens, G.R., and Miller, K.G., 1999, The source and fate of massive carbon input during the latest Paleocene thermal maximum: Science, v. 286, p. 1531–1533.

Kellough, G.R., 1965, Paleoecology of the Foraminiferida of the Wills Point Formation (Midway Group) in northeast Texas: Transactions of the Gulf Coast Association of Geological Societies, v. 15, p. 73–153.

Kelly, D.C., Bralower, T.J., Zachos, J.C., Premoli Silva, I., and Thomas, E., 1996, Rapid diversification of planktonic foraminifera in the tropical Pacific (ODP Site 865) during the late Paleocene thermal maximum: Geology, v. 24, p. 423–426.

Kennett, J.P., and Stott, L.D., 1991, Abrupt deep-sea warming, palaeoceanographic changes, and benthic extinctions at the end of the Paleocene: Nature, v. 353, p. 225–229.

Kouwenhoven, T.J., Speijer, R.P., van Oosterhout, C.W.M., and van der Zwaan, G.J., 1997, Benthic foraminiferal assemblages between two major extinction events: The Paleocene El Kef section, Tunisia: Marine Micropaleontology, v. 29, p. 105–127.

Kuss, J., Scheibner, C., and Gietl, R., 2000, Carbonate to platform transition along an Upper Cretaceous to lower Tertiary Syrian Arc uplift, Galala Plateaus, Eastern Desert of Egypt: GeoArabia, v. 5, p. 405–424.

Leckie, D.A., Morgans, H., Wilson, G.J., and Edwards, A.R., 1995, Mid-Paleocene dropstones in the Whangai Formation, New Zealand: Evidence of mid-Paleocene cold climate?: Sedimentary Geology, v. 97, p. 119–129.

Loeblich, A.R.J., and Tappan, H., 1987, Foraminiferal genera and their classification: New York, Van Nostrand Reinhold, 970 p.

Luger, P., 1985, Stratigraphie der marinen Oberkreide und des Alttertiärs im südwestlichen Obernil-Becken (southwest Ägypten) unter besonderer Berücksichtigung der Mikropaläontologie, Palökologie und Paläogeographie: Berliner Geowissenschaftliche Abhandlungen, Reihe A, Geologie und Paläontologie, v. 63, 151 p.

Luger, P., 1988, Maastrichtian to Paleocene facies evolution and Cretaceous–Tertiary boundary in middle and southern Egypt: Revista Espanola de Paleontologia, Numero extraordinario, p. 83–90.

Lüning, S., 1997, Late Cretaceous–early Tertiary sequence stratigraphy, paleoecology and geodynamics of eastern Sinai, Egypt [Ph.D. thesis]: Bremen University, Berichte aus dem Fachbereich Geowissenschaften der Universität Bremen, Nr. 98, 218 p.

Lüning, S., Marzouk, A.M., and Kuss, J., 1998, The Paleocene of central East Sinai, Egypt: "Sequence stratigraphy" in monotonous hemipelagites: Journal of Foraminiferal Research, v. 28, p. 19–39.

Mart, Y., 1991, Some Cretaceous and early Tertiary structures along the distal continental margin of the southeastern Mediterranean: Israel Journal of Earth Sciences, v. 40, p. 77–90.

Miller, K.G., Mountain, G.S., the Leg 150 Shipboard Party, and Members of the New Jersey Coastal Plain Drilling Project, 1996, Drilling and dating New Jersey Oligocene–Miocene sequences: Ice volumes, global sea level, and Exxon records: Science, v. 271, p. 1092–1095.

Nakkady, S.E., 1950, A new foraminiferal fauna from the Esna shales and Upper Cretaceous chalk of Egypt: Journal of Paleontology, v. 24, p. 675–692.

Norris, R.D., and Röhl, U., 1999, Carbon cycling and chronology of climate warming during the Paleocene–Eocene transition: Nature, v. 401, p. 775–778.

Oberhänsli, H., and Hsü, K.J., 1986, Paleocene–Eocene paleoceanography, *in* Hsü, K.J., ed., Mesozoic and Cenozoic oceans: Geodynamics Series 15, p. 85–100.

Olsson, R.K., Hemleben, C., Berggren, W.A., and Huber, B.T., 1999, Atlas of Paleocene planktonic foraminifera: Smithsonian Contributions to Paleobiology, v. 85, 252 p.

Posamentier, H.W., Jervey, M.T., and Vail, P.R., 1988, Eustatic controls on clastic deposition, I: Conceptual framework, *in* Wilgus, et al., eds., Sea-level changes: An integrated approach: Society of Economic Paleontologists and Mineralogists Special Publication 42, p. 109–124.

Reiss, Z., 1960, Structure of so-called Eponides and some other rotaliiform foraminifera: Israel Geological Survey, Bulletin, v. 29, p. 1–19.

Said, R., 1962, The geology of Egypt: Amsterdam, Elsevier, 377 p.

Said, R., 1990, Cenozoic, *in* Said, R., ed., The geology of Egypt: Rotterdam, A.A. Balkema, p. 451–486.

Scheibner, C., 2001, Architecture of a carbonate platform-to-basin transition on a structural high (Campanian–early Eocene, Eastern Desert, Egypt): Classical and modeling approaches combined [Ph.D. thesis]: Bremen University, Bremen, Berichte aus dem Fachbereich Geowissenschaften der Universität Bremen, Nr. 186, 174 p.

Scheibner, C., Kuss, J., and Marzouk, A.M., 2000, Slope sediments of a Paleocene ramp-to-basin transition in northeast Egypt: International Journal of Earth Sciences, v. 88, p. 708–724.

Scheibner, C., Marzouk, A.M., and Kuss, J., 2001, Maastrichtian–early Eocene litho-biostratigraphy and palaeogeography of the northern Gulf of Suez region, Egypt: Journal of African Earth Sciences, v. 32, p. 223–255.

Schmitz, B., Speijer, R.P., and Aubry, M.-P., 1996, Latest Paleocene benthic extinction event on the southern Tethyan shelf (Egypt): Foraminiferal stable isotopic ($\delta^{13}C$, $\delta^{18}O$) records: Geology, v. 24, p. 347–350.

Schmitz, B., Molina, E., and Von Salis, K., 1998, The Zumaya section in Spain: A possible global stratotype section for the Selandian and Thanetian Stages: Newsletters on Stratigraphy, v. 36, p. 35–42.

Schnack, K., 2000, Biostratigraphie und fazielle Entwicklung in der Oberkreide und im Alttertiär im Bereich der Kharga Schwelle, Westliche Wüste, southwest Ägypten [Ph.D. thesis]: Bremen University, Bremen, Berichte, Fachbereich Geowissenschaften, Universität Bremen, Nr. 151, 142 p.

Sen Gupta, B.K., and Machain-Castillo, M.L., 1993, Benthic foraminifera in oxygen-poor habitats: Marine Micropaleontology, v. 20, p. 183–201.

Shahar, J., 1994, The Syrian arc system: An overview: Palaeogeography, Palaeoclimatology, Palaeoecology, v. 112, p. 125–142.

Speijer, R.P., 1994, Extinction and recovery patterns in benthic foraminiferal paleocommunities across the Cretaceous–Paleogene and Paleocene–Eocene boundaries [Ph.D. thesis]: Geologica Ultraiectina, v. 124, p. 191 p.

Speijer, R.P., and Morsi, A.M., 2002, Ostracode turnover and sea-level changes associated with the Paleocene-Eocene thermal maximum: Geology, v. 30, p. 23–26.

Speijer, R.P., and Schmitz, B., 1998, A benthic foraminiferal record of Paleocene sea level and trophic/redox conditions at Gebel Aweina, Egypt: Palaeogeography, Palaeoclimatology, Palaeoecology, v. 137, p. 79–101.

Speijer, R.P., and Wagner, T., 2002, Sea-level changes and black shales associated with the late Paleocene thermal maximum: Organic-geochemical and micropaleontologic evidence from the southern Tethyan margin (Egypt-Israel), *in* Koeberl, C., and MacLeod, K.G., eds., Catastrophic events and mass extinctions: Impacts and beyond: Boulder, Colorado, Geological Society of America Special Paper 356, p. 533–549.

Speijer, R.P., Van der Zwaan, G.J., and Schmitz, B., 1996, The impact of Paleocene–Eocene boundary events on middle neritic benthic foraminiferal assemblages from Egypt: Marine Micropaleontology, v. 28, p. 99–132.

Speijer, R.P., Schmitz, B., and Luger, P., 2000, Stratigraphy of late Paleocene events in the Middle East: Implications for low- to middle-latitude successions and correlations: Journal of the Geological Society of London, v. 157, p. 37–47.

Stouge, S., Ferre, H.B., Rasmussen, J.A., Roncaglia, L., and Sheldon, E., 2000, Micro- and nannofossil biostratigraphy across the Danian-Selandian (Paleocene) stage boundary at Gemmas Allé, Copenhagen, Denmark: GFF, v. 122, p. 161–162.

Strougo, A., 1986, The Velascoensis event: A significant episode of tectonic activity in the Egyptian Paleogene: Neues Jahrbuch für Geologie und Paläontologie, Abhandlungen, v. 173, p. 253–269.

Tawadros, E.E., 2001, Geology of Egypt and Libya: Rotterdam, A.A. Balkema, 468 p.

Thomas, E., 1998, Biogeography of the late Paleocene benthic foraminiferal extinction, *in* Aubry, M.-P., et al., eds., Late Paleocene–early Eocene climatic and biotic events in the marine and terrestrial records: New York, Columbia University Press, p. 214–243.

Thomas, E., and Zachos, J.C., 2000, Was the late Paleocene thermal maximum a unique event?: GFF, v. 122, p. 169–170.

Thomas, E., Zachos, J.C., and Bralower, T.J., 2000, Deep-sea environments on a warm Earth: Latest Paleocene–early Eocene, *in* Huber, B.T., et al., eds., Warm climates in Earth history: Cambridge, UK, Cambridge University Press, p. 132–160.

Thomsen, E., 1994, Calcareous nannofossil stratigraphy across the Danian-Selandian boundary in Denmark: GFF, v. 116, p. 65–67.

Thomsen, E., and Heilmann-Clausen, C., 1985, The Danian-Selandian boundary at Svejstrup with remarks on the biostratigraphy of the boundary in western Denmark: Bulletin of the Geological Society of Denmark, v. 33, p. 341–632.

Tjalsma, R.C., and Lohmann, G.P., 1983, Paleocene–Eocene bathyal and abyssal benthic foraminifera from the Atlantic Ocean: Micropaleontology, Special Publication, v. 4, 90 p.

Toumarkine, M., and Luterbacher, H., 1985, Paleocene and Eocene planktic foraminifera, *in* Bolli, H.M., et al., eds., Plankton stratigraphy: Cambridge, UK, Cambridge University Press, p. 87–154.

Van Morkhoven, F.P.C.M., Berggren, W.A., and Edwards, A.S., 1986, Cenozoic cosmopolitan deep-water benthic foraminifera: Bulletin des Centres de Recherches Exploration-Production Elf-Aquitaine, Mémoire 11, 421 p.

Van Wagoner, J.C., Posamentier, H.W., Mitchum, R.M., Jr., Vail, P.R., Sarg, J.F., Loutit, T.S., and Hardenbol, J., 1988, An overview of the fundamentals of sequence stratigraphy and key definitions, *in* Wilgus, C.K., et al., eds., Sea-level changes: An integrated approach: Society of Economic Paleontologists and Mineralogists Special Publication 42, p. 39–45.

Zachos, J.C., Lohmann, K.C., Walker, J.C.G., and Wise, S.W., 1993, Abrupt climate changes and transient climates during the Paleogene: A marine perspective: Journal of Geology, v. 101, p. 191–213.

Zachos, J., Pagani, M., Sloan, L., Thomas, E., and Billups, K., 2001, Trends, rhythms, and aberrations in global climate 65 Ma to present: Science, v. 292, p. 686–693.

Manuscript Accepted by the Society August 13, 2002

Printed in the USA

Geological Society of America
Special Paper 369
2003

Palynology, paleoenvironments, and organic carbon isotope evolution in lagoonal Paleocene-Eocene boundary settings in North Belgium

E. Steurbaut
Royal Belgian Institute of Natural Sciences, Vautierstraat 29, B-1000 Brussels and KULeuven, Belgium
R. Magioncalda
C. Dupuis
Faculté Polytechnique de Mons, rue de Houdain 9, B-7000 Mons, Belgium
S. Van Simaeys
Katholieke Universiteit Leuven, Redingenstraat 16, B-3000 Leuven, Belgium
E. Roche
Royal Museum of Central Africa, B-3080 Tervuren, Belgium
M. Roche
Belgian Geological Survey, Jennerstraat 13, B-1000 Brussels, Belgium

ABSTRACT

Integration of new data in sedimentology, micropaleontology and carbon isotope analysis of the Paleocene-Eocene boundary sequence in northern Belgium enables differentiation of a series of biotic and tectonic events, generally associated with the brief episode of global warming, known as the Initial Eocene Thermal Maximum (IETM, formerly LPTM), and to establish their succession in time. The onset of the carbon isotope excursion (CIE), positioned at the base of the Tienen Formation, is shown to precisely coincide with the onset of a massive *Apectodinium* abundance, and to be coincident or very close to the development of deeply incised paleovalleys. The organic carbon isotope curve recorded in the lagoonal settings of northern Belgium, presents a complex structure, marked by multiple anomalies and, consequently, is not believed to result from a single instantaneous event. The most negative values (<–26‰) are recorded in the *Apectodinium*-dominated interval (>75%) in the lower part of the Tienen Formation. However, the range of the *Apectodinium* acme (>50%) suggests that the IETM probably lasted until the end of the deposition of the Tienen Formation. The kaolinite influxes at its base and within its middle part are proved to be unrelated to the CIE. The integration of the dinoflagellate cyst and the carbon isotope data has led to the establishment of a refined chronostratigraphic framework, allowing the depositional history of the southern North Sea Basin to be deciphered. During pre-IETM and post-IETM times widely uniform marine depositional regimes existed throughout the area because of tectonic quiescence. This strongly contrasted with the IETM interval, which is marked by restricted complex lagoonal conditions and differential subsidence. In spite of the extreme warm climate and the tectonic instability of the area, vegetation was relatively stable across the IETM and marked by low extinction and first-occurrence rates.

Steurbaut, E., Magioncalda, R., Dupuis, C., Van Simaeys, S., Roche, E., and Roche, M., 2003, Palynology, paleoenvironments, and organic carbon isotope evolution in lagoonal Paleocene-Eocene boundary settings in North Belgium, *in* Wing, S.L., Gingerich, P.D., Schmitz, B., and Thomas, E., eds., Causes and Consequences of Globally Warm Climates in the Early Paleogene: Boulder, Colorado, Geological Society of America Special Paper 369, p. 291–317.

INTRODUCTION

Since the work of Rea et al. (1990) and its continuation in International Geological Correlation Program Project 308 (Paleocene-Eocene [P-E] Boundary Events in Space and Time) it has become clear that the Paleocene-Eocene transition represents a critical period in earth history. The gradual warming trend that started in the late Paleocene and persisted into the early Eocene was interrupted by a short episode of greenhouse gas induced global warming, referred to as the Initial Eocene Thermal Maximum (IETM of Schmitz, 2000; also known as LPTM, Zachos et al., 1993). Several dramatic biotic and non-biotic changes are known to have occurred worldwide within this short time span of <0.1 m.y., among which are a land mammal turnover, a massive deep water benthic foraminiferal extinction, a major generic and family turnover among the calcareous nannoplankton, an anomalous negative shift in the global carbon isotope ($\delta^{13}C$) record, and a major influx of kaolinite (see Stott et al., 1996; Thomas and Shackleton, 1996; Aubry, 1998; and Aubry et al., 1998, for overviews and references). In order to unravel the temporal and spatial relationships of these P-E boundary events and understand the causes and consequences, numerous integrated high-resolution studies have been carried out worldwide recently (~90 short papers in Schmitz et al., 2000; ~100 communications during the Powell meeting, see abstracts volume edited by Ash and Wing, 2001).

The identification of the boundary events in northwest Europe is crucial, as historically, this area should be regarded as the standard for global Paleogene chronostratigraphy. However, recently obtained results from the southern North Sea Basin (Knox, 1996; Steurbaut, 1998) document that its late Paleocene and early Eocene depositional history is characterized by many periods of nondeposition and erosion, and is thus incomplete, implying that this area is inappropriate for global stratigraphic assignments (e.g., for the definition of Global Stratotype Sections and Points [GSSP]). However, its proximal position ensures that eustatically or tectonically driven changes, even the smallest, are documented in the sedimentological record, and thus are detectable and interpretable.

The present paper is intended to unravel the temporal relationships between the above-mentioned major events of global significance, including the $\delta^{13}C$ excursion and the mammal turnover, and the changes in sedimentological and palynological signature as recorded in Belgium, in a confined, predominantly lagoonal depocenter. The Kallo borehole (BGD 27E-148, map-sheet 15/2; x = 144.860, y = 217.840) and the Doel 1b borehole located 7.5 km northwest (map-sheet 7/5–6; x = 142.228, y = 224.452) are chosen as reference sections for the identification of these P-E boundary events (Figs. 1 and 2) because of the relative completeness of the P-E boundary interval, of its remarkable thickness (26 m), its richness in organic material and the almost continuous occurrence of well-preserved palynomorphs. Comparison of both sections will also lead to a better understanding of the lateral facies changes, the significance of the lithological trends and the omission surfaces and to clarify the differences in the $\delta^{13}C$ isotope record.

THE UPPER PALEOCENE–LOWER EOCENE SUCCESSION IN BELGIUM

In Belgium the P-E boundary has traditionally been placed at the base of the Ieper Clay (base of the Ypresian Stage), recently redefined as base of the Mont-Héribu Clay Member (Steurbaut, 1998; point 1 in Figure 3). In terms of dinoflagellate cyst biostratigraphy this boundary corresponds to or is very close to the first occurrence of representatives of the genus *Wetzeliella.* The underlying essentially fluvio-lagoonal succession, traditionally known as the Upper Landen Formation or "Upper (Late) Landenian" (Gulinck, 1973) and recently redefined as the Tienen Formation (Steurbaut, 1998), was completely included in the Paleocene. These traditional assignments were commented on by Steurbaut (1993; rediscussed in Steurbaut, 1998), who suggested that the P-E boundary should be lowered, and placed at the base of the locally preserved Zoute Silt Member (point 2 in Figure 3), or even lower, at the base of the underlying Tienen Formation (point 3 in Figure 3). Up to now the Zoute Silt Member has only been identified in the Knokke borehole in the extreme northwest of Belgium, occurring between the underlying Tienen Formation and the overlying Mont-Héribu Clay Member (Steurbaut, 1998). Recent investigations (Berggren and Aubry, 1996; Dupuis et al., 1998; Steurbaut et al., 1999; Steurbaut et al., 2000) have shown that the above-mentioned major events of global significance, which may be useful for the definition of the P-E boundary, apparently occur at the start of or within the time-span corresponding to the deposition of the Tienen Formation. The latter, which is widely distributed in subcrop in northern Belgium (Fig. 2), consists of a heterogeneous predominantly lagoonal succession. At Kallo and Doel, it rests unconformably on glauconitic fine sands of the Thanetian Hannut Formation (lower part of the Grandglise Sand Member; upper part of NP8, see Steurbaut, 1998; see also Figure 3), in both separated by a hiatus of ~0.8 m.y. (age estimates based on Berggren and Aubry, 1996, Figure 12).

THE P-E BOUNDARY DEFINITION

During the Paris meeting in September 1999 the members of the Working Group on the Paleocene-Eocene Boundary, headed by its chairman M.-P. Aubry, discussed several options concerning location of the P-E boundary (for an overview of the meeting see Aubry, 2000). It was decided by consensus that the series boundary should be lowered from the base of the Ypresian Stage, as defined by the base of the Mont-Héribu Clay Member in Belgium, to the carbon isotope excursion level (CIE). In May 2000 the voting members of the ISPS designated the base of the CIE as the criterion for the boundary (Luterbacher et al., 2000; Schmitz, 2000). In September 2002, it was agreed to place the GSSP for the P-E boundary at the base of a clay bed within the

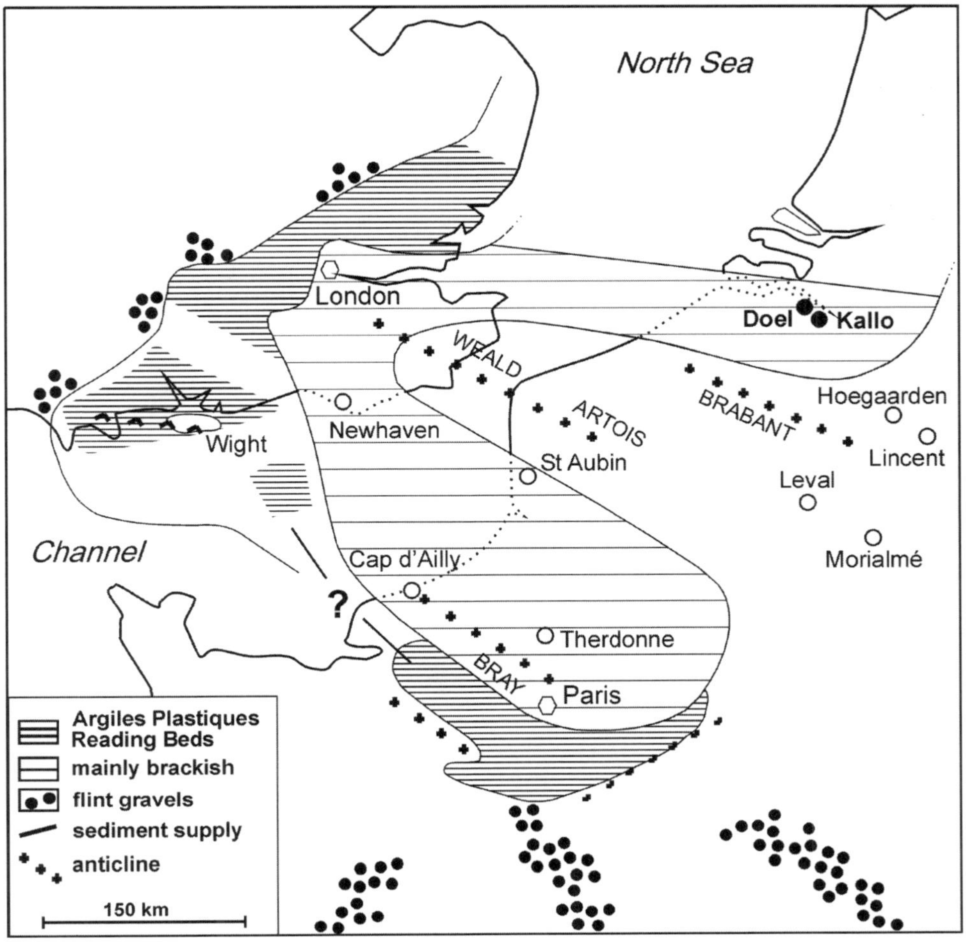

Figure 1. Paleogeography of northwest Europe during the Paleocene-Eocene boundary interval at ~55.5 Ma (after Leriche, 1928; Gulinck, 1973; Dupuis and Thiry, 1998) with locations of the studied Kallo and Doel boreholes.

DBH section at Dababiya (Upper Nile Valley, Egypt). This base is exactly coincident with the onset of the CIE.

MATERIAL AND METHODS

The Kallo borehole, drilled in the late 1960s, has not been stored as a series of consecutive cores, but split up into a series of separate closely spaced samples (at least a 10 cm large sample every 20 cm, sometimes every 10 cm). These cover ~90% of the initially cored interval. The samples studied here are collected from the stratigraphical interval between 404.5 m and 376 m depth. The Doel cores, which have been completely preserved, were sampled between 473 m and 442 m depth.

Mineralogy

The data on the clay mineralogy of the Kallo borehole result from the reinterpretation of analyses carried out by Mercier-Castiaux as part of a general study on the Tertiary clay sedimentation in northwest Europe (Mercier-Castiaux, 1986; Mercier-Castiaux et al., 1988; Mercier-Castiaux and Dupuis, 1991). It is based on 27 analyses, of which eight are new (Table DR1[1]). These, together with the 34 new analyses from the Doel borehole (Table DR2, see footnote 1) were all performed according to standard techniques. Carbonate content was measured with a Bernard calcimeter using standard procedures (Tables DR3 and DR4, see footnote 1). Most of the Doel samples have been directly treated with HCl, without calcite measurements. Analytical precision is ~5%. Determination of total organic carbon content (TOC) and of sulfur was performed with a standard LECO carbon and sulfur analyzer on 42 samples from Kallo and 60 samples from Doel. Results are listed in Tables DR3 and DR4 (see footnote 1). Mineralogical analyses were accomplished at the "Faculté Polytechnique de Mons" (Belgium).

Carbon isotopic measurements

The series of samples processed for determination of the TOC were used for carbon isotope analysis. Rock samples of ~20 g were dried and examined for purity. The cleaned samples

[1]GSA Data Repository item 2003050, Tables DR1–DR7, is available on request from Documents Secretary, GSA, P.O. Box 9140, Boulder, CO 80301-9140, USA, editing@geosociety.org, or at www.geosociety.org/pubs/ft2003.htm.

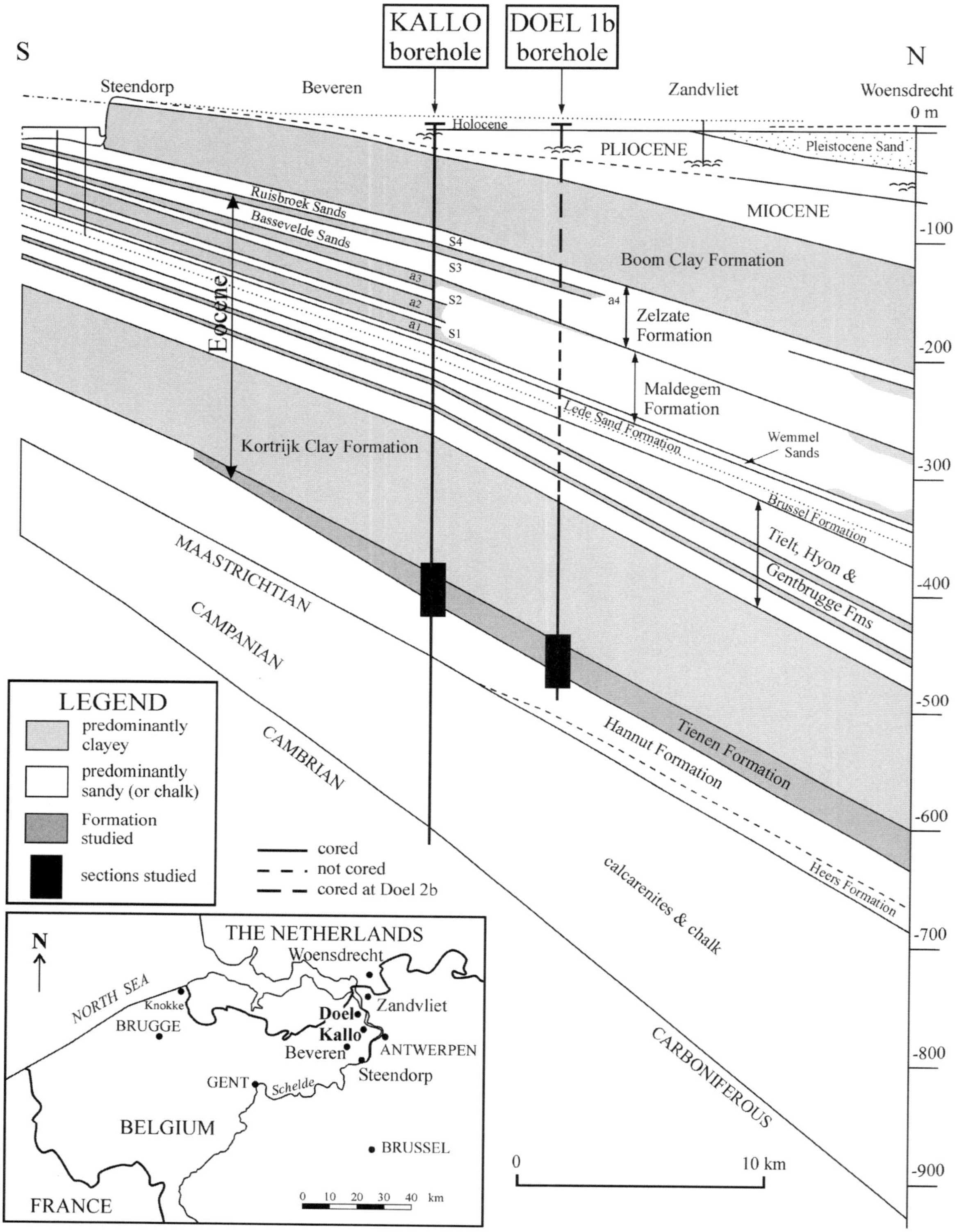

Figure 2. Geological cross section through the post-Paleozoic of northern Belgium with locations of the studied sections (modified after Gulinck, 1969).

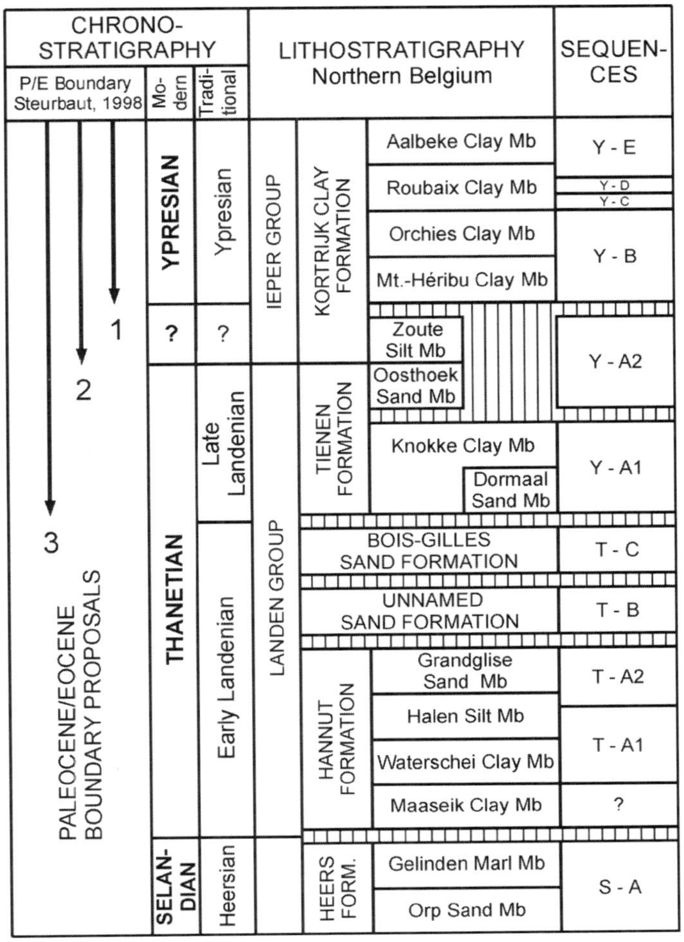

Figure 3. Lithostratigraphy of the Paleocene-Eocene boundary interval in northern Belgium and chronostratigraphic interpretations (after Steurbaut, 1998).

were powdered and treated with 10% HCl to remove carbonates. The soluble salts were removed by centrifuging. The residue was dried and powdered again. A small pellet of a few milligrams of each sample was fit into a small tin blank. The quantities of the material analyzed were determined on the basis of the TOC values. Sample quantities needed for accurate measurements vary between 0.05 mg (lignites) and 50 mg (sands). The isotopic measurements were carried out with a mass spectrometer with continuous flux Δ+ (Finnigan Mat), attached to a Carlo-Erba NA-1500 microanalyzer, at the "Laboratoire des Sciences du Climat et de l'Environnement du C.N.R.S." at Gif-sur-Yvette (France). Carbon dioxide for the mass spectrometric analysis was generated by O_2 and He flash-combustion. As CO_2 was recovered for mass spectrometry, signals corresponding to the masses 44, 45 and 46 were measured and the values of the carbon isotopic composition were calculated on the basis of these measurements. The $\delta^{13}C$ values listed in Tables DR3 and DR4 (see footnote 1) are reported relative to the international PDB standard. The precision of the measurement is <0.1‰, the reproducibility ~0.15‰.

Calcareous nannofossil, palynomorph, and palynofacies analyses

Closely spaced samples from the Grandglise Sand and from the Tienen Formation were processed for calcareous nannofossil investigations. Smear slides were made following standard procedures and examined with the light microscope at 1000 magnification. Each sample from the Grandglise Sand Member was also analyzed on a semiquantitative basis (magnification 1000 ×), based on the method described by Backman and Shackleton (1983). The samples from the Tienen Formation were not quantitatively analyzed because of the absence of in situ material (100% reworking).

Palynofacies and spores and pollen analyses were carried out only on the Kallo borehole samples. Thirty powdered samples of 30–40 g were processed at the Liège University according to standard laboratory procedure: Dissolution of carbonates and silicates through successive HCl and HF treatments, rinsing and filtration at 12 μ with a millipore system. The relative abundance of the kerogen constituents and the diversity of the aquatic palynomorphs were visually estimated (Fig. 5). Pollen, spores and aquatic palynomorphs (mainly dinoflagellate cysts) were separately counted (~300 palynomorph specimens per slide) in these and some extra samples, all together 53, and their percentages were calculated (Table DR5, see footnote 1). Some of these slides (K402, K398.5 and K389.15) were used for dinoflagellate cyst investigation (data are not quantitatively interpretable because of differences in preparation techniques, see Table DR6 [see footnote 1]). The dinoflagellate data from Kallo 377.35 m, 376.5 m and 376 m were taken from De Coninck (1975). Eleven extra samples from the Kallo borehole and 12 samples from the Doel borehole were processed using standard acid maceration techniques for dinoflagellate cyst analysis as described by Heilmann-Clausen (1985). The number of aquatic palynomorphs (mainly dinoflagellate cysts, but also acritarchs, green algae, Prasinophytes, etc.) counted generally exceeds 200 specimens per sample (Table DR7, see footnote 1). Slides and specimens are kept in the different institutions concerned. Details of the figured items are given in the explanation of the plates.

THE KALLO BOREHOLE

Sedimentology

The present study focuses on the stratigraphical interval between 404.5 m and 376 m depth, including in ascending order the Grandglise Sand Member, the entire Tienen Formation and the base of the Mont-Héribu Clay Member. The Tienen Formation is bounded by major omission surfaces at respectively 402.3 m and 377.35 m depth, representing major breaks in sedimentation, and corresponding to hiatuses of several 100 k.y. (age esti-

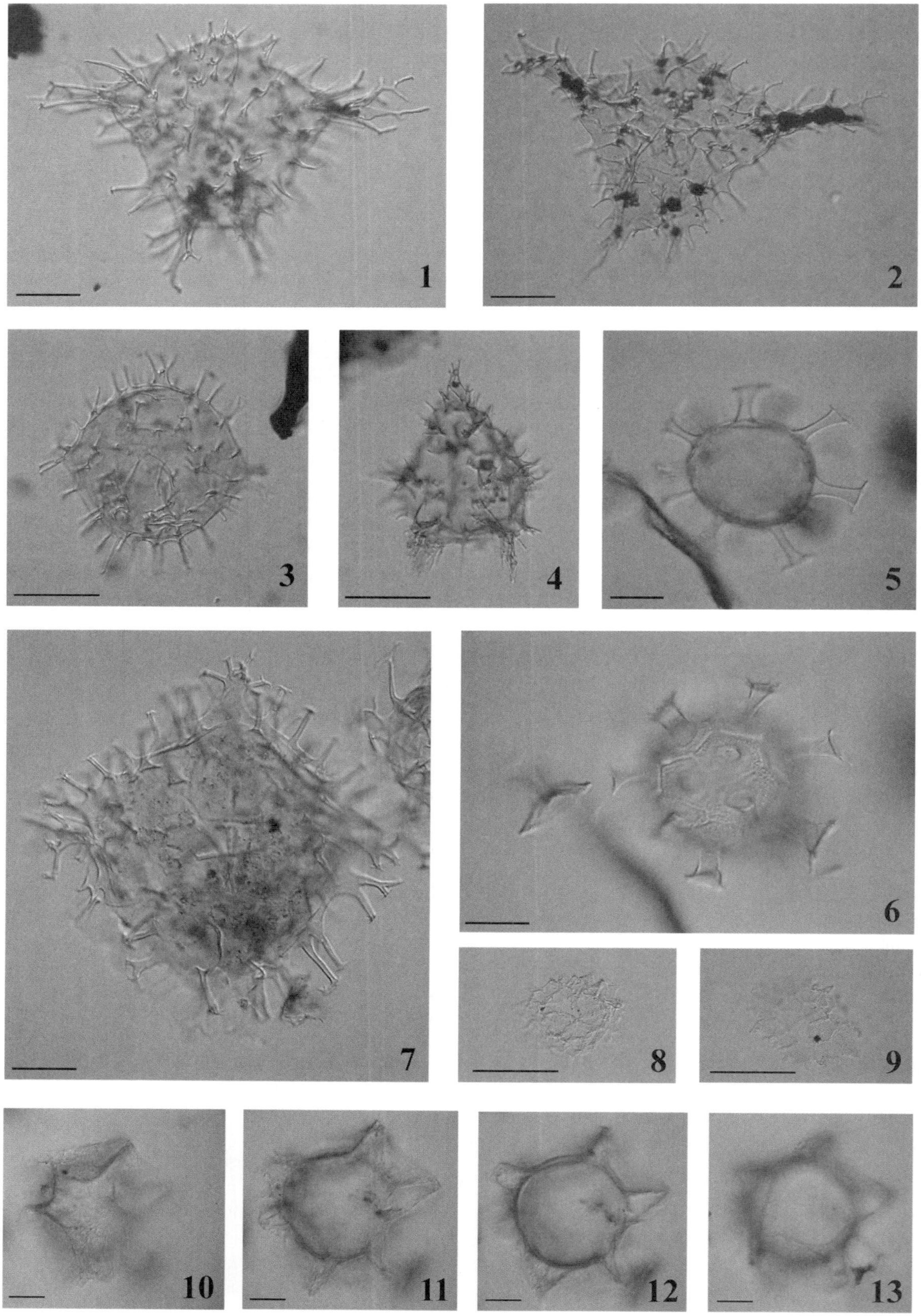
1
2
3
4
5
6
7
8
9
10
11
12
13

mates based on Berggren and Aubry, 1996; Bains et al., 2000). The underlying unit, representing the lower part of the Grandglise Sand Member (Hannut Formation), is made up of brownish green glauconitic bioturbated and decalcified very fine sand to sandy silt. It indicates fully marine environments below fair weather wave base. The base of the overlying Mont Héribu Clay consists of ~90 cm decalcified glauconitic clayey very fine sand, representing a transgressive deposit. The transition toward the overlying clays is sharp and indicates an abrupt deepening of the depositional environment.

Lithofacies analysis of the Tienen Formation. Eleven lithofacies are recognized within the Tienen Formation. They grade from stiff clays to fine sands, reflecting major changes in depositional hydrodynamics (Fig. 4).

Facies 1: Interbedded clay-silt. The beds are generally 0.5 cm thick, and often show wavy surfaces. This facies is generally found in thin intervals (from 402.2 m to 401.8 m; 400.70 m to 399.80 m; ~388.8 m), alternating with the laminated clayey facies 2. The silty beds in the lowermost interval contain reworked calcareous nannofossils. In the upper interval they consist of fine clastic mineral components. This facies represents lagoonal conditions with rapidly changing wave intensities and bottom turbulence, allowing episodic decantation of the silt fraction.

Facies 2: Dark grey laminated clay. The density of the silt laminae strongly varies throughout this facies. The surfaces between the laminae often show a wavy pattern. The silt laminae contain reworked calcareous nannofossils. Small in situ bivalves occur at certain levels. This facies predominates in the lower part of the Tienen Formation (401.8 m to 397.5 m depth). It is believed to represent shallow lagoonal low-energy environments, protected from high wave activity and strong currents.

Facies 3: Grey to purple sandy clay and silt. These partly leached clays and silts are found in two thin intervals (389.8–389.2 m and 388.6 m) below the lignitic horizons. They are penetrated by numerous tiny rootlets. This facies is related to soil development.

Facies 4: Pale grey calcareous stiff clay to clayey marl. This compacted clay contains abundant amorphous minute calcite particles of unknown origin (calcareous nannofossils are absent). It is furthermore marked by high percentages of the freshwater algae *Pediastrum* spp. and *Botryococcus* spp. (respectively 57% and 17.1% of the aquatic palynomorphs in sample 385.6 m), and relatively few dinoflagellate cysts (25.3%, almost exclusively *Apectodinium*). This facies is only known from a thin interval (385.6–385.2 m). It represents calm waters with high freshwater influx in a lagoonal setting, close to a river mouth.

Facies 5: Blackish laminated clay. The laminae are made up of dark organic-rich silt or very fine woody debris. This facies in found within the interval 389.1 m to 388.3 m. The organic material seems to be the result of disintegration of plant debris from a nearby source. Aquatic palynomorphs (dinoflagellates, *Botryococcus*) are generally present, although in low numbers (generally lower than 10%). This refers to swampy environments.

Facies 6: Lignitic horizons. Four levels made up of woody debris have been recognized in the interval 389.0–388.0 m. Two of these levels are devoid of aquatic palynomorphs (388.2 m and 388.6 m) reflecting subaerial conditions, plant colonization and the development of peat bogs. The others are marked by low numbers of aquatic palynomorphs (dinoflagellates ~7% of the palynomorphs), indicating aqueous, probably very low-salinity environments. The lignitic horizons are separated by sediments of facies 5, pointing to swamps and Everglades-like paleoenvironments.

Facies 7: Dark bioturbated sandy clay. The grey fine sand, sometimes containing shelly material (oyster fragments, *Cyrena*), is bioturbated in the clay matrix. Facies 7 is known to occur in a single interval between 381.3 m and 380.4 m. A rapidly changing environment is suggested, with high-energy pulses injecting coarser material into the lagoonal depocenter during storms, followed by a return to quiet sedimentation conditions and bioturbation. These sandy layers are interpreted as washover fans.

Facies 8: Interbedded dark clays with gastropod or bivalve beds. This sediment type only occurs at the top of the Tienen Formation (378.5–377.5 m). The shell beds are sometimes lithified. The gastropods dominate at the base of this interval (378.5 m). The mollusks occurring at the top are bivalved and concentrated in thin beds (from 378 m to 377.5 m). They present a very low diversity. It reflects in situ burial, without displacement, indicating very restricted wave and current activity. The dinoflagellate association indicates lagoonal environments.

Facies 9: Greyish bioturbated sandy silt. This facies is well represented in the lower part of the Tienen Formation. A lower unit of ~7.5 m thick is identified between 397.5 m and 389.8 m depth, of which the uppermost 3 m are leached and penetrated by tiny rootlets. This facies also predominates in the interval between 387.8 m and 386.2 m depth, just above the lignitic horizons. Bioturbation and homogenizing of the sediment indicates

Plate 1. All figures were made with differential interference contrast. Scale bar represents 20 µ. Slides, specimens, and negatives of micrographs figured on Plate 1 are stored at the "Afdeling Historische Geologie, Redingenstraat 16, B-3000 Leuven." Figures 1–2. *Apectodinium augustum* (Harland, 1979) Lentin and Williams, 1981; 1: Doel 468.7 m, slide 2; 2: Doel 468.7 m, slide 2. Figure 3. *Apectodinium homomorphum* (Deflandre and Cookson, 1955) Lentin and Williams, 1977; Doel 445 m, slide 2. Figure 4. *Apectodinium parvum* (Alberti, 1961) Lentin and Williams, 1977; emend. Harland, 1979; Doel 468.7 m, slide 2. Figures 5–6. *Homotryblium* spp.; 5: Doel 450.8 m, slide 3, cross section; 6: Same specimen, focused on disintegrating archeopyle. Figure 7. *Wetzeliella astra* Costa et al., 1978; Doel 442.49 m, slide 1. Figures 8–9. *Pediastrum* spp.; 8: Doel 444.09 m, slide 1; 9: Doel 445.6 m, slide 2. Figures 10–13. *Muratodinium fimbriatum* (Cookson and Eisenack, 1967) Drugg, 1970; Doel 450.8 m, slide 3, same specimen, from high to low focus.

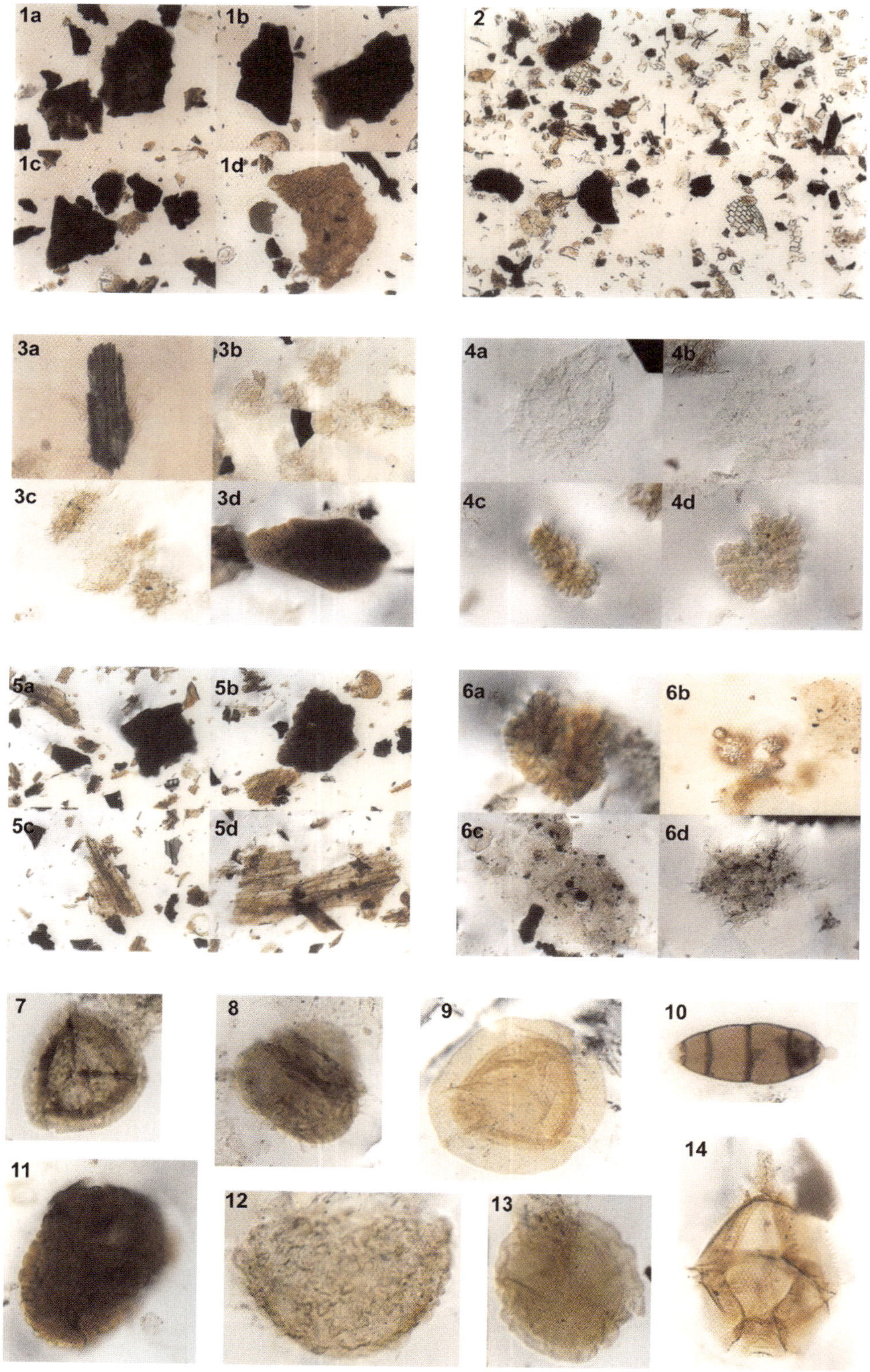
1a
1b
1c
1d
2
3a
3b
3c
3d
4a
4b
4c
4d
5a
5b
5c
5d
6a
6b
6c
6d
7
8
9
10
11
12
13
14

strong biological activity in a quiet aquatic environment. The dinoflagellate association points to lagoonal conditions.

Facies 10: Well-sorted bioturbated shelly fine sand. These coarser-grained beds that contain dispersed shells are only found in the upper part of the Tienen Formation (from 386.2 m upward), where they often alternate with coquina beds (facies 11). They occur as thin levels in the interval 386.2–386.0 m, and higher up as two thick units (385–381.6 m; 380.4–378.4 m). The very fine shell grit (1–2 mm) suggests intense shell abrasion and, thus, high-energy environments. However, wave activity must still have been minor at the deposition site, as neither channel erosion surfaces nor cross bedding have been observed. The sands are interpreted as washover fans, deposited in the inlets of the lagoons.

Facies 11: Lumachelle-coquina, locally cemented. These layers also occur in the upper part of the Tienen Formation (from 386.2 m upward). Some beds are made up of gastropods and oysters (386.0 m, 385.1 m), others by cemented oysters (381.5 m) or shell grit (379.3 m). The oysters seem to indicate an increase in salinity, the abraded shells high-energy environments (storms), both reflecting increased marine influence.

Omission surfaces. Thirteen surfaces, labeled K1 to K13, representing substantial breaks in sedimentation or important facies shifts are recognized (Fig. 4). Surface K1, which separates the Tienen Formation from the underlying Grandglise Sands, was not identified because of the discontinuous nature of the samples. K2 at 400.5 m depth is marked by shallow small burrows. It separates coarser sandy silt containing a substantial reworked calcareous nannofossil fraction from the underlying stiff clays, reflecting the installation of higher-energy environments. This might reflect a temporary landward shift of the lagoon. The burrowed contact at 397.5 m, representing K3, separates laminated clays from overlying bioturbated grey silt. It seems to indicate a renewed but now permanent shift of the depocenter toward higher-energy areas of the lagoon. This might be related to a landward shift of the lagoon because of a small sea-level rise. Surface K4 at 389.9 m announces the start of a clayey deposition regime, with several episodes of lignite development and soil formation, including several additional omission surfaces. The burrowed junction K5 at 387.8 m, separating sandy silt from the underlying lignite, represents a deepening of the depocenter from a temporary subaerial to a fully lagoonal paleoenvironment. It probably results from a small relative sea-level rise. The income of fine sand and oysters at 386.2 m (K6) reflects a substantial increase in salinity and representing a next step in a continuing sea-level rise. K7 at 385.6 m occurs at the base of calcareous stiff clay, rich in freshwater algae, and of restricted lagoonal origin. K8 at 385.1 m forms the base of an oyster-rich coquina, indicating higher salinities and a new sea-level rise. The bioturbated quartz sand occurring at 385.0 m is overlain by silts and oyster-rich sandstone. K9 at 381.4 m separates bioturbated sandy clay from the underlying oyster-rich sandstone bed. It indicates a return to quiet depositional conditions with accumulation of large quantities of organic material episodically interrupted by short influxes of sandy material, brought in by high-energy agents. K10 at 380.3 m marks the reintroduction of fine sand and oyster beds, reflecting high-energy conditions and relatively high salinities. K11 at 378.5 m separates shelly sand and sandstone from overlying dark shelly clays. It announces the return to restricted lagoonal environments. K12 at 377.35 m is marked by frequent burrows of glauconitic very fine sands in underlying black clays. It reflects a major landward facies shift, from lagoonal to shallow marine paleoenvironments, due to a major sea-level rise. Surface K13 at 376.20 m is sharp and marks the onset of a clayey deposition regime of marine origin. It represents a further deepening of the depocenter because of an abrupt and major sea-level rise.

Plate 2. Slides, specimens, and micrographs figured on Plate 2 can be consulted at the "Services associés de Paléontologie de l'Université de Liège—Paléobotanique et Paléopalynologie, Allée du 6 Août, Sart Tilman, B18, B-4000, Liège." Figure 1. Lignitic debris, sample Kallo 389.15 m: (a–c) large ± equidimensional black-brown coaly debris (up to 100 µ; slide 45580); (d) resinous debris showing ghostly cellular structure (~120 µ; slide 45580). Figure 2. Mixed continental (black-brown coaly debris, "fresh" phytoclasts, cuticles, resinous fragments, miospores and fungal remains)—aquatic (dinoflagellate cysts) palynofacies (largest coaly debris ~200 µ; slide 45683); sample Kallo 386.3 m. Figures 3–4. Palynofacies composition of sample Kallo 385.2 m. (3a) lath-shaped coaly fragment with well-preserved tracheidal and cellular structures (pyrofusinite sensu stricto) (100 µ, slide 45675); (3b–3c) dinoflagellate cyst assemblage dominated by *Apectodinium* spp. (slide 45675); (3d) resinous body (50 µ, slide 45675); 4a–4b: *Pediastrum* spp. (4a) 70 µ, slide 45675 EF N32; (4b) 80 µ, slide 45675 EF K35); (4c–4d) *Botryococcus* spp. (4c: 50 µ, slide 45675 EF S38; 4d: 50 µ, slide 45675 EF U44). Figure 5. Lignitic debris in sample Kallo 389.5 m. (5a, 5b, 5c, and 5d) 100% continental mixed palynofacies with large dark-brown coaly debris together with "fresh" phytoclasts (slide 45601; 5a–5b, coaly debris ~200 µ; 5c-5d, Phytoclasts length ~100 µ). Figure 6. Palynofacies composition of sample Kallo 378.5 m. (6a) *Botryococcus* spp. (50 µ, slide 45641 EF P36); (6b) framboïdal pyrite with some amorphous organic matter (viewed both in transmitted and reflected light; slide 45640); (6c) pyritous amorphous organic matter (slide 45640); (6d) A specimen of *Apectodinium* sp. embedded in amorphous organic matter (45 µ, slide 45641 EF C37). Figure 7. *Lycospora pusilla* (Ibrahim) Schopf, Wilson and Bentall, 1944 sensu Somers, 1971 (35 µ; sample Kallo 402 m; reworked; slide 45687 EF U37). Figure 8. *Heliosporites reissingeri* (Harris) Muir and Van Konijnenburg—Van Cittert, 1970 (55 µ; sample Kallo 402 m; reworked; slide 45687 EF S49). Figure 9. *Callialasporites turbatus* (Balme) Schulz, 1967 (70 µ; sample Kallo 402 m; reworked; slide 45687 EF G37). Figure 10. Fungal spore (40 µ; sample Kallo 402 m; slide 45687 EF U42). Figure 11. *Ricciisporites tuberculatus* (Lundblad) Lundblad, 1959 (tetrade; 85 µ; sample Kallo 402 m; reworked; slide 45687 EF M40). Figure 12. *Cerebropollenites macroverrucosus* (Thiergart) Schulz, 1967 (50 µ; Kallo 402 m; reworked; slide 45687 EF P35). Figure 13. *Callialasporites dampieri* (Balme) Dev, 1961 (45 µ; sample Kallo 402 m; reworked; slide 45687 EF V39). Figure 14. *Gonyaulacysta jurassica* Deflandre, 1938 (60 µ; sample Kallo 402 m; reworked; slide 45687 EF J39).

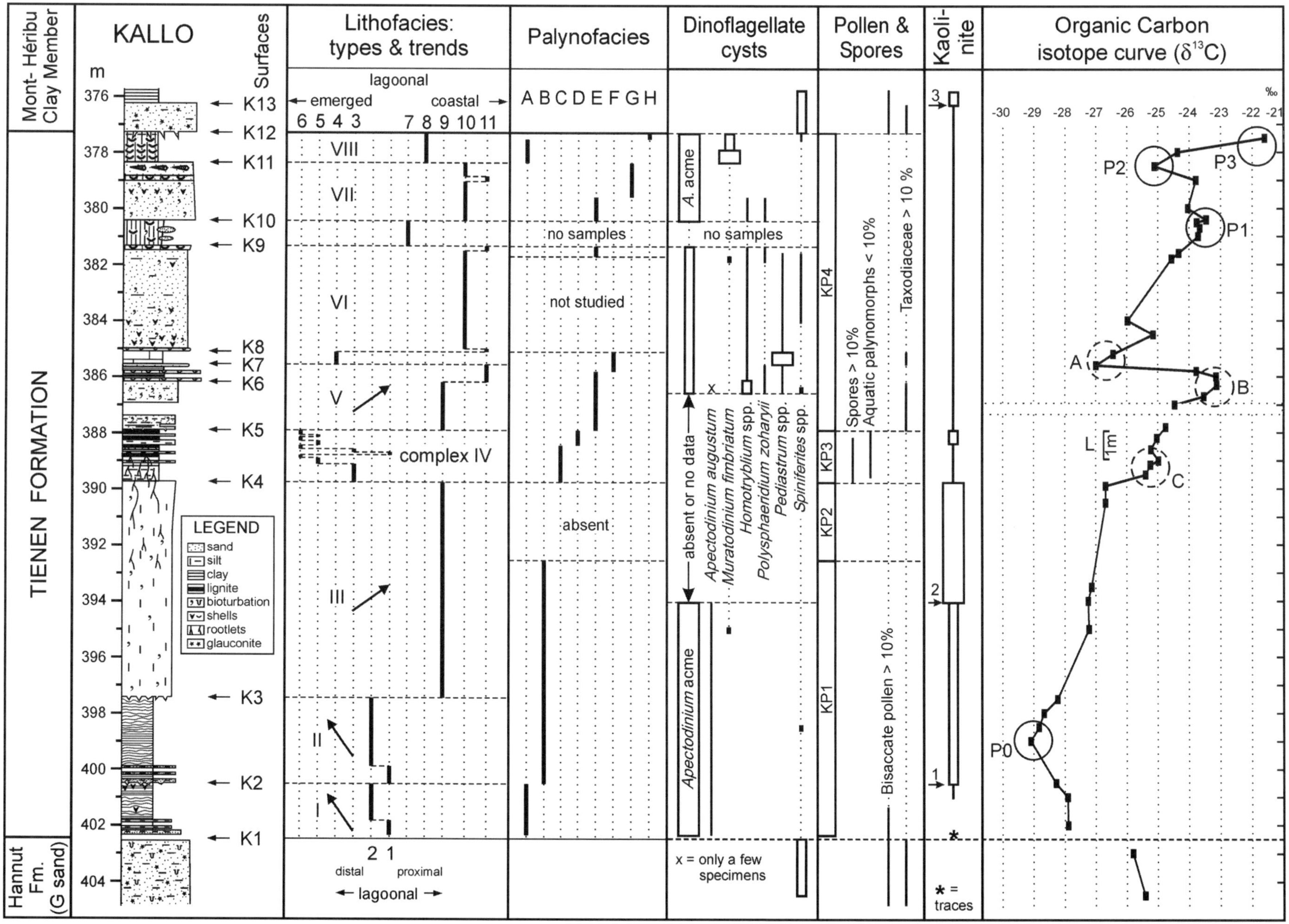

Figure 4. Lithofacies analysis, micropaleontology, and organic carbon isotope curve of the Paleocene-Eocene boundary interval in the Kallo borehole.

Clay mineralogy

The clay fraction (<2 μ), of which the distribution is reported in Table DR1 (see footnote 1), includes chlorite (C), illite (I), kaolinite (K) and an irregular interstratified illite-smectite (S-I). Chlorite is generally weakly represented (<5%), except for the topmost sample discussed (376 m) where it reaches 10%. The proportion of illite varies throughout the section, but never falls below 7%. It sometimes exceeds 40%. There is substantial decrease in illite content between 393.5 m and 393 m depth (from 23% to 13%). Kaolinite generally ranges between 5% and 15%. The highest values (>18%) are found in the interval 394–390 m, with a maximum peak of 36% at 393 m depth. The interstratified illite-smectite dominates the clayey fraction, although in variable proportions. The highest values are restricted to the lowermost (76% < x < 88% in 403.5–401 m) and the uppermost intervals studied (>70% in 386.5–377.5 m). There is a major decrease from 76% to 53% between 401 m and 400 m depth, and a major increase from 40% to 79% between 388 m and 386.5 m depth.

Calcareous nannofossils

A moderately diverse, but poorly preserved association, including *Heliolithus riedelii, Prinsius bisulcus, Toweius eminens, T. tovae* and *Hornibrookina arca,* was identified at 404 m depth in the lower part of the Grandglise Sands. The cooccurrence of these taxa indicates the upper part of nannofossil zone NP8 (codified as NP8b in Steurbaut, 1998) and to a mid-Thanetian age (~57.1 Ma). The associations found in the lower part of the Tienen Formation, between 401 m and 399.8 m depth are quantitatively richer, but entirely reworked. Their poor quality does not permit an exact biozonal attribution. However, the presence of *Ahmuellerella octoradiata, Eiffellithus eximius, E. turriseiffelii* s.str., in association with the absence of *Aspidolithus parcus* and *Arkhangelskiella* spp., suggests, according to species ranges in Perch-Nielsen (1985), an Early Late Cretaceous (probably pre-Campanian) age.

Palynofacies

The results of the semiquantitative palynofacies analysis of the Paleocene-Eocene boundary interval are outlined in Figure 5. The kerogen components discussed here are grouped into four major categories: (I) humic organic particles, essentially terrestrial plant debris, including black-brown lignite debris (BBD), lath-shaped opaque debris (pyrofusinites), fresh phytoclasts, cuticular matter and resinous matter, (II) nonaquatic palynomorphs, split up into miospores and fungal remains, (III) aquatic palynomorphs, of which some are of freshwater origin (*Pediastrum* spp., *Botryococcus* spp.) and others essentially marine (dinoflagellate cysts, acritarchs and foraminiferal inner tests or test-linings) and (IV) amorphous organic matter (AOM). Some of the palynomorphs have a pre-Cenozoic origin and are reworked.

The palynofacies is dominated throughout the interval by black-brown lignite debris (generally >30%, up to 75% in sample 388.60 m). Aquatic palynomorphs (strongly fluctuating percentages, in 1/3 of samples >30%) and miospores (generally between 10% and 20%) are well represented, although less consistently. The relative abundances of the different constituents of the organic matter allow the identification of several palynofacies types in the Tienen Formation. Its lowermost part (from 402 m to 393 m) presents a constantly mixed palynofacies composition: Mixture of comparable quantities of humic particles, miospores and aquatic palynomorphs. This palynofacies is pyritous and contains very little amorphous organic matter, cuticular matter and fresh phytoclasts. It can be split into a lower palynofacies A with relatively low frequencies of aquatic palynomorphs (<20% up to 400.5 m), and a palynofacies B with abundant aquatic palynomorphs (~40%). The Grandglise Sands and the base of the Mont-Héribu Clay also present a mixed palynofacies composition type A.

The interval from 392.5 m to 390.0 m is devoid of kerogen particles, because of complete oxidation during paleosol formation. The overlying interval (up to 388.5 m) presents very low values of aquatic palynomorphs, but a considerable amount (~5%) of fresh phytoclasts. Black-brown lignite debris is dominating in this palynofacies C (generally >50%; at 389.15 m and 388.60 m >90%). In the interval from 388.2 m to 388.0 m the composition of the organic matter is quite different (palynofacies D), because of the paucity of phytoclasts and the relatively high values of AOM (~11% at 388.0 m). Palynofacies E from 387.8 m to 386.0 m is marked by an increase in aquatic palynomorphs (~15%), high frequency of pyrofusinite (generally >5%, up to 20% at 387.5 m), referring to incomplete combustion of wood during natural fires, considerable amounts of phytoclasts (>5%) and cuticular material (~5%), but almost no amorphous material. This palynofacies also occurs in interval 381.8 m to 379.6 m, although with less high pyrofusinite values (2%–3%). Aquatic palynomorphs become dominant in the overlying 0.7 m thick interval (>60% of the palynomorphs in 385.8 m to 385.0 m), made up in the topmost 0.5 m of abundant *Pediastrum* spp. (57% of the aquatic palynomorphs). It characterizes palynofacies F. A comparable palynofacies G is encountered in samples 379 m and 378.5 m, although without *Pediastrum* spp. Sample 378.0 m contains a type A palynofacies. The top of the Tienen Formation (sample 377.5 m) is extremely rich in AOM (34%) and contains a relatively high amount of aquatic palynomorphs (25%). It defines palynofacies type H. The most common palynofacies types are illustrated in Plate 2.

Reworked Carboniferous miospores, Late Triassic miospores, Early to Middle Jurassic miospores and dinoflagellate cysts and Early Late Cretaceous calcareous nannofossils occur within the same levels at the base of the Tienen Formation (Figs. 5 and 9). The cooccurrence of these palynomorphs and nannofossils suggests that there might have been different sediment source areas and several phases of reworking. However, as Upper Triassic, Jurassic and Upper Cretaceous deposits are widely

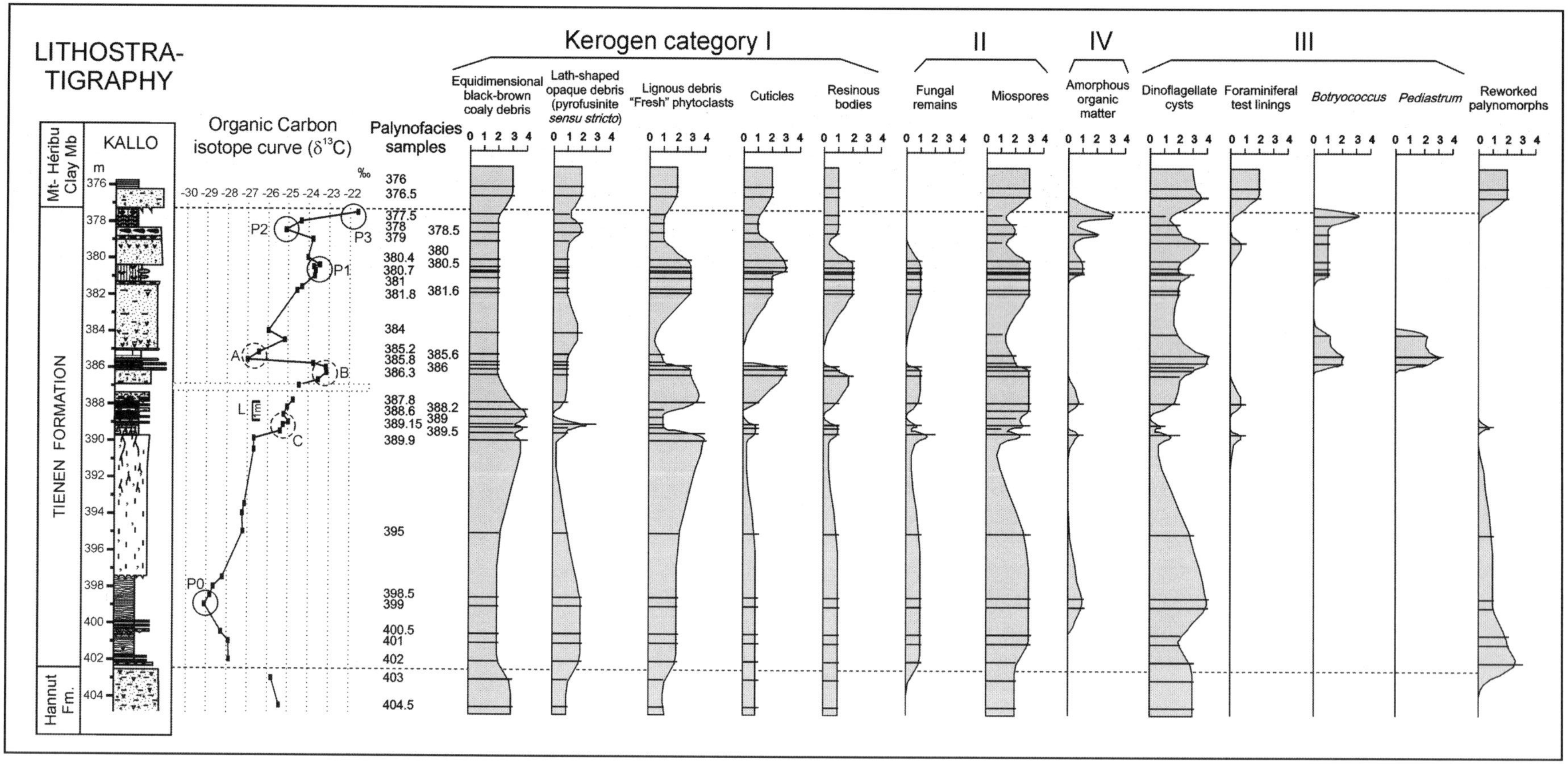

Figure 5. Semiquantitative palynofacies data of the Paleocene-Eocene boundary interval in the Kallo borehole (0—not observed, 1—present, 2—common, 3—abundant, 4—dominant).

distributed in Central England, and Jurassic palynomorphs were found in the Woolwich Beds in Kent (Gruas-Cavagnetto, 1970), in deposits that are coeval with the Tienen Formation, it seems that Central England, and probably the Gloucester district, was the main active sediment source area during the time of deposition of the lower part of the Tienen Formation. This is also true for the base of the Mont-Héribu Clay (Fig. 5), which is marked by a similar influx of reworked miospores, although without nannofossils. Southerly source areas are practically excluded in both cases in view of the fact that in the "Sparnacian" facies of the Paris Basin and northern France the reworked palynomorph fraction exclusively consists of Cretaceous forms (Gruas-Cavagnetto, 1968). Occurrences of Paleozoic and Mesozoic palynomorphs into Paleogene sediments have been discussed by Streel and Bless (1980).

Pollen and spores

The percentage distribution of pollen, spores and aquatic palynomorphs (essentially dinoflagellate cysts), including the individual percentages of several distinct pollen and spores groups, are given in Table DR5 (see footnote 1). The latter are expressed as fractions of the total amount of pollen and spores. The different groups discussed are bisaccate pollen, Palmae including *Monocolpopollenites parareolatus* and *M. tranquillus,* Myriacaceae including *Triatriopollenites aroboratus, T. belgicus, T. roboratus* and *T. rurensis,* Taxodiaceae including *Inaperturopollenites dubius* and *I. hiatus,* Restionaceae represented by *Milfordia hungarica,* Normapolles (*Basopollis atumescens* and *B. basilis, Nudopollis endangulatus, N. terminalis* and *N. thiergartii*), Juglandaceae including *Triatriopollenites engelhardtioides* and *T. platycaryoides,* Tiliaceae including *Intratripollenites microinstructus, I. microreticulatus* and *I. pseudinstructus,* Caryas represented by *Caryapollenites triangulus* and, finally, *Subtriporopollenites annulatus, S. constans, S. magnoporatus, S. spissoexinus* and *S. subporatus,* which are included in the taxon "Caryas ?."

The associations from the marine Grandglise Sands (samples 404 m and 402.5 m) substantially differ from these from the overlying Tienen Formation by the relative high percentages of aquatic palynomorphs (55% to 60% of the totality of the palynomorphs), by the high frequency of bisaccate pollen (17% to 21% of the totality of pollen and spores) and of Taxodiaceae (12% to 14%) and the low values of Juglandaceae (<4.5%), Restionaceae (<2.5%) and Myriacaceae pollen (<1%). The three last groups are much more abundant (respectively 11.7%, 10%, 6.5%) in the lowermost sample of the Tienen Formation (402 m), while Taxodiaceae are less frequent (6.5%). The high values of aquatic palynomorphs (63%) compared to these of the overlying associations might by due to reworking.

The palynomorph associations are quite similar throughout the interval from 402 m to 393 m, defining a distinct vegetation subzone KP1, with high frequencies of small pollen (<25 μ generally above 18%), Juglandaceae (>7%) and Palmae (>5% with a maximum of 11.6% at 393 m). There are several trends within KP1: A major decrease in bisaccate pollen (from 14.7% to 1.6%), a substantial increase in small pollen (from ~19% to 40%) and in aquatic palynomorphs (from 42% at 401.5 m to 88% at 393 m). The lowermost part up to 401 m, labeled KP1a, slightly differs from the upper KP1b by the lower values of aquatic palynomorphs (~44%) and higher values of spores (~8%) and Myriacaceae (9.9% at 402 m).

No palynomorphs have been encountered in the overlying samples up to 389.9 m (KP2), because of intense leaching of the sediments as the result of pedological processes. The overlying interval KP3 (from 389.5 m to 388 m) is marked by a major decrease in aquatic palynomorphs (generally below 14%) and a major increase in spores (generally above 10%) and in Myriacaceae, especially in its lower part (~7%). Important floral changes again occur at 387.8 m, including an increase in aquatic palynomorphs to moderate values (between 25% and 55%) and in Taxodiaceae (>10%). From this level up to the top of the Tienen Formation (387.8 m to 377.5 m) the associations, constituting unit KP4, are characterized by a moderately rich aquatic palynomorph fraction (generally between 20% and 35%). It can be subdivided into 5 subunits. KP4b (385.6 m to 385.2 m) differs from the lowermost KP4a subzone (387.8 m to 385.8 m) by the abundance of aquatic palynomorphs (~70% versus 20%–40%) and the high Taxodiaceae (up to 20%) and Palmae (up to 13%) contents. The assemblages from subzone KP 4c (382.0 m to 380.0 m) are somewhat similar to these of KP4a, although Juglandaceae and Tiliaceae are slightly frequent in the former. KP4d (379.0 m to 378.0 m) is marked by an increase in bisaccate pollen (from <3% to >7%) and in Restionaceae. Aquatic palynomorphs are increasing again in the uppermost sample 377.5 m, representing KP4e, whereas the numbers of bisaccate and Taxodiaceae pollen are decreasing.

The assemblages from the overlying Mont-Héribu Clay Member are globally similar to these of the Grandglise Sand Member, with comparable fractions of pollen, spores and aquatic forms and relatively high percentages of bisaccate pollen and Taxodiaceae. However, those from the Mont-Héribu Clay are clearly richer in Juglandaceae, Tiliaceae and Caryas.

Dinoflagellate cysts

The associations from the topmost part of the analyzed interval (377.35 m to 376 m depth) were studied in detail by De Coninck (1975, 1999b) as part of a general work on the Ypresian dinoflagellate cysts of northern Belgium. The underlying Tienen Formation has not been systematically investigated. A few levels were commented on by De Coninck (1975, 1999a, 1999b) and Steurbaut et al. (2000). All the previously investigated samples have been reviewed in the present paper (Table DR6, see footnote 1).

The lowermost sample, located 1 m below the base of the Tienen Formation (403.5 m) is dominated by *Areoligera senonensis* (66.2%) and *Spiniferites* spp. (23.8%), indicating fully marine deposition conditions. The presence of *Alisocysta margarita* (1.9%) and the abundance of *Areoligera* indicate subzone Ama 4 of Steurbaut et al. (personal commun.), which is dated as middle Biochron NP8 (NP8b of Steurbaut, 1998).

The overlying sample at the base of the Tienen Formation (402 m) is dominated by *Apectodinium homomorphum. Apectodinium* is dominating the assemblages throughout the Tienen Formation (Table DR6, see footnote 1). The highest percentages are registered in its lowermost third (83.3% at 395 m). Although dinoflagellate cysts are not continuously and abundantly represented in the interval studied, they allow the Tienen Formation in the Kallo borehole to be subdivided into 6 distinct paleoenvironmentally controlled dinoflagellate cyst intervals. Interval DB1, ranging from the base of the formation up to 398.5 m presents a very low species diversity, with *Apectodinium* >90% of the association. DB2, which includes sample 395 m, is somewhat more diversified, with a substantial influx of the marine taxa *Glaphyrocysta divaricata* (8.8%) and *G. pastielsii* (6.1%). The overlying interval DB3 is devoid of dinoflagellate cysts: The beds from 392.5 m up to 389.8 m because of leaching of the sediments, destroying most of the organic material, the overlying beds up to 387.8 m probably as the result of the installation of a subaerial nonaqueous environment. Interval DB4, ranging from 386.5 m up to 381.8 m, is marked by moderate values of *Apectodinium* (between 25% and 38%) and by the first occurrences of *Homotryblium* spp., *Polysphaeridium zoharyi* and of the green algae *Pediastrum* spp. The latter reaches up to 57% of the totality of the aquatic palynomorphs in sample 385.6 m. It decreases through the interval to ~3% at 381.8 m. There is a major influx of *Paralecaniella indentata* (40.7%) at 384 m. DB5 including samples 380 m and 379 m is marked by high percentages of *Apectodinium* (~65%) and of *Botryococcus* (20% to 30%). The last *P. zoharyi* are encountered at 380 m. DB6, encompassing the topmost part of the Tienen Formation (378–377.5 m), is characterized by high frequencies of *Muratodinium fimbriatum* (57.2% at 378 m, 21.6% at 377.5 m) and *Eocladopyxis peniculatum* (4.5% at 378 m, 20.8% at 377.5 m).

The glauconitic sandy base of the Mont-Héribu Clay is marked by the first occurrences of *Paucisphaeridium inversibuccinum* (2%) and *Surculosphaeridium oceaniae* (8%–12%), by high frequencies of *Spiniferites* spp. (24%) and very low values of *A. homomorphum* (1%). *Cerodinium wardenense* appears at the base of the clay facies within the Mont-Héribu Clay (376 m). The first representative of *Wetzeliella* (*W. lobisca*) occurs at 374.5 m, just above the studied interval.

The organic carbon isotope curve

The boundary between the Hannut Formation and the overlying Tienen Formation corresponds to an abrupt negative shift of –2‰ in $\delta^{13}C_{org}$ values from –25.8‰ to about –28‰ (Fig. 4). The isotope curve of the Tienen Formation presents a twofold general trend, including a lower short portion with decreasing isotope values to an absolute minimum of –29‰ at respectively 3 m above its base, and a long more complex shift toward more positive values with a maximum of –21.5% at the top of the interval studied. Superimposed on this general trend occurs a zigzag pattern, because of the occurrence of several major shifts in $\delta^{13}C_{org}$, referred to as peaks (Fig. 4). The shift of amplitude between the base of the Tienen Formation and the –29‰ minimum (= peak P0) is only –1‰. There is a return to more positive values in two steps, the lower, to about –25.4‰ at surface K4 (= C peak), the upper to about –23.1‰ at 386.30 m (= B peak). This is followed by a rapid return to very negative values of about –27‰ at 385.6 m (= A peak), and a prominent shift to positive values of about –23.5‰ at 380.4 m (= P1 peak). The most positive value of about –21.5‰, designated P3, is reached at the top (377.5 m). In between, at 378.5 m, there is a return to negative values of about –25.1‰ (= P2 peak).

THE DOEL BOREHOLE

Sedimentology

A close inspection of the cores led to the recognition of the lower junction of the Tienen Formation and of the lower junction of the Mont-Héribu Clay at respectively 468.94 m and 443.85 m depth. The sandy unit, underlying the Tienen Formation, corresponds to the Grandglise Sand Member. It is made up of brownish green glauconitic bioturbated and calcareous very fine sand with a few medium-grained sand lenses, indicating fully marine environments deposited below fair weather wave base. Between the top of the Tienen Formation at 444.13 m and the base of the Mont-Héribu Clay at 443.85 m occurs 30 cm thick stiff clay with fine sandy to silty bioturbations. This facies X is similar to facies recognized in the Zoute Silt Member of the Knokke borehole. The similarity of the dinoflagellate cyst assemblages and the presence of small pyritised diatoms in facies X and in the stratotype of the Zoute Silt Member (Moorkens et al., 2000) support this correlation. The lowermost 1 m of the Mont-Héribu Clay is made up of glauconite-rich clayey fine sand. At 442.85 m it passes rapidly into grey stiff clays.

Lithofacies analysis of the Tienen Formation. Several facies are identified (Fig. 6), of which the majority have been recognized in the Kallo borehole. The basal 35 cm of the Tienen Formation, labeled facies Y, consists of heterogeneous clayey sandy silt with shell fragments and phosphatic particles at the base. It has not been encountered at Kallo, probably because of core loss. Facies 1, made up of interbedded clay/silt and facies 2, represented by dark grey laminated clay dominate the interval from 468.65 m to 463.30 m. It is overlain by a predominantly greyish bioturbated clayey silt to silty clay (up to 461.0 m), corresponding to facies 9, although somewhat clayier than at Kallo. It is overlain, up to 459.70 m by mottled greyish silty clay with many rootlets, characterizing facies Z, and by a 30 cm thick greyish silt with some larger quartz grains (facies 9). The overlying interval from 459.40 m to 453.70 m is extremely heterogeneous, consisting of 5 thin lignitic horizons (facies 6), alter-

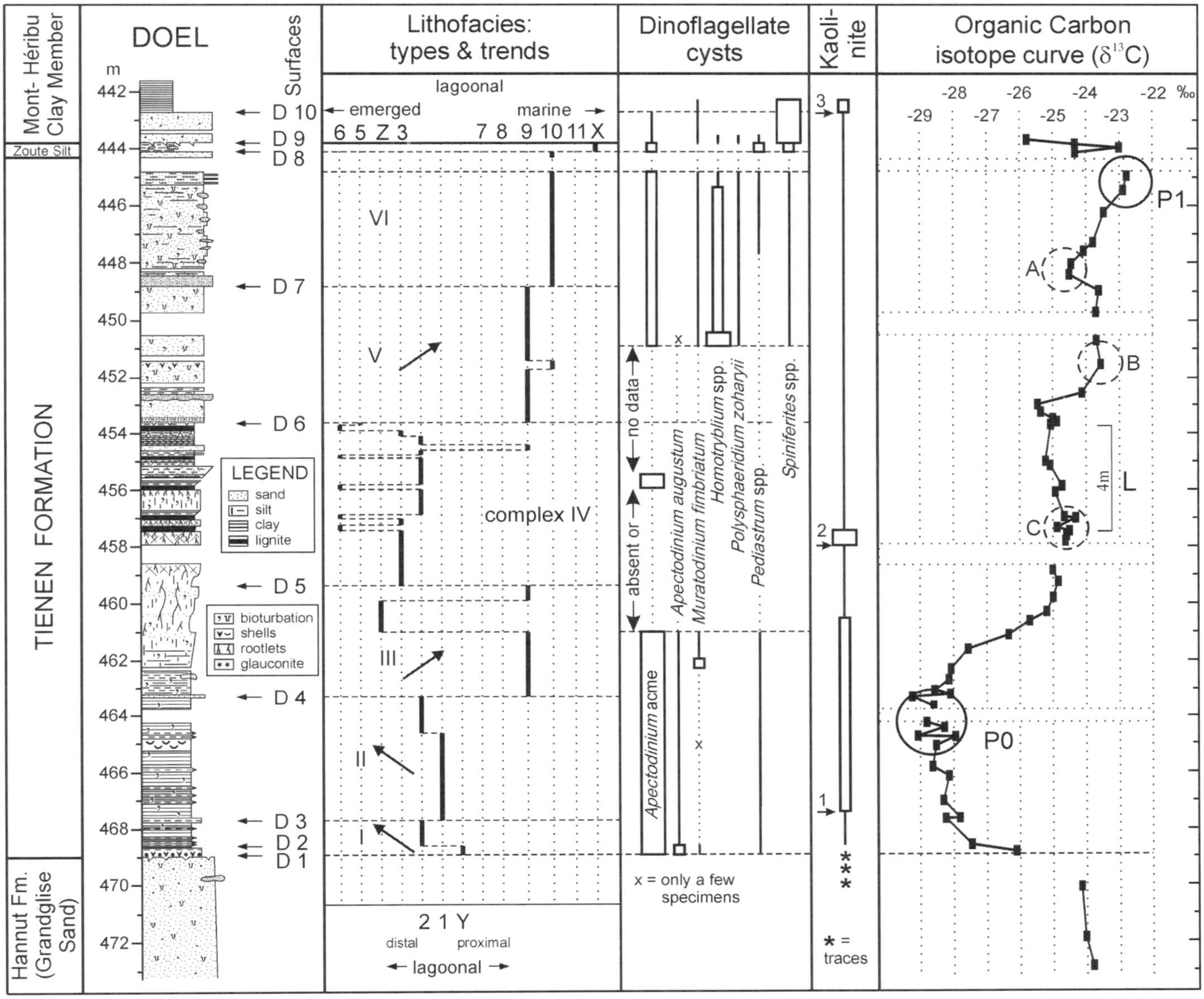

Figure 6. Lithofacies analysis, dinoflagellate cyst distribution, and organic carbon isotope curve of the Paleocene-Eocene boundary interval in the Doel borehole.

nating with purple silts with rootlets (facies 3 from 459.40 m to 457.05 m and several smaller units), bioturbated grey silt with rootlets (facies 9) and several units consisting of laminated grey-green sandy silt and darker-colored clay (facies 2). The upper part of the section from 453.7 m up to 444.85 m is more homogeneous and essentially consists of clayey bioturbated sand silt to fine sands, locally with shell fragments (facies 9 and 10). The loose very fine sands recovered between 444.33 m and 444.13 m might belong to the same facies.

Omission surfaces. Ten surfaces, labeled D1 to D10 are identified (Fig. 6). The bioturbated junction D1 at 468.95 m, corresponding to a major hiatus, separates the Tienen Formation from the underlying Grandglise Sands. Surface D2 at 468.65 m depth is sharp, separating heterogeneous clayey silt from laminated clay, reflecting a shift to low-energy environments. This might reflect a temporary landward shift of the lagoon. A return to higher-energy deposits, mainly interbedded clay/silt of facies 1 is noticed at 467.8 m (D3), followed again at 464.50 by laminated clay of facies 2. D4 at 463.30 m is sharp and marked by the income of coarser material (sandy silt). It seems to indicate a renewed but now permanent shift of the depocenter toward higher-energy areas of the lagoon. This might be related to a landward shift of the lagoon because of a small sea-level rise. Surface D5 at 459.4 m announces the start of a heterogeneous deposition regime, with several episodes of lignite development and soil formation, including several additional omission surfaces. The interburrowed junction D6 at 453.7 m, separating sandy silt from the underlying lignite, represents a deepening of the depocenter from a temporary subaerial to a fully lagoonal paleoenvironment. It probably results from a small sea-level rise. The sharp contact D7 at 448.8 m marks the income of somewhat coarser medium-grained loose sand units, which up to 444.85 m alternate with clayey fine sand. The latter contains large coarser sand bioturbations. D8 at 444.30 m is sharp and separates marine grey stiff clay with fine sandy to silty bioturbations from underlying grey loose fine sands. It represents a flooding surface, and deposition within a quiet marine bay environment, with substantial influx of fresh water. D9 at 443.85 m is marked by frequent burrows of glauconitic very fine sands in the underlying grey clays. It represents the onset of a marine transgression. Surface D10 at 442.75 m is sharp and marks the onset of an offshore clayey deposition regime. It represents a further deepening of the depocenter because of an abrupt and major sea-level rise.

Clay mineralogy

The clay fraction (<2 μ), of which the distribution is reported in Table DR2 (see footnote 1), presents almost the same mineral distribution as that from the Kallo borehole, although with slightly diverging percentages. Chlorite is generally weakly represented (<3%). The proportion of illite varies throughout the section, but never falls below 5%. It never exceeds 30%. There is substantial decrease in illite content between 460.60 m and 459.59 m depth (from 30% to 10%). Kaolinite generally ranges between 1% and 4%. The highest values (>7%) are found in the interval 467.60–460.60 m and at 457.61 m depth, where a maximum peak of 32% is reached. The interstratified illite-smectite dominates the clayey fraction (generally >85%). The lowest values are restricted to the middle part of the section (from 79% at 467.60 m to 54% at 457.61 m).

Calcareous nannofossils

The associations from the top of the Grandglise Sand in the Doel borehole (471 m to 469 m) are richer and better preserved than those from the Grandglise Sands at Kallo. The cooccurrence of *Heliolithus riedelii, Discoaster mohleri, Hornibrookina arca, Toweius eminens* and *T. tovae* also indicates nannofossil zone NP8b and a mid-Thanetian age. The associations from the lower part of the Tienen Formation (468 m to 465 m) are entirely reworked. They present a low species diversity, with *Watznaueria barnesae* as most frequent form. The presence of *Eiffellithus eximius, E. turriseiffelii* s.str., *Kamptnerius magnificus,* small *Aspidolithus* spp., *Corollithion* spp., in association with the absence of *Aspidolithus parcus* and *Arkhangelskiella* spp. suggests, according to species ranges in Perch-Nielsen (1985), an Early Late Cretaceous (probably pre-Campanian) age.

Dinoflagellate cysts

A few levels were commented on by Steurbaut et al. (2000). All these samples are reinterpreted here (Table DR7, see footnote 1). *Apectodinium* is abundantly represented throughout the Tienen Formation. The latter can be subdivided into four distinct paleoenvironmentally controlled dinoflagellate cyst intervals. Interval DB1, ranging from the base of the formation to ~463.30 m (surface D4) is dominated by *Apectodinium* (95% or more), with high frequencies of *A. augustum* (25% at the base to 6.5% at the top of DB1). The assemblage from the lowermost sample (468.70 m) is more diversified than that from the top of DB1 (465.05 m), probably because it originates from a lagoonal paleoenvironment with more pronounced marine influences. DB2, which includes sample 462.05 m, is again somewhat more diversified with a substantial influx of the marine taxa *Glaphyrocysta* (5.2%) and *Muratodinium fimbriatum* (12.6%). The overlying interval DB3 is generally devoid of dinoflagellate cysts: The beds from 461.0 m up to 457.6 m because of leaching of the sediments, destroying most of the organic material (TOC < 0.3%), most of the overlying beds up to 453.7 m probably as the result from the installation of a subaerial non aqueous environment. Interval DB4, ranging from 450.9 m up to 444.85 m, is marked by the first occurrences of *Homotryblium* spp., *Polysphaeridium zoharyi* and of *Spiniferites* spp. *Homotryblium* reaches 70.9% of the totality of the aquatic palynomorphs in sample 450.88 m. Species diversity is fairly high throughout DB4. It is furthermore marked by relatively low values of *Apectodinium* spp. (8% at the base, generally between 25% and 33%), and *Pediastrum* spp. (<6%), and by the presence of *Areoligera senonensis* (3%–4%) and *Phthanoperidinium crenulatum.* Sample 444.09 m substantially differs from the underlying samples by the first occurrence of *Thalassiphora delicata,* by the start of the acme of *Spiniferites* spp. (26.5%), by relatively high values of *Glaphyrocysta* (5.2%) and *Hystrichosphaeridium tubiferum* (6.2%) and by the presence of *Impagidinium* spp. The glauconitic sandy base of the Mont-Héribu Clay is marked by a major decrease in *Apectodinium,* by the first occurrences of *Tectatodinium pellitum* and by a further increase of *Spiniferites* spp. (50%). The first representative of *Wetzeliella* (*W. astra*) occurs within the clay facies at 442.49 m. The stratigraphically significant taxa are figured on Plate 1.

The organic carbon isotope curve

The isotope curve of the studied interval is complex because of the superposition of four distinct stratigraphic units. The values in the lowermost Grandglise Sand remain fairly constant, all close to –24‰. These in the Mont-Héribu Clay present a trend toward more negative values (from –24.31‰ to –25.74‰), these in the Zoute Silt to more positive values (from –24.30‰ to –23.00‰). The curve in the Tienen Formation has a twofold pattern, including a lower short portion with decreasing isotope values to an absolute minimum of –29.18‰ (= P0 peak) and a long overlying portion with a progressive shift toward more positive values, including a return to plateau values of –25‰, just below the lignitic package L (= C peak) and two small positive excursions, respectively of –23.54‰ at 451.58 m (= B peak) and of –22.77‰ at 445 m (= P1 peak). The most negative values are found at ~5.5 m above the base of the Tienen Formation (= P0 peak). The shift of amplitude from the base of the Tienen Formation to peak P0 is –3‰. The boundary between the Hannut Formation and the overlying Tienen Formation corresponds to an abrupt negative shift of –2‰ in $\delta^{13}C_{org}$ values from –24.13‰ to –26.07‰ (Figs. 6 and 8).

DISCUSSION

The kaolinite influx

The study of the clay-mineralogy of the Kallo and Doel boreholes forms part of a continuing program of research on the Paleocene-Eocene boundary sections in northwest Europe. At present, the information is insufficient to allow a comprehensive analysis, so that only brief details are given on the distribution of kaolinite.

Close inspection of the Kallo and Doel borehole data shows that there are three major influxes of kaolinite in the sections studied. The lowermost increase is located at ~1 m above the base of the Tienen Formation (event 1; see Figures 4 and 6 and Tables DR1 and DR2 [see footnote 1]). In the basal 1 m and in the underlying Grandglise Sands only traces of kaolinite were identified. High values continue to ~7 m higher up (to 394.0 m at Kallo and 460.6 m at Doel). The second and most conspicuous increase, up to >30%, is recognized within the middle of the formation (event 2: Kallo from 394.0 m to 390.0 m; Doel at 457.61 m). The third influx occurs at the base of the clay facies in the Mont-Héribu Member, at ~1 m above its base (event 3). Substantial increases of kaolinite have been reported by many authors in the Paleocene-Eocene boundary interval and directly linked to global climatic warming and increased humidity (Adatte et al., 2000; Bolle et al., 2000). Others (Knox, 1998; Thiry and Dupuis, 2000) pointed out that the arrival of kaolinite into marine basins is strongly influenced by sea-level changes and that this influx may occur long (several million years) after its formation on the continent.

Calibration of the changes in clay mineralogy and the organic carbon isotope values in the Kallo and Doel boreholes indicates no direct relation between the carbon isotope excursion (CIE) and the appearance of kaolinite, or with its maximum development. The increase in kaolinite in the lower and upper part of the sections (events 1 and 3) coincides with a substantial increase in illite (with percentages twice as high as those of kaolinite), and a small but distinct rise in chlorite. This covariation seems to be related to transgressive pulses at the base of depositional sequences (base Tienen Formation and base Mont-Héribu Member). It is interpreted as the result of reworking, related to erosion of paleo-kaolinite blankets on the illitic and chloritic shale of Paleozoic Massifs (Welsh Borderland?, Ardennes?) triggered by relative sea-level variations (Laenen, 1999). By contrast, the high kaolinite values identified within the top of sandy silt beds in the middle of the Tienen Formation (event 2) do not coincide with an increased development of illite, but are associated with a strong degradation of the crystallinity of the kaolinite (see Tables DR1 and DR2 [see footnote 1]). This poorly crystallized interstitial kaolinite may have formed during early diagenesis as a result of circulation of acidic meteoric pore waters. This kaolinite influx has also no causal relationship with the well-known Initial Eocene Thermal Maximum.

A refined dinoflagellate cyst zonation for the southern North Sea Basin

The vertical ranges and frequency distributions of the different dinoflagellate cyst taxa are generally consistent throughout the studied interval in the Kallo and Doel boreholes, allowing definition of a refined biozonation for the southern North Sea Basin. These DB intervals, which probably have only regional significance (Anglo-Franco-Belgian Basin), are bounded by distinct events (named Dib-events, see Figure 9), of which some were already discussed in Steurbaut et al. (2000).

The boundary between the Tienen Formation and the underlying Hannut Formation is marked by the start of the *Apectodinium* acme, including the first appearance of *A. augustum,* and by a strong decrease in *Areoligera* spp. (= Dib 1). The onset of the common occurrence of *Muratodinium fimbriatum* (= Dib 2) defines the base of zone DB2, the last consistent occurrence of *A. augustum* (= Dib 3) the base of DB3 (Fig. 9). The first occurrence of *Homotryblium* spp. (= Dib 4), which coincides in the southern North Sea Basin with the recurrence of *Polysphaeridium zoharyi,* marks the base of DB4. *Pediastrum* spp. is consistently represented throughout this interval, and dominates the assemblages within the middle of it. The lower boundary of DB5 corresponds to a major increase in *Apectodinium* and *Botryococcus,* the lower boundary of DB6 to the start of the acme of *Muratodinium fimbriatum* and *Eocladopyxis peniculatum.* The top of the *Apectodinium* acme coincides with the top of the Tienen Formation (= Dib 5).

From the foregoing it is clear that *Apectodinium* dominates the assemblages throughout the Tienen Formation. The highest percentages are registered in its lowermost third (>95% in DB1 and >75% in DB2) and, to a lesser degree, in the topmost 3 m, which are only preserved at Kallo (>50%). The disappearance of the extremely high frequencies >75% seems to coincide with the disappearance of *Apectodinium augustum.* Massive occurrences of *Apectodinium,* known to have occurred on a worldwide scale, are believed to be due to high sea surface temperatures and linked to high nutrient availability (Crouch et al., 2000; Brinkhuis and Crouch, 2001). The paleoenvironmental significance of the *M. fimbriatum* and *E. peniculatum* acmes are not understood up to now.

The assemblage from the thin silty clay unit identified between 444.30 m and 443.85 m at Doel substantially differs from these from the top of the Tienen Formation (Kallo 378 m to 377.5 m) by the increase in *Spiniferites* spp., the first appearance of *Thalassiphora delicata,* the incoming of *Impagidinium* spp., and the decrease in *Apectodinium* (from >50% to 28%). The same phenomena have been recorded within the stratotype of the Zoute Silt in the Knokke borehole (De Coninck, *in* Dupuis et al., 1990), allowing correlation of both units. They reflect fully marine conditions. The high amount of *Pediastrum* (25%) at Doel, compared to its total absence at Knokke, is believed to be paleoenvironmentally controlled (vicinity of river mouth or estuary). The increase in species diversity, in *Spiniferites* spp. and in *Glaphyrocysta* spp. in the overlying glauconitic sands at Kallo and Doel are reported in other boreholes at the base of the Mont-Héribu Clay (Dupuis et al., 1990; De Coninck, 1986, 1999b), evidencing correlation.

The interval from base DB1 to top DB6 corresponds to the lower part of the *A. augustum* Interval Biozone of Powell (1992), which is calibrated with the upper part of NP9. The assemblages from the Zoute Silt and the glauconitic base of the Mont-Héribu Member belong to Powell's *Glaphyrocysta ordinata* Interval Biozone, these of the overlying clay facies to Powell's *Wetzeliella astra* Interval Biozone.

Paleoenvironmental reconstruction

Calibration of the results from the lithofacies and palynofacies analyses reveals the presence of rapidly changing paleoenvironments in northern Belgium during the deposition of the Tienen Formation. Its paleoenvironmental history, which includes 7 major phases, can be summarized as followed (Fig. 9):

Phase 1 (from base Tienen Formation up to surface K4 and D5). At 55.50 Ma installation of an essentially lagoonal depositional regime, initially with a clayey sedimentation marked by a relatively high organic material content (TOC ~1%), with progressively decreasing percentage of spores and pollen, among which are bisaccate pollen, and increasing percentage of aquatic palynomorphs, reflecting a distal shift of the depocenter. *Apectodinium* dominates the aquatic assemblage (>75%), suggesting high surface water temperature and nutrient rich waters. Such environments occur in areas with high rates of continental runoff, e.g., in river mouths and estuaries (Crouch et al., 2000). Evidence therefore is also found in our study area, where several influxes of reworked marine coccolith material are recorded, resulting from coastal erosion or upstream river erosion. It was later on replaced by an essential silty sedimentation regime, with little organic material (TOC from 0.1% to 0.3%), dominated by dinoflagellate cysts (80%) of which a few are normal salinity forms, indicating an increase in salinity.

Phase 2 (K4 to K5, D5 to D6). Replacement by an essential paludal deposition regime (TOC between 1% and 3%), with several subaerial phases, including peat accumulation (TOC up to 28%) and soil formation (TOC of 0.7%–0.9% due to weathering and leaching). Black-brown lignite debris initially dominates the palynofacies, reflecting high influxes of terrestrial material from local sources. Stratified water masses and anoxic to dysoxic bottoms may have developed in a later stage, as shown by the considerable amounts of amorphous organic material (AOM ~15%). During this phase, occurring clearly subsequent to the CIE minimum record, sedimentation is very complex as the result of rapidly changing local conditions. As a whole, phase 2 corresponds to a stabilization of the depocenter and might have had a considerably long duration.

Phase 3 (K5 to K7, D6 to D7). Return to lagoonal paleoenvironments with an essentially sandy deposition regime, probably because of a minor sea-level rise. The organic material, generally between 0.1% and 0.9%, is essentially made up of pollen and spores (70%–80%). The dinoflagellate cyst assemblage is marked by a rather high species diversity, a high frequency of *Homotryblium* spp. (up to 70.9%) and by a strong decrease in *Apectodinium* (<10%). These changes, in association with the scarcity of *Pediastrum* and absence of *Botryococcus,* point to increasing marine influences. The first oysters accumulated in the depocenter during phase 3, providing additional evidence for increased salinity.

Phase 4 (only recorded at Kallo from K7 to K8). Installation of a marginal lagoonal depositional regime, with clayey, carbonate-rich (up to 50%) sedimentation and high organic material content (TOC between 1.3% and 1.8%) of palynofacies F type. The latter is dominated by aquatic palynomorphs (70%), essentially consisting of the freshwater algae *Pediastrum* spp. (57%) and *Botryococcus* spp. (17.1%). Sedimentation took place at the landward side of the lagoon, where small rivers enter, and which was probably bordered by dense Taxodiaceae and Palm forests, as witnessed by the high frequencies of their spores and pollen (respectively up to 11% and 20%).

Phase 5 (K8 to K10, D7 to D8). Return to less restricted lagoonal conditions with an essentially sandy sedimentation regime with episodically oyster accumulations and relatively low amounts of aquatic palynomorphs (20%–30%). The dinoflagellate cyst assemblages, which are moderately rich in *Apectodinium* (between 25% and 35%), present some substantial changes during this phase. Initially, they are marked by a relatively low to moderate species diversity (9 species) and by high frequencies of *Paralecaniella indentata* (40.7%) and freshwater *Pediastrum* spp. (17%). Later on, species diversity is increasing, up to 27 species. This trend is associated with decreases in *Pediastrum* (generally <3%) and in *P. indentata* (<3%) and substantial increases in *Homotryblium* spp. (from <1% to 10%–25%) and in *Lingulodinium machaerophorum* (>10%). These changes indicate a progressive increase in salinity (the highest of the entire Tienen Formation), due to shifting of the depocenter from the landward side of the lagoon, with substantial freshwater influx, to a more distal position.

Phase 6 (only recorded at Kallo, K10-K11). Installation of a carbonate-rich sandy depositional regime, initially with shell grit layers and moderately rich dinocyst assemblages (9 species), dominated by *Apectodinium* (68%), and to a lesser degree by *Botryococcus* (19.5%). It indicates high-energy environments with considerable freshwater influx, probably at the inlets of the lagoon, although close to a river mouth. The strong decrease in dinoflagellate cyst species diversity (4 species), and the increase in *Botryococcus* (31%), at the end of phase 6 point to a restricted setting, with substantial inflow of fresh water.

Phase 7 (only recorded at Kallo, K11 to top of Tienen Formation). Installation of a clayey depositional regime, consisting of organic-rich interbedded clays (TOC from 1% to 3%) and shell levels, essentially characterized by the abundances of the dinoflagellate cysts *Muratodinium fimbriatum* and *Eocladopyxis peniculata* (together 40%–60% of the aquatic palynomorphs), of which the exact significance is not understood up to now. The shell beds become almost monospecific in the uppermost levels, while the clays contain extremely high values of amorphous organic material (up to 34%). This probably points to water stratification and oxygen-depleted bottom waters.

The lagoonal paleoenvironments disappear in northern Belgium at the end of the deposition of the Tienen Formation. At the onset of the subsequent deposition of the Zoute Silt paleoenvironments drastically changed. A shallow marine epicontinental sea covered the area, with according to the dinoflagellate cysts almost normal salinities (the frequency of *Pediastrum* spp. at Doel is believed to be due to the inflow of fresh water from a nearby river mouth) and nutrient-rich, rather high-temperature surface waters (*Apectodinium* still reaching ~28%). The sea surface temperature might have decreased by the start of the deposition of the Mont-Héribu Clay (strong decrease in *Apectodinium*), while the depocenter was progressively shifting from offshore transition to outer shelf areas.

Vegetation history

Evaluation of the pollen and spore distribution in the Kallo borehole (Table DR5 [see footnote 1] and E. Roche, personal commun.) demonstrates that ~80% of the 54 taxa recognized in northern Belgium persisted without any noticeable changes over a period lasting 2.5 m.y., from ~1.5 m.y. before to 1 m.y. after the carbon isotope excursion (CIE) and the well-known global warming event (IETM). These pollen and spore assemblages are dominated by arboreal species, mainly Taxodiaceae, Palmae and Juglandaceae, and none is originating from aquatic plants. No evidence for the presence of mangroves was found (no *Nypa* pollen nor other mangrove pollen). Consequently, the assemblages reflect the dry land vegetation, surrounding the lagoonal depocenter, including that of swampy patches within the lagoon, of the nearby barrier islands as well as the hinterland (Fig. 7). This palynological-based flora indicates vegetation in a warm-temperate to subtropical climate.

Despite the overall similarity of the flora across the IETM in the southern North Sea Basin, several significant events have been identified in the pollen and spore assemblages: (1) A 20% higher species diversity in the Tienen Formation than in the underlying Grandglise Sand and overlying Mont-Héribu Clay. These differences mainly reflect changes in distance from de-

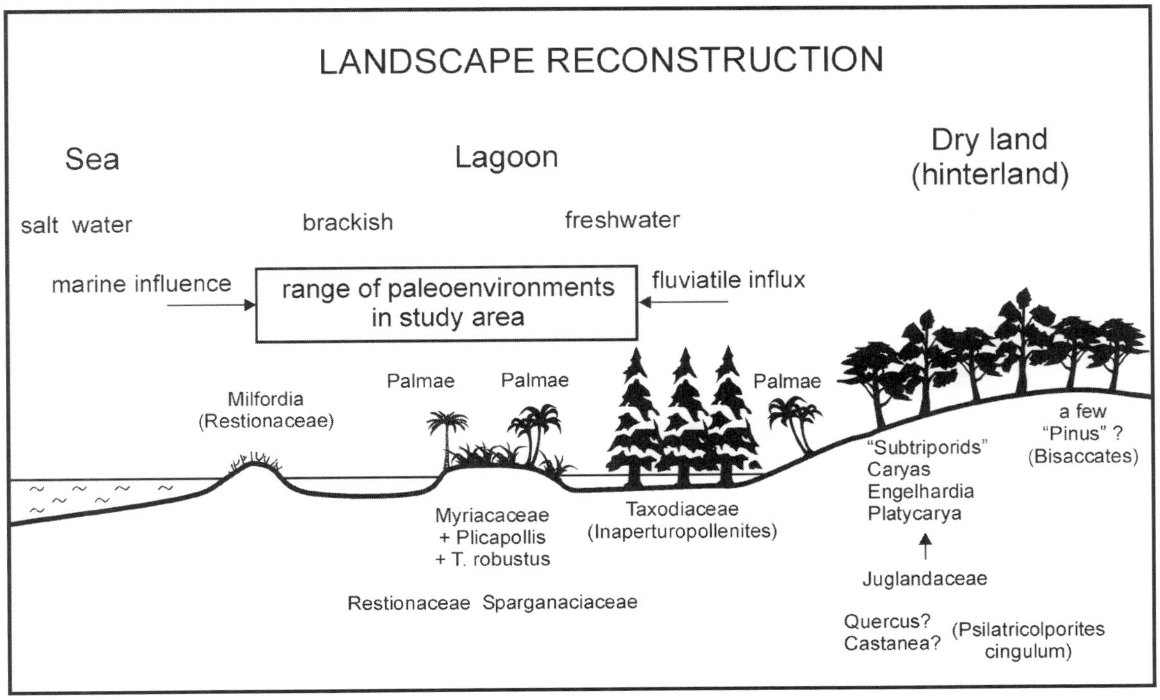

Figure 7. Landscape reconstruction of northern Belgium during the Paleocene-Eocene boundary interval at ~55.5 Ma.

pocenter to source area (the Grandglise Sands and the Mont-Héribu Clay are the most distal and consequently the poorest), and consequently have only local significance. (2) *Platycarya*-type pollen was already present at the beginning of the Thanetian, several million years before the IETM. (3) Several extinctions (8% of species recognized) and a few first occurrences (2%) occurred within Biochron NP8, ~2 m.y. before the IETM. (4) Four first occurrences, among which is the palm *Monocolpopollenites parareolatus,* and only one extinction are recognized at the onset of the CIE, which is further marked by a substantial increase in Juglandaceae and Restionaceae. (5) Very few first occurrences and extinctions are noted within the Tienen Formation. (6) Two taxa of Tiliaceae appear in the levels corresponding to the CIE minimum.

In conclusion, the changes in pollen and spore assemblages identified across the IETM in northern Belgium are essentially changes in numerical abundances, due to paleoenvironmental changes, probably driven by small sea-level changes, and do not originate from substantial changes in the overall vegetation. This vegetation is marked by low extinction and first-occurrence rates.

Sequential analysis

Integration of the Kallo and Doel borehole data into the updated sequence stratigraphical classification of Steurbaut (1998; Steurbaut et al., 2000) suggests that the Grandglise Sands belong to depositional sequence T-A2, dated as middle Thanetian, and that the upper Thanetian sequences T-B and T-C, normally occurring between the Grandglise Sands and the Tienen Formation (Fig. 3), are absent in the Kallo and Doel area. The overlying units have been interpreted somewhat differently in terms of sequence stratigraphy, as new data became available at the end of the 90s. The entire Tienen Formation and the overlying Zoute Silt Member were initially incorporated in a single third order depositional sequence, named TA 2.3 by Vandenberghe et al. (1998) and Y-A (Ypresian-A) by Steurbaut (1998). Both units were believed to represent respectively the Lowstand Systems Tract (LST) and the Transgressive Systems Tract (TST) of this sequence. Later on Steurbaut (in Steurbaut et al., 2000) subdivided sequence Y-A into Y-A1 and Y-A2, both believed to represent third order sequences (see also Figure 3). Sequence Y-A1 was introduced to include the lower and middle parts of the Tienen Formation (Dormaal Sand Member and Knokke Member), sequence Y-A2 to include the Oosthoek Sand Member, representing the upper part of the Tienen Formation, and the overlying Zoute Silt Member. The major criteria for this subdivision were (1) the presence of a substantial break in sedimentation at the base of the Oosthoek Sands (deeply burrowed contact, overlain by marine sand, probably representing a sandbank), (2) the sharp contact, the absence of pebbles (see also King, 1990) and the substantial paleoenvironmental deepening at the base of the Zoute Silt Member, suggesting a minor break in sedimentation, (3) the presence of small marine centric diatoms in both the Oosthoek Sands (Dupuis et al., 1990) and the Zoute Silt (Moorkens et al., 2000) and (4) the major break at the base of the overlying Mont-Héribu Member (burrowed contact and flint pebbles), representing the base of the overlying sequence Y-B.

The study of the Kallo and Doel boreholes does not add substantial new information on sequence Y-A2 because of the incomplete sedimentological record (absence of the Oosthoek Sands in both boreholes; the Zoute Silt only present at Doel and with extremely reduced thickness). On the contrary, the identification of the spatial and temporal relationships of the lithofacies in both boreholes (Figs. 4 and 6) allows the stratigraphic architecture of sequence Y-A1 be unraveled for the first time (Fig. 9). The two lowermost intervals (I and II) present fining upward trends and are bounded by omission surfaces. Interval III is marked by a slightly coarsening upward trend and somewhat higher salinities (higher dinoflagellate cyst species diversity, up to 13 species, and higher frequencies of *Glaphyrocysta*). Intervals I, II and III are believed to compose a distinct parasequence, labeled Y-A1a. Surface K4, at the base of the very complex interval IV, marks the onset of lignitic deposition and soil formation with episodically subaerial conditions. The dinoflagellate cyst assemblages, identified in a few levels within this interval, almost exclusively consist of *Apectodinium homomorphum* (99%). The base of interval V is marked by an abrupt change in dinoflagellate cyst assemblages, among which is the income of abundant *Homotryblium* spp. The assemblages of the intervals V and VI present a high species diversity, increasing from 14 species at the base of V to 24 at the top of VI (see Tables DR6 and DR7 [see footnote 1]), indicating an increasing marine influence. The overlying intervals VII and VIII are marked by a return to a restricted lagoonal setting with increasing freshwater influx (VII) and with oxygen-depleted bottoms (VIII). Intervals IV to VIII may constitute a second parasequence Y-A1b, with interval IV representing its lowstand depositional regime, intervals V and VI the transgressive pulses and intervals VII and VIII its late highstand conditions. However, it must be emphasized that the here presented sequence stratigraphic interpretation is only tentative, as the comparative study of the Cap d'Ailly section in northern France (Dupuis et al., 1998) reveals that the sedimentological record in the Kallo and Doel area is incomplete.

The organic carbon isotope evolution

The organic carbon isotope values ($\delta^{13}C$) of the Kallo and Doel boreholes show a rapid shift toward the most negative values in the lowermost part of the Tienen Formation, and a long return to the most positive value at the top of the formation. The same general trend, including an abrupt shift toward the most negative values at the onset of the CIE and a progressive return to pre-excursion values, is identified in the carbon isotope curves from marine sections, based on whole rock samples (Schmitz et al., 1996; Cramer et al., 1999; Speijer and Schmitz, 2000), and in the continental formations of the Bighorn Basin, based on paleosol carbonates (Koch et al., 1992, 1995; Koch and

Morrill, 2000; Norris and Röhl, 1999; Bowen et al., 2001). This general trend is also identified in the continental "Sparnacian" deposits from the Paris Basin (Stott et al., 1996; Thiry et al., 1998; Thiry and Dupuis, 1998) and in the lacustro-lagoonal deposits from the Mont-Bernon Group of Haute-Normandie, France (Magioncalda et al., 2001). However, the organic carbon isotopic records from these nonmarine sections and many others in northwest Europe (Therdonne, Newhaven, etc.; Magioncalda, personal commun.) display much more complex patterns. Superimposed on this general trend, the $\delta^{13}C$-record repeatedly shifts toward more negative and less negative values, although never as negative as the lowermost negative peak (P0). The significance of this superimposed zigzag pattern is not quite understood. It might be related to the nature of the organic matter analyzed, which in several settings consists of a mixture of continental plant debris (including pollen and spores), freshwater algae (e.g., *Pediastrum* spp.) and lagoonal dinoflagellate cysts, with potentially different magnitudes of photosynthetic fractionation. According to Stott et al. (1996) land plant communities seem to have utilized the same photosynthetic pathway during the IETM, but little is known about the significance of the freshwater algae and the dinoflagellate cyst assemblages.

The organic carbon isotope curves of the Kallo and Doel sections also present complex patterns. At first glance the isotope signature (Figs. 4, 6 and 8) displays marked similarities, but also a few discrepancies. The lower parts of the curves, up to the return to plateau values of –25‰ at peak C, are quite similar. In both sections the boundary between the Hannut Formation and the overlying Tienen Formation corresponds to a considerable negative shift of –2‰ in $\delta^{13}C_{org}$ values, although with different values, from respectively –24‰ to –26‰ at Doel and –25.5‰ to –28‰ at Kallo. It reveals that the onset of the carbon isotope excursion is located within the hiatus between both formations. The most negative values, slightly exceeding –29‰ (= P0 peak) occur within the lower part of the Tienen Formation, respectively at 5 m (Doel) and 3 m (Kallo) above its base. The return to more positive values includes three steps (Fig. 8), the lower to plateau values of about –25‰ just below the base of the lignitic package L (= peak C), the middle to values of about –23.5‰ (= B peak) within the middle of interval V, and the upper with values between –23.5‰ and –22.8‰ at peak P1. In both sections there is a shift toward more negative values between peak B and peak P1 (= A peak at the base of interval VI), although with strongly different values: A minimum of –24.5‰ at Doel and up to –27‰ at Kallo. The very negative values, only recorded at Kallo (–27‰ at 385.6 m, –26.4‰ at 385.2 m and –26‰ at 384 m) are characterized by high frequencies of the green algae *Pediastrum* spp. (respectively 57%, 37% and 17% of the aquatic palynomorphs). The latter is only poorly represented in the Tienen Formation at Doel (generally <3%).

The continuation of the isotope record is only registered in the Kallo borehole, due to local differences in depositional history. It includes a third, although less negative excursion to about –25.1‰ (= P2 peak) and a return to the highest positive values encountered in the Paleocene-Eocene boundary interval of northern Belgium, reaching –21.6‰ (= P3 peak).

From the foregoing it is clear that the succession of isotope trends and peaks is consistent throughout the Tienen Formation in northern Belgium, but that the absolute values of the excursions are influenced by the composition of the organic assemblages. Changes in the freshwater *Pediastrum* spp. content may cause considerable differences in carbon isotope values, up to a shift of –2‰ by very strong *Pediastrum* spp. increases. An identical succession of isotope trends and peaks, although with sometimes slightly different absolute values, has been identified in the coeval lacustro-lagoonal setting of Cap d'Ailly in northwest France (Magioncalda et al., 2001) and in exclusively continental settings in the Paris Basin (Thiry et al., 1998). Apparently, the observed zigzag pattern is not facies-related, but results from the complex nature of the $\delta^{13}C$ negative isotope shifts itself, as demonstrated on the basis of carbonate bulk sample analyses from several Ocean Drilling Program holes by Bains et al. (2000). These authors suggested that the complex $\delta^{13}C$-anomaly pattern represents multiple injections of light carbon into the ocean and atmosphere, as geographically separated hydrate reservoirs became unstable at slightly different times.

The depositional history

The integration of the organic carbon isotope, the lithofacies and the dinoflagellate cyst data clearly shows that the above-described signature of the carbon isotope curves provides a very refined, regional useful chronostratigraphic framework, allowing to identify sedimentary hiatus and to evaluate sedimentation rates.

The differences in isotope values at the base of the Tienen Formation and the location of peak P0 in the sections studied suggest that sedimentation started somewhat earlier at Doel and was marked by comparable to slightly higher sedimentation rates. The fluctuations in thickness of intervals III and IV point to differential subsidence. However, the total thickness of the lower part of the Tienen Formation, up to the top of the lignitic package L, appears to be rather consistent (14.45 m at Kallo and 15.10 m at Doel). The thinning of interval V and the presence of a major seaward facies shift at surface K7 at Kallo and its correlative surface D7 at Doel, may have a tectonic origin (uplift Brabant Massif). From surface K10 on, the sedimentological record in both boreholes is completely different: The top of the Tienen Formation has only been observed at Kallo, whereas the overlying Zoute Silt only at Doel. Around 54.8 Ma, depositional conditions became uniform again, marked by a clayey regime (Mont-Héribu Clay) with identical facies and thickness patterns.

Evaluation of the overall Kallo, Doel and Knokke borehole data (for the latter see Laga and Vandenberghe, 1990; Moorkens et al., 2000; Steurbaut, personal commun.) shows that, before and after the deposition of the Tienen Formation, northern Belgium formed part of a slowly subsiding basin with remarkable stability and uniform sedimentation. The absence of deposi-

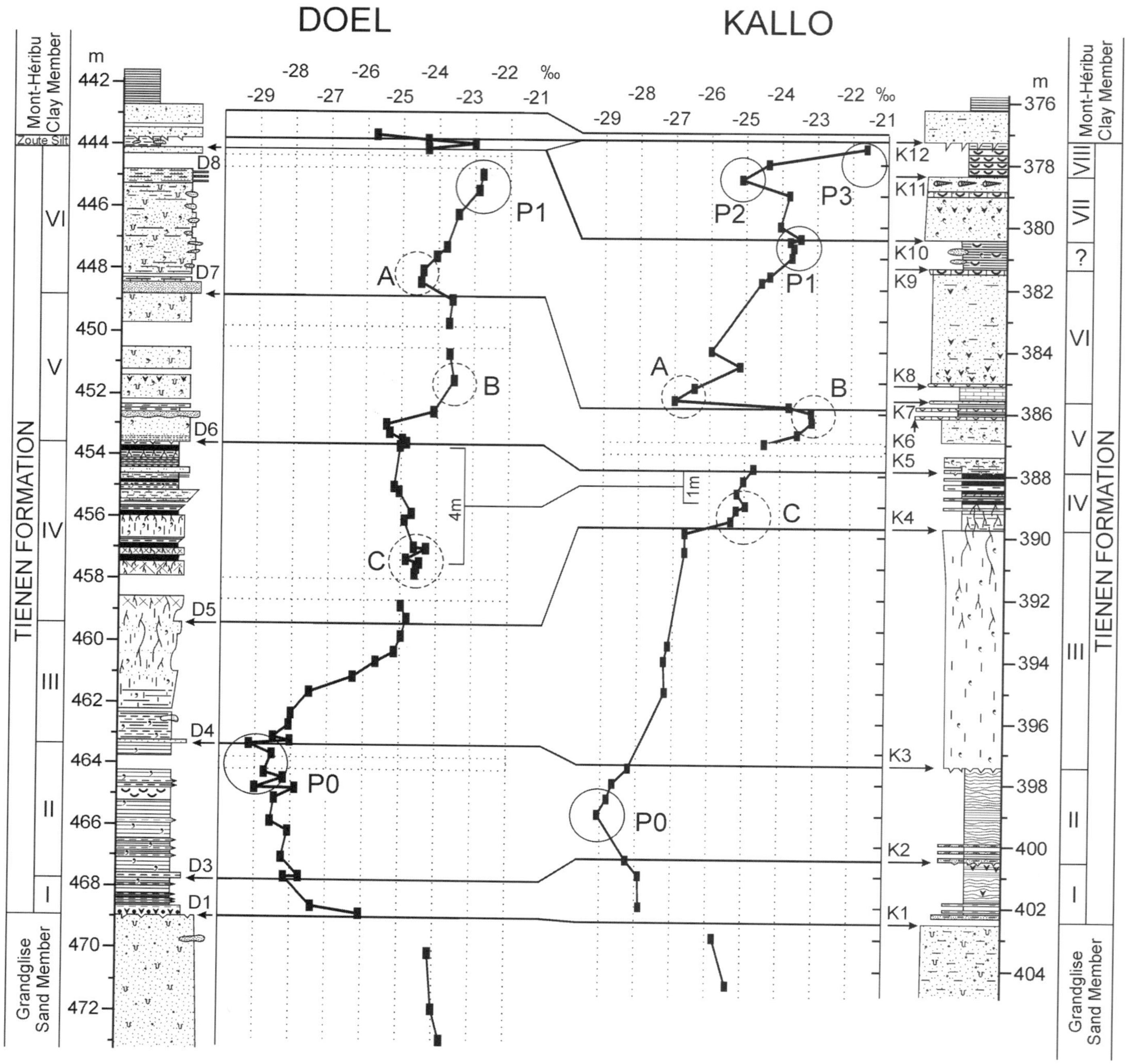

Figure 8. Correlation of the carbon isotope curves of the Kallo and Doel borehole sections based on lithofacies and dinoflagellate cyst analyses.

tional sequences T-B and T-C indicates a major erosion phase just before the start of deposition of the Tienen Formation. During the early phase of its deposition the area was tectonically very active and unstable (substantial subsidence in the Kallo and Doel area, but nondeposition at Knokke). The presence of a substantial facies shift in the Kallo and Doel area (K7 and D7) is interpreted as resulting of a distinct uplift phase of the Brabant Massif. The reasons and causes for the preservation of the upper part of the Tienen Formation at Kallo, but not at Doel, are far from understood. It could be related to subsequent erosion at the onset of the deposition of the Zoute Silt. The absence of the Zoute Silt at Kallo may be due to nondeposition. The high percentage of *Pediastrum* spp. at Doel compared to Knokke (none) might imply the presence of a nearby river mouth and coastline. Evidently, most of the evidence for understanding the complete scenario of the depositional and tectonic history of the area has been destroyed by the major erosive phase preceding the deposition of the Mont-Héribu Clay.

The δ^{13}C-record and its chronostratigraphic significance

Comparison of the organic carbon isotope curves from the Kallo and Doel borehole sections as well as many others reveals that the observed zigzag pattern is consistent throughout the southern North Sea Basin, with the same trends and turning points (8 major, called P-peaks), although sometimes with slightly different absolute values. As a consequence these parameters, essentially the isotope trends, have a chronostratigraphic connotation on a regional scale (Fig. 9). From a global point of view, it also suggests that the base or the inflection point of the carbon isotope excursion (CIE) falls within the hiatus between the Hannut Formation and the Tienen Formation. According to the decision of the voting members of the International Subcommission of Paleogene Stratigraphy (Luterbacher et al., 2000), this inflection point is the criterion for the recognition of the Paleocene-Eocene boundary. Consequently, the totality of the Tienen Formation, and probably also the famous Dormaal mammal fauna found at its base in more proximal continental settings (Steurbaut et al., 1999), have to be incorporated in the Eocene. The consistent high frequency of *Apectodinium* (>50%) throughout the Tienen Formation, indicating high sea-surface temperatures, suggests that the IETM probably lasted until the end of the deposition of the Tienen Formation. However, as shown by the range of the *Apectodinium*-dominated assemblages (>75%), the highest sea-surface temperatures seem to have occurred during the deposition of parasequence Y-A1a, before the first appearance of *Homotryblium* spp. in the North Sea Basin.

CONCLUSIONS

High-resolution sedimentological, micropaleontological and carbon isotope investigations have enabled to reconstruct the depositional history of northern Belgium across the Paleocene-Eocene boundary interval from ca. 57.1 Ma to 54.7 Ma. Although this history cannot be unraveled in its finest details because of the incomplete sedimentological record, the famous carbon isotope excursion (CIE) and the associated global warming (IETM, formerly LPTM), characterized by the *Apectodinium* acme, have for the first time been undeniably identified in this part of the southern North Sea Basin. The consistent high frequency of *Apectodinium* (>50%) throughout the Tienen Formation, indicating high sea-surface temperatures, suggests that the IETM probably lasted until the end of the deposition of the Tienen Formation. However, the highest sea-surface temperatures seem to have occurred during the deposition of its lower part, defined herein as parasequence Y-A1a (*Apectodinium* > 75%), before the first appearance of *Homotryblium* spp. Moreover, it is demonstrated that the carbon isotope excursion is complex, marked by multiple anomalies and, consequently, cannot be linked to a single instantaneous event. Its absolute minimum negative value (–29‰) is recorded within the lower part of the *Apectodinium* acme, its first return to plateau values of –25‰ just before the onset of the lignite accumulations and the first appearance of *Homotryblium* spp. Calibration of the dinoflagellate cyst events and the organic carbon isotopic signals demonstrates that the isotope trends provide excellent tools for precise stratigraphical correlation throughout the southern North Sea Basin. It also shows that changes in the freshwater *Pediastrum* spp. content may cause substantial differences in carbon isotope values in lagoonal settings, up to a shift of –2‰ by extremely high *Pediastrum* spp. values (>50% of the aquatic palynomorph assemblage).

Many substantial biotic and tectonic events have been recorded in the lagoonal settings of Belgium during the IETM, of which some are coincident and directly linked to the global warming (e.g., onset of the *Apectodinium* acme at the base of the CIE). Others traditionally linked to this global warming, such as the kaolinite influx, are proved to be unrelated and due to erosion phenomena on the paleocontinent (influx at the base of the Tienen Formation) or just to neomineralization (the poorly crystallized interstitial kaolinite in the middle of the Tienen Formation). The tectonic changes seem to have multiple causes, from global plate boundary rearrangements (the major sea-level fall and erosion just before the deposition of the Tienen Formation and the onset of the CIE), to uplift of the Brabant Massif (substantial facies shifts at surface K7 of Kallo and its correlative D7 at Doel). The reworking of Early Late Cretaceous calcareous nannofossils and Jurassic palynomorphs at the base of the Tienen Formation provides additional evidence for the magnitude of erosion and the provenance of the erosion products (probably Central England).

Most of the dinoflagellate cyst events appear to be caused by substantially fluctuating salinities and/or nutrient availability (*Apectodinium* dominance in the lower Tienen Formation by nutrient-rich waters; acme of *Homotryblium* in the middle of the Tienen Formation, and acme of *Spiniferites* in the Grandglise Sands and the Mont-Héribu Clay by increasing, almost normal salinities; the *Pediastrum* acme by high inflow of fresh water). Some of these are identified throughout the southern North Sea Basin (onset and end of the *Apectodinium* acme, the onset and the top of the *Homotryblium* spp. acme), and provide extra tie-points for calibrating the heterogeneous Paleocene-Eocene boundary sections in that area. They may represent the only tools for testing the correlation potential of the carbon isotope records. A few events, such as the *Muratodinium fimbriatum* acme at the top of the Tienen Formation, are reported for the first time and remain unclear, in terms of paleoenvironmental significance or correlation potential.

During the IETM the essentially lagoonal settings in northern Belgium were very unstable and marked by continuously changing paleoenvironments, as shown by the alternation of 13 distinct lithofacies, including paludal facies and blackish organic-rich clay facies, the presence of 8 palynofacies types (A to H)

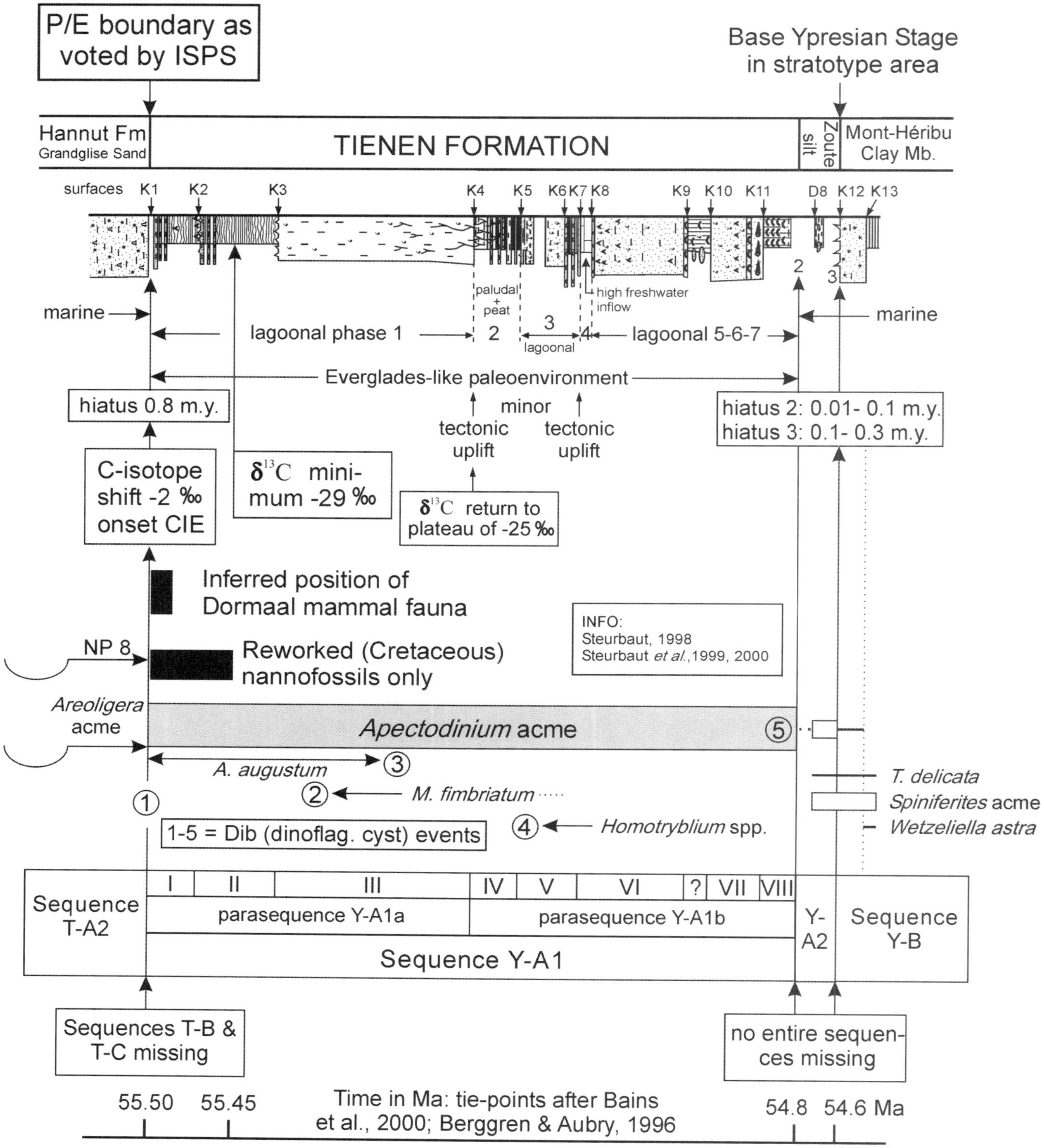

Figure 9. Compilation Kallo-Doel. Summary of the chronostratigraphic, biotic, tectonic, and paleoenvironmental history of the Paleocene-Eocene boundary interval in northern Belgium.

and by minor and major breaks in sedimentation. The stratigraphical architecture of the Tienen Formation is complex, consisting of at least 2 parasequences, together defining sequence Y-A1. The differences between the Kallo and Doel boreholes document substantial changes in accommodation space in the basin from the upper part of the Tienen Formation onward, up to the top of the Zoute Silt. Despite this instability, the vegetation remained fairly unchanged across the IETM (80% of the taxa persisted) and was dominated by arboreal species, mainly Taxodiaceae, Palmae and Juglandaceae. It was marked by low extinction and first-occurrence rates.

The tectonically, pale-oceanographically and climatologically unstable IETM period contrasted much with the pre-IETM and post-IETM history of the southern North Sea Basin. Before the IETM (deposition of the sandy Hannut Formation during Biochron NP8) and after the IETM (deposition of the Zoute Silt, during Biochron NP10 and the Mont-Héribu Clay, during transition of Biochron NP10 and Biochron NP11) distinct but widely uniform depositional regimes existed throughout Belgium because of the development of widespread marine offshore settings and tectonic quiescence. The Zoute Silt Member seems to be deposited during a transitional phase, marked by rather uniform shallow marine facies in relatively restricted areas and by diversified dinoflagellate cyst assemblages with moderately high *Apectodinium* values (28% at the base to 10% at the top). The return to tectonic stability at the base of the Mont-Héribu Member, and its lateral equivalent the Walton Member in England (King, 1981; Ellison et al., 1994), coincides with the disappearance of *Apectodinium*-containing dinoflagellate associations and the appearance of the genus *Wetzeliella.* These events, positioned at the base of the Ypresian s.str., together with the abrupt tectonic and paleoenvironmental changes at the onset of the CIE, are among the most crucial in the Paleogene depositional history of the southern North Sea Basin.

ACKNOWLEDGMENTS

The authors wish to thank Dr. D. Blamart (Lab. Sciences du Climat et de l'Environnement, Gif-sur-Yvette, France) for constructive comments on the methodology and the interpretation of the isotope data, Prof. Dr. N. Vandenberghe (KULeuven) and Dr. J. De Coninck (RUGent) for many valuable discussions concerning certain mineralogical or micropaleontological aspects. Dr. C. King (Greenwich University), Dr. R.W.O'B. Knox (British Geological Survey) and Prof. Dr. N. Vandenberghe (KULeuven) are greatly acknowledged for their critical reviews and useful suggestions. Dr. L. Wouters (Niras, Brussels), Dr. P. Laga (Geological Survey, Brussels), Mrs. I. Wemaere and Mr. S. Labat (SCK, Mol) are thanked for allowing access to the borehole cores, Mr. K. Belmili and Mr. D. Ostojski (Univ. Liège) for assistance in the pollen and spore analyses. Mr. H. De Potter (RBINS) provided graphical assistance. This study was supported by the Royal Belgian Institute for Natural Sciences (RBINS, Brussels), the "Faculté Polytechnique de Mons," the "Laboratoire des Sciences du Climat et de l'Environnement" (Gif-sur-Yvette, France) and the National Fund for Scientific Research (Belgium), Convention FRFC-IC 2.4540.99.

REFERENCES CITED

Adatte, T., Bolle, M.-P., De Kaenel, E., Gawenda, P., Winkler, W., and von Salis, K., 2000, Climatic evolution from Paleocene to earliest Eocene inferred from clay-minerals: A transect from northern Spain (Zumaya) to southern (Spain, Tunisia) and southeastern Tethys margins (Israel, Negev), *in* Schmitz, B., et al., eds., Early Paleogene warm climates and biosphere dynamics: GFF (Geologiska Föreningens i Stokholm Förhandlingar) (Journal of the Geological Society of Sweden), v. 122, p. 7–8.

Ash, A.W., and Wing, S.L., 2001, Climate and biota of the early Paleogene: International Meeting July 3–8, 2001, Northwest College, Powell, Wyoming, USA, Abstract volume, 104 p.

Aubry, M.-P., 1998, The upper Paleocene–lower Eocene stratigraphic puzzle, *in* La limite Paléocène-Eocène en Europe: Événements et correlations: Strata, v. 9, p. 7–9.

Aubry, M.-P., 2000, The vote on the Paleocene/Eocene criterion: The view of the Chairman of the P-E boundary Working Group: Newsletter of the International Subcommission on Paleogene Stratigraphy, Tübingen, November 2000, v. 9, p. 14–17.

Aubry, M.-P., Lucas, S., and Berggren, W.A., editors, 1998, Late Paleocene–early Eocene climatic and biotic events in the marine and terrestrial record: New York, Columbia University Press, 513 p.

Backman, J., and Shackleton, N.J., 1983, Quantitative biochronology of Pliocene and early Pleistocene calcareous nannofossils from the Atlantic, Indian, and Pacific Oceans: Micropaleontology, v. 8, p. 141–170.

Bains, S., Corfield, R., and Norris, R., 2000, Structure of the late Paleocene carbon isotope excursion, *in* Schmitz, B., et al., eds., Early Paleogene warm climates and biosphere dynamics: GFF (Geologiska Föreningens i Stokholm Förhandlingar) (Journal of the Geological Society of Sweden), v. 122, p. 19–20.

Berggren, W.A., and Aubry, M.-P., 1996, A late Paleocene–early Eocene northwest European and North Sea magnetobiochronological correlation network, *in* Knox, R.W.O'B., et al., eds., Correlation of the early Paleogene in northwest Europe: Geological Society [London] Special Publication 101, p. 309–352.

Bolle, M.-P., Pardo, A., Adatte, T., Aziz Tantawy, A., Hinrichs, K.-U., von Salis, K., and Burns, S., 2000, Climatic evolution on the southern and northern margins of the Tethys from the Paleocene to the early Eocene, *in* Schmitz, B., et al., eds., Early Paleogene warm climates and biosphere dynamics: GFF (Geologiska Föreningens i Stokholm Förhandlingar) (Journal of the Geological Society of Sweden), v. 122, p. 31–32.

Bowen, G.J., Koch, P.L., Gingerich, P.D., Norris, R.D., Bains, S., and Corfield, R.M., 2001, Refined isotope stratigraphy across the continental Paleocene/ Eocene boundary on Polecat Bench in the northern Bighorn Basin, *in* Gingerich, P.D., ed., Paleocene-Eocene stratigraphy and biotic change in the Bighorn and Clarks Fork Basins, Wyoming: University of Michigan Papers on Paleontology, v. 33, p. 73–88.

Brinkhuis, H., and Crouch, E.M., 2001, Cretaceous/Tertiary and Paleocene/ Eocene parallels: A dinoflagellate perspective, *in* Ash, A.W., and Wing, S.L., eds., Climate and biota of the early Paleogene: International Meeting July 3–8, 2001, Northwest College, Powell, Wyoming, USA, Abstract volume, p. 17.

Cramer, B.S., Aubry, M.-P., Miller, K.G., Olsson, R.K., Wright, J.D., and Kent, D.V., 1999, An exceptional chronologic, isotopic, and clay mineralogic record at the latest Paleocene thermal maximum, Bass River, NJ, ODP 174AX, *in* Thiry, M., and Aubry, M.-P., coor., Paleocene/Eocene boundary in Europe: Events and correlations: Bulletin de la Société géologique de France, v. 170, no. 6, p. 883–897.

Crouch, E.M., Bujak, J.P., and Brinkhuis, H., 2000, Southern and Northern Hemisphere dinoflagellate cyst assemblage changes in association with the late Paleocene thermal maximum, *in* Schmitz, B., et al., eds., Early Paleogene warm climates and biosphere dynamics: GFF (Geologiska Föreningens i Stokholm Förhandlingar) (Journal of the Geological Society of Sweden), v. 122, p. 40–41.

De Coninck, J., 1975, Microfossiles à paroi organique de l'Yprésien du Bassin belge: Professional Paper, Belgische Geologische Dienst (1975), v. 12, p. 1–151.

De Coninck, J., 1986, Microfossiles à paroi organique de l'Yprésien inférieur à Quenast: Professional Paper, Belgische Geologische Dienst (1986/1), no. 224, p. 1–59.

De Coninck, J., 1999a, Appearances of dinoflagellate species recorded in the Tienen Formation (Landen Group) and in the Kortrijk Formation (Ieper Group) in the Belgian Basin: Their correlation to transgressive phases in the southern part of the North Sea Basin, *in* Thiry, M., and Aubry, M.-P., coor., Paleocene/Eocene boundary in Europe: Events and correlations: Bulletin de la Société Géologique de France, v. 170, no. 1, p. 77–84.

De Coninck, J., 1999b, Phytoplancton à paroi organique et phases transgressives vers la transition paléocène-éocène dans la partie méridionale du Bassin de la Mer du Nord: Bulletin de la Société Belge de Géologie, v. 105(3–4) (1996), p. 139–169.

Dupuis, C., and Thiry, M., 1998, Geological frame of the "Sparnacian," *in* Thiry, M., and Dupuis, C., eds., The Paleocene/Eocene boundary in Paris Basin: The Sparnacian deposits: Field Trip Guide: Mémoires des Sciences de la Terre, Ecole des Mines de Paris, v. 34, no. 1, p. 3–12.

Dupuis, C., De Coninck, J., Guernet, C., and Roche, E., 1990, Biostratigraphic data: Ostracods and organic walled microfossils of the Landen Formation and the Base of the Ieper Formation in the Knokke borehole, *in* Laga, P., and Vandenberghe, N., eds., The Knokke well (11E/138) with a description of the Den Haan (22W/276) and Oostduinkerke (35E/142) wells: Mémoires pour servir à l'explication des Cartes géologiques et minières de la Belgique, v. 29, p. 33–45.

Dupuis, C., Steurbaut, E., De Coninck, J., and Riveline, J., 1998, The Western Argiles à lignite facies: Cap-d'Ailly sections, *in* Thiry, M., and Dupuis, C., eds., The Paleocene/Eocene boundary in Paris Basin: The Sparnacian deposits: Field Trip Guide: Mémoires des Sciences de la Terre, École des Mines de Paris, v. 34, no. 9, p. 60–71.

Ellison, R.A., Knox, R.W.O'B., Jolley, D.W., and King, C., 1994, A revision of the lithostratigraphical classification of the early Palaeogene strata of the London Basin and East Anglia: Proceedings of the Geologists' Association, v. 105, p. 187–197.

Gruas-Cavagnetto, C., 1968, Etude palynologique des divers gisements du Sparnacien du Bassin de Paris: Mémoires de la Société Géologique de France, N.S., XLVII, no. 110, p. 1–144.

Gruas-Cavagnetto, C., 1970, Microflore et microplancton des Woolwich Beds (Swanscombe, Kent): Pollen et Spores, v. 12, no. 1, p. 71–82.

Gulinck, M., 1969, Coupe résumée des terrains traversés au sondage de Kallo et profil géologique NS passant par Woensdrecht-Kallo-Halle: Mémoires pour servir à l'explication des Cartes géologiques et minières de la Belgique, v. 11, p. 1–7.

Gulinck, M., 1973, Note sur l'extension des faciès fluvio-lagunaires du Landénien de la Belgique: Mémoires pour servir à l'explication des Cartes géologiques et minières de la Belgique, v. 13, p. 1–12.

Heilmann-Clausen, C., 1985, Dinoflagellate stratigraphy of the uppermost Danian to Ypresian in the Viborg 1 borehole, central Jylland, Denmark: Danmarks Geologiske Undersøgelse A, v. 7, p. 1–69.

King, C., 1981, The stratigraphy of the London Clay and associated deposits: Tertiary Research Special Paper 6, p. 1–158.

King, C., 1990, Eocene stratigraphy of the Knokke borehole (Belgium), *in* Laga, P., and Vandenberghe, N., eds., The Knokke well (11E/138) with a description of the Den Haan (22W/276) and Oostduinkerke (35E/142) wells: Toelichtende Verhandelingen voor de Geologische en Mijnkaarten van België, no. 29, p. 67–102.

Knox, R.W.O'B., 1996, Tectonic controls on sequence development in the Paleocene and earliest Eocene of southeast England: Implications for North Sea stratigraphy, *in* Hesselbo, S.P., and Parkinson, D.N., eds., Sequence stratigraphy in British geology: Geological Society [London] Special Publication 103, p. 209–230.

Knox, R.W.O'B., 1998, Kaolinite influx within Paleocene/Eocene boundary strata of western Europe (Extended abstract): Newsletters on Stratigraphy, v. 36, no. 1, p. 49–53.

Koch, P.L., and Morrill, C., 2000, Paleocene-Eocene climate change in southwestern Wyoming: Isotopic records from paleosol carbonates and freshwater bivalves, *in* Schmitz, B., et al., eds., Early paleogene warm climates and biosphere dynamics: GFF (Geologiska Föreningens i Stokholm Förhandlingar) (Journal of the Geological Society of Sweden), v. 122, p. 89–91.

Koch, P.L., Zachos, J.C., and Gingerich, P.D., 1992, Correlation between isotope records in marine and continental carbon reservoirs near the Paleocene/Eocene boundary: Nature, v. 358, p. 319–322.

Koch, P.L., Zachos, J.C., and Dettman, D.L., 1995, Stable isotope stratigraphy and paleoclimatology of the Paleogene Bighorn Basin (Wyoming, USA): Palaeogeography, Palaeoclimatology, Palaeoecology, v. 115, p. 61–89.

Laenen, B., 1999, The geochemical signature of relative sea-level cycles recognized in the Boom Clay: Aardkundige Mededelingen, v. 9 (1998), p. 61–82.

Laga, P., and Vandenberghe, N., editors, 1990, The Knokke well (11E/138) with a description of the Den Haan (22W/276) and Oostduinkerke (35E/142) wells: Toelichtende Verhandelingen voor de Geologische en Mijnkaarten van België, no. 29, p. 1–117.

Leriche, M., 1928, Sur la répartition des facies lagunaires et fluviatiles du Landénien, dans les Bassins belges et parisiens: Bulletin de la Société Belge de Géologie, Paléontologie et Hydrologie, v. XXXVIII (1928), p. 69–91.

Luterbacher, H.P., Hardenbol, J., and Schmitz, B., 2000, Decision of the voting members of the International Subcommission on Paleogene Stratigraphy on the criterion for the recognition of the Paleocene/Eocene boundary: Newsletter of the International Subcommission on Paleogene Stratigraphy, Tübingen, November 2000, v. 9, p. 13.

Magioncalda, R., Dupuis, C., Blamart, D., Fairon-Demaret, M., Perreau, M., Renard, M., Riveline, J., Roche, M., and Keppens, E., 2001, L'excursion isotopique du carbone organique (δ^{13}Corg) dans les paléoenvironnements continentaux de l'intervalle Paléocène/Eocène de Varengeville (Haute Normandie): Bulletin de la Société Géologique de France, v. 172, no. 3, p. 349–358.

Mercier-Castiaux, M., 1986, Etude de la sédimentation argileuse tertiaire dans le Bassin belgo-franco-anglais: Comparaison avec l'Océan Atlantique Nord [Ph.D. thesis]: Lille, France, Université de Lille, 189 p.

Mercier-Castiaux, M., and Dupuis, C., 1991, Clay mineral associations in the Ypresian formations in the northwest European Basin: Time and geographical variations: Interpretations, *in* Dupuis, C., et al., eds., The Ypresian stratotype: Bulletin de la Société Belge de Géologie, v. 97, 3–4 (1988), p. 441–450.

Mercier-Castiaux, M., Chamley, H., and Dupuis, C., 1988, La sédimentation argileuse tertiaire dans le bassin belge et ses approches occidentaux: Annales de la Société Géologique du Nord, v. CVII, p. 139–154.

Moorkens, T., Steurbaut, E., Jutson, D., and Dupuis, C., 2000, The Knokke borehole of northwestern Belgium reanalyzed: New data on the Paleocene/Eocene transitional strata in the southern North Sea Basin, *in* Schmitz, B., et al., eds., Early Paleogene warm climates and biosphere dynamics: GFF (Geologiska Föreningens i Stokholm Förhandlingar) (Journal of the Geological Society of Sweden), v. 122, p. 111–114.

Norris, R.D., and Röhl, U., 1999, Carbon cycling and chronology of climate warming during the Paleocene-Eocene transition: Nature, v. 401, p. 775–778.

Perch-Nielsen, K., 1985, Cenozoic calcareous nannofossils, *in* Bolli, H.M., et al., eds., Plankton stratigraphy: Cambridge Earth Science Series, v. I, no. 11, p. 427–554.

Powell, A.J., 1992, Dinoflagellate cysts of the Tertiary system, *in* Powell, A.J., ed., A stratigraphic index of dinoflagellate cysts: London, Chapman and Hall, p. 155–251.

Rea, D.K., Zachos, J.C., Owen, R.M., and Gingerich, P.D., 1990, Global change at the Paleocene-Eocene boundary: Climatic and evolutionary consequences of tectonic events: Palaeogeography, Palaeoclimatology, Palaeoecology, v. 79, p. 117–128.

Schmitz, B., 2000, Is the CIE in the late Paleocene or the early Eocene?: Newsletter of the International Subcommission on Paleogene Stratigraphy, Tübingen, November 2000, v. 9, p. 19.

Schmitz, B., Speijer, R.P., and Aubry, M.-P., 1996, Latest Paleocene benthic extinction event on the southern Tethyan shelf (Egypt): Foraminiferal stable isotopic ($\delta^{13}C$, $\delta^{18}O$) records: Geology, v. 24, p. 347–350.

Schmitz, B., Sundquist, B., and Andreasson, F.P., editors, 2000, Early Paleogene warm climates and biosphere dynamics: GFF (Geologiska Föreningens i Stokholm Förhandlingar) (Journal of the Geological Society of Sweden), v. 122, 189 p.

Speijer, R.P., and Schmitz, B., 2000, A synthesis of biotic and stratigraphic data from the Middle East on late Paleocene global change, *in* Schmitz, B., et al., eds., Early Paleogene warm climates and biosphere dynamics: GFF (Geologiska Föreningens i Stokholm Förhandlingar) (Journal of the Geological Society of Sweden), v. 122, p. 152–153.

Steurbaut, E., 1993, High-resolution holostratigraphy of late Paleocene to early Eocene strata in Belgium, Southern North Sea Basin, *in* Knox, R.W.O'B., Corfield, R., and Dunay, R.E., convenors, Correlation of the early Paleogene in northwest Europe, International Meeting Geological Society of London, 1–2 December 1993, London: Programme and Abstracts volume, 2 p. (not numbered).

Steurbaut, E., 1998, High-resolution holostratigraphy of middle Paleocene to early Eocene strata in Belgium and adjacent areas: Palaeontographica Abt. A, v. 247, p. 91–156.

Steurbaut, E., De Coninck, J., Roche, E., and Smith, T., 1999, The Dormaal Sands and the Paleocene/Eocene boundary in Belgium, *in* Thiry, M., and Aubry, M.-P., coor., Paleocene/Eocene boundary in Europe: Events and correlations: Bulletin de la Société Géologique de France, v. 170, no. 2, p. 217–227.

Steurbaut, E., De Coninck, J., Dupuis, C., and King, C., 2000, Dinoflagellate cyst events and depositional history of the Paleocene/Eocene boundary interval in the southern North Sea Basin, *in* Schmitz, B., et al., eds., Early Paleogene warm climates and biosphere dynamics: GFF (Geologiska Föreningens i Stokholm Förhandlingar) (Journal of the Geological Society of Sweden), v. 122, p. 154–157.

Stott, L.D., Sinha, A., Thiry, M., Aubry, M.-P., and Berggren, W.A., 1996, Global $\delta^{13}C$ changes across the Paleocene-Eocene boundary: Criteria for terrestrial-marine correlations, *in* Knox, R.W.O'B., et al., eds., Correlation of the early Paleogene in northwest Europe: Geological Society [London] Special Publication 101, p. 381–399.

Streel, M., and Bless, M.J.M., 1980, Occurrence and significance of reworked palynomorphs: Mededelingen Rijks Geologische Dienst, no. 32, no. 10, p. 69–80.

Thiry, M., and Dupuis, C., 1998, Chemostratigraphy of the Sparnacian deposits and correlation to deep-sea record, *in* Thiry, M., and Dupuis, C., eds., The Paleocene/Eocene boundary in Paris Basin: The Sparnacian deposits: Field Trip Guide: Mémoires des Sciences de la Terre, Ecole des Mines de Paris, v. 34, no. 6, p. 39–47.

Thiry, M., and Dupuis, C., 2000, Use of clay minerals for paleoclimatic reconstructions: Limits of the method with special reference to the Paleocene–lower Eocene interval, *in* Schmitz, B., et al., eds., Early Paleogene warm climates and biosphere dynamics: GFF (Geologiska Föreningens i Stokholm Förhandlingar) (Journal of the Geological Society of Sweden), v. 122, p. 166–167.

Thiry, M., Dupuis, C., Sinha, A., Aubry, M.-P., Stott, L.D., and Berggren, W.A., 1998, Carbon isotopic signature of Sparnacian deposits, *in* Thiry, M., and Dupuis, C., eds., The Paleocene/Eocene boundary in Paris Basin: The Sparnacian deposits: Field Trip Guide: Mémoires des Sciences de la Terre, Ecole des Mines de Paris, v. 34, no. 5, p. 28–38.

Thomas, E., and Shackleton, N.J., 1996, The Paleocene-Eocene benthic foraminiferal extinction and stable isotope anomalies, *in* Knox, R.W.O'B., et al., eds., Correlation of the early Paleogene in northwest Europe: Geological Society [London] Special Publication 101, p. 401–441.

Vandenberghe, N., Laga, P., Steurbaut, E., Hardenbol, J., and Vail, P.R., 1998, Tertiary sequence stratigraphy at the southern border of the North Sea Basin in Belgium, *in* de Graciansky, P.-C., et al., eds., Mesozoic and Cenozoic sequence stratigraphy of European basins: SEPM (Society for Sedimentary Geology) Special Publication 60, p. 119–154.

Zachos, J.C., Lohmann, K.C, Walker, J.C.G., and Wise, S.W., 1993, Abrupt climate change and transient climates during the Paleogene: A marine perspective: Journal of Geology, v. 101, p. 191–213.

Manuscript Accepted by the Society August 13, 2002

Printed in the USA

Geological Society of America
Special Paper 369
2003

Extinction and food at the seafloor: A high-resolution benthic foraminiferal record across the Initial Eocene Thermal Maximum, Southern Ocean Site 690

Ellen Thomas*
Department of Earth and Environmental Sciences, Wesleyan University, Middletown, Connecticut 06459-0139, USA

ABSTRACT

A mass extinction of deep-sea benthic foraminifera has been documented globally, coeval with the negative carbon isotope excursion (CIE) at the Paleocene-Eocene boundary, which was probably caused by dissociation of methane hydrate. A detailed record of benthic foraminiferal faunal change over ~30 k.y. across the carbon isotopic excursion at Ocean Drilling Program Site 690 (Southern Ocean) shows that shortly before the CIE absolute benthic foraminiferal abundance at that site started to increase. "Doomed species" began to decrease in abundance at the CIE, and became smaller and more thin-walled. The main phase of extinction postdated the CIE by a few thousand years. After the extinction faunas were dominated by small species, which resemble opportunistic taxa under high-productivity regions in the present oceans. Calcareous nannofossils (primary producers), however, show a transition to more oligotrophic nannofloras exactly where the benthic faunas show the opposite. Plankton and benthos is thus decoupled. Possibly, a larger fraction of food particles reached the seafloor after the CIE, so that food for benthos increased although productivity declined. Enhanced organic preservation might have resulted from low-oxygen conditions caused by oxidation of methane. Alternatively, and speculatively, there was a food-source at the ocean-floor. Benthic foraminifera dominating the post-extinction fauna resemble living species that symbiotically use chemosynthetic bacteria at cold seeps. During increased, diffuse methane escape from hydrates, sulfate-reducing bacteria could have produced sulfide used by chemosynthetic bacteria, which in turn were used by the benthic foraminifera, causing extinction by a change in food supply.

INTRODUCTION

A major extinction of deep-sea benthic foraminifera in open ocean occurred at the Paleocene-Eocene boundary (as defined recently, Luterbacher et al., 2000), with 35%–50% of deep-sea species becoming extinct (e.g., Thomas, 1990, 1998; Pak and Miller, 1992; Coccioni et al., 1994; Katz et al., 1999). At this time, many long-lived bathyal to upper abyssal species, which had been in existence from the Maastrichtian or Campanian, became extinct within ~10,000 yr, as observed at locations where precise correlation is available (e.g., Thomas, 1998). Rapid extinction of many deep-sea benthic foraminiferal species at the same time is very unusual in earth history, and most faunal changes of deep-sea faunas occur gradually, over hundreds of

*Also at: Center for the Study of Global Change, Department of Geology and Geophysics, Yale University, New Haven, Connecticut 06520-8109, USA.

Thomas, E., 2003, Extinction and food at the seafloor: A high-resolution benthic foraminiferal record across the Initial Eocene Thermal Maximum, Southern Ocean Site 690, *in* Wing, S.L., Gingerich, P.D., Schmitz, B., and Thomas, E., eds., Causes and Consequences of Globally Warm Climates in the Early Paleogene: Boulder, Colorado, Geological Society of America Special Paper 369, p. 319–332.

thousands to a few million years (e.g., Thomas, 1992). The Maastrichtian-Paleocene bathyal and abyssal benthic foraminiferal faunas were globally rather uniform although relative abundances varied geographically (e.g., Widmark and Speijer, 1997), and had such common constituent species as *Stensioeina beccariiformis, Pullenia coryelli, Alabamina creta, Clavulinoides paleocenica, Clavulinoides havanensis, Gyroidinoides globosus, Paralabamina hillebrandti* and *Paralabamina lunata.*

The extinction has been recognized at many bathyal through upper abyssal sites worldwide (see review in Thomas, 1998). At lower abyssal depths below the Calcium carbonate Compensation Depth (CCD), agglutinated foraminiferal faunas show a probably coeval change in faunal composition although no major faunal turnover occurred (e.g., Kaminski et al., 1996; Galeotti et al., 2003). In agglutinated assemblages, the counterpart of the post-extinction faunas in calcium carbonate foraminifera is the "*Glomospira* assemblage," which has been recognized in lower Eocene sediments over a large area in the North Atlantic and western Tethys. Lesser extinctions or temporary changes in benthic foraminiferal faunal composition occurred in marginal and epicontinental basins (e.g., Speijer and Schmitz, 1998; Stupin and Muzylöv, 2001; Speijer and Wagner, 2002).

The extinction was coeval with rapid global warming and a negative excursion of at least 2.5‰ in $\delta^{13}C$ values in the oceanic and atmospheric carbon reservoirs (Kennett and Stott, 1991; Pak and Miller, 1992; Kaiho et al., 1996; Thomas and Shackleton, 1996; Koch et al., 1992, 1995; Stott et al., 1996; Fricke et al., 1998; Bowen et al., 2001). This carbon isotope excursion started rapidly, over less and possibly considerably <10,000 yr, with carbon isotope values returning to background in several hundred thousand years (Katz et al., 1999; Norris and Röhl, 1999; Röhl et al., 2000, this volume; Cramer, 2001). These authors documented that the initiation of the carbon isotope excursion (CIE) occurred over a period shorter than one precessional cycle.

The cause of the CIE probably was a rapid dissociation of sedimentary methane hydrates (e.g., Dickens et al., 1995; Dickens, 2000, 2001a), possibly in several steps (Bains et al., 1999). The cause of the dissociation is not clear, and might have been rapid warming of the intermediate ocean waters as a result of changing oceanic circulation patterns (e.g., Kennett and Stott, 1991; Kaiho et al., 1996; Thomas et al., 2000; Zachos et al., 2001; Dickens, 2001a; Bice and Marotzke, 2002), continental slope failure as the result of increased current strength in the North Atlantic Ocean (Katz et al., 1999, 2001), sea-level lowering (Speijer and Morse, 2002), the impact of a comet (Kent et al., 2001) or other extraterrestrial body (Dolenec et al., 2000), or a combination of several of these factors.

In the deep-sea, post-extinction faunas of carbonate foraminifera are mainly dominated by small individuals, with thin walls, and carbonate-cemented agglutinants are absent (e.g., Thomas, 1998). These faunas are low-diversity as compared with pre-extinction assemblages, in species richness as well as in such measures as alpha-diversity and rarefied number of species, because they are strongly dominated by few species. These post-extinction faunas were considerably more variable from site to site than the pre-extinction faunas, being at some sites (many in the South Atlantic) dominated by the oligotrophic indicator *Nuttallides truempyi* and possibly abyssaminids, at others by common eutrophic or low-oxygen indicators such as buliminid species (Thomas, 1998; Thomas and Röhl, 2002).

In marginal regions such as the Tethys and northeastern Peri-Tethys, faunal changes including extinction have been linked to low-oxygen conditions resulting from local high primary productivity of noncarbonate primary producers, as shown by the occurrence of the extinction at the base of laminated, dark-brown to black sediments with high concentrations of organic matter (Gavrilov et al., 1997, this volume; Stupin and Muzylöv, 2001; Speijer and Wagner, 2002). In these regions, faunal as well as geochemical evidence (Schmitz et al., 1997; Gavrilov et al., this volume) supports the theory that high productivity was the major cause of anoxia and extinction, possibly exacerbated by stratification of the water column and low ventilation. High productivity may also have occurred in shallow waters along the northeastern seaboard of America (Gibson et al., 1993). In an upper-middle bathyal section in New Zealand, laminated dark shales are slightly enriched in organic carbon; although surface productivity may have been high, benthic foraminiferal productivity dropped strongly at the extinction event, probably as the result of low-oxygen conditions (Kaiho et al., 1996).

In the open ocean, in contrast, the exact causes of the extinction are not so clear, although the extinction was more severe than in marginal basins. In land sections representing bathyal-abyssal depths (e.g., Spain, Coccioni et al., 1994; Ortiz, 1995) the benthic extinction commonly occurred in an interval with strong carbonate dissolution. Post-extinction faunas thus are enriched in agglutinated species, followed by faunas rich in thin-walled, small specimens of the oligotrophic indicator species *N. truempyi.* In some open-ocean sites (e.g., Site 999 in the Caribbean, and Sites 525 and 527 on Walvis Ridge, South Atlantic; site 929, Ceara Rise), there also is an interval of strong carbonate dissolution, where dark clays are not enriched in organic matter, and where neither benthic foraminifera nor planktonic organisms indicate high productivity (Thomas and Shackleton, 1996; D. Thomas et al., 1999; Thomas and Röhl, 2002).

At other locations (equatorial Pacific Site 865, Southern Ocean Sites 689, 690) there are sediment color variations (see Cramer, 2001), and variations in carbonate percentage (D. Thomas et al., 1999), but the percentage carbonate does not fall below ~65%. At these three sites, deep-sea benthic foraminiferal faunas suggestive of a high food supply or low oxygen levels cooccur with planktic foraminiferal and nannofossil assemblages indicative of low productivity (e.g., Thomas, 1998).

The benthic foraminiferal extinction occurred during a time characterized in general by decreasing oceanic productivity, after a high in productivity in the middle Paleocene (Aubry, 1998; Boersma et al., 1998; Corfield and Norris, 1998). Trace element (Ba) data from the Initial Eocene Thermal Maximum, however, have been interpreted as indicative of increased productivity in open ocean at such locations as Site 690 (Bains et al., 2000).

Possible causes of the benthic extinction event (BEE) at bathyal-abyssal depths in open ocean thus include low oxygenation, either as a result of increased deep sea temperatures or as a result of oxidation of methane in the water column, increased corrosivity of the waters for $CaCO_3$ as a result of methane oxidation, increased temperatures, globally increased or decreased productivity, locally increased and/or decreased productivity as a result of an expansion of the trophic resource continuum, or a combination of several of these factors (e.g., Boersma et al., 1998; Thomas, 1998; Thomas et al., 2000).

In this paper, the question of the cause(s) of the benthic foraminiferal extinction is addressed by a high-resolution study of benthic foraminifera at Ocean Drilling Program Site 690, which has been intensively studied for many parameters (e.g., Kennett and Stott, 1991; Thomas, 1990; Thomas and Shackleton, 1996; Thomas et al., 2000; Bains et al., 1999, 2000; Norris and Röhl, 1999; Röhl et al., 2000; Cramer, 2001; Bralower, 2002; D. Thomas et al., 2002). I intend to document the relative abundances of benthic foraminiferal species across the extinction in detail, and compare the benthic foraminiferal data with data on calcareous nannoplankton (Bralower, 2002) and geochemical data (Bains et al., 1999, 2000). Comparison of data on surface primary producers and bottom dwellers might help in trying to understand whether we have sufficient information to deduce whether productivity changed, and if it changed, whether it decreased or increased.

METHODS

Site 690 was drilled on Maud Rise, an aseismic ridge at the eastern entrance of the Weddell Sea in the Atlantic sector of the Southern Ocean (65°9.629′S, 1°12.296′E), at a present water depth of 2914 m, paleodepth during the Paleocene-Eocene ~1900–2000 m (Thomas, 1990). Site 690 is on the southwestern flank of this ridge. A continuous U-channel sample was collected from the archive half of Core 690B-19H, over the interval between 690B-19H-3, 50 cm to –121.5 cm (Bralower, 2002; D. Thomas et al., 2002). The U-channel was sectioned into 1 cm samples, which were processed (dried and sieved) at the University of North Carolina. The fraction >63 μ was used for the benthic foraminiferal study, from the interval between 690B-19H-3, 51–52 cm, and 690B-19H-3, 84–85 cm (170.42 and 170.75 mbsf), corresponding to ~31,000 yr in the time scale of Röhl et al. (2000), but less so in that by Cramer (2001).

In the samples located above the level of the benthic foraminiferal extinction, many specimens are small. The abundant specimens of biserial and triserial, small taxa may be washed through the sieve with 63 μ (230 mesh) openings and thus may be lost (see, for example, Thomas, 1985). It is therefore possible that samples washed more vigorously than others could have fewer of such specimens than other samples. Washing of the samples was not standardized, but the sediments were not strongly indurated, and have a very similar degree of induration over the short studied interval, so that the washing process was similar for all samples. This thus probably did not have a major effect on the observations on benthic foraminifera, especially because the samples with dominant small specimens were much richer in numbers of foraminifera than samples with larger specimens, suggesting that these faunas were not largely lost in sieving.

Benthic foraminifera were picked from 24 samples, at 1- or 2-cm intervals (Table DR1[1]). All samples contained sufficient specimens for study. Between 250 and 400 individuals were picked from a split of the samples (Table DR1). The weight of the sample split from which foraminifera were picked was determined, so that the number of foraminifera per gram of sediment (>63 μ) could be determined. Taxonomy follows Thomas (1990) and Thomas and Shackleton (1996), but modified in the use of generic names after Alegret and Thomas (2001) and as discussed in Appendix 1. In addition, species formerly classified in the genus *Stilostomella* now are placed in *Siphonodosaria* and *Myllostomella,* following Hayward (2002).

In order to estimate diversity, the species richness numbers were rarified following Sanders (1968), so that the number of species per 100 individuals was estimated.

RESULTS

The high-resolution data confirm that the benthic foraminiferal extinction event (BEE) occurred rapidly, over a few cm of sediment, with a rapid loss of diversity, extinction of species locally that also became extinct globally, and a change toward a strongly increased relative and absolute abundance of species from the Superfamily Buliminacea, generally thought to indicate a high food supply to the sediment and/or low-oxygen conditions (Figs. 1–4). At this high resolution, however, the BEE shows a more complex structure than earlier documented (Thomas, 1990; Thomas and Shackleton, 1996).

The sharp decline in species diversity occurred at 170.60 mbsf, slightly above the largest decrease in bulk sediment $\delta^{13}C$ at 170.63 mbsf (Bains et al., 1999). The species that became extinct at the event, however (e.g., *Stensioeina beccariiformis;* several calcareous agglutinated species such as *Clavulinoides havanensis, C. asper*a, *Paralabamina hillebrandti, P. lunata, Pullenia coryelli*) drastically declined in abundance exactly at the level of the initiation of the CIE at 170.63 mbsf (Figs. 1A, B). At this level these "extinction species" or "doomed species" decreased in test size, and between 170.63 and 170.60 mbsf they are represented by unusually small, thin-walled specimens only. There are a few large specimens of the "extinction species" present higher in the section, but all these specimens (marked with * in Table DR1) show signs of reworking: They are badly pre-

[1]GSA Data Repository item 2003051, Table DR1, is available on request from Documents Secretary, GSA, P.O. Box 9140, Boulder, CO 80301-9140, USA, editing@geosociety.org, or at www.geosociety.org/pubs/ft2003.htm.

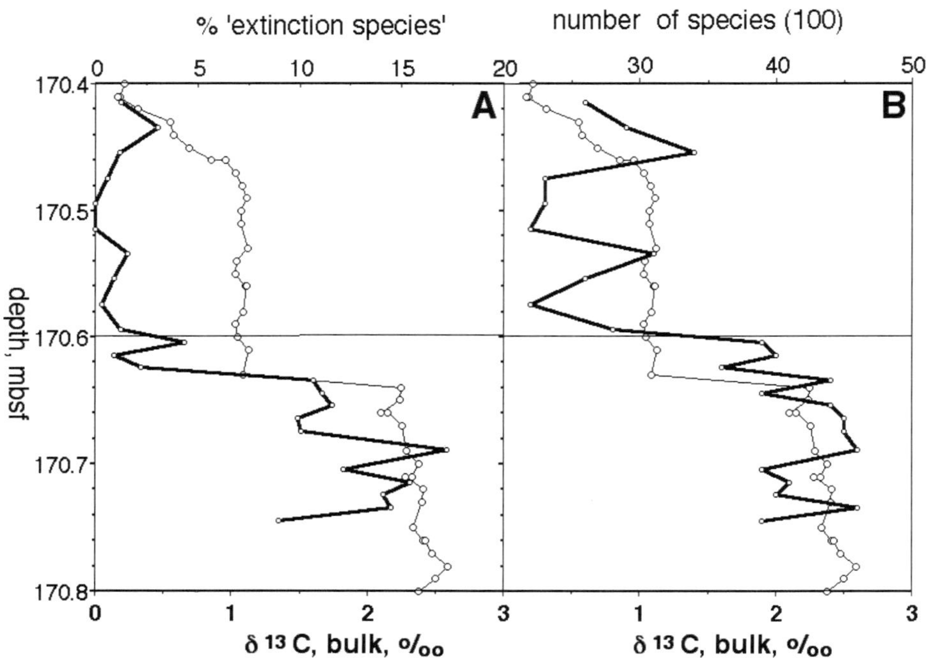

Figure 1. A: Percentage "extinction species" (*Stensioeina beccariiformis, Paralabamina* spp., *Clavulinoides* spp., *Alabamina creta, Pullenia coryelli*) shown by thick line, plotted over $\delta^{13}C$ in bulk sediment (Bains et al., 1999), shown in thin line with open circles. Horizontal line indicates the level of the main phase of benthic foraminiferal extinction. B: Number of species normalized to 100 specimens (rarefied, Sanders, 1968) shown by thick line, plotted over $\delta^{13}C$ in bulk sediment (Bains et al., 1999), shown in the thin line with open circles. Horizontal line indicates the level of the main phase of benthic foraminiferal extinction.

served, abraded, and riddled with fungal borings. Several were analyzed for carbon isotopes, and had pre-extinction values (Thomas and Shackleton, 1996).

At the BEE the species composition of the assemblages changed considerably at Site 690 (Thomas and Shackleton, 1996; Thomas, 1998), with a strong increase in relative abundance of species of the buliminid group ("infaunal species" in Thomas, 1990). These species morphologically resemble species that presently live infaunally, and are common in regions with a high, continuous food supply to the sediment, and/or where low-oxygen conditions may prevail (e.g., Sen Gupta and Machain-Castillo, 1993; Alve, 1995; Bernhard, 1996; Bernhard and Sen Gupta, 1999; Gooday and Rathburn, 1999; van der Zwaan et al., 1999; Gooday, 2002).

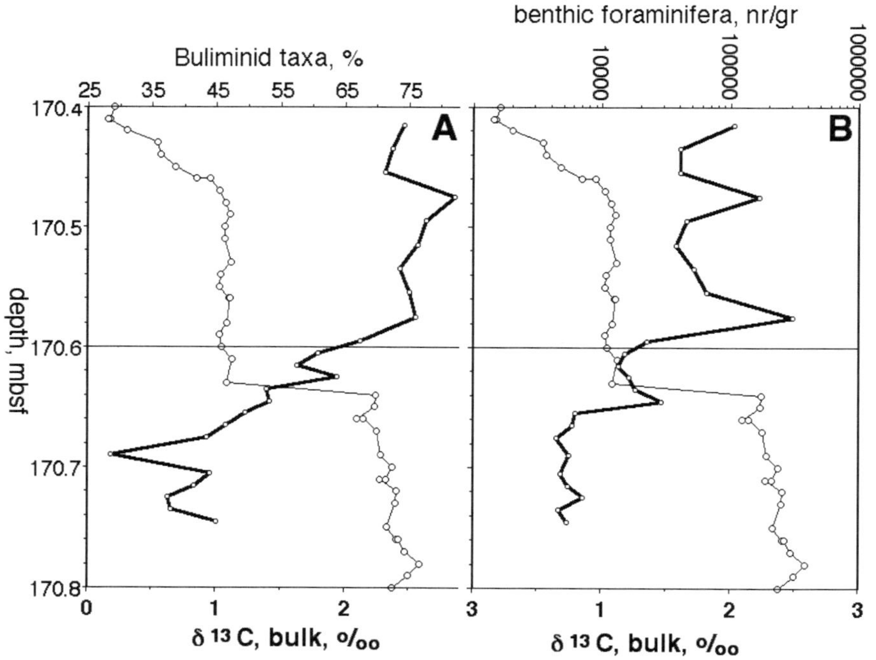

Figure 2. A: Percentage of buliminid taxa (*Bolivinoides* spp., *Bulimina* spp., *Buliminella beaumonti, Fursenkoina* spp., *Rectobulimina carpentierae, Siphogenerinoides* spp., *Tappanina selmensis* and *Turrilina brevispira*) shown by the thick line, plotted over $\delta^{13}C$ in bulk sediment (Bains et al., 1999), shown in the thin line with open circles. Horizontal line indicates the level of the main phase of benthic foraminiferal extinction. B: Number of benthic foraminifera per gram of sediment >63 μ, shown in thick line, plotted over $\delta^{13}C$ in bulk sediment (Bains et al., 1999), shown in the thin line with open circles. Horizontal line indicates the level of the main phase of benthic foraminiferal extinction.

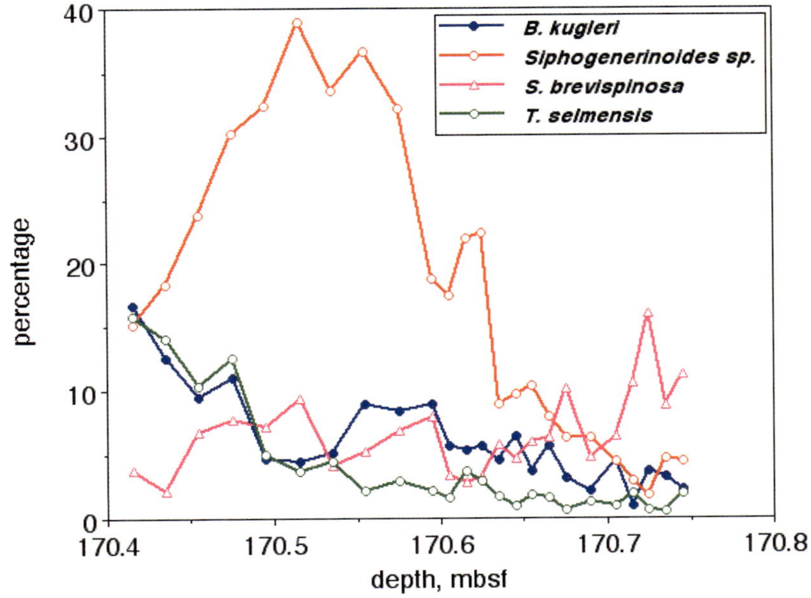

Figure 3. Relative abundances of common small, buliminid taxa.

This increased relative abundance of buliminid taxa after the BEE remains a feature in this high-resolution study (Fig. 2B), but the increase occurred gradually over the short distance between ~170.69 and 170.58 mbsf, and was not instantaneous. Another foraminiferal proxy that may indicate an increased supply of food to the benthic fauna is the absolute abundance of benthic foraminifera (nr/gr; Figure 2B) (Herguera and Berger, 1991; Jorissen et al., 1995), although this proxy may not be applicable under low-oxygen conditions (e.g., Den Dulk et al., 2000). This parameter (plotted on a logarithmic scale) shows a first rapid increase at 170.65 mbsf, just below the CIE. At equatorial Pacific Site 865 the absolute abundance of benthic foraminifera also increased after the BEE, but total numbers of benthic foraminifera per gram are lower by an order of magnitude at that site as compared to Site 690 (Thomas, 1998). I cannot say whether the total benthic foraminiferal biomass increased, however, because the abundant benthic foraminifera after the extinction are so much smaller than the common pre-extinction species that the

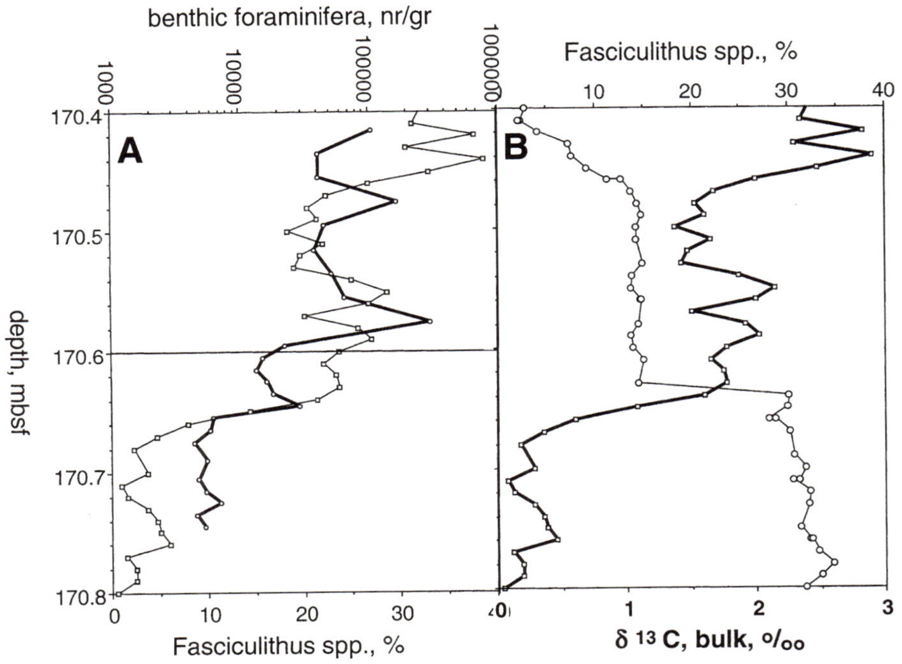

Figure 4. A: Number of benthic foraminifera per gram of sediment >63 μ, shown by the thick line, plotted over the percentage of *Fasciculithus* spp. in the calcareous nannofossil assemblage (Bralower, 2002), shown in the thin line with open circles. Horizontal line indicates the level of the main phase of benthic foraminiferal extinction. B: Percentage of *Fasciculithus* spp. in the calcareous nannofossil assemblage (Bralower, 2002), shown by the thick line, plotted over $\delta^{13}C$ in bulk sediment (Bains et al., 1999), shown in the thin line with open circles.

increased numbers may be balanced by the decreased size per individual. Note that the overall increase in numbers of foraminifera per gram combined with the increased relative abundance of buliminid taxa means that buliminids increased in numbers very strongly.

The taxa that are responsible for this increase in relative abundance of buliminids as well as the increased numbers of benthic foraminifera are somewhat different in this high-resolution report (Table DR1) than was documented earlier (Thomas, 1990; Thomas and Shackleton, 1996). These authors documented that abundant post-extinction taxa were mainly *Tappanina selmensis* and several small *Bolivinoides* and *Bulimina* species, including *Bulimina ovula* (now called *Bulimina kugleri,* see Appendix 1), followed by a short-lived increase in the relative abundance of *Aragonia aragonensis.* These species, however, begin their increase higher in the section, just above the interval studied here, as can be seen by their increasing abundance at the top of the studied interval (Fig. 3).

Between 170.63 mbsf and 170.45 mbsf, directly before and after the extinction, the faunas are dominated by a small taxon called *Siphogenerinoides* sp. (Plate 1; Appendix 1). This taxon is somewhat more flattened in overall test shape than the species *S. brevispinosa.* Both *S. brevispinosa* and *Siphogenerinoides* sp. are small species (diameter of ~20–40 µ, length ~100–150 µ), with *S. brevispinosa* in the lower part of that range in the post-extinction interval.

Siphogenerinoides sp. may be an ecophenotype present only in samples deposited during the unusual circumstances just before and after the extinction, similar to the occurrence of short-lived planktic taxa (Kelly et al., 1996). Other benthic foraminiferal species show similar morphological changes in populations after an extinction, such as across middle Cretaceous Oceanic Anoxic Event OAE1b (Holbourn et al., 2001).

At the BEE, the benthic foraminiferal faunas at Site 690 changed from highly diverse faunas with a rather equal distribution of specimens over species to a much less diverse fauna dominated by a few taxa (Figs. 1A, 1B). The pre-extinction faunas had common large, heavily calcified species with many chambered-tests. Such individuals were probably long-lived, and the assemblages can be interpreted as typical highly diversified faunas containing many specialist species (e.g., Valentine, 1973; Hallock, 1987). Such faunas, dominated by K-mode, specialist taxa, are common in oligotrophic regions (e.g., Boersma et al., 1998; Bralower, 2002).

In stark contrast, the post-extinction faunas are dominated by few species, for a short interval *Siphogenerinoides* sp., and all species are very small (<125 µ), consisting of few chambers (10–12 or fewer, in the case of *S. brevispinosa* and *Siphogenerinoides* sp.). Such faunas are usually common in regions with rapidly varying or disturbed environments, which are more eutrophic and more food is supplied to the benthos (e.g., Valentine, 1973; Hallock et al., 1991; Alve, 1995; Boersma et al., 1998; Bernhard and Sen Gupta, 1999). The overall appearance of the BEE at Site 690 is thus that of a diverse, oligotrophic community

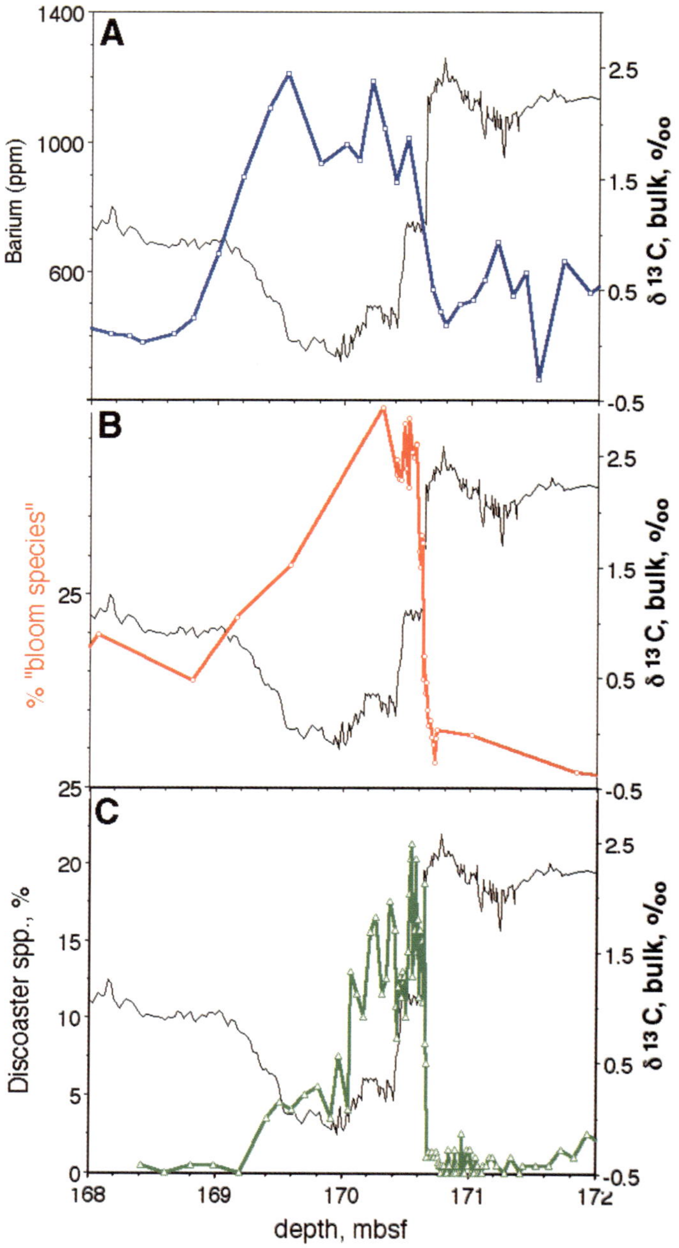

Figure 5. A: Ba content of the bulk sediment (in ppm) (Bains et al., 2000) shown in the thick line, plotted over the $\delta^{13}C$ in bulk sediment (Bains et al., 1999), shown in the thin line. B: Percentage of benthic foraminiferal opportunistic or "bloom" species (*Siphogenerinoides* spp., *Tappanina selmensis*) shown by the thick line (this paper; Thomas and Shackleton, 1996) plotted over the $\delta^{13}C$ in bulk sedi-ment (Bains et al., 1999), shown in the thin line. C: Percentage of *Discoaster* spp. in the calcareous nannoplankton assemblage (Bralower, 2002) shown in the thick line, plotted over the $\delta^{13}C$ in bulk sediment (Bains et al., 1999), shown in the thin line. Note that the scale for this figure differs from that in Figures 1–4, and covers the full extent of the carbon isotope excursion.

with highly specialized, K-selected species, which was disturbed and replaced by an opportunistic group of r-selected taxa, reflecting a larger food input to the benthos, as documented by the increased percentage of buliminid taxa as well as by the increase of absolute abundance of benthic foraminifera (Figs. 2A, 2B).

The increase in numbers of benthic foraminifera per gram at 170.65 mbsf coincided with a change in the calcareous nannoplankton, as exemplified by the increased abundance of *Fasciculithus* spp. and *Discoaster* spp. (Figs. 4, 5). The relative abundances of these taxa are shown as an example of floral change involving the complete nannoplankton community (Bralower, 2002). These floral changes reflect a change from more eutrophic, cool-water floras with dominant r-selected taxa to more warm-water floras dominated by various more oligotrophic, K-selected taxa such as *Discoaster* spp. and *Fasciculithus* spp. (Bralower, 2002). In benthic foraminifera, faunas changed at the same time from dominated by K-selected taxa to dominated by r-selected taxa, the reverse pattern.

This change in nannoplankton floras occurred slightly below the CIE, and at the same depth as the increased abundance of benthic foraminifera (Figs. 4A, 4B). There is thus a major discrepancy between the environmental implications of the observed floral change in the pelagic primary producers (calcareous nannoplankton) which indicates decreased productivity after the CIE, and the deep-sea benthos, which presumably received its food from the primary producers, and indicates an increase in food supply. A similar discrepancy was observed at equatorial Pacific Site 865 (Kelly et al., 1996; Thomas et al., 2000).

The apparent increased productivity indicated by the benthic foraminiferal assemblage composition (specifically the relative abundance of the "bloom" species *Siphogenerinoides* spp., *Aragonia aragonensis* and *Tappanina selmensis*) lasted for longer than the interval studied here, and remained throughout the CIE interval estimated to last ~220 k.y. (Röhl et al., 2000), as did the enhanced Ba-levels interpreted as indicating enhanced productivity (Bains et al., 2000), and the nannoplankton indicating decreased surface primary productivity (Bralower, 2002) (Figs. 5A–C). There is no information available on the quantitative nannoplankton floral assemblage or trace element composition from longer sections at Site 690, but the relative abundance of the "bloom species" of benthic foraminifera shows strong fluctuations within the whole interval between ~61.5 and 49 Ma, roughly corresponding to the late Paleocene through early Eocene (Berggren et al., 1995) (Fig. 6). At equatorial Pacific Site 865 these species have a similar distribution in time.

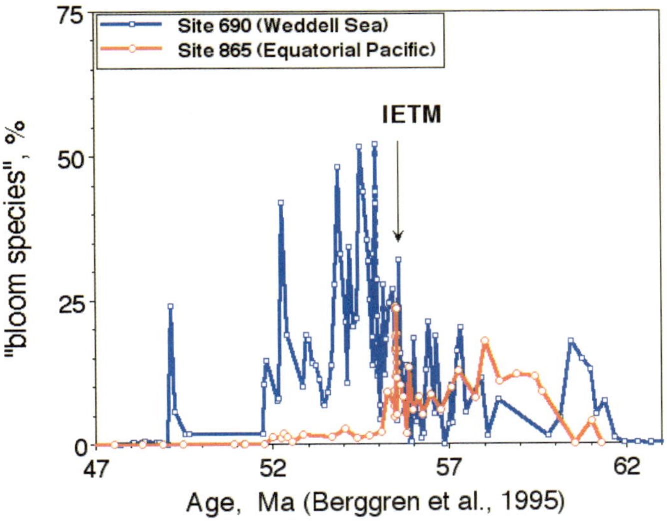

Figure 6. Percentage of bloom species (*Siphogenerinoides* spp., *Aragonia aragonensis,* and *Tappanina selmensis*) at Ocean Drilling Program Site 690 (Weddell Sea) and Site 865 (equatorial Pacific Ocean) plotted versus numerical time (Berggren et al., 1995). IETM—Initial Eocene Thermal Maximum.

DISCUSSION

Sequence of events

The benthic foraminiferal data, trace element data (Bains et al., 2000), nannofossil data (Bralower, 2002) and stable isotope data (D. Thomas et al., 2001) show considerable fluctuations at depth scales of a few cm. This is strong evidence that the record has not been fully homogenized by bioturbation on such scales, although individual benthic foraminifera and their isotopic signatures indicate that reworking must have occurred (Table DR1). In Pleistocene-Holocene sediment, sediment components in the size range >150 μ (such as foraminifera) are bioturbated differently from fine (<5 μ) components such as calcareous nannoplankton, as seen in different radiocarbon dates of fine fraction and foraminifera while sedimentation rates derived from the two different data sets are the same (Thompson et al., 1995). High-resolution isotope data on planktic foraminifera in the same samples as studied for benthic foraminifera show the CIE in planktic foraminifera 8 cm below the CIE in bulk sediment (Bains et al., 1999; D. Thomas et al., 2002). This offset thus might at least in part be explained by differential bioturbation.

At Site 690, the comparison of the nannofossil and foraminiferal carbon isotope records becomes even more difficult because the nannofloras show such a large change in floral composition. Different species of nannoplankton vary in isotopic signature, and a change in bulk $\delta^{13}C$ composition due to changes in nannoplankton floral composition may thus be superimposed on the isotope excursion resulting from gas hydrate dissociation (Bralower, 2002). It is thus not so simple to compare global events in the carbon cycle with local events in benthic foraminifera, and with events in nannoplankton, which has a different size and thus a different pattern of bioturbation, even when all these are studied in the same samples. The pattern may become even more complex because many foraminiferal species, and especially buliminid taxa, live at least a few cm in the sediments (e.g., Jorissen et al., 1995). The increase in relative abundance of small buliminid taxa a few cm somewhat below the CIE may thus have resulted from this infaunal life style.

At exactly the level in the sediment where the rapid, large decrease in bulk $\delta^{13}C$ occurs (Bains et al., 1999), however, the relative abundance of the "doomed species," as well as their size and wall thickness decreased, while there was a further increase in the relative abundance of small "bloom species," especially *Siphogenerinoides* sp. (Figs. 1, 2). About 2.5 cm above the decrease in bulk $\delta^{13}C$ the final extinction of "doomed species" occurred (all specimens at higher levels that were analyzed showed pre-excursion carbon isotope values), as well as a drop in species richness and evenness of distribution. These data thus might be interpreted as suggesting that at the time of the large change in bulk $\delta^{13}C$, which probably indicates large-scale dissociation of methane hydrates (e.g., Dickens, 2000, 2001a), the benthic foraminiferal association underwent a first stage of change, with large individuals no longer being able to survive.

Such a decrease in abundance of large, heavily calcified individuals occurred worldwide (Thomas, 1998), and could have been caused by increased solubility of calcite as a result of high total dissolved inorganic carbon levels resulting from oxidation of methane in the water column. At Site 690, dissolution of carbonate, as indicated by lower $CaCO_3$ values, started below the CIE and below this level of decrease in size of benthic foraminifera by several cm (D. Thomas et al., 1999), but this may well be the result of "burn-down," dissolution reaching down into the sediment.

The actual extinction (i.e., drop in diversity, local disappearance of species that become extinct globally, and evenness of distribution of species over specimens) apparently occurred a few cm higher in the sediment than the CIE, which would mean about a few thousand years after the initiation of the CIE (chronology as in Röhl et al., 2000 [see also D. Thomas et al., 2002]). The faunal changes in benthic foraminifera started earlier, however, because the increase in absolute abundance started ~2 cm below the decrease in bulk $\delta^{13}C$ signaling the initiation of the CIE, at the same level as a drop in cool, higher productivity, r-mode nannoplankton species (Bralower, 2002). The change in floral assemblages, which was observed in the same size fraction as the bulk $\delta^{13}C$ values, thus must have predated the CIE, and might have been coeval with the initial gradual warming of the surface waters observed in the oxygen isotope values of *Acarinina* spp. (D. Thomas et al., 2002). The possibility of differential bioturbation, however, makes it impossible to precisely correlate these events in planktic foraminifera and nannoplankton, and possibly between benthic foraminifera and nannoplankton.

Changes in oceanic productivity

The benthic foraminiferal data at Site 690 appear to suggest increased productivity after the BEE (as argued by Thomas and Shackleton, 1996), in agreement with an interpretation of the Ba data as resulting from increased productivity (Bains et al., 2000). This interpretation conflicts with the nannofossil evidence for decreased productivity (Bralower, 2002; see also Aubry, 1998). The Ba data, however, can be interpreted differently: During the dissociation of gas hydrates causing the CIE, significant dissolved Ba^{2+} was released to intermediate waters of the ocean. The dissolved Ba^{2+} concentrations in the deep ocean rose significantly, a smaller fraction of sinking barite particles dissolved, and "biogenic barite" accumulation increased (Dickens, 2001b; Dickens et al., this volume). In this model, no increased productivity is necessary to cause the increased Ba-concentrations during the CIE.

In marginal oceans and epicontinental basins there is strong evidence for increasing productivity at the time of the Initial Eocene Thermal Maximum (IETM) (e.g., Gibson et al., 1993; Gavrilov et al., 1997, this volume; Speijer et al., 1997; Speijer and Schmitz, 1998; Egger et al., this volume), in some regions leading to local anoxic conditions and deposition of laminated "black shales" which are enriched in organic carbon (Crouch, 2001; Crouch et al., 2001; Speijer and Wagner, 2002; Gavrilov et al., this volume). The record for open ocean settings, however, is not so clear.

At many pelagic locations, post-extinction benthic foraminiferal assemblages may be interpreted as indicative of a high food supply and/or low oxygenation because of the abundant occurrence of buliminid foraminifera (see Thomas, 1998, for a review). At increased temperatures the metabolic food requirements of foraminifera increase rapidly, and the larger number of foraminifera at these higher temperatures thus mean a strongly increased food supply (Boersma et al., 1998). At other locations (e.g., many sites in the Atlantic Ocean such as Walvis Ridge Sites 525 and 527 and Ceara Rise site 929), productivity appears to have been lowered, although the same faunal effect might have occurred at rising temperatures and a stable food supply (Thomas and Shackleton, 1996; Boersma et al., 1998; Thomas and Röhl, 2002). Not only at Site 690, but also at equatorial Pacific Site 865 the nannoplankton and benthic foraminiferal evidence is contradictory, with surface records suggesting lowered productivity, benthic foraminiferal records increased productivity.

This discrepancy in data on productivity of benthic and planktic organisms is rather unexpected. In the present oceans, changes in surface primary productivity at various time scales are expressed in deep-sea, open ocean benthic faunas because of bentho-pelagic coupling (e.g., Smart et al., 1994; Jorissen et al., 1995; Loubere and Fariduddin, 1999; Gooday and Rathburn, 1999; Van der Zwaan et al., 1999), which is present even in the warm waters (>13 °C) of the eastern Mediterranean (Duineveld et al., 2000).

There are indications, however, that such benthopelagic coupling was less tight in the warmer oceans of the Greenhouse world (Thomas et al., 2000). For instance, the occurrence of benthic foraminiferal species that rely on the rapid deposition of fresh phytodetritus to the seafloor is limited to the late Cenozoic, after the establishment of the Antarctic ice sheet, when the diversity gradient from low to high latitudes may have developed (Thomas and Gooday, 1996; Culver and Buzas, 2000). The mass extinction at the end of the Cretaceous was devastating for planktic primary producers such as nannoplankton, but benthic foraminifera in the deep oceans showed minor assemblage fluctuations only and extinction was not above background levels (e.g., Alegret et al., 2001). In my opinion this was not the result of the fact that oligotrophic faunas were adapted to low food levels and thus not sensitive to a major drop in productivity: In oligotrophic regions food is severely limiting, and large changes in such a limiting factor are highly likely to affect faunas significantly.

How can we explain such a lack of bentho-pelagic coupling while deep-sea faunas must depend on food supplied from the surface, and the deep-sea is a strongly food-limited environment? A first interpretation could be that the lack of bentho-pelagic coupling might have resulted from a higher transfer rate of organic matter from the surface water to the seafloor in the warmer, and thus possibly less oxygenated waters of a greenhouse world. Presently, only a few percent of organic material produced in surface waters makes it to the deep sea environment (e.g., Murray et al., 1996). Obviously, only a small increase in this fraction could have a major impact on the total food arriving at the seafloor at constant or even decreasing primary productivity (Thomas et al., 2000).

There is, however, not much evidence in the present ocean that variation in oxygenation leads to variation in the preservation of organic matter in the water column (e.g., Hedges and Keil, 1995). Delivery of food particles to the seafloor in warm, oligotrophic regions such as the Red Sea is extremely low (Thiel et al., 1987). At the higher temperatures, one would surmise that the metabolic ratios of foraminifera as well as bacteria would be higher, so that organic matter descending through the water column would end up being more refractory after more intense bacterial degradation (Thomas et al., 2000).

Therefore, I want to present the highly speculative idea that benthopelagic coupling might have been less pronounced in Greenhouse Oceans in general than in the present ones because there may have been a larger supply of food generated directly at the ocean floor, by chemosynthetic bacteria. In recent years it has become clear that in the present oceans benthic foraminifera live in cold-seep areas, where methane hydrate is present close to the seafloor (Jones, 1993; Akimoto et al., 1994; Sen Gupta and Aharon, 1994; Rathburn et al., 2000; Bernhard et al., 2000, 2001). Not all these studies may have successfully distinguished actually living foraminifera in the seeps, because Rose Bengal staining of living specimens is not always reliable, but authors using various methods of assessment documented that the foraminifera actually lived in the seeps (Bernhard et al., 2001).

Such cold seep faunas are not characterized by endemic seep-species that occur nowhere else. To the contrary, these associations vary considerably on a regional as well as local, patchiness scale, but they all have abundant species that elsewhere occur in high-productivity regions, specifically small, few-chambered taxa from the superfamiliy Buliminacea. Such foraminifera may use the sulfur-reducing bacterial partner active in a consortium together with methane-consuming Archaea (Boetius et al., 2000) symbiotically (Bernhard et al., 2000, 2001). The stable isotope signature of the cold seep faunas is not unequivocal, possibly because living individuals were not reliably distinguished, although at least in some cases specimens have unusually high values of $\delta^{18}O$ and low values of $\delta^{13}C$ (Rathburn et al., 2000).

There is another possible pathway of food supply to benthic foraminifera, based on bacteria and not on photosynthesis in the surface waters. Bacterial oxidation of methane in hydrothermal plumes in the present oceans contributes an amount of organic carbon up to 150% that of organic matter reaching the depth of the plume (2200 m) from the surface in the northeast Pacific (de Angelis et al., 1993). There is no evidence of possible plume-bacteria based foraminiferal associations in the present oceans, but this may be because no one searched for them. Such methane plumes might have been generated during dissociation of methane hydrates, and if they indeed moved through the Paleocene-Eocene oceans, they may have supported benthic foraminifera living on bacteria (see also Dickens, 2000).

I therefore suggest that benthic foraminiferal communities supported by chemo-autotrophic primary producers were more important in the overall food supply in the oceans at times when deep-water temperatures were generally above ~10 °C, such as the late Paleocene and early Eocene (Zachos et al., 2001). At these temperatures the metabolic rates of the Archaea and Bacteria involved would have been considerably higher than at the present temperatures close to 0 °C. Even in a warm ocean gas hydrates were probably commonly present along continental margins, wherever enough organic matter was deposited (Dickens et al., 1995). I speculate that under such conditions relatively small fluctuations in ocean temperatures could have led to relatively diffuse dissociation of gas hydrates, as occurs presently at the seafloor in regions where hydrates outcrop and downstream from hydrothermal areas. If the Paleogene climate fluctuated between warm and very warm intervals at Milankovich periodicity, as suggested by the occurrence of Milankovich periodicity in sediment parameters (Norris and Röhl, 1999; Röhl et al., 2000; Cramer, 2001), varying oceanic temperatures may have been expressed in varying, possibly diffuse, dissociation of methane hydrates (Dickens, 2001a).

If it is indeed true that the occurrence of common benthic foraminiferal "bloom" species indicates times of periodical, possibly diffuse dissociation of methane hydrates on the seafloor, then the warmest interval of the Cenozoic, the later part of the Paleocene through the early Eocene (Zachos et al., 2001) can possibly be seen as an interval during which such periodical dissociation occurred commonly (Fig. 6). During this entire interval the relative abundances of the bloom species varied strongly, and these species were present only in this time interval. The whole interval is characterized by overall low $\delta^{13}C$ values, although several short, more extreme excursions might, in fact, have occurred (Zachos et al., 2001). The isotope record has not been studied in sufficient detail to evaluate whether more than one large carbon isotope excursion (in contrast with more periodic, smaller fluctuations) did, in fact, occur, although this is suggested by the benthic foraminiferal records as well as by a few isotope data points (Thomas and Zachos, 2000).

The sudden increase in sulfur isotope values at the end of the early Eocene (Paytan et al., 1998) can then be interpreted as resulting from the end of a period during which periodic dissociation of gas hydrates occurred (Dickens, 2001b), with associated elevated levels of bacterial activity by sulfate reducing bacteria as fueled by methane consuming Archaea (Boetius et al., 2000). The drop in sulfur isotope values at the initiation of the CIE in New Zealand sections (Kaiho et al., 1996) suggests that an effect on sulfur isotopes can indeed be perceived at times of gas hydrate dissociation.

It is an intriguing idea that the warmest Greenhouse intervals may have been characterized by an oceanic carbon cycle that had a larger contribution of chemosynthetic organic matter. One might, for instance, speculate that the middle Maastrichtian extinction of the deep-sea clams *Inoceramus,* suggested to have harbored chemosynthetic symbionts, might have resulted from cooling below the threshold needed to support such a food flux (MacLeod and Hoppe, 1992).

The scenario for the IETM can then be speculated to have been as follows: The Earth warmed globally and gradually, possibly as a result of CO_2 emissions from the North Atlantic Volcanic Province (e.g., Eldholm and Thomas, 1993). This warming was more extreme at high latitudes, and caused changes in the nannoplankton floras (Bralower, 2002), as well as in the benthic faunas. The former changed to floras more typical of warmer, more oligotrophic waters, whereas the latter increased in abundance because bacteria on the seafloor increased their metabolic activity. At a threshold level of surface water warming, ocean circulation changed suddenly, possibly as a reaction to changing precipitation/evaporation patterns, with a switch from southern to northern high latitudes (Bice and Marotzke, 2002), or from high southern to lower latitudes (e.g., Kennett and Stott, 1991).

The changing circulation resulted in sudden higher temperatures of oceanic water masses and thus rapid methane dissociation, possibly also because the changing current patterns caused margin collapse (Katz et al., 2001). Methane dissociation was probably a very complex and spatially variable process, with one or more large slumps (Katz et al., 1999) generating large and sudden releases, whereas at the same time the temperature increase might have caused a much more widespread, diffuse dissociation (Dickens, 2001a). The methane may have been partially oxidized in the oceans, as suggested by the possibility of widespread low-oxygen conditions, and partially in the atmosphere, depending upon the way in which the methane was liberated. One would assume that sudden release of large amounts of methane as the result of a major slump (Katz et al., 1999, 2001) would have resulted in bubble formation and release into the atmosphere, whereas more diffuse dissociation might lead to more oxidation within the oceans.

Methane dissociation in the oceans may have triggered increased chemosynthetic activity and thus increased food for the benthos, even at a time of decreasing surface water primary productivity. Site 690 does not appear to be a prime candidate for vigorous local gas hydrate dissociation: At its paleodepth of 1900 m it was probably below the region of active dissociation (e.g., Dickens et al., 1995; Katz et al., 1999). These estimates are not very precise, however, and gas hydrate dissociation may have been a chaotic process, locally triggered by slumps or intense currents, or the presence of warmer currents than elsewhere. Interestingly, Kaminski et al. (1996) and Galeotti et al. (2003) mention that *Glomospira* spp., common in the early Eocene deep ocean basins, is an opportunistic taxon that in the Gulf of Mexico lives in areas of active petroleum seepage. In areas without methane dissociation and a chemosynthetic food supply the benthic foraminifera starved because their metabolic needs were not met at the suddenly increased temperatures.

Once the large-scale methane dissociation started, positive feedback would have been expected to cause a runaway greenhouse effect. How such an effect was prevented and why the Initial Eocene Thermal Maximum ended is not clear. I suggest that there is at present not sufficient evidence that primary production in open ocean surface waters increased, in contrast with Bains et al. (2000), and in agreement with Dickens et al. (this volume). There may, however, very well have been a large increase in carbon storage on land (Beerling, 2001), or in marginal ocean basins (Speijer and Wagner, 2002; Gavrilov et al., this volume) to take the added carbon dioxide produced from oxidized methane from gas hydrates out of the atmosphere-ocean system.

CONCLUSIONS

Detailed benthic foraminiferal records across the Paleocene-Eocene carbon isotope excursion suggest that highly diverse benthic foraminiferal faunas with an even distribution of specimens over species and with many specialized, large, K-selected taxa were rapidly replaced by low-diversity faunas, heavily dominated by small, opportunistic taxa, while the food supply to the benthos increased. The calcareous nannoplankton in the same samples, however, shows an exactly opposite trend, and suggests decreasing productivity (Bralower, 2002). I speculate that this contradiction might be solved if the benthic faunas in the Paleogene, warm oceans were less strictly coupled to surface water primary producers than benthos in the present oceans. Such coupling might have been much less efficient if a larger percentage of organic matter reached the ocean floor than does today, possibly because in warmer, less oxygenated waters less organic matter was oxidized.

But one could also speculate that in such oceans at least part of the faunas used food produced at the ocean floor by chemosynthetic bacteria. The benthic foraminiferal extinction might thus have been caused by decreasing oxygen levels or increasing $CaCO_3$ corrosivity of the deep waters, but a change in food supply may have been an important contributing factor. In locations where methane dissociation could occur, the benthos received considerably more food during the Initial Eocene Thermal Maximum, whereas in other regions (e.g., Walvis Ridge), the food supply to the benthos diminished as a result of decreasing surface water productivity, and starvation was exacerbated because foraminifera need more food to sustain life at higher temperatures. The increased overall patchiness of the deep ocean environments, resulting from the difference between regions with and without gas hydrates dissociation, may have been at least a contributing cause of the occurrence of biogeographically much more varied deep-sea faunas after the extinction (Thomas, 1998). Varying dependence upon chemosynthetic food supply by benthic foraminifera, with the supply depending on times of warming and/or cooling of the deep oceans may have been typical for the warmest period of the Cenozoic, the latest Paleocene through the early Eocene. We need to obtain detailed records of oxygen, carbon and sulfur isotopes over this whole time period in order to decipher the possible role of chemosynthesis in supporting ocean floor life, resulting in an oceanic carbon cycle very different from that in the present oceans.

ACKNOWLEDGMENTS

I want to thank Scott Wing for inviting me to attend the meeting in Powell, which was an intellectually stimulating experience as well as an introduction to a geologically spectacular area of the United States. I thank the Ocean Drilling Program for the samples from Site 690, and Tim Bralower and Debbie Thomas for asking me to participate in the study of these high-resolution samples. Tracy Krueger made the Scanning Electron Micrographs at the Wesleyan facility, and Daphne and Dylan Varekamp computer-processed the micrographs into a plate, for which my thanks. Lennart Bornmalm, Andy Gooday, Ann Holbourn, Mimi Katz, Robert Speijer, and an anonymous reviewer are all thanked for their well-reasoned and very helpful reviews and insightful comments. I gratefully acknowledge funding by grant NSF grant EAR0120727 to J.C. Zachos.

APPENDIX 1. TAXONOMIC NOTES

Siphogenerinoides sp.

Siphogenerinoides sp. has a strong resemblance to *Siphogenerinoides brevispinosa* (Plate 1), a species common in the upper Paleocene and lower Eocene at the Maud Rise drill sites (Thomas, 1990) as well as at equatorial Pacific Site 865 (Thomas, 1998). Both taxa have a test covered with fine spines and a short apertural neck. In contrast to *S. brevispinosa,* however, *Siphogenerinoides* sp. has a flattened rather than rounded early part of the test, with more chambers arranged biserially, and flattened rather than inflated in cross section. I have not noticed this morphotype in low-resolution samples from Site 689 (Maud Rise) and Sites 525 and 527 (Walvis Ridge, southeastern Atlantic Ocean), nor in high-resolution samples from Site 865 (equatorial Pacific). Both taxa cooccur just after the benthic

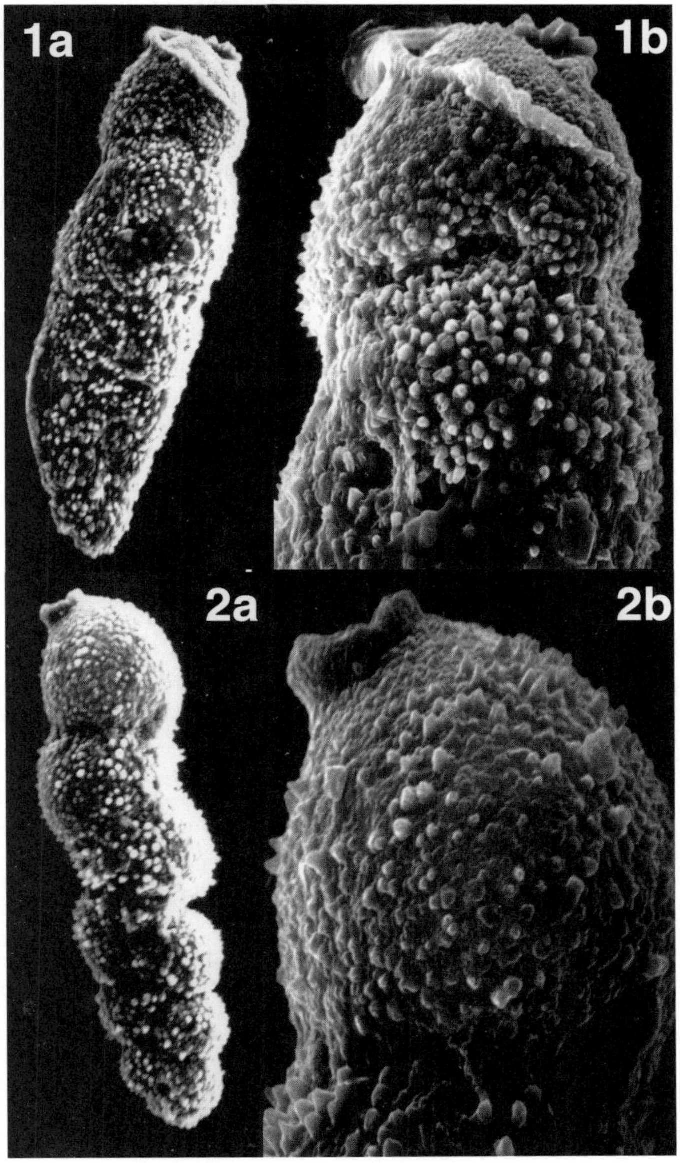

Plate 1. Figure 1: *Siphogenerinoides* sp. (a) Whole specimen; overall length 105 μ. (b) Last chamber (with later chamber broken off). Figure 2: *Siphogenerinoides brevispinosa* Cushman, overall length 125 μ. (a) Whole specimen. (b) Last chamber of 2a. Both specimens from Sample 690B-19H-3, 61–62 cm.

foraminiferal extinction at Site 690, but *Siphogenerinoides* sp. is not present over the rest of the section in Site 690. *Siphogenerinoides* sp. occurs in a very thin section of sediment only; higher-resolution studies are needed to document its biogeographic and stratigraphic range.

Bulimina kugleri Cushman and Renz, 1942

Bulimina kugleri Cushman and Renz, 1942, Cushman Laboratory for Foraminiferal Research, Contributions, v. 18, p. 9, Plate 2, Figure 9.

Bulimina ovula Reuss, Thomas, Thomas, E., 1990, Late Cretaceous through Neogene deep-sea benthic foraminifers (Maud Rise, Weddell Sea, Antarctica): Proceedings of the Ocean Drilling Program, Scientific Results: Texas A&M University, College Station, Texas, USA, v. 113, p. 571–594.

Bulimina ovula Reuss, Thomas, E., Zachos, J.C., and Bralower, T.J., 2000, Deep-sea environments on a warm earth: Latest Paleocene–early Eocene, *in* Huber, B., et al., eds., Warm climates in Earth history, Cambridge University Press, p. 32–160.

This is a small, fusiform species of *Bulimin*a, about twice as long as wide, greatest breadth at about its middle. The chambers are distinct and slightly inflated, with slightly depressed sutures. The wall is smooth, and the aperture a high and arched, curved opening at the base of the inner margin of the last chamber. I compared the material from Sites 689 and 690 with the holotype (collection number 38199) and several paratypes (collection number 38257) at the Smithsonian Institution. The Weddell Sea specimens are smaller, but very similar in shape and proportions to the type specimens.

I called this species *B. ovula* in the past (Thomas, 1990; Thomas et al., 2000), because its fusiform shape was very similar to that figured in White (1928). In a recent restudy of White's original collection (Alegret and Thomas, 2001), however, it became clear that White's rather unclear figure represented the large and fusiform *Praebulimina reussi* (Morrow), the *nomen novum* for *Bulimina ovula* Reuss, and thus was not cospecific with the small fusiform *Bulimina* specimens. There is considerable confusion as to the use of the name *Bulimina ovula,* which has been used for forms now called *Praebulimina reussi,* for *Bulimina ovula* d'Orbigny, with which *B. ovula* Reuss was homonymic, and which is a much more inflated, Recent form, and also with *Bulimina ovula* Terquem, which is now placed in the genus *Buliminella.* All three species are fusiform to some degree.

The specimens from Sites 689 and 690 closely resemble *B. kugleri* Cushman and Renz in overall fusiform shape of the test, shape and placement of aperture, smooth wall and slightly depressed sutures, but they are in general smaller than the type specimens. Some of the paratypes but not the holotype have a smaller breadth-to-length ratio than my specimens.

REFERENCES CITED

Akimoto, K., Tanaka, T., Hattori, M., and Hotta, H., 1994, Recent benthic foraminiferal assemblages from the cold seep communities: A contribution to the methane gas indicator, *in* Tsuchi, R., ed., Pacific Neogene events in time and space: Tokyo, University of Tokyo Press, p. 11–25.

Alegret, L., and Thomas, E., 2001, Upper Cretaceous and lower Paleogene benthic foraminifera from northeastern Mexico: Micropaleontology, v. 47, p. 269–316.

Alegret, L., Molina, E., and Thomas, E., 2001, Benthic foraminifera at the Cretaceous/Tertiary boundary around the Gulf of Mexico: Geology, v. 29, p. 891–894.

Alve, E., 1995, Benthic foraminiferal distribution and recolonization of formerly anoxic environments in Drammensfjord, southern Norway: Marine Micropaleontology, v. 25, p. 2–3.

Aubry, M.-P., 1998, Early Paleogene calcareous nannoplankton evolution: A tale of climatic amelioration, *in* Aubry, M.P., et al., eds., Late Paleocene–early Eocene biotic and climatic events in the marine and terrestrial records: New York, Columbia University Press, p. 158–201.

Bains, S., Corfield, R.M., and Norris, R.D., 1999, Mechanisms of climate warming at the end of the Paleocene: Science, v. 285, p. 724–727.

Bains, S., Norris, R.D., Corfield, R.M., and Faul, K., 2000, Termination of global warmth at the Paleocene/Eocene boundary through productivity feedback: Nature, v. 407, p. 171–174.

Beerling, D., 2000, Increased terrestrial carbon storage across the Paleocene Eocene boundary: Palaeogeography, Palaeoclimatology, Palaeoecology, v. 161, p. 395–405.

Berggren, W.A., Kent, D.V., Swisher, C.C., III, and Aubry, M. -P., 1995, A revised Cenozoic geochronology and chronostratigraphy, *in* Berggren, W.A., et al., eds., Geochronology, Time Scales and Global Stratigraphic Correlation: Society for Sedimentary Geology Special Publication, v. 54, p. 129–212.

Bernhard, J.M., 1996, Microaerophilic and facultative anaerobic benthic foraminifera: A review of experimental and ultrastructural evidence: Revue de Paléobiologie, v. 15, p. 261–275.

Bernhard, J.M., and Sen Gupta, B.K., 1999, Foraminifera of oxygen-depleted environments, *in* Sen Gupta, B.K., ed., Modern foraminifera: Dordrecht, Netherlands, Kluwer, p. 201–216.

Bernhard, J.M., Buck, K.R., Farmer, M.A., and Bowser, S.S., 2000, The Santa Barbara Basin is a symbiosis oasis: Nature, v. 406, p. 77–80.

Bernhard, J.M., Buck, K.R., and Barry, J.P., 2001, Monterey Bay cold-seep biota: Assemblages, abundance, and ultrastructure of living foraminifera: Deep-Sea Research I, v. 48, p. 2233–2249.

Bice, K., and Marotzke, J., 2002, Could changing ocean circulation have destabilized methane hydrate at the Paleocene/Eocene boundary?: Paleoceanography, 1029/2001PA000662.

Boersma, A., Premoli-Silva, I., and Hallock, P., 1998, Trophic models for the well-mixed and poorly mixed warm oceans across the Paleocene–Eocene Epoch boundary, *in* Aubry, M.P., et al., eds., Late Paleocene–early Eocene biotic and climatic events in the marine and terrestrial records: New York, Columbia University Press, p. 204–213.

Boetius, A., Ravenschlag, K., Schubert, C.J., Rickert, D., Widdel, F., Gieseke, A., Amann, R., Joergensen, B.B., Witte, U., and Pfannkuche, O., 2000, A marine microbial consortium apparently mediating anaerobic oxidation of methane: Nature, v. 407, p. 623–626.

Bowen, G.J., Koch, P.L., Gingerich, P.D., Norris, R.D., Bains, S., and Corfield, R.M., 2001, Refined isotope stratigraphy across the continental Paleocene-Eocene boundary on Polecat Bench in the northern Bighorn Basin: University of Michigan Papers on Paleontology, v. 33, p. 73–88.

Bralower, T.J., 2002, Evidence of surface water oligotrophy during the Paleocene-Eocene thermal maximum: Nannofossil assemblage data from Ocean Drilling Program Site 690, Maud Rise, Weddell Sea: Paleoceanography, 1029/2001PA000662.

Coccioni, R., Di Leo, R., Galeotti, S., and Monecchi, S., 1994, Integrated biostratigraphy and benthic foraminiferal faunal turnover across the Paleocene-Eocene boundary at Trabakua Pass section, Northern Spain: Palaeopelagos, v. 4, p. 87–100.

Corfield, R.M., and Norris, R.D., 1998, The oxygen and carbon isotopic context of the Paleocene/Eocene epoch boundary, *in* Aubry, M.P., et al., eds., Late Paleocene–early Eocene biotic and climatic events in the marine and terrestrial records: New York, Columbia University Press, p. 124–137.

Cramer, B.S., 2001, Latest Paleocene–earliest Eocene cyclostratigraphy: Using core photographs for reconnaissance geophysical logging: Earth and Planetary Science Letters, v. 186, p. 231–244.

Crouch, E.M., 2001, Environmental change at the time of the Paleocene-Eocene biotic turnover: Utrecht, Netherlands, Laboratory of Palaeobotany and Palynology Contributions Series, v. 14, 216 p.

Crouch, E., Heilmann-Clausen, C., Brinkhuis, H., Morgans, H.E., Rogers, K.M., Egger, H., and Schmitz, B., 2001, Global dinoflagellate event associated with the late Paleocene Thermal Maximum: Geology, v. 29, p. 315–318.

Culver, S.J., and Buzas, M.A., 2000, Global latitudinal species diversity gradient in deep-sea benthic foraminifera: Deep-Sea Research I, v. 47, p. 259–275.

Cushman, J.A., and Renz, R.H., 1942, Eocene Midway foraminifera from Soldado Rock, Trinidad: Cushman Laboratory for Foraminiferal Research, Contributions, v. 18, p. 7–15.

De Angelis, M.A., Lilley, M.D., Olson, E.J., and Baross, J.A., 1993, Methane oxidation in deep-sea hydrothermal plumes of the Endeavour Segment of the Juan de Fuca Ridge: Deep-Sea Research I, v. 40, p. 1169–1186.

Den Dulk, M., Reichart, G.J., van Heyst, S., Zachariasse, W.J., and van der Zwaan, G.J., 2000, Benthic foraminifera as proxies of organic matter flux and bottom water oxygenation? A case history from the northern Arabian Sea: Palaeogeography, Palaeoclimatology, Palaeoecology, v. 161, p. 337–359.

Dickens, G.R., 2000, Methane oxidation during the late Paleocene Thermal Maximum: Bulletin de la Société Géologique de France, v. 171, p. 37–49.

Dickens, G.R., 2001a, Modeling the global carbon cycle with a gas hydrate capacitator: Significance for the latest Paleocene thermal maximum: American Geophysical Union Geophysical Monograph Series, v. 124, p. 19–38.

Dickens, G.R., 2001b, Sulfate profiles and barium fronts in sediment on the Blake Ridge: Present and past methane fluxes through a large gas hydrate reservoir: Geochimica et Cosmochimica Acta, v. 65, p. 9–23.

Dickens, G.R., O'Neil, J.R., Rea, D.K., and Owen, R.M., 1995, Dissociation of oceanic methane hydrate as a cause of the carbon isotope excursion at the end of the Paleocene: Paleoceanography, v. 10, p. 965–971.

Dolenec, T., Pavsic, J., and Lojen, S., 2000, Ir anomalies and other elemental markers near the Paleocene-Eocene boundary in a flysch sequence from the Western Tethys (Slovenia): Terra Nova, v. 12, p. 199–204.

Duineveld, G.C.A., Tselepides, A., Witbaard, R., Bak, R.P.M., Berghuis, E.M., Nieuwland, G., van de Weele, J., and Kok, A., 2000, Benthopelagic coupling in the oligotrophic Cretan Sea: Progress in Oceanography, v. 46, p. 457–480.

Eldholm, E., and Thomas, E., 1993, Environmental impact of volcanic margin formation: Earth and Planetary Science Letters, v. 117, p. 319–329.

Fricke, H.C., Clyde, W.C., O'Neil, J.R., and Gingerich, P.D., 1998, Evidence for rapid climate change in North America during the latest Paleocene thermal maximum: Oxygen isotope compositions of biogenic phosphate from the Bighorn Basin (Wyoming): Earth and Planetary Science Letters, v. 160, p. 193–208.

Galleoti, S., Kaminski, M.A., Coccioni, R., and Speijer, R., 2003, High resolution deep water agglutinated foraminiferal record across the Paleocene/Eocene transition in the Contessa Road section (Italy), *in* Hart, M.B., et al., eds., Proceedings of the Sixth International Workshop on Agglutinated Foraminifera, in press.

Gavrilov, Y.O., Kodina, L.A., Lubchenko, I.Y., and Muzylöv, N.G., 1997, The late Paleocene anoxic event in epicontinental seas of Peri-Tethys and formation of the sapropelite unit: Sedimentology and geochemistry: Lithology and Mineral Resources, v. 32, p. 427–450.

Gibson, T.G., Bybell, L.M., and Owens, J.P., 1993, Latest Paleocene lithologic and biotic events in neritic deposits of southwestern New Jersey: Paleoceanography, v. 8, p. 495–515.

Gooday, A.J., 2002, Biological response to seasonally varying fluxes of organic matter to the ocean floor: Journal of Oceanography, v. 58, p. 305–332.

Gooday, A.J., and Rathburn, A.E., 1999, Temporal variability in living deep-sea benthic foraminifera: A review: Earth Science Reviews, v. 46, p. 187–212.

Hallock, P., 1987, Fluctuations in the trophic resource continuum: A factor in global diversity cycles?: Paleoceanography, v. 2, p. 457–471.

Hallock, P., Premoli-Silva, I., and Boersma, A., 1991, Similarities between planktonic and larger foraminiferal evolutionary trends through paleoceanographic change: Palaeogeography, Palaeoclimatology, Palaeoecology, v. 83, p. 49–64.

Hayward, B.W., 2002, Late Pliocene to middle Pleistocene extinction of deep-sea benthic foraminifera ("*Stilostomella* extinction") in the southwest Pacific: Journal of Foraminiferal Research. v. 32, p. 274–307.

Hedges, J.I., and Keil, R.G., 1995, Sedimentary organic matter preservation: An assessment and speculative synthesis: Marine Chemistry, v. 49, p. 81–115.

Herguera, J.C., and Berger, W.H., 1991, Paleoproductivity from benthic foraminiferal abundance: Glacial to postglacial changes in the west equatorial Pacific: Geology, v. 19, p. 1173–1176.

Holbourn, A., Kuhnt, W., and Erbacher, J., 2001, Benthic foraminifers from Lower Albian black shales (Site 1049, ODP Leg 171): Evidence for a non-"uniformitarian" record: Journal of Foraminiferal Research, v. 31, p. 60–74.

Jones, R.W., 1993, Preliminary observations on benthonic foraminifera associated with biogenic gas seep in the North Sea, *in* Jenkins, D.G., ed., Applied micropaleontology: Dordrecht, Netherlands, Kluwer Academic Publishers, p. 69–91.

Jorissen, F.J., de Stigter, H.C., and Widmark, J.G.V., 1995, A conceptual model explaining benthic foraminiferal microhabitats: Marine Micropaleontology, v. 26, p. 3–15.

Kaiho, K., Arinobu, T., Ishiwatari, R., Morgans, H.E., Okada, H., Takeda, N., Tazaki, K., Zhou, G., Kajiwara, Y., Matsumoto, R., Hirai, A., Niitsuma, N., and Wada, H., 1996, Latest Paleocene benthic foraminiferal extinction and environmental changes at Tawanui, New Zealand: Paleoceanography, v. 11, p. 447–465.

Kaminski, M.A., Kuhnt, W.A., and Radley, J.D., 1996, Paleocene–Eocene deep water agglutinated foraminifera from the Namibian Flysch (Rif, Northern Morocco): Their significance for the palaeoceanography of the Gibraltar Gateway: Journal of Micropalaeontology, v. 15, p. 1–19.

Katz, M.E., Pak, D.K., Dickens, G.R., Miller, K.G., 1999, The source and fate of massive carbon input during the latest Paleocene thermal maximum: Science, v. 286, p. 1531–1533.

Katz, M.E., Cramer, B.S., Mountain, G.S., Katz, S., and Miller, K.G., 2001, Uncorking the bottle: What triggered the Paleocene/Eocene thermal maximum methane release?: Paleoceanography, v. 16, p. 549–562.

Kelly, D.C., Bralower, T.J., Zachos, J.C., Premoli-Silva, I., and Thomas, E., 1996, Rapid diversification of planktonic foraminifera in the tropical Pacific (ODP Site 865) during the late Paleocene thermal maximum: Geology, v. 24, p. 423–426.

Kennett, J.P., and Stott, L.D., 1991, Abrupt deep-sea warming, palaeoceanographic changes, and benthic extinctions at the end of the Paleocene: Nature, v. 353, p. 225–229.

Kent, D.V., Cramer, B.S., and Lanci, L., Wang, D., Wright, J.D., and Van Der Voo, R., 2002, Evidence for a Comet Impact Trigger for the Paleocene/Eocene thermal maximum and carbon isotope excursion: GSA Annual Meeting, Abstracts, v. 34, p. 401.

Koch, P.L., Zachos, J.C., and Gingerich, P.D., 1992, Coupled isotopic changes in marine and continental carbon reservoirs at the Paleocene-Eocene boundary: Nature, v. 358, p. 319–322.

Koch, P.L., Zachos, J.C., and Dettman, D.L., 1995, Stable isotope stratigraphy and palaeoclimatology of the Palaeogene Bighorn Basin (Wyoming, USA): Palaeogeography, Palaeoclimatology, Palaeoecology, v. 115, p. 61–89.

Loubere, P., and Fariduddin, M., 1999, Quantitative estimation of global patterns of surface ocean biological productivity and its seasonal variation on timescales from centuries to millennia: Global Biogeochemical Cycles, v. 13, p. 115–133.

Luterbacher, H.P., Hardenbol, J., and Schmitz, B., 2000, Decision of the voting members of the International Subcommission on Paleogene Stratigraphy on the criterion for the recognition of the Paleocene/Eocene boundary: Newsletter of the International Subcommission on Paleogene Stratigraphy, Tübingen, November 2000, v. 9, p. 13.

Macleod, K.G., and Hoppe, K.A., 1992, Evidence that inoceramid bivalves were benthic and harbored chemosynthetic symbionts: Geology, v. 20, p. 117–120.

Murray, J.W., Young, J., Newton, J., Dunne, J., Chapin, T., Paul, B., and McCarthy, J.J., 1996, Export flux of particulate organic carbon from the central equatorial Pacific determined using a combined drifting trap–^{234}Th approach: Deep Sea Research II, v. 43, p. 1095–1132.

Norris, R.D., and Röhl, U., 1999, Carbon cycling and chronology of climate warming during the Paleocene/Eocene transition: Nature, v. 401, p. 775–778.

Ortiz, N., 1995, Differential patterns of benthic foraminiferal extinctions near the Paleocene/Eocene boundary in the North Atlantic and the western Tethys: Marine Micropaleontology, v. 26, p. 341–359.

Pak, D.K., and Miller, K.G., 1992, Paleocene to Eocene benthic foraminiferal isotopes and assemblages: Implications for deep water circulation: Paleoceanography, v. 7, p. 405–422.

Paytan, A., Kastner, M., Campbell, D., and Thiemens, M.H., 1998, Sulfur isotope composition of Cenozoic sea water sulfate: Science, v. 282, p. 1459–1462.

Rathburn, A.E., Levin, L.A., Held, Z., and Lohmann, K.C., 2000, Benthic foraminifera associated with cold methane seeps on the northern California margin: Ecology and stable isotope composition: Marine Micropaleontology, v. 38, p. 247–266.

Röhl, U., Bralower, T.J., Norris, R.D., and Wefer, G., 2000, A new chronology for the late Paleocene thermal maximum and its environmental implications: Geology, v. 28, p. 927–930.

Sanders, H.L., 1968, Marine benthic diversity: A comparative study: American Naturalist, v. 102, p. 243–282.

Schmitz, B., Charisi, S.D., Thompson, E.I., and Speijer, R.P., 1997, Barium, SiO_2 (excess) and P_2O_5 as proxies of biological productivity in the Middle East during the Paleocene and the latest Paleocene benthic extinction event: Terra Nova, v. 9, p. 95–99.

Sen Gupta, B.K., and Aharon, P., 1994, Benthic foraminifera of bathyal hydrocarbon vents of the Gulf of Mexico: Initial report on communities and stable isotopes: Geo-Marine Letters, v. 14, p. 88–96.

Sen Gupta, B.K., and Machain-Castillo, M.L., 1993, Benthic foraminifera in oxygen-poor habitats: Marine Micropaleontology, v. 20, p. 183–201.

Smart, C.W., King, S.C., Gooday, A.J., Murray, J.W., and Thomas, E., 1994, A benthic foraminiferal proxy of pulsed organic matter paleofluxes: Marine Micropaleontology, v. 23, p. 89–99.

Speijer, R.P., and Morsi, A.M., 2002, Ostracode turnover and sea-level changes associated with the Paleocene–Eocene thermal maximum: Geology, v. 30, p. 23–26.

Speijer, R.P., and Schmitz, B., 1998, A benthic foraminiferal record of Paleocene sea level and trophic/redox conditions at Gebel Aweina, Egypt: Palaeogeography, Palaeoclimatology, Palaeoecology, v. 137, p. 79–101.

Speijer, R.P., and Wagner, T., 2002, Sea-level changes and black shales associated with the late Paleocene thermal maximum: Organic-geochemical and micropaleontologic evidence from the southern Tethyam margin (Egypt-Israel), *in* Koeberl, C., and MacLeod, K., eds., Catastrophic events and mass extinctions: Impacts and beyond, Boulder, Colorado, Geological Society of America Special Paper 356, p. 533–549.

Speijer, R.P., Schmitz, B., and van der Zwaan, G.J., 1997, Benthic foraminiferal repopulation in response to latest Paleocene Tethyan anoxia: Geology, v. 25, p. 683–686.

Stott, L.D., Sinha, A., Thiry, M., Aubry, M.-P., and Berggren, W.A., 1996, Global ^{13}C changes across the Paleocene/Eocene boundary: Criteria for terrestrial-marine correlations: Geological Society [London] Special Publication 101, p. 381–399.

Stupin, S.I., and Muzylöv, N.G., 2001, The late Paleocene ecologic crisis in epicontinental basins of the eastern Peri-Tethys: Microbiota and accumulation conditions of sapropelitic bed: Stratigraphy and Geological Correlation,
v. 9, no. 5, p. 501–507.

Thiel, H., Pfannkuche, O., Theeg, R., and Schriever, G., 1987, Benthic metabolism and standing stock in the central and northern Red Sea: Marine Ecology, v. 8, p. 1–20.

Thomas, D.J., Bralower, T.J., and Zachos, J.C., 1999, New evidence for subtropical warming during the late Paleocene thermal maximum: Stable isotopes from Deep Sea Drilling Project 527, Walvis Ridge: Paleoceanography, v. 14, p. 561–570.

Thomas, D.J., Zachos, J.C., Bralower, T.J., Thomas, E., and Bohaty, S., 2002, Warming the fuel for the fire: Evidence for the thermal dissociation of methane hydrate during the Paleocene-Eocene thermal maximum: Geology, v., 30, p. 1067–1070.

Thomas, E., 1985, Late Eocene to Holocene deep-sea benthic foraminifers from the central equatorial Pacific Ocean, *in* Mayer, L., et al., eds., Initial reports of the Deep Sea Drilling Project, Washington, D.C., U.S. Government Printing Office, v. 85, p. 655–679.

Thomas, E., 1990, Late Cretaceous through Neogene deep-sea benthic foraminifers (Maud Rise, Weddell Sea, Antarctica), *in* Mayer, L., et al., eds., Proceedings of the Ocean Drilling Program, Scientific Results: Texas A & M University, College Station, Texas, USA, v. 113, p. 571–594.

Thomas, E., 1992, Cenozoic deep-sea circulation: Evidence from deep-sea benthic foraminifera: American Geophysical Union Antarctic Research Series, v. 56, p. 141–165.

Thomas, E., 1998, The biogeography of the late Paleocene benthic foraminiferal extinction, *in* Aubry, M.P., et al., eds., Late Paleocene–early Eocene biotic and climatic events in the marine and terrestrial records: New York, Columbia University Press, p. 214–243.

Thomas, E., and Gooday, A.J., 1996, Deep-sea benthic foraminifera: Tracers for Cenozoic changes in ocean productivity?: Geology, v. 24, p. 355–358.

Thomas, E., and Röhl, U., 2002, The Paleocene-Eocene Thermal Maximum in Ceara Rise Hole 929E, Ocean Drilling Program Leg 154: GSA Annual Meeting, Abstracts, v. 34, p. 462.

Thomas, E., and Shackleton, N.J., 1996, The Paleocene-Eocene benthic foraminiferal extinction and stable isotope anomalies: Geological Society [London] Special Publication 101, p. 401–441.

Thomas, E., and Zachos, J.C., 2000, Was the late Paleocene thermal maximum a unique event?: GFF, v. 122, p. 169–170.

Thomas, E., Zachos, J.C., and Bralower, T.J., 2000, Deep-sea environments on a warm Earth: Latest Paleocene–early Eocene, *in* Huber, B., et al., eds., Warm climates in Earth history: Cambridge, UK, Cambridge University Press, p. 132–160.

Thompson, J., Cook, G.T., Anderson, R., MacKenzie, A.B., Harkness, D.D., and McCave, I.N., 1995, Radiocarbon age offsets in different-sized carbonate components of deep-sea sediments: Radiocarbon, v. 37, p. 91–101.

Valentine, J.W., 1973, Evolutionary paleoecology of the marine biosphere: Englewood Cliffs, New Jersey, Prentice-Hall, 511 p.

Van der Zwaan, G.J., Duijnstee, I.A.P., den Dulk, M., Ernst, S.R., Jannink, N.T., and Kouwenhoven, T.J., 1999, Benthic foraminifers: Proxies or problems? A review of paleoecological concepts: Earth Science Reviews, v. 46, p. 213–236.

White, M.P., 1928, Some index foraminifera of the Tampico Embayment area of Mexico, Part II: Journal of Paleontology, v. 2, p. 280–317.

Widmark, J.G.V., and Speijer, R.P., 1997, Benthic foraminiferal faunas and trophic regimes at the terminal Cretaceous Tethyan seafloor: Palaios, v. 12, p. 354–371.

Zachos, J., Pagani, M., Sloan, L., Thomas, E., and Billups, K., 2001, Trends, rhythms, and aberrations in global climate change 65 Ma to present: Science, v. 292, p. 686–693.

Manuscript Accepted by the Society August 13, 2002

Printed in the USA

Geological Society of America
Special Paper 369
2003

Cobham Lignite Bed and penecontemporaneous macrofloras of southern England: A record of vegetation and fire across the Paleocene-Eocene Thermal Maximum

M.E. Collinson
Department of Geology, Royal Holloway University of London, Egham, Surrey, TW20 0EX, UK
J.J. Hooker
Department of Palaeontology, The Natural History Museum, Cromwell Road, London, SW7 5BD, UK
D.R. Gröcke
Department of Geology, Royal Holloway University of London, Egham, Surrey, TW20 0EX, UK

ABSTRACT

Carbon isotope, mesofossil, and qualitative palynological analyses have been undertaken through the Cobham Lignite Bed, Kent, England, which overlies the late Paleocene Upnor Formation and underlies the early Eocene lower Woolwich Shell Beds. Carbon isotope analysis has demonstrated a negative carbon isotope excursion similar to that seen in the lower part of the carbon isotope excursion that marks the onset of the Paleocene-Eocene Thermal Maximum (PETM) elsewhere in the marine and continental record. The carbon isotope excursion interval in the Cobham Lignite is shown to be characterized by repeated fires in a vegetation dominated by flowering plant trees and herbs and with many ferns. The abundance of ferns is likely to be a response to disturbance caused by fires. This detailed evidence from across the early PETM (as identified by stratigraphic context and a negative carbon isotope excursion), combined with data from less stratigraphically resolved, but almost coeval, macrofloras from other southern English sites, shows that there is no major change in vegetation at this time in southern England. In contrast, vegetation change occurs later with the emplacement of floras of the Eocene Thermal Maximum. Although there is no obvious vegetation response to the global warming at the PETM, the abundance and consistent occurrence of charcoal in this interval is unparalleled elsewhere in the southern English Paleogene. Therefore, repeated fires, at least on a local scale, may have been one environmental response to the events around the PETM.

INTRODUCTION

The Paleogene is an important interval for studies of global change as it contains three important climatic phenomena: A short-lived thermal maximum at the Paleocene-Eocene boundary, a longer-lived climatic optimum in the early Eocene, and a major cooling across the Eocene-Oligocene boundary (Zachos et al., 2001). The Paleogene sequence of southern England is renowned for its sequence of fossil floras (recently reviewed by Collinson and Cleal, 2001a, 2001b, 2001c), which provide a key resource to document floristic and vegetational responses to these climatic events. It is our long term research goal to document these responses and link them to faunal change, animal-plant coevolution and ultimately ecosystem evolution (e.g., Collinson, 2000b; Collinson and Hooker, 1987, 2000; Collinson and Cleal, 2001a, 2001b, 2001c, and references cited). The

Collinson, M.E., Hooker, J.J., and Gröcke, D.R., 2003, Cobham Lignite Bed and penecontemporaneous macrofloras of southern England: A record of vegetation and fire across the Paleocene-Eocene Thermal Maximum, *in* Wing, S.L., Gingerich, P.D., Schmitz, B., and Thomas, E., eds., Causes and Consequences of Globally Warm Climates in the Early Paleogene: Boulder, Colorado, Geological Society of America Special Paper 369, p. 333–349.

Paleocene-Eocene transitional interval is poorly documented by comparison with other parts of the succession and we have recently concentrated on obtaining new floras from this interval (Collinson, 2000a). However, most of these floras have not yet been precisely constrained temporally nor have they been located with respect to the carbon isotope excursion that marks the Paleocene-Eocene Thermal Maximum (PETM).

A rare temporary exposure (Fig. 1) of the Cobham Lignite (previously exposed only in 1923 and 1964) provided the opportunity to study floras from a key stratigraphic interval within the Paleocene-Eocene transition not represented by other samples. Furthermore, the recognition of the presence of a 2-m-thick, essentially continuous, lignite in the exposures offered the potential to use organic material to obtain a record of carbon isotope ratios in a context with little or no variation in lithology or the nature of the organic input. Stratigraphic correlation and chronological context indicate that the lower part of the PETM carbon isotope excursion, is likely to occur within the Cobham Lignite Bed.

The Cobham Lignite Bed only occurs in a small outlier of Paleogene strata in the eastern London Basin (county of Kent, southern England) (Figs. 2, 3). It underlies the lower Woolwich Shell Beds (Woolwich Formation) and overlies a nonmarine sand and mud unit that rests on the eroded top of the Upnor Formation (Fig. 3) whilst the London Clay Formation occurs higher up in the sequence. Following the Second World War, local companies attempted commercial mining of this bed (Anon., 1948) but, having proved uneconomical due to the limited thickness of lignite, the mine closed, probably in late 1948. The Cobham Lignite was exposed as a bed up to 1 m thick during construction of the A2 road (Chandler, 1923), and again by improvements to that road in 1964 (Martin, 1976). Four samples taken by Bill Chaloner in 1964, from a 60 cm thick exposure of this lignite, were subsequently studied by Martin (1976), who described *Salvinia* and *Azolla* megaspores and microspore massulae, and gave a brief overview of the palynology. Allen (1982) also studied palynology using 5 samples taken in 1964, which were made available by the Institute of Geological Sciences (now British Geological Survey); however, the exact sampling levels had not been recorded in the IGS register (Allen, 1982). These previous studies provide only minimal information on the Cobham Lignite.

The aims of this study are twofold. The first is to use the Cobham Lignite and penecontemporaneous macrofloras from other sites to document floral, vegetational and paleoenvironmental change across the Paleocene-Eocene transition in southern England. The second is to investigate whether carbon isotopes from the Cobham Lignite can help to constrain the location of the PETM in the onshore succession of southern England.

MATERIALS AND METHODS

Between late 1999 and early 2000 a 2 meter thickness of the Cobham Lignite Bed was temporarily exposed, by excavation of a cutting for the Channel Tunnel Rail Link (Fig. 1). Because of the short period of exposure we were only able to gain access on two days, which constrained our sampling strategy. Plaster-

Figure 1. The Paleogene section on the north side of the Channel Tunnel Rail Link (CTRL) cutting at Scalers Hill, Cobham, Kent, UK, showing the main units discussed in the text. Arrow points to thin pale clay separating laminated (below) from blocky (above) lignite. Heavy black line marks irregular junction between the sand and mud unit and underlying Upnor Formation. Spade and adjacent scale bar are ~1 m long.

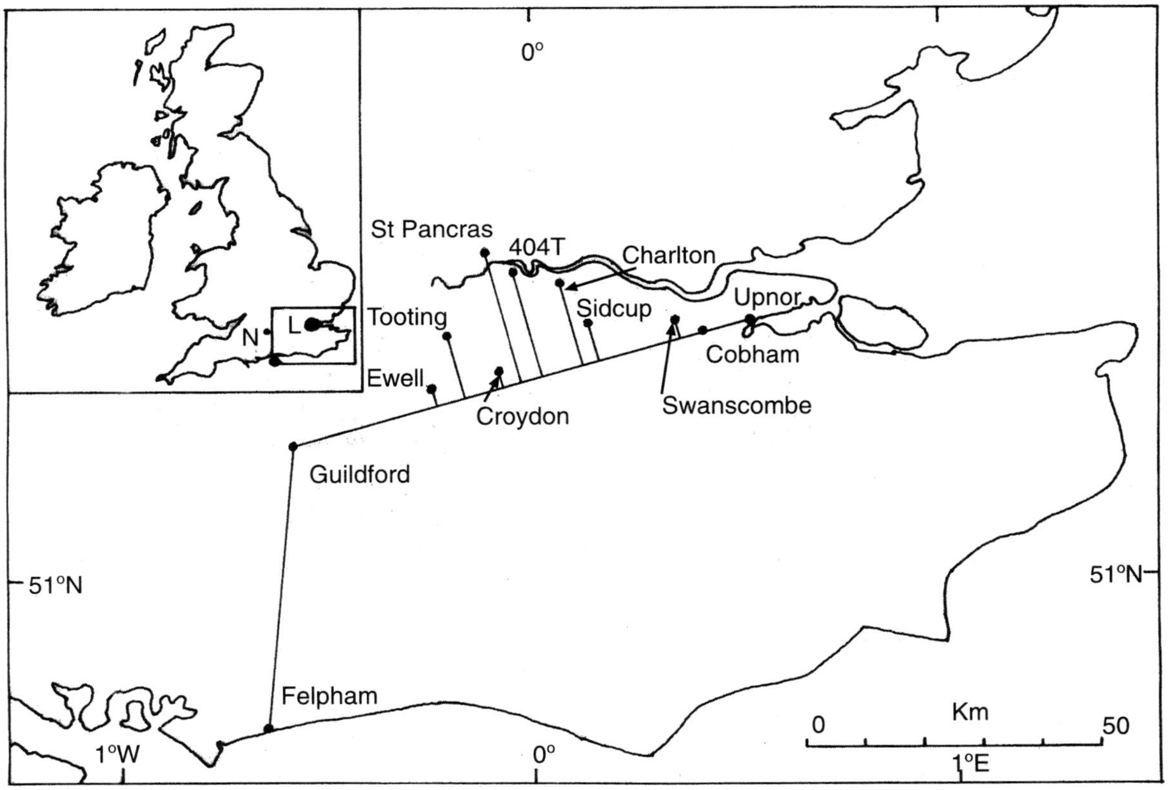

Figure 2. Map of southeast England, showing line of section in Figure 3 and how the localities named relate to it. Small inset map shows its location in the British Isles and the positions of London (L) and the plant fossil locality Coldash Quarry, Newbury (N).

jacketed box cores (columns between 10 × 10 and 20 × 20 cms in cross section) were made in the field to provide a continuous sample through the lignite that could be removed to the laboratory. These box cores were subsequently sampled in the laboratory for carbon isotope analysis of bulk lignite. An initial sample suite (set 1, Table 1), was taken at ~2–5 cm intervals to represent the entire sequence. 36 of these samples were run for carbon isotope analysis (Table 1), including two lots of two that represented the same stratigraphic interval (4.1 and 5.3; 4.2 and 5.2) in overlapping box cores (Table 1). Sampling was subsequently repeated at 3–5 mm intervals across the negative carbon isotope excursion (14 samples from 6.5 cm were run, set 2, samples SH23–38 on Table 1). The samples of bulk lignite were dried in an oven set at 60 °C overnight to remove any moisture. Weights of samples varied, depending on their total organic content, but were all adjusted to produce between 250 and 350 µg of CO_2 for carbon isotope analysis (Gröcke, 1998). The samples were sealed in 8 × 6 mm steel-foil cups and analyzed using a CF-IRMS with an EA1500 elemental analyzer on-line to an Optima MS. Isotopic results were calibrated to international (NBS-21) and internal standards. Carbon isotope ratios are expressed in the standard per mil (‰) notation against the international accepted standard, the Vienna Peedee belemnite (VPDB). Analytical machine reproducibility is better than ±0.1‰ based on analysis of internal standards. Five random replicate samples (see comments column on Table 1) gave an experimental variation between 0.06‰ and 0.25‰.

The qualitative mesofossil and palynological data reported here were obtained using sediment samples (Table 1) taken adjacent to the box cores from measured levels in the succession. This approach has enabled us to retain the box cores intact for future high-resolution, quantitative mesofossil, palynological and component-specific carbon isotope analysis directly linked to the carbon isotope results from bulk lignite and to detailed sedimentological and coal petrological variation. Lithological observations reported here are based on standard observations of hand specimens and field study of temporary exposures. The complete sequence of lithologies was studied on both sides of a 100 m long and 20 m wide cutting and parts of the sequence were visible at several other points in the works area.

Laboratory observations on mesofossils are based on four bulk samples of the lower laminated lignite (B4, 5, 6, 13), including two (B1, B13) replicate samples 50 m apart laterally; three bulk samples of upper blocky lignite (B8, 10, 15); two bulk samples of the basal pale clay (B3, B12), two bulk samples (B1, B2) of the underlying sand and mud unit (Fig. 1) and one sam-

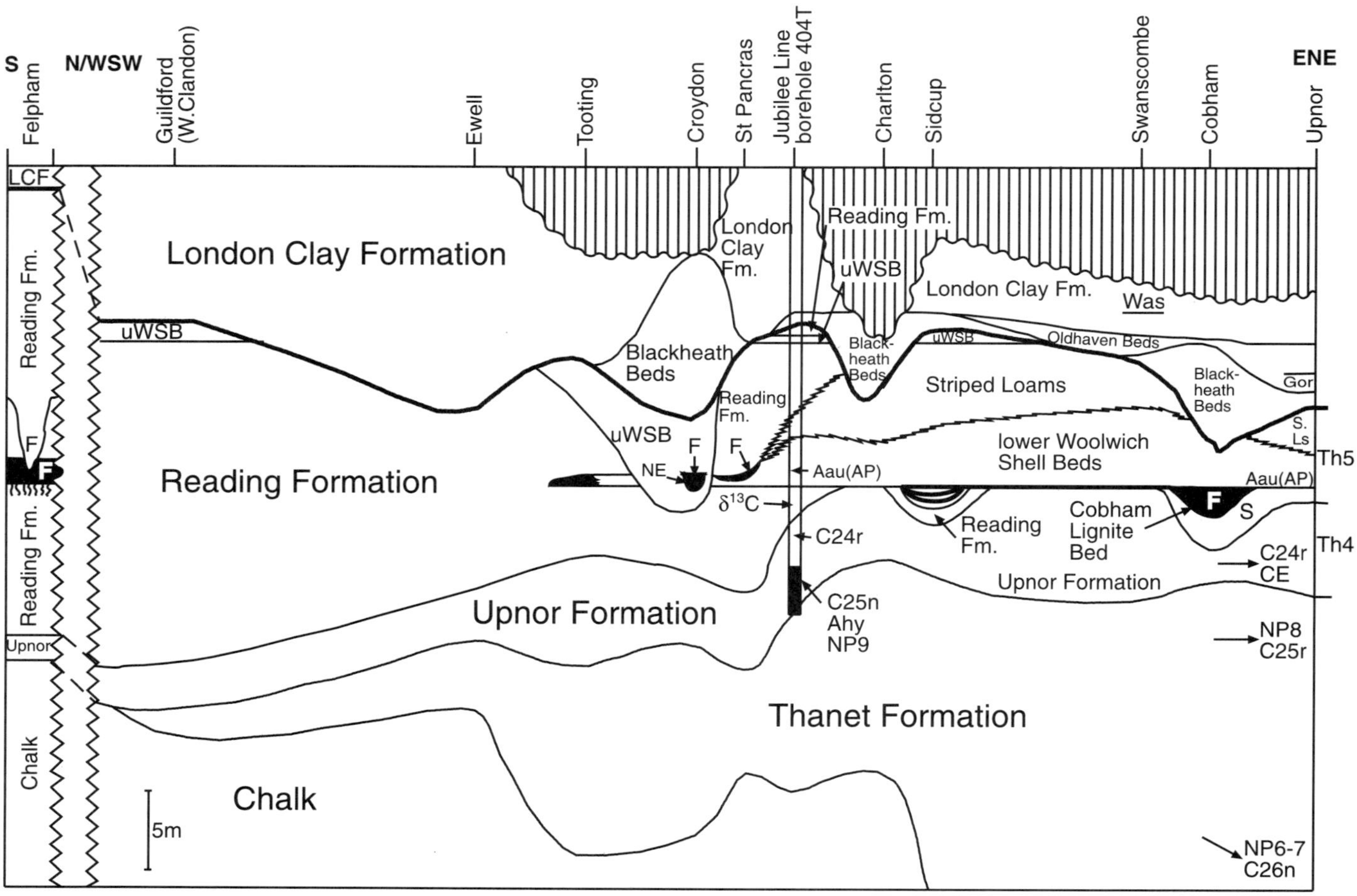

Figure 3. Section ENE to WSW through southeast England, plus log of the Felpham site (on the south coast). The section incorporates logs of key sites (see Figure 2 and text) with a succession of Thanet, Upnor, Woolwich (including lower Woolwich Shell Beds, Striped Loams and upper Woolwich Shell Beds), and Reading Formations, and the overlying Thames Group (including Blackheath Beds, Oldhaven Beds and London Clay Formation). This shows the regional setting and the stratigraphic relationships of the Cobham Lignite Bed and underlying unnamed sands and muds (S) (currently unattributed to a formation). F indicates the positions of fossil floras whilst lignitic lithologies are shown in solid black. The transgressive surface at the base of the lower Woolwich Shell Beds is considered an essentially synchronous horizon (Skipper, 1999) and is shown as a horizontal plane. The major unconformity at the base of the Thames Group is indicated by a thick line. The correlation indicators (magnetochron, NP, Dinocyst, ELMA, sequence) are defined and placed in a chronological framework in Figure 4. The explanation of the key correlation markers pertinent to the stratigraphic position of the Cobham Lignite is given in the text. Aau, Ahy, Gor and Was are dinocyst zones (Powell, 1992), the upper limit of Gor and lower limit of Was are indicated by short lines; AP—*Apectodinium* acme; CE—Cernaysian ELMA; C24-C26—magnetochrons (in the Jubilee Line borehole C24r pertains to all higher strata); $\delta^{13}C$—carbon isotope record of negative excursion; ELMA—European Land Mammal Age; F—flora; Fm—Formation; NE—Neustrian ELMA; NP—calcareous nannoplankton zone; S—sand and mud unit; S. Ls—Striped Loams; Th4 and Th5—sequences; uWSB—upper Woolwich Shell Beds. Arrows with magnetochrons and NP zones near the Upnor column indicate records further east. Sources for the site logs: Bone (1986), Brown and Wrigley (1925), Collinson (personal observation), Davis (1928), Dewey and Bromehead (1921), Dines et al. (1954), Ellison et al. (1996), Hooker (personal observation), Kennedy and Sellwood (1970), Klaassen (1883), Pitcher et al. (1958), Rundle (1972), Skipper (1999), Tracey (1986), Whitaker (1889, 1908, 1912).

ple of the overlying lower Woolwich Shell Beds (B16). Each sample weighed between 500 gms and 1 kg, and, within the lignite, samples represented between 15 and 20 cm of thickness. Palynological observations are based on 5 gm subsamples of these bulk samples.

The methods for mesofossil and palynological analysis followed standard procedures that are merely outlined here. Further details can be found in papers by Batten, Collinson, Pearson and Scott, Bruch and Pross, all in Jones and Rowe (1999). Samples for mesofossil analysis (mesofossils >125 µm) were disaggregated using hot water, hydrogen peroxide, or potassium hydroxide, the second and/or third methods only being applied if the first method failed to fully disaggregate the sample. Residues were sieved and size fractions from everything >4 mm down to 125 µm were retained and studied. Samples for palynological analysis (palynomorphs <125 µm) were treated with HCl, HF,

HCl, followed by 5% potassium hydroxide if necessary, then rinsing in water. In most cases potassium hydroxide treatment resulted in excellent lignite disaggregation as reported by Allen (1982) for most Cobham Lignite samples. If the disaggregation was incomplete, nitric acid and ammonia treatment was subsequently used. Palynological residues were sieved through 5 µm mesh to remove fine debris and 125 µm to remove larger particles. Slides were mounted by mixing a known aliquot volume of residue with PVA emulsion on a cover glass and mounting this on a slide with Petropoxy resin. The aliquot method retains quantitative relationships between palynomorph numbers in the residue on the slide and in the original rock sample.

Mesofossils were mounted for SEM using Bostick on a cover glass whilst palynomorphs were mounted from a water droplet onto negative film attached to Araldite (Moore et al., 1991). Samples were gold coated and photographed using a Philips 501B scanning electron microscope at 15 kv.

LITHOSTRATIGRAPHIC AND DEPOSITIONAL CONTEXT OF THE COBHAM LIGNITE BED

Figure 3 shows the lithostratigraphic context of the Cobham Lignite Bed. The shallow marine Upnor Formation was exhumed, folded and incised after deposition. Where succeeded by the fluvial Reading Formation it is pedogenically modified (Ellison et al., 1994). The irregular topography of the upper surface of the Upnor Formation was filled by the lower tongue of the Reading Formation plus laterally occurring lignitic and sandy units (which are unattributed to formations), prior to the Woolwich transgression. The Woolwich Formation (including lower Woolwich Shell Beds, Striped Loams and upper Woolwich Shell Beds) is the lateral equivalent of higher parts of the Reading Formation. Flood-plain deposits of the Reading Formation dominate in the west and north whilst tidal flat deposits of the Woolwich Formation dominate in the east, interdigitating in the central London and eastern Hampshire Basins (Ellison et al., 1994; Skipper, 1999). The observable sequence is complicated by minor incision below the upper Woolwich Shell Beds and by subsequent major incision at the base of the overlying Thames group (Blackheath Beds, Oldhaven Beds, London Clay Formation) (King, 1981; Ellison et al., 1994). In areas with least erosion below the Thames Group the Woolwich and Reading succession shows two tongues of Woolwich Shell Beds and two or three tongues of Reading Mottled Clays (Hester, 1965). Eastwards, where the Blackheath and Oldhaven overstep intensifies, only the lower tongue of Woolwich Shell Beds is preserved, and this is the case at Cobham (Chandler, 1923).

LITHOLOGICAL SUCCESSION AT COBHAM

Our samples were taken from temporary exposures at Scalers Hill, Cobham, Kent (National Grid Reference TQ 672 697) (Fig. 2). The succession (Figs. 1, 3) shows Upnor Formation at the base, followed by a sand and mud unit with an erosive junction between. The sand and mud unit is overlain, conformably in succession, by the Cobham Lignite Bed and the lower Woolwich Shell Beds. Above an unconformity these are succeeded by Blackheath Beds, then further east in the outlier by thin Oldhaven Beds and basal London Clay Formation (Chandler, 1923; Collinson and Hooker, personal observation). Since formalization of the interdigitating Woolwich and Reading Formations as distinct from the Upnor Formation, which were all once included in the "Woolwich and Reading Beds" (Ellison et al., 1994), the Cobham Lignite Bed and underlying sand and mud unit have remained unattributed to any formation. However, they occupy the same stratigraphic position in Kent as does the lower tongue of the Reading Formation (lower Reading Mottled Clay) further west in east and central London.

The Cobham Lignite Bed (Fig. 1) has a thin pale clay up to 3 cm thick at its base, which is succeeded by a strongly laminated (mm scale) lignite up to 55 cm thick. The laminated lignite is overlain by another thicker (up to 10 cm) pale clay. This is overlain by up to 132 cm of lignite that has a blocky fracture and little evidence of lamination in hand specimen, and that persists to the top of the Cobham Lignite Bed. The two pale clays were persistent across the extent of lignite exposure available to us (some 100 m in immediate lateral continuity) and were also recognizable wherever we saw partial exposures elsewhere through the works area. Other patchy and impersistent clay seams were also noticed within the blocky lignite. No large roots were observed in the field but fine rootlet-like structures were seen and sieved residues of the clays have yielded rootlet-like objects. A definitive statement concerning rooting, as well as the detail of the lamination of the lower lignite awaits preparation and study of polished blocks and thin sections. Originally, Chandler (1923) described the Cobham Lignite as varying in thickness from six inches (15 cm) to two feet six inches (75 cm) whilst Martin (1976) reported exposures with 60 cm and 120 cm of lignite. The 2 m section studied is, therefore, greatly expanded by comparison with previous studies and hence offers maximum opportunity for detecting floristic and environmental change.

CHRONOLOGICAL CONTEXT OF THE COBHAM LIGNITE BED AND POTENTIAL FOR RECORDING THE PETM CARBON ISOTOPE EXCURSION

In addition to the lithostratigraphy, Figure 3 shows the stratigraphic relationships of the Cobham Lignite Bed with the key biostratigraphic, magnetostratigraphic, and sequence stratigraphic markers discussed below in their geographical context (see also Figure 2) with respect to the Cobham site. Figure 4 is a chronological chart of the late Paleocene and early Eocene, which enables these key markers, shown on Figure 3, to be located in a global context.

Magnetochron C25n, dinocyst zone Ahy, and calcareous nannoplankton zone NP9 are present in the lower part of the Upnor Formation in the Jubilee Line 404T borehole, all higher

TABLE 1. LITHOLOGY AND CARBON ISOTOPIC MEASUREMENTS OF THE COBHAM LIGNITE

SEDIMENT SAMPLES				CARBON-ISOTOPE SAMPLES						
						Sample number				
Number*	Level[†] (cm)	Lithology	Lithostratigraphy[§]	Level (cm)	Lithology	set 2	set 1	δ³C	Carbon[#] wt%	Comments**
			IWSB	203.5–208.0	Silty clay		1.1			Excluded due to pyrite
B16	201–203	Lignitic clay	IWSB	201.0–203.5	Lignite		1.2	–26.4	47.6	
			IWSB	198.5–201.0	Shelly clay		1.3			Excluded shelly clay only
B15	192.5–198.5	Blocky lignite	CLB	196.5–198.5	Blocky lignite		1.4	–26.02	30.9	
B14	172–188	Blocky lignite	CLB	191.5–193.5	Blocky lignite		1.5	–26.09	38.0	Two Replicates,–26.21,–25.96
			CLB	184.0–186.5	Blocky lignite		1.6	–26.14	34.5	Two replicates,–26.07,–26.20
			CLB	175.5–178.0	Blocky lignite		3.14	–25.98	39.4	
			CLB	173.0–175.5	Blocky lignite		3.13			Excluded low yield
			CLB	170.0–173.0	Blocky lignite		3.12			Excluded due to pyrite
			CLB	167.0–170.0	Blocky lignite		3.11	–26.4	27.4	
			CLB	165.5–167.0	Blocky lignite		3.10			Not run
			CLB	162.5–165.5	Blocky lignite		3.9	–25.54	45.0	
			CLB	161.0–162.5	Blocky lignite		3.8			Result off scale
			CLB	158.0–161.0	Blocky lignite		3.7	–26.21	34.2	
B10	120–140	Blocky lignite	CLB	134.0–158.0	Blocky lignite		3.1–3.6			Excluded due to pyrite
			CLB	131.0–134.0	Blocky lignite		5.19	–26.28	47.8	
			CLB	127.0–131.0	Blocky lignite		5.18	–25.62	33.6	
			CLB	125.0–127.0	Blocky lignite		5.17	–26.61	28.6	
B9	100–120	Blocky lignite	CLB	123.0–125.0	Blocky lignite		5.16			Excluded due to pyrite
			CLB	114.5–118.0	Blocky lignite		5.15	–26.35	27.4	
			CLB	112.0–114.5	Blocky lignite		5.14	–26.19	31.8	
			CLB	107.0–111.0	Blocky lignite		5.13			Excluded due to pyrite
			CLB	102.0–104.0	Blocky lignite		5.12	–26.94	26.2	
B8	67–100	Blocky lignite	CLB	91.0–102.0	Blocky lignite		5.10–5.11			Excluded due to pyrite
			CLB	85.5–89.0	Blocky lignite		5.9	–25.96	23.6	
			CLB	82.0–85.5	Blocky lignite		5.8			Not run
			CLB	80.0–82.0	Blocky lignite		5.7	–25.93	44.2	
			CLB	77.0–80.0	Blocky lignite		5.6			Excluded due to pyrite
			CLB	72.5–75.0	Blocky lignite		5.5	–26.53	45.9	
			CLB	70.0–72.5	Blocky lignite		5.4			Not run
			CLB	67.0–70.0	Blocky lignite		4.1, 5.3	–26.08	66.8	–26.26,–25.90
B7	57.6–67.0	Pale clay	CLB (clay)	65.0–67.0	Lignite within clay		4.2, 5.2	–26.74	13.1	–26.87,–26.60
			CLB (clay)	60.0–63.0	Clay		4.3			Excluded clay only
			CLB (clay)	58.0–59.0	Clay		5.1			Excluded clay only
			CLB (clay)	57.6–67.0	Clay	SH1–SH22				Excluded clay only
B6	44–57.6	Laminated lignite	CLB	57.3–57.6	Laminated lignite	SH23		–26.18	50.2	
			CLB	56.9–57.3	Laminated lignite	SH24		–26.31	29.1	Two replicates,–26.21,–26.41
			CLB	56.6–56.9	Laminated lignite	SH25		–26.49	11.4	
			CLB	55.7–56.6	Thin dark clay	SH26–SH27				Excluded clay dominant

			CLB	55.3–55.7	Laminated lignite	SH28		–26.37	35.3	Two replicates,–26.34,–26.40
			CLB	55.0–55.3	Laminated lignite	SH29		–25.75	38.1	
			CLB	54.7–55.0	Laminated lignite	SH30		–26.96	35.8	
			CLB	54.4–54.7	Laminated lignite	SH31		–27.11	19.0	
			CLB	54.1–54.4	Laminated lignite	SH32		–27.5	17.3	MOST NEGATIVE
			CLB	53.0–54.0	Laminated lignite		4.4	–26.98	15.5	
			CLB	53.8–54.1	Laminated lignite	SH33		–27.1	21.6	
			CLB	53.5–53.8	Laminated lignite	SH34		–25.7	12.0	
			CLB	52.9–53.5	Laminated lignite	SH35		–25.6	19.0	
			CLB	52.0–53.0	Laminated lignite		4.5	–26.05	13.5	
			CLB	52.5–52.9	Laminated lignite	SH36		–26.42	25.7	
			CLB	52.0–52.5	Laminated lignite	SH37		–25.8	40.5	
			CLB	51.5–52.0	Laminated lignite	SH38		–26.25	13.65	
			CLB	46.0–48.0	Laminated lignite		4.6	–25.59	27.1	
			CLB	45.0–46.0	Laminated lignite		4.7	–25.41	56.3	
B5	23–44	Laminated	CLB	44.0–45.0	Laminated lignite		4.8			Not run
		lignite	CLB	41.0–42.0	Laminated lignite		4.9	–25.32	32.2	
			CLB	38.0–41.0	Laminated lignite		4.10–4.12			Not run
			CLB	36.5–38.0	Laminated lignite		4.13	–25.29	26.3	Two replicates–25.40,–25.17
			CLB	35.5–36.5	Laminated lignite		4.14			Not run
			CLB	34.0–35.5	Laminated lignite		4.15	–24.98	27.2	
			CLB	33.0–34.0	Laminated lignite		4.16			Not run
			CLB	32.0–33.0	Laminated lignite		4.17	–24.9	6.9	
			CLB	29.0–32.0	Laminated lignite		4.18–4.19			Not run
			CLB	28.0–29.0	Laminated lignite		4.20	–24.93	32.8	
B4; B13	2.5–23	Laminated	CLB	24.0–25.0	Laminated lignite		6.12	–24.62	45.4	
		lignite	CLB	18.5–20.0	Laminated lignite		6.11			Not run
			CLB	15.0–17.5	Laminated lignite		6.10	–24.47	46.2	MOST POSITIVE
			CLB	12.0–14.0	Laminated lignite		6.9			Not run
			CLB	6.0–7.0	Laminated lignite		6.8			Excluded low yield
			CLB	4.0–6.0	Laminated lignite		6.7			Not run
			CLB	2.5–4.0	Laminated lignite		6.6	–25.17	24.4	
B3; B12	0–2.5	Pale clay	CLB (basal clay)	0–2.5	Clay		6.4–6.5			Excluded clay dominant
B2; B11	–6 to 0	As below	Sand & mud unit	–3 to 0	Sandy lignite		6.3	–25.93	9.8	
B1	–10 to–6	Dark clayey	Sand & mud unit	–3 to–4	Sandy lignite		6.2	–26.26	8.5	
		sand with lignite	Sand & mud unit	–6 to–8	Sandy lignite		6.1	–26.37	7.4	

*Sediment samples (B1–B16) were collected for mesofossil and palynological analysis. Samples B1-6, 8,10,12,13,15,16 have been studied in this work. Samples B3 and B12, and B4 and B13, represent the same stratigraphic level in sites approximately 50 m apart.

†Sample point zero cm is taken at the base of the basal pale clay of the Cobham Lignite Bed.

§lWSB—lower Woolwich Shell Beds; CLB—Cobham Lignite Bed.

#Percent carbon is calculated from the peak area of CO_2 and is calibrated against international standards.

**Samples containing abundant pyrite were not run because of cost; "not run" indicates samples excluded in order to reduce cost. Samples with less than 5% carbon were excluded as un reliable.

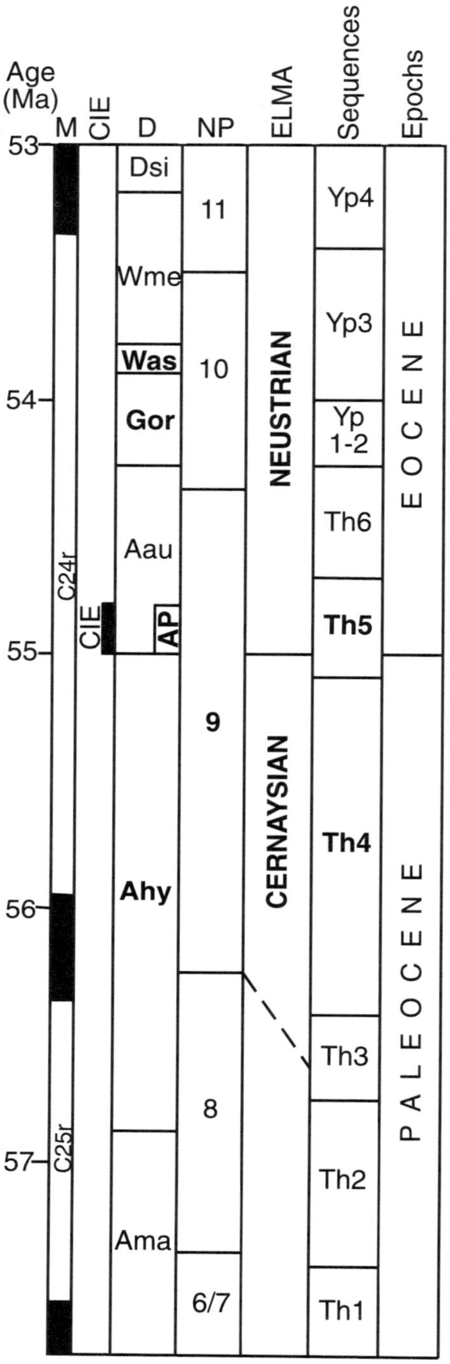

Figure 4. Chronological chart of the late Paleocene (Thanetian) and early Eocene (Ypresian) with time in Ma (Berggren et al., 1995; Norris and Röhl, 1999). It shows magnetochrons (M), biostratigraphic and biochronologic indicators (Powell, 1992), and sequences (Hardenbol et al., 1998) that have representation in the area under consideration and are relevant to recognition of the Paleocene-Eocene boundary. CIE indicates the carbon isotope excursion. Other abbreviations as in Figure 3. The indicators highlighted in bold are recorded from the central London Basin and, together with the magnetostratigraphic record, are shown in Figure 3 to date the Cobham Lignite Bed.

strata belonging to C24r (Ellison et al., 1996). The Cernaysian European Land Mammal Age (ELMA) is present in the upper reversed (C24r) interval of the Upnor Formation, 45 km east of Upnor at Herne Bay (Hooker and Millbank, 2001). The Th4-5 sequence boundary (Hardenbol et al., 1998) is interpreted as lying between the sand and mud unit (that underlies the Cobham Lignite) and the Upnor Formation (Fig. 3). This follows Powell et al.'s (1996) identification of the lower Woolwich Shell Beds as the transgressive systems tract of Th5 and Skipper's (1999) recognition of the unnamed underlying fluvial clastics (which include the sand and mud unit at Cobham) that fill incised valleys in the Upnor Formation, as lowstand or early transgressive deposits of the same sequence.

The *Apectodinium* acme occurs throughout the lower Woolwich Shell Beds at Upnor (Powell et al., 1996) and in the 404T borehole (Ellison et al., 1996). It is a proxy for the carbon isotope excursion at midpaleolatitudes and is taken as representative of the base of the Aau dinocyst zone (Powell, 1992; Bujak and Brinkhuis, 1998; Crouch et al., 2001). This is early Eocene according to the recent stabilization of the criterion for recognition of the Paleocene-Eocene boundary as the base of the carbon isotope excursion (Luterbacher et al., 2000). It should be noted that a second, stratigraphically higher, smaller *Apectodinium* acme is also recorded at Upnor (Powell et al., 1996) in Kennedy and Sellwood's (1970) bed 15. Kennedy and Sellwood (1970) identified bed 15 as Oldhaven Beds, whereas Powell et al. (1996, Figure 4) reinterpreted this without explanation as the local top of the Woolwich Formation. Bed 15 has an irregular contact with underlying strata and a molluscan assemblage typical of the Blackheath Beds (Hooker, personal observation) to which it is attributed here. The Blackheath Beds with their PE III mammal fauna (Hooker, 1996) calibrate to NP10 and Wasatchian North American Land Mammal Age (NALMA) zone Wa3 (Hooker, 1998). The bed 15 *Apectodinium* acme is thus too young to be part of the acme that is coeval with the carbon isotope excursion and is not relevant to our calibration of the Cobham Lignite Bed.

The lowest occurrence of the Neustrian ELMA is represented by *Coryphodon* in the upper Woolwich Shell Beds at Croydon (Klaassen, 1883; lithostratigraphic interpretation following Skipper, 1999). The Gor and Was dinocyst zones occur in the Oldhaven Beds (Powell et al., 1996) and lower London Clay (Costa et al., 1978) respectively, calibrating the basal Thames Group to NP10 (Powell, 1992; Hardenbol et al., 1998). From this evidence, the Upnor Formation is clearly Paleocene, whilst the lower Woolwich Shell Beds and higher strata are clearly Eocene. The intervening Cobham Lignite Bed and underlying sand and mud unit (Figs. 1, 3) at Cobham contain no biostratigraphic markers.

The Cobham Lignite Bed immediately underlies the lower Woolwich Shell Beds and hence immediately underlies the *Apectodinium* acme recorded 34 km west-northwest (404T borehole) and 8 km east-northeast (Upnor) (Fig. 3). The *Apectodinium* acme, at midpaleolatitudes, is coeval with the PETM carbon isotope excursion (see above). Therefore, the Cobham Lignite cannot postdate the carbon isotope excursion.

In order that the Cobham Lignite Bed should predate the carbon isotope excursion, either (1) the entire time-span of the *Apectodinium* acme would have to occur within the lower Woolwich Shell Beds, or (2) the Cobham Lignite Bed plus underlying sand and mud unit would have to occupy that part of sequence Th5 which precedes the carbon isotope excursion. Thiry et al. (1998) reported a carbon isotope excursion in soil carbonate nodules from the lower tongue of the Reading Formation, beneath lower Woolwich Shell Beds containing the *Apectodinium* acme, in the 404T borehole in an equivalent stratigraphic position to the Cobham Lignite (Fig. 3). This indicates that the first premise is not supported. The Th4-5 sequence boundary is estimated to precede the *Apectodinium* acme, and hence the carbon isotope excursion, by 80 k.y. (Hardenbol et al., 1998). This is a short interval of time, less than half the total length of the carbon isotope excursion (and by extrapolation the *Apectodinium* acme at midpaleolatitudes) which lasted 220 k.y. (Bujak and Brinkhuis, 1998; Crouch et al., 2001). To fit the Cobham Lignite Bed plus sand and mud unit into this 80 k.y. time interval, deposition would have to have been rapid following the clear hiatus early in Th5 represented by the incision of the Upnor Formation that occurred during the sea-level lowstand. In addition there would also have to be a significant hiatus between the Cobham Lignite Bed and the lower Woolwich Shell Beds because part of the carbon isotope excursion (see above) lies below the lower Woolwich Shell Beds. However, the sand and mud unit is interpreted as lowstand or early transgressive deposits (Skipper, 1999) and the lower Woolwich Shell Beds are interpreted as transgressive deposits (Powell et al., 1996). This suggests that the sand and mud unit, Cobham Lignite Bed and lower Woolwich Shell Beds form a logical succession of deposits reflecting sea-level rise in Th5 with no evidence for hiatus between any of them. There is thus no support for the second premise. Therefore, all stratigraphic data indicate that, in all likelihood, the lower part of the PETM carbon isotope excursion should be present in the Cobham Lignite Bed.

CARBON ISOTOPE ANALYSIS

Background

The modern carbon cycle is made up of many reservoirs, which can affect the ocean-atmosphere system on short-term and long-term time scales (Berner, 1990). The largest reservoirs consist of the Earth's crust, namely carbonate rocks, shales and fossil fuels. Owing to their size, the time required to affect the atmosphere is in the order of 100,000–1,000,000 yr and generally includes long-term weathering and burial, volcanic and diagenetic degassing and metamorphism. These long-term mechanisms generally have a major control on paleoatmospheric O_2 levels and CO_2 levels on the millions of years time-scale. The short-term carbon cycle (from days to 1000 yr) is related to air-sea exchange, photosynthesis, and respiration, which are dominated by the biosphere, surface, and deep ocean. The deep ocean is the largest of these three reservoirs and hence, the most effective in producing a carbon cycle perturbation. Therefore, a change in the carbon isotope composition of the oceanic carbon reservoir will be recorded in the atmosphere and biosphere owing to the short-term carbon cycle. The global record of the PETM negative carbon isotope excursion (lasting ~220 k.y.) observed in oceanic carbonates (Bains et al., 1999, 2000) will, therefore, be recorded in the atmosphere and biosphere, and consequently in terrestrial organic matter.

Carbon isotope values in the Cobham Lignite Bed

Carbon isotope results (Fig. 5) from terrestrial organic carbon derived from bulk lignite samples in the Cobham Lignite Bed indicate a background $\delta^{13}C$ level of ~ –26‰ and a pattern of fluctuations with a positive shift to –24.5‰, succeeded by a rapid negative shift to –27.5‰ followed by a return to background levels with minor variations. At the maximum negative shift, four contiguous samples in 10 mm of lignite (set 2, SH30-33), and the sample 4.4 from set 1 in the same 10 mm interval, give values –26.96‰ or greater (Table 1). The positive shift, the maximum negative shift, and the recovery to background levels all occur within the uniform lithology of the lower laminated lignite, containing only terrestrial organic matter as proven by mesofossil and palynological analyses (see below).

Comparison and discussion

The carbon isotope shifts documented in terrestrial lignite from the Cobham Lignite Bed show some similarity to those recorded for the Paleocene-Eocene transition in high-resolution deep sea successions such as Ocean Drilling Program (ODP) Site 690 (Fig. 6), which is judged to be the most expanded and complete marine section currently available (Bains et al., 1999, 2000; Röhl et al., 2000). The Site 690 carbonate carbon isotope excursion records a Paleocene background of +2‰ followed by a small positive shift to 2.5‰, before rapidly decreasing to ~0.5‰, via two successive rapid, negative shifts, with ~52 k.y. between the initiation and most negative carbon isotope values (Fig. 6). The entire excursion lasts for some 220 k.y. (Röhl et al., 2000). At ODP Site 1051 the negative shift is ~1.5‰ rather than 2‰ as at Site 690 (Bains et al., 1999). The excursion is thought to have been the result of pulsed releases of continental-margin marine methane gas hydrates (Dickens et al., 1995). The resultant interval of greenhouse warming (PETM), caused temperatures to rise by 4–8 °C in the mid to high paleolatitudes, thus reaching the warmest levels for the entire Cenozoic (Zachos et al., 2001).

Very similar carbon isotope fluctuations have been recorded in terrestrial paleosol carbonate nodules (Bowen et al., 2001a) (Fig. 6) and, though at much lower resolution, in mammalian tooth enamel (Koch et al., 1995), from Polecat Bench, Bighorn

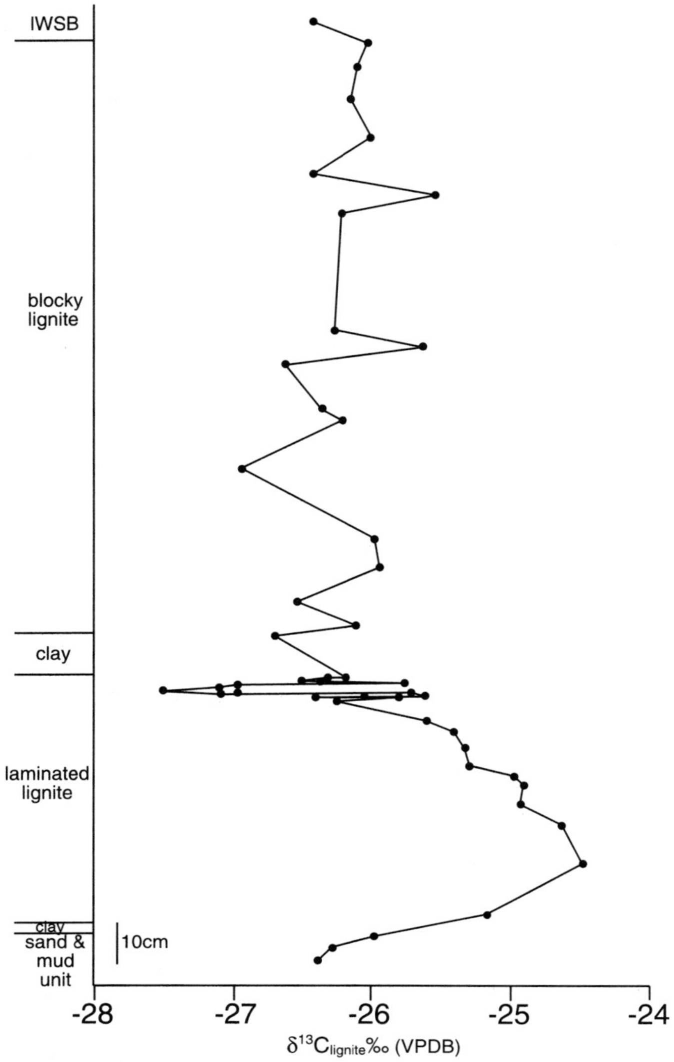

Figure 5. The carbon isotope record from bulk terrestrial lignite from the Cobham section. Details of samples and results can be found in Table 1. Where set 1 results (samples 4.4 and 4.5) and set 2 results (samples SH23–38) overlap, the more detailed set 2 results are joined to form the curve. lWSB—lower Woolwich Shell Beds.

Basin in Wyoming, USA. The Polecat Bench section is largely devoid of plant fossils and sections elsewhere in the Bighorn Basin, although rich in plant fossils, have a barren interval coincident with the earliest part of the warming event (Wing et al., 2000, p. 225; Wing and Harrington, 2001).

Given the magnitude of the carbon isotope fluctuations at ODP Site 690 the soil carbonate curve from Polecat Bench is not in isotopic equilibrium with expected fractionation factors between marine carbonate and soil carbonate (Koch et al., 1992). It has been suggested that the difference between expected and observed values in soil carbonates may be due to the complexity of the soil environment including such factors as the level of organic matter in the soil, carbon dioxide exchange rates, alkalinity and carbonate content (Cerling, 1984; Bowen et al., 2001b). The more negative carbon isotope values recorded from Polecat Bench soil carbonate nodules are probably the result of differential fractionation between soil biophysiology and increased photosynthetic fractionation invoked by high CO_2 levels (Bowen et al., 2001b). Fractionation factors of 1:1 or 1:2 between the carbonate reservoir and the organic reservoir (Koch et al., 1992) place the shift obtained from the Cobham Lignite Bed (1.5‰ from approximate background levels, 3‰ from most positive value, Figure 5) within reasonable isotopic equilibrium with that from ODP Sites 1051 and 690 (1.5‰ to maximum 2.5‰). Data from bulk lignite might mix subtle and small-scale changes. Therefore, it is not possible to judge from our data exactly how the major negative carbon isotope shift observed in the Cobham Lignite Bed corresponds to the steps in the negative carbon isotope excursion in the Site 690 sequence. Component-specific carbon isotope analyses will be needed to show the fine detail of the excursion in the Cobham Lignite Bed.

The effects leading to variation in the carbon isotope composition of ancient plant material have been discussed in detail in Gröcke (1998, 2001, 2002). These include, for example, original environmental effects, depositional environment, the type of plant material being preserved (such as lignin versus cellulosic components) and how the material is preserved (such as compression versus charcoalification). In spite of these effects, many studies on ancient terrestrial carbon isotope stratigraphy have shown that environmental effects are minor when considering a geological time sequence in the order of 1000–1,000,000 yr and greater (Gröcke, 1997, 2002; Hasegawa, 1997; Gröcke et al., 1999). In a lignite sequence it is possible that biological succession of a series of different plant communities, a change in the local depositional environment, or a change in the preservation state of the plant fossils could have caused the carbon isotope fluctuations that we have documented (Fig. 5). However, these possibilities are very unlikely due to the fact that the key isotopic shifts all occur within the lower laminated lignite lithology (Fig. 5). Furthermore, although a detailed quantitative assessment has not yet been undertaken the results presented here show that the types of dominant plant, and the preservational states of the dominant plant material, contributing to the lignite (see below) are uniform within the lower laminated lignite. Therefore, the pattern of carbon isotope values suggests that the Paleocene-Eocene boundary, and the onset of the PETM, in the eastern London Basin lies near the base of the Cobham Lignite Bed.

PLANT MESOFOSSIL ANALYSIS

Charcoal

Field and laboratory observations showed that every bedding surface (mm scale bedding) in the laminated lignite is completely covered with randomly oriented charcoalified plant axes,

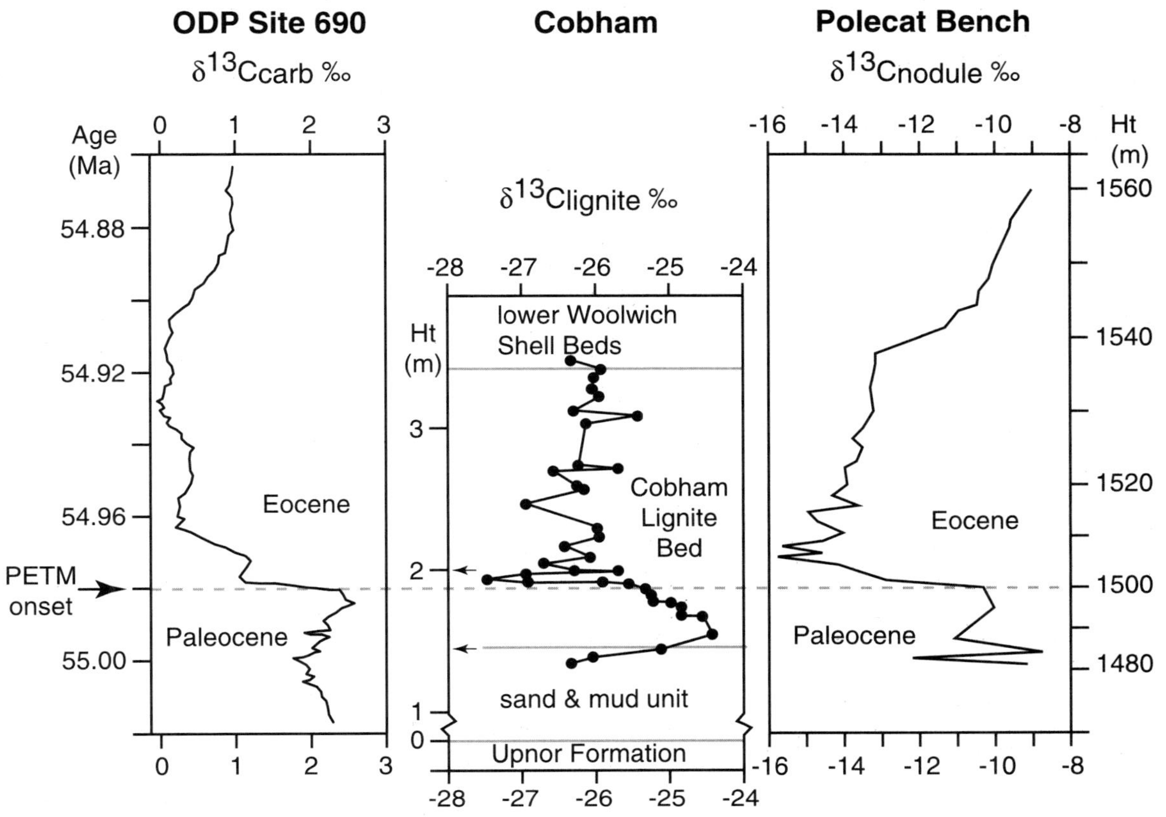

Figure 6. The carbon isotope record for the Cobham section and those from the Paleocene-Eocene Thermal maximum (PETM) in the marine (Ocean Drilling Program [ODP] Site 690; Bains et al., 1999) and continental (Polecat Bench; Bowen et al., 2001a) realms. ODP Site 690 is derived from bulk carbonate (Bains et al., 1999) recalibrated to a time scale (Röhl et al., 2000). Polecat Bench is derived from soil carbonate nodules (Bowen et al., 2001a), recalibrated to a time scale using Wing et al. (2000) model 1 with dated tie points and intervening assumed uniform sedimentation rate. Arrows point to the positions of thin clay bands in the Cobham Lignite, which are discussed in the text and evident in Figures 1 and 5.

from 1 mm to several cm in length (Fig. 7A). The axes were immediately evident in the field and their charcoal nature was confirmed in the laboratory using SEM observations (Fig. 7B). The following characteristics show that this material is fossil charcoal (= fusain): Black, shiny, brittle and/or friable, produces a black smear, has full three-dimensional preservation of cellular anatomy, and has homogenized cell walls (Fig. 7C; Scott, 2000; Collinson et al., 1999). Checking of cell walls (Fig. 7C) is also preserved (see discussion in Collinson et al., 1999). In the overlying blocky lignite, charcoal abundance decreases upwards. Higher in the succession, charcoal no longer covers every fractured or broken surface, rates as common rather than abundant, and its clast size decreases with many being 1 mm or smaller. The high abundance of charcoal is unparalleled elsewhere in the southern English succession, where usually only very rare (not noticeable without considerable search in a sieved residue), small (<250 μm), fragments of charcoal are found (Collinson, personal observations 1974–2002). Charcoal is also generally rarely reported in the Paleogene by comparison with other time intervals in the fossil record (Scott, 2000). Charcoal is absent, or present only as extremely rare, small (<125 μm) fragments, in the strata immediately underlying and overlying the Cobham Lignite.

In all lignite samples studied the charcoal includes both woody and herbaceous plant axes. The woody material is flowering plant wood, which contains vessels (Fig. 8A) with scalariform pitting (Fig. 7B) and scalariform perforation plates (Fig. 8B). No coniferous wood has yet been identified. The wood charcoal is rather fragmentary (for example no obvious twig fragments have been found), and most of the material seems to be radial slivers from larger axes. Some levels contain shredded, fibrous samples, which may reflect different charring temperatures or different wood anatomy, for example, reaction wood (see discussion in Collinson et al., 1999).

In contrast, the herbaceous axes are discrete, and epidermal layers are evident. Many of these herbaceous axes resemble fern rachides, having a U-shaped cross section and a sclerenchymatous to parenchymatous ground tissue, near the edge of which is a small, elongate-oval to comma-shaped, meristele of vascular tissue (Fig. 8D–F). Rare herbaceous axes have spiny external

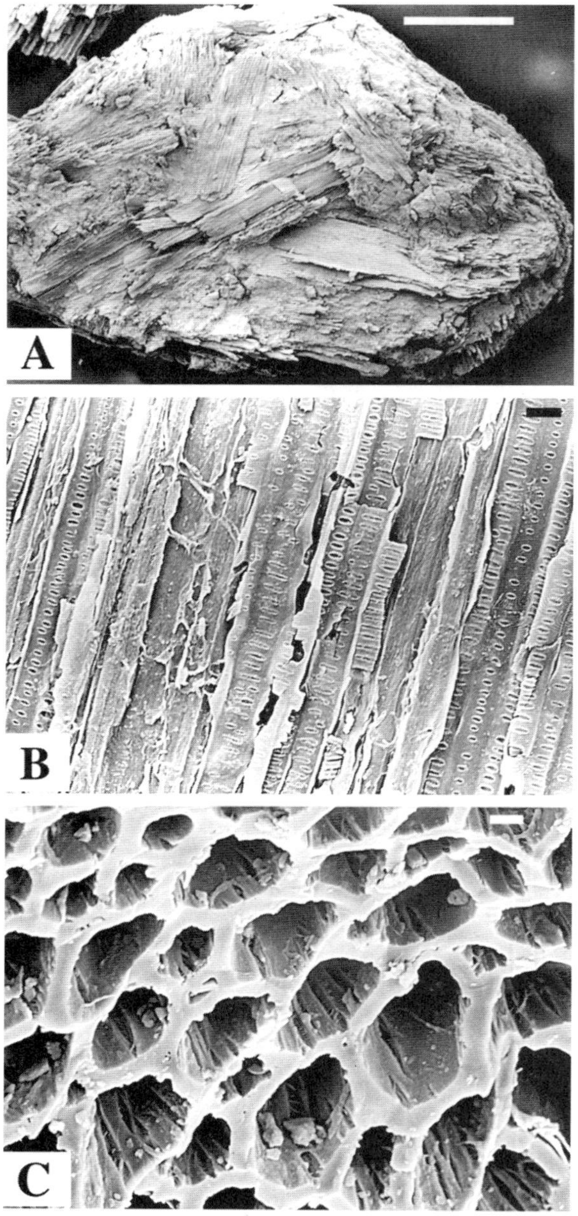

Figure 7. A: Representative fragment of lower laminated part of the Cobham Lignite Bed, showing bedding surface with numerous, randomly orientated charcoalified woody axes. B: Detail showing three-dimensional preservation of flowering plant wood structure in charcoalified fragment from A, and also showing evidence of the activity of fungi, probably prior to charcoalification, in the form of impressions of hyphae on the cell walls. C: Detail of sclerenchyma (transverse fracture) from herbaceous axis, showing homogenization of cell walls and checking. All scanning electron micrographs. Scale bars 1 mm in A, 10 µm in B and C.

ornament. "Fiber bundles," i.e., clumps of relatively thick-walled cells but lacking any primary or secondary vascular elements, are also common. These most probably derived from herbaceous axes and possibly from ferns (see also discussion of similar occurrences in the Cretaceous in Collinson et al., 1999).

Some of the charcoal contains evidence of fungal activity, which, on the basis of impressions of hyphae on cell walls (Fig. 7B) and apparent invasion of three-dimensionally preserved hyphae into cell walls (Fig. 8C), seems to indicate fungal activity prior to charring. This suggests that some, but certainly not all, of the charcoal was part of a litter layer.

We have not yet discovered any charcoalified leaves, sporangia or seeds, or animal material. However, we have not undertaken an exhaustive study of any of the sampled levels. The abundance, random orientation and consistent occurrence of large pieces of charcoal of both herbaceous and woody plants indicate that the local environment was characterized by repeated fires.

Noncharcoalified material

Samples from the base of the blocky lignite have yielded abundant specimens of microspore massulae of *Salvinia*. *Salvinia cobhamii* Martin was described by Martin (1976) from the Cobham Lignite Bed and the massulae conform to those of that species. However, *Salvinia* massulae lack specific characteristics (Batten and Collinson, 2001) and megaspores are needed to confirm the species determination. T. Martin has made his remaining Cobham Lignite residues available to us for study, which will facilitate comparisons. We have not yet found any *Azolla* material although this genus was reported by Martin (1976). Other megaspores, which might be expected in this time interval (e.g., Batten and Collinson, 2001), have also not been found. Martin (1976) emphasized the patchy distribution of Salviniaceae, which were only found in one (the basal) of his samples. Salviniaceae were not recorded by Allen (1982).

The *Salvinia* specimens we found occur only in the basal blocky lignite, immediately above the pale clay that overlies the laminated lignite. Thus, the occurrence of *Salvinia,* a free-floating water plant (see Collinson, 2001, 2002, for details of fossil record and ecology) is related to the occurrence of a clay layer within the lignite and to a change in lignite lithology. All these factors suggest a change in mire hydrology and they occur shortly above the maximum bulk lignite negative carbon isotope values suggesting a possible causal link. Bains et al. (1999) suggested that a change in the hydrological cycle might have occurred as a result of continental-margin methane gas-hydrate dissociation at the onset of the PETM.

PALYNOLOGICAL ANALYSIS

The palynology slides from the lower laminated lignite, and the lower part of the blocky lignite, contain very high proportions (for example >60%) of trilete fern spores of the *Cicatrico-*

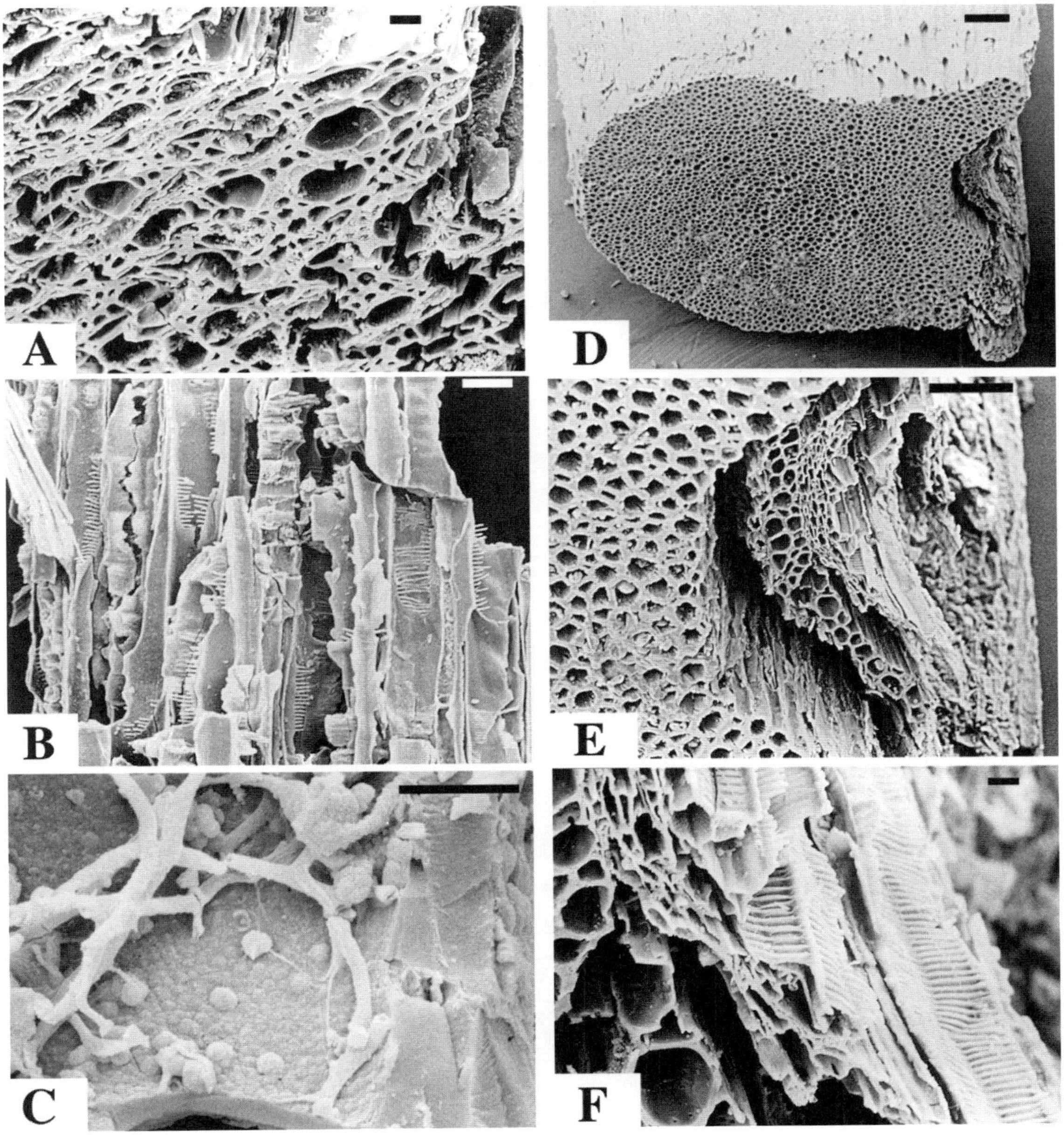

Figure 8. Charcoalified plant axes extracted from the Cobham Lignite. A, B: Flowering plant wood, A, in transverse fracture and B, longitudinal fracture with scalariform perforation plates on vessels. C: Detail of herbaceous axis showing three-dimensionally preserved fungal hyphae penetrating cell walls. D–F: herbaceous axis with fern-like anatomy and morphology, D, transverse fracture showing U-shape and meristele at right, E, detail of meristele, F, detail of vascular tissue within meristele. All scanning electron micrographs. Scale bars 20 μm in A, 50 μm in B, 10 μm in C and F, 200 μm in D, and 100 μm in E.

sisporites type. Proportions are reduced (for example <20%) in the blocky lignite higher in the sequence, at the same time as charcoal abundance is reduced. This spore type is known in situ in fossil ferns named *Ruffordia* which are closely related to the modern genus *Anemia* of the family Schizaeaceae. The ancient spore-producing plant was herbaceous with relatively delicately branching fronds >50 cm long from their first dichotomy above the main rachis. The plants are well known in the Paleogene including in peat-forming environments (see Collinson, 2001, 2002, for details). *Cicatricosisporites* spores in the Cobham Lignite occur in abundances comparable to those fern spore abundances that are termed "fern spikes" or "ferns spore abundance anomalies" at the Cretaceous-Tertiary (K-T) boundary (Collinson, 1996; Sweet et al., 1999; Vajda et al., 2001). Fern abundances at the K-T are taken to reflect the ease with which spore-bearing plants are able to colonize disturbed terrains and they may well reflect the same phenomenon in the Cobham Lignite Bed, where disturbance to the environment would have resulted from repeated fires. Occurrences of fossil charcoalified ferns and high fern abundances in association with disturbance are also documented by Collinson (2002).

Neither Martin (1976) nor Allen (1982) reported an abundance of spores and they lumped all trilete spore taxa in their charts. The samples studied by Allen (1982) were from the upper 50 cm of lignite in which we find reduced abundances of spores. The samples studied by Martin (1976) were from an area where the total lignite thickness was only 60 cm (in contrast to our 2 m) and it is not possible to judge which part of our lignite sequence his samples represent.

Other palynomorphs that occur in notable numbers include *Platycaryapollenites,* other Juglandaceae and Betulaceae pollen. Distinctive distributions of microechinae as seen with the SEM, characterize each family. *Platycaryapollenites,* and other Juglandaceae pollen, are known in situ from reconstructed ancient plants (e.g., Manchester, 1987, 1989) and Betulaceae pollen are also closely linked to other fossil organs of the Betulaceae (Crane, 1989). There is, therefore, little doubt that these ancient pollen were produced by trees of the families Juglandaceae and Betulaceae. Grains of *Pistillipollenites macgregorii* are also quite abundant in the Cobham Lignite. *Pistillipollenites* pollen has been reported in situ in at least two types of fossil flowers (Stockey and Manchester, 1988; Crepet and Daghlian, 1981). Some of the flowers occur in lacustrine settings. Possible affinities with Gentianaceae suggest that the parent plant was herbaceous, and other possible familial affinities also encompass the herbaceous habit (Euphorbiaceae). Allen (1982) and Martin (1976) also reported an abundance of *Platycarya* pollen (as *Platycaryapollenites platycaryoides* or *Triatriopollenites coryphaeus,* respectively) and the occurrence of *Pistillipollenites macgregorii* in the Cobham Lignite. Pollen of coniferous trees, e.g., bisaccate grains and inaperturate grains of the families Pinaceae/Podocarpaceae or Taxodiaceae respectively, are either rare or absent, a phenomenon also observed by Martin (1976) and Allen (1982). Other grains, such as a pentapantocolpate reticulate form, with enlarged reticulum at the poles, are not familiar Paleogene taxa and may represent very local, habitat-specific, flora. Preliminary observations indicate that a variety of subtle changes in palynomorph assemblages occur within the lignite. These include changes in relative abundance of the dominant taxa and in occurrence of rarer taxa, i.e. diversity of the assemblage.

PENECONTEMPORANEOUS MACROFLORAS

Collinson and Cleal (2001a) and Collinson (2000a) reviewed macrofloras of the Paleocene-Eocene transition in Britain. The fruit, seed and megaspore assemblages from Felpham (West Sussex), St. Pancras (central London), and Croydon (Surrey) (Figs. 2, 3) were recognized as containing a distinctive flora. This flora was originally termed the "near LPTM flora" (Collinson, 2000a) but the name is here revised to "near PETM flora." Near PETM floras differ from overlying Eocene Thermal Maximum (ETM) floras. The near PETM flora reflects a less tropical more mesothermal vegetation whilst the ETM flora reflects paratropical megathermal vegetation with coastal *Nypa*-dominated mangrove (Collinson, 2000a, 2000b). [Note that ETM as used here is equivalent to CGCO (Cenozoic Global Climatic Optimum) of Clyde et al., 2001, and EECO (early Eocene Climatic Optimum) of Zachos et al., 2001.] The Felpham and St. Pancras floras occur in lignites associated with, in at least the latter, the feather edge of the lower Woolwich Shell Beds, and are thus slightly (perhaps within 100 k.y.) younger than the Cobham Lignite (Figs. 3, 4). The Croydon flora comes from the stratigraphically higher upper Woolwich Shell Beds (Skipper, 1999), which could be several hundred thousand years younger.

In contrast to Felpham, St. Pancras, and Croydon, the flora from the Reading Formation at Cold Ash near Newbury (Fig. 2) is dominantly a leaf flora. The precise stratigraphic relationship of the Cold Ash floras to the Cobham succession or other sites with floras (Fig. 3) is uncertain because fluvial Reading Formation in the west (in this case in Berkshire) is remote from the zone of interdigitation with the Woolwich succession in the east. The Cold Ash leaf flora is more comparable with higher-latitude floras derived from polar deciduous forests than with the paratropical floras of the ETM in continental Europe (Collinson and Cleal, 2001a), further emphasizing the distinction between floras of the Paleocene-Eocene transition and the ETM interval.

The evidence from all four macrofloras supports the presence, across the Paleocene-Eocene transition, during deposition of the Woolwich and Reading Formations, of a vegetation that was distinct from that of later parts of the early Eocene. The global warming at the PETM may have been of insufficient duration (it was of sufficient magnitude—Clyde et al., 2001; Zachos et al., 2001) to enable establishment of the mangroves and paratropical forests that characterize the later Early and early middle Eocene in southern England (Collinson, 2000a, 2000b; Collinson and Cleal, 2001a, 2001b).

DISCUSSION AND CONCLUSIONS

The Cobham Lignite Bed (Kent, UK, Figures 1–3) and underlying sand and mud unit occur beneath lower Woolwich Shell Beds, dated as early Eocene, and above Upnor Formation, dated as late Paleocene (Figs. 3, 4). The Th4-5 sequence boundary is located between the sand and mud unit and the Upnor Formation and is estimated to be 80 k.y. earlier than the base of the *Apectodinium* acme (Hardenbol et al., 1998). The latter, an important global dinocyst event, is a proxy at midpaleolatitudes for the base of the carbon isotope excursion that indicates the PETM (Bujak and Brinkhuis, 1998; Crouch et al., 2001). The *Apectodinium* acme is recorded in the lower Woolwich Shell Beds within 8 km to the east and 34 km to the west of the section containing the Cobham Lignite Bed (Powell et al., 1996; Ellison et al., 1996) but is not in higher parts of the Woolwich Formation, or in the Upnor Formation that underlies the Cobham Lignite (Bujak and Brinkhuis, 1998). Thiry et al. (1998) reported a carbon isotope excursion in the Jubilee Line 404T borehole in London in lower Reading Mottled Clay underlying lower Woolwich Shell Beds, in the same stratigraphic position as the Cobham Lignite Bed (Fig. 3). All these data show that the lower part of the Paleocene-Eocene carbon isotope excursion should be expected to occur within the Cobham Lignite Bed. We have documented a rapid negative carbon isotope excursion in bulk lignite from the Cobham Lignite Bed (Fig. 5), the magnitude of which is in reasonable isotopic equilibrium with that observed in carbonate records at other sites (Fig. 6). The carbon isotope excursion occurs within a uniform laminated lignite lithology in which dominant plant types and preservational states remain constant. We therefore argue, on the basis of combined stratigraphic and carbon-isotopic evidence, that the Cobham Lignite Bed contains the lower part of the PETM carbon isotope excursion.

Plant mesofossils, especially charcoal, and qualitative palynological analysis of the Cobham Lignite Bed show that the vegetation was dominated by flowering plant trees (including Juglandaceae and Betulaceae) and herbaceous plants, including flowering plants (e.g., those producing *Pistillipollenites* pollen) and ferns. Ferns were abundant, especially those that produced *Cicatricosisporites* spores, attributed to *Ruffordia* of the Schizaeaceae on the basis of in situ spores from other sites. The fern abundances are likely to be related to the continuous disturbance to the environment caused by repeated fires, evidenced by high abundance and consistent occurrence of charcoal, especially of flowering plant trees and probable ferns.

The recognition of the onset of the PETM within the Cobham Lignite shows that the vegetation persisted with minimal change both with the onset and through the early part of the PETM. Penecontemporaneous macrofloras from Paleocene-Eocene transitional sites elsewhere in the London and Hampshire basins document a distinctive vegetation at this time, which is of less tropical aspect than that which existed during the ETM. These results from the Cobham Lignite and penecontemporaneous macrofloras are consistent with other studies from the Gulf Coast (Harrington, 2001; Harrington and Kemp, 2001) and the Bighorn Basin (Wing and Harrington, 2001), both of the United States, which also indicate little change in vegetation across the Paleocene-Eocene transition. However, the exact position of the PETM is not known in the Gulf Coast (Harrington, 2001; Harrington and Kemp, 2001) whilst in the Bighorn Basin the PETM is represented by an interval barren of plants (Wing et al., 2000; Wing and Harrington, 2001). In contrast, our study of the Cobham Lignite Bed shows that the vegetation described did indeed exist, and exhibit little change, at the PETM. A similar conclusion is drawn by Wing et al. (this volume) who record little change between latest Paleocene and earliest Eocene palynofloras, separated by strongly negative carbon isotope values from soil carbonates, in the Powder River Basin, Wyoming, United States. The Cobham Lignite contains not only a rich palynoflora but also a suite of plant mesofossils including charcoals in a lignite sequence within which a negative carbon isotope excursion has been documented. Therefore, the Cobham Lignite offers a unique opportunity to refine our knowledge with a high-resolution record of vegetation and environmental response to the rapid and short-lived global warming of the PETM.

ACKNOWLEDGMENTS

We would like to thank Alfred McAlpine, AMEC and Channel Tunnel Rail Link for access to the Cobham Lignite Bed, Sharon Rose for making arrangements, and Sharon and numerous staff for their considerable help on the site. We thank Jackie Skipper and Steve Tracey for help with sample collection and field discussions. Thanks to Dave Lowry and Nathalie Grassineau for assistance with, and the Department of Geology, Royal Holloway University of London for funding for the carbon isotope analyses, and Sharon Gibbons for preparing palynology slides and some mesofossil samples. We appreciate the efforts of two referees and the editor, Scott Wing, whose comments greatly improved this paper. This is a contribution to the Natural History Museum's Human Origins Programme, Project 298.

REFERENCES CITED

Allen, L.O., 1982, Palynology of the Paleocene and early Eocene of the London Basin [Ph.D. thesis]: University of London, 432 p.

Anonymous, 1948, Cobham miners "will carry on": Kent Messenger, August 1948.

Bains, S., Corfield, R.M., and Norris, R.D., 1999, Mechanisms of climate warming at the end of the Paleocene: Science, v. 285, p. 724–727.

Bains, S., Norris, R.D., Corfield, R.M., and Faul, K.L., 2000, Termination of global warmth at the Paleocene/Eocene boundary through productivity feedback: Nature, v. 407, p. 171–174.

Batten, D.J., and Collinson, M.E., 2001, Revision of species of *Minerisporites, Azolla* and associated plant microfossils from deposits of the upper Paleocene and Paleocene/Eocene transition in the Netherlands, Belgium, and the USA: Review of Palaeobotany and Palynology, v. 115, p. 1–32.

Berggren, W.A., Kent, D.V., Swisher, C.C., III, and Aubry, M.-P., 1995, A revised Cenozoic geochronology and chronostratigraphy, *in* Berggren, W.A., et al., eds., Geochronology, time scales, and global stratigraphic correlation: SEPM (Society for Sedimentary Geology) Special Publication 54, p. 129–212.

Berner, R.A., 1990, Atmospheric carbon dioxide over Phanerozoic time: Science, v. 249, p. 1382–1386.

Bone, D.A., 1986, The stratigraphy of the Reading Beds (Paleocene) at Felpham, West Sussex: Tertiary Research, v. 8, p. 17–32.

Bowen, G.J., Koch, P.L., Gingerich, P.D., Norris, R.D., Bains, S., and Corfield, R.M., 2001a, Refined isotope stratigraphy across the continental Paleocene-Eocene boundary on Polecat Bench in the northern Bighorn Basin: University of Michigan Papers on Paleontology, no. 33, p. 73–88.

Bowen, G.J., Koch, P.L., and Zachos, J.C., 2001b, Linking terrestrial carbon isotope records and atmospheric CO_2 response across the Paleocene/Eocene boundary: Eos (Transactions, American Geophysical Union), Washington (Supplement) 82, no. 47, p. F769.

Brown, E.E.S., and Wrigley, A., 1925, Notes on the sections exposed in the Sidcup by-pass road: Proceedings of the Geologists' Association, v. 36, p. 76–79.

Bujak, J.P., and Brinkhuis, H., 1998, Global warming and dinocyst changes across the Paleocene-Eocene epoch boundary, *in* Aubry, M.-P., et al., eds., Late Paleocene–early Eocene climatic and biotic events in the marine and terrestrial records: New York, Columbia University Press, p. 277–295.

Cerling, T.E., 1984, The stable isotopic composition of modern soil carbonate and its relationship to climate: Earth and Planetary Science Letters, v. 71, p. 229–240.

Chandler, R.H., 1923, The Tertiary section at Shorne Wood, Cobham, Kent: Proceedings of the Geologists' Association, v. 34, p. 137–141, plate 13.

Clyde, W.C., Sheldon, N.D., Koch, P.L., Gunnell, G.F., and Bartels, W.S., 2001, Linking the Wasatchian/Bridgerian boundary to the Cenozoic global climate optimum: New magnetostratigraphic and isotopic results from South Pass, Wyoming: Palaeogeography, Palaeoclimatology, Palaeoecology, v. 167, p. 175–199.

Collinson, M.E., 1996, What use are fossil ferns—20 yr on: With a brief review of the fossil history of extant pteridophyte families and genera, *in* Camus, J.M., et al., eds., Pteridology in perspective: Kew, UK, Royal Botanic Gardens, p. 349–394.

Collinson, M.E., 2000a, Fruit and seed floras from the Paleocene/Eocene transition and subsequent Eocene in southern England: Comparison and palaeoenvironmental implications: GFF, v. 122, p. 36–37.

Collinson, M.E., 2000b, Cenozoic evolution of modern plant communities and vegetation, *in* Culver, S.J., and Rawson, P.F., eds., Biotic response to global change: Cambridge, UK, Cambridge University Press, p. 223–243.

Collinson, M.E., 2001, Cainozoic ferns and their distribution: Brittonia, v. 53, p. 173–235.

Collinson, M.E., 2002, Cainozoic fern ecology: Review of Palaeobotany and Palynology, v. 119, p. 51–68.

Collinson, M.E., and Cleal, C.J., 2001a, The palaeobotany of the Palaeocene and Palaeocene/Eocene transitional strata in Great Britain, *in* Cleal, C.J., et al., eds., Mesozoic and Tertiary palaeobotany of Great Britain: Geological Conservation Review Series, no. 22, Peterborough, Joint Nature Conservation Committee, p. 157–184.

Collinson, M.E., and Cleal, C.J., 2001b, Early and early middle Eocene (Ypresian-Lutetian) palaeobotany of Great Britain, *in* Cleal, C.J., et al., eds., Mesozoic and Tertiary palaeobotany of Great Britain: Geological Conservation Review Series, no. 22, Peterborough, Joint Nature Conservation Committee, p. 187–226.

Collinson, M.E., and Cleal, C.J., 2001c, Late middle Eocene–early Oligocene (Bartonian-Rupelian) and Miocene palaeobotany of Great Britain, *in* Cleal, C.J., et al., eds., Mesozoic and Tertiary palaeobotany of Great Britain: Geological Conservation Review Series, no. 22, Peterborough, Joint Nature Conservation Committee, p. 229–274.

Collinson, M.E., and Hooker, J.J., 1987, Vegetational and mammalian faunal changes in the early Tertiary of southern England, *in* Friis, E.M., et al., eds., The origins of angiosperms and their biological consequences: Cambridge, UK, Cambridge University Press, p. 259–304.

Collinson, M.E., and Hooker, J.J., 2000, Gnaw marks on Eocene seeds: Evidence for early rodent behaviour: Palaeogeography, Palaeoclimatology, Palaeoecology, v. 157, p. 127–149.

Collinson, M.E., Featherstone, C., Cripps, J.A., Nichols, G.J., and Scott, A.C., 1999, Charcoal-rich plant debris accumulations in the Lower Cretaceous of the Isle of Wight, England: Acta Palaeobotanica, Supplement v. 2, p. 93–105.

Costa, L.I., Denison, C.N.D., and Downie, C., 1978, The Paleocene/Eocene boundary in the Anglo-Paris Basin: Journal of the Geological Society, London, v. 135, p. 261–264.

Crane, P.R., 1989, Early fossil history and evolution of the Betulaceae, *in* Crane, P.R., and Blackmore, S., eds., Evolution, systematics, and fossil history of the Hamamelidae, Volume 2: Higher Hamamelidae: Systematics Association Special Volume, no. 40B, p. 87–116.

Crepet, W.L., and Daghlian, C.P., 1981, Lower Eocene and Paleocene Gentianaceae: Floral and palynological evidence: Science, v. 214, p. 75–77.

Crouch, E.M., Heilmann-Clausen, C., Brinkhuis, H., Morgans, H.E.G., Rogers, K.M., Egger, H., and Schmitz, B., 2001, Global dinoflagellate event associated with the late Paleocene Thermal Maximum: Geology, v. 29, p. 315–318.

Davis, A.G., 1928, The geology of the city and south London railway Clapham-Morden extension: Proceedings of the Geologists' Association, v. 39, p. 339–352, plate 24.

Dewey, H., and Bromehead, C.E.N., 1921, The geology of South London: Memoirs of the Geological Survey of England and Wales, Explanation of sheet 270, London, HMSO, p. 1–92.

Dickens, G.R., O'Neil, J.R., Rea, D.K., and Owen, R.M., 1995, Dissociation of oceanic methane as a cause of the carbon isotope excursion at the end of the Paleocene: Paleoceanography, v. 10, p. 965–971.

Dines, H.G., Holmes, S.C.A., and Robbie, J.A., 1954, Geology of the country around Chatham: Memoirs of the Geological Survey of Great Britain, sheet 272, London, HMSO, p. 1–157.

Ellison, R.A., Knox, R.W.O'B., Jolley, D.W., and King, C., 1994, A revision of the lithostratigraphical classification of the early Palaeogene strata of the London Basin and East Anglia: Proceedings of the Geologists' Association, v. 105, p. 187–197.

Ellison, R.A., Ali, J.R., Hine, N.M., and Jolley, D.W., 1996, Recognition of Chron C25n in the upper Paleocene Upnor Formation of the London Basin, UK, *in* Knox, R.W.O'B., et al., eds., Correlation of the early Paleogene in northwest Europe: Geological Society [London] Special Publication 101, p. 185–193.

Gröcke, D.R., 1997, Carbon isotope stratigraphy of terrestrial plant fragments in the Early Cretaceous from southeastern Australia, *in* Wolberg, D.L., et al., eds., Dinofest(TM) International: Philadelphia, The Academy of Natural Sciences, p. 457–461.

Gröcke, D.R., 1998, Carbon isotope analyses of fossil plants as a chemostratigraphic and palaeoenvironmental tool: Lethaia, v. 31, p. 1–13.

Gröcke, D.R., 2001, Isotope stratigraphy and ocean-atmosphere interactions in the Jurassic and Early Cretaceous [Ph.D. thesis]: University of Oxford, England, 142 p.

Gröcke, D.R., 2002, The carbon isotope composition of ancient CO_2 based on higher-plant organic matter, *in* Gröcke, D.R., and Kucera, M., eds., Understanding climate change: Proxies, chronology, and ocean-atmosphere interactions: Philosophical Transactions of the Royal Society of London, series A, v. 360, p. 633–658.

Gröcke, D., Hesselbo, S.P., and Jenkyns, H.C., 1999, Carbon isotope composition of Lower Cretaceous fossil wood: Ocean-atmosphere chemistry and relation to sea-level change: Geology, v. 2, p. 155–158.

Hardenbol, J., Thierry, J., Farley, M.B., Jacquin, T., De Graciansky, P.-C., and Vail, P.R., 1998, Mesozoic and Cenozoic sequence chronostratigraphic framework of European basins: SEPM (Society for Sedimentary Geology) Special Publication 60, p. 3–13, 763–781.

Harrington, G.J., 2001, Impact of Paleocene-Eocene greenhouse warming on North American paratropical forests: Palaios, v. 16, p. 266–278.

Harrington, G.J., and Kemp, S.J., 2001, U.S. Gulf Coast vegetation dynamics during the latest Paleocene: Palaeogeography, Palaeoclimatology, Palaeoecology, v. 167, p. 1–21.

Hasegawa, T., 1997, Cenomanian-Turonian carbon isotope events recorded in terrestrial organic matter from northern Japan: Palaeogeography, Palaeoclimatology, Palaeoecology, v. 130, p. 251–273.

Hester, S.W., 1965, Stratigraphy and palaeogeography of the Woolwich and Reading Beds: Bulletin of the Geological Survey of Great Britain, v. 23, p. 117–137.

Hooker, J.J., 1996, Mammalian biostratigraphy across the Paleocene-Eocene boundary in the Paris, London, and Belgian basins, *in* Knox, R.W.O'B., et al., eds., Correlation of the early Paleogene in northwest Europe: Geological Society [London] Special Publication 101, p. 205–218.

Hooker, J.J., 1998, Mammalian faunal change across the Paleocene-Eocene transition in Europe, *in* Aubry, M.-P., et al., eds., Late Paleocene–early Eocene climatic and biotic events in the marine and terrestrial records: New York, Columbia University Press, p. 428–450.

Jones, T.P., and Rowe, N.P., editors, 1999, Fossil plants and spores: Modern techniques: London, The Geological Society, 396 p.

Kennedy, W.J., and Sellwood, R.W., 1970, *Ophiomorpha nodosa* Lundgren, a marine indicator from the Sparnacian of southeast England: Proceedings of the Geologists' Association, v. 81, p. 99–110.

King, C., 1981, The stratigraphy of the London Clay and associated deposits: Tertiary Research Special Papers, v. 6, p. 1–158.

Klaassen, H.M., 1883, On a section of the lower London Tertiaries at Park Hill, Croydon: Proceedings of the Geologists' Association, v. 8, p. 226–248.

Koch, P.L., Zachos, J.C., and Gingerich, P.D., 1992, Correlation between isotopic records near the Paleocene/Eocene transition: Nature, v. 401, p. 775–778.

Koch, P.L., Zachos, J.C., and Dettman, D.L., 1995, Stable isotope stratigraphy and paleoclimatology of the Paleogene Bighorn Basin (Wyoming, USA): Palaeogeography, Palaeoclimatology, Palaeoecology, v. 115, p. 61–89.

Luterbacher, H.P., Hardenbol, J., and Schmitz, B., 2000, Decision of the voting members of the International Subcommission on Paleogene Stratigraphy on the criterion for recognition of the Paleocene-Eocene boundary: Newsletter of the International Subcommission on Paleogene Stratigraphy, Tübingen, November 2000, v. 9, p. 13.

Manchester, S.R., 1987, The fossil history of the Juglandaceae: Monographs in Systematic Botany from the Missouri Botanical Garden, v. 21, p. 1–137.

Manchester, S.R., 1989, Early history of the Juglandaceae: Plant Systematics and Evolution, v. 162, p. 231–250.

Martin, A.R.H., 1976, Upper Paleocene Salviniaceae from the Woolwich/Reading Beds near Cobham, Kent: Palaeontology, v. 19, p. 173–184.

Moore, P.D., Webb, J.A., and Collinson, M.E., 1991, Pollen analysis (2nd edition): Oxford, Blackwell Scientific Publications, 216 p.

Norris, R.D., and Röhl, U., 1999, Carbon cycling and chronology of climate warming during the Paleocene/Eocene transition: Nature, v. 401, p. 775–778.

Pitcher, W.S., Peake, N.B., Carreck, J.N., Kirkaldy, J.F., Hester, S.W., and Hancock, J.M., 1958, The London region: Geologists' Association Guides No. 30, Colchester, Benham and Company, 41 p.

Powell, A.J., 1992, Dinoflagellate cysts of the Tertiary System, *in* Powell, A.J., ed., A stratigraphic index of dinoflagellate cysts: London, Chapman and Hall, p. 155–251.

Powell, A.J., Brinkhuis, H., and Bujak, J.P., 1996, Upper Paleocene–lower Eocene dinoflagellate cyst sequence biostratigraphy of southeast England, *in* Knox, R.W.O'B., et al., eds., Correlation of the early Paleogene in northwest Europe: Geological Society [London] Special Publication 101, p. 145–183.

Röhl, U., Bralower, T.J., Norris, R.D., and Wefer, G., 2000, New chronology for the late Paleocene thermal maximum and its environmental implications: Geology, v. 28, p. 927–930.

Rundle, A.J., 1972, Report of project meeting to Charlton, Kent: Tertiary Times, v. 1, p. 96–104.

Scott, A.C., 2000, The pre-Quaternary history of fire: Palaeogeography, Palaeoclimatology, Palaeoecology, v. 164, p. 281–330.

Skipper, J., 1999, The stratigraphy of the Lambeth Group (Palaeocene) of southeast England [Ph.D. thesis]: University of London, England, 221 p. + 65 p. of unpaginated figures.

Stockey, R.A., and Manchester, S.R., 1988, A fossil flower with *in situ Pistillipollenites* from the Eocene of British Columbia: Canadian Journal of Botany, v. 66, p. 313–318.

Sweet, A.R., Braman, D.R., and Lerbekmo, J.F., 1999, Sequential palynological changes across the composite Cretaceous-Tertiary (K-T) boundary claystone and contiguous strata, western Canada and Montana, USA: Canadian Journal of Earth Sciences, v. 36, p. 743–768.

Thiry, M., Dupuis, C., Aubry, M.-P., Berggren, W.A., Ellison, R.A., Knox, R.W.O'B., Sinha, A., and Stott, L., 1998, Tentative correlations between continental deposits of the Argiles Plastiques (Paris Basin) and Reading Beds (London Basin), based on chemostratigraphy: Strata, v. 9, p. 125–129.

Tracey, S., 1986, Lower Tertiary strata exposed in a temporary excavation at Well Hall, Eltham, southeast London: Tertiary Research, v. 7, p. 107–123.

Vajda, V., Raine, I.J., and Hollis, C.J., 2001, Indication of global deforestation at the Cretaceous-Tertiary boundary by New Zealand fern spike: Science, v. 294, p. 1700–1701.

Whitaker, W., 1889, The geology of London and of part of the Thames valley (explanation of sheets 1, 2, and 7): Memoirs of the Geological Survey of England and Wales, London, HMSO, Vol 1, Descriptive Geology, 556 p. Vol II, Appendices, 352 p.

Whitaker, W., 1908, The water supply of Kent, with records of sinkings and borings: Memoirs of the Geological Survey of England and Wales, London, HMSO, p. 1–399.

Whitaker, W., 1912, The water supply of Surrey, from underground sources, with records of sinkings and borings: Memoirs of the Geological Survey of England and Wales, London, HMSO, p. 1–352.

Wing, S.L., and Harrington, G.J., 2001, Floral response to rapid warming at the Paleocene-Eocene boundary and implications for concurrent faunal change: Paleobiology, v. 27, p. 539–563.

Wing, S.L., Bao, H., and Koch, P.L., 2000, An early Eocene cool period? Evidence for continental cooling during the warmest part of the Cenozoic, *in* Huber, B.T., et al., eds., Warm climates in Earth history: Cambridge, UK, Cambridge University Press, p. 197–237.

Zachos, J., Pagani, M., Sloan, L., Thomas, E., and Billups, K., 2001, Trends, rhythms, and aberrations in global climate 65 Ma to present: Science, v. 292, p. 686–693.

MANUSCRIPT ACCEPTED BY THE SOCIETY AUGUST 13, 2002

Printed in the USA

Geological Society of America
Special Paper 369
2003

Terrestrial vegetation record across the initial Eocene thermal maximum at the Tawanui marine section, New Zealand

Erica M. Crouch*
Henk Visscher
Laboratory of Palaeobotany and Palynology, Department of Geobiology, Utrecht University, Budapestlaan 4, 3584CD Utrecht, The Netherlands

ABSTRACT

The initial Eocene thermal maximum (IETM), an aberrant warming interval in the geological record (ca. 55 Ma), coincided with profound perturbations in global biogeochemical systems. To detect to what extent vegetation in a mid to high latitude Southern Hemisphere region responded to this intense warming interval, plant microfossils have been examined from a well-calibrated IETM marine sequence at Tawanui, New Zealand. Spore and pollen assemblages remained relatively stable across the IETM, with few notable changes. Pollen from an araucarian conifer, *Dilwynites granulatus* (Wollemi Pine), dominated the assemblages and the only notable local appearance near the beginning of the IETM was *Spinizonocolpites prominatus* (*Nypa* mangrove palm). The spore and pollen assemblage indicates that a mesothermal conifer-dominated multistratal rainforest, suggestive of a warm climate with high rainfall, was the predominant plant community. Megathermal coastal *Nypa* mangrove communities appear to have been established in New Zealand prior to the IETM, and the intense warming interval was not associated with significant immigrations and turnover events in plant communities. Rather, we suggest that the subtle variations recognized, primarily an increase in *D. granulatus* pollen at the beginning of the IETM, may reflect an increase in terrigenous inputs to the marine environment in the Tawanui region during the IETM. The lack of a clear response in plant communities to this aberrant event is in agreement with current observations from other mid to high latitude settings in both the Northern and Southern Hemispheres.

INTRODUCTION

The Paleocene-Eocene transition was characterized by an aberrant warming event in which deep ocean and high-latitude surface temperatures warmed by 6–8 °C (e.g., Kennett and Stott, 1991; Thomas and Shackleton, 1996; Katz et al., 1999). Beginning ca. 55 Ma (Norris and Röhl, 1999), this initial Eocene thermal maximum or IETM corresponds to a prominent negative carbon isotope excursion (CIE) of ≥ 2‰ in marine and terrestrial environments (Kennett and Stott, 1991; Zachos et al., 1993; Koch et al., 1995; Kaiho et al., 1996). The characteristics of the CIE indicate a massive injection of ^{12}C-rich carbon into the ocean-atmosphere system, followed by sequestering of excess carbon into carbonate and organic matter (Dickens et al., 1997; Dickens, 2000). The most probable source for the carbon input was methane released from seafloor gas hydrates (Katz et

*Present address: Institute of Geological and Nuclear Sciences, P.O. Box 30-368, Lower Hutt, New Zealand. E.Crouch@gns.cri.nz.

Crouch, E.M., and Visscher, H., 2003, Terrestrial vegetation record across the Initial Eocene Thermal Maximum at the Tawanui marine section, New Zealand, *in* Wing, S.L., Gingerich, P.D., Schmitz, B., and Thomas, E., eds., Causes and Consequences of Globally Warm Climates in the Early Paleogene: Boulder, Colorado, Geological Society of America Special Paper 369, p. 351–363.

al., 1999; Dickens, 2001). Well-documented changes that occurred in marine biota include a benthic foraminiferal extinction event (Thomas, 1998) and turnovers in planktonic organisms (Kelly et al., 1996; Aubry, 1998). Recent studies of organic-walled dinoflagellate cysts (dinocysts) indicate that coincident with the onset of the CIE representatives of one particular genus, *Apectodinium,* suddenly dominated dinocyst assemblages at locations around the world (Bujak and Brinkhuis, 1998; Heilmann-Clausen and Egger, 2000; Bybell et al., 2001; Crouch et al., 2001).

A distinct terrestrial biotic response to the IETM is a fundamental diversification of mammalian faunas (Beard, 1998; Clyde and Gingerich, 1998; Bowen et al., 2002). The response of terrestrial floras during the IETM is less well known and documented. Recent studies, predominantly from the Northern Hemisphere, encounter two main problems when attempting detailed reconstructions of vegetation during the IETM: (1) poor calibration of terrestrial sequences with the CIE and few well-recognized sequences that span the IETM (Koch et al., 1992; Frederiksen, 1994, 1998; Fricke et al., 1998; Beerling, 2000), and (2) a lack of well-preserved plant micro and macrofossils (Wing, 1998; Wing et al., 2000; Wing and Harrington, 2001). For similar reasons, records of plant micro and macrofossils from the IETM in the Southern Hemisphere are lacking (Askin, 1992; Greenwood and Wing, 1995; Greenwood et al., 2000).

To detect to what extent vegetation in a mid to high latitude Southern Hemisphere region responded to the intense warming and excess carbon of the IETM, we examined plant microfossils (spores and pollen) from a well-calibrated IETM marine sequence at Tawanui, New Zealand. Although plant microfossil records recovered from marine settings are biased toward well-represented sources in the regional vegetation, such as wind-pollinated gymnosperms, major changes within the composition of plant communities can still be recognized (e.g., Muller, 1959; MacPhail et al., 1994; Van der Kaars, 2001). Examination of coeval marine palynological samples allows direct correlation of terrestrial vegetation to marine faunas, through comparison with dinocyst assemblages.

TERRESTRIAL VEGETATION DURING THE INITIAL EOCENE THERMAL MAXIMUM

Although many studies of plant microfossils (e.g., spores and pollen) and macrofossils (e.g., leaves, seeds and fruits) span the late Paleocene to early Eocene, detailed documentation from the IETM is sparse. The most comprehensively studied continental upper Paleocene to lower Eocene succession is in the Bighorn Basin, northwestern Wyoming. Though the CIE is recognized (Koch et al., 1992; Fricke et al., 1998; Clyde and Gingerich, 1998), the IETM occurs within distinctive thick red paleosols in which both plant micro and macrofossils are absent (Koch et al., 1995; Wing, 1998; Harrington, 2000). Despite this, it appears that the composition of the flora did not drastically change from pre to post IETM, and a minimal number of extinctions and immigrations are recorded (Wing et al., 2000; Wing and Harrington, 2001). A plant microfossil study (Wing et al., this volume) from a recently discovered IETM section in the southwestern Powder River Basin, Wyoming, indicates that rapid, continental-scale range floral displacements were unlikely during the IETM. A study from the U.S. Gulf Coast did not recognize the exact position of the CIE but analysis of plant microfossils suggests there is little change in vegetation type across the Paleocene-Eocene transition, with no major extinctions or turnover events recorded (Harrington and Kemp, 2001).

Various studies in northwest Europe have examined plant micro and macrofossil records from the late Paleocene to early Eocene (Collinson and Hooker, 1987; Collinson, 1996, 2000a; Jolley, 1998) but nearly all studies lack precise recognition of the CIE. Recently, the CIE was apparently recorded from measuring $\delta^{13}C$ variations of near-pure taxon specific pollen in boreholes from northwest Europe (Beerling and Jolley, 1998). They suggested that the negative shift in $\delta^{13}C$ is coincident with a marked shift in dominance from mixed mesophytic to paratropical forests, although detailed plant microfossil information is lacking. Marine palynological analysis of a core from the central North Sea recorded an increase in paratropical plant forms (e.g., *Caryapollenites*) approximately coeval with an increase in the relative abundance of *Apectodinium* (King and Harding, 2000).

In the Southern Hemisphere one of the most studied regions of late Paleocene–early Eocene terrestrial vegetation is Australasia, although information from well-calibrated IETM sequences is lacking. Quantitative plant microfossil studies from the late Paleocene–early Eocene in New Zealand are limited (e.g., Raine and Wilson, 1988), and there are no high-resolution studies with which to compare the results from this study. New Zealand and southern Australia were situated in mid to high latitude positions during this time interval (Greenwood et al., 2000; King et al., 1999) and New Zealand was an isolated landmass (Ballance, 1993). During the Paleocene, vegetation in New Zealand and southern Australia was dominated by rainforest composed of plants with Gondwanan ancestry, including Araucariaceae, Podocarpaceae and *Nothofagus,* and more recently evolved and immigrant elements, such as Casuarinaceae (Harris, 1965; Mildenhall, 1980; Raine, 1984; MacPhail et al., 1994; Greenwood et al., 2000). Estimations of mean annual temperature (MAT) in the late Paleocene of southeast Australia range from 14 to 17 °C (Greenwood et al., 2000, this volume). Plant microfossils from the early Eocene are diverse, with angiosperms appearing to be more common and gymnosperms less abundant than in the late Paleocene (e.g., Martin, 1994). Some megathermal elements (preferring MAT >24 °C), such as *Nypa* mangroves, are well established, while coastal and more inland rainforest communities also contain megathermal taxa (MacPhail et al., 1994; Greenwood et al., 2000). MAT estimations in the early Eocene of southeast Australia range from 16 to 18 °C (Greenwood and Wing, 1995; Greenwood et al., 2000). Current evidence suggests that the general composition of vegetation in Australasia remained similar during the latest Paleocene to earliest Eocene (Martin, 1994; MacPhail et al., 1994; Greenwood et al., 2000).

THE INITIAL EOCENE THERMAL MAXIMUM AT TAWANUI

The Tawanui section is located along the upper Akitio River, on the southeastern margin of the Whangai Range, East Coast Basin (Fig. 1). The upper Paleocene to lower Eocene interval corresponds to part of the Wanstead Formation, a poorly bedded, calcareous, smectite-rich mudstone with scattered beds of glauconite sandstone (Moore and Morgans, 1987). Benthic foraminiferal assemblages suggest this formation was originally deposited in upper middle bathyal water depths (Moore and Morgans, 1987; Kaiho et al., 1996). The sequence at Tawanui was deposited in an environment relatively close to land, as the sediments contain abundant terrestrial components (Kaiho et al., 1996; Crouch, 2001).

The IETM at Tawanui is marked by a thin (~1.0 m) dark grey, calcareous siltstone within calcareous nannoplankton Zone NP9 (M.-P. Aubry, 2000, personal commun.) and planktonic foraminifera *Globigerina triloculinoides* Zone (Kaiho et al., 1996). The beginning of the IETM (marked as 0.00 m) is defined by a negative shift in the $\delta^{13}C$ of bulk carbonate and organic matter and n-C_{29} alkane, and by the disappearance of ~25% of benthic foraminiferal species (Kaiho et al., 1996; Crouch, 2001). The onset of *Apectodinium*-dominated assemblages at Tawanui also coincides with the beginning of the CIE (Crouch et al., 2001; Fig. 2). A thin glauconite horizon is present at ~0.02 m and a sandy siltstone glauconite lies between ~0.6 and 0.85 m (Fig. 2). A distinctive thin calcite layer, present at ~0.4–0.6 m (Fig. 2), runs parallel to bedding and consists primarily of recrystalized calcite (R. Cook, 2001, personal commun.).

MATERIAL AND METHODS

Samples, palynological processing, and quantitative examination

Twenty-one rock samples were collected across ~3.5 m of section, spanning the calcareous siltstone of the IETM (Fig. 2). Samples were taken at –1.45 m, –0.45 m, –0.06 m, 0.00 m, 0.04 m and 0.06 m; 10 samples were taken approximately every 10 cm between 0.10 m and 0.95 m; and 5 samples were taken approximately every 20 cm between 1.15 m and 1.95 m (Fig. 2; Appendix 2). The samples are assigned standard New Zealand Fossil Record File numbers, with the prefix U24/f (Fig. 2; Appendix 2).

All samples were prepared for palynological examination at the Laboratory of Palaeobotany and Palynology (LPP) using standard techniques (Van Steenbergen, 1997). Between 5 and 20 g of sediment were crushed, and the carbonate and siliceous components removed by adding 30% HCl and 38% HF, respectively. Samples were then sieved over a 10 μm mesh sieve. The >10 μm fraction was oxidized for 10 minutes using 30% HNO_3 and washed with 5% KOH to disaggregate amorphous and organic debris. Samples were sieved again over a 10 μm mesh sieve, and well-mixed representative fractions of the >10 μm residue were mounted on glass slides using a glycerine jelly medium.

Plant microfossil assemblages are generally well preserved and constitute between ~9% and 41% of the palynomorph association (Crouch, 2001). For all samples, ~200 spores and pollen were counted and the entire slide scanned for the presence of additional taxa (Appendix 2). In this paper, spore and pollen abun-

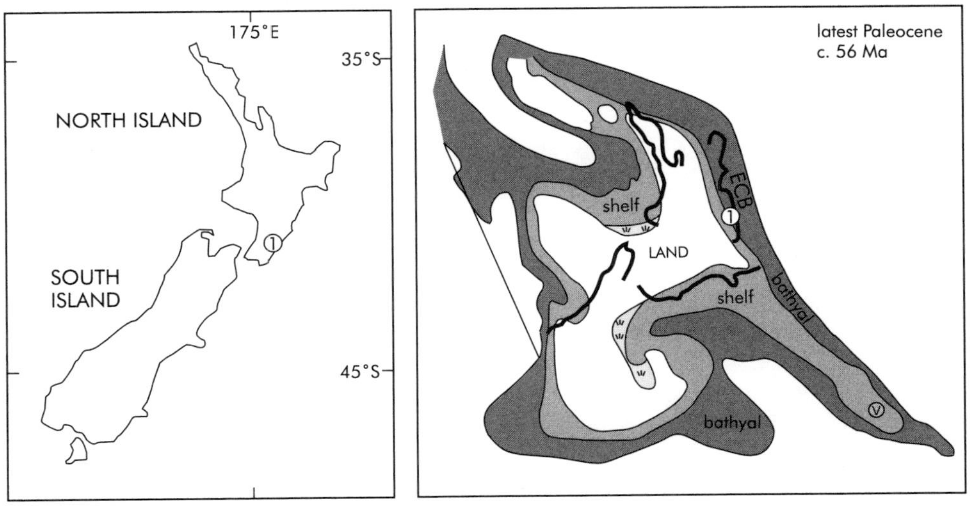

Figure 1. Map of present-day New Zealand landmass and paleogeographic map of New Zealand in the latest Paleocene, with the location of the Tawanui section. ECB—East Coast Basin, V—volcanics. The bold line on the paleogeographic map shows the approximate position of the present-day outline of New Zealand. Paleogeographic map modified from King et al. (1999).

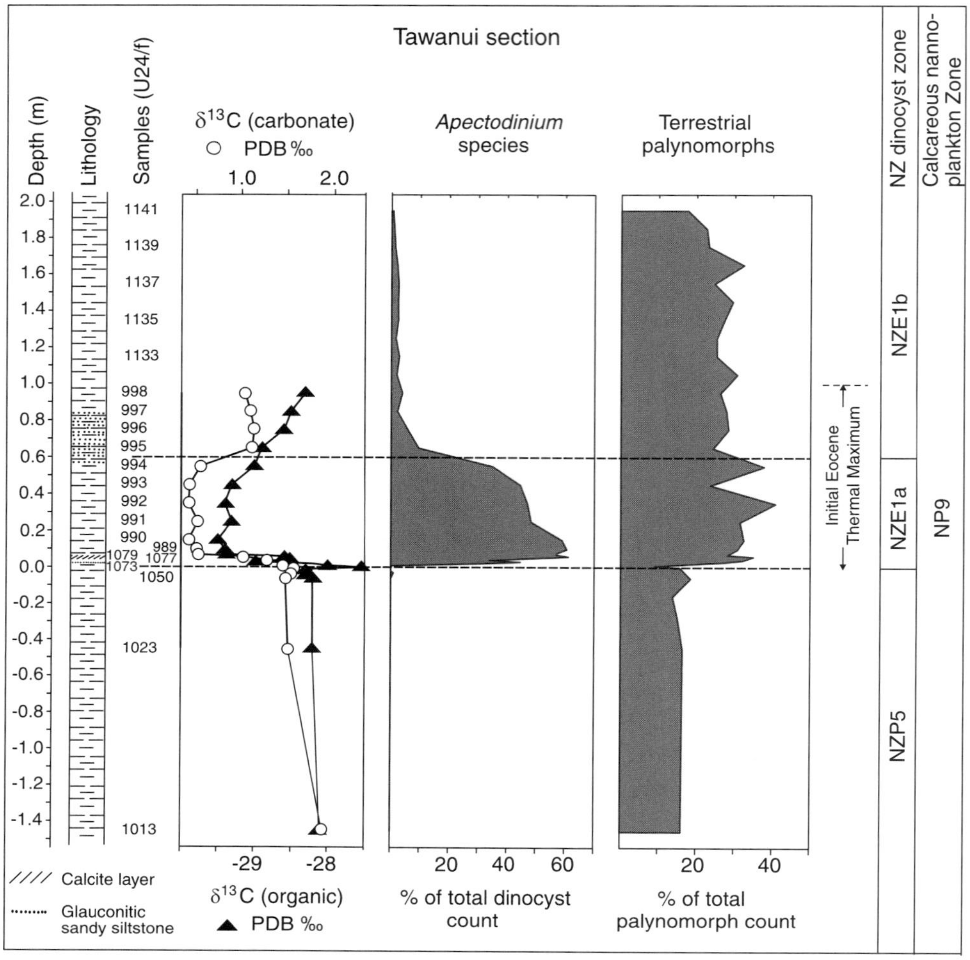

Figure 2. Profiles across the initial Eocene thermal maximum (IETM) at Tawanui of bulk organic and carbonate $\delta^{13}C$ isotope record, relative abundance of *Apectodinium* species, relative abundance of terrestrial palynomorphs, and calcareous nannoplankton (M.-P. Aubry, 2000, personal commun.) and dinocyst (Crouch, 2001) zonation. A summary lithological column and sample points are shown. The onset of the IETM is defined as 0.00 m.

dances are expressed as a percentage of the total count. The primary aim of this paper is to record major shifts in plant communities and, hence, most determinable spores and pollen were initially grouped into genera. Some taxa, principally those thought to be indicative of specific environmental conditions (e.g., *Spinizonocolpites prominatus*), were identified as individual species.

Correlation of the plant microfossil record

The spore and pollen analyses were completed on the same samples that have previously been examined for dinocyst assemblages (Crouch, 2001; Crouch et al., 2001). This allows direct correlation between plant microfossil and dinocyst records, in particular recognition of the *Apectodinium* acme. Moreover, splits taken from 16 of these samples have previously been analyzed for $\delta^{13}C$ of bulk carbonate and bulk organic matter (Crouch, 2001; Crouch et al., 2001). The positions of the samples in relation to profiles of $\delta^{13}C$, relative abundance of *Apectodinium* and percentage of terrestrial palynomorphs are shown in Figure 2 (Table 1).

ECOLOGICAL GROUPING OF SPORE AND POLLEN ASSEMBLAGES

It is generally accepted that changes within spore and pollen assemblages from both terrestrial and marine settings dominantly reflect changes in paleovegetation, rather than differences in transportation, preservation or sedimentation (e.g., Traverse, 1988; Tyson, 1995; Abbink, 1998; Collinson, 2000b; Van der Kaars, 2001). To best illustrate the plant microfossil record at Tawanui we have classified spore and pollen taxa in two distinct ways, further discussed below. The first grouping, into vegetation associations, is aimed at recognizing major changes in composition of plant communities across the IETM. The second grouping method aims to detect climate change by classifying taxa according to their temperature preferences. We appreciate that these two ecological groupings are based on the same underlying data and basic interpretations, however they provide alternative means of recognizing major plant community and temperature regime changes across the IETM.

Spores and pollen assigned to groups include quantitatively significant taxa (e.g., *Dilwynites granulatus*) and taxa that have

TABLE 1. RECORDS OF BULK ORGANIC AND CARBONATE $\delta^{13}C$, *APECTODINIUM* SPECIES ABUNDANCE, AND TERRESTRIAL FRACTION OF PALYNOMORPHS ACROSS THE INITIAL EOCENE THERMAL MAXIMUM AT TAWANUI

Sample depth (m)	Fossil record file number (U24/f)	$\delta^{13}C$ (organic) PDB (‰)	$\delta^{13}C$ (carbonate) PDB (‰)	*Apectodinium* spp. (percentage of total dinocyst count)	Relative terrestrial fraction of palynomorph assemblage (%)
1.95	1141	N.D.	N.D.	0.5	17.7
1.85	1140	N.D.	N.D.	1.0	22.6
1.75	1139	N.D.	N.D.	1.3	23.2
1.65	1138	N.D.	N.D.	2.1	32.4
1.55	1137	N.D.	N.D.	2.5	24.7
1.45	1136	N.D.	N.D.	2.3	29.5
1.35	1135	N.D.	N.D.	2.4	27.3
1.25	1134	N.D.	N.D.	1.6	25.3
1.15	1133	N.D.	N.D.	2.7	25.3
1.05	1132	N.D.	N.D.	2.3	30.7
0.95	998	–28.3	1.0	3.8	26.3
0.85	997	–28.5	1.1	2.2	27.9
0.75	996	–28.6	1.1	5.7	28.4
0.65	995	–28.9	1.1	9.5	24.4
0.55	994	–29.0	0.6	34.9	37.8
0.45	993	–29.3	0.4	44.5	23.7
0.35	992	–29.4	0.4	47.0	40.8
0.25	991	–29.3	0.5	48.3	31.5
0.15	990	–29.5	0.4	58.6	32.5
0.10	989	–29.4	0.5	60.3	30.8
0.08	1081	–29.3	0.5	57.2	28.2
0.07	1080	–29.4	0.5	56.9	N.D.
0.06	1078	–28.6	1.0	61.1	35.0
0.04	1077	–28.5	1.3	33.8	32.1
0.03	1076	–29.0	1.5	44.6	27.1
0.02	1075	–28.2	1.6	26.2	27.0
0.01	1074	–28.0	1.4	0.0	9.2
0.00	1073	–27.5	1.6	0.0	15.7
–0.02	1054	–28.2	1.5	0.0	14.9
–0.03	1053	–28.3	1.6	0.9	18.3
–0.06	1050	–28.2	1.5	0.4	18.5
–0.45	1023	–28.2	1.5	0.0	16.4
–1.45	1013	–28.1	1.9	0.0	16.1

known modern-day analogues and habitats (e.g., *Spinizonocolpites prominatus*). Knowledge of the ecological preferences of various taxa has been primarily collected from recent literature. The taxa included are listed in Appendix 1, along with information regarding their modern botanical affinities and the proposed plant community association and predominant temperature regime.

Plant community associations

Selected spore and pollen taxa recorded from Tawanui are placed within four broad plant community associations:

1. Conifer-dominated multistratal rainforest environment—most taxa at Tawanui, particularly those of Gondwanan ancestry, are considered characteristic of plants that lived within a diverse, multistratal rainforest. Conifers are the dominant component of this rainforest, represented by *Dilwynites granulatus, Dacrydiumites* and *Podocarpidites* spp. (MacPhail et al., 1994; Greenwood et al., 2000). Ferns were probably an important element of the understorey; included are *Cyathidites* spp. Another important component was broad-leaved angiosperm plants, although they comprise a minor proportion of the plant microfossil record at Tawanui. Taxa included are *Malvacipollis, Myrtaceidites, Nothofagus* and *Proteacidites* spp. (Christophel, 1989; MacPhail et al., 1994; Greenwood et al., 2000). This type of multistratal rainforest is suggestive of a warm climate with high rainfall (Pocknall, 1990; Greenwood et al., 2000).

2. Lowland rainforest environment—this association consists of pollen assigned to *Haloragacidites harrisii,* which appears to be related to an extant genus in the Casuarinaceae, *Gymnostoma* (Christophel, 1980; 1989). We consider *H. harrisii* to represent a distinct plant community because, like its modern

relative, it may have grown in lowlands (Christophel and Greenwood, 1989) and because it is also an important element of vegetation records from the early Eocene in New Zealand and southern Australia (Harris, 1965; Mildenhall, 1980; Raine, 1984; MacPhail et al., 1994; Greenwood et al., 2000).

3. Open heathland/disturbed environment—this association includes *Gleicheniidites* spores and *Ericipites* pollen.

4. *Nypa* mangrove coastal environment—this association consists of pollen assigned to *Spinizonocolpites prominatus.* This pollen is considered to have been produced by *Nypa* mangrove palms, which at present grow in coastal environments (Krutzsch, 1989; Collinson, 1993, 2000b; Pole and MacPhail, 1996).

Plant temperature regime

Climate is considered to be the most important of all environmental factors influencing the distribution of plants (e.g., Traverse, 1988; Martin, 1994). Following the method often used to describe temperature characteristics of living vegetation (Nix, 1982; MacPhail et al., 1994; Greenwood, 1994; Collinson, 2000b) we use the terms megathermal (MAT >24 °C), mesothermal (MAT >14 to 20 °C) and microthermal (MAT <12 °C). Three broad plant temperature regimes are established that provide a general overview of the temperature regime of vegetation at Tawanui.

1. Megathermal temperature regime—included is *Spinizonocolpites prominatus.*

2. Mesothermal to megathermal temperature regime—this includes *Haloragacidites harrisii* and *Proteacidites* spp.

3. Microthermal to mesothermal temperature regime—included are many components of the multistratal rainforest association; *Dilwynites granulatus, Dacrydiumites* and *Podocarpidites* spp., *Malvacipollis, Myrtaceidites* and *Nothofagus* spp.

RESULTS AND INTERPRETATION

The plant microfossil record across the IETM at Tawanui is characterized by rather stable spore and pollen assemblages with few notable changes (Figs. 3 and 4). Pollen derived from the araucarian conifer *Dilwynites granulatus* dominate the record, comprising between ~30% and 35% of the assemblage prior to the IETM and becoming more abundant (average ~50%) at the beginning of the IETM (Fig. 3). Other gymnosperms constitute between ~6% and 22% of the assemblage. Fern spores are common throughout (~20%–44%) and consist mainly of *Baculatisporites* spp., *Cibotiidites* spp., *Cyathidites* spp. and *Gleicheniidites* spp. The abundance of spores slightly decreases during the IETM and remains lower after the IETM. Pollen from angiosperms comprise between ~3% and 17% of the assemblage and are slightly more abundant from the beginning of the IETM. The most common angiosperms are *Haloragacidites harrisii, Malvacipollis* spp., *Proteacidites* spp. and small *Triorites* species. The only notable first appearance in this region is *Spinizonocolpites prominatus,* recorded near the beginning of the IETM (sample U24/f989).

As with the relatively stable plant microfossil assemblage, the broad plant community associations changed little during the interval studied (Fig. 4). Plants thought to be living within a multistratal rainforest community were dominant in the latest Paleocene and continued to dominate during and after the IETM. The majority of rainforest taxa appear to have lived within microthermal-mesothermal and mesothermal-megathermal temperature regimes (Fig. 4), suggesting a mesothermal climate was predominant. The multistratal rainforest consists primarily of plants of Gondwanan ancestry or those that established in Australasia during the Paleocene.

The megathermal mangrove palm *Nypa* appears to have been present in coastal environments during the IETM (Fig. 4). Extant *Nypa* is known to be a low producer of pollen and to have rather restricted pollen distribution, which can often result in the genus being underrepresented in the fossil pollen record (Collinson, 1993). Hence, it is difficult to determine if this record represents the first colonization of lower-latitude mangroves in this region. Recently, pollen assigned to *Spinizonocolpites prominatus* has been recorded from latest Paleocene sediments in southeastern New Zealand (Crouch, 2001), although most earliest records of *S. prominatus* in New Zealand are from sediments of early Eocene age (Raine, 1984; Raine and Wilson, 1988; Pocknall, 1990). In Australia the oldest record of *Nypa* mangroves is also from sediments of earliest Eocene age (Pole and MacPhail, 1996; Greenwood et al., 2001). The current results suggest that *Nypa* mangrove palms were already established in parts of New Zealand by the IETM, and that the intense warming of the IETM was not associated with significant floral immigrations into New Zealand.

As mentioned, the relative abundance of the araucarian conifer *Dilwynites granulatus* increases at the beginning of the IETM and the percentage remains elevated during and after the IETM (Fig. 3). This increase is also coincident with an elevated abundance of terrestrial palynomorphs (Fig. 2) and other factors (e.g., terrestrial organic carbon) that suggest the input of terrigenous material into the ocean was elevated in the Tawanui region during the IETM (Kaiho et al., 1996; Crouch et al., 2003). At present, the ecological preference of *D. granulatus* is not well established (e.g., MacPhail et al., 1994) and the species has been recorded in a wide variety of sediments, ranging from coastal to upland settings (Harris, 1965; Pocknall, 1990; MacPhail et al., 1995; D. Greenwood, 2001, personal commun.). We suggest that *D. granulatus* may have been a common element of upland vegetation in this region, and that the increase in abundance at the beginning of the IETM is a reflection of enhanced precipitation and terrigenous input into the marine environment. In marine palynological studies, a higher proportion of upland vegetation is frequently interpreted to reflect increased terrestrial runoff (e.g., Muller, 1959; Traverse, 1988).

Although the subtle variations recorded across the IETM at Tawanui (e.g., increase in *Dilwynites granulatus* and an-

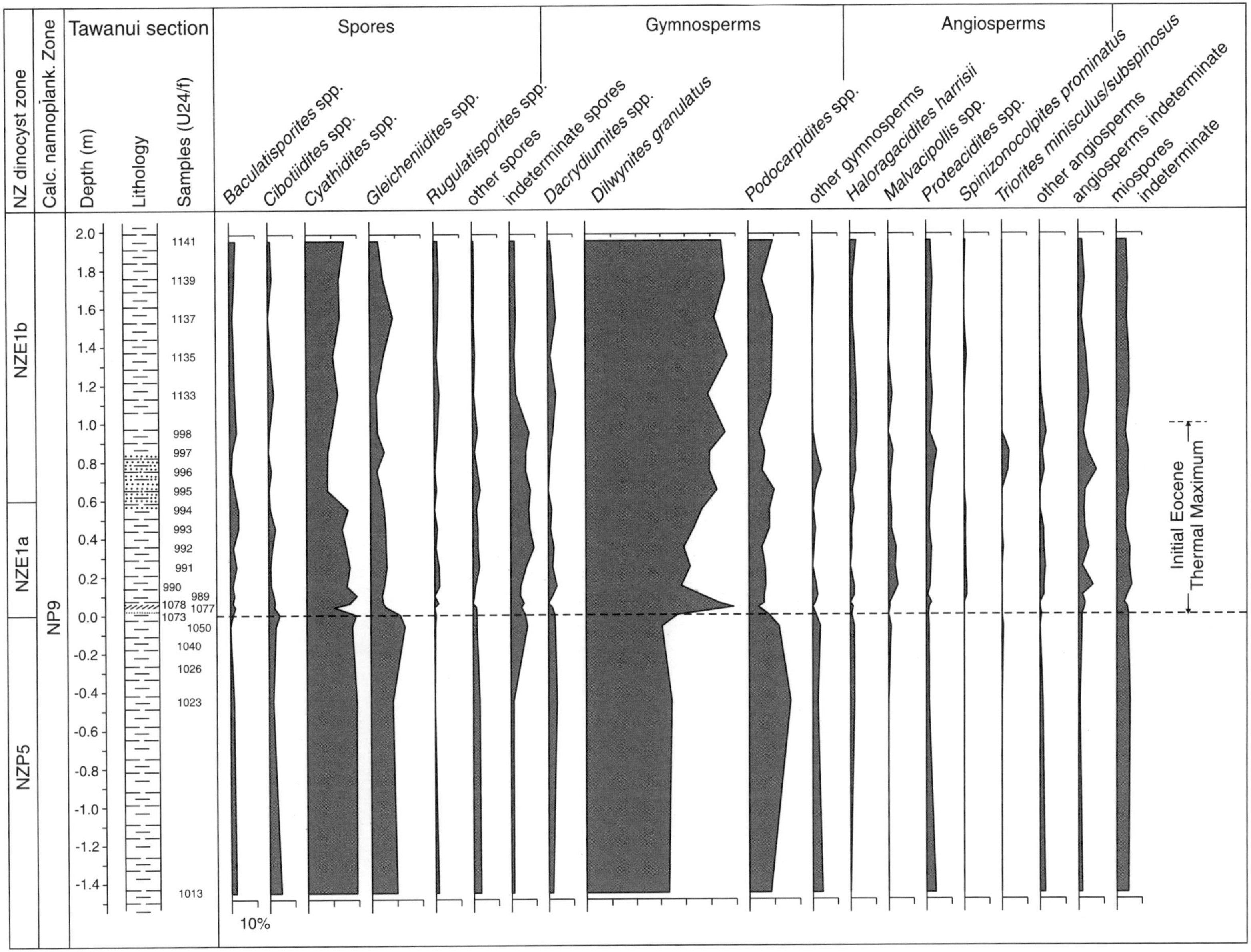

Figure 3. Relative abundance of the spore and pollen assemblage from across the initial Eocene thermal maximum (IETM) at Tawanui. The onset of the IETM is defined as 0.00 m.

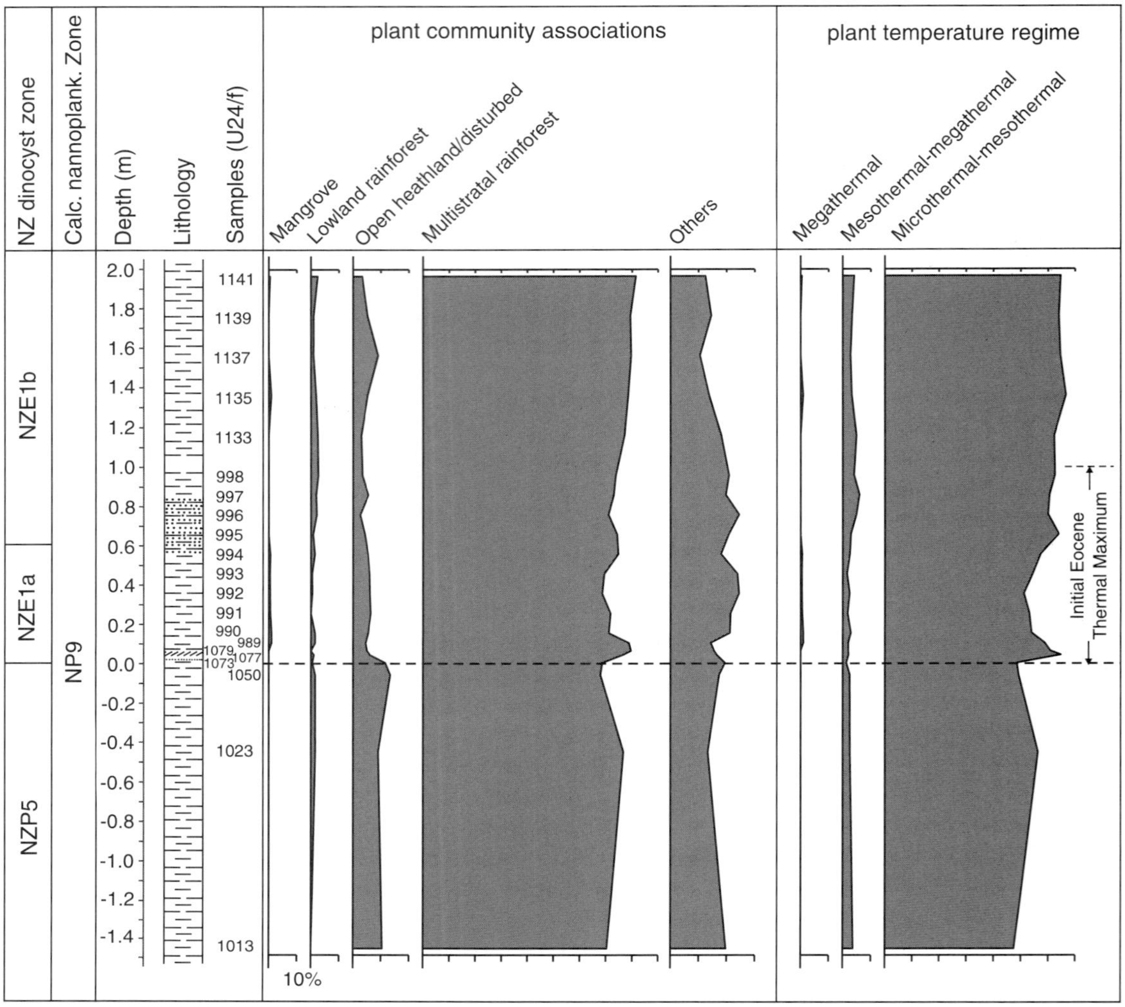

Figure 4. Record of broad plant community associations and plant temperature regimes across the initial Eocene thermal maximum (IETM) at Tawanui. The onset of the IETM is defined as 0.00 m.

giosperm pollen, decrease in *Podocarpidites, Cyathidites* and *Gleicheniidites* spp.) may indicate that changes occurred within the native plant communities, the record is difficult to interpret because it comes from a marine setting. Moreover, it is possible that due to this marine setting certain pollen groups may be under (e.g., angiosperms) or over (e.g., gymnosperms) represented in the plant microfossil record. It is, therefore, essential to identify and examine terrestrial sequences that span the IETM in order to determine to what extent the subtle changes recognized at Tawanui may signify vegetation change in response to marked climatic change.

DISCUSSION

Microfossil records of plant communities in both New Zealand and southeast Australia appear to show little change during the late Paleocene and earliest Eocene (Martin, 1994; Greenwood et al., 2001). This apparent lack of floral change suggests that vegetation communities during the intense warming of the IETM either (1) did not significantly change or (2) experienced a transient change in floral composition followed by a return to more Paleocene-like composition by the earliest Eocene (Wing and Harrington, 2001). The results from this study suggest that plant communities in the southern mid to high latitude setting of New Zealand did not significantly change in response to the intense warming of the IETM. This appears to agree with current observations from late Paleocene–early Eocene floral records in North America and northwest Europe (Wing et al., 2000; Collinson, 2000a; Wing and Harrington, 2001; Harrington and Kemp, 2001; Wing et al., this volume).

Two factors may explain why plant communities in New Zealand remained rather stable and were not significantly affected by the intense warming and excess carbon of the IETM. Firstly, the flora of New Zealand in the late Paleocene was dominated by conifers and ferns, with many of the plants having originated from the Gondwanan landmass in the Mesozoic (Mildenhall, 1980; Martin, 1994). Moreover, conifers can generally live within wide temperature regimes. This long-ranging and well-established flora appears to have been rather stable during the Paleocene and able to adapt to climatic cooling and warming trends in the late Paleocene and perhaps also to aberrant shifts in climate. Secondly, by the upper Paleocene–lower Eocene the New Zealand microcontinent was an isolated landmass and had reached maximum separation from Australia (Ballance, 1993). This geographical isolation made it more difficult for new floras to colonize New Zealand. However, this may not have been the most important controlling factor given the notable similarities between the New Zealand and southeast Australian floral records during this interval (Martin, 1994; Pole, 2001).

At present, the most significant change in vegetation in New Zealand during the late Paleocene–early Eocene does not appear to have been associated with the IETM, but instead occurred during the early Eocene. Vegetation change in the early Eocene is marked by a notable increase in the abundance of *Haloragacidites harrisii,* other angiosperms with mesothermal-megathermal affinities (e.g., *Anacolosidites* spp., *Spinizonocolpites prominatus*) and a decrease in the abundance of Podocarpaceae (Raine, 1984; Pocknall, 1990; Crouch, 2001; J. Raine, 2001, personal commun.). This observation of greater vegetation change in the early Eocene, rather than at the IETM, appears to be in agreement with current findings from North America and northwest Europe (Collinson, 2000a, 2000b; Wing et al., 2000; Wing and Harrington, 2001).

CONCLUSIONS

To detect to what extent vegetation in a mid to high latitude Southern Hemisphere region responded to the intense warming and excess carbon of the IETM, we examined terrestrial spores and pollen from a well-calibrated IETM marine sequence at Tawanui, New Zealand.

The plant microfossil record is characterized by relatively stable spore and pollen assemblages with few notable changes. Pollen from an araucarian conifer, *Dilwynites granulatus* (Wollemi Pine), dominates all samples and becomes slightly more abundant at the beginning of the IETM. Other gymnosperms and fern spores are relatively common, and angiosperm pollen comprises <20%. A mesothermal conifer-dominated multistratal rainforest, probably characteristic of a warm climate with high rainfall, was the dominant plant community. The only notable appearance is *Spinizonocolpites prominatus* (*Nypa* mangrove palm) near the onset of the IETM, although at present the oldest record of *S. prominatus* in New Zealand appears to be in the uppermost Paleocene. If megathermal *Nypa* mangrove coastal communities were already established in parts of New Zealand by the IETM, then the intense warming of the IETM does not appear to have been associated with significant immigrations from lower latitudes or major turnovers in plant communities in New Zealand.

Rather than indicating a significant response to marked climatic warming, we suggest that the subtle variations across the IETM may reflect a change in the terrestrial-runoff regime. An increase in *Dilwynites granulatus* pollen, possibly derived from upland vegetation, at the beginning of the IETM is in agreement with other evidence at Tawanui that suggests terrigenous inputs to the ocean were enhanced during the IETM. This may have been a result of intensification of biogeochemical feedbacks, such as precipitation and chemical weathering (Zachos and Dickens, 2000; Ravizza et al., 2001).

In contrast to the marine realm at mid to high latitudes, there appears to have been no clear response in vegetation communities to the profound perturbations in global biogeochemical systems associated with the IETM. This is further emphasized by an apparent lack of floral change documented from late Paleocene–earliest Eocene successions in Australasia and is in agreement with current observations from North America and northwest Europe.

ACKNOWLEDGMENTS

We are grateful to Henk Brinkhuis, David Greenwood, Guy Harrington, Johanna van Konijnenburg-van Cittert and Ian Raine for discussions and helpful comments on the manuscript. The reviewers, David Pocknall, Simon Ward and Scott Wing, are thanked for thought provoking comments and improving the manuscript. We thank Marie-Pierre Aubry for examination of calcareous nannoplankton and Richard Cook for mineralogical analyses. This is NSG publication number 20020405.

APPENDIX 1. BOTANICAL AFFINITIES OF SELECTED SPORE AND POLLEN TAXA

Selected spore and pollen taxa	Modern analogue	Modern distribution and habitat	Plant community association	Temperature regime
<u>Spores</u>				
Cyathidites spp.	Cyatheaceae	Widespread distribution, mainly in tropical and warm temperate regions. Tree ferns, primarily occur in the second layer of rainforests.	Forest understorey, multistratal rainforest	In regions with MAT ranging from ~11 to 23°C (Greenwood et al., 2001).
Gleicheniidites spp.	Gleicheniaceae	Widespread distribution, mainly temperate to tropical. Consists of thicket-forming ground ferns.	Open, disturbed areas, or alongside streams. Open heathland/disturbed environments. In mires	Wide distribution
<u>Gymnosperms</u>				
Araucariacites spp.	*Agathis/Araucaria*, Araucariaceae	Southwest Pacific (especially New Caledonia) and South America. Tall trees in forests.	Multistratal rainforest	Microthermal-mesothermal. In regions with MAT ranging from ~17 to 25°C (Greenwood et al., 2001).
Dacrydiumites spp.	*Dacrydium*, Podocarpaceae	Found in Malaysia, Tasmania, Chile, New Guinea, New Caledonia and New Zealand. Tall trees in forests.	Multistratal rainforest	Microthermal-mesothermal
Dilwynites granulatus	*Wollemia nobilis* (Wollemi Pine), Araucariaceae	Only known from two small populations, in sandstone gorges ~200km northwest of Sydney (MacPhail et al., 1995). In groves, at slightly higher altitude, may be favored by infertile soils that reduce competition from other plants (MacPhail et al., 1995). Tall trees.	Multistratal rainforest	Microthermal-mesothermal (D. Greenwood, pers. comm.)
Podocarpidites spp.	Podocarpaceae	Southern temperate regions through to tropical highlands. Tall trees in rainforests.	Multistratal rainforest	Microthermal-mesothermal. Macrofossil record from early Eocene, southeast Australia, suggests similar environments to the present tropical range of the genus (Christophel, 1989).
<u>Angiosperms</u>				
Ericipites spp.	Ericaceae/ Epacridaceae	Australia, New Zealand and New Caledonia. Widespread distribution; often in low, open vegetation types such as heath and sedgeland areas, and along the fringes of forests (MacPhail et al., 1994).	Open heathland/disturbed environments	Unknown
Haloragacidites harrisii	*Gymnostoma*, Casuarinaceae	New Guinea, New Caledonia and a small population in northern rainforests of Queensland (Christophel, 1989). Trees, often in low nutrient sites on the margins of various types of moist rainforests, tolerating disturbance (Prider & Christophel, 2000).	Lowland rainforest environment	Mesothermal-megathermal. In regions with MAT ranging from ~14 to 25°C (Greenwood et al., 2001).
Malvacipollis spp.	Euphorbiaceae	Widespread distribution, common in tropical regions. Shrubs and trees in forests.	Multistratal rainforest	Microthermal-mesothermal
Myrtaceidites spp.	Myrtaceae	Primarily in tropical to subtropical regions.	Multistratal rainforest	Microthermal-mesothermal
Nothofagus spp.	*Nothofagus* (Southern Beech)	New Guinea, New Caledonia, New Zealand, southern South America and southeastern Australia. Trees in cool temperate rainforests.	Multistratal rainforest	Microthermal-mesothermal
Proteacidites spp.	Proteaceae	Tropical-temperate regions, mainly Australia and South Africa. Shrublands and open forests, some genera form major elements of closed forests (Christophel, 1989).	Multistratal rainforest. Perhaps lived near forest edge or in open areas (Christophel, 1989).	Mesothermal-megathermal
Spinizonocolpites prominatus	*Nypa* (Arecaceae)	India to Indo-Malesia region. In protected brackish embayments or low-lying estuaries, not in shores exposed to much wave action (Krutzsch, 1989; Pole and MacPhail, 1996; Collinson, 1993, 2000b).	Nypa mangrove coastal environment	Megathermal

Note: MAT—mean annual temperature.

APPENDIX 2. QUANTITATIVE RECORD OF SPORE AND POLLEN ASSEMBLAGES ACROSS THE INITIAL EOCENE THERMAL MAXIMUM AT TAWANUI

Sample depth (m)	−1.45	−0.45	−0.06	0.0	0.04	0.06	0.1	0.15	0.25	0.35	0.45	0.55	0.65	0.75	0.85	0.95	1.15	1.35	1.55	1.75	1.95
Fossil Record File Number (U24/f)	1013	1023	1050	1073	1077	1078	989	990	991	992	993	994	995	996	997	998	1133	1135	1137	1139	1141
Spores																					
Baculatisporites spp.	4	2.5	0	2	3.75	2	3.25	2	4.25	2.5	7.25	7	3.5	1.25	2	5.5	4	3.5	2.5	4	5
Cibotiidites tuberculiformis	0	0	1	0	0	0	*	0	0	0	0	0	0	0	0	0	0	*	0	0	0
Cibotiidites spp.	9.5	3.25	4.5	8.5	4.25	5	4	2	1.25	3	6	1.5	0.5	2.5	0	1	4.5	2	0	3	2
Cyathidites spp. (w. pattern)	2.5	2	2	5.5	5	6	2.5	2	2	5	4.5	3	1	1	1.5	1	*	1	0	2	3
Cyathidites spp.	37	35.5	33.25	32.5	15.75	28	37	30.25	27.25	26.25	26.5	31.75	14.5	16.75	16.5	19.75	25.5	22	28.5	25	28.5
Foveosporites canalis	0	0	0	1	0	0	0	0	0	0	0	*	0	0	0	0	0	0	0	0	0
Foveosporites spp.	1	0	0	1	0	0	1	1	0	0	0	0	0	0	0	0	0	0	0	0	0
Gleicheniidites circinidites	5	6	5	5	1	1	1	1	1	0	2	2	1	0	1	0	0	2	5	2	1
Gleicheniidites spp.	15.5	16.5	26.5	22.75	11.5	10	9	11	11	12.25	13.25	11.5	8	5.5	10.75	6	5	9.5	14.5	9	6
Lycopodiumsporites spp.	1	0	2	0	3	0	0	0	1	0	0	1	1	0	0	1	0	0	0	0	0
Monolete sp. A	1	1	0	1	0	0	0	1	4	3	5	1	0	1	*	2	0	0	0	0	0
Neoristrickia truncatus	0	3	1	1	0	0	0	0	0	0	0	1	0	0	0	0	0	0	0	0	0
Rugulatisporites mallatus	0	0	0	0	0	1	*	1	1	0	0	0	0	0	0	*	0	0	0	*	*
Rugulatisporites spp.	2.5	0	0.5	1	0	2	0	3	2	1	2.5	0	1.5	0.5	1.5	2.75	4	2	3.5	4	3
Stereisporites spp.	0	1	0	0	0	0	0	0	0	1	0	0	3	1	1	1	0	*	0	0	*
Trilete sp. A	0	*	0	0	0	0	*	0	0	0	0	0	2	1	1	0	1	*	0	*	*
Spores indeterminate	5	2	13	11.25	9	10.5	7.5	8	10.5	18.5	17	15.25	14.75	13.75	13	15.5	4.5	5.5	5.75	5.5	5
Gymnosperms																					
Araucariacites spp.	2	2	2	7	9	7	2	1	3	5	5	6	3	4	4	9	4	7	3	8	4
Dacrydiumites spp.	3.5	6.5	5.5	5	3	0	3.5	7	3	4.5	2	3	0	1	2	4	6.5	2	7	4.5	2
Dilwynites granulatus	66.5	65	59	66	104.5	101.75	92.25	78	69.5	74.5	91.5	93.5	96.25	103.25	102.25	109.25	96.5	117.75	108.5	109.5	109.75
Phyllocladidites mawsonii	6	3	4	1.5	0	0	1	1	0	0	1.5	0.5	0	4.5	2	0.5	0	0	0	1	0
Phyllocladidites paleogenicus	2	0	0	0	0	0	0	2	0	0	1	0	0	1	0	0	0	0	0	0	0
Podocarpidites spp.	17.5	31.5	23.5	15.25	7	12	12	13.5	11	10.25	18	17.25	18.75	11.25	13.5	8.75	18	20	20.5	11	20
Microcachrydites antarcticus	0	1	2	1	0	2	3	0	0	1	0	0	2	2	1	0	0	0	0	0	0
Angiosperms																					
Ericipites spp.	1	1	0	0	0	0	0	0	0	0	0	0	0	0	0	1	1	0	0	0	0
Ilexpollenites spp.	0	0	0	0	0	0	0	1	0	0	0	0	0	0	0	0	0	0	0	0	0
Haloragacidites harrisii	0	3	3	1	2	0	3	3	0	1	1	3	1.5	4.5	4	5.5	5	4	2	2	5
Malvacipollis spp.	0	0	2	*	1	1	4	8	5	6	3	3	1	2	4	0	3	0	1	1	0
Myrtaceidites spp.	0	0	0	0	0	0	0	1	1	1	0	0	0	1	1	0	0	0	0	0	0
Nothofagus spp.	0	0	0	0	0	0	1	0	0	1	0	0	0	0	0	1	0	0	0	0	0
Proteacidites spp.	7.5	2	2	1.5	2	4	1	3	3	4	2.5	2.5	4	6	8.75	3	5	3	4	5	3.5
Spinizonocolpites prominatus	0	0	0	0	0	0	2	2	1	1	1	1.25	0	0	0	0	0	1	0	0	1
Tetracolporites oamaruensis	1	1	0	1	0	0	1	0	0	1	1	0	0	0	0	0	0	0	0	0	0
Tricolpites phillipsii	0	0	0	0	0	0	1	0	1	0	1	0	0	2	0	1	0	0	0	0	0
Tricolpites secarius	0	0	0	0	0	0	0	1	0	0	0	0	0	0.5	0	1	0	0	0	0	0
Triorites minisculus	0	0	0	0	0	0	0	0	0	1	0	0	0	4	4	0	0	0	0	0	0
Triorites subspinosus & *spinosus*	0	0	1	0	0	0	0	0	0	0	0	0	0	1	2	0	0	0	0	*	0
Triorites spp.	0	0	0	0	0	0	0	0	0	0	0	0	0	0	1	1	0	0	0	0	0
Tricolpate/tricolporate indeterminate	5	1	2.5	2.5	4.5	5	3	11.5	2.5	8	5.5	4.5	4.75	15	7.5	4	8.25	6.5	2.5	5	3
Miospores indeterminate	9.5	10	9	9	7.75	6	8.25	12	10	10.25	8.5	7.25	8.75	9.25	9.75	7.25	10	10.5	8.5	9	8
TOTAL COUNT	205.5	193.75	199.25	197.25	192.5	202.75	202.25	205.75	173.25	202	223.5	214.25	189.5	216	214.5	211.25	205.75	218.75	216.75	210	208.75

REFERENCES CITED

Abbink, O.A., 1998, Palynological investigations in the Jurassic of the North Sea region [Ph.D. thesis]: Utrecht, Netherlands, Utrecht University, Laboratory of Palaeobotany and Palynology Contributions Series, v. 8, 191 p.

Askin, R.A., 1992, Late Cretaceous–early Tertiary Antarctic outcrop evidence for past vegetation and climates, *in* Kennett, J.P., and Warnke, D.A., eds., The Antarctic paleoenvironment: A perspective on global change: Antarctic Research Series, v. 56, p. 61–73.

Aubry, M.-P., 1998, Early Paleogene calcareous nannoplankton evolution: A tale of climatic amelioration, *in* Aubry, M.-P., et al., eds., Late Paleocene–early Eocene climatic and biotic events in the marine and terrestrial records: New York, Columbia University Press, p. 158–203.

Ballance, P.F., 1993, The paleo-Pacific, postsubduction, passive margin thermal relaxation sequence (Late Cretaceous–Paleogene) of the drifting New Zealand continent, *in* Ballance, P.F., ed., South Pacific sedimentary basins of the world, 2: Elsevier Science Publishers, p. 93–110.

Beard, K.C., 1998, East of Eden: Asia as an important center of taxonomic origination in mammalian evolution, *in* Beard, K.C., and Dawson, M.R., eds., Dawn of the age of mammals in Asia: Carnegie Museum of Natural History Bulletin, v. 34, p. 5–39.

Beerling, D.J., 2000, Increased terrestrial carbon storage across the Paleocene-Eocene boundary: Palaeogeography, Palaeoclimatology, Palaeoecology, v. 161, p. 395–405.

Beerling, D.J., and Jolley, D.W., 1998, Fossil plants record an atmospheric $^{12}CO_2$ and temperature spike across the Paleocene-Eocene transition in northwest Europe: Journal of the Geological Society of London, v. 155, p. 591–594.

Bowen, G.J., Clyde, W.C., Koch, P.L., Ting, S.Y., Alroy, J., Tsubamoto, T., Wang, Y.Q., and Wang, Y., 2002, Mammalian dispersal at the Paleocene/Eocene boundary: Science, v. 295, p. 2062–2065.

Bujak, J., and Brinkhuis, H., 1998, Global warming and dinocyst changes across the Paleocene/Eocene Epoch boundary, *in* Aubry, M.-P., et al., eds., Late Paleocene–early Eocene climatic and biotic events in the marine and terrestrial records: New York, Columbia University Press, p. 277–295.

Bybell, L.M., Gibson, T.G., Heilmann-Clausen, C., and Zachos, J.C., 2001, Events associated with the late Paleocene thermal maximum in the northeastern U.S., *in* Ash, A.W., and Wing, S.L., eds., Climate and biota of the early Paleogene: International Meeting July 3–8, 2001, Northwest College, Powell, Wyoming, USA, Abstract volume, p. 18.

Christophel, D.C., 1980, Occurrence of *Casuarina* megafossils in the Tertiary of southeastern Australia: Australian Journal of Botany, v. 28, p. 249–259.

Christophel, D.C., 1989, Evolution of the Australian flora through the Tertiary: Plant Systematics and Evolution, v. 162, p. 63–78.

Christophel, D.C., and Greenwood, D.R., 1989, Changes in climate and vegetation in Australia during the Tertiary: Review of Palaeobotany and Palynology, v. 58, p. 95–109.

Clyde, W.C., and Gingerich, P.D., 1998, Mammalian community response to the latest Paleocene thermal maximum: An isotaphonomic study in the northern Bighorn Basin, Wyoming: Geology, v. 26, p. 1011–1014.

Collinson, M.E., 1993, Taphonomy and fruiting biology of recent and fossil *Nypa:* Special Papers in Paleontology, v. 49, p. 165–180.

Collinson, M.E., 1996, Plant macrofossils from the Bracklesham (early and middle Eocene), Bracklesham Bay, West Sussex, England: Review and significance in the context of coeval British Tertiary floras: Tertiary Research, v. 16, p. 175–202.

Collinson, M.E., 2000a, Fruit and seed floras from the Paleocene/Eocene transition and subsequent Eocene in southern England: Comparison and palaeoenvironmental implications: GFF, v. 122, p. 36–37.

Collinson, M.E., 2000b, Cenozoic evolution of modern plant communities and vegetation, *in* Culver, S.J., and Rawson, P.F., eds., Biotic response to global change: The last 145 million years: Cambridge, UK, Cambridge University Press, p. 223–243.

Collinson, M.E., and Hooker, J.J., 1987, Vegetation and mammalian faunal changes in the early Tertiary of southern England, *in* Fris, E.M., et al., eds., The origins of angiosperms and their biological consequences: Cambridge, UK, Cambridge University Press, p. 259–304.

Crouch, E.M., 2001, Environmental change at the time of the Paleocene-Eocene biotic turnover [Ph.D. thesis]: Utrecht, Netherlands, Utrecht University, Laboratory of Palaeobotany and Palynology Contributions Series, v. 14, 216 p.

Crouch, E.M., Heilmann-Clausen, C., Brinkhuis, H., Morgans, H.E.G., Rogers, K.M., Egger, H., and Schmitz, B., 2001, Global dinoflagellate event associated with the late Paleocene thermal maximum: Geology, v. 29, p. 315–318.

Crouch, E.M., Dickens, G.R., Brinkhuis, H., Aubry, M.-P., Hollis C.J., Rogers, K.M., and Visscher, H., 2003, The *Apectodinium* acme and terrestrial discharge during the Paleocene-Eocene thermal maximum: New palynological, geochemical and calcareous nannoplankton observations at Tawanui, New Zealand: Palaeogeography, Palaeoclimatology, Palaeoecology, in press.

Dickens, G.R., 2000, Methane oxidation during the late Paleocene thermal maximum: Bulletin de la Société Géologique de France, v. 171, p. 37–49.

Dickens, G.R., 2001, Carbon addition and removal during the late Paleocene Thermal Maximum: Basic theory with a preliminary treatment of the isotope record at Ocean Drilling Program Site 1051, Blake Nose, *in* Kroon, D., et al., eds., Western North Atlantic Paleogene and Cretaceous paleoceanography: Geological Society [London] Special Publication 183, p. 293–305.

Dickens, G.R., Castillo, M.M., and Walker, J.C.G., 1997, A blast of gas in the latest Paleocene: Simulating first-order effects of massive dissociation of methane hydrate: Geology, v. 25, p. 259–262.

Frederiksen, N.O., 1994, Paleocene floral diversities and turnover events in eastern North America and their relation to diversity models: Review of Palaeobotany and Palynology, v. 82, p. 225–238.

Fricke, H.C., Clyde, W.C., O'Neil, J.R., and Gingerich, P.D., 1998, Evidence for rapid climate change in North America during the latest Paleocene thermal maximum: Oxygen isotope compositions of biogenic phosphate from the Bighorn Basin (Wyoming): Earth and Planetary Science Letters, v. 160, p. 193–208.

Greenwood, D.R., 1994, Palaeobotanical evidence for Australian Tertiary climates, *in* Hill, R.S., ed., History of the Australian vegetation: Cretaceous to Holocene: Cambridge, UK, Cambridge University Press, p. 44–59.

Greenwood, D.R., and Wing, S.L., 1995, Eocene continental climates and latitudinal temperature gradients: Geology, v. 23, p. 1044–1048.

Greenwood, D.R., Vadala, A.J., and Banks, M.A., 2000, Climate change and vegetation responses during the Paleocene and Eocene in southeastern Australia: GFF, v. 122, p. 65–66.

Greenwood, D.R., Vadala, A.J., Anthony, J., Moss, P.T., and Keefe, R.L., 2001, Early Paleogene climates and vegetation in southeast Australia, *in* Ash, A.W., and Wing, S.L., eds., Climate and biota of the early Paleogene: International Meeting July 3–8, 2001, Northwest College, Powell, Wyoming, USA, Abstract volume, p. 38.

Harrington, G.J., 2000, Palynofloral dynamics of the late Paleocene to early Eocene from the Bighorn Basin, northwest Wyoming, USA: GFF, v. 122, p. 67–68.

Harrington, G.J., and Kemp, S.J., 2001, U.S. Gulf Coast vegetation dynamics during the latest Paleocene: Palaeogeography, Palaeoclimatology, Palaeoecology, v. 167, p. 1–21.

Harris, W.K., 1965, Basal Tertiary microfloras from the Princetown area, Victoria, Australia: Palaeontographica Abt. B, v. 115, p. 75–106.

Heilmann-Clausen, C., and Egger, H., 2000, The Anthering outcrop (Austria): A key-section for correlation between Tethys and northwestern Europe near the Paleocene/Eocene boundary: GFF, v. 122, p. 69.

Jolley, D.W., 1998, Palynostratigraphy and depositional history of the Paleocene Ormesby/Thanet depositional sequence set in southeastern England and its correlation with continental west Europe and the Lista Formation, North Sea: Review of Palaeobotany and Palynology, v. 99, p. 265–315.

Kaiho, K., Arinobu, T., Ishiwatari, R., Morgans, H.E.G., Okada, H., Takeda, N.,

Tazaki, K., Zhou, G., Kajiwara, Y., Matsumoto, R., Hirai, A., Niitsuma, N., and Wada, H., 1996, Latest Paleocene benthic foraminiferal extinction and environmental change at Tawanui, New Zealand: Paleoceanography, v. 11, p. 447–465.

Katz, M.E., Pak, D.K., Dickens, G.R., and Miller, K.G., 1999, The source and fate of massive carbon input during the latest Paleocene thermal maximum: Science, v. 286, p. 1531–1533.

Kelly, D.C., Bralower, T.J., Zachos, J.C., Premoli Silva, I., and Thomas, E., 1996, Rapid diversification of planktonic foraminifera in the tropical Pacific (ODP Site 865) during the late Paleocene thermal maximum: Geology, v. 24, p. 423–426.

Kennett, J.P., and Stott, L.D., 1991, Abrupt deep sea warming, paleoceanographic changes, and benthic extinctions at the end of the Paleocene: Nature, v. 353, p. 225–229.

King, A., and Harding, I.C., 2000, Temporal separation of Paleocene/Eocene boundary events in the North Sea Basin: GFF, v. 122, p. 88.

King, P.R., Naish, T.R., Browne, G.H., Field, B.D., and Edbrooke, S.W. (compilers), 1999, Cretaceous to Holocene sedimentation patterns in New Zealand: Institute of Geological and Nuclear Sciences folio series 1, version 1999.1., 35 p.

Koch, P.L., Zachos, J.C., and Gingerich, P.D., 1992, Correlation between isotope records in marine and continental carbon reservoirs near the Paleocene/Eocene boundary: Nature, v. 358, p. 319–322.

Koch, P.L., Zachos, J.C., and Dettmann, D.L., 1995, Stable isotope stratigraphy and paleoclimatology of the Paleogene Bighorn Basin (Wyoming, USA): Palaeogeography, Palaeoclimatology, Palaeoecology, v. 115, p. 61–89.

Krutzsch, W., 1989, Paleogeography and historical phytogeography (paleochronology) in the Neophyticum: Plant Systematics and Evolution, v. 162, p. 5–61.

MacPhail, M.K., Alley, N.F., Truswell, E.M., and Sluiter, I.R.K., 1994, Early Tertiary vegetation: Evidence from spores and pollen, *in* Hill, R.S., ed., History of the Australian vegetation: Cretaceous to Holocene: Cambridge, UK, Cambridge University Press, p. 189–261.

MacPhail, M., Hill, K., Partridge, A., Truswell, E., and Foster, C., 1995, Australia: "Wollemi Pine": Old pollen records for newly discovered genus of gymnosperm: Geology Today (March-April), p. 48–50.

Martin, H.A., 1994, Australian Tertiary phytogeography: Evidence from palynology, *in* Hill, R.S., ed., History of the Australian vegetation: Cretaceous to Holocene: Cambridge, UK, Cambridge University Press, p. 104–142.

Mildenhall, D.C., 1980, New Zealand Late Cretaceous and Cenozoic plant biogeography: A contribution: Palaeogeography, Palaeoclimatology, Palaeoecology, v. 31, p. 197–233.

Moore, P.R., and Morgans, H.E.G., 1987, Two new reference sections for the Wanstead Formation (Paleocene-Eocene) in southern Hawkes Bay: New Zealand Geological Survey Record, v. 20, p. 81–87.

Muller, J., 1959, Palynology of Holocene Orinoco delta and shelf sediments: Reports of the Orinoco Shelf Expedition: Micropaleontology, v. 5, p. 1–32.

Nix, H., 1982, Environmental determinants of biogeography and evolution in Terra Australia, *in* Barker, W.R., and Greenslade, P.J.M., eds., Evolution of the flora and fauna of arid Australia: Frewville, Peacock Publications, p. 47–66.

Norris, R.D., and Röhl, U., 1999, Carbon cycling and chronology of climate warming during the Paleocene/Eocene transition: Nature, v. 401, p. 775–778.

Pocknall, D.T., 1990, Palynological evidence for the early to middle Eocene vegetation and climate history of New Zealand: Review of Palaeobotany and Palynology, v. 65, p. 57–69.

Pole, M.S., 2001, Can long-distance dispersal be inferred from the New Zealand plant fossil record?: Australian Journal of Botany, v. 49, p. 357–366.

Pole, M.S., and MacPhail, M.K., 1996, Eocene *Nypa* from Regatta Point, Tasmania: Review of Palaeobotany and Palynology, v. 92, p. 55–67.

Prider, J.N., and Christophel, D.C., 2000, Distributional ecology of *Gymnostoma australianum* (Casuarinaceae), a putative palaeoendemic of Australian wet tropic forests: Australian Journal of Botany, v. 48, p. 427–434.

Raine, J.I., 1984, Outline of a palynological zonation of Cretaceous to Paleogene terrestrial sediments in West Coast region, South Island, New Zealand: New Zealand Geological Survey report, v. 109, p. 1–82.

Raine, J.I., and Wilson, G.J., 1988, Palynology of the Mt Somers (South Island, New Zealand) early Cenozoic sequence (Note): New Zealand Journal of Geology and Geophysics, v. 31, p. 385–390.

Ravizza, G., Norris, R.N., and Blusztajn, J., 2001, An osmium isotope excursion associated with the late Paleocene thermal maximum: Evidence of intensified chemical weathering: Paleoceanography, v. 16, p. 155–163.

Thomas, E., 1998, Biogeography of the late Paleocene benthic foraminiferal extinction, *in* Aubry, M.-P., et al., eds., Late Paleocene–early Eocene climatic and biotic events in the marine and terrestrial records: New York, Columbia University Press, p. 214–243.

Thomas, E., and Shackleton, N.J., 1996, The Paleocene-Eocene benthic foraminiferal extinction and stable isotope anomalies, *in* Knox, R.W.O'B., et al., eds., Correlation of the early Paleogene in northwest Europe: Geological Society [London] Special Publication 101, p. 401–444.

Traverse, A., 1988, Paleopalynology: London, Chapman and Hall, 600 p.

Tyson, R.V., 1995, Sedimentary organic matter: Organic facies and palynofacies: London, Chapman and Hall, 615 p.

Van der Kaars, S., 2001, Pollen distribution in marine sediments from the southeastern Indonesian waters: Palaeogeography, Palaeoclimatology, Palaeoecology, v. 171, p. 341–361.

Van Steenbergen, A., 1997, Palynological techniques for rock sampling: Utrecht, Netherlands, Laboratory of Palaeobotany and Palynology internal report, 25 p.

Wing, S.L., 1998, Late Paleocene–early Eocene floral and climatic change in the Bighorn Basin, Wyoming, *in* Aubry, M.-P., et al., eds., Late Paleocene–early Eocene climatic and biotic events in the marine and terrestrial records: New York, Columbia University Press, p. 380–400.

Wing, S.L., and Harrington, G.J., 2001, Floral response to rapid warming at the Paleocene/Eocene boundary and implications for concurrent faunal change: Paleobiology, v. 23, p. 539–563.

Wing, S.L., Bao, H., and Koch, P.L., 2000, An early Eocene cool period? Evidence for continental cooling during the warmest part of the Cenozoic, *in* Huber, B.T., et al., eds., Warm climates in Earth history: Cambridge, UK, Cambridge University Press, p. 197–237.

Zachos, J.C., and Dickens, G.R., 2000, An assessment of the biogeochemical feedback response to the climatic and chemical perturbations of the LPTM: GFF, v. 122, p. 188–189.

Zachos, J.C., Lohmann, K.C., Walker, J.C.G., and Wise, S.W., 1993, Abrupt climate change and transient climates during the Paleogene: A marine perspective: Journal of Geology, v. 101, p. 191–213.

Manuscript Accepted by the Society August 13, 2002

Printed in the USA

Geological Society of America
Special Paper 369
2003

Plant communities and climate change in southeastern Australia during the early Paleogene

David R. Greenwood*
School of Life Sciences and Technology (S008), Victoria University of Technology, PO Box 14428, Melbourne City MC, VIC 8001, Australia
Patrick T. Moss*
Department of Geography, University of Iowa, 316 Jessup Hall, Iowa City, Iowa 52242, USA
Andrew I. Rowett
Mineral Resources Group, Minerals and Energy Resources, SA, GPO Box 1671, Adelaide, SA 5001, Australia
Anthony J. Vadala
Rachael L. Keefe
School of Life Sciences and Technology (S008), Victoria University of Technology, PO Box 14428, Melbourne City MC, VIC 8001, Australia

ABSTRACT

In this study, data from fossil macrofloras and microfloras in southeastern Australia are used to reconstruct vegetation and climates for the early Paleogene. Our data show that for much of the late Paleocene to middle Eocene, complex, species-rich forests were predominant in southeastern Australia, under mesothermal humid climates (mean annual temperature 16–22 °C, coldest quarter mean temperature >10 °C, mean annual precipitation >150 cm/yr). A minor cooling episode may have occurred in the mid-early Eocene. Megathermal climates may have been present in lowlands in the latest early Eocene, during the Cenozoic Global Climatic Optimum. These forests were dominated by taxa characteristic of present-day mesothermal-megathermal high-rainfall multistratal forests; e.g., Cunoniaceae, Elaeocarpaceae, *Gymnostoma* (Casuarinaceae), Lauraceae (e.g., *Beilschmiedia, Cryptocarya* and *Endiandra*), and Proteaceae. A prominent treefern element (*Cyathea* and *Dicksonia* types) was present in the early Eocene. A number of megathermal taxa, including Cupanieae (Sapindaceae) and *Ilex* (Aquifoliaceae), were present through the early and middle Eocene. Taxa characteristic of modern-day microthermal to mesothermal forests were also present, e.g., *Nothofagus* (Nothofagaceae), *Eucryphia* (Eucryphiaceae), *Libocedrus* (Cupressaceae) and Podocarpaceae (*Acmopyle* and *Dacrycarpus*). The relictual araucarian conifer, *Wollemia,* and other Araucariaceae were present through the late Paleocene to early Eocene. There is limited physiognomic evidence to suggest the late Paleocene to early Eocene forests contained some deciduous canopy trees.

*David.Greenwood@vu.edu.au. Present address, Moss: Department of Geography, University of Wisconsin—Madison, 550 N. Park Street, Madison, Wisconsin 53706, USA.

Greenwood, D.R., Moss, P.T., Rowett, A.I., Vadala, A.J., and Keefe, R.L., 2003, Plant communities and climate change in southeastern Australia during the early Paleogene, *in* Wing, S.L., Gingerich, P.D., Schmitz, B., and Thomas, E., eds., Causes and Consequences of Globally Warm Climates in the Early Paleogene: Boulder, Colorado, Geological Society of America Special Paper 369, p. 365–380.

AUSTRALIAN EARLY PALEOGENE ENVIRONMENTS—SETTING AND PRIOR WORK

Worldwide, the warmest interval of the Cenozoic was the early Eocene, the latter part of which has been called the Cenozoic Global Thermal Optimum, or CGTO (Clyde et al., 2001). An additional warm episode in the early Paleogene, the Initial-Eocene Thermal Maximum, or IETM (formerly the late Paleocene Thermal Maximum or LPTM), was marked but of short duration (Zachos et al., 1993, 1994; Röhl et al., 2000). The warm character of the early Paleogene has led to the recognition of the early Paleogene as a test bed for understanding environments and factors controlling climates in a world much warmer than today. Most quantitative paleobotanical investigation of early Paleogene climates and their influence on regional or local vegetation has been based in North America (e.g., Wing and Greenwood, 1993; Wing et al., 1995, 2000; Wing, 1998; Wing and Harrington, 2001). Prior studies of the early Paleogene for southeastern Australia (e.g., Carpenter et al., 1994; Greenwood, 1994; Greenwood and Wing, 1995; Greenwood et al., 2000a) provided only limited analysis of Paleocene and early Eocene floras from the region. Climate was generally warm during this time period, although there may have been a possible cooling episode during the early Eocene (Fig. 1; Greenwood et al., 2000a; Wing et al., 2000; Greenwood and Christophel, 2003). Greater detail than was presented in these earlier studies is required to fully understand environmental change associated with the IETM and the early Eocene warm interval in high-latitude southeastern Australia.

The primary forcing factor for the warm early Paleogene is thought to be high concentrations of atmospheric CO_2. Recent studies however, suggest early Paleogene pCO_2 <500 ppm, concentrations similar to those of the last few decades when pCO_2 has been 340–360 ppm, but much lower than those determined using geochemical proxies and modeling (Royer et al., 2001). At these pCO_2 values, computer climate models only partially replicate the paleo-proxy climate estimates for continental interiors and high latitudes in the early Eocene (Shellito et al., 2003). However the model-proxy data discrepancies may reflect inaccurate specification of paleovegetation, mountain ranges in some cells and other boundary conditions in the models (Greenwood and Wing, 1995; Sewall et al., 2000; Shellito et al., 2003). A continuing problem however, with model and proxy data comparisons, is the lack of suitable proxy data in key areas of comparison such as Australia.

In this report we reconstruct in detail southeastern Australian latest Paleocene (56–55 Ma) to middle Eocene (49–37 Ma) environments from paleobotanical data. This report attempts to answer two questions: (1) what was the composition and character of vegetation in the late Paleocene to early Eocene, and (2) how did floristic composition change across the interval?

Macrofloras and microfloras from southeastern Australia were selected for four reasons. First, these floras are relatively well known. Second, southeastern Australia is a medium-size (7° × 10° lat and long) well-defined region at high latitudes in the early Paleogene (Fig. 1). Third, recent research has provided a detailed refinement of the palynostratigraphic schema based on sedimentary sequences from the Gippsland Basin (Partridge, 1999). Finally, paleoclimate can be reconstructed from leaf macrofloras using leaf physiognomy (Greenwood, 2001), and bioclimatic analysis using quantitative "climate profiles" (Kershaw and Nix, 1988; Kershaw, 1996) can be applied to both macrofloras and spore-pollen assemblages.

CORRELATIONS AND AGE OF EARLY PALEOGENE MACROFLORAS IN SOUTHEASTERN AUSTRALIA

The Cenozoic marine proxy record of sea surface temperatures (SST) for the Southern Ocean and for northeastern Australia show cycles of cooling and warming superimposed on the broad global trend of progressive cooling (Shackleton and Kennett, 1975; Feary et al., 1991). The SST record for northeastern Australia shows that the continent's northward movement at times negated or at least minimized the effects of the global climatic cooling during the Paleogene (Feary et al., 1991; Truswell, 1993). The greatest late Paleocene and early Eocene climate warming occurred at high latitudes. Southeastern Australia lay at high southern latitudes (~60°S) during the early Paleogene (Fig. 1), and topographic relief was in the order of 400–800 m in the southern sector of the Eastern Highlands (Truswell, 1993; Taylor, 1994). The pattern of Paleogene to early Neogene climate change in southeastern Australia corresponds to floristic changes and shifts in leaf physiognomic signatures over the same time (Truswell, 1993; Greenwood, 1994; Macphail et al., 1994; Greenwood and Christophel, 2003). The Gippsland, Otway and Murray Basins were the depositional areas for much of the sediment removed from the Southern Highlands during the Paleogene (Fig. 1). Stratigraphic assignment of macrofloras presented here is based on terrestrial spore and pollen assemblages, using palynological zones established for the Gippsland Basin (Macphail et al., 1994).

Correlation between the Gippsland Basin palynological zones and the international Cenozoic time-scale has developed over the last 30 yr, and is also based on marine microfossil and nannofossil data and isotopic dating of volcanic rocks (Macphail et al., 1994; Chaproniere et al., 1996; Truswell, 1997). Southeastern Australian Cenozoic palynological zones in Macphail et al. (1994) were correlated with the geochronometric scale of Berggren et al. (1985), and the Australian Tertiary time-scale in Chaproniere et al. (1996) was based on the Berggren et al. (1995) geochronometric scale. The biostratigraphic scheme used here has been updated to the geochronometric scale and European stages given in Hardenbol et al. (1998), and so shows some differences to earlier schemes. This scheme also has revisions and additional subdivisions to the spore-pollen zones based on recent reassessments (Partridge, 1999; 2001, personal commun.). Detailed magnetostratigraphy of terrestrial sediments in western North America by Clyde et al. (2001) indicated

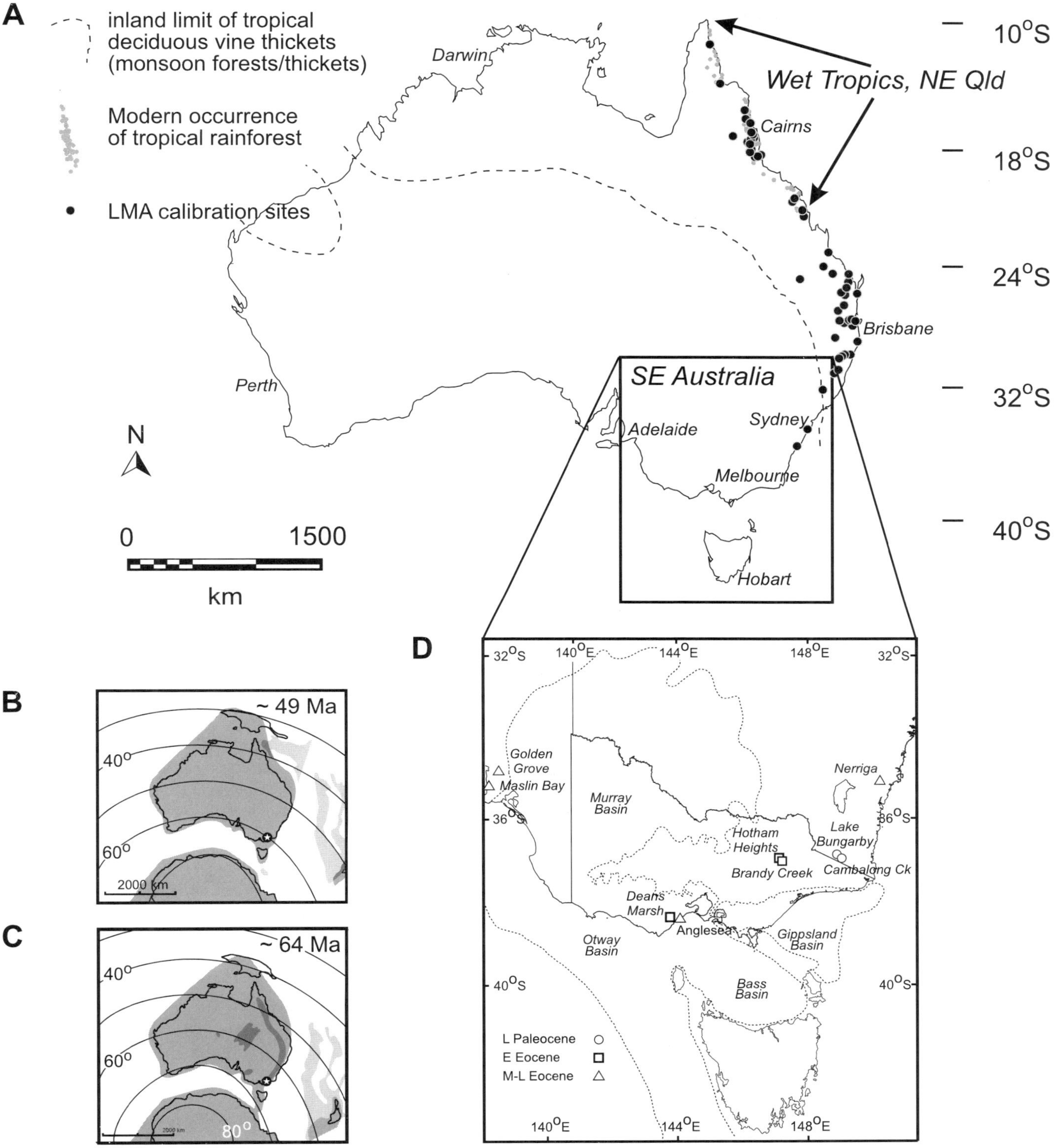

Figure 1. Location and paleogeographic setting for study area. A: Map of Australia showing location of southeastern Australia study area and other locations mentioned in the text, the modern distribution of tropical rainforests, and the location of the LMA calibration sites (Greenwood et al., 2001, personal commun.). B–C: Paleogeographic reconstructions of Australia in the middle Eocene (1B) and Paleocene (1C), adapted from Veevers et al. (1991). D: Detail of southeastern Australia showing the position of the Gippsland Basin and other major Cenozoic basins in the region, and the location of the Paleogene macrofloras discussed in the text (adapted in part from Greenwood et al., 2000b, and Greenwood and Christophel, 2003).

that the terrestrial onset of the CGCO falls within Chron C23r and lasts till Chron C22r. To facilitate comparisons with the North American terrestrial record, the stratigraphic assignment of floras has been related to the magnetostratigraphy in Hardenbol et al. (1998; Figure 2 herein) where possible.

Monaro Plains and region

Paleogene sediments crop out at several sites in southern New South Wales (Fig. 1), including Cambalong Creek and Nerriga (Hill, 1982; Taylor et al., 1990; Vadala and Drinnan, 1998). Outcrop at Cambalong Creek presents abundant well-preserved mummified leaves in well-defined sedimentary layers (Vadala and Drinnan, 1998). The microflora in the Cambalong Creek sediments correlates with the Upper *Lygistepollenites balmei* Zone (Stover and Partridge, 1973), consistent with a late Paleocene (upper Thanetian) age (Taylor et al., 1990). The Cambalong Creek flora probably sits just prior to the IETM (Fig. 2).

The Nerriga macroflora occurs in a lacustrine deposit (Titringo siltstone) that crops out to the north of the Monaro Plains (Hill, 1982). Microfloras in the Titringo siltstone were correlated with the Lower *Nothofagidites asperus* Zone, or the Upper *Malvacipollis diversus* to *Proteacidites asperopolus* Zones of the Gippsland Basin. Basalts associated with the Nerriga site give a corrected K/Ar age of 45.9–42.6 Ma (Hill, 1982), which falls within the lower part of the Lower *N. asperus* Zone (Fig. 2).

Mount Hotham and area (eastern Victorian Highlands)

The area around Mount Hotham in Victoria (Fig. 1) includes several peaks between 1700 and 1900 m. Early Cenozoic basalts are common in the area, capping an Oligocene to Eocene surface that preserves infilled stream systems containing fossiliferous sediments (Orr, 1999; Greenwood et al., 2000b). Outcrop at Hotham Heights and the abandoned Brandy Creek gold mine comprises Paleogene carbonaceous mudstones to silty sandstones that are capped by Eocene basalts (Greenwood et al., 2000b). Microfloras from both sites correlate to the uppermost part (E) of the *Malvacipollis diversus* Zone or the *Proteacidites asperopolus* Zone, indicating an early Eocene to basal middle Eocene age (Partridge, 1998a; 1998b; 1999, personal commun.). Here we treat Hotham Heights and Brandy Creek as being no younger than that portion of the *P. asperopolus* Zone that falls within the early Eocene (Fig. 2), giving greater weight to the early Eocene indicators (i.e., low counts of *Nothofagidites;* Partridge, 1998b). On available evidence these floras most likely fall within the CGCO (spanning Chrons C23r to C22r).

Western Victoria and South Australia

Two early Paleogene macrofloras from the Otway Basin (Fig. 1) are considered here; the Deans Marsh and Eocene Anglesea floras, from outcrop of the Eastern View Formation (Christophel et al., 1987; Greenwood et al., 2000b). The Deans Marsh macroflora was considered early Eocene based on palynological correlations (Christophel and Greenwood, 1989). However, Rowett and Sparrow (1994) indicated a lower middle Eocene age based on assignment of the microflora to the Lower *Nothofagidites asperus* Zone, and a biostratigraphic comparison of the dispersed cuticle flora of the Deans Marsh and other Australian Eocene macrofloras. Partridge (1998a), however considered the microflora correlated with his newly defined *Malvacipollis diversus* Zone C. Deans Marsh is thus considered here to be early Eocene (Fig. 2), potentially coinciding with the early Eocene cool interval detected in the western interior of North America (Wing et al., 2000). This age assignment differs from that used in Greenwood et al. (2000a, 2000b) and Greenwood and Christophel (2003).

A rich macroflora has been collected from the Eastern View Formation in a series of fluvio-lacustrine lenses in overburden from the Anglesea open-cut mine (Christophel et al., 1987; Greenwood et al., 2000b). The associated microflora was assigned by Christophel et al. (1987) to the base of the *Triorites magnificus* Zone of the Otway Basin (Harris, 1971), corresponding with the boundary of the Gippsland Basin Lower and Middle *Nothofagidites asperus* zones (equivalent to the P14-P15 boundary). The Anglesea macroflora is thus latest middle Eocene (late Bartonian) in age (Fig. 2), and is the youngest flora considered in this report.

The Golden Grove and Maslin Bay middle Eocene floras were uncovered in clay lenses in massive fluvial sand bodies of the North Maslin Sands in the St Vincent Basin, South Australia (Fig. 1). The Golden Grove clays (North Maslin Sands) are correlated with the Lower *Nothofagidites asperus* Zone of the Gippsland Basin (Alley, 1987; Christophel and Greenwood, 1987; Lindsay and Alley, 1995). Lindsay and Alley (1995) considered the Golden Grove flora to be no older than planktonic foraminifera zone P11. Partridge (1998a), offered evidence for assignment to either the *Proteacidites confragosus* Zone of the Otway Basin (Harris, 1971), which is a time equivalent of the *P. asperopolus* Zone of the Gippsland Basin (Stover and Partridge, 1973), or correlated it with his newly defined *Nothofagidites asperus* Zone A. The Golden Grove flora is here considered to be restricted within Partridge's (1999) *Nothofagidites asperus* Zone A, equivalent to zone P11 and that portion of P12 that correlates with the Lower *N. asperus* Zone A (Fig. 2). Uncertainty in our palynostratigraphic assignment means the flora spans the upper part of Chron C20r and the lower part of Chron C20n (Fig. 2).

FLORISTIC CHARACTER OF SOUTHEASTERN AUSTRALIAN EARLY PALEOGENE VEGETATION

Sampling, preparation and analysis of floristics

Preservation of leaf fossils at most sites was as compressions with the cuticular envelope preserved; this was liberated from the mudstone matrix by maceration with dilute H_2O_2. Ex-

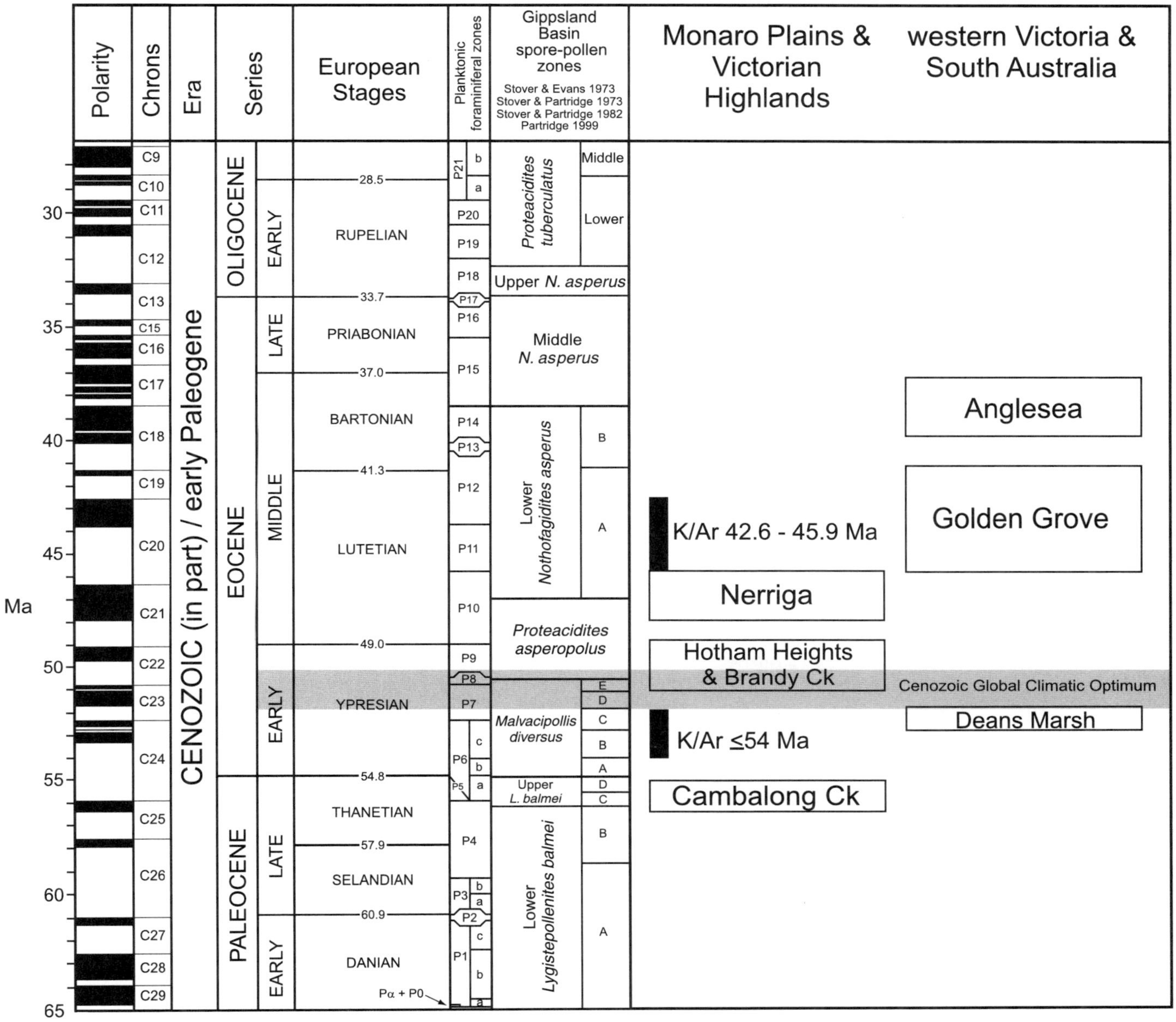

Figure 2. Early Paleogene palynostratigraphic schema for southeastern Australia showing the stratigraphic arrangement of the macrofloras discussed in the text (adapted from Greenwood et al., 2000b). Stratigraphy adapted from Macphail et al. (1994) and Chaproniere et al. (1996), to the geochronometric scale and European stages in Hardenbol et al. (1998), by Partridge (1998a; 1998b; 1999; 2001, personal commun.).

tracted leaves were cleaned with dilute aqueous chromic acid and/or hydrofluoric acid. Leaf floras preserve ecological information, such as dominance and diversity. Field census data from large laterally continuous fossiliferous sediments (e.g., Greenwood and Basinger, 1994; Wing, 1998), and modern leaf litter studies (Greenwood, 1991, 1992; Burnham et al., 1992), have shown that large collections (>600 specimens) will reflect the original floristic richness of the local forest. Anglesea and Golden Grove have been sampled for over two decades in a quantitative manner along bedding planes. Sampling at the Anglesea mine followed lateral transects within fluvial infill clay lenses as well as between several discrete lenses. Data presented here is from the site 2 transect (Christophel et al., 1987; Rowett and Christophel, 1990). Collections were examined from a single fluvio-lacustrine mudstone deposit at each of Hotham Heights, Brandy Creek, Nerriga, and Golden Grove (Hill, 1982; Christophel and Greenwood, 1987; Greenwood, 1991; Banks, 1999; Keefe, 2000). Sediment samples were collected from a spoil heap from the abandoned Deans Marsh mine (Christophel and Greenwood, 1989; Rowett and Sparrow, 1994).

Macrofloras were sorted into taxonomic units using leaf architecture and cuticle characters (Hickey, 1973). Identification of the fossil taxa was based on published diagnostic leaf and cuticle characters for key Australian plant families (e.g., Araucariaceae, Proteaceae, and Lauraceae) and member genera (Vadala and Greenwood, 2001, and references cited therein). Many of these taxa are climatically indicative (Kershaw and Nix, 1988). Unless otherwise stated, a "specimen" refers to a single leaf fragment, or whole leaf or leafy twig.

Spore-pollen counts used in this study were mainly derived from previous studies (e.g., Hill, 1982; Christophel et al., 1987; Taylor et al., 1990), but some data were collected specifically for the study (Partridge, 1998a, 1998b). Identification of nearest living modern analogs for the Tertiary palynomorphs follows the caveats and nomenclature provided in Macphail et al. (1994; Table 1 herein). Standard counts of >100 grains (typically 200 grains) were made in most instances. All data presented here for floras with multiple samples uses the average for each taxon for these samples.

Regional southeastern Australian early Paleogene spore-pollen record

Analysis of spore-pollen floras from terrestrial and marginal marine sediments in drill cores from the Gippsland and Otway Basins in southeastern Australia (Fig. 1), provide a general impression of changes in regional vegetation over the early Paleogene (Truswell, 1993; Macphail et al., 1994). According to Macphail et al. (1994) Araucariaceae (including the highly relictual conifer *Wollemia* as *Dilwynites granulatus;* Chambers et al., 1998), *Nothofagus* and some ferns (e.g., Cyatheaceae, Gleicheniaceae) fluctuated in importance during the Paleocene. Proteaceae were also prominent and diverse in microfloras throughout the Paleocene in southeastern Australia. Proteaceae today includes important canopy trees and understory shrubs in mesothermal-megathermal Australian rainforests (e.g., *Darlingia* and *Helicia*), as well as sclerophyllous forests and woodlands (e.g., *Banksia*). According to Macphail et al. (1994) Paleocene southeastern Australian forests were rich in araucarians and other conifers as emergents above an angiosperm (mostly *Nothofagus*) canopy. These forests included a diverse set of other

TABLE 1. LIST OF FOSSIL PALYNOMORPHS AND MACROFLORA AND THEIR NEAREST LIVING RELATIVES FOR FLORAS

Fossil palynomorph	Family	Suggested modern taxon/NLR (palynomorph or macrofossil)	1*	2	3	4	5	6	7
Cyathidites paleospora	Cyatheaceae	*Cyathea*		p†	p	p	p	p	p
Matonisporites ornamentalis	Dicksoniaceae	*Dicksonia*			p	p			
	Stangeriaceae	*Bowenia*					m§		m
Phyllocladidites mawsonii	"	*Lagarostrobos*	p	p			p	p	p
Podocarpidites spp.	"	*Podocarpus*	p	m	p		p	m	m
Araucariacites australis	Araucariaceae	*Agathis*	p		m	m	p	m	m
"	"	*Araucaria*	p	m			p		
Banksieaeidites arcuatus	Proteaceae	*Musgraveinae*						m	m
Proteacidites spp.	Proteaceae	*Orites*	m			m			
"	Proteaceae	*Helicia*	m			m		m	
	Elaeocarpaceae	*Aceratium*				m			
	"	*Elaeocarpus*	m		m	m			
	"	*Sloanea*						m	m
Not preserved	Lauraceae	*Beilschmiedia*	m		m	m		m	m
"	"	*Cryptocarya*	m	m	m	m	m	m	m
"	"	*Endiandra*	m	m	m	m	m	m	m
"	"	*Litsea*	m		m	m	m		m
Haloragacidites harrisii	Casuarinaceae	*Gymnostoma*	p	m	p	m	m	m	m
	Ebenaceae	*Diospyros*					m	m	m
	Grossulariaceae	*Quintinia*							m
Nothofagidities heterus-emarcidus	Nothofagaceae	*Nothofagus* sg *Brassospora*	p	p	p		p	p	p
N. asperus	"	*Nothofagus* sg *Lophozonia*						p	p
	Eucryphiaceae	*Eucryphia*	m						
Cupanieidites orthoteichus	Sapindaceae	Cupanieae (aff. *Cupaniopsis*)		p		p	p	p	p
Ilexpollenites sp.	Aquifoliaceae	*Ilex*	p	p	p	p			
	Sterculiaceae	*Brachychiton*					m	m	m

Note: Some taxa present in floras are omitted as they were not used in the bioclimatic analysis (i.e., not extant in the modern Australian flora so climate profiles could not be calculated). Palynomorph nearest living relatives (NLR) from Macphail et al. (1994).

*1—Cambalong Creek; 2—Deans Marsh; 3—Brandy Creek; 4—Hotham Heights; 5—Nerriga; 6—Golden Grove; 7—Anglesea.

§ p—palynomorphs; m—macrofloral taxa.

dicots, including *Ilex* (Aquifoliaceae, as *Ilexpollenites*). The importance of *Nothofagus* in the Paleocene forests may be overstated, however, as today trees in this genus are wind pollinated and can produce copious quantities of pollen, and the pollen of anemophilous taxa can swamp the output of zoophilous taxa in the same forest (Kershaw and Strickland, 1990).

According to Macphail et al. (1994) angiosperms with modern mesotherm-megathermal affinities make first appearances and/or increase significantly in abundance and diversity during the latest Paleocene. These taxa include palms (Arecaceae, primarily as *Arecipites* spp.), Olacaceae (*Anacolosidites* spp.), Proteaceae, Sapindaceae (e.g., *Cupaneidites* spp.), and Polygalaceae (*Polycolpites* spp.). By the early Eocene, conifer-*Nothofagus* forests of the early Paleocene had been replaced by diverse angiosperm dominated forests with emergent Araucariaceae, including *Wollemia* and *Araucaria* or *Agathis* (Macphail et al., 1994). A number of taxa made first appearances or became prominent in southeastern Australian microfloras by the early Eocene, including the climbing palm *Calamus* (Arecaceae), and *Gymnostoma* (Casuarinaceae, as *Haloragacidites harrisii*). *Nothofagus* and other mesothermal taxa were present, but likely as a minor component, perhaps reflecting compressed altitudinal and latitudinal vegetation zonation (Christophel and Greenwood, 1989; Truswell, 1993; Macphail et al., 1994). This view is reinforced by high counts of *Nothofagus* (*Brassospora*) pollen (34%–37%), the absence of megathermal pollen types, and macrofossil evidence of *Nothofagus* in northern Tasmanian early Eocene sites (Macphail et al., 1994).

The early–middle Eocene boundary in the Gippsland Basin was characterized by the rise to dominance of *Nothofagidites emarcidus-heterus* (av. 50%–60% of the pollen sum), representing *Nothofagus* subgenus *Brassospora.* Concomitantly, *Gymnostoma* pollen (as *Haloragacidites harrisii*) became consistently abundant (up to 30%), and gymnosperms declined in importance (<30%) compared to the early Eocene (Macphail et al., 1994). Overall angiosperm diversity increased in southeastern Australian microfloras in the middle Eocene compared to earlier times. The regional spore-pollen record is interpreted as reflecting middle to late Eocene vegetation in southeastern Australia as a mosaic of *Nothofagus*-dominated mesothermal rainforest associations, with an overstory of Podocarpaceae (Truswell, 1993; Macphail et al., 1994). This interpretation is not completely in accord with the macrofossil record. *Nothofagus* is scarce as a macrofossil and the southeastern Australian leaf record is dominated by Lauraceae and other common mesothermal-megathermal taxa typical of present-day tropical rainforests of northeastern Australia (Christophel and Greenwood, 1989; Vadala and Greenwood, 2001; Greenwood and Christophel, 2003). Podocarpaceae were diverse in some macrofloras (e.g., Anglesea, some sites in Tasmania), but macrofossils of these taxa are usually not abundant (Carpenter et al., 1994; Greenwood and Christophel, 2003).

The apparent discrepancy between the spore-pollen and macrofossil records for southeastern Australia reflect vegetational differences and relative transport potential of pollen and leaves and differential fossilization potential of various taxa (Christophel and Greenwood, 1989; Greenwood, 1991; Truswell, 1993). Wind pollinated taxa (e.g., conifers, and the angiosperms *Gymnostoma* and *Nothofagus*) are often over-represented relative to zoophilous taxa in multistratal mesotherm-megatherm forests (Kershaw and Strickland, 1990). Lauraceae leaves fossilize well, whereas their pollen does not, and the reverse situation may apply for *Nothofagus* (Macphail, 1980; Kershaw and Bulman, 1994; Vadala and Greenwood, 2001; Greenwood and Christophel, 2003). A satisfactory solution to this apparent discrepancy between the spore-pollen and macrofossil records is analyses of both types of plant organ assemblages (e.g., Blackburn and Sluiter, 1994), an approach that is applied here.

Local Paleogene environments from the spore-pollen record

Abundance data for the principal taxa from the spore-pollen sums and macrofloral records are summarized in Figure 3 for the principal taxa. A more comprehensive (but incomplete) floristic listing for each flora is given in Table 1. The microfloras show a consistent occurrence of a number of taxa at most sites through the early Paleogene (Fig. 3), namely *Nothofagus* (principally subgenus *Brassospora* as the palynomorphs *Nothofagidites emarcidus-N. heterus*), Casuarinaceae (*Haloragicidites harrisii*), diverse assemblages of Proteaceae (*Proteacidites* spp.), and the gymnosperm families Araucariaceae (principally *Araucariacites* spp., but also *Dilwynites granulatus*) and Podocarpaceae (principally *Lygistepollenites*/*Dacrydium* B, and *Podocarpidites* spp.). Fern and other free-sporing plants fluctuated in abundance and diversity through the early Paleogene, with treeferns such as *Cyathidites paleospora* (*Cyathea*) and *Matonisporites ornamentalis* (cf. *Dicksonia*), prominent in the latest early Eocene Brandy Creek and Hotham Heights floras. Gymnosperms and *Nothofagus* had the highest counts in the late Paleocene and middle Eocene (i.e. Cambalong Creek, Golden Grove and Anglesea). Casuarinaceae had low counts in the late Paleocene, and megathermal taxa such as *Ilexpollenites* (Aquifoliaceae) are largely characteristic of late Paleocene to early Eocene floras. Mesothermal-megathermal taxa such as *Cupaneidites* (Sapindaceae) and *Nothofagus* sg. *Brassospora* are characteristic of the middle Eocene.

The overall pattern demonstrated for the local sites (Fig. 3 and Table 1) is therefore generally consistent with the regional sequence described from cores sampled from within the Gippsland Basin of southeastern Australia (Macphail et al., 1994). The principal difference between the regional record and that reported here from local sites, is the lower counts of *Nothofagus* in the middle Eocene floras (i.e. <20%) than is typically reported for middle Eocene microfloras for the Gippsland Basin and other adjacent basins in southeastern Australia (i.e. av. 50%–60% of the pollen sum).

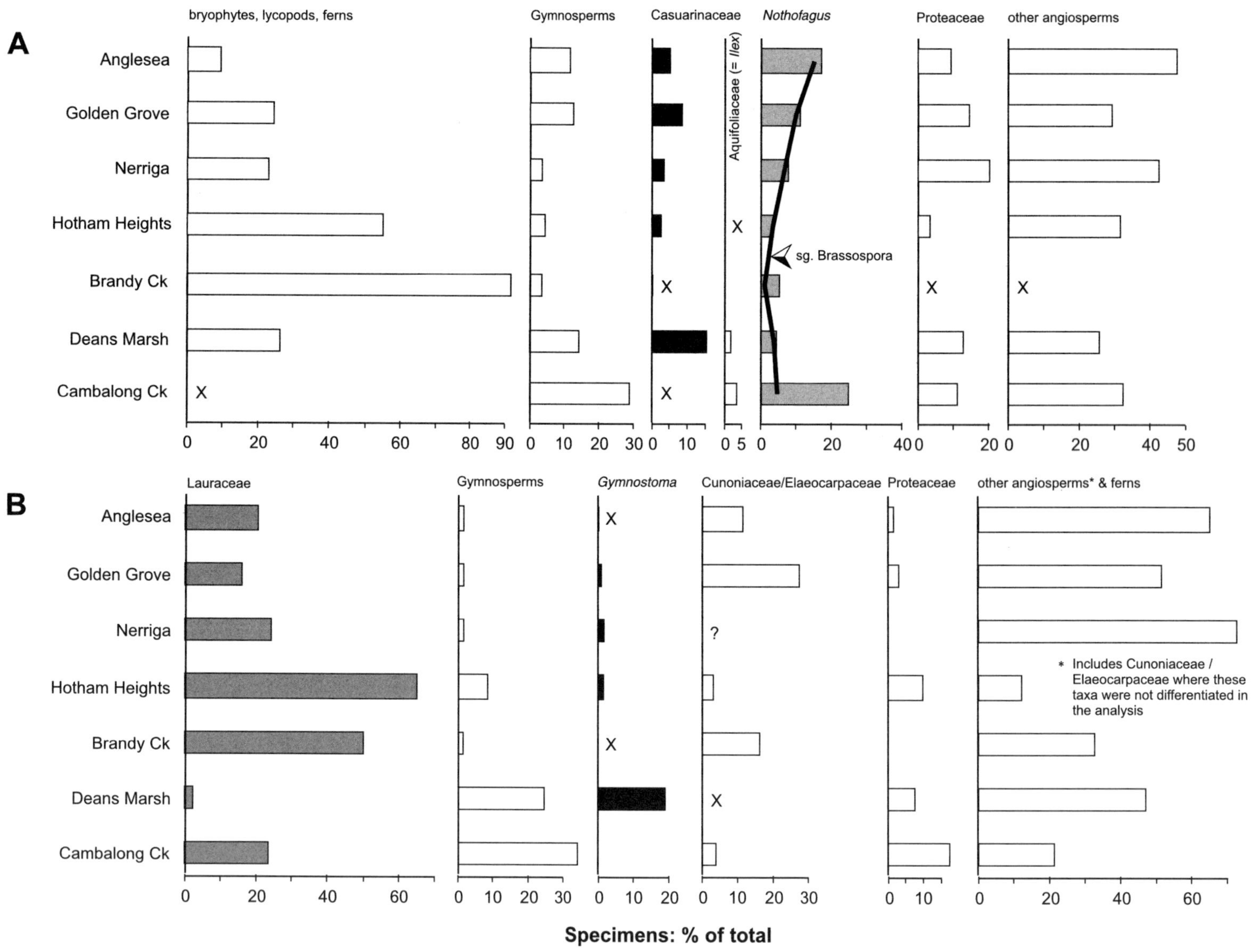

Figure 3. Changes in the relative abundance (spore-pollen grains, and macrofloral specimen counts) of the principal taxa for southeastern Australia over the late Paleocene to middle Eocene, based on microfloras (A) and macrofloras (B) from the fossil floras. *Nothofagus* sg. *Brassospora* differentiated within total *Nothofagus* for pollen (A) as a solid line. Data from Hill (1982), Christophel and Greenwood (1987), Christophel et al. (1987), Rowett and Christophel (1990), Taylor et al. (1990), Rowett and Sparrow (1994), Banks (1999), Partridge (1998a, 1998b), Keefe (2000), Vadala and Greenwood (2001). Where multiple samples were counted (spore-pollen and macrofossil), the average of these samples was used.

Local Paleogene environments from the macrofloral record

With the exception of the early Eocene Deans Marsh flora, leaves of Lauraceae are either abundant (>15%) or dominate (i.e. >40%) all of the early Paleogene macrofloras (Fig. 3). The same three genera of Lauraceae repeated in all floras (*Beilschmeidia, Cryptocarya* and *Endiandra*), with *Litsea* recorded only from Cambalong Creek and Brandy Creek (Table 1; see also Vadala and Greenwood, 2001). Gymnosperms were most abundant in the Cambalong Creek and Deans Marsh floras, but were present in low counts (<5%) in the middle Eocene macrofloras, although Anglesea had moderate diversity (i.e. 8 species of conifers and cycads). Cycads were recorded in the Nerriga and Anglesea floras (Table 1). Principal conifers in most floras were Podocarpaceae, typically one or more species of *Acmopyle, Dacrycarpus, Dacrydium, Phyllocladus,* and *Podocarpus* (e.g., Deans Marsh and Anglesea). Araucariaceae (e.g., *Agathis,* Hotham Heights, possible *Wollemia* at Cambalong Creek) and Cupressaceae (e.g., *Libocedrus,* Cambalong Creek) were rarely encountered. *Gymnostoma* was recorded as both foliage units (i.e. leafy twigs) and as infructescences (i.e. woody seed cones) in all of the Eocene floras (Fig. 4), but was not detected in the late Paleocene Cambalong Creek macroflora, although it occurs in the late Paleocene Lake Bungarby macroflora. The *Gymnostoma* seed cone records were recorded in the counts as a trace, and only where there were no other records for the flora (e.g., Brandy Creek). Cunoniaceae and Elaeocarpaceae can be difficult to discriminate between, and were grouped in the analysis, but included taxa such as *Aceratium* and *Elaeocarpus* (e.g., Hotham Heights and Cambalong Creek respectively), and *Sloanea* (e.g., Golden Grove and Anglesea). These two families were important elements in the Golden Grove and Anglesea middle Eocene floras, but significant counts of these taxa were also recorded for early Eocene floras. Macrofloral remains of *Nothofagus* were not found in any of the floras in this study (Table 1), but have been reported from Eocene floras from Tasmania, and from the Maslin Bay macroflora, which is coeval with the Golden Grove flora (Greenwood and Christophel, 2003).

The Proteaceae were moderately common (i.e., >10%) in only the Cambalong Creek and Hotham Heights floras (Fig. 4). This pattern masks the high diversity of Proteaceae in a number of the floras, with taxa such as the tribes Knightieae (e.g., *Darlingia*), Helicieae (e.g., *Helicia*), Banksieae (e.g., both *Musgravea*-type and *Banksia*-type), detected in a number of the floras (Table 1), and partly also reflects incomplete taxonomic knowledge of some floras (Vadala and Greenwood, 2001). The majority of angiosperm leaf taxa in these macrofloras remain unidentified, although a number of taxa not mentioned here are listed in Table 1 and by Greenwood et al. (2000b) and Greenwood and Christophel (2003).

The macrofloras show similar patterns to the microfloras, with a consistent set of taxa repeating throughout most of the early Paleogene sequence, and with some taxa most abundant in the late Paleocene and early Eocene (e.g., gymnosperms such as *Libocedrus,* and Proteaceae), while other taxa are most abundant in the early Eocene (Lauraceae) or middle Eocene (e.g., Cunoniaceae/Elaeocarpaceae). Some taxa were recorded in only a single flora (e.g., *Eucryphia;* Table 1). There are some notable similarities and differences between the two records. The pattern for the gymnosperms, *Gymnostoma* and Proteaceae largely matches the microfloral record. The absence of Lauraceae in the pollen record, as previously noted, is a taphonomic signature, and it seems likely that the apparent near absence of *Nothofagus* in the macrofloral record is also due to taphonomic bias, but also suggests that the role of *Nothofagus* in early Paleogene local vegetation may be less than is implied by pollen-based interpretations (Greenwood, 1991; Truswell, 1993).

SOUTHEASTERN AUSTRALIAN EARLY PALEOGENE CLIMATES

Methods of paleoclimate reconstruction

Bioclimatic analysis. This technique requires the development of "climatic profiles" (i.e., climatic parameters such as MAT and MAP) using bioclimatic analysis of distributions of modern plant genera. The first step of this approach is to identify as many "nearest living relatives" (NLRs) as possible in the fossil floras (micro- and macroflora; Table 1). Then a library of "climatic profiles" is produced for several key taxa based on climatic values such as mean annual temperature (MAT), warmest quarter mean temperature (WQMT), coldest quarter mean temperature (CQMT), mean annual precipitation (MAP), warmest quarter mean precipitation (WQMP) and coldest quarter mean precipitation (CQMP) (Kershaw and Nix, 1988; Kershaw, 1996; Moss and Kershaw, 2000). The climatic values were generated using a GIS based mathematical surface (incorporated in the BIOCLIM program) for present-day Australia, based on standard meteorological decadal means and a digital elevation model of Australia (Busby, 1991; Houlder et al., 1999). BIOCLIM summaries of climatic variables were generated for each taxa, and the maximum, minimum and percentiles (5, 25, 75 and 95) were calculated. These values constitute the climatic profiles for each taxon (Kershaw and Nix, 1988). Typically profiles were calculated for genera, but in some cases were generated for species groups within a genus, when this degree of taxonomic resolution was possible for fossil material. The zone of overlap for a set of NLRs defines the most likely climate space occupied for an individual early Paleogene flora (Kershaw and Nix, 1988; Kershaw, 1996; Moss and Kershaw, 2000).

For this study, climatic profiles were created for NLRs of 26 taxa found in the fossil spore-pollen and/or leaf floras of southeastern Australia (Table 1). Only a subset of these 26 taxa occurred in each flora. These profiles were based on the modern distribution of genera, which were determined from the electronic database of the Australian National Herbarium (CANB). These were supplemented in some cases using published climate profiles for taxa that are extinct in Australia today (i.e.

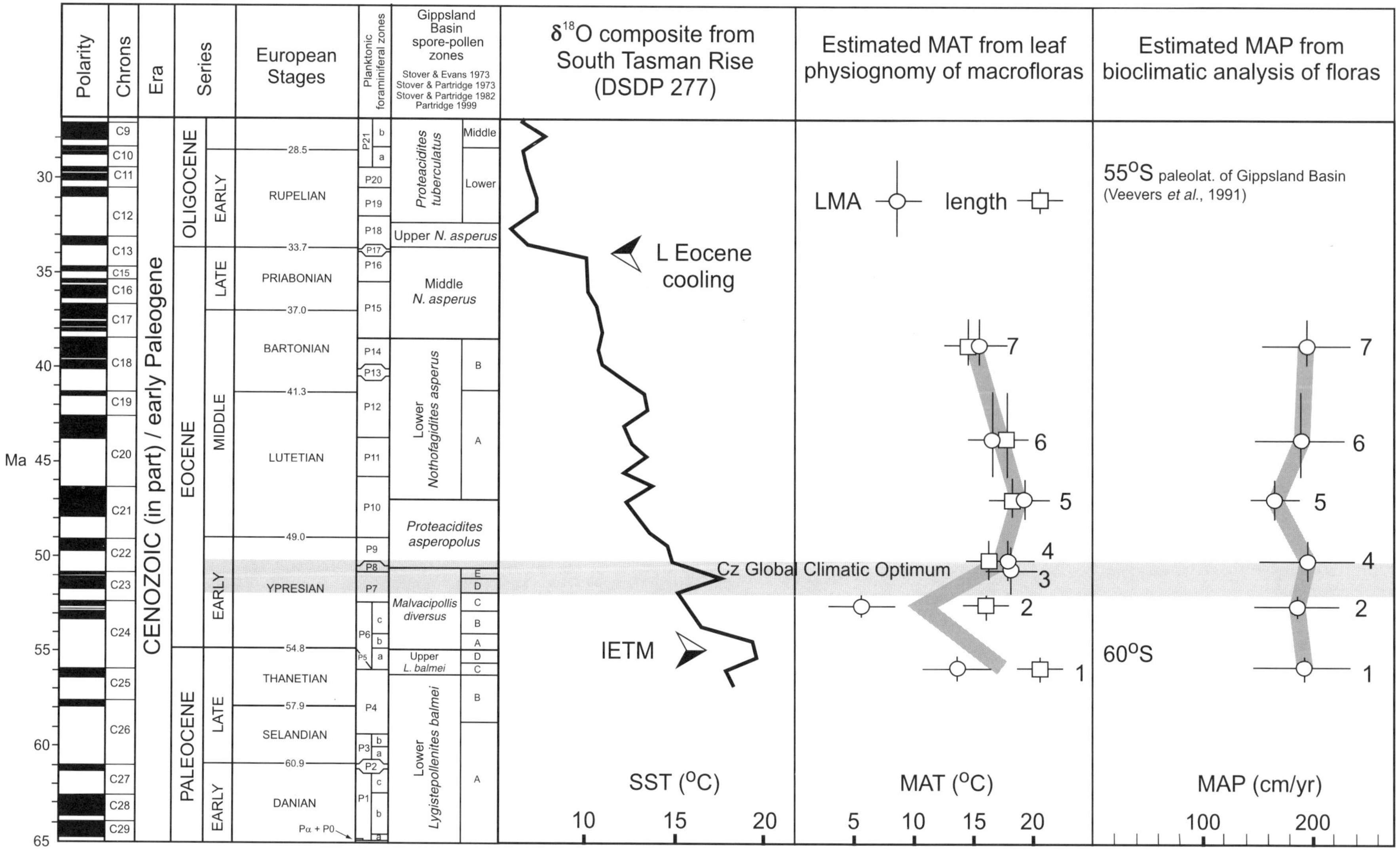

Figure 4. Summary chart of climate change over the early Paleogene for southeastern Australia. Sea surface temperatures for the South Tasman Rise (Shackleton and Kennett, 1975, diagrammatic only, adjusted to Berggren et al., 1995, time scale) are given to show the regional expression of the global Paleogene climate trend. Error bars for leaf margin analysis (LMA) are the binomial sampling error (Wilf, 1997) (equation 3), and for leaf size are the standard error of the estimate (±2.3 °C) from the regression equation for leaf size (equation 1). Fossil floras: 1 Cambalong Ck, 2 Deans Marsh, 3 Brandy Ck, 4 Hotham Heights, 5 Nerriga, 6 Golden Grove, 7 Anglesea, as per Table 3. IETM—Initial Eocene Thermal Maximum; MAP—mean annual precipitation; MAT—mean annual temperature; SST—sea surface temperature.

regional late Cenozoic extinctions), but still extant on neighboring landmasses (e.g., *Nothofagus* subgenus *Brassospora,* based on published data from Papua New Guinea and New Caledonia; Macphail, 1997). The mean climatic values for each site (Cambalong Creek, Deans Marsh, Hotham Heights, Nerriga, Golden Grove and Anglesea) was calculated by examining the climatic variables (e.g., MAT, WQMT, CQMT, MAP, WQMP and CQMP) for each NLR taxon found at that location. Mean climate values were then calculated for each site based on the range of the individual climatic variable (25–75 percentile values) for each of the NLR taxon and dividing these values by the total number of NLRs found at each site.

Leaf physiognomy. For the leaf physiognomy approach we apply two methods. Firstly, Greenwood's (1992; Greenwood et al., 1999) nontaxonomic method of modern forest floor litter studies, where the mean leaf size for the whole fossil leaf assemblage is the predictor of MAT

$$MAT_{size} = 0.1741\ ML + 4.3197, \text{ standard error} = 2.3\ ^\circ C \quad (1)$$

where *ML* is the mean length of ≥ 100 leaf specimens.

Secondly we apply leaf margin analysis (LMA; Wolfe, 1978; Wilf, 1997), which has been widely applied to North American Paleogene floras. Data from South and North America and east Asia, produce essentially the same statistical relationship between leaf margin proportion and MAT, however preliminary data from Australia had suggested a different relationship between these variables (Greenwood, 1992, 2001). Greenwood et al. (1999; 2001, personal commun.; Greenwood, 2001) completed an analysis of a comprehensive database of wet forest sites (mean annual rainfall >60 cm/yr) spanning a wide range of forest types along the east coast of Australia (Fig. 1). This analysis demonstrated that the regression of the proportion of toothed leaf margins versus MAT has the same slope for Australian floras as for those of other continents, although the intercept is different. The LMA equation is

$$MAT_{LMA} = 27.04\ P_{margin} - 2.1213, \text{ standard error} = 2.2\ ^\circ C \quad (2)$$

where P_{margin} is the proportion of dicot species in which the leaf margin lacks teeth ($0 < P_{margin} < 1$).

This new analysis provides a version of LMA calibrated for Australian conditions, providing more accurate estimates of MAT from Australian Paleogene leaf floras than was previously possible. Applying the method of Wilf (1997), the error of the estimate for LMA is expressed here as the binomial sampling error

$$\sigma[LMA] = c * (P_{margin}\ (1 - P_{margin})/r \quad (3)$$

where *c* is the slope from the LMA regression equation, P_{margin} as defined in (2), and *r* is the number of species scored for leaf margin type for the individual flora.

Paleoclimate from bioclimatic analysis of fossil spore-pollen and leaf floras

Climatic profiles for each of the seven sites suggest a mesothermal regime, with MAT ranging from 16 to 22 °C (average mean values 18–19 °C) throughout the early Paleogene (Table 2). Warmest quarter mean temperature (WQMT) was <25 °C (average mean values 22–23 °C) and the analysis suggests that coldest quarter mean temperatures (CQMT) were >13 °C (average mean values of 15 °C) for each of the seven sites. This bioclimatic analysis therefore suggests warm winter conditions, with temperatures in the region rarely falling below freezing at each of the sites throughout the early Paleogene (Table 2). Climatic profiles suggest MAP was at least >150 cm/yr (average mean values ~170–190 cm; Figure 4), consistent with modern rainforest climates of the Australian Wet Tropics (Fig. 1), and that precipitation was strongly seasonal, with the warmest quarter mean precipitation (WQMP) >50 cm (average mean values ~60–80 cm). This seasonality is further reinforced by precipitation values for the coldest quarter (<30 cm, average mean values ~20 cm; Table 2).

The magnitude of variation for all bioclimatic estimates between sites is within the measure of uncertainty for these estimates, however the mean values may reflect climatic change during the early Paleogene. Cambalong Creek (?IETM) and the early Eocene sites, Deans Marsh, Hotham Heights and Brandy Creek, all show warmer temperatures than the middle Eocene sites, Nerriga, Golden Grove and Anglesea. Our estimate for Cambalong Creek is at the upper end of the MAT 14–20 °C range suggested by Taylor et al. (1990). All sites reflect high mean annual rainfall, except Nerriga, which shows significantly lower annual and seasonal rainfall than the other sites. The warmer conditions of the late Paleocene and early Eocene sites is further reinforced by the presence of the megathermal taxa *Ilex* at all of these sites, which is absent at all of the middle Eocene sites (Table 1).

A caveat on the analysis is that three of the taxa, *Gymnostoma, Ilex* and *Lagarostrobos,* are represented by a sole species each in the modern Australian flora (and thus our calibration data set). Both *Gymnostoma* and *Ilex* are speciose outside of Australia; non-Australian *Ilex* includes temperate species (e.g., *I. aquifolium* [Europe] and *I. opaca* [southeast USA]), and *Gymnostoma* in Borneo can occur within seasonal swamp forests. It is possible that the modern Australian climatic range of these taxa represents a marked truncation of a broader climatic range for these taxa in the Paleogene. Each of these taxa were minor elements in most floras, and removing these taxa from the analysis had an insignificant effect.

Paleoclimate from leaf physiognomy of macrofloras

Greenwood and Christophel (2003) found that leaf size and the proportion of dicot species with entire leaf margins for southeastern Australia late Paleocene to middle Miocene floras varied in proportion to the northeastern Australian sea surface

TABLE 2. PALEOCLIMATE ESTIMATES BASED ON BIOCLIMATIC ANALYSIS OF SOUTHEASTERN AUSTRALIAN EARLY PALEOGENE FLORAS

Fossil flora (s)*	MAT† (°C)	WQMT (°C)	CQMT (°C)	MAP (cm/yr)	CQMP (cm)	WQMP (cm)
Cambalong Creek (16)*	16.3–20.7 (18.5)	20.4–24.3 (22.4)	12.6–17.6 (15.1)	155–240 (198)	14–31 (23)	51–93 (72)
Deans Marsh (11)	16.9–20.6 (18.8)	21.5–24.7 (23.1)	13.3–18.2 (15.7)	146–226 (186)	14–29 (21)	51–88 (70)
Hotham Heights (16)	16.8–21.3 (19.0)	20.9–25.1 (23.0)	13.1–18.3 (15.7)	150–195 (240)	13–28 (21)	54–101 (77)
Brandy Creek (12)	17.2–22.1 (19.6)	20.8–25.2 (23.0)	12.9–18.5 (15.7)	148–234 (191)	14–30 (22)	53–98 (76)
Nerriga (14)	16.8–19.4 (18.1)	21.3–23.4 (22.3)	12.9–15.9 (14.4)	143–190 (166)	13–23 (18)	54–75 (64)
Golden Grove (15)	16.8–21.1 (18.9)	21.0–24.9 (23.0)	12.7–18.1 (15.4)	148–231 (190)	13–30 (22)	54–95 (74)
Anglesea (17)	17.3–21.2 (19.3)	21.1–24.6 (22.8)	12.8–17.8 (15.3)	153–237 (195)	15–32 (23)	59–98 (78)

Note: Values are the mean of the 25th and 75th percentiles for all nearest living relatives in each flora shown as a range (mean in parentheses).
*The number of nearest living relative taxa used for each analysis (analyses).
†MAT—mean annual temperature; WQMT—warmest quarter mean temperature; CQMT—coldest quarter mean temperature; MAP—mean annual precipitation; CQMP—coldest quarter mean precipitation; WQMP—warmest quarter mean precipitation.

temperature curve (Feary et al., 1991), and for the early Paleogene, matched the Southern Ocean isotopic sea surface temperature record (Shackleton and Kennett, 1975). Significantly lower values for leaf size and proportion of entire margined leaf species occurred at times of cooling temperatures than for floras from warm intervals. Mean leaf size and proportion of species with entire leaf margins are correlated with mean annual temperature in modern Australian mesic forests, and are indicative of MAT for Paleogene floras (Greenwood, 1992, 2001; Greenwood and Wing, 1995).

A significant effect of the stratigraphy used here is to shift the position of the putative early Eocene cool interval proposed by Greenwood et al. (2000a) and Greenwood and Christophel (2003), from the latest early Eocene (i.e., after the Cenozoic Climatic Optimum) to prior to the CGCO. Mean annual temperatures remained in the mesothermal range for most of the early Paleogene (Fig. 4 and Table 3). Highest temperatures estimated from leaf physiognomy occur in the latest early Eocene to middle Eocene, whilst coolest temperatures were indicated in the mid-early Eocene (MAT <15 °C), although the 2 indices used (leaf size and leaf margin proportion) returned widely divergent estimates of MAT, and significantly, the bioclimatic analysis indicated MAT at Deans Marsh comparable to that at Hotham Heights (Table 2).

The leaf size based estimate for the late Paleocene is based on the published value for the coeval Lake Bungarby macroflora (Hill, 1992; Table 3). The LMA and leaf size estimated of MAT for the 2 floras from the southern Monaro are at the lower and upper ends respectively of the range (14–20 °C) estimated by Taylor et al. (1990). Greenwood and Christophel (2003) speculated that the divergence in the leaf size measures, the common occurrence of quite wide leaves in these floras, and the proportion of toothed leaf margins seen in the early Eocene and late Paleocene likely reflected the presence of a significant number of deciduous dicots in the forest canopy. Extant deciduous forest canopies in eastern North America have generally larger and broader leaves than extant Australian broad-leaved mesic evergreen forests growing under similar MAT (Basinger et al., 1994; Greenwood, 2001). Estimates of MAT based on mean leaf size of predominantly deciduous dicot fossil floras may therefore actually be estimates of mean growing season conditions, and so may potentially overestimate MAT (Basinger et al., 1994; Greenwood, 2001). Fossil wood with well-defined uniform growth rings have been reported from high latitude sites in Paleocene floras from Arctic Canada and the Southern Highlands of Australia (Basinger et al., 1994; Taylor et al., 1990), a wood character that is consistent with deciduous forests (Greenwood, 2001). Under the mild climates of the early Paleogene, deciduousness at high latitudes would likely be due to the effect of the seasonal light regime at high latitudes (Basinger et al., 1994).

SOUTHEASTERN AUSTRALIAN EARLY PALEOGENE ENVIRONMENTS

Leaf sizes were large throughout the early Paleogene in southeastern Australian macrofloras, but the percentage of dicot species with entire leaf margins was highest in the middle Eocene (Greenwood and Christophel, 2003; Table 3). This apparent discrepancy may reflect the presence of deciduous elements in the late Paleocene and early Eocene floras, and the lack of these elements in middle Eocene forests, as the dominance of species with toothed margins and broad lamina is characteristic of modern deciduous broad-leaved forest (Basinger et al., 1994; Greenwood, 2001). The magnitude of variation for all paleo-

TABLE 3. FOLIAR PHYSIOGNOMIC DATA AND CLIMATE ESTIMATES FOR SOUTHEASTERN AUSTRALIAN EARLY PALEOGENE FLORAS

				MAT (°C)	
Macroflora	Number of dicot spp. (*r*)	% spp. no teeth (*P*)	Mean length (*ML*, mm)	LMA (σ)	Length
1. Cambalong Ck (L Bungarby)*	21	58	No data	13.5 (2.9)	No data
	No data	No data	(93)	No data	(20.8)
2. Deans Marsh	34†	28	67	5.5 (2.9)	15.6
3. Brandy Ck	54§	75	No data	18.2 (2.2)	No data
4. Hotham Hts	26	74	78	17.9 (2.3)	17.8
5. Nerriga	24	79	80	19.2 (2.3)	18.2
6. Golden Grove	21	71	77	17.1 (2.7)	17.6
7. Anglesea	28	65	58	15.5 (2.4)	13.9

Note: MAT-LMA (mean annual temperature—leaf margin analysis), and MAT-length univariate regression are derived from modern Australian mesic vegetation (Greenwood, 1992; Greenwood et al. in prep.; equations 1 and 2). Measurements have been rounded here (but not in calculations) to the nearest whole number. Numbers in the first column refer to Figure 4.

* Data for both Lake Bungarby (leaf size, from Hill, 1992) and Cambalong Creek (LMA) were combined, as insufficient leaves were available to calculate mean length for Cambalong Creek.

† Only 18 of the 34 dicot taxa had a known margin type.

§ Only 28 of the 54 dicot taxa had a known margin type.

climate estimates between sites is within the measure of uncertainty for these estimates for a majority of sites, however climatic change during the early Paleogene is indicated. Estimates of terrestrial mean annual temperature (MAT) from foliar physiognomy and bioclimatic analysis over this interval generally track the northeastern Australian marine SST record (Feary et al., 1991; Macphail et al., 1994), with highest MAT values estimated for the middle and early Eocene, but not the regional high-latitude SST record (Fig. 4). The paleobotanical analyses indicate moist (i.e., MAP >150 cm/yr), mesothermal climates (i.e., MAT 16–22 °C, CQMT >10 °C) over the interval, but one line of physiognomic evidence points to a potential cooling episode in the mid-early Eocene. The cooling episode is not detected in our bioclimatic analysis. Uncertainty over the physiognomic MAT estimate for Deans Marsh prevent definitive statements on this putative cooling episode. The general low level of temperature change estimated for southeastern Australia during the early Paleogene appears to contradict the progressive cooling shown in the regional marine isotopic record (Fig. 4). The apparent mild cooling (2–3 °C) from the CGCO to the late middle Eocene in the terrestrial paleobotanical record versus the 6–8 °C drop in SST (Fig. 4), likely is a consequence of continental movement maintaining southeastern Australia in a mesothermal latitudinal zone (e.g., Feary et al., 1991; Truswell, 1993).

The magnitude of climate warmth in the late Paleocene and latest early Eocene, however may be "masked" by the paleoelevation of Cambalong Creek, Hotham Heights and Brandy Creek. The middle Eocene sites are at low altitudes (<200 m to ~550 m; Hill, 1982; Greenwood and Wing, 1995). Current elevation for the Paleocene site is ~800 m and the early Eocene sites are above 1400 m. Geomorphic evidence suggests that both areas were ~800 m above sea level during the Paleogene (Taylor, 1994; Orr, 1999). If these floras were deposited from local vegetation growing at ~800 m paleoelevation, then lowland sites at the same latitude would have experienced higher MAT. Applying the global average lapse rate (~0.59 °C/100 m; Meyer, 1992), 800 m elevation equates to MAT 4–5 °C higher than the estimates, namely 22–23 °C ±2 °C, and implies mesothermal to megathermal thermal regimes in southeastern Australia in the late Paleocene and latest early Eocene (e.g., Taylor et al., 1990; Macphail et al., 1994). The paleo-proxy estimates of MAT >20 °C for southeastern Australia is much higher than computer climate model output for southeastern Australia, which indicate early Eocene MAT <10 °C (Sewall et al., 2000; Shellito et al., 2003).

MAT estimates >20 °C are consistent with the observation by Macphail et al. (1994) that the early Eocene was the acme in development of lowland megathermal species-rich rainforest in southeastern Australia. Macphail et al. (1994) also suggested that mesothermal species-rich rainforests reached their maximum development in southeastern Australia during the middle-late Eocene (Figs. 1 and 2). Our paleoclimate analysis quantifies this hypothesis at the local scale, indicating slightly warmer conditions in the late-early Eocene than the middle Eocene. Macphail et al. (1994) stressed the continuity of many taxa throughout the Eocene, but noted a floristic shift in response to the establishment of mesothermal conditions by the middle Eocene expressed in part by a locally variable but very widespread expansion of *Nothofagus* subgenus *Brassospora* in microfloras (Fig. 3). Mesotherm-megatherm forests likely occupied parts of this landscape at periods of peak warmth, such as during the CGCO of the early Eocene.

These Paleogene forests contained many taxa closely related to extant taxa found in mesotherm-megatherm rainforests in northeastern Australia's "wet tropics" (Fig. 1). Other Paleogene taxa are not present in Australia but extant in the tropical rainforests of nearby landmasses (Vadala and Greenwood, 2001;

Greenwood and Christophel, 2003). Other taxa appear to persist today only in microtherm-mesotherm forests and other vegetation types. Present-day patterns reflect "sifting" of a regional biota during the vicissitudes of Cenozoic climate change, and particularly the climate oscillations of the Quaternary. What is apparent from the southeastern Australian record, is that the low level of temperature variation was reflected in the continuity of floristic composition in local vegetation, but that a transition from mesotherm-megatherm climates to mesotherm climates at the early–middle Eocene boundary in that region was a driving force for community reorganization (Fig. 3). Many plant taxa originated, or first migrated into the area during the interval (Truswell, 1993; Macphail et al., 1994; Vadala and Greenwood, 2001; Greenwood and Christophel, 2003). However, the greatest shifts in community composition during the early Paleogene are the replacement of mesotherm-megatherm taxa by mesotherm taxa, such as the rise to dominance of *Nothofagus* subgenus *Brassospora* in microfloras (Fig. 3).

ACKNOWLEDGMENTS

This work was funded by grants from the Australian Research Council to DRG and PTM (A39802019 and small grants scheme). RLK is in receipt of an APRA postgraduate scholarship. NSF and the conference organizers kindly provided financial assistance to RLK and DRG to participate in the Wyoming meeting. Analysis and interpretation of spore-pollen based stratigraphy was greatly facilitated at various stages by analyses and discussions with Alan Partridge (Biostrata Pty. Ltd.). We would also like to thank Andrew Drinnan, John Webb, Peter Kershaw and Stephen McLoughlin, for advice and practical assistance, and Mike Pole and Scott Wing for their valuable constructive criticism and advice on an earlier version of the paper. Permits to work in the Alpine National Park were provided by DNRE (Victoria). Larry Doyle and the staff of the Mount Hotham Management Authority kindly gave permission for access and provided material support.

REFERENCES CITED

Alley, N.F., 1987, Middle Eocene age of the megafossil flora at Golden Grove, South Australia: Preliminary report, and comparison with the Maslin Bay Flora: Transactions of the Royal Society of South Australia, v. 111, no. 3, p. 211–212.

Banks, M., 1999, The early Eocene macroflora of Hotham Heights [undergraduate thesis]: Melbourne, Victoria University of Technology, 209 p.

Basinger, J.F., Greenwood, D.R., and Sweda, T., 1994, Early Tertiary vegetation of Arctic Canada and its relevance to paleoclimatic interpretation, *in* Boulter, M.C., and Fisher, H.C., eds., Cenozoic plants and climates of the arctic: NATO ASI Series, v. 27 I: Berlin, Springer-Verlag, p. 176–198.

Berggren, W.A., Kent, D.V., Flynn, J.J., and Van Couvering, J.A., 1985, Cenozoic geochronology: Geological Society of America Bulletin, v. 96, p. 1407–1418.

Berggren, W.A., Kent, D.V., Swisher, C.C., III, and Aubry, M.-P., 1995, A revised Cenozoic geochronology and chronostratigraphy, *in* Berggren, W.A., et al., eds., Geochronology, time scales, and global stratigraphic correlation: SEPM (Society for Sedimentary Geology) Special Publication 54, p. 129–202.

Blackburn, D.T., and Sluiter, I.R.K., 1994, The Oligo-Miocene coal floras of southeastern Australia, *in* Hill, R.S., ed., History of the Australian vegetation: Cretaceous to Holocene: Cambridge, UK, Cambridge University Press, p. 328–367.

Burnham, R.J., Wing, S.L., and Parker, G.G., 1992, The reflection of deciduous forest communities in leaf litter: Implications for autochthonous litter assemblages from the fossil record: Paleobiology, v. 18, no. 1, p. 30–49.

Busby, J.R., 1991, BIOCLIM: A bioclimatic analysis and prediction system, *in* Margules, C.R., and Austin, M.P., eds., Nature conservation: Cost effective biological surveys and data analysis: Melbourne, CSIRO Publications, p. 64–68.

Carpenter, R.A., Hill, R.S., and Jordan, G., 1994, Cenozoic vegetation in Tasmania: Macrofossil evidence, *in* Hill, R.S., ed., History of the Australian vegetation: Cretaceous to Holocene: Cambridge, UK, Cambridge University Press, p. 276–298.

Chambers, T.C., Drinnan, A.N., and McLoughlin, S., 1998, Some morphological features of Wollemi Pine (*Wollemia nobilis:* Araucariaceae) and their comparison to Cretaceous plant fossils: International Journal of Plant Science, v. 159, no. 1, p. 160–171.

Chaproniere, G., Shafik, S., Truswell, E.M., MacPhail, M.K., and Partridge, A., 1996, Cainozoic (Chart 12), *in* Young, G.C., and Laurie, J.R., eds., An Australian Phanerozoic timescale: Melbourne, Oxford University Press, p. 175–186.

Christophel, D.C., and Greenwood, D.R., 1987, A megafossil flora from the Eocene of Golden Grove, South Australia: Transactions of the Royal Society of South Australia, v. 111, no. 3, p. 155–162.

Christophel, D.C., and Greenwood, D.R., 1989, Changes in climate and vegetation in Australia during the Tertiary: Review of Palaeobotany and Palynology, v. 58, p. 95–109.

Christophel, D.C., Harris, W.K., and Syber, A.K., 1987, The Eocene flora of the Anglesea locality, Victoria: Alcheringa, v. 11, p. 303–323.

Clyde, W.C., Sheldon, N.D., Koch, P.L., Gunnell, G.F., and Bartels, W.S., 2001, Linking the Wasatchian–Bridgerian boundary to the Cenozoic Global Climatic Optimum: New magnetostratigraphic and isotopic results from South Pass, Wyoming: Palaeogeography, Palaeoclimatology, Palaeoecology, v. 167, p. 175–199.

Feary, D.A., Davies, P.J., Pigram, C.J., and Symonds, P.A., 1991, Climatic evolution and control on carbonate deposition in northeast Australia: Palaeogeography, Palaeoclimatology, Palaeoecology (Global and Planetary Change Section), v. 89, p. 341–361.

Greenwood, D.R., 1991, The taphonomy of plant macrofossils, *in* Donovan, S.K., ed., The processes of fossilization: London, Belhaven Press, p. 141–169.

Greenwood, D.R., 1992, Taphonomic constraints on foliar physiognomic interpretations of Late Cretaceous and Tertiary palaeoclimates: Review of Palaeobotany and Palynology, v. 71, p. 142–96.

Greenwood, D.R., 1994, Palaeobotanical evidence for Tertiary climates, *in* Hill, R.S., ed., History of the Australian vegetation: Cretaceous to Holocene: Cambridge, UK, Cambridge University Press, p. 44–59.

Greenwood, D.R., 2001, Climate: Wood and leaves, *in* Briggs, D.E., and Crowther, P.R., eds., Palaeobiology II: London, Blackwell Scientific Publishers, p. 484–487.

Greenwood, D.R., and Basinger, J.F., 1994, The paleoecology of high-latitude Eocene swamp forests from Axel Heiberg Island, Canadian High Arctic: Review of Palaeobotany and Palynology, v. 81, p. 83–97.

Greenwood, D.R., and Christophel, D.C., 2003, Origins and Tertiary history of Australian "tropical" rainforests, *in* Moritz, C., et al., eds., Tropical rainforest diversity: Past and future: Chicago, Illinois, University of Chicago Press, in press.

Greenwood, D.R., and Wing, S.L., 1995, Eocene continental climates and latitudinal gradients: Geology, v. 23, p. 1044–1048.

Greenwood, D., Puchalski, B., and Baldyga, D., 1999, Leaf physiognomy and climate relationships in Australian mesic forests: International Botanical Congress, 14th, St. Louis, Missouri, Abstracts, p. 538.

Greenwood, D.R., Vadala, A.J., and Banks, M., 2000a, Climates and changes in forest floristics during the Paleocene and Eocene in southeastern Australia: GFF, v. 122, p. 65–66.

Greenwood, D.R., Vadala, A.J., and Douglas, J.G., 2000b, Victorian Paleogene and Neogene macrofloras: A conspectus: Proceedings of the Royal Society of Victoria, v. 112, p. 65–92.

Hardenbol, J., Thierry, J., Farley, M.B., Jacquin, T., de Graciansky, P.-C., and Vail, P.R., 1998, Mesozoic and Cenozoic sequence chronostratigraphic framework of European basins, *in* de Graciansky, P-C., et al., eds., Mesozoic and Cenozoic sequence stratigraphy of European basins: SEPM (Society for Sedimentary Geology) Special Publication 60, p. 3–14.

Harris, W.K., 1971, Tertiary stratigraphic palynology, Otway Basin, *in* Wopfner, H., and Douglas, J.G., eds., The Otway Basin of southeastern Australia: Melbourne, Special Bulletin of the Geological Surveys of South Australia and Victoria, p. 67–87.

Hickey, L.J., 1973, Classification of the architecture of dicotyledonous leaves: American Journal of Botany, v. 60, no. 1, p. 17–33.

Hill, R.S., 1982, The Eocene megafossil flora of Nerriga, New South Wales, Australia: Palaeontographica Abt. B, v. 181, p. 44–77.

Hill, R.S., 1992, Australian vegetation during the Tertiary: Macrofossil evidence: The Beagle, Records of the Northern Territory Museum of Arts and Sciences, v. 9, p. 1–10.

Houlder, D., Hutchinson, M., Nix, H., and McMahon, J., 1999, ANUCLIM version 5.0 user guide: Canberra, Centre for Resource and Environmental Studies, Australian National University, 73 p (paper copy) and CD-ROM.

Keefe, R.L., 2000, Windows on an ancient forest: The Palaeoecology of the early Eocene flora of Brandy Creek Mine, Eastern Highlands, Victoria [undergraduate thesis]: Melbourne, Victoria University of Technology, 98 p.

Kershaw, A.P., 1996, A bioclimatic analysis of Early to middle Miocene brown coal floras, Latrobe Valley, southeastern Australia: Australian Journal of Botany, v. 45, no. 3, p. 373–383.

Kershaw, A.P., and Bulman, D., 1994, The relationship between modern pollen samples and environment in the humid tropics region of northeastern Australia: Review of Palaeobotany and Palynology, v. 83, p. 83–96.

Kershaw, A.P., and Nix, H.A., 1988, Quantitative palaeoclimatic estimates from pollen data using bioclimatic profiles of extant taxa: Journal of Biogeography, v. 15, p. 589–602.

Kershaw, A.P., and Strickland, K.M., 1990, A 10 year pollen trapping record from rainforest in northeastern Queensland, Australia: Review of Palaeobotany and Palynology, v. 64, p. 281–288.

Lindsay, J.M., and Alley, N.F., 1995, St Vincent Basin, *in* Drexel, J.H., and Preiss, W.V., eds., The geology of South Australia, volume 2: The Phanerozoic: Adelaide, South Australian Geological Survey Bulletin 54, p. 208–217.

Macphail, M.K., 1980, Fossil and modern pollen of *Beilschmiedia* (Lauraceae) in New Zealand: New Zealand Journal of Botany, v. 18, p. 149–152.

Macphail, M.K., 1997, Late Neogene climates in Australia: Fossil pollen- and spore-based estimates in retrospect and prospect: Australian Journal of Botany, v. 45, p. 425–464.

Macphail, M.K., Alley, N.F., Truswell, E.M., and Sluiter, I.R.K., 1994, Early Tertiary vegetation: Evidence from spores and pollen, *in* Hill, R.S., ed., History of the Australian vegetation: Cretaceous to Holocene: Cambridge, UK, Cambridge University Press, p. 189–261.

Meyer, H.W., 1992, Lapse rates and other variables applied to estimating paleoaltitudes from fossil floras: Palaeogeography, Palaeoclimatology, Palaeoecology, v. 99, p. 71–99.

Moss, P.T., and Kershaw, A.P., 2000, The last glacial cycle from the humid tropics of northeastern Australia: Comparison of a terrestrial and a marine record: Palaeogeography, Palaeoclimatology, Palaeoecology, v. 155, p. 155–176.

Orr, M.L., 1999, Tectonic geomorphology of the Bogong and Dargo High Plains region, east Victorian highlands, Australia [Ph.D. thesis]: Melbourne, University of Melbourne, 155 p.

Partridge, A.D., 1998a, Miscellaneous palynological samples from macroplant fossil localities in Australia: Biostrata Report 1998/8, Biostrata Pty. Ltd., Melbourne, 19 p.

Partridge, A.D., 1998b, Palynological analysis of macroplant fossil locality, Hotham Heights, Victoria: Biostrata Report 1998/12, Biostrata Pty. Ltd., Melbourne, 3 p.

Partridge, A.D., 1999, Late Cretaceous to Tertiary geological evolution of the Gippsland Basin, Victoria [Ph.D. thesis]: Melbourne, LaTrobe University, 439 p.

Röhl, U., Bralower, T.J., Norris, R.D., and Wefer, G., 2000, New chronology for the late Paleocene thermal maximum and its environmental implications: Geology, v. 28, p. 927–930.

Rowett, A.I., and Christophel, D.C., 1990, The dispersed cuticle profile of the Eocene Anglesea clay lenses, *in* Douglas, J.G., and Christophel, D.C., eds., Proceedings of the 3rd IOP Conference, Melbourne, July 1988: Melbourne, A-Z Printers and International Organisation of Palaeobotanists, p. 115–21.

Rowett, A.I., and Sparrow, A.D., 1994, Multivariate analysis of Australian Eocene dispersed cuticle floras: Influence of age, geography, and taphonomy on biozonation: Review of Palaeobotany and Palynology, v. 81, p. 165–183.

Royer, D.L., Wing, S.L., Beerling, D.J., Jolley, D.W., Koch, P.L., Hickey, L.J., and Berner, R.A., 2001, Paleobotanical evidence for near present-day levels of atmospheric CO_2 during part of the Tertiary: Science, v. 292, p. 2310–2313.

Sewall, J.O., Sloan, L.C., Huber, M., and Wing, S., 2000, Climate sensitivity to changes in land surface characteristics: Global and Planetary Change, v. 26, p. 445–465.

Shackleton, N.J., and Kennett, J.P., 1975, Paleotemperature history of the Cenozoic and the initiation of Antarctic glaciation: Oxygen and carbon isotope analyses in DSDP sites 277, 279, and 281, *in* Kennett, J.P., et al., eds., Initial reports of the Deep Sea Drilling Project: Washington, D.C., U.S. Government Printing Office, v. 29, p. 743–755.

Shellito, C.J., Sloan, L.C., and Huber, M., 2003, Climate model sensitivity to atmospheric CO_2 levels in the early–middle Paleogene: Palaeogeography, Palaeoclimatology, Palaeoecology, v. 193, p. 113–123.

Stover, L.E., and Evans, P.R., 1973, Upper Cretaceous–Eocene spore-pollen zonation, offshore Gippsland Basin, Australia, *in* Glover, J.E., and Playford, G., eds., Mesozoic and Cainozoic palynology: Essays in honour of Isabel Cookson: Geological Society of Australia Special Publication 4, p. 55–72.

Stover, L.E., and Partridge, A.D., 1973, Tertiary and Late Cretaceous spores and pollen from the Gippsland Basin, southeastern Australia: Proceedings of the Royal Society of Victoria, v. 85, p. 237–286.

Taylor, G., 1994, Landscapes of Australia: Their nature and evolution, *in* Hill, R.S., ed., History of the Australian vegetation: Cretaceous to Holocene: Cambridge, UK, Cambridge University Press, p. 60–79.

Taylor, G., Truswell, E.M., McQueen, K.G., and Brown, M.C., 1990, Early Tertiary palaeogeography, landform evolution, and paleoclimates of the Southern Monaro, N.S.W., Australia: Palaeogeography, Palaeoclimatology, Palaeoecology, v. 78, p. 109–134.

Truswell, E.M., 1993, Vegetation changes in the Australian Tertiary in response to climatic and phytogeographic forcing factors: Australian Systematic Botany, v. 6, no. 5, p. 533–558.

Truswell, E.M., 1997, Palynomorph assemblages from marine Eocene sediments on the west Tasmanian continental margin and the South Tasman Rise: Australian Journal of Earth Science, v. 44, p. 633–654.

Vadala, A.J., and Drinnan, A.N., 1998, Elaborating the fossil history of Banksiinae: A new species of *Banksieaephyllum* (Proteaceae) from the late Paleocene of New South Wales: Australian Systematic Botany, v. 11, p. 439–463.

Vadala, A.J., and Greenwood, D.R., 2001, Australian Paleogene vegetation and environments: Evidence for palaeo-Gondwanic elements in the fossil records of Lauraceae and Proteaceae, *in* Metcalfe, I., et al., eds., Faunal and floral migrations and evolution in southeast Asia-Australsia: Lisse, Netherlands, Swets and Zeitlinger, p. 196–221.

Veevers, J.J., Powell, C.McA., and Roots, S.R., 1991, Review of seafloor spreading around Australia, 1: Synthesis of patterns of spreading: Australian Journal of Earth Sciences, v. 38, p. 373–389.

Wilf, P., 1997, When are leaves good thermometers? A new case for leaf margin analysis: Paleobiology, v. 23, p. 373–390.

Wing, S.L., 1998, Late Paleocene–early Eocene floral and climatic change in the Bighorn Basin, Wyoming, *in* Aubry, M.-P., et al., eds., Late Paleocene–early Eocene climatic and biotic events in the marine and terrestrial records: New York, Columbia University Press, p. 380–400.

Wing, S.L., and Greenwood, D.R., 1993, Fossils and fossil climates: The case for equable Eocene continental interiors: Philosophical Transactions of the Royal Society of London, series B, v. 341, p. 243–252.

Wing, S.L., and Harrington, G.J., 2001, Floral response to rapid warming in the earliest Eocene and implications for concurrent faunal change: Paleobiology, v. 27, p. 539–563.

Wing, S.L., Alroy, J., and Hickey, L.J., 1995, Plant and mammal diversity in the Paleocene to early Eocene of the Bighorn Basin: Palaeogeography, Palaeoclimatology, Palaeoecology, v. 115, p. 117–156.

Wing, S., Bao, H., and Koch, P., 2000, An early Eocene cool period? Evidence for continental cooling during the warmest part of the Cenozoic, *in* Huber, B., et al., eds., Warm climates in Earth history: Cambridge, UK, Cambridge University Press, p. 197–237.

Wolfe, J.A., 1978, A paleobotanical interpretation of Tertiary climates in the Northern Hemisphere: American Science, v. 66, p. 694–703.

Zachos, J.C., Lohmann, K.C., Walker, J.C.G., and Wise, S.W., 1993, Abrupt climate change and transient climates during the Paleogene: A marine perspective: Journal of Geology, v. 101, p. 191–213.

Zachos, J., Stott, L., and Lohmann, K., 1994, Evolution of early Cenozoic marine temperatures: Paleoceanography, v. 9, p. 353–387.

Manuscript Accepted by the Society August 13, 2002

Printed in the USA

Geological Society of America
Special Paper 369
2003

Geographic patterns in the floral response to Paleocene-Eocene warming

Guy J. Harrington*
Department of Geology, National University of Ireland Cork, Cork, Ireland

ABSTRACT

Studies on vegetation change from North America show that floral composition in the early Eocene was very similar to that in the late Paleocene. This interpretation is counterintuitive with expectations that rapid global warming, which marks the Paleocene-Eocene boundary, would have led to intense vegetation turnover. In this review, data on Paleocene-Eocene pollen and spores are taken from published sources from four main regions in North America: (1) The U.S. Gulf Coast (paleolatitude 32 °N) (2) the eastern seaboard of Virginia (paleolatitude 39 °N) (3) the Bighorn and Powder River basins (paleolatitude 47 °N) and (4) the Canadian Arctic (paleolatitude 78 °N). This data set shows that there is no apparent latitudinal pattern to vegetation turnover in the early Eocene, but rather a strong distinction between the continental interior and the coastal areas. No one immigrant taxon characterizes the whole of North America and the coasts have a greater number of Eocene first occurrences than the continental interior. Immigrants can be classified as those that migrated from other continents, of which 40% have European Paleocene records, and those that immigrated from other regions in North America. Intracontinental immigration was exclusively from south to north and was restricted to coastal areas. The consequence of global warming was a complex vegetational response caused by the probable inability of many plants to disperse into North America.

INTRODUCTION

Climatic events of the Paleocene-Eocene boundary have attracted much attention because of the fundamental affect they had on oceanic and some terrestrial biota (Kennett and Stott, 1991; Kelly et al., 1996; Speijer et al., 1997; Clyde and Gingerich, 1998; Crouch et al., 2001; Gingerich, 2001). The Initial Eocene Thermal Maximum (IETM) is central to our understanding of community change because it represented an interval of global warming. The IETM lasted between 85 k.y. and 220 k.y. (Röhl et al., 2000; Bowen et al., 2001) and, at high latitudes, led to an increase in mean annual land temperature of ~3–4 °C (Fricke et al., 1998). This event marks the position of the Paleocene-Eocene boundary and lies totally within the Eocene. With this significant change in mean annual temperature (MAT), concomitantly high plant community change is expected because plants are sensitive to temperature. However, results from regions where fossil plants have been studied in detail show that the early Eocene flora is highly similar to the late Paleocene (Wing, 1998; Rull, 1999; McIver and Basinger, 1999; Wing et al., 2000; Crouch, 2001; Harrington, 2001a; Wing and Harrington, 2001; Crouch and Visscher, this volume). The plant record shows a gradual change from typically Paleocene to typically Eocene floras (Wing, 1998; Rull, 1999). Furthermore, a small set of pollen samples from within the IETM in North America (Wing et al., this volume) and a more extensive record of ter-

*Present address: Department of Paleobiology, Smithsonian Institution, Washington, D.C. 20560-0121, USA.

Harrington, G.J., 2003, Geographic patterns in the floral response to Paleocene-Eocene warming, *in* Wing, S.L., Gingerich, P.D., Schmitz, B., and Thomas, E., eds., Causes and Consequences of Globally Warm Climates in the Early Paleogene: Boulder, Colorado, Geological Society of America Special Paper 369, p. 381–393.

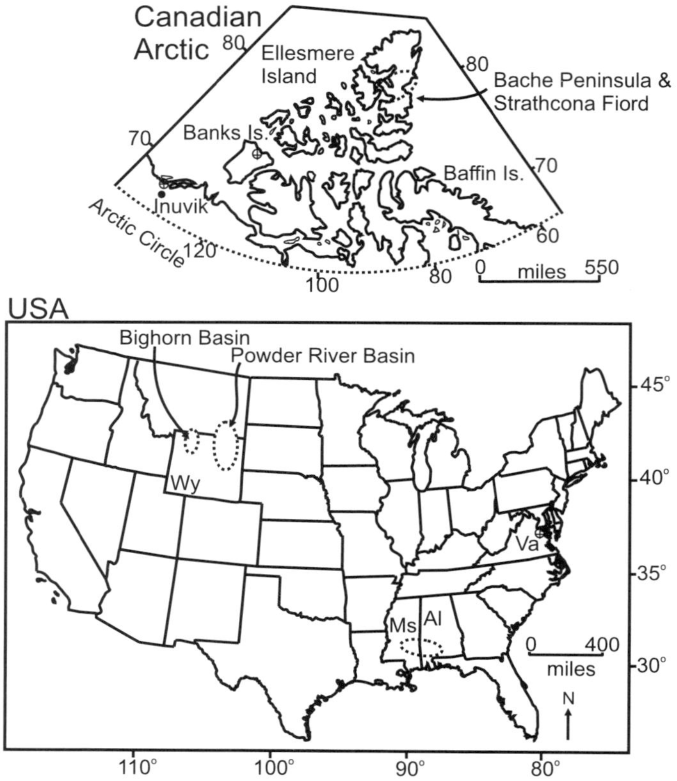

Figure 1. Location map of North America showing studied regions and current latitude. Wy—Wyoming; Ms—Mississippi; Al—Alabama. Pollen data from Ellesmere Island is taken from McIntyre (1991, 1994), Kalkreuth et al. (1996). Data from Bains Island and Inuvik from Doerenkamp et al. (1976). Data from the Bighorn Basin taken from Harrington (1999, 2000, 2001b) and Wing and Harrington (2001). Powder River Basin data from Tschudy (1976), Pocknall (1986, 1987a, 1987b), Pocknall and Nichols (1996), and Wing et al. (this volume). Virginia data from Frederiksen (1979, 1998), and Frederiksen in Gibson et al. (1980). Pollen data from the eastern U.S. Gulf Coast from Elsik (1968, 1974), Tschudy (1973), Frederiksen (1980a, 1980b, 1991, 1998), Harrington and Kemp (2001), and Harrington (2001a).

restrial vegetation from the IETM from New Zealand (Crouch, 2001; Crouch and Visscher, this volume) document little vegetation change during the IETM.

In North America, the early Eocene flora had some new taxa that were not present in Paleocene strata but the distribution of first occurrences is unknown in a latitudinal context. The best continent to undertake any study on the pattern of early Eocene floral occurrences is North America because the landmass has geologic sections that span the late Paleocene–early Eocene from different latitudes and regions that have good records of pollen and spores. Hence, in this paper Eocene first occurrences are reviewed from four main areas (Figs. 1 and 2): (1) The U.S. Gulf Coast (paleolatitude 32 °N) (2) the Bighorn and Powder River basins (paleolatitude 47 °N) (3) the eastern seaboard of Virginia (paleolatitude 39 °N) and (4) the Canadian Arctic (paleolatitude 78 °N). These areas represent a range of paleo–vegetation types from the paratropical forests of the U.S. Gulf Coast to the Arctic deciduous forests of Ellesmere Island and provide a sound framework for studying the potential latitudinal and regional components of floral immigration across the Paleocene-Eocene boundary.

CONSIDERATIONS

Chronology

Currently there is no Paleocene-Eocene boundary stratotype (Berggren and Aubry, 1998; Aubry et al., 1999) but the negative carbon isotope excursion in mid to upper NP9, which marks the base of the IETM, is used widely because it is found in both marine and terrestrial carbon reservoirs (Koch et al., 1992; Bowen et al., 2001). Unfortunately, sediments preserving sporomorphs tend to be barren of calcareous fossils or vertebrate remains that can be used for stable isotope analysis and independent age determination. In North America the best pollen records lie in deltaic sediments that provide high-quality pollen data but are difficult to correlate with marine faunal or phytoplanktonic biostratigraphic zones. These sections include the U.S. Gulf Coast, the Atlantic seaboard (e.g., Virginia), and the Canadian Arctic. In many cases it is not possible to look directly

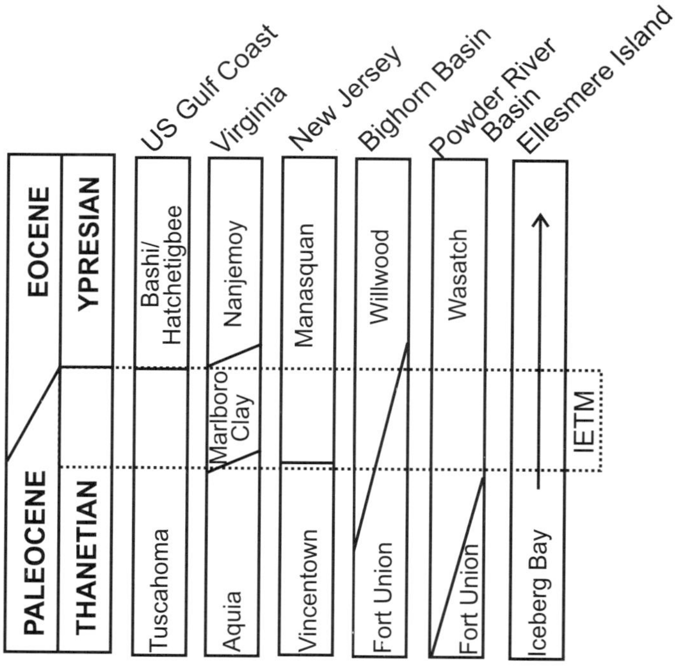

Figure 2. Schematic comparison of formations discussed in this review from the various regions. In many cases, the actual position of the Initial Eocene Thermal Maximum (IETM) is unknown but all sections are unequivocally late Paleocene–early Eocene.

at change across the IETM sensu stricto because of correlation problems. Hence, in the following discussion, I have incorporated only regions with pollen records that span unequivocally the late Paleocene to early Eocene. This has been ascertained either by direct isotope analysis, or through the presence of other fossil groups that have been correlated directly with the isotope excursion. These include vertebrate faunas, calcareous microfossils, and the dinoflagellate *Apectodinium.*

First occurrences of sporomorph taxa can be attributed either to evolutionary change in established plant clades or to immigration into new regions. Any discussion that centers on immigration invariably needs to address the direction of plant movement (Tiffney, 1985a, 1985b). This is fraught with chronostratigraphic correlation problems between host and recipient regions of immigration. Specifically, the east Asian floral record is difficult to date with precision. However, the correlation between strata in North America and those in Europe is more secure because the carbon isotope excursion has been identified in the UK and in France. In the UK, the –2‰ to –3‰ $\delta^{13}C$ excursion falls in the lower Reading Beds of the Lambeth Group (reported by Aubry et al., 1999, Figure 4, and Beerling and Jolley, 1998) which correlates lithostratigraphically with the "Sparnacian" facies in the Paris Basin (Ali and Jolley, 1996; Berggren and Aubry, 1998; Aubry et al., 1999). Some localities in the Paris Basin contain the negative carbon isotope excursion in the lower part of the "Sparnacian" (Stott et al., 1996; Sinha et al., 1996).

Immigration pathways

Early Paleogene biotic interchange between North America and other Holarctic continents has received much attention (e.g., Tschudy, 1975; Tiffney, 1985a, 1985b; McKenna, 1983; Krause and Maas, 1990; Frederiksen, 1993; Beard and Dawson, 1999). Paleogeographic reconstructions indicate that animals and plants could have moved over land bridges in arctic regions during times of climate optima. The major distinction between animal and plant migration is that autotrophs are susceptible to light regimes and migrate at slower rates than heterotrophs. In particular, plants that lack dormancy mechanisms, and evergreens especially, are unsuited to habitation at high latitudes because during the winter months plants need to survive in prolonged darkness (Tiffney, 1985a; Read and Francis, 1992); in the late Paleocene–early Eocene, plants were exclusively deciduous in the Canadian arctic (Hickey et al., 1983; McIver and Basinger, 1999). Climate was not particularly limiting in the arctic because seasonality was low to moderate (Tripati et al., 2001) but rare frosts may have characterized the coldest months in the early Eocene (Tiffney, 1994; McIver and Basinger, 1999).

The Bering land connection was a route that could be used by deciduous taxa only because it was affected by day length and polar night (Tiffney, 1985a). This route is probably significant for the dispersal of many mammals in the earliest Eocene (Beard, 1998; Beard and Dawson, 1999). Other high-latitude land bridges included a Scandinavian-Greenland route that connected to a Greenland-Canadian Arctic bridge, which had the same day length problems as the Bering connection (Tiffney, 1985a). A lower-latitude connection, which potentially avoided the polar night problem, linked Scotland–southern Greenland to Baffin Island in the Canadian Arctic. Evergreen taxa might have exploited this latter land bridge (Tiffney, 1985a, 1985b). Alternatively, seeds of some plants may have been dispersed by ocean currents but large-scale plant migration is associated more strongly with animals and wind rather than water transport (Pitelka et al., 1997). No land bridges existed between South and North America. Thus, little floral similarity existed between northern South America and the U.S. Gulf Coast in the late Paleocene (estimated at 0.7% similarity; Jaramillo and Dilcher, 2001a).

STUDY SECTIONS

U.S. Gulf Coast

Stratigraphy. The exact stratigraphic position of the IETM is unknown on the eastern U.S. Gulf Coast (Gibson et al., 2000; Harrington, 2001a; Harrington and Kemp, 2001). However, biostratigraphic studies show that: (1) The boundary between the Tuscahoma and Bashi/Hatchetigbee formations is usually the boundary between nannoplankton zones NP9 and NP10 (Frederiksen et al., 1982; Gibson and Bybell, 1994); (2) the upper half of the Tuscahoma Formation is contained within part of P5 foram zone of Berggren et al. (1995); (3) the uppermost Tuscahoma Formation contains a fauna that belongs to the Graybullian subdivision of the Wasatchian (early Eocene) land mammal age (Beard et al., 1995); and (4) sporomorph first occurrences are noted in the upper Tuscahoma Formation that include *Celtispollenites, Brosipollis, Interpollis microsupplingensis, Nuxpollenites psilatus,* and *Platycarya* spp. (Harrington, 2003). These pollen taxa were observed previously only from the Bashi/Hatchetigbee formations and distinguish "Eocene" from "Paleocene" sporomorph assemblages (Tschudy, 1973; Elsik, 1974; Frederiksen, 1979, 1980a, 1980b; Frederiksen et al., 1982; Harrington, 2001a).

Palynofloral composition. The pollen record from ancient swamps and estuarine sediments preserves a diverse palynoflora of >200 taxa including families such as Anacardiaceae, Annonaceae, Apocynaceae, Bombacaceae, Burseraceae, Loranthaceae, Nyssaceae, Sparganiaceae, Sterculiaceae, and Symplocaceae. Anemophilous taxa are represented most abundantly by Myricaceae-Betulaceae, Juglandaceae, Taxodiaceae, and Ulmaceae. The composition of pollen assemblages from the U.S. Gulf Coast is well known but these assemblages can include pollen and spores transported from the continental interior. Unfortunately, megafloras are not known in detail in the late Paleocene–early Eocene, although megafloral data from the Mississippi embayment show that between 57 ma and 55 ma, plants grew in a tropical to paratropical climate with a MAT of ~27 °C (Wolfe, 1978; Wolfe and Dilcher, 2000).

First occurrences. First occurrences represent a diverse group of plants (Table 1). The biogeographic origins of many Gulf Coast taxa are unknown, although five taxa are known from the Paleocene of northwest Europe. The probable liana *Dicolpopollis* (Palmae, *Calamus*) is present in Thanetian III of the Paris Basin but the pollen is more abundant in the "Sparnacian" and the later early Eocene (Gruas-Cavagnetto, 1968, 1978). The Normapolles taxa *Basopollis* and *Interpollis microsupplingensis* are present from at least Thanetian II in the Paris Basin (Gruas-Cavagnetto, 1978). *Basopollis* sp. is also recorded in the Thanetian of Belgium at Gelinden-Overbroek (Schumacker-Lambry, 1978) but it appears to be absent from British sections (e.g., Jolley, 1997, 1998). *Interpollis* spp. are documented from the North Sea prior to the *Apectodinium* acme (Schröder, 1992), but the species of *Interpollis* are not identified. *Platycarya* spp. (Fig. 3) are noted from the late Paleocene from Scotland (Jolley, 1997), the North Sea (Schröder, 1992), southern England from at least Chron 26n (Jolley, 1998), Belgium (Schumaker-Lambry, 1978), and the Paris Basin from at least Thanetian II (Gruas-Cavagnetto, 1978). Jolley (1998) recognizes *Platycarya platycaryoides, P. anticyclus* and *P. swasticoidus* from the Thanet Sand-Ormeby Clay formations in southeast England that are all found only in the Eocene of the U.S. Gulf Coast (e.g., Elsik, 1974; Frederiksen, 1998). *Intratriporopollenites instructus*

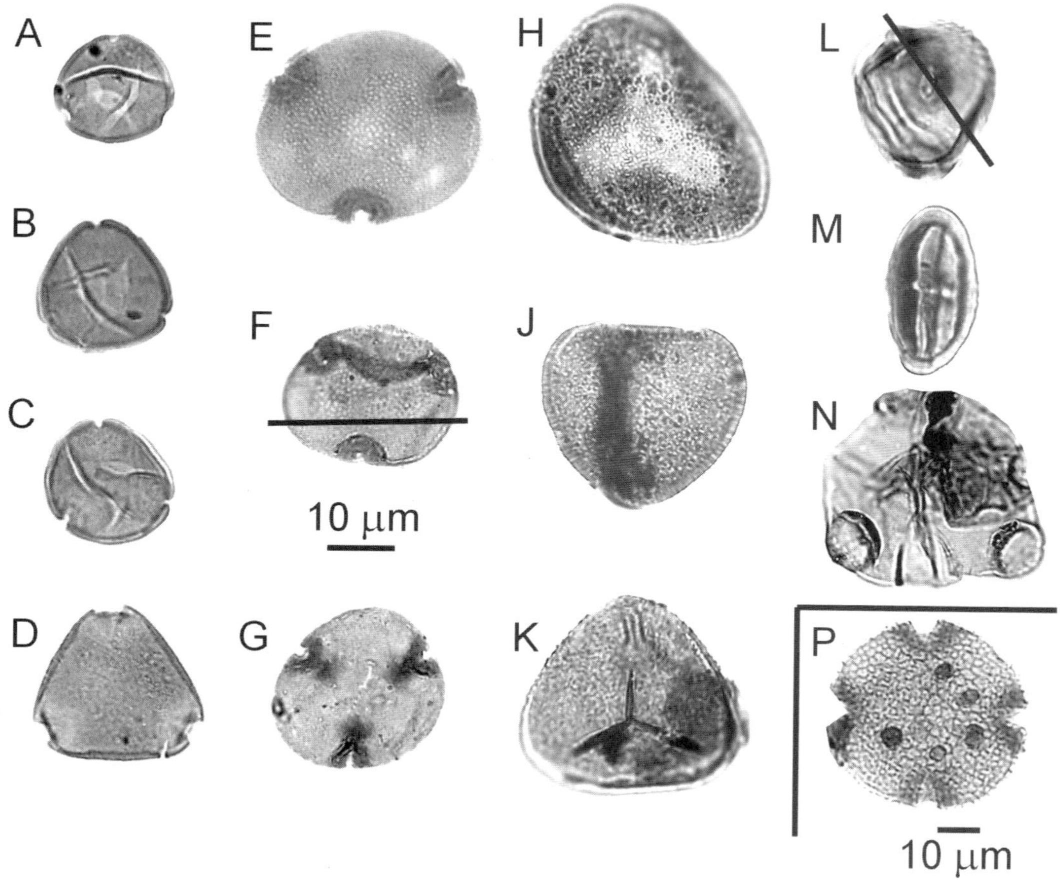

Figure 3. Plate of selected taxa from North America including biostratigraphically significant sporomorphs and those that are currently undescribed. The scale bars represent 10 μm. Notes include the sample number after the species or genus name and the England finder reference taken on Leica microscope LIS 253 housed in the Department of Geology, University College Cork. (A) *Platycarya platycaryoides,* HB15a, O28/center. (B) *Platycarya* sp. similar to *P. anticyclus,* Rh05c b, F34/center. (C) *Platycarya* sp. similar to *P. swasticoidus,* Rh05c a, J35/2. (D) *Symplocos ? contracta,* Wa252.05a, P34/2. (E) *Intratriporopollenites instructus,* PS0004, O34/4, (F) *Intratriporopollenites instructus,* high and low focus, AL620 a, P37/3. (G) *Intratriporopollenites* cf. *instructus,* this specimen is similar to *I. microreticulatus* because the exine is not coarsely reticulated, GH97/1.28a, O21/2. (H) *Punctatisporites* sp. in dorsal view, GH97/1.28c, P27/3. (J) *Punctatisporites* sp. in equatorial view, GH97/1.28a, K40/center. (K) *Granulatisporites* sp., ventral view, Ha35 454.02c, J14/1. (L) *Brosipollis* sp. in high and low focus, Ht/39 265.0a, C22/1. (M) *Nuxpollenites psilatus,* Ht/01a 295.09, B18/1. (N) *Corsinipollenites psilatus,* Ha/05 389.06, C25/4. (P) *Retistephanocolporites* sp., note the different scale bar, RH99–44b, E28/center.

TABLE 1. PALYNOFLORAL LIST OF TAXA NEW TO NORTH AMERICA IN THE EARLY EOCENE

	Origin	U.S. Gulf Coast	Virginia	Continental Interior	Canadian Arctic
SPORES					
LYCOPSID					
Lycopodiaceae					
Punctatisporites sp.	U	—	—	X	—
FERN					
Pteridaceae					
Granulatisporites sp.	U	X	—	—	—
GYMNOSPERM POLLEN					
Picea					
**Tsugeapollenites* sp.†	U	—	—	—	X
MONOCOT POLLEN					
Palmae					
**Dicolpopollis* sp.§(†)	E	X	—	—	—
DICOT POLLEN					
?Anacardiaceae/Simaroubaceae/Sapindaceae					
Ailanthipites berryi	P	P	X	P	—
Aquifoliaceae					
**Ilexpollenites* sp.§	P	P	—	—	X
?Araliaceae					
Bagelopollis verrucatus	P	P	X	—	—
Bombacaceae					
#*Bombacacidites* sp.§**	P	P	P	—	X
#*Bombacacidites naciementoenensis*§**	P	P	X	—	—
Bursuraceae					
Brosipollis spp.§	U	X	—	—	—
Juglandaceae					
#*Caryapollenites* sp. cf. *C. circulus*†	E	—	—	—	X
#*Juglanspollenites*†	U	—	—	—	X
#*Platycarya* spp.†	E	X	X	X	—
#*Polyatriopollenites vermontensis*†	P	P	—	P	X
?Loranthaceae					
Nuxpollenites psilatus§**	E	X	X	—	—
Nyssaceae					
#*Nyssapollenites kruschii*	P	P	—	P	X
?Onagraceae					
Corsinipollenites sp.§**	P	P	—	—	X
Corsinipollenites psilatus§**	SA	X	—	—	X
Sterculiaceae/Bombacaceae					
Intratriporopollenites instructus§**	E	X	—	X	X
Symplocaceae					
**Symplocos ? contracta* group§	U	X	X	—	—
Ulmaceae					
#*Celtispollenites tschudyi*§	C	X	—	—	—
Unknown					
Retistephanocolporites sp.††	U	X	—	—	—
Subtriporopollenites nanus††	P	P	X	—	—
INCERTAE SEDIS					
Normapolles					
Basopollis sp.††	P	—	—	[X]	—
Basopollis atumescens††	E	—	—	[?]	X
Basopollis obscurocostatus††	P	P	X	[?]	—
Interpollis microsupplingensis††	E	X	X	—	—
Kyandopollenites anneratus††	P	P	X	—	—
Thomsonipollis magnificus††	P	P	X	—	—

Note: Taxa prefixed with * are tentative evergreens, with # are deciduous. Those left blank are unknown or the family is not dominated by one particular condition. Dominant pollinator vectors are approximate and represent the dominant type for the family. Origins are E for Europe, SA for South America, C for China, P for already present in North America, and U for unknown. Paleocene presence is marked by P, early Eocene presence by X, uncertainty regarding species identification by [?], and an absence marked by a long dash. Continental interior pollen records combine the Bighorn and Powder River basins. Canadian Arctic pollen records include Inuvik, and Bains and Ellemere Islands combined.

†Anemophilous.

§Entomophilous.

**Vertebrate animal.

††Unknown.

(Fig. 3) also occurs first in the Eocene of the U.S. Gulf Coast (Frederiksen, 1980b; Frederiksen et al., 1982; Harrington, 2001a). *Intratriporopollenites instructus* sensu stricto is not recorded in the Paleocene from Europe, but similar species such as *Intratriporopollenites microreticulatus* are recorded from the late Paleocene (Thanetian) (Schumacker-Lambry, 1978; Gruas-Cavagnetto, 1978; Jolley, 1997, 1998). The distinction between *Intratriporopollenites instructus* and *I. microreticulatus* is small (Fig. 3, specimens F and G) and here they are treated synonymously. Together with *Platycarya,* this taxon has the widest geographic range in North America and probably came from an Holarctic region in the Paleocene because it is not observed in tropical floras of that age in South America (Jaramillo and Dilcher, 2001a, 2001b; Rull, 1999; Muller et al., 1987; C. Jaramillo, 2001, personal commun.). There are no Chinese records of tiliaceous pollen in the Paleocene (Wang et al., 1990), which suggests an European origin in the Paleocene. A South American origin for *Corsinipollenites psilatus* (Fig. 3, N) is probable because this taxon is observed in the Paleocene of Colombia (Jaramillo and Dilcher, 2001a) but this species represents the only known immigrant from South America.

The remaining taxa with a first occurrence in the Eocene (Fig. 3) are of unknown origin. A species of *Granulatisporites* occurs also in the "Sparnacian" of the Paris Basin together with a species of *Brosipollis* (Gruas-Cavagnetto, 1978) but both these are different from species observed on the U.S. Gulf Coast. *Nuxpollenites psilatus* (Fig. 3) is rare on the U.S. Gulf Coast but is more typical of the early Eocene in South Carolina and Virginia (Frederiksen, 1979, 1980a). Schumacker-Lambry (1978, Plate 14, Figure 28) illustrates similar pollen to *N. psilatus* that she refers to as *Tricolporopollenites* sp. 10 and documents it as being rare throughout her Paleocene Belgian sections. It is difficult to determine if this is the same taxon as *N. psilatus* because the features used to distinguish species in tricolporate grains tend to be very fine and require resolution under higher magnification. *Retistephanocolporites* sp. (Fig. 3) also has a first occurrence in the early Eocene on the U.S. Gulf Coast and was erroneously described as *Bombacacidites fereparilis* by Harrington (2001a). No other author has described this taxon either from North America, South America (C. Jaramillo, 2001, personal commun.), or Europe. The only remaining taxon with a potential first occurrence in the Eocene is the *Symplocos ? contracta* group (Fig. 3) of Frederiksen (1988). However, it apparently has one occurrence in the Paleocene Nanafalia Formation (~NP8) (Frederiksen, 1998, Figure 4), although this could be the result either of misidentification or core hole contamination (N. Frederiksen, 2001, personal commun.). A second Paleocene occurrence is reported by Frederiksen (1991) in the Porters Creek Clay of the Midwayan (~NP4) although this too is regarded as misidentification or contamination (N. Frederiksen, 2001, personal commun.). A further Gulf Coast occurrence of this species is reported by Christopher et al. (1980) from poorly time constrained strata in Warm Springs district, Meriwether County, Georgia. The Warm Springs pollen assemblages are consistent with a late Paleocene–early Eocene age. The first consistent occurrence of *Symplocos ? contracta* is in the upper Tuscahoma Formation in both the Wahalak #2 Core from Choctaw County, Alabama, and from the Harrell Core, Lauderdale County, Mississippi (Harrington, 2003). There are no known Paleocene occurrences of the *Symplocos ? contracta* group outside North America.

Virginia

Stratigraphy. The late Paleocene–early Eocene is represented by the upper Aquia Formation and the Marlboro Clay (Fig. 2). Calcareous nannofossils place the upper Aquia in NP9 and the Marlboro Clay in NP9 to lower NP10 (Gibson et al., 1980; Gibson et al., 1993, 2000; Gibson and Bybell, 1994). Abundant kaolinite is frequently found in earliest Eocene strata (Gibson et al., 2000; Bolle and Adatte, 2001) and accounts for 20%–60% of the clay mineral suite in the Marlboro Clay (Gibson et al., 2000). *Apectodinium* occurs in the upper Aquia and throughout the Marlboro Clay (Edwards, 1996). In down basin sections, *Apectodinium augustum* is found (Edwards, 1996) which has a first occurrence in strata in the upper half of the *Apectodinium* acme in European sections (Heilmann-Clausen, 1985; Powell, 1992; Bujak and Brinkhuis, 1998). The Oak Grove core has high kaolinite abundance (Gibson et al., 2000) and several first occurrences of "Eocene" pollen taxa (Frederiksen, *in* Gibson et al., 1980; Frederiksen, 1980a, 1998) including *Platycarya* (Frederiksen, 1980a) that first occurs after the IETM in North America (Harrington, 2000, 2001b; Wing and Harrington, 2001; Wing et al., this volume). The position of the carbon isotope excursion is unknown in the Aquia-Marlboro transition but these units correlate with sections in New Jersey (Gibson and Bybell, 1994; Gibson et al., 2000) that have been analyzed isotopically (Fig. 2). The IETM lies in the topmost Vincentown Formation and extends through to the lowermost Manasquan Formation (Thomas et al., 1997). Hence, pollen changes observed in the Marlboro Clay should correspond to changes across the IETM.

Palynofloral composition. Except for clay mineral interpretations, the paleoclimate of Virginia is unknown. However, the palynoflora is similar to the U.S. Gulf Coast (Frederiksen, 1979) although it lacks some numerically important Gulf Coast taxa such as *Holkopollenites charmendensis, Faguspollenites, Lanagiopollis cribellatus,* and *L. lihokus* (Frederiksen, 1998). Normapolles pollen are well represented but palms are not (Frederiksen, 1979, 1993). In general, the palynofloral composition has considerably greater floristic similarity to the Gulf Coast than to regions in the continental interior.

First occurrences. First occurrences in Virginia can be split into two types: (1) Those that immigrate from other continents and (2) those that migrate from other regions of North America (Table 1). Intercontinental immigrants include plants that produced *Interpollis microsupplingensis, Nuxpollenites psilatus* and *Platycarya* from Europe. *Symplocos ? contracta* also first occurs in the Marlboro Clay. The remaining taxa with first occurrences in the Marlboro Clay expanded their range from the south. *Ailanthipites berryi* is present from at least mid NP9 from the Tuscahoma Formation (Harrington and Kemp, 2001), and occurs in the Clarkforkian (late Paleocene) in the Bighorn Basin (Harrington, 1999; Wing and Harrington, 2001). *Bombacacidites nacimientoensis* and *Bagelopollis verrucatus* both have definite Paleocene occurrences on the U.S. Gulf Coast (Frederiksen, 1991, 1998; Harrington and Kemp, 2001) although *Bagelopollis verrucatus* first occurs in the middle Tuscahoma Formation (Frederiksen, 1988, 1998; Carroll, 1999; Harrington and Kemp, 2001). *Subtriporopollenites nanus* also probably expanded range from the south where it has an upper Midwayan (~NP4–5) occurrence in Mississippi-Alabama (Frederiksen, 1991, 1998). Other Paleocene taxa that make their first Virginia occurrences in the Marlboro Clay include *Thomsonipollis magnificus,* which is present in Alabama from at least NP1, and *Basopollis obscurocostatus* that is present in the Naheola Formation on the U.S. Gulf Coast from ~NP5 (Frederiksen, 1991). *Kyandopollenites anneratus* is observed in lower–mid NP9 on the U.S. Gulf Coast (Frederiksen, 1998; Harrington and Kemp, 2001) before its first occurrence in the Marlboro Clay.

Bighorn Basin

Stratigraphy. The IETM is stratigraphically tightly constrained in the Bighorn Basin and is known from Polecat Bench and the Clarks Fork Basin (Koch et al., 1992, 1995; Fricke et al., 1998; Wing et al., 2000; Bowen et al., 2001) where it correlates with a faunal immigration event (Clyde and Gingerich, 1998; Gingerich, 2001). This event marks the base of the Wasatchian land mammal age (Gingerich, 1989). Unfortunately, pollen and spores are not preserved in the red beds that typify the IETM at Polecat Bench (Harrington, 2001b). Pollen floras are known from further south in the Bighorn Basin from the Antelope/Elk Creek section (Wing, 1984; Farley, 1990; Harrington, 2000; Wing and Harrington, 2001) and from the North Butte area where the carbon isotope excursion has been identified near the base of the Willwood Formation (Koch et al., 1992, 1995). Even in the southern Bighorn Basin, where the lower Willwood Formation preserves plant fossils, the IETM is barren of both megafloras and palynomorphs (Wing, 1998; Wing et al., 2000; Harrington, 2001b; Wing and Harrington, 2001).

Palynofloral composition. Megafloras and palynofloras of the Bighorn Basin have been well studied and represent a subtropical–warm-temperate vegetation type (e.g., Hickey, 1980; Farley, 1990; Wing, 1998; Wing et al., 2000; Harrington, 1999, 2001b; Wing and Harrington, 2001). The pollen assemblages are dominated by Juglandaceae (*Caryapollenites* spp., *Momipites* spp., and *Polyatriopollenites vermontensis*) and Taxodiaceae as well as other predominantly anemophilous families such as Betulaceae and Ulmaceae (Harrington, 2001b). Mean annual temperatures in the latest Paleocene reached ~16–17 °C and returned to this level after the IETM (Wing, 1998; Wing et al., 2000).

First occurrences. The first observed taxon new to the Eocene is *Intratriporopollenites instructus* found one meter above the top of the IETM (Wing and Harrington, 2001) together with *Punctatisporites* sp. (Fig. 3). The biogeographic origin of *Punctatisporites* sp. is completely unknown and it occurs solely in the Bighorn Basin. *Platycarya* spp. (specifically *Platycarya platycaryoides* and *P. anticyclus*) had a delayed entry into the Eocene flora and first occur in low numbers 38 m (>100 k.y.) above the top of the IETM in the Antelope/Elk Creek section. Some of the first occurrences listed in Table 1 of Wing and Harrington (2001), such as *Cyrillaceaepollenites,* and *Bombacacidites,* migrated into the Bighorn Basin in response to the early Eocene climatic optimum. Other taxa with an Eocene first occurrence are rare and were potentially present in the Paleocene. Therefore, only *Platycarya* spp., *Intratriporopollenites instructus,* and *Punctatisporites* can be confidently identified as IETM immigrants and probably there are no intracontinental migrants (Table 1, Figure 4).

Powder River Basin

Stratigraphy. The position of the IETM is only known at a few localities characterized by the Wa0 fauna in the Powder River Basin (Wing et al., this volume). Pollen records are known from the Powder River Basin (e.g., Tschudy, 1976; Pocknall, 1986, 1987a, 1987b) but until recently, the correlation between mammalian faunas and pollen floras was poorly understood. Recent work has located the carbon isotope excursion in the central Powder River Basin that can be traced to a Wa0 fauna and carbonaceous beds favorable for pollen preservation in the lower Wasatch Formation (Wing et al., this volume).

Palynofloral composition. The composition of the palynoflora is almost identical to that of the Bighorn Basin, although Taxodiaceae is more abundant in the Powder River Basin and *Aesculiidites wyomingensis, Nyssapollenites explanata, Retitrescolpites catenatus, Striatopollis tectatus,* and *Talisiipites* spp. are recorded in the Powder River Basin (Pocknall and Nichols, 1996) but they are absent from the Bighorn Basin (Harrington, 1999; appendix 1 of Wing and Harrington, 2001). The subtle differences in palynofloral composition between the Powder River and Bighorn basins probably reflect the generally swampier conditions of the Powder River Basin in the late Paleocene–early Eocene.

First occurrences. The only first occurrences noted in the Powder River Basin are *Intratriporopollenites instructus* and *Platycarya* spp. (Tschudy, 1976; Pocknall, 1986, 1987b). *Platycarya* pollen is first recorded in low abundance (Wing et al., this volume) which is also the case in the Bighorn Basin (Wing and Harrington, 2001). There appear to be no intracontinental immigrants in the early Eocene palynoflora of the Powder River Basin (Table 1) with the possible exception of *Basopollis* sp. which is recorded, but not illustrated, by Tschudy (1976). It is possible that *Basopollis* in the Powder River Basin could either belong to the North American *B. obscurocostatus,* or to *B. atumescens,*

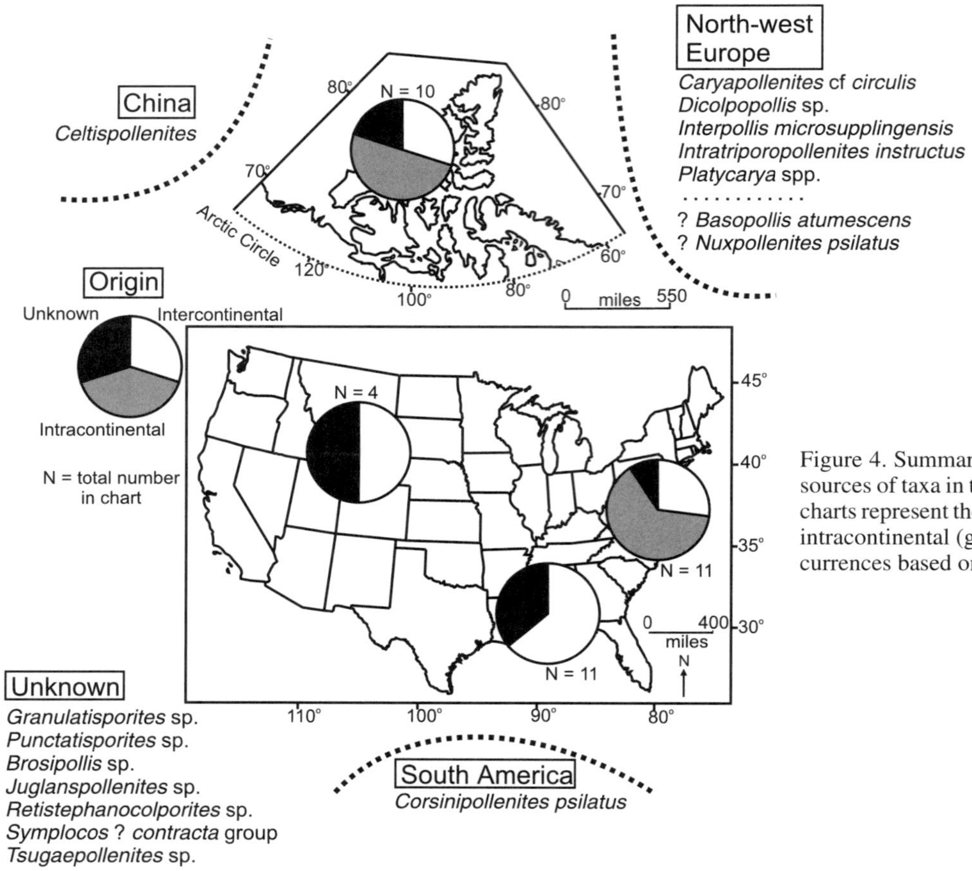

Figure 4. Summary diagram depicting possible immigration sources of taxa in the earliest Eocene into North America. Pie charts represent the proportion of intercontinental (white fill), intracontinental (gray fill), and unknown (black fill) first occurrences based on data presented in Table 1.

which may have immigrated from another Holarctic continent. The European record of *B. atumescens* is sketchy but it is present in the "Sparnacian" of the Paris Basin (Gruas-Cavagnetto, 1968; 1978, Plate 7, Figure 12). Schumacker-Lambry (1978) illustrates a species of *Basopollis* from the Paleocene of Belgium (Schumaker-Lambry, 1978, Plate 11, Figure 11) that resembles Gruas-Cavagnetto's (1968) illustration of *B. atumescens* from the "Sparnacian" of the Paris Basin.

Canadian Arctic

Stratigraphy. The Eureka Sound Group in the Canadian Arctic represents a thick unit of deltaic sediments that span much of the Paleocene to early Eocene. The Paleocene-Eocene transition is present in the Iceberg Bay Formation of Ricketts (1986). On Ellesmere Island, the Iceberg Bay Formation is divided into four main members and there is a variety of terrestrial fossils preserved within them. Member two is Paleocene because it contains the freshwater unionid pelecypod *Plesielliptio priscus* (Kalkreuth et al., 1996). The Margaret Formation (terminology of Miall, 1986 = Coal member, or upper division of the Iceberg Bay Formation) contains a Graybullian (Wasatchian, early Eocene) fauna (West et al., 1977; McKenna, 1980; Tauxe and Clark, 1987; Lucas, 1998). Sediments from both Banks Island and near Inuvik on the northern Canadian mainland (Fig. 1) contain some dinoflagellates including *Apectodinium* (Doerenkamp et al., 1976) which at high latitudes first occurs with the IETM (Bujak and Brinkhuis, 1998; Crouch, 2001; Crouch et al., 2001). On Inuvik and Banks Island, the presence of *Apectodinium* immediately precedes an early Eocene palynoflora (Doerenkamp et al., 1976).

Palynofloral composition. The pollen and spore composition of the Canadian Arctic is similar in some ways to the continental interior basins of the Rocky Mountains. Juglandaceae and Taxodiaceae are abundant, as are Betulaceae, Buxaceae, and Ulmaceae (McIntyre, 1991; Kalkreuth et al., 1993, 1996). Late Paleocene–early Eocene megafloras show the presence of Taxodiaceae, *Ginko,* Ulmaceae, Cercidiphylloid species, and *Nordenskioldia borealis* (McIver and Basinger, 1999). The climate in the early Eocene was temperate (Tiffney, 1994; McIver and Basinger, 1999).

First occurrences. Even at high latitudes Paleocene taxa dominate early Eocene floras (McIntyre, 1991; McIver and Basinger, 1999). First occurrences are more numerous, however, in the palynoflora of the Canadian high Arctic than in the continental interior basins (Fig. 4), although the same taxa are not observed in both regions (Table 1). *Tsugaepollenites* first occurs in the Canadian Arctic in the early Eocene (Doerenkamp et al.,

1976; McIntyre, 1991, 1994) although its biogeographic origin is unknown. Other taxa present in the early Eocene of Ellesmere Island, Bains Island and Inuvik include *Intratriporopollenites* sp. resembling *Intratriporopollenites instructus* (Doerenkamp et al., 1976, Plate 1, Figure 15; McIntyre, 1991, Plate 3, Figure 12), *Basopollis atumescens,* and *Caryapollenites* sp. The Normapolles species *Basopollis atumescens* probably migrated from another Holarctic continent (see above) and the species of *Caryapollenites* that appears in the early Eocene is comparable to *C. circulis,* which is present in the Paleocene and Eocene of Europe (Jolley, 1997; Jolley and Spinner, 1991, Plate 1, Figures 2, 4, 6). Doerenkamp et al. (1976) also illustrate a specimen of *Juglanspollenites* (Doerenkamp et al., 1976, Plate 1, Figure 26) that makes its first occurrence with *Intratriporopollenites* sp. cf. *I. instructus* at Inuvik. There are no other records of *Juglanspollenites* in North America until the middle Eocene (e.g., Frederiksen, 1988; McIntyre, 1991, 1994) when the appearance of the pollen corresponds with the first unambiguous megafossil plant organs of *Juglans* in North America (Manchester, 1987, 1999).

Other taxa probably expanded their ranges from Paleocene populations established already in North America. These include *Bombacacidites* and *Corsinipollenites* sp. (recorded as *Jussiaea champlinensis,* Doerenkamp et al., 1976, Plate 1, Figures 13 and 14), which are present at Inuvik (Doerenkamp et al., 1976) and both have short ranges coinciding with the first occurrence of *Intratriporopollenites* sp. cf. *I. instructus.* The illustrated specimen of *Corsinipollenites* sp. is highly similar to thick-walled species such as *Corsinipollenites oculus-noctis* that are present in the late Paleocene (NP9) on the U.S. Gulf Coast (Harrington and Kemp, 2001). *Ilexpollenites* has a late Paleocene record on the U.S. Gulf Coast (Harrington and Kemp, 2001), and *Nyssapollenites kruschii* is present on the U.S. Gulf Coast and in the Bighorn Basin in the late Paleocene (Harrington, 2001b; Harrington and Kemp, 2001; Wing and Harrington, 2001). *Polyatriopollenites vermontensis* (recorded as *Pterocarya* by Doerenkamp et al., 1976) is present in the continental interior basins, where it is especially abundant (Pocknall, 1986; Pocknall and Nichols, 1996; Wing and Harrington, 2001).

POTENTIAL BIAS IN THE POLLEN AND SPORE RECORD

There are several potential problems that can affect the recognition of floral turnover that are specifically related to pollen taphonomy, depositional environment, collection strategy, and long distance pollen transport. Here, I concentrate on pollination vectors because an important taphonomic problem arises from zoophily (animal-dispersed pollen). Zoophilous plants usually produce less pollen than anemophilous plants (wind-dispersed pollen) and pollen that is produced usually does not disperse far from the parent plant without the appropriate pollinator (Proctor et al., 1996). The zoophilous syndrome is almost universal in high-diversity wet-tropical forests where only 2%–2.5% of taxa are anemophilous (Bawa et al., 1985; Bush, 1995). Subtropical to paratropical floras of the U.S. Gulf Coast should have had the highest proportion of zoophily among Paleogene North American floras. This has been suggested for both the late Paleocene (Harrington and Kemp, 2001; Harrington, 2001a) and the early-middle Eocene floras (Frederiksen, 1988). By contrast, in the Canadian high Arctic, anemophily should have been important because temperate vegetation is predominantly anemophilous (e.g., Whitehead, 1983; Ritchie, 1995). If community change is predominantly restricted to zoophilous plants on the U.S. Gulf Coast, then plant immigration across the IETM may be underestimated. However, the pollen of zoophilous plants can be present in high abundance and high diversity in sedimentary environments conducive for pollen preservation (Colinvaux et al., 1999; Colinvaux and De Oliveira, 2001). Pollen from plants pollinated by small generalist insects and bees, together with plants with large dioecious flowers can distribute pollen in significant quantities without their animal vector (Bush and Rivera, 1998). A review of the probable pollinator vectors of taxa with Eocene first appearances in North America (Table 1) shows that only four taxa (*Tsugaepollenites, Caryapollenites* sp. cf. *C. circulus, Juglanspollenites,* and *Platycarya* spp.) in two families (Picea and Juglandaceae) can be tied strongly to families that are predominantly anemophilous. This excludes free-sporing plants, which presumably had efficacious dispersal mechanisms. The majority of immigrant taxa belong to families that today are predominantly zoophilous, and the remainder (*Basopollis* spp., *Interpollis microsupplingensis,* and *Retistephanocolporites* sp.) are of completely unknown pollinating type. This suggests that mode of pollination does not have an overwhelming impact upon the assessment of turnover at middle latitudes.

PATTERNS OF TURNOVER

This preliminary review of the pollen changes across the Paleocene-Eocene boundary from North America shows that the pattern of change is complex and not clearly delineated on a latitudinal gradient (Fig. 4). Specifically this data set shows:

1. All regions record some intercontinental immigration, with the greatest absolute frequency on the coastal margins (Fig. 4). However, even if high-latitude land bridges were a major route enabling plants to enter North America, the majority of immigrants did not establish themselves in the high Arctic in the early Eocene after the IETM because most immigrants have no record in the Canadian Arctic regions (Table 1). One explanation is that the IETM has not been sampled there, but the presence of the *Apectodinium* acme suggests that some pollen records, such as that near Inuvik (Doerenkamp et al., 1976), do go through the IETM. The only unusual palynofloral elements picked out by Doerenkamp et al. (1976) are *Bombacacidites* and *Corsinipollenites* sp. which both are short-ranging and more typical of lower latitude pollen records (Table 1). The pollen record would imply that the arctic region was not necessarily a

gateway into North America. Hence, sea routes may have been significant for the introduction of some first occurrences into North America. If water transport was important for the dispersal of plants into and around North America, this may explain why there is a relatively high level of intra- and intercontinental immigration in the coastal pollen records and a concomitantly low incidence of first occurences in the continental interior. There is also some geological evidence for oceanic dispersal because during the Paleocene and early Eocene, the limited interchange of taxa between South America, the Caribbean and Central America, and the U.S. Gulf Coast appears to be restricted to coastal plain environments (Graham, 1992; Frederiksen, 1998; Jaramillo and Dilcher, 2001a).

2. The range expansions of taxa already established in North America were also significant in the early Eocene. Pollen records from Virginia show that intracontinental immigration is actually more significant than intercontinental immigration (Fig. 4) because 64% (n = 7) of taxa have Paleocene occurrences elsewhere in North America (Table 1). However, intracontinental immigration across the Paleocene-Eocene boundary is strictly from south to north because the Gulf Coast shows no intracontinental immigration (Fig. 4). If the species of *Basopollis* found in the Powder River Basin belongs to the intercontinental immigrant *B. atumescens,* then intracontinental plant range expansions are totally restricted to coastal regions. However, the continental interior was probably the source area for some of the intracontinental immigrants such as *Polyatriopollenites vermontensis* that migrated north (Table 1). Presumably, plants growing in coastal settings were ecologically unsuited to inhabiting environments typical of the continental interior.

3. If the dominant directions of plant migration trend east to west, as also noted by Frederiksen (1993), they run counter to the main direction of mammal migration. Establishing the direction of mammal immigration into the different Holarctic continents has proved difficult (e.g., Krause and Maas, 1990; Hooker, 1998; Beard, 1998; Beard and Dawson, 1999). However, improvements in the dating and compositional identification of Asian faunas has led to a model of latest Paleocene–earliest Eocene immigration that starts in Asia, goes through North America and into Europe (Beard, 1998; Beard and Dawson, 1999). Some palynomorph taxa with unknown areas of origin (Table 1, Fig. 4) may have come from eastern Asia, but the high latitude and polar night in Beringia may have presented problems for some plant groups. Unfortunately, with the exception of *Dicolpopollis,* we do not know the Paleocene geographic origin of the pollen taxa likely to belong to families that are today predominantly evergreen that have Eocene first occurrences on the U.S. Gulf Coast. It is unlikely that plants migrated from Asia to Europe via Russia and then into North America using lower-latitude land bridges unaffected by polar night regimes because the Turgay Strait in Siberia would have been a formidable obstacle for biotic dispersal from Asia in the IETM (Iakovleva et al., 2001).

ACKNOWLEDGMENTS

I am grateful to Norman Frederiksen and Carlos Jaramillo for correspondence. This paper benefited greatly from constructive reviews by Norman Frederiksen, Fred Rich and Scott Wing. This work is supported by Enterprise Ireland award SC/00/251.

REFERENCES CITED

Ali, J.R., and Jolley, D.W., 1996, Chronostratigraphic framework for the Thanetian and lower Ypresian deposits of southern England, *in* Knox, R.W.O'B., et al., eds., Correlation of the early Paleogene in northwest Europe: Geological Society [London] Special Publication 101, p. 130–144.

Aubry, M.-P., Berggren, W.A., Van Couvering, J.A., and Steininger, F., 1999, Problems in chronostratigraphy: Stages, series, unit and boundary stratotypes, global stratotype section, and point and tarnished golden spikes: Earth Science Reviews, v. 46, p. 99–148.

Bawa, K.S., Bullock, S.H., Perry, D.R., Coville, R.E., and Grayum, M.H., 1985, Reproductive biology of tropical lowland rain forest trees, II: Pollination systems: American Journal of Botany, v. 72, p. 346–356.

Beard, K.C., 1998, East of Eden: Asia as an important center of taxonomic origination in mammalian evolution: Bulletin of the Carnegie Museum of Natural History, v. 34, p. 5–39.

Beard, K.C., and Dawson, M.R., 1999, Intercontinental dispersal of holarctic land mammals near the Paleocene-Eocene boundary: Paleogeographic, paleoclimatic and biostratigraphic implications: Bulletin de la Société Géologique de France, v. 170, p. 697–706.

Beard, K.C., Dawson, M.R., and Tabrum, A.R., 1995, First diverse land mammal fauna from the early Cenozoic of the southeastern United States: The Red Hot local fauna, Lauderdale County, Mississippi: Geological Society of America Abstracts with Programs, v. 27, no. 6, p. A453.

Beerling, D.J., and Jolley, D.W., 1998, Fossil plants record an atmospheric $^{12}CO_2$ and temperature spike across the Paleocene-Eocene transition in northwest Europe: Journal of the Geological Society of London, v. 155, p. 591–594.

Berggren, W.A., and Aubry, M.-P., 1998, The Paleocene-Eocene Epoch/Series boundary: Chronostratigraphic framework and estimated geochronology, *in* Aubry, M.-P., et al., eds., Late Paleocene–early Eocene climatic and biotic events in the marine and terrestrial records: New York, Columbia University Press, p. 18–36.

Berggren, W.A., Kent, D.V., Swisher, C.C., III, and Aubry, M.-P., 1995, A revised Cenozoic geochronology and chronostratigraphy, *in* Berggren, W.A., et al., eds., Geochronology, time scales, and global stratigraphic correlations: SEPM (Society for Sedimentary Geology) Special Publication 54, p. 126–212.

Bolle, M.-P., and Adatte, T., 2001, Paleocene–early Eocene climatic evolution in the Tethyan realm: Clay mineral evidence: Clay Minerals, v. 36, p. 249–261.

Bowen, G.J., Koch, P.L., Gingerich, P.D., Norris, R.D., Bains, S., and Corfield, R.M., 2001, Refined isotope stratigraphy across the continental Paleocene-Eocene boundary on Polecat Bench in the northern Bighorn Basin, *in* Gingerich, P.D., ed., Paleocene-Eocene stratigraphy and biotic change in the Bighorn and Clarks Fork Basins, Wyoming: University of Michigan Papers on Paleontology, v. 33, p. 73–88.

Bujak, J.P., and Brinkhuis, H., 1998, Global warming and dinocyst changes across the Paleocene-Eocene epoch boundary, *in* Aubry, M.-P., et al., eds., Late Paleocene–early Eocene climatic and biotic events in the marine and terrestrial records: New York, Columbia University Press, p. 277–295.

Bush, M.B., 1995, Neotropical plant reproductive strategies and fossil pollen representation: The American Naturalist, v. 145, p. 595–609.

Bush, M.B., and Rivera, R., 1998, Pollen dispersal and representation in a neotropical rain forest: Global Ecology and Biogeography Letters, v. 7, p. 379–392.

Carroll, R.E., 1999, Morphology and distribution of *Bagelopollis verrucatus* (Frederiksen 1988) gen. et. comb. nov. from lower Tertiary coastal plain deposits of the southeastern USA: Palynology, v. 23, p. 55–66.

Christopher, R.A., Prowell, D.C., Reinhardt, J., and Markewich, H.W., 1980, The stratigraphic and structural significance of Paleocene pollen from Warm Springs, Georgia: Palynology, v. 4, p. 105–124.

Clyde, W.C., and Gingerich, P.D., 1998, Mammalian community response to the latest Paleocene thermal maximum: An isotaphonomic study in the northern Bighorn Basin, Wyoming: Geology, v. 26, p. 1011–1014.

Colinvaux, P.A., and De Oliveira, P.E., 2001, Amazon plant diversity and climate through the Cenozoic: Palaeogeography, Palaeoclimatology, Palaeoecology, v. 166, p. 51–63.

Colinvaux, P.A., De Oliveira, P.E., and Moreno Patiño, J.E., 1999, Amazon Pollen Manual and Atlas: Amsterdam, Harwood Academic Publishers, 332 p.

Crouch, E.M., 2001, Environmental change at the time of the Paleocene-Eocene biotic turnover: LPP Foundation, Utrecht, Netherlands, Laboratory of Palaeobotany and Palynology Contributions Series, v. 14, p. 1–216.

Crouch, E.M., Heilmann-Clausen, C., Brinkhuis, H., Morgans, H.E.G., Rogers, K.M., Egger, H., and Schmitz, B., 2001, Global dinoflagellate event associated with the late Paleocene thermal maximum: Geology, v. 29, p. 315–318.

Doerenkamp, A., Jardine, S., and Moreau, P., 1976, Cretaceous and Tertiary palynomorph assemblages from Banks Island and adjacent areas (N.W.T.): Bulletin of Canadian Petroleum Geology, v. 24, p. 372–417.

Edwards, L.E., 1996, Graphic correlation of the Marlboro Clay and Nanjemoy Formation (uppermost Paleocene and lower Eocene) of Virginia and Maryland, *in* Jansonius, J., and McGregor, D.C., eds., Palynology: Principles and applications: American Association of Stratigraphic Palynologists Foundation, v. 3, p. 989–999.

Elsik, W.C., 1968, Palynology of a Paleocene Rockdale Lignite, Milam County, Texas, II: Morphology and taxonomy (end): Pollen et Spores, v. 10, p. 599–664.

Elsik, W.C., 1974, Characteristic Eocene palynomorphs in the Gulf Coast, USA: Palaeontographica Abt. B, v. 149, p. 90–111.

Farley, M.B., 1990, Vegetation distribution across the early Eocene depositional landscape from palynological analysis: Palaeogeography, Palaeoclimatology, Palaeoecology, v. 79, p. 11–27.

Frederiksen, N.O., 1979, Paleogene sporomorph biostratigraphy, northeastern Virginia: Palynology, v. 3, p. 129–167.

Frederiksen, N.O., 1980a, Paleogene sporomorphs from South Carolina and quantitative correlations with the Gulf Coast: Palynology, v. 4, p. 125–179.

Frederiksen, N.O., 1980b, Sporomorphs from the Jackson Group (upper Eocene) and adjacent strata of Mississippi and western Alabama: U.S. Geological Survey Professional Paper 1084, p. 1–75.

Frederiksen, N.O., 1988, Sporomorph biostratigraphy, floral changes, and paleoclimatology, Eocene and earliest Oligocene of the eastern Gulf Coast: U.S. Geological Survey Professional Paper 1448, p. 1–68.

Frederiksen, N.O., 1991, Midway (Paleocene) pollen correlations in the eastern United States: Micropaleontology, v. 37, p. 101–137.

Frederiksen, N.O., 1993, Paleocene and Eocene floral migrations between Europe and North America as indicators of climatic change: SEPM (Society for Sedimentary Geology) Meeting, Stratigraphic Record of Global Change, Pennsylvania State University, Abstract with Program, p. 44.

Frederiksen, N.O., 1998, Upper Paleocene and lowermost Eocene angiosperm pollen biostratigraphy of the eastern Gulf Coast and Virginia: Micropaleontology, v. 44, p. 45–68.

Frederiksen, N.O., Gibson, T.G., and Bybell, L.M., 1982, Paleocene-Eocene boundary in the eastern Gulf Coast: Transactions Gulf Coast Association of Geological Societies, v. 32, p. 289–294.

Fricke, H.C., Clyde, W.C., O'Neil, J.R., and Gingerich, P.D., 1998, Evidence for rapid climate change in North America during the late Paleocene thermal maximum: Oxygen isotope composition of biogenic phosphate from the Bighorn Basin (Wyoming): Earth and Planetary Science Letters, v. 160, p. 193–208.

Gibson, T.G., and Bybell, L.M., 1994, Sedimentary patterns across the Paleocene-Eocene boundary in the Atlantic and Gulf Coastal plains of the United States: Bulletin de la Société Belge de Géologie, v. 103, p. 237–265.

Gibson, T.G., Andrews, G.W., Bybell, L.M., Frederiksen, N.O., Hansen, T., Hazel, J.E., McLean, D.M., Witmer, R.J., and Van Nieuwenhuise, D.S., 1980, Biostratigraphy of the Tertiary strata of the core, *in* Geology of the Oak Grove core: Virginia Division of Mineral Resources Publication 20, part 2, p. 88.

Gibson, T.G., Bybell, L.M., and Owens, J.P., 1993, Latest Paleocene lithologic and biotic events in neritic deposits from southwestern New Jersey: Paleoceanography, v. 8, p. 495–514.

Gibson, T.G., Bybell, L.M., and Mason, D.B., 2000, Stratigraphic and climatic implications of clay mineral changes around the Paleocene-Eocene boundary of the northeastern U.S. margin: Sedimentary Geology, v. 134, p. 65–92.

Gingerich, P.D., 1989, New earliest Wasatchian mammalian fauna from the Eocene of northwestern Wyoming: Composition and diversity in a rarely sampled high–flood plain assemblage: University of Michigan Papers on Paleontology, v. 28, p. 1–97.

Gingerich, P.D., 2001, Biostratigraphy of the continental Paleocene-Eocene boundary interval on Polecat Bench in the northern Bighorn Basin, *in* Gingerich, P.D., ed., Paleocene-Eocene stratigraphy and biotic change in the Bighorn and Clarks Fork Basins, Wyoming: University of Michigan Papers on Paleontology, v. 33, p. 37–71.

Graham, A., 1992, Utilization of the isthmian land bridge during the Cenozoic: Paleobotanic evidence for timing, and the selective influence of altitudes and climate: Review of Palaeobotany and Palynology, v. 72, p. 119–128.

Gruas-Cavagnetto, C., 1968, Étude palynologique des divers gisements du Sparnacien du Bassin de Paris: Mémoires de la Société Géologique de France, v. 110, p. 144.

Gruas-Cavagnetto, C., 1978, Étude palynologique de L'Éocène du Bassin Anglo-Parisien: Mémoires de la Société Géologique de France, v. 131, p. 63.

Harrington, G.J., 1999, North American palynofloral dynamics in the late Paleocene to early Eocene [Ph.D. thesis]: University of Sheffield, UK, p. 1–321.

Harrington, G.J., 2000, Palynofloral dynamics of the late Paleocene to early Eocene from the Bighorn Basin, northwest Wyoming, USA, *in* Schmitz, B., et al., eds., Early Paleogene warm climates and biosphere dynamics: GFF (Geologiska Föreningens i Stokholm Förhandlingar) (Journal of the Geological Society of Sweden), v. 122, p. 67–68.

Harrington, G.J., 2001a, Impact of Paleocene-Eocene greenhouse warming on North American paratropical forests: Palaios, v. 16, p. 266–278.

Harrington, G.J., 2001b, Pollen assemblages and Paleocene-Eocene stratigraphy in the Bighorn and Clarks Fork Basins, *in* Gingerich, P.D., ed., Paleocene-Eocene stratigraphy and biotic change in the Bighorn and Clarks Fork Basins, Wyoming: University of Michigan Papers on Paleontology, v. 33, p. 89–96.

Harrington, G.J., 2003, Wasatchian (Early Eocene) pollen floras from the Red Hot Truck Stop, Mississippi, USA: Paleontology, v. 46 (in press).

Harrington, G.J., and Kemp, S.J., 2001, U.S. Gulf Coast vegetation dynamics in the latest Paleocene: Palaeogeography, Palaeoclimatology, Palaeoecology, v. 167, p. 1–21.

Heilmann-Clausen, C., 1985, Dinoflagellate stratigraphy of the uppermost Danian to Ypresian in the Viborg 1 borehole, central Jylland, Denmark: Geological Survey of Denmark, Series A, v. 7, p. 1–69.

Hickey, L.J., 1980, Paleocene stratigraphy and flora of the Clarks Fork Basin, *in* Gingerich, P.D., ed., Early Cenozoic paleontology and stratigraphy of the Bighorn Basin: University of Michigan Papers on Paleontology, v. 24, p. 33–49.

Hickey, L.J., West, R.M., Dawson, M.R., and Choi, D.K., 1983, Arctic terrestrial biota: Paleomagnetic evidence of age disparity with midnorthern latitudes during the Late Cretaceous and early Tertiary: Science, v. 221, p. 1153–1156.

Hooker, J.J., 1998, Mammalian faunal change across the Paleocene-Eocene transition in Europe, *in* Aubry, M.-P., et al., eds., Late Paleocene–early Eocene climatic and biotic events in the marine and terrestrial records: New York, Columbia University Press, p. 428–450.

Iakovleva, A.I., Brinkhuis, H., and Cavagnetto, C., 2001, Late Paleocene–early Eocene dinoflagellate cysts from the Turgay Strait, Kazakhstan: Correlations across ancient seaways: Palaeogeography, Palaeoclimatology, Palaeoecology, v. 172, p. 243–268.

Jaramillo, C.A., and Dilcher, D.L., 2001a, Middle Paleogene palynology of central Colombia, South America: A study of pollen and spores from tropical latitudes: Palaeontographica Abt. B, v. 258, p. 87–213.

Jaramillo, C.A., and Dilcher, D.L., 2001b, Microfloral diversity patterns of the late Paleocene-Eocene interval in Colombia, northern South America: Geology, v. 28, p. 815–818.

Jolley, D.W., 1997, Palaeosurface palynofloras of the Skye lava field and age of the British Tertiary volcanic province, *in* Widdowson, M., ed., Palaeosurfaces: Recognition, reconstruction, and palaeoenvironmental interpretation: Geological Society [London] Special Publication 120, p. 67–94.

Jolley, D.W., 1998, Palynostratigraphy and depositional history of the Paleocene Ormesby/Thanet depositional sequence set in southeastern England and its correlation with continental West Europe and the Lista Formation, North Sea: Review of Palaeobotany and Palynology, v. 99, p. 265–315.

Jolley, D.W., and Spinner, E., 1991, Sporepollen associations from the lower London Clay (Eocene), East Anglia, England: Tertiary Research, v. 13, p. 11–25.

Kalkreuth, W.D., McIntyre, D.J., and Richardson, R.J.H., 1993, The geology, petrography, and palynology of Tertiary coals from the Eureka Sound Group at Strathcona Fjord and Bache Peninsula, Ellesmere Island, Arctic Canada: International Journal of Coal Geology, v. 24, p. 75–111.

Kalkreuth, W.D., Riediger, C.L., McIntyre, D.J., Richardson, R.J.H., Fowler, M.G., and Marchioni, D., 1996, Petrological, palynological, and geochemical characteristics of Eureka Sound Group coals, Stenkul Fiord, southern Ellesmere Island, Arctic Canada: International Journal of Coal Geology, v. 30, p. 151–182.

Kelly, D.C., Bralower, T.J., Zachos, J.C., Premoli Silva, I., and Thomas, E., 1996, Rapid diversification of planktonic foraminifera in the tropical Pacific (ODP Site 865) during the late Paleocene thermal maximum: Geology, v. 24, p. 423–426.

Kennett, J.P., and Stott, L.D., 1991, Abrupt deep-sea warming, palaeoceanographic changes, and benthic extinctions at the end of the Paleocene: Nature, v. 353, p. 225–229.

Koch, P.L., Zachos, J.C., and Gingerich, P.D., 1992, Correlation between isotope records in marine and continental carbon reservoirs near the Paleocene-Eocene boundary: Nature, v. 358, p. 319–322.

Koch, P.L., Zachos, J.C., and Dettman, D.L., 1995, Stable isotope stratigraphy and paleoclimatology of the Paleogene Bighorn Basin (Wyoming, USA): Palaeogeography, Palaeoclimatology, Palaeoecology, v. 115, p. 61–89.

Krause, D.W., and Maas, M.C., 1990, The biogeographic origins of late Paleocene–early Eocene mammalian immigrants to the western interior of North America, *in* Bown, T.M., and Rose, K.D., eds., Dawn of the Age of Mammals in the northern part of the Rocky Mountain interior, North America: Boulder, Colorado, Geological Society of America Special Paper 234, p. 71–105.

Lucas, S.G., 1998, Fossil mammals and the Paleocene-Eocene series boundary in Europe, North America, and Asia, *in* Aubry, M.-P., et al., eds., Late Paleocene–early Eocene climatic and biotic events in the marine and terrestrial records: New York, Columbia University Press, p. 451–500.

Manchester, S.R., 1987, The fossil history of the Juglandaceae: Monographs in Systematic Botany from the Missouri Botanical Garden, v. 21, p. 1–137.

Manchester, S.R., 1999, Biogeographical relationships of North American Tertiary floras: Annals of the Missouri Botanical Garden, v. 86, p. 472–522.

McIntyre, D.J., 1991, Pollen and spore flora of an Eocene forest, eastern Axel Heiberg Island, N.W.T., *in* Christie, R.L., and McMillan, N.J., eds., Tertiary fossil forests of the Geodetic Hills, Axel Heiberg Island, Arctic Archipelago: Geological Survey of Canada Bulletin, v. 403, p. 83–97.

McIntyre, D.J., 1994, Appendix 1: Palynology, *in* Ricketts, B.D., ed., Basin analysis, Eureka Sound Group, Axel Heiberg and Ellesmere Islands, Canadian Arctic Archipelago: Geological Survey of Canada Bulletin, v. 439, p. 85–101.

McIver, E.E., and Basinger, J.F., 1999, Early Tertiary floral evolution in the Canadian high arctic: Annals of the Missouri Botanical Garden, v. 86, p. 523–545.

McKenna, M.C., 1980, Eocene paleolatitude, climate, and mammals of Ellesmere Island: Palaeogeography, Palaeoclimatology, Palaeoecology, v. 30, p. 349–362.

McKenna, M.C., 1983, Cenozoic paleogeography of North Atlantic land bridges, *in* Bott, et al., eds., Structure and development of the Greenland-Scotland Ridge: New York, Plenum Press, p. 351–399.

Miall, A.D., 1986, The Eureka Sound Group (Upper Cretaceous–Oligocene), Canadian Arctic Islands: Bulletin of Canadian Petroleum Geology, v. 34, p. 240–270.

Muller, J., De Di Giacoma, E., and van Erve, A.W., 1987, A palynological zonation for the Cretaceous, Tertiary, and Quaternary of northern South America: American Association of Stratigraphic Palynologists Contributions Series, v. 19, p. 7–76.

Pitelka, L.F., and the Plant Migration Workshop Group, 1997, Plant migration and climate change: American Scientist, v. 85, p. 464–473.

Pocknall, D.T., 1986, Palynological data from the Fort Union and Wasatch formations, Powder River Basin, Wyoming and Montana: U.S. Geological Survey Open-File Report 86-117, p. 58.

Pocknall, D.T., 1987a, Paleoenvironments and age of the Wasatch Formation (Eocene), Powder River Basin, Wyoming: Palaios, v. 2, p. 368–376.

Pocknall, D.T., 1987b, Palynomorph biozones for the Fort Union and Wasatch Formations (upper Paleocene–lower Eocene), Powder River Basin, Wyoming and Montana, USA: Palynology, v. 11, p. 23–35.

Pocknall, D.T., and Nichols, D.J., 1996, Palynology of coal zones of the Tongue River Member (upper Paleocene) of the Fort Union Formation, Powder River Basin, Montana and Wyoming: American Association of Stratigraphic Palynologists Contributions Series, number 32, p. 1–58.

Powell, A.J., 1992, Dinoflagellate cysts of the Tertiary System, *in* Powell, A.J., ed., A stratigraphic index of dinoflagellate cysts: London, Chapman and Hall, p. 155–252.

Proctor, M., Yeo, P., and Lack, A., 1996, The natural history of pollination: Portland, Oregon, Timber Press Inc., 487 p.

Read, J., and Francis, J.E., 1992, Responses of some Southern Hemisphere tree species to a prolonged dark period and their phytogeographic and palaeoecological implications for high-latitude Cretaceous and Tertiary floras: Palaeogeography, Palaeoclimatology, Palaeoecology, v. 99, p. 271–290.

Ricketts, B.D., 1986, New formations in the Eureka Sound Group, Canadian Arctic Islands: Current Research of the Geological Survey of Canada, v. 86-1B, p. 363–374.

Ritchie, J.C., 1995, Current trends in studies of long-term plant community dynamics: New Phytologist, v. 130, p. 469–494.

Röhl, U., Bralower, T.J., Norris, R.D., and Wefer, G., 2000, New chronology for the late Paleocene thermal maximum and its environmental implications: Geology, v. 28, p. 927–930.

Rull, V., 1999, Paleofloristics and palaeovegetational changes across the Paleocene-Eocene boundary in northern South America: Review of Palaeobotany and Palynology, v. 107, p. 83–95.

Schröder, T., 1992, A palynological zonation for the Paleocene of the North Sea Basin: Journal of Micropalaeontology, v. 11, p. 113–126.

Schumacker-Lambry, J., 1978, Palynologie du Landenien inférieur (Paléocène) à Gelinden-Overbroek/Belgique: Relations entre les microfossiles et le sédiment: Laboratoire de Paléobotanique et de Paléopalynologie, Liège, Belgium, Université de Liège, 157 p.

Sinha, A., Aubry, M.-P., Stott, L.D., Thiry, M., and Berggren, W.A., 1996, Chemostratigraphy of the "lower" Sparnacian deposits (Argiles plastique bariolées) of the Paris Basin, *in* Aubry, M.-P., and Benjamini, C., eds., Paleocene-Eocene boundary events in space and time: Israel Journal of Earth Sciences, v. 44, p. 223–237.

Speijer, R.P., Schmitz, B., and van der Zwaan, G., 1997, Benthic foraminiferal extinction and repopulation in response to latest Paleocene Tethyan anoxia: Geology, v. 25, p. 683–686.

Stott, L.D., Sinha, A., Thiry, M., Aubry, M.-P., and Berggren, W.A., 1996, Global $\delta^{13}C$ changes across the Paleocene-Eocene boundary: Criteria for terrestrial-marine correlations, *in* Knox, R.W.O'B., et al., eds., Correlation of the early Paleogene in northwest Europe: Geological Society [London] Special Publication 101, p. 381–399.

Tauxe, L., and Clark, D.R., 1987, New paleomagnetic results from the Eureka Sound Group: Implications for the age of the early Tertiary Arctic biota: Geological Society of America Bulletin, v. 99, p. 739–747.

Thomas, E., Gibson, T.G., Bybell, L.M., and Zachos, J.C., 1997, The late Paleocene Thermal Maximum in New Jersey (USA): Eos (Transactions, American Geophysical Union), v. 78, p. F363.

Tiffney, B.H., 1985a, The Eocene North Atlantic land bridge: Its importance in Tertiary and modern phytogeography of the Northern Hemisphere: Journal of the Arnold Arboretum, v. 66, p. 243–273.

Tiffney, B.H., 1985b, Perspectives on the origin of the floristic similarity between eastern Asia and eastern North America: Journal of the Arnold Arboretum, v. 66, p. 73–94.

Tiffney, B.H., 1994, An estimate of the early Tertiary paleoclimate of the southern Arctic, *in* Boulter, M.C., and Fisher, H.C., eds., Cenozoic plants and climates of the Arctic, NATO ASI Series 1: Global Environmental Change, v. 27, Springer-Verlag, Berlin, p. 265–280.

Tripati, A., Zachos, J.C., Marincovich, L., and Bice, K., 2001, Late Paleocene arctic coastal climate inferred from molluscan stable and radiogenic isotope ratios: Palaeogeography, Palaeoclimatology, Palaeoecology, v. 170, p. 101–113.

Tschudy, R.H., 1973, Stratigraphic distribution of significant Eocene palynomorphs of the Mississippi Embayment: U.S. Geological Survey Professional Paper 743-B, p. 1–24.

Tschudy, R.H., 1975, Normapolles pollen from the Mississippi Embayment: U.S. Geological Survey Professional Paper 865, p. 1–42.

Tschudy, R.H., 1976, Pollen changes near the Fort Union–Wasatch boundary, Powder River Basin: Wyoming Geological Association, 28th Annual Field Conference, Guidebook, p. 73–81.

Wang, D.-N., Sun, X.-Y., and Zhao, Y.-N., 1990, Late Cretaceous to Tertiary palynofloras in Xinjiang and Qinghai, China: Review of Palaeobotany and Palynology, v. 65, p. 95–104.

West, R.M., Dawson, M.R., and Hutchinson, J.H., 1977, Fossils from the Paleogene Eureka Sound Formation, N.W.T., Canada: Occurrence, climatic, and paleogeographic implications: Milwaukee Public Museum Special Publications in Biology and Geology, v. 2, p. 77–93.

Whitehead, D.R., 1983, Wind pollination: Some ecological and evolutionary perspectives, *in* Real, L., ed., Pollination biology: London, Academic Press, p. 97–108.

Wing, S.L., 1984, A new basis for recognizing the Paleocene-Eocene boundary in western interior North America: Science, v. 226, p. 439–441.

Wing, S.L., 1998, Late Paleocene–early Eocene floral and climatic change in the Bighorn Basin, Wyoming, *in* Aubry, M.-P., et al., eds., Late Paleocene–early Eocene climatic and biotic events in the marine and terrestrial records: New York, Columbia University Press, p. 380–400.

Wing, S.L., and Harrington, G.J., 2001, Floral response to rapid warming in the earliest Eocene and implications for concurrent faunal change: Paleobiology, v. 27, p. 539–563.

Wing, S.L., Bao, H., and Koch, P.L., 2000, An early Eocene cool period? Evidence for continental cooling during the warmest part of the Cenozoic, *in* Huber, B.T., et al., eds., Warm climates in Earth history: Cambridge, UK, Cambridge University Press, p. 197–237.

Wolfe, J.A., 1978, A paleobotanical interpretation of Tertiary climates in the Northern Hemisphere: American Scientist, v. 66, p. 694–703.

Wolfe, J.A., and Dilcher, D.L., 2000, Late Paleocene through middle Eocene climates in lowlands North America, *in* Schmitz, B., et al., eds., Early Paleogene warm climates and biosphere dynamics: GFF (Geologiska Föreningens i Stokholm Förhandlingar) (Journal of the Geological Society of Sweden), v. 122, p. 184–185.

Manuscript Accepted by the Society August 13, 2002

Printed in the USA

Geological Society of America
Special Paper 369
2003

Paleogene West Antarctic climate and vegetation history in light of new data from King George Island

Richard J. Hunt*
*School of Earth Sciences, University of Leeds, Leeds LS2 9JT, UK,
and British Antarctic Survey, Natural Environmental Research Council,
Madingley Road, High Cross, Cambridge CB3 0ET, UK*
Imogen Poole
*Wood Anatomy Section, National Herbarium of the Netherlands Utrecht University Branch,
and Department of Geochemistry, Faculty of Earth Science, Utrecht University,
P.O. Box 80021, 3508 TA Utrecht, The Netherlands*

ABSTRACT

Early Tertiary sediments of the Antarctic Peninsula region continue to yield rich assemblages of well-preserved fossil plants. This paper presents detailed sedimentological background and paleoecological (i.e., environmental, climatic and vegetational) inferences drawn from the Dragon Glacier flora from Point Hennequin, King George Island. This leaf assemblage was originally thought to be an impoverished, postglacial Oligo-Miocene flora but now, based on new dating and botanical evidence, is considered to be a taxonomically diverse, middle Eocene preglacial flora that can be compared with other contemporaneous but relatively species depauperate middle Eocene floras from the same region. Eocene vegetation on King George Island grew in a sedimentary paleoenvironment characterized by fluvio-lacustrine deposition with a strong pyroclastic component providing evidence of volcanic activity set against a backdrop of gradual climatic change. The floral composition is comparable with modern South American rain forest vegetation growing under similar environmental conditions today and thus supports the Valdivian Model of disturbance-controlled successional vegetation.

INTRODUCTION

Fossil floras from the Antarctic Peninsula region play an integral role in reconstructing the biodiversity and ecology of high-latitude forests during the geological past. Our understanding of the floristic composition of the Antarctic Peninsula macro- and microfloras during the Late Cretaceous and early Tertiary has increased dramatically in recent years with well-documented macrofloras known from the Coniacian-Maastrichtian (Hayes, 1999; Poole and Cantrill, 2001; Poole and Francis, 1999, 2000; Dutra and Batten, 2000; Poole and Gottwald, 2001; Poole et al., 2000a, 2000b, 2000c), Paleocene (Poole et al., 2000b; Poole and Gottwald, 2001; Birkenmajer and Zastawniak, 1988, 1989) and preglacial Eocene strata (Birkenmajer and Zastawniak, 1988, 1989; Li, 1994; Poole et al., 2001). In addition, Late Cretaceous and Paleogene microfloras have been recorded (Stuchlik, 1981; Askin, 1988, 1989, 1990, 1992, 1994, 1997; Askin and Baldoni, 1998; Cao, 1992, 1994; Dettmann, 1989; Dettmann and Thomson, 1987; Truswell, 1990). Moreover, Antarctic Paleogene plant records are important in tracking paleoclimate trends during the

*richard.hunt@arup.com

Hunt, R.J., and Poole, I., 2003, Paleogene West Antarctic climate and vegetation history in light of new data from King George Island, *in* Wing, S.L., Gingerich, P.D., Schmitz, B., and Thomas, E., eds., Causes and Consequences of Globally Warm Climates in the Early Paleogene: Boulder, Colorado, Geological Society of America Special Paper 369, p. 395–412.

final phases of Gondwana breakup and the progressive deterioration of climate that culminated in major Cenozoic glaciation (Cantrill and Poole, 2002; Truswell, 1991; Hill and Scriven, 1995; Zachos et al., 2001).

The abundance of fossil floras in terrestrial Antarctic Peninsula strata is not reflected in the post-Oligocene glaciation record, which until recently comprised only two floras, the Dragon Glacier flora (DGF) and the slightly younger Mount Wawel flora (MWF)—from Point Hennequin on King George Island in the South Shetland Islands, West Antarctica (Figs. 1 and 2). Zastawniak et al. (1985) concluded these floras to be *Nothofagus*-Podocarpaceae dominated vegetation and interpreted their low apparent diversity as an indication of prevailing cool-temperate climates. The floras were dated as late Oligocene (~24.5 Ma) based on K-Ar data, with the MWF occupying a slightly younger stratigraphic position than the DGF (Birkenmajer et al., 1983; Zastawniak et al., 1985) (Fig. 2B). According to Birkenmajer (1997) the floras represented vegetation that had recolonized the area from outlying refugia during a milder climate phase that succeeded peninsula-wide West Antarctic glaciation and extinction at ca. 30 Ma (Dingle and Lavelle, 1998). However, K-metasomatism and Ar-loss is thought to have affected many of the K-Ar data for King George Island (Birkenmajer, 1990; Soliani and Bonhomme, 1994; Willan and Kelley, 1999). Reassessment of the age data using $^{40}Ar/^{39}Ar$ geochronology indicates a much older preglacial, middle Eocene age (~44–49 Ma) for the fossil floras from Point Hennequin (Hunt, 2001).

Recent British Antarctic Survey (BAS) field collections from the Dragon Glacier locality suggest that the flora was much more diverse than previously realized, probably representing vegetation growing under a mild climate, a finding that Hunt (2001) interpreted as consistent with the new preglacial, rather than the previous postglacial age. However, the discovery of the new, monotypic Smok Hill flora, contemporary with the Dragon Glacier flora, as well as further collections from the Mount Wawel locality, continue to reveal only low-diversity *Nothofagus*-dominated assemblages.

Consequently the climatic and paleoenvironmental setting of the floras at Point Hennequin has been in need of reassessment, with focus on underlying reasons for the apparent major variation in diversity between the three closely spaced, similar aged horizons. Leaf physiognomic analyses and comparisons with contemporaneous macrofossil floras have been undertaken to more fully understand the paleoecology (i.e., environment, climate and vegetation) of King George Island during the early Tertiary.

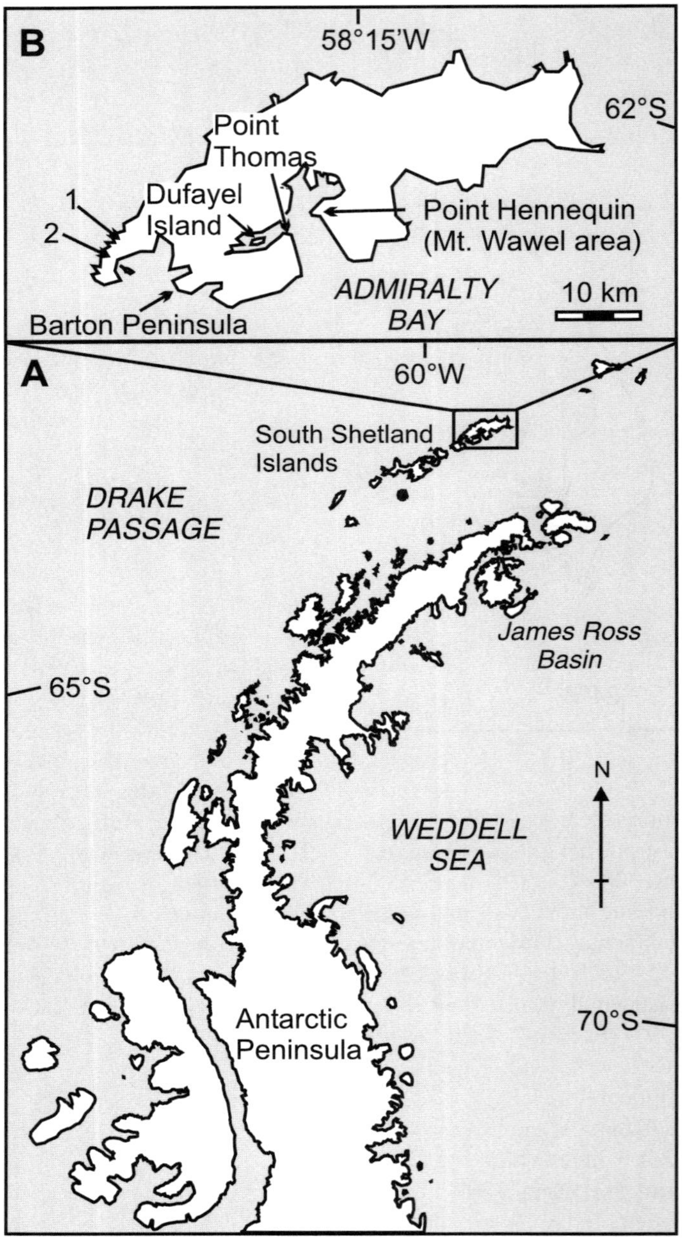

Figure 1. Map of the Antarctic Peninsula region showing the position of King George Island. A: Antarctic Peninsula region. B: King George Island showing the position of Point Hennequin and other localities mentioned in the text, 1 Collins Glacier, 2 Fossil Hill.

GEOLOGICAL SETTING AND SYSTEM DYNAMICS

King George Island (KGI), situated at a latitude of 62°S (approximately coincident with that in the early Tertiary; Lawver et al., 1992) (Fig. 1) has been one of the main foci for Antarctic paleobotany for over a century. During the early Tertiary Pacific Ocean floor was being subducted eastwards beneath the Antarctic Peninsula, resulting in uplift and stratovolcanic activity (Pankhurst and Smellie, 1983; Smellie et al., 1984; McCarron and Larter, 1998). A mixed effusive-explosive style of volcanism resulting in agglomerate and fossiliferous tuff deposits, intercalated with lavas, characterizes Late Cretaceous to early Tertiary strata on KGI (Barton, 1965; Smellie et al., 1984; Shen, 1994; Xue et al., 1996; Dutra and Batten, 2000).

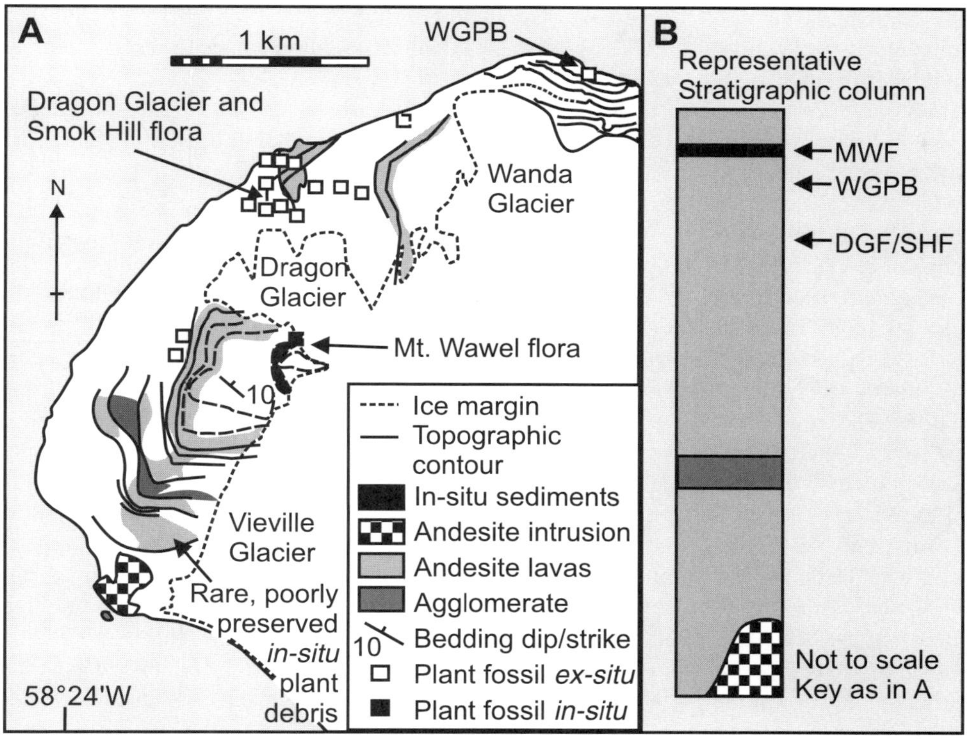

Figure 2. Map of Point Hennequin. A: Point Hennequin, showing distribution of exotic fossiliferous blocks and in situ floras. Geology modified from Smellie et al. (1984). WGPB—Wanda Glacier Plant Beds, this flora known as a secondary deposit in Wanda Glacier moraines was collected but not officially described by Birkenmajer (1981). B: Representative stratigraphic column showing relative age of plant beds (not to scale). MWF—Mount Wawel flora, DGF—Dragon Glacier flora, SHF—Smok Hill flora.

This is clearly demonstrated at Point Hennequin, situated on the east side of Admiralty Bay (62°7′S, 58°24′W), which comprises an ~300 m thick east-dipping sequence of basaltic-andesite and andesite lavas, with interbedded thin tuffaceous plant-bearing sedimentary horizons (Jardine, 1950; Barton, 1959, 1961, 1964a and 1964b, 1965; Birkenmajer, 1980, 1981; Zastawniak, 1981; Zastawniak et al., 1985; Smellie et al., 1984; Duan and Cao, 1998) (Fig. 2). Volcanism therefore played an important role in the development, dynamics and preservation of local and regional floras. Poole et al. (2001) highlighted the impact of terrain modification and widespread vegetation disturbance on KGI, which in turn led to successional changes in the vegetation. This scenario also occurs in areas controlled by volcanic activity today (Burnham and Spicer, 1986; Spicer, 1989; Burnham, 1994), including tectonically and floristically similar areas such as Valdivia, Chile in southern South America (Veblen et al., 1977, 1981; Markgraf et al., 1996). Indeed, the Valdivian Model (Poole et al., 2001) for West Antarctic floras uses modern southern South American forest dynamics to characterize the influence of volcanism on the paleovegetation. In the modern vegetation high levels of disturbance result in low-diversity *Nothofagus*-dominated assemblages, whilst low levels of disturbance give rise to relatively species-rich, climax forest assemblages (Veblen et al., 1977, 1981, 1996). Such trends can also be recognized in the fossil record of King George Island (Poole et al., 2001), but they require careful resolution from the overall trend of Tertiary cooling and associated reduction in diversity observed in Antarctic floras (Hill and Scriven, 1995).

MATERIAL AND METHODS

Over 200 individual angiosperm leaves and leaf mats as well as conifer shoots, fruits, seeds and an inflorescence, were collected from transported blocks of tuffaceous sandstone and siltstone (clasts varying in size from 0.1 to 2 m^3) in terminal moraines to the north of Dragon Glacier (Fig. 2A). This flora is known as the Dragon Glacier flora (Birkenmajer, 1981). The plant material is preserved as carbonized films that lack cuticle and generally are oriented parallel or subparallel to the bedding. A smaller collection of 24 mineralized angiosperm leaf impressions preserved in a pale green coarse air fall tuff, and representing a separate but apparently coeval sedimentary unit, was also collected from this locality and is here named the Smok Hill flora (Figs. 2A and B). A small and highly fragmented plant assemblage was also collected from the in situ Mount Wawel flora horizon (Figs. 2A and B; Locality 232 in Zastawniak et al., 1985).

Plant macrofossil specimens were cleaned and prepared using traditional methods (Fairon-Demaret et al., 1999). The plant macrofossils were examined microscopically under ethanol using low-angle lighting to enhance surface texture. Leaf specimens were grouped into morphotypes using the terminology of Hickey (1973, 1979) and Hickey and Wolfe (1975) with modifications from the Manual of Leaf Architecture (Leaf Architecture Working Group, 1999). Leaf size classes are based on the categories of Raunkiaer (1934) as modified by Webb (1959). Leaf area was estimated using the method of Cain and de Castro (1959). Counts of total leaf populations are based on fragmented as well as whole leaves. Only the high morphotype diversity of the Dragon Glacier flora meets the requirements for mean annual temperature (Wolfe, 1971) and mean annual precipitation analysis (Wilf et al., 1998), while the low diversity Mount Wawel and Smok Hill floras support nearest living comparisons and offer evidence for paleoecological changes through the stratigraphic section. Estimates for mean annual temperature (MAT) were based on leaf margin analysis (Wolfe, 1979; Greenwood, 1992; Wing and Greenwood, 1993; Wilf, 1997; Wiemann et al., 1998; see Gregory [1996] for a discussion of this technique). An estimate of mean annual precipitation (MAP) was determined using the leaf area analysis method of Wilf et al. (1998). Other indicators of paleoenvironmental conditions were determined including leaf size index (LSI), leaf shape and leaf apex style as described by Wolfe and Upchurch (1987) and Upchurch and Wolfe (1987).

Thin section and field descriptions were used to define sedimentary facies and follow standard sedimentary (Pettijohn, 1975) and volcaniclastic terminology (Fisher and Schmincke, 1984; Cas and Wright, 1987). The term "volcaniclastic" follows the definition of Fisher (1966) where it is used to refer to any clastic sediment with a volcanic component that has been transported and deposited by any mechanism in any environment. Volcaniclastic sediments that have experienced little or no reworking are termed "primary." Tuffaceous refers to sediment that has both pyroclastic and reworked elements. Sedimentary thin sections were studied using transmitted light microscopy. All sedimentological material, data, and the majority of the leaf macrofossil material were collected during the British Antarctic Survey 1999 field season (BAS sample reference numbers prefixed by P.). Leaf material was supplemented by collections from previous BAS expeditions. All material is deposited at the British Antarctic Survey, Cambridge, UK.

RESULTS

Sedimentology

Less than 40 m of thin volcaniclastic sediment intercalations, predominantly agglomerate with subordinate reworked tuffaceous material (Birkenmajer, 1981; Smellie et al., 1984), are present in the 300 m sequence at Point Hennequin, which is otherwise lava dominated (Jardine, 1950; Smellie et al., 1984). Exotic fossiliferous blocks, glacially eroded from this sequence, are found in highest concentration in terminal moraines along the northern margin of Dragon Glacier. The paucity of fossiliferous material along the margins of the Vieville and Wanda glaciers (Fig. 2A), to the south and north respectively, suggests that the eroded source beds originated at, or close to, the "neck" of Dragon Glacier—a view shared by other workers of this area (Birkenmajer, 1981; J. Smellie, 2001, personal commun.). Certainly the source beds are <5 km distant, coincident with the present-day ice divide.

Sedimentological logs and facies descriptions for sediments producing the Dragon Glacier and Smok Hill floras are given in Table 1 and Figure 3. Thin section descriptions from the dominant plant bearing lithologies are summarized in Table 2. This provides important information for determining whether the sediments are primary or reworked.

The exotic plant-bearing blocks from the Dragon Glacier flora are subdivided into six facies (DGF-1 to DGF-6), which are grouped as two facies associations (DGF-a, DGF-b). The DGF facies are commonly gradational in nature, suggesting that the exotic blocks sample through a complete fluvial-lacustrine section, from terrestrial fluvial/shoreline lacustrine sequences (facies association DGF-a) to relatively deeper water lacustrine facies (DGF-b; Fig. 5C). The subaqueously deposited tuffaceous sandstones of these two facies associations (Table 1) are interpreted as lacustrine, rather than brackish or marine deposits, due to the lack of any preserved marine organisms and the abundance of paleosols and plant material. A microflora comprising only terrestrial palynomorphs also implies a terrestrial setting (Duan and Cao, 1998; G. Harrington, 2001, personal commun.). Cyclical development of paleosols, defined by alternating rootlet beds and symmetrically wave rippled horizons (facies DGF-2), suggests fluctuating water levels possibly resulting from draining and refilling of an ephemeral lacustrine basin. This interpretation is also supported by the absence of a well-developed lacustrine fauna except for rare invertebrate trackways and burrows (facies DGF-2; Birkenmajer, 1981; Hunt, 2001).

Compositionally the DGF sedimentary rocks are very similar to one another, with varying proportions of crystals (plagioclase, clinopyroxene, and opaques that become increasingly common with reworking), scoria or glass shards (which are more abundant in primary or minimally reworked sediments), carbonized organic material and rare charcoalified wood fragments (Tables 1 and 2). The dominance of basic-intermediate grains in the sediments is consistent with their origin from a compositionally limited source, the abundant basic to intermediate volcanics and pyroclastic lithologies that form much of the Eocene strata (Birkenmajer, 1981; Smellie et al., 1984). The composition, fine grain size and sorting of most of the DGF sedimentary rocks suggests moderate reworking and a relatively distal volcaniclastic source. This excludes the coarser, scoriaceous deposits of DGF-3, which are considered to be more proximal (Vessel and Davies, 1981; Carey, 1991). Given the close sedimentological similarity of the DGF facies, these lithologies are assumed to represent a thin stratigraphic interval.

Previous interpretations of the Dragon Glacier sedimentary rocks have not included a facies description. Birkenmajer (1980, 1981) described pelitic and psammitic tuffs, shaley sandstones and shale, including fine-grained sandstone, siltstone and clay-shale triplets from the moraine samples and tentatively suggested that the latter were varves. No other evidence has been found for glacial sedimentation in the Dragon Glacier or Mount Wawel sedimentary rocks but the possibility of seasonal rhythmic sedimentation cannot be excluded. "Suncracked" surfaces were described by both Barton (1964a) and Birkenmajer (1981) in the Dragon Glacier sedimentary rocks and used as evidence in support of ephemeral lake conditions. However the supposed cracks are extremely regular and show no evidence of post-

TABLE 1. SEDIMENTARY FACIES DESCRIPTIONS FOR THE DRAGON GLACIER AND SMOK HILL SEDIMENTS FROM WHICH THE FOSSIL MATERIAL WAS DERIVED

Facies	Description	Interpretation
DGF-1	Blue gray to gray-brown coarse grained tuffaceous sandstone to tuffaceous conglomerate with interbedded tuffaceous siltstone unidirectional flow structures (microscours, rare flute casts, asymmetric current ripples) Massive to normally graded, with common internal erosion surfaces. Plant material is generally rare except for rootlets, which often arise from the upper part of the sandstone beds.	Fluvial deposits and/or possible overbank sediments (cf. Smith, 1987; Nakayama and Yoshikawa, 1997).
DGF-2	Dark gray to light brown, well sorted, normally graded, fine grained tuffaceous siltstone, interbedded with coarse grained tuffaceous sandstone, specimens capped by massive mudstone. Fine siltstone drape. Laminated to bedded. Some samples show cyclic development of symmetrical wave ripples and rootlets on a centimeter scale. Vertical (length 1–4 mm) and horizontal roots (length <120 mm, width <4 mm). Trackways and burrows present on the draped surfaces. Horizontal invertebrate trackways are frequently associated with symmetrically wave-rippled horizons. Leaf fossil remains are occasionally preserved in the troughs of wave ripples. Plant material is rare in rootlet horizons.	Marginal lacustrine and shoreline paleosol deposits with mudstone drapes and bioturbation (cf. Smith, 1987; Nakayama and Yoshikawa, 1997). Oxygenated conditions are implied by the scarcity of plant fossils and abundant bioturbation. Samples lacking rootlets are interpreted as slightly deeper water, distal shoreline deposits. Cyclic development of rootbeds and wave-rippled horizons suggests rapid fluctuations in water depth (cf. Spicer, 1989; Burnham, 1994).
DGF-3	Scoria dominated breccia with rare plant compressions. Poorly sorted granule to small pebble, matrix supported scoria breccia. Clasts angular to rounded. Normally graded to massive. Cryptocrystalline groundmass with fine glass shards and microcryst plagioclase and pyroxene. Long axis clast alignment. Rootlets are sometimes present in the upper portion of deposits. Fine to coarse grained laminated tuffs with impact sags sometimes overlie scoria deposits. Plant fossils comprise leaf fragments, carbonaceous layers, a single small fruit and rare charcoalified wood.	Subaerial primary or slightly reworked, scoria flow (pyroclastic) deposit (cf. Carey, 1991). Suggests small-scale strombolian type eruption or collapse of an eruption column. The laminated tuffs may be the upper section of the flow unit but are more likely to represent an airfall ash and scoria deposit (cf. Carey, 1991). The development of paleosol characteristics (i.e., rootlets and lack of subaqueous textures) is interpreted as subaerial deposition.
DGF-4	Light brown and gray laminated tuffaceous siltstones to coarse tuffaceous sandstones with soft sediment deformation. Moderate to well sorted. Laminated to medium bedded. Soft sediment deformation includes dish structures, sandstone balls, small-scale compaction faults and convolute bedding. Dewatering and ?slump folds vary in scale from <0.1 to 0.5 m. Plant fossils rare.	Reworked tuffaceous sandstones. The scale of discrete folds up to 0.5 m long suggests a basin slope setting with loading of unstable lacustrine sediments caused by rapid sediment accumulation and/or possible syn-sedimentary volcanotectonic activity (cf. Breitkreutz, 1991).
DGF-5	Light brown, gray-green, pink and gray, normally graded tuffaceous siltstones to coarse grained tuffaceous sandstones. Laminated to bedded. Sometimes rhythmically laminated defined by changes in color and grain size. Laminations defined by alternations of massive tuffaceous sandstone, siltstone and massive or laminated siltstone/mudstone. The coarser laminae often truncate underlying laminated siltstones and grade normally into laminated siltstones. Occasional asymmetric current ripples. Rare rounded pebbles of cryptocrystalline brown, red and green plagioclase-rich scoriaceous clasts (<63 mm) occur as isolated poorly sorted clasts. Plants are most abundant in these sediments and occur in discrete planes, sometimes as dense mats but generally as scattered leaves. Alternating angiosperm rich leaf horizons and debris-dominated horizons with abundant conifer shoots are common in many samples.	Low to moderate energy lacustrine sediments. Rhythmically laminated sediments may indicate seasonal cyclicity or cyclic deposition of normally graded microturbidites (cf., e.g., Nakayama and Yoshikawa, 1997).
DGF-6	Pale yellow brown tuffs, dominated by low vesicularity and highly fragmented glass shards with diffuse contacts between laminae. Moderate to excellent preservation of abundant leaf and stem compressions.	Probable hydroclastic deposits suggested by high fragmentation of glass shards, lacustrine conditions indicated by diffuse contacts due to bioturbation (Fisher and Schmincke, 1984; Cas and Wright, 1987).
Smok-1	Pale green, moderate to well sorted, angular, medium to coarse, massive to weakly laminated sideromelane-rich tuff. Plant fossils comprise a mineralized monotypic leaf assemblage preserved in non bed-parallel orientations, with a single possible stem fragment.	Distal primary or slightly reworked airfall ash deposit. The low vesicularity of the sideromelane fragments suggests a hydroclastic origin.

Note: DGF-1 to DGF-3 form the facies association DGF-a (subaerial to shallow lacustrine volcaniclastic sediments). DGF-4 to DGF-6 form the facies association DGF-b (nonmarginal lacustrine sediments). The Smok-1 facies forms a separate Smok Hill flora facies.

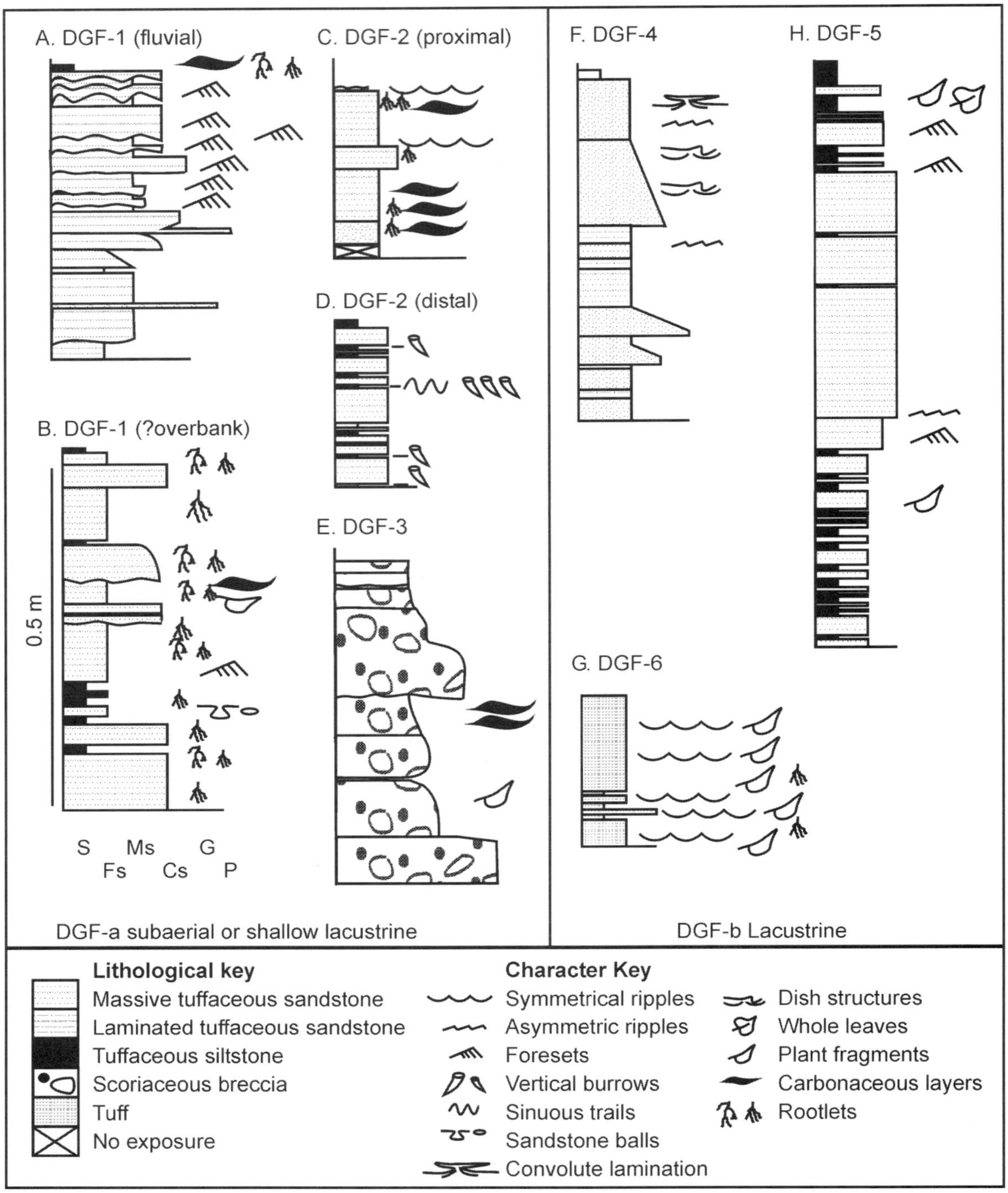

Figure 3. Sedimentological logs of Point Hennequin facies associations. Scale given for DGF-1 applies to all logs. (DGF—Dragon Glacier flora. Grain size: S—siltstone, Fs/Ms/Cs—fine/medium/coarse sandstone, G—granule conglomerate, P—pebble conglomerate). A: Fluvial deposits. B: Possible overbank or flood-plain deposits. C: Shoreline lacustrine deposits. D: Distal shoreline deposits. E: Coarse subaerial scoria flow (base) to scoria fall (top) deposit. F: Basin slope to delta plain sediments with soft-sediment deformation. G: Fine-grained subaqueous ash fall deposit. H: Laminated basin plain sediments with good plant preservation.

TABLE 2. SELECTED THIN SECTION DESCRIPTIONS FOR THE MAIN PLANT-BEARING FACIES AT POINT HENNEQUIN

Facies	Nongenetic sedimentary description	Grain size (mm)	Sorting and support	Angularity		Composition (%)	Interpretation
DGF-5 (P.2810)	Laminated to bedded, laminae massive or normally graded, pink and gray siltstone and fine sandstone, rarely medium sandstone	<0.19 <0.38	Poor to well sorted, matrix supported	Angular—subrounded	PVC*	<2% glass shards	Reworked, crystal dominated lacustrine sandstone
					Crystals	<30% plagioclase <10% clinopyroxene	
					Matrix	<60% microcryst plagioclase, clinopyroxene and opaque grains in very fine grained groundmass	
DGF-6 (P.3007)	Diffusely laminated light brown fine siltstone to sandstone. Occasionally massive with extensive recrystallization of matrix	0.19† <0.5	Well sorted, variable matrix to clast support	Angular	PVC	<60% glass shards (<20% vesicularity)	Primary or slightly reworked subaqueously deposited airfall tuff
					Crystals	10–25% plagioclase <1% opaque grains	
					Matrix	<40% cryptocrystalline groundmass <2% organic material	
DGF-3 (P.236.21)	Massive granule to small pebble conglomerate	0.5–9 1.0†	Poorly sorted, matrix supported	Angular—subrounded	PVC	<60% scoria <20% glass shards	Primary scoriaceous breccia
					Crystals	N.A.	
					Matrix	<10–20% plagioclase <5% clinopyroxene <1% opaque grains	
Smok-1 (P.1404.2b)	Pale green, massive medium to coarse sandstone	0.5–0.75 0.6†	Moderate to well sorted, clast supported	Angular—subangular	PVC	50–70% low vesicularity sideromelane	Primary ?subaqueous airfall tuff ?formed by hydroclastic activity
					Crystals	5–10% plagioclase; possible altered orthopyroxene <2% opaques	
					Matrix	<20% Chlorite infilled porosity	

Note: The bracketed numbers indicate British Antarctic Survey specimen numbers for thin section slides.
*Primary volcanic clasts.
†Average.

formation infilling by other sediments. Red-orange iron oxide staining is concentrated along the cracks suggesting that they were related to fluid flow, possibly as a result of diagenesis.

Sedimentary rocks preserving the Smok Hill flora are placed in a separate facies (Smok-1) on the basis of their distinctive green color and composition. They are dominated by highly fragmented, low-vesicularity sideromelane (up to 70%) suggestive of hydrovolcanic activity (J. Smellie, 2001, personal commun.). These sediments, compositionally distinct from the predominantly reworked, crystal-rich Dragon Glacier sedimentary rocks, are primary volcaniclastics, the weak planar bedding and abundant chlorite infill of the porosity (Table 2) may also indicate lacustrine deposition.

Paleofloral composition

Dragon Glacier flora. This flora comprises 37 dicotyledonous (dicot) angiosperm morphotypes, seven conifer morphotypes (belonging to the Araucariaceae, Podocarpaceae and Cupressaceae), four fern morphotypes and a single form of *Equisetum* (Fig. 4A and B). This extends the findings of Zastawniak (1981), who described the flora as a *Nothofagus*-Podocarpaceae-pteridophyte assemblage. The vegetation supported an arthropod community as evidenced from the abundant traces on leaves (Hunt, 2001).

Smok Hill flora. This new flora comprises 24 leaves of only a single *Nothofagus* species (morphotype 31; Fig. 4B). Therefore the flora is more consistent with the interpretation of Zastawniak et al. (1985) of low-diversity vegetation at this time.

Mount Wawel flora. According to Zastawniak et al. (1985), the Mount Wawel flora is dominated by *Nothofagus* leaves both in terms of specimens (158 impressions) and species. These fossil leaves are most similar to the modern South American species of *Nothofagus* and to the fossil species *N. ledenfeldi* Ettingshausen, from the Upper Cretaceous of New Zealand, and *N. simplicidens* Dusén and *N. ulmifolia* Ettingshausen from the lower Tertiary of South Patagonia. Two fragmentary angiosperm leaves with affinities to the Rhamnaceae and Cochlospermaceae were also present (Zastawniak et al., 1985). Seventy seeds and several fragments of podocarpaceous shoots represent the gymnosperm component of the flora (Zastawniak et al., 1985). The new collection described here contains a single cone scale belonging to the *Araucaria* section *Eutacta* and a small clump of moss axes in addition to poorly preserved Nothofagaceae leaf fragments and rare, indeterminate fragments of other angiosperms (Hunt, 2001).

Wood, often abundant in modern volcanic environments (Burnham, 1994), is extremely rare at Point Hennequin but abundant elsewhere on King George Island (Lucas and Lacey, 1981; Jagmin, 1987; Torres, 1984; Torres et al., 1984; Torres and Lemoigne, 1988; Hee and Soon-Keun, 1991; Zhang and Wang, 1994; Poole, 2002). Hydraulic sorting by rivers or floods may have been responsible for the sorting of organic material, some of which (fruits, seeds, and leaves) became trapped in the (?)ephemeral lakes (Jardine, 1950; Birkenmajer, 1980, 1981), whereas the wood may have been transported greater distances.

Paleoclimate analysis

Although the leaves from the Dragon Glacier flora are seemingly well preserved, with up to five orders of venation recognizable, no cuticular material was observed by ordinary light or fluorescence microscopy. Consequently paleoclimate inferences were determined using predominantly taxon-independent methods focusing on leaf morphological characters. The 37 dicotyledonous angiosperm morphotypes found in the Dragon Glacier flora are illustrated in Figures 4A and B. Thirty four of these morphotypes were used in the paleoclimatic analyses, which excluded possible herbaceous forms (Fig. 4A; morphotype 4) and suspected juvenile taxa (Figs. 4B; morphotypes 36 and 37).

The relationship between mean annual temperature and the proportion of dicotyledonous species with entire margined leaves has been recognized since the early twentieth century (Bailey and Sinnott, 1915, 1916). This relationship has been increasingly studied as more attention is focused on the potential of leaves as a paleoclimatic indicator (Wolfe, 1979; Wing and Greenwood, 1993; Boyd, 1994; Jordan, 1997; Wilf, 1997; Wilf et al., 1998; Wiemann et al., 1998).

The results from mean annual temperature (MAT) and mean annual precipitation (MAP) analyses are summarized in Table 3. Leaf margin analysis using a variety of predictor data sets indicates an average MAT of between 10 and 11 °C. The most divergent estimate (7.4–8.8 °C) is derived from the Southern Hemisphere regression equation (LMA[6] in Table 3). This estimate is likely to be incorrect because the predictor data set is derived from modern Southern Hemisphere floras, which tend to be dominated by evergreen taxa with entire margined leaves, whereas Paleogene Antarctic floras, like modern Northern Hemisphere floras, probably had a high proportion of deciduous taxa (Wolfe, 1979; Spicer and Parrish, 1990). Therefore the slope of the relationship between MAT and leaf margin percent in modern Northern Hemisphere vegetation is probably more appropriate for inferring the temperature for the Eocene KGI floras. An LSI based MAT value (~16 °C) for the DGF is also derived from regressions based on modern Southern Hemisphere vegetation (Greenwood, 1992, 1994). Jordan (1997), however, found that modern Australasian floras often produced biased MAT results due to overrepresentation by evergreen taxa and by certain families, for example the Myrtaceae, which are dominantly entire margined. Consequently the modern Southern Hemisphere vegetation with a predominance of evergreen entire margined leaves is considered to be less representative of the KGI vegetation with its high proportion of toothed margined leaves and a potentially high proportion of deciduous taxa. This is seemingly confirmed by the higher estimated MAT value predicted from LSI-MAT correlation of the Dragon Glacier flora (Table 3) (cf. Greenwood, 1992, 1994).

Of the total dicot morphotypes ~11% are stenophylls, suggesting that a component of the flora grew in habitats at or near the edge of water (Richards, 1952), implying a possible bias in the flora toward toothed margined leaves, which are overrepresented in these settings (Burnham et al., 2001). MAT estimates derived here should therefore be considered as minimum estimates (Burnham et al., 2001). Lobed and compound leaves characterize the understorey of forests, and successional vegetation (Upchurch and Wolfe, 1987) but correlations have been found between MAT and the proportion of lobed leaves within a flora (Wolfe, 1995; Spicer and Wolfe, 1999). 16% of the Dragon Glacier flora morphotypes are pinnately or palmately lobed (such as morphotype 4 in Figure 4A), which supports warm microthermal MAT estimates (cf. Figure 44.2 in Spicer and Wolfe, 1999).

Leaf size is strongly coupled to available moisture, thus leaf area has been used as a proxy for mean annual precipitation (Dolph and Dilcher, 1979, 1980; Givnish, 1984; Greenwood, 1992, 1996; Wing and Greenwood, 1993; Wilf, 1997; Wilf et al., 1998; Jacobs, 1999; Hayes, 1999). Wing and Greenwood (1993) derived MAP estimates based on the proportion of fossil leaf species that possessed drip tips and the percentage of large leaves within the flora (i.e. ≥ mesophyll I). Large leaves such as those falling in the mesophyll and greater size classes are rare in the floras, thus precluding the use of MAP regressions derived from the CLAMP (climate leaf analysis multivariate program) data set, which are based on the percentage of large leaves (Wing and Greenwood, 1993; Wolfe, 1993). Large leaves are likely to have been present in the original flora since fragments of leaves up to 14 cm in length and 10 cm in width are present in the fossil assemblages. This implies that the larger leaf size classes may have been present in the paleovegetation but were preferentially removed by fragmentation and hydraulic sorting (Ferguson, 1985). The leaf area analysis, which suggests MAP of ~1039 mm for the Dragon Glacier flora is therefore considered to be a minimum value (Table 3). Drip tips are absent from the Dragon Glacier flora (and Smok Hill and Mount Wawel floras) however ~10% of leaves from DGF have more attenuated apices suggesting moderate availability of moisture, an observation supported by the derived MAP values.

The leaf size index (LSI) is 23.53, which equates to that found in modern, predominantly microphyllous, vegetation (Burnham, 1989) and is equivalent to samples from modern simple notophyll vine forest in Australia (Webb, 1959; Greenwood, 1994).

DISCUSSION

Sedimentary paleoenvironment

The volcaniclastic lithologies outlined above highlight the importance of volcanic activity in the paleoenvironment of King George Island. The exact location and distance to the volcanic center that sourced both the lavas and the pyroclastic sediments deposited at Point Hennequin remains unknown. Birkenmajer (1981) and Zastawniak et al. (1985) proposed that a vent complex was situated at Point Hennequin itself, however the new middle Eocene age of ~44–49 Ma for the lava sequence at Point Hennequin (Hunt, 2001) conflicts with this interpretation, instead suggesting that the lithologies inferred to be vent core strata by Birkenmajer (1981) and Zastawniak et al. (1985) may actually be younger intrusive bodies. This would imply a more distal source vent for the volcanic and volcaniclastic deposits at Point Hennequin than previously supposed. The easterly dip of the volcanic pile at Point Hennequin suggests a source vent to the west, possibly between Point Hennequin and Point Thomas 5–6 km distant (Fig. 1B; J. Smellie, 2001, personal commun.).

The Dragon Glacier and Smok Hill sedimentary rocks are characterized in terms of (i) the degree of reworking—whether the sediments are primary pyroclastic, or reworked volcaniclastics (Fisher and Schmincke, 1984; Cas and Wright, 1987); and (ii) vent proximity—the relative distance to the vent that sourced these sediments (Vessel and Davies, 1981; Smith, 1991). These two factors may be used as proxies for disturbance levels (Poole et al., 2001), and thus have implications for interpretations of floral composition, since vegetation growing proximal to an active volcano is likely to experience higher levels of disturbance and consequent floral turnover. Abundant plant material and low proportions of primary volcaniclastics such as scoria and glass within the predominantly reworked Dragon Glacier facies, suggest relatively low levels of volcanic disturbance during the growth of this flora (see Poole et al., 2001, and the references therein) and the generally fine grain size of the sediments, a medial to distal depositional setting. However, the presence of charcoalified wood fragments found associated with both this flora (facies DGF-3) and that at Fossil Hill (Poole et al., 2001; Fig. 1B) may provide evidence for blast effects or forest fire, similar to that seen in areas associated with volcanic disturbance today (Burnham, 1994). Rare waterlain tuffs (DGF-6) also imply volcanic activity and elevated disturbance levels. This might reflect the initial phases of volcanic disturbance of the climax vegetation, since the reworked DGF facies are otherwise indicative of low levels of disturbance. Increased disturbance is implied by the primary tuff composition of the Smok Hill facies (Smok-1), while the fine grain size supports a relatively distal setting.

From observations of the Mount Wawel sequence and the distribution of exotic moraine blocks, the Smok Hill and Dragon Glacier floras are interpreted as originating from a single small source area close to the neck of Dragon Glacier. Therefore the difference in sedimentary composition of the respective plant-bearing lithologies is considered here to imply a stratigraphic, rather than lateral separation.

In summary, the available sedimentological evidence derived from fossiliferous exotic blocks from the Dragon Glacier and Smok Hill localities suggests a medial to distal volcanic source. Reworked volcaniclastic sediments suggesting low levels of volcanic disturbance and deposition in an ephemeral lacustrine basin, dominate the Dragon Glacier sedimentary rocks. In contrast the Smok Hill sediments are a primary ash fall deposit, possibly derived from hydrovolcanic activity and imply high levels of disturbance. The low morphotype diversity of the Mount Wawel flora (Zastawniak et al., 1985) also tends to suggest high levels of disturbance.

Paleoenvironment

Growing at a paleolatitude of ~62°S, these southern high-latitude forests would have experienced a growth regime with

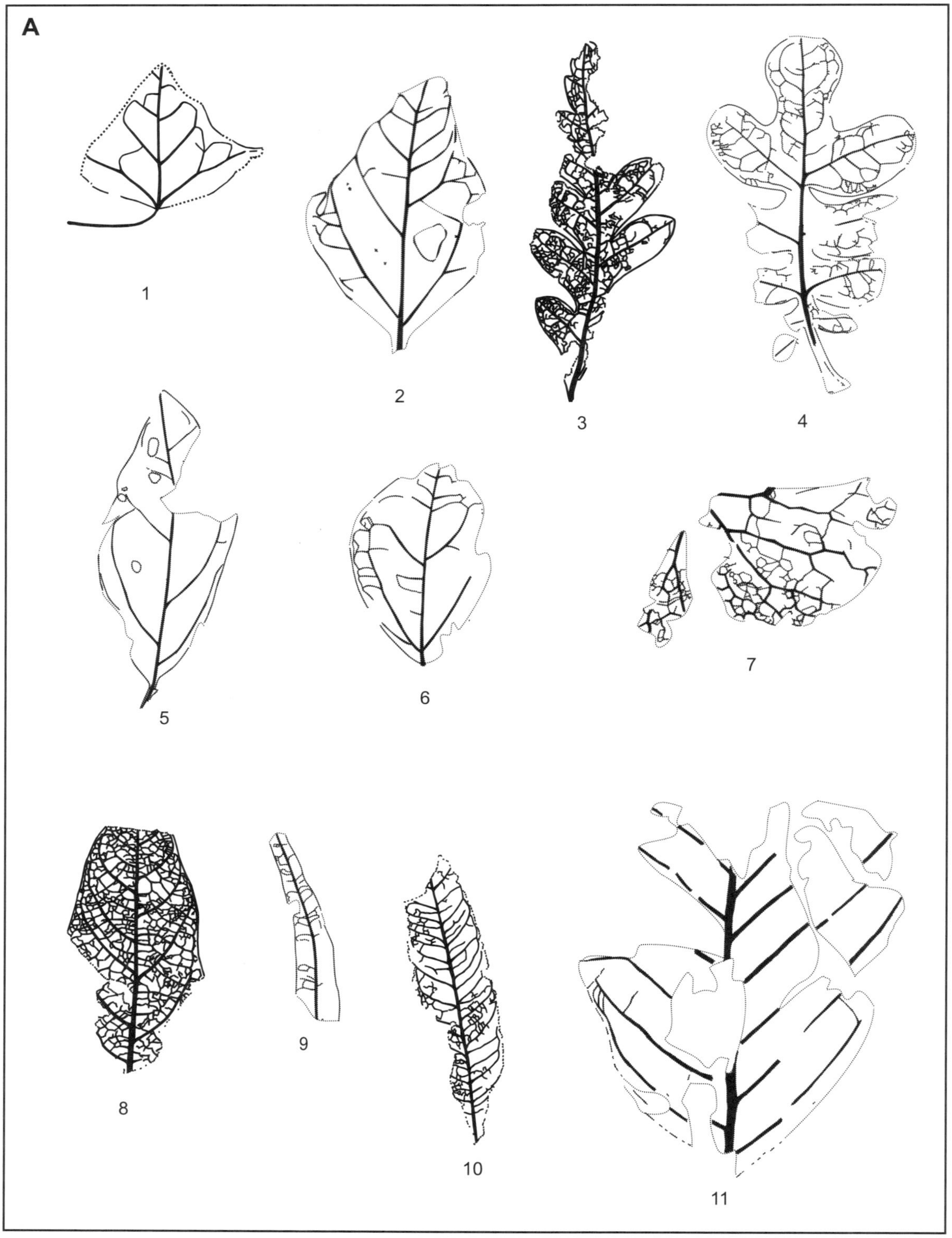

Figure 4 (on this and facing page). A: Entire margined dicotyledonous angiosperm morphotypes from Point Hennequin numbered arbitrarily with a magnification × 0.75 except for morphotypes 3 and 8 (× 1). Morphotype 4 is a probable herbaceous morphotype. B: Toothed margined morphotypes from Point Hennequin numbered arbitrarily with a magnification at × 0.75 except for morphotypes 20, 25, 36, 37 (× 2). Morphotype 12 is of compound leaf morphology; morphotypes 36 and 37 are probable juvenile leaves; lamina of morphotypes 26, 29, 31, 33, and 35 have plicate vernation suggesting deciduousness.

B
12
13
14
15
16
17
18
19
20
21
22
23
24
25
26
27
28
29
30
31
32
33
34
35
36
37

TABLE 3. SUMMARY OF THE PALEOCLIMATE DATA DERIVED FROM THE DRAGON GLACIER FLORA AT POINT HENNEQUIN

Leaf margin analysis (LMA) regression	Mean annual temperature (MAT) (°C)	SE* (°C)	σMAT† (°C)	Mean annual precipitation (mm)	SE (mm)	Leaf size index
LMA[1]	10.14	0.8	2.4	N.A	N.A	N.A
LMA[2]	10.87	Not stated	1.7	N.A	N.A	N.A
LMA[3]	10.65	2	2.2	N.A	N.A	N.A
LMA[4]	10.80	3.4	2.3	N.A	N.A	N.A
LMA[5]	10.4	2.1	1.9	N.A	N.A	N.A
Average MAT	10.58	N.A	2.1	N.A	N.A	N.A
LMA[6]	7.35[§]/8.8[#]	N.A	N.A	N.A	N.A	N.A
N.A	N.A	N.A	N.A	N.A	N.A	23.53
N.A	N.A	N.A	N.A	1039.1	+448.8 –313.4	N.A

Note: These analyses are based on 34 dicot leaf morphotypes, 26 of which are toothed margined and 11 that are entire margined. The proportion of entire margined leaves is 0.294.

*SE—Standard error.

†σMAT—Binomial sampling error of Wilf (1997) a minimum standard error of ±2 °C is applied where the binomial sampling error is less than this value.

[1]Wing and Greenwood (1993) based on Wolfe (1979).

[2]Wiemann et al. (1998).

[3]Wilf (1997) based on temperate and tropical floras of North and South America.

[4]Wilf (1997) based on the CLAMP data set of Wolfe (1993).

[5]Wilf (1997) with the 32 coldest months removed.

[6]Southern Hemisphere LMA not included in average MAT value (see text for details).

[§]Using an increase in 4% entire margined species to equate to an increase in 1°C (Wolfe, 1979).

[#]Plot given by Spicer (1990) and Spicer and Parrish (1990) which does not intercept with the origin.

lengthy periods of winter darkness and prolonged low-angle sunlight during the summer months. The forests contained evergreen and deciduous elements, as evidenced by *Nothofagus* leaves with characteristic deciduous plicate vernation (Birkenmajer and Zastawniak, 1986, 1988, 1989; Hill, 1991; Hill and Read, 1991; Hunt, 2001).

Dragon Glacier flora. Although systematic descriptions and identifications of the leaf material found in the Dragon Glacier flora are not presented here, it is clear from the floral assemblage (Figs. 4A and B) that the vegetation during the middle Eocene was more diverse than the impoverished flora envisaged by Zastawniak et al. (1985). The angiosperm leaves comprise elements such as Nothofagaceae, Cunoniaceae, Proteaceae and possible Eucryphiaceae and Sterculiaceae (Hunt, 2001). The coniferous component of the vegetation is also diverse with organs of both Cupressaceae and Araucariaceae in the flora (Hunt, 2001) supplementing the previously recognized Podocarpaceae (Zastawniak et al., 1985). Ferns and *Equisetum* are also present (Zastawniak et al., 1985). Perhaps not surprisingly, a recent study of an Eocene wood flora from King George Island described similar elements (Poole et al., 2001). The relatively diverse composition of this vegetation and inferred low disturbance levels suggest that this was a climax vegetation controlled only by small scale, successional dynamics (e.g., Veblen et al., 1996; Poole et al., 2001).

Smok Hill and Mount Wawel floras. In contrast to the DGF, both the in situ Mount Wawel and Smok Hill floras are low in diversity and *Nothofagus*-dominated, in accordance with previous descriptions (Zastawniak et al., 1985). The only addition to this flora is Araucariaceae found in the MWF. Low-diversity floras are more characteristic of the vegetation subsequent to periods of active volcanism and therefore suggest a raised level of disturbance in this immediate area (Poole et al., 2001)—an interpretation supported by the primary nature of the Smok Hill tuffs.

Since the three floras are geographically close, compositional variations are likely to reflect temporal, rather than spatial, variations in disturbance. In this respect the transition from a higher-diversity flora (i.e., from the Dragon Glacier flora to the lower-diversity Mount Wawel and Smok Hill floras) could reflect different phases in the disturbance and succession cycle (see Poole et al., 2001). At the modern El Chichón volcano, Mexico, certain sections of the canopy and subcanopy vegetation were preserved in different lithologies at different intervals in the volcaniclastic sequence and could be interpreted as separate stratigraphic horizons if preserved in the fossil record (Burnham and Spicer, 1986; Burnham, 1994)—a possible scenario for the Point Hennequin floras. However, the absence of known canopy components other than *Nothofagus* in the Smok Hill flora, such as Araucariaceae, argue against this interpretation.

The impact of volcanic disturbance and its effects on the diversity and composition of the Point Hennequin floras is represented schematically in Figure 5. The inferred paleoenvironmental setting of the Dragon Glacier flora is a fluvial-lacustrine environment on the lower flanks of a local stratovolcano (Fig.

5A). In modern environments, active pyroclastic volcanism and resulting high sedimentation rates leads to damming of local drainage networks and the formation of ephemeral lacustrine basins (Fig. 5B; Spicer, 1989). Local subsidence and basin formation due to construction of the volcanic edifice is also common. In either case, local, often ephemeral, depositional basins are created that preserve organic material produced by both pyroclastic stripping of the surrounding vegetation or by normal taphonomic processes. Due to the enhanced sedimentation rates associated with the erosion of loosely consolidated volcaniclastic deposits, these plants have a high preservation potential (e.g., Spicer, 1989; Burnham, 1994). Volcanic activity (Figs. 5B and C) causes major disturbance of the local vegetation, possibly damming local drainage networks and forming ephemeral lacustrine basins, in which further plant and sediment deposition occurs. During this phase, shade intolerant plants such as *Nothofagus* quickly come to dominate the vegetation and further volcanism leads to the preservation of low-diversity *Nothofagus*-dominated fossil assemblages such as the Smok Hill and Mount Wawel floras. Over time, and in the absence of further volcanic activity, the vegetation will return to its predisturbance composition (Fig. 5A).

Paleoclimate. Calculated minimum MATs from the Dragon Glacier flora (Table 3) range between 10 and 11 °C with an average of ~10.5 °C that supports the warm microthermal climate inferences published in other paleovegetational studies (summarized in Figure 6). Cooler MAT values (~7–9 °C; Table 3) are derived from the modern Southern Hemisphere regression suggested by Wolfe (1979). Warm microthermal climates are also inferred from the proportion of lobed morphotypes in the flora.

The new minimum MAP estimate for the Dragon Glacier flora overlaps with modern MAP ranges of 900–2500 mm from the Valdivian rainforests (Ohga, 1987), which are considered to have a composition analogous to the Paleogene vegetation of West Antarctica (Birkenmajer and Zastawniak, 1986, 1989; Poole et al., 2001).

The DGF LSI value (23.53) is similar to LSI values (20–25) derived from the coeval middle Eocene, Australian, Anglesea, Golden Grove, Nerriga and Nelly Creek floras with estimated MATs of 15–21 °C (Greenwood, 1994) rather than MATs of ~10 °C as indicated by the DGF. However, these coeval middle Eocene floras are dominated by entire margined taxa (66%–75%) suggesting a taxon related bias of the LSI-MAT relationship. An LSI of 23.53 suggests a predominantly microphyllous flora tending toward notophyly (Burnham, 1989).

Comparative studies of the wood flora from the James Ross Basin (Fig. 1) tentatively suggest late Paleocene–early Eocene MAT of between 7 and 15 °C (Francis and Poole, 2002; Fig. 6) whereas oxygen isotope studies of shallow marine Eocene faunas (which are interpreted here as a proxy for minimum terrestrial paleotemperatures) from the same region suggest temperatures of 8.6–11.7 °C (Ditchfield et al., 1994). The Mount Wawel flora and Smok Hill flora, although not morphologically diverse enough to be used for MAT analyses, are dominated by nothofagaceous leaves that also suggest cool to warm temperate climates (Zastawniak et al., 1985; Hill, 1991).

General paleoclimatic inferences based on fossil plant material from the Antarctic Peninsula region (e.g., Birkenmajer and Zastawniak, 1986, 1988, 1989; Askin, 1992; Hayes, 1999; Francis and Poole, 2002) provide evidence for the climate becoming cooler and somewhat drier from the Late Cretaceous into the Tertiary with a temporary increase in temperature during the Maastrichtian–earliest Paleocene (Fig. 6). The minimum MAT and MAP estimates presented here reflect a cool to warm temperate climate (cf. Holdridge, 1967; Whittaker, 1975) prevailing across the northern Antarctic Peninsula region during the mid Eocene. This seems to have preceded a relatively warm, moist phase during the late Eocene (Fig. 6) prior to the onset of glaciation in the Oligocene. The overall trend in MAT derived from fossil plant material is similar to that from oxygen isotope studies of marine macrofossils and foraminifera from the James Ross Basin (Barrera et al., 1987; Pirrie and Marshall, 1990; Ditchfield et al., 1994) and complements the general trends in the global $\delta^{18}O$ curve given by Zachos et al. (2001; redrawn in Figure 6). However, both the terrestrial MAT and MAP values are relatively poorly constrained during the late Eocene–earliest Oligocene, highlighting the need for further research on Antarctic floras to help define terrestrial climatic trends and vegetational responses to the changing environment during the onset of cooler climates during the late Paleogene.

CONCLUSIONS

The results from this study provide new evidence for Antarctic paleoecology prior to the onset of the widespread Cenozoic glaciation. Three roughly contemporaneous floras from Point Hennequin on King George Island, have been discussed. The major findings can be summarized as:

1. The Dragon Glacier flora, originally described as an impoverished postglacial upper Oligocene *Nothofagus*-Podocarpaceae-pteridophyte assemblage (Zastawniak et al., 1985), has been reinvestigated and found to be a comparatively species-rich middle Eocene flora (Hunt, 2001) with a strong bias toward *Nothofagus* species.
2. The Mount Wawel flora (Zastawniak et al., 1985) and the newly described Smok Hill flora, are contemporaneous with the Dragon Glacier flora and also *Nothofagus*-dominated but are species-poor relative to the Dragon Glacier flora.
3. These three middle Eocene floras are considered to reflect successional changes in the paleovegetation that was strongly controlled by volcanic disturbance. This supports the Valdivian Model proposed by Poole et al. (2001). On this basis, interpretations of climate derived from Antarctic fossil floras, have to be set in context of the vegetation dynamics in an environment largely controlled by volcanic activity.
4. Given the high morphotype diversity of the Dragon Glacier flora and the inferred low levels of disturbance, taxon-independent physiognomic analyses are considered appropriate for this flora. A prevailing warm microthermal, temperate climate with MAT of between 10 and 11 °C and minimum MAP of ~1039 mm is inferred.

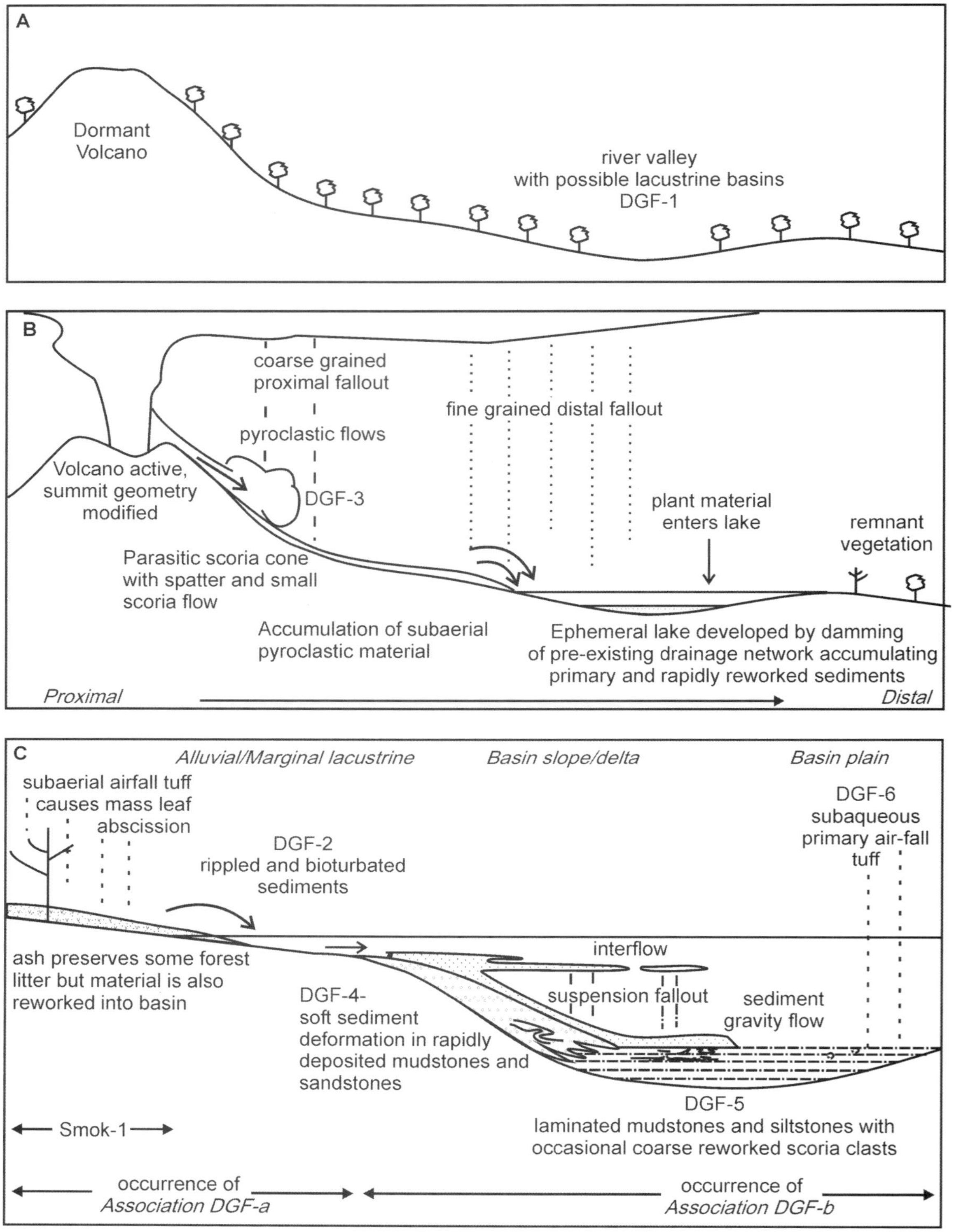

Figure 5. Paleoenvironmental reconstruction of Point Hennequin. Schematic representations of facies distribution and the likely mechanism for sedimentation at Point Hennequin. DGF—Dragon Glacier flora. A: Hypothetical forest environment prior to volcanism. B: Syn-volcanic and postvolcanic phase, rapid primary pyroclastic sedimentation deposits large volumes of material that is quickly resedimented into the existing and/or newly formed ephemeral lacustrine basin(s). C: Schematic cross section of the proposed lacustrine setting relevant to the syn-volcanic phase and for the early post-volcanic phase (modified from clastic sedimentation model proposed for oligotrophic lakes in Talbot and Allen, 1996). See text for further details on facies and numbering.

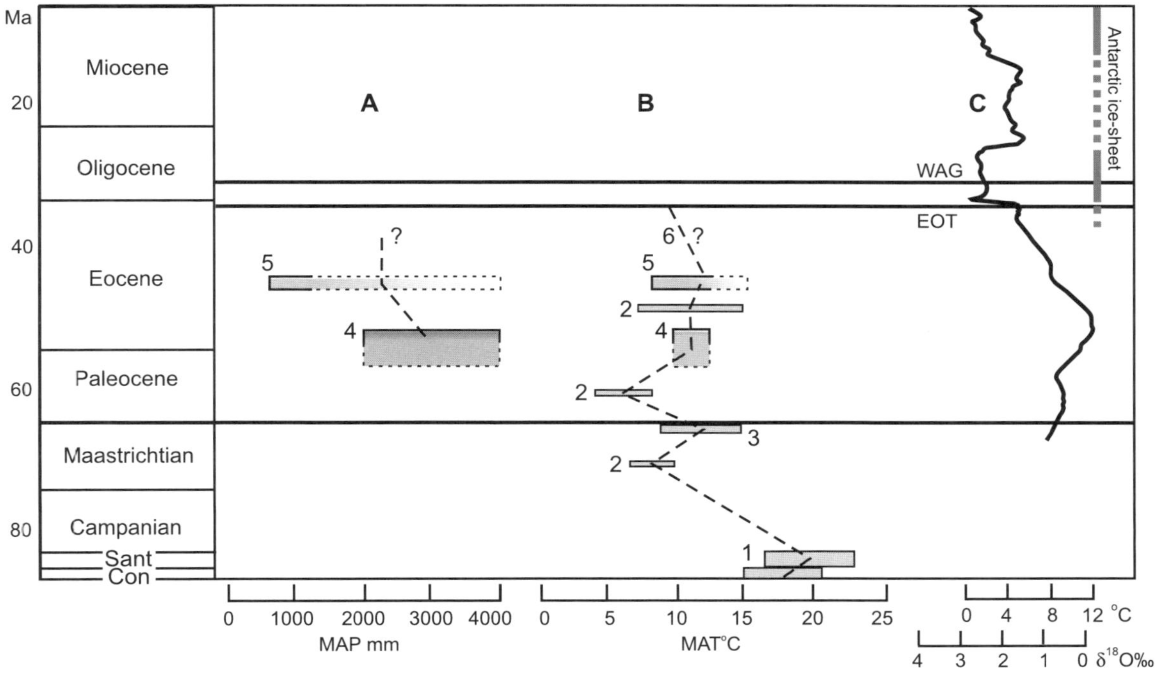

Figure 6. Trend (black dashed line) in paleoclimate for the Late Cretaceous through Miocene in the Antarctic Peninsula region as derived from fossil plant assemblages (redrawn and updated from Francis and Poole, 2002). A: Estimated mean annual precipitation (MAP, mm). B: Estimated mean annual temperature (MAT, °C) derived from leaf and wood macrofossils and floras. C: Global $\delta^{18}O$ isotope curve for the latest Cretaceous and Cenozoic (from Zachos et al., 2001) included here for comparisons. Data sources: 1—Hayes (1999) for the Coniacian (Con) Hidden Lake Formation (15–20 °C), and the early Santonian (Sant)–early Campanian Santa Marta Formation (16–23 °C); 2—Francis and Poole (2002); 3—Askin (1992); 4—Birkenmajer and Zastawniak (1986); 5—herein; 6—Askin (1997). EOT—Eocene-Oligocene Transition (Wilson et al., 1998), WAG—West Antarctic glaciation (Dingle and Lavelle, 1998). Dashed boxes and gradation in shading indicate range of possible geological age due to uncertainties in dating of some sedimentary sequences, or total range in climatic parameter (see text). "?" denotes uncertainty as to the directionality of MAT during the late Eocene as datum point 6 provides no absolute value being derived from Nearest Living Relative extrapolations rather than from morphological and/or anatomical characters. Relative extent of the Antarctic ice sheet is denoted by the gray line with partial or ephemeral state denoted by the broken line, and full scale and permanent state represented by the solid line.

Further detailed studies of the early Tertiary fossil floras such as the leaf, pollen and wood assemblages in the South Shetlands and Antarctic Peninsula region will no doubt continue to provide further evidence of the complex interplay between environmental disturbance and vegetational dynamics set against a backdrop of gradual climate deterioration prior to the onset of late Tertiary glaciation in these high southern latitudes.

ACKNOWLEDGMENTS

We thank the British Antarctic Survey for the loan of the fossil material and the invitation to collect further plant material. Mr. N. Cundall, Leeds University, is thanked for preparing the sedimentary thin sections. Dr. L. Amy is thanked for critical feedback on the sedimentary interpretations. Dr. J.L. Smellie is thanked for discussion of the Point Hennequin stratigraphy and access to sedimentary field data. Drs. D.J. Cantrill and J. Francis are thanked for their discussions, support and encouragement throughout the course of this work. RJH undertook this research whilst was in receipt of an NERC CASE studentship, which is here gratefully acknowledged.

REFERENCES CITED

Askin, R.A., 1988, The palynological record across the Cretaceous-Tertiary transition on Seymour Island, Antarctica: Boulder, Colorado, Geological Society of America Memoir 169, p. 155–162.

Askin, R.A., 1989, Endemism and heterochroneity in the Late Cretaceous (Campanian) to Paleocene palynofloras of Seymour Island, Antarctica: Implications for origins, dispersal, and palaeoclimates of southern floras, *in* Crame, J.A., ed., Origins and evolution of the Antarctic biota: London, Geological Society Special Publication 47, p. 217–226.

Askin, R.A., 1990, Cryptogam spores from the upper Campanian and Maastrichtian of Seymour Island, Antarctica: Micropaleontology, v. 36, p. 141–156.

Askin, R.A., 1992, Late Cretaceous–early Tertiary Antarctic outcrop evidence for past vegetation and climates: Antarctic Research Series, v. 56, p. 61–73.

Askin, R.A., 1994, Monosulcate angiosperm pollen from the Lopez de Bertodano Formation (upper Campanian-Maastrichtian-Danian) of Seymour Island, Antarctica: Review of Palaeobotany and Palynology, v. 81, p. 151–164.

Askin, R.A., 1997, Eocene–?Earliest Oligocene terrestrial palynology of Seymour Island, Antarctica, *in* Ricci, C.A., ed., The Antarctic Region: Geological evolution and processes: Proceedings of the VII International Symposium of Antarctic Earth Sciences: Siena, Italy, Terra Antartica Publication, p. 993–996.

Askin, R.A., and Baldoni, A.M., 1998, The Santonian through Palaeogene record of Proteaceae in the southern South America–Antarctic Peninsula region: Australian Systematic Botany, v. 11, p. 373–390.

Bailey, I.W., and Sinnott, E.W., 1915, A botanical index of Cretaceous and Tertiary climates: Science, v. 41, p. 831–834.

Bailey, I.W., and Sinnott, E.W., 1916, The climatic distribution of certain types of angiosperm leaves: American Journal of Botany, v. 3, p. 24–39.

Barrera, E., Huber, B., Savin, S., and Webb, P., 1987, Antarctic marine temperatures: Late Campanian through early Paleocene: Paleoceanography, v. 2, p. 21–47.

Barton, C.M., 1959, The east coast of Admiralty Bay from Point Hennequin to Ternyck Glacier: F.I.D.S. Base G. Admiralty Bay Report (1949). Geology Survey Archive No AD6/2G/1959/GI (paper copy, 19 p.).

Barton, C.M., 1961, The geology of King George Island, South Shetland Islands: Preliminary Report of the Falkland Islands Dependencies Survey Number 12 (paper copy, 18 p.).

Barton, C.M., 1964a, Significance of the Tertiary fossil floras of King George Island, South Shetland Island, *in* Adie, R.J., ed., Antarctic geology: Amsterdam, North Holland Publishing Company, p. 603–608.

Barton, C.M., 1964b, The geology of King George Island, South Shetland Islands [Ph.D. thesis]: Birmingham, UK, University of Birmingham, 167 p.

Barton, C.M., 1965, The Geology of the South Shetland Islands, Part 3: The stratigraphy of King George Island: British Antarctic Survey Scientific Reports, v. 44, p. 1–33.

Birkenmajer, K., 1980, Tertiary volcanic-sedimentary succession at Admiralty Bay, King George Island (South Shetland Islands, Antarctica): Studia Geologica Polonica, v. 64, p. 7–65.

Birkenmajer, K., 1981, Lithostratigraphy of the Point Hennequin Group (Miocene volcanics and sediments) at King George Island (South Shetland Islands, Antarctica): Studia Geologica Polonica, v. 74, p. 175–197.

Birkenmajer, K., 1990, Geochronology and climatostratigraphy of Tertiary glacial and interglacial successions on King George Island, South Shetland Islands (West Antarctica): Zentralblatt für Paläontologie, v. 1, p. 41–151.

Birkenmajer, K., 1997, Tertiary glacial/interglacial palaeoenvironments and sea-level changes, King George Island, West Antarctica: An overview: Bulletin of the Polish Academy of Sciences, Earth Sciences, v. 44, p. 157–181.

Birkenmajer, K., and Zastawniak, E., 1986, Plant remains of the Dufayel Island Group (early Tertiary?), King George Island, South Shetland Islands (West Antarctica): Acta Palaeobotanica, v. 26, p. 33–49.

Birkenmajer, K., and Zastawniak, E., 1988, Late Cretaceous–early Neogene vegetation history of the Antarctic Peninsula Sector: Gondwana breakup and Tertiary glaciations: Bulletin of the Polish Academy of Sciences, Earth Sciences, v. 37, p. 63–88.

Birkenmajer, K., and Zastawniak, E., 1989, Late Cretaceous–early Tertiary floras of King George Island, West Antarctica: Their stratigraphic distribution and palaeoclimatic significance, *in* Crame, J.A., ed., Origins and evolution of the Antarctic biota: London, Geological Society Special Publication 47, p. 17–226.

Birkenmajer, K., Narêbski, W., Nicoletti, M., and Petrucciani, C., 1983, Late Cretaceous through Oligocene K-Ar ages of the King George Island Supergroup volcanics, South Shetland Islands (West Antarctica): Bulletin de l'Académie Polonaise des Sciences Série des Sciences de la Terre, v. 30, p. 133–143.

Boyd, A., 1994, Some limitations in using leaf physiognomic data as a precise method for determining paleoclimates with an example from the Late Cretaceous Pautût flora of West Greenland: Palaeogeography, Palaeoclimatology, Palaeoecology, v. 112, p. 261–278.

Breitkreutz, C., 1991, Fluvio-lacustrine sedimentation and volcanism in a Late Carboniferous tensional intraarc basin, northern Chile: Sedimentary Geology, v. 4, p. 173–187.

Burnham, R.J., 1989, Relationships between standing vegetation and leaf litter in paratropical forest: Implications for palaeobotany: Review of Palaeobotany and Palynology, v. 58, p. 5–32.

Burnham, R.J., 1994, Plant deposition in modern volcanic environments: Transactions of the Royal Society of Edinburgh Earth Sciences, v. 84, p. 275–281.

Burnham, R.J., and Spicer, R.A., 1986, Forest litter preserved by volcanic activity at El Chichón, Mexico: A potentially accurate record of the preeruption vegetation: Palaios, v. 1, p. 558–161.

Burnham, R.J., Pitman, N.C.A., Johnson, K.R., and Wilf, P.W., 2001, Habitat-related error in estimating temperatures from leaf margins in a humid tropical forest: American Journal of Botany, v. 88, p. 1096–1102.

Cain, S.A., and de Castro, O.G.M., 1959, Manual of vegetation analysis: New York, Harper and Row, 325 p.

Cantrill, D.J., and Poole, I., 2002, Cretaceous patterns of floristic change in the Antarctic Peninsula, *in* Crame, J.A., and Owen, A.W., eds., Palaeobiology and biodiversity change: A comparison of the Ordovician and Mesozoic-Cenozoic radiations: Geological Society [London] Special Publication 194, p. 141–152.

Cao, L., 1992, Late Cretaceous and Eocene palynofloras from Fildes Peninsula, King George Island (South Shetland Islands), Antarctica, *in* Yoshida, Y., ed., Recent progress in Antarctic Earth science: Tokyo, Terra Scientific Publishing Company (Terrapus), p. 363–369.

Cao, L., 1994, Late Cretaceous palynoflora in King George Island of Antarctica with reference to its palaeoclimatic significance, *in* Shen, Y., ed., Stratigraphy and palaeontology of Fildes Peninsula, King George Island: State Antarctic Committee Monograph 3: Beijing, China, Science Press, p. 85–96.

Carey, S.N., 1991, Transport and deposition of tephra by pyroclastic flows and surges, *in* Fisher, R.V., and Smith, G.A., eds., Sedimentation in volcanic settings: SEPM (Society for Sedimentary Geology) Special Publication 45, p. 39–57.

Cas, R.A.F., and Wright, J.V., 1987, Volcanic successions, modern and ancient: London, Allen and Unwin, 528 p.

Dettmann, M.E., 1989, Antarctica: Cretaceous cradle of austral temperate rainforests?, *in* Crame, J.A., ed., Origins and evolution of the Antarctic biota: London, Geological Society Special Publication 47, p. 89–105.

Dettmann, M.E., and Thomson, M.R.A., 1987, Cretaceous palynomorphs from the James Ross Island area, Antarctica: A pilot study: British Antarctic Survey Bulletin, v. 77, p. 13–59.

Dingle, R.V., and Lavelle, M., 1998, Late Cretaceous–Cenozoic climatic variations of the northern Antarctic Peninsula: New geochemical evidence and review: Palaeogeography, Palaeoclimatology and Palaeoecology, v. 141, p. 215–232.

Ditchfield, P.W., Marshall, J.D., and Pirrie, D., 1994, High-latitude palaeotemperature variation: New data from the Tithonian to Eocene of James Ross Island, Antarctica: Palaeogeography, Palaeoclimatology, Palaeoecology, v. 107, p. 79–101.

Dolph, G.E., and Dilcher, D.L., 1979, Foliar physiognomy as an aid in determining palaeoclimate: Palaeontographica Abteilung B, v. 170, p. 151–172.

Dolph, G.E., and Dilcher, D.L., 1980, Variation in leaf size with respect to climate in the tropics of the Western Hemisphere: Bulletin of the Torrey Botanical Club, v. 107, p. 154–162.

Duan, W., and Cao, L., 1998, Late Palaeogene palynoflora from Point Hennequin of the Admiralty Bay, King George Island, Antarctica, with reference to its stratigraphical significance: Chinese Journal of Polar Science, v. 9, p. 125–132.

Dutra, T., and Batten, D., 2000, Upper Cretaceous floras of King George Island, West Antarctica, and their palaeoenvironmental and phytogeographic implications: Cretaceous Research, v. 21, p. 181–209.

Fairon-Demaret, M., Hilton, J., and Berry, C.M., 1999, Fossil leaf character states: Multivariate analyses, *in* Jones, T.P., and Rowe, N.P., eds., Fossil plants and spores: Modern techniques: London, The Geological Society, p. 33–35.

Ferguson, D.K., 1985, The origin of leaf assemblages: New light on an old problem: Review of Palaeobotany and Palynology, v. 46, p. 117–188.

Fisher, R.V., 1966, Rocks composed of volcanic fragments and their classification: Earth Science Reviews, v. 1, p. 287–298.

Fisher, R.V., and Schmincke, H.U., 1984, Pyroclastic rocks: Berlin, Springer-Verlag, 472 p.

Francis, J.E., and Poole, I., 2002, Cretaceous and Tertiary climates of Antarctica: Evidence from fossil wood: Paleogeography, Palaeoclimatology, Palaeoecology, v. 182, p. 47–64.

Givnish, T.J., 1984, Leaf and canopy adaptations in tropical forests, *in* Medina, E., et al., eds., Physiological ecology of plants of the wet tropics: Tasks for Vegetation Science 12: The Hague, Junk, p. 51–84.

Greenwood, D.R., 1992, Taphonomic constraints on foliar physiognomic interpretations of Late Cretaceous and Tertiary palaeoclimates: Review of Palaeobotany and Palynology, v. 71, p. 149–190.

Greenwood, D.R., 1994, Palaebotanical evidence for Tertiary climates, *in* Hill, R.S., ed., History of the Australian vegetation: Cretaceous to Holocene: Cambridge, UK, Cambridge University Press, p. 44–59.

Greenwood, D.R., 1996, Eocene monsoon forests in central Australia?: Australian Systematic Botany, v. 9, p. 95–112.

Gregory, K.M., 1996, Are paleoclimate estimates biased by foliar physiognomic responses to increased atmospheric CO_2?: Palaeogeography, Palaeoclimatology, Palaeoecology, v. 124, p. 39–51.

Hayes, P., 1999, Cretaceous angiosperm floras of Antarctica [Ph.D. thesis]: Leeds, UK, University of Leeds, 310 p.

Hee, Y.C., and Soon-Keun, C., 1991, Study on the gymnospermous fossil woods from the King George Island: Korean Journal of Polar Research Special Issue, v. 2, p. 179–185.

Hickey, L.J., 1973, Classification of the architecture of dicotyledonous leaves: American Journal of Botany, v. 60, p. 17–33.

Hickey, L.J., 1979, A revised classification of the architecture of dicotyledonous leaves, *in* Metcalfe, C.R., and Chalk, L., eds., Anatomy of the dicotyledons (2nd edition): Oxford, Clarendon, p. 25–39.

Hickey, L.J., and Wolfe, J.A., 1975, The bases of angiosperm phylogeny: Vegetative morphology: Annals of the Missouri Botanical Garden, v. 62, p. 538–589.

Hill, R.S., 1991, Tertiary *Nothofagus* (Fagaceae) macrofossils from Tasmania and Antarctica and their bearing on the evolution of the genus: Botanical Journal of the Linnean Society, v. 105, p. 73–112.

Hill, R.S., and Read, J., 1991, A revised infrageneric classification of *Nothofagus* (Fagaceae): Botanical Journal of the Linnean Society, v. 105, p. 37–72.

Hill, R.S., and Scriven, L.J., 1995, The angiosperm dominated woody vegetation of Antarctica: A review: Review of Palaeobotany and Palynology, v. 86, p. 175–198.

Holdridge, L.R., 1967, Life zone ecology: San Jose, Tropical Science Center, p. 506.

Hunt, R.H., 2001, Biodiversity and palaeoecology of Tertiary fossil floras in Antarctica [Ph.D. thesis]: Leeds, UK, University of Leeds, 400 p.

Jacobs, B.F., 1999, Estimation of rainfall variables from leaf characters in tropical Africa: Palaegeography, Palaeoclimatology, Palaeoecology, v. 145, p. 231–250.

Jagmin, N.I.B., 1987, Estudo anatômico dos troncos fósseis de Admiralty Bay, King George Island (Península Antártica): Acta Biologica Leopoldensia, v. 9, p. 81–98.

Jardine, D.J., 1950, Base G. Admiralty Bay Report (1949). Geology Report F.I.D.Sc Bureau No. 83/50 (paper copy, 31 p.).

Jordan, G.J., 1997, Uncertainty in palaeoclimatic reconstructions based on leaf physiognomy: Australian Journal of Botany, v. 45, p. 527–547.

Lawver, L.A., Gahagan, L.M., and Coffin, M.F., 1992, The development of paleoseaways around Antarctica: Antarctic Research Series, v. 56, p. 7–30.

Leaf Architecture Working Group, 1999, Manual of leaf architecture: Morphological description and categorization of dicotyledonous and netveined monocotyledonous angiosperms: Washington, D.C., Smithsonian Institution, 67 p.

Li, H.M., 1994, Early Tertiary Fossil Hill flora from Fildes Peninsula of King George Island, Antarctica, *in* Shen, Y., ed., Stratigraphy and palaeontology of Fildes Peninsula, King George Island: State Antarctic Committee Monograph 3: Beijing, China, Science Press, p. 133–171.

Lucas, R.C., and Lacey, W.S., 1981, A permineralized wood flora of probable Tertiary age from King George Island, South Shetland Islands: British Antarctic Survey Bulletin, v. 53, p. 147–151.

Markgraf, V., Romero, E., and Villagrán, C., 1996, History and palaeoecology of South American *Nothofagus* forests, *in* Veblen, T.T., et al., eds., The ecology and biogeography of *Nothofagus* forests: New Haven and London, Yale University Press, p. 354–386.

McCarron, J.J., and Larter, R.D., 1998, Late Cretaceous to early Tertiary subduction history of the Antarctic Peninsula: Journal of the Geological Society of London, v. 155, p. 255–268.

Nakayama, K., and Yoshikawa, S., 1997, Depositional processes of primary to reworked volcaniclastics on an alluvial plain: An example from the lower Pliocene Ohta tephra bed of the Tokai Group, central Japan: Sedimentary Geology, v. 107, p. 211–229.

Ohga, N., 1987, The plant ecology of the cloud forests of the eastern and western slopes of the Andes, I: Microclimatic conditions of the forests of Bolivia and Chile, *in* Nishida M., ed., Contributions to the botany of the Andes, II: Academia Scientific Book Inc., Tokyo, p. 12–22.

Pankhurst, R.J., and Smellie, J.L., 1983, K-Ar geochronology of the South Shetland Islands, Lesser Antarctica: Apparent lateral migration of Jurassic to Quaternary island-arc volcanism: Earth and Planetary Science Letters, v. 66, p. 214–222.

Pettijohn, F.J., 1975, Sedimentary rocks: New York, Harper and Row, 628 p.

Pirrie, D., and Marshall, J.D., 1990, High-paleolatitude Late Cretaceous paleotemperatures: New data from James Ross Island, Antarctica: Geology, v. 18, p. 31–34.

Poole, I., 2002, Systematics of Cretaceous and Tertiary *Nothofagoxylon:* Implications for Southern Hemisphere biogeography and evolution of the Nothofagaceae: Australian Systematic Botany, v. 15, no. 2, p. 247–276.

Poole, I., and Cantrill, D.J., 2001, Fossil woods from Williams Point Beds, Livingston Island, Antarctica: A Late Cretaceous southern high-latitude flora: Palaeontology, v. 44, p. 1081–1112.

Poole, I., and Francis, J.E., 1999, The first record of fossil atherospermataceous wood from the Upper Cretaceous of Antarctica: Review of Palaeobotany and Palynology, v. 107, p. 97–107.

Poole, I., and Francis, J.E., 2000, The first record of Winteraceae wood from the Cretaceous of Antarctica: Annals of Botany, v. 85, p. 307–315.

Poole, I., and Gottwald, H., 2001, Monimiaceae sensu lato, an element of Gondwanan polar forests: Evidence from the Late Cretaceous–early Tertiary wood flora of Antarctica: Australian Systematic Botany, v. 14, p. 207–230.

Poole, I., Cantrill, D.J., Hayes, P., and Francis, J.E., 2000a, The fossil record of Cunoniaceae: New evidence from Late Cretaceous fossil wood of Antarctica: Review of Palaeobotany and Palynology, v. 111, p. 127–144.

Poole, I., Gottwald, H., and Francis, J.E., 2000b, Illiciaceae, an element of Gondwanan polar forests? Late Cretaceous and early Tertiary woods of Antarctica: Annals of Botany, v. 86, p. 421–432.

Poole, I., Richter, H.G., and Francis, J.E., 2000c, Gondwanan origins for Sassafras (Lauraceae): Evidence from Late Cretaceous fossil wood of Antarctica: International Association of Wood Anatomists Journal, v. 21, p. 463–475.

Poole, I., Hunt, R.J., and Cantrill, D.J., 2001, A fossil wood flora from King George Island: Ecological implications for an Antarctic Eocene vegetation: Annals of Botany, v. 88, p. 33–54.

Raunkiaer, C., 1934, The life forms of plants and statistical plant geography: Oxford, Clarendon Press, 632 p.

Richards, P.W., 1952, The tropical rain forest: Cambridge, UK, Cambridge University Press, 450 p.

Shen, Y., 1994, Subdivision and correlation of Cretaceous to Paleogene volcano-sedimentary sequence from Fildes Peninsula, King George Island, Antarctica, *in* Shen, Y., ed., Stratigraphy and palaeontology of Fildes Peninsula,

King George Island: State Antarctic Committee Monograph 3: Beijing, China, Science Press, p. 1–36.

Smellie, J.L., Pankhurst, R.J., Thomson, M.R.A., and Davies, R.E.S., 1984, The geology of the South Shetland Islands, VI: Stratigraphy, geochemistry, and evolution: British Antarctic Survey Scientific Report 87 (paper copy, p. 1–85).

Smith, G.A., 1987, The influence of explosive volcanism on fluvial sedimentation: The Deschutes Formation (Neogene) in central Oregon: Journal of Sedimentary Petrology, v. 57, p. 613–629.

Smith, G.A., 1991, Sedimentology of proximal to distal volcaniclastics dispersed across an active foldbelt: Ellensburg Formation (late Miocene), central Washington: Sedimentology, v. 35, p. 935–977.

Soliani, E., and Bonhomme, M.G., 1994, New evidence for Cenozoic resetting of K-Ar ages in volcanic rocks of the northern portion of the Admiralty Bay, King George Island, Antarctica: Journal of South American Earth Sciences, v. 7, p. 85–94.

Spicer, R.A., 1989, The formation and interpretation of plant fossil assemblages: Advances in Botanical Research, v. 16, p. 96–191.

Spicer, R.A., 1990, Climate from plants, *in* Briggs, D.E.G., and Crowther, P.R., eds., Palaeobiology: A synthesis: Oxford, Blackwell Science, p. 401–403.

Spicer, R.A., and Parrish, J.T., 1990, Late Cretaceous–early Tertiary palaeoclimates of northern high latitudes: A quantitative view: Journal of the Geological Society of London, v. 147, p. 329–341.

Spicer, R.A., and Wolfe, J.A., 1999, Fossil leaf character states: Multivariate analyses, *in* Jones, T.P., and Rowe, N.P., eds., Fossil plants and spores: Modern techniques: London, The Geological Society, p. 233–239.

Stuchlik, L., 1981, Tertiary pollen spectra from the Ezcurra Inlet Group of Admiralty Bay, King George Island (South Shetland Islands, Antarctica): Studia Geologica Polonica, v. 72, p. 109–132.

Talbot, M.R., and Allen, P.A., 1996, Lakes, *in* Reading, H.G., ed., Sedimentary environments: Processes, facies, and stratigraphy: Oxford, Blackwell Science, p. 83–124.

Torres, T., 1984, *Nothofagoxylon antarcticus* n. sp., madera fósils del Terciario de la isla Rey Jorge, islas Shetland del Sur, Antártica: Serie Cìentifica Instituto Àntartico Chileno, v. 31, p. 39–52.

Torres, T., and Lemoigne, Y., 1988, Maderas fósiles terciarias de la Formación Caleta Arctowski, Isla Rey Jorge, Antártica: Serie Cìentifica Instituto Àntartico Chileno, v. 37, p. 69–107.

Torres, T., Román, A., Deza, A., and Rivera, C., 1984, Anatomiá, mineralogía y termoluminiscencia de madera fósil del Terciario de la isla Rey Jorge, Islas Shetland del Sur, *in* Ma del Carmen Perrilliat, ed., III Congreso Latinoamericano de Paleontologia Brasil 2, p. 566–574.

Truswell, E.M., 1990, Cretaceous and Tertiary vegetation of Antarctica: A palynological perspective, *in* Taylor, T.N., and Taylor, E.L., eds., Antarctic palaeobiology: Its role in the reconstruction of Gondwana: Berlin, Springer-Verlag, p. 71–88.

Truswell, E.M., 1991, Antarctica: A history of terrestrial vegetation, *in* Tingey, R.J., ed., The geology of Antarctica: Oxford, Clarendon Press, p. 499–537.

Upchurch, G.R., and Wolfe, J.A., 1987, Mid-Cretaceous to early Tertiary vegetation and climate: Evidence from fossil leaves and woods, *in* Friis, E.M., et al., eds., The origin of angiosperms and their biological consequences: Cambridge, UK, Cambridge University Press, p. 75–105.

Veblen, T.T., Ashton, D.H., Schlegel, F.M., and Veblen, A.T., 1977, Plant succession in a timberline depressed by vulcanism in south-central Chile: Journal of Biogeography, v. 4, p. 275–294.

Veblen, T.T., Donoso, C.Z., Schlegel, F.M., and Escobar, B.R., 1981, Forest dynamics in south-central Chile: Journal of Biogeography, v. 8, p. 211–247.

Veblen, T.T., Donoso, C., Kitzberger, T., and Rebertus, A.J., 1996, Ecology of southern Chilean and Argentinean *Nothofagus* forests, *in* Veblen, T.T., et al., eds., The ecology and biogeography of *Nothofagus* forests: New Haven and London, Yale University Press, p. 293–353.

Vessel, R.K., and Davies, D.K., 1981, Nonmarine sedimentation in an active fore-arc basin: Society of Economic Palaeontologists and Mineralogists Special Publication 31 (paper copy), p. 31–45.

Webb, L.J., 1959, A physiognomic classification of Australian rainforests: Journal of Ecology, v. 47, p. 551–570.

Whittaker, R.H., 1975, Communities and ecosystems: New York, Macmillan Publishers, 385 p.

Wiemann, M.C., Manchester, S.R., Dilcher, D.L., Hinojosa, L.F., and Wheeler, E.A., 1998, Estimation of temperature and precipitation from morphological characters of dicotyledonous leaves: American Journal of Botany, v. 85, p. 1796–1802.

Wilf, P., 1997, When are leaves good thermometers? A new case for leaf margin analysis: Palaeobiology, v. 23, p. 373–390.

Wilf, P., Wing, S.L., Greenwood, D.R., and Greenwood, C.L., 1998, Using fossil leaves as paleoprecipitation indicators: An Eocene example: Geology, v. 26, p. 203–206.

Willan, R.C.R., and Kelley, S.P., 1999, Mafic dike swarms in the South Shetland Islands volcanic arc: Unravelling multiepisodic magmatism related to subduction and continental rifting: Journal of Geophysical Research, v. 104, p. 23,051–23,068.

Wilson, G.S., Roberts, A.P., Verosub, K.L., Florindo, F., and Sagnotti, L., 1998, Magnetobiostratigraphic chronology of the Eocene-Oligocene transition in the CIROS-1 core, Victoria Land margin, Antarctica: Implications for Antarctic glacial history: Geological Society of America Bulletin, v. 110, p. 35–47.

Wing, S.L., and Greenwood, D.R., 1993, Fossils and fossil climate: The case for equable continental interiors in the Eocene: Philosophical Transactions of the Royal Society of London, series B, v. 341, p. 243–252.

Wolfe, J.A., 1971, Tertiary climatic fluctuations and methods of analysis of Tertiary floras: Palaeogeography, Palaeoclimatology, Palaeoecology, v. 9, p. 27–57.

Wolfe, J.A., 1979, Temperature parameters of humid to mesic forests of eastern Asia and relation to forests of other regions in the Northern Hemisphere and Australasia: U.S. Geological Survey Professional Paper 1106, p. 1–37.

Wolfe, J.A., 1993, A method of obtaining climatic parameters from leaf assemblages: U.S. Geological Survey Bulletin, v. 2040, p. 1–71.

Wolfe, J.A., 1995, Palaeoclimatic estimates from Tertiary leaf assemblages: Annual Review of Earth and Planetary Sciences, v. 23, p. 119–142.

Wolfe, J.A., and Upchurch, G.R., 1987, North American nonmarine climates and vegetation during the Late Cretaceous: Palaeogeography, Palaeoclimatology, Palaeoecology, v. 33, p. 1–77.

Xue, Y., Shen, Y., and Zhuo, E., 1996, Petrological characteristics of the sedimentary volcaniclastic rocks of the Fossil Hill Formation (Eocene) in King George Island, West Antarctica: Antarctic Research, v. 7, p. 99–117.

Zachos, J., Pagani, M., Sloan, L., Thomas, E., Billups, K., 2001, Trends, rhythms, and aberrations in global climate 65 Ma to present: Science, v. 292, p. 686–693.

Zastawniak, E., 1981, Tertiary leaf flora from the Point Hennequin Group of King George Island (South Shetland Islands), Antarctica: Preliminary report: Studia Geologica Polonica, v. 72, p. 97–112.

Zastawniak, E., Wrona, R., Gazdzicki, A., and Birkenmajer, K., 1985, Plant remains from the top part of the Point Hennequin Group (upper Oligocene), King George Island (South Shetland Islands, Antarctica): Studia Geologica Polonica, v. 81, p. 143–164.

Zhang, S., and Wang, Q., 1994, Paleocene petrified wood on the west side of Collins glacier in the King George Island, Antarctica, *in* Shen, Y., ed., Stratigraphy and palaeontology of Fildes Peninsula, King George Island, Antarctica: State Antarctic Committee Monograph 3: Beijing, China, Science Press, p. 231–238.

Manuscript Accepted by the Society August 13, 2002

Printed in the USA

Geological Society of America
Special Paper 369
2003

Status of plant megafossils during the early Paleogene in India

R.C. Mehrotra*
Birbal Sahni Institute of Palaeobotany, 53 University Road, Lucknow-226007, India

ABSTRACT

Plant megafossil assemblages from the early Paleogene of India are primarily known from the Maastrichtian–Danian Deccan Intertrappean beds, upper Paleocene sediments of Meghalaya (northeast India), upper Paleocene–lower Eocene sediments of Kutch of Gujarat and middle Eocene Fuller's Earth deposits of Rajasthan (western India). Among them the most important and well studied is the Deccan Intertrappean flora. This flora is very interesting because it is marked by the presence of certain elements typical of Gondwana continents. Although the megafloral assemblages of Rajasthan and Kutch are not very rich, the presence of calcareous red algae and some coastal elements makes the floras important. The most notable feature of the Paleogene flora of India is the absence of Dipterocarpaceae, which is quite abundant during the Neogene. The general climate during the early Paleogene in India was warm and humid with plenty of rainfall. The land was covered with tropical evergreen to moist deciduous forests. Paleogeographers should take paleobotanical data into account in reconstructing the location of the fast moving Indian plate during the early Paleogene.

INTRODUCTION

One of the basic aims of many paleobotanical studies is to reconstruct the paleoenvironment of the area. Vegetation is sensitive to climate so knowing the paleovegetation helps in deducing the paleoclimate. Correct identification of the fossils is very important in order to avoid wrong conclusions therefore, in the present paper many forms of uncertain affinities have been excluded while listing the fossil flora.

The study of plant megafossils from the early Paleogene of India is very important, as during the period India was on its way to north after being separated from the other Gondwana continents. The flora serves as an important tool in supporting or rejecting any existing paleogeographic map showing the position of India during the early Paleogene.

The important megafossil assemblages known from India during the early Paleogene (Fig. 1) are given below. Though rare reports of plant megafossils are also known from Ladakh and a few other places in northeast India (Tripathi et al., 1981; Lakhanpal et al., 1983; Mehrotra, 2000b), they are not treated in this paper.

1. The Deccan Intertrappean flora—The exact age of the Deccan Traps has been a matter of controversy (Bande et al., 1988) but according to the latest views based on radiometric dating the age may range from upper Maastrichtian to Danian (Venkatesan et al., 1993; Shukla et al., 1997; Salis and Saxena, 1998; Khosla, 1999). As the samples for the dating were not taken from many of the fossil localities, it becomes difficult to say whether the fossils are of upper Maastrichtian or Danian in age.

2. The floras of the Tura Formation of the Garo Hills and the Lakadong Sandstone Formation of the Khasi Hills, Meghalaya, are considered upper Paleocene on the basis of various parameters discussed in detail by Rajarao (1981).

3. In Kutch, Gujarat Province, two lignite mines, Panandhro and Rajpardi, have produced Paleogene floras. The age of the former was earlier thought to be lower Eocene (Ypre-

*E-Mail: rcmehrotra@yahoo.com

Mehrotra, R.C., 2003, Status of plant megafossils during the early Paleogene in India, *in* Wing, S.L., Gingerich, P.D., Schmitz, B., and Thomas, E., eds., Causes and Consequences of Globally Warm Climates in the Early Paleogene: Boulder, Colorado, Geological Society of America Special Paper 369, p. 413–423.

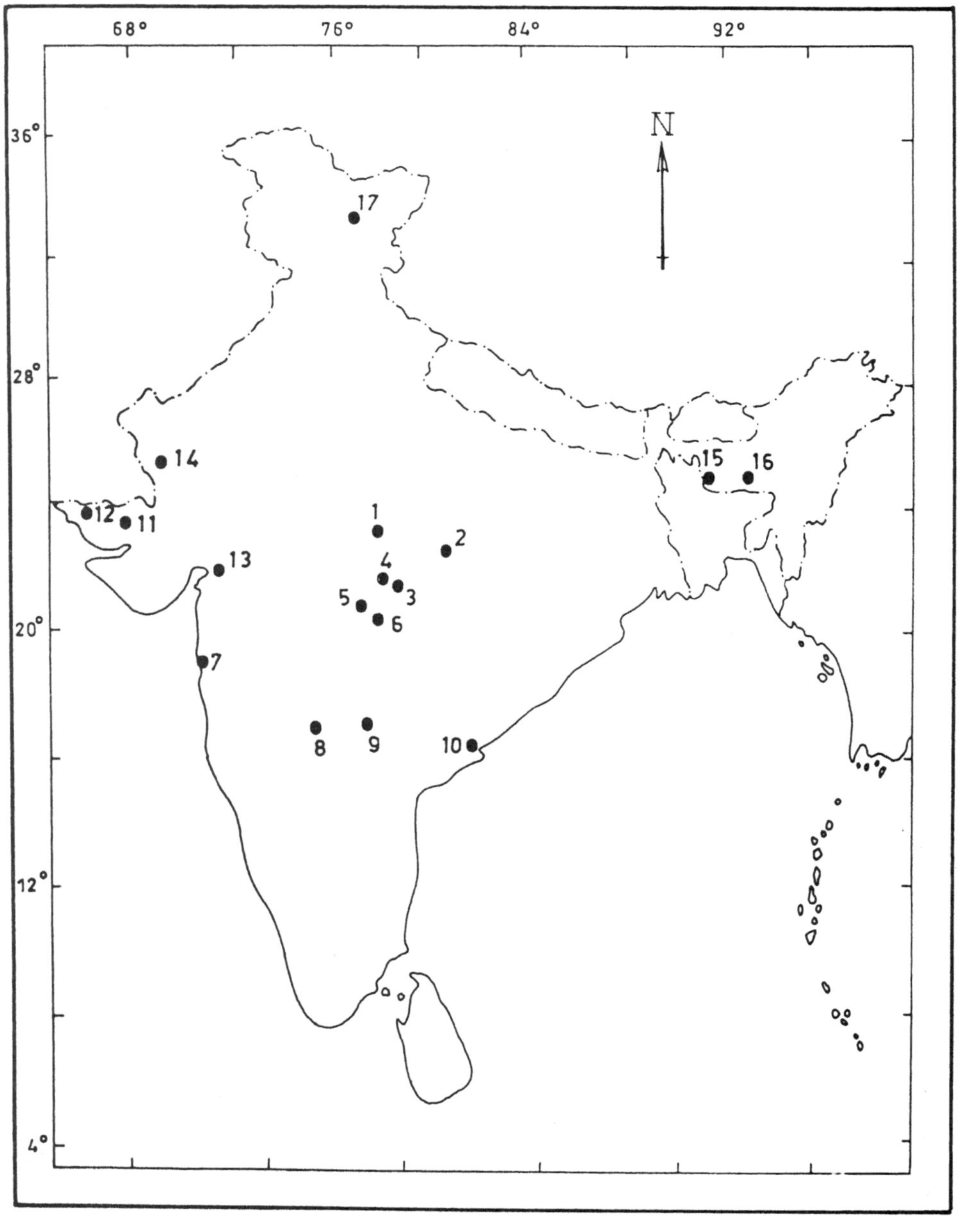

Figure 1. Map showing important early Paleogene localities of India (1. Sagar 2. Shahpura 3. Seoni 4. Mohgaon Kalan 5. Mahurzari 6. Nawargaon 7. Mumbai 8. Gulbarga 9. Vicarabad 10. Rajahmundary and 11. Anjar—all from the Deccan Intertrappean beds 12. Panandhro and 13. Rajpardi—both from Kutch 14. Kapurdi—from Fuller's Earth deposits of Rajasthan 15. Tura and 16. Cherrapunji—from upper Paleocene sediments of Meghalaya and 17. Ladakh—from Hemis conglomerate horizon).

sian) (Biswas and Raju, 1971, 1973) but it is now considered to be upper Paleocene (Biswas, 1986, 1992). Rajpardi is considered to be lower Eocene on the basis of palynological data (Kar and Bhattacharyya, 1992).

4. A small florule from the Fuller's Earth deposits of Barmer, Rajasthan is little known, but is mentioned as middle Eocene by Bande (1992).

The Deccan Intertrappean flora is the most important because it is very rich and diverse, comprising all the major divisions of the Plant Kingdom from algae to angiosperms, and all the possible plant parts from roots to flowers. The flora is also very interesting as it is marked by the presence of certain elements of Gondwana continents.

DECCAN INTERTRAPPEAN FLORA

The Deccan Intertrappean flora is mostly known from central India, with important localities at Mohgaon Kalan in Chindwara District, Nawargaon in Wardha District and Shahpura and Parapani in the Mandla District (Bande et al., 1988). The other localities are Mahurzari and Mumbai in Maharashtra, Sagar and Seoni in Madhya Pradesh, Vicarabad and Rajahmundry in Andhra Pradesh, Gurmatkal in Karnataka and Anjar in Kutch (Guleria and Srivastava, 2001). Important and common plant taxa described from these localities are listed in Table 1.

Angiosperms are the most dominant group in the flora and are equally represented by both monocotyledonous and

TABLE 1. PLANT MEGAFOSSILS FROM THE DECCAN INTERTRAPPEAN BEDS OF INDIA

Fossil species	Comparable extant forms	Organ	Reference
Algae			
Mougeotiates deccani	*Mougeotia*	Filament	Barlinge and Paradkar, 1982
Ulothrix	*Ulothrix*	Filament	Dwivedi, 1959; Mahadevan and Sarma, 1948; Sahni and H.S. Rao, 1943
Peyssonnelia antiqua	Corallinaceae	Thallus	Bande et al., 1981
Distichoplax raoi	Corallinaceae	Thallus	Bande et al., 1981
Westiellopsis mahabalei	Stigonemataceae	Filament	Biradar, 1977
Anabaena	*Anabaena*	Filament	Trivedi et al., 1986
Microcystiopsis mohgaonkalaensis	*Microcystis*	Filament	Mishra and Maithy, 1984
Palaeohydrococcus raoi	*Hydrococcus*	Filament	Mishra and Maithy, 1984
Chara sausari	*Chara*	Filament with gyrogonite	Sahni and S.R.N. Rao, 1943
Chara spp.	*Chara*	Gyrogonite	Bande et al., 1988
Fungi			
Palaeopthora mohgaonensis	Pythiaceae	Filament with spores	Singhai, 1978
Palaeosordaria lagena	Sordariaceae	Filament with spores	Sahni and H.S. Rao, 1943
Ustilago deccani	*Ustilago*	Filament with spores	Chitaley and Yawale, 1978
Chlamydosporites gramineum	*Tilletia*	Filament with spores	Paradkar, 1975
Helminthosporites mohgaoense	*Helminthosporium*	Spores	Chitaley and Sheikh, 1971
Diplodia rodei	*Diplodia*	Spores	Mahabale, 1969
Helicominites salvinitis	*Helicomina*	Spores	Barlinge and Paradkar, 1982
Phomites ebenoxyloni	*Phoma*	Fruiting body	Chitaley and Patil, 1972
Dactylosporites dicotylophylli	*Dactylosporium*	Spores	Paradkar, 1974
Bryophyta			
Shuklanites deccanii	Anthocerotales	Sporogonium	Singhai, 1964, 1975
Notothylites nirulai	Anthocerotales, Mosses	Sporogonium	Chitaley and Yawale, 1980
Riccia chitaleyi	*Riccia*	Thallus	Sheikh and Kapgate, 1982
Pteridophyta			
Selaginella	*Selaginella*	Strobilus	Singh and Patil, 1979
Azolla intertrappea	*Azolla*	Leaf, root, stem, micro and megasporocarp	Sahni, 1941; Sahni and H.S. Rao, 1943
Marsilea intertrappea	*Marsilea*	Rhizome, petiole	Upadhye and Patil, 1979
Salvinia intertrappea	*Salvinia*	Sporocarps	Mahabale, 1950a; Paradkar and Barlinge, 1979
Rodeites dakshini	*Regnellidium*	Sporocarp	Sahni, 1943; Chitaley and Paradkar, 1973
Gymnosperms			
Gymnovulites	Cycadales	Ovule	Paradkar, 1976
Takliostrobus alatus	Coniferales	Cone	Sahni, 1931
Indostrobus bifidolepis	Coniferales	Cone	Sahni, 1931
Pityostrobus crassitesta	Coniferales	Cone	Sahni, 1931
Mohgaostrobus sahnii	Coniferales	Cone	Prakash, 1957
Harrisostrobus intertrappea	Coniferales	Cone	Chitaley and Sheikh, 1973
Podocarpoxylon deccanensis	Podocarpaceae	Wood	Trivedi and Srivastava, 1989
Podocarpoovulites chitaleyi	Podocarpaceae	Ovule	Sheikh and Kolhe, 1982
Araucarioxylon keriense	*Araucaria-Agathis*	Wood	Trivedi and Srivastava, 1989
Angiosperms			
Monocotyledons			
Pandanusocarpon umariense	Pandanaceae	Fruit	Bonde, 1990
Palmoxylon ghuguensis	*Chrysalidocarpus*	Wood	Ambwani and Prakash, 1983
P. sundaram	*Cocos*	Wood	Sahni, 1947; Rao and Menon, 1964
Palmocarpon cocoides	*Cocos*	Fruit	Mehrotra, 1988
Phoenicites lakhanpalii	*Sclerosperma*	Leaf	Guleria and Mehrotra, 1999
Hyphaeneocarpon indicum	*Hypaene*	Fruit	Bande et al., 1982
Nypa	*Nypa*	Fruit	Sahni and Rode, 1937; Nambudiri, 1966
Cyclanthodendron sahnii	*Cyclanthus*	Floral axis	Sahni and Surange, 1944, 1953; Mahabale, 1950b
Sahnipushpam shuklai	N.A.	Flower	Verma, 1956; Chitaley, 1964
Musa cardiosperma	Zingiberaceae	Fruit	Jain, 1964a

continued

TABLE 1. ***Continued***

Fossil species	Comparable extant forms	Organ	Reference
Musocaulon indicum	Musaceae	Petiole, pseudostem	Rao and Menon, 1963; Jain, 1964b
Musophyllum indicum	*Musa*	Leaf	Prakash et al., 1979
Callistemonites indicus	*Musa*	Fruit and seeds	Bande et al., 1993
Heliconiaites mohgaoensis	*Heliconia*	Petiole and lamina	Trivedi and Verma, 1972
Dicotyledons			
Polyalthioxylon parapaniense	*Polyalthia*	Wood	Bande, 1973; Mehrotra, 1990
Nymphaeocaulon intertrappeum	Nymphaeaceae	Wood	Trivedi and Ambwani, 1971
Homalioxylon mandlaense	*Homalium*	Wood	Bande, 1974
Hydnocarpoxylon indicun	*Hydnocarpus*	Wood	Bande and Khatri, 1980; Guleria and Mehrotra, 1999
Garcinioxylon tertiarum	*Garcinia*	Wood	Bande and Khatri, 1980
Sterculinium spp.	*Sterculia* spp.	Wood	Lakhanpal et al., 1978; Bande and Prakash, 1984a
Grewioxylon indicum	*Grewia*	Wood	Prakash and Dayal, 1965; Khare et al., 2001
Elaeocarpoxylon spp.	*Elaeocarpus-Echinocarpus*	Wood	Prakash and Dayal, 1964; Khare et al., 2001
Atalantioxylon indicum	*Atalantia-Limonia*	Wood	Lakhanpal et al., 1978
Evodinium indicum	*Evodia*	Wood	Bande and Prakash, 1984b
Ailanthoxylon indicum	*Ailanthus*	Wood	Prakash et al., 1967; Mehrotra, 1990
Simarouboxylon indicum	*Simarouba*	Wood	Shallom, 1960b
Burseroxylon preserratum	*Bursera*	Wood	Bande and Prakash, 1984a
Canarioxylon shahpuraensis	*Canarium*	Wood	Trivedi and Srivastava, 1985
Turraeanthus deccanensis	*Turraeanthus*	Wood	Awasthi et al., 1996
Heyneoxylon tertiarum	*Heynea*	Wood	Bande and Prakash, 1984a
Amooroxylon deccanensis	*Amoora*	Wood	Bande and Prakash, 1984b
Walsura deccanensis	*Walsura*	Wood	Mehrotra, 1989b
Gomphandroxylon samnapurensis	*Stemonurus*	Wood	Bande and Khatri, 1980
Lophopetalumoxylon indicum	*Lophopetalum*	Wood	Mehrotra et al., 1984
Leeoxylon multiseriatum	*Leea*	Wood	Prakash and Dayal, 1964
Euphorioxylon deccanense	*Otonephelium*	Wood	Mehrotra, 1987
Dracontomelumoxylon mangiferumoides	*Dracontomelum*	Wood	Bande and Prakash, 1984a
Eucalyptus dharmendrae	*Eucalyptus*	Wood	Bande et al., 1986
Tristania confertoides	*Tristania conferta*	Wood	Bande et al., 1986
Callistemonoxylon deccanensis	*Callistemon-Melaleuca*	Wood	Bande et al., 1986
Syzygioxylon mandlaense	*Syzygium cumini*	Wood	Ingle, 1973
Barringtonioxylon spp.	*Barringtonia*	Wood	Shallom, 1960a; Prakash and Dayal, 1965; Bande and Khatri., 1980
Enigmocarpon parijai	*Sonneratia*	Fruit	Sahni and Rode, 1937; Mahabale and Deshpande, 1957
Sahnianthus parijai	*Sonneratia*	Flower	Shukla, 1944; Chitaley, 1955
Sonneratioxylon preapetalum	*Sonneratia*	Wood	Mehrotra, 1989a
Euphorbiocarpon drypeteoides	Euphorbiaceae	Fruit	Mehrotra et al., 1983
Chitaleypushpam mohgaoense	Lythraceae	Flower	Paradkar, 1973
Tetramelioxylon prenudiflora	*Tetrameles*	Wood	Lakhanpal and Verma, 1966
Gmelina tertiara	*Gmelina*	Wood	Bande, 1987
Aristolochioxylon prakashii	*Aristolochia*	Wood	Kulkarni and Patil, 1977
Bischofia palaeojavanica	*Bischofia*	Wood	Guleria and Srivastava, 2001
Mallotoxylon keriense	*Mallotus*	Wood	Lakhanpal and Dayal, 1964; Guleria and Srivastava, 2001
Artocarpoxylon deccanensis	*Artocarpus*	Wood	Mehrotra et al., 1984
Harrisocarpon sahnii	N.A.	Fruit	Chitaley and Nambudiri, 1973
Sahniocarpon harrisii	N.A	Fruit	Chitaley and Patil, 1973

Note: N.A.—Not applicable

dicotyledonous taxa. Though a large number of palm woods occur in the flora, their precise identification is very difficult and, therefore, most of them are described under the organ genus *Palmoxylon* Schenk. Other noteworthy genera are *Cyclanthodendron, Simarouboxylon* and *Rodeites,* which possess affinities with the South American taxa *Cyclanthus, Simarouba* and *Regnellidium* respectively. *Musa* is another interesting genus represented by its fruit, petiole and leaf. The fruits are known as *Musa cardiosperma* and *Callistemonites indicus* while petiole and leaf are described as *Musocaulon* and *Musophyllum* respectively. Manchester and Kress (1993) while describing a fossil Musaceae from the Eocene of western North America as *Ensete oregonense,* raised a doubt about the affinities of *Musa cardiosperma,* which they considered to be Zingiberaceae. They pointed out certain anatomical features that could easily differentiate the seeds of these two families. Unfortunately, being a cast, these features could not be confirmed in *Callistemonites indicus.* They also advocated a pantropical origin for the Musaceae. However, the authors did not make any comments on the described taxa *Musocaulon* and *Musophyllum,* which support an old world origin of the family. A comprehensive study is required in this direction to confirm the facts.

UPPER PALEOCENE FLORA OF MEGHALAYA

The fossiliferous exposures of the upper Paleocene sediments occur mostly in Meghalaya, northeast India and floras are known from the Tura Formation and the Lakadong Sandstone Formation.

The Tura Formation is exposed in the Garo Hills District and the well-known localities are Nangwalbibra, Damalgiri and Tura, whereas the Lakadong Sandstone Formation is exposed in Cherrapunji and Laitryngew of the Khasi Hills District. The important fossils known from these two formations are listed in Table 2. The flora is primarily based on leaf impressions and/or compressions though a few fruits and/or seeds are also known from there.

LIGNITE MINE FLORA OF KUTCH

The Kutch region lies in the western part of India and the flora comprises a few red algae and a variety of angiospermous leaf impressions and woods. The commonly known taxa of the flora are presented in Table 3. The algae are known from Babbia Hills, leaf impressions from near Panandhro, Bhuj District and mostly woods from Rajpardi lignite, Bharuch District, Gujarat.

FULLER'S EARTH FLORA OF RAJASTHAN

The assemblage presented in Table 4 is not well known and a few clusiaceous leaf remains along with a fruit of Arecaceae are known from Kapurdi, Barmer District of Rajasthan, western India.

TABLE 2. PLANT MEGAFOSSILS FROM THE UPPER PALEOCENE SEDIMENTS OF NORTHEAST INDIA

Fossil species	Modern comparable forms	Organ	Reference
Tura Formation			
Nypa	*Nypa*	Fruit	Bhattacharyya, 1983
Nelumbo nagalensis	*Nelumbo nucifera*	Leaf	Bhattacharyya, 1983
Grewia spp.	*Grewia*	Leaf	Lakhanpal, 1955
Bombacites orientalis	*Bombax*	Leaf	Lakhanpal, 1955
Atalantia palaeomonophylla	*Atalantia*	Leaf	Mehrotra, 2000a
Schleichera palaeotrijuga	*Schleichera trijuga*	Leaf	Mehrotra, 2000a
Eomangiferophyllum damalgiriensis	*Mangifera*	Leaf	Mehrotra et al., 1998
Mangifera someshwarica	*Mangifera*	Leaf	Mehrotra, 2000a
Leguminocarpon spp.	Fabaceae	Fruit	Bhattacharyya, 1985
Terminalia palaeocatappa	*Terminalia catappa*	Leaf	Mehrotra, 2000a
Syzygium nangalbibrensis	*Syzygium*	Leaf	Mehrotra, 2000a
Chrysophyllum tertiarum	*Chrysophyllum*	Leaf	Mehrotra, 2000a
Phoebe sublanceolata	*Phoebe*	Leaf	Bhattacharyya, 1983
Artocarpus garoensis	*Artocarpus*	Leaf	Bhattacharyya, 1983
Lakadong Sandstone Formation			
Nypa	*Nypa*	Fruit	Bhattacharyya, 1967
Calophyllum spp.	*Calophyllum*	Leaf	Bhattacharyya, 1967; Ambwani, 1992
Bauhinia siwalika	*Bauhinia*	Leaf	Ambwani, 1992
Pterocarpus eocenicus	*Pterocarpus*	Leaf	Ambwani, 1992
Sonneratia meghalayensis	*Sonneratia*	Leaf	Ambwani, 1992
Terminalia panandhroensis	*Terminalia*	Leaf	Ambwani, 1992
Ficus cherrapunjiensis	*Ficus mysorensis*	Leaf	Ambwani, 1992

TABLE 3. PLANT MEGAFOSSILS FROM THE UPPER PALEOCENE—LOWER EOCENE SEDIMENTS OF KUTCH, WESTERN INDIA

Fossil species	Modern comparable forms	Organ	Reference
Algae			
Lithophyllum sp.	Corallinaceae	Thallus with reproductive structures	Lakhanpal et al., 1984
Lithothamnium spp.	Corallinaceae	Thallus with reproductive structures	Lakhanpal et al., 1984
Angiosperms			
Pandanus	*Pandanus*	Leaf	Lakhanpal et al., 1984
Terminalia panandhroensis	*Terminalia*	Leaf	Guleria, 1994
Terminalioxylon palaeocalamansanai	*Terminalia*	Wood	Guleria, 1991
Syzygium	*Syzygium*	Leaf	Lakhanpal et al., 1984
Lagerstroemia	*Lagerstroemia*	Leaf	Lakhanpal et al., 1984
Cinnamomum	*Cinnamomum*	Leaf	Lakhanpal et al., 1984
Ficus	*Ficus*	Leaf	Lakhanpal et al., 1984
Sonneratioxylon preapetalum	*Sonneratia*	Wood	Guleria, 1991

DISCUSSION AND CONCLUSIONS

A cursory look at the megaflora of the early Paleogene of India gives an indication that the flora was typically tropical. Champion and Seth (1968) have classified the modern tropical Indian forests into 7 types: (i) wet evergreen (ii) semievergreen (iii) moist deciduous (iv) littoral and swamp (v) dry deciduous (vi) thorn and (vii) dry evergreen forest. Modern plants comparable to the fossils occur mostly in tropical evergreen and moist deciduous forests, suggesting this type of vegetation was widespread in India during the early Paleogene. Most of the taxa found as fossils are presently growing in the Western Ghats and northeast India where the climate is tropical with a uniform temperature throughout the year, annual rainfall >2000 mm (Bande and Prakash, 1982), and a long rainy season. The presence of Tethys Sea around India during the early Paleogene could be one of the factors responsible for such a climate.

The occurrence of the marine algae *Lithophyllum* and *Lithothamnium* indicates that there was a transgression in the Kutch region during the upper Paleocene–lower Eocene. The presence of *Lagerstroemia, Syzygium,* and *Pandanus* suggests the prevalence of marshes around the area in which the vegetation got buried and later formed lignite. The finding of *Cocos* from the middle Eocene of Rajasthan is very interesting because the area in Rajasthan from which this fossil was collected is now a desert, but living *Cocos* are coastal.

On the basis of *Nypa* and *Terminalia catappa,* estuarine condition can easily be suggested during the upper Paleocene in Meghalaya as *Nypa* is an extant genus of palms known only from mangrove swamps today (Mehrotra, 2000a), while *Terminalia catappa* occurs in the beach forests of Andamans and Malayan Peninsula (Awasthi and Mehrotra, 1995; Mehrotra, 2000a).

On the basis of some marine algal forms such as *Peyssonnelia* and *Distichoplax* and some coastal forms like *Cocos, Nypa,* and *Sonneratia* from the Deccan Intertrappean sediments, Lakhanpal (1974b) and Bande and Prakash (1982) have envisaged an arm of the Tethys Sea in central India during the early Paleogene. However, this theory can not be sustained in the absence of other supportive evidence. In India during the early Paleogene the inland areas were marked by freshwater lakes and ponds, as indicated by freshwater algae, water ferns, *Barringtonia, Nelumbo* and *Syzygium.*

Most existing paleocontinental maps (Smith and Briden, 1979; Smith et al., 1994) show the Indian Subcontinent as an isolated island from the Aptian to the Eocene. Paleogeographers have not noted the geologically sudden appearance of angiosperms

TABLE 4. PLANT MEGAFOSSILS FROM THE FULLER'S EARTH DEPOSITS OF BARMER, RAJASTHAN, WESTERN INDIA

Fossil species	Modern comparable forms	Organ	References
Mesua tertiara	*Mesua*	Leaf	Lakhanpal and Bose, 1951; Lakhanpal, 1964
Garcinia barooabii	*Garcinia*	Leaf	Lakhanpal and Bose, 1951; Lakhanpal, 1964
Cocos sahnii	*Cocos*	Fruit	Kaul, 1951

in huge quantities in India during Maastrichtian-Danian. No reliable older megafossils of flowering plants are known from India, though Sharma (1997) and Banerji (2000) have described a few angiospermous fruits of uncertain affinities and a questionable flower from the Lower Cretaceous sediments of Rajmahal Hills, India. These records should be seriously questioned before they are accepted as reliable records of Lower Cretaceous angiosperms in India.

It is interesting and possibly significant that in India the rich plant megafossil assemblages from the Barremian to Albian do not have any definitive angiosperms (Borkar, 1993), especially since Albian deposits in Europe, North America, South America and Australia have angiosperms in them. However, the absence of angiosperms in India after the Albian and before the Maastrichtian may be due to the presence of marine deposits (Raju and Mishra, 1996) which are devoid of plant fossils. Preliminarily, it appears that the entry of angiosperms into India might have been relatively late (post-Albian) compared with their appearance on other continents, but it is hard to tell how late because of the lack of good floras in the Cenomanian through Campanian interval.

Only a few fragmentary angiospermous megafossils have been recorded from the Lameta Formation of central India that is considered as Maastrichtian in age (Mohabey, 1996). The Lameta beds are considered as infratrappeans as they are found below the Deccan Intertrappean beds.

The sudden presence of angiosperms in huge quantities in the Deccan Intertrappean flora was earlier thought to be due to their poor fossil records especially from the Cretaceous. But the situation changed when a few years ago Chatterjee and Scotese (1999) proposed several paleogeographic maps showing the evolution of the Indian plate from its origin to its union with Asia. The authors have given due importance to the animal fossils especially the dinosaur remains from India. According to them, "The lack of endemism among Indian Cretaceous terrestrial vertebrates is clearly inconsistent with the island continent hypothesis."

The current knowledge of the paleobotanical record in India is consistent with their paleogeographic interpretations. In their model an unrecognized land area named Greater Somalia occupied the position between eastern Arabia and northwest India. During the Late Jurassic, India became separated from this landmass and remained so till the Upper Cretaceous. According to them Greater India established its contact with Africa via Greater Somalia during the Campanian–early Maastrichtian. This may have paved the way for the influx of angiosperms to India (Fig. 2).

The possibility that Africa was a source area for the initial angiosperms found in India is supported by the presence of some African elements such as *Hyphaene* (Bande et al., 1982), *Chrysalidocarpus* (Ambwani and Prakash, 1983), *Turraeanthus* (Awasthi et al., 1996) and *Sclerosperma* (Guleria and Mehrotra, 1999) in the Deccan Intertrappean flora. India maintained contact with Greater Somalia until the Eocene when it collided with Asia, resulting in the formation of Himalayas (Chatterjee and Scotese, 1999). Greater Somalia may have been subducted beneath Eurasia after the collision.

The initial collision between Greater India and Asia took place during the early Eocene along the northeastern corner of India (Chatterjee and Scotese, 1999). The collision ceased the marine sedimentation and started the terrestrial deposition in the suture zone, allowing a subaerial link between the two landmasses. The suturing was complete by the end of Oligocene, as Dipterocarpaceae, a typical Southeast Asian family, is found in India during Miocene. It is generally believed that this family originated in western Malaysia (Lakhanpal, 1974a) and entered the Indian Subcontinent only after the Indian plate joined the Asian plate.

The formation of Himalayas took place in three phases. The first was in the middle Eocene, the second about the middle Miocene and the third took place in the post Pliocene. The first phase of elevation culminating in the Oligocene resulted in the continuity of land from western Malaysia to eastern India, facilitating a way for the migration of dipterocarps northwards into India. Though Paradkar (1972) recorded a dipterocarpaceous wood of *Shorea* from the Deccan Intertrappean sediments, its identification has already been questioned (Bande et al., 1988) due to the absence of vasicentric tracheids and vertical gum canals in it. This family is considered a marker of the Neogene in India and is known mainly by wood and leaves from various horizons. The detailed listing of various fossil taxa of the family has been made by Guleria (1992) from the peninsular part of India and Awasthi (1992) from the Siwalik sediments.

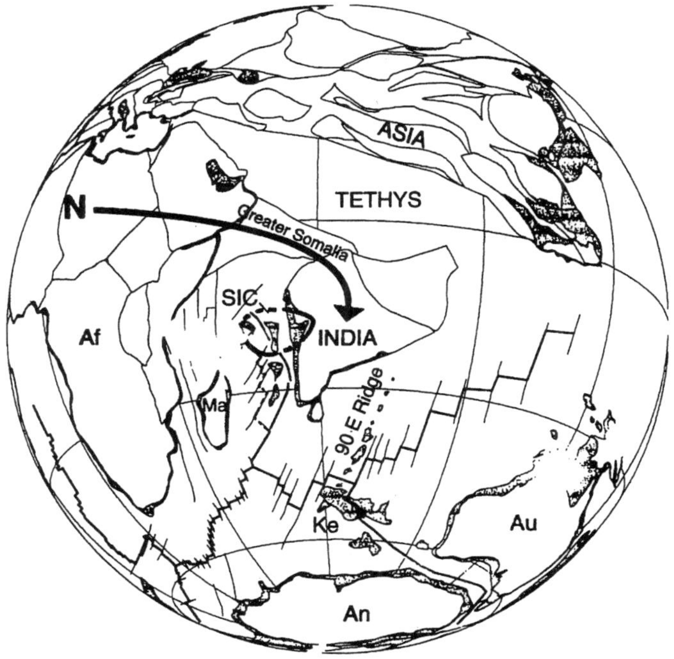

Figure 2. A paleogeographic map showing the positions of Gondwana continents during K-T boundary (after Chatterjee and Scotese, 1999).

The presence of *Eucalyptus* along with some other Australian taxa during the Maastrichtian-Danian is another riddle that needs to be solved, as almost all the species of *Eucalyptus* are found in Australia. As its identification is based on a fossil wood, which has unique anatomy characterized by the echelon arrangement of small to medium vessels profusely plugged with tyloses, uniseriate homocellular xylem rays, vasicentric tracheids, simple perforation plates, and thick-walled fibres (Kribs, 1959), it becomes necessary to rethink the position of India in relation to Australia in the past. There may be two possibilities: Either India was attached to Australia through a chain of islands even in the Upper Cretaceous, or *Eucalyptus* is not indigenous to Australia. As there is no evidence at present of any land connections between these two landmasses during the Upper Cretaceous, the second possibility seems to be more plausible. This can be further supported by the fact that there is no reliable macrofossil record of *Eucalyptus* in Australia before the Miocene (Hill, 1994). It shows that the genus might have migrated to Australia from India at the end of Oligocene.

ACKNOWLEDGMENTS

The author is thankful to the Director, Birbal Sahni Institute of Palaebotany, Lucknow, India for his constant encouragement throughout the progress of this work and for providing the necessary facilities. He is also grateful to Profs. David Dilcher and Steven Manchester of the Florida Museum of Natural History, USA for their constructive suggestions while reviewing the manuscript. The author is indebted to Prof. Scott Wing, Smithsonian National Museum of Natural History, USA, not only for his suggestions in improving the manuscript but also for his help in various ways.

REFERENCES CITED

Ambwani, K., 1992, Leaf impressions belonging to the Tertiary age of northeast India: Phytomorphology, v. 41, no. 1–2, p. 139–146.

Ambwani, K., and Prakash, U., 1983, *Palmoxylon ghuguensis* sp. nov. resembling *Chrysalidocarpus* from the Deccan Intertrappean beds of Mandla District in Madhya Pradesh: Palaeobotanist, v. 31, no. 1, p. 76–81.

Awasthi, N., 1992, Changing patterns of vegetation through Siwalik succession: Palaeobotanist, v. 40, p. 312–327.

Awasthi, N., and Mehrotra, R.C., 1995, Oligocene flora from Makum coal field, Assam, India: Palaeobotanist, v. 44, p. 157–188.

Awasthi, N., Mehrotra, R.C., and Srivastava, R., 1996, Fossil woods from the Deccan Intertrappean beds of Madhya Pradesh: Geophytology, v. 25, p. 113–118.

Bande, M.B., 1973, A petrified dicotyledonous wood from the Deccan Intertrappean beds of Mandla District, Madhya Pradesh: Botanique, v. 4, no. 1, p. 41–47.

Bande, M.B., 1974, Two fossil woods from the Deccan Intertrappean beds of Mandla District, Madhya Pradesh: Geophytology, v. 4, no. 2, p. 189–195.

Bande, M.B., 1987, Fossil wood of *Gmelina* Linn. (Verbenaceae) from the Deccan Intertrappean beds of Nawargaon with comments on the nomenclature of Tertiary woods: Palaeobotanist, v. 35, no. 2, p. 165–170.

Bande, M.B., 1992, The Palaeogene vegetation of Peninsular India (megafossil evidences): Palaeobotanist, v. 40, p. 275–284.

Bande, M.B., and Khatri, S.K., 1980, Some more fossil woods from the Deccan Intertrappean beds near Shahpura, Mandla District, Madhya Pradesh: Geophytology, v. 10, no. 2, p. 268–271.

Bande, M.B., and Prakash, U., 1982, Palaeoclimate and palaeoecology of central India during the early Tertiary: Geophytology, v. 12, no. 2, p. 152–165.

Bande, M.B., and Prakash, U., 1984a, Fossil dicotyledonous woods from the Deccan Intertrappean beds near Shahpura, Mandla District, Madhya Pradesh: Palaeobotanist, v. 31, no. 1, p. 13–29.

Bande, M.B., and Prakash, U., 1984b, Occurrence of *Evodia, Amoora,* and *Sonneratia* from the Palaeogene of India, *in* Sharma, A.K., et al., eds., Evolutionary botany and biostratigraphy, Prof. Ghosh Commemoration Volume: New Delhi, Today and Tomorrow's Publishers, p. 97–114.

Bande, M.B., Prakash, U., and Bonde, S.D., 1981, Occurrence of *Peyssonnelia* and *Distichoplax* in the Deccan Intertrappeans, with remarks on the age of Chhindwara Traps and palaeogeography of the region: Geophytology, v. 11, no. 2, p. 182–188.

Bande, M.B., Prakash, U., and Ambwani, K., 1982, A fossil palm fruit *Hyphaeneocarpon indicum* gen. et sp. nov. from the Deccan Intertrappean beds, India: Palaeobotanist, v. 30, no. 3, p. 303–309.

Bande, M.B., Mehrotra, R.C., and Prakash, U., 1986, Occurrence of Australian element in the Deccan Intertrappean flora of India: Palaeobotanist, v. 35, no. 1, p. 1–12.

Bande, M.B., Chandra, A., Venkatachala, B.S., and Mehrotra, R.C., 1988, Deccan Intertrappean floristics and its stratigraphic implications, *in* Maheshwari, H.K., ed., Paleocene of India: Lucknow, Indian Association of Palynostratigraphers, p. 83–123.

Bande, M.B., Mehrotra, R.C., and Awasthi, N., 1993, Revision of *Callistemonites indicus* Bande, Mehrotra, and Prakash from the Deccan Intertrappean beds of Mandla District, Madhya Pradesh: Palaeobotanist, v. 42, no. 1, p. 66–69.

Banerji, J., 2000, Occurrence of angiosperm remains in an Early Cretaceous Intertrappean bed, Rajmahal Basin, India: Cretaceous Research, v. 21, p. 781–784.

Barlinge, S.G., and Paradkar, S.A., 1982, Record of new fossil algal and fungal forms from the Deccan Intertrappean of Mohgaon-Kalan, M.P., India: Botanique, v. 10, no. 1–4, p. 163–174.

Bhattacharyya, B., 1967, Tertiary plant fossils from Cherrapunji and Laitryngew in Khasi and Jaintia Hills, Assam: Quarterly Journal of Geological Mining and Metallurgical Society of India, v. 39, no. 2, p. 131–134.

Bhattacharyya, B., 1983, Fossil plants from the Tura Formation (Eocene) in the Garo Hills, Meghalaya: Indian Journal of Earth Sciences, v. 10, no. 1, p. 1–10.

Bhattacharyya, B., 1985, Leguminous fruits from the Eocene of Garo Hills, Meghalaya: Quarterly Journal of Geological Mining and Metallurgical Society of India, v. 57, no. 4, p. 215–225.

Biradar, N.V., 1977, On the occurrence of cyanophycean member *Westiellopsis* in the Deccan Intertrappean Series, M.P., India: Geophytology, v. 7, no. 2, p. 204–207.

Biswas, S.K., 1986, Paleocene of Kutch: A rejoinder: Indian Journal of Earth Sciences, v. 13, no. 4, p. 343–353.

Biswas, S.K., 1992, Tertiary stratigraphy of Kutch: Journal of the Palaeontological Society of India, v. 37, p. 1–29.

Biswas, S.K., and Raju, D.S.N., 1971, Note on the rock stratigraphic correlation of the Tertiary sediments of Kutch: Quarterly Journal of Geological Mining and Metallurgical Society of India, v. 43, no. 3, p. 177–180.

Biswas, S.K., and Raju, D.S.N., 1973, The rock stratigraphic classification of the Tertiary sediments of Kutch: Bulletin of the Oil and Natural Gas Commission, v. 10, p. 37–45.

Bonde, S.D., 1990, A new palm peduncle *Palmostroboxylon umariense* (Arecaceae) and a fruit *Pandanusocarpon umariense* (Pandanaceae) from the Deccan Intertrappean beds of India, *in* Douglas, J.G., and Christophel, D.C., eds., Proceedings of the 3rd IOP Conference, Melbourne, July 1988: Melbourne, A-Z Printers and International Organisation of Palaeobotanists, p. 59–65.

Borkar, V.D., 1993, Palaeoflora of Wealden Gondwana basins of India and their significance, *in* Dutta, K.K., ed., Birbal Sahni Centenary National Symposium on Gondwana of India: Nagpur, Gondwana Geological Society, Gondwana Geological Magazine Special Volume, p. 256–258.

Champion, H.G., and Seth, S.K., 1968, A revised survey of the forest types of India: Delhi, The Manager of Publications, 404 p.

Chatterjee, S., and Scotese, C.R., 1999, The breakup of Gondwana and the evolution and biogeography of the Indian plate: Proceedings of the Indian National Science Academy, v. 65A, no. 3, p. 397–425.

Chitaley, S.D., 1955, A further contribution to the knowledge of *Sahnianthus:* Journal of the Indian Botanical Society, v. 34, no. 2, p. 121–129.

Chitaley, S.D., 1964, Further observations on *Sahnipushpam:* Journal of the Indian Botanical Society, v. 43, no. 1, p. 69–74.

Chitaley, S.D., and Nambudiri, E.M.V., 1973, *Harrisocarpon sahnii* gen. et sp. nov. from the Deccan Intertrappean beds of Mohgaonkalan, district Chhindwara: Geophytology, v. 3, no. 1, p. 36–41.

Chitaley, S.D., and Paradkar, S.A., 1973, *Rodeites* Sahni reinvestigated, II: Palaeobotanist, v. 20, no. 3, p. 293–296.

Chitaley, S.D., and Patil, G.V., 1972, Ebenaceous fossil wood infected with deuteromycetous fungus from the Deccan Intertrappean beds of India: Botanique, v. 3, no. 2, p. 99–106.

Chitaley, S.D., and Patil, G.V., 1973, *Sahniocarpon harrisii* gen. et sp. nov. from the Mohgaon Kalan beds of India: Palaeobotanist, v. 20, no. 3, p. 288–292.

Chitaley, S.D., and Sheikh, M.T., 1971, An infected grain from the Deccan Intertrappean cherts of Mohgaonkalan: Journal of the Indian Botanical Society, v. 50, no. 2, p. 137–142.

Chitaley, S.D., and Sheikh, M.T., 1973, *Harrisocarpon intertrappea* gen. et sp. nov., a petrified gymnospermous cone from the Deccan Intertrappean beds of India: Palaeontographica, v. 144B, no. 1/2, p. 25–30.

Chitaley, S.D., and Yawale, N.R., 1978, Fungal remains from the Deccan Intertrappean beds of Mohgaonkalan, India: Botanique, v. 7, no. 4, p. 189–194.

Chitaley, S.D., and Yawale, N.R., 1980, On *Notothylites nirulai* gen. et sp. nov., a petrified sporogonium from the Deccan Intertrappean beds of Mohgaonkalan, M.P., (India): Botanique, v. 9, no. 1–4, p. 111–118.

Dwivedi, J.N., 1959, Occurrence of fossil alga probably *Ulothrix* in Tertiary beds of Mohgaonkalan locality (M.P.), India: Current Science, v. 28, no. 11, p. 456–457.

Guleria, J.S., 1991, On the occurrence of carbonised woods resembling *Terminalia* and *Sonneratia* in Palaeogene deposits of Gujarat, western India: Palaeobotanist, v. 39, no. 1, p. 1–8.

Guleria, J.S., 1992, Neogene vegetation of peninsular India: Palaeobotanist, v. 40, p. 285–311.

Guleria, J.S., 1994, Fossil leaf of *Terminalia* from the Rajpardi Lignite mine, district Bharuch, Gujarat, India: Geophytology, v. 24, no. 1, p. 55–58.

Guleria, J.S., and Mehrotra, R.C., 1999, On some plant remains from Deccan Intertrappean localities of Seoni and Mandla districts of Madhya Pradesh, India: Palaeobotanist, v. 47, p. 68–87.

Guleria, J.S., and Srivastava, R., 2001, Fossil dicotyledonous woods from the Deccan Intertrappean beds of Kachchh, Gujarat, western India: Palaeontographica, v. 257B, no. 1–6, p. 17–33.

Hill, R.S., 1994, The history of selected Australian taxa, *in* Hill, R.S., ed., History of Australian vegetation: Cretaceous to Holocene: Cambridge, UK, Cambridge University Press, p. 390–419.

Ingle, S.R., 1973, *Syzygioxylon mandlaense* gen. et sp. nov., a fossil dicotyledonous wood from the Deccan Intertrappean Series, India: Botanique, v. 4, no. 1, p. 71–76.

Jain, R.K., 1964a, Studies in Musaceae, I: *Musa cardiosperma* sp. nov., a fossil banana fruit from the Deccan Intertrappean Series, India: Palaeobotanist, v. 12, no. 1, p. 45–58.

Jain, R.K., 1964b, Studies in Musaceae, II: *Musocaulon indicum* gen. et sp. nov., a petrified pseudostem from the Deccan Intertrappean Series, India: Palaeobotanist, v. 12, no. 1, p. 115–120.

Kar, R.K., and Bhattacharyya, M., 1992, Palynology of Rajpardi Lignite, Cambay Basin and Gujra Dam and Akri lignite, Kutch Basin: Palaeobotanist, v. 39, no. 2, p. 250–263.

Kaul, K.N., 1951, A palm fruit from Kapurdi (Jodhpur, Rajasthan Desert): *Cocos sahnii* sp. nov.: Current Science, v. 29, no. 5, p. 138.

Khare, E.G., Prasad, M., and Awasthi, N., 2001, Contributions to the Deccan Intertrappean flora of Nawargaon, Wardha District, Maharashtra, India: Palaeobotanist, v. 49, no. 3, p. 443–460.

Khosla, S.C., 1999, *Costabuntonia,* a new genus of Ostracoda from the Intertrappean beds (Paleocene) of east coast of India: Micropaleontology, v. 45, p. 319–323.

Kribs, D.A., 1959, Commercial foreign woods on the American market: Pennsylvania State University, Pennsylvania, 203 p.

Kulkarni, A.R., and Patil, K.S., 1977, *Aristolochioxylon prakashii* from the Deccan Intertrappean beds of Wardha District, Maharashtra: Geophytology, v. 7, no. 1, p. 44–49.

Lakhanpal, R.N., 1955, Recognizable species of Tertiary plants from Damalgiri in the Garo Hills, Assam: Palaeobotanist, v. 3, p. 27–31.

Lakhanpal, R.N., 1964, Specific identification of the guttiferous leaves from the Tertiary of Rajasthan: Palaeobotanist, v. 12, no. 3, p. 265–266.

Lakhanpal, R.N., 1974a, Geological history of the Dipterocarpaceae, *in* Lakhanpal, R.N., ed., Symposium on origin and phytogeography of angiosperms: Lucknow, Birbal Sahni Institute of Palaeobotany, p. 30–39.

Lakhanpal, R.N., 1974b, Physical conditions of the Indian Tertiary in the light of palaeobotanical evidence, *in* Surange, K.R., et al., eds., Aspects and appraisal of Indian palaeobotany: Lucknow, Birbal Sahni Institute of Palaeobotany, p. 516–524.

Lakhanpal, R.N., and Bose, M.N., 1951, Some Tertiary leaves and fruits of the Guttiferae from Rajasthan: Journal of the Indian Botanical Society, v. 30, no. 1–4, p. 132–136.

Lakhanpal, R.N., and Dayal, R., 1964, *Mallotoxylon keriense* gen. et sp. nov., a fossil dicotyledonous wood from the Deccan Intertrappean Series, India: Palaeobotanist, v. 11, no. 3, p. 149–153.

Lakhanpal, R.N., and Verma, J.K., 1966, Fossil wood of *Tetrameles* from the Deccan Intertrappean beds of Mohgaon Kalan, Madhya Pradesh: Palaeobotanist, v. 14, no. 1–3, p. 209–213.

Lakhanpal, R.N., Prakash, U., and Bande, M.B., 1978, Fossil dicotyledonous woods from the Deccan Intertrappean beds of Mandla District, Madhya Pradesh: Palaeobotanist, v. 25, p. 190–204.

Lakhanpal, R.N., Sah, S.C.D., Sharma, K.K., and Guleria, J.S., 1983, Occurrence of *Livistona* in the Hemis conglomerate horizon of Ladakh, *in* Thakur, V.C., and Sharma, K.K., eds., Geology of Indus Suture Zone of Ladakh: Dehra Dun, Wadia Institute of Himalayan Geology, p. 179–185.

Lakhanpal, R.N., Guleria, J.S., and Awasthi, N., 1984, The fossil floras of Kachchh, III: Tertiary megafossils: Palaeobotanist, v. 33, p. 228–319.

Mahabale, T.S., 1950a, A species of fossil *Salvinia* from the Deccan Intertrappean Series, India: Nature, v. 165, p. 410.

Mahabale, T.S., 1950b, Some new fossil plants from the Deccan Intertrappeans: Journal of the Indian Botanical Society, v. 29, no. 1, p. 31–33.

Mahabale, T.S., 1969, On a fossil species *Diplodia* from the Deccan Intertrappean Series, M.P., India: Palaeobotanist, v. 17, no. 3, p. 295–297.

Mahabale, T.S., and Deshpande, S.R., 1957, The genus *Sonneratia* and its fossil allies: Palaeobotanist, v. 6, no. 2, p. 51–64.

Mahadevan, C., and Sarma, S.R., 1948, Charophyta from the Deccan: Journal of the Indian Botanical Society, v. 26, no. 4, p. 260.

Manchester, S.R., and Kress, W.J., 1993, Fossil bananas (Musaceae): *Ensete oregonense* sp. nov. from the Eocene of western North America and its phytogeographic significance: American Journal of Botany, v. 80, no. 11, p. 1264–1272.

Mehrotra, R.C., 1987, A new fossil dicot wood from the Deccan Intertrappean beds of Mandla District, Madhya Pradesh: Palaeobotanist, v. 35, no. 2, p. 146–149.

Mehrotra, R.C., 1988, Some new palm fruits from the Deccan Intertrappean beds of Mandla District, Madhya Pradesh: Geophytology, v. 17, no. 2, p. 204–208.

Mehrotra, R.C., 1989a, Fossil wood of *Sonneratia* from the Deccan Intertrappean beds of Mandla District, Madhya Pradesh: Geophytology, v. 18, no. 2, p. 129–134.

Mehrotra, R.C., 1989b, Fossil wood of *Walsura* from the Deccan Intertrappean beds of Mandla District, with a review of the Intertrappean flora of the district: Review of Palaeobotany and Palynology, v. 58, p. 205–213.

Mehrotra, R.C., 1990, Further observations on some fossil woods from the Deccan Intertrappean beds of central India: Phytomorphology, v. 40, no. 1–2, p. 169–174.

Mehrotra, R.C., 2000a, Study of plant megafossils from the Tura Formation of Nangwalbibra, Garo Hills, Meghalaya: Palaeobotanist, v. 49, no., 2, p. 225–237.

Mehrotra, R.C., 2000b, A new rhizome-like structure from near Jowai, Jaintia Hills, Meghalaya: Palaeobotanist, v. 49, no. 2, p. 329–331.

Mehrotra, R.C., Prakash, U., and Bande, M.B., 1983, *Euphorbiocarpon drypeteoides,* a new euphorbiaceous fruit from the Deccan Intertrappean beds of Mandla District, Madhya Pradesh: Geophytology, v. 13, no. 1, p. 127–135.

Mehrotra, R.C., Prakash, U., and Bande, M.B., 1984, Fossil wood of *Lophopetalum* and *Artocarpus* from the Deccan Intertrappean beds of Mandla District, Madhya Pradesh, India: Palaeobotanist, v. 32, no. 3, p. 310–320.

Mehrotra, R.C., Dilcher, D.L., and Awasthi, N., 1998, A Paleocene *Mangifera*-like leaf fossil from India: Phytomorphology, v. 48, no. 1, p. 91–100.

Mishra, P.K., and Maithy, P.K., 1984, Occurrence of cyanophycean remains from the Deccan Intertrappean beds, Madhya Pradesh: Palaeobotanist, v. 31, no. 2, p. 120–125.

Mohabey, D.N., 1996, Depositional environment of Lameta Formation (Late Cretaceous) of Nand-Dongargaon Inland Basin, Maharashtra: The fossil and lithological evidences: Geological Society of India Memoir 37, p. 1–33.

Nambudiri, E.M.V., 1966, More *Nypa* fruits from the Deccan Intertrappean beds of Mohgaon Kalan: Current Science, v. 35, no. 16, p. 421–422.

Paradkar, S.A., 1972, *Shoreoxylon mahurzarii* sp. nov. from Deccan Intertrappean beds of India: Journal of Biological Sciences, v. 15, p. 19–25.

Paradkar, S.A., 1973, *Chitaleypushpam mohgaoense* gen. et sp. nov. from the Deccan Intertrappean beds of India: Palaeobotanist, v. 20, no. 3, p. 334–338.

Paradkar, S.A., 1974, Pollen and fungal spore association on a fossil leaf from the Deccan Intertrappean beds of India: Journal of Palynology, v. 10, no. 2, p. 119–125.

Paradkar, S.A., 1975, On a new monocot axis with pathogenic fungi from the Deccan Intertrappean beds of India: Geophytology, v. 5, no. 1, p. 94–97.

Paradkar, S.A., 1976, Pollination in a fossil ovule from the Deccan Intertrappean of Mohgaon Kalan: Botanique, v. 7, no. 2–3, p. 95–100.

Paradkar, S.A., and Barlinge, S.G., 1979, *Salvinia intertrappea* Mahabale reinvestigated, *in* Bhardwaj, D.C., et al., eds., Proceedings 4th International Palynological Conference: Lucknow, Birbal Sahni Institute of Palaeobotany, v. 1, p. 494–499.

Prakash, U., 1957, Studies in the Deccan Intertrappean flora, I: On a petrified ovuliferous cone from Mohgaon cherts in the Deccan: Palaeobotanist, v. 5, no. 2, p. 91–94.

Prakash, U., and Dayal, R., 1964, Fossil woods resembling *Elaeocarpus* and *Leea* from the Deccan Intertrappean beds of Mahurzari near Nagpur: Palaeobotanist, v. 12, no. 2, p. 121–127.

Prakash, U., and Dayal, R., 1965, *Barringtonioxylon eopterocarpum* sp. nov., a fossil wood of Lecythidaceae from the Deccan Intertrappean beds of Mahurzari: Palaeobotanist, v. 13, no. 1, p. 25–29.

Prakash, U., Verma, J.K., and Dayal, R., 1967, Revision of the genus *Gondwanoxylon* Saksena with critical remarks on the fossil woods of *Ailanthoxylon:* Palaeobotanist, v. 15, no. 3, p. 294–307.

Prakash, U., Bande, M.B., and Ambwani, K., 1979, *Musophyllum indicum* sp. nov.: A leaf impression resembling banana leaf from the Deccan Intertrappean Series, India: Palaeobotanist, v. 26, no. 2, p. 175–179.

Rajarao, C.S., 1981, Coalfield of northeastern India, I: Bulletin of the Geological Survey of India, no. 45A, p. 1–76.

Raju, D.S.N., and Mishra, P.K., 1996, Cretaceous Stratigraphy of India: A review: Geological Society of India Memoir 37, p. 363–386.

Rao, A.R., and Menon, V.K., 1963, Further contribution to our knowledge of *Musocaulon indicum* Jain: Proceedings Indian Academy of Sciences, v. 57B, no. 6, p. 389–399.

Rao, A.R., and Menon, V.K., 1964, On a new specimen probably of *Palmoxylon sundaram* Sahni from Mohgaon Kalan, Madhya Pradesh: Proceedings Indian Academy of Sciences, v. 59B, no. 3, p. 137–144.

Sahni, B., 1931, Revisions of Indian fossil plants, part II: Coniferales (b. Petrifactions): Memoirs Geological Survey of India, Palaeontologica indica, new series, v. 11, p. 51–124.

Sahni, B., 1941, Indian silicified plants, I: *Azolla intertrappea* Sahni and H.S., Rao: Proceedings Indian Academy of Sciences, v. 14B, no. 6, p. 489–501

Sahni, B., 1943, Palaeobotany in India, IV: Journal of the Indian Botanical Society, v. 22, no. 1, p. 180–181.

Sahni, B., 1947, A silicified *Cocos*-like palm stem, *Palmoxylon (Cocos) sundaram,* from the Deccan Intertrappean beds: Journal of the Indian Botanical Society, M.O.P., Iyengar Commemoration Volume, p. 361–374.

Sahni, B., and Rao, H.S., 1943, A silicified flora from the Intertrappean cherts round Sausar in the Deccan: Proceedings National Academy of Sciences, India, v. 13, no. 1, p. 36–75.

Sahni, B., and Rao, S.R.N., 1943, On *Chara sausari* sp. nov., a *Chara* (sensu stricto) from the Intertrappean cherts at Sausar in the Deccan: Proceedings National Academy of Sciences, India, v. 13, no. 3., p. 215–223.

Sahni, B., and Rode, K.P., 1937, Fossil plants from the Intertrappean beds of Mohgaon Kalan in the Deccan with a sketch of geology of the Chhindwara District: Proceedings National Academy of Sciences, India, v. 7, no. 3, p. 165–174.

Sahni, B., and Surange, K.R., 1944, On the structure and affinities of *Palmoxylon sahnii* (Rode): Proceedings National Academy of Sciences, India, v. 14, no. 1–2, p. 83–85.

Sahni, B., and Surange, K.R., 1953, On the structure and affinities of *Cyclanthodendron sahnii* (Rode) Sahni and Surange from the Deccan Intertrappean Series: Palaeobotanist, v. 2, p. 83–100.

Salis, K.V., and Saksena, R.K., 1998, Calcareous nannofossils across the K/T boundary and the age of Deccan Trap volcanism in southern India: Journal of the Geological Society of India, v. 51, p. 183–192.

Shallom, L.J., 1960a, Fossil dicotyledonous wood of Lecythidaceae from the Deccan Intertrappean beds of Mahurzari: Journal of the Indian Botanical Society, v. 39, no. 2, p. 198–203.

Shallom, L.J., 1960b, A new simaroubaceous fossil dicotyledonous wood from the Deccan Intertrappean beds of Mohgaon Kalan, India: Bulletin Botanical Society of Nagpur, v. 1, p. 37–42.

Sharma, B.D., 1997, An early angiosperm fructification resembling *Lesqueria* Crane and Dilcher from the Rajmahal Hills, India: Phytomorphology, v. 47, no. 3, p. 305–310.

Sheikh, M.T., and Kapgate, D.K., 1982, First petrified fossil *Riccia* L. from Indian beds: Current Science, v. 51, no. 25, p. 241–242.

Sheikh, M.T., and Kolhe, P.D., 1982, A new petrified ovule *Podocarpoovulites chitaleyi* from a new locality of Nagpur, Deccan, Maharashtra (India): Botanique, v. 10, no. 1–4, p. 99–108.

Shukla, V.B., 1944, On *Sahnianthus,* a new genus of petrified flowers from the intertrappean beds of Mohgaonkalan in the Deccan, and its relation with the fruit *Enigmocarpon parijai* Sahni from the same locality: Proceedings National Academy of Sciences, India, v. 14, no. 1–2, p. 1–39.

Shukla, P.N., Shukla, A.D., and Bhandari, N., 1997, Geochemical characterisation of the Cretaceous-Tertiary sediments at Anjar, India: Palaeobotanist, v. 46, no. 1–2, p. 127–132.

Singh, R.B., and Patil, G.V., 1979, A petrified strobilus *Selaginella:* Current Science, v. 48, no. 15, p. 684–685.

Singhai, L.C., 1964, On a fossil bryophytic sporogonium from the Deccan Intertrappean beds: Current Science, v. 33, no. 4, p. 117–119.

Singhai, L.C., 1975, *Shuklanites deccanii* Singhai, an anthocerotaceous sporogonium from the Deccan Intertrappean beds of Mohgaon Kalan: Palaeobotanist, v. 22, no. 2, p. 171–175.

Singhai, L.C., 1978, *Palaeopthora mohgaoensis* Singhai—a fossil fungus from the Deccan Intertrappean beds of Mohgaon Kalan, Chhindwara District, M.P., India: Palaeobotanist, v. 25, p. 481–485.

Smith, A.G., and Briden, J.C., 1979, Mesozoic and Cenozoic palaeocontinental maps: Cambridge, London, New York, Melbourne, Cambridge University Press, 63 p.

Smith, A.G., Smith, D.G., and Funnell, M., 1994, Atlas of Mesozoic and Cenozoic coastlines: Cambridge, UK, Cambridge University Press, 99 p.

Tripathi, C., Chowdhury, J.R., and Das, D.P., 1981, Discovery of Tertiary plants from Geku Formation of Dihang Valley, Siang District, Arunachal Pradesh, *in* Sinha, A.K., ed., Contemporary geoscientific researches in Himalaya: Dehra Dun, Wadia Institute of Himalayan Geology, v. 1, p. 225–230.

Trivedi, B.S., and Ambwani, K., 1971, On the structure of *Nymphaeocaulon intertrappeum* gen. et sp. nov. from the Deccan Intertrappean Series of Mohgaon Kalan (M.P.), India: Palaeontographica, v. 133B, no. 4–6, p. 129–136.

Trivedi, B.S., and Srivastava, K., 1985, *Canarioxylon shahpuraensis* sp. nov. from the Deccan Intertrappean beds of Shahpura, District Mandla (M.P.), India: Geophytology, v. 15, no. 1, p. 27–32.

Trivedi, B.S., and Srivastava, R., 1989, Gymnospermous woods from early Tertiary of Chhindwara District of Madhya Pradesh: Phytomorphology, v. 39, no. 1, p. 61–68.

Trivedi, B.S., and Verma, C.L., 1972, Occurrence of *Heliconiaites mohgaonensis* gen. et sp. nov. from the early Eocene of Deccan Intertrappean Series, M.P., India: Palaeontographica, v. 139B, no. 5/6, p. 73–82.

Trivedi, B.S., Bajpai, S.K., and Trivedi, G.K., 1986, A new cyanophycean fossil from the Deccan Intertrappean beds of Mohgaon Kalan, Chhindwara District (M.P.), India: Journal of the Indian Botanical Society, v. 64, no. 4, p. 393–394.

Upadhye, E.V., and Patil, G.V., 1979, On *Marsilea*-like fossil rhizome and petiole from the Deccan Intertrappean beds of Mohgaon Kalan, India: Journal of the Indian Botanical Society, v. 58, no. 1, p. 63–69.

Venkatesan, T.R., Pande, K., and Gopalan, K., 1993, Did Deccan volcanism predate the Cretaceous–Tertiary transition?: Earth and Planetary Science Letters, v. 119, p. 181–189.

Verma, J.K., 1956, On a petrified flower, *Sahnipushpam shuklai* sp. nov., from the Intertrappean beds of Mohgaon Kalan in the Deccan: Journal of the Palaeontological Society of India, v. 1, no. 1, p. 131–141.

Manuscript Accepted by the Society August 13, 2002

Printed in the USA

Geological Society of America
Special Paper 369
2003

Floral change during the Initial Eocene Thermal Maximum in the Powder River Basin, Wyoming

Scott L. Wing
Department of Paleobiology, Smithsonian Institution, Washington, D.C. 20560, USA
Guy J. Harrington
Department of Geology, University College Cork, Cork, Ireland
Gabriel J. Bowen
Paul L. Koch
Department of Earth Sciences, University of California, Santa Cruz, California 95062, USA

ABSTRACT

Rapid warming at the beginning of the Eocene (the Initial Eocene Thermal Maximum, or IETM) has been associated with modest changes in floral composition, mostly shifts in the relative abundances of taxa rather than large numbers of first or last appearances. Although floral change across the Paleocene-Eocene transition has been studied in many areas, few fossils demonstrably come from the ~200 k.y.-long IETM. The rarity of fossils from the IETM permits two end-member hypotheses: (1) IETM floras were similar to, or intermediate in composition between, Paleocene and Eocene floras, or (2) they were distinct from both Paleocene and Eocene assemblages in having a high proportion of taxa that were temporary, thermophilic immigrants. The latter hypothesis is consistent with expectations developed from the study of late Quaternary floras, which demonstrate rapid northward range displacements in the wake of retreating continental glaciers. Here we report lithological, paleontological and isotopic evidence for a Paleocene-Eocene boundary section in the lower Wasatch Formation of the southwestern Powder River Basin, Wyoming. Pollen samples from within the IETM interval do not record immigrants from the south, making it unlikely that there were rapid, continental-scale range displacements during the IETM. Floral response consisted of shifts in the relative abundances of native taxa at the beginning of, or even prior to, the IETM, followed by immigration of taxa from outside of North America, probably Europe, near the end of, or possibly after, the IETM.

INTRODUCTION

The Initial Eocene Thermal Maximum (formerly LPTM or latest Paleocene Thermal Maximum) was a geologically brief interval of global warming at the beginning of the Eocene epoch (Kennett and Stott, 1991; Zachos et al., 2001). During the IETM mid- to high-latitude temperatures increased by ~4–8 °C over a period of 10–20 k.y., then returned to the warm climatic conditions characteristic of most of the Paleogene over the succeeding ~200 k.y. This warming was probably caused in part by greenhouse forcing following the release of carbon from seafloor methane clathrates at the Paleocene-Eocene boundary (Dickens et al., 1997).

Previous work on floral change across the Paleocene-Eocene boundary has revealed relatively modest changes in composition and diversity. This is true at low latitudes in northern South America (Rull, 1999; Jaramillo and Dilcher, 2000), throughout North America (Pocknall, 1987; Frederiksen, 1994;

Wing, S.L., Harrington, G.J., Bowen, G.J., and Koch, P.L., 2003, Floral change during the Initial Eocene Thermal Maximum in the Powder River Basin, Wyoming, *in* Wing, S.L., Gingerich, P.D., Schmitz, B., and Thomas, E., eds., Causes and Consequences of Globally Warm Climates in the Early Paleogene: Boulder, Colorado, Geological Society of America Special Paper 369, p. 425–440.

Harrington, 2001a, 2001b; Harrington and Kemp, 2001; Harrington, this volume), and in Europe (Gruas-Cavagnetto, 1978; Schumacker-Lambry, 1978; Jolley, 1998). In these studies the stratigraphic position of the IETM was not identified precisely, so transient changes in floras might have gone undetected.

Three other studies have examined palynofloras from the IETM. Beerling and Jolley (1998) worked with material from a set of cores from the North Sea and England. They found an increase in the abundance of tropical taxa (particularly pollen of the mangrove palm, *Nypa*) coincident with the interval inferred to represent the IETM. The identification of the IETM remains uncertain, however, because the best carbon isotope evidence for the IETM does not come from the same cores as the increase in tropical taxa. Crouch (2001) and Crouch and Visscher (this volume) have examined terrestrial palynomorphs from the marine Paleocene-Eocene section at Tawanui on the north island of New Zealand. They identified only modest changes in floral composition across this time interval, and no indication of a distinctive and transient flora during the IETM, although there was an increase in the proportion of terrestrial palynomorphs. Collinson et al. (this volume) present stratigraphic and isotopic evidence for the presence of the IETM within the ~3 m thick Cobham Lignite in southeastern England. Palynofloral composition changes little within this unit.

In the Bighorn Basin of northwestern Wyoming the IETM has been identified in outcrops spanning >100 km, and both megafloras and palynofloras have been used to describe change across the IETM (Wing et al., 1995; Wing, 1998; Wing and Harrington, 2001). These studies are consistent with work done elsewhere in that they show relatively minor changes in composition and diversity across the IETM; four megafossil and two palynofloral taxa have first appearances in the earliest Eocene, and diversity increases either not at all (megafloral data) or by ~20% (palynofloral data). There are significant shifts in relative abundances of several taxa. Even in the Bighorn Basin sections however, there is a 40–50-m-thick interval representing the IETM that has yielded neither pollen nor megafossils. Although the average temporal separation of pollen samples before and after the boundary interval in the Bighorn Basin is ~30 k.y. (10 m), the sample gap at the IETM probably represents ~200 k.y. during an interval marked by extreme climatic change.

Decades of work on floral response to Quaternary climate change has led us to expect that rapid warming of 5–10 °C in the midlatitudes will result in rapid, continental-scale shifts in the ranges of many plant species (Davis, 1976; Overpeck et al., 1992; Jackson and Overpeck, 2000). Pollen records that extend back into the Pleistocene show multiple, substantial fluctuations in floral composition that correspond to Milankovitch orbital cycles (Hooghiemstra and Melice, 1994; Mommersteeg et al., 1995). Clearly the ~200 k.y. sampling gap at the IETM in the Bighorn Basin could conceal a significant transient response to global warming.

By analogy with Quaternary floral change we expect that the IETM flora in the northern Rockies should differ from both latest Paleocene and earliest Eocene floras in having a higher abundance of thermophilic taxa that extended their ranges northward during the IETM. The stratigraphic ranges of these tropical elements might be confined entirely to the IETM if their migration into the northern Rockies was strongly controlled by temperature or other climatic features that changed rapidly but temporarily during the IETM. In both surface ocean and terrestrial vertebrate communities, transient faunas and floras are characteristic of the IETM (Gingerich, 1989; Kelly et al., 1998; Crouch et al., 2001, this volume).

In this paper we report geological, paleontological, and geochemical data indicating that the IETM is represented by sediments in the lower Wasatch Formation of the Powder River Basin, Wyoming. Palynological samples from within the IETM help characterize floral response to this geologically brief interval of global warming.

PREVIOUS WORK

The field area is near the southwestern margin of the Powder River Basin, Wyoming (Fig. 1). In this area, Upper Cretaceous, Paleocene and Eocene strata dip off the northern side of the Casper Arch and the regional strike is nearly east-west. Resistant strata of the Lance and Fort Union Formations form the east-west trending Pine Ridge, in which beds dip at 20–40 degrees north into the basin. Dips shallow rapidly to the north, and at Chalk Butte, a north-south oriented salient 2–3 km north of Pine Ridge, the lower part of the Wasatch Formation dips only 4–7 degrees.

The lower part of the Wasatch Formation in this area is drab, contains many thick carbonaceous shales and thin lignites, and rests on the Fort Union Formation with a minor angular unconformity (Robinson, 2000, personal commun.). In parts of the eastern and southern Powder River Basin the lower Wasatch Formation has produced Paleocene vertebrates such as *Champsosaurus gigas* (Robinson, 2000, personal commun.).

The transition from drab mudstones, carbonaceous shales and lignites to pedogenically altered red and variegated mudstones is exposed on the lower to middle elevations of Chalk Butte (Fig. 2). The lowest laterally extensive red-variegated mudstone bed is ~135 m above the base of the Wasatch Formation (Robinson, 2000, personal commun.). Vertebrate fossils collected from sandstone and variegated mudstone layers ~175 m above the base of the formation include mammals, such as *Cantius torresi* and *Copecion davisii* that are characteristic of the basal Eocene (Wasatchian 0, or Wa 0) faunal zone (Gingerich, 1989; Thewissen, 1990; Robinson and Williams, 1997). In the Bighorn Basin, these taxa have been found only in paleosols that are correlated with the IETM via carbon isotope stratigraphy (Koch et al., 1995; Bowen et al., 2001).

LITHOSTRATIGRAPHY

We measured three local sections in the vicinity of Chalk Butte (Fig. 3). The base of each section is a single, laterally ex-

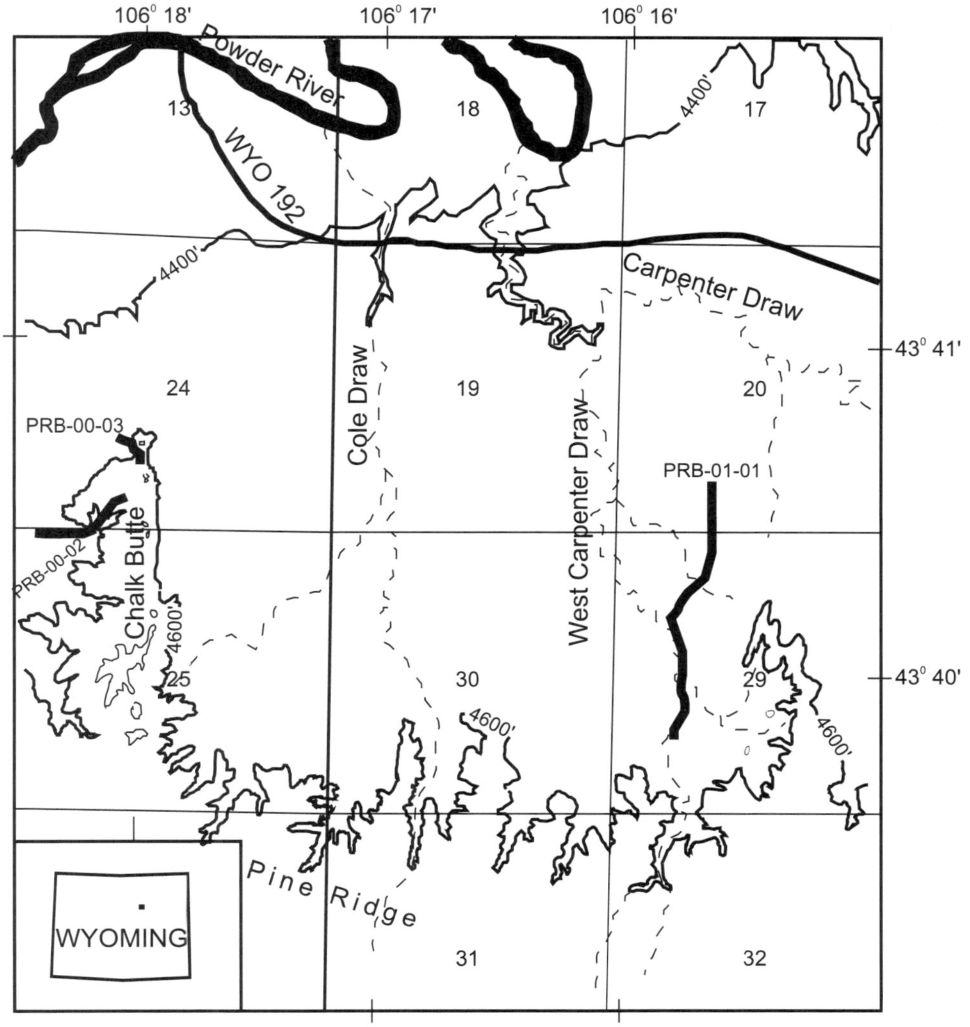

Figure 1. Map of the Chalk Butte study area showing topography, local landmarks, and the three lines of section. Based on U.S. Geological Survey Sussex Quadrangle (1:24,000).

tensive carbonaceous shale with lignitic zones informally called the Chalk Butte lignite (Figs. 2 and 3; Robinson, 2000, personal commun.). The base of the southern Chalk Butte section (PRB-00-02) is in the northern part of section 25, T. 43 N, R. 79 W, and extends 80 m up section to the top of the butte (Figs. 1, 3). We collected paleosol carbonate nodules for stable isotope analysis from this section because it offers the best local exposures of pedogenically altered mudstones.

The northern Chalk Butte section (PRB-00-03) extends through the same stratigraphic interval, but has its base in the southern part of section 24, T. 43 N, R. 79 W (Figs. 1, 3). Near the northern tip of Chalk Butte there are at least two channel-form bodies that diagonally truncate previously deposited mudstones. These channel-form bodies consist of gray, organic rich mudstones that preserve leaves and palynomorphs, as well as thin, red-variegated mudstones that probably represent paleosols. It is difficult to determine the depth of these channel fills because the diagonally truncated lower surfaces are eroded away at the top of the butte. Most of the palynological samples were taken from this section, and leaf fossils are also present at the base and top (Fig. 3).

The Loehse Ranch section (PRB-01-01) is approximately three km east of Chalk Butte on the west side of West Carpenter Draw (sections 20 and 29, T. 43 N, R. 78 W; Figs. 1, 3). We traced the Chalk Butte lignite along nearly continuous exposures in the intervening drainages in order to establish a stratigraphic correlation between the two areas. The Loehse section begins just below the base of the Chalk Butte lignite and records 85 m of strata. Two University of Colorado vertebrate localities with Wa 0 faunas (UCM 83117 and 86101), as well as a pollen sample and a leaf fossil site, occur near the top of this section.

With the exception of the Chalk Butte lignite itself, it is difficult to make bed-for-bed correlations between the Loehse Ranch section and the sections on Chalk Butte. There are consistencies in the order of lithologies, however. In both areas the Chalk Butte lignite is overlain by a thick sandstone unit. A distinctive, white, poorly cemented, sandy or silty mudstone with a leached appearance occurs above the sandstone interval in

Figure 2. Photograph of the west face of Chalk Butte showing the primary marker strata referred to in Figure 3. Photograph by Henry Fricke.

both areas (Figs. 2 and 3). On Chalk Butte the unit is a single bed ~1 m thick, whereas in the Loehse Ranch section there appear to be two thinner white mudstone units separated by ~2 m of weakly developed orange-brown paleosols. In both areas the first bright and laterally extensive red bed occurs ~12–15 m above the white mudstone horizons. In both areas the upper part of the section contains a number of purple-red and variegated mudstone paleosol units. The channel cut and fill structures that are clearly visible on the upper part of the Chalk Butte were not seen in upper part of the Loehse Ranch section, but much of this interval is covered (Fig. 3).

In general, the stratigraphic sequence in the Chalk Butte area is similar to Paleocene-Eocene sequences in the southeastern part of the Bighorn Basin near North Butte, ~100 km to the west. There the IETM falls at the base of the Willwood Formation (Koch et al., 1995; Wing, 1998; Strait, 2001), which is defined as beginning with the lowest laterally extensive red-variegated mudstone beds (Bown, 1979). Although unusually thick, red paleosols characterize the lowest part of all known IETM sequences in the Bighorn Basin (in at least five separate locations), paleosols in the upper portions of those same IETM sequences are similar to overlying typical early Eocene paleosols. Thus, the occurrence of Wa 0 mammals above the thick red paleosols at Chalk Butte is entirely consistent with observations in the Bighorn Basin.

STABLE ISOTOPES

We collected paleosol carbonate nodules, mammal teeth, and organic matter from a number of stratigraphic levels in the Chalk Butte area. Fossil soils (paleosols) were identified by their vertical color horizonation and the presence of features such as root traces, mottling, and absence of primary sedimentary structure. We collected nodules from trenches dug through the lower parts of paleosols. Carbonate nodules were polished flat on a lapidary wheel, and subsamples (~100 μg) of micritic carbonate were drilled from polished surfaces, then roasted at 400 °C for 1 hr to remove organic contaminants. Mammalian tooth enamel was drilled from fossil molars and prepared for analysis of structural carbonate by treatment with 2% NaOCl to remove organic contaminants and leaching with 1 N acetic acid to dissolve diagenetic carbonate. All carbonate samples were analyzed using a VG Optima or Prism mass spectrometer following reaction with 100% phosphoric acid at 90 °C in an online carbonate preparation device.

Seven samples were taken from carbonaceous shale and mudstone beds in the northern Chalk Butte section (PRB-00-03) and treated both for stable carbon isotope and palynological analyses. At each site we dug away weathered material until we obtained coherent pieces of rock ≥3 cm on a side that were not obviously penetrated by modern roots. Samples for isotopic

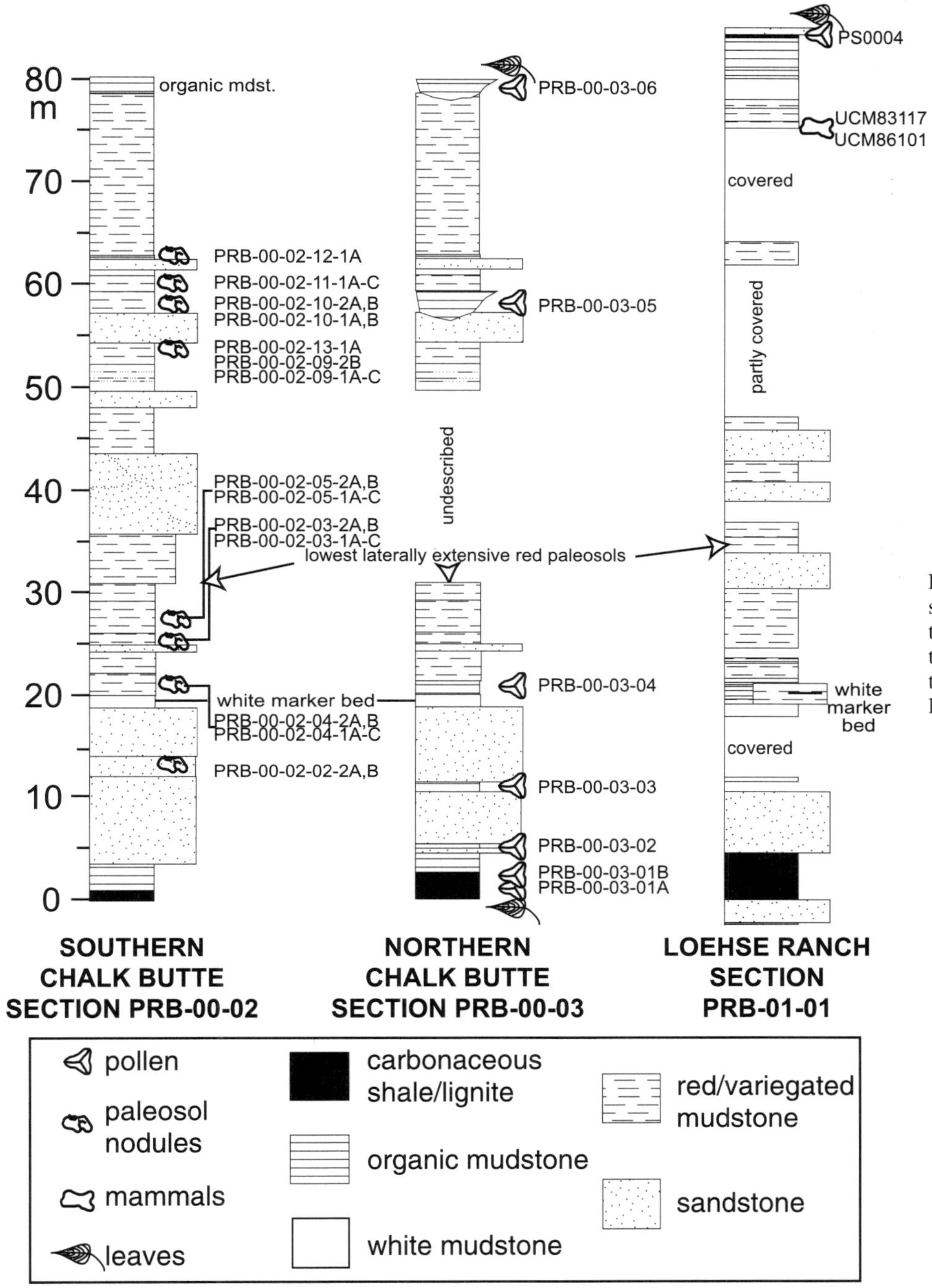

Figure 3. Chalk Butte and Loehse Sections showing lithology, and stratigraphic positions of samples. Correlation between sections is based on the Chalk Butte "lignite" at the base. See Figure 1 for map of section lines.

analysis of bulk organic matter were prepared by leaching with 0.5 N HCl to remove carbonate, followed by extraction of soluble organic compounds with dichloromethane for 1 hour. The extracted fraction was discarded, and the remaining solid sample dried and analyzed using a Carlo Erba Elemental Analyzer coupled to a VG Optima mass spectrometer.

All isotopic results are presented in per mil units and δ notation relative to the V-PDB standard, where $\delta^{13}C = 1000 * (R_{sample} - R_{standard})/R_{standard}$ and R is the ratio $^{13}C/^{12}C$. Analytical precision was ~00.1‰ for carbonates and 0.3‰ for organic carbon based on repeated analysis of in-house standards (Carerra marble, PU gel). Results of stable isotope analyses are shown in Tables 1–3.

No single substrate for isotopic analyses is present at numerous levels throughout the Chalk Butte section, therefore we present a composite carbon isotope stratigraphy incorporating measurements made on paleosol carbonate, organic matter, and fossil enamel. These values are plotted by stratigraphic position

TABLE 1. ISOTOPIC MEASUREMENTS OF PALEOSOL CARBONATE NODULES

Sample	Level (m)	$\delta^{13}C$ (‰)	$\delta^{18}O$ (‰)
PRB-00-02-02-2A	13.20	–14.599	–7.524
PRB-00-02-02-2B	13.20	–14.707	–7.722
PRB-00-02-04-1A	21.20	–14.031	–7.573
PRB-00-02-04-1B	21.20	–14.142	–8.133
PRB-00-02-04-1C	21.20	–14.310	–7.870
PRB-00-02-04-2A	21.20	–13.841	–8.146
PRB-00-02-04-2B	21.20	–13.180	–8.179
PRB-00-02-03-1A	25.65	–14.581	–7.391
PRB-00-02-03-1B	25.65	–14.929	–7.564
PRB-00-02-03-1C	25.65	–14.607	–7.287
PRB-00-02-03-2A	25.65	–14.767	–7.592
PRB-00-02-03-2B	25.65	–15.071	–7.811
PRB-00-02-05-1A	27.85	–13.989	–7.801
PRB-00-02-05-1B	27.85	–13.886	–7.760
PRB-00-02-05-1C	27.85	–13.835	–7.849
PRB-00-02-05-2A	27.85	–14.058	–7.574
PRB-00-02-05-2B	27.85	–14.383	–8.131
PRB-00-02-09-1A	53.50	–14.052	–7.361
PRB-00-02-09-1B	53.50	–13.953	–7.619
PRB-00-02-09-1C	53.50	–13.908	–7.851
PRB-00-02-09-2B	53.50	–13.983	–7.935
PRB-00-02-13-1A	53.50	–15.053	–7.762
PRB-00-02-13-1A	53.50	–15.176	–7.082
PRB-00-02-13-1B	53.50	–15.279	–7.822
PRB-00-02-13-2A	53.50	–15.322	–7.658
PRB-00-02-13-2B	53.50	–15.444	–7.764
PRB-00-02-10-1A	58.50	–13.825	–7.504
PRB-00-02-10-1B	58.50	–13.727	–7.858
PRB-00-02-10-2A	58.50	–12.814	–8.682
PRB-00-02-10-2B	58.50	–12.522	–8.411
PRB-00-02-11-1A	60.30	–14.350	–8.385
PRB-00-02-11-1B	60.30	–14.040	–7.805
PRB-00-02-11-1C	60.30	–14.306	–7.392
PRB-00-02-12-1A	62.95	–11.920	–8.272
PRB-00-02-12-1B	62.95	–11.603	–8.228
PRB-00-02-12-1C	62.95	–12.342	–8.094
PRB-00-02-12-2A	62.95	–12.174	–8.241
PRB-00-02-12-2B	62.95	–11.878	–8.188

in Figure 4, with axes for different substrates aligned to reflect the typical isotopic fractionation between these substrates in the modern (Cerling, 1984) or in Bighorn Basin fossil localities (Koch et al., 1995). The $\delta^{13}C$ values of organic matter from near the base of the section range between –24.6‰ and –27.0‰. This is near the expected values given the average carbon isotope fractionation between modern C3 plants and the atmosphere (–19.6‰, Passey et al., 2002) and estimated late Paleocene to early Eocene atmospheric $\delta^{13}C$ values between –6.5‰ and –4.5‰ (Bowen and Bloch, 2002). A single sample (PRB-00-03-05) from the upper part of the Chalk Butte section, however, has a lower $\delta^{13}C$ (–29.4‰), suggesting an atmospheric $\delta^{13}C$ value of ~ –9.8‰. Paleosol carbonates from the middle of the Chalk Butte section have $\delta^{13}C$ values between –14.7‰ and –13.2‰ (averages for single stratigraphic levels), with one level high in the section producing a value of –12.0‰. Using the soil gas diffusion model of Cerling (1984), the paleosol carbonate $\delta^{13}C$ values from the middle of the section would have been generated from decomposing and/or respiring plants with values no higher than –29.7‰ to –28.2‰. Thus, most of the soil carbonate data suggest that co-occurring organic matter had $\delta^{13}C$ values lower than those measured at the base of the Chalk Butte section, but similar to that measured for the organic matter sample at 58 m.

TABLE 2. ISOTOPIC MEASUREMENTS OF TOOTH ENAMEL

Sample	Level (m or zone)	$\delta^{13}C$ (‰)	$\delta^{18}O$ (‰)
Coryphodon Loc. 94025	Wa 2	–12.1	–8.2
Coryphodon Loc. 94024	Wa 2	–13.7	–8.6
Phenacodus—Chalk Butte	21.2	–16.1	–8.0
Phenacodus B—Chalk Butte	21.2	–15.7	–7.4

TABLE 3. ISOTOPIC MEASUREMENTS OF ORGANIC CARBON

Sample	Level	% C†	C/N†	$\delta^{13}C$ (‰)
PRB-00-03-06	79.0	0.86	12.0	–27.3
PRB-00-03-05	58.0	0.62	08.7	–29.4
PRB-00-03-04	20.5	0.06	02.2	–24.6
PRB-00-03-03	10.9	1.99	18.1	–27.0
PRB-00-03-02	5.0	0.62	08.4	–25.4
PRB-00-03-01A	1.0	2.28	18.4	–26.9
PRB-00-03-01A (duplicate)	1.0	2.93	18.4	–26.3
PRB-00-03-01B	2.0	1.15	10.1	–26.6

†Estimated from ion source strength calibrated using standards of known size.

Finally, the specimens of *Phenacodus* from the 21.2 m level had $\delta^{13}C$ values of –15.7‰ and –16.1‰, slightly lower than the value for Wa 0 *Phenacodus* in the Bighorn Basin (–15.3‰), but much lower than values for non-Wa 0 *Phenacodus* specimens in the Bighorn Basin. (The mean value for non-Wa 0 Wasatchian specimens is –11.9 ± 0.7‰ [±1σ], the minimum is –13.9‰, and the maximum is –11.0‰ [N = 18]. The values for Tiffanian-Clarkforkian *Phenacodus* are: mean, –10.2 ± 1.3‰; minimum, –13.1‰; maximum, –7.4‰; *N*, 22.) In contrast, specimens of the pantodont *Coryphodon* from Wa 2 fossil localities at Pumpkin Buttes, ~35 km east of Chalk Butte in the Powder River Basin, produced $\delta^{13}C$ values of –12.1‰ and –13.7‰. This overlaps the range observed for non-Wa 0 *Coryphodon* from the Bighorn Basin. (The mean value for non-Wa 0 Wasatchian specimens is –12.6 ± 1.3‰ [±1σ], the minimum is –14.3‰, and the maximum is –9.3‰ [N = 18]. Comparable values for Clarkforkian *Coryphodon* are: –12.4 ± 1.0, –14.5‰, –11.1‰, 9.) Based on the average carbon isotope offset between Bighorn Basin *Phenacodus* or *Coryphodon* and co-occurring paleosol carbonate (Koch et al., 1995), the $\delta^{13}C$ values of Chalk Butte *Phenacodus* should correspond to paleosol carbonate with $\delta^{13}C$

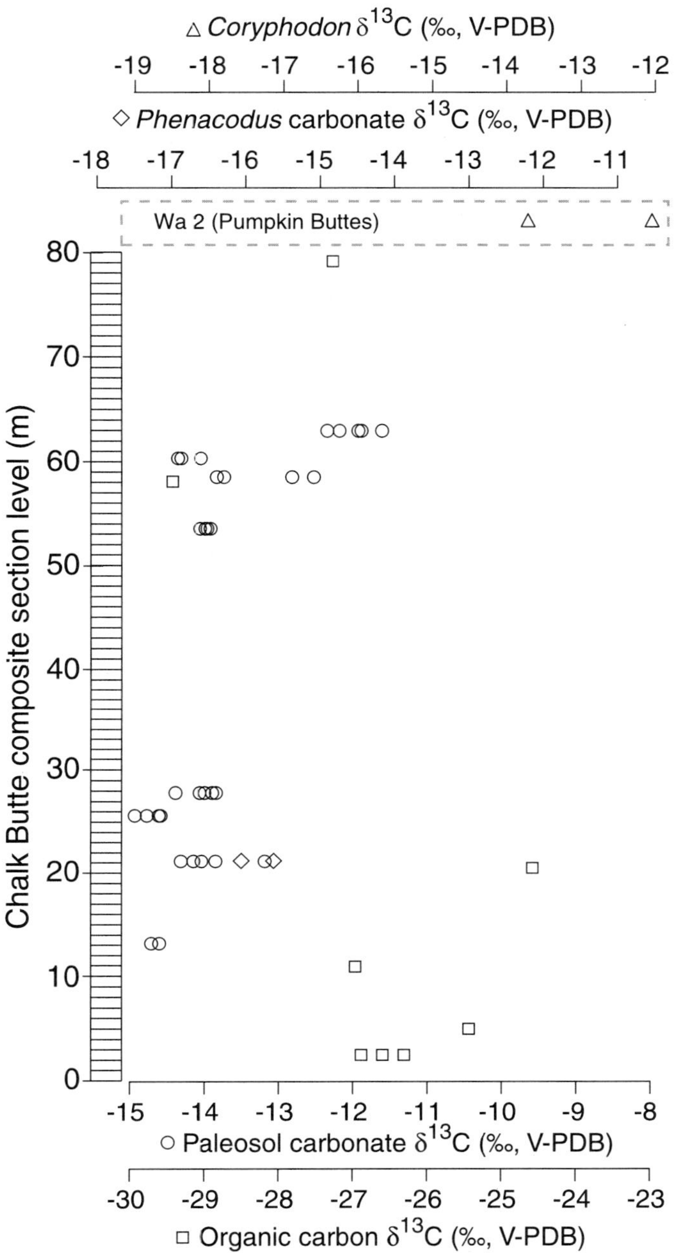

Figure 4. Plot of carbon isotope values of pedogenic calcium carbonate nodules, bulk paleosol organic matter, and fossil tooth enamel carbonate by stratigraphic level. Axes are aligned according to "average" carbon isotope fractions between sample substrates; these are: $\delta^{13}C_{\text{soil carbonate}} - \delta^{13}C_{\text{organic carbon}}$ = ~15‰ or greater for modern soils (Cerling, 1984), and $\delta^{13}C_{\text{soil carbonate}} - \delta^{13}C_{Phenacodus}$ = 2.6‰ and $\delta^{13}C_{\text{soil carbonate}} - \delta^{13}C_{Coryphodon}$ = 4.1‰ based on the fossil data given by Koch et al. (1995). Symbols: Circles—pedogenic carbonate nodules, squares—organic carbon, diamonds—*Phenacodus* tooth enamel, triangles—*Coryphodon* tooth enamel. V-PDB—Vienna Peedee belemnite.

values of –13.1‰ to –13.5‰ (similar to the value measured for this level, –13.8‰). The *Coryphodon* enamel values suggest Powder River Basin Wa 2 paleosol carbonate with $\delta^{13}C$ values of about –8.0‰ to –9.6‰, much higher than any values for paleosol carbonates from the Chalk Butte section.

The combined carbon isotope data (Fig. 4) suggest that the middle of the Chalk Butte section contains an extended interval characterized by low $\delta^{13}C$ values. Most of the evidence for low $\delta^{13}C$ values in this interval comes from paleosol carbonates, which produce values lower than those for any non-IETM Paleogene paleosol carbonates nodules from the Bighorn, Wind River, or Green River Basins (Koch and Morrill, 2000; Clyde et al., 2001; Koch et al., this volume). Because we did not measure any Powder River Basin paleosol carbonates with "normal" late Paleocene $\delta^{13}C$ values, we must consider the possibility that the low values reported here reflect either diagenetic alteration or the contribution of a secondary source of isotopically light carbon to Powder River Basin soils, rather than the global $\delta^{13}C$ signal marking the IETM. Burial diagenesis should affect both the $\delta^{18}O$ and $\delta^{13}C$ values of carbonate, and typically decreases $\delta^{18}O$ values due to mineral formation at elevated temperatures. The range of $\delta^{18}O$ values measured in Chalk Butte paleosol carbonates is small and $\delta^{18}O$ and $\delta^{13}C$ values are negatively correlated ($r^2 = 0.41$, $p < 0.0001$). This suggests that if the paleosol carbonate isotope values were affected by diagenesis, this modification probably increased $\delta^{13}C$ and can not account for the very low $\delta^{13}C$ values. A second scenario for the very low paleosol carbonate $\delta^{13}C$ values, influx of light carbon from outside the soils, would most likely imply the upward migration and oxidation of isotopically light biogenic methane produced in underlying organic-rich sediments (e.g., Bowen and Bloch, 2002). This scenario is unlikely, however, as the carbonate-bearing paleosols from the Chalk Butte sections are well oxidized and any methane in the Chalk Butte sediment stack would have been produced and oxidized deep in the subsurface. Furthermore, since mammals obtain their carbon directly from plants, the low $\delta^{13}C$ values for *Phenacodus* tooth enamel suggest that plants, rather than subsurface methane, were the source of low $\delta^{13}C$ values at Chalk Butte. This hypothesis is supported by the relatively low $\delta^{13}C$ value of organic matter sample PRB-00-03-05. Finally, the *Coryphodon* data from Pumpkin Butte demonstrate that "normal" Wasatchian $\delta^{13}C$ values are preserved in fossil tooth enamel from the Powder River Basin. Thus, we suggest that the low $\delta^{13}C$ values from the middle of the Chalk Butte section record the low $\delta^{13}C$ values for atmospheric CO_2 that characterized the IETM.

High-resolution measurements of carbon isotope values through continental IETM sections show a characteristic stratigraphic pattern: A rapid, possibly stepped, decrease of 6‰–8‰ at the onset of the IETM, succeeded by an interval of stable low values, then an asymptotic rise of carbon isotope values back to pre-IETM levels (Bowen et al., 2001; Bains et al., this volume). Continental sections sampled at lower resolution across the IETM also show an abrupt shift in $\delta^{13}C$ values, and may preserve a gradual recovery to pre-excursion values (Koch et al., 1992, 1995; Bowen et al., 2002; Ting et al., this volume). When organic matter, enamel carbonate, and soil carbonate isotopic data from Chalk Butte are considered together, the pattern of values suggests an excursion of similar form to that recorded elsewhere, but this record lacks the resolution necessary to distinguish the more subtle features of the event (Fig. 4). The "normal" isotope values for organic matter near the base of the section suggest that these rocks predate the IETM, whereas comparison with the record from Polecat Bench (Bowen et al., 2001) suggests that the low, relatively stable $\delta^{13}C$ values of paleosol carbonates from the Chalk Butte section may represent the middle portion of the carbon isotope excursion. The precise stratigraphic level of the base of the carbon isotope excursion at Chalk Butte is uncertain, as organic matter from the 20.5 m level (PRB-00-03-04) has a "normal" $\delta^{13}C$ value, while stratigraphically lower paleosol carbonates (PRB-00-02-02) produce IETM-like $\delta^{13}C$ values. The organic carbon content of sample PRB-00-03-04 is an order of magnitude less than for the other samples (Table 3). Given the questionable quality of that isotopic datum, we tentatively place the onset of the isotope excursion between the 5 and 13 m levels of the Chalk Butte section.

PALYNOLOGY

Nine samples were used in the palynological study. Seven pollen samples from the northern Chalk Butte section (PRB-00-03) are splits of the samples used for organic matter stable isotopic measurements discussed in the previous section. One additional sample (PS0004) comes from a carbonaceous shale 10 m above UCM83117 at the top of the Loehse section (PRB-01-01), and the other (PS0003) from a laterally extensive carbonaceous shale above the top of all measured sections (Fig. 3). This highest sample comes from the informally designated Loehse lignite (Robinson, 2000, personal commun.) which is exposed on the north side of the highway due north of the Loehse section. We estimate that the Loehse lignite is ~220 m above the base of the Wasatch Formation, or 120 m above the Chalk Butte lignite. Sample PS0003 is thus ~35 m above the highest sample in any measured section (PS0004).

Palynological samples were processed by Global Geolabs of Drumheller, Alberta using standard HF maceration followed by sieving with a true 10 μ screen, oxidation of organics, and staining. Slides were scanned under a transmitted light microscope. GJH identified and counted a mean of 406 grains in each sample (SD = 253). The high variability in sample size reflects two samples with high counts (PRB-00-03-01B, N = 826 and PS0003, N = 834) and two with low counts (PRB-00-03-02, N = 170 and PRB-00-03-04, N = 143). The high counts were made because of strong dominance of the sample by a single taxon. To ensure that rarer species would be detected grains were counted until the sum of the nondominant taxa reached 200. The low counts are for two samples that produced few palynomorphs, possibly because of poor preservation, low pollen production in the environment of deposition, or because the samples were weathered. (Sample PRB-00-03-04, which was the least productive pollen sample, contained only 0.06% organic carbon by weight.) Species lists and abundances for each sample are given in Appendix 1.

Important elements of the palynoflora from the Chalk Butte area include fern spores (*Cyathidites, Deltoidospora,* and *Laevigatosporites haardtii* together are 1%–41% of grains), taxodiaceous conifers (9%–76%), *Caryapollenites* (2%–30%), *Polyatriopollenites vermontensis* (0%–75%), *Ulmipollenites* (1%–6%) and Betulaceae/Myricaceae (2%–35%). Although the overall floral similarity with Bighorn Basin samples is high, a few moderately common pollen and/or spore taxa are unique to each basin. For example, *Cycadopites scabratus* is present in 7 of the 9 Chalk Butte samples at abundances of 0.5%–11.9%, but is unknown in 52 samples from the latest Paleocene and earliest Eocene of the Bighorn Basin (Harrington, 1999; Wing and Harrington, 2001). Other taxa differ consistently in their relative abundances between the Chalk Butte and Bighorn Basin samples. *Momipites* is moderately common in the Paleocene (mean of 9.5%) and Eocene (4.6%) of the Bighorn Basin, but is <1% of the flora in all but two samples from Chalk Butte. *Alnipollenites,* which has a mean abundance of 1% in the Paleocene of the Bighorn Basin and 8% in the Eocene, is <2% in all of the Chalk Butte samples. Both of these taxa were probably less important in the local vegetation around Chalk Butte than in coeval vegetation of the Bighorn Basin, just 100 km away.

The mean richness of the Chalk Butte samples is 24.1 taxa, as compared with 20.5 taxa in the set of 52 late Paleocene and early Eocene samples from the Bighorn Basin. Even with only 9 samples from Chalk Butte the difference is moderately significant (t-test with separate variances $p \leq 0.07$). There is also a stratigraphic pattern in richness of the Chalk Butte samples: The five lowest samples (0–22 m) have 18–23 species, whereas the upper four samples (56–120 m) have 24–31 species. This upward increase in diversity is consistent with the stratigraphic pattern described in the Bighorn Basin, where Eocene samples had higher average diversity than Paleocene samples (Wing and Harrington, 2001). The within-sample diversity differences are unlikely to result from sample size because the latter is poorly correlated with richness ($r^2 = 0.1$ for a logarithmic regression of richness on sample size).

Two stratigraphically important taxa have local first appearances in the Chalk Butte sections. *Intratriporopollenites instructus,* a *Tilia*-like pollen, is found in the three highest samples at Chalk Butte (79–120 m), and is moderately common in two of them (3.3% and 8.4%). In the Bighorn Basin *I. instructus* is moderately common from the lowest to the highest Eocene sample, but is absent in the Paleocene. One grain of *Platycarya* pollen is present in sample PRB-00-03-06 (79 m), and another in PS0003 (120 m). Although the form is locally so rare that its absence from lower samples could be due to vagaries of sampling, this taxon is also rare in the first half of the early Eocene in the Bighorn Basin, where it is absent from a well-sampled interval ~40 m thick immediately above the IETM (Wing and Harrington, 2001). The shift to palynofloras with Eocene index taxa at Chalk Butte occurs between sample PRB-00-03-05 at 58 m and sample PRB-00-03-06 at 79 m. The Wa 0 fauna of Robinson is in this interval, at ~74 m.

IDENTIFICATION OF THE IETM IN THE POWDER RIVER BASIN

In the western interior of the United States the Paleocene-Eocene boundary has been recognized by three criteria: A major immigration of mammalian taxa (Rose, 1981; Gingerich, 1989, this volume), a shift in palynofloral composition (Nichols and Ott, 1978; Wing, 1984; Harrington, 2001b), and a negative excursion in carbon isotope values (Koch et al., 1995; Bowen et al., 2001). In many areas the Paleocene-Eocene boundary is also stratigraphically close to the boundary between drab, lignitic strata assigned to the Fort Union Formation and red or variegated strata belonging to the Wasatch Formation, Willwood Formation, and other units, although the transition from drab to variegated rocks has been shown to be time transgressive over short distances in some areas (Wing and Bown, 1985). Over an extensive area in the Bighorn Basin where the IETM has been identified from faunal, floral and/or isotopic evidence, its onset is characterized by pedogenically altered red mudstones of unusual thickness and intense color (Gingerich, 1989; Wing et al., 2000). These prominent paleosols are evident even in areas where the transition from Fort Union to Willwood deposition predates the IETM (Gingerich, 2001).

Four lines of evidence suggest that exposures of the lower Wasatch Formation in the middle part of Chalk Butte and stratigraphically correlated areas nearby were deposited during the IETM. (1) There are prominent, laterally extensive red paleosols immediately above drab, lignitic strata, just as there are in IETM sequences in the eastern and southern Bighorn Basin. (2) Vertebrate faunas from ~40 m above the lowest red beds have a characteristic earliest Wasatchian (Wa 0) fauna that has been shown to be restricted to the isotopically defined IETM in the Bighorn Basin (Koch et al., 1995; Gingerich, this volume). (3) Palynofloras 5 m above the Wa 0 fauna document first appearances of Eocene index taxa and some of the shifts in diversity and relative abundances that are typical of the Paleocene-Eocene boundary in the Bighorn Basin. (4) $\delta^{13}C$ measurements made on a variety of substrates suggest a shift in the $\delta^{13}C$ value of terrestrial carbon stores similar to that seen in other terrestrial IETM sections.

Although the evidence for the IETM being represented at Chalk Butte is strong, the stratigraphic extent of the IETM is not clear. The first occurrence of *I. instructus* at 79 m suggests that this level is within or above the IETM. The presence of *Copecion davisii* and *Cantius torresi* at 74 m suggests that this level is within the IETM. The first laterally extensive red paleosol (at ~26 m) is probably near the base of the IETM, if the paleosols in Bighorn Basin sections can be used as a guide. The isotopic data are sparse, but imply that the interval between 13 and 63 m is within the IETM, with the base of the IETM lying between 5 and 13 m. Although these sources of data do not pinpoint the stratigraphic level for the base or top of the IETM, our best estimate is that the base of the IETM is between 5 and 13.2 m, and that the top of the IETM is above 74 m. If this is correct, then it implies that depositional rates at Chalk Butte were higher than they were at Polecat Bench, because low carbon isotope values at Chalk Butte span 50 m of section, whereas low values persist for ~40 m at Polecat Bench. The possibility of high rates of sediment deposition during the IETM has also been noted at other sites. Preliminary isotopic results (Koch et al., 1995), and the distribution of prominent red beds (personal observation), on the Sand Creek Divide in the eastern Bighorn Basin suggest that the IETM may be represented by 60–80 m of sediment, although this is far from certain. Ting et al. (this volume) have noted that the IETM in the Hengyang Basin of China is probably represented by a similar thickness of section. Unfortunately we lack other lines of evidence to test the idea that depositional rates increased during the IETM, although Crouch and Visscher (this volume) argue for an increase in terrestrial runoff during the IETM based on the abundance of terrestrial palynomorphs in marine environments.

FLORAL RESPONSE TO THE IETM

Lag in first appearances

Information on floral response to the IETM comes from just five pollen samples in the Chalk Butte area: PRB-00-03-03, 04, 05, and 06 at 10.9, 20.5, 58, and 79 m respectively in the northern Chalk Butte section, and PS0004 at 84 m in the Loehse Ranch section (Fig. 3). The lowest of these samples (10.9 m) is below isotope values characteristic of the excursion, but above the highest normal values, so it is unclear if it is within the IETM. The samples at 20.5 and 58 m are within the zone where pedogenic carbonate nodules have low carbon isotope values. The sample from 79 m is 5 m above the Wa 0 fauna, but could be in the upper part of the IETM.

The absence of Eocene index palynomorphs in samples at 20.5 and 58 m indicates that the immigration of *I. instructus* and *Platycarya* spp. in this area postdates the onset of the IETM by at least the amount of time represented by the ~45 m of sediments that lie between the lowest excursion isotopic value and the highest sample lacking the Eocene index palynomorphs. Although numerical age control is lacking in the Chalk Butte region, the deposition of ~45 m of fluvial sediment and paleosols is likely to represent many tens-of-thousands of years at rates of rock accumulation typical of intermontane basins. If the carbon isotope excursion begins below 13.2 m, and/or if the pollen sample from 58 m (which is in the base of a channel cut-and-fill deposit) represents a somewhat later time than its stratigraphic level indicates, then the lag between the beginning of the IETM and the first appearances of *I. instructus* and *Platycarya* is even greater. Outside limits are placed on the length of the lag period by the observation that Wa 0 mammals occur at 74 m. In the Bighorn Basin Wa 0 mammals are associated with only the most negative $\delta^{13}C$ excursion values, and these are thought to occur only in the first ~85 k.y. of the IETM (Röhl et al., 2000; Bowen et al., 2001; Gingerich, this volume). Consequently, the maximum period of lag between the onset of the IETM and the immigration of *I. instructus* and *Platycarya* into the Chalk Butte area would be slightly >85 k.y.

To summarize, the first appearances of *I. instructus* and *Platycarya* appear to lag the onset of the IETM by tens of thousands of years. The lag between the beginning of the IETM and the FADs for these two taxa is long enough that their appearance is associated with the later, cooling phase of the IETM rather than with the initial, geologically brief, warming. Both *I. instructus* and *Platycarya* are known from the late Paleocene of northwestern Europe. Thus, a reasonable hypothesis is that these taxa crossed to North America over high-latitude land bridges during the IETM, but did not expand their ranges southward in the continental interior until climate cooled in the latter part of the IETM. The absence of Asian records of these taxa during the Paleocene-Eocene interval reflects lack of study.

Absence of southern immigrants

A simplistic extrapolation of Quaternary patterns of floral response to the IETM would lead to the prediction that samples from within the IETM should contain a substantial number of immigrants from the Gulf Coastal Plain or other more southerly parts of North America. The relevant samples (at 84, 79, 58, 20.5, and possibly at 10.9 m) do not contain any of the southern taxa that might be expected, such as the Normapolles, which are numerically important constituents of Gulf Coastal Plain palynofloras of this age (Harrington and Kemp, 2001; Harrington, this volume). It is still possible that southern immigrants could have moved north into the area for a very short period of time, but evidence for such floral immigrants would have to occur in the gaps between samples low in the section. It seems more likely that the absence of evidence is evidence of absence.

Local changes in the native flora

Stratigraphic changes in the relative abundance of palynomorphs in the Chalk Butte samples also reveal aspects of the local floral response to the IETM. In the shift from Paleocene to Eocene palynofloras in the Bighorn Basin, changes in the relative abundances of taxa played a larger role than changes in presence and/or absence of taxa (Wing and Harrington, 2001). Considering only taxa that cross the boundary, Paleocene samples are consistently distinguished by higher abundance of *Polyatriopollenites vermontensis, Momipites,* and *Caryapollenites;* Eocene samples consistently have higher abundance of *Alnipollenites,* Betulaceae/Myricaceae, and *Ulmipollenites* (Wing and Harrington, 2001; Harrington, 2001b). Similar shifts in relative abundances are seen in the Chalk Butte samples, except for the overall rarity of *Alnipollenites* and *Momipites.*

We conducted a detrended correspondence analysis with down-weighting of rare taxa on a combined data set of Bighorn Basin and Chalk Butte samples to illuminate overall changes in floral composition through time. The first two axes of the ordination capture 28.2% of the variation in the data set, but they discriminate very well between Paleocene and Eocene samples (Fig. 5). The line separating the two groups is diagonal, but scores on the first axis only overlap in a narrow zone. Only two Bighorn Basin samples do not group as expected in the ordination. GH97-1.16 from the Paleocene of the Bighorn Basin groups with the Eocene samples even though it lacks Eocene indicator taxa because it is low in *Polyatriopollenites* and high in *Ulmipollenites.* PS9413 from the Eocene of the Bighorn Basin groups with Paleocene samples because it lacks Eocene indicators, is low in *Ulmipollenites* and high in *Polyatriopollenites.* Taxa are also plotted on the ordination, and they occupy positions that are consistent with their tendency to be more abundant either in the Paleocene or Eocene (Fig. 5).

If samples from within the IETM (marked by stars in Figure 5) had distinctive abundances of species, they would be expected to plot separate from the rest of the samples; they do not. Sample PRB-00-03-03 (either lowest IETM or just prior to the IETM) and sample PRB-00-03-04 (IETM), have relative abundances of taxa typical of Paleocene samples (Fig. 5). Sample PRB-00-03-05 (IETM) and samples PRB-00-03-06 and PS0004 (uppermost IETM or just post IETM) have relative abundances more typical of the Eocene, even though PRB-00-03-05 lacks Eocene index taxa. PRB-00-03-05 plots with the Eocene samples because of the low abundance of *Polyatriopollenites vermontensis.*

We can show shifts in floral composition through time by plotting the first axis score of each sample against its stratigraphic level (Fig. 6). Plotting samples from the Powder River and Bighorn Basin on the same y-axis makes the implicit assumption that depositional rates were the same in both basins, which cannot be independently verified. We are confident, however, that with the exception of the three Chalk Butte samples discussed above, we have correctly assigned the samples to pre-IETM, IETM, and post-IETM time. Figure 6 confirms that Paleocene and Eocene palynofloras have distinct compositions, and that the permanent shift to typical Eocene relative composition takes place in the latter part of the IETM. Sample PRB-00-03-02 plots with Eocene samples even though it is below the IETM as defined by low carbon isotope values, and in an interval with other pollen samples that have more typical "Paleocene" composition. The main cause of its similarity to Eocene samples is its low abundance of *Polyatriopollenites vermontensis, Caryapollenites,* and *Momipites.* We do not know if the low abundances of these taxa reflects climatic shifts prior to the onset of the IETM proper, or just chance fluctuations in pollen preservation and depositional environment. Major fluctuations in the abundance of *P. vermontensis* are also seen in the last part

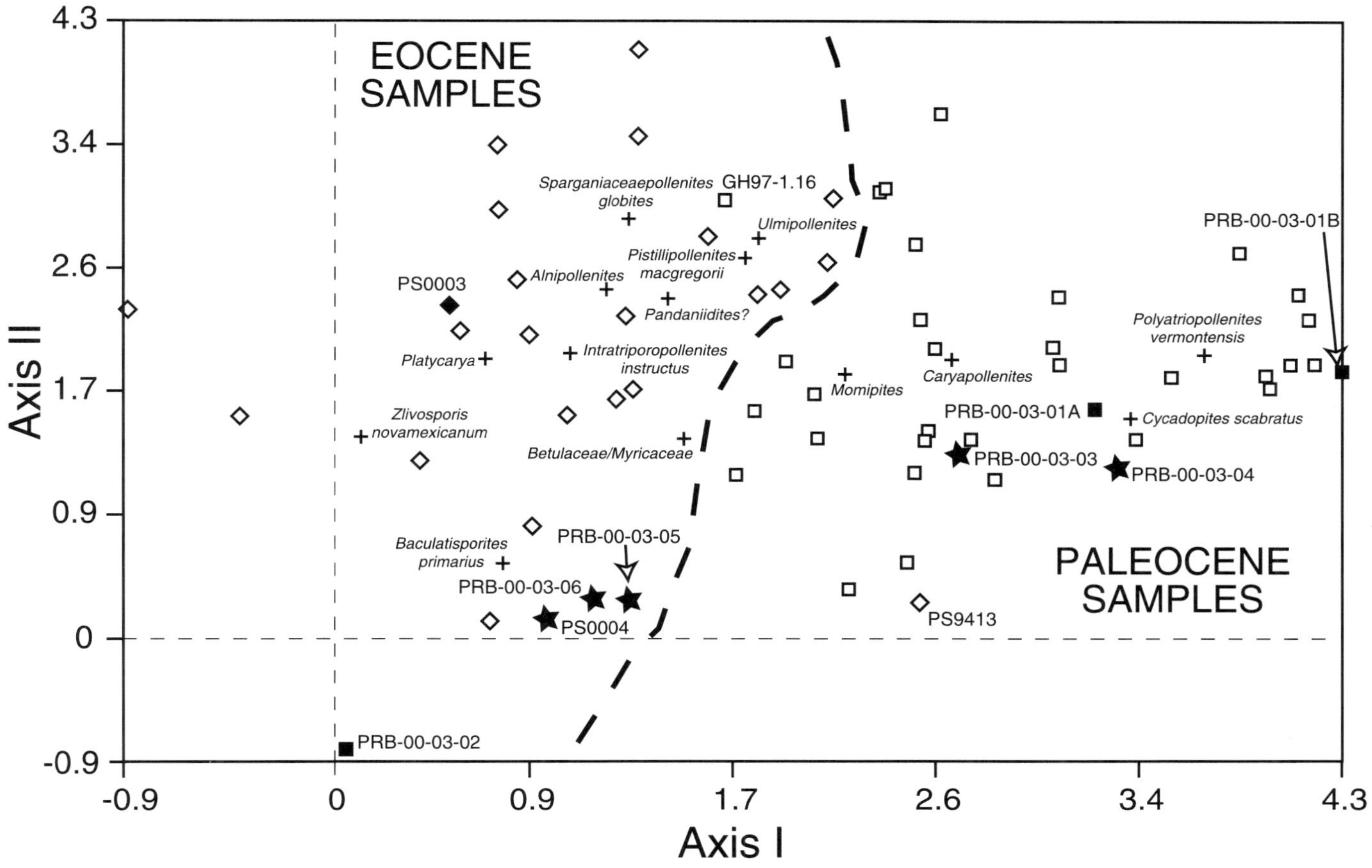

Figure 5. Detrended correspondence analysis of latest Paleocene and earliest Eocene pollen samples from the Bighorn and Powder River Basins. Axis I expresses 22% of the variance and axis II 6%. Filled symbols are samples from Chalk Butte, empty symbols are samples from the Bighorn Basin. Diamonds are Eocene samples, squares are Paleocene samples, and stars are samples that are possibly or definitely from the Initial Eocene Thermal Maximum. (See text discussion.) Crosses show the positions of selected taxa in the ordination space.

of the Paleocene in the Bighorn Basin (Wing and Harrington, 2001). If PRB-00-03-02 does reflect change in vegetation near Chalk Butte, then it suggests that relative abundances of taxa that were already present in the region (specifically *P. vermontensis*) began to shift in response to climate change well before the arrival of immigrant taxa late in the IETM.

DISCUSSION

Pollen samples from the IETM in the Powder River Basin do not appear to record the pattern of rapid, continental-scale range change that would be expected from a simple extrapolation of Quaternary plant response to this Paleogene example of rapid global warming. There is no indication of rapid northerly migration of plant populations, and the small number of immigrants that did appear probably lagged the onset of warming by at least tens of thousands of years. This slow response is in sharp contrast to that in mammalian communities, which show a pulse of first appearances of entirely new lineages during the IETM event as well as suspected rapid evolutionary transitions within pre-existing lineages (Gingerich, 1989; Clyde and Gingerich, 1998; Gingerich, this volume). As yet we do not know why the pace of floral change during the IETM was so slow, but at least three types of explanations should be considered: Ecological limitations on range shifts, physiological limitations on range shifts, and historical development of the flora.

Ecological limitations on movement of plant populations might have been very different during the IETM than during Quaternary deglaciations. During deglaciation, plant populations expanded into areas undergoing primary succession in the wake of retreating glaciers. Models and data show that rare, long-distance dispersal and successful establishment are necessary to account for the high rate of plant migration in the Holocene (Clark et al., 1998), and that successful invasion in the present day is associated with disturbance (Richardson and Bond, 1991; Rejmánek and Richardson, 1996). The probability of successful establishment might well have been much higher during deglaciation than during the IETM, because Paleogene

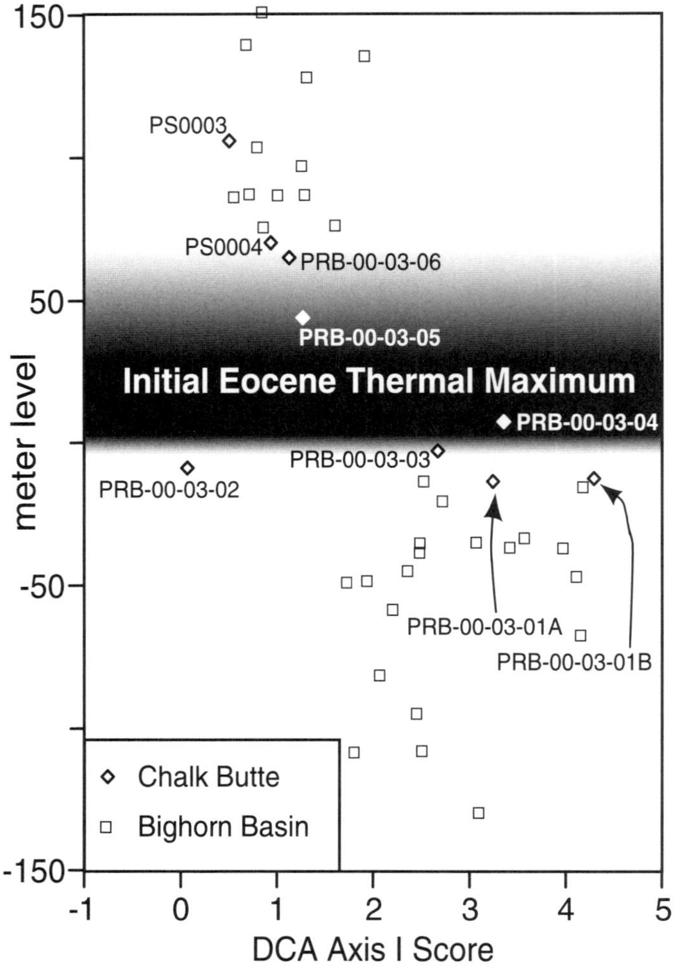

Figure 6. Ordination axis I scores of pollen samples against stratigraphic level, including Powder River Basin (PRB) and Bighorn Basin samples. Diamonds are samples from Chalk Butte area, empty squares are samples from the Bighorn Basin. DCA—detrended correspondence analysis.

climates permitted forests to grow from the Gulf of Mexico (Wolfe, 1985) to the shores of the Arctic Ocean (Hickey et al., 1983; McIver and Basinger, 1999). During the IETM potential long-distance dispersers would have had to establish new populations under an existing forest canopy.

Physiological limitations might also have slowed plant migration during the IETM. The southern part of North America was probably occupied by broad-leaved evergreen forest during the late Paleocene and early Eocene (Wolfe, 1985; Wolfe and Dilcher, 2000). Broad-leaved, evergreen species with poorly developed physiological mechanisms for leaf abscission or dormancy may not have grown well in the warm, dim, winter conditions that probably existed in the middle to high latitudes during the IETM (Peters and Sloan, 2000). Plants that do not drop their leaves during the winter continue to lose water through transpiration, and respiration in the leaves may use stored resources (Read and Francis, 1992). Warm winters with low light levels might therefore favor deciduous plants, as has been proposed for high-latitude Cretaceous floras (Falcon-Lang and Cantrill, 2001). Similar arguments presumably apply in a less absolute form to floras living between 45 and 55 degrees during the IETM. Maladaptation to long, dim but warm, winters may have prevented many broad-leaved evergreen taxa from shifting ranges during the IETM. Longer intervals of warm climate, such as the one that occurred at the end of the early Eocene and persisted for at least a million years, might have been long enough for evergreen taxa to evolve the deciduous habit that would permit colonization and dispersal across high-latitude corridors.

A final factor that may explain the slow floral response to rapid warming during the IETM and the rapid floral response to late Quaternary warming is the difference in climatic regimes that preceded the two events. The IETM was a sudden climate shift in the midst of a geologically long period when the amplitude of orbitally forced temperature change was probably low. In contrast, the late Quaternary deglaciation was merely the last of many rapid, high-amplitude climatic changes that severely affected northern midlatitude floras over the last 2–3 million years. The late Quaternary flora of the midnorthern latitudes thus consists of plants that survived many glacial-interglacial fluctuations, and it may be composed of an unusually high proportion of species that disperse well. The latest Paleocene floras that experienced the IETM may have consisted of a more typical mix of species including many that were not particularly good dispersers.

CONCLUSIONS

1. The lower Wasatch Formation in the area of Chalk Butte in the southwestern Powder River Basin preserves a sequence of fossiliferous, pedogenically altered, fluvial sediments that represent the IETM. This is the first record of a terrestrial IETM sequence in North America outside of the Bighorn Basin. Isotopic evidence, mammalian biostratigraphy, and palynostratigraphy are in agreement with expectations developed in the Bighorn Basin.

2. Pollen samples from within the IETM give no evidence of a transient flora containing tropical or subtropical taxa, thus there is no evidence for northward range changes during the IETM.

3. The initial floral response to the IETM may have been a shift in relative abundances of taxa that were already present in the area, specifically declining abundance of *Polyatriopollenites vermontensis*. This is consistent with the idea that warming and/or other climatic and atmospheric changes during the IETM rapidly affected the relative success of different types of native plants.

4. First appearances of the Eocene index palynomorphs *Intratriporopollenites instructus* and *Platycarya* spp. probably occurred during the cooling late in the IETM, not with the initial warming. These Eocene floral immigrants to the northern

Rockies were cool adapted species that probably came across high-latitude land bridges.

5. Floral changes during the IETM are much slower than would be expected from a model based on floral response to Quaternary glacial cycles. This probably reflects limits on the dispersal and colonization rates of Paleogene plants.

ACKNOWLEDGMENTS

We thank Peter Robinson for guidance and orientation to the field area, the Meike and Loehse families for permission to work on their land, Henry Fricke, Brenda Beitler, Amanda Ash, Kira Lawrence, David Blackburn, and Anjali Goswami for assistance in the field, and Jason Newton and Beth Zotter for their help with isotopic analyses. An earlier version of this manuscript was substantially improved by comments from Bruce Tiffney and an anonymous, constructive skeptic. SLW was supported by a grant from the Evolution of Terrestrial Ecosystems Program of the Smithsonian Institution, GJH by Enterprise Ireland award SC/00/251, and GJB by the National Science Foundation Graduate Research Fellowship Program. This is ETE contribution 84.

APPENDIX 1. ABUNDANCES OF POLLEN AND SPORE TAXA—CHALK BUTTE SAMPLES

Taxon name	Sample numbers and stratigraphic levels								
	PS0003 (~120 m)	PS0004 (84 m)	PRB-00-03-06 (79 m)	PRB-00-03-05 (58 m)	PRB-00-03-04 (20.5 m)	PRB-00-03-03 (10.9 m)	PRB-00-03-02 (5 m)	PRB-00-03-01B (2 m)	PRB-00-03-01A (1 m)
Spores									
Appendicisporites spp.	0	0	0	0	0	0	1	0	0
Baculatisporites primarius	1	0	0	1	0	0	2	0	0
Camarozonosporites	1	1	3	13	0	0	8	0	0
Cicatricosisporites dorogensis	0	1	1	1	0	0	9	0	0
Cyathidites triangulus	1	2	3	5	0	0	7	0	1
Deltoidospora spp.	13	7	12	25	9	3	35	2	0
Gleicheniidites	0	6	6	7	0	0	7	1	0
Laevigatosporites haardtii	57	9	22	40	25	26	21	0	6
Microfoveolatasporis pseudodentata	0	0	0	0	1	0	0	0	0
Reticuloidosporites pseudomurii	1	0	0	0	0	0	0	0	0
Stereisporites steriodes	0	0	1	1	0	0	0	0	0
Zlivosporis novamexicanum	0	0	0	0	0	0	1	0	0
Gymnosperms									
Bisaccate	0	2	3	19	0	10	15	3	6
Cupressacites hiatipites	616	70	73	125	19	83	19	64	49
Erdtmanipollis cretaceus	1	3	1	0	2	0	0	0	0
Sequoia	15	9	12	41	4	20	23	8	19
Monocots									
Arecipites spp.	3	0	0	15	0	11	0	1	1
Cycadopites scabratus	0	2	0	3	2	2	2	29	43
Cycadopites spp.	4	4	0	4	3	1	0	0	11
Liliacidites spp.	0	0	1	0	0	0	0	0	0
Monocolpopollenites tranquillus	2	1	2	0	1	1	0	0	2
Pandaniidites?	0	0	0	1	0	0	0	0	0
Sparganiaceaepollenites globites	2	1	1	0	0	0	0	1	0
Dicots (Tricolpate + Tricolporate)									
Ailanthipites berryi	0	0	1	0	1	0	0	0	0
Caprifoliipites spp.	0	0	1	0	0	0	0	0	0
Cercidiphyllites sp.	0	0	0	1	0	1	0	0	0
Ericipites spp.	0	0	0	0	0	1	0	0	0
Eucommia sp. A	3	0	1	1	0	1	0	3	0
Fraxinoipollenites spp.	0	1	1	1	0	0	2	3	2
Fraxinoipollenites varabilis	3	1	1	0	0	9	0	0	0

continued

APPENDIX 1. ***Continued***

Taxon name	Sample numbers and stratigraphic levels								
	PS0003 (~120 m)	PS0004 (84 m)	PRB-00-03-06 (79 m)	PRB-00-03-05 (58 m)	PRB-00-03-04 (20.5 m)	PRB-00-03-03 (10.9 m)	PRB-00-03-02 (5 m)	PRB-00-03-01B (2 m)	PRB-00-03-01A (1 m)
Malvasipollis?	0	0	1	0	1	0	0	0	0
Retitrescolpites anguloluminosus	0	19	3	0	1	8	0	0	1
Retitrescolpites catenatus	0	1	0	0	0	0	0	0	0
Rousea spp.	0	2	0	0	0	0	0	1	0
Rousea crassimurina	0	1	0	0	0	2	0	0	1
Striatopollis sp. (Pocknall & Nichols 1996)	0	5	0	0	1	0	0	0	0
Tricolpites hians	3	2	7	1	1	2	2	1	1
Ilexpollenites spp.	0	0	0	0	0	0	1	0	0
Insulapollenites rugulatus	0	0	0	0	0	0	0	0	1
Intratriporopollenites spp.	0	2	0	0	0	0	0	0	0
Intratriporopollenites instructus	5	26	11	0	0	0	0	0	0
Intratriporopollenites sp. cf. *I. tetraforaminipites*	2	2	2	0	0	1	3	5	13
Intratriporopollenites vescipites?	0	0	0	0	0	0	0	2	5
Nyssa kruschii	4	2	0	3	0	0	1	1	1
Nyssapollenites explanata	4	2	3	3	2	0	0	0	0
Rhoipites bradleyi	0	0	0	2	0	0	0	0	0
Siltaria hanleyi	0	0	0	1	0	0	0	0	0
Dicots (Triporate—Periporate)									
Alnipollenites spp.	0	1	0	0	0	0	0	1	0
Alnipollenites scoticus	1	0	0	0	0	0	0	0	0
Alnipollenites verus	0	4	1	1	0	3	0	4	3
Caryapollenites spp.	4	0	8	20	2	6	3	26	30
Caryapollenites imparalis group	0	0	0	1	0	1	0	0	0
Caryapollenites veripites	14	5	7	18	2	18	1	49	81
Caryapollenites wodehousei	0	0	2	0	1	3	1	1	1
Momipites spp.	2	1	3	0	0	1	0	3	5
Momipites ventifluminis	1	8	9	0	0	2	0	0	0
Momipites wyomingensis	0	0	0	0	0	0	0	0	0
Periporopollenites (Chenopod type)	0	0	1	1	0	0	0	0	0
Periporopollenites (*Liquidambar* type)	0	1	0	0	0	0	0	0	1
Pistillipollenites macgregorii	0	0	0	0	0	0	0	0	0
Platycarya platycaryoides	1	0	1	0	0	0	0	0	0
Polyatriopollenites vermontensis	0	2	6	5	56	42	1	602	57
Paraalnipollenites confusus	0	1	7	0	0	2	0	0	1
Betulaceae/Myricaceae undiff.	7	19	67	16	6	25	4	5	17
Triatriopollenites subtriangulus	0	0	2	0	1	2	0	2	0
Triporopollenites granulatus	12	80	39	1	0	2	0	5	0
Ulmipollenites spp.	17	1	3	6	2	0	1	1	0
Ulmipollenites krempii	26	2	3	3	0	5	0	1	2
Ulmipollenites tricostatus	8	0	0	1	0	0	0	1	2
Number of Grains (sum)	834	309	332	387	143	294	170	826	363
Number of Taxa (richness)	31	39	40	34	22	30	24	28	28
Reworked and Aquatic Taxa									
Dinoflagellates	4	0	7	22	0	0	2	0	0
Acritarchs	0	0	2	0	0	0	0	0	0
R *Classopollis*	0	0	1	0	0	0	0	0	0
Pterospumella?	0	0	0	0	18	0	1	0	0
R *Tricolpites foeveolatus*	0	0	0	0	0	0	2	0	0
R ?*Complexipollis*	0	0	0	0	0	0	2	0	0
K spores undiff.	0	0	0	0	0	0	4	0	0
Pediastrum	0	0	0	0	0	0	0	0	0

REFERENCES CITED

Beerling, D.J., and Jolley, D.W., 1998, Fossil plants record an atmospheric $^{12}CO_2$ and temperature spike across the Paleocene-Eocene transition in northwest Europe: Journal of the Geological Society of London, v. 155, p. 591–594.

Bowen, G.J., and Bloch, J.I., 2002, Petrography and geochemistry of flood-plain limestones from the Clarks Fork Basin, Wyoming, USA: Carbonate deposition and fossil accumulation on a Paleocene-Eocene flood plain: Journal of Sedimentary Research, v. 72, p. 46–58.

Bowen, G.J., Koch, P.L., Gingerich, P.D., Norris, R.D., Bains, S., and Corfield, R.M., 2001, Refined isotope stratigraphy across the continental Paleocene-Eocene boundary on Polecat Bench in the northern Bighorn Basin: University of Michigan Papers on Paleontology, v. 33, p. 73–88.

Bowen, G.J., Clyde, W.C., Koch, P.L., Ting, Suyin, Alroy, J., Tsubamoto, Takehisa, Wang, Yuanqing, and Wang, Yuan, 2002, Mammalian dispersal at the Paleocene/Eocene boundary: Science, v. 295, p. 2062–2065.

Bown, T.M., 1979, Geology and mammalian paleontology of the Sand Creek facies, lower Willwood Formation (lower Eocene), Washakie County, Wyoming: Geological Survey of Wyoming Memoir, v. 2, p. 1–151.

Cerling, T.E., 1984, The stable isotopic composition of modern soil carbonate and its relationship to climate: Earth and Planetary Science Letters, v. 71, p. 229–240.

Clark, J.S., Fastie, C., Hurtt, G., Jackson, S.T., Johnson, C., King, G.A., Lewis, M., Lynch, J., Pacala, S., Prentice, C., Schupp, E.W., Webb, T.I., and Wyckoff, P., 1998, Reid's paradox of rapid plant migration: BioScience, v. 48, p. 13–24.

Clyde, W.C., and Gingerich, P.D., 1998, Mammalian community response to the latest Paleocene thermal maximum: An isotaphonomic study in the northern Bighorn Basin, Wyoming: Geology, v. 26, p. 1011–1014.

Clyde, W.C., Sheldon, N.D., Koch, P.L., Gunnell, G.F., and Bartels, W.S., 2001, Linking the Wasatchian/Bridgerian boundary to the Cenozoic Global Climate Optimum: New magnetostratigraphic and isotopic results from South Pass, Wyoming: Palaeogeography, Palaeoclimatology, Palaeoecology, v. 167, p. 175–199.

Crouch, E.M., 2001, Environmental change at the time of the Paleocene-Eocene biotic turnover: Utrecht, Netherlands, Laboratory of Palaeobotany and Palynology Contributions Series, v. 14, p. 1–216.

Crouch, E.M., Heilmann-Clausen, C., Brinkhuis, H., Morgans, H.E.G., Rogers, K.M., Egger, H., and Schmitz, B., 2001, Global dinoflagellate event associated with the late Paleocene thermal maximum: Geology, v. 29, p. 315–318.

Davis, M.B., 1976, Pleistocene biogeography of temperate deciduous forests: Geoscience and Man, v. 13, p. 13–26.

Dickens, G.R., Castillo, M.M., and Walker, J.C.G., 1997, A blast of gas in the latest Paleocene: Simulating first-order effects of massive dissociation of oceanic methane hydrate: Geology, v. 25, p. 259–262.

Falcon-Lang, H.J., and Cantrill, D.J., 2001, Leaf phenology of some mid-Cretaceous polar forests, Alexander Island, Antarctica: Geological Magazine, v. 138, p. 39–52.

Frederiksen, N.O., 1994, Paleocene floral diversities and turnover events in eastern North America and their relation to diversity models: Review of Palaeobotany and Palynology, v. 82, p. 225–238.

Gingerich, P.D., 1989, New earliest Wasatchian mammalian fauna from the Eocene of northwestern Wyoming: Composition and diversity in a rarely sampled high–flood plain assemblage: University of Michigan Papers on Paleontology, v. 28, p. 1–97.

Gingerich, P.D., 2001, Biostratigraphy of the continental Paleocene-Eocene boundary interval on Polecat Bench in the northern Bighorn Basin: University of Michigan Papers on Paleontology, v. 33, p. 37–72.

Gruas-Cavagnetto, C., 1978, Étude palynologique de L'Éocène du bassin Anglo-Parisien: Mémoire de la Société Géologique de France, v. 131, p. 1–60.

Harrington, G.J., 1999, North American palynofloral dynamics in the late Paleocene to early Eocene [Ph.D. thesis]: Sheffield, University of Sheffield, Department of Earth Sciences, 320 p.

Harrington, G.J., 2001a, Impact of Paleocene/Eocene greenhouse warming on North American paratropical forests: Palaios, v. 16, p. 266–278.

Harrington, G.J., 2001b, Pollen assemblages and Paleocene-Eocene stratigraphy in the Bighorn and Clarks Fork Basins: University of Michigan Papers on Paleontology, v. 33, p. 89–96.

Harrington, G.J., and Kemp, S.J., 2001, U.S. Gulf Coast vegetation dynamics during the latest Paleocene: Palaeogeography, Palaeoclimatology, Palaeoecology, v. 167, p. 1–21.

Hickey, L.J., West, R.M., Dawson, M.R., and Choi, D.K., 1983, Arctic terrestrial biota: Paleomagnetic evidence of age disparity with midnorthern latitudes during the Late Cretaceous and early Tertiary: Science, v. 221, p. 1153–1156.

Hooghiemstra, H., and Melice, J.L., 1994, Pleistocene evolution of orbital periodicities in the high-resolution pollen record Funza I, Eastern Cordillera, Colombia: Special Publication of the International Association of Sedimentology, v. 19, p. 117–126.

Jackson, S.T., and Overpeck, J.T., 2000, Responses of plant populations and communities to environmental changes of the late Quaternary: Paleobiology, v. 26, p. 194–220.

Jaramillo, C.A., and Dilcher, D.L., 2000, Microfloral diversity patterns of the late Paleocene-Eocene interval in Colombia: Geology, v. 28, p. 815–818.

Jolley, D.W., 1998, Palynostratigraphy and depositional history of the Paleocene Ormesby/Thanet depositional sequence set in southeastern England and its correlation with continental West Europe: Review of Palaeobotany and Palynology, v. 99, p. 265–315.

Kelly, D.C., Bralower, T.J., and Zachos, J.C., 1998, Evolutionary consequences of the latest Paleocene thermal maximum for tropical planktonic foraminifera: Palaeogeography, Palaeoclimatology, Palaeoecology, v. 141, p. 139–161.

Kennett, J.P., and Stott, L.D., 1991, Abrupt deep-sea warming, palaeoceanographic changes and benthic extinctions at the end of the Paleocene: Nature, v. 353, p. 225–229.

Koch, P.L., and Morrill, C., 2000, Paleocene/Eocene climate change in southwestern Wyoming: Isotopic records from paleosol carbonates and freshwater bivalves: GFF, v. 122, p. 89–91.

Koch, P.L., Zachos, J.C., and Gingerich, P.D., 1992, Correlation between isotope records in marine and continental carbon reservoirs near the Paleocene/Eocene boundary: Nature, v. 358, p. 319–322.

Koch, P.L., Zachos, J.C., and Dettmann, D.L., 1995, Stable isotope stratigraphy and paleoclimatology of the Paleogene Bighorn Basin (Wyoming, USA): Palaeogeography, Palaeoclimatology, Palaeoecology, v. 115, p. 61–89.

McIver, E.E., and Basinger, J.F., 1999, Early Tertiary floral evolution in the Canadian high Arctic, *in* Crane, P.R., and Herendeen, P.S., eds., The origin of modern terrestrial ecosystems: Annals of the Missouri Botanical Garden, v. 86, p. 523–545.

Mommersteeg, H.J.P.M., Loutre, M.F., Young, R., Wijmstra, T.A., and Hooghiemstra, H., 1995, Orbital forced frequencies in the 975,000 year pollen record from Tenaghi Phillipon (Greece): Climate Dynamics, v. 11, p. 4–24.

Nichols, D.J., and Ott, H.L., 1978, Biostratigraphy and evolution of the *Momipites-Caryapollenites* lineage in the early Tertiary in the Wind River Basin, Wyoming: Palynology, v. 2, p. 37–48.

Overpeck, J.T., Webb, R.S., and Webb, T.I., 1992, Mapping eastern North American vegetation change of the past 18 ka: No-analogs and the future: Geology, v. 20, p. 1071–1074.

Passey, B.H., Cerling, T.E., Perkins, M.E., Voorhies, M.R., Harris, J.M., and Tucker, S.T., 2002, Environmental change in the Great Plains: An isotopic record from fossil horses: Journal of Geology, v. 110, p. 123–140.

Peters, R.B., and Sloan, L.C., 2000, High concentrations of greenhouse gases and polar stratospheric clouds: A possible solution to high-latitude faunal migration at the latest Paleocene thermal maximum: Geology, v. 28, p. 979–982.

Pocknall, D.T., 1987, Palynomorph biozones for the Fort Union and Wasatch Formations (upper Paleocene–lower Eocene), Powder River Basin, Wyoming and Montana, USA: Palynology, v. 11, p. 23–35.

Read, J., and Francis, J., 1992, Responses of some Southern Hemisphere tree species to a prolonged dark period and their implications for high-latitude Cretaceous and Tertiary floras: Palaeogeography, Palaeoclimatology, Palaeoecology, v. 99, p. 271–290.

Rejmánek, M., and Richardson, D.M., 1996, What attributes make some plant species more invasive?: Ecology, v. 77, p. 1655–1661.

Richardson, D.M., and Bond, W.J., 1991, Determinants of plant distribution: Evidence from pine invasions: The American Naturalist, v. 137, p. 639–668.

Robinson, P., and Williams, B., 1997, Species diversity, tooth size, and shape of *Haplomylus* (Condylarthra, Hyopsodontidae) from the Powder River Basin, northeastern Wyoming: University of Wyoming, Contributions to Geology, v. 31, p. 59–78.

Röhl, U., Bralower, T.J., Norris, R.D., and Wefer, G., 2000, New chronology for the late Paleocene thermal maximum and its environmental implications: Geology, v. 28, p. 927–930.

Rose, K.D., 1981, The Clarkforkian land-mammal age and mammalian faunal composition across the Paleocene-Eocene boundary: University of Michigan Papers on Paleontology, v. 26, p. 1–197.

Rull, V., 1999, Palaeofloristic and palaeovegetational changes across the Paleocene/Eocene boundary in northern South America: Review of Palaeobotany and Palynology, v. 107, p. 83–95.

Schumacker-Lambry, J., 1978, Palynologie du Landenien inférieur (Paléocène) à Gelinden-Overbroek/Belgique: Relations entre les microfossiles et le sédiment [Ph.D. thesis]: Liège, Belgium, Laboratoire de Paléobotanique et de Paléopalynologie, Universite de Liège, 158 p.

Strait, S.G., 2001, New Wa 0 mammalian fauna from Castle Gardens in the southeastern Bighorn Basin: University of Michigan Papers on Paleontology, v. 33, p. 127–144.

Thewissen, J.G.M., 1990, Evolution of Paleocene and Eocene Phenacodontidae (Mammalia, Condylarthra): University of Michigan Papers on Paleontology, v. 29, p. 1–107.

Wing, S.L., 1984, A new basis for recognizing the Paleocene/Eocene boundary in western interior North America: Science, v. 226, p. 439–441.

Wing, S.L., 1998, Late Paleocene–early Eocene floral and climatic change in the Bighorn Basin, Wyoming, *in* Berggren, W., et al., eds., Late Paleocene–early Eocene biotic and climatic events: New York, Columbia University Press, p. 380–400.

Wing, S.L., and Bown, T.M., 1985, Fine scale reconstruction of late Paleocene–early Eocene paleogeography in the Bighorn Basin of northern Wyoming, *in* Flores, R., and Kaplan, S., eds., Cenozoic paleogeography of the west-central United States: Rocky Mountain Section, Society of Economic Paleontologists and Mineralogists, p. 93–105.

Wing, S.L., and Harrington, G.J., 2001, Floral response to rapid warming in the earliest Eocene and implications for concurrent faunal change: Paleobiology, v. 27, p. 539–562.

Wing, S.L., Alroy, J., and Hickey, L.J., 1995, Plant and mammal diversity in the Paleocene to early Eocene of the Bighorn Basin: Palaeogeography, Palaeoclimatology, Palaeoecology, v. 115, p. 117–156.

Wing, S.L., Bao, H., and Koch, P.L., 2000, An early Eocene cool period? Evidence for continental cooling during the warmest part of the Cenozoic, *in* Huber, B., et al., eds., Warm climates in Earth history: Cambridge, UK, Cambridge University Press, p. 197–237.

Wolfe, J.A., 1985, Distribution of major vegetational types during the Tertiary: American Geophysical Union Geophysical Monograph Series, v. 32, p. 357–375.

Wolfe, J.A., and Dilcher, D.L., 2000, Late Paleocene through middle Eocene climate in lowland North America: GFF, v. 122, p. 184–185.

Zachos, J., Pagani, M., Sloan, L., Thomas, E., and Billups, K., 2001, Trends, rhythms, and aberrations in global climate 65 Ma to present: Science, v. 292, p. 686–693.

Manuscript Accepted by the Society August 13, 2002

Printed in the USA

Geological Society of America
Special Paper 369
2003

Lacertilian faunal change across the Paleocene-Eocene boundary in Europe

Marc Augé
Laboratoire de Paléontologie, Muséum National d'Histoire Naturelle, 8 rue Buffon, 75005 Paris, France

ABSTRACT

The Franco-Belgian Basin contains numerous early Eocene localities that have yielded a large sample of lacertilian fossil specimens. Paleocene faunal successions are far less documented in Europe. A major lacertilian faunal turnover occurs across the Paleocene-Eocene boundary, with a strong increase in diversity during the early Eocene. Families represented during the Paleocene are still present in the early Eocene, and new taxa (Agamidae, Iguanidae*, Varanidae, Gekkonidae, Glyptosaurinae) appear in the locality of Dormaal (earliest Eocene). Some important changes in ecology (diet, size) seem to affect the lizard community during the Paleocene-Eocene events. An increase in richness of insect eaters and carnivorous lizards between Paleocene and early Eocene faunas is noteworthy. New carnivorous (Varanidae) and insectivorous lizards (Iguanidae*, Agamidae) enter the intermediate size class during the transition. More equable climate may have allowed immigration through high-latitude routes. The area of origin of these newcomers has not been identified.

INTRODUCTION

Faunal succession of reptiles across the Paleocene-Eocene boundary is more poorly documented than the mammalian turnover. In Europe and North America there is a sharp distinction between Paleocene and Eocene mammal faunas, with the Eocene appearance of the modern Orders Perissodactyla, Artiodactyla, Primates and family Hyaenodontidae (Gingerich, 1989).

Among the few studies concerning reptiles, Alifanov (1993) reported an impoverished lizard fauna from the upper Paleocene of Central Asia that included only two families (Necrosauridae and Agamidae). Hou (1974) placed the genus *Qianshanosaurus* (Paleocene, China) in the Iguanidae* but this "aberrant" lizard was later regarded as Lacertilia incertae sedis by Estes (1983). Lizard jaw fragments belonging to agamid lizards and large varanid lizard (?*Saniwa*) vertebrae have been reported from the early Eocene of Kyrgyzstan (Averianov and Danilov, 1996, 1997). The middle Eocene fauna of this area seems more diversified and includes new genera of agamid and varanid lizards. According to Alifanov (1993), Iguanidae* reappear after their Late Cretaceous local extinction.

In western North America, Hutchison (1998) and Holroyd and Hutchison (2000) reported the appearance of several new turtle taxa across the Paleocene-Eocene transition.

The Bighorn Basin (Wyoming) glyptosaurine lizards underwent a significant radiation (Bartels, 1983). The Paleocene-Eocene boundary is also marked by the disappearance of choristoderan reptiles (champsosaurs) in both Europe and North America (Gervais, 1873; Bartels, 1983; Gunnell et al., 1993).

In Europe, according to Rage and Augé (1993), the Paleocene squamate fauna appears depauperate, containing only lacertid and necrosaurid lizards and boid snakes.

Conversely, late Paleocene–early Eocene lacertilian assemblages are abundant and diversified (Hecht and Hoffstetter, 1962; Augé, 1990a, 1990b). In Europe, snakes are very rare in the Paleocene but they became numerous and rather diversified at the beginning of the Eocene (Rage, 1987a). The nonmarine turtles underwent an increase in diversity and modernization of

Augé, M., 2003, Lacertilian faunal change across the Paleocene-Eocene boundary in Europe, *in* Wing, S.L., Gingerich, P.D., Schmitz, B., and Thomas, E., eds., Causes and Consequences of Globally Warm Climates in the Early Paleogene: Boulder, Colorado, Geological Society of America Special Paper 369, p. 441–453.

the fauna through the boundary interval (Broin, 1977). Buffetaut (1985) found no important change in the taxonomic diversity of crocodilians across the limit. In studies of general crocodilian diversity during the Cenozoic (Markwick, 1998; Vasse, 1993; Vasse and Hua, 1998), the record of diversity from the late Paleocene to the early Eocene is steady. However, these authors note an increase both in rate of origination and extinction during the period.

The record of South American and African fossil lizards is poor compared to that of Europe and North America. It is not possible to develop an adequate interpretation about lizard diversity in Gondwanan territories during the Paleogene. The Paleocene lizards in South America include Iguania, Gekkonidae, Teiidae and Scincomorpha (Albino, 1996). The Eocene record of South American lizards is poorly represented by a teiid genus (Donadio, 1985). According to Albino (1993), the snakes from the late Paleocene of Patagonia (Argentina) were larger than those of the Late Cretaceous. During the early Eocene in Patagonia a more diverse snake fauna is present and, during this period, the snakes were the most important predators, together with ziphodont crocodiles, phorusrhacoid birds and borhyaenid mammals. The richness of snakes probably reflects the diversification of mammals, providing abundant prey of "large" size.

This study focuses on the faunal turnover that affected Lacertilia (lizards) as more material and studies become available from Paleocene and Eocene localities, mainly situated in the same geographic area, the Paleogene Franco-Belgian basin. Few Paleocene localities have yielded lacertilian fossils in Europe. Among them, Le Hainin (Belgium) is early–middle Paleocene in age (Mammalian reference level for the Paleogene MP1–5, Schmidt-Kittler, 1987; Aguilar et al., 1997) and there is universal agreement that the fauna from Cernay is late Paleocene (MP6) (Russell, 1964). The Paleocene lacertilian fauna from Walbeck (Germany; Kuhn, 1940a; Estes, 1983) and from Menat (Central France; Piton, 1940) are not currently available. Hence, it is emphasised that the estimate of Lacertilian diversity during the Paleocene in Europe is provisional. Additional records are required for a better knowledge of paleobiodiversity.

In contrast, abundant lacertilian assemblages are known from several localities close to the Paleocene-Eocene boundary or from the early Eocene in western Europe: Le Quesnoy (MP7, Nel et al., 1999); Meudon (MP7, Russell et al., 1990); Avenay (MP8 + 9); Mutigny (MP8 + 9); Condé-en-Brie (MP8 + 9; Augé, 1990b); Prémontré (MP10; Augé et al., 1997). Escarguel (1999), Hooker (1996), and Louis (1996), among others, provided a detailed framework for the chronology of these early Eocene localities. The Dormaal Sand Member deserves special attention in regard to the richness of its lacertilian fauna and its situation close to the Paleocene–Eocene boundary (Smith and Smith, 1996; Smith, 2000; Steurbaut et al., 1999). These authors assumed that the burial of the vertebrate fauna preceded the kaolinite influx and the negative $\delta^{13}C$ excursion that marks the Initial Eocene Thermal Maximum (IETM).

DIVERSITY

Taxonomic composition is rather homogeneous within the Paleocene localities of Hainin (mammalian standard level MP1–5; Van Dyck, 1983) and Cernay (MP6), and within the set of early Eocene localities. In contrast, there is a strong difference in the relative abundance of lacertilian families between the two groups of localities, which represent two stratigraphic levels (Figs. 1–2).

During the Paleocene, the maximum familial (and subfamilial) richness is documented at Cernay, which includes 5 families. Van Dyck (1983) notes the presence of a single varanid dorsal vertebra at Hainin, but her description casts doubt on this taxonomic assignment. There is a strong increase in familial diversity across the Paleocene-Eocene boundary, since 10 lacertilian families are present in the standard level MP7 (Dormaal, Le Quesnoy). The lacertilian fauna at the family level in the early Eocene localities is the most diverse in the Cenozoic record of Europe. Paleocene families persist and several new taxa make their first appearance in the Tertiary at Dormaal: Iguanidae*, Agamidae, Gekkonidae, Glyptosaurinae and Helodermatidae. The possible presence of the Superfamily Scincoidea (Scincidae + Cordylidae) is only documented by an incomplete premaxilla. The number of families remains high during the entire early Eocene with 10 families present in the standard level MP7 + 8 (locality of Condé-en-Brie; Augé, 1990b) and 11 families in the level MP10 (locality of Prémontré; Augé et al., 1997). Apparently, there is a decrease in diversity after the early Eocene in Europe, around MP11 + 12 with 6–7 families in the German localities of Messel and Geiseltal (Rossmann, 2000a). This relatively low diversity is in part an artefact, because many fossil lizards from Messel are currently under study.

On the whole, the main event affecting the species diversity of the Lacertilia during the period occurs between the localities of Cernay (MP6) and Dormaal (MP7, 55.5 Ma, Smith, 2000). The turnover of species is pronounced: Four out of seven species

Figure 1. A: Amphisbaenidae, left dentary, Paleocene, Cernay (MP6). B: ?Scincidae, right dentary, Paleocene, Cernay (MP6). C: Lacertidae, left dentary, Paleocene, Cernay (MP6). D: Iguanidae, *Geiseltaliellus* cf. *longicaudus,* left dentary, early Eocene, Prémontré (MP7). E: Glyptosaurinae, left dentary, early Eocene, Dormaal (MP7). F: Agamidae, *Tinosaurus europeocaenus,* left dentary, early Eocene, Dormaal (MP7). G: Amphisbaenidae, left dentary, early Eocene, Dormal (MP7). H: Varanidae, ?*Saniwa orsmaelensis,* left dentary, early Eocene, Dormaal (MP7). I: Necrosauridae, cf. *Necrosaurus,* left dentary (detail of teeth), early Eocene, Dormaal (MP7). J: Glyptosaurinae, osteoscute, early Eocene, Dormaal (MP7). K: Necrosauridae, osteoscute, Paleocene, Cernay (MP6). L: Gekkonidae, anterior part of frontal (ventral view), early Eocene, Dormaal (MP7). M: Lacertidae, right dentary, early Eocene, Dormaal (MP7). N: Anguidae, caudal vertebra with fracture plane (ventral view), early Eocene, Prémontré (MP10). All dentaries in lingual view, scale bar 1 mm.

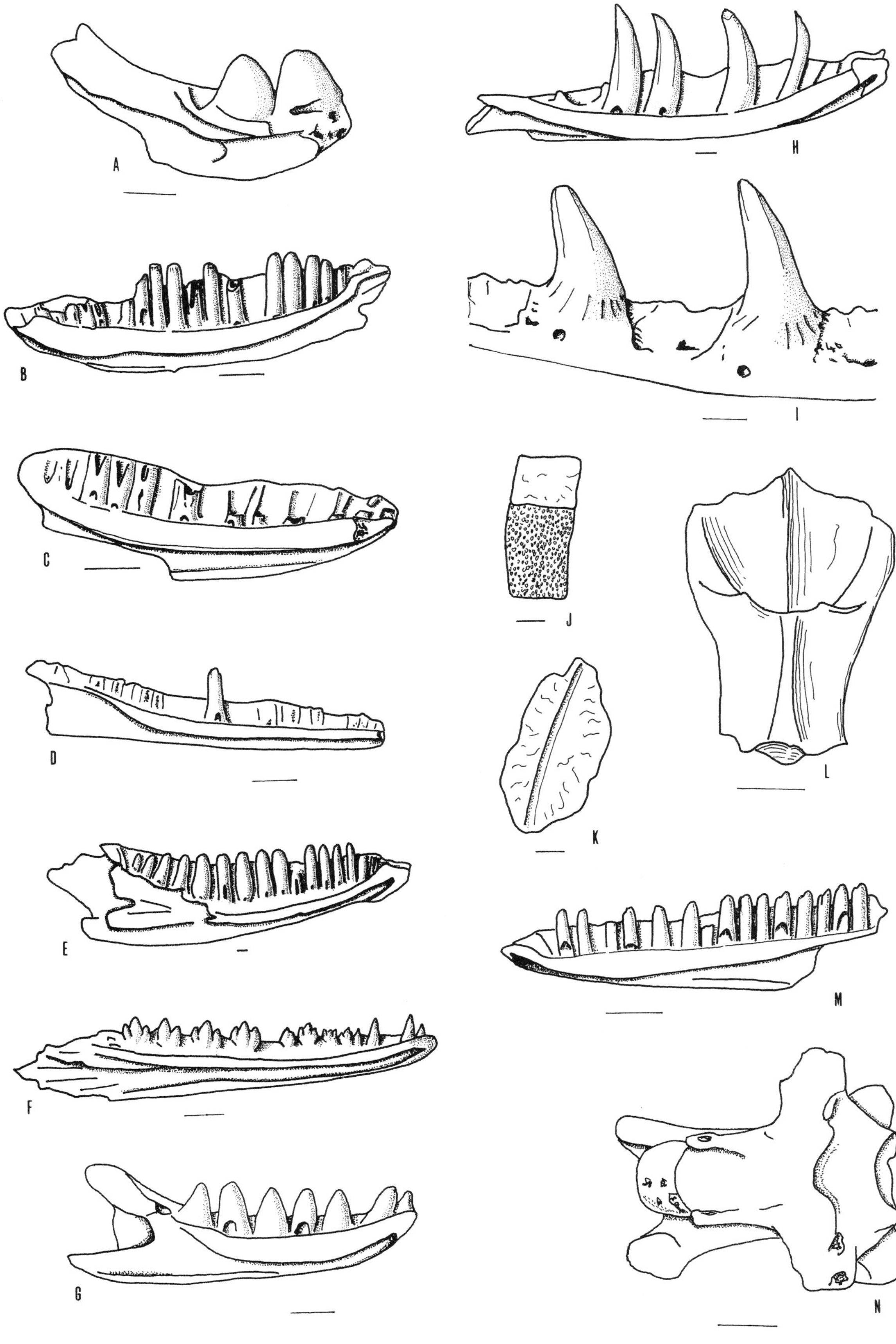
A
B
C
D
E
F
G
H
I
J
K
L
M
N

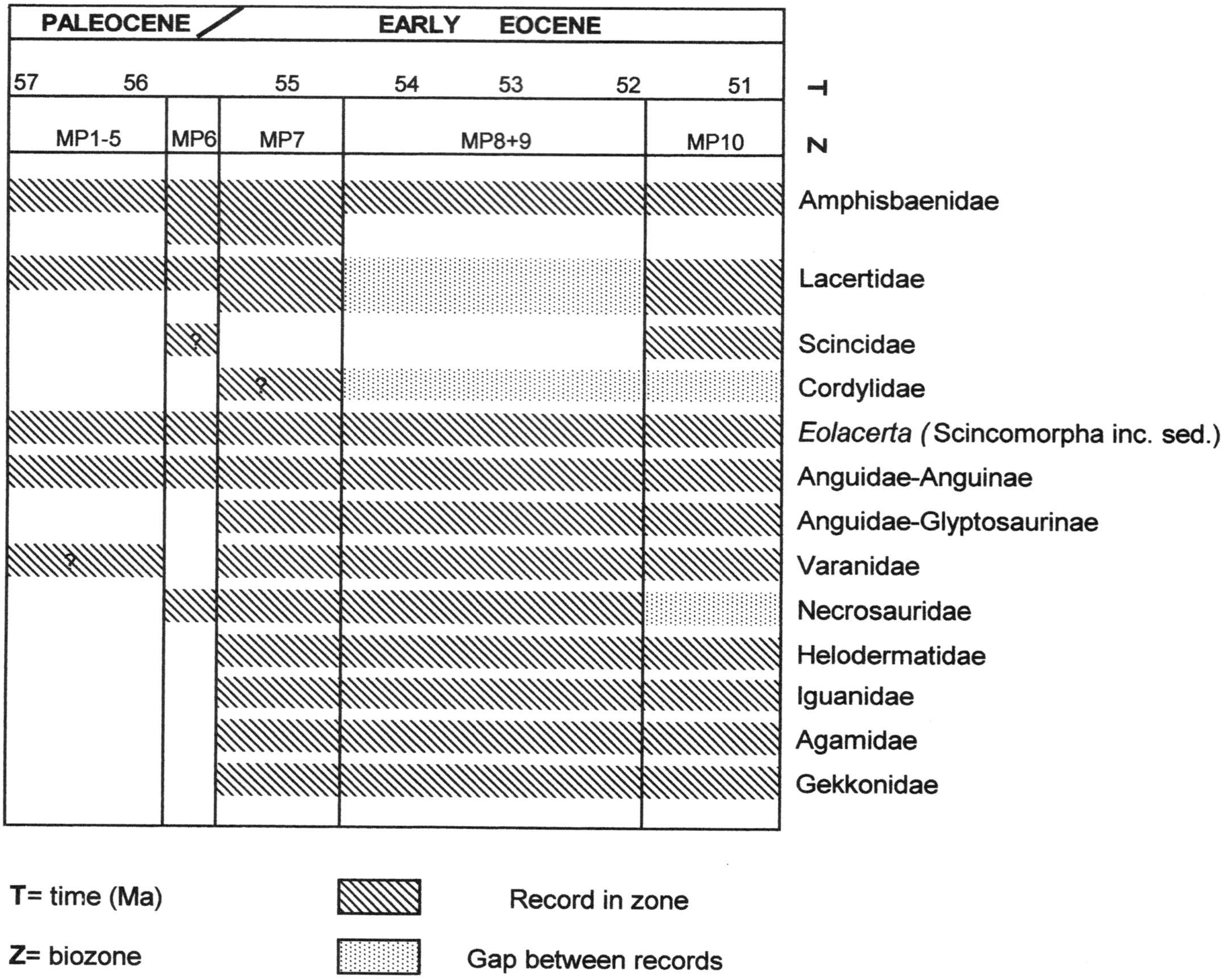

Figure 2. Record of Paleocene–early Eocene lizards in Europe. Z—biozone (mammalian standard levels); T—time (Ma). Localities: MP1–5, Hainin, Belgium. MP6: Cernay, France. MP7: Dormaal, Belgium; Le Quesnoy, Meudon, France. MP8+9: Avenay, Mutigny, Condé-en-Brie, France. MP10: Prémontré, France.

have their last appearance in the level MP6, and the level MP7 has nine first appearances. Six species have no close relatives in the level MP6 faunas and certainly represent dispersals from outside. The remainder certainly evolved from taxa of the MP6 level in the area. However, these numbers of species are still preliminary. More systematic work at the species level needs to be done. Thus, a clear picture of the faunal turnover across the Paleocene-Eocene and through the whole Eocene can be given for the families and not yet for the species (Table 1). For the period, both the percentage and the probability of origination reached a peak in the standard level MP7 (0.6 and 0.4, respectively).

The relative abundance of each family, as reflected by the number of specimens, is shown on Figure 3. A striking fact is the dominance of the Amphisbaenidae during the Paleocene. They remain dominant in the Dormaal fauna, however they are represented by species having a different dental adaptation (see below). The Scincidae obviously diminished. They have not been recognized in the Dormaal fauna until now, however the family still persists in Europe (see below).

Among the immigrants, the Iguanidae* are abundant, being the second dominant group as early as the Dormaal level.

Iguanidae Gray, 1827*

The family Iguanidae* is a metataxon indicated by an asterisk (Gauthier et al., 1988) which can only be defined provi-

TABLE 1. DIVERSITY, EXTINCTION, AND ORIGINATION OF EUROPEAN LACERTILIAN FAMILIES DURING THE PALEOCENE AND THE EOCENE

	MP1–5	MP 6	MP 7	MP8–9	MP 10	MP 11	MP12–13	MP14–15	MP 16	MP 17	MP 18	MP19–20
Diversity (D) (families)	4	5	10	10	11	6	7	8	9	9	9	10
Extinction (E)	N.D.	?1	1	0	0	5	0	0	1	0	0	0
Origination (O)	N.D.	2	6	0	1	0	1	1	2	0	0	1
Percentage extinction (E/D)	N.A.	?0.16	0.1	N.A.	N.A.	0.83	N.A.	N.A.	0.11	N.A.	N.A.	N.A.
Percentage origination (O/D)	—	0.33	0.6	—	0.09	N.A.	0.16	0.12	0.22	—	—	0.1
Probability extinction (E/D/t)	—	0.1	0.06	—	—	0.55	—	—	0.07	—	—	—
Probability origination (O/D/t)	—	0.22	0.4	—	0.06	—	0.05	0.04	0.14	—	—	0.03

Note: N.D.—no data; N.A.—not applicable. Percentages of extinction and origination are measured according to Sepkoski and Koch (1995) (i.e., number of origination or extinction divided by diversity). Probability of extinction and origination are measured according to Sepkoski and Koch (1995) (i.e., percentage of origination or extinction divided by estimated duration of the interval, with 1 MP = 1.5Ma).

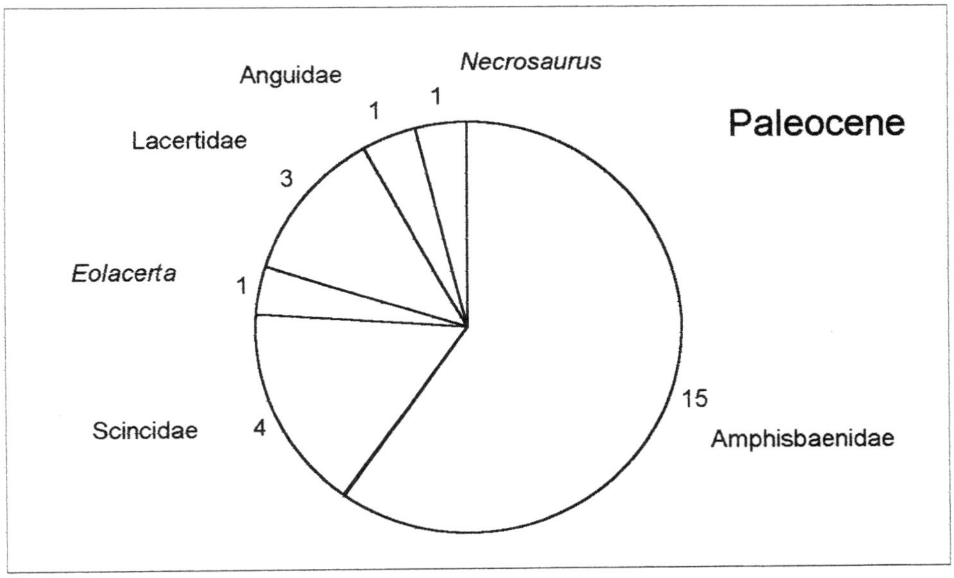

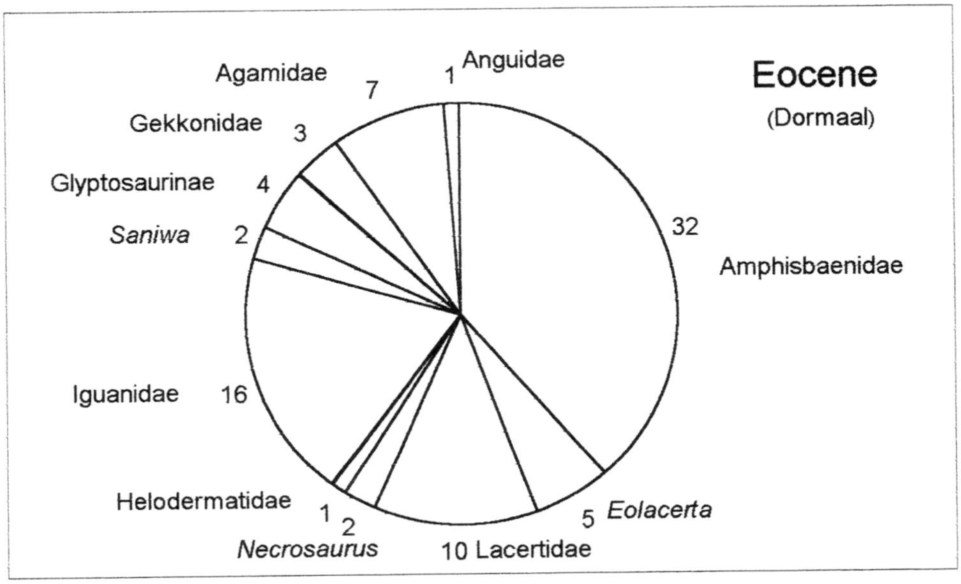

Figure 3. Record of the relative abundance of lizard families in Europe during the Paleocene–early Eocene. Here, the number of dentaries (right or left) has been used as an estimate of the abundance of each family. Numbers of specimens are shown on the graph.

sionally as iguanians that lack acrodontan synapomorphies. One iguanid genus, *Geiseltaliellus* (Fig. 1D) is present during the early and middle Eocene in Europe (Kuhn, 1944; Augé, 1990a; Rossmann, 2000a). The family is represented by *Cadurciguana hoffstetteri* (Augé, 1987) through the middle–late Eocene (standard level MP16-MP19). The last European iguanids are recorded in the locality of Escamps, just before the "Grande Coupure," i.e., the Eocene/Oligocene boundary.

Agamidae Gray, 1827

Agamid lizards are characterized by acrodont teeth, marginal teeth with apical implantation on the dentary and the maxilla. They made their first appearance in the European early Eocene (MP7, locality of Dormaal), in the form of a single genus and species, *Tinosaurus europeocaenus* (Augé and Smith, 1997; Fig. 1F). *Tinosaurus* becomes progressively less abundant during the early Eocene and its last record in Europe appears to be in the middle Eocene (MP13; Duffaud and Rage, 1997). This genus has a large geographic distribution, it is known in Asia, both in the Paleocene (*Tinosaurus doumuensis* Hou, 1974) and in the Eocene (*T. lushihensis* Dong, 1965 and *T. yuanquensis* Li, 1991). *Tinosaurus* is the only agamid lizard ever found in North America, with *Tinosaurus pristinus* and *T. stenodon* described from the middle Eocene of Wyoming (Gilmore, 1928; Hecht, 1959; Estes, 1983).

Gekkonidae Gray, 1825

Gekkonid remains are easily identified owing to their Meckelian canal fused as a tube and their unicuspid and numerous marginal teeth. A possible early record of Gekkota (fragmentary remains of vertebrae) was reported from the Middle Jurassic of England (Evans, 1998). The first true Gekkonid fossil in Europe was recovered from the Dormaal sands (Fig. 1L). Hoffstetter (1946) erected two taxa, *Rhodanogekko vireti,* from the middle Eocene of Lissieu (France) and *Cadurcogekko piveteaui* from the late Eocene of the Phosphorites du Quercy, France. The family is present throughout the Cenozoic. Recent Gekkonid species in Europe are restricted to the Mediterranean part of Europe and the coasts of the Black and Caspian seas. (Gasc et al., 1997).

Lacertidae Bonaparte, 1831

The locality of Hainin (Paleocene, MP1–5) contains the first confirmed lacertid fossils (Van Dyck, 1983), although Alifanov (1993) suggests a long "cryptic" history of the family during the Late Cretaceous in Eurasia. Lacertid lizards are well-represented in the fossil record during the whole Cenozoic in Europe (Fig. 1C, M). Two small species are known in the early Eocene from Dormaal and Prémontré, *Dormaalisaurus girardoti* and *Dormaalisaurus rossmanni* (Augé and Smith, 2002). *Succinilacerta succinea* (Böhme and Weitschat, 1998; Borsuk-Bialynicka et al., 1999) is another small species preserved in the Baltic amber (middle–late Eocene). The present distribution of the lacertid lizards and that of the iguanids are mutually exclusive, which suggests some negative interactions between the two taxa. Both families are represented in the same localities during the European Eocene but the disappearance in Europe of the Iguanidae* at the Eocene/Oligocene boundary seems to be an abrupt event, without any ascertained decline in diversity during the late Eocene.

Scincoidea Gray, 1825 (Scincidae Gray, 1825 + Cordylidae Fitzinger, 1826)

Scincoidea are easily recognized on the basis of their cranial morphology (frontal, parietal, maxilla) but, within the Superfamily, families Scincidae and Cordylidae are often difficult to distinguish. An undescribed scincid species with unicuspid teeth (Fig. 1B) is certainly present in the Paleocene of Cernay and a different, undescribed scincid lizard appears in the late early Eocene of Prémontré. At Dormaal, Hecht and Hoffstetter (1962) reported two forms of cordylid lizards, one with tricuspid teeth. However, comparisons of the material from Dormaal with complete skeletons from Messel (early middle Eocene) demonstrate that these fossils unequivocally belong to the iguanid genus *Geiseltaliellus* (Rossmann, 2000a; Augé, 1990a). A premaxilla from Dormaal and an undescribed dentary from Condé-en-Brie (early Eocene, MP8 + 9) possibly belong to the Scincoidea. The first confirmed record of a cordylid lizard in Europe is reported from the middle Eocene (MP16) of Le Bretou (Rage, 1988).

Anguidae Gray, 1825

Anguid lizards are a diverse group with broad past and present geographic distributions. They include four subfamilies, among which Anguinae Gray, 1825, and Glyptosaurinae Marsh, 1872, are represented in the European Cenozoic. Several anguinine species have been named from the early middle Eocene (Kuhn, 1940b) but the only valid species is *Ophisauriscus quadrupes* (Sullivan et al., 1999). Numerous Paleocene and early Eocene localities have yielded anguinine vertebrae and osteoderms, uninformative at the species or genus level (Fig. 1N).

Glyptosaurinae range in size from small to rather large taxa and are characterized by subrectangular trunk osteoderms with tubercular sculpture (Fig. 1J). The large genus *Melanosaurus,* described in the early Eocene from Wyoming (Gilmore, 1928), may be present in the early Eocene from Dormaal (Hecht and Hoffstetter, 1962; Fig. 1E). Glyptosaurine lizards are represented during the whole Eocene in Europe by the Eurasian genus *Placosaurus,* which bears a striking resemblance to the North American genus *Glyptosaurus.* The subfamily probably becomes extinct in Europe across the Eocene/Oligocene boundary, although Rocek (1984) referred a premaxillary from the early Miocene of Dolnice to the Glyptosaurinae.

Necrosauridae Hoffstetter, 1943

This extinct family is a member of the Platynota, which includes several lizard families (Necrosauridae, Helodermatidae, Varanidae + snakes and mosasaurs, McDowell and Bogert, 1954; Pregill et al., 1986; Lee, 1997; Norell and Gao, 1997) having fang-like teeth (Fig. 1I) that feed on a variety of insects and vertebrates. Oval osteoderms with a keel (Fig. 1K), a characteristic of the family, are relatively common in the locality of Cernay (Paleocene, MP6). The genus *Necrosaurus* is known throughout the Eocene and the first part of the early Oligocene in Europe (Hecht and Hoffstetter, 1962; Rage, 1978; Estes, 1983; Augé, 2000).

Helodermatidae Gray, 1837

The two extant species are the only venomous lizards in the world. Today the family is restricted to North America but it enjoyed a wider distribution during the Cretaceous (Asia) and the Eocene (Europe). Hoffstetter (1954, 1957) described *Eurheloderma gallicum* from the late Eocene of the Phosphorites du Quercy (France). Moreover, several dorsal vertebrae from Dormaal (Augé, 1995) and other Eocene localities demonstrate the presence of the family in Europe throughout the Eocene.

Varanidae Gray, 1827

The only varanid species described in the European Eocene is *Saniwa orsmaelensis* from Dormaal (Dollo, 1923; Fig. 1H). Hecht and Hoffstetter (1962) compare *S. orsmaelensis* with the North American species *Saniwa ensidens* and note the close resemblance between the two forms. Godinot et al. (1978) state that *Saniwa orsmaelensis* is problematic because no figure has been given nor a type chosen.

However, Hoffstetter (1969) figured a vertebra from the original Dollo series and Augé (1990b) reported some significant differences between the axial skeleton of the two species. To date, the last positive identification of the family in the European Eocene is from the locality of Condé-en-Brie (MP8 + 9).

Amphisbaenidae Gray, 1865

The largest and most diverse family of the Amphisbaenia, Amphisbaenidae includes limbless lizards extremely modified for a burrowing existence. Amphisbaenid skulls are heavily ossified and the mandibles are short. In the fossil record, the reduced length of the maxillae and the dentaries demonstrates that ancient amphisbaenid lizards were also fossorial (Rieppel, 1984). This family may have a long stratigraphic range in Europe, from the Late Cretaceous (Gheerbrant et al., 1997; Rage, 1999) to the recent species *Blanus cinereus* present in the Iberian Peninsula. Hecht and Hoffstetter (1962) note the presence of small amphisbaenid dentaries and vertebrae that are very similar to those of the living genus *Blanus* at Dormaal (Fig. 1G). Numerous localities throughout the European Cenozoic (Rage, 1978, 1988; Augé and Rage, 1995; Bailon, 1989; Schleich, 1985; Böhme, 1999; Delfino, 1997, among others) have yielded amphisbaenid fossils, some of them closely resembling the genus *Blanus*. However, assignment to *Blanus* is certainly premature if not dubious, as Gans (1990) envisioned that the genus is possibly paraphyletic. Rossmann (2000b) presented a preliminary account of a new species from the late Eocene of Le Bretou (Phosphorites du Quercy, France).

Two undescribed species are present in the Paleocene of Cernay (Fig. 1A). They disappear before or near the Paleocene-Eocene boundary and are replaced by two species at Dormaal.

ECOLOGICAL PARAMETERS

Taxon-free parameters can be used to assess lizard community response to the Paleocene-Eocene boundary events. This approach may provide information about ancient ecosystem function and environmental conditions. Morphology based ecological inferences for fossil members of small reptiles are far less common in the literature than are those for larger reptiles and mammals. Such inferences can be made only in view of sound knowledge of extant communities and species. Luckily, many ancient lizards show morphological specializations similar to those of their recent counterparts.

Diet

The dentigerous bones of the jaws (dentary, premaxilla and maxilla) are rather common in the fossil record. Lacertilians exhibit two main types of marginal teeth: Acrodont (teeth with apical implantation, restricted to Agamidae and some amphisbaenians) and pleurodont (teeth with mesial implantation, widespread among the other families). It is commonplace that the shape of teeth and pattern of dentition reflect particular dietary habits. It has been suggested that there is a fairly tight correlation between diet and crown shape in lizards (Hotton, 1955; Montanucci, 1968). However, on a broad scale, there appears to be little diet-related variation in crown form and the vast majority of pleurodont squamates have numerous, relatively small, unicuspid to tricuspid teeth. These tooth forms are associated with a variety of invertebrate prey types of food (arthropod-insect eating lizards) as well as some percentage of plant food. Three dietary types, however, seem to promote modification in the teeth (Schwenk, 2000). Here, fossils presenting such dental shapes are considered as arthropod eaters (insectivorous): Iguanidae*, some Lacertidae, *Eolacerta,* Gekkonidae and some Amphisbaenidae.

In many herbivorous lizards the teeth are laterally compressed and multicuspate (Hotton, 1955; Montanucci, 1968; Throckmorton, 1976) but this dental pattern is not represented in our fossil record. Species that consume hard food items (durophagy) such as shelled molluscs, tend to form molariform teeth (amblyodont; Hoffstetter, 1944) at the back of the tooth row (Rieppel, 1979; Sullivan, 1979; Estes and Williams, 1984; Augé, 2001; Rage, 1987b). These teeth are typically broad based with blunted or no cusps (Fig. 1A). Amblyodont lizards are well represented in the European fossil record (Amphisbaenia, Lacertidae and Glyptosaurinae). Here, this adaptation is not taken as a unique specialization in durophagy, but rather corresponds to a diversified diet including arthropods, shelled molluscs and some vertebrates. Gauthier (1982) and Shea (1990) consider these lizards as omnivorous. Most varanoid lizards and some anguid taxa (*Dopasia*) typically have very sharp unicuspid teeth (Fig. 1H–I) that are slightly recurved and with an expanded base (McDowell and Bogert, 1954). Such a tooth form is usually connected to carnivory, and, often, varanoid lizards include vertebrates in their diet. Moreover, some varanoids have a cutting edge along the anterior and posterior margins of their teeth and some anguid lizards (*Ophisauriscus, Dopasia*) exhibit such blades. According to Rieppel (1979), the formation of plicidentine in varanoid lizards may represent an adaptation to handle large prey. Here, Necrosauridae, Varanidae (*Saniwa*) and Helodermatidae are considered as carnivorous. The acrodont dentition produces a precise dental occlusion and a cutting mechanism (Borsuk-Bialynicka, 1996; Schwenk, 2000) but this dentition is not related to a particular diet as demonstrated by the large dietary types found in agamid lizards.

The most notable change in trophic structure between Paleocene and early Eocene faunas is a dramatic increase in the abundance of arthropod eaters (Fig. 4A) which pertains to five instead of two families. Insectivorous lizards make up around one third of the Paleocene lizard community whereas they become the largely dominant group in the early Eocene. Holroyd (2001) reports an increase in insectivorous mammals from Wyoming, between Clarkforkian and Wasatchian faunas, certainly related to an increase in insect diversity. A high insect diversity is also found in the early Eocene (MP7) locality of Le Quesnoy (France, Bassin de Paris; Nel et al., 1999). Carnivorous lizards also increased in abundance (Fig. 4A). This is partly due to the appearance of the varanid species *Saniwa orsmaelensis* in the localities of Dormaal and Le Quesnoy.

Size

Distribution of body sizes among the vertebrates is a significant feature of community structure and may indicate something about environmental conditions. Inferences from the distribution of body sizes concern almost exclusively ancient mammalian and large reptilian communities (Fleming, 1973; Legendre, 1989; Legendre and Hartenberger, 1992; Hooker, 1992, among others for mammalian faunas). To my knowledge, no such studies are available in the paleontological literature for small reptiles. Similarly, there are no body mass estimation techniques for fossil lizards. Here, dentary size has been used as an estimate of size in fossil lizard taxa.

It allowed a comparison of size classes in lizard faunas across the Paleocene-Eocene boundary (Fig. 4B). The intermediate size class does not seem to have been filled during the Paleocene. New carnivore (Varanidae) and insectivore lizards (Iguanidae*, Agamidae) enter this size class during the Paleocene-Eocene transition. In parallel, Alroy et al. (2000) reported the filling of the middle body size range by members of new mammalian groups during the Clarkforkian–Wasatchian boundary in North America.

Missing data and inadequate sampling during the Paleocene could explain the gap in the body size distribution. According to Blackburn and Gaston (1998), unsampled species will tend to have low abundance and small body size. In a variety of studies, including extant and fossil species, taxa described earlier tend to have large body size (Gaston, 1991; Patterson, 1994; Gaston et al., 1995; Peczkis, 1994). However, in the size frequency distribution for Paleocene and early Eocene lizards, the smaller class is the more abundant. Hence, inadequate sampling seems less likely.

LACERTILIAN FAUNAL DYNAMIC AND CLIMATE

Can lizard faunas give clues about past climates? Can these faunas contribute to the understanding of Paleocene-Eocene boundary events?

Most living lizards are restricted to tropical or warm-temperate regions, as are members of the families Agamidae, Varanidae, Helodermatidae and Amphisbaenidae. Due to their physiological requirements, vertebrate ectotherms are often considered good climatic indicators. As the name ectotherm implies, these vertebrates largely rely on heat coming from outside of the body. The use of cold-blooded vertebrates in determining paleoclimates has so far been almost restricted to tortoises and crocodilians (Hibbard, 1960; Brattstrom, 1961; Cheylan, 1981; Markwick, 1998; and many others). Snakes have been used by Brattstrom (1967) who showed that the proportion of natricids to colubrids and crotalids correlates well with the glacial and interglacial phases of the High Plains in North America.

Another use of a fauna to indicate paleoclimates is to map out their present ranges and find the region of greatest overlap. The climate in this area should approximate that experienced by the fossil fauna when it was extant.

Many lizard groups are associated with particular climatic regimes, which suggests that the species richness could be directly related to climate. Schall and Pianka (1978) compared the species richness of lizards in the United States and Australia. Despite difference between the total number of species, the species richness of lizards is highly and positively correlated with mean annual sunshine and negatively correlated with mean annual precipitation on these two continents, and the species richness of snakes is positively related to both mean annual rainfall and mean annual temperature.

However, several shortcomings affect this simple picture.

Reptiles, more specifically lizards, can regulate their body temperature with considerable precision, and the level at which the temperature is regulated is characteristic of a species. Ther-

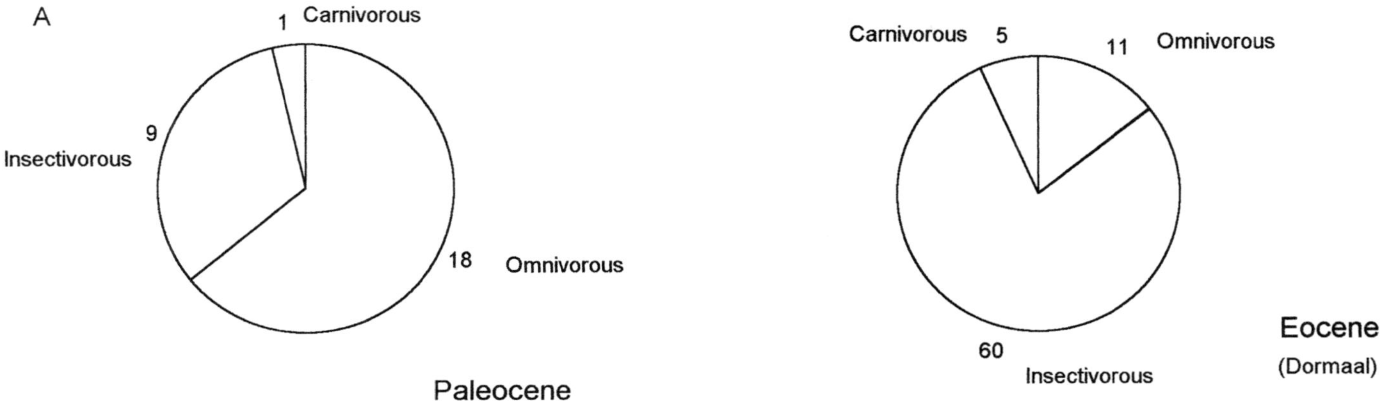

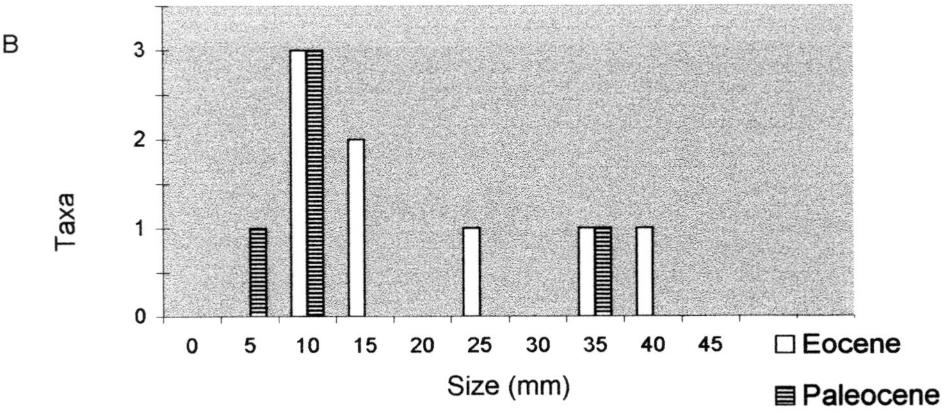

Figure 4. Ecological diversity for Paleocene–early Eocene lizard faunas in Europe. A: Diet, abundance of lizards in each trophic category. Numbers of specimens are shown on the graph. B: Size = length of dentaries (mm).

mal homeostasis is achieved by behavioral (external source of energy) as well as physiological adjustments (Cowles and Bogert, 1944; Huey and Slatkin, 1976; Pough et al., 1998).

A brief glance at the present distribution of lizards in Europe (Gasc et al., 1997) shows that the ranges of *Anguis fragilis* (Anguidae) and *Lacerta vivipara* (Lacertidae) extend from Spain to North Sweden and beyond the Arctic circle for *Lacerta vivipara.* Hence, they can endure long and cold winters (Grenot et al., 1996), and lizards have evolved many ways of coping with low temperatures. For example, various lizards, like the lacertid *Podarcis* endure some freezing (Storey and Storey, 1996).

Some past climatic deteriorations did not affect lizard faunas dramatically. During the North American Pleistocene, the herpetofauna (Squamata + Amphibia) was quite stable, compared to the avian and mammalian faunas (Kurten and Anderson, 1980). No extinct families or genera of Pleistocene reptiles are currently recognized in the North American continental Pleistocene. Lizards had only one unquestioned extinct species among 46 identified taxa. However, beyond general taxonomic stability, wide geographical changes and range adjustments occur in lacertilian populations (Holman, 1995).

In the European Pleistocene, only three species of oceanic island lizards become unquestionably extinct, and no mainland form is considered to be an extinct species (Holman, 1998). There is still no doubt that the herpetofaunas of Europe and North America have been strikingly more stable since the beginning of the Pleistocene than the endothermic faunas. However, range adjustments of the herpetofauna in response to climatic conditions during the Pleistocene are recorded throughout Europe (Bailon and Rage, 1992; Barbadillo et al., 1997; Holman, 1998). Formerly glaciated areas were reinvaded by species from the south during warming cycles and during climatic optima, several southern species existed well north of their present ranges.

Hence, it seems that climatic changes cause wide geographic displacements of lacertilian populations. On the other hand, no major extinction of lacertilian family is recorded during those events. The history of the lacertilian faunas during the Paleocene–early Eocene confirms this pattern: Few, if any, extinction of families are recorded during that time and dispersal of higher-level taxa added new families to the lacertilian faunas typical of the European Paleocene. Presumably, the late Paleocene–early Eocene warming trend, which culminated in the IEMT, made intercontinental dispersal across high-latitude land bridges increasingly feasible for a broad range of Lacertilia. The early Eocene reptilian fauna of the Ellesmere Island locality

(Estes and Hutchison, 1980) is indicative of pan-Arctic faunal continuity. The presence of alligators, large tortoises, agamid and varanid lizards implies a relatively mild winter climate in an area that lay well inside the Arctic Circle. However, a low sea level occured just before the IEMT (Hardenbol et al., 1998; Steurbaut et al., 1999) and this suggests the presence of land bridges linking Asia to Europe or/and North Africa to Europe (Hooker, 2000; Iakovleva et al., 2001).

Once an invader has arrived it must manage to establish. The success of an invader is often linked to climatic matching between the region of origin and the new range (Williamson, 1996). Hence, the establishment of warm and equable climates close to the early Eocene may have allowed the success of immigrants from more tropical areas. Another critical factor is the availability of accessible resources. The data on Lacertilian suggests a marked increase in insect availability. This in turn may be correlated with changes in vegetation and productivity during the IETM (Collinson, 2000; Wilf and Labandeira, 2000).

CONCLUSION

Based on the lacertilian data, it appears that a fundamental reorganization of the European lizard faunas occurs across the Paleocene-Eocene boundary. There is a strong increase in diversity in the early Eocene (or at the Paleocene-Eocene boundary, Standard Level of Dormaal, MP7). The number of lacertilian families found at this time in Europe is among the highest in the whole Cenozoic. The primary mechanism that affected lacertilian richness is certainly by opening of new routes to immigrant taxa, like agamid, gekkonid, and varanid lizards. This event is regarded as a dispersal, however the area of origin has not been identified. High-latitude warming may explain this dispersal but because of our very restricted knowledge of the Paleocene Asian and African faunas, this scenario cannot be confirmed for all families. There were also marked changes in lacertilian communities: An increase in arthropod eaters and carnivorous lizards is clearly documented.

In future studies, the "mutual climatic range" method could be applied to lizard taxa. This method is based on climatic conditions found across the modern range of particular taxa, which defines their tolerance limits. The climate of localities where several fossil taxa coexisted is defined by the overlapping range of climatic conditions that is compatible with their occurrence together in one place (Bradley, 1999). Although initially applied to Pleistocene coleopteran faunas (Atkinson et al., 1987), attempts could be made to use this approach for lacertilian families found at Dormaal. The two extant families Agamidae and Varanidae seem good candidates for such studies. Their present and past distribution demonstrate their great sensibility to climatic conditions. In the Eocene of Europe, their cooccurence is restricted to the climatic optimum, from the early Eocene to the middle Eocene. Later, both families occur together sporadically, during the Miocene and the Pliocene. They certainly immigrated from Asia or Africa during climatic warmings. Macey et al. (2000) suggest that Afro-Arabian acrodont lizards are basal groups, some species of which have dispersed into Laurasia after the collision of Afro-Arabia with Laurasia in the early Miocene (18 Ma). However, climatic warmings may also explain the repeated pattern of acrodont lizards dispersion in Europe, after the collision.

ACKNOWLEDGMENTS

This work was supported by the Comité National Français de Géologie. I thank J.C. Rage, M. Godinot, C. Beard, J.A. Wilson and W. Bartels for constructive reviews of the manuscript. I would like to acknowledge the help and advice from workers in the French-Belgian Basin Paleogene, particulary R. Smith, F. Duchaussois, P. Louis and D.E. Russell.

REFERENCES CITED

Aguilar, J.P., Legendre, S., and Michaux, J., editors, 1997, Actes du Congrès Biochrom'97: Mémoires Travaux de l'Ecole Pratique des Hautes Etudes: Montpellier, France, Institut de Montpellier, v. 21, 817 p.

Albino, A.M., 1993, Snakes from the Paleocene and Eocene of Patagonia (Argentina): Paleoecology and coevolution with mammals: Historical Biology, v. 7, p. 51–69.

Albino, A.M., 1996, The South American fossil Squamata (Reptilia: Lepidosauria): Münchner Geowissenschaft Abhandlung, v. A30, p. 185–202.

Alifanov, V.R., 1993, Some peculiarities of the Cretaceous and Palaeogene lizard faunas of the Mongolian people's republic: Kaupia, v. 3, p. 9–13.

Alroy, J., Koch, P.L., and Zachos, J., 2000, Global climate change and North American mammalian evolution, *in* Erwin, D.H., and Wing, S.L., eds., Deep time, paleobiology's perspective: Paleobiology, supplement, v. 26, no. 4, p. 259–288.

Atkinson, T.C., Briffa, K.R., and Coope, G.R., 1987, Seasonal temperatures in Britain during the last 22,000 years, reconstructed using beetle remains: Nature, v. 325, p. 587–592.

Augé, M., 1987, Confirmation de la présence d'Iguanidae (Reptilia, Lacertilia) dans l'Eocène européen: Comptes Rendus de l'Académie des Sciences, Paris, Série II, v. 305, p. 633–636.

Augé, M., 1990a, La faune de lézards et d'Amphisbaenes de l'Eocène inférieur de Condé-en-Brie (France): Bulletin du Muséum National d'Histoire Naturelle, Paris, 4e série, section C, v. 12, p. 111–141.

Augé, M., 1990b, La faune de lézards et d'amphisbènes (Reptilia, Squamata) du gisement de Dormaal (Belgique, Eocène inférieur): Bulletin de l'Institut royal des Sciences naturelles de Belgique, Sciences de la Terre, v. 60, p. 161–173.

Augé, M., 1995, Un helodermatidé (Reptilia, Lacertilia) dans l'Eocène inférieur de Dormaal, Belgique: Bulletin de l'Institut royal des Sciences naturelles de Belgique, v. 65, p. 277–281.

Augé, M., 2000, Diversité des faunes de lézards du Tertiaire en Europe de l'Ouest: Bulletin de la Société Herpétologique de France, v. 96, p. 5–14.

Augé, M., 2001, Les Lacertiliens (Reptilia, Squamata) du Paléogène d'Europe de l'Ouest: Paléobiodiversité, Evolution, Paléobiogéographie [Ph.D. thesis]: Muséum National d'Histoire Naturelle, Laboratoire de Paléontologie, Paris, 325 p.

Augé, M., and Rage, J.C., 1995, Le Garouillas et les sites contemporains (Oligocène, MP25) des Phosphorites du Quercy (Lot, Tarn-et-Garonne, France) et leurs faunes de vertébrés: Amphibiens et Squamates: Palaeontographica Abt. A, v. 236, p. 11–32.

Augé, M., and Smith, R., 1997, Les Agamidae (Reptilia, Squamata) du Paléogène d'Europe occidentale: Belgian Journal of Zoology, v. 127, no. 2, p. 123–138.

Augé, M., and Smith, R., 2002, Nouveaux Lacertidae (Reptilia, Squamata) de l'Eocène inférieur européen: Belgian Journal of Zoology, v. 131, no. 1, p. 3–15.

Augé, M., Duffaud, S., Lapparent de Broin, F., Rage, J.C., and Vasse, D., 1997, Les Amphibiens et les reptiles de Prémontré (Cuisien, Bassin parisien): Une herpétofaune de référence pour l'Eocène inférieur: Géologie de la France, v. 1, p. 23–33.

Averianov, A., and Danilov, I., 1996, Agamid lizards (Reptilia, Sauria, Agamidae) from the Early Eocene of Kyrgyzstan: Neues Jahrbuch für Geologie und Paläontologie, Monatshefte, v. H12, p. 739–750.

Averianov, A., and Danilov, I., 1997, A varanid lizard (Squamata: Varanidae) from the early Eocene of Kirghizia: Russian Journal of Herpetology, v. 4, p. 143–147.

Bailon, S., 1989, Les amphibiens et les reptiles du pliocène supérieur de Balaruc II (Herault, France): Palaeovertebrata, v. 19(1), p. 7–28.

Bailon, S., and Rage, J.C., 1992, Amphibiens et Reptiles du Quaternaire: Relations avec l'Homme: Mémoires de la Société Géologique de France, v. 160, p. 95–100.

Barbadillo, L.J., Garcia-Paris, M., and Sanchiz, B., 1997, Origenes y relaciones evolutivas de la Herpetofauna Iberica, *in* Pleguezuelos, J.M., ed., Distribucion y Biogeographica de los Anfibios y reptiles en Espana y Portugal: Granada, University of Granada, p. 47–100.

Bartels, W.S., 1983, A transitional Paleocene–Eocene reptile fauna from the Bighorn Basin, Wyoming: Herpetologica, v. 39, p. 359–374.

Blackburn, T.M., and Gaston, K.J., 1998, Some methodological issues in macroecology: The American Naturalist, v. 151, p. 68–83.

Böhme, M., 1999, Doppelschleichen (Sauria, Amphisbaenidae) aus dem Untermiozän von Stubersheim 3 (Süddeutschland): Mitteilungen des Bayerische Staatsslag Paläontogie und Historische Geologie, v. 39, p. 85–90.

Böhme, W., and Weitschat, W., 1998, Redescription of the Eocene lacertid lizard *Nucras succinea* Boulenger, 1917, from Baltic amber and its allocation to *Succinilacerta* n. gen.: Mitteilungen des Geologisch-Paläontologischen Instituts der Universität Hamburg, v. 81, p. 203–222.

Borsuk-Bialynicka, M., 1996, The Late Cretaceous lizard *Pleurodontagama* and the origin of tooth permanency in Lepidosauria: Acta Palaeontologica Polonica, v. 41(3), p. 231–252.

Borsuk-Bialynicka, M., Lubka, M., and Böhme, W., 1999, A lizard from baltic amber (Eocene) and the ancestry of the crown group lacertids: Acta Palaeontologica Polonica, v. 44(4), p. 349–382.

Bradley, R.S., 1999, Paleoclimatology (2nd edition): International Geophysics Series, v. 64: San Diego, Academic Press, 613 p.

Brattstrom, B.H., 1961, Some new fossil tortoises from western North America with remarks on zoogeography and paleoecology of tortoises: Journal of Paleontology, v. 35, p. 543–560.

Brattstrom, B.H., 1967, A succession of Pliocene and Pleistocene snake faunas from the High Plains of the United States: Copeia, no. 1, p. 188–202.

Broin, F. de, 1977, Contribution à l'étude des chéloniens: Chéloniens continentaux du Crétacé supérieur et du Tertiaire de France: Mémoires du Muséum National d'Histoire Naturelle, série C, v. 38, 366 p.

Buffetaut, E., 1985, Les crocodiliens de l'Eocène inférieur de Dormaal (Brabant, Belgique): Bulletin de la Société Belge de Géologie, v. 94(1), p. 51–59.

Cheylan, M., 1981, Biologie et Ecologie de la Tortue d'Hermann, *Testudo hermanni* Gmelin, 1789: Contribution de l'espèce à la connaissance des climats quaternaires de la France: Mémoires et Travaux de l'Institut de Montpellier de l'Ecole Pratique des Hautes Etudes, v. 13, p. 1–383.

Collinson, M.E., 2000, Fruit and seed floras from the Palaeocene–Eocene transition and subsequent Eocene in southern England: Comparison and palaeoenvironmental implications: GFF, v. 122, p. 36–37.

Cowles, R.B., and Bogert, C.M., 1944, A preliminary study of the thermal requirements of desert Reptiles: Bulletin of the American Museum of Natural History, v. 83, p. 261–296.

Delfino, M., 1997, *Blanus* from the Early Pleistocene of Southern Italy *in* Böhme, W., et al., eds., Another small tessera from a big mosaic: Herpetologica bonnensis, p. 89–97.

Dollo, L., 1923, *Saniwa orsmaelensis,* varanidé nouveau du Landénien supérieur d'Orsmael (Brabant): Bulletin de la Société Belge de Géologie, Paléontologie et Hydrologie, v. 33, p. 76–82.

Donadio, O.E., 1985, Un nuevo Lacertilio (Squamata, Sauria, Teiidae) de la Formacion Lumbrera (Eoceno temprano), Provincia de Salta, Argentina: Ameghiniana, v. 22, p. 221–228.

Dong, Z.M., 1965, A new species of *Tinosaurus* from Lushih, Honan: Vertebrata Palasiatica, v. 9, p. 79–82.

Duffaud, S., and Rage, J.C., 1997, Les remplissages karstiques polyphasés (Eocène, Oligocène, Pliocène) de Saint-Maximin (Phosphorites du Gard) et leur apport à la connaissance des faunes européennes, notamment pour l'Eocène moyen (MP13): 2-Systématique: Amphibiens et Reptiles, *in* Aguilar, J.P., et al., eds., Actes du congrès Biochrom'97: Mémoires et Travaux de l'Ecole Pratique des Hautes Etudes: Montpellier, France, Institut de Montpellier, v. 21, p. 729–735.

Escarguel, G., 1999, Les Rongeurs de l'Eocène inférieur et moyen d'Europe Occidentale: Systématique, phylogénie, biochronologie et paléobiogéographie des Niveaux-repères MP7 à MP14: Palaeovertebrata, v. 28, p. 89–345.

Estes, R., 1983, Sauria terrestria, Amphisbaenia: Handbuch der Paläoherpetologie: Teil 10A: Stuttgart, Germany, Gustav Fischer Verlag, 249 p.

Estes, R., and Hutchinson, J., 1980, Eocene lower vertebrates from Ellesmere island, Canadian Arctic archipelago: Palaeogeography, Palaeoclimatology, Palaeoecology, v. 30, p. 325–347.

Estes, R., and Williams, E., 1984, Ontogenetic variations in the molariform teeth of lizards: Journal of Vertebrate Paleontology, v. 4, p. 96–107.

Evans, S., 1998, Crown Group lizards (Reptilia, Squamata) from the Middle Jurassic of the British Isles: Palaeontographica Abt. A, v. 250, p. 123–154.

Fleming, T.H., 1973, Numbers of mammal species in North and Central American forest communities: Ecology, v. 54, p. 555–563.

Gans, C., 1990, Patterns in amphisbaenian biogeography: A preliminary analysis, *in* Peters, G., and Hutterer, R., eds., Vertebrates in the tropics: Bonn, Museum Alexander Koenig, p. 133–143.

Gasc, J.P., Cabela, A., Crnobrnja-Isailovic, J., Dolmen, D., Grossenbacher, K., Haffner, P., Lescure, J., Martens, H., Martinez Rica, J.P., Maurin, H., Oliveira, M.E., Sofianidou, T.S., Veith, M., and Zuiderwijk, A., 1997, Atlas of amphibians and reptiles in Europe: Societas Europaea Herpetologica, Muséum National d'Histoire Naturelle, 494 p.

Gaston, K.J., 1991, Body size and probability of description: The beetle fauna of Britain: Ecological Entomology, v. 16, p. 505–508.

Gaston, K.J., Blackburn, T.M., and Loder, N., 1995, Who gets described first? The case of North American butterflies: Biodiversity and Conservation, v. 4, p. 119–127.

Gauthier, J., 1982, Fossil xenosaurid and anguid lizards from the early Eocene formation Wasatch, southeast Wyoming, and a revision of the Anguioidea: Contribution Geology University Wyoming, v. 21, p. 7–54.

Gauthier, J., Estes, R., and de Queiroz, K., 1988, A phylogenetic analysis of Lepidosauromorpha, *in* Estes, R., and Pregill, G., eds., Phylogenetic relationships of the lizard families: Palo Alto, California, Stanford University Press, p. 15–98.

Gervais, P., 1873, Enumération de quelques ossements d'animaux vertébrés recueillis aux environs de Reims par M. Lemoine: Journal de Zoologie, Paris, v. 2, p. 351–355.

Gheerbrant E., Abrial, C., and Cappetta, H., 1997, Nouveaux sites à microvertébrés continentaux du Crétacé terminal des Petites Pyrénées (Haute-Garonne et Ariège, France): Geobios, v. 20, p. 257–269.

Gilmore, C.W., 1928, Fossil lizards of North America: Memoir National Academy of Sciences, v. 22, p. 1–201.

Gingerich, P.D., 1989, New Earliest Wasatchian mammalian fauna from the Eocene of northwestern Wyoming: Composition and diversity in a rarely sampled high–flood plain assemblage: University of Michigan Papers on Paleontology, v. 28, p. 1–97.

Godinot, M., Broin, F. de, Buffetaut, E., Rage, J.C., and Russell, D., 1978, Dormaal, une des plus anciennes faunes éocènes d'Europe: Comptes Rendus Académie des Sciences, Paris, v. 287, p. 1273–1276.

Grenot, C.J., Garcin, L., and Tsere-Pages, H., 1996, Cold-hardiness and behavior of the European common lizard, from French populations in winter, *in* Geiser, F., et al., eds., Adaptations to the cold, tenth international hibernation symposium: Armidale, University New England Press, p. 115–121.

Gunnell, G.F., Bartels, W., and Gingerich, P.D., 1993, Paleocene–Eocene boundary in continental North America: Biostratigraphy and geochronology, northern Bighorn Basin, Wyoming: Vertebrate paleontology in New Mexico: New Mexico Museum of Natural History and Science, Bulletin 2, p. 137–144.

Hardenbol, J., Thierry, J., Farley, M., Jacquin, T., de Graciansky, P.C., and Vail, P.R., 1998, Mesozoic and Cenozoic sequence chronostratigraphic framework of European basins: SEPM (Society for Sedimentary Geology) Special Publication 60, p. 1–13.

Hecht, M.K., 1959, Amphibians and reptiles, *in* McGrew, P., et al., eds., The geology and paleontology of the Elk Mountain and Tabernacle Butte area, Wyoming: Bulletin of the American Museum of Natural History, v. 117, p. 130–146.

Hecht, M.K., and Hoffstetter, R., 1962, Note préliminaire sur les Amphibiens et les Squamates du Landenien supérieur et du Tongrien de Belgique: Bulletin de l'Institut royal des Sciences naturelles de Belgique, v. 39, p. 1–30.

Hibbard, C.W., 1960, An interpretation of Pliocene and Pleistocene climates in North America: President's Address, 62nd Annual Report of the Michigan Academy of Science, Arts and Letters, p. 5–30.

Hoffstetter, R., 1944, Sur les Scincidae fossiles, I: Formes européennes et nord-Américaines: Bulletin du Muséum National d'Histoire Naturelle, Paris, v. 16, p. 547–553.

Hoffstetter, R., 1946, Sur les Gekkonidae fossiles: Bulletin du Muséum National d'Histoire Naturelle, Paris, v. 18, p. 195–203.

Hoffstetter, R., 1954, Présence d'un Saurien helodermatidé dans la faune fossile des Phosphorites du Quercy: Comptes Rendus de l'Académie des Sciences, Paris, v. 239, p. 1826–1827.

Hoffstetter, R., 1957, Un Saurien helodermatidé (*Eurheloderma gallicum* nov. gen. et sp.) dans la faune fossile des Phosphorites du Quercy: Bulletin de la Société Géologique de France, v. 7, p. 775–786.

Hoffstetter, R., 1969, Présence de Varanidae (Reptilia, Sauria) dans le Miocène de Catalogne: Considérations sur l'histoire de la famille: Bulletin du Muséum National d'Histoire Naturelle, v. 40, p. 1051–1064.

Holman, J.A., 1995, Pleistocene amphibians and reptiles in North America: Oxford University Press, Oxford Monographs on Geology and Geophysics, v. 32, 243 p.

Holman, J.A., 1998, Pleistocene amphibians and reptiles in Britain and Europe: Oxford University Press, Oxford Monographs on Geology and Geophysics, v. 38, 254 p.

Holroyd, P.A., 2001, Mammalian turnover and trophic structure in the early Paleogene of the Washakie Basin, southwestern Wyoming, *in* Ash, A.W., and Wing, S.L., eds., Climate and biota of the early Paleogene: International Meeting July 3–8, 2001, Northwest College, Powell, Wyoming, USA, Abstract volume, p. 44.

Holroyd, P.A., and Hutchison, J.H., 2000, Proximate causes for changes in vertebrate diversity in the early Paleogene: An example from turtles in the western interior of North America: GFF, v. 122, p. 75–76.

Hooker, J.J., 1992, British mammalian paleocommunities across the Eocene-Oligocene transition and their environmental implications, *in* Prothero, D., and Berggren, W., eds., Eocene–Oligocene climatic and biotic evolution: Princeton, New Jersey, Princeton University Press, p. 494–511.

Hooker, J.J., 1996, Mammalian biostratigraphy across the Paleocene-Eocene boundary in the Paris, London and Belgian Basins, *in* Knox, R.W.O'B., et al., eds., Correlation of the early Paleogene in North West Europe: The Geological Society [London] Special Publication 101, p. 205–218.

Hooker, J.J., 2000, Ecological response of mammals to global warming in the late Paleogene and early Eocene: GFF, v. 122, p. 77–79.

Hotton, N., 1955, A survey of adaptive relationships of dentition to diet in the North American Iguanidae: American Midland Naturalist, v. 53, p. 88–114.

Hou, L.H., 1974, Paleocene lizards from Anhui, China: Vertebrata Palasiatica, v. 12, p. 193–200.

Huey, R.B., and Slatkin, M., 1976, Cost and benefits of lizard thermoregulation: The Quarterly Review of Biology, v. 51(3), p. 363–383.

Hutchison, J.H., 1998, Turtles across the Paleocene/Eocene Epoch boundary in West-Central North America, *in* Aubry, M.P., et al., eds., Late Paleocene–early Eocene climatic and biotic events in the marine and terrestrial records: New York, Columbia University Press, p. 401–408.

Iakovleva, A.I., Brinkhuis, H., and Cavagnetto, C., 2001, Late Palaeocene–early Eocene dinoflagellate cysts from the Turgay Strait, Kazakhstan: Correlations across ancient seaways: Palaeogeography, Palaeoclimatology, Palaeoecology, v. 172, p. 243–268.

Kuhn, O., 1940a, Crocodilier- und Squamatenreste aus dem oberen Paläozän von Walbeck: Sonder-Abdruck aus dem Zentralblatt für Mineralogie und Paläontologie, Abteilung B(1), p. 21–25.

Kuhn, O., 1940b, Die Placosauriden und Anguiden aus dem Mittleren Eozän des Geiseltales: Nova Acta Academia Leopoldina-Carolinska, v. 53(8), p. 461–486.

Kuhn, O., 1944, Weitere Lacertilier, insbesondere Iguaniden aus dem Eozän des Geiseltales: Paläontologische Zeitschrift, v. 23, p. 360–366.

Kurten, B., and Anderson, E., 1980, Pleistocene mammals of North America: New York, Columbia University Press, 442 p.

Lee, M.S., 1997, The phylogeny of varanoid lizards and the affinities of Snakes: Philosophical Transactions of the Royal Society of London, series B, v. 352, p. 53–91.

Legendre, S., 1989, Les communautés de mammifères du Paléogène (Eocène supérieur et Oligocène) d'Europe occidentale: Structure, milieux et Evolution: Münchner Geowissenschaft Abhandlung (A), v. 16, p. 1–110.

Legendre, S., and Hartenberger, J.L., 1992, Evolution of mammalian faunas in Europe during the Eocene and Oligocene, *in* Prothero, D., and Berggren, W., eds., Eocene–Oligocene climatic and biotic evolution, Princeton, New Jersey, Princeton University Press, p. 516–528.

Louis, P., 1996, Recherches de mammifères Paléogènes dans les Départements de l'Aisne et de la Marne pendant la deuxième moitié du vingtième siècle: Palaeovertebrata, vol. jubil., D.E. Russell (Godinot, M., and Gingerich, P.D., eds.), v. 25(2-4), p. 83–113.

Macey, J.R., Schulte, J.A., II, Larson, A., Ananjeva, N.B., Wang, Y., Pethiyagoda, R., Rastegar-Pouyani, N., and Papenfuss, T., 2000, Evaluating Trans-Tethys migration: An example using Acrodont Lizard phylogenetics: Systematic Biology, v. 49(2), p. 233–256.

Markwick, P.J., 1998, Crocodilian diversity in space and time: The role of climate in paleoecology and its implication for understanding K/T extinctions: Paleobiology, v. 24, p. 470–497.

McDowell, S.B., and Bogert, C.M., 1954, The systematic position of *Lanthanotus* and the affinities of the anguinomorphan lizards: Bulletin American Museum Natural History, v. 105, p. 1–145.

Montanucci, R.R., 1968, Comparative dentition in four iguanid lizards: Herpetologica, v. 24, p. 305–315.

Nel, A., Plöeg G. de, Dutheil, D., Franceschi D. de, Gheerbrant, E., Godinot, M., Hervet, S., Menier, J.J., Augé, M., Bignot, G., Cavagnatto, C., Duffaud, S., Gaudant, J., Hua, S., Jossang, A., Lapparent de Broin, F., Pozzi, J.P., Paicheler, J.C., Beuchet, F., and Rage, J.C., 1999, Un gisement sparnacien exceptionnel à plantes, arthropodes et vertébrés (Eocène basal, MP7): Le Quesnoy (Oise, France): Comptes Rendus de l'Académie des Sciences, Paris, v. 329 (IIa), 1, p. 65–72.

Norell, M.A., and Gao, K., 1997, Braincase and phylogenetic relationships of *Estesia mongoliensis* from the Late Cretaceous of the Gobi desert and the recognition of a new clade of lizards*: American Museum Novitates, v. 3211, p. 1–25.

Patterson, B.D., 1994, Accumulating knowledge on the dimensions of biodiversity: Systematic perspectives on neotropical mammals: Biodiversity Letters, v. 2, p. 79–86.

Peczkis, J., 1994, Implications of body-mass estimates for dinosaurs: Journal of Vertebrate Paleontology, v. 14, p. 520–533.

Piton, L., 1940, Paléontologie du gisement éocène de Menat (Puy-de-Dôme) [Ph.D. thesis]: Faculté des Sciences de Clermont, Paul Vallier, Clermont. 303 p.

Pough, F.H., Andrews, R.M., Cadle, J.E., Crump, M.L., Savitzky, A.H., and Wells, K.D., 1998, Herpetology: Englewood Cliffs, New Jersey, Prentice-Hall, 577 p.

Pregill, G.K., Gauthier, J.A., and Greene, H.W., 1986, The evolution of helodermatid squamates, with description of a new taxon and an overview of Varanoidea: Transactions of the San Diego Society of Natural History, v. 21, p. 167–202.

Rage, J.C., 1978, Squamates, *in* Geze, B., et al., eds., La Poche à Phosphate de Ste-Néboule (Lot) et sa faune de vertébrés du Ludien supérieur: Palaeovertebrata, v. 8, p. 167–326.

Rage, J.C., 1987a, Fossil history, *in* Seigel, et al., eds., Snakes, ecology and evolutionary biology: New York, Macmillan, p. 51–76.

Rage, J.C., 1987b, Extinction chez les Squamates (Reptiles) à la fin de l'Oligocène en France: Adaptations et modifications de l'environnement: Mémoires de la Société Géologique de France, N.S., v. 150, p. 145–150.

Rage, J.C., 1988, Le gisement du Bretou (Phosphorites du Quercy, Tarn-et-Garonne, France) et sa faune de vertébrés de l'Eocène supérieur, I: Amphibiens et Reptiles: Palaeontographica Abt. A, v. 205, p. 3–27.

Rage, J.C., 1999, Squamates (Reptilia) from the Upper Cretaceous of Lano (Basque Country, Spain): Estudios del Museo de Ciencias Naturales de Alava, v. 14 (Numero especial 1), p. 121–133.

Rage, J.C., and Augé, M., 1993, Squamates from the Cainozoic of the western part of Europe: A review: Revue de Paléobiologie, volume spécial, v. 7, p. 199–216.

Rieppel, O., 1979, A functional interpretation of the Varanid dentition (Reptilia, Lacertilia, Varanidae): Gegenbaurs Morphologisches Jahrbuch, Leipzig, v. 125, p. 797–817.

Rieppel, O., 1984, Miniaturization of the lizard skull: Its functional and evolutionary implications, *in* Ferguson, M., ed., The structure, development and evolution of reptiles: Symposium Zoological Society, London, p. 503–520.

Rocek, Z., 1984, Lizards (Reptilia: Sauria) from the lower Miocene locality Dolnice (Bohemia, Czechoslovakia): Rozpravy Ceskoslovenske akademie ved Rada matematickych a prirodnich ved, v. 94(1), p. 1–69.

Rossmann, T., 2000a, Osteologische Beschreibung von *Geiseltaliellus longicaudus* Kuhn, 1944 (Squamata: Iguanoidea) aus dem Mittleren Eozän der Fossillagerstätten Geiseltal und Grube Messel (Deutschland), mit einer Revision der Gattung *Geiseltaliellus:* Palaeontographica Abt. A, v. 258, p. 117–158.

Rossmann, T., 2000b, A new Amphisbaenian (Squamata) from the Phosphorites du Quercy, France: 5th European Workshop on Vertebrate Paleontology, Karlsruhe, 27/06, p. 66.

Russell, D.E., 1964, Les mammifères Paléocènes d'Europe: Mémoires du Muséum National d'Histoire Naturelle, Paris, Série C 13, p. 1–324.

Russell, D.E., Broin, F.D., Galoyer, A., Gaudant, J., Gingerich, P.D., and Rage, J.C., 1990, Les Vertébrés du Sparnacien de Meudon: Bulletin d'Information des Géologues du Bassin de Paris, v. 27, p. 21–31.

Schall, J.J., and Pianka, E.R., 1978, Geographical trends in numbers of species: Science, v. 201, p. 679–686.

Schleich, H.H., 1985, Neue Reptilienfunde aus dem Tertiär Deutschlands: 3 Erstnachweis von Doppelschleichen (*Blanus antiquus* sp. nov.) aus dem Mittelmiozän Süddeutschlands: Münchner Geowissenschaft Abhandlung A, v. 4, p. 1–16.

Schmidt-Kittler, N., editor, 1987, European reference levels and correlation tables: Münchner Geowissenschaft Abhandlung, A, v. 10, p. 15–31.

Schwenk, K., editor, 2000, Feeding, form, function and evolution in Tetrapod Vertebrates: San Diego, Academic Press, 537 p.

Sepkoski, J.J., and Koch, C.F., 1995, Evaluating paleontologic data relating to bio-events, *in* Walliser, O.H., ed., Global events and Event Stratigraphy in the Phanerozoic: Berlin, Springer-Verlag, p. 21–34.

Shea, G.M., 1990, Ontogenetic changes and interspecific variation in the dentition of blue-tongue skinks (Lacertilia: Scincidae): Anatomical Society of Australia and New Zealand, 28th Annual Conference, University of Sydney, p. 26.

Smith, T., 2000, Mammals from the Paleocene–Eocene transition in Belgium (Tienen Formation, MP7): Palaeobiological and biostratigraphical implications: GFF, v. 122, p. 148–149.

Smith, T., and Smith, R., 1996, Synthèse des données actuelles sur les vertébrés de la transition Paléocène-Eocène de Dormaal (Belgique): Bulletin de la Société Belge de Géologie, v. 104, p. 119–131.

Steurbaut, E., De Coninck, J., Roche, E., and Smith, T., 1999, The Dormaal Sands and the Paleocene–Eocene boundary in Belgium: Bulletin de la Société Géologique de France, v. 170, p. 17–227.

Storey, K.B., and Storey, J.M., 1996, Natural freezing survival in animals: Annual review ecology and systematic, v. 27, p. 365–386.

Sullivan, R.M., 1979, Revision of the Paleogene Genus *Glyptosaurus* (Reptilia, Anguidae): Bulletin of the American Museum of Natural History, v. 163, p. 1–72.

Sullivan, R.M., Keller, T., and Habersetzer, J., 1999, Middle Eocene (Geiseltalian) anguid lizards from Geiseltal and Messel, Germany, I: *Ophisauriscus quadrupes* Kuhn 1940: Courier Forschungsinstitut Senckenberg, v. 216, p. 97–129.

Throckmorton, G.S., 1976, Oral food processing in two herbivorous lizards, *Iguana iguana* (Iguanidae) and *Uromastix aegyptius* (Agamidae): Journal of Morphology, v. 148, p. 363–390.

Van Dyck, M-C., 1983, Etude de la faune herpétologique du "Montien" continental de Hainin (Hainaut, Belgique) et d'autres gisements paléogènes du Nord-Ouest de l'Europe [Ph.D. thesis]: Université catholique de Louvain, Faculté des Sciences, 199 p.

Vasse, D., 1993, Les crocodiles du Crétacé supérieur et du Paléogène d'Europe: Aspects paléobiogéographiques et paléoécologiques [Ph.D. thesis]: Université Paris 6, Paléontologie, 173 p.

Vasse, D., and Hua, S., 1998, Diversité des Crocodiliens du Crétacé supérieur et du Paléogène: Influences et limites de la crise Maastrichtien-Paléocène et des "Terminal Eocene Events": Oryctos, v. 1, p. 65–77.

Wilf, P., and Labandeira, C.C., 2000, Effects of Paleocene–Eocene warming on insect herbivory: GFF, v. 122, p. 178–179.

Williamson, M., 1996, Biological invasions: London, Chapman & Hall, 244 p.

Manuscript Accepted by the Society August 13, 2002

Geological Society of America
Special Paper 369
2003

Mammalian faunal turnover in the Eocene of central Europe

Jens Lorenz Franzen
Jakobistrasse 10, D-79822 Titisee-Neustadt, Germany

ABSTRACT

There are two major mammal turnovers on the Central European Island during the middle Eocene, one at 47–44 Ma, and one at 42–36 Ma. The first immigration brought the first adapines that replaced cercamoniine prosimians, and the first palaeotheriids that replaced, step by step, equids. It is also at this time that the first bunoselenodont and selenodont artiodactyls appear and replace the brachyo- and bunodont forms of the early Eocene and the early middle Eocene. The second immigration brought the first anoplotheriid artiodactyls, and led to a considerable replacement of taxa among the palaeotheriids. There is no indication that the two immigrations originated from outside of Europe although their origin is still enigmatic. The appearance of certain paleobiologic adaptations indicates that, except for the appearance of land bridges, paleoclimatic development may have played a role in these migrations.

INTRODUCTION

It is well known that the lower and the upper boundaries of the European Eocene are characterized by important immigration events. Although the first occurs on all continents of the Northern Hemisphere, that at the Eocene-Oligocene boundary is restricted to Eurasia ("grande coupure"; Stehlin, 1909). Both are intercontinental.

This paper deals with two other faunal turnovers that occurred within the Eocene of central Europe. Europe was then an archipelago of larger and smaller islands (Fig. 1). One of the largest was that of central Europe. It is this island to which the following statements refer, and it is mainly the ungulates they are focused on, because they belong to the most frequently recorded fossil mammals.

THE FIRST INTRA-EOCENE FAUNAL TURNOVER

The first faunal turnover occurred in the middle of the middle Eocene at the Geiseltalian-Robiacian boundary of European land mammal ages (Figs. 2 and 3). It began with immigrants like the palaeothere *Paloplotherium cartieri,* the artiodactyl *Anthracobunodon,* and anchomomyine primates that first appeared in the European mammal unit MP 12 (Hellmund, 2000; Erfurt, 1993; Godinot, 1988). Others, like the first *Palaeotherium, Palaeotherium eocaenum,* the tapiromorph *Anchilophus,* the artiodactyls *Meniscodon* and *Cebochoerus* as well as bunoselenodont forms, like *Dacrytherium* and *Tapirulus,* and among primates, the first adapines arrived in MP 13 (Franzen, 1968, 1987, 1994; Franzen and Haubold, 1986; Godinot, 1998, p. 238). Some immigrants appeared still somewhat later, at the base of the Robiacian, in MP 14, among them the first selenodont artiodactyls, the Pseudamphimerycidae, and the Xiphodontidae (Sudre, 1978, p. 128–139, 162).

On the other side, the last cercamoniine primate of the *Protoadapis-Europolemur* group, *Europolemur collinsonae,* disappeared a little later in MP 15 (Hooker, 1986, p. 373), while the anchomomyins survived into the Early Headonian, MP 17 (Godinot, 1998, p. 228).

Estimated on the basis of a recent numerical dating of the Eckfeld locality (Mertz et al., 2000), the whole turnover spanned the time range of ~47–44 Ma, corresponding with the early to

* jlfranzen@t-online.de

Franzen, J.L., 2003, Mammalian faunal turnover in the Eocene of central Europe, *in* Wing, S.L., Gingerich, P.D., Schmitz, B., and Thomas, E., eds., Causes and Consequences of Globally Warm Climates in the Early Paleogene: Boulder, Colorado, Geological Society of America Special Paper 369, p. 455–461.

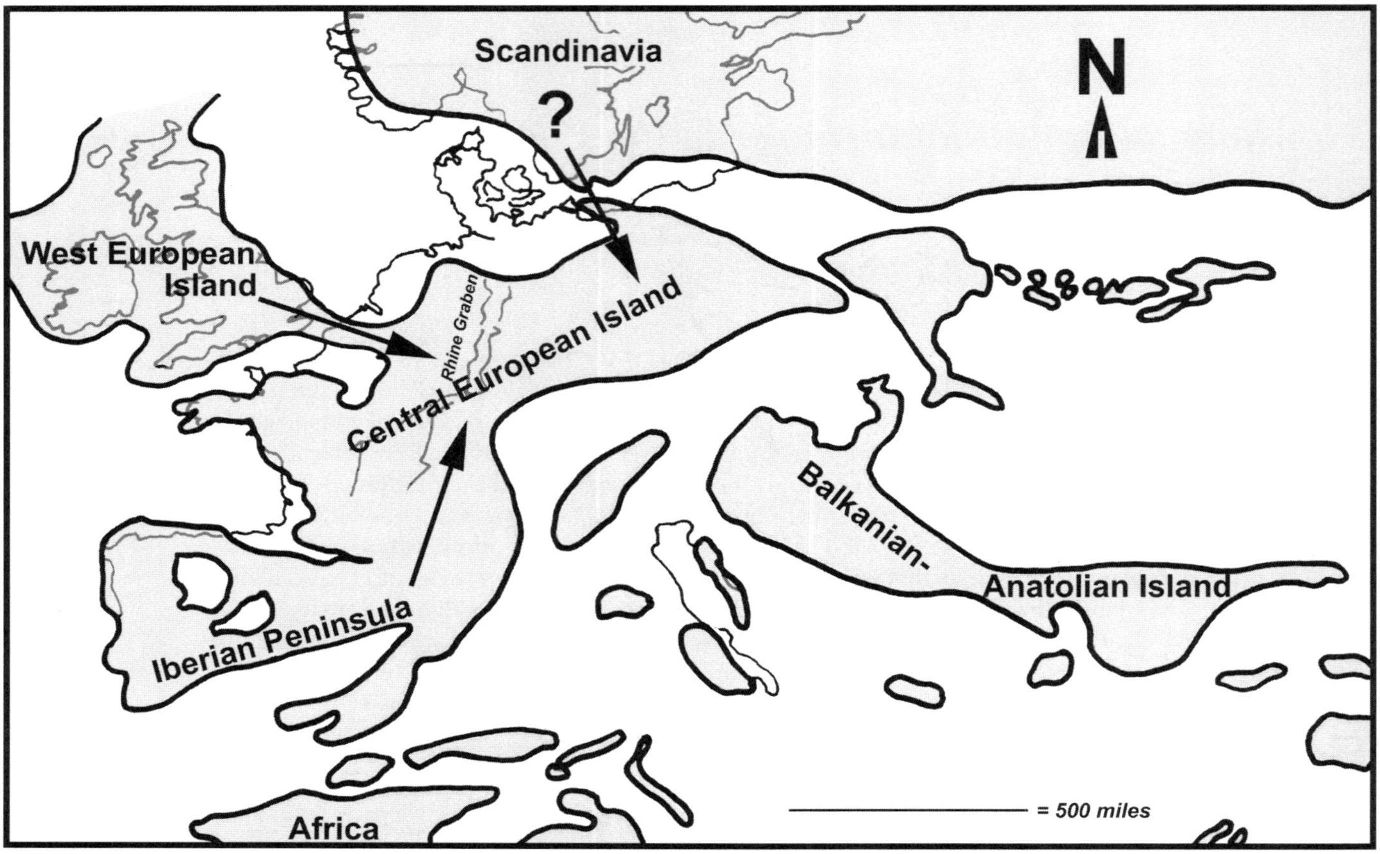

Figure 1. Paleogeography of Eocene Europe during the late Lutetian (44–41 Ma) and migrations of mammals (arrows) into the Central European Island. Drawing: J.L. Franzen (after Crasquin, 2000; Meulenkamp et al., 2000).

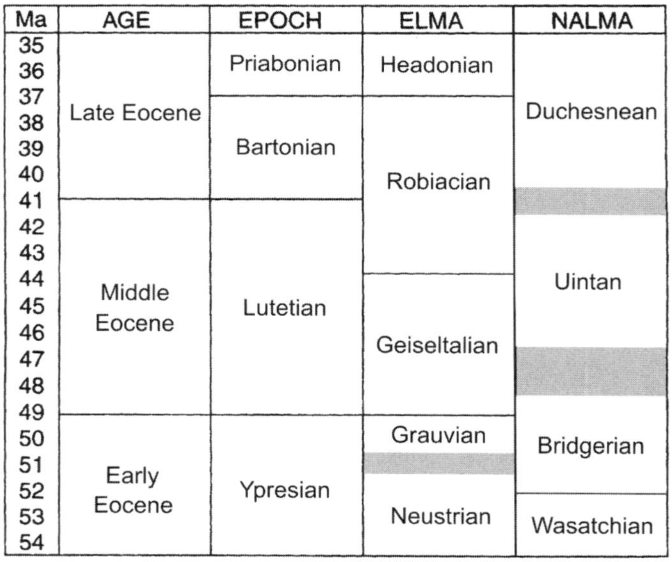

Figure 2. Time table showing a correlation of European and North American land mammal ages (ELMAs versus NALMAs). Boundaries of uncertain age are indicated by gray bars instead of black lines. Drawing: J.L. Franzen (after Mertz et al., 2000; Franzen, 2001; Clyde et al., 2001).

middle Uintan of North American stratigraphy (Fig. 2; Stucky, 1992, p. 465).

It is an interesting aspect that some biogeographic differentiation is observed between Europe west of the Rhine Graben—a big rift valley that came into existence at the beginning of the middle Eocene—and east of that (Franzen, 1995, p. 43). Although the appearance of the first bunoselenodont artiodactyls and of the first adapine primates is restricted to the west, the first palaeothere, *Paloplotherium cartieri,* occurred in the east. On the other hand, the palaeotheres *Paraplagiolophus codiciensis, Palaeotherium ruetimeyeri,* and *Palaeotherium castrense* are only known from the west. It is a question, however, if this differentiation depends on the Rhine Graben beginning to operate as a biogeographic barrier at this time or if it is due to paleoecological differentiation of the few fossil localities available.

Although Messel, Eckfeld, or the Geiseltal are spectacular in their preservational quality, they represent rare and isolated spots of information compared with the vast and highly fossiliferous areas of the North American West. So there is always the pitfall of generalizing information too much based on small areas that may represent very special paleoecological conditions. This holds particularly true for the Geiseltal, being the only

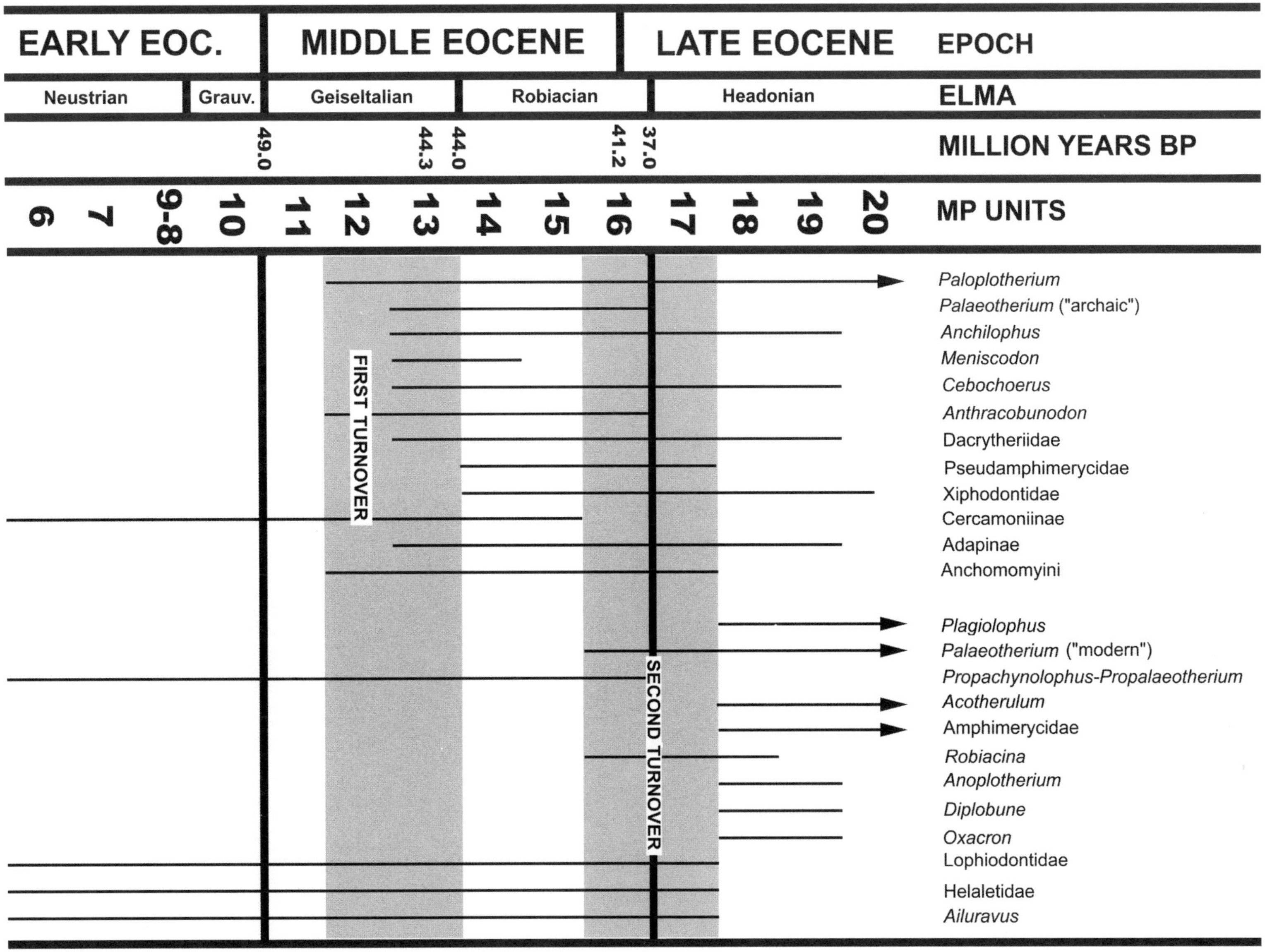

Figure 3. Turnovers of land mammal faunas on the Eocene Central European Island. Gray bars indicate the time range of each turnover. Arrows point to taxa existing longer than the time range expressed in the diagram. Drawing: J.L. Franzen.

mammal locality of that age on the Central European Island east of the Rhine Graben.

THE SECOND INTRA-EOCENE FAUNAL TURNOVER

The second intra-Eocene faunal turnover occurred mainly at the middle-late Eocene boundary, between the European land mammal ages of the Robiacian below and the Headonian above, ~42–36 Ma, corresponding with the early to middle Duchesnean (Figs. 2 and 3). It led to considerable replacement of taxa particularly among palaeotheres and artiodactyls (Franzen, 1968, p. 144, 1995, p. 34; Franzen and Haubold, 1986; Hooker, 1986, p. 450; Sudre, 1978). Somewhat archaic species, like *Palaeotherium ruetimeyeri* or *Palaeotherium castrense,* disappear. On the other hand, taxa of modern aspect with molarized premolars and cheek teeth becoming buccally progressively hypsodont, such as *Palaeotherium magnum, P. medium, P. muehlbergi* and *P. curtum,* appear suddenly. At the same time, lophiodons, once dominating in Europe, as well as the paramyid *Ailuravus* and the helaletids, vanish completely.

A few immigrants, however, like the first anoplothere, *Robiacina,* and *Palaeotherium duvali* arrived earlier, within the Robiacian in MP 16 (Sudre, 1978, p. 127; Hooker and Weidmann, 2000, p. 5, 114, 124, 128), while some late comers, like the first *Plagiolophus, Pl. minor,* the cebochoerid *Acotherulum,* and the anoplotheres s.s., *Anoplotherium, Diplobune,* and *Oxacron,* appeared not earlier than in MP 18 (Brunet and Jehenne, 1989; Sudre, 1978, p. 163).

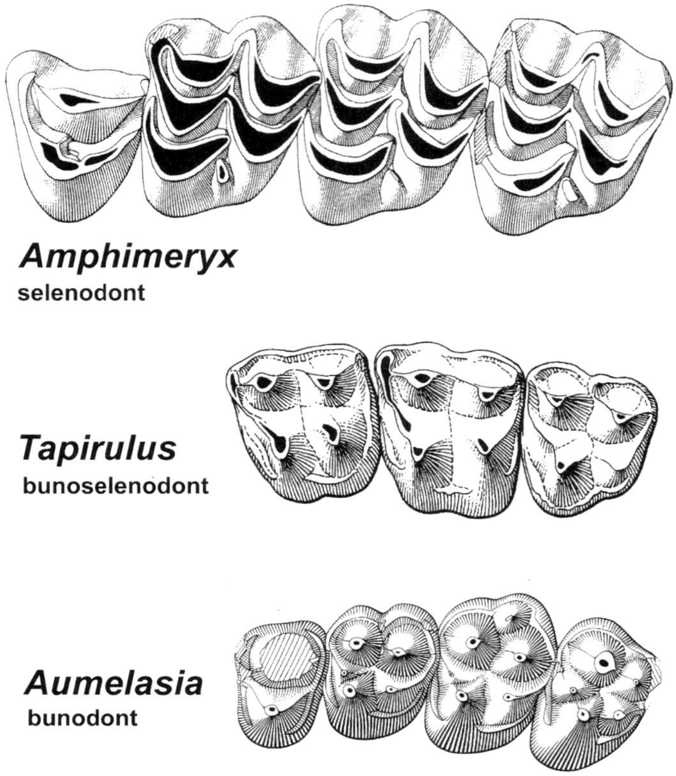

Figure 4. European Eocene artiodactyls: bunodont (e.g., *Aumelasia*) versus bunoselenodont (e.g., *Tapirulus*) and selenodont upper cheek teeth (e.g., *Amphimeryx*). Not to scale. Drawing: J.L. Franzen (after Sudre, 1972, 1978, 1980).

DISCUSSION

What may be the reasons for these intra-Eocene faunal turnovers in Europe? Do both reflect a similar paleogeographic and/or paleoclimatic scenario, or is each caused by different factors?

Three facts seem to be revealing in this context: First, in both cases the new taxa arrive over a considerable time of 2–3 mammal units or 4–6 million years. Secondly, some of the immigrant taxa display certain paleobiological adaptations, e.g., semihypsodonty, bunoseleno- and selenodonty, and thirdly they replace, step by step, brachyodont and bunodont species.

The appearance of bunoselenodonty and selenodonty refers to the artiodactyls (Fig. 4: bunoselenodont are the Dacrytheriidae, selenodont the Xiphodontidae and Amphimerycidae). Among the perissodactyls, it is the equid brachyodont hyracotheres, like *Propalaeotherium,* that are replaced by the increasingly semihypsodont palaeotheres, like *Paloplotherium, Plagiolophus,* and *Palaeotherium,* and the precociously hypsodont *Leptolophus* (Fig. 5). An additional paleobiological development is the appearance of cementum on the cheek teeth of *Leptolophus* and *Plagiolophus* on the one hand, and the extinction of the low-crowned lophiodons and helaletids on the other. Correspondingly, among primates, it is the slightly more hypsodont adapines with their sharper molar crests and their molarized fourth premolars that replace the cercamoniines (Fig. 6; Franzen, 1987, p. 174–175).

Hypsodonty as well as molarization of premolars means an increase of masticatory surface per time while cementum reinforces the structure of the tooth crown. Increasing selenodonty and sharper crests stand for increasing efficiency of the shearing edges of the cheek teeth. Therefore, all these developments are interpreted as adaptations for higher-abrasive food, typical for more arid environments. Consequently, the faunal turnover in the ungulates and primates of the Eocene of central Europe seems to indicate that a shift of climatic and vegetational zones may be responsible for corresponding changes in mammal assemblages. This hypothesis is corroborated by the mammals of the Geiseltal, who remain relatively conservative (no bunoselenodont or selenodont artiodactyls, no adapines, and almost no palaeotheres in MP 12–14) corresponding to persisting swamp forest habitats.

The geographic origin of the immigrants could have been in the south, on the Iberian peninsula (Fig. 1). This is suggested by paleogeography—Iberia was an island during the early Eocene that did not become connected to the Central European Island until the late Lutetian (Meulenkamp et al., 2000, Maps

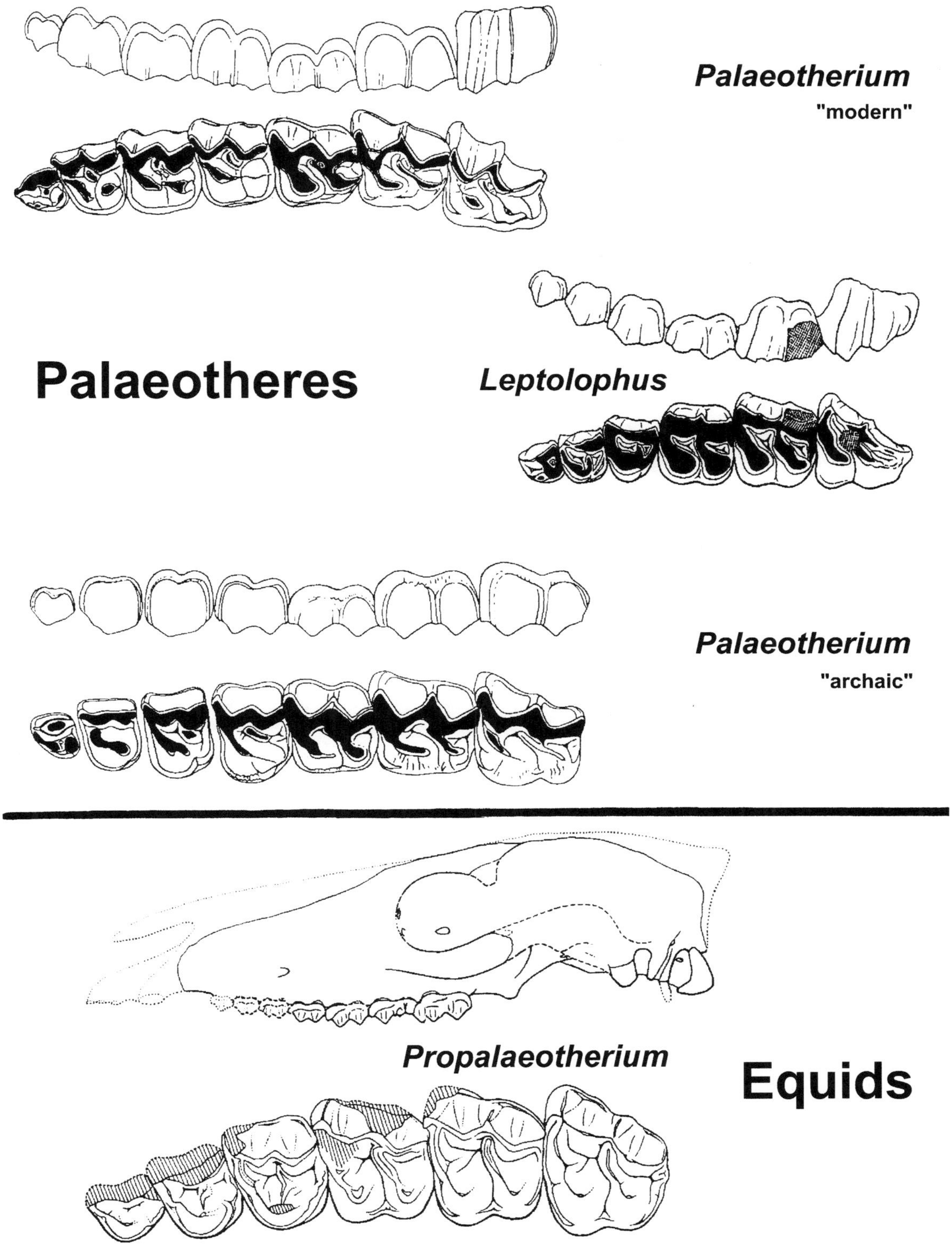

Figure 5. European Eocene equoids: brachyodont and bunolophodont upper cheek teeth of equids (e.g., *Propalaeotherium)* versus semihypsodont and lophodont palaeotheres (e.g., *Palaeotherium, Leptolophus*). Not to scale. Drawing: J.L. Franzen (after Franzen, 1968; Remy, 1998, 2001).

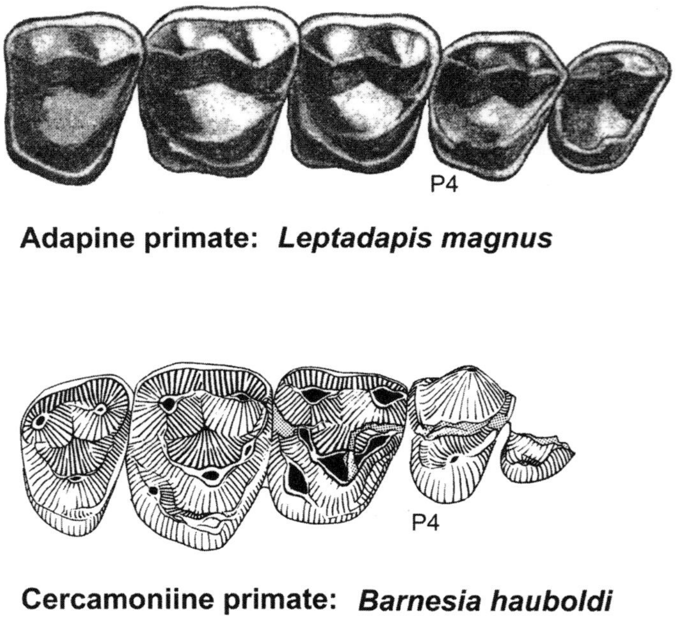

Figure 6. European prosimian primates: upper cheek teeth of cercamoniines (e.g., *Barnesia hauboldi*) versus adapines (e.g., *Leptadapis magnus*). Note the sharper crests and the molarized P4 in adapines. Not to scale. Drawing: J.L. Franzen (after Stehlin, 1912; Thalmann, 1994).

17–18). A southern origin is also indicated by the earlier appearance of some immigrants already in the early Eocene of Iberia, such as the artiodactyls *Cebochoerus* and *Mouillacitherium* (Golpe Posse, 1974), although according to Sudre (1978, p. 194) the evidence is not very strong.

Looking into the east, the scarce mammal remains from the Balkanian Island indicate Asian affinities. It is only in the late Eocene that the occurrence of the palaeothere *Plagiolophus minor* points to some immigration from the Central European Island (Fig. 1; Nikolov and Heissig, 1985). Considering the Balkanian-Anatolian Island, a very peculiar ungulate fauna was recently described from the middle Eocene that excludes any closer relationship with the European faunas of that time (Maas et al., 2001).

For the second faunal turnover there is some evidence that the immigrants into Central Europe came not only from the south but also from the west (Hooker, 1986, p. 450). The primitive rodents *Treposciurus* and *Suevosciurus,* later on common to all provinces of the Headonian in Europe, as well as *Palaeotherium muehlbergi* and the primate *Leptadapis magnus,* are known on the West European Island already from the late Robiacian (MP 16) of Creechbarrow, and Barton bed F, England (Hooker and Weidmann, 2000, p. 124). Such an origin of at least some of the immigrants appearing at the Robiacian-Headonian boundary on the Central European Island is also indicated by the northern distribution of newcomers, like *Palaeotherium duvali, P. muehlbergi,* and *P. curtum* (Franzen, 1968, p. 158–159). A western origin of those mammals is corroborated by the development of a land bridge between the Western and the Central European Island during the late Lutetian (Meulenkamp et al., 2000, Maps 17–18).

A general northward shift of the mammal faunas of Eocene Europe may be related to a tendency toward increasing aridity. Such a climatic development corresponds very well with evaporate deposits appearing in northern Iberia already "close to or at the P-E boundary" (Schmitz and Pujalte, 2001) while gypsum evaporites emerge during the late Eocene of southern France (Basin d'Aquitaine, Plateau Central, Vaucluse), the Paris Basin, and the southern part of the Rhine Graben (Franzen, 1968, p. 162). This climatic development is not related, however, to higher temperatures because oxygen isotopes from the shales of Tertiary mollusks from the North Sea Basin (Buchardt, 1978), as well as the marine record in general (Zachos et al., 1994) indicate decreasing temperatures during the late Eocene.

Therefore, an alternative or supplementary hypothesis would be a northern origin of the immigrants, pursuing cooler environments as they moved southward. The role, however, that Scandinavia played in this context remains enigmatic as long as there is no fossil evidence from that area at all.

CONCLUSIONS

Being aware of all we do not know yet sufficiently, some very cautious conclusions can be drawn:

There are two intra-Eocene turnovers of mammal faunas, both characterized by taxa that have no ancestors on the Central European Island, hence they are immigrants. One occurs around the Geiseltalian-Robiacian boundary, ~47–44 Ma. The other focuses around the Robiacian-Headonian boundary, ~42–36 Ma (Hooker and Weidmann, 2000, p. 128, 130).

Both turnovers may have resulted from land bridges appearing between Iberia and the Central European Island at the Ypresian-Lutetian boundary, and between the West and the Central European Island at the end of the Lutetian.

Both turnovers are characterized by immigrants showing adaptations, including increasing hypsodonty and development of cement in perissodactyls as well as bunoseleno- and selenodonty in artiodactyls, that are typically associated with more abrasive foods.

This corresponds with a climatic change to increasing aridity indicated by other evidence. So both turnovers were probably driven by climatic conditions shifting in a northward direction.

Alternatively or supplementary, decreasing paleotemperature during the late Eocene resulted in a southward movement of climatic zones together with their faunas.

Finally, looking at the rather close relationships between the immigrants and the preexisting fauna on the Central European Island, and taking into consideration that no closer relatives of the immigrants are known from outside of Europe, it appears rather improbable that they came from other continents. It was more the archipelago of islands and the climatic shift that made Europe a center of taxonomic differentiation during the Eocene.

ACKNOWLEDGMENTS

The author thanks Scott Wing for inviting him to the Paleogene Climate and Biota Conference as well as for financial support of his travel expenses. Many thanks go also to Philip Gingerich and Holly Smith for all their hospitality during the conference as well as to Gregg Gunnell and Ken Rose for helpful comments on an earlier version of this article.

REFERENCES CITED

Brunet, M., and Jehenne, Y., 1989, Révision des genres *Plagiolophus* POMEL, 1847, et *Paloplotherium* OWEN, 1848, Mammalia, Palaeotheriidae du Paléogène d'Europe-intérêt biochronologique: Annales de Paléontologie (Vertébrés-Invertébrés), v. 75, p. 23–52.

Buchardt, N., 1978, Oxygen isotope palaeotemperatures from the Tertiary period in the North Sea area: Nature, v. 275, p. 121–123.

Clyde, W.C., Sheldon, N.D., Koch, P.L., Gunnell, G.F., and Bartels, W.S., 2001, Linking the Wasatchian/Bridgerian boundary to the Cenozoic Global Climate Optimum: New magnetostratigraphic and isotopic results from South Pass, Wyoming: Palaeogeography, Palaeoclimatology, Palaeoecology, v. 167, p. 175–199.

Crasquin, S., coordinator, 2000, Explanatory notes, *in* Dercourt, J., et al., eds., Atlas Peri-Tethys: Paleogeographical maps: Paris, Commission de la Carte Géologique du Monde, p. 155–170.

Erfurt, J., 1993, Interim report on taxonomy and biostratigraphy of Artiodactyla from the middle Eocene locality Geiseltal near Halle (Germany): Kaupia, Darmstadt, v. 3, p. 131–136.

Franzen, J.L., 1968, Revision der Gattung *Palaeotherium* CUVIER 1804 (Palaeotheriidae, Perissodactyla, Mammalia) [Doctoral thesis]: Freiburg i.Br., Germany, Albert-Ludwigs-Universität, 2 vols.: 181 p.

Franzen, J.L., 1987, Ein neuer Primat aus dem Mitteleozän der Grube Messel (Deutschland, S-Hessen): Courier Forschungsinstitut Senckenberg, v. 91, p. 151–187.

Franzen, J.L., 1994, The Messel primates and anthropoid origins, *in* Fleagle, J.F., and Kay, R.F., eds., Anthropoid origins: New York, Plenum Press, p. 99–122.

Franzen, J.L., 1995, Die Equoidea des europäischen Mitteleozäns (Geiseltalium): Hallesches Jahrbuch der Geowissenschaften, v. B17, p. 31–45.

Franzen, J.L., 2001, Taphonomic analysis of the Messel Formation (Germany), *in* Gunnell, G.F., ed., Eocene biodiversity: Unusual occurrences and rarely sampled habitats: New York, Kluwer Academic/Plenum Press, p. 197–214.

Franzen, J.L., and Haubold, H., 1986, The middle Eocene of European mammalian stratigraphy: Definition of the Geiseltalian: Modern Geology, v. 10, p. 159–170.

Godinot, M., 1988, Les primates adapidés de Bouxwiller (Eocène Moyen, Alsace) et leur apport à la compréhension de la faune de Messel et à l'évolution des Anchomomyini: Courier Forschungsinstitut Senckenberg, v. 107, p. 383–407.

Godinot, M., 1998, A Summary of Adapiform systematics and phylogeny: Folia primatologica, v. 69 (supplement 1), p. 218–249.

Golpe Posse, J.M., 1974, Faunas de yacimientos con Suiformes en el Terciario español: Paleontología y Evolución, v. 8, p. 1–87.

Hellmund, M., 2000, Erstnachweis von *Plagiolophus cartieri* STEHLIN (Palaeotheriidae, Perissodactyla) in der unteren Mittelkohle (uMK, MP 12) des Geiseltales bei Halle (Sachsen-Anhalt, Deutschland): Neues Jahrbuch für Geologie und Paläontologie, Monatshefte, v. 2000(4), p. 205–216.

Hooker, J.J., 1986, Mammals from the Bartonian (middle/late Eocene) of the Hampshire Basin, southern England: Bulletin of the British Museum (Natural History), Geology Series, v. 39(4), p. 191–478.

Hooker, J.J., and Weidmann, M., 2000, The Eocene mammal faunas of Mormont, Switzerland: Systematic revision and resolution of dating problems: Schweizerische Paläontologische Abhandlungen, v. 120, p. 1–141.

Maas, M.C., Thewissen, J.G.M., Sen, S., Kazanci, N., and Kappelman, J., 2001, Enigmatic new ungulates from the early middle Eocene of central Anatolia, Turkey: Journal of Vertebrate Paleontology, v. 21, no. 3, p. 578–590.

Mertz, D.F., Swisher, C.C., Franzen, J.L., Neuffer, F.O., and Lutz, H., 2000, Numerical dating of the Eckfeld maar fossil site, Eifel, Germany: A calibration mark for the Eocene time scale: Naturwissenschaften, v. 87, p. 270–274.

Meulenkamp, J.E., Sissingh, W., Calvo, J.P., Daams, R., Londeix, L., Cahuzac, B., Kovac, M., Nagymarosy, A., Badescu, D., Rusu, A., Studencka, B., Beniamovskii, V.N., Scherba, I.G., Roger, J., Platel, J.-P., Hirsch, F., Sadek, A., Abdel-Gawad, G.I., Zaghbib-Turki, D., Ben Ismail-Lattrache, K., Bouaziz, S., Karoui-Yaakoub, N., and Yaich, C., 2000, Map 17: Early to Middle Ypresian (55–51 Ma), Map 18: Late Lutetian (44–41 Ma), *in* Dercourt, J., et al., eds., Atlas Peri-Tethys: Paleogeographical maps: Paris, Commission de la Carte Géologique du Monde, maps 17–18.

Nikolov, I., and Heissig, K., 1985, Fossie Säugetiere aus dem Obereozän und Unteroligozän Bulgariens und ihre Bedeutung für die Palaeogeographie: Mitteilungen der Bayerischen Staatssammlung für Paläontologie und Historische Geologie, v. 25, p. 61–79.

Remy, J.A., 1998, Le genre *Leptolophus* (Perissodactyla, Mammalia): Morphologie et histologie dentaires, anatomie crânienne, implications fonctionelles: Palaeovertebrata, v. 27, no. 1–2, p. 45–108.

Remy, J.A., 2001, Sur le crâne de *Propalaeotherium isselanum* (Mammalia, Palaeotheriidae) de Pépieux (Minervois, Sud de la France): Geodiversitas Paris, v. 23, no. 1, p. 105–127.

Schmitz, B., and Pujalte, V., 2001, Correlating the Paleocene-Eocene boundary in the Pyrenees, Spain, by carbon isotopes: A transect from deep marine to continental environments, *in* Ash, A.W., and Wing, S.L., eds., Climate and biota of the early Paleogene: International Meeting July 3–8, 2001, Northwest College, Powell, Wyoming, USA, Abstract volume, p. 81.

Stehlin, H.G., 1909, Remarques sur les faunules de mammifères des couches éocènes et oligocènes du bassin de Paris: Bulletin de la Société Géologique de France, v. (4), no. 9, p. 488–520.

Stehlin, H.G., 1912, Die Säugetiere des schweizerischen Eocaens: Critischer Catalog der Materialien: Siebter Teil, erste Hälfte: *Adapis:* Abhandlungen der schweizerischen paläontologischen Gesellschaft, v. 38, p. 1163–1298.

Stucky, R.K., 1992, Mammalian faunas in North America of Bridgerian to Early Arikareean "Ages" (Eocene and Oligocene), *in* Prothero, D.R., and Berggren, W.A., eds., Eocene-Oligocene climatic and biotic evolution: Princeton, New Jersey, Princeton University Press, p. 464–493.

Sudre, J., 1972, Révision des Artiodactyles de l'Eocène moyen de Lissieu (Rhône): Palaeovertebrata, v. 5, no. 4, p. 111–156.

Sudre, J., 1978, Les artiodactyles de l'Éocène moyen et supérieur d'Europe Occidentale (Systématique et évolution): Ecole Pratique des Hautes Etudes, Mémoires et Travaux de l'Institut de Montpellier, v. 7, 229 p.

Sudre, J., 1980, *Aumelasia gabineaudi* n.g. n.sp. Nouveau Dichobunidae (Artiodactyla, Mammalia) du gisement d'Aumelas (Herault) d'âge Lutetien terminal: Palaeovertebrata, Mémoire Jubilée R. Lavocat, p. 197–211.

Thalmann, U., 1994, Die Primaten aus dem eozänen Geiseltal bei Halle/Saale (Deutschland): Courier Forschungsinstitut Senckenberg, v. 175, p. 1–161.

Zachos, J.C., Stott, L.D., and Lohmann, K.C., 1994, Evolution of early Cenozoic marine temperatures: Paleoceanography, v. 9, p. 353–387.

MANUSCRIPT ACCEPTED BY THE SOCIETY AUGUST 13, 2002

Printed in the USA

Geological Society of America
Special Paper 369
2003

Mammalian responses to climate change at the Paleocene-Eocene boundary: Polecat Bench record in the northern Bighorn Basin, Wyoming

Philip D. Gingerich
Department of Geological Sciences and Museum of Paleontology, University of Michigan, Ann Arbor, Michigan 48109-1079, USA

ABSTRACT

The Paleocene-Eocene transition in North American land-mammal faunas is well documented on the south side of Polecat Bench and in surrounding badlands of northwestern Wyoming. Here a rich fossil record is known from a stratigraphic section with an established geomagnetic polarity time scale and medium- to high-resolution carbon and oxygen isotope records. The Paleocene-Eocene carbon isotope excursion (CIE) and temperature maximum (PETM) associated with greenhouse warming are both represented, starting at ca. 55 Ma and lasting ~86 k.y., during the transition from Paleocene to Eocene epochs and from Clarkforkian to Wasatchian land-mammal ages. Attention is focused on transient dwarfing of earliest Wasatchian (Wa-0) mammals during the CIE-PETM interval and on more lasting changes in generic composition and diversity. Faunal changes lagged initiation of the CIE by some 13 and 22 k.y., consistent with the idea of a stepped and pulsed response to climate change. Body size dwarfing in Wa-0 mammals cannot be explained as a response to temperature alone, and elevated atmospheric CO_2 probably played an important role.

INTRODUCTION

Evolution is an expected response to environmental change, and, similarly, climatic change and other environmental shifts are widely acknowledged to be important mechanisms driving evolution. Why then is it so difficult to show correlations in time between mammalian evolution and changing climates?

For every study claiming to see some correlation of mammalian evolution with climate change, there is another questioning such an interpretation. Sloan (1970), Gingerich (1976, 1985), and Rose (1981b) studied late Paleocene and early Eocene mammals in North America and regarded mammalian faunal change through this interval as a consequence of climatic cooling followed by warming, a pattern questioned by Krause and Maas (1990). Stucky (1990), Janis (1993), and Webb and Opdike (1995) studied broad changes in Cenozoic mammalian faunas and related these to climate change, but subsequent studies of the North American record by Prothero (1999) and Alroy et al. (2000) failed to find relationships to climate and instead emphasized intrinsic biotic controls on faunal change. Vrba (1985) studied mammalian faunal change through the Plio-Pleistocene record in southern Africa and developed a "turnover pulse" hypothesis of climatically driven rapid speciation and faunal evolution, but evidence for the principal pulse was queried by Behrensmeyer et al. (1997).

The Paleocene-Eocene boundary is one case where the correlation of a turnover pulse of mammalian faunal change with climate change appears unequivocal. Here I summarize what is known of the North American Paleocene-Eocene transition based on the stratigraphic record of mammals and climate in the vicinity of Polecat Bench in the northern Bighorn Basin of northwestern Wyoming.

Gingerich, P.D., 2003, Mammalian responses to climate change at the Paleocene-Eocene boundary: Polecat Bench record in the northern Bighorn Basin, Wyoming, *in* Wing, S.L., Gingerich, P.D., Schmitz, B., and Thomas, E., eds., Causes and Consequences of Globally Warm Climates in the Early Paleogene: Boulder, Colorado, Geological Society of America Special Paper 369, p. 463–478.

HISTORY OF STUDY

The Thanetian stage/age of the Paleocene and the Ypresian stage/age of the Eocene in Europe have long been known to contain different mammalian faunas. Similarly, the Tiffanian and Clarkforkian land-mammal ages of the Paleocene and the Wasatchian land-mammal age of the Eocene in North America have long been known to differ faunally. The Thanetian and Tiffanian-Clarkforkian on both continents are dominated by archaic mammalian orders such as Multituberculata, plesiadapiform Proprimates, and Condylarthra, while the younger Ypresian and Wasatchian on both continents are dominated by modern orders such as Artiodactyla, Perissodactyla, and true Primates (see, for example, Gervais, 1877; Lemoine, 1891; Russell, 1964; Cope, 1888; Granger, 1914; Jepsen, 1930; Rose, 1981a). From the 1870s and 1880s, these differences were understood to indicate a profound reorganization of early Cenozoic mammalian faunas. Sometime between 1910 and 1912, vertebrate paleontologists accepted the Paleocene as representing a distinct epoch of mammalian evolution. First appearances of Artiodactyla, Perissodactyla, and true Primates were taken to mark the beginning of the succeeding Eocene epoch. Sinclair (1912) and Scott (1913) were at the forefront of modern acceptance of the Paleocene as a distinct epoch.

What does the transition from Paleocene to Eocene mammalian faunas look like? This is a more difficult question that requires high-resolution biostratigraphy. It is a question best studied in a sharply eroding badland setting like the Bighorn Basin of Wyoming, and the contiguous Clarks Fork Basin of Wyoming and Montana (Fig. 1), where accumulation of Paleocene-Eocene sediments was virtually continuous and the resulting strata are thick, well exposed, and richly fossiliferous. This area has special interest as the type area of the Clarkforkian land-mammal age.

Granger (1914), Jepsen (1930), Wood et al. (1941), and others, considered Clarkforkian and Wasatchian faunas to be distinct. However, in the 1960s, two studies cast doubt on this, and indeed on our whole understanding of the Paleocene-Eocene transition in North America. Wood (1967) studied published evidence of the Clarkforkian land-mammal age fauna and concluded that it resulted from an intermixture of late Paleocene Tiffanian land-mammal age faunal elements and early Eocene Wasatchian elements. This meant, at best, that the Paleocene-Eocene transition was slow and gradual, and, at worst, that very little was known about the transition. Jepsen and Woodburne (1969) published a record of the "dawn horse" perissodactyl *Hyracotherium* that was said to come from the Paleocene of the Clarks Fork Basin. This extension of temporal range reinforced the idea of a phased faunal transition, and led to temporary confusion of intercontinental faunal correlation.

GENERAL RESULTS

With this background, resolution could only come from new, larger, stratigraphically controlled collections of mammals from the Clarks Fork and northern Bighorn Basins. I initiated a project of new collections in 1975, which has since gone through several phases with substantial studies completed on the Clarkforkian land-mammal age (Rose, 1979, 1980, 1981a, 1981b), paleomagnetic stratigraphy (Butler et al., 1981, 1987; Clyde et al., 1994), mammalian biostratigraphy (Gingerich, 1983, 1991, 2001a), the Clarkforkian-to-Wasatchian faunal transition (Gingerich, 1989; Clyde and Gingerich, 1998; Gunnell, 1998; Gingerich and Clyde, 2001), carbon and oxygen isotope stratigraphy (Koch et al., 1992, 1995; Fricke et al., 1998; Bowen et al., 2001), and the Wasatchian land-mammal age (Badgley and Gingerich, 1988; Clyde, 1997, 2001). A dissertation on the Tiffanian land-mammal age is in progress by R. Secord. Parallel studies of sedimentology, paleobotany, and mollusks have been carried out by Kraus (1980, 1987, 1996, 2001), Hickey (1980), Wing (1998), Hartman and Roth (1998), Wing et al. (2000), and Wing and Harrington (2001).

There is no longer any doubt that the Clarkforkian land-mammal fauna in the type area is distinct from both the preceding Tiffanian-age fauna and the succeeding Wasatchian-age fauna (Fig. 2). The Clarkforkian fauna contains Rodentia, Tillodontia, the condylarth *Haplomylus,* and the pantodont *Coryphodon* not found in the preceding Tiffanian, and lacks Artiodactyla, Perissodactyla, Primates, and hyaenodontid Creodonta that first appear in the succeeding Wasatchian fauna. The beginning of the Clarkforkian land-mammal age in the type area is defined by the first appearance of Rodentia, which is constrained within about ±50 m stratigraphically and within about ±100 k.y. temporally. The end of the Clarkforkian land-mammal age coincides with the beginning of the Wasatchian land-mammal age defined by the first appearance of Perissodactyla, and this is even more tightly constrained stratigraphically and temporally (Fig. 2). Note that the Wa-0 faunal interval at the beginning of the Wasatchian, of particular interest here, coincides with the carbon isotope excursion (CIE) and Paleocene-Eocene Thermal Maximum (PETM) at the beginning of the Eocene (see below).

The Tiffanian, Clarkforkian, and Wasatchian land-mammal ages of the Clarks Fork and northern Bighorn basins can be further subdivided, respectively, into six, three, and seven zones and their corresponding biochrons (Fig. 2). Stratigraphic levels and interpolated ages and durations of the biochrons are shown and explained in Table 1. Briefly, age and duration were calculated by interpolation, using a cubic spline to describe the fit of stratigraphic thickness to radiometric age. The spline was calibrated using the stratigraphic positions and accepted ages of paleomagnetic anomalies. Accumulation of Paleocene and lower Eocene strata on Polecat Bench was not linear with time (although through much of the Clarkforkian and Wasatchian this was approximately linear), and a spline provides a model-free description of the pattern. Faunal samples are generally aggregated from 10 to 20 m of section in any particular locality, with the position of each locality in the master section rounded to the closest 5 m level (meaning that precision and accuracy are both

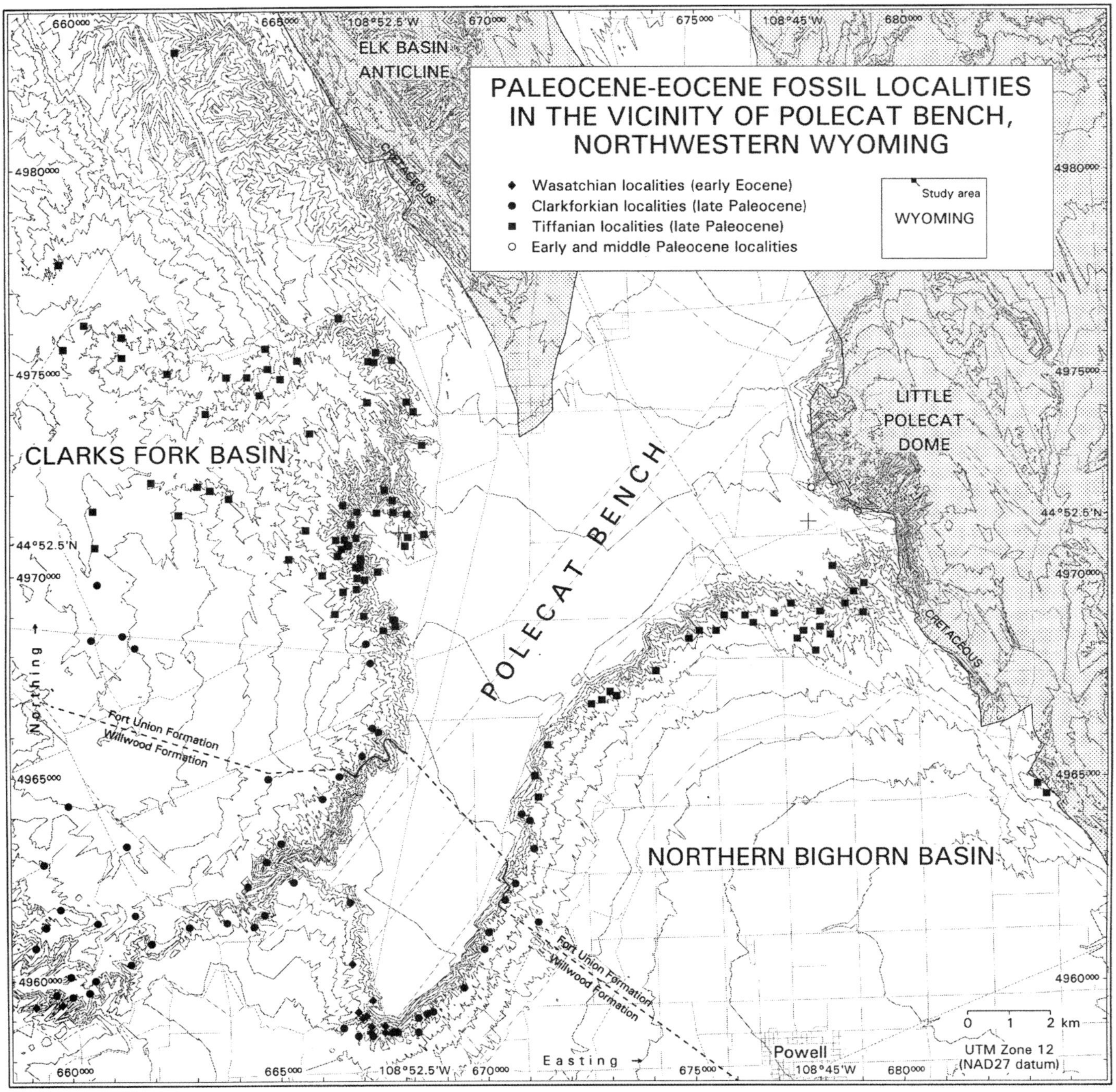

Figure 1. Map of vertebrate fossil localities in the vicinity of Polecat Bench in the Clarks Fork and northern Bighorn basins of northwestern Wyoming. This 20 × 20 km area includes Cretaceous and older strata in the northeast (stippled) overlain by Paleogene strata that dip shallowly to the southwest. Polecat Bench is a Pleistocene river terrace surrounded by sharply weathering badlands exposing Paleocene and lower Eocene strata and making them accessible for fossil collecting. The Clarks Fork Basin is the type area of Clarkforkian land-mammal age (Granger, 1914; Wood et al., 1941). The area immediately west of Polecat Bench is called Sand Coulee, which includes the type area of the overlying Sandcouleean land-mammal subage (Granger, 1914). The town of Powell is located in the lower right at the bottom of the map. Most localities shown here are Tiffanian late Paleocene (solid squares) and Clarkforkian late Paleocene (solid circles). Wasatchian localities (solid diamonds) are exposed at the southwestern tip of Polecat Bench and in the head of Big Sand Coulee. Vertebrate-bearing Paleocene and lower Eocene strata extend some 20 km farther to the west across the Clarks Fork Basin, and some 120 km farther to the southeast through the entire length of the Bighorn Basin.

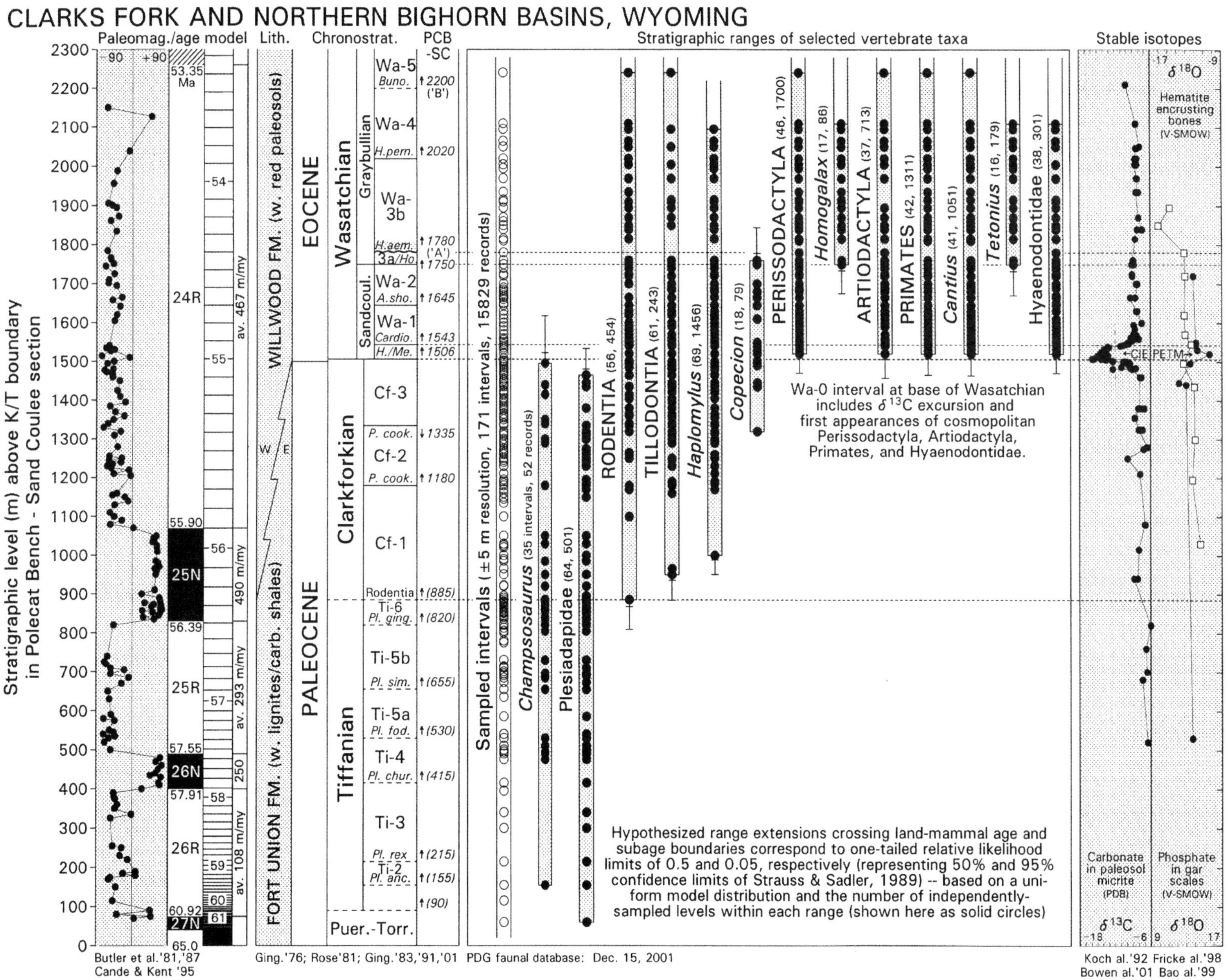

Figure 2. Stratigraphic ranges of selected vertebrate taxa in the Polecat Bench–Sand Coulee stratigraphic section of the Clarks Fork and northern Bighorn basins, Wyoming. These are shown in the context of paleomagnetic stratigraphy, interpolated age, lithostratigraphy, chronostratigraphy, and stable isotope stratigraphy, based on diverse studies of the Polecat Bench–Sand Coulee stratigraphic section by Butler et al. (1981, 1987; calibrated following Cande and Kent, 1995); by Gingerich (1976, 1983, 1991, 2001a) and Rose (1981a); and by Koch et al. (1992), Fricke et al. (1998), Bao et al. (1999), and Bowen et al. (2001). Note first appearances of cosmopolitan Perissodactyla, Artiodactyla, Primates, and Hyaenodontidae in the Wa-0 faunal zone coincident with the carbon isotope excursion (CIE) and oxygen isotope excursion representing the Paleocene-Eocene Thermal Maximum (PETM), interpolated to be near or slightly younger than 55.0 Ma (see Tables 1 and 2). Confidence limits on taxon ranges are calculated following Strauss and Sadler (1989) to emphasize the quality of the record. One-tailed limits are employed in each case because they are larger and hence more conservative. Chart updated from Gingerich (2000). V-SMOW—Vienna standard mean ocean water; PDB—Peedee belemnite.

TABLE 1. AGE AND DURATION ESTIMATES FOR BIOSTRATIGRAPHIC ZONES ABOVE AND BELOW THE PALEOCENE-EOCENE BOUNDARY ON POLECAT BENCH

Zone name	Zone bottom (m)	Zone top (m)	Zone bottom (Ma)	Zone top (Ma)	Zone duration (k.y.)
Wa-5	2200	—	–53.48	—	—
Wa-4	2020	2200	–53.87	–53.48	393.2
Wa-3b	1780	2020	–54.40	–53.87	530.4
Wa-3a	1750	1780	–54.47	–54.40	66.2
Wa-2	1645	1750	–54.70	–54.47	230.8
Wa-1	1543	1645	–54.92	–54.70	221.9
Wa-0	1506	1543	–55.00	–54.92	79.8
Cf-3	1335	1506	–55.36	–55.00	362.3
Cf-2	1180	1335	–55.68	–55.36	317.8
Cf-1	885	1180	–56.27	–55.68	587.6
Ti-6	820	885	–56.41	–56.27	145.0
Ti-5b	655	820	–56.89	–56.41	478.5
Ti-5a	530	655	–57.39	–56.89	492.9
Ti-4	415	530	–57.85	–57.39	463.4
Ti-3	215	415	–58.93	–57.85	1079.7
Ti-2	155	215	–59.48	–58.93	551.7

Note: These are calculated by interpolation, using a cubic spline describing the fit of stratigraphic thickness to radiometric age. Spline calibration points are the bottoms and tops of magnetic anomalies 26R, 26N, 25R, 25N, and 24R in the Polecat Bench–Sand Coulee stratigraphic section (Fig. 2). See Table 2 for more detailed interpolation in the vicinity of Wa-0.

on the order of 20 k.y.). Faunal samples analyzed here include all specimens from all localities recorded at a particular 5 m stratigraphic level.

The Clarkforkian-Wasatchian boundary is of special interest because it coincides closely with the Paleocene-Eocene epoch boundary, and both boundaries lie within paleomagnetic anomaly 24R (Fig. 2). The average long-term sediment accumulation rate through this interval is ~467 m/m.y. Spline and linear interpolations yield ages for the Clarkforkian-Wasatchian boundary of 55.00 and 54.97 Ma respectively (base of Wa-0?), or 54.99 and 54.96 Ma respectively (base of Wa-0), and ages for the Paleocene-Eocene boundary of 55.01 and 54.98 Ma respectively (Table 2). Note that both interpolations yield a Clarkforkian-Wasatchian boundary recognized by the first appearance of perissodactyls that is at most 21.5 k.y. younger than the Paleocene-Eocene boundary at the base of the carbon isotope excursion (CIE; see below).

Stable isotopes in soil nodules, gar scales, and hematite encrusting bones document: (1) a profound 4–5 per mil (‰) negative excursion in $\delta^{13}C$ at the time of the Clarkforkian-Wasatchian transition (Koch et al., 1992; Bowen et al., 2001; Bains et al., this volume), reflecting the global carbon isotope excursion, or CIE, first recognized in the deep sea (the base of the CIE is now taken to mark the Paleocene-Eocene boundary here and elsewhere; Luterbacher et al., 2000; Fig. 2); (2) a substantial ~2‰ positive excursion in $\delta^{18}O$ at the time of the Clarkforkian-Wasatchian transition (Fricke et al., 1998), possibly reflecting a brief interval of warm climate during the Paleocene-Eocene Thermal Maximum (PETM) associated with the CIE (Fig. 2); and (3) a longer-term 4‰ negative trend in $\delta^{18}O$ through much of Clarkforkian and early Wasatchian time (Bao et al., 1999), consistent with a background pattern of slight climatic cooling across the Paleocene-Eocene boundary (Fig. 2).

HIGH-RESOLUTION STRATIGRAPHY

High-resolution carbon isotope stratigraphy shows the CIE to span the 1500 through 1540 m interval in the Polecat Bench section, with an interpolated duration of ~86,000 yr (Fig. 3; Table 2). Stratigraphically, the Wa-0 mammalian fauna at the beginning of the Wasatchian land-mammal age falls entirely within the CIE. Clarkforkian mammals are found as high as 1505 m, and the onset of the CIE evidently happened during latest Clarkforkian time.

The interval from 1506 to 1511 m contains unusual brown paleosols, abundant *Celtis* (hackberry) endocarps, and the only specimens of the rare condylarth *Meniscotherium* from the Bighorn Basin for which precise stratigraphic information is available. *Meniscotherium* is unusual for a condylarth in being a highly specialized selenodont herbivore. It is common in later Wasatchian faunas in more southern localities (e.g., New Mexico). The *Meniscotherium*-bearing interval in the Polecat Bench section coincides with maximum excursion values of the CIE. The remainder of the CIE, from 1512 through ~1540 m, contains numerous bright red and purple paleosols and yields the classic Wa-0 mammalian fauna (Gingerich, 1989).

TABLE 2. AGE AND DURATION ESTIMATES FOR STRATIGRAPHIC EVENTS AND INTERVALS IN THE PALEOCENE-EOCENE AND CLARKFORKIAN (CF)-WASATCHIAN (WA) TRANSITIONS ON POLECAT BENCH

		Spline interpolation			Linear interpolation		
Stratigraphic level (m)	Position in 24R	Estimated age (Ma)	Time in 24R (m.y.)	CIE duration (k.y.)	Estimated age (Ma)	Time in 24R (m.y.)	CIE duration (k.y.)
2260	1190	−53.35	2.55	—	−53.35	2.55	—
1543	473	−54.92	0.98	—	−54.89	1.01	—
1540	**470**	**−54.93**	**0.97**	**86.2**	**−54.89**	**1.01**	**85.7**
1510	440	−54.99	0.91	21.5	−54.96	0.94	21.4
1506	436	−55.00	0.90	12.9	−54.97	0.93	12.9
1500	**430**	**−55.01**	**0.89**	**0.0**	**−54.98**	**0.92**	**0.0**
1070	0	−55.90	0.00	—	−55.90	0.00	—

Note: The stratigraphic interval from 1500 to 1540 m (in bold) is the interval of the carbon isotope excursion (CIE) marking the beginning of the Eocene. The earliest Eocene interval from 1500 to 1506 m has a Cf-3 mammalian fauna. The interval from 1506 to 1510 m is the *Meniscotherium priscum* zone (Wa-0?), and the interval from 1510 to 1540 m is in the *Hyracotherium sandrae* zone (Wa-0). Ages and durations were calculated using cubic spline interpolation (columns 3–5) and, for comparison, linear interpolation (columns 6–8). Spline calibration points are the bottoms and tops of magnetic anomalies 26R, 26N, 25R, 25N, and 24R in the Polecat Bench–Sand Coulee stratigraphic section (Fig. 2). Linear interpolation is calibrated using the bottom and top of anomaly 24R at 1070 and 2260 m, respectively.

Discovery that there are three successive, distinctly different, mammalian faunas, a late Clarkforkian fauna, a *Meniscotherium* zone fauna, and the classic Wa-0 fauna, all within an 86-k.y. interval of the CIE, means that the Paleocene-Eocene boundary interval on Polecat Bench requires more intensive study. As the vertebrate record improves it should be possible to study mammalian body size changes through the Wa-0 interval at higher resolution. It should also be possible to develop an enhanced paleotemperature record for the Wa-0 interval using $\delta^{18}O$ in the phosphate of gar scales and mammalian teeth. There is a high-resolution $\delta^{18}O$ record in paleosol carbonate showing a ~1‰ positive excursion consistent with warming, but interpretation is confounded by effects of temperature, precipitation, evaporation, and large-scale atmospheric circulation (Bowen et al., 2001).

MAMMALIAN BODY SIZE ACROSS THE PALEOCENE-EOCENE BOUNDARY

The earliest Wasatchian or Wa-0 mammalian fauna known from University of Michigan locality SC-67 at the south end of Polecat Bench was first recognized to be unusual because of the abundance of small and otherwise-rare specimens identified as *Ectocion parvus* Granger, 1915 (see Gingerich, 1982). Granger (1915) stated that the type and only previously known specimen of *Ectocion parvus* had come from "Clark Fork" beds at the head of Big Sand Coulee, just north and west of Polecat Bench, but it now appears more likely that the specimen came from the Wa-0 stratigraphic interval at the south end of Polecat Bench (Gingerich, 1989, p. 51). Specimens initially identified as *Ectocion parvus* were later found to belong to two different species, one being *Ectocion parvus* and the other being *Copecion davisi.*

Ectocion is one of the more common genera of mammals in strata of Clarkforkian and early Wasatchian age. It is represented by a long lineage, stationary or unchanging through time, comprising a single species *Ectocion osbornianus,* and a contemporary sample or very short parallel lineage of smaller *E. parvus.* This pattern is shown in Figure 4, where the tooth size of individual specimens is plotted against their geological age. *Ectocion parvus* was first distinguished from *E. osbornianus* by its smaller size (Granger, 1915), and no further differences have been discovered subsequently. The body weight of *E. osbornianus,* estimated from molar size, averaged ~8–10 kg in life, while the body weight of *E. parvus* with smaller molars was only ~4–5 kg (Fig. 4).

Differences in mean values of samples of *Ectocion osbornianus* from successive stratigraphic levels are all small (circles in the range of 0–4 standard deviations plotted in Figure 4A), and these differences fall within a $1 - \alpha = 99.9\%$ confidence band (open area) for samples of a given size drawn at random from a population with the previous mean (calculated as $\pm t_{\alpha[n-1]}/\sqrt{n}$; Sokal and Rohlf, 1981, p. 148). In contrast, differences involving Wa-0 *Ectocion parvus* are larger (diamonds in stippled area of Figure 4A).

Ectocion parvus appears to be a distinctly smaller congeneric species that replaced *E. osbornianus* ecologically for a brief interval at the beginning of Wasatchian time, but it could conceivably be an evolutionarily dwarfed member of the *E. osbornianus* lineage (current taxonomy implies but does not demonstrate ecological replacement; the taxonomy may well need to be changed when the evidence is clear). The phrase of "evolutionary dwarfing" is used to indicate that the observed case is unusual, that it involves small size, and to suggest inherited adaptive change (as opposed to temporary ecological stunting).

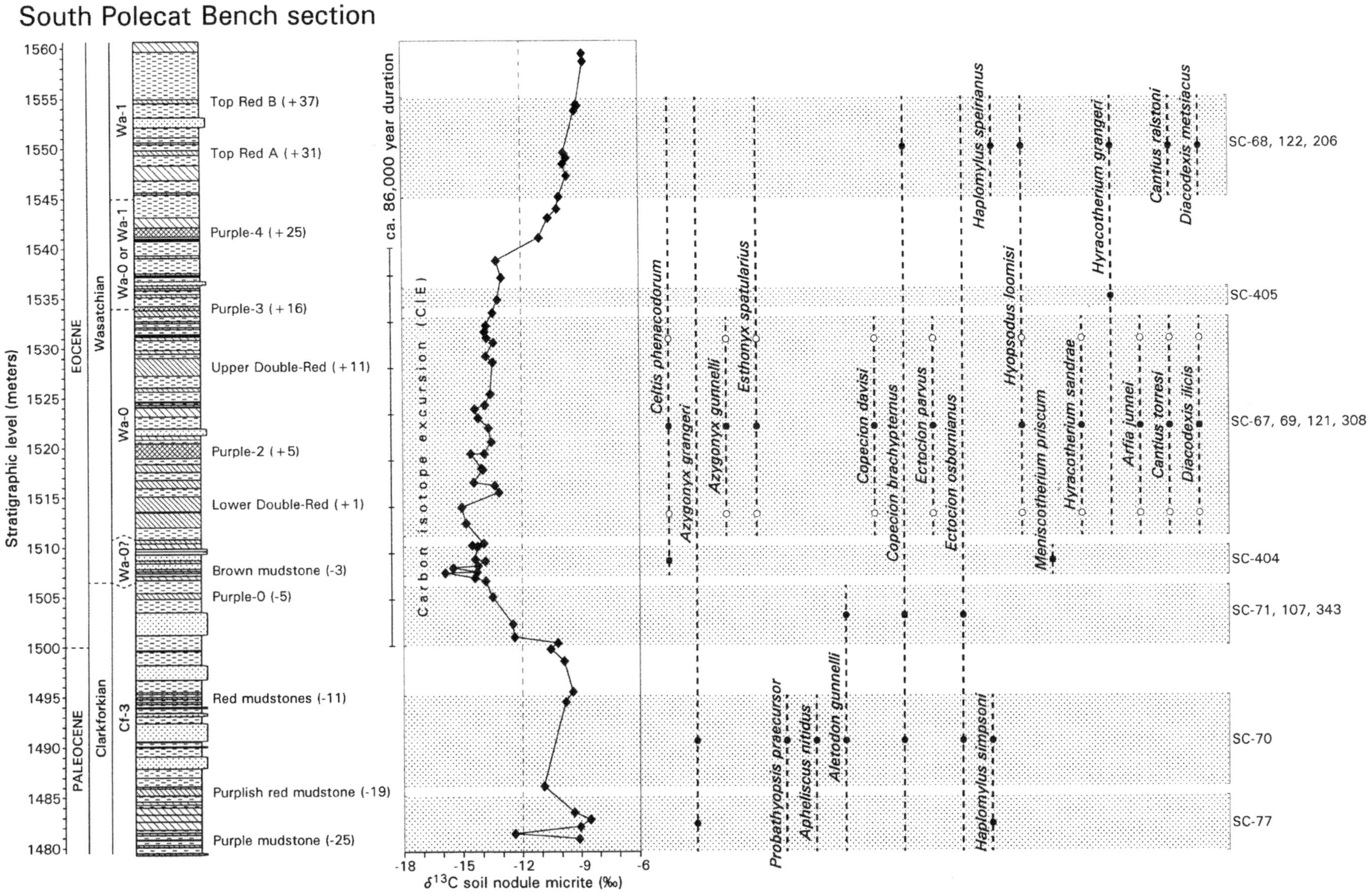

Figure 3. High-resolution carbon isotope stratigraphy and biotic change at the south end of Polecat Bench. Lithostratigraphy and stratigraphic ranges of the elm-related dicot *Celtis* (hackberry) and important mammalian taxa crossing the Paleocene-Eocene boundary are from Gingerich (2001a). Fourteen mammal-bearing fossil localities are known (SC-77, etc.), representing seven distinct stratigraphic intervals (shaded). These can be grouped into four time-successive associations based on their biota (Cf-3, Wa-0?, Wa-0, and Wa-1 on the ordinate). The interval labeled "Wa-0 or Wa-1" on the ordinate is poorly sampled and could be either Wa-0 or Wa-1 in age. The 86,000-year time scale shown here is consistent with age interpolations in Table 2. Note that the beginning of the carbon isotope excursion is in the late Clarkforkian, overlapping the stratigraphic interval of localities SC-71, etc. The most negative $\delta^{13}C$ values (~15‰) are in the earliest zone of the Wasatchian (Wa-0?), represented by SC-404 with the first hackberry *Celtis* and rare condylarth *Meniscotherium.* The remainder of the carbon isotope excursion spans the interval of the classic Wa-0 fauna with small *Ectocion* and *Copecion* from localities SC-67, etc. *Ectocion* and *Copecion* of standard size are known from localities SC-68, etc., in zone Wa-1 after carbon isotope values returned to normal (here ~9‰). Reproduced with modifications from Figure 6 in Bowen et al. (2001).

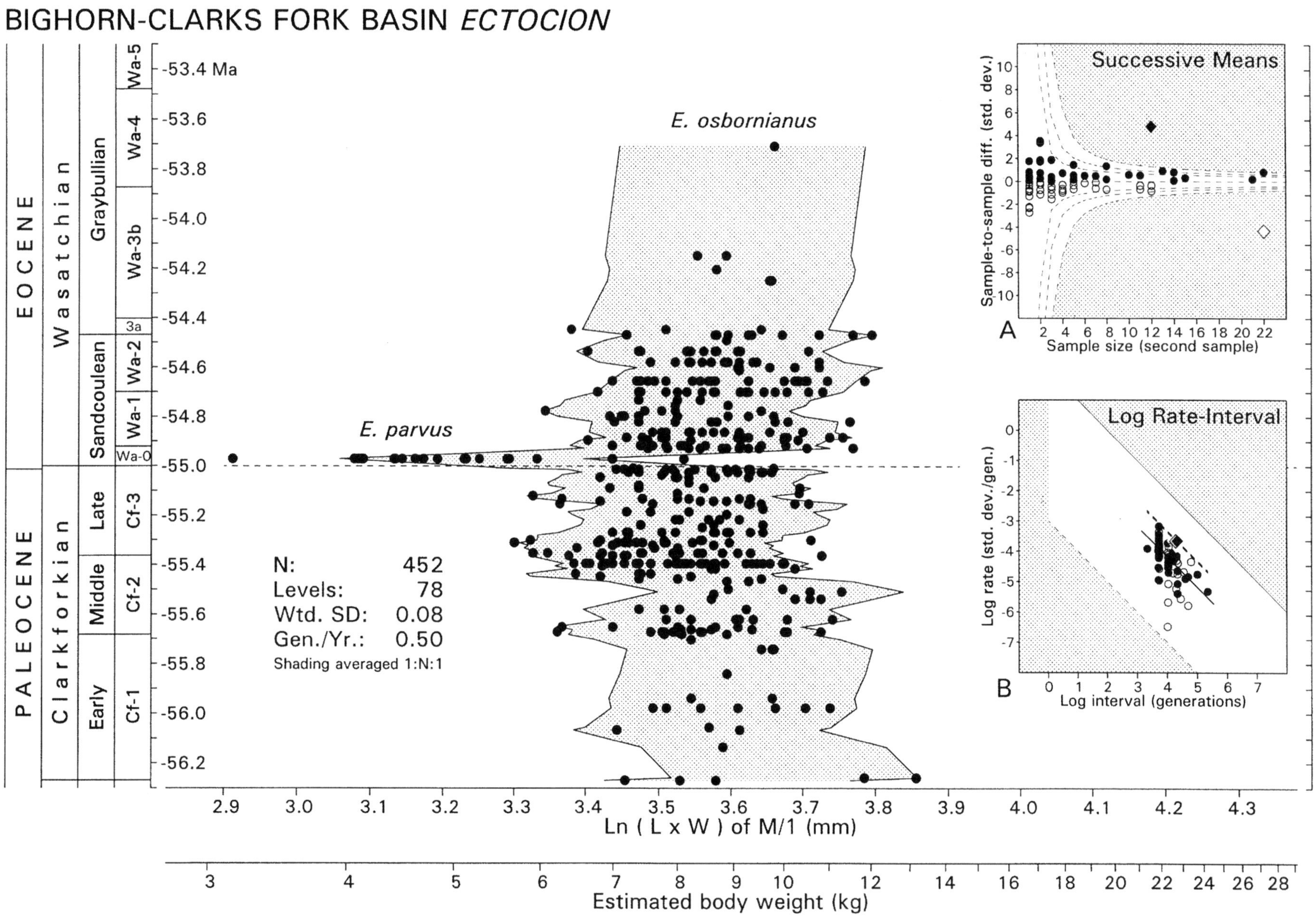

Figure 4. Stratigraphic record of tooth size and body size for the mammalian genus *Ectocion* through Clarkforkian and early Wasatchian time in the northern Bighorn and Clarks Fork basins. Most specimens represent the common species *Ectocion osbornianus* of ~8–10 kg body weight, but smaller specimens of *Ectocion parvus* of ~4–5 kg body weight predominate in the Wa-0 stratigraphic interval coinciding with the CIE and PETM (cf. Figure 2). Vertical axis is time, based on cubic spline interpolation of Figure 2 and Table 1. Inset A shows that the Wa-0 sample (diamonds) differs from adjacent samples more than expected by chance (symbols are open or closed reflecting negative and positive changes in tooth size, respectively). Dashed lines in A are confidence limits for $\alpha = 0.05$, 0.01, and 0.001. Inset B shows rates for Wa-0 samples to be higher than all other rates calculated on the same time scale (dashed lines of slope 1 are fit to diamond rates involving Wa-0). Body weight is estimated from tooth size following ungulate regression of Legendre (1989, Table 1), and generation time is estimated from body weight (Eisenberg, 1981). *Ectocion osbornianus* in the Bighorn and Clarks Fork basins was chosen by Simpson (1943) as the featured example when he introduced the concept of an evolving chronocline, but improved sampling and stratigraphic control shown here indicates that *E. osbornianus* changed little through its known evolutionary history.

Rates required for dwarfing and recovery, plotted as diamonds in inset Figure 4B, are high but at the margin of rates calculated for the rest of the time series (and again within the range expected for evolutionary change, including island dwarfing, on these time scales; Gingerich, 2001b). There are two Wa-0 specimens the size of ordinary *E. osbornianus* in the Wa-0 fauna, which supports the idea of replacement rather than dwarfing, but these could be contaminants (this idea can be tested isotopically). Whether *E. parvus* represents a parallel lineage or a brief evolutionary dwarfing event in the longer *E. osbornianus* lineage, it is a conspicuous outlier with a body weight estimated to have been only ~50% the weight of its more common and longer-ranging congener.

Copecion too is represented by a long stationary lineage comprising a single species *Copecion brachypternus,* and a contemporary sample or very short parallel lineage of *C. davisi.* This pattern is shown in Figure 5. *Copecion davisi* was first distinguished from *C. brachypternus* by its smaller size (Gingerich, 1989), and no further differences have been discovered subsequently. The body weight of *C. brachypternus* estimated from molar size, like that of *E. osbornianus,* averaged ~8–10 kg in life, while the body weight of *C. davisi* with smaller molars was only ~5–6 kg.

Here again, differences in mean values of successive samples of *Copecion brachypternus* are all small (circles in the range of 0–4 standard deviations plotted in Figure 5A), and these differences again fall within a $1 - \alpha = 99.9\%$ confidence band (open area) for samples of a given size drawn at random from a population with the previous mean. In contrast, differences involving Wa-0 *Copecion davisi* are larger (diamonds in Figure 5A).

C. davisi appears to be a distinctly smaller congeneric species that replaced *C. brachypternus* ecologically for a brief interval at the beginning of Wasatchian time, but it could conceivably be an evolutionarily dwarfed member of the *C. brachypternus* lineage (again, current taxonomy implies but does not demonstrate ecological replacement).

Rates required for dwarfing and recovery, plotted as diamonds in inset Figure 5B, are high but again at the margin of rates calculated for the rest of the time series (and again within the range expected for evolutionary change on these time scales; Gingerich, 2001b). There are no Wa-0 specimens the size of ordinary *C. brachypternus.* Whether *C. davisi* represents a parallel lineage or a transient evolutionary dwarfing event in the longer *C. brachypternus* lineage, it is a conspicuous outlier with a body weight estimated to have been only ~60% the weight of its more common and longer-ranging congener.

Perissodactyls make their first appearance in the North American fossil record in the Wasatchian land-mammal age. The only common perissodactyl from Wa-0 is the small 5–6 kg, short-lived species *Hyracotherium sandrae* (Fig. 6). There is limited evidence that the larger 8–10 kg *Hyracotherium grangeri* comes from the Wa-0 interval too, but it is much rarer than *H. sandrae,* at least in northern Bighorn Basin localities (*H. grangeri* is the only species reported from the Castle Gardens Wa-0 fauna in the southeastern Bighorn Basin; Strait, 2001). Thus, the body size pattern for *Hyracotherium* through the Wa-0 to Wa-2 part of the record is similar to that for *Ectocion* and *Copecion,* with *H. sandrae* being a conspicuous outlier in having a body weight estimated to have been only some 60% the weight of its more common and longer-ranging congener *H. grangeri.* There are no differences known to distinguish *H. sandrae* from *H. grangeri* except size.

More examples could be shown (see Gingerich, 1989), but Figures 4–6 convey the idea that unusually small body size is a salient characteristic of Wa-0 species from the CIE and PETM interval of the earliest Wasatchian. Some species, such as *Hyopsodus loomisi,* seem to have made the transition from Wa-0 to Wa-1 with no substantial change in body size. No species are known to exhibit larger body size in response to the elevated mean annual temperatures in the Wa-0 interval.

EXPLANATION FOR EVOLUTIONARY DWARFING

How can we explain the rapid replacement or evolution of a common, successful Paleocene species by a congener only 50%–60% of its body weight? Requisite rates of change lie within the range of evolutionary rates observed on 10^4–10^5 generation time scales in other settings, so the replacement is not rate limited. The problem is rather to identify the selective forcing factor or factors. We know that the CIE-PETM interval on land is unusual for its isotopically light carbon, its isotopically heavy oxygen, its ambient temperature elevated by up to 8 °C, and its elevated levels of greenhouse gases such as CO_2 in the atmosphere (Koch et al., 1992; Fricke et al., 1998; Dickens, 2000; Royer et al., 2001; Zachos et al., 2001). Elevated temperature and elevated CO_2 are both possible forcing factors.

Bergmann's rule is one of the best documented and widely known ecological and biogeographic generalizations (Bergmann, 1847; Rensch, 1929; Ashton et al., 2000): As a rule, the body size of a species increases in cooler climates and decreases in warm climates. Koch (1986) studied geographic clines in extant North American mammals and showed that today 0.02 ln units of tooth crown area are approximately equivalent to 1.0 °C of temperature change. The difference in tooth crown area distinguishing, for example, *Ectocion parvus* in the earliest Wasatchian from *E. parvus* in the Clarkforkian is about –0.3 ln units, which would require a temperature rise of some 15 °C (compared to ~4 °C estimated by Fricke et al., 1998). Thus, a simple Bergmann response to temperature elevation cannot be the whole explanation for the dwarfing observed Wa-0 mammals.

Body size in mammals is also influenced by food quality (Case, 1979). Here new and ongoing experiments with herbivores fed on plants grown in CO_2-enriched atmospheres are particularly interesting (e.g., Fajer et al., 1989; Kinney et al., 1997; Tuchman et al., 2002). Experiments show that "high" CO_2 partial pressures (~700 ppmv, compared to present-day pressures of ~350 ppmv) are sufficient to increase photosynthesis, and to increase carbon and decrease nitrogen in plant tissues. The extra

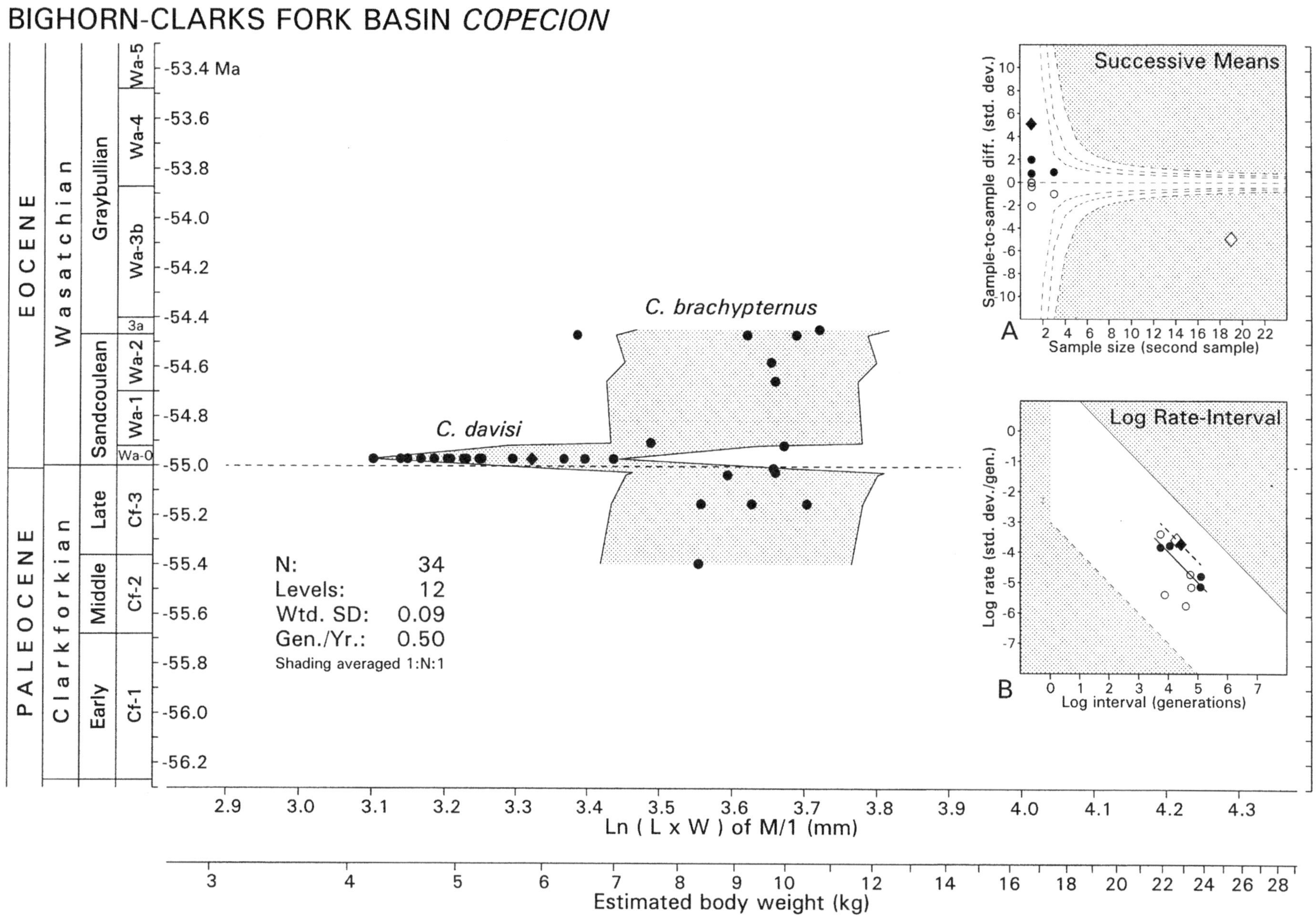

Figure 5. Stratigraphic record of tooth size and body size for the mammalian genus *Copecion* through Clarkforkian and early Wasatchian time in the northern Bighorn and Clarks Fork basins. Most specimens represent the common species *Copecion brachypternus* of ~8–10 kg body weight, but smaller specimens of *Copecion davisi* of ~5–6 kg body weight predominate in the Wa-0 stratigraphic interval coinciding with the CIE and PETM (cf. Figure 2). Vertical axis is time based on cubic spline interpolation. Insets A and B show that the Wa-0 sample (diamonds) differs from adjacent samples more than expected by chance (symbols are open or closed reflecting negative and positive changes in tooth size, respectively). Dashed lines in A are confidence limits for $\alpha = 0.05$, 0.01, and 0.001. Distribution in B shows rates for Wa-0 samples to be higher than all other rates calculated on the same time scale (dashed lines of slope 1 are fit to diamond rates involving Wa-0). Body weight is estimated from tooth size following ungulate regression of Legendre (1989, Table 1), and generation time is estimated from body weight (Eisenberg, 1981).

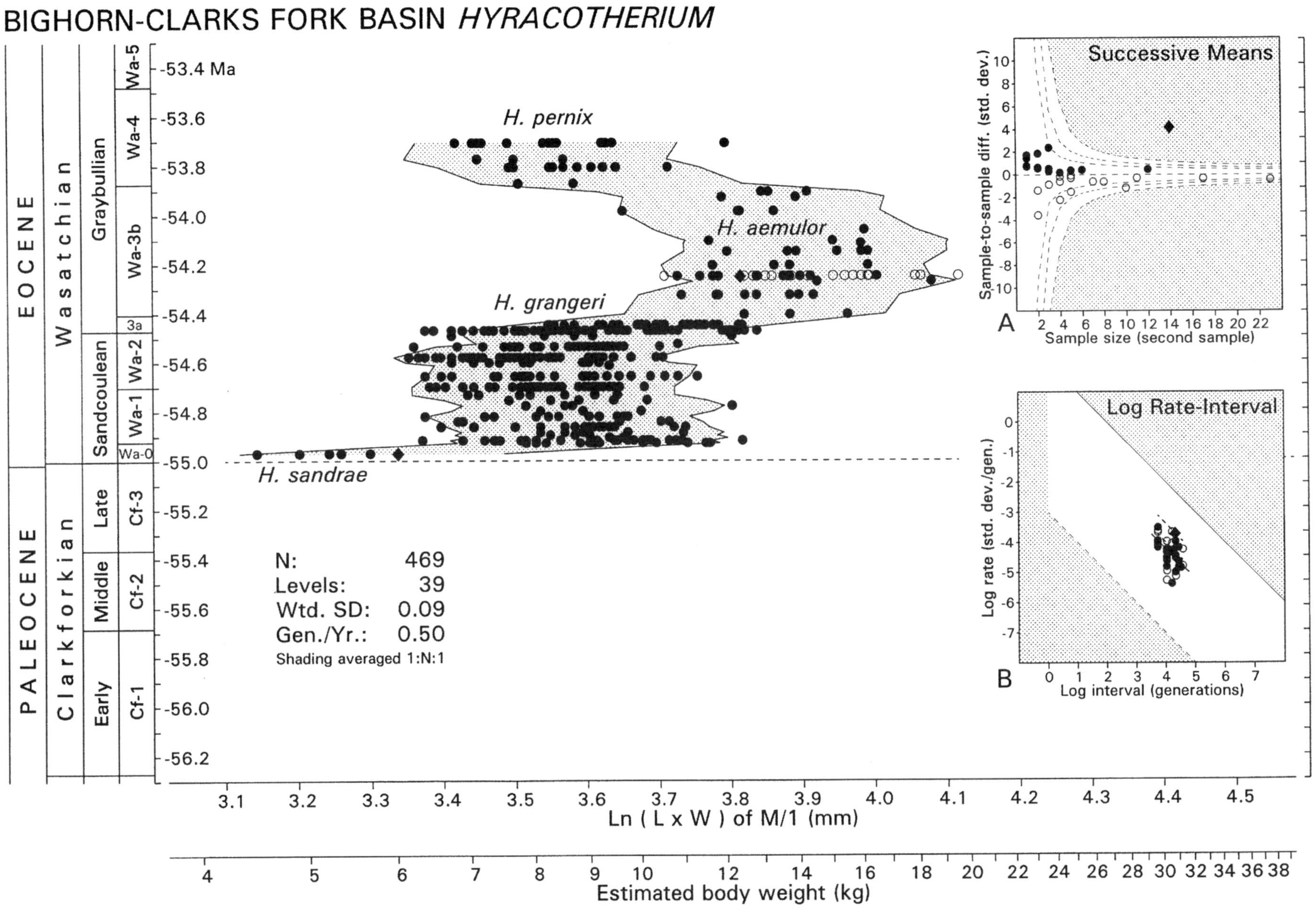

Figure 6. Stratigraphic record of tooth size and body size for the mammalian genus *Hyracotherium* through Clarkforkian and early Wasatchian time in the northern Bighorn and Clarks Fork basins. Most specimens represent the common species *Hyracotherium grangeri* of ~8–10 kg body weight, but smaller specimens of *Hyracotherium sandrae* of ~5–6 kg body weight predominate in the Wa-0 stratigraphic interval coinciding with the CIE and PETM (cf. Figure 2). Vertical axis is time based on cubic spline interpolation. Inset figures A and B show that the Wa-0 sample (diamonds) differs from adjacent samples more than expected by chance (symbols are open or closed reflecting negative and positive changes in tooth size, respectively). Dashed lines in A are confidence limits for $\alpha = 0.05$, 0.01, and 0.001. Distribution in B shows rates for Wa-0 samples to be higher than all other rates calculated on the same time scale (dashed line of slope 1 is fit to diamond rate involving Wa-0). Body weight is estimated from tooth size following ungulate regression of Legendre (1989, Table 1), and generation time is estimated from body weight (Eisenberg, 1981).

carbon is used to enhance secondary compounds, inhibiting digestibility for herbivores. CO_2 enrichment also leads to reduction in rubisco enzyme concentrations in foliage, thus decreasing leaf protein and lowering nutritional value for herbivores. Experimentally, insect herbivores are observed to grow more slowly on high-CO_2 foliage. No experiments have been done with mammalian herbivores, but it is logical to expect that mammals too would experience slower growth. Slower growth, coupled with any of the usual seasonal temperature, rainfall, or day-length effects controlling reproductive cycles, would lead to reproduction at smaller body size rather than at the same or larger size. This would be, simply and directly, environmental selection favoring evolutionary dwarfing. Alternatively, under some conditions, we might expect "normal"-sized species to be replaced by smaller congeners or other species.

We do not know very much about the partial pressure of CO_2 during the PETM or Wa-0 interval of dwarfing in mammals. Royer et al. (2001) published one estimate of 800 ppmv based on the stomatal index of *Ginkgo gardneri* from Ardtun Head on the Isle of Mull, which is correlated over some distance to an interval with CIE and PETM indicators. Further, we do not yet know anything experimentally about the effects that a diet of plants grown in CO_2-enhanced environments might have on the body size of mammalian herbivores. Nevertheless, the coincidental transience of both the elevated partial pressure of CO_2 postulated during the PETM and the transience of the dwarfed body size of mammals during this interval suggest a cause and effect relationship. The temperature rise during the PETM interval is not sufficient by itself to explain all of the dwarfing observed through any simple Bergmann response to warming, but temperature plus an elevated partial pressure of CO_2 might explain this.

OTHER RESPONSES TO CLIMATE CHANGE

Differences distinguishing the Wa-0 mammalian fauna of the CIE-PETM interval from faunas that preceded or followed in time can be divided into two categories: Transient differences and permanent differences. The size changes in *Ectocion* and *Copecion* illustrated here are transient in the sense that the sizes of species observed following the CIE-PETM event returned to the sizes observed before the event. Unusually small species appeared during the PETM interval, and smaller size is plausibly a response to elevated temperature and elevated CO_2. A second transient change concerns the proportion of digging mammals (palaeanodonts and taeniodonts) in the Wa-0 fauna (Gingerich, 1989). Digging mammals typically constitute <1% of all specimens in a Clarkforkian or Wasatchian fauna, but rise to 6% during Wa-0 time (of which ~3% are palaeanodonts and ~3% are taeniodonts). It is not clear that digging would be favored by a warmer climate, but a better-drained landscape promoting development of especially mature paleosols, like those in the Wa-0 interval, could plausibly favor digging mammals. Both of these transient faunal changes are so closely linked to the CIE-PETM interval, occurring in precisely the same strata, that there can be little question that they are, directly or indirectly, responses to climate change.

A more permanent faunal change occurring at the time of the CIE-PETM event is evident in a range chart showing the relative timing of first and last appearances of mammalian genera from the northern Bighorn and Clarks Fork basins (Fig. 7). Integration of appearances and disappearances shows a plateau of diversity of ~45–50 genera in the middle and late Clarkforkian, which then rises to a higher plateau of ~55–60 genera in the early Wasatchian. This rise coincides with and results from a peak of 14 first appearances of new genera in the Wa-0 fauna. Lesser peaks of first and last appearances are scattered through Clarkforkian and Wasatchian time, but the most conspicuous one is the peak of first appearances coinciding with the CIE-PETM event.

Other lasting changes at the CIE-PETM event are an increase in mean species body size, and an increase in the relative importance of herbivores and frugivores (see Clyde and Gingerich, 1998). These equally rapid but more permanent changes almost certainly reflect immigration of new taxa. The cosmopolitanism of the new taxa (Artiodactyla, Perissodactyla, Primates, etc.) is itself evidence of adaptive advantages enabling the immigrants to invade and live side by side with endemic taxa.

DISCUSSION

Webb and Opdike (1995, Tables 11.1 and 11.2) listed 15 mammalian genera as North American immigrants at the beginning of Wasatchian time. Nine of the genera listed by Webb and Opdike not only make their first appearance in the Wasatchian land-mammal age, but also make their first appearance in the Wa-0 fauna (as do five genera not listed by Webb and Opdike as Wasatchian immigrants). The remaining six genera listed by Webb and Opdike appear at different times later in the

Figure 7. Stratigraphic ranges of late Paleocene and early Eocene mammalian genera in the Polecat Bench–Sand Coulee area of the Clarks Fork and northern Bighorn basins. Dashed lines in the range chart separate zones of the Clarkforkian and Wasatchain land mammal ages (labeled Cf-1, etc.). Stippled column is Wa-0. Graph at the bottom of the figure shows the cumulative number of genera inferred to be present through the Paleocene-Eocene boundary interval based on first and last appearances of genera in the range chart. Numbers of first and last appearances are shown as solid and dashed lines, respectively, in the histogram at the bottom of the graph. Note the peak of first appearances in Wa-0 and the change from a plateau of diversity in the middle and late Clarkforkian to a plateau of higher diversity in the early Wasatchian. Interpretation is complicated by dependency of diversity on both sample size and number of contributing localities. Rarefaction curves for species show Wasatchian assemblages to be more diverse for a given number of specimens (Clyde and Gingerich, 1998). At the same time, the average number of localities per stratigraphic level is about two for Clarkforkian levels and about three for Wasatchian levels, which might explain the higher generic diversity seen here. Range chart is abstracted from Figure 2 of Gingerich and Clyde (2001).

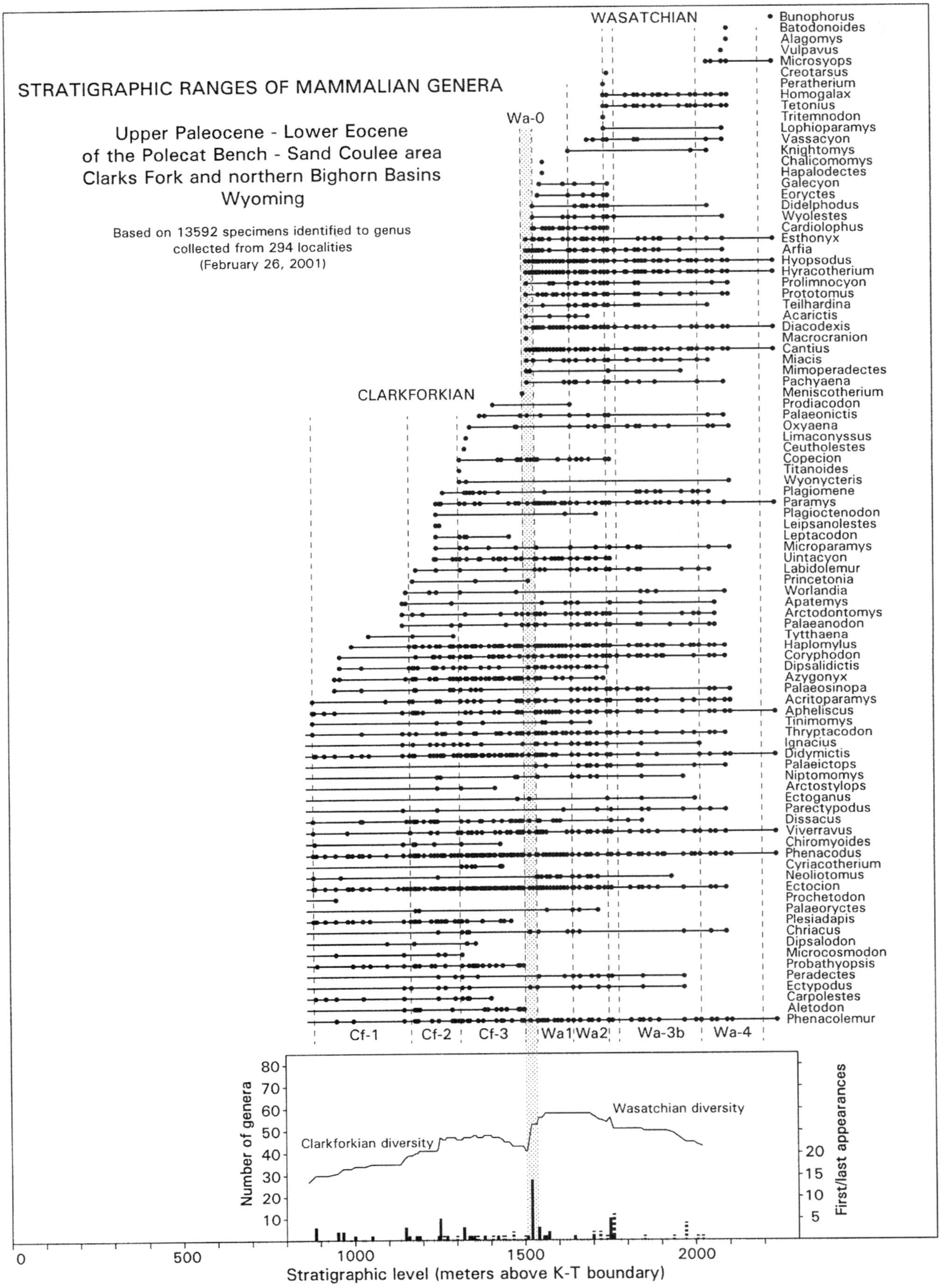
STRATIGRAPHIC RANGES OF MAMMALIAN GENERA
Upper Paleocene - Lower Eocene
of the Polecat Bench - Sand Coulee area
Clarks Fork and northern Bighorn Basins
Wyoming
Based on 13592 specimens identified to genus
collected from 294 localities
(February 26, 2001)
WASATCHIAN
CLARKFORKIAN
Wa-0
Bunophorus
Batodonoides
Alagomys
Vulpavus
Microsyops
Creotarsus
Peratherium
Homogalax
Tetonius
Tritemnodon
Lophioparamys
Vassacyon
Knightomys
Chalicomomys
Hapalodectes
Galecyon
Eoryctes
Didelphodus
Wyolestes
Cardiolophus
Esthonyx
Arfia
Hyopsodus
Hyracotherium
Prolimnocyon
Prototomus
Teilhardina
Acarictis
Diacodexis
Macrocranion
Cantius
Miacis
Mimoperadectes
Pachyaena
Meniscotherium
Prodiacodon
Palaeonictis
Oxyaena
Limaconyssus
Ceutholestes
Copecion
Titanoides
Wyonycteris
Plagiomene
Paramys
Plagioctenodon
Leipsanolestes
Leptacodon
Microparamys
Uintacyon
Labidolemur
Princetonia
Worlandia
Apatemys
Arctodontomys
Palaeanodon
Tytthaena
Haplomylus
Coryphodon
Dipsalidictis
Azygonyx
Palaeosinopa
Acritoparamys
Apheliscus
Tinimomys
Thryptacodon
Ignacius
Didymictis
Palaeictops
Niptomomys
Arctostylops
Ectoganus
Parectypodus
Dissacus
Viverravus
Chiromyoides
Phenacodus
Cyriacotherium
Neoliotomus
Ectocion
Prochetodon
Palaeoryctes
Plesiadapis
Chriacus
Dipsalodon
Microcosmodon
Probathyopsis
Peradectes
Ectypodus
Carpolestes
Aletodon
Phenacolemur
Cf-1
Cf-2
Cf-3
Wa1
Wa2
Wa-3b
Wa-4
Number of genera
80
70
60
50
40
30
20
10
Clarkforkian diversity
Wasatchian diversity
First/last appearances
20
15
10
5
0
500
1000
1500
2000
Stratigraphic level (meters above K-T boundary)

Wasatchian. Appearance of so many mammals in such a restricted interval of time coinciding with the CIE-PETM event reinforces the arguments by Stucky (1990), Janis (1993), and Webb and Opdike (1995) that climate has an important effect on Cenozoic mammalian faunas. This is shown too by the transient changes mentioned above. Whether climate has any direct effect on speciation remains to be substantiated, but mammalian turnover is pulsed and the Wa-0 pulse discussed here clearly coincides with climate change (Vrba, 1985).

Prothero (1999) and Alroy et al. (2000) studied Cenozoic climate and mammals on million-year time scales and failed to find any substantial biotic response to climate change. Prothero (1999) excluded the Paleocene-Eocene transition from his study because it involved immigrant genera, while Alroy et al. (2000) included it and found evidence associating climate and biotic change in this brief transition. One reason the relationship of faunal change to climate is clear at the Paleocene-Eocene boundary is because the isotopic record being compared comes from the very same stratigraphic section, standardizing the spatial scale and removing uncertainties inherent in correlation of distant sections like those on land and in the deep sea. Another reason the relationship is clear here may be that important climatic events operate on short time scales, and any direct forcing of mammalian change will be blurred if mammals being compared are not sampled on comparably short time scales. Aggregation of mammalian records on million-year time scales may not be precise enough to reveal patterns on the time scale of climate change (just as study of evolutionary change on million-year time scales is inadequate to characterize rates on the time scale of the evolutionary process; Gingerich, 2001b).

Climate plays a large role in structuring living mammalian faunas both geographically and ecologically (Badgley and Fox, 2000). The links between Wa-0 faunal change and climate change demonstrated here cannot be generalized to the whole faunal history of Cenozoic mammals, but at the same time it is clear that climate plays an important role in faunal change and such relationships deserve careful investigation.

ACKNOWLEDGMENTS

I thank Dr. James A. Teeri for information on experiments with plants grown in environments of elevated carbon dioxide; Daniel C. Fisher for discussion; and Catherine E. Badgley, David L. Fox, Gregg F. Gunnell, and John P. Hunter for reviews improving the manuscript. Research on mammalian faunal evolution across the Paleocene-Eocene boundary has been supported by a series of grants from the National Science Foundation, most recently EAR-0125502.

REFERENCES CITED

Alroy, J., Koch, P.L., and Zachos, J.C., 2000, Global climate change and North American mammalian evolution, *in* Erwin, D.H., and Wing, S.L., eds., Deep time: Paleobiology's perspective: Paleobiology, v. 26 (supplement), p. 259–288.

Ashton, K.G., Tracy, M.C., and de Queiroz, A., 2000, Is Bergmann's Rule valid for mammals?: American Naturalist, v. 156, p. 390–415.

Badgley, C.E., and Fox, D.L., 2000, Ecological biogeography of North American mammals: Species density and ecological structure in relation to environmental gradients: Journal of Biogeography, v. 27, p. 1437–1467.

Badgley, C.E., and Gingerich, P.D., 1988, Sampling and faunal turnover in early Eocene mammals: Palaeogeography, Palaeoclimatology, Palaeoecology, v. 63, p. 141–157.

Bao, H., Koch, P.L., and Rumble, D., 1999, Paleocene-Eocene climatic variation in western North America: Evidence from the $\delta^{18}O$ of pedogenic hematite: Geological Society of America Bulletin, v. 111, p. 1405–1415.

Behrensmeyer, A.K., Todd, N.E., Potts, R., and McBrinn, G.E., 1997, Late Pliocene faunal turnover in the Turkana Basin, Kenya and Ethiopia: Science, v. 278, p. 1589–1594.

Bergmann, C., 1847, Über die Verhältnisse der Wärmeökonomie der Thiere zu ihrer Grösse: Göttinger Studien, v. 1, p. 595–798.

Bowen, G.J., Koch, P.L., Gingerich, P.D., Norris, R.D., Bains, S., and Corfield, R.M., 2001, Refined isotope stratigraphy across the continental Paleocene-Eocene boundary on Polecat Bench in the northern Bighorn Basin, *in* Gingerich, P.D., ed., Paleocene-Eocene stratigraphy and biotic change in the Bighorn and Clarks Fork basins, Wyoming: University of Michigan Papers on Paleontology, v. 33, p. 73–88.

Butler, R.F., Gingerich, P.D., and Lindsay, E.H., 1981, Magnetic polarity stratigraphy and biostratigraphy of Paleocene and lower Eocene continental deposits, Clarks Fork Basin, Wyoming: Journal of Geology, v. 89, p. 299–316.

Butler, R.F., Krause, D.W., and Gingerich, P.D., 1987, Magnetic polarity stratigraphy and biostratigraphy of middle-late Paleocene continental deposits of south-central Montana: Journal of Geology, v. 95, p. 647–657.

Cande, S.C., and Kent, D.V., 1995, Revised calibration of the geomagnetic polarity timescale for the Late Cretaceous and Cenozoic: Journal of Geophysical Research, v. 100, p. 6093–6095.

Case, T.J., 1979, Optimal body size and an animal's diet: Acta Biotheoretica, v. 28, p. 54–69.

Clyde, W.C., 1997, Stratigraphy and mammalian paleontology of the McCullough Peaks, northern Bighorn Basin, Wyoming: Implications for biochronology, basin development, and community reorganization across the Paleocene-Eocene boundary [Ph.D. dissertation]: Ann Arbor, Michigan, University of Michigan, 270 p.

Clyde, W.C., 2001, Mammalian biostratigraphy of the McCullough Peaks area in the northern Bighorn Basin, *in* Gingerich, P.D., ed., Paleocene-Eocene stratigraphy and biotic change in the Bighorn and Clarks Fork basins, Wyoming: University of Michigan Papers on Paleontology, v. 33, p. 109–126.

Clyde, W.C., and Gingerich, P.D., 1998, Mammalian community response to the latest Paleocene thermal maximum: An isotaphonomic study in the northern Bighorn Basin, Wyoming: Geology, v. 26, p. 1011–1014.

Clyde, W.C., Stamatakos, J., and Gingerich, P.D., 1994, Chronology of the Wasatchian land-mammal age (early Eocene): Magnetostratigraphic results from the McCullough Peaks section, northern Bighorn Basin, Wyoming: Journal of Geology, v. 102, p. 367–377.

Cope, E.D., 1888, Synopsis of the vertebrate fauna of the Puerco series: Transactions of the American Philosophical Society, v. 16, p. 298–361.

Dickens, G.R., 2000, Methane oxidation during the late Paleocene thermal maximum: Bulletin de la Société Géologique de France, v. 177, p. 37–49.

Eisenberg, J.F., 1981, The mammalian radiations: An analysis of trends in evolution, adaptation, and behavior: Chicago, Illinois, University of Chicago Press, 610 p.

Fajer, E.D., Bowers, M.D., and Bazzaz, F.A., 1989, The effects of enriched carbon dioxide atmospheres on plant-insect herbivore interactions: Science, v. 243, p. 1198–1200.

Fricke, H.C., Clyde, W.C., O'Neil, J.R., and Gingerich, P.D., 1998, Evidence for rapid climate change in North America during the latest Paleocene thermal maximum: Oxygen isotope compositions of biogenic phosphate from the Bighorn Basin (Wyoming): Earth and Planetary Science Letters, v. 160, p. 193–208.

Gervais, P., 1877, Enumération de quelques ossements d'animaux vertébrés recueillis aux environs de Reims par M. Lemoine: Journal de Zoologie, Paris, v. 6, p. 74–79.

Gingerich, P.D., 1976, Cranial anatomy and evolution of early Tertiary Plesiadapidae (Mammalia, Primates): University of Michigan Papers on Paleontology, v. 15, p. 1–140.

Gingerich, P.D., 1982, Paleocene "*Meniscotherium semicingulatum*" and the first appearance of Meniscotheriidae (Condylarthra) in North America: Journal of Mammalogy, v. 63, p. 488–491.

Gingerich, P.D., 1983, Paleocene-Eocene faunal zones and a preliminary analysis of Laramide structural deformation in the Clarks Fork Basin, Wyoming: Wyoming Geological Association, 34th Annual Field Conference, Guidebook, p. 185–195.

Gingerich, P.D., 1985, South American mammals in the Paleocene of North America, *in* Stehli, F.G., and Webb, S.D., eds., The great American biotic interchange: New York, Plenum Press, p. 123–137.

Gingerich, P.D., 1989, New earliest Wasatchian mammalian fauna from the Eocene of northwestern Wyoming: Composition and diversity in a rarely sampled high-floodplain assemblage: University of Michigan Papers on Paleontology, v. 28, p. 1–97.

Gingerich, P.D., 1991, Systematics and evolution of early Eocene Perissodactyla (Mammalia) in the Clarks Fork Basin, Wyoming: Contributions from the Museum of Paleontology, University of Michigan, v. 28, p. 181–213.

Gingerich, P.D., 2000, Paleocene-Eocene boundary and continental vertebrate faunas of Europe and North America, *in* Schmitz, B., et al., eds., Early Paleogene warm climates and biosphere dynamics: GFF, v. 122, p. 57–59.

Gingerich, P.D., 2001a, Biostratigraphy of the continental Paleocene-Eocene boundary interval on Polecat Bench in the northern Bighorn Basin, *in* Gingerich, P.D., ed., Paleocene-Eocene stratigraphy and biotic change in the Bighorn and Clarks Fork basins, Wyoming: University of Michigan Papers on Paleontology, v. 33, p. 37–71.

Gingerich, P.D., 2001b, Rates of evolution on the time scale of the evolutionary process, *in* Hendry, A., et al., eds., Contemporary microevolution—rate, pattern, and process: Genetica, v. 112/113, p. 127–144.

Gingerich, P.D., and Clyde, W.C., 2001, Overview of mammalian biostratigraphy in the Paleocene-Eocene Fort Union and Willwood formations of the Bighorn and Clarks Fork basins, *in* Gingerich, P.D., ed., Paleocene-Eocene stratigraphy and biotic change in the Bighorn and Clarks Fork basins, Wyoming: University of Michigan Papers on Paleontology, v. 33, p. 1–14.

Granger, W., 1914, On the names of lower Eocene faunal horizons of Wyoming and New Mexico: Bulletin of the American Museum of Natural History, v. 33, p. 201–207.

Granger, W., 1915, A revision of the lower Eocene Wasatch and Wind River faunas, Part III: Order Condylarthra, families Phenacodontidae and Meniscotheriidae: Bulletin of the American Museum of Natural History, v. 34, p. 329–361.

Gunnell, G.F., 1998, Mammalian faunal composition and the Paleocene-Eocene epoch/series boundary: Evidence from the northern Bighorn Basin, Wyoming, *in* Aubry, M.-P., et al., eds., Late Paleocene–early Eocene climatic and biotic events in the marine and terrestrial records: New York, Columbia University Press, p. 409–427.

Hartman, J.H., and Roth, B., 1998, Late Paleocene and early Eocene nonmarine molluscan faunal change in the Bighorn Basin, northwestern Wyoming and south-central Montana, *in* Aubry, M.-P., et al., eds., Late Paleocene–early Eocene climatic and biotic events in the marine and terrestrial records: New York, Columbia University Press, p. 323–379.

Hickey, L.J., 1980, Paleocene stratigraphy and flora of the Clarks Fork basin, *in* Gingerich, P.D., ed., Early Cenozoic paleontology and stratigraphy of the Bighorn Basin, Wyoming: University of Michigan Papers on Paleontology, v. 24, p. 33–49.

Janis, C.M., 1993, Tertiary mammal evolution in the context of changing climates, vegetation, and tectonic events: Annual Review of Ecology and Systematics, v. 24, p. 467–500.

Jepsen, G.L., 1930, Stratigraphy and paleontology of the Paleocene of northeastern Park County, Wyoming: Proceedings of the American Philosophical Society, v. 69, p. 463–528.

Jepsen, G.L., and Woodburne, M.O., 1969, Paleocene hyracothere from Polecat Bench Formation, Wyoming: Science, v. 164, p. 543–547.

Kinney, K.K., Lindroth, R.L., Jung, S.M., and Nordheim, E.V., 1997, Effects of CO_2 and NO_3 availability on deciduous trees: Phytochemistry and insect performance: Ecology, v. 78, p. 215–230.

Koch, P.L., 1986, Clinal geographic variation in mammals: Implications for the study of chronoclines: Paleobiology, v. 12, p. 269–281.

Koch, P.L., Zachos, J.C., and Gingerich, P.D., 1992, Correlation between isotope records in marine and continental carbon reservoirs near the Paleocene-Eocene boundary: Nature, v. 358, p. 319–322.

Koch, P.L., Zachos, J.C., and Dettman, D.L., 1995, Stable isotope stratigraphy and paleoclimatology of the Paleogene Bighorn Basin (Wyoming, USA): Palaeogeography, Palaeoclimatology, Palaeoecology, v. 115, p. 61–89.

Kraus, M.J., 1980, Genesis of a fluvial sheet sandstone, Willwood Formation, northwest Wyoming, *in* Gingerich, P.D., ed., Early Cenozoic paleontology and stratigraphy of the Bighorn Basin, Wyoming: University of Michigan Papers on Paleontology, v. 24, p. 87–94.

Kraus, M.J., 1987, Integration of channel and flood-plain suites, II: Vertical relations of alluvial paleosols: Journal of Sedimentary Petrology, v. 57, p. 602–612.

Kraus, M.J., 1996, Avulsion deposits in lower Eocene alluvial rocks, Bighorn Basin, Wyoming: Journal of Sedimentary Research, v. 66, p. 354–363.

Kraus, M.J., 2001, Sedimentology and depositional setting of the Willwood Formation in the Bighorn and Clarks Fork basins, *in* Gingerich, P.D., ed., Paleocene-Eocene stratigraphy and biotic change in the Bighorn and Clarks Fork basins, Wyoming: University of Michigan Papers on Paleontology, v. 33, p. 15–28.

Krause, D.W., and Maas, M.C., 1990, The biogeographic origins of late Paleocene–early Eocene mammalian immigrants to the western interior of North America, *in* Bown, T.M., and Rose, K.D., eds., Dawn of the age of mammals in the northern part of the Rocky Mountain interior, North America: Boulder, Colorado, Geological Society of America Special Paper 243, p. 71–105.

Legendre, S., 1989, Les communautés de mammifères du Paléogène (Eocène supérieur et Oligocène) d'Europe occidentale: Structures, milieux et évolution: Münchner Geowissenschaftliche Abhandlungen, Reihe A, Geologie und Paläontologie, v. 16, p. 1–110.

Lemoine, V., 1891, Étude d'ensemble sur les dents des mammifères fossiles des environs de Reims: Bulletin de la Société Géologique de France, v. 19, p. 263–290.

Luterbacher, H., Hardenbol, J.A., and Schmitz, B., 2000, Decision of the voting members of the International Subcommission on Paleogene Stratigraphy on the criterion for the recognition of the Paleocene-Eocene boundary: Newsletter of the International Subcommission on Paleogene Stratigraphy, Tübingen, November 2000, v. 9, p. 13.

Prothero, D.R., 1999, Does climatic change drive mammalian evolution?: GSA Today, v. 9, no. 9, p. 1–7.

Rensch, B., 1929, Das Prinzip geographischer Rassenkreise und das Problem der Artbildung: Berlin, Borntraeger, 206 p.

Rose, K.D., 1979, The Clarkforkian land-mammal "age" and mammalian faunal composition across the Paleocene-Eocene boundary [Ph.D. thesis]: Ann Arbor, Michigan, University of Michigan, 628 p.

Rose, K.D., 1980, Clarkforkian land-mammal age: Revised definition, zonation, and tentative intercontinental correlations: Science, v. 208, p. 744–746.

Rose, K.D., 1981a, The Clarkforkian land-mammal age and mammalian faunal composition across the Paleocene-Eocene boundary: University of Michigan Papers on Paleontology, v. 26, p. 1–197.

Rose, K.D., 1981b, Composition and species diversity in Paleocene and Eocene mammal assemblages: An empirical study: Journal of Vertebrate Paleontology, v. 1, p. 367–388.

Royer, D.L., Wing, S.L., Beerling, D.J., Jolley, D.W., Koch, P.L., Hickey, L.J., and Berner, R.A., 2001, Paleobotanical evidence for near present-day levels of atmospheric CO_2 during part of the Tertiary: Science, v. 292, p. 2310–2313.

Russell, D.E., 1964, Les Mammifères Paléocènes d'Europe: Mémoires du Muséum National d'Histoire Naturelle, Paris, Série C, v. 13, p. 1–324.

Scott, W.B., 1913, A history of land mammals in the Western Hemisphere: New York, Macmillan Company, 693 p.

Simpson, G.G., 1943, Criteria for genera, species, and subspecies in zoology and paleontology: Annals of the New York Academy of Sciences, v. 44, p. 145–178.

Sinclair, W.J., 1912, Contributions to geologic theory and method by American workers in vertebrate paleontology: Geological Society of America Bulletin, v. 23, p. 262–266.

Sloan, R.E., 1970, Cretaceous and Paleocene terrestrial communities of western North America, *in* Yochelson, E.L., ed., Proceedings of the North American Paleontological Convention, Two Volumes: Lawrence, Kansas, Allen Press, p. 427–453.

Sokal, R.R., and Rohlf, F.J., 1981, Biometry (2nd edition): San Francisco, W.H. Freeman, 859 p.

Strait, S.G., 2001, New Wa-0 mammalian fauna from Castle Gardens in the southeastern Bighorn Basin, *in* Gingerich, P.D., ed., Paleocene-Eocene stratigraphy and biotic change in the Bighorn and Clarks Fork basins, Wyoming: University of Michigan Papers on Paleontology, v. 33, p. 127–143.

Strauss, D.J., and Sadler, P.M., 1989, Classical confidence intervals and Bayesian probability estimates for ends of local taxon ranges: Mathematical Geology, v. 21, p. 411–427.

Stucky, R.K., 1990, Evolution of land mammal diversity in North America during the Cenozoic, *in* Genoways, H.H., ed., Current mammalogy, Volume 2: Plenum, New York, p. 375–432.

Tuchman, N.C., Wetzel, R.G., Rier, S.T., Wahtera, K.A., and Teeri, J.A., 2002, Elevated atmospheric CO_2 lowers leaf litter nutritional quality for stream ecosystem food webs: Global Change Biology, v. 8, p. 163–170.

Vrba, E.S., 1985, Environment and evolution: Alternative causes of the temporal distribution of evolutionary events: South African Journal of Science, v. 81, p. 229–236.

Webb, S.D., and Opdike, N.D., 1995, Global climatic influences on Cenozoic land mammal faunas, *in* Stanley, S.M., et al., eds., Effects of past global change on life: Washington, D.C., National Academies Press, p. 184–208.

Wing, S.L., 1998, Late Paleocene–early Eocene floral and climatic change in the Bighorn Basin, Wyoming, *in* Aubry, M.-P., et al., eds., Late Paleocene–early Eocene climatic and biotic events in the marine and terrestrial records: New York, Columbia University Press, p. 380–400.

Wing, S.L., and Harrington, G.J., 2001, Floral response to rapid warming in the earliest Eocene and implications for concurrent faunal change: Paleobiology, v. 27, p. 539–563.

Wing, S.L., Bao, H., and Koch, P.L., 2000, An early Eocene cool period? Evidence for continental cooling during the warmest part of the Cenozoic, *in* Huber, B.T., et al., eds., Warm climates in Earth history: Cambridge, UK, Cambridge University Press, p. 197–237.

Wood, H.E., Chaney, R.W., Clark, J., Colbert, E.H., Jepsen, G.L., Reeside, J.B., and Stock, C., 1941, Nomenclature and correlation of the North American continental Tertiary: Geological Society of America Bulletin, v. 52, p. 1–48.

Wood, R.C., 1967, A review of the Clark Fork vertebrate fauna: Breviora, Museum of Comparative Zoology, Harvard University, v. 257, p. 1–30.

Zachos, J.C., Pagani, M., Sloan, L.C., Thomas, E., and Billups, K., 2001, Trends, rhythms, and aberrations in global climate 65 Ma to present: Science, v. 292, p. 686–693.

Manuscript Accepted by the Society August 13, 2002

Printed in the USA

Geological Society of America
Special Paper 369
2003

Evidence for direct mammalian faunal interchange between Europe and Asia near the Paleocene-Eocene boundary

J.J. Hooker
Department of Palaeontology, The Natural History Museum, Cromwell Road, London, SW7 5BD, UK
D. Dashzeveg
Geological Institute, Mongolian Academy of Sciences, Peace Avenue 63, Ulaanbaatar, Mongolia

ABSTRACT

Dispersal of mammals into Europe at the beginning of the Eocene is conventionally interpreted as being solely from North America via the Greenland land bridge, as other routes were apparently unavailable because of the intervention of major seaways. The Tethys separated Europe from Africa to the south, and the West Siberian Sea and Turgai Straits separated Europe from Asia to the east. Using cladistic analysis and paleogeographic reconstructions, three cases of multiple land mammal dispersal across the Turgai Straits from Asia to Europe in the late Paleocene and early Eocene are recognized. Two of these involve the hyopsodontid condylarths *Lessnessina* (with its synonym *Midiagnus,* and including the species *L. khushuensis* sp. nov.) and a *Hyopsodus* clade (*H. orientalis + H. itinerans*), which most likely entered Europe early in the Eocene at ~54.5 Ma. The third case is of the order Perissodactyla, whose analysis provides an early history of global dispersal for the group. It indicates that their entry into Europe was earlier than for the hyopsodontids, in the latest Paleocene, perhaps at ~56.5 Ma. As the Turgai Straits are at mid latitudes, not high latitudes like the Bering and Greenland land bridges, it is unlikely that ameliorating climate was the driving force behind these three dispersals. The cause was most likely low sea levels, providing either a narrow land bridge or a greatly narrowed strait, allowing rafting.

INTRODUCTION

At the beginning of the Eocene, dispersal was a major phenomenon, radically affecting mammal faunas in all three Northern Hemisphere continents. This event is known as the Mammalian Dispersal Event (MDE; e.g., Hooker, 2000). Reconstructing dispersal routes and directions is relevant to understanding centers of origin of newly appearing groups. It is thought that the warming event known as the Paleocene-Eocene Thermal Maximum (PETM), associated with the Carbon Isotope Excursion (CIE), was the driving force behind these dispersals. This is because dispersal directions were northwards on northern continents, whereupon the high-latitude Bering and Greenland land bridges became available for intercontinental exchange (e.g., Gingerich, 1989; Clyde and Gingerich, 1998).

Most intercontinental exchange is thought to be via these land bridges, allowing movement between Asia and North America on the one hand and between North America and Europe on the other. However, for much of the Eocene direct land connection between Europe and Asia was interrupted by a major epicontinental seaway, extending from the Arctic Ocean to the Peritethys, known as the West Siberian Sea (Iakovleva et al., 2001) or Obik Sea (McKenna, 1975). This sea narrowed at its southern end at the Turgai Straits. It is widely accepted that no land dispersal took place across this barrier until the Oligocene. In fact, the majority (probably about two thirds) of mammalian

Hooker, J.J., and Dashzeveg, D., 2003, Evidence for direct mammalian faunal interchange between Europe and Asia near the Paleocene-Eocene boundary, *in* Wing, S.L., Gingerich, P.D., Schmitz, B., and Thomas, E., eds., Causes and Consequences of Globally Warm Climates in the Early Paleogene: Boulder, Colorado, Geological Society of America Special Paper 369, p. 479–500.

taxa appearing in Europe at the MDE can be readily traced to a North American origin (Hooker, 1998; Beard and Dawson, 1999). Their closest relatives occur only in North America and have in most instances a record in that continent extending back into the Paleocene. Moreover, they have no unequivocal Paleocene record in any other continent. Taxa falling into this category are: the multituberculate *Ectypodus,* peradectid and herpetotheriid marsupials, paramyid rodents, nyctitheriids, oxyaenid creodonts, apatotheres, the tillodont *Esthonyx,* and the "condylarth" genera *Phenacodus* and *Hyopsodus.*

A number of other mammalian taxa have no authenticated Paleocene records anywhere, yet appear around the MDE in North America, Europe and much of Asia. Taxa in this category are: Perissodactyls, artiodactyls, primates, and bats. Hyaenodontid creodonts were formerly also included until their occurrence in the Gashatan Asian Land Mammal Age (ALMA) was demonstrated to be Paleocene (Bowen et al., 2002). The centers of origin for taxa in this category are unknown, although hypothesized at different times to be Africa, Central America, India or other parts of Asia (Gingerich, 1976b, 1989; Hooker, 1998; Krause and Maas, 1990). Although some of these taxa are thought to have dispersed to Europe before moving on to North America, the concept of a direct African or Asian source has been inhibited by the knowledge that the Tethys Ocean and the Turgai Straits respectively formed major long-term barriers to dispersal of land mammals from these continents to Europe.

Critical to Beard and Dawson's (1999) hypothesis that all taxa entering Europe at the MDE came from North America via the Greenland land bridge was the acceptance that no occurrences of these taxa were shared between Europe and Asia exclusive of North America. The present communication documents three such shared occurrences between Europe and Asia, which suggest multiple cases of Asia-Europe interchange with a westerly dispersal direction in the latest Paleocene and earliest Eocene. The method used is cladistic analysis of the taxa concerned, combined with paleogeography (Forey et al., 1992).

Institutional abbreviations

BMNH, Natural History Museum, London; MNHN, Muséum National d'Histoire Naturelle, Paris; UM, University of Michigan Museum of Paleontology; USTL, Université des Sciences et Techniques du Languedoc (Montpellier II).

LESSNESSINA

Lessnessina is an enigmatic mammal genus that was described originally based on upper teeth, from the early Eocene, Neustrian European Land Mammal Age (ELMA), zone PE III site of Abbey Wood, London, UK (Hooker, 1979). Until now it has not been recorded from any other site. However, new finds of a P_4 and an M_2 from Abbey Wood, which occlude well with the uppers, show that the genus *Midiagnus,* described on a lower dentition from the Neustrian PE IV site of Palette, France (Godinot et al., 1987), should be synonymized with *Lessnessina* (Fig. 1). This extends its range to two sites in Europe (see below). A new species of *Lessnessina, L. khushuensis* sp. nov., from the early Eocene Bumbanian site of Tsagan Khushu, Mongolia, extends the range of *Lessnessina* farther from Europe to Asia. The systematics of the three species and the synonymy of *Midiagnus* with *Lessnessina* is established below as a necessary basis for consideration of its dispersal history. Dental terminology follows Hooker (1986).

Systematics

Order "Condylarthra" Cope, 1881
Family Hyopsodontidae Trouessart, 1879
Genus *Lessnessina* Hooker, 1979

Type species: *Lessnessina packmani* Hooker, 1979.

Included species: *Lessnessina praecipuus* (Russell, 1987) comb. nov.; *Lessnessina khushuensis* sp. nov.

Emended diagnosis: Small hyopsodontids, M_2 measuring 2.20–2.55 mm in length. Tooth formula ?/? 1/1 4/4 3/3. P^3-M^2 with "transverse" long axis running slightly distobuccally-mesiolingually. Upper molars with: Centrocrista slightly flexed buccally, meeting mesostyle on M^{1-2}; parastyle not salient buccally. Upper preultimate molars (known only for the type species) with: Protocone massive and with long lingual slope; large metaconule; metacone smaller than paracone. P^{3-4} (known only for the type species) without metacone, massive mesiodistally compressed protocone, taller than paracone, with tall, buccolingually elongate crestiform parastyle and with weak ectocingulum. Lower molars with tall approximated or fused paraconid and metaconid; low entoconid; paracristid with sharp mesiobuccal angle; cristid obliqua joins trigonid at protoconid. P_4 premolariform, without paraconid or metaconid, and with unbasined talonid with single tall terminal cusp.

All other hyopsodontids (sensu herein—see below) have: P^3-M^2 long axis more or less transverse; upper molars with straight centrocrista, smaller metaconule, and equal sized paracone and metacone; lower molars with lower paraconid-metaconid complex, paracristid with rounded mesiobuccal angle, and cristid obliqua with more lingual attachment positions;

Figure 1. Scanning electron micrographs of teeth of species of *Lessnessina,* early Eocene. A–D, F–G, J–K, M–N, *Lessnessina packmani,* Blackheath Beds, Abbey Wood, UK. A, holotype left maxilla with P^2-M^3 (BMNH.M29632); B, paratype right P^4 (reversed) (BMNH.M29756); C, paratype right M^1 (reversed) (BMNH.M44944); D, paratype right M^3 (reversed) (BMNH.M29637); F, J, N, right P_4 (reversed in J and N); G, K, M, right M_2 (reversed in K and M). E, I, O, *Lessnessina khushuensis,* holotype left M_{2-3} (reversed in E) (Geological Institute, Ulaanbaatar, PSS.20–225), Bumban Member, Tsagan Khushu, Mongolia. H, L, P, *Lessnessina praecipuus,* holotype C_1–M_3 (composite of right and left dentitions, shown as right in H, left in L and P) (USTL.PAT7), lignitic marls, Palette, France. Views are occlusal (A–H), buccal (I–L) and lingual (M–P). All images are of gold-coated epoxy casts.

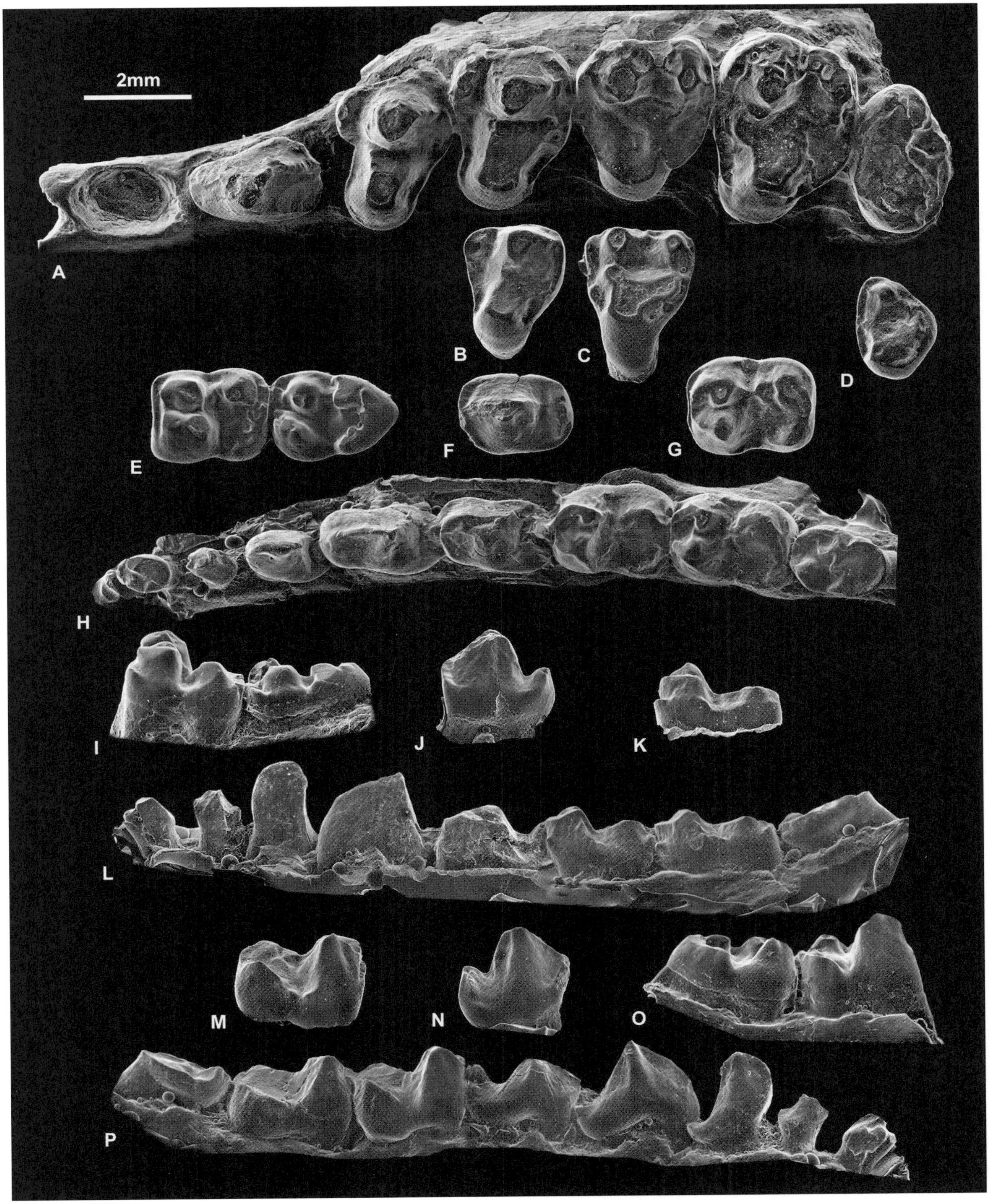
2mm
A
B
C
D
E
F
G
H
I
J
K
L
M
N
O
P

P^{3-4} with smaller protocones and less buccolingually extensive parastyles; and P_4 with paraconid, metaconid and basined talonid.

All other hyopsodontids apart from *Protoselene* lack an upper molar mesostyle. All except *Haplomylus* and *Hyopsodus* have a buccally salient upper molar parastyle. All have lower upper molar protocones and taller lower molar entoconids except *Litaletes,* where only M^2 has a massive protocone and M_2 a low entoconid. All except *Litaletes* have a lower P_4 talonid. *Haplomylus, Dorraletes,* and *Protoselene* have: Semimolariform fourth premolars with small but distinct P^4 metacone and subterminal P_4 hypoconid; and lower molars with low or absent paraconid. *Hyopsodus* has a much smaller P^4 parastyle.

Synonymy: *Midiagnus* Russell, 1987 is here considered to represent the lower dentition of *Lessnessina* Hooker, 1979, which was until now only known from upper teeth. The former is synonymized with the latter and the reasons given under *L. praecipuus.*

Discussion of family and higher relationships: *Lessnessina* was originally placed in the periptychid "condylarth" subfamily Anisonchinae because of similarity to certain anisonchines in the long lingual slope of the upper molar protocone, upper molar metacone smaller than paracone and large P^3 protocone, the last resembling only *Hemithlaeus* and *Anisonchus* (Hooker, 1979). Subsequent suggestions of relationships have been with the Pseudictopidae or other Anagalida (Archibald et al., 1983, p. 55) or with the order Taeniodonta (Estravis and Russell, 1992).

Now that both upper and lower dentitions of *Lessnessina* are known, new evidence can be brought to bear on this issue. On the basis of the lower teeth, *Lessnessina* is clearly not an anisonchine. Members of the Periptychidae retain the lower molar paraconid quite distinct from the metaconid (Matthew, 1937). Moreover, the entoconid is always a prominent subterminal cusp, and the upper molar precingulum extends to the lingual margin of the tooth. The upper premolar and molar protocone enlargements are thus likely to be convergent with the Anisonchinae.

Regarding possible anagalid relationships, it is true that the strong upper molar pre- and postcingula of *Pseudictops* and *Hsiuannania* superficially resemble those of *Lessnessina* (Xu, 1976; Sulimski, 1969), but here the similarity ends. Pseudictopids have transversely elongate upper molar trigons and lower molars with short low talonids to receive them. There is a strong bilophodonty and this is superimposed on a fundamentally tribosphenic plan. There is no "squaring off" of the lower molars that one finds in *Lessnessina* and other "condylarths." Moreover, the lower molar paraconid has undergone significant reduction with no sign of fusion to the metaconid. Anagalids also show important trends in hypsodonty unlike *Lessnessina.*

With the stem taeniodont *Onychodectes,* there are also superficial similarities to *Lessnessina* in the upper dentition, although major differences are the complete absence of pre- and postcingula and strong reduction of the paraconule and metaconule in the former (Schoch, 1986). Once again the greatest differences are in the lower molars. In *Onychodectes,* there is no sign of a shift of the paraconid toward the metaconid. Like the Anagalida, the lower molars are fundamentally tribosphenic, on which plan increase in height of the talonid and general inflation of cusps are superimposed. There is no "squaring off" as found in "condylarths."

The greatest similarities are in fact between *Lessnessina* and certain members of the Hyopsodontidae, thus following Russell in Godinot et al.'s (1987) familial attribution for *"Midiagnus."* Unfortunately, there is little consensus on content of the family. Various Paleocene genera are usually separated at subfamily or family level as Mioclaeninae or Mioclaenidae (e.g., Simpson, 1937; Williamson, 1996). However, it has been argued that the hyopsodontines *Litomylus* and *Haplaletes* are not even ungulates, and that some mioclaenines are ancestral to some hyopsodontines (Rigby, 1980), raising the issues of paraphyly and polyphyly. The family is clearly in need of thorough revision. Given these problems, to try and establish the relationships of *Lessnessina,* a cladistic analysis is conducted here of *Hyopsodus* plus genera that share with it characters that are unique within "condylarths." These are taxa that are here construed to constitute the family Hyopsodontidae. The characters observed are: A, lower molars with tall paraconid close to or fused with metaconid; B, upper first molar precingulum with protostyle and extending only two thirds to three quarters the distance to the lingual tooth wall; C, P_4 with strong paracristid; D, P_4 with strong distal protoconid crest. *Lessnessina* plus two other genera share all of these characters with *Hyopsodus.* All three species of *Lessnessina* and three species of *Hyopsodus* (see later) have been included in the analysis. Three genera that share some but not all of these characters have also been included. *Protungulatum* is chosen as the outgroup taxon as being the generally accepted most primitive ungulate. The character descriptions are in Appendix 1, the data matrix in Appendix 2. The analysis using PAUP 3.1 (Swofford, 1990) produced two maximum parsimony cladograms of 50 steps, a consistency index excluding uninformative characters of 0.644, and a retention index of 0.758. One result shows an apomorphic *Promioclaenus* branching off below a clade dichotomizing into *Lessnessina* + *Litaletes* and *Hyopsodus.* The other shows a plesiomorphic *Promioclaenus* in a trichotomy with the two clades *Lessnessina* + *Litaletes* and *Hyopsodus.* The strict and majority rule consensus results have the same topology as the second maximum parsimony cladogram, which is shown in Figure 2. It demonstrates paraphyly for the Mioclaenidae, some of the genera included by Archibald (1998) in the family having sister group relationships to different members of the Hyopsodontidae. *Litaletes* is sister taxon to *Lessnessina.* It shares with the latter tall P_4 talonid cusps and a tall M^2 protocone that occludes with an M_2 talonid basin with a reduced entoconid. *Lessnessina* has extended this adaptation to all three molars and raised the height of the paraconid-metaconid complex, thus emphasizing both puncture crush function and the late buccal phase of occlusion (e.g., Butler, 1972). Similar heightening of the protocone on P^{3-4} as well

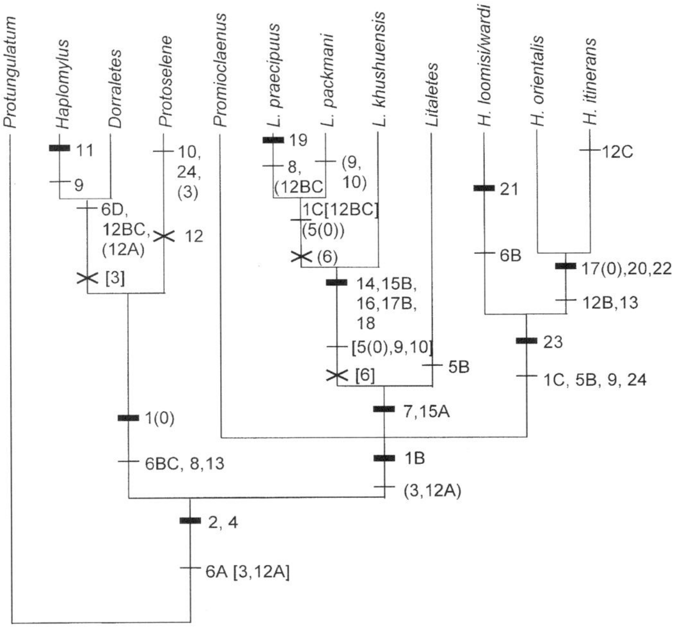

Figure 2. Cladogram of species of *Hyopsodus, Lessnessina,* and their close relatives in the Hyopsodontidae. This is one of two maximum parsimony results, whose topology resembles the strict and majority rule consensuses. Generated by PAUP 3.1 from data matrix in Appendix 2. Broad bar, synapomorphy; narrow bar, normal polarity homoplasy; X, reversal; change to 0 state, where derived, of multistate character indicated by (0). Characters that vary with different optimizations are enclosed between () for DELTRAN and between [] for ACCTRAN. *L.—Lessnessina; H.—Hyopsodus.*

as simple shortening of these teeth is accommodated in the lowers by a simplification, i.e. reduction of the metaconid to a rib on the lingual side of the protoconid and loss of the basin of the talonid.

Lessnessina packmani Hooker, 1979
(Figs. 1A–D, F–G, J–K, M–N, and 3)

v*1979 *Lessnessina packmani* Hooker, p. 50–54, Figure 15.

Holotype: Left maxilla with P^2-M^3, alveolus for P^1 and distal margin of alveolus for canine (BMNH.M29632).

Paratypes: Right P^4 (BMNH.M29756) and right M^1 (BMNH.M44944) possibly associated; right M^3 (BMNH. M29637).

New material: Right dentary fragment with P_4, distal root and mesial alveolus of P_3 and part of alveolus for P_2 (BMNH.M44456); right M_2 (BMNH.M44927).

Horizon and locality: All specimens are from the Lessness Shell Bed, Blackheath Beds, Zone PE III, Neustrian ELMA, early Eocene, Abbey Wood, London, UK.

Emended diagnosis. Lower molars with mesoconid and weak, notched, protocristid; P_4 with mesially sloping paracristid. Like *L. khushuensis* but unlike *L. praecipuus,* lower molars weakly crested, with cuspate entoconid, strong mesial cingulum. Like *L. praecipuus,* but unlike *L. khushuensis,* lower preultimate molar with hypoconulid subsumed by strong posthypocristid, talonid more than half tooth length, postcingulid reduced, and paraconid fused to metaconid; third molar much smaller than second.

***Comparison of new material with* L. praecipuus:** The dentary fragment with P_4 (BMNH.M44456) has a distinctly convex lateral surface and is about twice as deep as the P_4 crown is tall. It shows no sign of shallowing anteriorly to where P_3 was once rooted, and there is a tiny mental foramen below the junction that would have existed between these two teeth, as in *L. praecipuus* (Godinot et al., 1987, pl. 2, figs j–k). There is a slight medial curvature anteriorly but no sign of a symphysis. The P_4 is 2.20 mm long by 1.65 mm wide, thus slightly shorter and broader than that of *L. praecipuus.* It is a simple tooth dominated by a large protoconid with a lingual rib that is slightly less salient than in *L. praecipuus* (Fig. 1F, J, and N). The unicuspid talonid is just over half the height of the slightly tip-worn protoconid and is not as tall as that of *L. praecipuus,* unless the latter is enhanced by the evident damage in this area (Godinot et al., 1987, pl.2, figs h–i). A very faint crest bridges the valley between the talonid and the distal crest of the protoconid. There is a strong precingulid represented on both buccal and lingual sides of the tooth as in *L. praecipuus.* The clearest difference from the latter species is in the angle of slope of the paracristid. This is ~45° in BMNH.M44456, but estimated at ~20°–30° in *L. praecipuus.* The latter has the tip of the protoconid broken and both teeth are slightly worn. However, the slightly greater height of the front of the tooth in *L. praecipuus* versus *L. packmani,* despite heavier wear of the former, testifies to the stronger cresting of the paracristid in *L. praecipuus.*

The M_2 (BMNH.M44927) is 2.55 mm in length, 2.10 mm in trigonid width and 1.95 mm in talonid width, thus relatively very slightly broader than the same tooth in *L. praecipuus.* It is judged to be an M_2 rather than an M_1 because the talonid is slightly narrower than the trigonid. It is very similar morphologically to M_2 in *L. praecipuus,* but differs in the following ways (Fig. 1G, K, and M). The precingulid is slightly stronger. The entoconid is unworn and difficult to compare with the worn one in *L. praecipuus,* but forms a distinct cusp with an entocristid, whereas in the latter there is no obvious buccal salient interrupting the low crest bordering the tooth distolingually. Despite slightly greater tip wear on the protoconid than on the fused paraconid-metaconid, the latter is clearly the taller cusp with a shallower slope lingually than that of the protoconid buccally, so that the paraconid-metaconid tip is nearly at a median point on the tooth when viewed occlusally. Viewed distally, the weak protocristid describes a deep V, meeting the weak buccal cristid obliqua at the base of the distal slope of the protoconid. The talonid basin has a low bump in the middle adjacent to and virtually joining both the mesoconid and hypoconid; the bump bears a lingual phase wear facet on its surface. *L. praecipuus* appears not to have such a bump. It is not clear how significant it is taxonomically, although a similar but slightly more lingually situated one exists in *L. khushuensis* sp. nov.

Lessnessina praecipuus (Russell, 1987) comb. nov.
(Fig. 1H, L, and P)

v*1987 *Midiagnus praecipuus* Russell in Godinot et al., p. 281, pl. 2, Figures g–k.

Holotype: Associated left and right dentaries with left P_2-M_3 and right canine—M_2 (USTL.PAT7), from lignitic marls between the Calcaire de Langesse and Calcaire de Montaiguet, Zone PE IV, Neustrian ELMA, early Eocene, Palette, Bouches-du-Rhône, France.

Emended diagnosis: Lower molars strongly crested, with crestiform entoconid, strong notched protocristid, and weak mesial cingulum; P_4 with high near horizontal paracristid. Like *L. packmani,* but unlike *L. khushuensis,* lower preultimate molars with paraconid fused to metaconid, hypoconulid subsumed by strong posthypocristid, talonid more than half tooth length, postcingulid reduced; third molar much smaller than second. Like *L. khushuensis,* but unlike *L. packmani,* mesoconid absent.

Reasons for synonymy of Midiagnus with Lessnessina: The new M_2 from Abbey Wood attributed to *Lessnessina packmani* occludes well with the M^1 and M^2 in a unique way (Fig. 3). In the following discussion, wear facets are numbered according to Butler (1973).

Normally in hyopsodontids, the upper and lower molar cusps interlock only in early wear stages. Abrasion quickly produces nearly flat attritional surfaces. In *L. packmani,* the increased height of the upper molar protocone and of the lower molar paraconid-metaconid means that interlocking lasts for more of the life of the teeth (e.g., there is still much relief on M^1 of the holotype even though the enamel pattern of the trigon has been completely worn away). The entoconid, although low, has retreated distally leaving a wide open talonid notch lingually for reception of the protocone. In fact there are large and deep facets on the mesial face of the protocone and distal face of the metaconid, testifying to the majority of translatory wear being late in the buccal phase and predominantly prevallum-postvallid (see e.g., Butler, 1973; Crompton and Kielan-Jaworowska, 1978). In contrast, the wear in early buccal phase is minor, the facets on buccal crests being narrow and difficult to discern except on the upper molar precingulum. Evidence of lingual phase wear is largely obliterated by the heavy cusp tip and crest edge abrasion caused by puncture crush. However, it can be best seen on the M_2, where the protocone of M^2 has produced lingual phase facets (10) on the lingual edges of the mesoconid, hypoconid, posthypocristid and raised bump in the talonid basin. It is the postcingulum of M^1 on the other hand that has produced the lingual phase facets on the lingual edge of the M_2 paracristid (5). A group of facets have an angle intermediate between the buccal and lingual phase ones. They are labeled here with Greek letters. They imply a short intermediate phase between the buccal and lingual.

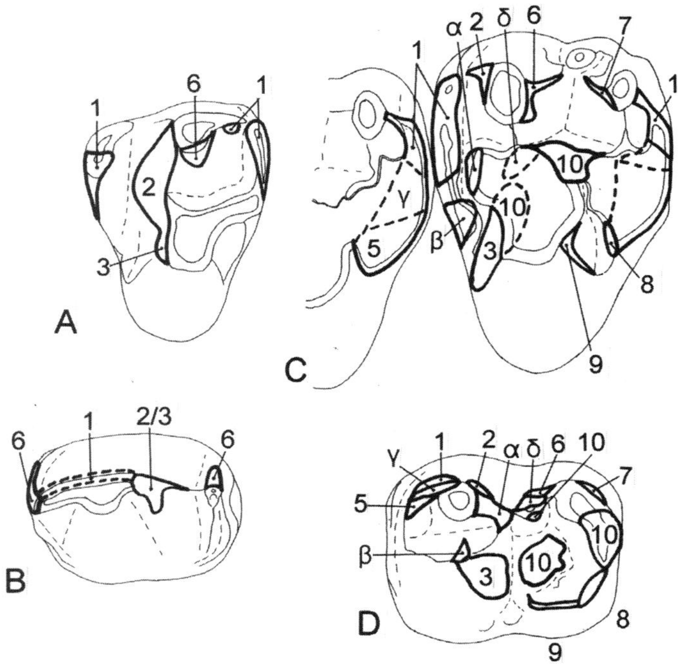

Figure 3. Wear facets of upper (A) and lower (B) fourth premolars and upper first (part) and second (C) and lower second (D) molars of *Lessnessina packmani,* demonstrating matching occlusion of "*Midiagnus,*" based on lowers, with *Lessnessina,* based on uppers. Numbering system for buccal and lingual phase facets follows Butler (1973). Greek letters used for upper and lower homologous molar facets of intermediate angle between buccal and lingual as discussed in text.

Occlusion between P_4 and P^4 is best seen on *L. packmani* specimens M44456 and M44944 respectively. On the P^4, a large facet extends on the mesial face from the paracone to the protocone, via the preprotocrista that links the two cusps (2–3). The large parastyle wraps over the protoconid tip and bears an extensive facet (1). The postparacrista, metacingulum and possibly the short postprotocrista of P^3 make contact with the P_4 paracristid, whilst the P^3 paracone makes a very small area of contact with the P_4 precingulid (6). These P^{3-4} facets make contact only with the longitudinal crest edge of the P_4 trigonid. The lingual margin of the P^4 paracone bears a facet, which occludes with one on the buccal face of the P_4 talonid cusp (6). There is no evidence of lingual phase wear on the premolars. The main function of the massive P^{3-4} protocone is thus mainly puncture crushing.

The P_4 and M_2 of *Midiagnus praecipuus* are so similar to those of *Lessnessina packmani* as to leave little doubt as to the congeneric status of the two, *Lessnessina* having priority. The entire lower tooth rows of *Midiagnus* provide a pattern of cusps, crests and facets that make for accurate occlusal matching with *Lessnessina.* The only differences are minor ones of degree of cresting etc., as borne out in the specific diagnoses. The upper molars of *L. praecipuus,* if found, would be expected to have a more buccally flexed centrocrista, which, together with the postmetacrista and preprotocrista, would also be more strongly crested.

Direct manipulated occlusion between the holotype dentitions of *L. packmani* and *L. praecipuus* is not strictly possible because of slight size differences and because of diagenetic foreshortening and offsetting of some of the teeth of the latter. However, in addition to the distal premolar and preultimate molar occlusion already discussed, it can be seen that the long high P_3 paracristid of *L. praecipuus* matches well with the long P^2 postparacrista of *L. packmani*. Also, the reduced M_3 of *L. praecipuus* fits well against the similarly reduced M^3 of *L. packmani.*

Lessnessina khushuensis sp. nov.
(Fig. 1E, I, and O)

Holotype: Left dentary fragment with M_{2-3} (Geological Institute, Ulaanbaatar, PSS.20–225) from the base of the Bumban Member, Naran Bulak Formation, Bumbanian, early Eocene, Tsagan Khushu, near Naran Bulak, Mongolia.

Etymology: Named for the type locality Tsagan Khushu.

Diagnosis: Lower molars with paraconid separate from metaconid; lower preultimate molars with distinct hypoconulid and weak posthypocristid; talonid less than half tooth length; postcingulid strong, joining hypoconulid, protocristid absent; third molar subequal in size to second molar. Like *L. packmani,* but unlike *L. praecipuus,* lower molars weakly crested, with cuspate entoconid, and strong mesial cingulum. Like *L. praecipuus,* but unlike *L. packmani,* lower molar mesoconid absent.

Description and comparisons: M_2 is 2.20 mm in length, 1.80 mm in trigonid width and 1.75 mm in talonid width. M_3 is 2.30 mm in length, 1.70 mm in trigonid width and 1.50 mm in talonid width. Most of the morphological features are covered in the diagnosis. Additional points are made here. The teeth are moderately worn, but the trigonid cusps seem to be taller than those of either *L. packmani* or *L. praecipuus.* Certainly, there is less overhang of the protoconid by the paraconid-metaconid complex. In M_2 the weak cristid obliqua is bowed buccally, notched in the middle and joins the trigonid at the back of the protoconid. There is an obliquely orientated low bump in the talonid basin, behind the metaconid but close to the entoconid, similar to *L. packmani.*

M_3 is difficult to compare with the other species. The tooth type is otherwise only known in *L. praecipuus,* where heavy wear has removed most features, and is most easily interpreted by comparison with *L. khushuensis.* The cusps are all lower than on M_2. The protoconid and paraconid-metaconid complex are more widely separated than on M_2 and the deep valley separating them passes uninterrupted into the talonid basin. The weak paracristid descends steeply to the mesial tooth margin to meet a strong precingulid. The paraconid and metaconid are only separated at their tips. A bump in the talonid basin is situated between the hypoconid and entoconid and has a narrow link to the latter. The hypoconulid is a weak cusp on the distally bowed postcristid, which forms only a slightly bulging hypoconulid lobe. The *L. praecipuus* M_3 is similar in structure, but differs mainly in having an accessory crest trending toward but stopping short of the metaconid.

Dispersal route and direction

There is no record of the genus *Lessnessina* from the well-sampled early Eocene deposits of North America. This indicates direct interchange between Europe and Asia without the intervention of North America, implying that the Turgai Straits were the dispersal route.

To establish direction, it is necessary to orient the relevant part of the cladogram in Figure 2 on a paleogeographic map. The nearest sister taxon to *Lessnessina* is the North American Paleocene genus *Litaletes. Litaletes* has M_3 and M_2 subequal and a paraconid on M_2 and M_3. Therefore, *L. packmani* and *L. praecipuus* are more derived than *L. khushuensis* sp. nov. in having fused the lower molar paraconid and metaconid, and reduced the size of M_3 (Fig. 2). *L. khushuensis* may have some autapomorphies preventing it from being a candidate for direct ancestry of the European species. However, individual variation cannot be assessed and principal character polarity suggests dispersal from east to west (Fig. 4).

HYOPSODUS

Our second example involves the hyopsodontid *Hyopsodus* Leidy, 1870. This genus occurs in the early Eocene in North America, Asia and Europe. It was cited in the Introduction as an example of an immigration into Europe from North America, on the basis of a first occurrence in the late Clarkforkian North American Land Mammal Age (NALMA) of the Bighorn Basin, Wyoming, USA (Rose, 1981). Europe contains two species, *H. wardi* Hooker, 1979 from the London and Paris Basins and *H. itinerans* Godinot, 1978 from the southern French site of Rians. Both are similarly small, but morphologically rather different from one another as noted by Godinot (1981). Compared to *H. wardi, H. itinerans* is more strongly crested, has an M_3 smaller than M_2, M_{1-2} hypoconulid slightly closer to the entoconid, smaller P_4 with shorter, narrower talonid, and no lower molar entocristid. It shares these characters with *H. orientalis* Dashzeveg, 1977 from the Bumbanian ALMA of Tsagan Khushu, near Naran Bulak, Mongolia (Figs. 6–9). In contrast, *H. wardi* is morphologically very similar to the slightly larger, earliest and most primitive North American species, *H. loomisi* McKenna, 1960 (Figs. 5–8). Upper teeth are less well known for *H. itinerans,* but *H. orientalis* has a similarly reduced M^3 and a mesiodistally shorter P^4 with projecting parastyle, which occlusally matches the short talonid of P_4 (Figs. 5E–G, 6G). The considerable transverse width of M^2 in *H. itinerans* may be an autapomorphy of this species (Fig. 5F).

It should be noted that species of *Hyopsodus* are notoriously variable intraspecifically in morphology (Redline, 1997), and contemporaneous North American species are usually distinguished on size (e.g., Gingerich, 1976a). In fact, in the large assemblages of North American species, it is possible to find variants with one or several of the characters of the Eurasian *H. itinerans* and *H. orientalis,* although these are uncommon. *H. itinerans* and *H. orientalis* are consistent for their defining characters (Godinot, 1981; Kondrashov and Agadjanian, 1999), and, even though assemblage sizes are small, the character states represented are likely to be in the vicinity of the mode, not at the

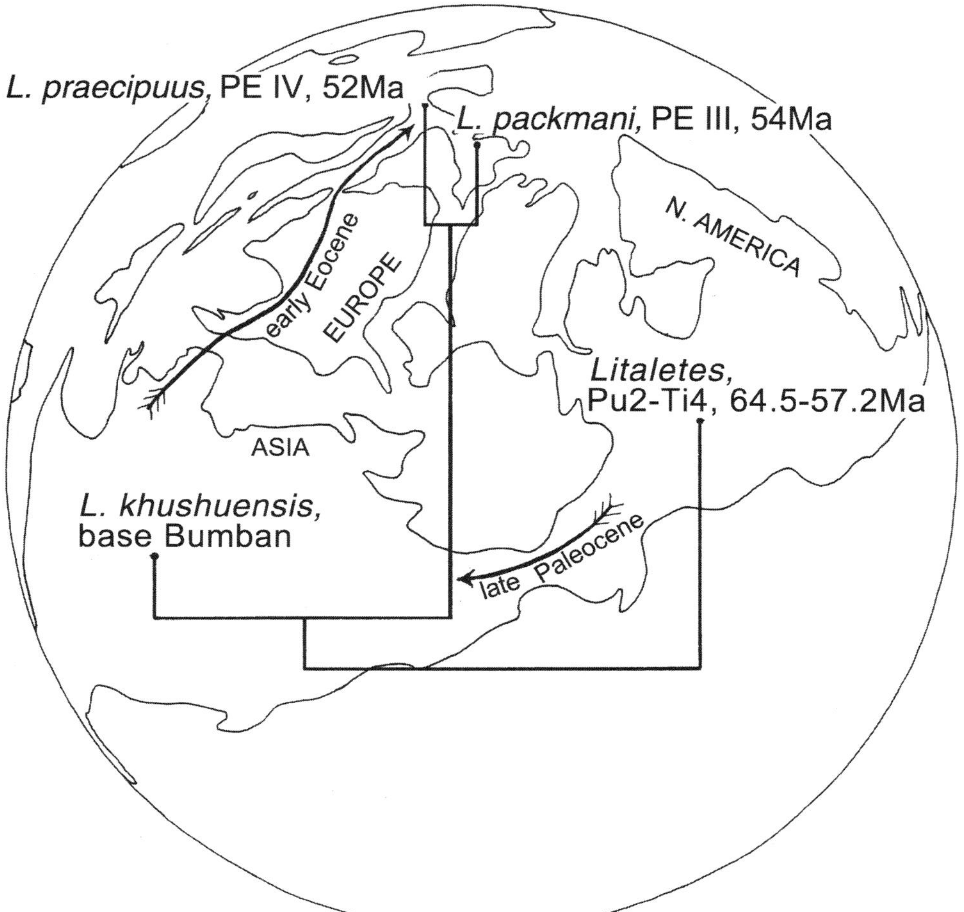

Figure 4. Early Eocene polar projection Northern Hemisphere paleogeographic shoreline map, with superimposed cladogram of *Lessnessina* species from Figure 2 with zones, time ranges in Ma where known and proposed dispersal directions (arrows) (map from unpublished work by Paul Markwick, see Markwick et al., 2000, with minor modifications to North Sea–Arctic Ocean link following Iakovleva et al., 2001).

margins of variation. Figure 9 shows lengths and widths of M_2 and M_3 in the four species concerned. The values for M_3 of *H. itinerans* and *H. orientalis* lie well outside the range of the large assemblage of *H. loomisi* from Wa0 (Gingerich, 1989), whereas those for M_2 nearly coincide. Therefore, *H. itinerans* and *H. orientalis* consistently have relatively smaller lower third molars than do *H. loomisi* or *H. wardi.*

The many other species of North American *Hyopsodus* are not considered here as they are recognized to be derivatives of *H. loomisi* (Gingerich, 1974, 1976a, 1985; Redline, 1997). All are more derived than *H. wardi, H. itinerans* or *H. orientalis* in features such as large M^{1-2} hypocone, hypocone on M^3, and fusion of M_{1-2} paraconid to metaconid. Their relationships are thus judged to be relatively remote from *H. itinerans* or *H. orientalis.*

The cladogram in Figure 2 shows southern French *H. itinerans* to be more closely related to Asian *H. orientalis* than to southern English *H. wardi* and North American *H. loomisi,* which were combined for the analysis. In fact, *H. wardi* is only distinguishable from *H. loomisi* on size (Fig. 9; Hooker, 1979; Gingerich, 1989). On the basis of both an extensive and diverse North American Paleocene record of the family Hyopsodontidae and of the earliest record of the genus *Hyopsodus* being from that continent, one can envisage initial dispersal from North America, probably at the beginning of the Eocene, in two directions. One of these was to Asia via the Bering Straits and the other to northern Europe via Greenland. Speciation followed with subsequent dispersal from Asia to central Europe via the Turgai Straits a little later in the early Eocene and at least by Zone PE III time (Fig. 10).

PERISSODACTYLS

The first two examples discussed here are members of a family that has a known record back through most of the Paleocene. Our third example relates to an order that is not certainly known before the Eocene and has no very close relatives in the Paleocene and thus no clear center of origin. A cladistic analysis of the most primitive members of the order and of their closest nonperissodactyl relatives is carried out here. Their relationships in terms of paleobiogeography provide a hypothesis on the nature and place of their origins and subsequent dispersal.

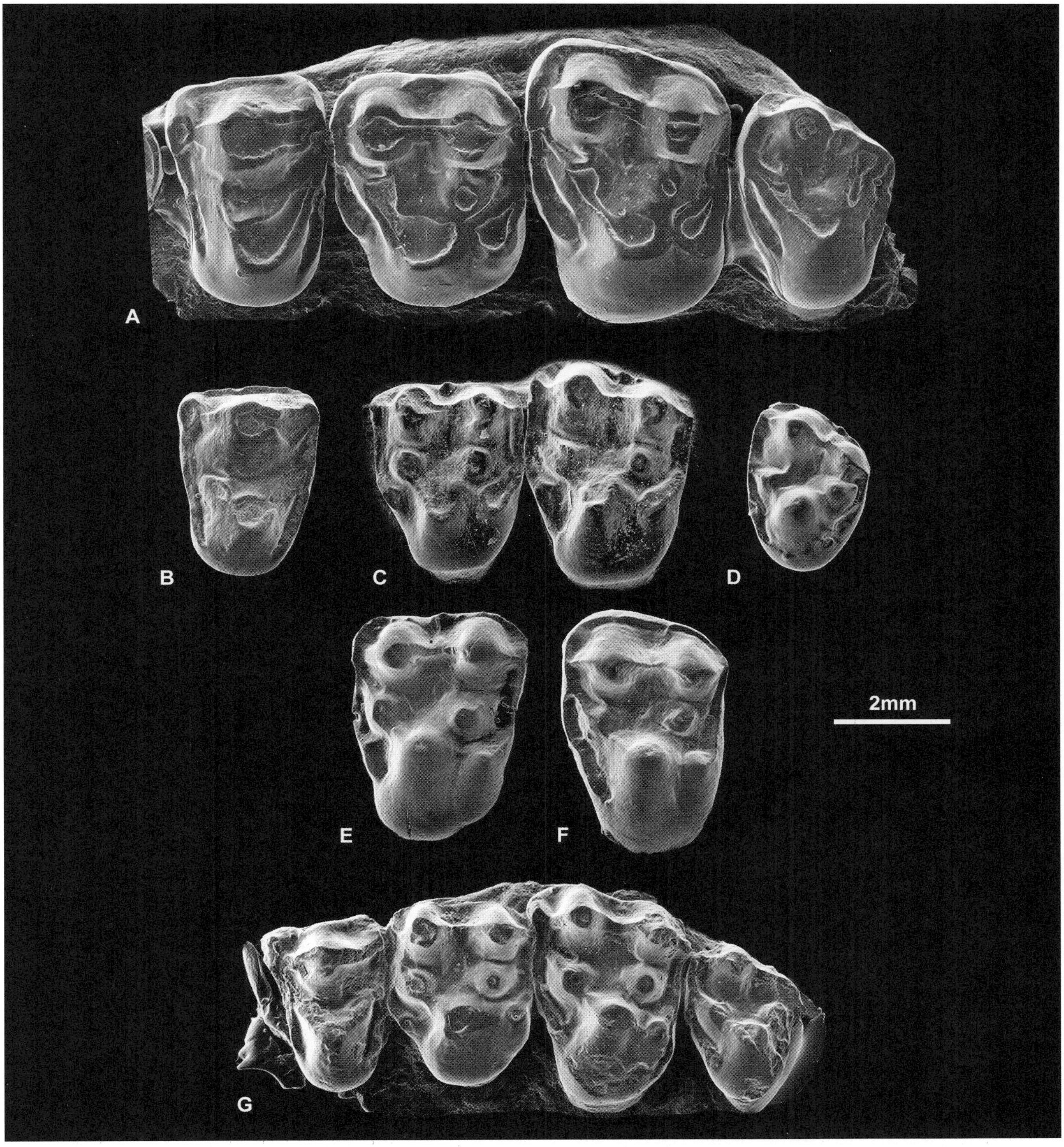

Figure 5. Scanning electron micrographs of occlusal views of upper cheek teeth of species of *Hyopsodus,* early Eocene. A, *Hyopsodus loomisi,* left P^4-M^3 (UM.63620), Willwood Formation, locality 383, Bighorn Basin, Wyoming, USA. B–D, *Hyopsodus wardi,* Blackheath Beds Abbey Wood, UK; B, paratype left P^4 (BMNH.M44946); C, holotype right M^{1-2} (reversed) (BMNH.M29762); D, paratype right M^3 (reversed) (BMNH.M29644). E–F, *Hyopsodus itinerans,* Sables Bleutés, Rians, France; E, holotype left M^1 (MNHN.RI 279); F, right M^2 (reversed) (MNHN.RI 157). G, *Hyopsodus orientalis,* left P^4-M^3 (Geological Institute, Ulaanbaatar, PSS 20–226), Bumban Member, Tsagan Khushu, Mongolia. All images are of gold-coated epoxy casts.

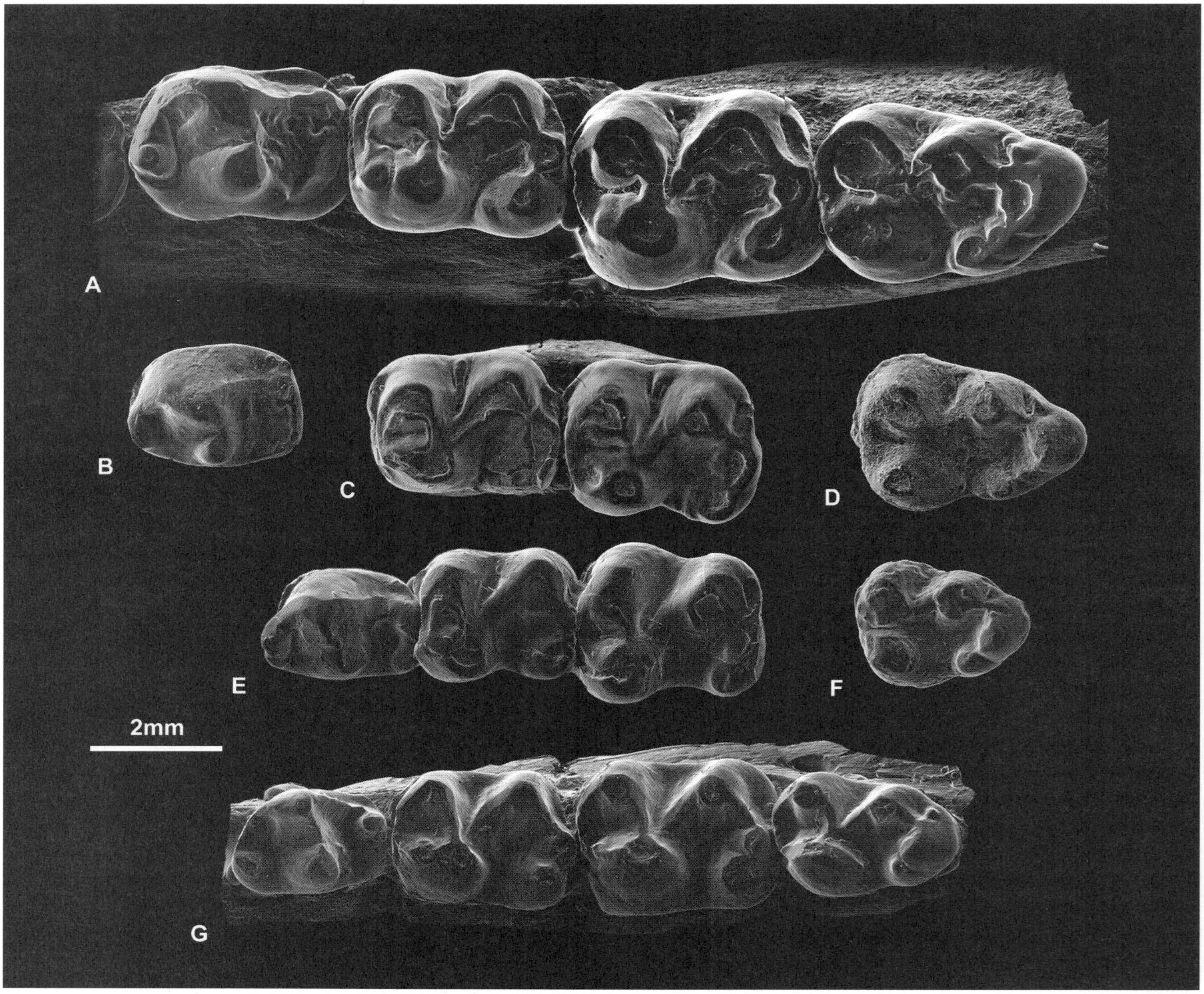

Figure 6. Scanning electron micrographs of occlusal views of lower cheek teeth of species of *Hyopsodus,* early Eocene. A, *Hyopsodus loomisi,* right P_4-M_3 (UM.63621), Willwood Formation, locality 383, Bighorn Basin, Wyoming, USA. B–D, *Hyopsodus wardi,* Blackheath Beds, Abbey Wood, UK; B, paratype right P_4 (BMNH.M29757); C, paratype right M_{1-2} (BMNH.M15146); D, paratype right M_3 (BMNH.M15132). E–F, *Hyopsodus itinerans,* Sables Bleutés, Rians, France; E, left P_4-M_2 (reversed) (original Loggia Collection, cast BMNH.M42629); F, right M_3 (MNHN.RI 280). G, *Hyopsodus orientalis,* paratype left P_4-M_3 (reversed) (Geological Institute, Ulaanbaatar, PSS.20–46), Bumban Member, Tsagan Khushu, Mongolia. All images are of gold-coated epoxy casts.

Methods

The list of taxa includes several relatively recently described Asian genera. Three phenacodontid genera, *Ectocion, Lophocion* and *Phenacodus,* are also included in the analysis (Appendix 4). The last of these is used as the outgroup, for which some justification needs to be made.

The Chinese genus *Radinskya* has been suggested as an outgroup for perissodactyls because of its incipient perissodactyl-like bilophodonty (McKenna et al., 1989; Beard, 1998; Froehlich, 1999). Although only upper teeth are known for *Radinskya,* these show more similarity to phenacolophids than to perissodactyls. Examples of the relevant derived characters involve the upper molars: Crestiform parastyle; and tall metaconule close to the hypocone, lying exactly on a line drawn between the metacone and hypocone. The perissodactyl-like bilophodonty is, however, more similar to derived members of the suborder Ceratomorpha, than to primitive perissodactyls

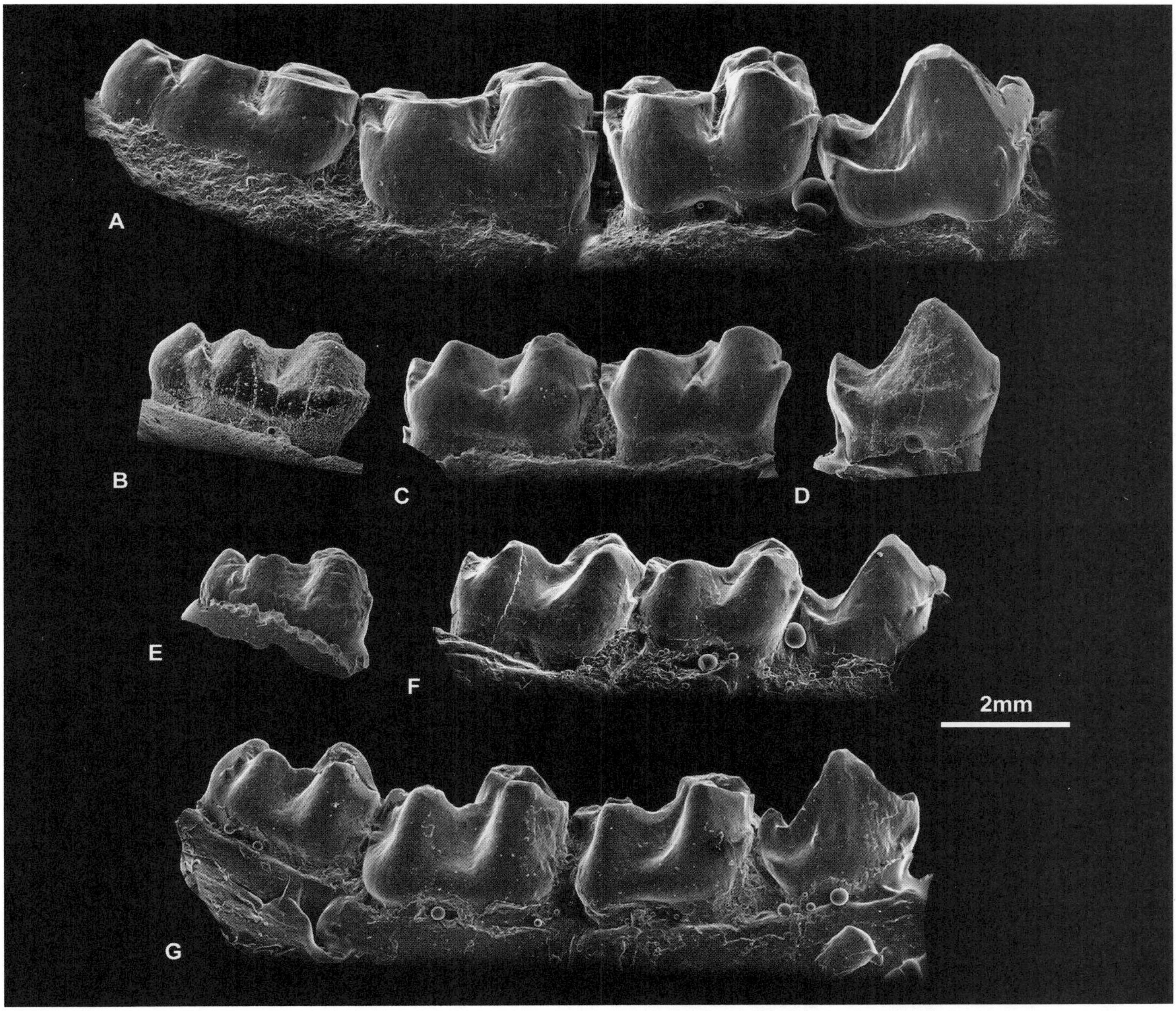

Figure 7. Scanning electron micrographs of buccal views of lower cheek teeth of species of *Hyopsodus,* early Eocene. A, *Hyopsodus loomisi,* right P_4-M_3 (UM.63621), Willwood Formation, locality 383, Bighorn Basin, Wyoming, USA. B–D, *Hyopsodus wardi,* Blackheath Beds, Abbey Wood, UK; B, paratype right M_3 (BMNH.M15132); C, paratype right M_{1-2} (BMNH.M15146); D, paratype right P_4 (BMNH.M29757). E–F, *Hyopsodus itinerans,* Sables Bleutés, Rians, France; E, right M_3 (MNHN.RI 280); F, left P_4-M_2 (reversed) (original Loggia Collection, cast BMNH.M42629). G, *Hyopsodus orientalis,* paratype left P_4-M_3 (reversed) (Geological Institute, Ulaanbaatar, PSS.20–46), Bumban Member, Tsagan Khushu, Mongolia. All images are of gold-coated epoxy casts.

where the metaloph is based around a small, mesially shifted metaconule. Phenacodontids, on the other hand, have long been regarded as close relatives of perissodactyls because of phenetic similarity (e.g., Radinsky, 1966), although also to other pantomesaxonians (Thewissen and Domning, 1992). *Phenacodus* is a relatively generalized member of the family and for that reason is chosen here as outgroup.

The 39 characters and the descriptions of their states are listed in Appendix 3. They are slightly modified from those used solely for equoids by Hooker (1994). As in this last work, they frequently combine characters of upper and lower teeth that are linked by occlusal relationships in an attempt to reduce the incidence of inadvertent weighting. Characters that are judged to form transformation series are treated as ordered and

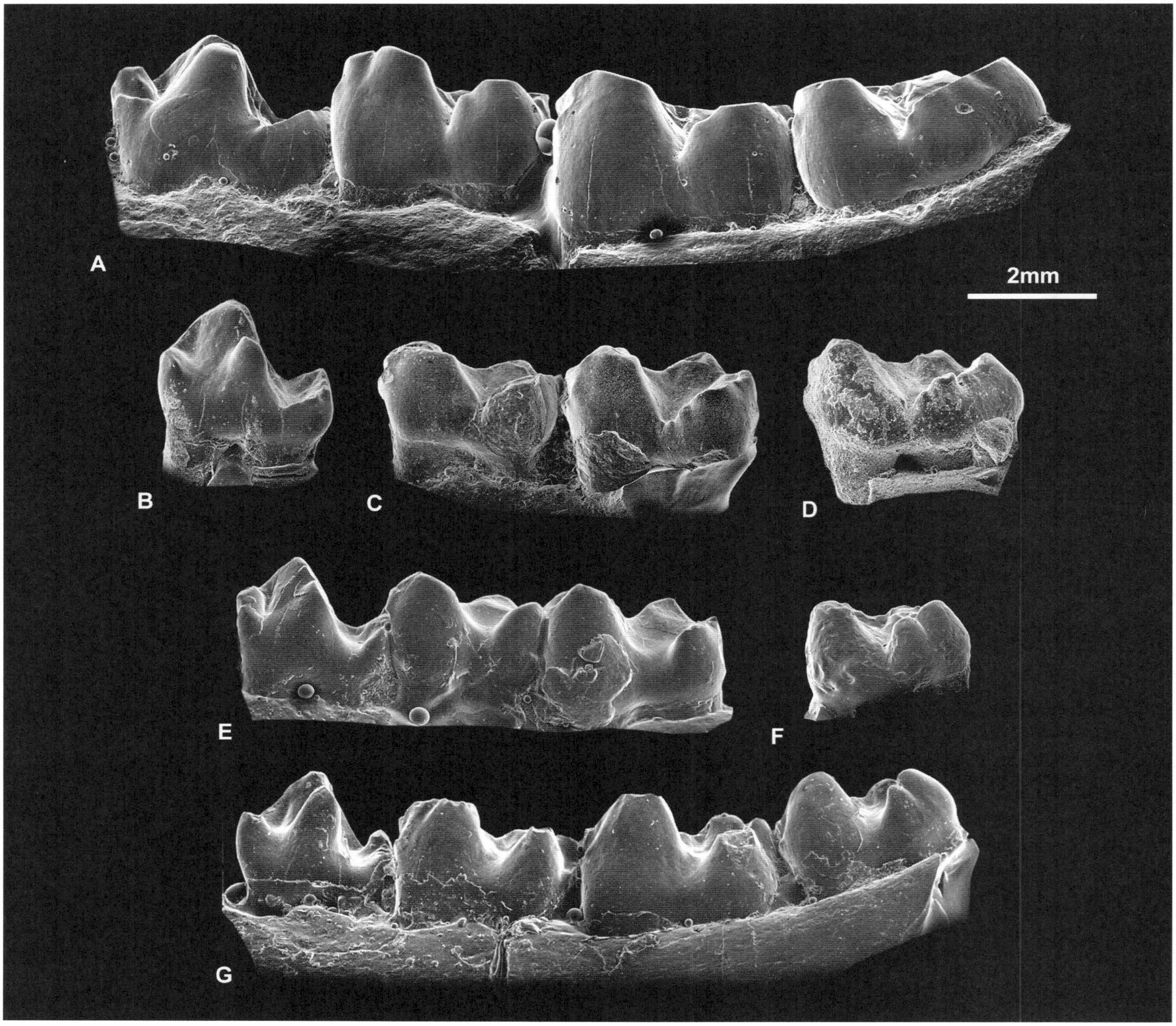

Figure 8. Scanning electron micrographs of lingual views of lower cheek teeth of species of *Hyopsodus,* early Eocene. A, *Hyopsodus loomisi,* right P_4-M_3 (UM.63621), Willwood Formation, locality 383, Bighorn Basin, Wyoming, USA. B–D, *Hyopsodus wardi,* Blackheath Beds, Abbey Wood, UK; B, paratype right P_4 (BMNH.M29757); C, paratype right M_{1-2} (BMNH.M15146); D, paratype right M_3 (BMNH.M15132). E–F, *Hyopsodus itinerans,* Sables Bleutés, Rians, France; E, left P_4-M_2 (reversed) (original Loggia Collection, cast BMNH.M42629); F, right M_3 (MNHN.RI 280). G, *Hyopsodus orientalis,* paratype left P_4-M_3 (reversed) (Geological Institute, Ulaanbaatar, PSS.20–46), Bumban Member, Tsagan Khushu, Mongolia. All images are of gold-coated epoxy casts.

some are entered as stepmatrices to reduce or avoid duplication and consequent potential weighting of primitive states (Lipscomb, 1992). Characters are dominantly dental because of the limited representation of other parts of the anatomy in these early perissodactyls.

Results

The data matrix in Appendix 4 was analyzed using PAUP 3.1 (Swofford, 1990) and resulted in three maximum parsimony cladograms of 100 steps, with a consistency index excluding un-

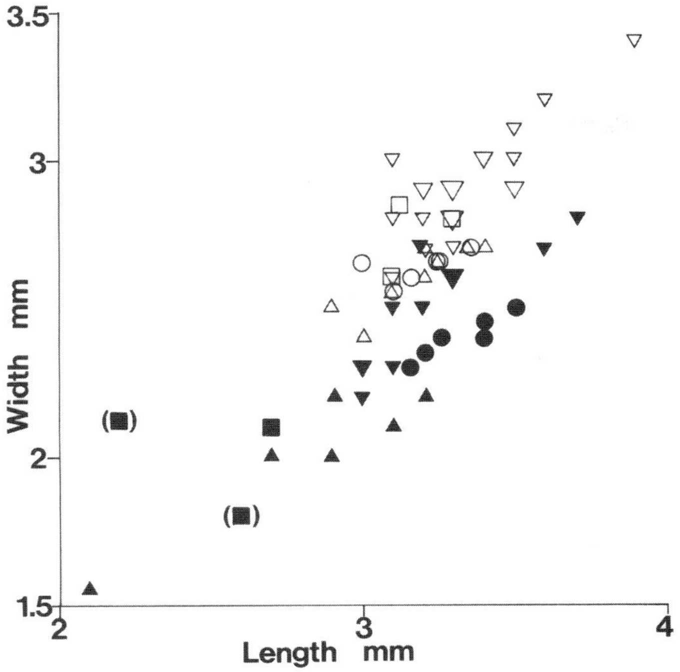

Figure 9. Scatter diagram of length versus width in millimeters of second and third lower molars of species of *Hyopsodus*. ○—*H. wardi*, □—*H. itinerans*, △—*H. orientalis*, ▽—*H. loomisi*. Hollow symbol—second molar, solid symbol—third molar. Three successive increasing sizes of inverted triangle symbol indicate 1, 2, and 3 coincident plots respectively. Bracketed symbols indicate estimated measurements. Plots of *H. orientalis* partly taken from measurements in Kondrashov and Agadjanian (1999). Those of *H. loomisi* from measurements provided by P.D. Gingerich.

informative characters of 0.600 and a retention index of 0.746 (Fig. 11). All three topologies show *Ectocion* and *Lophocion* branching off successively from the outgroup. *Lophocion,* with its incipient metaloph and reduced metaconule on upper molars is a step nearer perissodactyls than is *Ectocion.* Both these phenacodontids have bulbous upper molar parastyles like those of perissodactyls, and ones that are enlarged over those of *Phenacodus.* They are morphologically unlike those of *Radinskya.*

All three topologies also show *Lambdotherium* and its sister taxon *Danjiangia* branching off next. *Danjiangia* was described as a primitive chalicothere by Wang (1995). *Danjiangia* lacks the complete high upper premolar and upper molar metaloph joining the ectoloph, and the absence of upper molar mesostyle of primitive chalicotheres such as *Paleomoropus* and *Lophiaspis* (Radinsky, 1964; Depéret, 1910; Savage et al., 1966). In contrast, it has the strong dilambdodonty (especially the buccally deflected upper molar preparacrista) of *Lambdotherium* (e.g., Osborn, 1929).

Above this node, topologies fall into two main categories, those that show equoids to be paraphyletic (Fig. 11B–C) and that which shows them to be monophyletic (Fig. 11A). The paraphyletic alternatives differ in the relationships of *Karagalax* plus *Pachynolophus* to *Orientolophus* and *Cardiolophus.* When time ranges of taxa that can be accurately calibrated (as in Europe and North America) are considered, these fit better with the order of nodes in the monophyletic equoid alternative. The paraphyletic equoid alternatives show equoid ghost lineages inversely proportional in length to their closeness of relationships. Other lines of evidence also favor the monophyletic equoid alternative. The functionless vestigial upper molar mesostyle of the otherwise bilophodont *Cardiolophus* indicates a rapid shift from dilambdodonty to bilophodonty early in perissodactyl evolution (Hooker, 1989). Moreover, *Hyracotherium,* with its posterior optic foramen and incipiently deepening narial incision is best viewed as the initiation of the family Palaeotheriidae (Hooker, 1994) rather than as the first step away from dilambdodonty. Froehlich's (1999) analysis showed *Cymbalophus* as a tapiromorph, not an equoid, and contrasts with the results obtained here. This may be a shortcoming of not combining clearly linked upper and lower tooth characters, so inadvertently weighting certain features.

The maximum parsimony cladogram that best fits the stratigraphic ranges of taxa, by having the shortest ghost ranges, is shown with character state changes in Figure 12. It shows the next branch to split off above the brontotheres to be the family Isectolophidae, comprising *Cardiolophus* and *Orientolophus.* Above this is a clade consisting of equoids (with *Cymbalophus* at its stem) and *Pachynolophus* plus *Karagalax.* The main characters defining this entire clade are: Reduction of the upper molar parastyle, lengthening of upper molar postmetacrista and of occlusally related lower molar paracristid, and lengthening of the postcanine diastema. These characters are shared by both equoids and ceratomorphs. *Pachynolophus* and *Karagalax* lack equoid characters but share a tall P_3 paraconid and are interpreted here as very primitive ceratomorphs. This contrasts with Maas et al. (2001) who placed *Karagalax* in the Isectolophidae.

Figure 13 shows this cladogram placed on an early Eocene paleogeographic map, and Figure 14 the postulated dispersal events that are listed below:

1. A crossing of the Bering Straits by *Ectocion* in the late Paleocene leading, via *Lophocion,* to origin and initial radiation of the perissodactyls.
2. Dispersal to Europe, via the Turgai Straits in the latest Paleocene.
3. Origin of equoids in Europe before the beginning of the Eocene.
4. Dispersal of equoids to N. America in two phases early in the Eocene.
5. Subsequent colonizations of North America by isectolophids (*Cardiolophus*), ceratomorphs (*Heptodon, Hyrachyus*) and brontotheriids (*Lambdotherium, Eotitanops*), in that order, via the Bering Straits in the course of the early Eocene.
6. Dispersal to the Indian subcontinent sometime in the early Eocene.

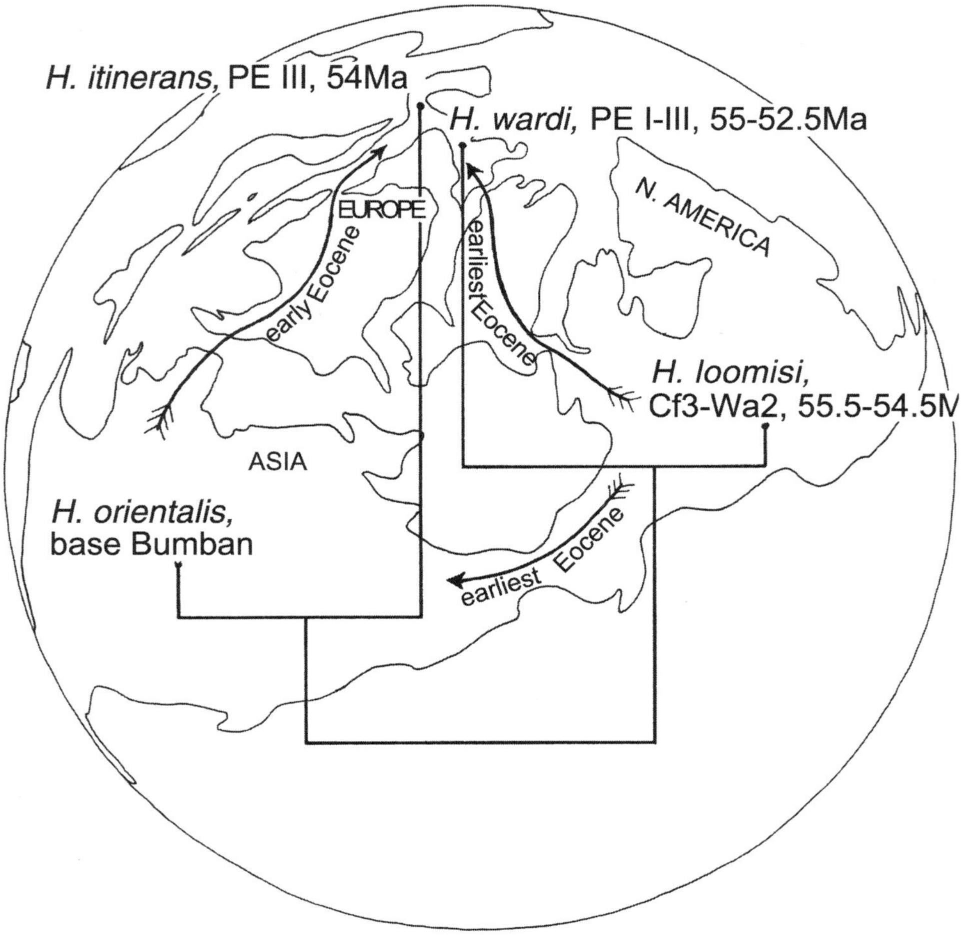

Figure 10. Early Eocene polar projection Northern Hemisphere paleogeographic shoreline map, with superimposed cladogram of early *Hyopsodus* species from Figure 2 with zones, time ranges in Ma where known and proposed dispersal directions (arrows) (map from unpublished work by Paul Markwick, see Markwick et al., 2000, with minor modifications to North Sea–Arctic Ocean link following Iakovleva et al., 2001).

Until *Lophocion* was described (Wang and Tong, 1997), the phenacodontids that are phenetically closest to perissodactyls were restricted to North America, an unlikely center of origin for the order Perissodactyla, in view of the sudden appearance there of only equoids and isectolophids at the beginning of the Eocene. The discovery of *Lophocion,* although young in age (contemporaneous with perissodactyls in the early Eocene Wutu Formation) makes the origin of perissodactyls in Asia a more tenable hypothesis. The fragmentary perissodactyl teeth from the Gashatan of Bayan Ulan, China (Meng et al., 1998), if really Paleocene (see Bowen et al., 2002), would represent the oldest perissodactyl. Although difficult to interpret because of breakage and heavy wear, they give the appearance of a *Lambdotherium*-like perissodactyl as Meng et al. (1998) proposed, but less dilambdodont. Thus, the upper molar centrocrista is slightly less buccally flexed and the lower preultimate molar hypoconulid is central rather than lingually displaced. Also, the entoconid seems not to be isolated as in *Orientolophus* and the more derived primitive equoids like *Pliolophus* and *Hyracotherium.* In these respects it is close to what would be expected of the basal perissodactyl.

THE TURGAI ROUTE AND TIMING OF DISPERSALS

The three examples above provide evidence of east to west dispersal between Asia and Europe as well as a model for origin and early evolutionary history of the order Perissodactyla. Similar treatments could help elucidate the origins of other enigmatic Eocene newcomers such as primates and artiodactyls. According to paleogeographic reconstructions (e.g., McKenna, 1975), the most likely crossing point was the Turgai Straits.

Iakovleva et al. (2001), in a study of combined dinocyst and sequence stratigraphy in the Turgai area, have indicated four hiatuses in the succession, relating to intervals of low sea level during the late Paleocene (Thanetian) and earliest Eocene (early Ypresian). These could have bridged the Turgai Straits, thus allowing terrestrial dispersal. They are (1) between 57.2 and 57.4 Ma (early Thanetian); (2) between 56.4 and 57.1 Ma (mid Thanetian); (3) between 55 and 55.1 Ma (late Thanetian, just before the CIE); and (4) between 54 and 54.5 Ma (early Ypresian). The time scale is that of Berggren et al. (1995) calibrated to dinocyst zonations and sequence stratigraphy via Hardenbol et al. (1998), Mudge and Bujak (1996) and Powell

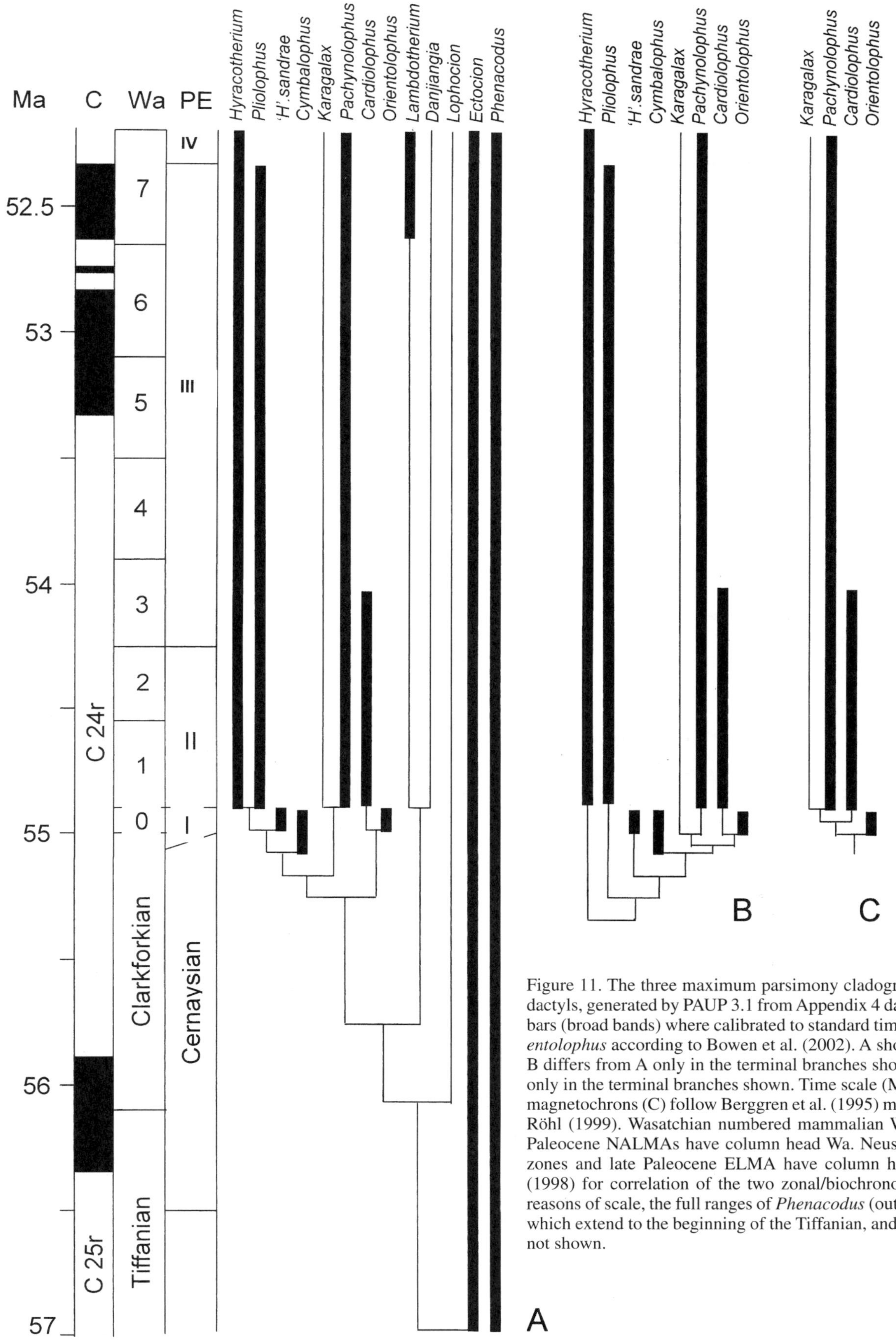

Figure 11. The three maximum parsimony cladograms of stem perissodactyls, generated by PAUP 3.1 from Appendix 4 data matrix, with range bars (broad bands) where calibrated to standard time scale; range of *Orientolophus* according to Bowen et al. (2002). A shows a full cladogram; B differs from A only in the terminal branches shown; C differs from B only in the terminal branches shown. Time scale (Ma) and calibration of magnetochrons (C) follow Berggren et al. (1995) modified by Norris and Röhl (1999). Wasatchian numbered mammalian Wa biozones and late Paleocene NALMAs have column head Wa. Neustrian mammalian PE zones and late Paleocene ELMA have column head PE. See Hooker (1998) for correlation of the two zonal/biochronological systems. For reasons of scale, the full ranges of *Phenacodus* (outgroup) and *Ectocion,* which extend to the beginning of the Tiffanian, and the basal branch, are not shown.

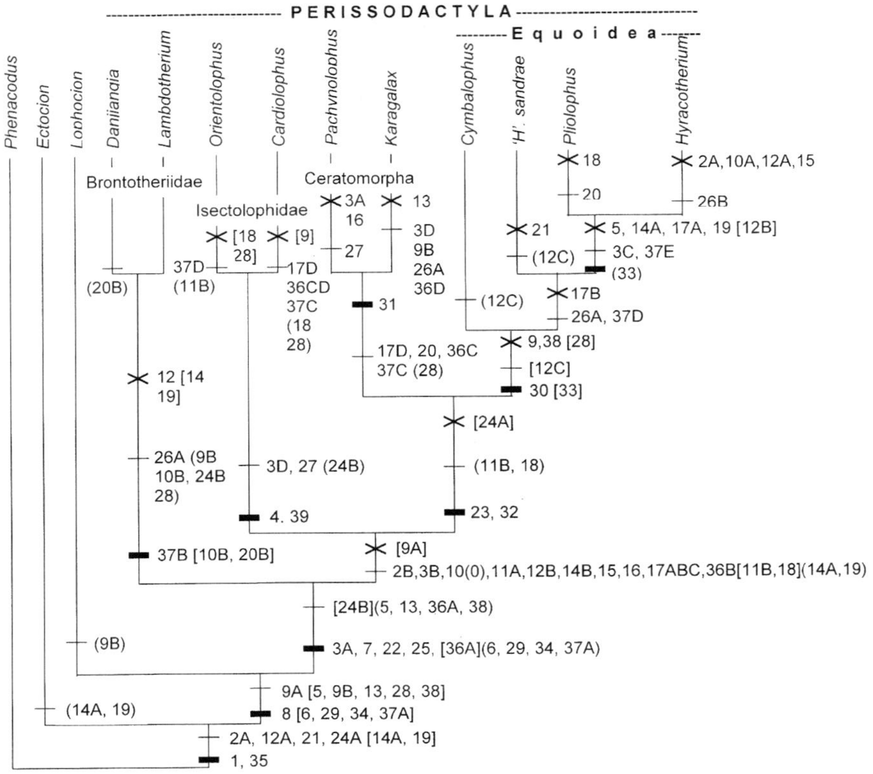

Figure 12. The maximum parsimony cladogram from Figure 11A, showing character state changes. See Appendix 3 for description of numbered characters, Appendix 4 for data matrix, and Figure 2 for explanation of notation.

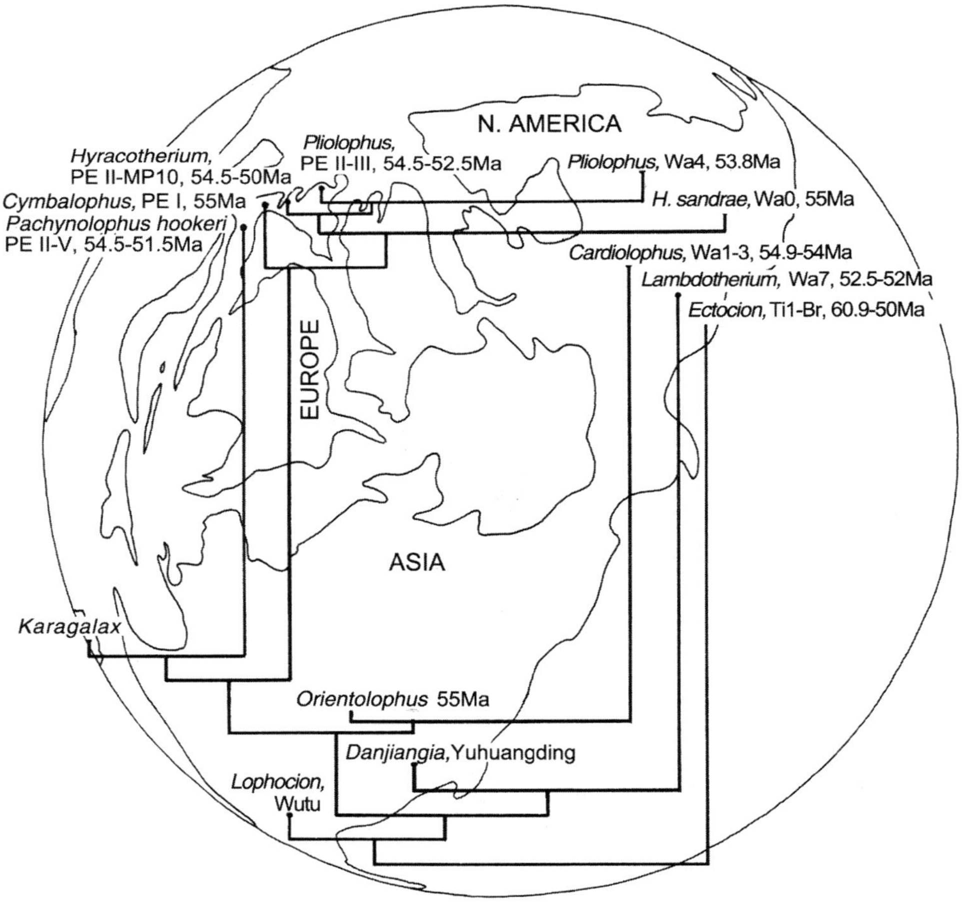

Figure 13. Early Eocene polar projection Northern Hemisphere paleogeographic shoreline map, with superimposed cladogram of stem perissodactyl and phenacodontid taxa from Figure 12 (map from unpublished work by Paul Markwick, see Markwick et al., 2000, with minor modifications to North Sea–Arctic Ocean link following Iakovleva et al., 2001). Zones and time ranges in Ma given where known, formational names given where dating uncertain.

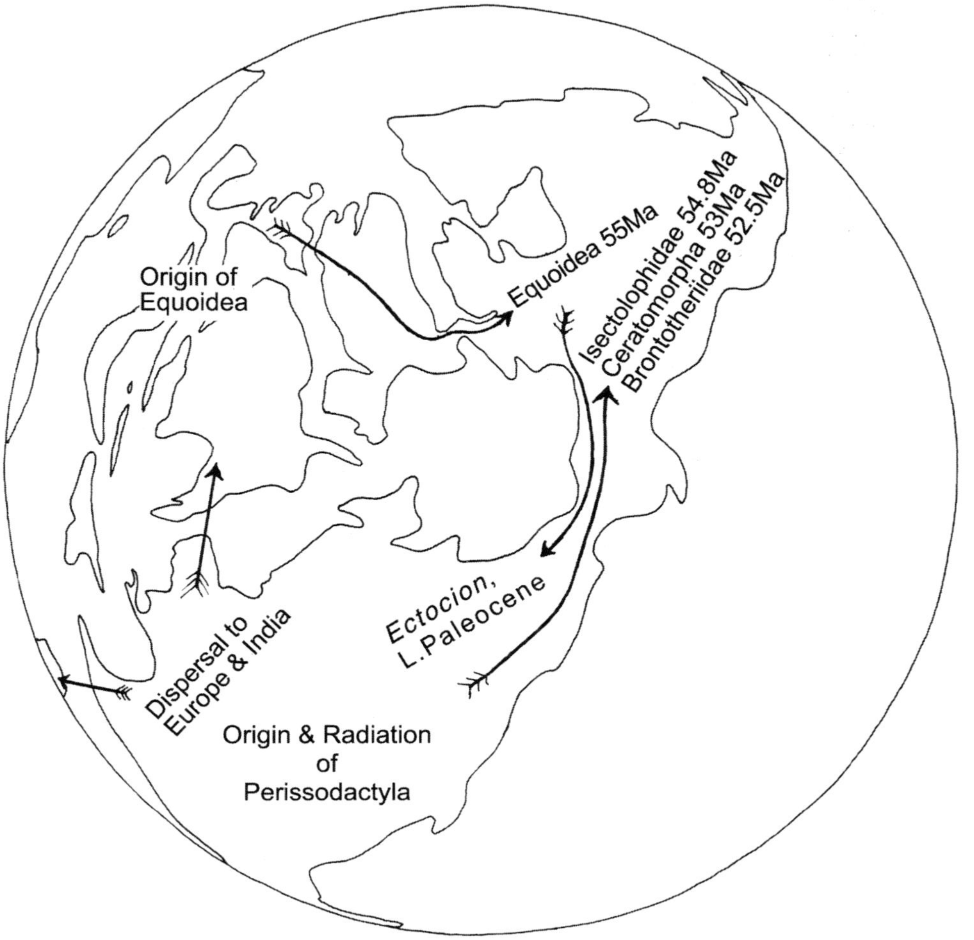

Figure 14. Same map as in Figure 13, with proposed dispersal directions and timings (arrows) for phenacodontids and stem perissodactyls.

et al. (1996), modified around the CIE according to Norris and Röhl (1999).

If the Bumbanian ALMA indeed extends no earlier than early Eocene (Bowen et al., 2002) and if the occurrence at Tsagan Khushu of *Lessnessina khushuensis* approximates the appearance of the genus in Asia, then hiatuses 1–3 are too early to be the ones allowing dispersal of *Lessnessina* to Europe. The earliest occurrence of the genus in Europe in mammal zone PE III (Hooker, 1998), however, fits closely with hiatus 4. It is therefore likely that *Lessnessina* crossed the Turgai Straits to Europe ~54.5 Ma, shortly before its appearance at Abbey Wood.

The same arguments apply to the Mongolian (Bumbanian) and southern French (PE III) occurrences of *Hyopsodus. Hyopsodus* is thus also likely to have crossed the Turgai Straits early in the timespan of hiatus 4, perhaps at ~54.5 Ma.

Perissodactyls including equoids have their first appearance in northwest Europe at around the MDE (Hooker, 1994, 1998). Differentiation of equoids must have taken place within Europe (although apparently not in the northwest) prior to this, but there is currently no record of it. Hiatus 3, which immediately predates the CIE, hardly allows time for this differentiation prior to the equoid first appearance in PE I (Hooker, 1994, 1998). Late in the long hiatus 2 (e.g., ~56.5 Ma) thus is the most likely interval for a perissodactyl crossing of the Turgai Straits. This has the feeling of being "too early," bearing in mind our near total lack of knowledge of this order before the beginning of the Eocene at 55 Ma. However, it provides a working hypothesis and a prediction that is testable by future discovery of the elusive late Paleocene perissodactyl record, which the cladistic analysis predicts to exist somewhere.

As the Turgai Straits are at mid latitudes, not high latitudes like the Bering and Greenland land bridges, it is unlikely that ameliorating climate was the driving force behind these three dispersals. Low sea level providing either a narrow land bridge or a greatly narrowed strait, allowing relatively easy rafting, is the scenario that seems best to fit the current evidence.

ACKNOWLEDGMENTS

We thank Dr. Paul Markwick for allowing us to use outlines of his unpublished paleogeographic maps; Dr. Philip Gingerich for providing measurements of the *Hyopsodus loomisi* teeth used herein and, together with Drs. Don Russell, Marc Godinot, Spencer Lucas, and Hans Thewissen, casts of relevant specimens; and Dr. Adrian Rundle, Messrs. David Bone, Brian Gasson, Allan Lawson, Bill Morris and David Ward, for presenting important Abbey Wood mammals to the Natural History Museum, London. Ms. Pat Hart and Messrs. Harry Taylor and Kevin Webb of the NHM Photostudio ably composed the photographic illustrations. Prof. Bill Clemens provided helpful discussion on hyopsodontids. Comments by Spencer Lucas and an anonymous reviewer improved the manuscript. We are grateful to The Royal Society for funding visits to Mongolia and England. This is a contribution to the NHM Human Origins Programme, project 298.

APPENDIX 1. HYOPSODONTID CHARACTER DESCRIPTIONS

Description of characters used in the hyopsodontid cladistic analysis (see text and data matrix in Appendix 2). Multistate characters are ordered, representing linear transformation series. 0 is primitive state except for characters 1, 5, and 17, where A is primitive state and 0 is derived. 1 is derived state for binary characters. A, B, C, and D are otherwise derived states of multistate characters.

Character 1: Lower molar paraconid, small, distinctly separate from metaconid (0); tall, distinctly separate from metaconid (A) PRIMITIVE; tall, close to metaconid (B); fused to metaconid on M_{2-3} (C).

Character 2: Upper molars with precingulum extending to lingual wall of tooth (0); precingulum with protostyle extending on M^1 two thirds to three quarters the distance to the lingual wall (1).

Character 3: P_4 with paracristid weak (0); strong (1).

Character 4: P_4 with distal protoconid crest weak (0); strong (1).

Character 5: P_4 without metaconid (0); with small metaconid (A) PRIMITIVE; with large metaconid (B).

Character 6: P^4 without metacone and with short protocone, P_4 without basined talonid (0); P^4 with faint metacone and short protocone, P_4 with short basined talonid (A); P^4 with faint metacone and long protocone, P_4 with long basined talonid (B); P^4 with small distinct metacone and long protocone, P_4 with long basined talonid with subterminal hypoconid (C); P^4 with small distinct metacone and long protocone, P_4 with long basined talonid with subterminal hypoconid and distinct entoconid (D).

Character 7: P_4 with distal talonid cusp(s) low (0); tall (1).

Character 8: P_4 with paraconid (or its position on the mesial end of the paracristid) half the height of the protoconid or less (0); taller (1).

Character 9: Upper molars with parastyle and ectocingulum strong (0); weak (1).

Character 10: Upper molar mesostyle absent (0); present (1).

Character 11: Upper molar postprotocingulum absent (0); present (1).

Character 12: M_3 larger than M_2 (0); M_3 and M_2 subequal (A); M_3 slightly smaller than M_2 (B); M_3 much smaller than M_2 (C). E.g., Figure 9.

Character 13: Lower preultimate molar hypoconulid nearly median (0); distinctly displaced lingually (1).

Character 14: Lower molar paracristid with rounded mesiobuccal corner (0); with sharp mesiobuccal corner (1).

Character 15: Upper molars with low protocone, lower molars with tall entoconid (0); M^2 with tall protocone, M_2 with low entoconid (A); all upper molars with tall protocone, all lower molars with low entoconid (B).

Character 16: Lower molar paraconid-metaconid complex and protoconid subequal in height (0); paraconid-metaconid complex distinctly taller (1).

Character 17: Lower molar cristid obliqua contact with back of trigonid lingual of midline (0); more or less median (A) PRIMITIVE; more buccal (B).

Character 18: Lower molar protocristid shallowly notched (0); deeply notched or missing (1).

Character 19: Molar cresting weak (0); strong (1).

Character 20: Crown height low (0) (e.g., Figure 7B–C); higher (1) (e.g., Figure 7G).

Character 21: P^4 with mesial parastyle projection strong (0); weak (1).

Character 22: Lower molar entocristid present (0); absent (1).

Character 23: Mesial and distal contact walls of adjacent lower molars transverse (0); oblique (1).

Character 24: Lower preultimate molar hypoconulid poorly differentiated from postcristid (0) (e.g., Figure 1G); well differentiated from postcristid (1) (e.g., Figure 6A).

APPENDIX 2. HYOPSODONTID DATA MATRIX

Data matrix for hyopsodontids derived from characters in Appendix 1. *Protungulatum* is outgroup taxon.

Taxon	111111111122222 123456789012345678901234
Protungulatum	A000A00000000000A0000000
Protoselene	0111AC0101001000A0000001
Dorraletes	0101AD01000C1000A0000000
Haplomylus	0101AD01101C1000A0000000
Promioclaenus	B111AA00000A0000A0000000
Litaletes	B111BA10000A00A0A0000000
Lessnessina khushuensis	B??????????A01B1B100?000
Lessnessina packmani	C1110010110?01B1B1000000
Lessnessina praecipuus	C?110011???C01B1B110?000
Hyopsodus loomisi/wardi	C111BB00100A0000A0001011
Hyopsodus orientalis	C111BA00100B100000010111
Hyopsodus itinerans	C111BA00100C10000001?111

APPENDIX 3. PERISSODACTYL CHARACTER DESCRIPTIONS

Description of characters used in the perissodactyl cladistic analysis (see text and data matrix in Appendix 4). These follow, with modifications and additions, those used by Hooker (1994). Multistate characters are fully ordered, except for 3 and 37, which are partially ordered and coded as stepmatrices. 0 is primitive state except for characters 10 and 20, where A is primitive state and 0 is derived. 1 is derived state for binary characters. A, B, C, D, and E are otherwise derived states of multistate characters.

Character 1: Cheek teeth generally bunodont with cresting weak (0); more strongly crested, with tendency to form ectoloph on upper molars (1).

Character 2: Upper molar preparaconule crista directed toward parastyle (0); toward preparacrista, variably joining it (A); toward preparacrista, constantly joining it (B).

Character 3: Upper molar paraconule not mesially distanced from the protocone, lacking facet 2A, preprotocrista mesiobuccally directed toward the paraconule, lower molar metaconid single, bearing facets 2 and 3 (0); upper molar paraconule, with facet 2A, mesially distanced from protocone, preprotocrista mesiobuccally directed toward the paraconule, lower molar metaconid widely twinned, bearing facets 2, 2A and 3 (A); upper molar paraconule, with facet 2A, not mesially distanced from protocone, preprotocrista nearly buccally directed toward the paraconule, lower molar metaconid narrowly twinned, bearing facets 2, 2A and 3 (B); upper molar paraconule, with facet 2A, mesially distanced from protocone, preprotocrista buccally directed to just behind the paraconule, lower molar metaconid widely twinned, bearing facets 2, 2A and 3 (C); upper molar paraconule nearly in line with protocone, lacking facet 2A and facets 2 and 3 nearly merged, lower molar metaconid single, undivided facets 2 and 3 forming distal bevel (D). Stepmatrix:

	0	A	B	C	D
0	-	1	2	3	3
A	1	-	1	2	2
B	2	1	-	1	1
C	3	2	1	-	2
D	3	2	1	2	-

Character 4: Lower molar cristid obliqua straight (0); bowed buccally (1).

Character 5: Lower molar protolophid notched (0); shallowly indented, lophoid (1).

Character 6: Lower molar metastylid a prominent cuspule (0); weak to lacking (1).

Character 7: Upper molar metaconule situated on a line drawn between the metacone and hypocone, lower molar hypoconulid an integral part of the postcristid (0); upper molar metaconule (or its position if subsumed by metaloph) distinctly mesial of a line drawn between the metacone and hypocone, lower molar hypoconulid separated from and distal of crest joining hypoconid and entoconid (hypolophid) (1).

Character 8: Upper molar metaconule not joined to hypocone by crest (0); joined to hypocone by crest (1).

Character 9: Upper molar metaconule large (0); size variable, some large, some small (A); all small (B).

Character 10: Upper molar centrocrista straight (0); slightly flexed buccally (A) PRIMITIVE; sharply flexed buccally (B).

Character 11: Upper molar mesostyle large (0); small or variably developed (A); lacking (B).

Character 12: Lower molar buccal and lingual cusp outer walls converge at ~45° (0); ~20° (A); ~10° (B); ~5° (C).

Character 13: Lower molar metaconid butress (Hooker, 1994, Figure 2E–F) weak or absent (0); strong (1).

Character 14: Lower molar cristid obliqua attaches to trigonid nearer to metaconid than to protoconid (0); midway between protoconid and metaconid (A); nearer to protoconid than to metaconid (B).

Character 15: Lower molar cristid obliqua attachment on back wall of trigonid, high (0); low (1).

Character 16: Lower molar entoconulid a prominent cuspule (0); weak to lacking (1).

Character 17: Lower molar paracristid, when lightly worn, makes angle to tooth long axis of 50° (0); 40° (A); 30° (B); 20° (C); 10° (D).

Character 18: Lower molar paracristid with mesiobuccal angle rounded (0); sharp or bulging (1).

Character 19: Lower molar trigonid back wall shallow (0); steep (1).

Character 20: P_{1-2} diastema absent (0); short (A) PRIMITIVE; long (B).

Character 21: M^{1-2} as long as broad (0); broader than long (1).

Character 22: M^3 without hypocone (0); with hypocone (1).

Character 23: Upper molar postmetacrista and lower molar paracristid buccal segment short (0); long (1).

Character 24: Upper molar parastyle small (0); medium (A); large (B).

Character 25: M^3 smaller than M^2, M_3 not larger than M_2 (0); M^3 not smaller than M^2, M_3 larger than M_2 (1).

Character 26: M_3 hypoconulid as close to hypoconid as this is to entoconid (0); more distant but closer to hypoconid than this is to protoconid (A); as far or further from hypoconid than this is from protoconid (B).

Character 27: M_3 hypoconulid forming distal margin of posttalonid lobe (0); bearing more distal lobe (1).

Character 28: M_3 hypolophid incomplete (0); complete (1).

Character 29: P^3 metacone smaller than paracone (0); as large as paracone (1).

Character 30: P^3 with trigon relatively narrow, protoloph weak, paraconule and P_3 metaconid very small and poorly defined (0); trigon broader, protoloph stronger, paraconule and metaconid much larger and better defined, but smaller than protocone and protoconid respectively (1).

Character 31: P_3 paraconid weak and much lower than protoconid (0); strong and approaching height of protoconid (1).

Character 32: Lower canine to P_1 diastema short (0); long (1).

Character 33: Optic foramen of orbit significantly in front of (0) or close to (1) anterior lacerate foramen (1). [N.B. *Ectocion* recoded following Holbrook (2001, p. 16–17).]

Character 34: Navicular facet of astragalus convex (0); saddle-shaped (1).

Character 35: Astragalar canal present (0); absent (1).

Character 36: Upper molars with no metaloph (0); with metaconal fold (Hooker, 1994, Figure 2A–C) not joined to metaconule (A); some joined to metaconule, some not (B); consistently joined to metaconule forming complete metaloph (C); buccal end of metaloph (homologue of metaconal fold) shifted mesially from metacone (D).

Character 37: Lower molar without hypolophid (0); hypolophid complete, comprising equal buccal and lingual segments of former postcristid joining in middle at notch in front of hypoconulid (A); buccal segment lengthened at expense of lingual segment with lingual hypoconulid (B); equal buccal and lingual segments joined into strong unnotched loph, hypoconulid median (C); hypolophid notched but complete like state A, some with lingual segment broken (D); hypolophid with lingual segment consistently broken (E). Stepmatrix:

	0	A	B	C	D	E
0	-	1	2	2	2	3
A	1	-	1	1	1	2
B	2	1	-	2	2	3
C	2	1	2	-	2	3
D	2	1	2	2	-	1
E	3	2	3	3	1	-

Character 38: Lower M_{1-2} distal cingulum lingual of hypoconulid absent (0); present (1).

Character 39: Upper molar protoloph and metaloph and lower molar protolophid and hypolophid transverse (0); oblique (1).

APPENDIX 4. PERISSODACTYL DATA MATRIX

Data matrix for perissodactyls and related phenacodontids derived from characters in Appendix 3. N.B. *Hyracotherium* indicates only the type species *H. leporinum* (see Hooker, 1994). *Pachynolophus* involves mainly the most primitive species (*P. hookeri* Godinot in Godinot et al., 1987), but also others for skull characters (Remy, 1972; Savage et al., 1965) and other pachynolophids for postcranial characters (Depéret, 1917). Data from original material, casts and, in the case of *Danjiangia* and *Lophocion,* from the literature (Wang, 1995; Wang and Tong, 1997). Relevant additional sources are: Gingerich (1989, 1991), Thewissen (1990) and Ting (1993).

Taxon	111111111122222222223333333333 123456789012345678901234567890123456789
Hyracotherium	1AC001110ABA1A01A10A111A1B0011?11??BE00
Cymbalophus	1BB0111100BC1B11C11A111A10001101???BA00
"H." sandrae	1BB0111100BC1B11B11A011A1A001101?11BD00
Cardiolophus	1BD11111A0AB1B11D11A110B1011100 0?11DC11
Orientolophus	1BD11111A0BB1B11C01?1?0B1010???????BD11
Lambdotherium	1AA01111BB001000000?110B1A01100??11AB10
Danjiangia	1AA01111BB001000000B110B1A011000???AB?0
Lophocion	1A0???01BA0A????????100A0??????????0??0
Ectocion	1A0000000A0A0A00001A100A000000000010000
Karagalax	1BD01111B0BB0B11D110111A1A011011?11DC10
Pachynolophus	1BA01111A0BB1B10D110111A10111011011CC10
Pliolophus	1BC0011100BB1A11A000111A1A00110 1111BE00
Phenacodus	000000000A000000000A0000000000000000000

REFERENCES CITED

Archibald, J.D., 1998, Archaic ungulates ("Condylarthra"), *in* Janis, C.M., et al., eds., Evolution of Tertiary mammals of North America, volume 1: Terrestrial carnivores, ungulates, and ungulatelike mammals: Cambridge, UK, Cambridge University Press, p. 292–331.

Archibald, J.D., Rigby, J.K., Jr., and Robison, S.F., 1983, Systematic revision of *Oxyacodon* (Condylarthra, Periptychidae) and a description of *O. ferronensis* n. sp.: Journal of Paleontology, v. 57, p. 53–72.

Beard, K.C., 1998, East of Eden: Asia as an important center of taxonomic origination in mammalian evolution: Bulletin of Carnegie Museum of Natural History, v. 34, p. 5–39.

Beard, K.C., and Dawson, M.R., 1999, Intercontinental dispersal of Holarctic land mammals near the Paleocene-Eocene boundary: Paleogeographic, paleoclimatic and biostratigraphic implications: Bulletin de la Société Géologique de France, v. 170, p. 697–706.

Berggren, W.A., Kent, D.V., Swisher, C.C., III, and Aubry, M.-P., 1995, A revised Cenozoic geochronology and chronostratigraphy, *in* Berggren, W.A., et al., eds., Geochronology, time scales, and global stratigraphic correlation: SEPM (Society for Sedimentary Geology) Special Publication 54, p. 129–212.

Bowen, G.B., Clyde, W.C., Koch, P.L., Ting Suyin, Alroy, J., Tsubamoto, Takehisa, Wang Yuanqing, and Wang Yuan, 2002, Mammalian dispersal at the Paleocene–Eocene boundary: Science, v. 295, p. 2062–2065.

Butler, P.M., 1972, Some functional aspects of molar evolution: Evolution, v. 26, p. 474–483.

Butler, P.M., 1973, Molar wear facets of early Tertiary North American primates, *in* Zingeser, M.R., ed., Craniofacial biology of primates: Symposia of the Fourth International Congress of Primatology, Portland, Oregon, August 15–18, 1972, v. 3, p. 1–27.

Clyde, W.C., and Gingerich, P.D., 1998, Mammalian community response to the latest Paleocene thermal maximum: An isotaphonomic study in the northern Bighorn Basin, Wyoming: Geology, v. 26, p. 1011–1014.

Cope, E.D., 1881, A new type of Perissodactyla: American Naturalist, v. 15, p. 1017–1018.

Crompton, A.W., and Kielan-Jaworowska, Z., 1978, Molar structure and occlusion in Cretaceous therian mammals, *in* Butler, P.M., and Joysey, K.A., eds., Studies in the development, function and evolution of teeth: London, Academic Press, p. 249–287.

Dashzeveg, D., 1977, On the discovery of *Hyopsodus* Leidy, 1870 (Mammalia, Condylarthra) in the Mongolian People's Republic, *in* Barsbold, R., and Vorobjeva, E.L., eds., Mesozoic and Cenozoic faunas, floras and biostratigraphy of Mongolia: Trudy Sovmestnaya Sovetsko-Mongol'skaya Paleontologicheskaya Ekspeditsiya, v. 4, p. 7–13.

Depéret, C., 1910, Etudes sur la famille des lophiodontidés: Bulletin de la Société Géologique de France, ser. 4, v. 10, p. 558–577.

Depéret, C., 1917, Monographie de la faune de mammifères fossiles du Ludien inférieur d'Euzet-les-Bains (Gard): Annales de l'Université de Lyon, new series, I Sciences, Médecine, v. 40, p. 1–290.

Estravis, C., and Russell, D.E., 1992, The presence of Taeniodonta (Mammalia) in the early Eocene of Europe: Ciências da Terra (Universidade Nova de Lisboa), v. 11, p. 191–201.

Forey, P.L., Humphries, C.J., Kitching, I.J., Scotland, R.W., Siebert, D.J., and Williams, D.M., 1992, Cladistics, a practical course in systematics: Systematics Association Publication No. 10: Oxford, Clarendon Press, 191 p.

Froehlich, D.J., 1999, Phylogenetic systematics of basal perissodactyls: Journal of Vertebrate Paleontology, v. 19, p. 140–159.

Gingerich, P.D., 1974, Stratigraphic record of early Eocene *Hyopsodus* and the geometry of mammalian phylogeny: Nature, v. 248, p. 107–109.

Gingerich, P.D., 1976a, Paleontology and phylogeny: Patterns of evolution at the species level in early Tertiary mammals: American Journal of Science, v. 276, p. 1–28.

Gingerich, P.D., 1976b, Cranial anatomy and evolution of early Tertiary Plesiadapidae (Mammalia, Primates): University of Michigan Papers on Paleontology, v. 15, p. 1–141.

Gingerich, P.D., 1985, Species in the fossil record: Concepts, trends, and transitions: Paleobiology, v. 11, p. 27–41.

Gingerich, P.D., 1989, New earliest Wasatchian mammalian fauna from the Eocene of northwestern Wyoming: Composition and diversity in a rarely sampled high-floodplain assemblage: University of Michigan Papers on Paleontology, v. 28, p. 1–97.

Gingerich, P.D., 1991, Systematics and evolution of early Eocene Perissodactyla (Mammalia) in the Clarks Fork Basin, Wyoming: Contributions from the Museum of Paleontology, University of Michigan, v. 28, p. 181–213.

Godinot, M., 1978, Diagnoses de trois nouvelles espèces de mammifères du Sparnacien de Provence: Compte Rendu Sommaire des Séances de la Société Géologique de France, v. 1978, p. 286–288.

Godinot, M., 1981, Les mammifères de Rians (Eocène inférieur, Provence): Palaeovertebrata, v. 10, p. 43–126.

Godinot, M., Crochet, J.-Y., Hartenberger, J.-L., Lange-Badré, B., Russell, D.E., and Sigé, B., 1987, Nouvelles données sur les mammifères de Palette (Eocène inférieur, Provence): Münchner Geowissenschaftliche Abhandlungen, ser. A, v. 10, p. 273–288.

Hardenbol, J., Thierry, J., Farley, M.B., Jacquin, T., De Graciansky, P.-C., and Vail, P.R., 1998, Mesozoic and Cenozoic sequence chronostratigraphic framework of European basins: SEPM (Society for Sedimentary Geology) Special Publication 60, p. 3–13, 763–781.

Holbrook, L.T., 2001, Comparative osteology of early Tertiary tapiromorphs (Mammalia, Perissodactyla): Zoological Journal of the Linnean Society, v. 132, p. 1–54.

Hooker, J.J., 1979, Two new condylarths (Mammalia) from the early Eocene of southern England: Bulletin of the British Museum (Natural History), Geology series, v. 32, p. 43–56.

Hooker, J.J., 1986, Mammals from the Bartonian (middle/late Eocene) of the Hampshire Basin, southern England: Bulletin of the British Museum (Natural History), Geology series, v. 39, p. 191–478.

Hooker, J.J., 1989, Character polarities in early Eocene perissodactyls and their significance for *Hyracotherium* and infraordinal relationships, *in* Prothero, D.R., and Schoch, R.M., eds., The evolution of perissodactyls: New York, Oxford University Press, p. 79–101.

Hooker, J.J., 1994, The beginning of the equoid radiation: Zoological Journal of the Linnean Society, v. 112, p. 29–63.

Hooker, J.J., 1998, Mammalian faunal change across the Paleocene-Eocene transition in Europe, *in* Aubry, M.-P., et al., eds., Late Paleocene–early Eocene climatic and biotic events in the marine and terrestrial records: New York, Columbia University Press, p. 428–450.

Hooker, J.J., 2000, Ecological response of mammals to global warming in the late Paleocene and early Eocene: GFF, v. 122, p. 77–79.

Iakovleva, A.I., Brinkhuis, H., and Cavagnetto, C., 2001, Late Paleocene–early Eocene dinoflagellate cysts from the Turgay Strait, Kazakhstan: Correlations across ancient seaways: Palaeogeography, Palaeoclimatology, Palaeoecology, v. 172, p. 243–268.

Kondrashov, P.E., and Agadjanian, A.K., 1999, New material on the genus *Hyopsodus* (Mammalia, Condylarthra) from the Eocene of Mongolia: Morphological variability and taxonomic position: Paleontological Journal, v. 33, p. 667–676.

Krause, D.W., and Maas, M.C., 1990, The biogeographic origins of late Paleocene–early Eocene mammalian immigrants to the western interior of North America: Boulder, Colorado, Geological Society of America Special Paper 243, p. 71–105.

Leidy, J., 1870, Remarks on a collection of fossils from the western territories: Proceedings of the Academy of Natural Sciences, Philadelphia, v. 22, p. 109–110.

Lipscomb, D.L., 1992, Parsimony, homology and the analysis of multistate characters: Cladistics, v. 8, p. 45–65.

Maas, M.C., Hussain, S.T., Leinders, J.J.M., and Thewissen, J.G.M., 2001, A new isectolophid tapiromorph (Perissodactyla, Mammalia) from the early Eocene of Pakistan: Journal of Paleontology, v. 75, p. 407–417.

Markwick, P.J., Rowley, D.B., Ziegler, A.M., Hulver, M.L., Valdes, P.J., and Selwood, B.W., 2000, Late Cretaceous and Cenozoic global palaeogeographies: Mapping the transition from a "hot-house" to an "ice-house" world: GFF, v. 122, p. 103.

Matthew, W.D., 1937, Paleocene faunas of the San Juan Basin, New Mexico: Transactions of the American Philosophical Society, v. 30, p. 1–510.

McKenna, M.C., 1960, Fossil mammalia from the early Wasatchian Four Mile fauna, Eocene of northwest Colorado: Berkeley, California, University of California Publications in Geological Science, v. 37, p. 1–130.

McKenna, M.C., 1975, Fossil mammals and early Eocene North Atlantic land continuity: Annals of the Missouri Botanical Garden, v. 62, p. 335–353.

McKenna, M.C., Chow Minchen, Ting Suyin, and Luo Zhexi, 1989, *Radinskya yupingae,* a perissodactyl-like mammal from the late Paleocene of China, *in* Prothero, D.R., and Schoch, R.M., eds., The evolution of Perissodactyls: New York, Oxford University Press, p. 24–36.

Meng Jin, Zhai Renjie, and Wyss, A.R., 1998, The late Paleocene Bayan Ulan fauna of Inner Mongolia, China: Bulletin of Carnegie Museum of Natural History, v. 34, p. 148–185.

Mudge, D.C., and Bujak, J.P., 1996, An integrated stratigraphy for the Paleocene and Eocene of the North Sea, *in* Knox, R.W.O'B., et al., eds., Correlation of the early Paleogene in northwest Europe: London, Geological Society Special Publication 101, p. 91–113.

Norris, R.D., and Röhl, U., 1999, Carbon cycling and chronology of climate warming during the Paleocene-Eocene transition: Nature, v. 401, p. 775–778.

Osborn, H.F., 1929, The titanotheres of ancient Wyoming, Dakota and Nebraska: Monographs of the U.S. Geological Survey, v. 55, p. 1–953.

Powell, A.J., Brinkhuis, H., and Bujak, J.P., 1996, Upper Paleocene–lower Eocene dinoflagellate cyst sequence biostratigraphy of southeast England, *in* Knox, R.W.O'B., et al., eds., Correlation of the early Paleogene in northwest Europe: London, Geological Society Special Publication 101, p. 145–183.

Radinsky, L.B., 1964, *Paleomoropus,* a new early Eocene chalicothere (Mammalia, Perissodactyla), and a revision of Eocene chalicotheres: American Museum Novitates, no. 2179, p. 1–28.

Radinsky, L.B., 1966, The adaptive radiation of the phenacodontid condylarths and the origin of the Perissodactyla: Evolution, v. 20, p. 408–417.

Redline, A.D., 1997, Revision of the Wind River faunas, early Eocene of central Wyoming, Part 13: Systematics and phylogenetic pattern of early Eocene *Hyopsodus* (Mammalia: Condylarthra): Annals of Carnegie Museum, v. 66, p. 1–81.

Remy, J.-A., 1972, Etude du crâne de *Pachynolophus lavocati* n. sp. (Perissodactyla, Palaeotheriidae) des Phosphorites du Quercy: Palaeovertebrata, v. 5, p. 45–78.

Rigby, J.K., Jr., 1980, Swain Quarry of the Fort Union Formation, middle Paleocene (Torrejonian), Carbon County, Wyoming: Geologic setting and mammalian fauna: Evolutionary Monographs, v. 3, p. 1–179.

Rose, K.D., 1981, The Clarkforkian Land-Mammal Age and mammalian faunal composition across the Paleocene-Eocene boundary: University of Michigan Papers on Paleontology, v. 26, p. 1–197.

Savage, D.E., Russell, D.E., and Louis, P., 1965, European Eocene Equidae (Perissodactyla): Berkeley, California, University of California Publications in Geological Science, v. 56, p. 1–94.

Savage, D.E., Russell, D.E., and Louis, P., 1966, Ceratomorpha and Ancylopoda (Perissodactyla) from the lower Eocene Paris Basin, France: Berkeley, California, University of California Publications in Geological Science, v. 66, p. 1–38.

Schoch, R.M., 1986, Systematics, functional morphology, and macroevolution of the extinct mammalian order Taeniodonta: Bulletin of the Peabody Museum of Natural History, v. 42, p. 1–307.

Simpson, G.G., 1937, The Fort Union of the Crazy Mountain Field, Montana and its mammalian faunas: Bulletin of the United States National Museum, v. 169, p. 1–287.

Sulimski, A., 1969, Paleocene genus *Pseudictops* Matthew, Granger and Simpson, 1929 (Mammalia) and its revision: Palaeontologia Polonica, v. 19, p. 101–131.

Swofford, D.L., 1990, PAUP: Phylogenetic Analysis Using Parsimony, Version 3.0: Computer program distributed by the Illinois Natural History Survey, Champaign, Illinois, 162 p.

Thewissen, J.G.M., 1990, Evolution of Paleocene and Eocene Phenacodontidae (Mammalia, Condylarthra): University of Michigan Papers on Paleontology, v. 29, p. 1–107.

Thewissen, J.G.M., and Domning, D.P., 1992, The role of phenacodontids in the origin of the modern orders of ungulate mammals: Journal of Vertebrate Paleontology, v. 12, p. 494–504.

Ting Suyin, 1993, A preliminary report on an early Eocene mammalian fauna from Hengdong, Hunan Province, China: Kaupia, v. 3, p. 201–207.

Trouessart, E.L., 1879, Catalogue des mammifères vivants et fossiles: Revue et Magazine de Zoologie, v. 7, p. 219–285.

Wang Jingwen, and Tong Yongsheng, 1997, A new phenacodontid condylarth (Mammalia) from the early Eocene of the Wutu Basin, Shandong: Vertebrata PalAsiatica, v. 35, p. 283–289.

Wang Yuan, 1995, A new primitive chalicothere (Perissodactyla, Mammalia) from the early Eocene of Hubei, China: Vertebrata PalAsiatica, v. 33, p. 138–159, 1 plate.

Williamson, T.E., 1996, The beginning of the Age of Mammals in the San Juan Basin, New Mexico: Biostratigraphy and evolution of Paleocene mammals of the Nacimiento Formation: Bulletin of the New Mexico Museum of Natural History and Science, v. 8, p. 1–141.

Xu Qinqi, 1976, New materials of Anagalidae from the Paleocene of Anhui (A): Vertebrata PalAsiatica, v. 14, p. 174–184.

Manuscript Accepted by the Society August 13, 2002

Printed in the USA

Geological Society of America
Special Paper 369
2003

Rare taxa, biostratigraphy, and the Wasatchian-Bridgerian boundary in North America

Krister T. Smith*
Department of Geology and Geophysics, Yale University, P.O. Box 208109, New Haven, Connecticut 06520-8109, USA
Patricia A. Holroyd*
University of California Museum of Paleontology, 1101 Valley Life Sciences Building, University of California, Berkeley, California 94720, USA

ABSTRACT

The boundary between the Wasatchian and Bridgerian North American land-mammal "ages" has received considerable recent attention and is widely recognized as a time of significant faunal turnover in the terrestrial realm. However, it is also less well understood than other such transitions. Here, we report the discovery of *Eotitanops borealis,* long considered the principal defining taxon for the beginning of the Bridgerian, from a classic late Wasatchian (Lostcabinian) fauna near the center of the northern Green River Basin. We argue that such a range extension should not be surprising given the relative rarity of this taxon in other geographic areas. This discovery implies that the beginning of the Bridgerian is in need of redefinition, but the selection of a different principal taxon (taxa) for this purpose is complicated by our poor understanding of the true first appearances of many traditionally Bridgerian taxa. We defer such redefinition and instead suggest several different migrational scenarios that might be considered as we continue to refine our understanding of this period of comparatively gradual climatic warming.

INTRODUCTION

Questions of sampling bedevil the analysis of paleontological data (e.g., Schankler, 1981; Badgley and Gingerich, 1988; Kidwell, 2001), because sampling may control what one can and cannot confidently assert about the fossil record. It is widely appreciated that fossil "communities" are assembled by taphonomic processes and only in the rarest of cases or in certain respects represent actual past communities. Within paleontology, sampling issues are particularly relevant to biostratigraphy, because one of the foremost goals of that discipline is the identification of isochronous events, such as the first appearances of fossil species, which may then be used to demarcate and correlate units of geologic time.

The boundary between the Wasatchian and Bridgerian North American land-mammal "ages" (NALMAs) has been accorded much recent attention (Gunnell et al., 1992; Clyde et al., 1997; Zonneveld et al., 2000; Clyde et al., 2001) and is widely recognized as a time of considerable, if not remarkable, faunal turnover (Krishtalka et al., 1987; Woodburne and Swisher, 1995). In this paper, we first review the biostratigraphic record spanning the Wasatchian-Bridgerian boundary, focusing on rare taxa and the extent to which stability is promoted by the choice of indicator taxa. We then discuss new data from a classic Lostcabinian (late

*E-mails: Krister.smith@yale.edu; Pholroyd@uclink4.berkeley.edu.

Smith, K.T., and Holroyd, P.A., 2003, Rare taxa, biostratigraphy, and the Wasatchian-Bridgerian boundary in North America, *in* Wing, S.L., Gingerich, P.D., Schmitz, B., and Thomas, E., eds., Causes and Consequences of Globally Warm Climates in the Early Paleogene: Boulder, Colorado, Geological Society of America Special Paper 369, p. 501–511.

Wasatchian) fauna in the northwestern Greater Green River Basin. Finally, we present an argument for a reformulation of the Gardnerbuttean "subage" of the Bridgerian and a reevaluation of intercontinental migration scenarios at this boundary.

RARE TAXA IN MAMMALIAN BIOSTRATIGRAPHY

The development of provincial land-mammal "ages" (LMAs) reflects the difficulty of correlating terrestrial and marine sediments (e.g., Wood et al., 1941). Various criticisms have been levied against LMAs (see, e.g., Tedford, 1970; Walsh, 1998), despite their widely recognized utility. Some of these criticisms are legalistic in nature (LMAs do not or cannot fulfill the requirements of Age status established by the North American Commission on Stratigraphic Nomenclature [1983] or the International Subcommission on Stratigraphic Classification [Salvador, 1994]). The continued usage of the NALMA system in the literature reflects both its practicality and the hope that it will one day become more formally established (e.g., Savage, 1977; Woodburne, 1987).

Sample size is an especially important issue in biostratigraphic studies. It has long been recognized that the most useful organisms for temporal discrimination—that is, those which best lend themselves to biochronology—are those that (1) are widely distributed (to which one might add eurytopic) (2) are relatively common, and (3) evolve rapidly (e.g., Schindewolf, 1950; Boggs, 1995).

Mammals may be useful for biochronology because they are thought to evolve relatively rapidly. For example, Stanley (1985) estimated for them a mean species duration of 1–2 m.y. Gingerich and Gunnell (1995) calculated a median duration of ~0.8 m.y. for mammals in the Paleocene and early Eocene of the Clarks Fork Basin, and Flynn et al. (1995), based on work in the Neogene Siwalik sequence in Pakistan, estimated median species durations of 1–3 m.y., depending on the time period examined.

However, mammals sometimes fail to fulfill the first two requirements for biostratigraphic utility. Many taxa, for example, are not widely distributed. Limited geographic distribution potentially is a problem because LMAs are supposed to be useful throughout the province (continent) for which they are created. In many cases, however, narrow geographic distribution of taxa is directly related to narrow geographic distribution of sediments of the correct age. Since one can only correlate strata that exist, this is less a problem for biostratigraphy than for paleobiogeographic and evolutionary studies.

A more serious problem in mammalian biostratigraphy is the use of rare taxa for establishing zonal boundaries. (The debate over range or interval zones versus assemblage zones is a complicated one and will not be entertained here.) Walsh (1998, 2000) in particular has forcefully advocated a revised and more precise terminology for biostratigraphy and biochronology in which a distinction is made between data-oriented terms in biostratigraphy and theory-laden terms in biochronology. In his 2000 paper, he recommends the use of lowest occurrence (LO) for the lowest known stratigraphic occurrence of a taxon in a section (instead of first appearance datum, or FAD), and first historic appearance (FHA) for the true first appearance of a taxon in the same region. In other words, "occurrence" refers to the known, physical record and "appearance" to historical truth: A distinction is made between reality and our knowledge, which is hoped to approximate it. (Some mammalian biostratigraphers have appended a subscript to FAD for "known," i.e., FAD_k, which is basically synonymous with LO.)

The distinction between true first appearance and lowest occurrence is particularly important when dealing with relatively rare taxa. It is almost certain that a taxon's true stratigraphic and temporal range is greater than its known range. How well the known range can be expected to approximate the true range depends on the sampling density of the taxon in local sections (Strauss and Sadler, 1989). Taxa whose stratigraphic ranges consist of few occurrences and many large gaps have poorly constrained first and last appearances, other things being equal. It is preferable to use taxa that are relatively common because (1) their first appearances in a given area can be expected to be near their first occurrences, which permits (2) testing of isochroneity and (3) the potential for establishing chronostratigraphic (not just biostratigraphic) units. The Wasatchian-Bridgerian boundary will be discussed in light of these principles.

A "GARDNERBUTTEAN" MAMMAL IN THE LOSTCABINIAN

Since Stucky's (1984a, 1984b) work on the biostratigraphy of the Wind River Basin and broader regional application of biozones developed there, the brontotheriid *Eotitanops borealis* has been used as the principal taxon defining the beginning of the Gardnerbuttean "subage" (and hence, the Bridgerian NALMA). (In Stucky's taxonomy, *Eotitanops* was a junior synonym of *Palaeosyops.* We follow the recent revision of Gunnell and Yarborough [2000] in recognizing *Eotitanops* as a valid taxon.) Stucky (1984b) placed the Wasatchian-Bridgerian boundary at "the first appearance in the fossil record of *Palaeosyops* (including *Eotitanops*), *Hyrachyus, Trogosus, Helaletes, Omomys, Helohyus, Pantolestes, Homacodon, Washakius,* and/or *Smilodectes*" (p. 347), essentially at the base of his *Palaeosyops borealis* Assemblage Zone. Thus, the beginning of the Bridgerian biochronologic interval is defined by the "penecontemporaneous" immigration (Krishtalka et al., 1987, p. 87) of multiple taxa, and its recognition is tied to the *Eotitanops borealis* Assemblage Zone (cf. Walsh, 1998). The statement also constitutes a redefinition of the Gardnerbuttean, the first "subage" of the Bridgerian, from Robinson's (1966) original treatment of it. This new concept of the Bridgerian has been followed by a number of workers (e.g., Gunnell and Yarborough, 2000; Zonneveld et al., 2000; Gunnell and Bartels, 2001) as they have recognized and described an *Eotitanops borealis* Assemblage Zone and Bridgerian-age deposits elsewhere in the western interior of the United States.

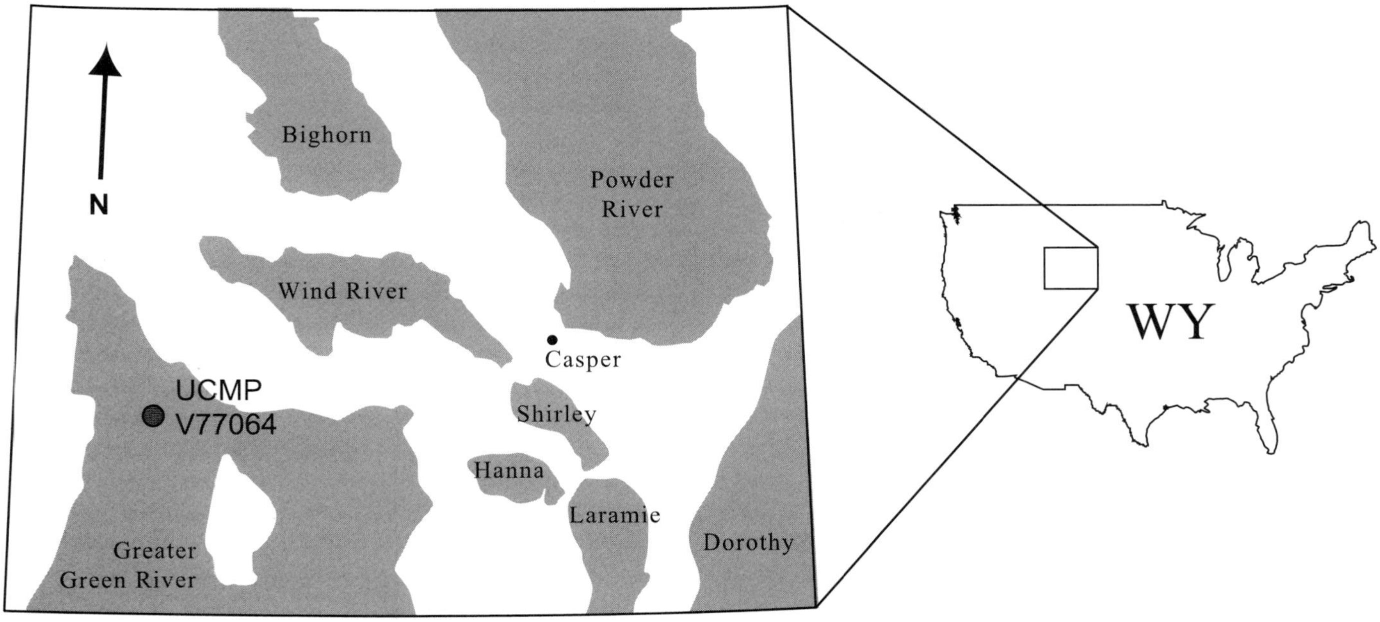

Figure 1. Map showing the locations of Laramide structural basins within the state of Wyoming (WY), USA. UCMP locality V77064 in the Greater Green River Basin is marked by filled circle. The city of Casper, indicated with a dot, is included for reference.

The La Barge–Big Piney area, Green River Basin

The Green River Basin in southwestern Wyoming (Fig. 1) is one area known to preserve strata from near the Wasatchian-Bridgerian boundary. In particular, the area around the towns of La Barge and Big Piney in the northern Green River Basin has long been known for its rich late Wasatchian (Lostcabinian) faunas (Wood et al., 1941; Gazin, 1952, 1962, 1965; West, 1973). In recent years, University of California Museum of Paleontology (UCMP) field parties have recollected a productive locality ~12 miles north of Big Piney (UCMP locality V77064) in the Main Body (formerly Knight Member) of the Wasatch Formation. V77064 is near the geographic center of the northern part of the basin and occurs stratigraphically ~35 m below the contact of the Wasatch Formation with the Scheggs Bed of the Tipton Shale Member of the Green River Formation, which documents a great expansion of paleo–Lake Gosiute (Roehler, 1993). The fossils are preserved in a drab, gray, tan-weathering mudstone and were collected by surface prospecting and quarrying of associated material.

Locality V77064 preserves a diversity of vertebrate and invertebrate taxa, including fish, turtles, crocodylians, lizards, mammals, and freshwater molluscs. Much of the vertebrate material consists of isolated teeth, jaw fragments, and bones (including abundant postcrania), although a few partial skeletons also have been found. A list of the 20 mammalian species thus far recovered and identified appears in Table 1; all these species are of medium to small size (but not very small). These mammals are for the most part Lostcabinian species (see, e.g., Schankler, 1980; Stucky, 1984a), consistent with previous assessments of the age of faunas in this area of the Green River Basin (e.g., Wood et al., 1941; Gazin, 1952; Stucky, 1984b). Although Stucky (1984a) has noted that some Wasatchian mammals are known from the lower parts of the *Eotitanops borealis* Assemblage Zone in the Wind River Basin, the predominance of Lostcabinian forms at V77064, especially the presence of the brontotheriid *Lambdotherium popoagicum*—whose first appearance defines the beginning of the Lostcabinian "subage" (Savage, 1977; Krishtalka et al., 1987; but see Smith, 2001)—indicates that this fauna should be assigned to the Lostcabinian. (For further discussion of age interpretation, see below.)

Two other perissodactyl taxa are of interest. The first is *Orohippus proteros* Korth and Evander, 1982, represented by a left maxilla with P^2 and P^4–M^3 (UCMP 154440; Fig. 2A). This species was described on the basis of a partial right maxilla with P^4–M^3 and represents a transitional form between a *Hyracotherium* ancestor and later *Orohippus* (e.g., Kitts, 1957). UCMP 154440 is from an adult individual (P^4 and M^3 have erupted) and is comparable in size to *Hyracotherium index*[1]

[1]Froehlich (2002) has recently revised the genus-level taxonomy of certain taxa formerly placed in *Hyracotherium, Orohippus,* and *Protorohippus* based on a phylogenetic analysis of selected species. Under his taxonomy, *H. index* is transferred to *Minippus*, and *H. venticolum* to *Protorohippus*. *O. proteros* was not discussed in his analysis, but it possesses the features diagnostic of the genus. Pending further testing of the phylogenetic hypothesis upon which the new genera were erected and a more comprehensive consideration of described species, we conservatively retain the genus-name *Hyracotherium* herein and provide reference to the new taxonomy where possible.

TABLE 1. MAMMALIAN FAUNAL LIST FOR UCMP LOCALITY V77064

Order	Species	N	MNI
Apatotheria	*Apatemys* sp.	1	1
Artiodactyla	*Bunophorus macropternus*	1	1
Artiodactyla	*Diacodexis* sp.	2	1
Artiodactyla	*Hexacodus pelodes*	1	1
Carnivora	*Viverravus lutosus*	3	3
Condylarthra	*Hyopsodus wortmani*	26	11
Creodonta	*Gazinocyon vulpeculus*	2	2
Creodonta	*Prolimnocyon elizabethae*	2	2
Perissodactyla	*Eotitanops borealis*	1	1
Perissodactyla	*Heptodon calciculus*	13	6
Perissodactyla	*Hyracotherium index*	21	6
Perissodactyla	*Hyracotherium venticolum*	24	7
Perissodactyla	*Lambdotherium popoagicum*	21	7
Perissodactyla	*Orohippus proteros*	1	1
Primates	*Absarokius abbotti*	4	4
Primates	*Cantius frugivorus*	6	5
Primates	*Cantius* sp. (large)	1	1
Primates	*Microsyops knightensis*	12	7
Rodentia	*Paramys copei*	3	3
Tillodontia	*Esthonyx* sp.	1	1

Note: N is number of specimens, and MNI is minimum number of individuals. Totals are N = 146 and MNI = 71. Counts based on UCMP collections made in 1976–1977, 1980, 1996–1997, and 2001. Hyracotherium index = Minippus index, and Hyracotherium venticolum = Protorohippus cf. P. venticolum, of Froehlich (2002).

(which also occurs at V77064). P^2 has two labial cusps with no trace of a lingual cusp. P^4 is subtriangular in outline, with the protocone shifted mesially relative to primitive species of *Hyracotherium.* The tooth is well worn, especially in the region of the metaloph, and there are two distinct patches there where dentine has been exposed (Fig. 2A). One is in the place of the metaconule, and the second, smaller patch is in the place of a hypocone (h). M^{1-3} are similar to those of most *Hyracotherium,* except that complete lingual cingula are present. More significantly, there are faint mesostyles (m).

The only taxon that shares these characters (size, small P^4 hypocone with corresponding subtriangular tooth shape, and incipient mesostyles on the upper molars) is *Orohippus proteros.* Our referral of UCMP 154440 to this species allows us to expand the morphological description of the species as well as confirm that, if it had a single origin, *Orohippus* diverged from *Hyracotherium* (= *Protorohippus* of Froehlich, 2002) prior to the end of the Wasatchian. UCMP 154440 is the stratigraphically lowest record of *Orohippus* in the Green River Basin (occurring below the Scheggs Bed of the Tipton Shale Member), and it also extends the geographic range of *O. proteros* to the south. *Orohippus* has previously been recognized in Lostcabinian deposits (e.g., Stucky, 1984a), so although its presence is usually associated with the Bridgerian faunas (e.g., Froehlich, 2002), that association has never been formalized in a biostratigraphic scheme that would require revision.

The second remarkable perissodactyl taxon is *Eotitanops borealis* (Cope, 1880), represented by UCMP 170392, a left maxilla fragment with M^{2-3} (Fig. 2B). Both teeth bear the characteristically brontotheriid W-shaped ectoloph with a prominent mesostyle. On M^2, the paraconule is rudimentary and is connected by a low crest to the base of the paracone. There is a large hypocone, and the metaconule (as a distinct cusp) is scarcely visible. On M^3, the paraconule is larger, and the metaconule is larger and more distinct. A very small hypocone is present. These morphologic characteristics and its size constitute the basis for assigning UCMP 170392 to *E. borealis* (although it falls at the smaller end of the *E. borealis* size spectrum). Given the nature of the rest of the fauna, we argue that this specimen represents an occurrence of *E. borealis* in the Lostcabinian.

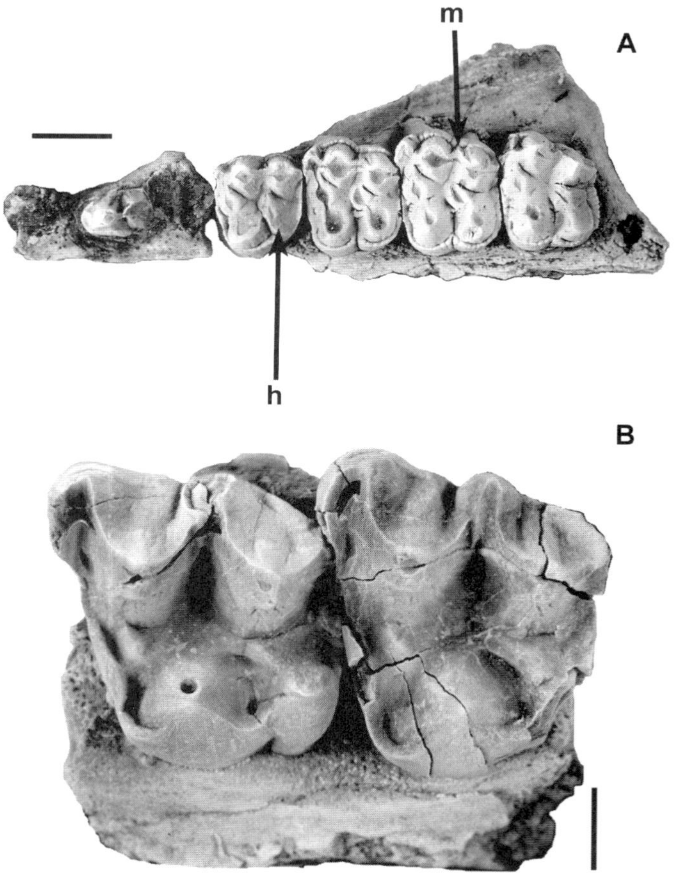

Figure 2. Perissodactyl specimens from UCMP locality V77064. A: *Orohippus proteros,* left maxilla with P^2 and P^4–M^3, UCMP 154440. h—Small hypocone, m—Mesostyle. Scale bar is 5 mm. B: *Eotitanops borealis,* left maxilla fragment with M^{2-3}, UCMP 170392. Scale bar is 5 mm.

Mammalian species richness at V77064

A relevant question to consider is whether the recovery of *Orohippus* and *Eotitanops* is simply a function of sample size. Is it possible that the recovery of these rare taxa is merely the result of greater sampling at V77064 than at other places, where these taxa are not known to cooccur with *Lambdotherium?* The sample size for V77064 (146 dental specimens) is larger than that of many Lostcabinian localities. Gazin (1952) reported 41 species from the Wasatch Formation in the La Barge–Big Piney area. Of these, 20 are currently known from V77064. (Virtually all others are represented in localities and stratigraphic levels immediately above V77064.) Table 1 shows counts and minimum numbers of individuals of all mammalian taxa recovered from V77064 by UCMP crews. Despite continued collecting, species richness at V77064 or even in the composite "La Barge fauna" is lower than that previously reported by Stucky (1984a) for the Lostcabinian of the Wind River Basin. At his single richest locality in the Wind River Basin, University of Colorado Museum (UCM) locality 80062 (Gardnerbuttean in age) in the Deadman Butte section, Stucky reported 278 specimens representing 59 species. Even if specimens belonging to very small species (notably underrepresented in our samples) are excluded, UCM locality 80062 still has 151 specimens representing 33 species.

Rarefaction techniques can be used to compare species richness in samples of different size (e.g., Tipper, 1979). We calculated rarefaction curves for these two productive localities (UCM locality 80062, UCMP V77064) using the program Analytic Rarefaction 1.3 (Holland, 2001). Using all species listed by Stucky (1984a), the expected diversity for subsampling 140 specimens is 45 ± 5 species, or 32 ± 2 species (with very small taxa excluded), as compared with the 20 species and 146 specimens documented from V77064 (Fig. 3).

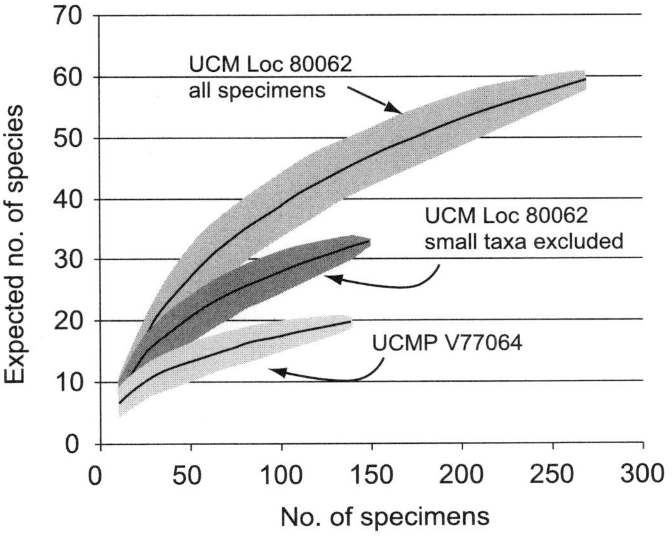

Figure 3. Rarefaction curves for three localities showing the expected number of species recovered from subsampling *x* specimens. 95% confidence intervals shaded for each curve. Calculations made using Analytic Rarefaction 1.3 (Holland, 2001). Data for UCM locality 80062 are from Stucky (1984a). "Small taxa" excluded from the middle curve are *Ectypodus* sp., *Peratherium* spp., *Peradectes chesteri,* Leptictidae sp. indet., *Apatemys* sp. A, Proteutheria sp. indet., *Scenopagus* spp., *Macrocranion nitens,* Adapisoricidae sp., *Centetodon* spp., *Nyctitherium* sp., *Microsyops* sp. cf. *M. scottianus, Uintasorex* sp., *Alveojunctus* sp., cf. *Ignacius* sp., cf. Anemorhysis sp., *Absarokius* sp. cf. *A. noctivagus, Shoshonius cooperi,* Omomyidae sp. indet., *Acritoparamys francesi, Microparamys* sp., and *Mattimys kalicola.*

Apparent lower overall diversity cannot be used as any kind of probability estimate for recovering particular taxa, but this comparison of relatively species-rich faunas highlights the problem of accurately recording the first historic appearance of a taxon if it is rare within a fauna or if the fauna is less rich due to taphonomic factors or real local or temporal differences among faunas. If the real abundance of *Eotitanops* or *Orohippus* in the Wasatchian is <1% of the fauna (as can be inferred from our samples; see also Gunnell and Yarborough, 2000, p. 363), it is unlikely that it will be found, except at particularly rich sites or due to chance.

Gunnell and Bartels (2001) have advanced the idea that "anachronistic" pairings of otherwise biostratigraphically informative taxa may represent a real phenomenon, resulting from the greater heterogeneity of basin-margin environments and hence an increased frequency of speciation events there. Although this explanation may fit for some faunas (e.g., at South Pass, Wyoming; Gunnell and Yarborough, 2000), it is not a good explanation for cooccurrences such as *Lambdotherium-Eotitanops* at V77064, because its paleogeographic position (equidistant between the Tetons and the Wind River Range) is far from the basin margins. Further, we regard such "anachronisms" as tests of biostratigraphic hypotheses, ones which may demonstrate that biostratigraphic schemes developed in one local section may not be applicable on a regional scale.

The stratigraphic range of Eotitanops

Nearly all specimens of *Eotitanops* derive from localities that preserve other characteristic Bridgerian taxa. However, Gunnell and Yarborough (2000) discussed three other potential occurrences of *Eotitanops* in Lostcabinian deposits, and their discussion can be extended in light of the new data presented here. Although we can add little regarding two of the occurrences to which they referred (Baja California: Novacek et al., 1991; and New Fork, Wyoming: West, 1973), we should note that the third locality, Yale Peabody Museum (YPM) locality 8 in the Willwood Formation of the Bighorn Basin, preserves an apparently Lostcabinian mammalian fauna. Based on examination of the collections and records at YPM, Lostcabinian forms at this locality include *Lambdotherium popoagicum, Coryphodon* sp., and *Heptodon calciculus.*[2] Other localities at

[2]The specimen of *Heptodon calciculus* is currently missing from the YPM collections, although it was measured, identified, and catalogued by D. M. Schankler two decades ago (personal communication to K.T.S., 2001).

the same stratigraphic level (591 m in the Elk Creek section; Bown et al., 1994) preserve more Lostcabinian mammals, and none of them preserves Bridgerian index taxa (besides *E. borealis* mentioned above). YPM locality 8 is not, geographically speaking, a basin-margin area. The new data reported here confirm that *Eotitanops* was present alongside traditional Lostcabinian species in some non–basin margin areas.

One might argue that the problem of anomalous occurrence is resolved simply by extending recognition of the Bridgerian down into deposits that have traditionally been called Wasatchian, because the Bridgerian is defined (in part) by the first appearance of *Eotitanops borealis.* There are two problems with this course of action. First, it would decouple the recognition of the Bridgerian from the known ranges of other traditional index taxa, none of which are present at V77064 or at the 591-m level of the southern Bighorn Basin sections (Bown et al., 1994). This disrupts the traditional use of the biozone. Second, if read literally, the known record suggests that the *Eotitanops borealis* FHA is diachronous. In the southern Bighorn Basin, the LO of this taxon is at the 591-m level of Bown et al.'s (1994) stratigraphic section, within Chron 24n.1r at ~52.95 Ma, using Age Model 1 of Wing et al. (2000). By contrast, in the Green River Basin at South Pass, the LO of *E. borealis* would fall within Chron 23r, at ca. 52 Ma based on the occurrence of *Eotitanops* and other typical Bridgerian taxa (Clyde et al., 2001). Given the seemingly diachronous nature of the *E. borealis* FHA, and the lack of any other Bridgerian taxa at V77064, we conservatively retain it within the Lostcabinian.

Anomalously low occurrences for rare taxa (e.g., *Eotitanops borealis*) may not be surprising. Biostratigraphic zones, especially in mammalian paleontology, are frequently defined by the first occurrences of particular taxa. Because essentially all lowest occurrences will be younger than first historic appearances (Strauss and Sadler, 1989), lowest occurrences are conservative estimates of first appearances. How badly these estimates "miss their mark" depends on the density of stratigraphic occurrences. Bearing this in mind, confidence intervals placed on the lowest occurrences of particular taxa are relevant. A confidence interval estimator for the lower or upper boundary of a taxon's range is provided by the equation

$$\alpha = (1 - C)^{-1(H-1)} - 1$$

where α is the confidence interval (as a proportion of the known stratigraphic range), C is the confidence level (here, we will use 0.95), and H is the number of fossiliferous levels (Strauss and Sadler, 1989; Marshall, 1990). Here, we assume that known fossil horizons are randomly distributed: Fossils are not differentially preserved in different facies (or facies are randomly distributed) and collecting has been reasonably homogeneous. Any deviation from these assumptions must temper our conclusions, but they seem reasonable for the cases we will examine.

In the Wapiti Valley sections in the Willwood Formation of the western Bighorn Basin (Fig. 1; Gunnell et al., 1992), *Eotitanops borealis* is known from two horizons, spaced ~10 m apart. With only this data, a 95% confidence interval for the first appearance of that taxon is 19 times its known stratigraphic range, or 190 m. In other words, its true range might begin as low as –80 m in the section, or well below the Willwood Formation's unconformable contact with the Cody Shale. Confidence interval estimation becomes less precise with fewer occurrences, and it is not really meaningful to speak of a confidence interval that extends across such an unconformity; however, the magnitude of the confidence interval is what we would like to emphasize. The first appearance of that taxon—because of its scarcity—cannot be confidently placed even within the Lostcabinian.

The Wind River Formation of the Wind River Basin (Fig. 1) has become a classic area for examining the Wasatchian-Bridgerian boundary, especially since the work of Stucky (1984a, 1984b). We shall return to this formation later, and here we shall simply note that *Eotitanops borealis,* in the area where the *E. borealis* Assemblage Zone was defined, has two occurrences each in two different sections. In the Red Creek section, the upper *E. borealis* specimen comes from UCM locality 80065, for which a meter-level was not provided. Thus, there is not enough (published) data to place a confidence interval on the range of *E. borealis* in this section. In the Deadman Butte section, the two occurrences are spaced ~60 m apart, implying a vast, >1100-m 95% confidence interval on its lowest occurrence. Since this interval is greater than the length of the section (as in Wapiti Valley), one cannot take the figure at face value. One can, however, conclude that the first historic appearance of *E. borealis* is poorly constrained in these sections. As these are the best documented first appearances of *E. borealis,* it is clear that the record is not sufficiently dense in these areas and that this species is too rare to say with confidence when the first historic appearance of this species might have been.

Given the poor confidence limits that can be placed on the known occurrences, we should not be surprised to find taxa such as *Eotitanops borealis* considerably lower than currently documented. The precise values on confidence intervals that we give here are not important; what is quite pertinent is their magnitude. The lowest occurrence of *E. borealis* is probably a poor estimator of its first historic appearance, and therefore it is probably not a useful taxon for establishing biostratigraphic or biochronologic boundaries. Stability in biochronology and accuracy of correlation are not promoted by using such taxa in the definitions of stratigraphic units.

THE WASATCHIAN-BRIDGERIAN BOUNDARY

Evolution of the Gardnerbuttean

The Gardnerbuttean has had a checkered history. It was originally proposed by Robinson (1966), who characterized it as an assemblage zone typified by the presence of a mix of Lostcabinian taxa, taxa unique to the zone, and taxa also known from the younger Bridgerian Blacksforkian "subage." Further, Robinson tied this zone to his Huerfano localities I, II, III, and V in the upper facies of the Huerfano Formation in northeastern Colorado. As originally defined, the Gardnerbuttean was functionally an assemblage zone characterized by taxa without regard to their range limits, because as Robinson explicitly discussed, some of these taxa may occur earlier or later in other areas. When Robinson proposed the Gardnerbuttean, he considered it a "substage" of the Wasatchian, and he did not recognize *Eotitanops borealis* in his list of characteristic taxa. In fact, the only locality at which he reported *E. borealis* (Huerfano VII) was specifically excluded from either the Gardnerbuttean or Lostcabinian. However, the Gardnerbuttean has come to mean something else (Stucky, 1984a, 1984b; Krishtalka et al., 1987; Gunnell and Yarborough, 2000; Zonneveld et al., 2000). If we follow Robinson's original conception of the Gardnerbuttean, then there is less difficulty—in light of the data presented here—in recognizing that "subage," again, as the final "subage" of the Wasatchian. But given that the last 20 years of precedent has held the Gardnerbuttean as a "subage" of the Bridgerian, we suspect that this solution would be distasteful to most workers.

If the term "*Eotitanops borealis* Assemblage Zone" is intended to denote a biozone of early Bridgerian (Gardnerbuttean) age, then it is inappropriate to apply it outside the area in which it was first defined (the Wind River Basin), and perhaps even there it may be less useful than originally thought. Because *E. borealis* is now known to occur in at least two, non–basin margin, Lostcabinian deposits and because of the apparent diachronous occurrence of *Eotitanops borealis* in different basins, we suggest that the use of the *E. borealis* Assemblage Zone (and its proposed subzones; e.g., Gunnell and Yarborough, 2000) be dropped outside the area in which it was originally defined. This course of action, however, begs the larger question of which taxon (taxa) to use in the definition and recognition of the Lostcabinian-Gardnerbuttean (hence, Wasatchian-Bridgerian) boundary. The first occurrences of immigrant taxa have long been a mainstay for mammalian paleontologists when establishing biozones, and several taxa have been argued to immigrate around the Lostcabinian-Gardnerbuttean boundary (e.g., Stucky, 1984b; Krishtalka et al., 1987). But are the first appearances of these other immigrants any better constrained than that of *Eotitanops borealis?*

First occurrences of Bridgerian taxa in the western interior

One solution to this biostratigraphic problem is predicated on the supposed nature of the Lostcabinian-Gardnerbuttean boundary, which is generally thought to reflect a pulse-like immigration event of several taxa. Krishtalka et al. (1987, p. 87) state that the "onset of the Bridgerian (Wasatchian-Bridgerian boundary) is best defined by the first, penecontemporaneous, and abrupt appearance of a number of exotic and new taxa." The taxa they list are the genera *Palaeosyops* (including *Eotitanops*), *Trogosus, Hyrachyus, Homacodon, Helohyus, Smilodectes, Omomys, Washakius,* and *Pantolestes.* This assemblage-zone concept may only reflect the poverty of our knowledge of the Lostcabinian-Gardnerbuttean boundary interval. On the other hand, if each geographic area studied is thought of as an independent sample of the boundary interval, then a similar sequence of first occurrences in each area may lend credence to the "multiple-immigrations" conception of the Wasatchian-Bridgerian boundary. Here, we briefly review some of the classic or stratigraphically better-documented boundary intervals.

The stratigraphic distributions of various Lostcabinian and Gardnerbuttean index taxa (based on the literature) and correlation of the base of the *Eotitanops borealis* Assemblage Zone between different geographic regions are shown in Figure 4. In Stucky's Red Creek section in the Wind River Formation (Fig. 4A), most of the first occurrences of Gardnerbuttean taxa are approximately at the 84-m level (UCM locality 80062). The rarity of occurrences below this level suggests that the coincidence of first occurrences may be a preservational artifact: A particularly productive locality has provided specimens of many species that may already have been present. The coincidence does not necessarily imply the simultaneous immigration of multiple taxa. In Stucky's Deadman Butte section, also in the Wind River Formation (Fig. 4A), the known record is even less dense. In Robinson's Huerfano Park study area (Fig. 4B), the records of most of the important Bridgerian taxa are known from only one, or possibly two, "levels" (if taxa such as *Helaletes, Hyrachyus, Palaeosyops,* and *Trogosus* occur in the upper part of level I; Robinson's "levels" are collection intervals thicker than a single bed). Gunnell and Yarborough (2000, p. 366) note that, in the Huerfano Park area, *Eotitanops borealis* does not overlap stratigraphically with *Lambdotherium popoagicum, Palaeosyops fontinalis,* or *Eotitanops minimus,* but we would add that *E. borealis* is known there from only one specimen. In the southwestern Green River Basin (Zonneveld et al., 2000), only *Hyrachyus* sp. and *Palaeosyops fontinalis* occur at more than two horizons (4 and 3 horizons, respectively; Fig. 4C). Note, however, that *Eotitanops borealis* and *Trogosus* sp. do not occur at all. Additionally, some Bridgerian index taxa (*Washakius* sp., *Smilodectes* sp.) first occur hundreds of meters above the first occurrence of *Hyrachyus* sp., which was used by Zonneveld et al. (2000) to recognize the base of the *Eotitanops borealis* Assemblage Zone. In Wapiti Valley, on the western margin of the Bighorn Basin, *E. borealis* first occurs 130 m below other Bridgerian taxa and does not overlap with them (Fig. 4D). *Trogosus* spp. and *Hyrachyus modestus* have LOs coincident with one another but also with *Palaeosyops fontinalis,* a later Bridgerian taxon.

In summary, the relative first occurrences of traditionally Bridgerian taxa are not consistent in different geographic areas. Inconsistent LOs do not necessarily imply that FHAs were not synchronous. As discussed above, LOs may be poor estimators of FHAs when sample sizes are small. However, this exercise does highlight the relative poverty of our current understanding of the Wasatchian-Bridgerian boundary. If this boundary is to be

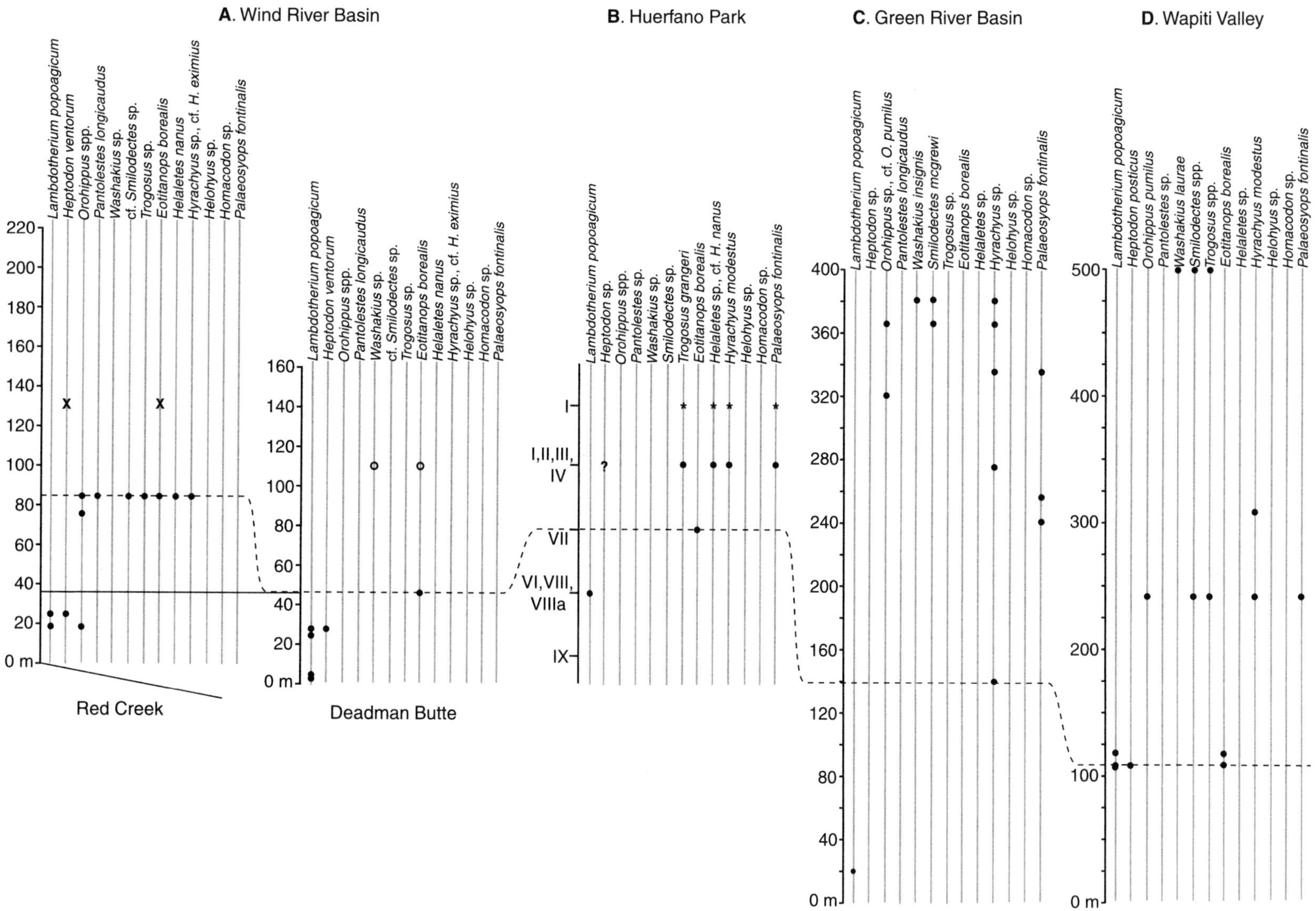

Figure 4. Stratigraphic ranges of late Wasatchian and early Bridgerian index taxa in four geographic areas of the western interior. A solid dot indicates the occurrence, at a given level, of at least one specimen of a taxon. The dashed line indicates correlation between sections of the base of the *Eotitanops borealis* Assemblage Zone (operationally, the beginning of the Bridgerian). A: Wind River Basin, Wyoming (see Figure 1). Data for the Red Creek and Deadman Butte sections are from Stucky (1984a). Solid lines indicate lithologic correlations. X indicates an occurrence at an unknown stratigraphic level above ~84 m. Open circle indicates an occurrence at an approximate stratigraphic level. B: Huerfano Park, Colorado. Data are from Robinson (1966). ? indicates questionable taxonomic assignment of the *Heptodon* specimen. * indicates that at least one specimen is from level I, which spans a significant vertical range; the specimen may or may not be known from the upper part of the range. Our correlation, though different from Robinson's, is consistent with current usage (e.g., Gunnell and Yarborough, 2000). C: Southwestern Green River Basin, Wyoming (see Figure 1). Data are from Zonneveld et al. (2000). D: Wapiti Valley, western margin of Bighorn Basin (see Figure 1). Data are from Gunnell et al. (1992). Note that the base of the *Eotitanops borealis* Assemblage Zone, as shown here, differs from the proposed correlation of Gunnell et al. (1992). However, since *Palaeosyops fontinalis* (generally regarded to be a later Bridgerian species; e.g., Gunnell and Yarborough, 2000) occurs at the 240-m level, and *Eotitanops borealis* first occurs at the 110-m level, we feel that our correlation is accurate insofar as the relevant biozone is currently defined.

redefined, as the data presented and reviewed here suggest, then it is not entirely clear how to do so. We should note that *Hyrachyus* appears to be more common than other traditionally Bridgerian taxa and less patchy in its geographic occurrence, but it would be premature, without much more extensive review, to redefine the boundary using the first historic appearance (or, in functional terms, the lowest occurrence) of *Hyrachyus.*

Nature of the Wasatchian-Bridgerian boundary

The migration of taxa along high-latitude corridors is a well-documented phenomenon. Dispersals to North America during the Paleogene (Krause and Maas, 1990), in particular from Asia (Beard, 1998), are thought to have been important factors effecting the reorganization of North American faunas (Webb and Opdike, 1995). The Clarkforkian-Wasatchian boundary marks a particularly well-documented migration episode that was pulse-like in nature (Rose, 1981; Gingerich, 1989) and coincident with geologically rapid global warming (e.g., Koch et al., 1995). Perissodactyls, artiodactyls, hyaenodontid creodonts, and omomyid and adapid primates all have their first North American appearances at this time.

As noted above, the Wasatchian-Bridgerian boundary is also thought to reflect a pulse-like migration event. However, given our uncertain knowledge of the boundary interval, several other possibilities should be considered. These alternative hypotheses might be grouped into (1) those postulating pulse-like migration (i.e., with multiple, synchronous first appearances) (2) those postulating staggered filtering of taxa across high-latitude corridors, and (3) evolution of taxa in poorly sampled areas, with subsequent migration into areas where fossil faunas are well represented. The abundance of particular taxa may also be important.

First, there may have been a pulse-like dispersal event that occurred at or somewhat before the currently recognized Wasatchian-Bridgerian boundary, but the first occurrences of many of the immigrant taxa have been smeared forward in time in a manner analogous to the classic Signor-Lipps effect (Signor and Lipps, 1982), in which truly synchronous last appearances are smeared backward in time due to incomplete preservation. This possibility for immigration events in general was suggested by Springer (1990) and may be termed the Sppil-Rongis effect, although its implications for biostratigraphy have not been adequately addressed. Second, there may have been a pulse-like dispersal event prior to the currently recognized Wasatchian-Bridgerian boundary, but the abundances of immigrant taxa remained very low until some time after their first appearances. In this scenario, first occurrences are recorded when taxa become more abundant, or they may be recorded earlier if sample sizes are large enough. As noted above, sample sizes from V77064 and UCM locality 80062 are considerably larger than in Wapiti Valley (Bighorn Basin margin), although they are smaller than at South Pass (Greater Green River Basin). Both of these suggestions imply that as temperatures increased, some ecologic threshold was attained that suddenly permitted numerous taxa to emigrate from elsewhere simultaneously.

Third, immigrant taxa may have filtered across high-latitude corridors in a staggered fashion. In this case, *Eotitanops* would simply represent a taxon that came earlier than has been thought (i.e., prior to Bridgerian taxa such as *Hyrachyus* and *Trogosus*). Among perissodactyls, *Heptodon* and *Lambdotherium* are other examples of taxa that appear at irregular intervals during the latter part of the Wasatchian. It is important to recognize here that, in constrast to the Clarkforkian-Wasatchian boundary (e.g., Koch et al., 1995), global temperatures were increasing far more gradually near the Wasatchian-Bridgerian boundary (see, e.g., Zachos et al., 2001), and therefore one may even expect that immigration events are staggered through this interval. This scenario also implies that factors other than just climate—e.g., paleoautecology—are responsible for controlling the immigration of taxa during optimal periods.

Finally, taxa may have immigrated either in concert or disjunctly, but initially resided in geographic regions not well sampled prior to the Wasatchian-Bridgerian boundary. Areas of origin or initial residence could lie elsewhere in North America or perhaps within the same basin (i.e., basin margin faunas; Gunnell and Bartels, 2001). The record of early to middle Eocene faunas is extremely poor outside the Laramide intermontane basins of Wyoming and Colorado, and in areas where individual faunas from this interval are well known (e.g., in the San Juan Basin of New Mexico, Baja California), the overall record of the early to middle Eocene is discontinuous.

CONCLUSIONS

Real and objective differences exist between faunas traditionally called late Wasatchian (Lostcabinian) and early Bridgerian. However, the nature of the boundary between these land-mammal "ages" is less clear. This lack of clarity stems in part from the use of rare taxa to define the boundary. The discovery of a Bridgerian index taxon (*Eotitanops borealis*) from the same locality as traditional Lostcabinian taxa highlights the problems associated with the definition of this boundary and associated biozones. Given the low relative abundance of *E. borealis* in the Bridgerian faunas from which it is best known, it is difficult to place reliable confidence intervals on its lowest occurrences in local sections. Its rarity makes it less useful for biostratigraphy and the identification of isochronous events in the fossil record. Our results lead us to affirm that rare taxa should not be used as the sole biostratigraphic indicators, since they are unlikely to be reliable chronostratigraphic proxies. More importantly, the uncertain nature of the Wasatchian-Bridgerian boundary may also reflect the possibility that it does not represent a short-term faunal turnover (as seen at the Paleocene-Eocene boundary). Anomalous occurrences, as reported here, emphasize the poverty of our knowledge of the Wasatchian-Bridgerian boundary interval (in comparison with the much better-studied Clarkforkian-Wasatchian boundary) and suggest that other scenarios be considered as we continue to refine our understanding of this important transition.

ACKNOWLEDGMENTS

Howard Hutchison (UCMP) has been invaluable during fieldwork in the Green River Basin. William Clyde, Gregg Gunnell, Peter Robinson, and John-Paul Zonneveld provided much useful discussion of and information on the (bio)stratigraphic problems associated with the Wasatchian-Bridgerian boundary, and Gregg Gunnell and Richard Stucky thoughtfully reviewed the manuscript. We gratefully acknowledge Anne Weil (Duke University) for having suggested the term "Sppil-Rongis effect." Bob Purdy (USNM) provided access to collections and assistance in archival work. Fieldwork has been carried out with the cooperation of the Bureau of Land Management, and we especially thank the staff at the Pinedale Field Office for their help and enthusiasm. Daniel Brinkman, Lyndon Murray, and Mary Ann Turner provided the faunal data for YPM locality 8, and David Schankler kindly provided measurements and specific identification for the *Heptodon* specimen from that locality. Funding for fieldwork has come from the Annie M. Alexander endowment to UCMP. This work is also supported by a NSF Graduate Research Fellowship to Krister T. Smith. This is UCMP publication number 1761.

REFERENCES CITED

Badgley, C., and Gingerich, P.D., 1988, Sampling and faunal turnover in early Eocene mammals: Palaeogeography, Palaeoclimatology, Palaeoecology, v. 63, p. 141–157.

Beard, K.C., 1998, East of Eden: Asia as an important center of taxonomic origination and mammalian evolution, *in* Beard, K.C., and Dawson, M.R., eds., Dawn of the Age of Mammals in Asia: Bulletin of Carnegie Museum of Natural History 34, p. 5–39.

Boggs, S., Jr., 1995, Principles of sedimentology and stratigraphy: Upper Saddle River, New Jersey, Prentice-Hall, 774 p.

Bown, T.M., Rose, K.D., Simons, E.L., and Wing, S.L., 1994, Distribution and stratigraphic correlation of upper Paleocene and lower Eocene fossil mammal and plant localities of the Fort Union, Willwood, and Tatman Formations, southern Bighorn Basin, Wyoming: U.S. Geological Survey Professional Paper 1540, p. 1–103.

Clyde, W.C., Zonneveld, J.-P., Stamatakos, J., Gunnell, G.F., and Bartels, W.S., 1997, Magnetostratigraphy across the Wasatchian/Bridgerian NALMA boundary (early to middle Eocene) in the western Green River Basin, Wyoming: Journal of Geology, v. 105, p. 657–669.

Clyde, W.C., Sheldon, N.D., Koch, P.L., Gunnell, G.F., and Bartels, W.S., 2001, Linking the Wasatchian/Bridgerian boundary to the Cenozoic Global Climate Optimum: New magnetostratigraphic and isotopic results from South Pass, Wyoming: Palaeogeography, Palaeoclimatology, Palaeoecology, v. 167, p. 175–199.

Cope, E.D., 1880, Geology and palaeontology: American Naturalist, v. 14, p. 745–748.

Flynn, L.J., Barry, J.C., Morgan, M.E., Pilbeam, D., Jacobs, L.L., and Lindsay, E.H., 1995, Neogene Siwalik mammalian lineages: Species longevities, rates of change, and modes of speciation: Palaeogeography, Palaeoclimatology, Palaeoecology, v. 115, p. 249–264.

Froehlich, D.J., 2002, Quo vadis eohippus? The systematics and taxonomy of early Eocene equids (Perissodactyla): Zoological Journal of the Linnean Society, v. 134, p. 141–256.

Gazin, C.L., 1952, The lower Eocene Knight Formation of western Wyoming and its mammalian faunas: Smithsonian Miscellaneous Collections, v. 117, no. 18, p. 1–82.

Gazin, C.L., 1962, A further study of the lower Eocene mammalian faunas of southwestern Wyoming: Smithsonian Miscellaneous Collections, v. 144, no. 1, p. 1–98.

Gazin, C.L., 1965, Early Eocene mammalian faunas and their environment in the vicinity of the Rock Springs uplift, Wyoming: Wyoming Geological Association, 19th Annual Field Conference, Guidebook, p. 171–180.

Gingerich, P.D., 1989, New earliest Wasatchian mammalian fauna from the Eocene of northwestern Wyoming: Composition and diversity in a rarely sampled high-floodplain assemblage: University of Michigan Papers on Paleontology, v. 28, p. 1–97.

Gingerich, P.D., and Gunnell, G.F., 1995, Rates of evolution in Paleocene-Eocene mammals of the Clarks Fork Basin, Wyoming, and a comparison with Neogene Siwalik lineages of Pakistan: Palaeogeography, Palaeoclimatology, Palaeoecology, v. 115, p. 227–247.

Gunnell, G.F., and Bartels, W.S., 2001, Basin margins, biodiversity, evolutionary innovation, and the origin of new taxa, *in* Gunnell, G.F., ed., Eocene vertebrates: Unusual occurrences and rarely sampled habitats: New York, Kluwer Academic/Plenum Press, p. 403–432.

Gunnell, G.F., and Yarborough, V.L., 2000, Brontotheriidae (Perissodactyla) from the late early and middle Eocene (Bridgerian), Wasatch and Bridger formations, southern Green River Basin, southwestern Wyoming: Journal of Vertebrate Paleontology, v. 20, p. 349–368.

Gunnell, G.F., Bartels, W.S., Gingerich, P.D., and Torres, V., 1992, Wapiti Valley faunas: Early and middle Eocene fossil vertebrates from the North Fork of the Shoshone River, Park County, Wyoming: Contributions from the Museum of Paleontology, University of Michigan, v. 28, p. 247–287.

Holland, S.M., 2001, Analytic Rarefaction 1.3: http://www.uga.edu/~strata/software (October 2001).

Kidwell, S.M., 2001, Preservation of species abundance in marine death assemblages: Science, v. 294, p. 1091–1094.

Kitts, D.B., 1957, A revision of the genus *Orohippus* (Perissodactyla, Equidae): American Museum Novitates, no. 1864, 40 p.

Koch, P.L., Zachos, J.C., and Dettman, D.L., 1995, Stable isotope stratigraphy and paleoclimatology of the Paleogene Bighorn Basin (Wyoming, USA): Palaeogeography, Palaeoclimatology, Palaeoecology, v. 115, p. 61–89.

Korth, W.W., and Evander, R.L., 1982, A new species of *Orohippus* (Perissodatyla, Equidae) from the early Eocene of Wyoming: Journal of Vertebrate Paleontology, v. 2, p. 167–171.

Krause, D.W., and Maas, M.C., 1990, The biogeographic origins of late Paleocene–early Eocene mammalian immigrants to the western interior of North America, *in* Bown, T.M., and Rose, K.D., eds., Dawn of the Age of Mammals in the northern part of the Rocky Mountain interior, North America: Boulder, Colorado, Geological Society of America Special Paper 243, p. 71–105.

Krishtalka, L., West, R.M., Black, C.C., Dawson, M.R., Flynn, J.J., Turnbull, W.D., Stucky, R.K., McKenna, M.C., Bown, T.M., Golz, D.J., and Lillegraven, J.A., 1987, Eocene (Wasatchian through Duchesnean) biochronology of North America, *in* Woodburne, M.O., ed., Cenozoic mammals of North America: Geochronology and biostratigraphy: Berkeley, California, University of California Press, p. 77–117.

Marshall, C.R., 1990, Confidence intervals on stratigraphic ranges: Paleobiology, v. 16, p. 1–10.

North American Commission on Stratigraphic Nomenclature, 1983, North American stratigraphic code: AAPG (American Association of Petroleum Geologists) Bulletin, v. 67, p. 841–875.

Novacek, M.J., Ferrusquia-Villafranca, I., Flynn, J.J., Wyss, A.R., and Norell, M.A., 1991, Wasatchian (early Eocene) mammals and other vertebrates from Baja California, Mexico: The Lomas las Tetas de Cabra fauna: Bulletin of the American Museum of Natural History, v. 208, p. 1–88.

Robinson, P., 1966, Fossil mammalia of the Huerfano Formation, Eocene, of Colorado: Peabody Museum of Natural History Bulletin 21, p. 1–95.

Roehler, H.W., 1993, Eocene climates, depositional environments, and geography, Greater Green River Basin, Wyoming, Utah, and Colorado: U.S. Geological Survey Professional Paper 1506F, p. 1–74.

Rose, K.D., 1981, The Clarkforkian Land-Mammal Age and mammalian faunal composition across the Paleocene-Eocene boundary: University of Michigan Papers on Paleontology, v. 26, p. 1–197.

Salvador, A., editor, 1994, International stratigraphic guide: A guide to stratigraphic classification, terminology, and procedure: Boulder, Colorado, Geological Society of America, 214 p.

Savage, D.E., 1977, Aspects of vertebrate paleontological stratigraphy and geochronology, *in* Kauffman, E.G., and Hazel, J.E., eds., Concepts and methods of biostratigraphy: Stroudsburg, Pennsylvania, Dowden, Hutchison, and Ross, p. 427–442.

Schankler, D.M., 1980, Faunal zonation of the Willwood Formation in the central Bighorn Basin, Wyoming: University of Michigan Papers on Paleontology, v. 24, p. 99–114.

Schankler, D.M., 1981, Local extinction and ecological reentry of early Eocene mammals: Nature, v. 293, p. 135–138.

Schindewolf, O., 1950 [1993], Basic questions in paleontology: Geologic time, organic evolution, and biological systematics [translated by Judith Schaefer]: Chicago, Illinois, University of Chicago Press, 467 p.

Signor, P.W., III, and Lipps, J.W., 1982, Sampling bias, gradual extinction patterns, and catastrophes in the fossil record, *in* Silver, L.T., and Schultz, P.H., eds., Geological implications of impacts of large asteroids and comets on the Earth: Boulder, Colorado, Geological Society of America Special Paper 190, p. 291–296.

Smith, K.T., 2001, Reassessing the *Lambdotherium* first appearance datum (Wasatchian, early Eocene) in the Bighorn Basin, Wyoming: PaleoBios, v. 21, p. 1–11.

Springer, M.S., 1990, The effect of random range truncations on patterns of evolution in the fossil record: Paleobiology, v. 16, p. 512–520.

Stanley, S.M., 1985, Rates of evolution: Paleobiology, v. 11, p. 13–26.

Strauss, D., and Sadler, P.M., 1989, Classical confidence intervals and Bayesian probability estimates for ends of local taxon ranges: Mathematical Geology, v. 21, p. 411–427.

Stucky, R.K., 1984a, Revision of the Wind River faunas, early Eocene of central Wyoming, Part 5: Geology and biostratigraphy of the upper part of the Wind River Formation, northeastern Wind River Basin: Annals of Carnegie Museum, v. 53, p. 231–294.

Stucky, R.K., 1984b, The Wasatchian-Bridgerian land mammal age boundary (early to middle Eocene) in western North America: Annals of Carnegie Museum, v. 53, p. 347–382.

Tedford, R.H., 1970, Principles and practices of mammalian geochronology in North America: Proceedings of the North American Paleontological Convention, Part F, p. 666–703.

Tipper, J.C., 1979, Rarefaction and rarefiction: The use and abuse of a method in paleontology: Paleobiology, v. 5, p. 423–434.

Walsh, S.L., 1998, Fossil datum and paleobiological event terms, paleontostratigraphy, chronostratigraphy, and the definition of Land Mammal "Age" boundaries: Journal of Vertebrate Paleontology, v. 18, p. 150–179.

Walsh, S.L., 2000, Eubiostratigraphic units, quasibiostratigraphic units, and "assemblage zones": Journal of Vertebrate Paleontology, v. 20, p. 761–775.

Webb, S.D., and Opdike, N.D., 1995, Global climatic influence on Cenozoic land mammal faunas, *in* National Research Council, reviewers, Effects of past global change on life: Washington, D.C., National Academies Press, p. 184–208.

West, R.M., 1973, Geology and mammalian paleontology of the New Fork–Big Sandy area, Sublette County, Wyoming: Fieldiana, Geology, v. 29, p. 1–193.

Wing, S.L., Bao, H., and Koch, P.L., 2000, An early Eocene cool period? Evidence for continental cooling during the warmest part of the Cenozoic, *in* Huber, B.T., et al., eds., Warm climates in Earth history: Cambridge, UK, Cambridge University Press, p. 197–237.

Wood, H.E., II, Chaney, R.W., Clark, J., Colbert, E.H., Jepsen, G.L., Reeside, J.B., and Stock, C., 1941, Nomenclature and correlation of the North American continental Tertiary: Geological Society of America Bulletin, v. 52, p. 1–48.

Woodburne, M.O., 1987, Principles, classification, and recommendations, *in* Woodburne, M.O., ed., Cenozoic mammals of North America: Geochronology and biostratigraphy: Berkeley, California, University of California Press, p. 9–17.

Woodburne, M.O., and Swisher, C.C., 1995, Land mammal high-resolution geochronology, intercontinental overland dispersals, sea level, climate, and vicariance, *in* Berggren, W.A., et al., eds., Geochronology, time scales, and global stratigraphic correlations: Unified temporal framework for an historical geology: SEPM (Society for Sedimentary Geology) Special Publication 54, p. 335–364.

Zachos, J., Pagani, M., Sloan, L., Thomas, E., and Billups, K., 2001, Trends, rhythms, and aberrations in global climate 65 Ma to present: Science, v. 292, p. 686–693.

Zonneveld, J.-P., Gunnell, G.F., and Bartels, W.S., 2000, Early Eocene fossil vertebrates from the southwestern Green River Basin, Lincoln and Uinta Counties, Wyoming: Journal of Vertebrate Paleontology, v. 20, p. 369–386.

Manuscript Accepted by the Society August 13, 2002

Geological Society of America
Special Paper 369
2003

Terrestrial mammals as biostratigraphic indicators in upper Paleocene–lower Eocene marine deposits of the southern North Sea Basin

Thierry Smith*
Département de Paléontologie, Institut royal des Sciences naturelles de Belgique, 29 rue Vautier, B-1000 Brussels, Belgium
Richard Smith
Laekenveld 6, B-1780 Wemmel, Belgium

ABSTRACT

Teeth of terrestrial mammals found in shallow marine deposits of the late Paleocene and early Eocene in the southern North Sea Basin (Belgium, northern France and southeastern England) have been used as biostratigraphic indicators. Analyses indicate that the age of the continental Walbeck mammal fauna (Germany) is close to that of the Upper Selandian Heers Formation of Belgium (NP4-5). The MP6 reference-level of Cernay (France) is probably correlated with the lower part of NP9 (late Thanetian). The MP7 – MP8 + 9 intermediate faunas of Meudon and Pourcy could be partly equivalent in age to Biochron NP10. The MP8 + 9 reference-level of Avenay corresponds to the upper part of the London Clay and Kortrijk Formations, which are of late middle Ypresian age (lower NP12), or to the lower part of the Wittering and Tielt Formations, which are dated early late Ypresian (middle NP12). The MP10 Grauves and Prémontré faunas (France) are correlated with the NP13 Upper Wittering Formation. The taphonomy of terrestrial mammals discovered in marine deposits indicates several origins of the material such as reworking, action of predators or fluvial transport.

INTRODUCTION

The correlation between marine and continental domains has always been a problem of compatibility. Nearly all biostratigraphic markers are different and, so far, only palynomorphs have been used efficiently. Chemostratigraphic correlation of $\delta^{13}C$ variation allows a better precision and may be the solution for a good correlation between terrestrial and marine deposits. However, this depends on some sedimentological factors such as the presence of carbonate or organic matter in the sediments. Also, sufficiently thick and complete outcrops are needed in order to obtain a significant $\delta^{13}C$ curve. A series of successive values are indeed necessary to indicate the tendencies of the correlated curves. Contrary to some very thick early Paleogene continental outcrops of North America (e.g., Wasatchian deposits in the Bighorn Basin of Wyoming), most of the continental outcrops in Northwestern Europe are small and alternating with marine deposits. This is why only a few outcrops or boreholes in the southern North Sea Basin give convincing $\delta^{13}C$ variation results (see Steurbaut et al., this volume).

The discovery of new fossils of terrestrial mammals in several shallow marine deposits of the late Paleocene and early

*E-mail: Thierry.Smith@naturalsciences.be

Smith, T., and Smith, R., 2003, Terrestrial mammals as biostratigraphic indicators in upper Paleocene–lower Eocene marine deposits of the southern North Sea Basin, *in* Wing, S.L., Gingerich, P.D., Schmitz, B., and Thomas, E., eds., Causes and Consequences of Globally Warm Climates in the Early Paleogene: Boulder, Colorado, Geological Society of America Special Paper 369, p. 513–520.

Eocene in the southern North Sea Basin (Belgium, northern France and southeastern England; Fig. 1) enable us to use them as biostratigraphic indicators. Previously published discoveries have also been included.

MICROMAMMALS FROM MARINE DEPOSITS

New discoveries

The small European microsyopoid plesiadapiformes *Sarnacius* and *Berruvius,* so far only known from fluviatile deposits of Cernay and Berru (MP6, Marne, France), are present in the middle Thanetian Tuffeau de St-Omer of Templeuve (NP8, Nord, France). Five teeth and fragments of teeth, discovered in a level very rich in shark teeth (see Moreau and Mathis, 2000), are referred to *Sarnacius* sp. (Fig. 2). They resemble *S. gingerichi* (Russell, 1981) from Cernay but the cusps are more slender and higher, in particular on the lower molar, giving an insectivorous aspect to the teeth. This character attests to a more primitive condition of *Sarnacius* sp. from Templeuve than in all other toliapinids. *S. gingerichi* and other toliapinids show lower cusps, more prominent crests encircling a concave talonid, and a more reduced hypoconulid than in *Sarnacius* sp. from Templeuve. The most derived toliapinids such as *Avenius* Russell et al., 1992 and *Toliapina* Hooker et al., 1999 from the middle Ypresian have the lower molar talonid surrounded by strong

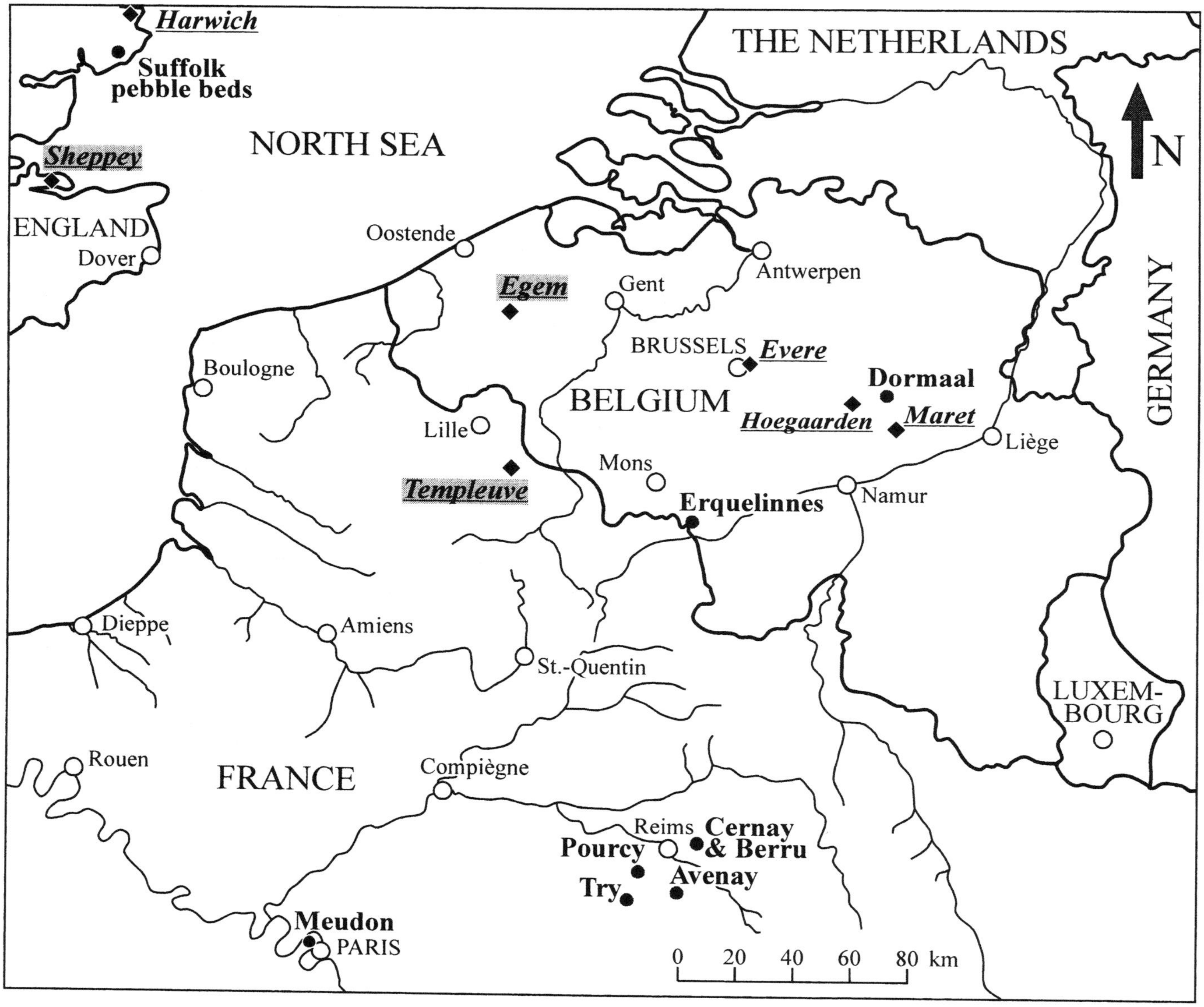

Figure 1. Location of mammal sites (in bold) from the upper Paleocene and lower Eocene in Belgium, northern France and southeastern England. The marine localities are underlined and the three sites that have yielded new mammal material are in grey.

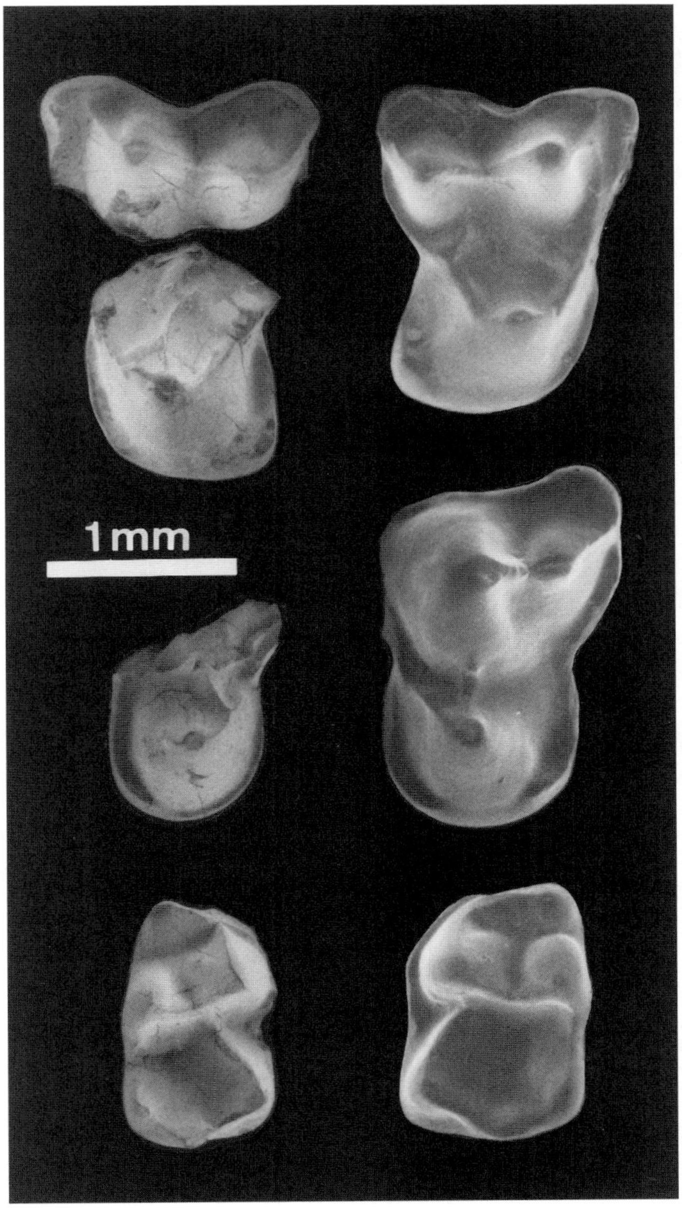

Figure 2. M^1 (above), P^4 (middle) and m_1 (below) of *Sarnacius* sp. from the NP8 Thanetian Tuffeau de St-Omer of Templeuve (at left) and *Sarnacius gingerichi* from the MP6 reference-level of Cernay (at right). All in occlusal view.

crests in which cusps are difficult to recognize (Hooker et al., 1999). One left M^2 of *Berruvius* cf. *lasseroni* Russell, 1964 (Fig. 3A) found in the same level as *Sarnacius* sp. is similar in size and general morphology to *B. lasseroni* from Cernay. Additional material is needed to confirm the species identification.

A left m_1 of the nyctitheriid insectivore *Leptacodon* sp. (Fig. 3B) has been found from late middle Ypresian of the Isle of Sheppey (NP12, London Clay Formation, Kent, England). It is very similar to the lower molars of the undescribed *Leptacodon* n. sp. from the "Falun d'Avenay" (MP8 + 9, "Sparnacian," Marne, France) by the morphology of the talonid. In both taxa, the hypoflexid is not well developed (the anterior part of the oblique crest does not go toward the centre of the posterior wall of the trigonid but is relatively labial). The hypoconulid and the entoconid are relatively small and a small crest is present between the hypoconid and the hypoconulid. *Leptacodon* is not present in the Paleocene of Europe and seems to be originating from North America at the Paleocene-Eocene boundary (Smith, 1999). The oldest European species such as *L. dormaalensis* (Quinet, 1964) from Dormaal (MP7, Tienen Formation, Belgium) and *L. nascimentoi* Estravis, 1996 from Silveirinha (MP7, lower "Sparnacian," Baixo Mondego, Portugal) have a larger and deeper hypoflexid, and more developed entoconid and hypoconulid than in *Leptacodon* sp. from Sheppey and Avenay.

The Euramerican erinaceomorph insectivore *Macrocranion* cf. *nitens* (Matthew, 1918), described from the middle Wasatchian continental deposits of the Willwood Formation (Bighorn Basin, Wyoming) and also from Avenay and Mutigny (MP8 + 9, Marne, France; Russell et al., 1975), is recognized on the basis of a right M^1 (Fig. 4) in the early late Ypresian locality of Egem (NP12 according to Steurbaut, 1998; Tielt Formation, West Vlaanderen, Belgium). A partial P^4 (Fig. 3C) of the multituberculate *Neoplagiaulax* aff. *sylvani* Vianey-Liaud, 1986 known from Mutigny (Vianey-Liaud, 1986, p. 160, Figure 57) and a right mandible fragment with m_{1-3} of an indeterminate nyctitheriid-type insectivore have also been discovered in Egem.

Other known micromammals

Among the oldest Paleocene micromammals known from the southern North Sea Basin, the erinaceomorph insectivore *Adapisorex* sp. has been described on the basis of a right DP^4 from the upper Selandian Orp Sand Member at Maret (Heers Formation, NP4-5, Brabant) in Belgium (Herman and Sigé, 1975). It is closely related to *A. abundans* Russell, 1964 from Walbeck (MP5?, Germany). The type species *A. gaudryi* Lemoine, 1883 (see Russell, 1964) from Cernay (MP6, upper Thanetian, Paris Basin, France) is probably the most derived species because it has a larger and more molariform DP^4 than *Adapisorex* sp. and *A. abundans*. A fragment of a right M^2 of the arctocyonid archaic ungulate *Arctocyonides* cf. *weigelti* Russell, 1964 has been found in the basal gravel of the "Tuffeau de Lincent," also in Maret. It is very similar to the M^2 of *Arctocyonides weigelti* only known from Walbeck. Nearly all material from this gravel, which also yielded many shark teeth, is reworked from the underlying Heers Formation (Herman and Sigé, 1975, p. 232).

The typical MP6 *Arctocyonides arenae* Russell 1964 described originally from Cernay and Berru (MP6, upper Thanetian, Paris Basin, France) has been recently identified on the basis of a DP^4 in the Beltinge Fish Bed at the base of the late Thanetian Upnor Formation (NP9, Herne Bay, Kent, England) by Hooker and Millbank (2001). No DP^4 of *A. arenae* have been formally described from Cernay or Berru and the determination

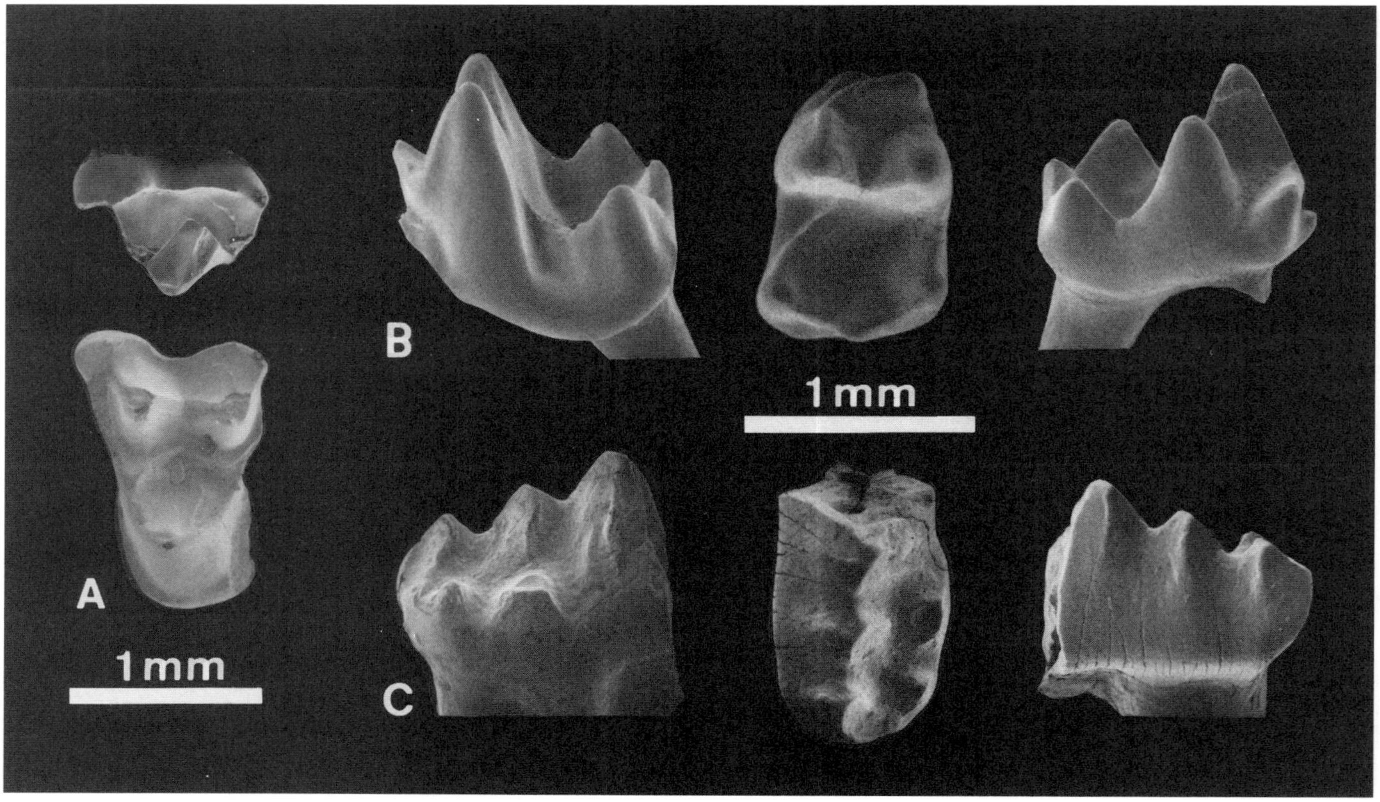

Figure 3. A: Left M^2 of *Berruvius* cf. *lasseroni* from the NP8 Thanetian Tuffeau de St-Omer of Templeuve. Occlusal view (below) and labial view (above). B: Left m_1 of *Leptacodon* sp. from the NP12 Ypresian London Clay Formation of the Isle of Sheppey. Labial view (left), occlusal view (middle) and lingual view (right). C: Partial P^4 of the multituberculate *Neoplagiaulax aff. sylvani* from the NP12 early late Ypresian of Egem. Lateral views (left and right) and occlusal view (middle).

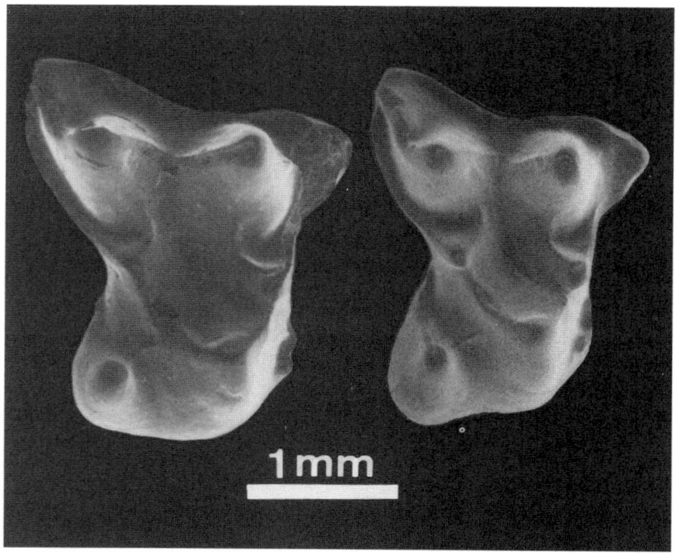

Figure 4. Right M^1 of *Macrocranion* cf. *nitens* from the NP12 early late Ypresian of Egem (at left) and from the MP8 + 9 reference-level of Avenay (at right). Both in occlusal view.

of the species from the Upnor Formation needs to be confirmed. Nevertheless, this species is closely related to *A. arenae.*

The equoid perissodactyl *Pliolophus vulpiceps* Owen, 1858 from the Harwich Stone band at Harwich (Harwich Formation, "NP10", Essex, England) has also been described (14 lower and upper jugal teeth) from the "falun de Pourcy" (MP7?, Paris Basin, Marne, France) by Hooker (1994).

The presence of the MP8 + 9 rodent *Paramys ageiensis* (Michaux, 1964) is mentioned (one tooth not figured) in the basal Swanscombe Member (London Clay Formation, "NP10," early Ypresian) of Upnor (Kent) by Hooker (1991). If the identification is confirmed, this could be the first occurrence of this species.

The equoid perissodactyl *Hyracotherium leporinum* Owen, 1841 and the plesiadapid plesiadapiformes *Platychoerops richardsonii* Charlesworth, 1855 have been described from Studd Hill (London Clay Formation, NP11, Kent, England). *Hyracotherium leporinum* could be derived from *H.* aff. *leporinum* from Meudon (MP7?, Paris Basin, France) and Pourcy (Hooker, 1994).

The holotype of the microsyopoid plesiadapiformes *Toliapina lawsoni* Hooker et al., 1999 was described from the late middle Ypresian of Sheppey (London Clay Formation, NP12) and one specimen comes from the fluviatile MP8 + 9 deposit of Mutigny (Hooker et al., 1999).

A left m_{1-2} of the MP8 + 9 didelphid marsupial *Peratherium matronense* Crochet, 1979 is known from the late middle Ypresian of Evere (Brussels, Brabant) in Belgium (Kortrijk Clay Formation, NP12; Smith and Russell, 1992). The holotype (a right mandible fragment with p_4-m_1) of the bat *Hassianycteris joeli* Smith and Russell, 1992 was also found in this level.

Hooker (1996) described mammals from the late Ypresian to Lutetian Bracklesham Group from southern England. Among the Ypresian material, a tooth fragment of the rodent *Pseudoparamys* sp. comes from Peel Common (bed PC6) in the Lower Wittering Formation (NP12, Hampshire, England). It compares closely in size and morphology with *Pseudoparamys teilhardi* (Wood, 1962) from Mutigny and Avenay (Hooker, 1996).

Three localities in the Upper Wittering Formation (NP13) have yielded mammals. The didelphid marsupials *Peratherium* sp. and *Amphiperatherium maximum* Crochet, 1979, the paramyid rodent *Pantrogna* sp., the pantolestid Cimolesta *Buxolestes* sp., the nyctitheriid insectivore *Saturninia* sp. 1 and the equoid perissodactyl *Hyracotherium leporinum* Owen, 1841 have been found in Dummer's Copse (Hampshire, England). Another nyctitheriid, *Saturninia* sp. 2, comes from Shore Road Berd (Hampshire). The rodent *Ailuravus michauxi* Hartenberger, 1975 known from Grauves (MP10, Marne, France) has been described in Bracklesham Bay (West Sussex, England). The omomyid primate *Nannopithex zuccolae* Godinot, Russell and Louis, 1992 described originally from Prémontré (MP10, Aisne, France) has been found in the three English localities (Hooker, 1996).

The mammal taxa useful as indicators for the marine-continental correlation are summarized in Figure 5.

Age		Taxa from marine deposits	Marine localities	Nannofossil levels	Closest related taxa from continental deposits	Continental localities	Mammal levels
EOCENE	Latest Ypresian	*Nannopithex zuccolae*	Upper Wittering Formation	NP13	*Nannopithex zuccolae*	Prémontré	MP10
		Ailuravus michauxi			*Ailuravus michauxi*	Grauves	
		Amphiperatherium maximum			*Amphiperatherium maximum*	Grauves	
	Early Late Ypresian	*Pseudoparamys* sp.	Peel Common	NP12	*Pseudoparamys teilhardi*	Avenay	MP8+9
		Neoplagiaulax aff. *sylvani*	Egem	NP12	*Neoplagiaulax* aff. *sylvani*	Mutigny	MP8+9
		Macrocranion cf. *nitens*			*Macrocranion* cf. *nitens*	Avenay	
	Late Middle Ypresian	*Peratherium matronense*	Evere	NP12	*Peratherium matronense*	Avenay	MP8+9
		Toliapina lawsoni	Sheppey	NP12	*Toliapina lawsoni*	Mutigny	MP8+9
		Leptacodon sp.			*Leptacodon* n. sp.	Avenay	
	Early Ypresian	*Paramys ageiensis*	Swamscombe	NP10	*Paramys ageiensis*	Avenay	MP8+9
		Pliolophus vulpiceps	Harwich	NP10	*Pliolophus vulpiceps*	Pourcy	MP7?
PALEOCENE	Late Thanetian	*Arctocyonides arenae*	Beltinge	NP9	*Arctocyonides arenae*	Cernay	MP6
	Middle Thanetian	*Sarnacius* sp.	Templeuve	NP8	*Sarnacius gingerichi*	Cernay	MP6
		Berruvius cf. *lasseroni*			*Berruvius lasseroni*	Cernay	
	Upper Selandian	*Adapisorex* sp.	Maret	NP4-5	*Adapisorex abundans*	Walbeck	MP5?
		Arctocyonides cf. *weigelti*			*Arctocyonides weigelti*	Walbeck	

Figure 5. Mammal taxa used as indicators for marine-continental correlation.

CORRELATION MARINE-CONTINENTAL BY TERRESTRIAL MAMMAL TEETH

Following the biostratigraphic correlation (see Figure 6), the Belgian taxa *Adapisorex* sp. and *Arctocyonides* cf. *weigelti* suggest that the continental Walbeck mammal fauna is older than the MP6 reference-level of Cernay, and is probably coeval or close in age with the Selandian Heers Formation (NP4-5). The MP6 reference-level of Cernay seems to be somewhat younger than zone NP8 according to the evolutionary stage of the micromammal *Sarnacius,* which is confirmed by the dating of the underlying NP8 "Sables de Châlons-sur-Vesle" (see Steurbaut, 1998). The presence of *Arctocyonides arenae* at the base of the Upnor Formation indicates a correlation between Cernay and the lower part of NP9 (late Thanetian). No typical MP7 mammal species has been found yet from the upper NP9 marine deposit. Evidently, most of deposits from this time interval are continental or brackish in the southern North Sea Basin, confirming the presence of a major sea-level lowstand as discussed by Steurbaut et al. (1999). The species *Pliolophus vulpiceps* in the Harwich Formation suggests a correlation between Pourcy and the zone "NP10". *Paramys ageiensis* at the base of the London Clay Formation (Hooker, 1991) is the first occurrence of the species. *Paramys ageiensis* occurs already in the MP7 – MP8 + 9 intermediate fauna of Meudon (Smith, 1999). The MP8 + 9 reference-level of Avenay seems to be correlated with zone NP12 of late middle Ypresian age by the presence of *Toliapina lawsoni* and *Leptacodon* sp. at Sheppey, and *Peratherium matronense* at Evere, and to early late Ypresian age with *Macrocranion* cf. *nitens* and *Neoplagiaulax* aff. *sylvani* at Egem and *Pseudo-paramys* sp. at Peel Common. Finally, the taxa *Amphiperatherium maximum, Ailuravus michauxi* and *Nannopithex zuccolae* of the upper Wittering Formation indicate a correlation between the MP10 reference-level of Grauves and zone NP13.

ORIGIN OF TERRESTRIAL MAMMALS IN MARINE DEPOSITS

Micromammals as biostratigraphic markers in marine deposits are useful biostratigraphic indicators, especially where no

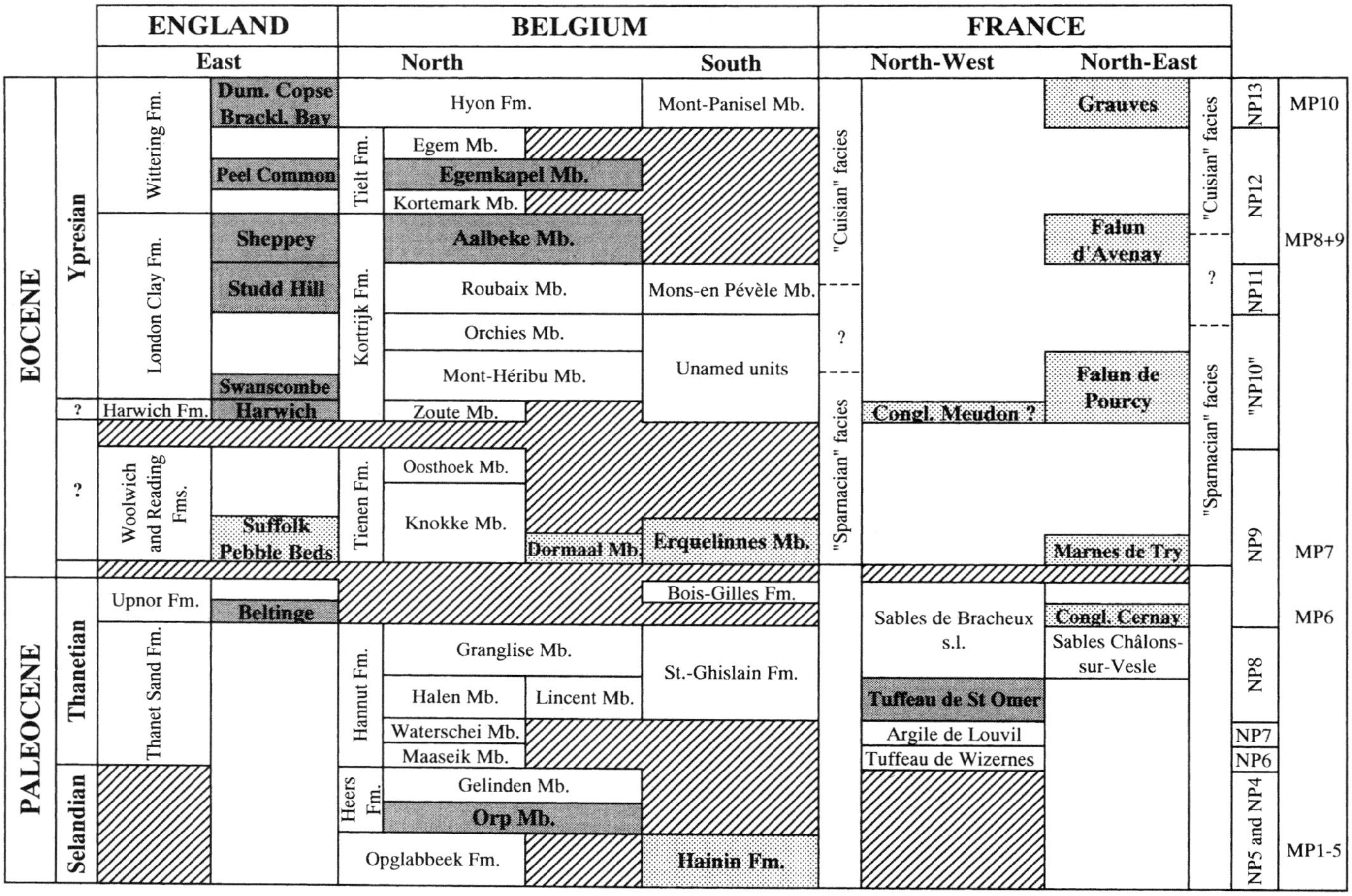

Figure 6. Relative position of the upper Paleocene–lower Eocene stratigraphical units in the English-French-Belgian Basin (modified from Steurbaut, 1998, and Smith, 2000).

$\delta^{13}C$ records are available. However, terrestrial mammal remains are generally rare in marine deposits and most of them have been discovered by chance during excavations looking principally for shark teeth. They are practically always found in coarse sand and gravel beds at the base of sequences, where contemporaneous vertebrate remains deposited on the bottom of shallow seas were concentrated. This phenomenon can be explained by the transgressive nature of most of these basal beds. However, sometimes, levels can also contain reworked faunas. For example, Smith and Hooker (1996) described two mammal taxa reworked in the Brussels Formation at Hoegaarden (middle Eocene, Belgium). In this case, not only the mammal teeth but also about half of the shark teeth (18 species out of 31) were reworked from the local underlying continental Tienen Formation (Paleocene-Eocene transition) and the marine Hannut Formation (upper Thanetian), respectively. Therefore, it is necessary to verify independently that mammal remains are contemporaneous with the marine deposits. The new mammal remains from the three marine localities here described (Templeuve, Sheppey, Egem) are not reworked because they all come from marine concentration levels which, according to the nanofossils and microfossils, are only slightly younger than the underlying beds (see Fig. 6).

Several taphonomical hypotheses could be proposed to explain the presence of terrestrial mammals in shallow marine deposits. The small size of some mammals suggests that they were brought into the association as indigestible material regurgitated as pellets by birds, e.g., in the case of the Evere, Egem and Sheppey finds. Indeed, mammals occur in the diet of living coastal birds of prey such as Sea-eagle (*Haliaeetus* spp.) and of marine gulls (Laridae) such as Herring Gull, Lesser Black-backed Gull, Greater Black-backed Gull and Common Gull (Cramp and Simmons, 1983). Their bones and teeth are regurgitated as pellets (Del Hoyo et al., 1996, p. 579). However, other large vertebrate remains of champsosaurs, crocodiles, turtles and birds in Templeuve may suggest fluvial transport toward a shallow marine depocenter. In any case, the presence of a large quantity of shark teeth discovered together with the terrestrial mammal teeth in all these marine deposits indicates a concentration effect of the material on the bottom of the sea.

ACKNOWLEDGMENTS

We express our gratitude to Fabrice Moreau and Didier Dutheil for providing the specimens from Templeuve, and to Jacques Herman and Ben D'Haeze for the specimens from Egem. Special thanks are also due to Etienne Steurbaut for documentation and discussions about the stratigraphy of the Belgian Basin, to Jerry Hooker for discussions about the toliapinid plesiadapiform, to Jan Veen and René-Marie Lafontaine for discussions about the diet of living birds. At the IRSNB, Julien Cillis and Wilfried Miseur produced the photographs and Véronique Arkosi checked the English version of the manuscript. Suggestions from an anonymous reviewer improved this paper significantly. This work has benefited from financial support by the grant FRFC-IC 2.4540.99 of the Belgian National Fund of Scientific Research.

REFERENCES CITED

Charlesworth, E., 1855, On several new vertebrate fossils: Report from the British Association for the Advancement of Science, Liverpool, v. 24, p. 80.

Cramp, S., and Simmons, K.E.L., 1983, Handbook of the birds of Europe, the Middle East, and North Africa (The birds of the Western Palearctic), Vol. III: Oxford, Oxford University Press, 913 p.

Crochet, J.-Y., 1979, Diversité systématique des Didelphidae (Marsupialia) européens tertiaires: Géobios, v. 12, no. 3, p. 365–378.

Del Hoyo, J., Elliott, A., and Sargatal, J., 1996, Handbook of the birds of the world: Lynx Edicions, v. 3, 752 p.

Estravis, C., 1996, *Leptacodon nascimentoi* n. sp., un nouveau Nyctitheriidae (Mammalia, Lipotyphla) de l'Eocène inférieur de Silveirinha (Baixo Mondego, Portugal): Palaeovertebrata, v. 25, no. 2–4, p. 279–286.

Godinot, M., Russell, D.E., and Louis, P., 1992, Oldest known *Nannopithex* (Primates, Omomyiformes) from the early Eocene of France: Folia Primatologica, v. 58, p. 32–40.

Hartenberger, J.-L., 1975, Evolution des rongeurs primitifs de l'Ancien Monde, *in* Problèmes actuels de Paléontologie: Évolution des vertébrés: Colloques internationaux du Centre National de la Recherche Scientifique, Paris, v. 218, p. 777–790.

Herman, J., and Sigé, B., 1975, Présence du genre paléocène *Adapisorex* (Lipotyphla, Mammalia) dans les Sables d'Orp-le-Grand (Heersien) à Maret en Brabant (Belgique): Géobios, v. 8, no. 4, p. 231–239.

Hooker, J.J., 1991, The sequence of mammals in the Thanetian and Ypresian of the London and Belgian Basins: Location of the Paleocene-Eocene boundary: Newsletter on Stratigraphy, v. 25, no. 2, p. 75–90.

Hooker, J.J., 1994, The beginning of the equoid radiation: Zoological Journal of the Linnean Society, v. 112, p. 29–63.

Hooker, J.J., 1996, Mammals from the early (late Ypresian) to middle (Lutetian) Eocene Bracklesham Group, southern England: Tertiary Research, v. 16, no. 1–4, p. 141–174.

Hooker, J.J., and Millbank, C., 2001, A Cernaysian mammal from the Upnor Formation (late Paleocene, Herne Bay, UK) and its implications for correlation: Proceedings of the Geologists' Association, v. 112, p. 331–338.

Hooker, J.J., Russell, D.E., and Phélizon, A., 1999, A new family of Plesiadapiformes (Mammalia) from the old lower Paleogene: Palaeontology, v. 42, p. 377–407.

Lemoine, V., 1883, Etude sur le *Neoplagiolax* de la faune Éocène inférieur des environs de Reims: Bulletin de la Société Géologique de France, 3ème série, v. 11, p. 249–271.

Matthew, W.D., 1918, A revision of the lower Eocene Wasatch and Wind River faunas, Part V: Insectivora (continued), Glires, Edentata: Bulletin of the American Museum of Natural History, v. 38, p. 565–657.

Michaux, J., 1964, Diagnose de quelques Paramyidés de l'Eocène inférieur de France: Comptes rendus sommaires des séances de la Société Géologique de France, avril 1964, no. 4, p. 153–154.

Moreau, F., and Mathis, S., 2000, Les élasmobranches du Thanétien (Paléogène) du nord de la France, des carrières de Templeuve et de Leforest: Cossmanniana, v. 7, no. 1–4, p. 1–18.

Owen, R., 1841, Description of the fossil remains of a mammal (*Hyracotherium leporinum*) and of a bird (*Lithornis vulturinus*) from the London Clay: Transactions of the Geological Society of London, v. 6, p. 203–208.

Owen, R., 1858, Description of a small lophiodont mammal (*Pliolophus vulpiceps*, Owen), from the London Clay, near Harwich: Quarterly Journal of the Geological Society of London, v. 14, p. 54–71.

Quinet, G.E., 1964, Morphologie dentaire des mammifères éocènes de Dormaal: Bulletin du Groupement international pour la Recherche scientifique en Stomatologie, v. 7, p. 272–294.

Russell, D.E., 1964, Les mammifères paléocènes d'Europe: Mémoires du Muséum National d'Histoire Naturelle, série C, v. 13, 324 p.

Russell, D.E., 1981, Un primate nouveau du Paléocène supérieur de France: Géobios, Mémoire spécial, v. 14, no. 3, p. 399–405.

Russell, D.E., Louis, P., and Savage, D.E., 1975, Les Adapisoricidae de l'Eocène Inférieur de France: Réévaluation des formes considérées affines: Bulletin du Muséum National d'Histoire Naturelle, 3ième série, no. 327, Sciences de la Terre, v. 45, p. 129–194.

Russell, D.E., Phélizon, A., and Louis, P., 1992, *Avenius* n. gen. (Mammalia, Primates?, Microsyopidae) de l'Eocène inférieur de France: Comptes rendus de l'Académie des Sciences, Paris, Série II, v. 314, p. 243–250.

Smith, R., and Hooker, J.J., 1996, Sur la présence de dents de mammifères (Creodonta, Perissodactyla) près de la Limite Paléocène-Eocène à Hoegaarden, Belgique: Palaeovertebrata, v. 25, no. 2–4, p. 115–124.

Smith, R., and Russell, D.E., 1992, Mammifères (Marsupialia, Chiroptera) de l'Yprésien de la Belgique: Bulletin de l'Institut royal des Sciences naturelles de Belgique, v. 62, p. 223–227.

Smith, T., 1999, Les mammifères de la transition Paléocène-Eocène de Belgique (Formation de Tienen, MP7): Systématique, phylogénie, paléoécologie, paléobiogéographie et biostratigraphie [Ph.D. thesis]: Louvain-la Neuve, Belgium, University of Louvain, 311 p.

Smith, T., 2000, Mammals from the Paleocene-Eocene transition in Belgium (Tienen Formation, MP7): Palaeobiogeographical and biostratigraphical implications: GFF, v. 122, no.1, p. 148–149.

Steurbaut, E., 1998, High-resolution holostratigraphy of middle Paleocene to early Eocene strata in Belgium and adjacent areas: Palaeontographica Abt. A, v. 247, p. 91–156.

Steurbaut, E., De Coninck, J., Roche, E., and Smith, T., 1999, The Dormaal Sands and the Paleocene-Eocene boundary in Belgium: Bulletin de la Société Géologique de France, v. 170, no. 2, p. 217–227.

Vianey-Liaud, M., 1986, Les multituberculés thanétiens de France, et leurs rapports avec les multituberculés nord-américains: Palaeontographica Abt. A, v. 191, p. 85–171.

Wood, A.E., 1962, The early Tertiary rodents of the family Paramyidae: Transactions of the American Philosophical Society, New Series, v. 52, no. 1, p. 1–261.

Manuscript Accepted by the Society August 13, 2002

Printed in the USA

Geological Society of America
Special Paper 369
2003

Biostratigraphic, chemostratigraphic, and magnetostratigraphic study across the Paleocene-Eocene boundary in the Hengyang Basin, Hunan, China

Suyin Ting
LSU Museum of Natural Science, Baton Rouge, Louisiana 70803, USA, and the Institute of Vertebrate Paleontology and Paleoanthropology, Chinese Academy of Sciences, P.O. Box 643, Beijing 100044, China
Gabriel J. Bowen
Paul L. Koch
Department of Earth Sciences, University of California, Santa Cruz, California 95064, USA
William C. Clyde
Department of Earth Sciences, University of New Hampshire, Durham, New Hampshire 03824, USA
Yuanqing Wang
Yuan Wang
Institute of Vertebrate Paleontology and Paleoanthropology, Chinese Academy of Sciences, P.O. Box, 643, Beijing 100044, China
Malcolm C. McKenna
Department of Vertebrate Paleontology, American Museum of Natural History, Central Park W. at 79th Street, New York, New York 10024, USA

ABSTRACT

The Lingcha Formation of the Hengyang Basin, Hunan, China, is one of two currently known sites thought to preserve sedimentary rocks spanning the Paleocene-Eocene boundary in Asia. The formation includes two fossiliferous intervals: the upper, yielding about 12 mammalian species, and the lower, yielding one mammalian and one reptilian species. To constrain the temporal correlation of the Lingcha Formation faunas, we collected samples for isotopic and paleomagnetic analysis in three measured stratigraphic sections tied to the fossil sites. The isotopic and paleomagnetic results demonstrate that the transient carbon isotope excursion that marks the Paleocene-Eocene boundary as currently advocated is present in the upper Lingcha Formation. The fauna from the upper Lingcha Formation occurs within the excursion interval, and is thus of earliest Eocene age. It is correlative with the Wasatchian-0 faunal zone in the Bighorn Basin, Wyoming, to within ~10^4 yr. Cluster analysis demonstrates compositional ties between the upper Lingcha fauna, the fauna from the Bumban Member of the Naran Bulak Formation, southern Mongolia, and the fauna from the Wutu Formation, northeastern China. These faunas form a cohort separated from earlier faunas by turnover at the ordinal and familial level, and from subsequent faunas by generic-level turnover. The discovery of a hyaenodontid in the Nomogen Formation of Inner Mongolia indicates that hyaenodontid creodonts may have appeared first in Asia and then spread to North America at the Paleocene-Eocene transition.

Ting, Suyin, Bowen, G.J., Koch, P.L., Clyde, W.C., Wang, Yuanqing, Wang, Yuan, McKenna, M.C., 2003, Biostratigraphic, chemostratigraphic, and magnetostratigraphic study across the Paleocene-Eocene boundary in the Hengyang Basin, Hunan, China, *in* Wing, S.L., Gingerich, P.D., Schmitz, B., and Thomas, E., eds., Causes and Consequences of Globally Warm Climates in the Early Paleogene: Boulder, Colorado, Geological Society of America Special Paper 369, p. 521–535.

INTRODUCTION

The late Paleocene through early Eocene has been identified as a period of significant global climate and biotic change. Paleoceanographic records from deep-sea cores (late Paleocene through early Eocene) indicate the existence of a short-term (100–200 k.y.) climatic anomaly superimposed on a longer-term (5–7 m.y.) change in global climate (Kennett and Stott, 1991, among many others). Both long- and short-term changes involved the warming of deep sea and high-latitude surface water, as well as negative shifts in $\delta^{13}C$ values of marine carbonates (Thomas and Shackleton, 1996). Coincident changes in temperature (Koch et al., 1995; Wing et al., 1991; Wolfe, 1994), equability (Wing and Greenwood, 1994), and carbon isotopes (Koch et al., 1992, 1995) in the continental record suggest significant climatic change across the entire globe through the late Paleocene-early Eocene (Initial Eocene Thermal Maximum, [IETM]). The unique character, global synchronicity, and preservation of the short-term carbon isotope shift in terrestrial and marine sections has led to its use as a chronostratigraphic marker (e.g., Koch et al., 1992; Cojan et al., 2000) and to the proposition that the isotope excursion be used to define the Paleocene-Eocene boundary (Koch et al., 1992). In the Holarctic continents, the modern mammalian orders, Perissodactyla, Artiodactyla, and Primates *sensu stricto,* made their first appearances in the fossil record during the Paleocene-Eocene (Clarkforkian-Wasatchian) transition. The abrupt appearance of these new orders of mammals in North America, Europe, and Asia (Beard, 1998; Clyde and Gingerich, 1998; Gingerich, 1989; Gunnell, 1999; Hooker, 1996, 1998; Meng and McKenna, 1998; Ting, 1998), and a large-scale benthic marine faunal turnover (Thomas and Shackleton, 1996), suggest a significant reorganization in the biosphere (McKenna, 1998).

Due to a series of important discoveries of Paleocene and early Eocene fossil mammals, Asia has increasingly been considered the source of immigrants of several mammalian orders to North America and Europe (Beard, 1998; Beard and Dawson, 1999; Dashzeveg, 1988; Krause and Maas, 1990; Sloan, 1987; South China "Red Beds" Research Group, IVPP, 1977; Zhai, 1978). However, the Asian record across Paleocene-Eocene boundary is poorly understood when compared to records from its Holarctic neighbors, North America and Europe. Asian Cenozoic deposits are virtually restricted to continental rocks, and the age of most mammal-bearing strata is unconstrained radioisotopically. As such, fossil mammals have long been the primary tool for stratigraphic correlation and subdivision of Cenozoic strata. The Asian Paleocene and early Eocene faunas have been divided into a series of land mammal ages from oldest to youngest: The Shanghuan, Nongshanian, Gashatan, and Bumbanian Asian Land Mammal Ages (ALMAs) (Li and Ting, 1983; Russell and Zhai, 1987; Ting, 1998; Tong et al., 1995). The faunas thought to represent the latest Paleocene and early Eocene have been grouped within the Gashatan and Bumbanian respectively, but precise biostratigraphic correlation of the postulated Asian Paleocene and early Eocene faunas to those of North America and Europe has not been achieved. The Gashatan is suggested to correlate to the Clarkforkian North American Land Mammal Age (NALMA) by Meng et al. (1998) and Ting (1998), and to correlate with the latter half of the Tiffanian NALMA and early part of the Clarkforkian NALMA by Wang et al. (1998) and Beard (1998). The Bumbanian is suggested to correlate to the Wasatchian NALMA by Dashzeveg (1988), Russell and Zhai (1987), and Ting (1998), and to correlate to the pre-Wasatchian, based on the Wutu fauna, Wutu Basin, Shangdong, by Beard (1998). To provide better evidence for intercontinental correlation, we conducted a chemostratigraphic and magnetostratigraphic study of the Lingcha Formation, Hengyang Basin, Hunan, China, one of only two currently known Asian localities with a continuous Paleocene-Eocene sedimentary record (the other being the Nemegt Basin, Mongolia). Here we report results that demonstrate the presence of the Paleocene-Eocene boundary carbon isotope excursion in the Lingcha Formation, use quantitative biostratigraphic methods to place this isotope excursion within the broader context of Asian Paleocene-Eocene mammalian faunal turnover, and explore the implications of these results.

HISTORY OF WORK IN THE HENGYANG BASIN

The Hengyang Basin is located in the south-central part of Hunan Province, China (27°05′N, 112°57′E, Fig. 1). It is roughly square, with an area of ~5200 km^2 (Liu and Fu, 1986). The basin is bounded by the Heng Shan mountain on the north, and by hills of late Paleozoic rocks on the south near Laiyang city (Young et al., 1938). To the east are ridges of early Paleozoic rocks, and westward lie the mountain ranges near the cities of Hengyang, Shaoyang, and Changning. The basin is drained by the Hsiang River and its two major tributaries, the Laishui on the east and the Chengshui on the west. The basement of the Hengyang Basin is composed of Proterozoic and early Paleozoic low-grade metamorphic rocks and late Paleozoic limestone, which crop out around the basin (417 Geological Team, 1979). Cretaceous rocks are widely distributed in the basin, whereas the early Tertiary rocks are mainly in the northeastern part of the basin.

The discoveries of early Paleogene mammalian fossils in Hengyang, Hunan, come from extensive geological investigations of the "Red Beds," the widely distributed late Mesozoic and early Cenozoic terrestrial rocks of southern China. The red beds of Hunan were carefully investigated by C.C. Tien, H.C. Wang, and Y.T. Hsu in 1932 (Tien et al., 1933). They distinguished two sets of red beds in the surveyed areas. The older set was named the "Hengyang Red Beds" or "Hengyang Sandstone" and considered to be Eocene in age; the younger was named the "Tanshih Red Beds" and suggested to be Miocene. By 1936, however, Tien had come to consider that the "Tanshih Red Beds" were most probably Eocene, based on the "plants in association with fishes," and that the "Hengyang Red Beds"

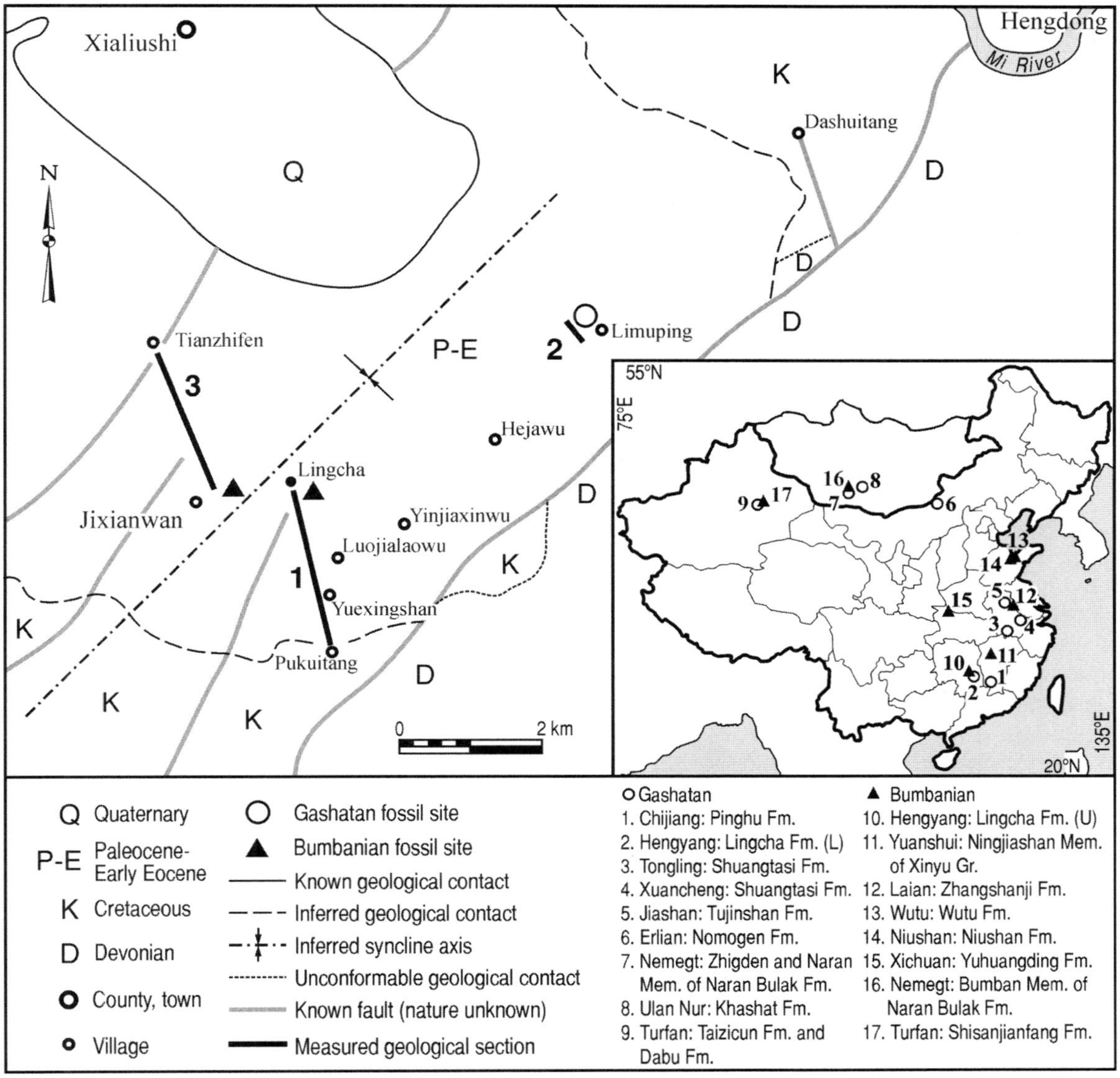

Figure 1. Geological sketch map near Hengdong County, Hengyang Basin (27°05′N, 112°57′E; modified from a Geological Map by the Hengyang Geological Survey, Hunan, China). Inset is a map of Central Asia showing the location of key Gashatan and Bumbanian faunas. (Modified with permission from Bowen et al., 2002. Copyright 2002, American Association for the Advancement of Science).

were Cretaceous (Tien, 1936, p. 462). To prove the age of the two red beds of Hunan, C.C. Young, M.N. Bien, and Y.Y. Lee took three trips in 1938 to several basins of central Hunan (Young et al., 1938). The first fossil mammal, a left dentary with m/3, was found in the "Hengyang Red Beds" from the Hengyang Basin. The identity of this specimen is discussed below with reference to new materials. Young (1944) reported this single find and proposed that the "Hengyang Red Beds" from which the fossil came were middle Eocene in age. However, he also pointed out that this "does not necessarily mean that the whole formation is middle Eocene in age" (Young, 1944, p. 3).

Increasingly detailed studies of the red beds in the Hengyang Basin have been made since the1960s. Survey of industrial mineral resources of the red beds called for a reliable stratigraphic framework. This was achieved mainly through biostratigrapic study of invertebrate, vertebrate, and pollen fossils. The invertebrate paleontologic and palynologic research was carried out by the Hubei Geological Institute, Hubei Geological Bureau, Hubei Petroleum Team, Hunan Petroleum Team, and Nanjing Institute of Geology and Paleontology, Chinese Academy of Sciences. Research on ostracodes by S.Z. Guan (1979, 1989) and Liang and Zhang (1984), on Cretaceous pollen by Z.L. Zhang (1979), on Cretaceous freshwater lamellibranchiata by R.J. Zhang (1979), on Eocene plants by H.M. Li (1965), and on stratigraphy by J.M. Hu and D.M. Zeng (1980), J.J. Zhang (1982), and C. Chu (1986) has all been important for the development of a basic stratigraphic framework. In addition, paleomagnetic analysis has been carried out on Cretaceous rocks, but only a brief report has been published (Hsu et al., 1990).

Study of fossil vertebrates has led to major progress in defining the early Tertiary strata of the Hengyang Basin. This research has been carried out by the Institute of Vertebrate Paleontology and Paleoanthropology, Chinese Academy of Sciences (IVPP), since the early 1970s. In 1974 and 1976, a team led by Z.X. Qiu and C.K. Li found five new species of fossil mammals from the localities ~15 km southwest of Hengdong, a small town in the northeastern part of the Hengyang Basin (Fig. 1). The name "Lingcha Formation" is given to the fossil-bearing strata by Li et al. (1979, p. 77). They further subdivided the fossil-bearing strata into an upper fossil layer, yielding six mammalian species and representing early Eocene (no later than the Cuisian European Land Mammal Age [ELMA], probably equivalent to the Sparnacian ELMA), and a lower fossil layer, yielding one mammalian species and probably representing the earliest Eocene (perhaps correlative with the fauna from the Naran Bulak Formation of southern Mongolia) (Li et al., 1979, p. 78). J.J. Zheng, X.S. Huang, and others investigated the fossil localities later in 1978, and proposed that both upper and lower fossil layers are probably earlier Eocene in age and may be comparable to the Graybullian subage of the Wasatchian NALMA in North America or Sparnacian ELMA in Europe (Zheng and Huang, 1984, p. 202). However, Li and Ting (1983, p. 40) considered that the lower fossil layer of the Lingcha Formation may be Paleocene based on occurrence of *Archaeolambda* sp. To further enlarge the collections and establish the age of the fauna, C.K. Li, S.Y. Ting, and others worked there from 1982 to 1986. S.Y. Ting, J.L. Li, and others collected more fossils, including new species, in these localities in 1987. These six years of field work in the 1980s, as well as work by Y.Q. Wang and Y.M. Hu during the early 1990s, have tremendously improved both the quality and quantity of collections from the Hengyang Basin (Li and Ting, 1983, 1985; Li et al., 1989; Ting, 1993, 1995; Ting and Li, 1984, 1987).

SYSTEMATIC PALEONTOLOGY

New material bearing on the identity of the specimen reported by Young (1944) is described here:

Class Mammalia Linnaeus, 1758
Order Perissodactyla Owen, 1848
Family incertae sedis
Propachynolophus hengyangensis
(Young, 1944; Li et al., 1979)

This species is represented by the type specimen, a fragmentary mandible with a left m/3 (IVPP V-214), the first mammalian fossil found in the basin. The species was first identified as ?*Adapidium* sp. (Young et al., 1938, p. 268) and then assigned to *Propalaeotherium,* a European equid (Young, 1944). It was reassigned to the genus *Propachynolophus,* another European equid, when additional material, a partial left mandible with the alveoli of p/2 and deeply worn p/3-m/2 (IVPP V-5349), was found (Li et al., 1979, p. 74).

The new material described here is a single left m/3 (IVPP V-7453; Fig. 2), ~12.5 mm long and 5.5 mm wide. This specimen is slightly longer and narrower than the type specimen, but similar to the latter in basic morphology; both m/3s are rectangular. The protoconid and the metaconid are conical and connected by a crest. The metastylid is well developed, but not separated from the metaconid. The metalophid (= cristid obliqua, Hooker, 1989) extends from the middle of the posterior wall of the protolophid (= protocristid, Hooker, 1989), labial to the metaconid. The third lobe of the m/3 in both the type specimen and IVPP V-7453 is characterized by a closed, circular basin. The hypoconulid is not distinct. The new specimen differs from the type in having higher, more acute protolophid, hypolophid, and metalophid. The new specimen is referred to this species mainly based on morphological similarity, especially the characteristic third lobe of the m/3.

The species *Propachynolophus hengyangensis,* previously identified as an equid, has long been thought to be related to a counterpart taxon in Europe. The type specimen of this species was referred to the European equid genus *Propalaeotherium,* based mainly on "the general structure of the tooth (which) fits well with that of *Palaeotherium sinensis*" (Young, 1944, p. 2). It was reassigned to the genus *Propachynolophus,* based on similarities in the size, structure of the lower molars, and degree of molarization of the premolars to those of *Propachynolophus maldani* (Li et al., 1979, p. 74). These authors also pointed out that the larger size and the weaker cingulum of *P. hengyangensis* represent more advanced features, and that the lesser molarization of p/3 represents a primitive feature (Li et al., 1979, p. 74).

Both the type and new specimen are about a fourth larger than *P. maldani* and smaller than *Propachynolophus gaudryi.* The most noticeable differences distinguishing *P. hengyangensis* from *P. maldani* and *P. gaudryi* are in the third lobe of m/3. The third lobe of the m/3 in P. *hengyangensis* is characterized by a closed circular basin; however, the third lobe of m/3 in *P. maldani* and *P. gaudryi* is composed mainly of a crest connecting the middle of the posterior wall of the hypolophid to the hypoconulid. The crest then turns lingually to meet the entoconid, which is the condition commonly seen in the primitive equids. *P. hengyangensis* also differs from *P. maldani* and *P. gaudryi* in having a more lophodont protolophid and metalophid, and metalophid more labially connected with the posterior wall of the protolophid, instead of with the metastylid as in the latter species.

Compared with early tapiroids, *P. hengyangensis* is about the same size as *Cardiolophus radinskyi* and smaller than *Homo-*

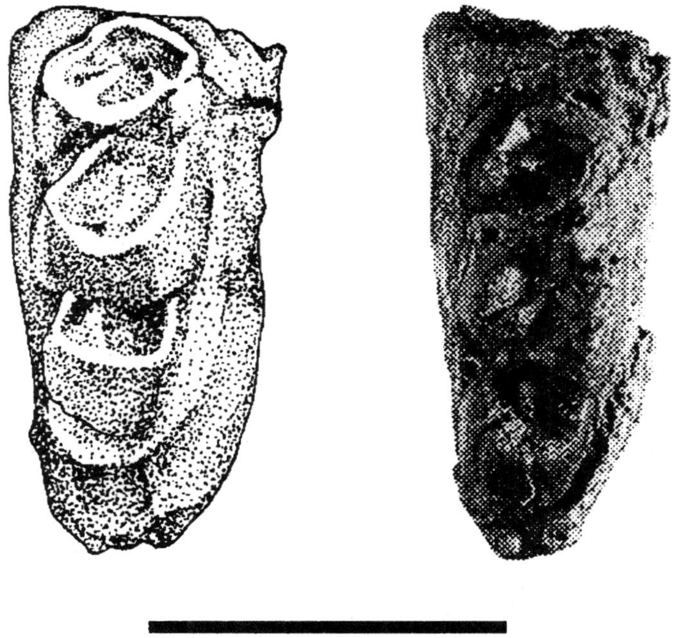

Figure 2. *Propachynolophus hengyangensis,* left m/3 (IVPP V-7453), occlusal view, scale = 1 cm.

galax wutuensis. However, its lower molars are much narrower than those of *C. radinskyi.* The lophodont protolophid and hypolophid, and the labially connected metalophid in *P. hengyangensis* are features similar to those of the primitive tapiroids, such as *H. wutuensis* and *C. radinskyi. P. hengyangensis* differs from these primitive tapiroids in having the metalophid connected with the posterior wall of the protolophid more lingually.

Early chalicothere materials are very rare. The type specimen of the species *Paleomoropus jepseni* from the early Wasatchian lower Gray Bull beds (Radinsky, 1964) includes left M1/ and M3/ and right M2/. Unfortunately, no information about the lower molars has been reported. Comparison of *P. hengyangensis* with the late Eocene chalicothere *Litolophus gobiensis* reveals a striking similarity of dental morphology in their lower molars. The metalophid of *P. hengyangensis,* extending from the posterior wall of the protolophid more lingually than that of tapiroids, is quite similar to that of the late Eocene chalicothere. The closed, circular, basinlike third lobe on the m/3 is also similar to that of *L. gobiensis.* Li et al. (1979, p. 74) emphasized that one of the most interesting features in *P. hengyangensis* is either lack of p/1 (most likely) or p/1 very small, which occurred in other representatives of the genus *Propachynolophus.* This feature lead them to consider that *P. hengyangensis* belongs to *Propachynolophus.* In the early chalicotheres *L. gobiensis* and *Schizotherium avitum* the absence of p/1 is common and characteristic (Colbert, 1934; Lucas and Schoch, 1989). *L. gobiensis* also possesses a long diastema between P1/and P2/. The recently reported Asian early Eocene species, *Danjiangia pingi* (Wang, 1995), clearly shows the cranial and dental features of early chalicotheres, and indicates that a long diastema between p/1 and p/2 occurred in early chalicothere forms. The long diastema before p/2 and either lack of p/1 or having reduced p/1 in *P. hengyangensis* allies this species with early chalicothere. Comparing *P. hengyangensis* and *ropalaeotherium sinense* with *D. pingi,* we find further striking similarities in dental morphology. *P. hengyangensis* is similar to *D. pingi* in basic dental features, especially the closed lobe of m/3, but differs from the latter in having a slightly smaller metastylid. The lower m/3 of *P. hengyangensis* is more similar to that of *Propalaeotherium sinense. Propalaeotherium sinense* is about the same size as *D. pingi* and *P. hengyangensis.* It is similar to *D. pingi* in basic dental morphology, but differs from the latter in having a more developed mesostyle on the upper molar and advanced lower p/3. We consider that *Propachynolophus hengyangensis, Danjiangia pingi,* and *Propalaeotherium sinense* may represent a new group closely related to the basal chalicotheres.

HENGYANG BASIN FAUNAS

Collections from the Hengyang Basin now include (classification based on McKenna and Bell, 1997):

Upper Lingcha Formation

Class Mammalia Linnaeus, 1758
- Order Cimolesta McKenna, 1975
  - Family ?Micropternodontidae Stirton and Rensberger, 1964
    - *Hsiangolestes* Zheng and Huang, 1984
      - *Hsiangolestes youngi* Zheng and Huang, 1984 (see Ting, 1995; Ting in preparation)
  - Family Cimolestidae Marsh, 1889
    - *Naranius* Russell and Dashzeveg, 1985
      - cf. *Naranius infrequens* Russell and Dashzeveg, 1985 (see Ting, 1995; Ting in preparation)
- Order Insectivora gen. et sp. nov. (see Ting, 1995; Ting in preparation)
- Order Mixodontia Sych, 1971
  - Family Eurymylidae Matthew, Granger and Simpson, 1929
    - *Matutinia* Li, Chiu, Yan, and Hsieh, 1979
      - *Matutinia nitidulus* Li, Chiu, Yan, and Hsieh, 1979 (see Li et al., 1979; Ting, et al., 2002)
- Order Acreodi Matthew, 1909
  - Family Hapalodectidae (Szalay and Gould, 1966) Ting and Li, 1987
    - *Hapalodectes* Matthew, 1909
      - *Hapalodectes hetangensis* Ting and Li, 1987 (see Ting and Li, 1987)
  - Family Mesonychidae Cope, 1875
    - *Dissacus* Cope, 1881
      - *Dissacus* sp. n. (Ting and Wang, in preparation)
- Order Pantodonta Cope, 1873
  - Family Coryphodontidae Marsh, 1876

?*Asiocoryphodon* Xu, 1976
?*Asiocoryphodon* sp.
(see Li et al., 1979)
Order Primates Linnaeus, 1758
(Li and Wang, in preparation)
Order Rodentia Bowdich, 1821
Family Chapattimyidae Hussain, de Bruijn, and Leinders, 1978
Cocomys Dawson, Li, and Qi, 1984
Cocomys lingchaensis Li, Chiu, Yan, and Hsieh, 1979
(see Li et al., 1979)
cf. *Tsagamys subitus* Dashzeveg, 1990
(see Ting, 1995; Dawson, Ting, and Li, in preparation)
Order Perissodactyla Owen, 1848
Family Isectolophidae Peterson, 1919
Orientolophus Ting, 1993
Orientolophus hengdongensis Ting, 1993
(see Ting, 1993)
Family incertae sedis
Propachynolophus hengyangensis (Young, 1944)
Li, Chiu, Yan, and Hsieh, 1979
(see Li et al., 1979)
Superorder Leptictida McKenna, 1975
Family Didymoconidae Kretzoi, 1943
Hunanictis Li, Chiu, Yan, and Hsieh, 1979
Hunanictis inexpectatus Li, Chiu, Yan, and Hsieh, 1979
(see Li et al., 1979)
Lower Lingcha Formation
Class Reptilia Laurenti, 1768
Order Crocodilia, Gmelin, 1788
Family Crocodylidae Curier, 1807
Planocrania, Li, 1976
Planocrania hengdongensis Li, 1984
(see Li, 1984)
Class Mammalia Linnaeus, 1758
Order Pantodonta Cope, 1873
Family Archaeolambdidae Flerov, 1952
Archaeolambda, Flerov, 1952
Archaeolambda sp.
(see Li et al., 1979)
This species was mistakenly listed in the Bumbanian in Ting (1998).

TEMPORALLY CONSTRAINED BIOSTRATIGRAPHY

Samples for isotopic and paleomagnetic analysis were taken along three measured stratigraphic sections southwest of Hengdong (Fig. 1). Section 1 and 2 are on the south limb of the basin-forming syncline, while section 3 is on the north. Section 1 (555.4 m thick) is between the villages of Pukuitang and Lingcha, composed of sandstone intercalated with red mudstone. The upper Lingcha fauna is found between the 530 and 540 m levels of this section. Section 2 (16.8 m thick) is west of Limuping, and is composed of red mudstone intercalated with sandstone. The lower Lingcha fauna is found in the middle of the section. Although section 2 can not be directly correlated into section 1, bedding orientations along the south limb of the fold are broadly consistent and local topography is precisely known, so a simple trigonometric projection can be used to make an approximate correlation between these two sections. Such a projection indicates that section 2 correlates to the ~350 m level in section 1. Section 3 (284.4 m thick) is from Tianzhifen to near Jixianwan, and is also composed of sandstone intercalated with red mudstone. Mammalian fossils, found between the 255 and 257 m levels, are considered equivalent to those found at the top of section 1 and grouped with the upper Lingcha fauna. An unidentified turtle was found at 204 m within section 3.

Magnetostratigraphy

Paleomagnetic samples were collected from all three of the local stratigraphic sections measured in the Lingcha Formation in an effort to provide independent chronostratigraphic constraints on fossil and isotopic samples collected from this area. One hundred and fifty oriented samples of red mudstone and sandstone were extracted from 31 sites. Sampling resolution was necessarily coarse due to patchy exposures. Samples were analyzed at the paleomagnetics laboratory at University of California, Santa Cruz, using thermal demagnetization. Most samples showed very to moderately stable demagnetization behavior and were characterized by 2–3 components of magnetization. All specimens carried a low-temperature component (<300 °C unblocking temperature) that exhibits random directions suggesting that it is a viscous overprint. A handful of samples also exhibited an intermediate temperature component (300–600 °C unblocking temperature) of unknown origin. Most specimens exhibited a strong high-temperature component characterized by unblocking temperatures of 670–690 °C, indicating a hematite remanence carrier. This high-temperature characteristic component exhibits antipodal directions with an average declination/inclination of 10.4/26.2 ($N = 118$, $\alpha = 4.6°$) when all reversed samples are inverted. This component also passes the reversal test at 99% confidence limit and is not significantly different from the expected Paleocene-Eocene direction for this location on the South China Block (Dec = 3.3°, Inc = 27.5°, $\alpha = 10.2°$, reference pole from McElhinny and McFadden, 2000). We interpret this component to be an early chemical remanent magnetization associated with syndepositional oxidation of these fine-grained terrigenous sediments. Although the beds are gently folded into a syncline, the fold test is equivocal with the best sites ($N \geq 5$, $K \geq 50$) showing maximum clustering at ~70% unfolding but a 95% confidence interval that includes 100% unfolding (and excludes 0% unfolding) using the parametric bootstrap (Tauxe and Watson, 1994). This result could indicate a synfolding magnetization but is more likely due to the small difference in bedding orientation between limbs and/or the existence of a small unaccounted vertical axis rotation.

Sites were split into two groups (alpha and beta) depending on their reliability and plotted against stratigraphic position (Fig. 3A). Alpha sites (shown as closed symbols in Figures 3 and 4) are the most reliable sites and have 4 or more stable samples that are significantly clustered at $p < 0.01$ (using the Watson, 1956, test for randomness). Beta sites (shown as open symbols in Figures 3 and 4) are characterized by 3 or more stable samples significantly clustered at $p < 0.05$ or two samples significantly clustered at $p < 0.01$. Samples that exhibited stable demagnetization but were not included in these sites are also used to help confirm the pattern of polarity changes. Section 1, which is the longest section, shows a total of 4 polarity intervals, with the upper Lingcha faunal horizon falling in a long reversed zone at the top of the section. Section 2 and 3 each exhibit two polarity zones, with rocks of reversed polarity overlying normal polarity rocks (Fig. 3A).

Carbon isotope stratigraphy

The carbon isotope composition of authigenic carbonate precipitated within ancient soils (paleosols) is indirectly determined by the isotopic composition of atmospheric CO_2. During

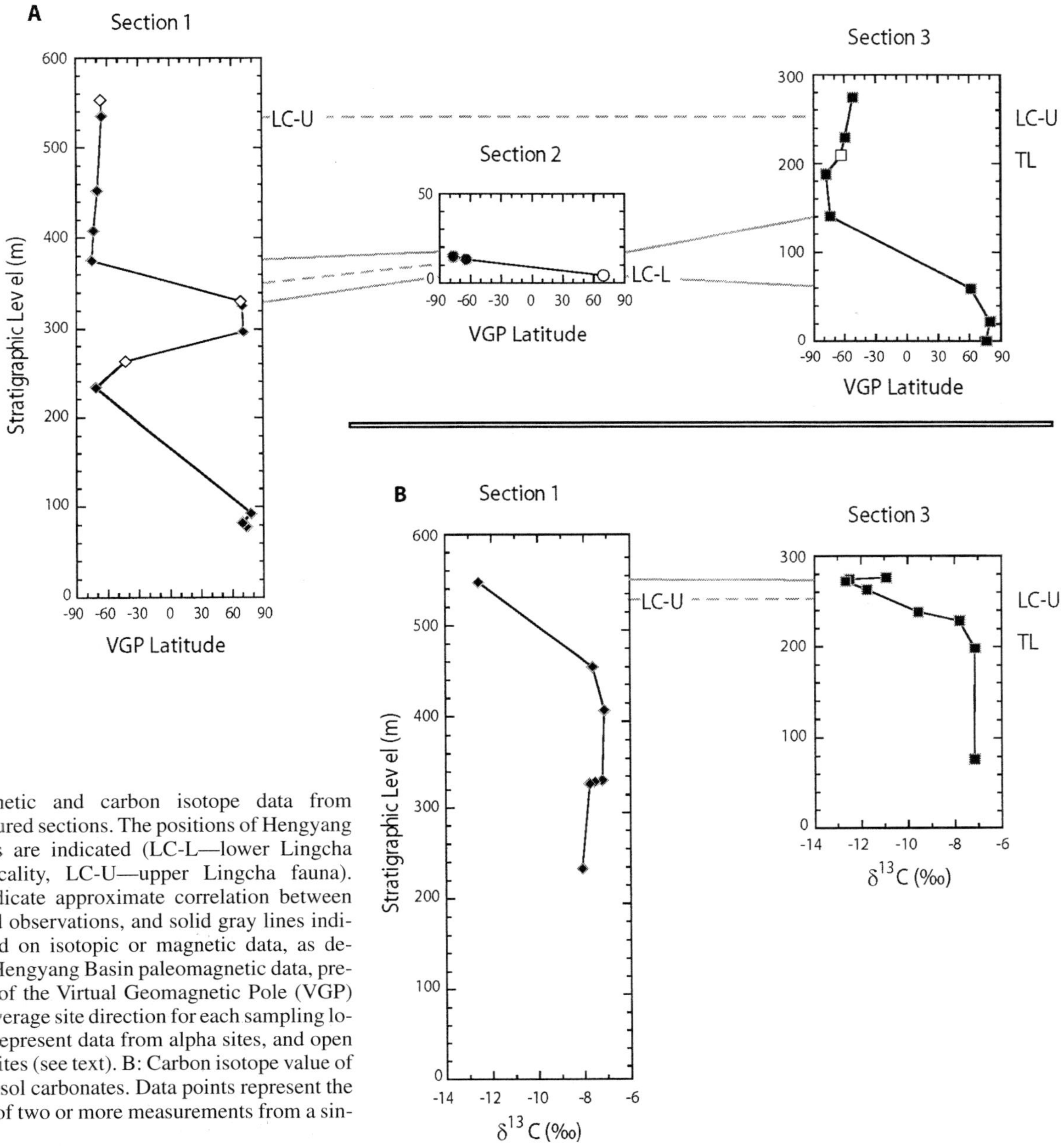

Figure 3. Paleomagnetic and carbon isotope data from Hengyang Basin measured sections. The positions of Hengyang Basin faunal horizons are indicated (LC-L—lower Lingcha fauna, TL—turtle locality, LC-U—upper Lingcha fauna). Dashed gray lines indicate approximate correlation between sections based on field observations, and solid gray lines indicate correlations based on isotopic or magnetic data, as described in the text. A: Hengyang Basin paleomagnetic data, presented as the latitude of the Virtual Geomagnetic Pole (VGP) corresponding to the average site direction for each sampling locality. Solid symbols represent data from alpha sites, and open circles represent beta sites (see text). B: Carbon isotope value of Hengyang Basin paleosol carbonates. Data points represent the average isotope value of two or more measurements from a single sampling level.

photosynthesis, plants assimilate carbon from the atmosphere with an approximately consistent fractionation relative to atmospheric CO_2 (e.g., Bocherens et al., 1993; Arens et al., 2000). Within soils, gaseous CO_2 is typically a mixture of atmospheric carbon dioxide and CO_2 produced during respiration in plant roots and microbial degradation (two nonfractionating processes). The mixing of CO_2 within soils and fractionation of carbon isotopes during soil carbonate precipitation have been investigated (Cerling et al., 1991), and the $\delta^{13}C$ values of soil carbonate precipitating more than ~30 cm below the soil surface typically reflects the carbon isotope composition of soil vegetation, and thus that of atmospheric CO_2.

We collected and determined the stable isotope composition of 32 paleosol carbonates from 15 levels within sections 1 and 3. In the laboratory, carbonate nodules were polished flat on a lapidary wheel using 600-grit silica carbide powder, washed, then dried in a low-temperature oven. Samples (~100 µg) were drilled from the polished surfaces under a binocular microscope using a mounted dental drill. Micritic carbonate was sampled exclusively; care was taken to avoid sampling sparry phases that might represent diagenetic, rather that pedogenic, conditions. Where possible, two samples were drilled from each of two nodules from every carbonate-bearing horizon in each measured section. Samples were roasted in vacuo at 400 °C for 1 hour to remove organic contaminants. These were analyzed using a Micromass Optima or Prism gas-source mass spectrometer, following reaction with 100% phosphoric acid at 90 °C, with the aid of an automated Isocarb device. Isotope ratios are reported in per mil (‰) units as $\delta^{13}C$ or $\delta^{18}O$, the deviation of the isotope ratio of a sample from that for a standard (V-PDB). $\delta = (R_{sample}/R_{standard} - 1) \times 1000$, where $R = {}^{13}C/{}^{12}C$ or ${}^{18}O/{}^{16}O$. Analytical precision, based on repeated analysis of an in-house standard, was better than 0.1‰ for both carbon and oxygen. All data are presented in Appendix 1 (Tables A1 and A2).

The $\delta^{13}C$ values of multiple samples from individual nodules and of multiple nodules from a single sampling level are tightly clustered. The average range of values from multiple samples of a single nodule is 0.3‰ (n = 26), and the average range of multiple nodules from a single level is 0.6‰ (n = 13). The average carbon isotope composition of paleosol carbonates from each sampling level in local sections 1 and 3 is shown in Figure 3B. Values from both sections are nearly constant at approximately –7‰ through most of the sampled interval. Each section, however, includes one or more levels characterized by significantly lower $\delta^{13}C$ values. In section 1, a single level at 548 m contains nodules with an average $\delta^{13}C$ value of –12.5‰. In section 3, the carbonate nodule $\delta^{13}C$ values decrease from 238 m to 272 m, reaching a minimum of –12.7‰, and then rebound slightly at the top of the section.

Composite section

We use a combination of field observations, isotopic data, and paleomagnetic data to correlate the three measured sections into a single composite section (Fig. 4). The similar composition and lithostratigraphic association of the upper Lingcha fauna from sections 1 and 3 suggests that these horizons may be correlative. This correlation is supported by the dramatic decrease in paleosol carbonate $\delta^{13}C$ values associated with the upper Lingcha faunal horizon in these two sections. Both of these sections also preserve a normal to reverse polarity transition that lies ~180 m below the level characterized by the upper Lingcha fauna and carbon isotope excursion. This suggests very similar sediment accumulation rates for these two sections. Based on this evidence, we correlate the 548 m level of section 1 and the 272 m level of section 3 and correlate the remainder of these sections by assuming equal sedimentation rates. The short stratigraphic interval represented by section 2 also appears to preserve a reversal from normal to reversed polarity, although sampling is relatively poor. Section 2 is estimated to correlate to the ~350 m level in section 1 using trigonometric projection. This lies within the interval of the uppermost normal to reversed polarity transition in section 1, suggesting that the reversals in these two sections are correlative. This defines a ~40 m interval within the composite of sections 1 and 3, within which section 2 likely correlates. In the absence of other constraints, we correlate section 2 to the middle of this interval.

Figure 4 presents our composite carbon isotope and paleomagnetic stratigraphy, with all values projected on the section 1 stratigraphic scale. The composite section shows two well-delineated magnetic polarity zones and a dramatic shift in paleosol carbonate $\delta^{13}C$ values. The only known decrease in the carbon isotope composition of pedogenic carbonate during the Cenozoic that approximates the magnitude and abruptness of that seen in the record from Hengyang occurs during the Paleocene-Eocene boundary carbon isotope excursion, and we correlate the Hengyang excursion with this event. This correlation is supported by the association of the isotopic excursion with a fauna having early Eocene affinities, the existence of a typical Paleocene taxon from a level 180 m below the excursion, and the stratigraphic position of the excursion within an interval of reversed polarity. We place the Paleocene-Eocene boundary (54.97 Ma, Wing et al., 1999) at 516 m in the Hengyang Basin composite section, the first level producing unusually low $\delta^{13}C$ values. Given this identification, and with reference to correlated marine Paleocene-Eocene boundary sections (Kennett and Stott, 1991), we correlate the uppermost reversed polarity interval within the Hengyang composite section to chron C24r of the most recent Geomagnetic Polarity Time Scale (Cande and Kent, 1995), and the underlying normal polarity interval to chron C25n (Fig. 4). The age of the C25n/C24r boundary has been estimated to be 55.7 Ma in marine sections (Cande and Kent, 1995). Below the 275 m level, the pattern of polarity reversals is too poorly resolved to allow confident identification of magnetostratigraphic units.

Our chemostratigraphic and magnetostratigraphic results allow us to correlate the Hengyang Basin faunal horizons directly to the early Paleogene biostratigraphy of North America

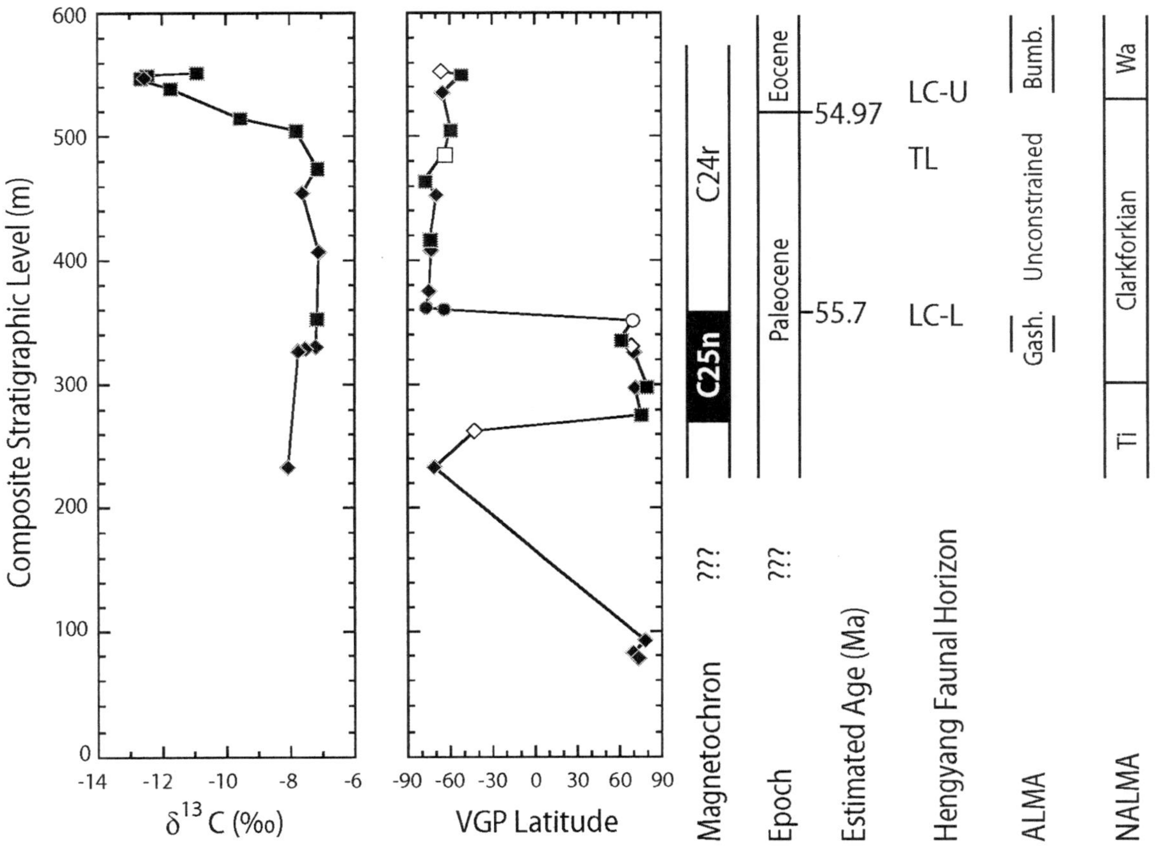

Figure 4. Composite carbon isotope and magnetic stratigraphy of the Hengyang Basin Paleocene-Eocene boundary interval. Composite stratigraphy represents data from local sections 1 (diamonds), 2 (circles), and 3 (squares) projected on the section 1 stratigraphic scale. Open and closed symbols for paleomagnetic data are as in Figure 3A. Magnetochron assignments are made with reference to marine Paleocene-Eocene boundary sections (e.g., Kennett and Stott, 1991). The Paleocene-Eocene boundary is placed at the base of the carbon isotope excursion, and assigned an age of 54.97 Ma after Wing et al. (1999). The age assignment for the C25n/C24r magnetic reversal is an estimated age for this reversal in marine sections. The stratigraphic and temporal range of the Gashatan (Gash.) and Bumbanian (Bumb.) Asian Land Mammal Ages (ALMAs) are constrained by the lower and upper Lingcha faunal horizons (LC-L and LC-U, respectively); the Gashatan-Bumbanian boundary lies within the interval labeled "Unconstrained." Position of the Tiffanian (Ti), Clarkforkian, and Wasatchian (Wa) North American Land Mammal Ages (NALMAs) are after Bowen et al. (2001) and the compilation of Gingerich (2001). (Reprinted with permission from Bowen et al., 2002. Copyright 2002, American Association for the Advancement of Science).

(Bowen et al., 2002). The upper Lingcha fauna occurs between ~15 m above the Paleocene-Eocene boundary and the carbon isotope minimum at 548 m. In the Bighorn Basin of Wyoming, the first occurrence of Wasatchian mammals is ~2 m above the carbon isotope minimum (Bowen et al., 2001). The Hengyang Basin $\delta^{13}C$ record is poorly resolved relative to that from the Bighorn Basin, however, and we present the conservative interpretation that the upper Lingcha fauna is of earliest Eocene age and is time-correlative with the earliest Wasatchian equivalent on a scale of $\sim 10^4$ yr (Fig. 4). The lower Lingcha fauna, found within meters of the C25n/C24r reversal, is of late Paleocene age and can be correlated to the earliest subzone of the Clarkforkian North American Land Mammal Age (Cf-1, Figure 4).

Faunal analysis

In order to place the chemostratigraphic and magnetostratigraphic correlation points in the Hengyang Basin within the broader context of Asian early Paleogene mammalian faunal turnover, we correlate the lower and upper Lingcha fauna in the Hengyang Basin to other Asian Gashatan-Bumbanian faunas (see Appendix 2). Faunal similarity was determined using a commonly employed clustering technique (Meng and McKenna, 1998). A similarity matrix (19 localities × 91 taxa) was built from presence/absence data at the generic level using Dice coefficients. We then clustered the faunas by reordering the similarity matrix using unweighted pair-group clustering (arithmetic means) within the SAHN (Sneath and Sokal, 1973) module of the NTSYSpc computer program (Rohlf, 2000). The program can find alternative trees when there are ties in the input matrix.

The results show that the upper Lingcha fauna most closely resembles the fauna from the Bumban Member of the Naran Bulak Formation, and shares secondary compositional affiliation with the fauna from the Wutu Formation (Fig. 5). The Wutu fauna (or faunas) includes both advanced taxa, such as *Homogalax wutuensis* (Chow and Li, 1963), and primitive taxa, such as *Chronolestes simul* and *Carpocristes oriens* (Beard and Wang, 1995), and the stratigraphy of this locality must be further examined to ensure that this collection is not significantly time-averaged. Correlation of this fauna must further be considered preliminary, because only ~1/3 of the taxa (11 of 31 taxa) in the Wutu fauna have been studied and published so far (Tong and Wang, 1998). Given our current level of knowledge, however, these three faunas share only broad compositional ties with Gashatan and other Bumbanian faunas. We suggest that they might represent a cohort of earliest Bumbanian faunas of transitional composition. The presence of a transitional fauna near the Paleocene-Eocene boundary in North America has also been demonstrated (the Wa-0 fauna of Gingerich, 1989). Faunas traditionally assigned to the Gashatan ALMA cluster within a discrete group, indicating significant compositional similarities between them. As expected, the lower Lingcha fauna clusters within the Gashatan cluster, where it is most closely allied with another low-diversity fauna containing *Archaeolambda* (a fauna from the Shuangtasi Fm, Tongling Basin, Anhui Province, China).

The association of the upper Lingcha fauna with the Paleocene-Eocene boundary carbon isotope excursion in the Hengyang Basin allows us to correlate the Paleocene-Eocene boundary below (or within) the basal Bumbanian faunal cohort. Thus, the analysis suggests that most or all Bumbanian faunas are of Eocene age. Further, if any Bumbanian assemblages prove to be of late Paleocene age, the stratigraphic coincidence of the lower Lingcha fauna with the C25n/C24r magnetochron boundary suggests that no Bumbanian fauna is older than the 55.7 Ma age of that reversal, and that no Bumbanian fauna correlates before the latter part of the Cf-1 subzone of the Clarkforkian NALMA (Fig. 4).

The cluster analysis highlights the substantial faunal turnover separating the Gashatan and Bumbanian ALMAs. Except for a single mixodont species, *Eomylus zhigdenensis,* we have not found any Gashatan species that occurred in the Bum-

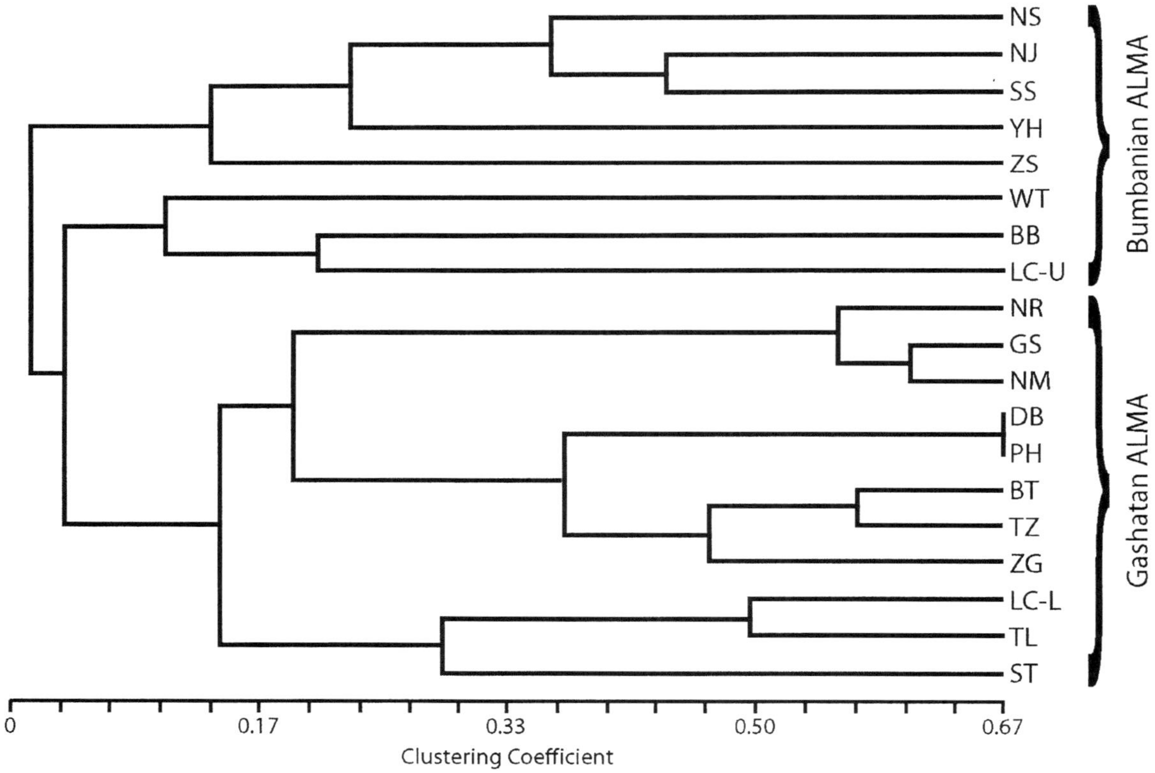

Figure 5. Cluster analysis of Gashatan-Bumbanian faunas (based on genera). Clustering coefficient provides an index of the compositional similarity of any two faunas (higher coefficent indicates greater similarity). Faunas discussed in the text include the lower (LC-L) and upper (LC-U) Lingcha faunas, the Bumban fauna (BB) from the Bumban Member of Naran Bulak Formation, Nemegt Basin, Mongolia, the Wutu fauna (if a single entity) (WT), Shangdong Province, China, and the Nomogen + Bayan Ulan fauna (NM), near Erlian, Inner Mongolia, China. See Appendix 2 for the names of other faunas.

banian. Only four Gashatan genera, belonging to Arctostylopida, Mesonychidae, Pantodonta, and Insectivora, survived at the beginning of the Bumbanian. Additionally, the families Isectolophidae (Perissodactyla), Chapattimyidae (Rodentia), Omomyidae (Primates), Hapalodectidae[1] (Acreodi), and Hyopsodontidae ("Condylarthra") made their first appearance at the beginning of the Bumbanian. The cluster analysis also indicates a substantial faunal turnover between the basal Bumbanian cohort and subsequent Bumbanian faunas. Faunal turnover across this interval is almost complete at the generic level, but does not include the substantial replacement at higher taxonomic levels that characterizes the Gashatan-Bumbanian boundary.

The extant mammalian orders Perissodactyla, Primates, and Artiodactyla, and the extinct family Hyaenodontidae (Creodonta) first appeared in North America at the beginning of the Wasatchian NALMA. Because our results indicate that all Gashatan faunas are of Paleocene age, the recent discovery of a new hyaenodontid species, *Prolimnocyon chowi,* in the Gashatan Bayan Ulan fauna, Inner Mongolia (Meng et al., 1998), indicates that Hyaenodontidae was present in Asia during the Paleocene, and most likely spread to North America during the Paleocene-Eocene transition (also see Bowen et al., 2002). Reports of a possible perissodactyl in the Bayan Ulan fauna (Meng et al., 1998) may also indicate that the order Perissodactyla (or at least their sister-group) occurred first in Asia, though the stratigraphic position of this specimen relative to the rest of the Bayan Ulan fauna needs to be further examined (Jin Meng, 2002, personal commun.).

ACKNOWLEDGMENTS

This project was funded by National Geographic Society Grant 6528-99. We greatly appreciate the support of the Society. Our special thanks goes to the Institute of Vertebrate Paleontology and Paleoanthropology, Chinese Academy of Sciences, and the Hengdong County Office, Hunan, China, for their arrangements and aid in the field. Yuanqing Wang and Yuan Wang's work was also supported by the Major Basic Research Project of Ministry of Science and Technology, China (G2000077700). G.J.B. was supported by the National Science Foundation Graduate Research Fellowship program. We thank Dr. Judith A. Schiebout of Louisiana State University for her help and support, especially in providing the NTSYS computer program for cluster analysis, and Profs. Chuankuei Li, Jiajian Zheng, Yongsheng Tong, Xueshi Huang, Jingwen Wang, and Jinling Li of IVPP for discussion. Dr. John Flynn of Field Museum of Natural History helped to organize the project during a 1997 Penrose Conference in New Mexico. Shuhua Xie of IVPP assisted the field work, Rob Coe of the University of California, Santa Cruz, facilitated paleomagnetic analysis, Paul White of Louisiana State University helped with the cluster analysis, and Mary Eggart of Louisiana State University helped draft the figures. Priscilla McKenna helped with the paleontological collecting. Ruth Hubert of LSU Geoscience Associates read the manuscript. This paper benefited from reviews by Kenneth Rose and Jonathan Bloch. We express our sincere thanks to all of them.

[1] Zhang et al. (1979) reported ?*Hapalodectes* sp. from the Lannikeng Member of Chijiang Formation (Nongshanian ALMA), Chijiang Basin, Jiangxi, China. The specimen is a fragmentary mandible with left p/3 and root of p/4 (IVPP V-5038). The authors assigned it to *Hapalodectes* because of its small size and laterally compressed teeth with acute cusps. Ting and Li (1987) suggested that this species might be a mesonychid. We do not include this potential occurrence of *Hapalodectes* in our analysis, as the evidence to support the classification of this specimen is questionable.

APPENDIX 1. STABLE ISOTOPE DATA

TABLE A1. STABLE ISOTOPE DATA FROM HENGYANG BASIN SECTION 1

Sample ID	Level (m)	$\delta^{13}C$ (‰)	$\delta^{18}O$ (‰)
7.5+NOD-1-1	547.2	–13.9	–6.6
17.5+NOD-1-2	547.2	–12.9	–6.6
17.5+NOD-2-1	547.2	–12.0	–6.6
17.5+NOD-2-2	547.2	–11.4	–5.6
15U-NOD-2-1	454.4	–7.5	–9.9
15U-NOD-1-1	454.4	–7.4	–8.8
15U-NOD-1-2	454.4	–8.1	–7.9
14L-NOD-1-1	406.5	–7.1	–8.4
14L-NOD-1-2	406.5	–7.5	–8.6
14L-NOD-2-1	406.5	–6.9	–7.7
14L-NOD-2-2	406.5	–7.0	–6.9
11.6-NOD-1-1	330.2	–7.2	–8.1
11.6-NOD-2-1	330.2	–7.2	–7.7
11.5-NOD1-1-1	328.4	–7.6	–7.3
11.5-NOD1-2-1	328.4	–7.4	–8.8
011.1-NOD-1-1	326.3	–7.7	–6.7
011.1-NOD-1-2	326.3	–7.8	–6.7
008-NOD-1-1	232.7	–8.1	–6.7

TABLE A2. STABLE ISOTOPE DATA FROM HENGYANG BASIN SECTION 3

Sample ID	Level (m)	$\delta^{13}C$ (‰)	$\delta^{18}O$ (‰)
ROOTCASTS-1-1	276.0	–10.6	–6.3
ROOTCASTS-1-2	276.0	–10.5	–6.9
ROOTCASTS-2-1	276.0	–11.3	–6.6
ROOTCASTS-2-2	276.0	–11.2	–7.0
031-NOD-1-1	274.2	–12.1	–7.1
031-NOD-1-2	274.2	–12.5	–6.3
031-NOD-2-1	274.2	–12.7	–6.8
031-NOD-2-2	274.2	–12.6	–6.4
SS#24-NOD-1-1	271.8	–12.7	–7.7
SS#24-NOD-1-2	271.8	–12.6	–7.7
SS#24-NOD-2-1	271.8	–12.5	–7.4
SS#24-NOD-2-2	271.8	–13.0	–6.9
SS#23+NOD1-1-1	262.8	–12.5	–6.8
SS#23+NOD1-1-3	262.8	–12.1	–6.3
SS#23+NOD1-1-4	262.8	–12.6	–6.5
SS#23+NOD1-2-1	262.8	–12.1	–6.6

continued

TABLE A2. *Continued*

Sample ID	Level (m)	$\delta^{13}C$ (‰)	$\delta^{18}O$ (‰)
SS#23+NOD1-2-2	262.8	−12.6	−6.6
SS#23+NOD2-1-1	262.8	−10.8	−6.3
SS#23+NOD2-1-2	262.8	−10.8	−6.6
SS#23+NOD2-2-1	262.8	−11.6	−6.0
SS#23+NOD2-2-2	262.8	−11.2	−5.9
SS#20U-NOD1-1-1	238.4	−8.6	−7.8
SS#20U-NOD1-1-2	238.4	−8.8	−7.2
SS#20U-NOD1-2-1	238.4	−9.3	−7.8
SS#20U-NOD1-2-2	238.4	−9.7	−7.2
SS#20U-NOD2-1-1	238.4	−9.9	−7.5
SS#20U-NOD2-1-2	238.4	−10.5	−7.0
SS#20U-NOD2-2-1	238.4	−9.8	−7.8
SS#20U-NOD2-2-2	238.4	−9.8	−7.7
SS#19-NOD-1-1	228.6	−8.0	−7.7
SS#19-NOD-1-2	228.6	−8.0	−7.9
SS#19-NOD-2-1	228.6	−7.6	−7.3
SS#19-NOD-2-2	228.6	−7.6	−7.6
SS#11B-NOD-1-1	198.0	−7.1	−8.4
SS#11B-NOD-1-2	198.0	−7.2	−8.2
SS#11B-NOD-2-1	198.0	−7.0	−7.9
SS#11B-NOD-2-2	198.0	−7.2	−8.0
SS#6-NOD-1-1	76.8	−6.9	−5.6
SS#6-NOD-1-2	76.8	−7.1	−7.0
SS#6-NOD-2-1	76.8	−7.5	−6.6
SS#6-NOD-2-2	76.8	−7.2	−7.1

APPENDIX 2. FAUNAL LISTS

Gashatan Land Mammal Age

ST (Shuangtasi Group, Xuancheng Basin, Anhui)
Hsiuannania maguensis, Dissacus magushanensis, Bothriostylops progressus, Archaeolambda yangtzeensis, Wanotherium xuanchengensis, Harpyodus progressus, Palaeostylops iturus, Wanostylops youngi

PH (Pinghu Fm., Chijiang Basin, Jiangxi)
Prodinoceras lacustris

LC-L (Lower part of Lingcha Fm., Hengyang Basin, Hunan)
Archaeolambda cf. *planicanina*

TL (Shuangtasi Fm., Tongling Basin, Anhui)
Bothriostylops progressus, Archaeolambda sp. cf. *A. yangtzeensis, Guichilambda zhaii*

TS (Tujinshan Fm., Jiashan Basin, Anhui)
Sinonyx jiashanensis, Sinomylus zhaii

TZ (Taizicun Fm., Turfan Basin, Xinjiang)
Multituberculata indet., Eurymylidae gen. et. sp. indet., *Pseudictops chaii,* Mesonychidae indet., *Archaeolambda speciosa, Archaeolambda* sp. 2, *Tienshanilophus subashiensis, Tienshanilophus lianmuqinensis, Tienshanilophus shengjinkouensis, Prodinoceras turfanensis, Prodinoceras diconicus, Prodinoceras primigenum, Prodinoceras simplum*

DB (Dabu Fm., Turfan Basin, Xinjiang)
Coryphodon dabuensis, Prodinoceras xinjiangensis

NM (Nomogen Fm.; former Bayan Ulan plus Nomogen Fm., Erlian, Inner Mongolia)
Lambdopsalis bulla, Prionessus lucifer, Prionessus cf. *P. lucifer, Sphenopsalis nobilis, Sarcodon minor, ?Sarcodon pygmaeus, Hyracolestes ermineus, Bayanulanius tenuis,* Lepictidae gen. et. sp. n., Palaeoryctoidea gen. et. sp. indet., *Pseudictops lophiodon, Eomylus borealis, ?Khaychina elongata, Tribosphenomys minutus, Prolimnocyon chowi,* cf. *Viverravus* sp., *Pachyaena* sp., ?Creodonta indet., *Dissacus serratus, Pastoralodon lacustris, Pastoralodon haliutensis, Convallisodon convexus, Prodinoceras xinjiangensis, Palaeostylops iturus, Gashatostylops macrodon,* Perissodactyla (Family indeterminate)

BT (Bugin Mem. of Naran Bulak Fm., Bugin Tsav Basin, Mongolia)
Khaychina elongata, Archaeolambda trofimovi, Prodinoceras sp.

ZG (Zhigden Mem. of Naran Bulak Fm., Nemegt Basin, Mongolia)
Ernanodon sp., *Amar eleator, Archaeolambda planicanina, Prodinoceras martyr, Gashatostylops macrodon, ?Arctostylops* sp.

NR (Naran Mem. of Naran Bulak Fm., Nemegt Basin, Mongolia)
Prionessus lucifer, Prionessus sp., cf. *Sinopa* sp., *Eurymylus laticeps, Eomylus zhigdenensis, Oxyaena* sp., *Archaeolambda planicanina, Coryphodon tsaganensis, Prodinoceras martyr, Gashatostylops macrodon, Pseudictops lophiodon, Dissacus indigenus, Pachyaena nemegetica, Palaeostylops iturus, Khaychina elongata*

GS (Member I of Gashato Fm., Ulan Nur Basin, Mongolia)
Sarcodon pygmaeus, Hyracolestes ermineus, Praolestes nanus, Khashanagale zofiae, ?Khashanagale unamed n. sp., *Pseudictops lophiodon, Eurymylus laticeps, Eomylus zhigdenensis, Prodinoceras martyr, Palaeostylops iturus, Gashatostylops macrodon, Prionessus lucifer, Sphenopsalis nobilis, Phenacolophus fallax, ?Dissacus* sp. n.

Bumbanian Land Mammal Age

LC-U (Upper part of Lingcha Fm., Hengyang Basin, Hunan)
Hsiangolestes youngi, cf. *Naranius infrequens,* a primate, *Matutinia nitidulus, Cocomys lingchaensis,* cf. *Tsagamys* sp., *Dissacus* sp., *Hapalodectes hetangensis, Asiocoryphodon* sp., *Hunanictis inexpectatus, Orientolophus hengdongensis, Propachynolophus hengyangensis*

BB (Bumban Mem. of Naran Bulak Fm., Nemegt Basin, Mongolia)
Naranius infrequens, Tsaganius ambiguus, cf. *Hyracolestes* sp., Pantolestidae gen. indet., *Bumbanius rarus, Oedolius perexiguus,* Nyctitheriidae gen. indet., Lipotyphla indet., *Altanius orlovi, Zagmys insolitus, Gomphos elkema, Decipomys mongoliensis, Rhombomylus* cf. *turpanensis, Alagomys inopinatus, Sharomys singularis, Sharomys parvus, Kharomys mirandus, Kharomys gracilis, Tsagamys subitus, Ulanomys*

mirificus, Orogomys obscurus, Ivanantonia efremovi, Adolomys ameristus, Tsagankhushumys deriphatus, Bumbanomys edestus, ?Esesempeomys centralasiae, Eomylus zhigdenensis, Hapalodectes sp., *Hyopsodus orientalis, Archaeolambda* sp. n., *Arctostylops* sp., *?Orientolophus* (*Homogalax,* Dashzeveg, 1988) *namadicus, ?Orientolophus* (*Hyracotherium,* Dashzeveg, 1988) *gabuniai*

WT (Wutu Fm., Wutu Basin, Shandong)
Mesodmops dawsonae, Auroratherium sinense, ?Palaeoryctidae indeterminate, *Changlelestes dissetiformis,* Erinaceidae gen. et sp. n., ?Nyctitheriidae gen. et sp. n., Pseudictopidae gen. et sp. n., *Chronolestes simul, Carpocristes oriens,* cf. *Ignacius* sp. n., Micromomyidae gen. et sp. n., *Bandaomys zhonghuaensis, Alagomys oriensis, Acritoparamys? wutui, Taishanomys changlensis, Oxyaena* sp. n., cf. *Protictis* sp. n., *Dissacus* sp., *Hapalodectes* sp. n., Hyopsodontidae gen.et sp. n., *Lophocion asiaticus, Pastoralodon* sp. n., *Heterocoryphodon* sp., Arctostylopidae gen. et sp. n., *Homogalax wutuensis, Ampholophus luensis,* Eomoropidae gen. et sp. n., Entelodontidae gen. et sp. n., Isectolophidae gen. et sp. n., Esthonychidae gen. et sp. n, Equoidea (gen. et sp. n.)

NJ (Ningjiashan Mem., of Xinyu Group, Yuanshui Basin, Jiangxi)
Xinyuictis tenuis, Coryphodon ninchiashanensis, ?Probathyopsis sinyuensis, ?Heptodon sp.

ZS (Zhangshanji Fm., Laian Basin, Anhui)
Rhombomylus laianensis

YH (Yuhuangding Fm., Xichuan Basin, Henan)
Rhombomylus sp., *Advenimus hupeiensis, Asiocoryphodon conicus, Heterocoryphodon flerowi,* cf. *Heptodon* sp., *Danjiangia pingi, Hannanomys lini, Hohomys lii, Dinocerata, indet.,*

NS (Niushan Mem. of Wutu Fm., Niushan Basin, Shandong)
Heptodon niushanensis

SS (Shisanjianfang Fm., Turfan Basin, Xinjiang)
Rhombomylus turpanensis, Anatolostylops dubius, Hyopsodus sp., *Coryphodon* sp., *Heptodon tianshanensis*

REFERENCES CITED

417 Geological Team, Hunan Geological Bureau, 1979, Preliminary report on Cretaceous–early Tertiary lithofacies and paleogeography of Hengyang basin, *in* Chang, Y., ed., Cretaceous–early Tertiary lithofacies and paleogeography of south-central region: Beijing, Geology Publishing House Press, p. 140–150.

Arens, N.C., Jahren, A.H., and Amundson, R., 2000, Can C3 plants faithfully record the carbon isotopic composition of atmospheric carbon dioxide?: Paleobiology, v. 26, p. 137–164.

Beard, K.C., 1998, East of Eden: Asia as an important center of taxonomic origination in mammalian evolution, *in* Beard, K.C., and Dawson, M.R., eds., Dawn of the age of mammals in Asia: Bulletin of the Carnegie Museum of Natural History, v. 34, p. 5–39.

Beard, K.C., and Dawson, M.R., 1999, Intercontinental dispersal of Holarctic land mammals near the Paleocene-Eocene boundary: Paleogeographic, paleoclimatic and biostratigraphic implications: Bulletin de la Société Géologique de France, v. 170, p. 697–706.

Beard, K.C., and Wang, Jing Wen, 1995, The first Asian plesiadapoids (Mammalia: Primatomorpha): Annals of Carnegie Museum, v. 64, p. 1–33.

Bocherens, H., Friis, E.M., Mariotti, A., and Pedersen, K.R., 1993, Carbon isotopic abundances in Mesozoic and Cenozoic fossil plants: Palaeoecological implications: Lethaia, v. 26, p. 347–358.

Bowen, G.J., Koch, P.L., Gingerich, P.D., Norris, R.D., Bains, S., and Corfield, R.M., 2001, Refined isotope stratigraphy across the continental Paleocene-Eocene boundary on Polecat Bench in the northern Bighorn Basin, *in* Gingerich, P.D., ed., Paleocene-Eocene stratigraphy and biotic change in the Bighorn and Clarks Fork Basins, Wyoming: University of Michigan Papers on Paleontology, v. 33, p. 73–88.

Bowen, G.J., Clyde, W.C., Koch, P.L., Ting, Su Yin, Alroy, J., Tsubamoto, Takehisa, Wang, Yuan Qing, and Wang, Yuan, 2002, Mammalian dispersal at the Paleocene/Eocene boundary: Science, v. 295, p. 2062–2065.

Cande, S.C., and Kent, D.V., 1995, Revised calibration of the geomagnetic polarity timescale for the Late Cretaceous and Cenozoic: Journal of Geophysical Research, v. 100, p. 6093–6095.

Cerling, T.E., Solomon, D.K., Quade, J., and Bowman, J.R., 1991, On the isotopic composition of carbon in soil carbon dioxide: Geochimica et Cosmochimica Acta, v. 55, p. 3403–3405.

Chow, Min Chen, and Li, Chuan Kuei, 1963, A fossil of *Homogalax* from the Eocene of Shantung: Scientia Sinica, v. 12, p. 1411–1412.

Chu, C., 1986, Late Late Cretaceous red beds of Hunan: Journal of Stratigraphy, v. 10, p. 54–59.

Clyde, W.C., and Gingerich, P.D., 1998, Mammalian community response to the latest Paleocene thermal maximum: An isotaphonomic study in the northern Bighorn Basin, Wyoming: Geology, v. 26, p. 1011–1014.

Cojan, I., Moreau, M.G., and Stott, L.E., 2000, Stable carbon isotope stratigraphy of the Paleogene pedogenic series of southern France as a basis for continental-marine correlation: Geology, v. 28, p. 259–262.

Colbert, E.H., 1934, Chalicotheres from Mongolia and China in the American Museum: Bulletin of the American Museum of Natural History, v. 67, p. 353–387.

Dashzeveg, D., 1988, Holarctic correlation of nonmarine Paleocene-Eocene boundary strata using mammals: Journal of the Geological Society of London, v. 145, p. 473–478.

Gingerich, P.D., 1989, New earliest Wasatchian mammalian fauna from the Eocene of northwestern Wyoming: Composition and diversity in a rarely sampled high-floodplain assemblage: University of Michigan Papers on Paleontology, v. 28, p. 1–97.

Gingerich, P.D., 2001, Biostratigraphy of the continental Paleocene-Eocene boundary interval on Polecat Bench in the northern Bighorn Basin, *in* Gingerich, P.D., ed., Paleocene-Eocene stratigraphy and biotic change in the Bighorn and Clarks Fork Basins, Wyoming: University of Michigan Papers on Paleontology, v. 33, p. 37–71.

Guan, Shao Zeng, 1979, Cretaceous and early Tertiary ostracode assemblage from the south-central region and related stratigraphic subdivision and correlation, *in* Institute of Vertebrate Paleonotology and Paleoanthropology, and Nanjing Institute of Geology and Paleontology, Chinese Academy of Sciences, eds., The Mesozoic and Cenozoic Red Beds of South China: Beijing, Science Press, p. 121–131.

Guan, Shao Zeng, 1989, The age of the Dongtang Formation in the Hengyang Basin of Hunan, and the Cretaceous and Tertiary boundary: Acta Geologica Sinica, v. 63, p. 59–72.

Gunnell, G.F., 1999, Mammalian faunal composition and the Paleocene/Eocene epoch/series boundary: Evidence from the northern Bighorn Basin, Wyoming, *in* Aubry, M.-P., et al., eds., Late Paleocene–early Eocene climatic and biotic events in the marine and terrestrial records: New York, Columbia University Press, p. 409–427.

Hooker, J.J., 1989, Character polarities in early perissodactyls and their significance for *Hyracotherium* and infraordinal relationships, *in* Prothero, D.R., and Schoch, R.M., eds., The evolution of perissodactyls: Oxford University Press, Oxford Monographs on Geology and Geophysics, v. 15, p. 79–101.

Hooker, J.J., 1996, Mammalian biostratigraphy across the Paleocene-Eocene boundary in the Paris, London and Belgian basins: London, Geological Society Special Publication 101, p. 205–218.

Hooker, J.J., 1998, Mammalian faunal change across the Paleocene-Eocene transition in Europe, *in* Aubry, M.-P., et al., eds., Late Paleocene–early Eocene climatic and biotic events in the marine and terrestrial records: New York, Columbia University Press, p. 428–450.

Hsu, Vindell, Ge, Tong Ming, Baksi, Ajoy K., Fan, Li Min, and Liu, Jan, 1990, Late Cretaceous magnetostratigraphy from a red basin in central China and a preliminary radiometric date for the K-N (Cretaceous Normal) superchron: American Geophysical Union 1990 fall meeting, Abstracts, v. 71, p. 1298.

Hu, Ji Min, and Zeng, De Min, 1980, On the age of Chejiang Formation of Hengyang basin: Petroleum and Natural Gas Geology, v. 1, p. 159–165.

Kennett, J.P., and Stott, L.D., 1991, Abrupt deep-sea warming, palaeoceanographic changes, and benthic extinctions at the end of the Paleocene: Nature, v. 353, p. 225–229.

Koch, P.L., Zachos, J.C., and Gingerich, P.D., 1992, Correlation between isotope records in marine and continental carbon reservoirs near the Paleocene/Eocene boundary: Nature, v. 358, p. 319–322.

Koch, P.L., Zachos, J.C., and Dettman, D.L., 1995, Stable isotope stratigraphy and paleoclimatology of the Paleogene Bighorn Basin (Wyoming, USA): Palaeogeography, Palaeoclimatology, Palaeoecology, v. 115, p. 61–89.

Krause, D.W., and Maas, M.D., 1990, The biogeographic origins of late Paleocene–early Eocene mammalian immigrants to the western interior of North America, *in* Bown, T.M., and Rose, K.D., eds., Dawn of the age of mammals in the northern part of the Rocky Mountain interior, North America: Boulder, Colorado, Geological Society of America Special Paper 243, p. 71–105.

Li, Chuan Kuei, and Ting, Su Yin, 1983, The Paleogene mammals of China: Bulletin of Carnegie Museum of Natural History, v. 21, 93 p.

Li, Chuan Kuei, and Ting, Su Yin, 1985, Possible phylogenetic relationship of Asiatic eurymylids and rodents, with comments on mimotonids, *in* Luckett, W.P., and Hartenberger, J.-L., eds., Evolutionary relationships among rodents: A multidisciplinary analysis: New York, Plenum Press, p. 35–58.

Li, Chuan Kuei, Chiu, Chan Siang, Yan, De Fa, and Hsieh, Shu Hua, 1979, Notes on some early Eocene mammalian fossils from Hengdong County, Hunan Province: Vertebrata PalAsiatica, v. 17, p. 71–80.

Li, Chuan Kuei, Zheng, Jia Jian, and Ting, Su Yin, 1989, The skull of *Cocomys lingchaensis,* an early Eocene ctenodactyloid rodent of Asia: Contributions in Science, Natural History Museum of Los Angeles County, v. 33, p. 179–192.

Li, Hao Min, 1965, The early Tertiary fossil plants from Chashanao, Hengyang basin, Hunan: Acta Palaeontologica Sinica, v. 13, p. 540–547.

Li, Jin Ling, 1984, A new species of *Planocrania* from Hengdong, Hunan: Vertebrata PalAsiatica, v. 22, p. 123–133.

Liang, Ming Sheng, and Zhang, Wei, 1984, The Tertiary ostracods of China, *in* Li, Yun Tong, ed., The Tertiary system of China: Beijing, Geological Publishing, p. 297–305.

Liu, Xun, and Fu, De Yung, 1986, Sedimentary facies and the tectonic development of the Hengyang Basin, Hunan Province: Bulletin of the Chinese Academy of Geological Sciences, v. 13, p. 13–36.

Lucas, S.G., and Schoch, R.M., 1989, Taxonomy and biochronology of *Eomoropus* and *Grangeria,* Eocene chalicotheres from the western United States and China, *in* Prothero, D.R., and Schoch, R.M., eds., The evolution of perissodactyls: Oxford, UK, Oxford University Press, p. 422–437.

McElhinny, M.W., and McFadden, P.L., 2000, Paleomagnetism: Continents and oceans: San Diego, Academic Press, 386 p.

McKenna, M.C., 1998, Semiisolation of the Arctic Ocean in the late Thanetian to earliest Ypresian time: Geological Society of America Abstracts with Programs, v. 30, no. 7, p. 118.

McKenna, M.C., and Bell, S.K., 1997, Classification of mammals above the species level: New York, Columbia University Press, 547 p.

Meng, Jin, and McKenna, M.C., 1998, Faunal turnovers of Paleogene mammals from the Mongolian Plateau: Nature, v. 394, p. 364–367.

Meng, Jin, Zhai, Ren Jie, and Wyss, A.R., 1998, The late Paleocene Bayan Ulan fauna of Inner Mongolia, China, *in* Beard, K.C., and Dawson, M.R., eds., Dawn of the age of mammals in Asia: Bulletin of the Carnegie Museum of Natural History, v. 34, p. 148–185.

Radinsky, L.B., 1964, *Paleomoropus,* a new early Eocene chalicothere (Mammalia, Perissodactyla), and a revision of Eocene chalicotheres: American Museum Novitates, v. 2179, p. 1–28.

Rohlf, J.F., 2000, NTSYS numerical taxonomy and multivariate analysis system: Version 2.1: Exeter Software, 47 Route 25A, Suite 2, Setauket, New York, 38 p.

Russell, D.E., and Zhai, Ren Jie, 1987, The Paleogene of Asia: Mammals and stratigraphy: Mémoires du Muséum National d'Histoire Naturelle, v. 52, 488 p.

Sloan, R.E., 1987, Paleocene and latest Cretaceous mammal ages, biozones, magnetozones, rates of sedimentation, and evolution, *in* Fassett, J.E., and Rigby, J.K., Jr., eds., The Cretaceous-Tertiary boundary in the San Juan and Raton Basins, New Mexico and Colorado: Boulder, Colorado, Geological Society of America Special Paper 209, p. 165–200.

Sneath, P.H.A., and Sokal, R.R., 1973, Numerical taxonomy: San Francisco, Freeman, 573 p.

South China "Red Beds" Research Group, IVPP, 1977, Paleocene vertebrate horizons and mammalian faunas of south China: Scientia Sinica, v. 3, p. 232–244.

Tauxe, L., and Watson, G.S., 1994, The fold test: An Eigen analysis approach: Earth and Planetary Science Letters, v. 122, p. 331–341.

Thomas, E., and Shackleton, N.J., 1996, The Paleocene-Eocene benthic foraminiferal extinction and stable isotope anomalies: London, Geological Society Special Publication 101, p. 401–441.

Tien, Chi Chun, 1936, Orogenic movements in Hunan: Bulletin of Geology Society of China, v. 15, p. 453–465.

Tien, Chi Chun, Wang, Hsiao Ching, and Hsu, Tiao Yuan, 1933, The geology of Changsha, Hsiangtan, Hengshan, Hsianghsiang, Hengyang, and Shaoyang districts, central Hunan: Bulletin 15, Geology 2, Geological Survey of Hunan, p. 29–31.

Ting, Su Yin, 1993, A preliminary report on an early Eocene mammalian fauna from Hengdong, Hunan Province, China: Kaupia, Darmstädter Beiträge zur Naturgeschichte, v. 3, p. 201–207.

Ting, Su Yin, 1995, An early Eocene mammalian fauna from Hengdong, Hunan Province, China [doctoral dissertation]: Baton Rouge, Louisiana State University, 200 p.

Ting, Su Yin, 1998, Paleocene and early Eocene land mammal ages of Asia, *in* Beard, K.C., and Dawson, M.R., eds., Dawn of the age of mammals in Asia: Bulletin of the Carnegie Museum of Natural History, v. 34, p. 124–147.

Ting, Su Yin, and Li, Chuan Kuei, 1984, The structure of the ear region of *Rhombomylus* (Anagalida, Mammalia): Vertebrata PalAsiatica, v. 22, p. 92–102.

Ting, Su Yin, and Li, Chuan Kuei, 1987, The skull of *Hapalodectes* (?Acreodi, Mammalia), with notes on some Chinese Paleocene mesonychids: Vertebrata PalAsiatica, v. 25, p. 161–186.

Ting, Su Yin, Meng, Jin, McKenna, M.C., and Li, Chuan Kuei, 2002, The osteology of *Matutinia* (Simplicidentata, Mammalia) and its relationship to *Rhombomylus:* American Museum Novitates, v. 3371, p. 1–33.

Tong, Yong Sheng, and Wang, Jing Wen, 1998, A preliminary report on the early Eocene mammals of the Wutu fauna, Shandong Province, China, *in* Beard, K.C., and Dawson, M.R., eds., Dawn of the age of mammals in Asia: Bulletin of the Carnegie Museum of Natural History, v. 34, p. 186–193.

Tong, Yong Sheng, Zheng, Shao Hua, and Qiu, Zhu Ding, 1995, Cenozoic mammal ages of China: Vertebrata PalAsiatica, v. 33, p. 290–314.

Wang, Yuan, 1995, A new primitive Chalicothere (Perissodactyla, Mammalia) from the early Eocene of Hubei, China: Vertebrata PalAsiatica, v. 33, p. 138–159.

Wang, Yuan Qing, Hu, Yao Ming, Chow, Min Chen, and Li, Chuan Kuei, 1998, Chinese Paleocene mammal faunas and their correlation, *in* Beard, K.C., and Dawson, M.R., eds., Dawn of the age of mammals in Asia: Bulletin of the Carnegie Museum of Natural History, v. 34, p. 89–123.

Watson, G.S., 1956, A test for randomness of directions: Monthly Notices of the Royal Astronomical Society Geophysical Supplement, v. 7, p. 160–161.

Wing, S.L., and Greenwood, D.R., 1994, Fossils and fossil climate: The case for equable continental interiors in the Eocene: Philosophical Transactions of the Royal Society of London, series B, v. 341, p. 243–252.

Wing, S.L., Brown, T.M., and Obradovich, J.D., 1991, Early Eocene biotic and climatic change in interior western North America: Geology, v. 19, p. 1189–1192.

Wing, S.L., Bao, Hui Ming, and Koch, P.L., 1999, An early Eocene cool period? Evidence for continental cooling during the warmest part of the Cenozoic, *in* Huber, B.T., et al., eds., Warm climates in Earth history: Cambridge, UK, Cambridge University Press, p. 197–237.

Wolfe, J.A., 1994, Tertiary climatic changes at middle latitudes of western North America: Palaeogeography, Palaeoclimatology, Palaeoecology, v. 108, p. 195–205.

Young, Chung Chien, 1944, Note on the first Eocene mammal from south China: American Museum Novitates, v. 1268, p. 1–3.

Young, Chung Chien, Bien, Mei Nian, and Lee, Yue Yan, 1938, "Red Beds" of Hunan: Bulletin of Geology Society of China, v. 18, p. 259–300.

Zhai, Ren Jie, 1978, More fossil evidences favouring an early Eocene connection between Asia and Neoarctic: Memoir of Institute of Vertebrate Paleontology and Paleoanthropology, v. 13, p. 107–115.

Zhang, Jin Jian, 1982, The age of the Chejiang Formation in Hengyang Basin: Journal of Stratigraphy, v. 6, p. 140–143.

Zhang, Ren Jie, 1979, The Cretaceous freshwater lamellibranchiates from the middle southern region of China: Acta Palaeontologica Sinica, v. 18, p. 189–196.

Zhang, Yu Ping, Zheng, Jia Jian, and Ting, Su Yin, 1979, On Paleocene condylarths from Jiangxi, *in* Institute of Vertebrate Paleonotology and Paleoanthropology, and Nanjing Institute of Geology and Paleontology, Chinese Academy of Sciences, eds., The Mesozoic and Cenozoic Red Beds of South China: Beijing, Science Press, p. 382–386.

Zhang, Zhen Lai, 1979, The Cretaceous pollen assemblage of middle-south region of China, *in* Institute of Vertebrate Paleonotology and Paleoanthropology, and Nanjing Institute of Geology and Paleontology, Chinese Academy of Sciences, eds., The Mesozoic and Cenozoic Red Beds of South China: Beijing, Science Press, p. 132–140.

Zheng, Jia Jian, and Huang, Xue Shi, 1984, A new didymoconid (Mammalia) from the early Eocene of Hunan: Vertebrata PalAsiatica, v. 22, p. 198–207.

Manuscript Accepted by the Society August 13, 2002

Printed in the USA

Geological Society of America
Special Paper 369
2003

Paleogene paleosols of Big Bend National Park, Texas

Paul D. White*
Department of Geology and Geophysics, Louisiana State University, Baton Rouge, Louisiana 70803, USA
Judith A. Schiebout
LSU Museum of Natural Science, and Department of Geology and Geophysics, Louisiana State University, Baton Rouge, Louisiana 70803, USA

ABSTRACT

A chemostratigraphic section using the stable isotopes of carbon from pedogenic carbonate nodules has been developed in Big Bend National Park, Texas, with the goal of pinpointing the Paleocene-Eocene Thermal Maximum and thus the Paleocene-Eocene boundary. A negative carbon excursion has been identified within C24r and is bracketed by a late Tiffanian or Clarkforkian fossil site and an early Wasatchian (Wa-1) site. The range in δ^{13}C values is from –14.08‰ to –8.14‰.

Outcrops in Big Bend National Park are North America's most southern exposure of early Paleogene continental deposits in which the Paleocene-Eocene boundary is constrained by terrestrial mammal localities. The exposures are mostly red and black mudstones, the colors of which result from pedogenic processes. Common pedogenic features of the red paleosols are slickensides, sub angular, blocky peds, carbonate nodules, redoximorphic features, and Gilgai microrelief. Smectite is the most abundant clay mineral present. Quartz, calcite and plagioclase make up the remaining bulk of their mineralogy. These ancient soils are interpreted to have formed mainly under dry conditions with seasonal wetting and a fluctuating water table. Black paleosols are less common than red paleosols in the Paleocene and absent from the Eocene. Slickensides and sub angular blocky peds are common pedogenic features in them. The smectite content of black paleosols is much higher compared to that of red paleosols, and they contain significantly less calcite and plagioclase. Black paleosols formed under very wet and poorly drained conditions in which calcite and plagioclase were leached and authigenic smectite was produced.

Due to the condensed nature of the section as well as pedogenic overprinting, the interpretation of the negative carbon excursion is complex, as is its effects on pedogenic processes. During this global warming event, changes in pedogenesis did occur and resulted in a significant increase in the translocation of clays and the leaching of calcite and plagioclase. This increase in the weathering profile may have been caused by an increase in mean annual temperature and rainfall.

*pwhite1@lsu.edu

White, P.D., and Schiebout, J.A., 2003, Paleogene paleosols of Big Bend National Park, Texas, *in* Wing, S.L., Gingerich, P.D., Schmitz, B., and Thomas, E., eds., Causes and Consequences of Globally Warm Climates in the Early Paleogene: Boulder, Colorado, Geological Society of America Special Paper 369, p. 537–550.

INTRODUCTION

This paper presents new data on southern North American continental deposits spanning the Paleocene-Eocene boundary. The outcrop in the study area (Fig. 1) is the southernmost exposure of continental deposits in North America that record the terrestrial Paleocene-Eocene transition constrained by fossil mammal occurrences. Unlike the Bighorn and Clarks Fork basins in Wyoming, which also record the Paleocene-Eocene transition, the exposures in Big Bend National Park are condensed, and mammalian fossils are relatively rare. The early Paleogene deposits of the Bighorn basin in Wyoming are an order of magnitude thicker for the same period than those exposed in Big Bend National Park (Schiebout, 1995)

The Paleocene-Eocene transition is punctuated by a well-documented global warming event, the Paleocene-Eocene Thermal Maximum (PETM). Evidence for the PETM using stable isotopes of carbon and oxygen has been found in both continental and marine deposits (Bowen et al., 2001; Norris and Rohl, 1999; Fricke et al., 1998; Koch et al., 1992; Kennet and Stott, 1991). A goal of this research was to pinpoint the PETM using stable isotope stratigraphy on the deposits in Big Bend National Park. Combining a chemostratigraphic section with a detailed study of Paleogene paleosols in Big Bend National Park is aimed at recognizing effects of the PETM on pedogenesis.

GEOLOGICAL SETTING

The Paleocene-Eocene transition in the study area lies within the Tornillo Formation (Schiebout et al., 1987) (Fig. 2). The Paleocene–Eocene boundary has been considered to correspond roughly to the contact between the Black Peaks Member and the Hannold Hill Member (Schiebout et al., 1987). The rocks crop out in typical badlands exposures (Fig. 3) in a basin

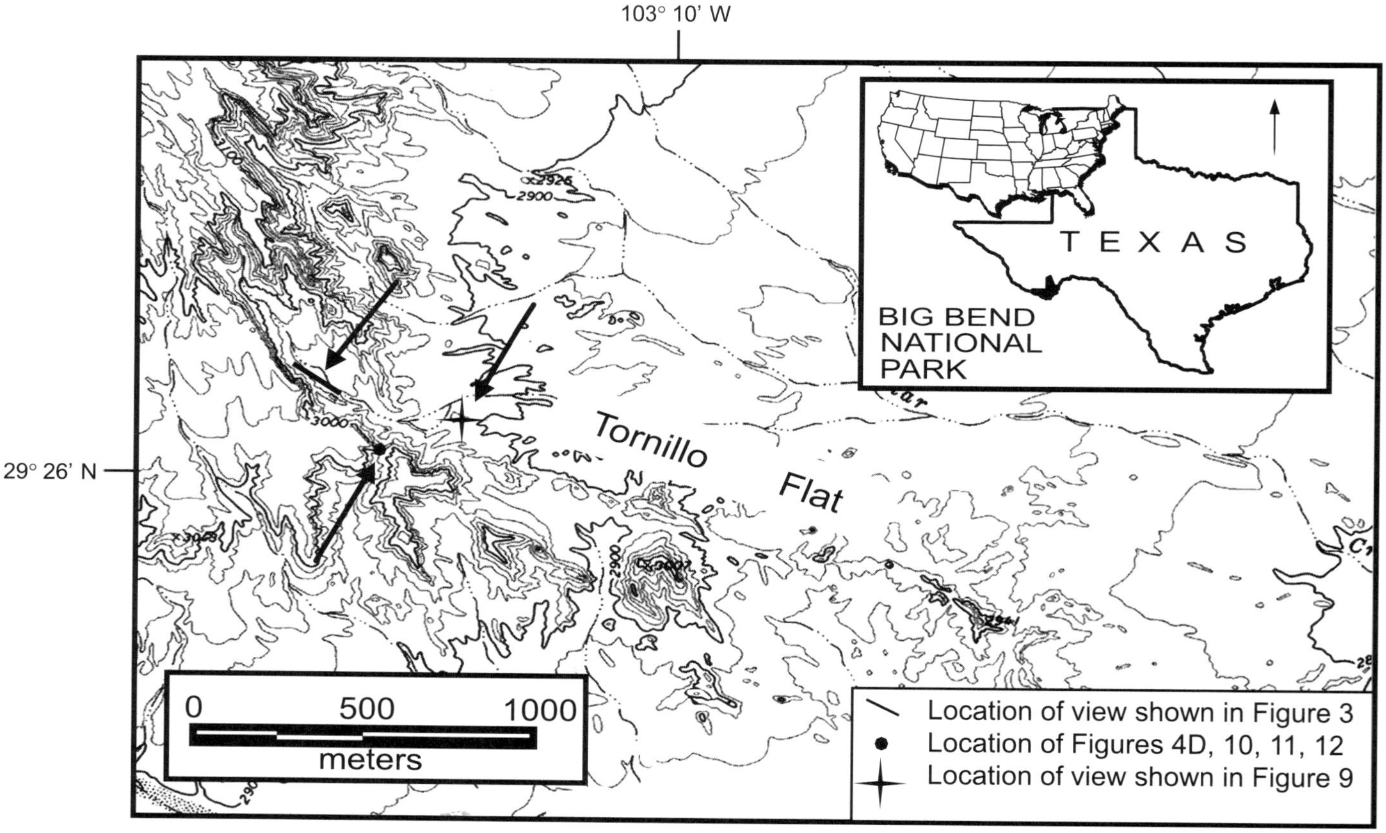

Figure 1. The study area is located on western Tornillo Flat in Big Bend National Park, Texas.

Figure 2. Stratigraphic control and stable isotope data for the study area. Columns 1 and 2 are the chronology and polarity history after Berggren et al. (1995). Column 3 is the magnetic polarity data for the study area after Rapp et al. (1983). Column 4 is formation and members in the study area after Schiebout et al. (1987). Column 5 shows the fossil mammal locations (Ti—Tiffanian, Cf—Clarkforkian, Wa—Wasatchian) and elevations of black paleosols. Column 6 is the chemostratigraphic section. Four data points from bulk samples have been added to complete the lower part of the section.

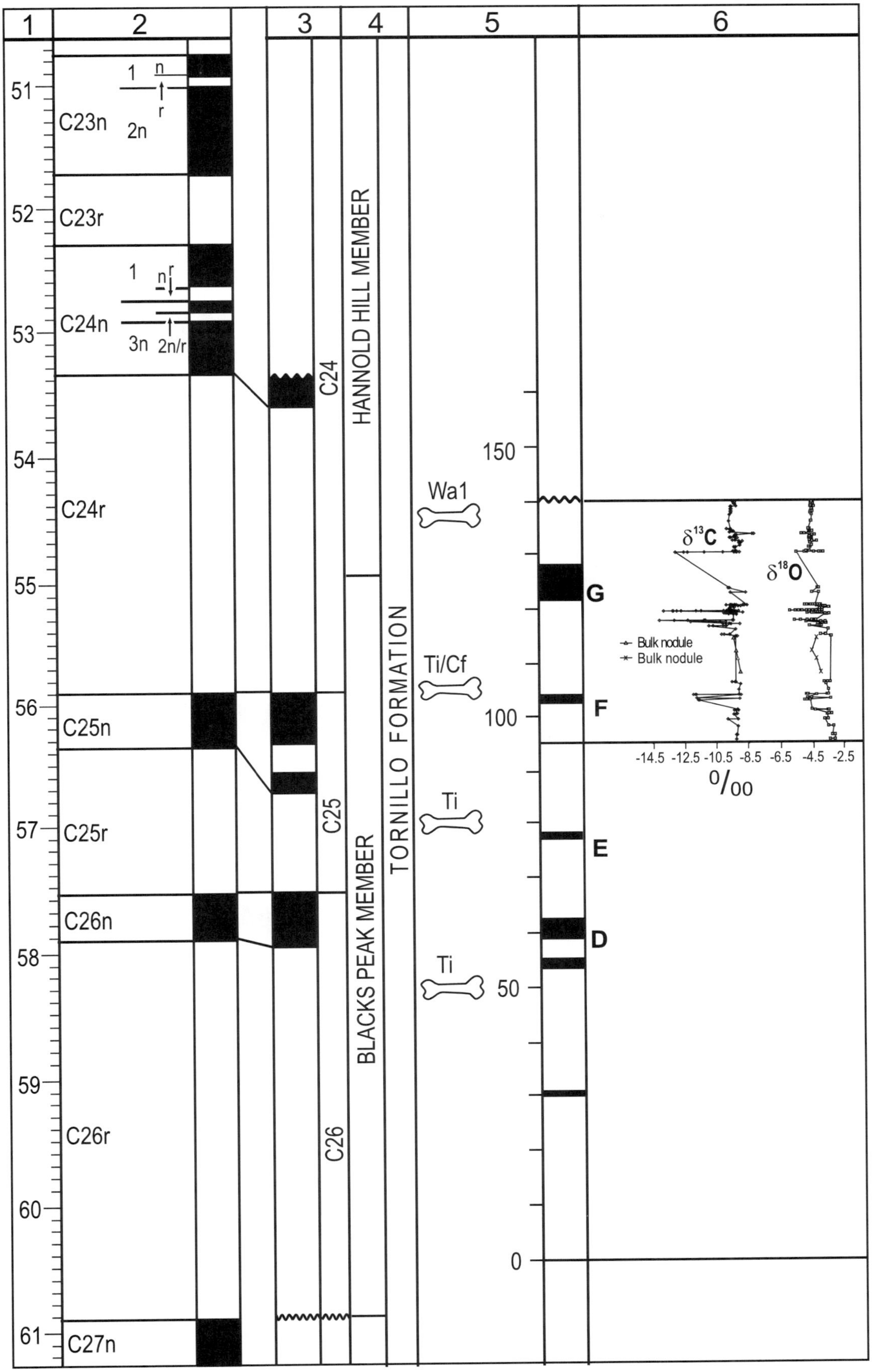

1
2
3
4
5
6
51
52
53
54
55
56
57
58
59
60
61
C23n
1
2n
n
r
C23r
C24n
1
3n
2n/r
C24r
C25n
C25r
C26n
C26r
C27n
C24
C25
C26
HANNOLD HILL MEMBER
BLACKS PEAK MEMBER
TORNILLO FORMATION
Wa1
Ti/Cf
Ti
Ti
150
100
50
0
G
F
E
D
δ13C
δ18O
Bulk nodule
Bulk nodule
-14.5 -12.5 -10.5 -8.5 -6.5 -4.5 -2.5
0/00

Figure 3. Typical badlands exposure in Big Bend National Park with a black paleosol and a series of stacked red paleosols.

and range graben (Maxwell et al., 1967; Schiebout et al., 1987). The fluvial Black Peaks and Hannold Hill Members of the Tornillo Formation in the study area are made up of 60% mudstone, which occur mainly as stacked red and black layers that have been interpreted as paleosols (Schiebout, 1974, 1995). This cumulative stacking of paleosols results from the addition of sediment by successive overbank flooding events (Kraus and Gwinn, 1997).

METHODS

Seventeen detailed columnar sections were measured to characterize the morphology of Big Bend paleosols. Sections were measured using a Jacob's staff and a Brunton compass. Each section was tied to a marker bed so a composite section could be constructed. Marker beds are prominent black paleosols that crop out laterally for hundreds of meters (Fig. 3).

Carbonate nodules were sampled from ~60 paleosols to develop a chemostratigraphic section. The chemostratigraphic section is a composite of eight individual sections. The carbon isotopic composition of authigenic soil carbonate is controlled by the type of plant cover, and the formation of authigenic soil carbonate has been described and its fractionation pathways under various types of plant cover have been documented (Cerling, 1984; Quade et al., 1989; Cerling et al., 1991). Since C4 plants where not a significant part of the flora until the Miocene (Quade and Cerling, 1995), and the $\delta^{13}C$ values of a C3 cover are between –14‰ and –8‰, any change in the atmospheric $^{13}C/^{12}C$ ratio should be recorded in pedogenic carbonate (Cerling, 1984; Quade et al., 1989; Cerling et al., 1991).

Carbonate nodules were collected from B horizons of paleosols. Isotopic analyses were performed on samples (n = 178) taken from polished surfaces of nodules under magnification to insure the sampling of micrite and not spar. In most cases, at least two nodules from each paleosol were analyzed. The number of analyses per paleosol was increased to at least three at the approximate expected level of the Paleocene-Eocene transition based on fossil and magnetic polarity data. Approximately 150 µg of sample was roasted in a vacuum at 400 °C to remove organics. A Micromass Optima or Prism gas-source mass spectrometer was used to analyze the sample after reaction with 100% phosphoric acid. Results are reported relative to the V-PDB standard in per mil units. $\delta^{18}O$ and $\delta^{13}C$ values were calculated using the following equation:

$$\delta^{18}O \text{ or } \delta^{13}C = 1000 \times (R_{sample} - R_{standard})R_{standard}$$

where R = $^{18}O/^{16}O$ or $^{13}C/^{12}C$.

The mineralogy of paleosols (n = 27) and carbonate nodules (n = 8) was determined by X-ray diffraction (XRD) analysis of randomly oriented powders on a Phillips APD 3720 system using Cu Kα radiation. Samples where powdered in a mortar and pestle and then micronized to a maximum grain size of 35 μm.

Stable isotope stratigraphy

A major goal of developing a chemostratigraphic section for the Paleogene of Big Bend National Park is to identify the PETM using the stable isotopes of carbon and to tie it to the sequence of paleosols. The PETM and its negative carbon excursion in the Bighorn Basin occur in magnetic polarity chron 24R2 (Bowen et al., 2001; Koch et al., 1992, 1995). Magnetic polarity data places the Paleocene-Eocene boundary in the study area at ~125 m above the base of the Black Peaks Member of the Tornillo Formation (Rapp et al., 1983; Schiebout et al., 1987) (Fig. 2). The results of the chemostratigraphic section for the study area are shown in Figures 2 and 4. The range in carbon and oxygen values are between –14.08‰ to –8.14‰ and –5.93‰ to –3.10‰ respectively. Figure 5 shows that the data have two significant groupings, excursion and nonexcursion values. Carbon excursion values range from –14.08‰ to –10.70‰ with a mean oxygen value of –4.93‰. Nonexcursion carbon values range from –10.18‰ to –8.14‰ and have an average $\delta^{18}O$ of –4.20‰. Variation within individual paleosol $\delta^{13}C$ values can be as high as 4.63‰. Variations in $\delta^{18}O$ within paleosols is low and never exceeds 1.70‰. The average $\delta^{18}O$ variation within nodules is 0.44‰. The difference in mean excursion and mean nonexcursion $\delta^{18}O$ values is low (<1‰). This small difference suggests that the isotopic data presented have not been affected by diagenetic alteration. The small $\delta^{18}O$ variation within paleosols and within nodules variation also supports this conclusion.

Interpretation of stable isotope stratigraphy

The carbon excursion reported by Bowen et al. (2001) for the Bighorn Basin is different in pattern from the excursion data reported here. Rates of sedimentation in the Bighorn Basin during the Paleogene were much higher than those in Big Bend (Schiebout, 1995). The result for the Bighorn Basin is an apparently smooth and continuous excursion starting in the Clarkforkian, continuing into the Wasatchian, and ending early in Wasatchian (Wa1) time. The condensed early Paleogene section in Big Bend results in four apparently separate $\delta^{13}C$ excursion events punctuating the section (Figs. 2, 4). This is a result of $\delta^{13}C$ excursion and nonexcursion values occurring within single paleosols, interspersed with paleosols that cannot be sampled at all. Big Bend soils are interpreted as having been active at the surface for longer periods of time than those in the Bighorn Basin and therefore a single paleosol can record pre-PETM and PETM nodule formation or PETM and post PETM nodule formation. Gaps in the data set result from paleosols that have not produced carbonate nodules. The most significant gap

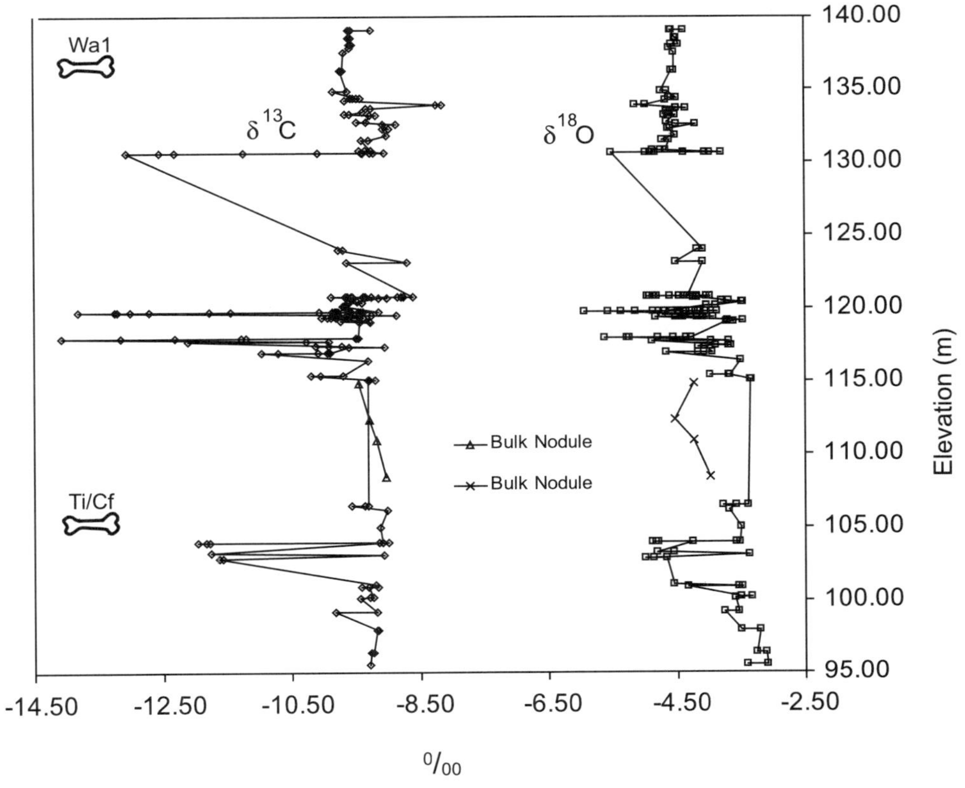

Figure 4. Chemostratigraphic section for the study area. The three upper spikes fall within C24r and are bracketed by fossil mammal localities. The gaps between spikes in the section are a result of pedogenic overprinting and the occurrence of interspersed soils that do not contain carbonate nodules.

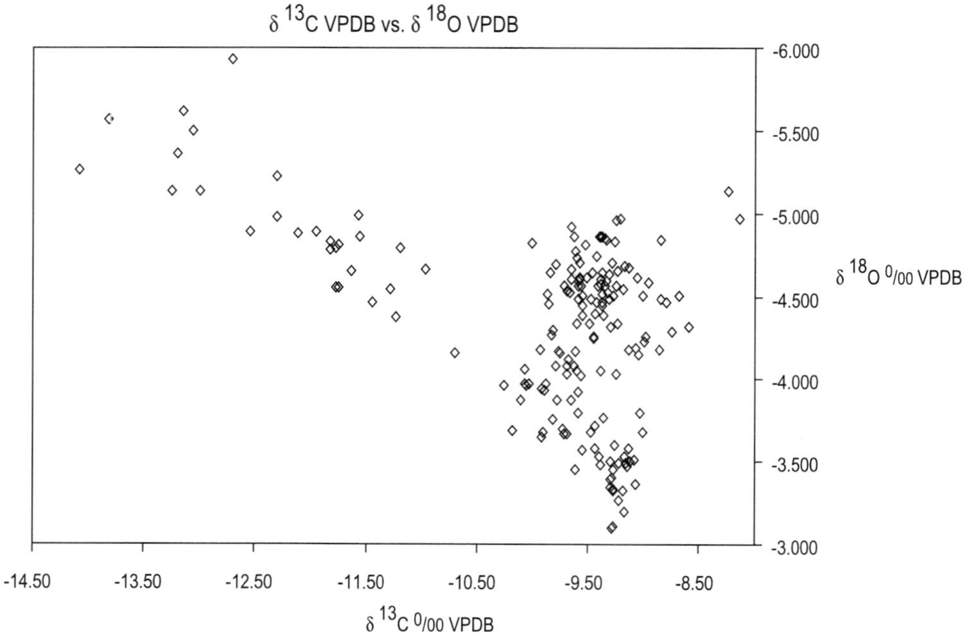

Figure 5. $\delta^{13}C$ vs. $\delta^{18}O$ plot showing two groupings of the data. Excursion values plot on the left and nonexcursion values on the right.

is from ~124.00 m to 130.6 m. See the paleosol section below for factors controlling the presence of carbonate nodules.

The first strong shift in $\delta^{13}C$ values is at ~103 m in the section, has a $\delta^{13}C$ of −11.95‰ and is interpreted as too early to equate to the PETM based on magnetic polarity data (Fig. 2). However, Bowen et al. (2001) report a similar excursion in their data set that occurs early in the Clarkforkian. It is interpreted that the excursion at 103 m in the Big Bend section correlates to the early Clarkforkian excursion in the Bighorn Basin.

The next three negative $\delta^{13}C$ spikes lie within C24r and are bracketed by a latest Tiffanian or Clarkforkian fossil location and a Wasatchian (Wa1) fossil locality. Based on magnetic polarity and fossil data the three spikes fall within the predicted stratigraphic level for the PETM. The apparent punctuation is interpreted as a result of interspersed soils barren of nodules, ones in which no nodules recording the PETM have yet been sampled, and pedogenic overprinting.

Developing a chemostratigraphic section for the study area is difficult because of ancient pedogenic processes and the condensed nature of the section. However, all excursions in the Big Bend section are comparable in magnitude to those reported for the Bighorn Basin (Bowen et al., 2001) and their stratigraphic occurrence is constrained by both mammalian fossil data and magnetic polarity data.

PALEOSOL FEATURES AND MINERALOGY

Fastovsky and McSweeney (1987, 1991), Bown and Kraus (1981), Kraus (2001), and Retallack (1997, 1999) are just a few of the many authors that have provided the basis for interpreting paleosols and changes in ancient pedogenic process and what they may tell us with regard to climate change, sea-level change, or basin dynamics. This study follows the approach of Fastovsky and McSweeney (1987) in that it does not try to identify paleosols using the modern soil classification systems, because the processes that formed these ancient soils may no longer be active or the soil types themselves may be extinct (Retallack, 1981; Lehman, 1989). It looks at stratigraphic changes in paleosol morphology and mineralogy and attempts to link these changes to a possible cause. Below are descriptions of common pedogenic features found in Big Bend paleosols and their interpretation, as well as paleosol mineralogy.

Color

Soil color is either a reflection of the original colors of the soil parent material or a result of the reduction and oxidation of iron, in part or all of the soil. Once reduction has taken place, it is followed by oxidation and the precipitation of iron oxides (Fanning and Fanning, 1989). Iron as Fe (II) is soluble and leaches or diffuses within the soil profile or possibly leaches through the soil. This process results in soil redoximorphic features or mottles. The area of the soil where iron has been removed is commonly a low chroma gray, and where the iron has accumulated is typically a high chroma rust color, such as red. The cause of redoximorphic features such as mottling in modern soils and paleosols has been attributed to endosaturation in response to a fluctuating groundwater table (Duchaufour, 1982; Fanning and Fanning, 1989; Pipujol and Buurman, 1994, 1997; Therrien and Fastovsky, 2000). Other constituents that color soils are manganese oxides (black), jarosite (yellow) and colored clay minerals.

Red mudstones in both the Black Peaks and Hannold Hill Members have colors with hues that range from 2.5YR to 5Y/R with values from 2.5 to 3 and a chroma of 3 on the Munsell Soil Color Chart (Munsell, 1992). Redoximorphic features are common in these soils and areas of the soil that have been depleted of iron are gray (Fig. 6A). They range in size from a few millimeters to tens of centimeters and occur as root halos and large mottles. Redox features occur in B horizons and, where preserved, also occur in E horizons. Most paleosols in Big Bend have abundant redoximorphic features and they are interpreted to result from ground water saturation in response to a fluctuating water table. The dark red color of paleosols also suggests that they were predominately well drained during pedogenesis (Kraus, 2001; Duchaufour, 1982).

Six stratigraphically distinct black paleosols (hue 5YR; value 2.5; chroma 1) occur in the Paleocene Black Peaks Member on Tornillo Flat. They are absent in the overlying Eocene Hannold Hill Member. The black coloration of these paleosols results from a high manganese oxide content. Two field tests using the application of hydrogen peroxide and sodium citrate-bicarbonate-dithionite resulted in a vigorous reaction, indicating the presence of manganese oxide (Fanning and Fanning, 1989).

Paleosol mineralogy

Paleosol mineralogy was determined using X-ray diffraction methods on bulk samples. Figure 7A is an X-ray pattern from a typical B-horizon of a red paleosol. The clay mineralogy is dominated by smectite, and occasionally a small kaolinite peak is evident. Other major minerals in red paleosols are quartz, plagioclase and calcite. Qualitative soil mineralogy is consistent in stacked B and E horizons in red paleosols. This is demonstrated in Figure 8, which is a series of X-ray patterns from stacked B and E horizons. Figure 9 shows the stacked horizons. Figure 7B is an X-ray pattern from a typical Big Bend black paleosol. The most abundant clay mineral present is smectite, and a small kaolinite peak is also recorded. The clay mineralogy of the black paleosols is similar to the red paleosols, but the smectite peak of the black paleosol is more pronounced, indicating a higher clay content. In the X-ray pattern for the black paleosol (Fig. 7B), peaks for calcite and plagioclase are absent. This suggests that the black paleosols formed under very wet to saturated conditions, resulting in the leaching and removal of calcite and plagioclase. The dominance of smectite, an expandable clay, is reflected in the soil structures that are common in Big Bend paleosols.

Soil structures

Common soil structures in Big Bend paleosols are sub angular blocky peds, slickensides and Gilgai microrelief (Fig. 6D). In modern soils these features form as a result of shrinking and swelling in response to cycles of wetting and drying (Fanning and Fanning, 1989; Dudal and Eswaran, 1988). The development of these features is also dependent on the clay content and clay mineralogy of the soil (Dudal and Eswaran, 1988). In modern soils with a high smectite content, features such as microhighs and microlows (Gilgai microrelief), and slickensides are common (Dudal and Eswaran, 1988). Rare nodular peds are found in Big Bend black paleosols and show a concretionary growth pattern (Fig. 6E). These peds are interpreted as desiccation peds like those observed in modern Louisiana soils (personal com., W. Hudnall).

Carbonate nodules

Carbonate nodules are abundant in most red paleosols in Big Bend and formed under dry conditions as shown by the red oxidized color of the paleosols and ferric iron preserved within the nodules (Kleinert and Strecker, 2001; Therrien and Fastovsky, 2000). Carbonate nodules have been found in only one of the six black paleosols and are rare in it. Nodules range in size from 0.50 cm to several centimeters. Micrite, spar, and quartz are the major components of the Big Bend nodules, with rare occurrences of barite and plagioclase. Figure 6F is a photomicrograph of a typical carbonate nodule from a red paleosol. In the center of the picture, plagioclase is being replaced by calcite. Figure 6G illustrates the preservation of oxidized iron in a carbonate nodule. X-ray diffraction methods were also used to identify the mineralogy of carbonate nodules. Results are shown in Figure 7C.

Soil horizons

The most common soil horizons identified in the study area are B and E horizons (Soil Survey Staff, 1999). C horizons are less common and no A horizons have been identified. Carbonate nodules are abundant in most red paleosols and are interpreted as Bk horizons. Paleosol horizons with redoximorphic features are Bg horizons, and where slickensides are present, they are designated as Bss horizons. In addition, many horizons show more than one pedogenic feature. For example, a soil horizon with carbonate nodules and slickensides would be a Bkss horizon. Coarser grained horizons, which are lighter in color and lie above B horizons are interpreted as E-horizons. Their coarser texture suggests that clays have been translocated to the underlying B horizon. In addition, mottles from the underlying B horizon extend up into the layers interpreted as E horizons. Figure 9 illustrates the stacking of B and E horizons in red paleosols. Black paleosols do not show multiple horizons and are interpreted as Bss horizons due to the presence of large slickensides (Fig. 6C).

INTERPRETATIONS OF BIG BEND PALEOSOLS

Red paleosols

Slickensides and peds develop in response to the shrinking and swelling of clays due to cycles of wetting and drying

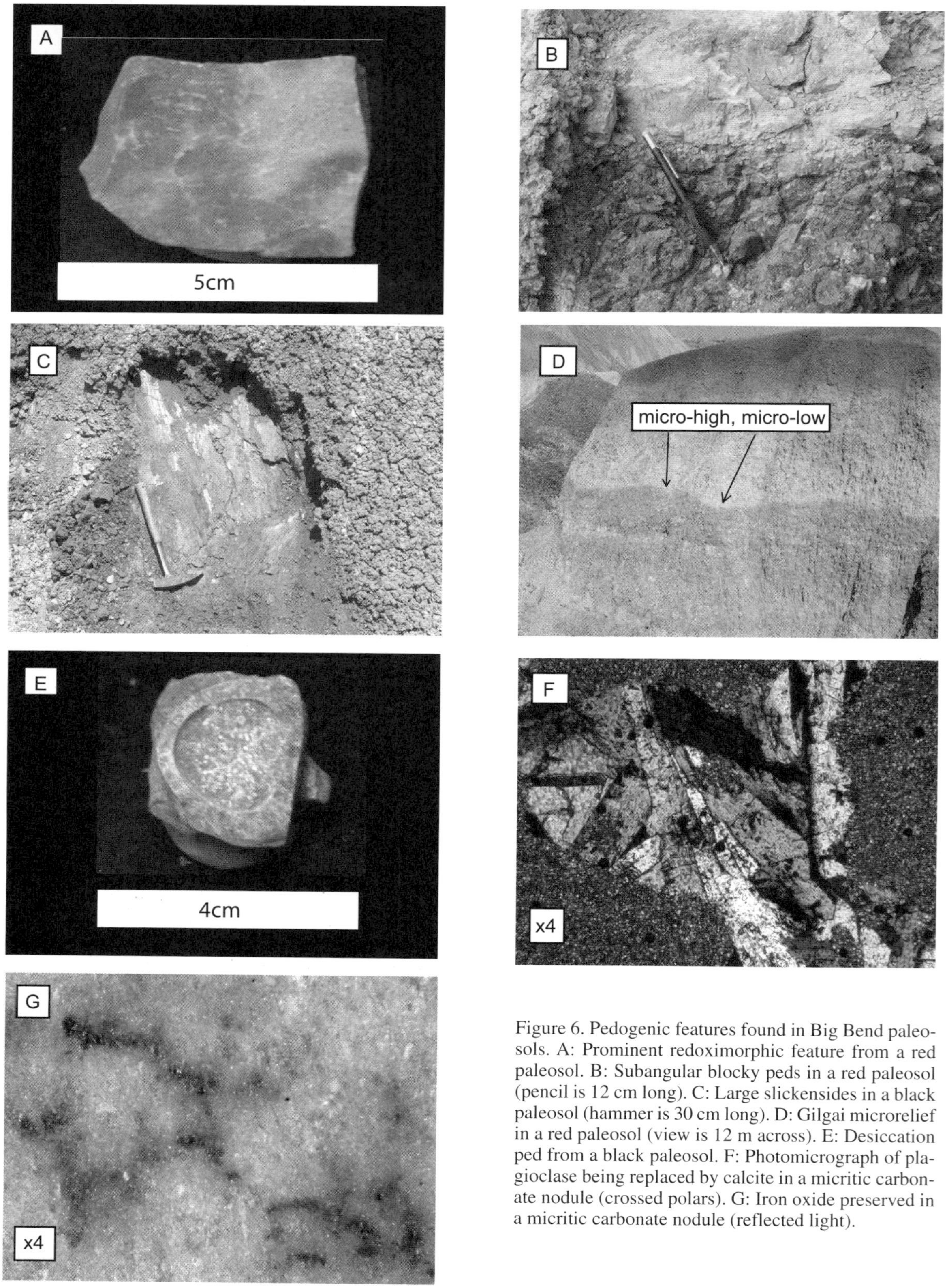

Figure 6. Pedogenic features found in Big Bend paleosols. A: Prominent redoximorphic feature from a red paleosol. B: Subangular blocky peds in a red paleosol (pencil is 12 cm long). C: Large slickensides in a black paleosol (hammer is 30 cm long). D: Gilgai microrelief in a red paleosol (view is 12 m across). E: Desiccation ped from a black paleosol. F: Photomicrograph of plagioclase being replaced by calcite in a micritic carbonate nodule (crossed polars). G: Iron oxide preserved in a micritic carbonate nodule (reflected light).

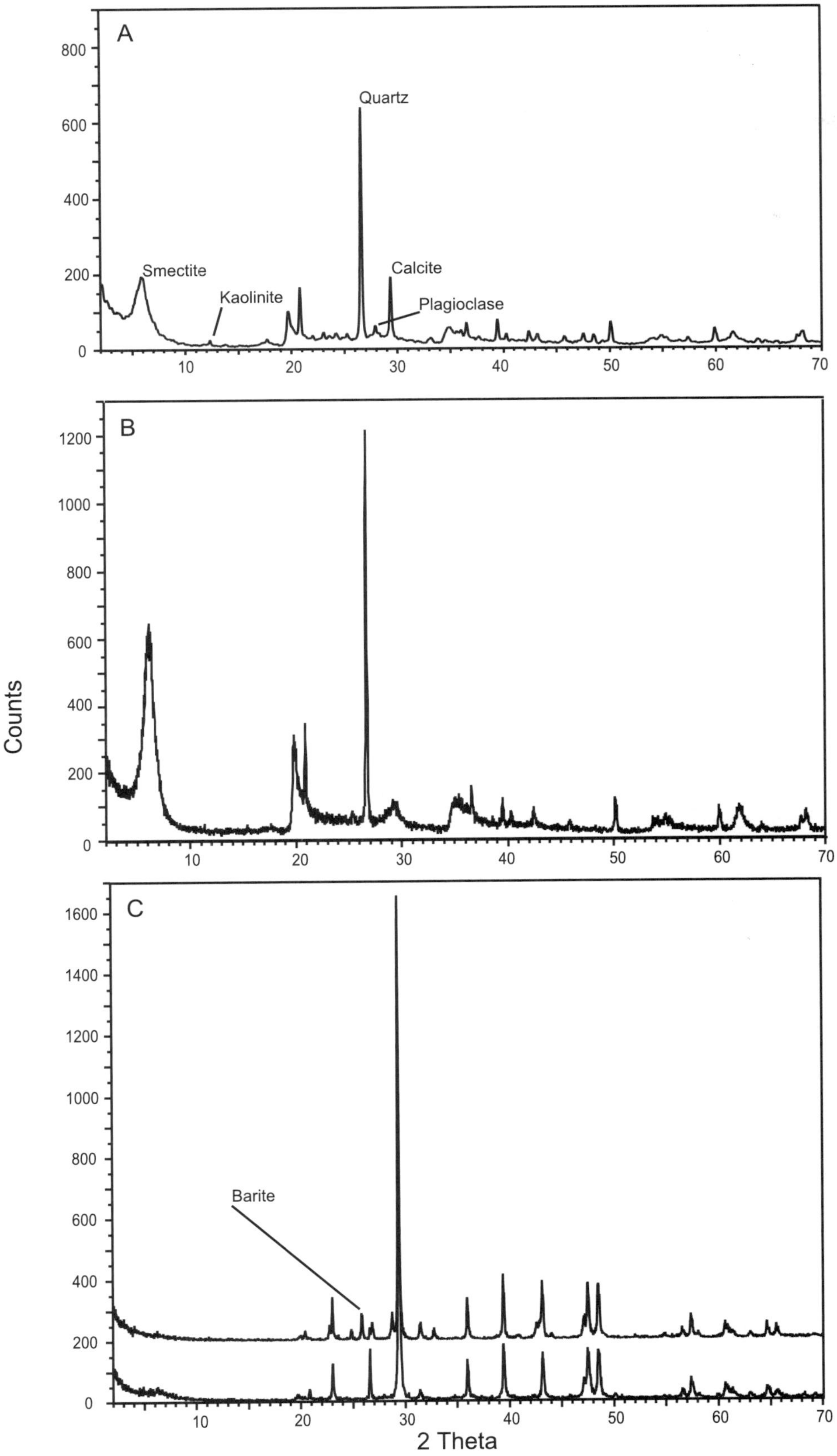

Figure 7. XRD patterns of bulk randomly oriented samples. A: From a red paleosol. B: From a black paleosol. C: From carbonate nodules. The upper pattern shows a sample that has barite.

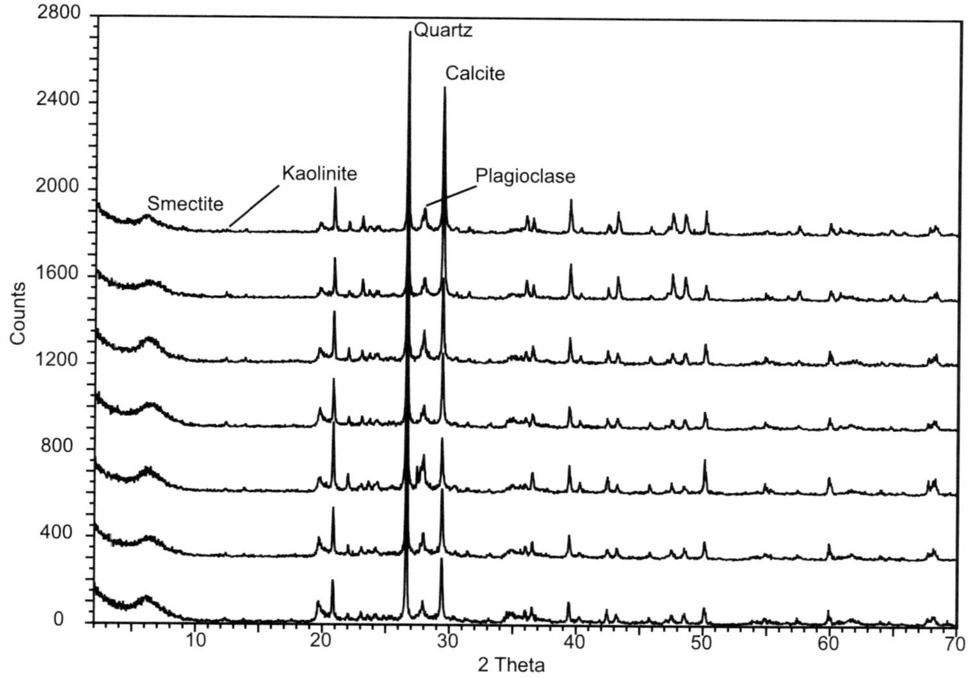

Figure 8. XRD patterns of bulk randomly oriented samples from a series of stacked B and E horizons. See Figure 9 for a photograph and descriptions of the seven horizons.

(Dudal and Eswaran, 1988), but can form under climates with a pronounced dry season with as little precipitation as 18 cm/yr (Retallack, 1990). It seems probable that the pedogenesis of Big Bend red paleosols took place mainly under dry conditions resulting in their red color and the presence of carbonate nodules. In addition to a mainly dry climate, seasonal rains must have occurred to develop slickensides. Moreover, it is interpreted that these ancient soils formed in a topographic setting that was well drained. Prominent redoximorphic features occur in B horizons and E horizons and suggest the presence of a high fluctuating water table (Kleinert and Strecker, 2001; Fastovsky, 1987).

Black paleosols

The absence of calcite and plagioclase (Fig. 7B) suggests that these ancient soils formed under conditions in which they were strongly leached. The larger smectite peak seen in black paleosols (Fig. 7B) indicates an increase in authigenic clay compared to the red paleosols. This is a direct result of the leaching of plagioclase. Smectite has been reported as the first weathering product of plagioclase (Glassmann, 1982). Big Bend black paleosols are interpreted to have formed under very wet to saturated conditions punctuated by short dry periods that would have produced the slickensides. It is possible that during these short time periods, the rare carbonate nodules may have formed, along with the desiccation peds. It is also probable that carbonate nodules and desiccation peds formed as the soil became well drained.

Stratigraphic changes in paleosols

Red paleosols are the most common ancient soils in the Black Peaks and Hannold Hill Members and formed under mainly dry and well-drained conditions but experienced some degree of seasonal wetting. The most dramatic stratigraphic change in pedogenesis in the study area is the repetitive replacement of these red paleosols with black paleosols that formed under poorly drained and very wet conditions. An increase in rainfall, a change in basin dynamics affecting drainage, or a change in sea level also affecting drainage, are all possible causes for this significant change in pedogenesis.

The PETM and pedogenesis

Based on mineralogical and field observations, it appears that the PETM affected pedogenesis in the study area. The first set of paleosols that record both pre-PETM and PETM $\delta^{13}C$ values occurs between 118 m and 125 m in the section and is capped by a black paleosol that does not contain carbonate nodules (Fig. 10). Figure 11 is a columnar section illustrating the soils affected by the PETM. XRD analyses indicate there has been intense translocation of clay minerals (Fig. 12) through the section and textural observations in the field also support this

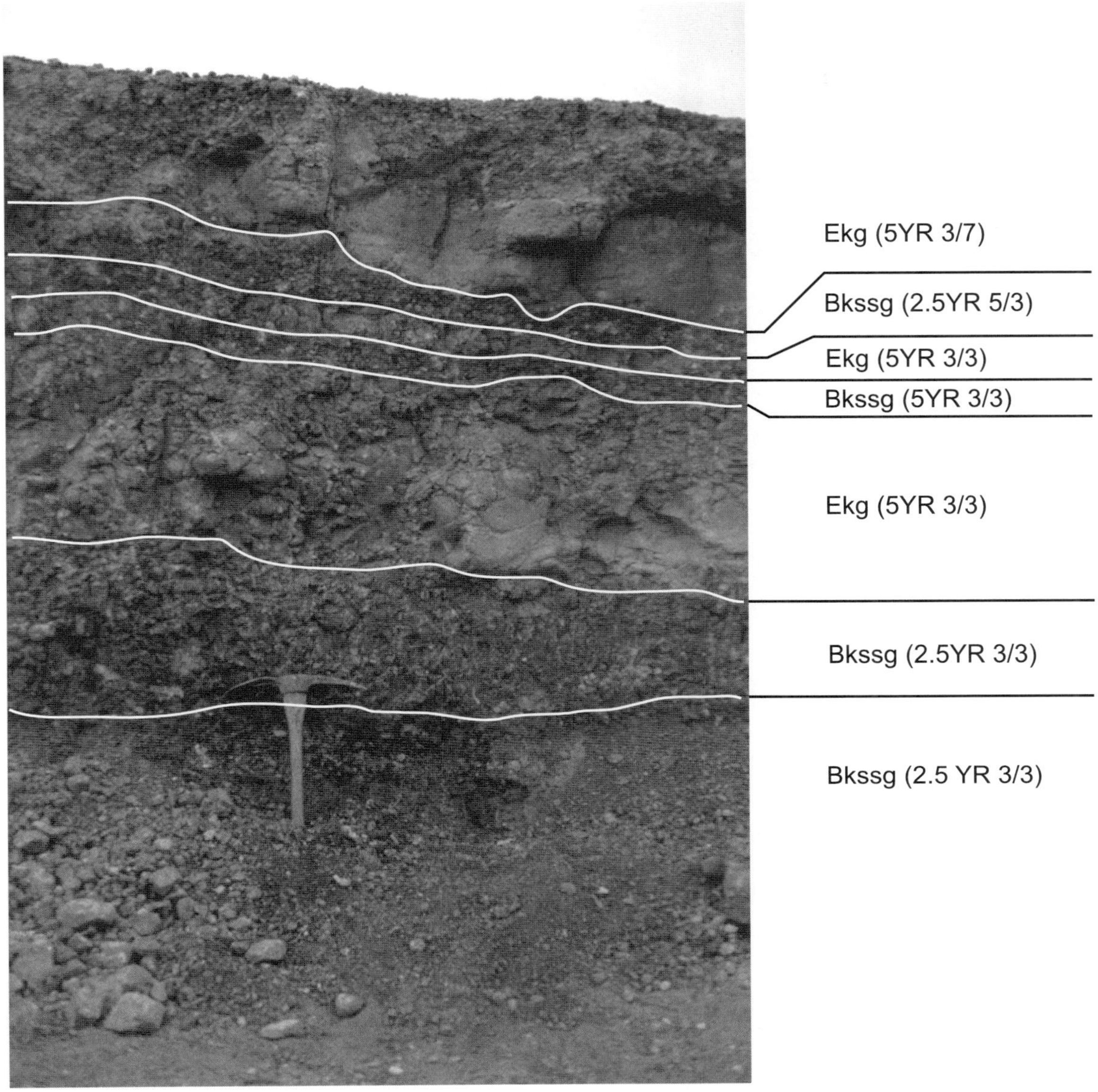

Figure 9. Horizonation of a series of stacked B and E horizons (pick is 35 cm long). k—carbonate nodules, ss—slickensides, g—gleyed (mottled).

conclusion. The XRD patterns also indicate that these soils have been leached of calcite and plagioclase relative to non-PETM soils. It is interpreted that the stacked red B and E horizons from 118 m to 125 m were deposited before the PETM and were similar in mineralogy and texture to those lower in the section (Fig. 8, 9). In addition, E horizons typically do not exceed 30 cm in thickness prior to the PETM. During the PETM, the weathering profile depth increased, possibly due to an increase in temperature and rainfall, causing the translocation of clay minerals and leaching of calcite and plagioclase through seven meters of section.

SUMMARY AND CONCLUSIONS

The PETM has been identified in the study area using chemostratigraphic methods, occurs within C24r, and is bracketed by mammalian fossil sites. The range in $\delta^{13}C$ values is from −14.08‰ to −8.14‰. Small differences in mean $\delta^{18}O$ values within individual nodules and within single paleosol horizons, as well as the small difference in mean excursion and nonexcursion $\delta^{18}O$ values indicate that the data presented here have not been affected by diagenesis. The pattern of the excursion is different than that reported for

Figure 10. Lower part of the section that records both pre-PETM (Paleocene-Eocene Thermal Maximum) and PETM $\delta^{13}C$ values (Jacob staff = 2 m).

the PETM in the Bighorn Basin. Instead of a smooth continuous curve, the interval is punctuated by three individual spikes. This results from single paleosol horizons recording both nonexcursion and excursion $\delta^{13}C$ values (i.e., pedogenic overprinting), as well as the presence of interspersed paleosols that do not contain carbonate nodules.

Red paleosols dominate North America's most southern exposure of early Paleogene continental rocks. These soils formed mainly under well-drained dry conditions with periods of seasonal wetting as evidenced by their red color, the presence of carbonate nodules and slickensides. Abundant and prominent redoximorphic mottling throughout B horizons and occasionally extending up into E horizons suggests a high fluctuating water table.

Black paleosols occur in the Paleocene, Black Peaks Member but not the overlying Hannold Hill Member. Their color is attributed to manganese oxide. X-ray patterns indicate that these paleosols have minor amounts of plagioclase and calcite. The reduction of these two minerals indicates leaching conditions and the additional smectite is interpreted to be authigenic. It is most likely that the black paleosols formed under very wet and poorly drained conditions.

Mineralogical and field observations suggest that the PETM affected pedogenesis in the study area. A suite of stacked red B and E horizons record an increase in the depth of the weathering profile compared to non-PETM soils. This suite of red paleosols is characterized by an intense translocation of clay minerals and leaching of plagioclase and calcite through seven meters of previously developed soils.

This study has characterized early Paleogene paleosols and has identified the PETM using chemostratigraphic methods in North America's most southern exposure of terrestrial rocks spanning the boundary that is constrained by fossil mammals. In addition, it has identified changes in pedogenesis caused by the PETM. Moreover, it has demonstrated the complexities of developing a chemostratigraphic section in a condensed record of stacked paleosols that were affected by pedogenic overprinting.

ACKNOWLEDGMENTS

We thank Ray Ferrell, Wayne Hudnall, Brooks Ellwood, Paul Koch, Jason Newton, Brian Lareau, Matt Hackworth, Huiming Bao, Suyin Ting, and David Fastovsky for their comments and suggestions. We also thank the research staff at Big Bend National Park. A Geological Society of America Graduate Student Research Grant to White supported this work. We also thank the Department of Geology and Geophysics at Louisiana State University and the Louisiana State University Museum of Natural Science for their support. Special thanks are extended to Bruce MacFadden and an anonymous reviewer for their comments and suggestions that have greatly improved this manuscript.

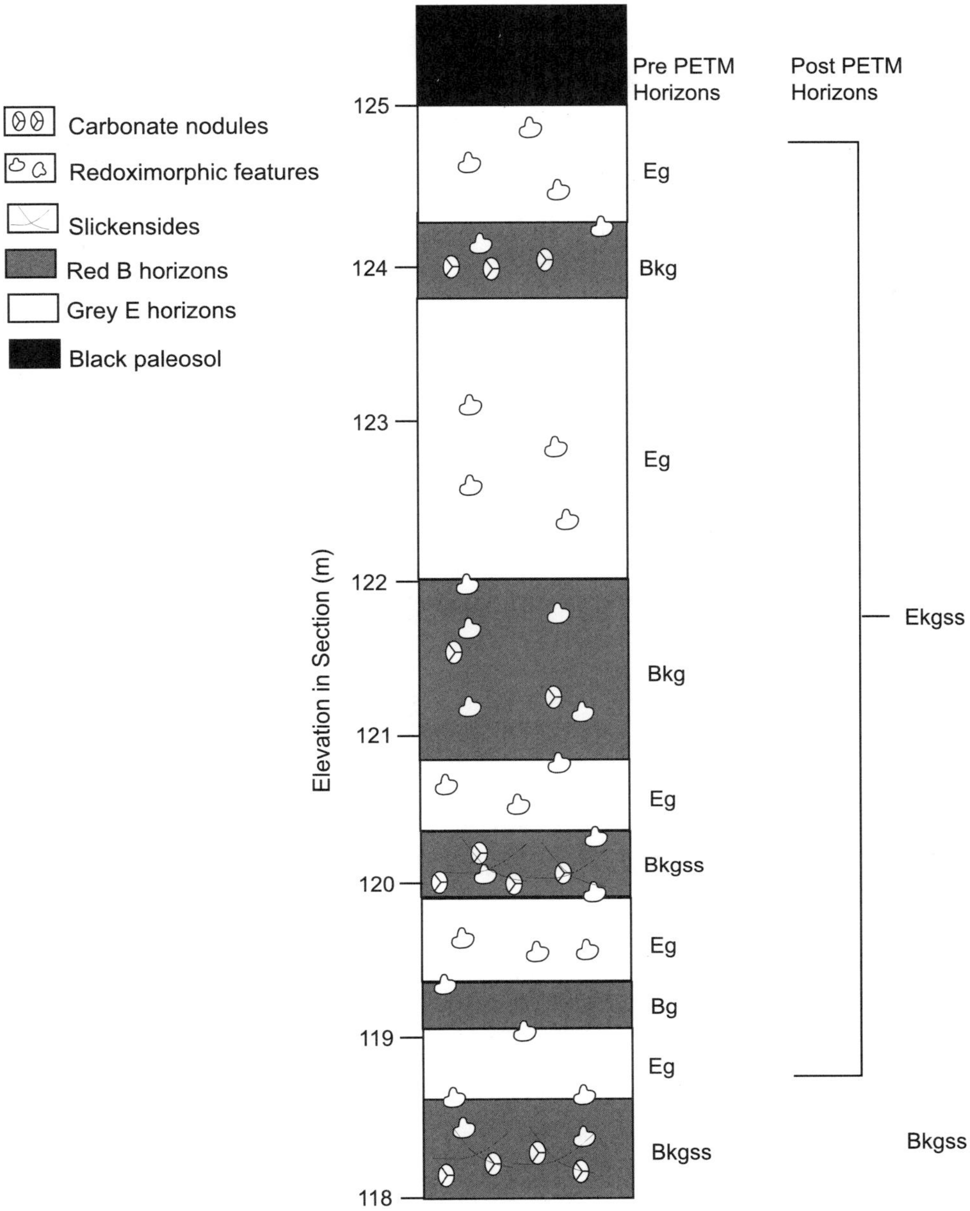

Figure 11. Columnar section of the area shown in Figure 10. Interpreted paleosol horizons on the right of the column describe pre-PETM (Paleocene-Eocene Thermal Maximum) soil conditions and the changes in soil horizons that occurred during PETM time.

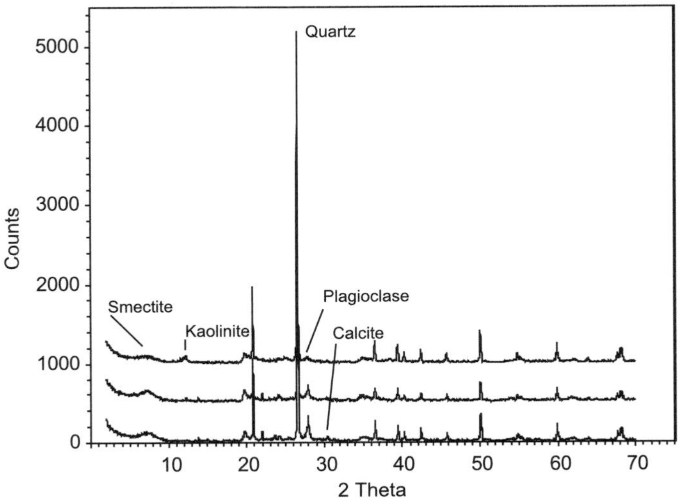

Figure 12. XRD patterns of bulk randomly oriented samples from a series of stacked B and E horizons that were affected by the PETM (Paleocene-Eocene Thermal Maximum). Note the significant reduction in the smectite, calcite and plagioclase peaks as compared to XRD patterns from soils that were not affected by the PETM (Fig. 8). See Figure 9 for a photograph of the paleosols.

REFERENCES CITED

Berggren, W.A., Kent, D.V., Swisher, C.C., III, and Aubry, M.-P., 1995, A revised Cenozoic geochronology and chronostratigraphy, *in* Berggren, W.A., et al., eds., Geochronology time scales and global stratigraphic correlation: Tulsa, Oklahoma, SEPM (Society for Sedimentary Geology) Special Publication No. 54, p. 129–212.

Bowen, G.J., Koch, P.L., Gingerich, P.D., Norris, R.D., Santo, B., and Corflield, R.M., 2001, Refined isotope stratigraphy across the continental Paleocene-Eocene boundary on Polecat Bench in the northern Bighorn Basin, *in* Gingerich, P.D., ed., Paleocene-Eocene stratigraphy and the biotic change in the Bighorn and Clarks Fork basins, Wyoming: University of Michigan Papers on Paleontology, v. 33, p. 73–88.

Bown, T.M., and Kraus, M.J., 1981, Vertebrate fossil-bearing paleosol units (Willwood Formation, lower Eocene, northwest Wyoming, USA): Implications for taphonomy, biostratigraphy, and assemblage analysis: Palaeogeography, Palaeoclimatology, Palaeoecology, v. 34, p. 31–56.

Cerling, T.E., 1984, The stable isotope composition of modern soil carbonates and its relation to climate: Earth and Planetary Science, v. 71, p. 229–240.

Cerling, T.E., Solomon, D.K., Quade, J., and Bowman, J.R., 1991, On the isotopic composition of carbon in soil carbon dioxide: Geochimica et Cosmochimica Acta, v. 55, p. 3403–3405.

Duchaufour, P., 1982, Pedology: London, Allen and Unwin, 448 p.

Dudal, R., and Eswaran, H., 1988, Distribution, properties, and classification of Vertisols, *in* Wilding, L.P., and Puentes, R., eds., Vertisols: Their distribution properties, classification, and management: College Station, Texas, Texas A&M University Printing Center, p. 1–22.

Fanning, D.S., and Fanning, M.C.B., 1989, Soil: Morphology, genesis and classification: New York, Wiley & Sons, 395 p.

Fastovsky, D.E., 1987, Paleoenvironments of vertebrate bearing strata during the Cretaceous-Paleogene transition: Palaios, v. 2, p. 282–295.

Fastovsky, D.E., and McSweeney, K., 1987, Paleosols spanning the Cretaceous-Paleogene transition, eastern Montana and western North Dakota: Geological Society of America Bulletin, v. 99, p. 66–77.

Fastovsky, D.E., and McSweeney, K., 1991, Paleocene paleosols of the petrified forests of Theodore Roosevelt National Park, North Dakota: A natural experiment in compound pedogenesis: Geological Society of America Bulletin, v. 6, p. 67–80.

Fricke, H.C., Clyde, W.C., O'Neil, J.R., and Gingerich, P.D., 1998, Evidence for rapid climate change in North America during the latest Paleocene thermal maximum: Oxygen isotope compositions of biogenic phosphate from the Bighorn Basin (Wyoming): Earth and Planetary Science Letters, v. 160, p. 193–208.

Glassmann, J.R., 1982, Alteration of andesite in wet, unstable soils of Oregon's western Cascades: Clays and Clay Minerals, v. 30, p. 253–263.

Kennet, J.P., and Stott, L.D., 1991, Abrupt deep-sea warming, paleoceanographic changes, and benthic extinctions at the end of the Paleocene: Nature, v. 353, p. 225–229.

Kleinert, K., and Strecker, M.R., 2001, Climate change in response to orographic barrier uplift: Paleosol and stable isotope evidence from the late Neogene Santa Maria Basin, northwestern Argentina: Geological Society of America Bulletin, v. 113, p. 728–742.

Koch, P.L., Zachos, J.C., and Gingerich, P.D., 1992, Correlation between isotopic records in the marine and continental carbon reservoirs near the Paleocene/Eocene boundary: Nature, v. 358, p. 319–322.

Koch, P.L., Zachos, J.C., and Dettman, D.L., 1995, Stable isotope stratigraphy and paleoclimatology of the Paleogene Bighorn Basin (Wyoming, USA): Palaeogeography, Palaeoclimatology, Palaeoecology, v. 115, p. 61–89.

Kraus, M.J., 2001, Sedimentology and depositional setting of the Willwood Formation in the Bighorn and Clarks Fork Basins, *in* Gingerich, P.D., ed., Paleocene-Eocene stratigraphy and biotic change in the Bighorn and Clarks Fork Basins, Wyoming: University of Michigan Papers on Paleontology, v. 33, p. 15–28.

Kraus, M.J., and Gwinn, B.M., 1997, Controls on the development of early Eocene avulsion deposits and flood-plain paleosols, Willwood Formation, Bighorn Basin: Sedimentary Geology, v. 114, p. 33–54.

Lehman, T.M., 1989, Upper Cretaceous (Maastrichtian) paleosols in Trans-Pecos Texas: Geological Society of America Bulletin, v. 101, p. 188–203.

Maxwell, R.A., Lonsdale, J.T., Hazzard, R.T., and Wilson, J.A., 1967, Geology of the Big Bend National Park, Brewster County, Texas: University of Texas Bureau of Economic Geology, publication 6711, p. 23–156.

Munsell Color, 1992, Munsell soil color charts (revised edition): Newburgh, New York, Macbeth Division of the Kollmorgen Corporation.

Norris, R.D., and Rohl, U., 1999, Carbon cycling and chronology of climate warming during the Paleocene/Eocene transition: Nature, v. 401, p. 775–778.

Pipujol, M.D., and Buurman, P., 1994, The distinction between ground-water gley and surface water gley phenomena in Tertiary paleosols of the Ebro Basin, northeast Spain: Palaeogeography, Palaeoclimatology, Palaeoecology, v. 110, p. 103–113.

Pipujol, M.D., and Buurman, P., 1997, Dynamics of iron and calcium carbonate redistribution and paleohydrology in middle Eocene alluvial paleosols of the southeast Ebro Basin Margin (Catalonia, northeast Spain): Palaeogeography, Palaeoclimatology, Palaeoecology, v. 134, p. 87–107.

Quade, J., and Cerling, T.E., 1995, Expansion of C4 grasses in the late Miocene of Northern Pakistan: Evidence from stable isotopes in paleosols: Palaeogeography, Palaeoclimatology, Palaeoecology, v. 115, p. 91–116.

Quade, J., Cerling, T.E., and Bowman, J., 1989, Development of the Asian monsoon revealed by marked ecological shift in the latest Miocene of northern Pakistan: Nature, v. 342, p. 163–166.

Rapp, S.D., MacFadden, B.J., and Schiebout, J.A., 1983, Magnetic polarity stratigraphy of the early Tertiary Black Peaks, Big Bend National Park, Texas: Journal of Geology, v. 95, p. 555–572.

Retallack, G.J., 1981, Fossil soils: Indicators of ancient terrestrial environments, *in* Niklas, K.J., ed., Paleobotany, paleoecology, and evolution, Volume 1: New York, Praeger Publishers, p. 55–102.

Retallack, G.J., 1990, Soils of the past: London, Allen and Unwin, 520 p.

Retallack, G.J., 1997, Paleosols of the upper Narrabeen Group of New South Wales as evidence of Early Triassic paleoenvironments without modern analogues: Australian Journal of Earth Sciences, v. 44, p. 185–201.

Retallack, G.J., 1999, Postapocalyptic greenhouse paleoclimate revealed by earliest Triassic paleosols in the Sydney Basin, Australia: Geological Society of America Bulletin, v. 111, p. 52–70.

Schiebout, J.A., 1974, Vertebrate paleontology and paleoecology of the Paleocene Black Peaks Formation, Big Bend National Park, Texas: Bulletin of the Texas Memorial Museum, v. 24, 88 p.

Schiebout, J.A., 1995, The Paleocene–Eocene transition on the Tornillo Flat in Big Bend National Park, Texas, *in* Santucci, V.L., and McLelland, L., eds., National Park Service Paleontological Research: Denver, National Park Service, p. 40–45.

Schiebout, J.A., Rigsby, C.A., Rapp, S.D., Hartnell, J.A., and Standhardt, B.R., 1987, Stratigraphy of the Cretaceous-Tertiary and Paleocene-Eocene transition rocks of Big Bend National Park, Texas: Journal of Geology, v. 95, p. 359–375.

Soil Survey Staff, 1999, Soil taxonomy, 2nd edition (U.S. Department of Agriculture Handbook 436): Washington, D.C., U.S. Government Printing Office, 869 p.

Therrien, F., and Fastovsky, D.E., 2000, Paleoenvironments of the early theropds, Chinle Formation (Late Triassic), Petrified Forest National Park, Arizona: Palaios, v. 16, p. 194–211.

Manuscript Accepted by the Society August 13, 2002

Printed in the USA

Geological Society of America
Special Paper 369
2003

Chronostratigraphic terminology at the Paleocene/Eocene boundary

Marie-Pierre Aubry
Department of Geological Sciences, Rutgers University, 610 Taylor Road, Piscataway, New Jersey 08554, USA
William A. Berggren
Department of Geology and Geophysics, Woods Hole Oceanographic Institution, Woods Hole, Massachussetts 02543, USA, and Department of Geological Sciences, Rutgers University, 610 Taylor Road, Piscataway, New Jersey 08554, USA
John A. Van Couvering
Micropaleontology Press, American Museum of Natural History, New York, New York 10024, USA
Jason Ali
Department of Earth Sciences, University of Hong Kong, Pokfulam Road, Hong Kong, China
Henk Brinkhuis
Laboratory of Palaeobotany and Palynology, University of Utrecht, Budapestlaan 4, 3584 CD, Utrecht, The Netherlands
Benjamin Cramer
Dennis V. Kent*
Carl C. Swisher III
Department of Geological Sciences, Rutgers University, 610 Taylor Road, Piscataway, New Jersey 08554, USA
Christian Dupuis
Géologie Fondamentale et Appliquée, Mons Polytechnics, Rue de Houdain, 9, B-7000 Mons, Belgium
Philip D. Gingerich
Department of Geological Sciences and Museum of Paleontology, University of Michigan, Ann Arbor, Michigan 48109–1079, USA
Claus Heilmann-Clausen
Geologisk Institut, Aarhus Universitet, DK-8000 Aarhus C, Denmark
Chris King
David J. Ward
Department of Earth Sciences, University of Greenwich, Chatham, Kent, UK
Robert W. O'B. Knox
British Geological Survey, Keyworth, Nottingham NG12 5GG, UK
Khaled Ouda
Department of Geology, University of Assiut, Assiut, 71516 Egypt
Lowell D. Stott
Department of Geology, University of Southern California, Los Angeles, California 90089–0740, USA
Medard Thiry
Ecole des Mines de Paris, 35, rue Saint Honoré, 77305 Fontainebleau, France

*Also at: Lamont-Doherty Earth Observatory, Palisades, New York 10964, USA

Aubry, M.-P., Berggren, W.A., Van Couvering, J.A., Ali, J., Brinkhuis, H., Cramer, B., Kent, D.V., Swisher, C.C., III, Dupuis, C., Gingerich, P.D., Heilmann-Clausen, C., King, C., Ward, D.J., Knox, R.W.O'B., Ouda, K., Stott, L.D., and Thiry, M., 2003, Chronostratigraphic terminology at the Paleocene/Eocene boundary, *in* Wing, S.L., Gingerich, P.D., Schmitz, B., and Thomas, E., eds., Causes and Consequences of Globally Warm Climates in the Early Paleogene: Boulder, Colorado, Geological Society of America Special Paper 369, p. 551–566.

ABSTRACT

Integrated research over the past decade has led to the recognition of a short 150–200 k.y.) interval of Paleogene time within Chron C24r at ~55.5 Ma, formerly termed the late Paleocene Thermal Maximum (LPTM) but more recently the Paleocene-Eocene Thermal Maximum (PETM), that was crucial in the climatic, paleoceanographic, and biotic evolution of our planet. Stable isotope analysis of marine carbonates indicates that there were transient changes in surface and deep-water temperatures. These climatic changes coincided with a negative 3‰–4‰ carbon isotope excursion (CIE), which is recorded in both marine and terrestrial deposits. It was soon realized that the CIE not only constitutes a powerful tool for long distance ("global") isochronous correlation, but even more importantly that it is coeval with notable biotic events in both marine and continental fossil records that have long been taken as criteria for the beginning of the Eocene in North America and more recently in deep sea cores. On the other hand, the conventional Paleocene/Eocene boundary level at the Thanetian/Ypresian boundary in Belgium and the London Basin has been found to be ~1 m.y. younger than the CIE, based on the association of the First Appearance Datum (FAD) of the (calcareous nannoplankton) *Tribrachiatus digitalis* (at ~54.4 Ma) with the base of the Ypresian in the London Basin. Although the Ypresian definition would take priority under normal circumstances, a consensus has been reached to redefine the Eocene in recognition of the worldwide significance and correlatibility of stratigraphic features associated with the PETM.

Redefinition of the Eocene, however desirable, nevertheless cannot proceed in a stratigraphic vacuum, and this paper is concerned with resolving the consequences of this action. To be made coincident with the CIE at ~55.5 Ma, the Ypresian/Thanetian boundary would have to be lowered by ~1 m.y., resulting in the inflation of the span of the Ypresian by 20% and a reduction of the span of the Thanetian by 30%. At the same time, the terminology of the strata in the leapfrogged interval would be thrown into total conflict with the literature, with the substitution of one widely used stage name for the other in the conflicted interval. On the other hand, to relocate the Paleocene/Eocene boundary without moving the stage boundaries would result in the upper third of the Thanetian falling within the Eocene, demolishing a century-old consensus. We propose that the destabilizing effect of the new boundary in the classic chronostratigraphy of western Europe can best be minimized with the introduction of a pre-Ypresian Stage, to encompass the orphaned upper Thanetian interval as the basal unit of the Eocene under a separate name. To this end, we suggest the reintroduction of the Sparnacian Stage, now that its original concept has been shown to correlate essentially with the interval between the CIE and the FAD of *T. digitalis*.

INTRODUCTION

The redefinition of the Paleocene/Eocene boundary to coincide with the distinctive carbon isotope excursion (CIE) in Chron C24r at ~55.5 Ma leaves the stratigraphic community with the essential task of reconciling the relocated series/epoch within the principles of chronostratigraphy. It is now commonly accepted that chronostratigraphy is a hierarchical discipline in which the stage is the lowest, basic unit (Hedberg, 1976; Cowie et al., 1986). We have reviewed elsewhere (Aubry et al., 1999) the conflicting treatment of chronostratigraphic hierarchy in the two primary authorities, the International Stratigraphic Guide (Hedberg, 1976; Salvador, 1994) and the Revised Guidelines of the International Commission on Stratigraphy (ICS) (Cowie et al., 1986; Remane et al., 1996). In regard to the role of the stage in defining chronostratigraphic boundaries, or Global Standard Stratotype-section and Points (GSSPs), we have argued that the approach taken by the Hedberg Committee is preferable, where these authorities conflict (Aubry, 2000; Aubry and Berggren, 2000a, 2000b; Aubry et al., 2000a, 2002).

In chronostratigraphy of the Cenozoic, far more than in earlier parts of the geologic continuum, age correlation is controlled by multiple independent criteria, not only in biostratigraphy, but also in terms of cyclostratigraphy, magnetostratigraphy, radioisotopic age determinations, stable isotope stratigraphy, and regional or global signals of large scale environmental changes. Using these overlapping controls, the relative age of well-characterized stratigraphic levels in the Neogene and

Paleogene can normally be resolved to within 100 k.y., and frequently better than that. This means that the difference in age (~1 m.y.) between the CIE, a worldwide geochemical marker that reflects a sharp transient spike in global temperature at ~55.5 Ma (Berggren and Aubry, 1998) and the base of the Ypresian Stage at ~54.4 Ma (Aubry, 1996; Aubry et al., 1996) (both ages used here from Berggren and Aubry, 1998, for consistency with the time scale of Berggren et al., 1995), is significant. (The relationships between the lithostratigraphic units in northwestern Europe and the conventional chronostratigaphic units upon which they are based have been thoroughly discussed in Aubry [2000 and references therein] are therefore not repeated here.) In this paper we discuss the options for reconciling the lack of synchrony between the base of the Ypresian in the London-Paris Basin, historically and for many workers still the undisputed criterion for the base of the Eocene, and the level of the CIE, which the Paleocene/Eocene Working Group of the International Subcommission on Paleogene Stratigraphy has recently agreed to make the guide for selecting a new GSSP for the basal stage of the Eocene Series. Our aim is to suggest how the new concept of the Eocene can be established in accordance with the hierarchical standards recommended under the International Guidelines, and in consistency with principles outlined in the ICS Guidelines.

HISTORICAL BACKGROUND

> We have little doubt that it will be necessary hereafter to intercalate other periods, and that many of the deposits, now referred to a single era, will be found to have been formed at very distinct periods of time . . . [. . .] . . . we might have three divisions of the Eocene epoch,—the older, middle, and newer; [. . .] In that case, the formations of the middle period must be considered as the types from which the assemblages of organic remains in the groups immediately antecedent or subsequent will diverge. (Lyell, 1833, p. 57)

Despite Lyell's foresight, the introduction of the Paleocene Epoch by Schimper (1874)—which represented a fulfillment of Lyell's vision of "the older" Eocene—met with unrelenting resistance. It took >65 yr for the Paleocene to become accepted by stratigraphic bodies in North America and almost 100 yr for continental Europe to follow. Whereas, in retrospect, Lyell's insight and Schimper's initiative can only be praised, we should be aware of the train of thought underlying their approach, which is espoused to this day by a large part of the scientific community, although it is divorced from the philosophy of modern chronostratigraphy. Current controversies (e.g., Aubry et al., 1999, 2000a; Aubry and Berggren, 2000a, 2000b; Remane, 2000; Walsh, 2001) in chronostratigraphic procedures can be understood as a consequence of this dual thinking.

Lyell built a chronologic framework that recognized time before rock. His subdivisions are based on a succession of fossil assemblages arranged according to internal evidence of their antiquity, and not on rock units. Formations, grouped in *Series,* are fitted to *Epochs* on the basis of their molluscs and echinoderms. For Lyell, the Eocene Epoch was known before the Eocene Series. Applying modern language, Lyell saw chronology as synonymous with biochronology. His *Classification of Tertiary Formations in Chronological Order* is alien to chronostratigraphy, a concept that was to be developed later.

Schimper (1874) followed precisely in Lyell's steps when he introduced the Paleocene Epoch, and so did the paleontologists and stratigraphers who first saw the appropriateness of an older Eocene interval. Schimper characterized the Paleocene not by its stratigraphic units, but by its flora "directly related to the Heersian flora, itself very close to the Cretaceous flora, and even more to the flora of the Eocene Epoch, but with a distinctive aspect which characterises it at once" (translated from Schimper, 1874, p. 464). Similarly, the Wood Committee (Wood et al., 1941) characterized the Paleocene in terms of its North American Land Mammal Ages (NALMA) and many vertebrate paleontologists have followed this path to this day. For example, Archibald et al. (1987, p. 25) stated, "We consider mammal age and mammal zones to be a type of biochronological unit, a unit that is not used by the NASC and is only briefly considered by the ISG. Such units are characterized by faunal content."

The mammalian turnover recorded between the upper Clarkforkian and lower Wasatchian beds in North America and the Cernay and Meudon Conglomerates in Europe (Granger, 1914; Wood et al., 1941; Russell, 1967, 1968; Gingerich, 1989) was instrumental in convincing European stratigraphers, who had worked hard for almost a century to reduce Schimper's Paleocene concept to a non sequitur, to see the Paleocene as a useful concept after all. Pomerol (1969, 1977) found in this prominent turnover a justification for proposing the base of the Sparnacian Stage as the definition for a revived Paleocene/Eocene boundary (Fig. 1, column H in association with columns E and G). Although Pomerol's proposal could have constituted the ultimate reconciliation between different schools of thought, it proved only to increase discord. According to Pomerol, the base of the Sparnacian was appropriate for the base of the Eocene primarily because the Conglomérat de Meudon—the mammal-bearing formation interpreted as immediately younger than the faunal turnover—was understood to lie at the base of the Sparnacian Stage. This point of view, reinforced by the subsequent correlation of the basal Zone Wa0 of the Wasatchian NALMA to the Conglomérat de Meudon (Gingerich, 1989), was essentially a validation of the Lyellian geohistorical approach, with its assumption that the Paleocene/Eocene boundary should be defined by a paleobiologic event. Unfortunately, this assumption was inconsistent with the advanced chronostratigraphic principles that had more recently become accepted by the general stratigraphic community.

In the late 1940s the International Subcommission on Stratigraphic Classification (ISSC), chaired by Hollis Hedberg, began to outline principles of definition and practice that were eventually brought together in the Guide to Stratigraphic Classification (Hedberg, 1976). This work elevated the earlier efforts in various national groups to a new level of logical philosophy that established chronostratigraphy as *the* independent means of definition for temporal reference. Notably, the ISSC proposed the stage as the basic unit in the hierarchy of chronostratigraphy (Hedberg, 1972, 1976; Salvador, 1994). Following the guiding principle of field geology that observation precedes interpretation, the Hedbergian school built on the logical premise that recognized rock before time. The subdivisions of geological time, in this view, are directly based on rock units rather than on indirect inferences about relative antiquity, or on the age of inferred events that were derived from paleontologic changes such as those observed by Lyell. The definition of geological ages thus follows from the definition of stages as the basic units of chronostratigraphy. This is in accord with the original definition by d' Orbigny (cf. Aubry et al., 1999), which clearly understands the stage as a unit of geological time defined in strata that record events of the same age. As marine stratigraphers attempted to unravel correlations between formations on various continents and deep-sea deposits recovered by the nascent Deep Sea Drilling Project, much concerted effort was devoted to the definition of chronostratigraphic boundaries. The standard chronostratigraphic scale was born as the boundaries between series became tied to those between stages (Berggren, 1971). The names of Lyell's subdivisions were retained but the concept and content of series were altered to fit modern (chrono)stratigraphic concepts. The base of the Eocene thus became tied to the base of the Ypresian Stage, following the pioneering work of von Koenen (1885; Fig. 1, columns A, D and F; see discussion below).

Two definitions of the Paleocene/Eocene boundary had thus emerged. One (Fig. 1, column H in association with columns E and G), identified the beginning of the Eocene with a major paleobiologic event observed at the base of a continental lithologic unit (e.g., Wood et al., 1941; Pomerol, 1969, 1977; Archibald et al., 1987; Gingerich, 1989) referred to by some as Sparnacian Stage (e.g., Russell, 1968; Pomerol, 1969, 1977; Russell et al., 1982; Savage and Russell, 1983), much in the tradition of Lyell's thought. The other, in which the base of the Eocene is inherent in the definition of the base of the Ypresian Stage (Fig. 1, columns A and F), was followed in observance of the hierarchical principles espoused by the international stratigraphic community (Hedberg, 1976; Salvador, 1994).

CURRENT (PRE-GSSP) UPPER PALEOCENE–LOWER EOCENE STANDARD CHRONOSTRATIGRAPHIC FRAMEWORK

Intense research and debate, continuing to the present generation, supports the consensus that two stages typified in the Paris-London Basin, Thanetian and Ypresian, constitute the primary reference for upper Paleocene–lower Eocene chronostratigraphy. In the hierarchical logic of chronostratigraphy, the Thanetian/Ypresian boundary simultaneously determined the Paleocene/Eocene boundary. A discussion of the role of these stages in the changed concept of the Eocene Series recommended by the International Commission on Stratigraphy (ICS) requires a fresh evaluation of the consensus. Rehabilitation of the Sparnacian Stage, which has suffered from inconsistent and preconceived treatment, appears to be crucial in reconciling the new concept of the Eocene with stability in global chronostratigraphy.

The introductions of the Ypresian, Sparnacian and Thanetian Stages are part of a late nineteenth century effort at describing stratigraphic successions in northwestern Europe and at sorting out correlations between disjunct stratigraphies in the Belgium, Paris and London basins (Figs. 1, 2). An extensive historical review of this effort is given in Aubry (2000).

The Ypresian Stage

The Ypresian Stage was defined by Dumont (1839) in Belgium, near the town of Ypres. Numerous studies over the succeeding one hundred and 50 yr established a firm framework of correlation between the Thanetian and Ypresian successions of Belgium, the London-Hampshire Basin(s), and the Paris Basin, such that the Ypresian Stage has acquired a strong regional significance (see Knox, 1994). The commonly accepted base of the Ypresian Stage is the base of the Mont Héribu Member of the Ieper Formation, Belgium (De Coninck et al., 1983). This level is correlative with the Walton Member at the base of the London Clay Formation, England, and the base of the Sables de Cuise s.l., Paris Basin (Fig. 2; Aubry, 2000, Figure 4). Because it is precisely correlated to the Thanetian-Ypresian succession in Belgium (Ali et al., 1993; King, 1991), the more marine—and thus more amenable to correlation with the deep sea—Thanetian-Ypresian succession in England has become the reference for upper Paleocene-lower Eocene chronostratigraphy. This succession in the London Basin has been formalized to comprise the Thanet Formation, the Lambeth (Upnor Formation and Woolwich-Reading Formation) Group and the Thames (Harwich Formation and London Clay Formation) Group (Ellison et al., 1994). Following the chronostratigraphic principles of the *International Stratigraphic Guide,* the base of the Eocene is thus tied to the base of the London Clay Formation, where it has in fact been identified for well >120 yr, in agreement with von Koenen (1885) who placed the Paleocene/Eocene boundary in England between the Thanetian deposits and the London Clay based on analysis of molluscan faunas (Fig. 1, column F). Importantly the base of the London Clay Formation has been indirectly tied to the First Appearance Datum of *Tribrachiatus digitalis* (Aubry, 1996) whose estimated age of 54.4 Ma (Aubry et al., 1996; Berggren and Aubry, 1996) was proposed to approximate the age of the base of the Ypresian Stage. This proposal has been validated with the recovery of *T. digitalis* in the Basement Bed (= upper Harwich Formation) that lies immediately below the Walton Member (Fig. 2).

The term Ypresian has also been used in a different—and incorrect—chronostratigraphic context. Below the base of the stratotypic Ypresian in northwest Europe, the Argiles Plastiques in the Paris Basin and the Sables d'Erquelinnes in Belgium cor-

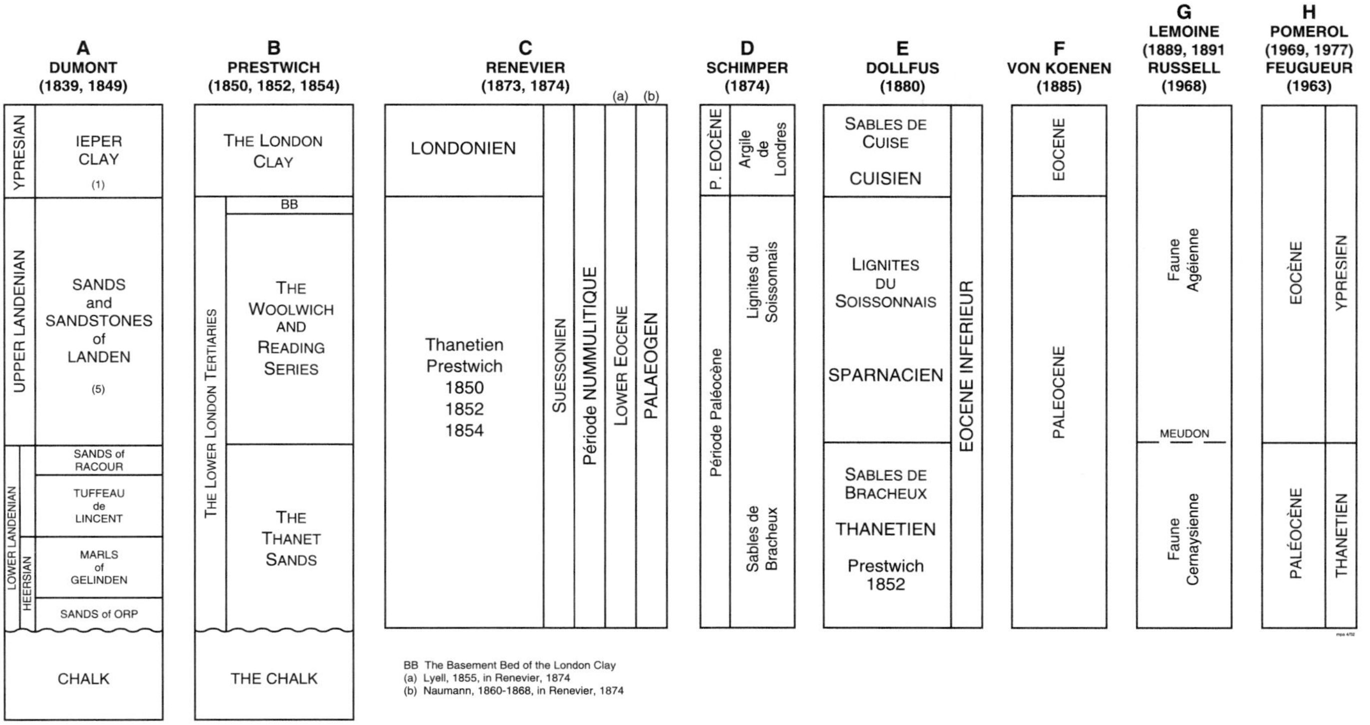

Figure 1. Historical review of the lithostratigraphic units of northwest Europe that have served as references for upper Paleocene–lower Eocene chronostratigraphy until the recent designation of a GSSP for the Paleocene/Eocene boundary. Note that the terms are presented in the format they were introduced. Two schemes of references were established in parallel, one in which lithostratigraphic horizons and units served as basis for the designation of stages (columns A–C, E), another in which paleontologic changes were used to designate epochs (column D) and their boundary (column H). (Modified after Aubry, 2000).

relate with the Woolwich-Reading Formation in the London Basin (Fig. 2). In contrast with most of his contemporaries, Feugueur (1963) believed in age equivalence between the Argiles plastiques, the London Clay and the (lower) Ypresian clays, based on facies similarity and apparent faunal equivalence (Fig. 1, column H in association with columns E, B and A and Figure 2). This miscorrelation implied that the Sparnacian Stage (Dollfus, 1880), represented in the Argiles plastiques, was equivalent with the lower part of the Ypresian Stage of Dumont (Fig. 1, columns A and E; Fig. 2). Because the Sparnacian had long been identified with the upper part of the Thanetian, Feugueur's view meant that the lower part of the Ypresian Stage would overlap with the upper part of the Thanetian Stage sensu Renevier (Fig. 1, H and C). When placing the Paleocene/Eocene boundary at the base of the Sparnacian Stage on the basis of the mammalian turnover event, Pomerol (1969, 1977) perpetuated Feugueur's miscorrelation by essentially accepting the view that the base of Dollfus' Sparnacian was correlative with the base of Dumont's Ypresian (Fig. 1, compare columns E and H with columns E and A). Two concepts of Ypresian Stage had now emerged, although few stratigraphers followed Pomerol's view. Nevertheless this created a further complication (see below).

The Sparnacian Stage

Dollfus (1880) introduced three stages: Thanetian, Sparnacian and Cuisian (Fig. 2) (unaware that a Thanetian Stage had already been established by Renevier, 1873; Figure 1, compare columns C and E). His main objective was to substitute the Cuisian for Dumont's Ypresian, whose stratotype he saw as unsatisfactory. Dollfus did not immediately propose a stratotype for his Sparnacian, but he clearly delineated its lithostratigraphic extent, in particular with reference to the London Basin succession. Importantly, Dollfus used sequence stratigraphy much before its time, recognizing that the Lignites du Soissonnais and the Reading-Woolwich Beds on the one hand, and the Sables de Sinceny and Oldhaven/Basement Bed on the other hand represented correlative packages separated by erosion surfaces at the same time as the correlative packages rested unconformably over, respectively, the Thanet Sands and their French equivalents (Sables de Bracheux s.l.) and below the Sables de Cuise and the London Clay (Fig. 2). From its introduction, the Sparnacian Stage thus contained the criteria required for sound chronostratigraphic units with one exception (acknowledged by Dollfus, 1905) that it was based primarily on continental and brack-

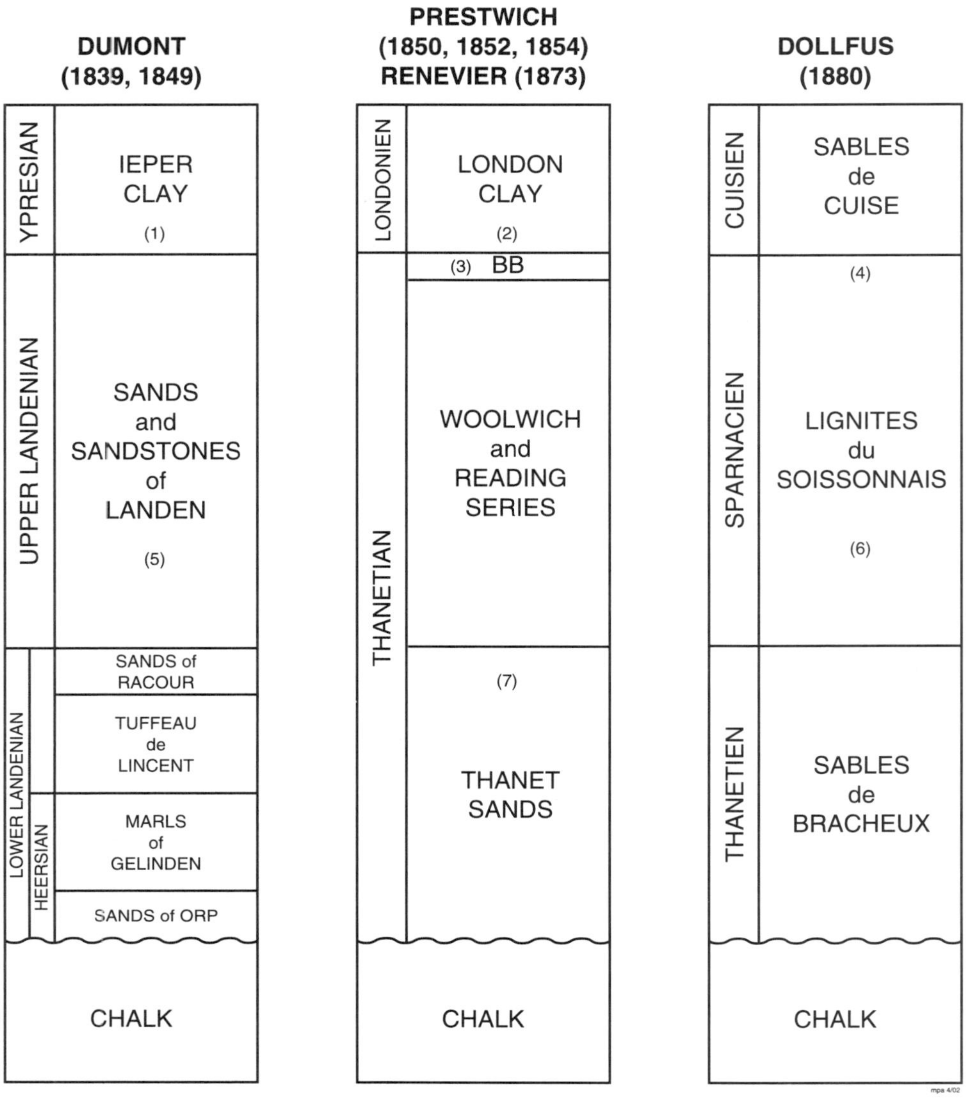

Figure 2. Relationships between the stages defined in the Belgian (Dumont, 1839, 1849), London (Renevier, 1873) and Paris (Dollfus, 1880) basins that have served in the standard chronostratigraphic scale (e.g., Berggren et al., 1985, 1995; Harland et al., 1990, i.al.) until formal definition of chronostratigraphic boundaries in the form of Global Standard-stratotype Section and Points become mandatory. Note that the terms are presented in the format they were introduced, as in Figure 1. We have added the lithostratigraphic units in the Landenian Stage given in Laga (1981) and we show the location of stratigraphic units discussed in the text: (1) Mont Héribu Member; (2) Walton Member; (3) Oldhaven Beds = Basement Bed of the London Clay (BB) = upper part of the Harwich Formation; (4) Sables de Sinceny; (5) Sables d'Erquelinnes; (6) Argiles Plastiques; (7) Upnor Formation (= Woolwich Bottom Bed). Discussion of the content of the stages can be found in Berggren (1971), Cavelier and Roger, (1980) and Pomerol (1981) (modified after Aubry et al., 1999).

ish deposits (although the shallow marine Sables de Sinceny are part of the definition).

What Dollfus did not recognize, when he later (1905) designated the Mont Bernon section (near Epernay) as the stratotype of the Sparnacian Stage, was that his type designation and his stage definition (1880) were irreconcilable. The type-section is much younger than the original definition, being of Cuisian (Ypresian) Age (Laurain et al., 1983). This inconsistency largely contributed to the dismissal of the stage by some (e.g., Hooker, 1996, 1998), regardless of 1) Dollfus' clear definition and sound correlations and 2) the extreme difficulty in establishing a sound lithostratigraphic framework for the lignite-bearing clays of the Paris Basin, including the Mont Bernon section (Thiry et al., 1998). Cavelier and Pomerol (1986) proposed to use the term "Sparnacian s.s." in a regional sense, according to the definition of Dollfus, while other authors (e.g., Laurain et al., 1983; Hooker, 1998) chose to restrict the Sparnacian to its stratotype. Aubry (2000) recognized the clarity of Dollfus' definition and the validity of his correlations between the London and Paris Basin stratigraphies, which have been well corroborated by numerous studies (e.g., Curry et al., 1978), and even more importantly, the significance of his early descriptions of sequences.

The Thanetian Stage

The Thanetian Stage, as it is used today, essentially corresponds to the Thanetian Stage erected by Renevier (1873; Fig. 2), to encompass Prestwich's description of the "Lower London Tertiaries" (Fig. 1, columns B and C). Intercalated between the Chalk and the London Clay, these strata consisted of the Thanet Sands (Prestwich, 1852), the Woolwich and Reading Series (Prestwich, 1854) and the "Basement Bed of the London Clay" (Prestwich, 1850; now the upper part of the Harwich Formation of Ellison et al., 1994). On the other hand, the Thanetian Stage of Dollfus was restricted to the Thanet Sands, while his Sparnacian Stage included the Woolwich and Reading Series and Basement Bed of the London Clay (Fig. 1, compare columns B and E). Thus, broader and narrower concepts of the Thanetian Stage have been utilized, the narrower concept being applied by those who accepted the Sparnacian Stage. The broader concept of Thanetian, heretofore in somewhat wider use than the narrower, is inconsistent with the new concept of Eocene, in which the definitions of Ypresian and Thanetian as lowermost Eocene and uppermost Paleocene stages, respectively, are inconsistent with a series boundary linked to the CIE.

Discussion

The upper Paleocene–lower Eocene chronostratigraphy in the Paris-London Basin, as described above, is marked by inconsistency—both conceptual and terminological—from very early on, due to unawareness of published material and genuine miscorrelation. It is important to note, however, that the shallow-marine and epicontinental nature of the stratigraphy under scrutiny makes any attempt at regional correlation difficult under the best circumstances.

One major conflict has been the placement of the Paleocene/Eocene boundary at the base of the Ypresian Stage by most marine stratigraphers, and at the base of the Sparnacian Stage by most vertebrate paleontologists. Another example has been the identification of the beginning of the Eocene by some as corresponding to the moment when a lithostratigraphic horizon was being deposited, while others have seen it as the moment when a paleobiologic event occurred. Finally, there has been a basic dualism in definition of the stages themselves, wherein two parallel concepts of Thanetian Stage are in active use, and the definition of the Sparnacian Stage is disconnected from the designation of its stratotype. These problems with the definition and characterization of chronostratigraphic units are quite apart from the further problems of their global correlation (see Berggren and Aubry, 1996, 1998; Aubry, 2000).

The definition of a GSSP for the Paleocene/Eocene boundary is thus a welcome endeavor, a chance to clarify definition and means of correlation in this important area of classical stratigraphy for the benefit of the entire geological community. In recognizing the value of the CIE as a clear and unambiguous criterion for the new GSSP we hope that it also provides an opportunity for reconciliation of the various conflicts identified above. For example, even though the new GSSP level serendipitously coincides with the Lyellian (i.e., paleobiologically defined) Paleocene/Eocene boundary favored by vertebrate paleontologists and some deep sea workers, such alternative methods of definition will henceforth cease to be relevant.

With regard to the relationship of the Paleocene/Eocene GSSP to the chronostratigraphy of the Paris-London basin, in which Eocene and Paleocene series were originally recognized, and which has become a de facto global standard, we now have the benefit of such tools as carbon isotope stratigraphy in particular, but also an improved biozonal subdivision based on microfossils (planktonic and benthic foraminifera, calcareous nannoplankton and dinoflagellates) associated with the isotopic excursion and finally, radioisotopic dating and astrochronology. In applying this information, we have taken a position that we believe is consistent with the hierarchical logic of chronostratigraphy, if not the requirements of the current ICS Guidelines (Aubry et al., 1999, 2000a), which is that the newly proposed series boundary should be defined in the context of a framework of globally correlatable stages. To this end, we have suggested a solution that we hope will be seen as a satisfactory reconciliation of practical goals and fundamental principles.

RECONCILED CONCEPT OF THE PALEOCENE/EOCENE BOUNDARY

The late Paleocene–early Eocene transition was characterized by a marked global warming, particularly at high latitudes, reflected in evolutionary turnovers, diversification, and migrations as well as extinctions that are recorded over a span of ~2 million years in both marine and terrestrial environments. This warming, in turn, was punctuated by a brief (<0.1 m.y.) extreme ocean warming event (formerly known as the late Paleocene Thermal Maximum and now termed the Paleocene-Eocene Thermal Maximum (PETM), and was associated with the CIE, a large (~3%–4%) negative excursion in $\delta^{13}C$ first recognized in planktonic and benthic foraminifera (Kennett and Stott, 1991).

The marine biotic response to the brief climatic warming is seen in the form of a series of events that have been recognized in the calcareous and organic walled plankton and calcareous and agglutinated benthos (Berggren et al., 1998) at/near the PETM. These are enumerated below.

Calcareous nannoplankton. The onset of the CIE is marked by the sudden appearance of species (e.g., *Discoaster anartios, D. araneus, Rhomboaster* spp.) with unusual morphology and structure that for a short time dominated all PETM/CIE assemblages in the Tethys and Atlantic area (Cramer et al., 1999; Aubry et al., 2000b). Superimposed on this short-term event, a long-term turnover was initiated with the onset of the CIE, leading to the extinction of taxa that radiated during the Paleocene and the evolution of several modern structural groups. Of particular significance for the Chron C24r stratigraphy/chronology is the evolution of the *Rhomboaster-Tribrachiatus* lineage, which permits the subdivision of the NP9-NP10 zonal interval into six subzones.

Planktonic foraminifera. Muricate, nonspinose (sub)tropical morozovellids and their geographically more widespread nonspinose, bluntly muricate acarininid cousins increased in abundance and expanded into high southern (Kerguelen Plateau, Maud Rise, South Atlantic) as well as northern (London Basin, North Sea) latitudes, while a distinct association of small acarininids (*Acarinina africana* and *A. sibaiyaensis*) and a lone morozovellid (*Morozovella allisonensis*) taxon—the so-called "excursion fauna"—characterized the PETM.

Benthic foraminifera. Deep-sea (bathyal and abyssal) benthic foraminifera underwent the largest (>50%) essentially global extinction/taxonomic turnover in >70 million years (since the mid-Cretaceous Cenomanian/Turonian boundary, Thomas, 1998). This event is referred to as the Benthic Foraminiferal Extinction Event. A global, cosmopolitan, taxonomically diverse fauna (characterized by the taxon *Stensioina beccariiformis*) was replaced by a fauna dominated by *Nuttallides truempyi,* commonly in mid-to-lower bathyal and abyssal Atlantic Ocean (and specifically South Atlantic) sites. At many other oceanic sites, and in Tethyan bathyal sections in Spain, postextinction faunas are dominated by various buliminid taxa. Postextinction faunas vary widely, are all of low diversity, and contain small, thin-walled individuals (if calcareous). A lesser, but nevertheless recognizable, event occurred in neritic/shelf environments as well (Speijer, 1994; Cramer et al., 1999).

Dinoflagellates. Associated with the CIE interval was the first global increase in *Apectodinium*-dominated assemblages (Bujak and Brinkhuis, 1998; Crouch et al., 2000). Subsequently, close to the beginning of the Ypresian Age (as determined by the base of the conventional (pre-GSSP Ypresian Stage) there was an increase in new dinocyst taxa, including the *Deflandrea phosphorica* group, and a concomitant increase in diversity within the Wetzelielloideae (*Wetzeliella, Charlesdowniea* and *Dracodinium*) (Bujak and Brinkhuis, 1998; Egger et al., 2000; Heilmann-Clausen and Egger, 2000).

The terrestrial Paleocene-Eocene transition coincides with the CIE and is marked by significant turnovers in mammalian faunas. Turnovers in the plant floras occurred more progressively.

Mammals. A significant and relatively rapid appearance/incursion (mammal dispersion event, MDE) of the earliest representatives of perissodactyls (odd-toed ungulates), artiodactyls (even-toed ungulates) and euprimates and rodents (Rose, 1981; Gingerich, 1989, 2001) as well as extinction of the larger plesiadapids in the Bighorn Basin is seen also in contemporary levels in Europe (Paris-London Basins) and Asia (Rose, 1981; Russell and Zhai, 1987).

Plants. The Paleocene-Eocene transition in North America witnessed a ~30% diminution in the number of plant species (Frederiksen, 1994; Wing et al., 1995, 2000; Wing, 1997) over an interval of ~1 m.y., followed by a rapid rebound to levels well in excess of pre-PETM floral richness in the mid-early Eocene, reflecting the trend toward higher global temperatures associated with the Cenozoic Global Climatic Optimum. The Paleocene-Eocene turnover consisted of a replacement of largely deciduous groups with modern north temperate distribution (Betulaceae, Cercidiphyllaceae, Juglandaceae, Hamamelidaceae, *Metasequoia*) by largely subtropical evergreen groups (Annonaceae, Lauraceae, Leguminosae, Myristicaceae, Palmae, Zamiaceae, i.al.) (Wing, 1997). Similar patterns have been recorded in Gulf Coastal Plain (Frederiksen, 1994), North Sea (Jolley, 1996) and New Zealand (Crouch, 2001) palynofloras, attesting to the essentially global nature of these changes.

Of particular interest in the context of our discussion is the climatic implications of leaf margin analysis (Wing et al., 2000) on Bighorn Basin assemblages: (1) mean annual temperatures on the order of 12.9–15.3 °C during the Clarkforkian, (2) earliest Wasatchian mean annual temperatures of 18.2 °C followed by (3) decline to 16.4 °C and then 10.8 °C at an (interpolated) age of 53.56 Ma (youngest part of Chron C24r), and (4) mid-Wasatchian rise to 15.8 °C, eventually rising to ~22.2 °C by 52.8 Ma (Chron C24n.1n) near the beginning of the Lostcabinian Subage of the Wasatchian NALMA (≅mid-Ypresian).

These then are the bio- and chemostratigraphic events that have shaped the discussion and debate within the Paleocene/Eocene Boundary Working Group over the past decade regarding criteria suitable for correlating the Paleocene/Eocene boundary. The recognition of a succession of distinct, closely juxtaposed events associated with an unmistakable isotopic signal in marine and nonmarine contexts alike justifies the recommendation that the Paleocene/Eocene boundary should be defined at a level coincident with the CIE, as the most widely applicable criterion for global correlation. In consequence, however, one is confronted by the problem of what to do with the boundaries of the stages that formerly defined the Paleocene/Eocene boundary in its original context. We consider this point in further detail below.

THE PALEOCENE/EOCENE BOUNDARY GSSP

The place given to the stage by ICS revised guidelines is currently inconsistent (Aubry et al., 1999, 2000a). On the one hand, the stage is the basic unit of chronostratigraphy; but on the other hand, in use the stage has been subservient to series, in the sense that stage boundaries have been defined by series boundaries, and not the reverse. For instance, the base of the Rupelian Stage was adjusted after the fact to correspond to the base of the Oligocene Series (Brinkhuis and Visscher, 1995; Aubry et al., 1999). In like manner the base of the Aquitanian Stage was adjusted after the fact to correspond to the base of the Miocene Series, itself adjusted to coincide with the base of the Neogene System (Aubry et al., 1999). This is in contrast with the ISSC International Guide, which recommends that series boundaries be based on existing stage boundaries (Hedberg, 1976, p. 25, 71, 73; Aubry et al., 1999, p. 110).

A large part of the marine stratigraphic community has long recognized that the base of the Eocene Series corresponds to the base of the Ypresian Stage (Fig. 1, columns A and F; Fig. 3), itself defined by the base of the Mont Héribu Member in Belgium (or its correlative base of the Walton Member of the London Clay Formation in England; see above). With the revision of the concept of the Eocene, this relationship could be maintained only by lowering the base of the Ypresian Stage to a level that is ~0.8–1 m.y. older than the historically and presently recognized base of the stage (Fig. 3, option 1). This would have a number of unsuitable consequences. First, it would create further terminologic and conceptual confusion, blurring even more the unfortunate situation described above. Second, the base of the GSSP-defined Ypresian Stage would lie within the Thanetian Stage, according to the almost universally accepted definition of Renevier. Third, the expanded Ypresian Stage would be defined in a manner that would disconnect it from the unit-stratotype in Belgium, and indeed, there would be no stratigraphic relevance between the GSSP boundary and the unit-stratotype. And fourth, and most regrettably, we would lose the possibility to enhance the stratigraphic resolution of this vitally important, highly informative interval with an identity of its own (Fig. 4). In recognition of this fact, the same meeting of the Working Group of the Paleocene/Eocene Boundary in Paris (1998) that favored lowering the Paleocene/Eocene boundary to the level of the CIE, also recommended retention of the historical definition of the Ypresian Stage and insertion of a new stage for the (chrono)stratigraphic interval between the CIE and the base of the Ypresian, in order to reconcile the redefined concept of Eocene with existing stage hierarchy (Fig. 3, options 2 and 3).

Although the current ICS revised guidelines are not primarily concerned with historical precedence and, furthermore, have largely eliminated the role of the unit stratotype in favor of the boundary stratotype, we do not believe that chronostratigraphic standards should be erected *de novo*. Instead, we believe that chronostratigraphy is best served when new standards are defined in harmony with old ones. The advantages and disadvantages of the different options for reconciliation in this particular instance are discussed below.

Chronostratigraphic options

It is now recognized that there is a distinct stratigraphic gap at the base of the unit stratotype of the Ypresian Stage in northwest Europe (Pomerol, 1988). This hiatus in the type area has generated the argument that the base of the Ypresian Stage could logically be located older than the level currently used (Remane, 2000), and could in fact be equally well defined at a level corresponding to the PETM, thus maintaining the present termi-

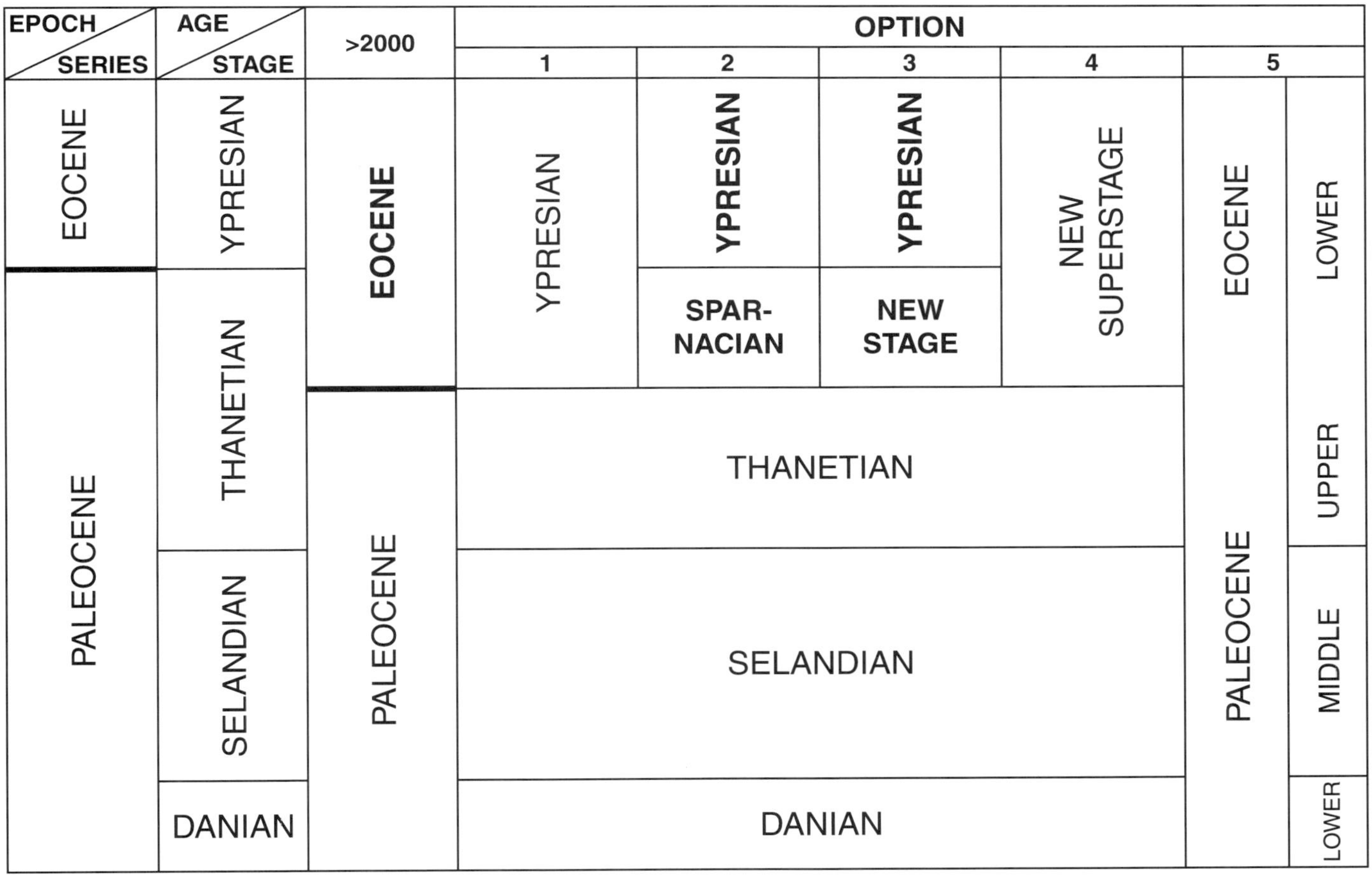

Figure 3. Options for accommodating a newly Global Standard Stratotype-section and Point–defined Paleocene/Eocene boundary. The authors favor Option 2, i.e., the reintroduction of Sparnacian as the lowest standard stage of the newly modified Eocene. Alternatively, a new name (option 3) might be introduced if Sparnacian should prove undesirable in use.

nology in which the Ypresian is the basal stage of the Eocene. This view, however, ignores monographic studies of the Ypresian stratotype (Dupuis et al., 1991, i.al.) and currently accepted practice (see studies in Knox et al., 1996; de Graciansky et al., 1998), in which the physical stratigraphic limits of the Ypresian synthem have been translated into regional (and global) chronostratigraphy. Furthermore, the hiatus below the unit stratotype does not represent "unoccupied territory" into which the boundary could be moved without repercussions.

If the base Ypresian Stage is lowered to the level of the CIE (Fig. 3, option 1), it will in fact be at a stratigraphic level well within the Thanetian Stage as commonly understood (i.e., that of Renevier, *non* Dollfus), requiring the Woolwich and Reading Series and Basement Bed of the London Clay, not to mention the Lignites du Soissonnais and Sables de Sinceny, to henceforth bear the name "Ypresian." In the stratigraphic literature, collections, maps and other documents dating back over a century, the integrity of the Thanetian with regard to the Ypresian has been

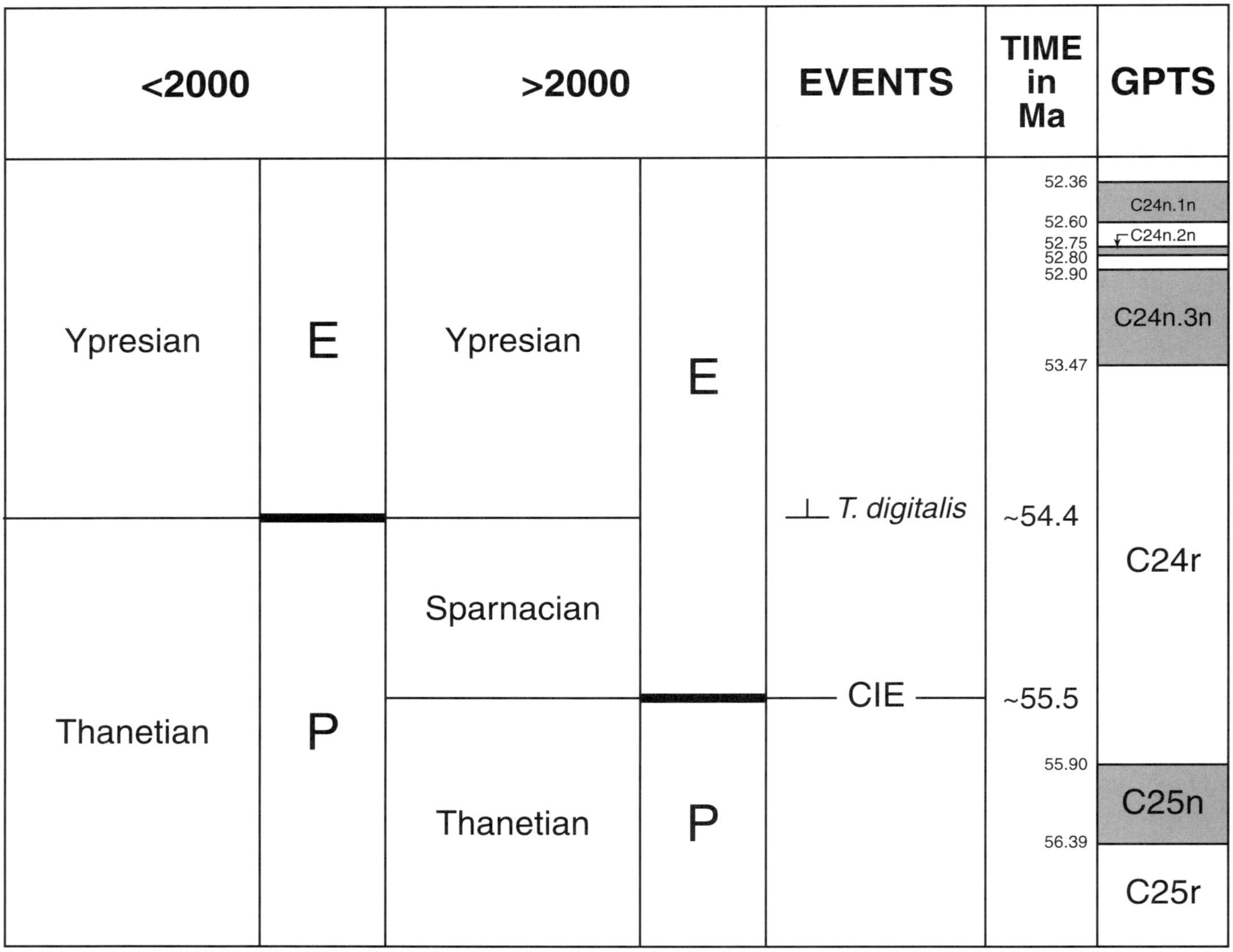

Figure 4. Lower Paleogene chronostratigraphy if the Sparnacian is inserted as the lowest Eocene stage (between Thanetian and Ypresian) as required by newly defined Paleocene/Eocene boundary. The Sparnacian Stage is essentially the (chrono)stratigraphic interval bracketed by the CIE (carbon isotope excursion) and the FAD (First Appearance Datum) of the calcareous nannoplankton *Tribrachiatus digitalis.* The age estimates of 55.5 Ma and 54.4 Ma for the CIE and FAD of *Tribrachiatus digitalis,* respectively, are based on the time scale of Berggren et al. (1995). Recent astrochronologic estimates suggest that the CIE may actually be closer to 55 Ma and the FAD of *T. digitalis* (~ base Ypresian) may be 0.8–1.0 m.y. younger (Cramer, 2002). *Tribrachiatus digitalis* occurs in the upper part of the Harwich Formation just below the Walton Member at the base of the London Clay Formation. The FAD of this species thus very closely approximates the base of the Ypresian stage in its stratotypic area of northwest Europe. GPTS—geomagnetic polarity time scale.

axiomatic. The confusion that would result from simply substituting one familiar and strongly characterized stage name for the other in the dense literature of this interval would be unacceptable to most European stratigraphers.

In addition, the lowering of the base of the Ypresian Stage into the upper levels of the previous system seriously conflicts with the strong understanding of the terms Ypresian and Thanetian Age/Stages as corresponding to marine transgressions of geohistorical significance (Aubry et al., 1999). Of all the possible compromises required by the new concept of the Eocene, the temporal distortion of the Ypresian Age by almost 20%, to absorb >30% of the Thanetian seems to be the least desirable in principle and on geohistorical grounds.

On the other hand, as we have discussed above, the original definition of the Sparnacian (Dollfus, 1880) is as a chronostratigraphic unit between a more restricted Thanetian (sensu Dollfus) and the normal Ypresian Stage (Fig. 1, columns A, B and E). Dollfus explicitly and correctly correlated the Paris Basin deposits that embodied his 1880 Sparnacian to the (marginal marine) successions of the Woolwich-Reading Beds in the London Basin (Aubry, 2000). The original base of the Sparnacian, as exposed in outcrops in the western Paris Basin, is closely correlated with the CIE (Fig. 4; Sinha and Stott, 1993a, 1993b; Stott et al., 1996; Sinha et al., 1996). This usage unites marine and terrestrial chronostratigraphy, even though the initial correlation of base Wasatchian NALMA with the Conglomérat de Meudon was incorrect (Sinha et al., 1996; see Aubry, 2000).

We note that the term Neustrian, in place of the continental version of the Sparnacian, has been recently reintroduced by vertebrate paleontologists (Hooker, 1996, 1998; Lucas, 1998) as a European Land-Mammal Age to which the Conglomérat de Meudon fauna is assigned. The Sparnacian is thus released from collateral service as a land mammal age and becomes fully available as a global chronostratigraphic term (Fig. 3, option 2). Despite its somewhat checkered past, in which the Sparnacian has suffered from inconsistent definitions, as well as an uncertain identity ("stage," "lithofacies," "biochronologic unit," "continental stage"), the original concept (Dollfus, 1880) of a stage that encompassed shallow-marine strata is clear, and can be readily restored with reference to unit and boundary stratotypes that conform to the modern guidelines (Hedberg, 1976; Cowie et al., 1986).

The Sparnacian has been criticized in discussions as being too short (c. 1 m.y.) to be useful in global chronostratigraphy. In fact, shorter ages, reflecting the improved resolution in such geochronometric tools as stable isotope curves, magnetostratigraphy, and cyclostratigraphy, are becoming essential for worldwide recognition of distinctive periods (e.g., Gelasian, Ionian and Sicilian stages, cf. Van Couvering, 1997). In this instance, the base of the Sparnacian, in a GSSP-linked to the onset of the CIE, could be resolved to within ~10–20 k.y. in any part of the world, and cyclostratigraphy and stable isotope stratigraphy allow internal chronostratigraphic resolution of similar order in marine (Bains et al., 1999; Röhl et al., 2000; Cramer, 2001) as well as in terrestrial systems (Bowen et al., 2001). If the test of useful duration of a stage is how far its time span exceeds the error in dating its boundary, the Sparnacian rates higher than most other stages. We note that if for any reason the name Sparnacian, for the orphaned upper Thanetian strata above the new Paleocene/Eocene boundary, proves unacceptable to the stratigraphic community, another new name for this interval would be preferable to moving the Thanetian/Ypresian boundary and thereby increasing the terminological confusion of the past decades (Fig. 3, option 3). In Table 1 we show that the stratigraphic interval encompassing the Sparnacian Stage varies from several meters to several hundred meters in marine depositional environments to ~1000 m in some terrestrial settings.

Substitution of a new superstage for the entire interval between the CIE and the middle Eocene (Lutetian Stage) (Fig. 3, option 4) might, at first glance, be seen as a suitable means of providing a clear and unambiguous break from the terminological confusion of the past four decades while at the same time allowing the term Ypresian to be retained without departing from its historical definition by keeping it as a regional stage (along with the Sparnacian). However, it will be readily seen that this approach merely coarsens resolution at the stage level rather than refining it (since it would span an interval of ~6.5 m.y.), while at the same time requiring the (re)introduction/(re)insertion of the Sparnacian as a local/regional stage to fill the gap between the CIE and the Ypresian. This approach would furthermore require eventual modification/reduction of all other Paleogene and Neogene "stages" to local/regional status. Paleogene and Neogene Stratigraphic Subcommissions have recently introduced GSSPs for numerous global stage units and this modification of the connotation of a fundamental chronostratigraphic unit is not likely to be met with favor.

Legitimacy of stage/ages as the basis of chronostratigraphy

In the continuing discussions about rectifying the relationship of the classical European stages and the new Paleocene/Eocene boundary, we have encountered a strongly held opinion among some colleagues that ages/stages have no correlation value outside of their local domains, even though their names may be applied for essentially sentimental reasons to the global time scale. In this view, the effort to work out resolution of the stage problem is pointless, and the appropriate solution is simply to retrofit the lower Eocene subseries to the enlarged interval between the new boundary and the base of the middle Eocene (currently the base of the Lutetian). Within this framework there would be no need to clarify the stratigraphic relationship of the Ypresian, Sparnacian or Thanetian with regard to the base of the Eocene, since only their names would be required to fill in arbitrary blanks.

In effect, this approach accepts the series, not the stage, as the basic unit in chronostratigraphy (Fig. 3, option 5). Hence, the subseries is a *post hoc* subdivision of series, without independent calibration beyond its internal middle and upper di-

TABLE 1. APPROXIMATE THICKNESS OF THE STRATIGRAPHIC INTERVAL BETWEEN CARBON ISOTOPE EXCURSION AND BASE YPRESIAN STAGE* IN DIFFERENT GEOLOGICAL AND/OR SEDIMENTOLOGICAL SETTINGS

Location	Geological setting	Thickness (m)
Atlantic Coastal Plain		
1. Bass River Corehole	Marls, continental shelf	~12
2. Ancora corehole	Marls, continental shelf	~9
Gulf Coastal Plain†		
Alabama	Tuscahoma Sands (nearshore sands)	~23
North Sea Basin§		
1. Sele Formation to base Balder Formation (ash series)	Marine sands	~30–50
2. Outer Moray Firth	Substatial deltaic progradation; siltstones with subordinate; sandstones	~350–400
London-Hampshire Basin		
1. Whitecliff Bay, Isle of Wight	Woolwich-Reading Formation, brown silty clays	~45–47
2. London Basin	Woolwich-Reading Beds, fluvio-deltaic	~5–15
Paris Basin#		
1. Limay Quarry (West Paris Basin)	Argile plastique, paludal	~15
2. Mount Bernon (Epernay East Paris Basin)	Argile plastique	~20
3. Bougival corehole (East Paris Basin)	Marnes blanches	~15
4. Cap d'Ailly	Shallow marine sands, clays	~12
Belgium Basin**		
1. Kallo borehole	Shallow marine sands, clays	~23
Austria††		
1. Salzburg	Anthering Formation, Gosau Basin (flysch basin)	~100
Spain		
1. Zumaya	Turbidite sands	~12
2. Alamedilla	Deep-sea marls	~7
Israel		
1. Ben Gurion	Marine marls (shelf)	~11
2. Zonet Telalim	Same	~2.3
Egypt (Upper Nile Valley)		
1. Dababiya	outer shelf-upper slope; marls	~42
2. Qeriya	Same	~22
3. Gebel Gurnah/Oweina	Same	>36
Deep Sea sections		
1. ODP Site 690 (South Atlantic)	Nannooze/chalks	~30
2. DSDP 550 (North Atlantic)	Marls and volcanic ashes	~25
Bighorn Basin, Wyoming§§		
	Continental-fluvial-alluvial beds	~1000

*Denoted by First Appearance Datum of *Tribrachiatus digitalis,* or proxy. This interval (about 0.8 to 1 m.y. in duration) would correspond to the reintroduced Sparnacian Stage. Data sources (work done by authors of this paper indicated by initials). No footnote indicates personal observation by MPA. ODP—Ocean Drilling Program; DSDP—Deep Sea Drilling Project.

†E. Mancini, personal communication, 9/2001.

§R.W.O'B. Knox.

MT.

**CD.

††Egger et al., 1998.

§§Wing et al., 2000.

vision boundaries, and the stage in turn has no global application, except as a name given to a *post hoc* subdivision of a subseries.

Apart from the violence that this does to the hierarchical logic recommended in both the International Guide (Hedberg, 1976) and, ostensibly, in the ICS revised guidelines (Cowie et al., 1986) this option would abandon the advantages of a chronostratigraphy based on stages/ages of relatively short duration, with boundaries made ever more useful by continuing advances in the precision and accuracy of correlation tools, in favor of a relatively coarse system in which only series boundaries would be independently defined and dated. Assumptions that series boundaries are more precise than stage boundaries, simply because it is easier to distinguish elements of one series from another, or possess greater significance because series boundaries are supposedly based on more important criteria than stage boundaries, are illusory, at least in the Cenozoic. It is only realistic to note that any stratigraphic feature that reflects a globally significant event, appropriate for a series GSSP, is also certain to be the basis of a stage in a regional section. There is thus nothing to be gained from claiming that only series boundaries based on such features have worldwide correlatibility, and much to be lost because there are clearly more globally correlatable boundary criteria than there are series to fit them.

THE INTRODUCTION OF A NEW CHRONOSTRATIGRAPHIC UNIT

In recommending the term Sparnacian as a standard stage in the Paleogene chronostratigraphic hagiography, we must acknowledge that the inconsistent usage of the term has caused it to be generally (but not completely) ignored as a chronostratigraphic term, in favor of extending the Thanetian up to the base of the Ypresian London Clay Formation (Berggren, 1971; Berggren and Aubry, 1996, 1998). However, it is clear that the decision to lower the Paleocene/Eocene boundary to be coincident with the negative $\delta^{13}C$ excursion makes the upper third of Renevier's expanded Thanetian early Eocene in age. Reassigning this amputated interval to Ypresian has the undesirable effect of concealing, rather than illuminating, the fact that the boundary had been moved to a significantly older level. The least confusion, and the greatest respect for the stability of the literature, requires the insertion of a formal stage between the Thanetian (sensu limito) and the Ypresian (s.s.) to accommodate the redesignated interval.

The restoration of the Sparnacian Stage, which in its original definition is essentially equivalent to the redesignated interval, seems to be the most appropriate, if not the only possible, choice for the identification of this crucial interval (Fig. 4). Furthermore, it must be recognized that this, the basal stage of the Eocene in its conceptual home of the Paris-London Basin (Lyell, 1833), would necessarily be regarded as having the special status of a "global standard stage" or defining hierarchical component of the Eocene Series. In accord with the hierarchical logic that requires that the base of this unit defines the base of the higher unit (Hedberg, 1976), and following the precedent of the Calabrian Stage and the Pliocene/Pleistocene boundary (Van Couvering, 1997), we consider that the boundary-stratotype of this basal stage will automatically be created by the adoption of a GSSP for the Paleocene/Eocene boundary, without further action.

The authors of this paper most of whom are members of the Working Group on the Paleocene/Eocene boundary that voted in favor of the revised Eocene concept, endorse the recommendation that a new stage, preferably the Sparnacian Stage restored to the original sense of Dollfus (1880), should be adopted for that former part of the classical Paleocene that now becomes the lowest part of the Eocene (Fig. 4). In fact, if we accept hierarchical logic and precedent, the approval of a GSSP for the base of the Eocene Series must simultanously establish the GSSP for the Sparnacian Stage in the worldwide chronostratigraphic hierarchy.

ACKNOWLEDGMENTS

The information and ideas presented here represent the outgrowth of over a decade of discussions amongst the members of the Working Group of the Paleocene/Eocene Boundary of the International Subcommission on Paleogene Stratigraphy as well as with other members of the stratigraphic community. The ideas expressed here were presented at the meeting on Paleogene Climates in Powell, Wyoming, in July 2001, organized by the Smithsonian Institution National Museum of Natural History. We wish to express our heartfelt thanks to Scott Wing, coordinator of the meeting, and to the many members of the stratigraphic community at large who, over the years, have engaged us in an ongoing dialogue regarding chronostratigraphic philosophy and methodology and the terminology and nomenclature associated with the placement of the Paleocene/Eocene boundary. We warmly thank Mike Woodburne for thorough discussion on the content and presentation of this manuscript. We thank him, Ellen Thomas and an anonymous reviewer for their critical and constructive review of the manuscript. In addition, Steven Walsh's comments on an incomplete draft contained some stimulating points. Our field research and the participation of several of us (MPA, WAB and KO) at the Powell meeting was supported, in part, by a grant from the National Geographic Society. This is Woods Hole Oceanographic Institution Contribution Number 10710.

REFERENCES CITED

Ali, J.R., King, C., and Hailwood, E.A., 1993, Magnetostratigraphic calibration of early Eocene depositional sequences in the southern North Sea Basin, *in* Hailwood, E.A., and Kidd, R.B., eds., High-resolution stratigraphy: Geological Society [London] Special Publication 70, p. 99–125.

Archibald, J.D., Gingerich, P.D., Lindsay, E.H., Clemens, W.A., Krause, D.W., and Rose, K.D., 1987, First North American Land Mammal Ages of the Cenozoic era, *in* Woodburne, M.O., ed., Cenozoic mammals of North America: Berkeley, California, University of California Press, p. 24–76.

Aubry, M.-P., 1996, Toward an upper Paleocene–lower Eocene high-resolution stratigraphy, *in* Aubry, M.-P., and Benjamini, C., eds., Paleocene/Eocene boundary events in space and time: Israel Bulletin of Earth Sciences, v. 44, p. 239–253.

Aubry, M.-P., 2000, Where should the global stratotype section and point (GSSP) for the Paleocene/Eocene boundary be located?: Bulletin de la Société Géologique de France, v. 171, no. 4, p. 461–476.

Aubry, M.-P., and Berggren, W.A., 2000a, When conflict arises between principles and rules: The case of the GSSP for the Paleocene/Eocene boundary: GFF (Geologiska Föreningens i Stockholm Förhandlingar) (Journal of the Geological Society of Sweden), v. 122, p. 13–14.

Aubry, M.-P., and Berggren, W.A., 2000b, The homeless GSSP: The dilemma of the Paleocene/Eocene boundary: Tertiary Research, v. 20, no. 1–4, p. 107–112.

Aubry, M.-P., Berggren, W.A., Stott, L., and Sinha, A., 1996, The upper Paleocene–lower Eocene stratigraphic record and the Paleocene-Eocene boundary carbon isotopre excursion: Implications for geochronology, *in* Knox, R.W.O'B., et al., eds., Correlation of the early Paleogene in northwestern Europe: London, Geological Society Special Publication 101, p. 353–380.

Aubry, M.-P., Berggren, W.A., Van Couvering, J.A., and Steininger, F., 1999, Problems in chronostratigraphy: Stages, series, unit and boundary strato-

types, global stratotype section and point and tarnished golden spikes: Earth Science Reviews, v. 46, p. 99–148.
Aubry, M.-P., Van Couvering, J.A., Berggren, W.A., and Steininger, F., 2000a, Should the golden spike glitter?: Episodes, v. 23, no. 3, p. 203–210.
Aubry, M.-P., Cramer, B., Miller, K.G., Wright, J., Kent, D.V., and Olsson, R.D., 2000b, Late Paleocene event chronology: Unconformities not diachrony: Bulletin de la Société Géologique de France, v. 171, p. 367–378.
Aubry, M.-P., and the Working Group on the Paleocene/Eocene Boundary, 2002, The Paleocene/Eocene boundary global standard stratotype-section and point (GSSP): Criteria for characterisation and correlation: Tertiary Research, v. 21, no. 1–4, p. 57–70.
Bains, S., Corfield, R.M., and Norris, R.D., 1999, Mechanisms of climate warming at the end of the Paleocene: Science, v. 285, p. 724–727.
Berggren, W.A., 1971, Tertiary Boundaries, *in* Funnel, B.M., and Riedel, W.R., eds., Micropaleontology of the oceans: Cambridge, UK, Cambridge University Press, p. 693–808.
Berggren, W.A., and Aubry, M.-P., 1996, A late Paleocene–early Eocene northwest European and North Sea magnetobiochronological correlation network, *in* Knox, R.W.O'B., et al., eds., Correlation of the early Paleogene in northwestern Europe: London, Geological Society Special Publication 101, p. 309–352.
Berggren, W.A., and Aubry, M.-P., 1998, The Paleocene/Eocene Epoch/Series boundary: Chronostratigraphic framework and estimated geochronology, *in* Aubry, M.-P., et al., eds., Late Paleocene–early Eocene climatic and biotic events in the marine and terrestrial records: New York, Columbia University Press, p. 18–36.
Berggren, W.A., Kent, D.V., and Flynn, J.J., 1985, Paleogene geochronology and chronostratigraphy, *in* Snelling, N.J., ed., The chronology of the geological record: London, Geological Society Memoir 10, p. 141–195.
Berggren, W.A., Kent, D.V., Swisher, C.C., III, and Aubry, M.-P., 1995, A revised Cenozoic geochronology and chronostratigraphy, *in* Berggren, W.A., et al., eds., Geochronology, time-scales, and global stratigraphic correlation: SEPM (Society for Sedimentary Geology) Special Publication 54, p. 129–212.
Berggren, W.A., Lucas, S., and Aubry, M.-P., 1998, Late Paleocene–early Eocene climatic and biotic evolution, *in* Aubry, M.-P., et al., eds., Late Paleocene–early Eocene climatic and biotic events in the marine and terrestrial records: New York, Columbia University Press, p. 1–17.
Bowen, G.J., Koch, P.L., Gingerich, P.D., Norris, R.D., Bains, S., and Corfield, R.M., 2001, Refined isotope stratigraphy across the continental Paleocene-Eocene boundary on Polecat Bench in the northern Bighorn Basin, *in* Gingerich, P.D., ed., Paleocene-Eocene stratigraphy and biotic change in the Bighorn and Clarks Fork Basins, Wyoming: Ann Arbor, Michigan, University of Michigan Papers on Paleontology, v. 33, p. 73–88.
Brinkhuis, H., and Visscher, H., 1995, The upper boundary of the Eocene Series: A reappraisal based on dinoflagellate cyst biostratigraphy and sequence stratigraphy, *in* Berggren, W.A., et al., eds., Geochronology, time scales, and global stratigraphic correlations: SEPM (Society for Sedimentary Geology) Special Publication 54, p. 295–304.
Bujak, J.P., and Brinkhuis, H., 1998, Global warming and dinocyst changes across the Paleocene/Eocene boundary, *in* Aubry, M.-P., et al., eds., Late Paleocene–early Eocene climatic and biotic events in the marine and terrestrial records: New York, Columbia University Press, p. 277–295.
Cavelier, C., and Pomerol, C., 1986, Stratigraphy of the Paleocene: Bulletin de la Société Géologique de France, ser. 8, v. 2, no. 2, p. 255–265.
Cavelier, C., and Roger, J., editors, 1980, Les Etages Français et leurs Stratotypes: Mémoires du Bureau de Recherches Géologiques et Minières, no. 109, 295 p.
Cowie, J.W., Ziegler, W., Boucot, A.J., Bassett, M.G., and Remane, J., 1986, Guidelines and statutes of the International Commission on Stratigraphy: Courier Forschungsinstitut Senckenbergiana, v. 83, p. 1–9.
Cramer, B.S., 2001, Latest Paleocene–earliest Eocene cyclostratigraphy using core photographs for reconnaisance in geophysical logging: Earth and Planetary Science letters, v. 186, p. 231–244.
Cramer, B.S., 2002, Milankovitch cycles and comet catastrophes: Carbon isotope variations in the Late Paleocene–eartly Eocene: Ph.D. thesis, Rutgers University, 116 p.
Cramer, B.S., Aubry, M.-P., Miller, K.G., Olsson, R.K., Wright, J.D., and Kent, D.V., 1999, An exceptional chronologic, isotopic, and clay mineralogic record of the latest Paleocene thermal maximum, Bass River, NJ, ODP 174AX: Bulletin de la Société Géologique de France, v. 170, p. 883–897.
Crouch, E.M., 2001, Environmental change at the time of the Paleocene-Eocene biotic turnover: Utrecht, Netherlands, Laboratory of Palaeobotany and Palynology Contributions Series, v. 14, 216 p.
Crouch, E., Heilmann-Clausen, C., Brinkhuis, H., Morgans, H.E.G., Rogers, K.M., Egger, H., and Schmitz, B., 2000, Global dinoflagellate event associated with the late Paleocene thermal maximum: Geology, v. 29, p. 315–318.
Curry, D., Adams, C.G., Boulter, M.C., Dilley, F.C., Eames, F.E., Funnell, B.M., and Wells, M.K., 1978, A correlation of Tertiary rocks in the British Isles: Geological Society [London] Special Report 12, p. 1–72.
De Coninck, J., Geets, S., and Willems, W., 1983, The Mont Héribu Member: Base of the Ieper Formation in the Belgian Basin: Tertiary Research, v. 5, no. 2, p. 83–104.
de Graciansky, P.-C., Hardenbol, J., and Vail, P.R., editors, 1998, Mesozoic and Cenozoic sequence stratigraphy of European basins: SEPM (Society for Sedimentary Geology) Special Publication 60, 768 p.
Dollfus, G., 1880, Essai sur l'extension des terrains tertiaires dans le Bassin anglo-parisien: Bulletin de la Société Géologique de Normandie, Le Havre, v. VI (1879), p. 584–605.
Dollfus, G., 1905, Critique de la classification de l'Eocène inférieur (Lettre de M.G. Dollfus à M. Mce Leriche): Annales de la Société Géologique du Nord, v. 34, p. 373–382.
Dumont, A.H., 1839, Rapport sur les travaux de la carte géologique en 1839, avec une carte géologique des environs de Bruxelles: Bulletin de l' Académie du Royaume de Belgique, Classe Sciences, v. 6, p. 464–485.
Dumont, A.H., 1849, Rapport sur la carte géologique du Royaume: Bulletin de l'Académie du Royaume de Belgique, Classe Sciences, v. 16, p. 351–373.
Dupuis, C., De Coninck, J., and Steurbaut, E., editors, 1991, The Ypresian stratotype: Bulletin de la Société Belge de Géologie, v. 97, no. 3/4, 481 p.
Egger, H., Heilmann-Clausen, C., and Schmitz, B., 2000, The Paleocene/Eocene boundary interval of a Tethyan deep-sea section (Austria) and its correlation with the North Sea basin: Bulletin de la Société Géologique de France, v. 171, no. 2, p. 207–216.
Ellison, R.A., Knox, R.W.O'B., Jolley, D.W., and King, C., 1994, A revision of the lithostratigraphical classification of the early Palaeogene strata of the London Basin and East Anglia: Proceedings of the Geologists' Association, v. 105, p. 187–197.
Feugueur, L., 1963, L'Yprésien du Bassin de Paris: Mémoire explicatif de la Carte géologique détaillée de la France: Orléans, France, Bureau de Recherches Géologiques et Minières, 568 p.
Frederiksen, N.O., 1994, Paleocene floral diversities and turnover events in eastern North America and their relation to diversity models: Review of Palaeobotany and Palynology, v. 82, p. 225–238.
Gingerich, P.D., 1989, New earliest Wasatchian mammalian fauna from the Eocene of northwestern Wyoming: Composition and diversity in a rarely sampled high-floodplain assemblage: University of Michigan Papers on Paleontology, v. 28, p. 1–97.
Gingerich, P.D., 2001, Biostratigraphy of the continental Paleocene-Eocene boundary interval on Polecat Bench in the northern Bighorn Basin, *in* Gingerich, P.D., ed., Paleocene-Eocene stratigraphy and biotic change in the Bighorn and Clarks Fork Basins, Wyoming: University of Michigan Papers on Paleontology, v. 33, p. 37–71.
Granger, W., 1914, On the names of lower Eocene faunal horizons of Wyoming and New Mexico: Bulletin of the American Museum of Natural History, v. 33, p. 201–207.
Harland, W.B., Armstrong, R.L., Cox, A.V., Craig, L.E., Smith, A.G., and Smith, D.G., 1990, A geologic time scale 1989 (revised edition): Cambridge, UK, Cambridge University Press, 263 p.

Hedberg, H.D., editor, 1972, An international guide to stratigraphic classification, terminology, and usage: Lethaia, v. 5, p. 283–295 (reprinted from International Subcommission on Stratigraphic Classification, Report No. 7).

Hedberg, H.D., editor, 1976, International stratigraphic guide: A guide to stratigraphic classification, terminology, and procedure: New York, John Wiley and Sons, 200 p.

Heilmann-Clausen, C., and Egger, H., 2000, The Anthering outcrop (Austria): A key-section for correlation between Tethys and northern Europe near the Paleocene/Eocene boundary: GFF (Geologiska Föreningens i Stockholm Förhandlingar) (Journal of the Geological Society of Sweden), v. 22, p. 69.

Hooker, J.J., 1996, Mammalian biostratigraphy across the Paleocene-Eocene boundary in the Paris, London and Belgian Basins, *in* Knox, R.W.O'B., et al., eds., Correlation of the early Palaeogene in northwest Europe: London, Geological Society Special Publication 101, p. 219–254.

Hooker, J.J., 1998, Mammalian faunal change across the Paleocene–Eocene transition in Europe, *in* Aubry, M.-P., et al., eds., Late Paleocene–early Eocene climatic and biotic events in the marine and terrestrial records: New York, Columbia University Press, p. 428–450.

Jolley, D.W., 1996, The earliest Eocene sediments of eastern England: An ultrahigh-resolution palynological correlation, *in* Knox, R.W.O'B., et al., eds., Correlation of the early Palaeogene in northwest Europe: London, Geological Society Special Publication 101, p. 219–254.

Kennett, J.P., and Stott, L.D., 1991, Abrupt deep-sea warming, palaeoceanographic changes and benthic extinctions at the end of the Palaeocene: Nature, v. 353, p. 225–229.

King, C., 1991, Stratigraphy of the Ieper Formation and Argile des Flandres (early Eocene) in western Belgium and southern France: Bulletin of the Belgium Geological Society, v. 97, p. 349–372.

Knox, R.W.O'B., 1994, From regional to standard stage: Implications for the historical Paleogene stratotypes of northwest Europe: GFF (Geologiska Föreningens i Stockholm Förhandlingar) (Journal of the Geological Society of Sweden), Meeting Proceedings "Stratigraphy of the Paleocene," v. 116, p. 56–57.

Knox, R.W.O'B., Corfield, R.M., and Dunay, R.E., editors, 1996, Correlation of the early Palaeogene in northwest Europe: London, Geological Society Special Publication 101, 480 p.

Laga, P., 1981, Landenian, *in* Pomerol, C., ed., Stratotypes of Paleogene stages: Bulletin d' Information des Géologues du Bassin de Paris, p. 123–145.

Laurain, M., Barta, L., Bolin, C., Guernet, C., Gruas Cavagnetto, C., Louis, P., Perreau, M., Riveline, J., and Thiry, M., 1983, Le sondage et la coupe du Mont Bernon à Epernay (Marne): Etude sédimentologique et paléontologique du stratotype du Sparnacien et de la série éocène: Géologie de la France, v. 3, p. 235–254.

Lemoine, M., 1891, Etude d' ensemble sur les dents des mammifères fossiles des environs de Reims: Bulletin de la Société Géologique de France, Sér. 3, p. 263–290.

Lemoine, V., 1889, Considérations généraux sur les vertébrés fossiles des environs de Reims et spécialement sur les mammifères de la faune cernaysienne: Comptes Rendus des Séances du Congrès International Zoologique, Paris, 1889, p. 233–279.

Lucas, S.G., 1998, Fossil mammals and the Paleocene/Eocene Series boundary in Europe, North America, and Asia, *in* Aubry, M.-P., et al., eds., Late Paleocene–early Eocene climatic and biotic events in the marine and terrestrial records: New York, Columbia University Press, p. 451–500.

Lyell, C., 1833, Principles of geology: Being an attempt to explain the former changes of the Earth's surface by reference to causes now in operation: London, John Murray, v. 3, 398 p.

Pomerol, C., 1969, Limite Paléocène-Eocène: Mémoires du Bureau de Recherches Géologiques et Minières, v. 69, p. 447–449.

Pomerol, C., 1977, La limite Paléocène-Eocène en Europe occidentale: Comptes rendus sommaires de la Société Géologique de France, fasc. 4, p. 199–202.

Pomerol, C., editor, 1981, Stratotypes of Paleogene Stages: Bulletin d'Information des Géologues du Bassin de Paris, Mém. hors série, no. 2, 301 p.

Pomerol, C., 1988, Stratigraphy of the Palaeogene: Hiatuses and transitions: Proceedings of the Geologists' Association, v. 100, no. 3, p. 313–324.

Prestwich, J., 1850, On the structure of the strata between the London Clay and the Chalk in the London and Hampshire Tertiary Systems, Part I: Quarterly Journal of the Geological Society [London], v. 6, p. 252–281.

Prestwich, J., 1852, On the structure of the strata between the London Clay and the Chalk in the London and Hampshire Tertiary Systems, Part III: The Thanet Sands: Quarterly Journal of the Geological Society [London], v. 8, p. 235–264.

Prestwich, J., 1854, On the structure of the strata between the London Clay and the Chalk in the London and Hampshire Tertiary Systems, Part II: The Woolwich and Reading Series: Quarterly Journal of the Geological Society [London], v. 10, p. 75–170.

Remane, J., 2000, Comments on the paper of "Should the Golden Spike Glitter?" by M.-P. Aubry et al.: Episodes, v. 23, no. 3, p. 211–213.

Remane, J., Bassett, M.G., Cowie, J.W., Gohrbandt, K.H., Lane, H.R., Michelsen, O., and Naiwen, W., 1996, Revised guidelines for the establishment of global chronostratigraphic standards by the International Commission on Stratigraphy (ICS): Episodes, v. 19, p. 77–81.

Renevier, E., 1873, Tableau des terrains sédimentaires formés pendant les époques de la phase organique du globe terrestre: Bulletin de la Société Vaudoise de Sciences Naturelles, v. 12, no. 70, Table III.

Renevier, E., 1874, Tableau des terrains sédimentaires formés pendant les époques de la phase organique du globe terrestre: Bulletin de la Société Vaudoise de Sciences Naturelles, v. 13, no. 72, p. 218–252.

Röhl, U., Bralower, T.J., Norris, R.D., and Wefer, G., 2000, New chronology for the late Paleocene thermal maximum and its environmental implications: Geology, v. 28, p. 927–930.

Rose, K.D., 1981, The Clarkforkian land-mammal age and mammalian faunal composition across the Paleocene/Eocene boundary: University of Michigan Papers on Paleontology, v. 26, p. 1–197.

Russell, D.E., 1967, Le Paléocène continental d'Amérique du Nord: Mémoires du Muséum National d'Histoire Naturelle, Sér. C, v. 16, p. 1–99.

Russell, D.E., 1968, Succession, en Europe, des faunes mammaliennes au début du Tertiaire: Mémoires du Bureau de Recherches Géologiques et Minières, v. 58, p. 291–296.

Russell, D.E., and Zhai, R., 1987, The Paleogene of Asia: Mammals and stratigraphy: Mémoires du Muséum National d'Histoire Naturelle, Série C, Sciences de la Terre, v. 52, 488 p.

Russell, D.E., Hartenberger, J.L., Pomerol, C., Sen, S., Schmidt-Kittler, N., and Vianey Liaud, M., 1982, Mammals and stratigraphy: The Paleocene of Europe: Palaeovertebrata, Mémoire extraordinaire, p. 1–77.

Salvador, A., editor, 1994, International Stratigraphic Guide: A guide to stratigraphic classification, terminology, and procedure (2nd edition): International Union of Geological Sciences and Geological Society of America, Trondheim, Norway, and Boulder, Colorado, iii–xix, 214 p.

Savage, D.E., and Russell, D.E., 1983, Mammalian paleofaunas of the world: London, Addison-Wesley Publishing Company, 432 p.

Schimper, W.Ph., 1874, Traité de Paléontologie Végétale ou la Flore du Monde Primitif dans ses Rapports avec les Formations Géologiques, v. 3: Paris, J.B. Ballière et Fils, 836 p.

Sinha, A., and Stott, L.D., 1993a, Recognition of the Paleocene/Eocene boundary in the northwestern European stratotype sections on the basis of carbon isotope stratigraphy: Journal of Vertebrate Paleontology, v. 13(3) (Supplement, p. 57A, Abstracts of papers, 53d Annual Meeting of Society of Vertebrate Paleontology).

Sinha, A., and Stott, L.D., 1993b, Recognition of the Paleocene/Eocene boundary carbon isotope excursion in the Paris Basin, France: Symposium on correlation of the early Paleogene in northwest Europe, Geological Society of London Program and Abstracts (1–2 December 1993), 1 p.

Sinha, A., Aubry, M.-P., Stott, L.D., Thiry, M., and Berggren, W.A., 1996, Chemostratigraphy of the "lower" Sparnacian deposits (Argiles plastiques bariolées) of the Paris Basin, *in* Aubry, M.-P., and Benjamini, C., eds., Paleocene-Eocene boundary events in space and time: Israel Journal of Earth Sciences, v. 44, p. 223–237.

Speijer, R.P., 1994, Extinction and recovery patterns in benthic foraminiferal paleocommunities across the Cretaceous/Paleogene and Paleocene/Eocene boundaries: Geologica Ultraiectina, Mededelingen van de Faculteit Aardwetenschappen Universiteit Utrecht, v. 124, 191 p.

Stott, L.D., Sinha, A., Thiry, M., Aubry, M.-P., and Berggren, W.A., 1996, Global $\delta^{13}C$ changes across the Paleocene-Eocene boundary: Criteria for terrestrial-marine correlations, *in* Knox, R.W.O'B., et al., eds., Correlation of the early Paleogene in northwest Europe: London, Geological Society Special Publication 101, p. 388–399.

Thiry, M., Dupuis, C., Aubry, M.-P., Berggren, W.A., Ellison, R.A., Knox, R.W.O'B., Sinha, A., and Stott, L., 1998, Tentative correlations between continental deposits of the Argiles Plastiques (Paris Basin) and Reading Beds (London Basin), based on chemostratigraphy: Strata, v. 1, p. 125–129.

Thomas, E., 1998, Biogeography of the late Paleocene benthic foraminiferal extinction. *in* Aubry, M.-P., et al., eds., Late Paleocene–early Eocene climatic and biotic events in the marine and terrestrial records: New York, Columbia University Press, p. 214–243.

Van Couvering, J.A., 1997, Preface: The new Pleistocene, *in* Van Couvering, J.A., ed., The base of the Pleistocene and the beginning of the Quaternary: New York, Cambridge University Press, p. iii–x.

von Koenen, A., 1885, Über eine Palaocäne Fauna von Kopenhagen: Abhandlungen Köngliche Gesellschaft Wissenschaften, Göttingen, v. 32, p. 1–128.

Walsh, S.L., 2001, Notes on geochronological and chronostratigraphic units: Geological Society of America Bulletin, v. 113, p. 704–713.

Wing, S.L., 1997, Global warming and land species richness: A case study of the Paleocene/Eocene boundary, *in* Reska-Kudia, M.L., et al., eds., Biodiversity II: Understanding and protecting our biological resources: Washington, D.C., Joseph Henry Press, p. 163–185.

Wing, S.L., Alroy, J., and Hickey, L.J., 1995, Plant and mammal diversity in the Paleocene to early Eocene of the Bighorm Basin: Palaeogeography, Palaeoclimatology, Palaeoecology, v. 115, p. 117–155.

Wing, S.L., Bao, H., and Koch, P., 2000, An early Eocene cool period? Evidence for continental cooling during the warmest part of the Cenozoic, *in* Huber, B.T., et al., eds., Warm climates in Earth history: New York, Cambridge University Press, p. 197–237.

Wood, H.E., II, Chaney, R.W., Clark, J., Colbert, E.H., Jepsen, G.L., Reeside, J.B., Jr., and Stock, C., 1941, Nomenclature and correlation of the North American continental Tertiary: Geological Society of America Bulletin, v. 52, p. 1–48.

Manuscript Accepted by the Society August 13, 2002

Geological Society of America
Special Paper 369
2003

Cyclostratigraphy of upper Paleocene and lower Eocene sediments at Blake Nose Site 1051 (western North Atlantic)

Ursula Röhl*
Geosciences Department, DFG Research Center for Ocean Margins, Bremen University, 28334 Bremen, Germany
Richard D. Norris
308 Vaughan Hall, Scripps Institution of Oceanography, University of California San Diego, La Jolla, California 92093-0244, USA
James G. Ogg
Department of Earth and Atmospheric Sciences, Purdue University, West Lafayette, Indiana 47907-1397, USA

ABSTRACT

Upper Paleocene to lower Eocene sediments drilled at Ocean Drilling Program (ODP) Site 1051 (Blake Nose, off Florida) display well-defined orbital cycles, a detailed magnetic stratigraphy, and a suite of planktonic foraminiferal datums. We derived a cyclostratigraphy by using spectral analysis of high-resolution records of elemental concentrations obtained by an X-ray fluorescence (XRF) Core Scanner. XRF counts of iron serve as a proxy for the relative amount of terrestrial material. Sliding-window spectral analysis, bandpass filtering, and direct counting of precession and obliquity cycles yield minimum durations for magnetic polarity chrons C22 to C26 (~49 to ~61 Ma), calculations of sediment accumulation rates, as well as constraints on the timing of biostratigraphic and climatological events in the vicinity of the Initial Eocene Thermal Maximum (IETM).

Durations of polarity chrons as represented in sediments drilled at Site 1051 were estimated using a conservative assignment of 41 k.y. for obliquity cycles and 21 k.y. for precession cycles. Combined polarity chrons C26r and C26n span 3.61 m.y., and chron C25r spans 1.07 m.y. Polarity chron C24r is estimated as 2.877 m.y. The interpretation of polarity chron C24n is ambiguous, but its duration is probably <1.23 m.y. Polarity chron C23r spans 0.53 m.y., chron C23n is 0.74 m.y., and chron C22r is 0.9 m.y.

Spectral analysis through this interval indicates that spectral peaks shift through time and are related to changes in sedimentation rate at Site 1051. The sedimentation rates dramatically increased ~200 k.y. after the IETM and remained high for most of chron C24r.

*Uroehl@allgeo.uni-bremen.de

Röhl, U., Norris, R.D., and Ogg, J.G., 2003, Cyclostratigraphy of upper Paleocene and lower Eocene sediments at Blake Nose Site 1051 (western North Atlantic), *in* Wing, S.L., Gingerich, P.D., Schmitz, B., and Thomas, E., eds., Causes and Consequences of Globally Warm Climates in the Early Paleogene: Boulder, Colorado, Geological Society of America Special Paper 369, p. 567–588.

INTRODUCTION

Climatic variance in the "Milankovitch" band offers a unique record of climate sensitivity to external forcing. Milankovitch climate cycles are created by oscillations in the annual and seasonal distribution of insolation due to periodic changes in the obliquity or tilt of the Earth's spin axis, in the precession or relative tilt of the Earth upon closest approach to the Sun, and in the eccentricity of the Earth's orbit. Orbital-climatic oscillations recorded in deep-sea sediments enabled the development of a high-resolution astronomically calibrated time scale of geomagnetic polarity reversals and biostratigraphy from the late Cenozoic (e.g., Hilgen, 1991; Hilgen et al., 1995; Shackleton et al., 1990, 1995; Shackleton and Crowhurst, 1997) to the base of the late Eocene (Shackleton et al., 1999, 2000; Pälike et al., 2001). Because analysis of sedimentological and physical properties data from Paleogene sequences documents the sensitivity of the ocean to Milankovitch-band forcing and the rapidity of threshold events in Paleogene paleoceanographic and paleoclimatic history, "floating" astronomical time scales could be based on the number of precession or obliquity-controlled cycles multiplied by the period off the corresponding astronomical cycle at that time (Norris and Röhl, 1999; Röhl et al., 2000, 2001; Cramer, 2001).

Analysis of orbital-climate cycles and associated astronomical tuning provides the best practical method for measuring the duration of individual magnetochrons. We have concentrated on analysis of cyclostratigraphy in Ocean Drilling Program (ODP) Site 1051 (Atlantic margin off northern Florida, 30°N, 76°W, water depth of 1983 m) (Fig. 1) where an expanded

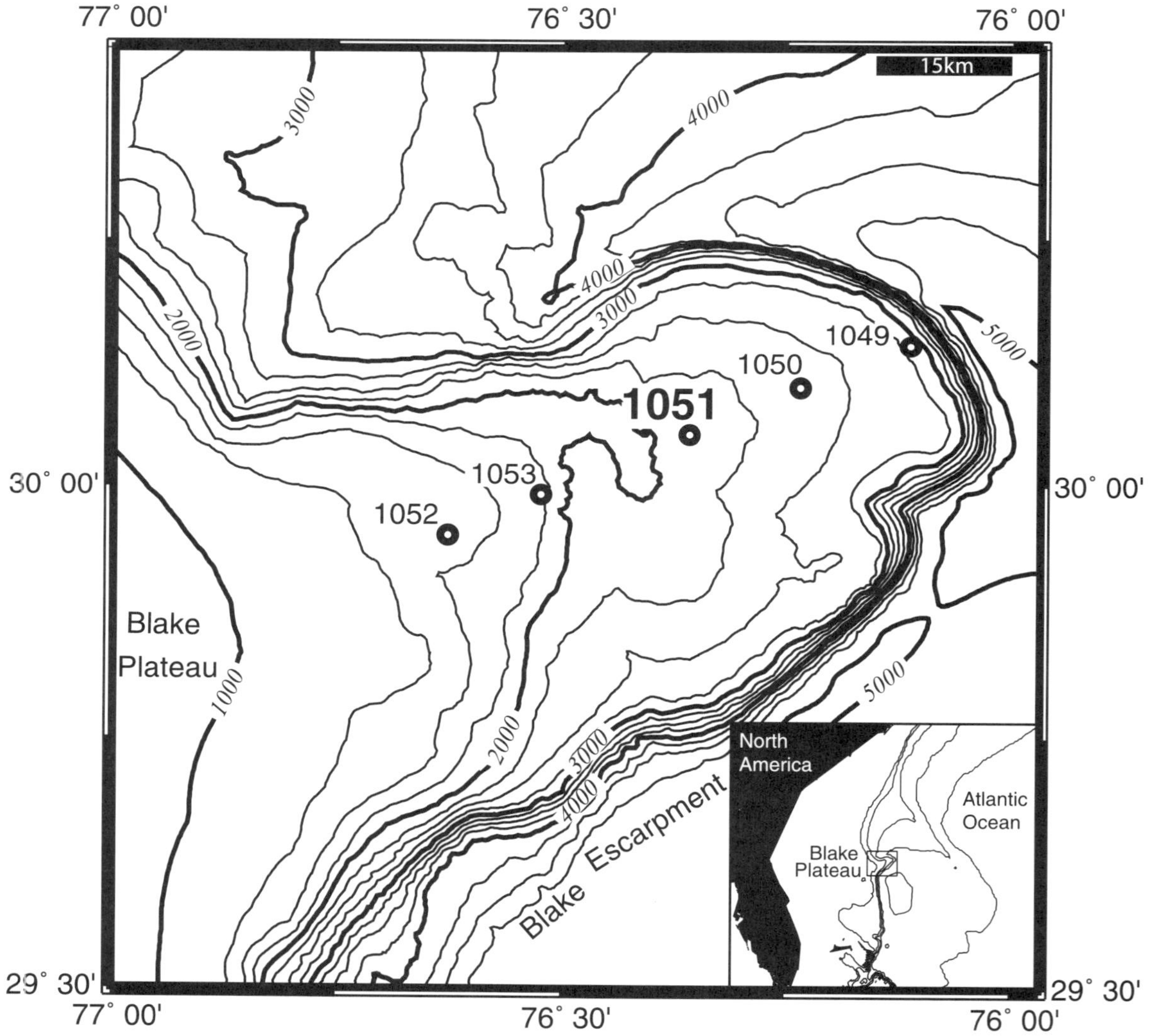

Figure 1. Ocean Drilling Program Leg 171B drilling transect and location of Site 1051 on Blake Nose in the western North Atlantic. Bathymetry in meters.

record of upper Paleocene and lower Eocene strata was recovered that preserves both an interpretable magnetostratigraphy and well-defined cycles in sediment color and geochemistry. The record at Site 1051 is one of the most expanded sequences presently available through the Paleocene-Eocene transition with a magnetostratigraphy. Although the sequence is cut by several unconformities it is sufficiently complete to provide estimates of the minimum duration of polarity reversals between chrons C26 and C22, as well as constraints on the timing of biostratigraphic and climatological events in the vicinity of the Initial Eocene Thermal Maximum (IETM).

STRATIGRAPHY AND ANALYTICAL METHODS

Paleocene-Eocene of ODP Site 1051

Ocean Drilling Program (ODP) Site 1051 on the Blake Nose (Fig. 1) was drilled down to the lower Paleocene and was double cored as low as the IETM. The studied interval (~380–630 m below seafloor (mbsf), cores 1051A-42X to -68X and 1051B-46X to -61X) spans the early Paleocene through early Eocene with an average sedimentation rate of 1.8 cm/kyr (Shipboard Scientific Party, 1998). The Paleocene-Eocene sediments range from greenish-gray clayey siliceous chalk, slightly darker nannofossil-rich claystone to light siliceous nannofossil chalk (Shipboard Scientific Party, 1998). Decimeter to one-meter-scale oscillations in $CaCO_3$ and clay content are expressed as white to pale brown or olive green banding and are pronounced features in the iron (Fe) intensity core logs (Fig. 2).

Polarity chron C24r is of special interest as its early portion includes the Paleocene-Eocene boundary. This episode in Earth history is characterized by a brief, but intense interval of global warming, by changes in ocean circulation and chemistry, and by a massive perturbation of the global carbon cycle (e.g., Kennett and Stott, 1991; Zachos et al., 1993). A leading hypothesis for the carbon isotope excursion (CIE) is a large-scale input of methane into the ocean and atmosphere from the dissociation of methane hydrates in continental margin sediments (e.g., Matsumoto, 1995; Dickens et al., 1995, 1997). This climatic excursion was formerly called the Late Paleocene Thermal Maximum (LPTM); but was renamed the Initial Eocene Thermal Maximum (IETM), after the Paleocene-Eocene boundary was redefined to be marked by the base of the (CIE) (Luterbacher et al., 2000), thus placing the thermal maximum in the initial Eocene.

Polarity chron C24r in Site 1051 spans 76 m of siliceous nannofossil chalk, the thickest known sedimentary record of this magnetic polarity chron in the deep sea (Norris and Röhl, 1999; Cramer, 2001). The IETM record at Site 1051 contains a minor unconformity spanning ~50 k.y. (~2.5 precession cycles) at the base of the carbon isotope excursion (CIE) if compared to the record at the more complete IETM interval at Site 690 in the Southern Ocean (Röhl et al., 2000). This minor unconformity may be related to a postulated slope failure caused by methane release (Katz et al., 1999, 2001), or to changes in bottom water circulation. With the exception of this minor disruption at the IETM, the expanded sedimentary record of polarity chron C24r is apparently complete.

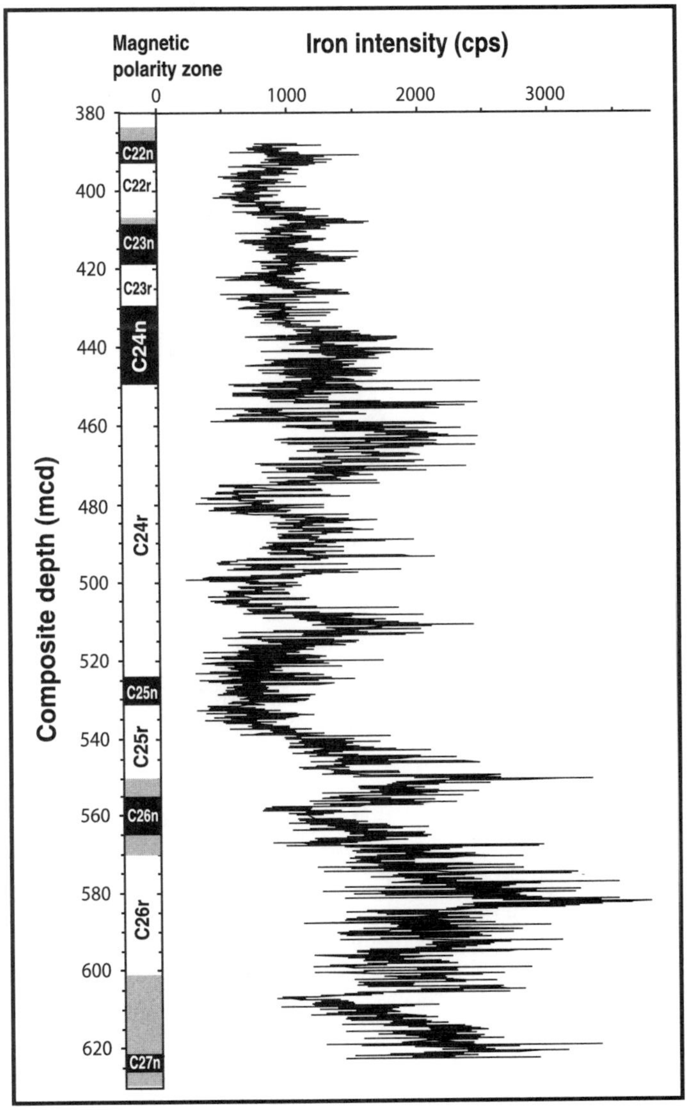

Figure 2. Composite Fe intensity record versus meters composite depth (mcd) at Site 1051 spanning magnetochrons C22n through C26r (49–61 Ma) (black—normal; white—reverse magnetic polarity zone; gray—error bar; cps—counts per second).

Magnetostratigraphy

A reference geomagnetic polarity time scale for the latest Cretaceous and early Cenozoic was derived from an age model for magnetic polarity chrons derived from a cubic-spline fit of the marine magnetic anomaly pattern in the South Atlantic to selected radiometric ages (Cande and Kent, 1992, 1995). Magnetobiochronological correlations allow projection of ages from this "CK95" model to biostratigraphic datums and associated geological stages (Berggren et al., 1995). The Paleocene-Eocene portion of the CK95 model was calibrated by four age control points—at 65 Ma (C29r.7 base of Cenozoic), 55 Ma (C24r.34; lowermost Eocene), 46.8 Ma (C21n.67; lower middle Eocene), and 33.7 Ma (C13r.86; base of Oligocene). The polarity chron (time) and polarity zone (stratigraphy) nomenclature in this paper is the system of Cande and Kent (1992), with suffix n denoting normal polarity and r denoting the underlying reversed polarity interval. The relative timing (position) of an event (level) within a polarity chron (zone) is "defined as the relative position in time or distance between the younger and older chronal boundaries" (Hallam et al., 1985, p. 126). In this proportional stratigraphic convention, the location of the Cretaceous-Paleogene boundary at Gubbio (Alvarez et al., 1977) occurs at C29r.75, indicating that 75% of reversed-polarity zone C29r is below the event. Cande and Kent (1992) used an inverted stratigraphic placement relative to present; therefore, C29r.3 in their notation indicates that 30% of reversed-polarity chron C29r follows the event. This system mirrors the convention of measuring geological time and the numbering magnetic anomalies backwards from the present.

Magnetic polarity zones are identified by paleomagnetic measurements on discrete minicores from Holes 1051A and 1050A (Ogg and Bardot, 2001) (Fig. 2). Polarity chron assignments are based upon the biostratigraphic constraints coupled with the distinctive pattern of normal- and reversed-polarity zones. Paleomagnetic reversals are moderately well defined in Hole 1051A but the relatively broad spacing of minicores (~1 m to 3 m) and sometimes indeterminate paleomagnetic interpretation leaves the exact placement of reversal boundaries uncertain in some cases (e.g., Ogg and Bardot, 2001). In Hole 1051A, the uncertainty in the placement of C23n/C22r boundary is ~4 m and that for the base of C23n is 2.55 m. Other reversals show a similar range of uncertainty if we consider only the results from minicores; the range of placement in the chron boundary is 3.16 m for the base of C24n, 63 cm for the top of C25n, 2.05 m for the top of C26n, and 3.15 m for the base of C26n (Fig. 2). In turn, these sampling gaps contribute to further uncertainties in the number of physical property and geochemical cycles that can be identified with polarity chrons. Paleomagnetic resolution is improved somewhat by use of pass-through magnetometer data from both Holes 1051A and 1051B (Shipboard Scientific Party, 1998; Ogg and Bardot, 2001). Pass-through magnetometer measurements show considerable scatter but intervals of reversed polarity can generally be interpreted and provide additional support for our placement of reversal boundaries.

We have been able to improve the resolution of some polarity boundaries by correlating the paleomagnetic records from Site 1051 with those from Site 1050 (Figs. 3 and 4). In general, the resolution of discrete paleomagnetic sampling was higher in Hole 1050A (50 cm to 1 m) than in Hole 1051A and the resulting paleomagnetic data contained fewer indeterminate polarity assignments. For example, comparison of the physical property data for the two sites allows us to constrain the location of the

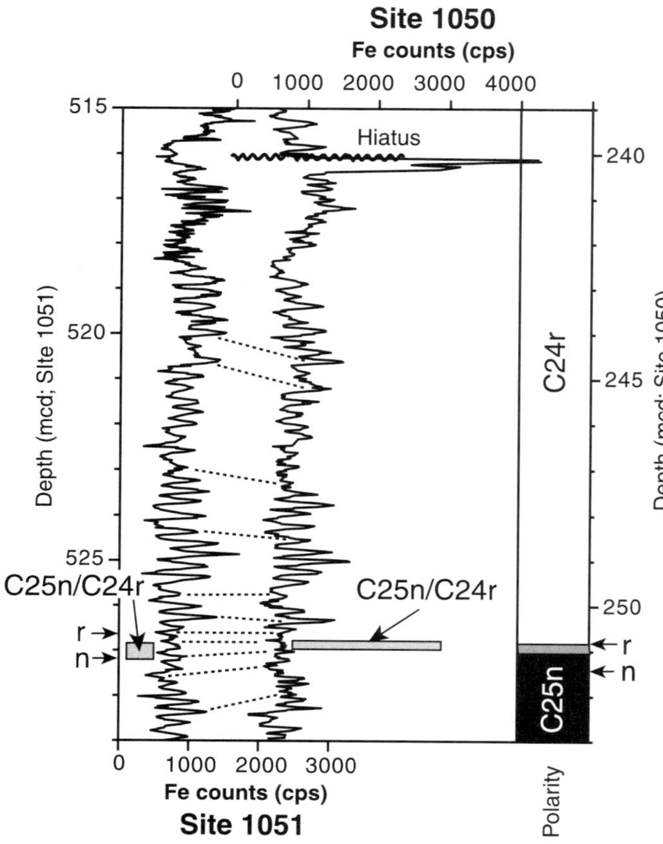

Figure 3. Correlation of the Fe cycles in Sites 1050 and 1051 in the vicinity of the C24r/C25n boundary. The cycle-to-cycle correlation of discrete magnetopolarity measurements restricts the placement of the reversal boundary to a single precession cycle between 526.92 and 527.06 mcd (meters composite depth) in Site 1051 (n—normal; r—reversed polarity interpretation for discrete samples in both holes; cps—counts per second).

C25n/C24r and C24r/C24n reversals to within a single physical property cycle in each case. The correlation of the two sites using XRF scanner Fe counts places the C25n/C24r boundary between 1051B-61X-5, 106 cm and -61X-5, 120 cm (523.96–524.1 mbsf = 526.92–527.06 mcd [meters composite depth]; Fig. 3). In the case of the C25n/C24r boundary, the peak-to-peak correlation between the Fe count records allows us to reduce the estimated error of the cycle count by a full precession cycle (~21 k.y.) between the chron boundary and the start of the IETM compared to that reported by Norris and Röhl (1999). For the C24r/C24n boundary, our correlation with Hole 1050A (based upon shipboard measurements of color reflectance) places the boundary between the levels 1051A-48X-2, 60 cm and -48X-2, 100 cm (449.6–450.0 mbsf, 447.59–447.99 mcd; Fig. 4).

Correlation between Sites 1050 and 1051 using color reflectance data also improves the resolution of the magnetic reversals for other chron boundaries. In Hole 1051A, we place the top of C26n at ~548.85–549.15 mbsf (549.86–550.16 mcd) based upon correlation to Hole 1050A using the Munsell hue

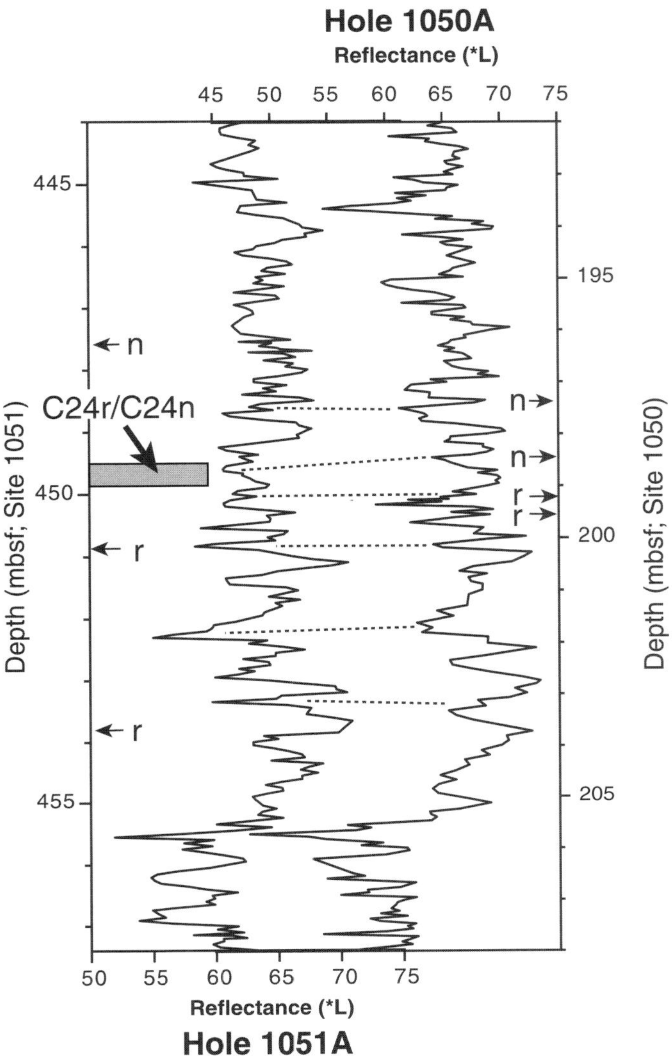

Figure 4. Correlation of color reflectance profiles ("lightness," L*) in Holes 1050A and 1051A across the C24r/C24n reversal boundary. Peak-to-peak correlation of color reflectance suggests that the polarity reversal defined by analysis of minicores occurs between 449.60 and 450.0 mbsf in Hole 1051A (n—normal; and r—reversed polarity interpretation for discrete samples in both holes; mbsf—meters below seafloor).

and value scales. However, attempts to refine the location of the C23n/C22r boundary by correlation between the two sites were unsuccessful since no unambiguous correlation could be established using either color reflectance data or magnetic susceptibility. The base of C25n is difficult to constrain using intersite correlation owing to a core gap in Hole 1050A. However, we follow Norris and Röhl (1999) who used the pass through shipboard magnetometer data to place this boundary at ~530.5 mcd at Site 1051. The remaining polarity zone boundaries were assigned to midway between the constraining paleomagnetic samples of opposing polarities.

X-ray fluorescence scanning

We analyzed the cyclicity of Fe variations in the sediments encompassing polarity zones C22 through C26 (spanning ~12 m.y.) in cores from adjacent Holes 1051A and 1051B with the X-ray fluorescence (XRF) Core Scanner at Bremen University (Röhl and Abrams, 2000; Röhl et al., 2000). The XRF scanner allows closely spaced (or even continuous), nondestructive analyses of major and minor elements at the surfaces of the split cores (Jansen et al., 1998). XRF measurements of the element ranging in atomic number between potassium (K) and strontium (Sr) were recorded at 2 cm intervals (e.g., Figures 2, 3, 5, 8A, 9A, 10A, and 11A). Where necessary, the sampling plan was adjusted to avoid core gaps, drilling mud intervals between single core biscuit pieces, and other disturbances.

Our high-resolution data sets reveal pronounced periodicity. Fe intensity (counts per second, cps) shows the clearest signal and was selected as a geochemical proxy for climate-paleoceanographic cycles. This major element mirrors changes in carbonate/clay ratios, and occurs at sufficiently high concentrations throughout the sediments. We did not convert cps values to concentrations because we are interested in the variation in elemental concentrations rather than absolute values. The intensities in counts per second (cps) were calibrated between each core by measuring a suite of lithologic standards.

Construction of a "meters composite depth" scale

We constructed a composite record for the Fe counts determined by XRF on the meters composite depth (mcd) scale (Fig. 2). For the expanded interval of polarity zone C24r, we constructed a spliced record from adjacent Holes 1051A and 1051B, which overlap from the seafloor down to the base of polarity zone C25. The sediment in the lower Eocene and uppermost Paleocene portion of Site 1051 is a stiff siliceous nannofossil chalk. With only a few exceptions, these rocks exhibit nearly complete recovery in both holes and very little core disturbance. Above polarity zone C24r, the excellent core recovery from Hole 1051A was adequate for a near-continuous geochemical record. Only Hole 1051A penetrated below polarity zone C24r. A composite section and associated mcd scale was constructed by aligning prominent features, sediment color data and magnetic susceptibility variations between these parallel holes (Shipboard Scientific Party, 1998). Because the chalk is quite stiff and displays relatively little differential stretching or compaction within cores, we were able to apply a constant depth offset to cores in the two holes to create the composite record. The hardness of the rocks also contributes to nearly complete recovery of adjacent cores in most cases (e.g., Fig. 5).

Initially, the composite depth scale was constructed using shipboard magnetic susceptibility and color reflectance records that were each measured at ~5 cm spacing. These records are adequate to construct a spliced record above the IETM, where the prominent color cycles are 70 cm to 1 m in wavelength. However, below the IETM, the physical property logs are too aliased to provide a reliable splice. We enhanced this splice by utilizing

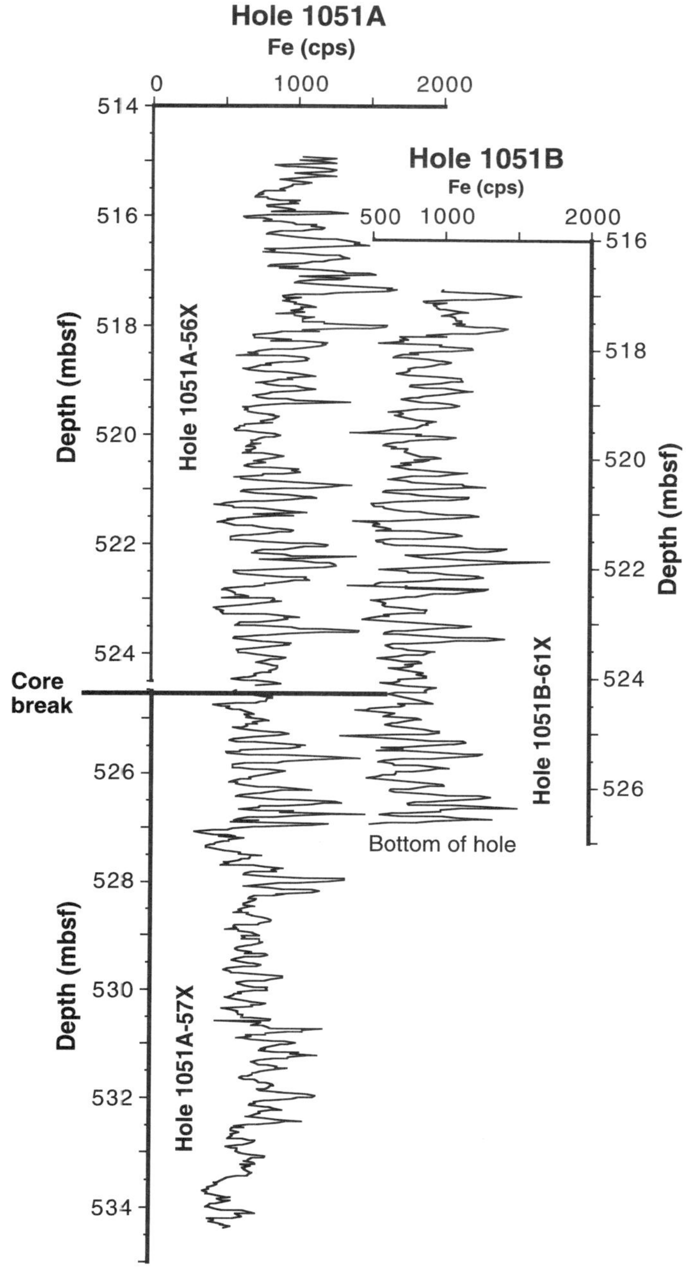

Figure 5. Example of the excellent hole-to-hole correlation using XRF Fe counts (cps—counts per second) in Site 1051. A meters composite depth scale (mbsf is meters below seafloor) was constructed by splicing records across core gaps. The correlation suggests that there is almost no sediment missing between successive cores in Hole 1051A.

the high-resolution (2 cm measurement spacing) geochemical fluctuations obtained from the cores of both holes with the XRF scanner. The excellent quality of the resulting splice is illustrated in Figure 5. The resulting meters composite depth scale was compared with downhole logs of gamma ray and Formation Microscanner Resistivity to check the overall continuity of the cored record.

Calculation of cycle periods in the depth domain

The interval we have chosen for study between magnetochron C22 and C26 is ~12 million years in duration. We have used a combination of several strategies to calculate the number of physical property cycles observed in sediments spanning this interval and interpret their temporal significance. Our first step was to calculate an evolutive spectrum in the depth domain to identify intervals where there are marked changes in cycle periods. We then analyzed specific intervals of core where the evolutive spectrum suggested that the cyclicity had relatively constant wavelengths. Two different approaches were used—spectral analysis of sliding windows and direct cycle identification—to verify the presence and relative importance of different cycles within the Fe intensity measurements through the upper Paleocene and lower Eocene sediments. We applied spectral analysis using the AnalySeries 1.2 PPC software package (Paillard et al., 1996) to a series of windows through the Fe record to identify the continuity and ratios of significant peaks in cyclicity with depth (Fig. 6). The XRF Fe intensity measurements within each window were linear detrended and normalized to unit variance before spectral analysis by the Blackman-Tukey method. Windows of 20–40 meters were applied to monitor the continuity and relative importance of spectral peaks and the variations in sedimentation rate.

Figure 7 shows the results obtained from an evolutive spectral analysis for the Fe intensities from our XRF measurements of the full intervals of polarity zones C22 to C26. The evolutive spectral analysis of the raw XRF Fe data was performed with a window length of 10 m and a step size of 5 m (500 values spaced at 0.02 m, Figure 7B). A Gaussian filter was used to remove variations with wavelength >10 m prior to calculation of evolutive spectra. As each window was plotted at its central position the figure does not extend to the ends of the full data set. Variations in cycle period produce patches rather than continuous bands across the spectra. One distinct "band" varies ~0.3–0.1 cycles/meter (3.5–10 m wavelength) and is most visible in the interval above 510 mcd and below 544 mcd (e.g., Figure 7B). A secondary band with a period of 1–1.5 cycles/meter (~1 m to 60 cm wavelength) appears in parts of the record between 460 and 490 mcd and below 560 mcd. A distinct shift in spectral peaks occurs between 544 mcd and 510 mcd where the high frequency is 3–4.5 cycles/meter (~29–33 cm wavelength). Another interval between 430 and 450 mcd also has a relatively high frequency of ~1–2 cycles/meter (1 m to 50 cm wavelength).

Continuous downhole geophysical logging of Hole 1051A (e.g., the measurement of natural gamma radiation at 15 cm intervals) provided an independent record to demonstrate that the high-resolution Fe-cycle record was not significantly distorted by core recovery. Many of the broad major oscillations in Fe have corresponding features in the downhole logs of natural gamma radiation, and these probably indicate relative concentrations of nonbiogenic clastics (e.g., Figures 8C and 9C). Even some of the individual natural gamma-ray excursions in Hole 1051A could be directly matched to intensity peaks in the XRF

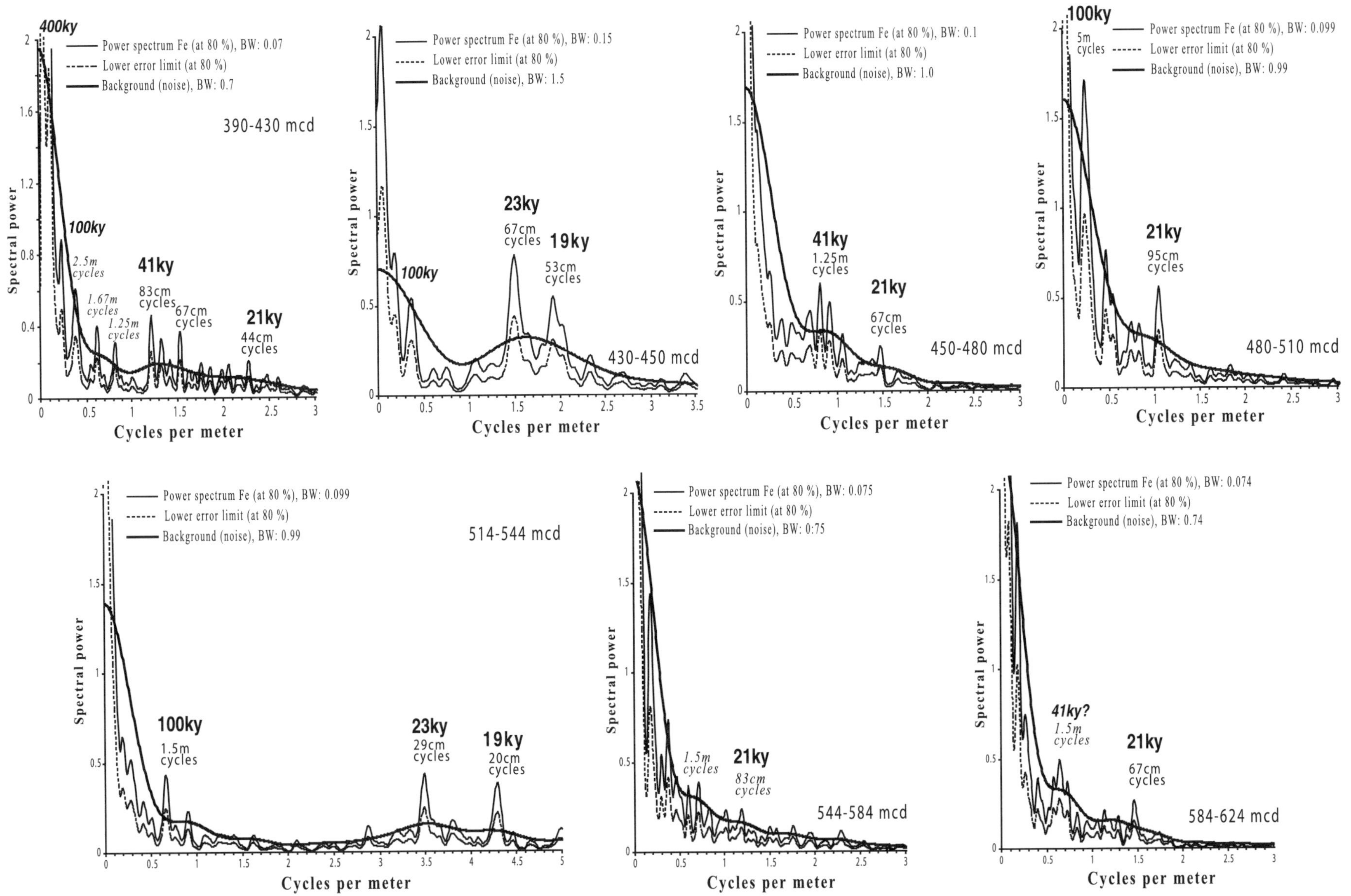

Figure 6. Selected interval-spectra covering magnetochron C22 to C26 from Site 1051. Spectral windows span 20 m to 40 m intervals. Precession and obliquity cycles can be identified in most cases; mcd—meters composite depth.

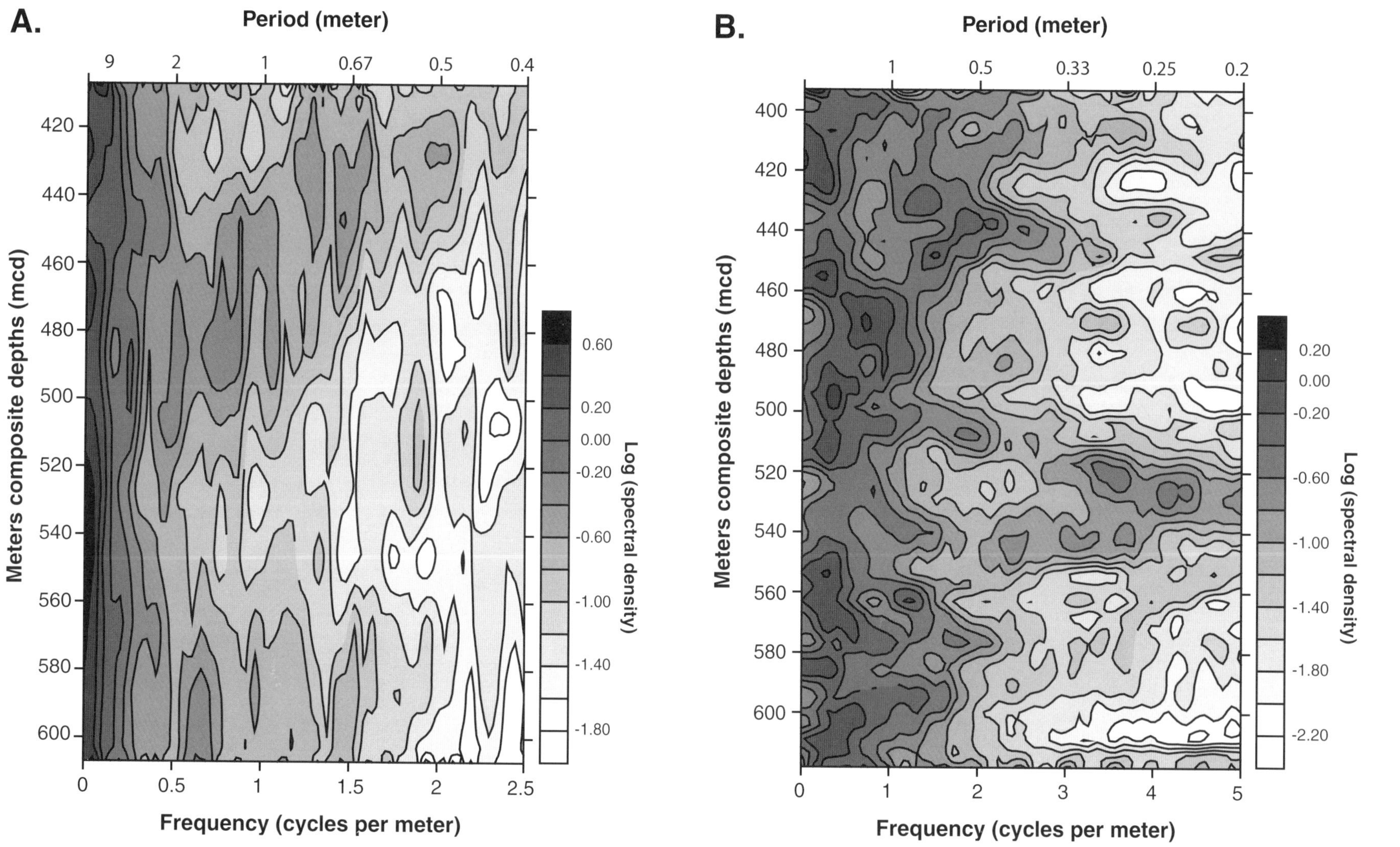

Figure 7. Evolutive spectral analysis in the depth domain for the Fe intensities. We used the Blackman-Tukey method with (A) a window length of 40 m and a step size of 20 m, and (B) a window length of 10 m and a step size of 5 m. Changes in sedimentation rates produces patches rather than continuous bands across the spectra. Note the spectacular shift in spectral peaks below 544 mcd and 510 mcd.

Fe record. Therefore we believe that the major oscillations in cycle period are not data processing artifacts but reflect some combination of: (1) a constant climate-driven cycle that is modulated by changes in sedimentation rate, (2) long period variations in the dominance of different climate cycles, and (3) variations in sedimentation rate unrelated to orbital periodicity.

Age interpretation of sedimentary cycles

To interpret the record in terms of time, we made a first-order estimate of the duration of the dominant cycles using published age determinations for the magnetochrons. We also looked for frequency modulation of the various spectral peaks in the depth domain to try and verify the presence of the expected modulation of orbital precession by the long and short eccentricity cycles. This last strategy was recently employed by Olsen and Kent (1996) and Pälike et al. (2001) who used the 400 k.y. modulation of precession as a tuning target.

We converted Milankovitch cyclicity to elapsed time by applying the modern average obliquity period of 41 k.y. and modern average precession period of 21 k.y. (Herbert et al., 1995). Herbert et al. (1995) have shown that 21 k.y. is a good estimate for the average of the modern 23.7 and 19.4 k.y. precession periods for cycle counts of more than ~7 cycles. The actual orbital period may be slightly less than modern. Berger et al. (1992) have estimated that the obliquity and precessional bands should be shorter than their modern values by 2.8% and 5.7%, respectively at 100 m.y. ago. Therefore, we would expect the central obliquity period to be ~40.4 k.y. and the central precession cycle to have a period of ~20.7 k.y. at 55 Ma. However, in view of the relatively small decrease predicted for orbital periods in the early Paleogene, the error due to assuming a constant (modern) rate of tidal friction is not likely to be significant. Olsen and Kent (1996) analyzed cycle periods in Triassic lake deposits and found a close correspondence between the ratios of the modern 400 k.y. long eccentricity cycle and precession with those estimated for the Triassic. The 400 k.y. cycle is generally estimated to be a stable cycle on long time scales (e.g., Laskar, 1990, 1999; Berger et al., 1992) so the results of Olsen and Kent (1996) suggest that the various bands of the precession cycle have not changed appreciably in the last 250 m.y. Accordingly, we have used the modern periods of the orbital cycles in calculating the duration of events between the late Paleocene and the early Eocene.

ASTRONOMICAL CALIBRATION OF THE MAGNETIC POLARITY TIME SCALE

Estimated durations of polarity chrons C25 and C26

The power spectrum of polarity zones C25 and C26 (529–622 mcd) shows a set of several peaks; the major ones (above confidence level) occur between 0.8 and 1.2 cycles/meter (Fig. 8B). An estimated duration of 4.53 m.y. (Cande and Kent, 1995) for the 93-m-thick section covering magnetochrons C25r through C26r indicates that an 0.9-m cyclicity would have an average period of ~43.9 k.y. and could correspond to variations in the Earth's obliquity with a period of ~41 k.y. This is supported by the presence of additional spectral peaks between 1.4 and 1.6 cycles/meter (70–55 cm), which can be interpreted as variations linked to climatic precession. Similarly, the longer eccentricity frequency of ~400 k.y. seems to be clearly present at ~9- to 10-m cyclicity, and a weak spectral peak at ~2.5 m may correspond to the shorter eccentricity peak (~100 k.y.). We applied bandpass filters to isolate the components of cyclicity with wavelength 0.7–1.4 m (1.43–0.7 cycles/m), 50–80 cm (3.6–1.5 cycles/m), 1.5–6.7 m (0.67–0.15 cycles/m), and 6.7–20 m (0.15–0.05 cycles/m) (Fig. 8A). Amplitude maxima in the 60-cm cycles (21 k.y.) seem to correspond to amplitude minima of the 2.5-m filter (~100 k.y.?), especially between 569 and 600 mcd, but generally show a weak pattern indicating that the 100 k.y. cyclicity is less pronounced in this interval. The amplitude modulation of the 60-cm filter is well pronounced and fits to the 10-m filter. The 1-m filter (41 k.y.) shows some amplitude modulation but does not fit to either 2.5 m or 10 m cyclicity.

The 1-m filter displays 26 cycles within polarity zone C25r, whereas the 60-cm filter suggests 51 cycles within C25r. Therefore, we assigned polarity zone C25r a duration of either 26 obliquity cycles of 41 k.y. (1.07 m.y.) or 51 precession cycles of 21 k.y. (1.07 m.y.). As both the spectral analysis and the filters are applied to the Fe data in the depth domain, any major variation in sedimentation rate would distort these interpretations. However, the value that we calculated is only slightly less than the estimate derived from a spreading rate model by Cande and Kent (1995) (1.163 m.y.).

The interpretation of cyclicity in underlying polarity zone C26 is more problematic. Only a single hole spans polarity zone C26, and correlation of the Fe intensity data with the natural gamma-ray log of the continuous downhole measurements indicates that there are only a few minor gaps in the recovery (Fig. 8C). An unconformity at ~564 mcd in the lower part of polarity zone C26n had been tentatively inferred from biostratigraphic markers (Shipboard Scientific Party, 1998), even though the relative thickness of polarity zone C26n relative to adjacent C25r and C26r resembles the estimated magnetic anomaly pattern (Fig. 8A).

The biostratigraphic data do not provide a firm constraint on the completeness of the record in 1051. It is likely that there is a hiatus in C26r. The absence of calcareous nannofossil zone CP6 suggests an unconformity in C25r at ~545 mbsf. The interval between zone CP4–7 is condensed relative to zones CP3 and CP9a. The relatively condensed nature of the nannofossil zones suggests that the interval down to ~560 mbsf either contains one or more unconformities or the whole section accumulated at a slower sedimentation rate than for the lower Paleocene or the lower Eocene sequence. Therefore, our estimated durations for chrons C26r and C26n should be considered as minimum values (Fig. 8C, Table 2).

We combined the cycle analysis of narrow C26n and the wider C26r into a single polarity zone C26, due to the un-

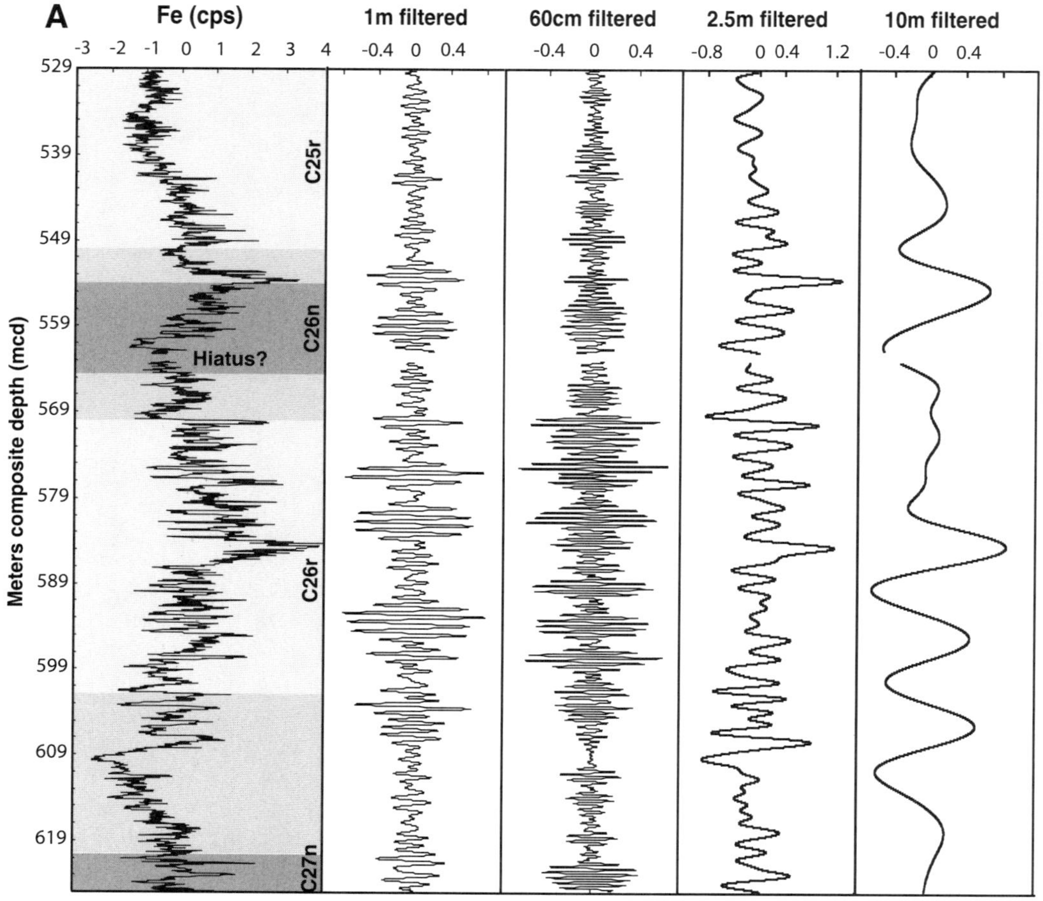

529-622 mcd (top C25r - base C26r)

4.529 myr (Cande and Kent, 1995)

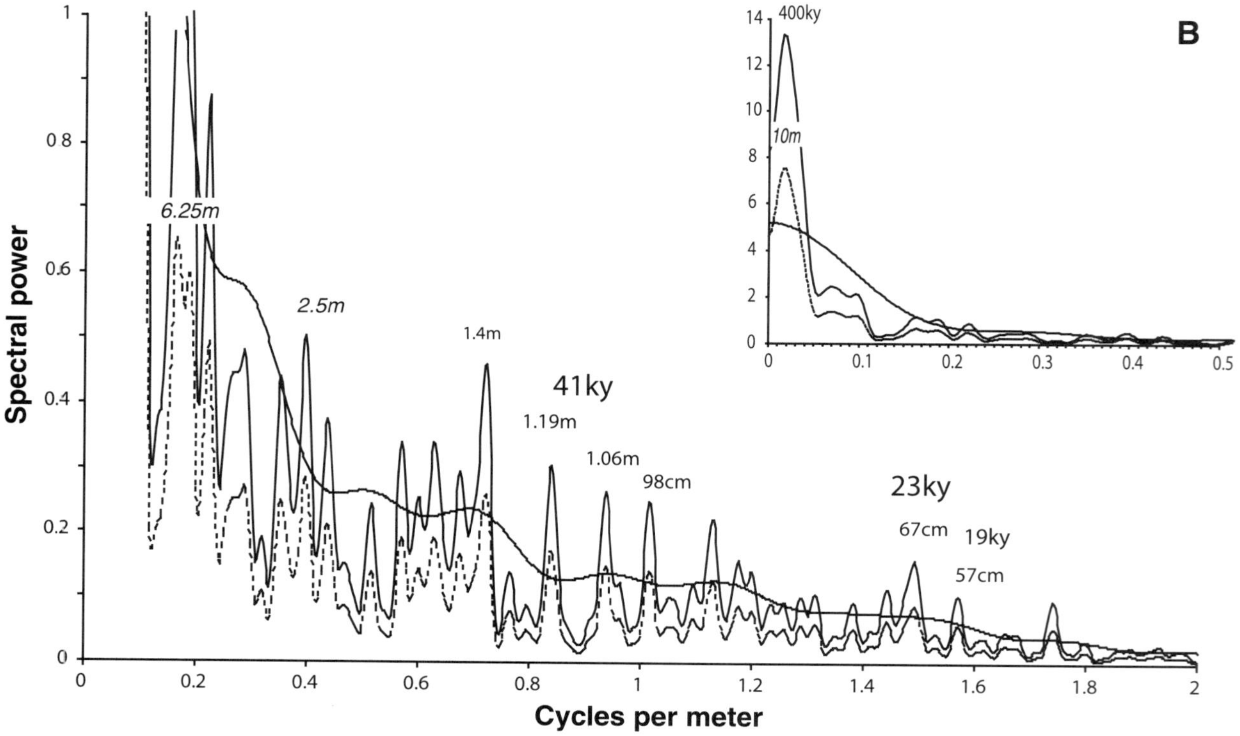

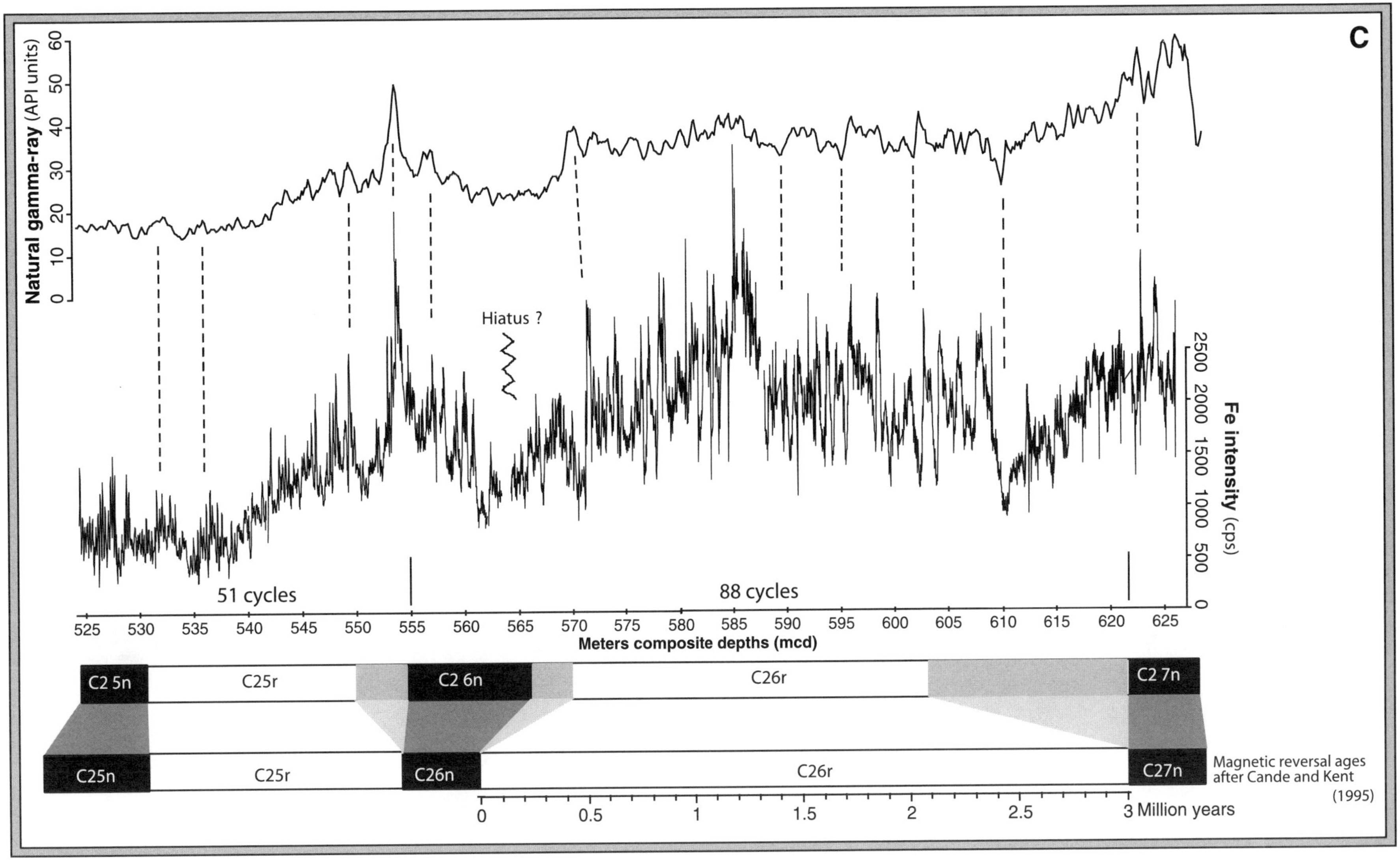

Figure 8 (here and on facing page). A: Magnetostratigraphy, Fe intensity curves for magnetochrons C25r through C27n, and bandpass filters for the Fe intensity data in the depth domain. (Black—normal; white—reversed polarity zone; gray—error bar; cps—counts per second.) B: Power spectrum (thin solid line) (at 80%, bandwidth 0.031) covering magnetochrons C25r through C26r (529–622 mcd). Broken line—lower error limit (at 80%, bandwidth 0.31); thick solid line—background noise. C: Composite of Fe intensity curves, magnetostratigraphy, natural gamma-ray downhole log, and cycle counts for magnetochrons C25n through C26r. (Black—normal; white—reverse; gray—error bar.)

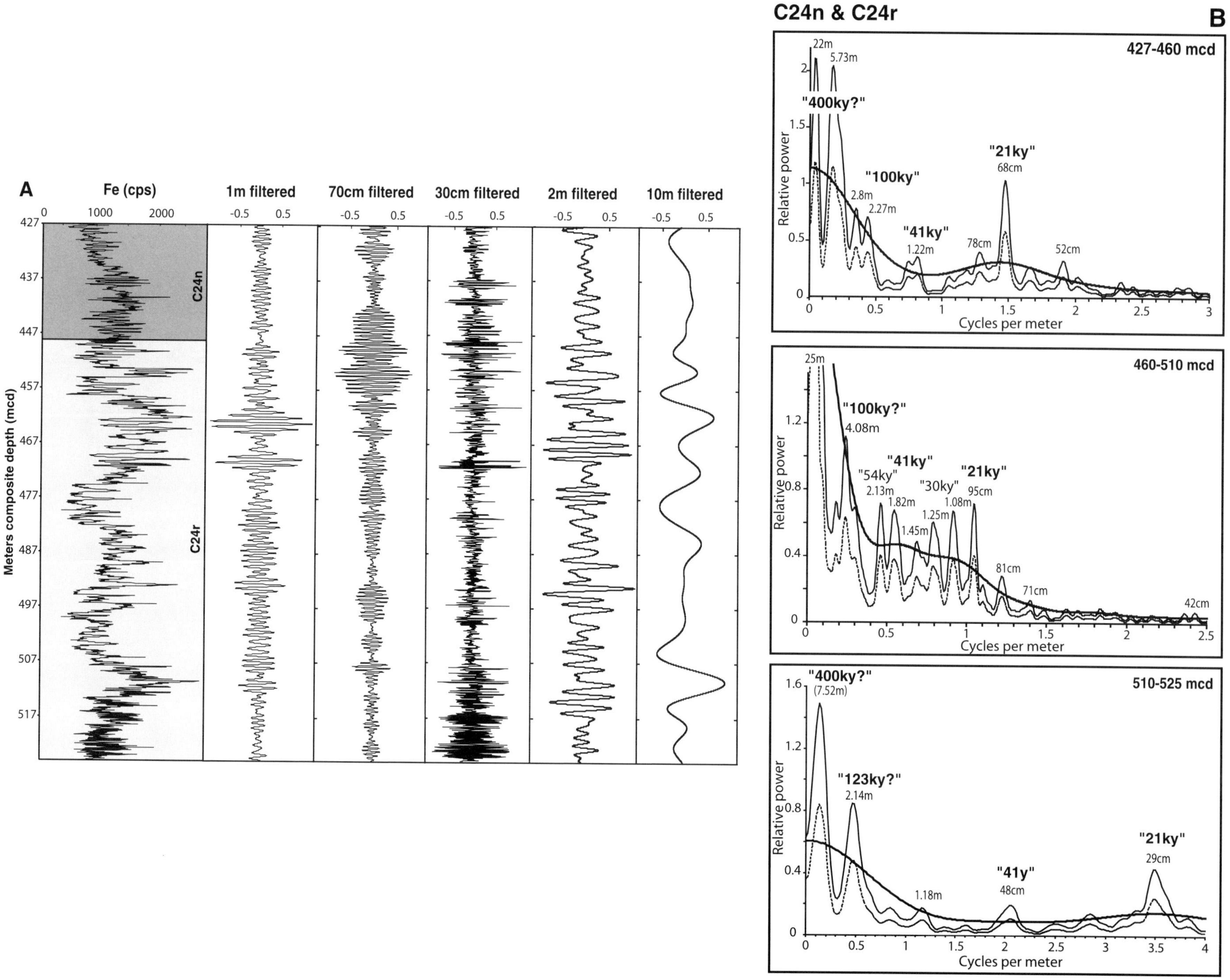

A
Fe (cps)
1m filtered
70cm filtered
30cm filtered
2m filtered
10m filtered
C24n
C24r
Meters composite depth (mcd)
C24n & C24r
B
427-460 mcd
460-510 mcd
510-525 mcd
"400ky?"
"100ky"
"41ky"
"21ky"
"100ky?"
"54ky"
"30ky"
"123ky?"
"41y"
Relative power
Cycles per meter

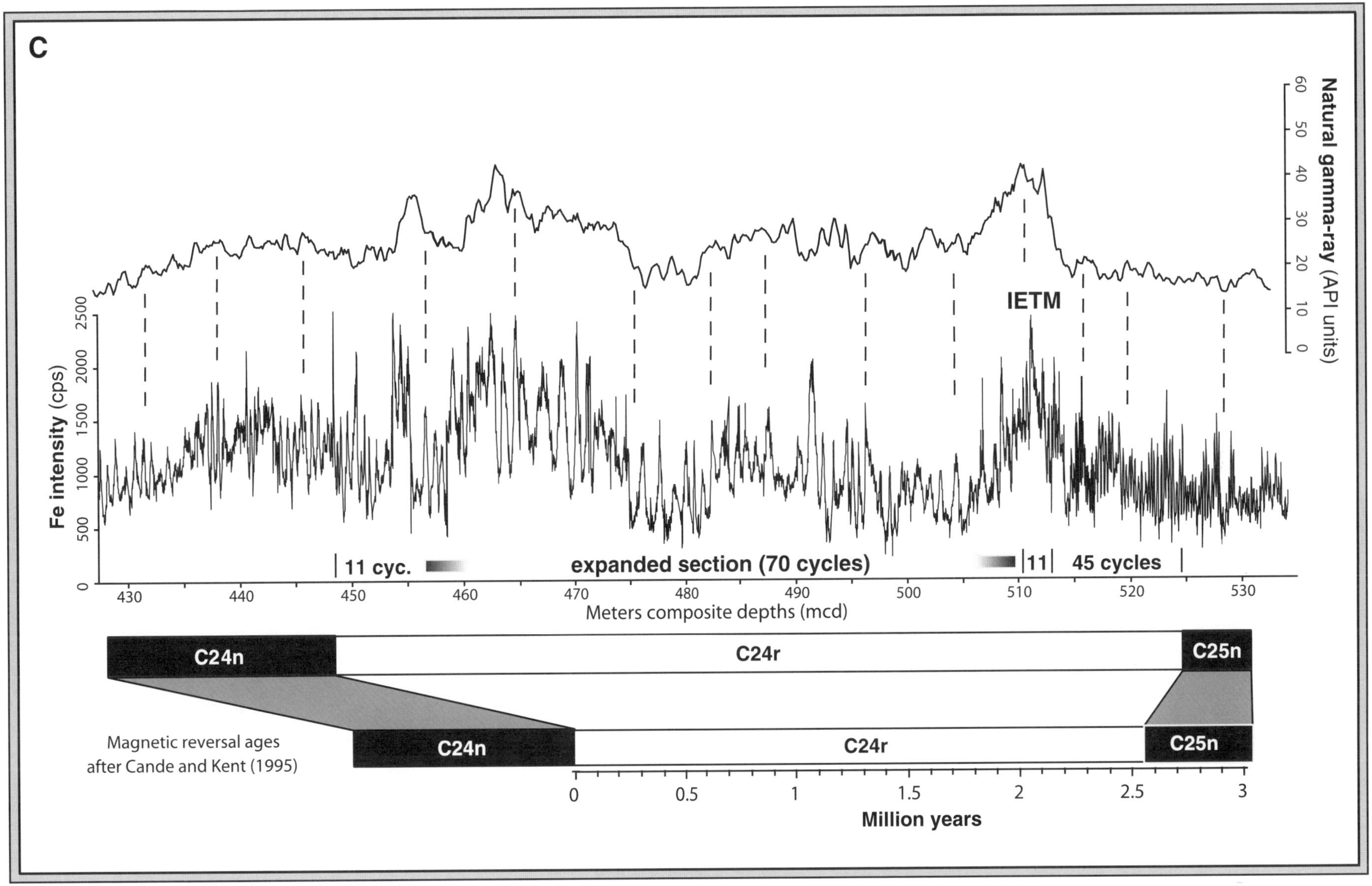

Figure 9 (here and on facing page). A: Magnetostratigraphy, Fe intensity curves for magnetochrons C24n and 24r and bandpass filters for the Fe intensity data in the depth domain (cps—counts per second). B: Power spectra for three selected windows (representing different amounts of sediment accumulation) covering magnetochrons C24n and C24r (427–460 mcd, 460–510 mcd, 510–525 mcd). Power spectrum (thin solid line) at 80%, bandwidth 0.091 (427–460 mcd), 0.06 (460–510 mcd), and 0.2 (510–525 mcd); broken line—lower error limit (at 80%, bandwidth 0.91 (427–460 mcd), 0.6 (460–510 mcd), and 2.02 (510–525 mcd); thick solid line—background noise. C: Composite of Fe intensity curves, magnetostratigraphy, natural gamma-ray downhole log, and cycle counts for chron C24r. Cycle counts for C24r portion below the Initial Eocene Thermal Maximum (IETM) are from Norris and Röhl (1999); cycle counts for the IETM portion of C24r are from Röhl et al. (2000).

certainty in delimiting the reversal boundary between them. The basal reversal boundary of C26r to C27n was assigned just above the level of the first definite normal-polarity minicore, following the interpretation of Ogg and Bardot (2001). Altogether we counted 88 cycles in the raw data for the C26n+r interval, the 1 m bandpass filter picked 72 cycles. Minor changes in sedimentation rate and therefore cycle thickness are probably responsible for the difference in estimated cycle numbers. The cycle counting indicates that the ~1 m cycles must be obliquity: If they were precessional it would require about half of chron C26 to be missing in Site 1051 and imply the existence of one or more substantial unconformities in the section.

Duration of polarity chrons C24n and C24r

The cyclicity for the interval of polarity zone C24n and C24r (427–525 mcd) is very pronounced (Figs. 9A to 9C). This magnetochron is characterized by changes in cycle thickness through time (Fig. 9A). If we compare the Fe intensity data with the natural gamma-ray log of the continuous downhole measurements it seems that the complete interval was recovered. Röhl et al. (2000) compared the cycle record at Site 1051 with that at ODP Site 690 and showed that there is a minor unconformity spanning ~50 k.y. (~2.5 precession cycles) at the base of the IETM; the section in Site 1051 includes the base of polarity zone C24r and the top of zone C24n (Fig. 9C).

There are two pronounced shifts in the typical thickness of the Fe oscillations within polarity zone C24r (Figs. 9A and 9C). The lower 15 m, including the IETM interval, contains closely spaced peaks in a pattern similar to the precession-dominated spectra of underlying polarity zone C25n. Because of the varying cycle thickness (which we hypothesize results from changes in sedimentation rate) we defined three subunits: 427–460 mcd, 460–510 mcd, and 510–525 mcd, for each of which we conducted a spectral analysis in the depth domain (Fig. 9B). The wavelength of the primary high-frequency cycle increases from 29 cm below the IETM, to 95 cm above it and up to 460 mcd, then slightly decreases to 68 cm in the 427–460 mcd interval (Fig. 9B).

The power spectrum of the 510–525 mcd interval shows a set of several peaks, the major ones (above confidence level) occur at 3.5 cycles/meter (29 cm), 2.05 cycles/meter (48 cm), 0.5 cycles/meter (2.14 m), and 0.14 cycles/meter (~7.5 m) (Fig. 9B, lower panel). Direct counting of the pronounced ~29 cm oscillations in the Fe record have yielded estimates of 45 cycles from the base of C24r to the onset of the IETM (Norris and Röhl, 1999) and 11 cycles covering the IETM (Röhl et al., 2000). Norris and Röhl (1999) interpreted the high-frequency (~23–29 cm) cycle between the base of C25n and the base of the CIE as precession based upon (1) a close correspondence between the duration of C25n estimated from cycle counts and the geomagnetic time scale; and (2) the similarity in peak height and spacing between the modern precession bands and those produced from spectral analysis of Fe cycles in the core. Inspection of the Fe record between 520 and 530 mcd (Fig. 5) also reveals an impressive modulation of the high-frequency cycle by low-frequency cycles in a 5:1 ratio, as is expected for modulation of the precession band by the short eccentricity cycle. We interpret the cycle with ~1.5 m wavelength to be the short eccentricity band while the 29 cm and 20-cm cycles are probably the precession bands (e.g., Figure 6, 514–544 mcd). The longer eccentricity cycle (400 k.y.) might be also present, but the window length is too short to detect cycles of this wavelength.

The power spectrum for the overlying 460–510 mcd interval shows several spectral peaks; the most prominent is the peak at 1.05 cycles/meter (95 cm) (Fig. 9B, central panel). We can make a first-order estimate of the cycle period by estimating the average number of these cycles within the various magnetochrons. For example, there is ~60 m of section between the base of C24n [nominally 53.347 Ma in the Cande and Kent (1995) time scale] and the top of the IETM (~54.76 Ma, Röhl et al., 2000). Hence the 60 m of the upper part of C24r in Site 1051 represents some 1.4 m.y. and the cycles with ~1 m wavelength would have an average duration of ~23.6 k.y. In places, the average cycle wavelength is somewhat shorter (~80 cm) and has an average cycle duration of ~18.9 k.y. These estimates are compatible with the 19.4, and 23.7 k.y. bands in modern precession. The ~1-m cycle cannot be the obliquity band, since the ~60–70 cycles above the top of the IETM and below the base of C24n would require the upper part of C24r to be at least 2460 k.y. in duration and the entire duration of the revered polarity zone to be nearly 3.5 m.y.—nearly 1 m.y. longer than that estimated by Cande and Kent (1995).

The power spectrum for the overlying 427–460 mcd interval exhibits a set of several peaks that can be assigned to Milankovitch frequencies: At 1.5 cycles/meter (68 cm, 21 k.y.), 0.8 cycles/meter (1.22 m, 41 k.y.), 0.3–0.4 cycles/meter (2.2–2.8 m, ~100 k.y.?), and 0.2 cycles/meter (~5.7 m, 400 k.y.) (Fig. 9B, upper panel).

We applied bandpass filters to the Fe record of magnetochron C24r (427–525 mcd) to isolate the components of cyclicity with wavelength 0.8–1.33 m (1.23–0.8 cycles/meter), 60–83 cm (1.66–1.23 cycles/meter), 20–63 cm (5–1.62 cycles/meter), 1.33–4 m (0.75–0.25 cycles/meter), and 6.7–20 m (0.148–0.05 cycles/meter) (Fig. 9A). The bandpass filtering in the depth domain means that there was no correction for variation in sedimentation rate applied. Therefore, the different filters cannot be directly transferred into the Milankovitch frequencies valid throughout the interval. Nevertheless, it is quite probable that the 70-cm filter in the upper part (C24n and the upper ~7 m of C24r), the 1-m filter in the central (expanded) part, and the 30-cm filter in the lowermost part, picked the 21 k.y. cyclicity. Because of the lack of any compensation for the obvious changes in sedimentation rates a detailed interpretation of amplitude modulations for the different filters and a correlation of amplitude maxima or minima between different filters is not possible.

The filters were also used to verify our cycle counts. The interval between 455 mcd and the top of the IETM contains 70 main oscillations, which display an approximate one-meter wavelength through the central part. At ~455 mcd is a return to closer-spaced peaks (e.g., shift in spectra peaks exhibited between 450 and 460 mcd in Figures 7 and 9B), of which 11 are at the top of polarity zone C24r (between 455 mcd and 448 mcd). Accordingly, the duration of polarity chron C24r is equivalent to 137 precession cycles or 2.877 m.y. (Fig. 9C). We counted 30 variations in the 70-cm filter for polarity zone C24n, assuming 41 k.y. duration for these 70-cm intervals. The count result into a 1.23 m.y. duration for this zone.

There are large differences in the number of cycles counted directly from the raw Fe record versus the counts from the filtered record, particularly in the interval between the top of the CIE and the base of C24n. Use of the filtered record is the most conservative method (since it implies relatively modest changes in sedimentation rate) whereas the direct counts allow for more variation in sedimentation rate, including relatively abrupt decreases in sedimentation rate within sequences that otherwise were deposited at relatively fast rates. There are some partial or truncated cycles in the interval where mostly ~1-m cycles prevail, which are not picked by the 1 m filter. The interval between the base of C24n (448 mcd) down to the top of the IETM at 508 mcd is represented by 81 cycles (direct counting) or 59 cycles (counting of filtered records). We have previously shown that the IETM lasted about ~220 k.y. and the interval between the start of the IETM and the top of C25n was ~945 k.y. in duration. Accordingly the total duration of C24r is between 2414 k.y. (from the filtered records) and 2877 k.y. (from direct counts of Fe cycles).

We examined two alternative interpretations for the cyclostratigraphy of the 460- to 510-mcd interval ("expanded section," Figure 9C). The first hypothesis is that the interval of the relatively monotonous ~95-cm-wavelength oscillations is dominated by obliquity with a partial superposition of precession influences. The second hypothesis is that this interval represents an expanded section of precession-dominated oscillations at more rapid sediment accumulation rates. The "obliquity hypothesis" assumes that a relatively steady sediment accumulation rate was maintained during and following the IETM, and would imply that the total duration of polarity chron C24r was ~3.5 m.y. (~54 obliquity cycles plus 65 precession cycles).

In contrast, the "precession hypothesis" (137 cycles in the raw Fe data) yields a total duration of 2.877 m.y. for chron C24r. We consider this "precession hypothesis" to be the best interpretation of the cyclostratigraphy for the following reasons:

1. The computed duration (2.877 m.y.) is only 11% longer than the 2.56 m.y. estimate by Cande and Kent (1995). The "obliquity hypothesis" would have implied an at least 35% deviation, which is also incompatible with our estimates of the age deviation of adjacent chrons C25 and C23 from the Cande and Kent (1995) time scale.

2. Counts of precession cycles in color logs from Hole 1051A resulted in a very similar estimate of the duration of 2.856 m.y. ±0.441 m.y. (Cramer, 2001).

3. The implied increase in sedimentation rate after the IETM is consistent with seismic and biostratigraphic interpretations of an enhanced downslope drift of sediment. Deposits of uppermost Paleocene on the more landward slope (e.g., Site 1052) underwent partial exhumation and/or slumping during and after the IETM episode (Katz et al., 1999, 2001).

4. Tuning the XRF Fe signal to a fixed precession (based on the direct counts in the raw Fe record) beat enables resolution of eccentricity components. The resulting age model for chron C24r is shown in Figure 10A, with the implied shifts in mean sedimentation rates. The raw Fe data are shown to the left and the right four columns show filters that extract those variations in the data that fall within astronomical frequency bands for obliquity (41 k.y.), climatic precession (~21 k.y.), and the short and long eccentricity (~100 k.y. and ~400 k.y.). The filter for the climatic precession component shows a strong amplitude modulation that covaries with the short eccentricity orbital component (~100 k.y.) in sediments older than ~54.6 m.y. In the record younger than 54.6 m.y. the maxima in amplitude modulation of the proposed 21-kyr cycle does not always fit to the maxima of the 100-kyr cycle: Sometimes these maxima correlate to the minima of the shorter eccentricity cycles. The 100 k.y. cycle may be less developed in the time interval younger ~54.6 m.y., which would fit to the relative weak 100 k.y. spectral peak in the spectrum (Fig. 10B). The tuned Fe data also show a direct contribution of variability at long eccentricity frequencies; the ~400-kyr cycle is rather obvious between 55.5 and 53.5 m.y. (Fig. 10A). A very pronounced 400-kyr cycle (~55.1–54.7 m.y.) includes the IETM. The CIE occurred ~150 k.y. after the 400 k.y. cycle minimum in agreement with the results of Cramer and Wright (2001).

Estimated durations of polarity chrons C22r through C23r

The cyclicity in the interval of polarity zones C22r through C23r (393–428 mcd) is very well pronounced. As our data for these polarity zones are derived only for a single hole, our estimated values of the duration for chrons C22r to C23r should be considered as minimum values. The correlation of the Fe intensity data with the natural gamma-ray log of the downhole measurements indicates that there are only a few minor gaps in the recovery (Fig. 11C). The sedimentation rates clearly slow down above C24r and the record exhibits both higher-frequency cycles and longer cycles that nicely modulated the shorter variations (Figs. 11A to 11C).

It is difficult to determine the orbital periodicity associated with the section above C24n. The resolution of the foraminifer and nannofossil biostratigraphy is not particularly good above C24n owing to poor preservation and the relatively widely spaced datum levels (Shipboard Scientific Party, 1998). The cycle wavelength is typically only 60–80 cm in C22r, C23n, C23r and C24n. An estimated duration of 3.622 m.y. (Cande and Kent, 1995) for the 56 m-thick section covering magnetochrons C22r through C24n indicates that these cyclicities would have average durations of 39–52 k.y. The cycles do not clearly correspond to known orbital periods except possibly the 41 k.y. obliquity cycle.

Again we examined two alternative interpretations for the cyclostratigraphy of the 393- to 428-mcd interval (Figs. 11A and

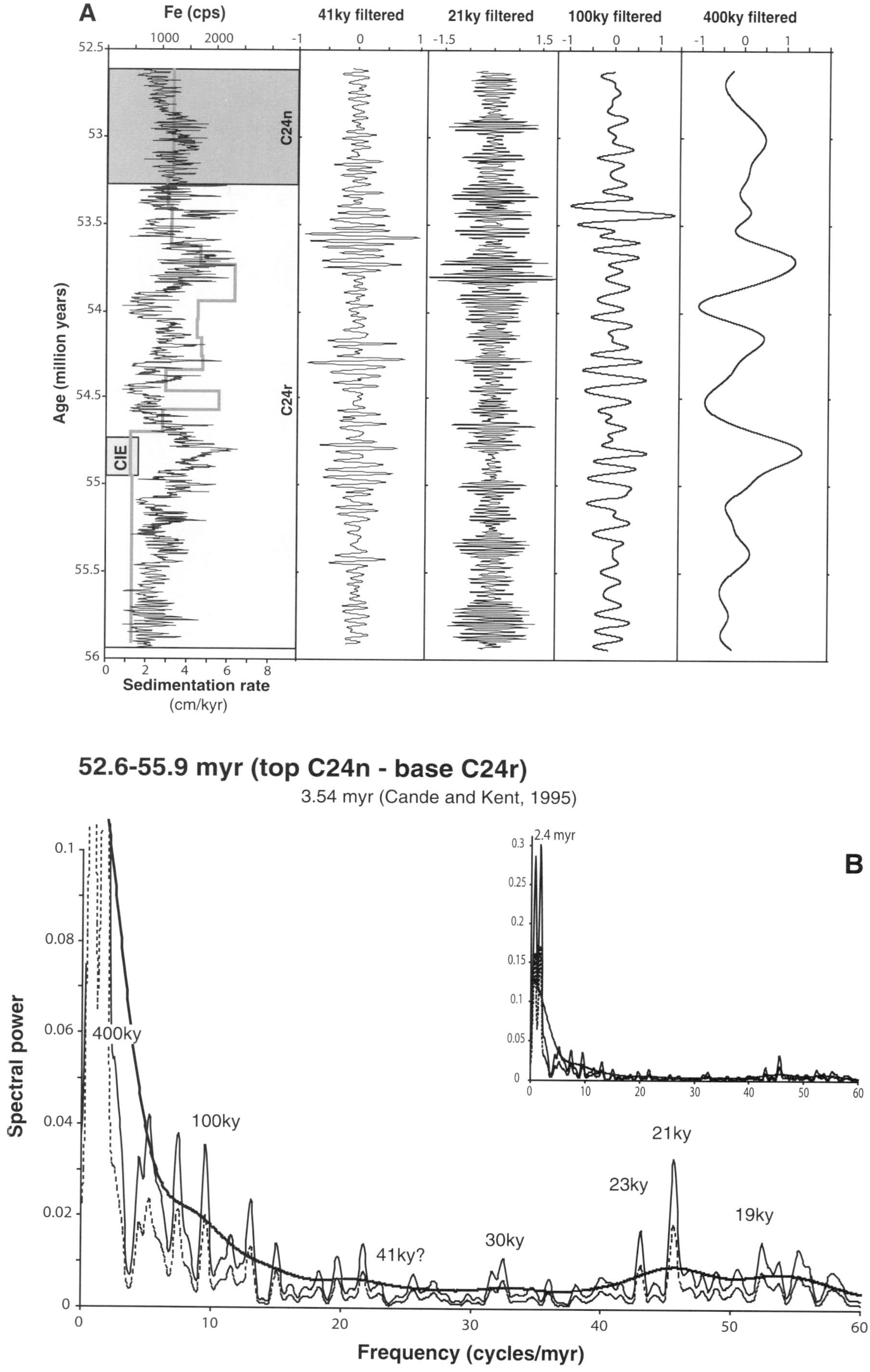
A
Fe (cps)
1000
2000
41ky filtered
21ky filtered
100ky filtered
400ky filtered
Age (million years)
C24n
C24r
CIE
Sedimentation rate
(cm/kyr)
52.6-55.9 myr (top C24n - base C24r)
3.54 myr (Cande and Kent, 1995)
B
2.4 myr
Spectral power
400ky
100ky
41ky?
30ky
23ky
21ky
19ky
Frequency (cycles/myr)

11C). The first hypothesis is that the interval of the relatively monotonous 60- to 80-cm-wavelength oscillations is dominated by obliquity with a partial superposition of precession influences. The "obliquity hypothesis" assumes that a relatively steady sediment accumulation rate was maintained during polarity zones C23r, C23n, and C22r, and would imply that the durations of polarity chrons are quite similar to the durations estimated by Cande and Kent (1995) (Table 2). Alternatively, these magnetochrons may have been truncated by erosion or nondeposition of varying duration that led to overestimates of the cycle duration. Indeed, if the 60-cm to 80-cm cycle corresponds to the precession band, then there is likely to be as much as a third to half of the interval between C22r and C24n missing in unrecognized unconformities.

To test which of the hypotheses is the best interpretation of the cyclostratigraphy we investigated the data in more detail. The power spectrum of this interval shows a set of several peaks, with the major ones occur between 1.2 and 1.5 cycles/meter (82 cm, 75 cm, and 67 cm) (Fig. 11B), but others at 2.35 cycles/meter (0.43 m), 0.6 cycles/meter (1.67 m), and 0.4 cycles/meter (2.5 m). An estimated duration of 2.65 m.y. (Cande and Kent, 1995) for the 35-m-thick section covering magnetochrons C22r to C23r indicates that the 75-cm cyclicity with an average duration of ~56.8 k.y. could correspond to variations in the Earth's obliquity with a period of ~41 k.y. If we assume the record is complete, this interpretation is supported by the presence of an additional spectral peak at 2.35 cycles/meter (42.5-cm wavelength with an average duration of ~31.2 k.y.) interpreted as variation linked to climatic precession (Figs. 11B and 7). Similarly, the longer eccentricity frequency of ~400 k.y. seems to be clearly present at ~9-m cyclicity, the shorter eccentricity peak (~100 k.y.) might be represented by the weak spectral peak at ~0.4 cycles/meter. Altogether the relative position of the spectral peaks broadly agrees with the expected orbital bands. The longer-than-expected period may be due to slight variations in cycle thickness in the C22r through C23r record (Fig. 7) that is also expressed by the occurrence of two spectral peaks at frequencies of 1.2 and 1.35 cycles per meter.

We applied bandpass filters to isolate the components of cyclicity with wavelength 0.5–1.4 m (2–0.7 cycles/meter), 28–67 cm (3.6–1.5 cycles/meter), 1.5–6.7 m (0.67–0.15 cycles/meter), and 6.7–20 m (0.15–0.05 cycles/meter). We also used the resulting filtered data to test for characteristic amplitude modulations. If we assign the 40-cm filter to precession, the 75-cm filter to obliquity, and the 2.5-m filter to eccentricity, we obtain a typical relationship: Amplitude maxima in the 40-cm cycles ("21 k.y.") correspond to amplitude maxima of the 2.5-m filter (~"100 k.y.") in many cases (e.g., at 395, 397, 404, 408, 415, 417 mcd). At 412, 420, 423 mcd, for instance, amplitude maxima in the 40 cm filter coincide with minima of the 2.5-m filter, indicating that the 100 k.y. cyclicity is less pronounced in this interval or that slight changes in sedimentation rates occurred. The amplitude of the 75-cm filter is not modulated by neither the 2.5-m nor the 10-m filters indicating that the 75-cm filter indeed is most readily interpreted as obliquity and not precession.

As both the spectral analysis and the filters are applied to the Fe data in the depth domain, any major variation in sedimentation rate would distort these interpretations. Polarity zone C22r spans the equivalent of 22 cycles in the 75-cm filter, C23r spans 13 cycles in the 75-cm filter, C22n 18 cycles in the 75-cm filter (Figs. 11A and 11C, Table 2). An assignment of 41 k.y. for each of the 75-cm filter variations would be consistent with the estimated total duration of 2.65 m.y. for these three chrons (Cande and Kent, 1995). Indeed, our calculations derived from counting variations in the 75-cm filter (0.902 m.y. (C22r), 0.738 m.y. (C23n), 0.533 m.y. (C23r) are very similar to the estimate derived from a spreading rate model by Cande and Kent (1995) [1.064 m.y. (C22r), 0.965 m.y. (C23n), 0.621 m.y. (C23r)].

RELATIVE TIMING OF MAJOR EARLY EOCENE PLANKTONIC FORAMINIFER DATUMS

The siliceous nannofossil chalk from Holes 1051A and 1051B contains a diverse, subtropical planktonic foraminifer assemblage. The pronounced cyclicity in sediment color and geochemistry is reflected in percentage of carbonate and therefore in foraminifer preservation. Preservation is highly variable, generally poor to moderate in light-colored carbonate-rich intervals, and moderate to good in dark greenish gray, clay-rich intervals. Some intervals, most notably within the IETM are well indurated and yielded poorly preserved foraminifera.

Reworking of Paleocene foraminifera is endemic through much of the lower Eocene section. For example, *Globanomalina pseudomenardii,* whose last occurrence marks the top of biozone P4 (Table 1), is a frequent to common constituent of assemblages as high as biozone P6b. Occasional specimens of *Morozovella occlusa* (nominal last occurrence at the end of P5) are present in sediments referable to Biozone P7. Calcareous nannofossils display similar evidence for extensive reworking (M.-P. Aubry, 2001, personal commun.).

This extensive reworking poses problems for interpretation of the foraminiferal biozonation at Site 1051. The accurate placement of lowest occurrences (LOs) should be unaffected by reworking, but highest occurrences (HOs) may be placed too high in the sequence as illustrated by the unusually young HOs for *G. pseudomenardii* and *M. occlusa.* We have attempted to partly rectify this problem by picking HOs based on the last common occurrence, a strategy that provides datums for *M. occlusa* and *Morozovella velascoensis* that are broadly in accordance with previous records. However, the common occurrence

Figure 10. A: Magnetostratigraphy, Fe intensity curves and sedimentation rates for magnetochrons C24r after astronomical tuning, and bandpass filters for the Fe intensity data for the main orbital frequencies. CIE—carbon isotope excursion; cps—counts per second. B: Power spectrum (thin solid line) (at 80%, bandwidth 0.9), lower error limit (at 80%, bandwidth 9.04), and background noise (thick solid line) covering magnetochrons C24n and C24r (427–525 mcd) after astronomical tuning.

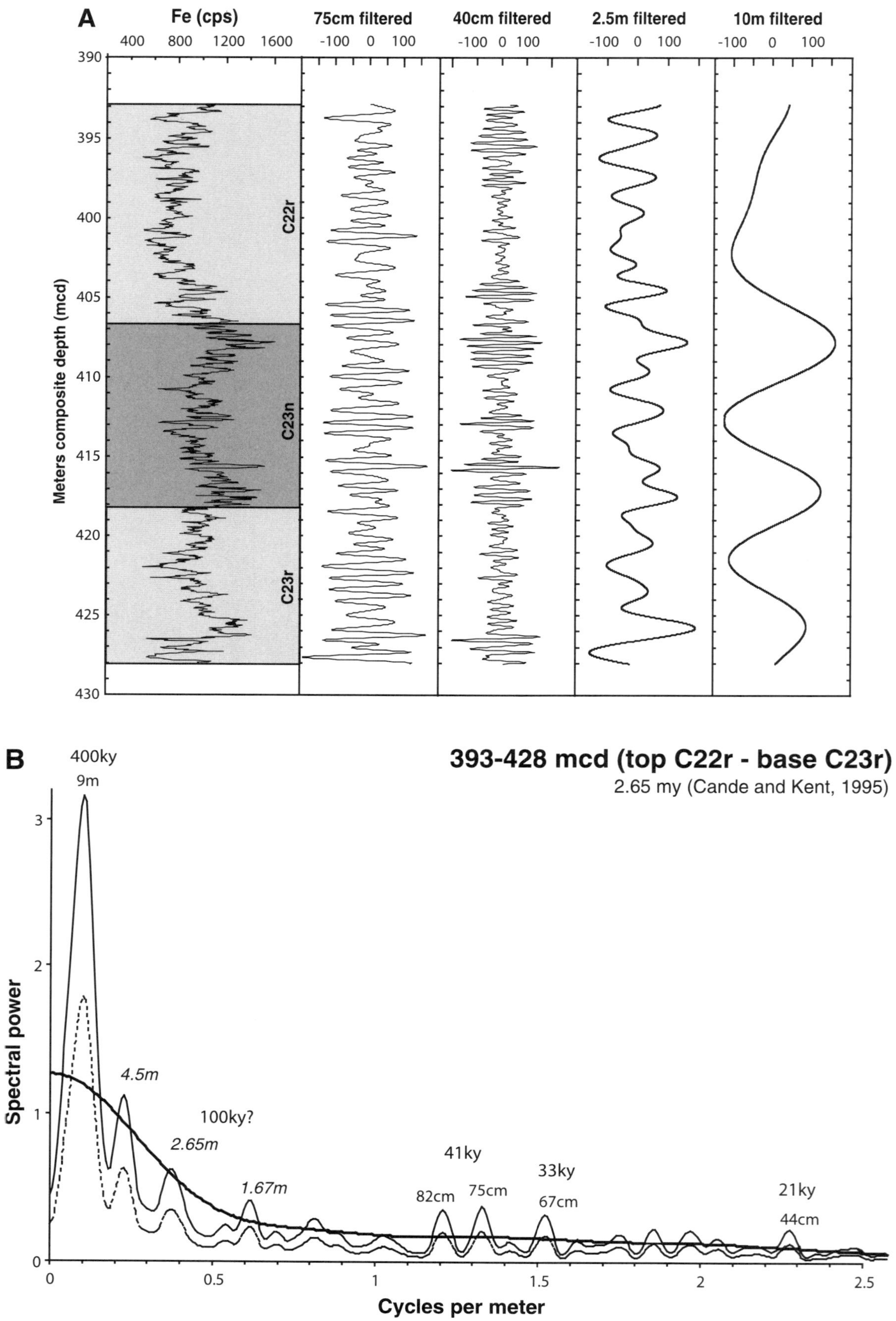
A
Fe (cps)
400 800 1200 1600
75cm filtered
-100 0 100
40cm filtered
-100 0 100
2.5m filtered
-100 0 100
10m filtered
-100 0 100
Meters composite depth (mcd)
390
395
400
405
410
415
420
425
430
C22r
C23n
C23r
B
393-428 mcd (top C22r - base C23r)
2.65 my (Cande and Kent, 1995)
400ky
9m
4.5m
100ky?
2.65m
1.67m
41ky
82cm
75cm
33ky
67cm
21ky
44cm
Spectral power
0
1
2
3
Cycles per meter
0
0.5
1
1.5
2
2.5

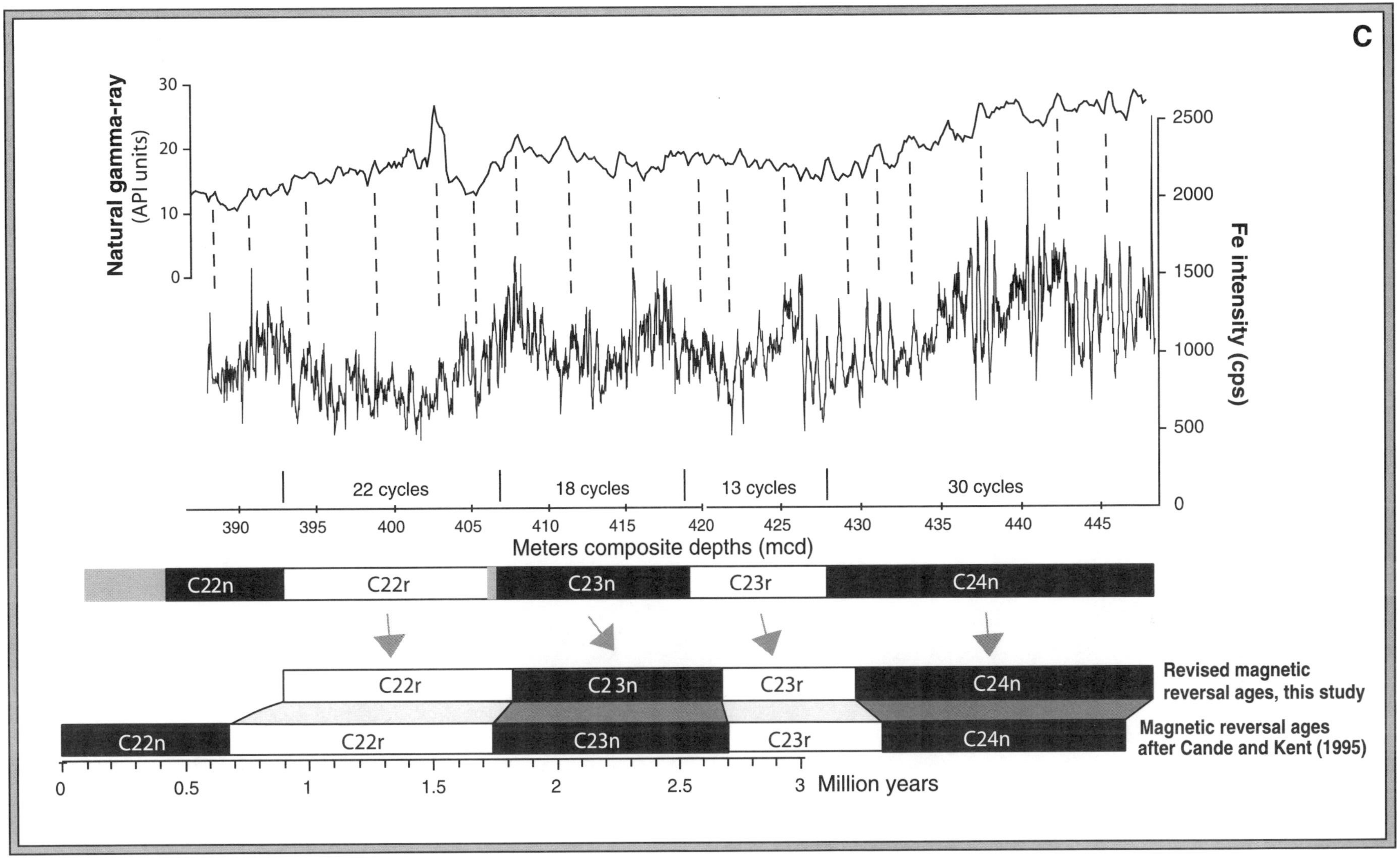

Figure 11 (here and on facing page). A: Magnetostratigraphy, Fe intensity curves, and bandpass filters for magnetochrons C22r through C23r in the depth domain (cps—counts per second). B: Power spectrum (thin solid line) (at 80%, bandwidth 0.085) covering magnetochrons C22r through C23r (393–428 mcd). Broken line—lower error limit (at 80%, bandwidth 0.85); thick solid line—background noise. C: Composite of Fe intensity curves, magnetostratigraphy, natural gamma-ray downhole log, and cycle counts for magnetochrons C22n through C24n.

TABLE 1. REVISED ESTIMATES FOR PLANKTONIC FORAMINIFERA BIOSTRATIGRAPHIC DATUM LEVELS RELATIVE TO THE AGE OF THE INITIAL EOCENE THERMAL MAXIMUM

Datum	Species	Core, section, interval	Depth (mbsf)	Depth range (m)	Depth (mcd)	Age range (Ma)	Mean age (Ma)
		1051A-					
LO	*Morozovella aragonensis*	46X-4, 62 cm to 46X-5, 67 cm	434.19	±0.75	432.18	52.730–52.780	52.760
LO	*Acarinina angulosa*	46X-6, 67 cm to 46X-6, 69 cm	435.01	±0.02	433.00	52.781–52.783	52.782
HO	*Morozovella aequa*	47X-1, 58 cm to 47X-1, 68 cm	438.55	±0.05	436.54	52.889–52.892	52.891
HO	*Globanomalina pseudoimitata*	47X-1, 68 cm to 47X-1, 58 cm	438.55	±0.05	436.54	52.889–52.892	52.891
HO	*Acarinina berggreni*	48X-1, 56 cm to 48X-1, 66 cm	448.13	±0.05	446.12	53.183–53.186	53.184
LO	*Morozovella dolobrata*	48X-1, 66 cm to 48X-2, 68 cm	448.91	±0.75	446.89	53.185–53.230	53.208
LO	*Acarinina quetra*	49X-1, 68 cm to 49X-1, 78 cm	458.50	±0.05	456.49	53.517–53.553	53.520
LO	*Morozovella lensiformis*	50X-2, 66 cm to 50X-4, 66 cm	463.87	±1.50	461.86	53.658–53.700	53.680
HO	*Subbotina triangularis*	52X-7, 32 cm to 52X-7, 42 cm	485.89	±0.05	483.88	54.071–54.075	54.073
HO	*Subbotina velascoensis*	53X-3, 20 cm to 53X-4, 50 cm	490.22	±0.90	488.21	54.150–54.190	54.170
HO	*Morozovella occlusa*	53X-6, 17 cm to 53X-6, 66 cm	494.93	±0.25	492.91	54.262–54.273	54.268
HO	*Morozovella velascoensis*	54X-1, 110 cm to 54X-1, 120 cm	496.89	±0.05	494.88	54.309–54.311	54.310
LO	*Acarinina berggreni*	54X-2, 30 cm to 54X-2, 40 cm	497.58	±0.05	495.57	54.324–54.326	54.333
LO	*Morozovella marginodentata*	54X-2, 110 cm to 54X-2, 120 cm	498.38	±0.05	496.37	54.340–54.344	54.342
LO	*Morozovella gracilis*	54X-5, 94 cm to 54X-5, 84 cm	502.53	±0.05	499.89	54.455–54.464	54.462
		1051B-					
LO	*Chiloguembelina wilcoxensis*	60X-2, 20 cm to 60X-3, 50 cm	512.77	±0.15	512.43	54.933–54.957	54.944
LO	*Pseudohastigerina wilcoxensis*	60X-2, 20 cm to 60X-3, 50 cm	512.77	±0.15	512.43	54.933–54.957	54.944

Note: HO—Highest occurrence; LO—Lowest occurrence.

of *G. pseudomenardii* at several levels in the lower Eocene suggests that the HO of this species cannot be determined reliably. The same is probably true for locally rare species such as *Igorina albeari* and *Acarinina mckannai.* Species with first occurrences in the Eocene are probably not affected by substantial reworking.

We have calibrated the LOs and HOs of planktonic foraminifera in the lower Eocene using our cyclostratigraphy. Ages were determined by taking the age of the onset of the IETM (54.98 Ma) as a reference point and counting precessional cycles with assumed average durations of 21 k.y. The assumption of an average 21 k.y. periodicity for the cycles has an intrinsic error of ~5 k.y. for counts of more than seven cycles owing to the averaging of the 23 and 19 k.y. precession bands (Herbert et al., 1995; Norris and Röhl, 1999; Röhl et al., 2000). Errors accumulate in any cycle count with more cycles included, because of problems associated with the recognition of condensed or incomplete cycles and unconformities.

A major uncertainty is the numerical age of the IETM (e.g., Cramer, 2001). As discussed by Norris and Röhl (1999), our 54.98 Ma estimate is based on a cycle count between the IETM and the top of C25n in ODP Site 1051, calibrated to the 55.9 Ma calculated age for the C25n/C24r boundary. However, the age of the magnetochron boundary was determined using a 55 Ma age of the calcareous nannofossil biozone NP9/10 boundary estimated by Berggren et al. (1995). This boundary is estimated to have occurred ~500 k.y. after the IETM, and thus the 54.98 Ma date cannot be accurate if the NP9/10 boundary is indeed at 55.0 Ma. It is now apparent that the age of the NP9/10 boundary determined by Berggren et al. (1995) is in error owing to unconformities in the sections they utilized (e.g., Norris and Röhl, 1999; Wing et al., 2000). Therefore, the absolute age of the C25n/C24r boundary is probably too young, and the IETM as determined by Norris and Röhl (1999) and Röhl et al. (2000) is likely to be too young by several 100 k.y. Because our calibrated ages are determined relative to magnetostratigraphy in Site 1051, it will be a relatively straightforward exercise to adjust the ages of the datums when a firm absolute chronology is finally put into place.

Wing et al. (2000) used a recent Ar-Ar date on volcanic ashes in the Bighorn basin to estimate the age of the top and base of C24n and work out the age of the IETM. Their age model suggests that the base of C24n.3n is 53.566 Ma, which is ~200 k.y. older than estimated by Cande and Kent (1995). Our cycle count of ~92 precession cycles (counted directly from the Site 1051 Fe record) or 70 cycles (counted from the filtered records) between the base of C24n and the base of the CIE, suggests that it has an age—defined relative to the age of the base of C24n—of 55.498 or 55.036 k.y., respectively. Our estimates straddle the 55.234 Ma age determined for the IETM by Wing et al. (2000). The location of the NP9/10 boundary has not been identified at Site 1051. However the top of the planktonic foraminifer P5 biozone (which is slightly younger than the NP9/10 boundary) is at 494 mcd or ~25 precession cycles above the IETM. Therefore the top of biozone P5 has an age of 54.97–54.51 based on direct Fe cycle counts and counts of filtered 1-m cycles, respectively.

CONCLUSIONS

High-resolution geochemical profiles through upper Paleocene and lower Eocene (representing the Selandian, Thanetian, and Ypresian stages) pelagic sediments from the Blake Nose

TABLE 2. REVISED ESTIMATES OF NUMERICAL AGES OF MAGNETOCHRONS C22 THROUGH C26 AS REPRESENTED AT SITE 1051 IN COMPARISON TO THE NUMERICAL AGES OF CANDE AND KENT (1995)

Chron	Top* (Ma)	Base* (Ma)	Duration* (Ma)	Obliquity cycles	Precession cycles	Duration cycles (ka)	Difference to Cande and Kent (1995) (ka)
22r	49.714	50.778	1.064	22		902	–162
23n	50.778	51.743	0.965	18		738	–227
23r	51.743	52.364	0.621	13		533	–88
24n	52.364	53.347	0.983	30		1230	+247
24r	53.347	55.904	2.557		137	2877	+320
25n	55.904	56.391	0.487		25	525	+38
25r	56.391	57.554	1.163	26	51	1066	–92
26n and r	57.554	60.920	3.366	88		3608	+242

*Cande and Kent (1995).

margin of Florida display regular variations in Fe abundance associated with Milankovitch orbital-climate cycles of obliquity, precession, and eccentricity. The collection of X-ray fluorescence elemental concentration data greatly helps enhancing the signal-to-noise ratio and the hole-to-hole integration of marine sediment cores compared to the use of shipboard physical properties data. The data enabled us to create an astronomically calibrated relative age-scale for the late Paleocene to early Eocene magnetic polarity zones C22 through C26 drilled at Site 1051. Spectral analysis of overlapping windows through this interval indicates that shifting of spectral peaks through time is related to changes in sedimentation rate. The Selandian portion (most of chron C26r) is characterized by moderate sedimentation rates. Lower sedimentation rates occurred in the latest Paleocene (Thanetian, C25r and C25n, lower third of C24r), which is characterized by pronounced high-frequency cycles and amplitude modulation of climatic precession by the short (~100 k.y.) eccentricity cycle.

The sedimentation rates dramatically increased ~200 k.y. after the Initial Eocene thermal maximum (IETM) and remained high for most of chron C24r. In the uppermost portion of C24r a trend toward slower accumulation occurred, which continued in chrons C23 and C22 where both precession and obliquity cycles can be identified.

Our new age model mainly relies on the recognition of precession cycles in the record. Bandpass filters extracted variations in the data that fall within astronomical frequency bands for climatic precession (~21 k.y.), obliquity (41 k.y.), and the short and long eccentricity (~100 k.y. and ~400 k.y.). The Fe data also show a direct contribution of variability at long eccentricity frequencies, and include an obvious ~400-kyr cycle. The filter for the climatic precession component also shows a strong amplitude modulation that covaries with the short eccentricity orbital component (~100 k.y.) in some intervals, namely C25n and the lower portion of C24r.

Estimates for the duration of chrons C26, C24n, C23r and C23n, and C22r are based on the variation in eccentricity, obliquity, and precession. Age estimates for chron C26 are less well constrained due to the inferred presence of an unconformity and the fact that this interval was covered by data from one hole only. Hence our calculated duration for chrons C26r and C26n should be interpreted as a minimum estimate.

Our estimates for the relative duration of magnetochrons C22 to C26 as represented in cores from Site 1051 broadly agree with the durations estimated by Cande and Kent (1995) (Table 2). The relative duration of these chrons is the same in both models with the exception of the relative durations of chrons C24r, C24n and the combined C26n+r. These seem to be longer, by ~320 k.y., ~250 k.y., and ~240 k.y. respectively. We are confident of the new estimates for chron C24, because the precession signal is strong. In upper Paleocene and especially lower Eocene sediments from Site 1051 planktonic foraminiferal datums were recognized and we introduce revised age estimates for several of these.

ACKNOWLEDGMENTS

F.G. Hilgen, E. Thomas, G.P. Weedon, and an unnamed reviewer significantly improved an earlier version of the manuscript. We thank H. Pälike and T. Westerhold for discussions. Funding for this research was provided by the Deutsche Forschungsgemeinschaft (DFG) to U. Röhl, and by JOI-USSAC and the Marine Geology and Geophysics Program at the U.S. National Science Foundation to R.D. Norris and J.G. Ogg. This research used samples and data provided by the Ocean Drilling Program (ODP). The ODP is sponsored by the U.S. National Science Foundation (NSF) and participating countries under management of Joint Oceanographic Institutions (JOI) Inc. The complete data set presented in this paper is available in electronic form in the World Data Center for Marine Environmental Sciences (WDC-MARE) PANGAEA database under www.pangaea.de.

REFERENCES CITED

Alvarez, W., Arthur, M.A., Fischer, A.G., Lowrie, W., Napoleone, G., Premoli Silva, I., and Roggenthen, M.W., 1977, Upper Cretaceous–Paleocene magnetic stratigraphy at Gubbio, Italy, V: Type section for the Late Cretaceous–Paleocene geomagnetic reversal time scale: Geological Society of America Bulletin, v. 88, p. 383–389.

Berger, A., Loutre, M.F., and Laskar, J., 1992, Stability of the astronomical frequencies over the Earth's history for paleoclimate studies: Science, v. 255, p. 560–566.

Berggren, W.A., Kent, D.V., Swisher, C.C., III, and Aubry, M.-P., 1995, A revised Cenozoic geochronology and chronostratigraphy, *in* Berggren, W.A., et al., eds., Geochronology, time scales, and global stratigraphic correlations: A unified temporal framework for a historical geology: SEPM (Society for Sedimentary Geology) Special Publication 54, p. 129–212.

Cande, S.C., and Kent, D.V., 1992, A new geomagnetic polarity time scale for the Late Cretaceous and Cenozoic: Journal of Geophysical Research, v. 97, p. 13917–13951.

Cande, S.C., and Kent, D.V., 1995, Revised calibration of the geomagnetic polarity time scale for the Late Cretaceous and Cenozoic: Journal of Geophysical Research, v. 100, p. 6093–6095.

Cramer, B.S , 2001, Latest Paleocene–earliest Eocene cyclostratigraphy: Using core photographs for reconnaissance geophysical logging: Earth and Planetary Science Letters, v. 186, p. 231–244.

Cramer, B.S., and Wright, J.D., 2001, Bringing Milankovitch to bear in the Paleocene/Eocene carbon isotope excursion: Eos (Transactions, American Geophysical Union), v. 82, no. 47, Fall Meet. Suppl., Abstract PP31B-08, 2001.

Dickens, G.R., O'Neil, J.R., Rea, D.K., and Owen, R.M., 1995, Dissociation of oceanic methane as a cause of the carbon isotope excursion at the end of the Paleocene: Paleoceanography, v. 10, p. 965–971.

Dickens, G.R., Castillo, M.M., and Walker, J.G.C., 1997, A blast of gas in the latest Paleocene: Simulating first-order effects of massive dissociation of oceanic methane hydrate: Geology, v. 25, p. 259–262.

Hallam, A., Hancock, J.M., LaBrecque, J.L., Lowrie, W., and Channell, J.E.T., 1985, Jurassic and Cretaceous geochronology and Jurassic to Palaeogene magnetostratigraphy, *in* Snelling, N.J., ed., The chronology of the geological record: Geological Society [London] Memoir 10, p. 118–140.

Herbert, T.D., D'Hondt, S.L., Premoli-Silva, I., Erba, E., and Fischer, A.G., 1995, Orbital chronology of Cretaceous–early Paleocene marine sediments, *in* Berggren, W.A., et al., eds., Geochronology, time scales, and global stratigraphic correlations: A unified temporal framework for a historical geology: SEPM (Society for Sedimentary Geology) Special Publication 54, p. 81–94.

Hilgen, F.J., 1991, Astronomical calibration of Gauss to Matuyama sapropels in the Mediterranean and implication for the geomagnetic polarity time scale: Earth and Planetary Science Letters, v. 104, p. 226–244.

Hilgen, F.J., Krijgsman, W., Langereis, C.G., Lourens, L.J., Santarelli, A., and Zachariasse, W.J., 1995, Extending the astronomical (polarity) time scale into the Miocene: Earth and Planetary Science Letters, v. 136, p. 495–510.

Jansen, J.H.F, Van der Gaast, S.J., Koster, B., and Vaars, A.J., 1998, CORTEX, a shipboard XRF-scanner for element analyses in split sediment cores: Marine Geology, v. 151, p. 143–153.

Katz, M.E., Pak, D.K., Dickens, G.R., and Miller, K.G., 1999, The source and fate of massive carbon input during the latest Paleocene thermal maximum: Science, v. 286, p. 1531–1533.

Katz, M.E., Cramer, B.S., Mountain, G.S., Katz, S., and Miller, K.G., 2001, Uncorking the bottle: What triggered the latest Paleocene methane release?: Paleoceanography, v. 16., p. 549–562.

Kennett, J.P., and Stott, L.D., 1991, Abrupt deep-sea warming, palaeoceanographic changes and benthic extinctions at the end of the Paleocene: Nature, v. 353, p. 225–229.

Laskar, J., 1990, The chaotic motion of the Solar System: A numerical estimate of the size of the chaotic zones: Icarus, v. 88, p. 266–291.

Laskar, J., 1999, The limits of Earth's orbital calculations for geological timescale use: Philosophical Transactions of the Royal Astronomical Society of London, series A, v. 357, p. 1735–1759.

Luterbacher, H.P., Hardenbol, J., and Schmitz, B., 2000, Decision of the voting members of the International Subcommission on Paleogene Stratigraphy on the criterion of recognition of the Paleocene/Eocene boundary: Newsletter of the International Subcommission on Paleogene Stratigraphy, Tübingen, November 2000, v. 9, p. 13.

Matsumoto, R., 1995, Causes of the $\delta^{13}C$ anomalies of carbonates and a new paradigm "'Gas-Hydrate Hypothesis": Journal of the Geological Society of Japan, v. 11, p. 902–924.

Norris, R.D., and Röhl, U., 1999, Carbon cycling and chronology of climate warming during the Paleocene/Eocene transition: Nature, v. 401, p. 775–778.

Ogg, J.G., and Bardot, L., 2001, Aptian through Eocene magnetostratigraphic correlation of the Blake Nose transect (ODP Leg 171), Florida continental margin, *in* Kroon, D., et al., eds., Proceedings of the Ocean Drilling Program, Scientific Results: Texas A&M University, College Station, Texas, USA, v. 171B, p. 2–58.

Olsen, P.E., and Kent, D.V., 1996, Milankovitch climate forcing in the tropics of Pangaea during the Late Triassic: Palaeogeography, Palaeoclimatology, Palaeoecology, v. 122, p. 1–26.

Paillard, D., Labeyrie, L., and Yiou, P., 1996, Macintosh program performs timeseries analysis: Eos (Transactions, American Geophysical Union), v. 77, p. 379.

Pälike, H., Shackleton, N.J., and Röhl, U., 2001, Astronomical forcing on late Eocene marine sediments: Earth and Planetary Science Letters, v. 193, p. 589–602.

Röhl, U., and Abrams, L.J., 2000, High-resolution, downhole and nondestructive core measurements from Sites 999 and 1001 in the Caribbean Sea: Application to the late Paleocene Thermal Maximum, *in* Leckie, R.M., et al., eds., Proceedings of the Ocean Drilling Program, Scientific Results: Texas A&M University, College Station, Texas, USA, v. 165, p. 191–203.

Röhl, U., Bralower, T.J., Norris, R.D., and Wefer, G., 2000, A new chronology for the late Paleocene Thermal Maximum and its environmental implications: Geology, v. 28, p. 927–930.

Röhl, U., Ogg, J., Geib, T., and Wefer, G., 2001, Astronomical calibration of the Danian time scale, *in* Norris, R.D., et al., eds., Western North Atlantic Palaeogene and Cretaceous palaeoceanography: London, Geological Society Special Publication 183, p. 163–183.

Shackleton, N.J., and Crowhurst, S., 1997, Sediment fluxes based on an orbitally tuned time scale 5–14 Ma, Site 926, *in* Curry, W.B. et al., eds., Proceedings of the Ocean Drilling Program, Scientific Results: Texas A&M University, College Station, Texas, USA, v. 154, p. 69–82.

Shackleton, N.J., Berger, A., and Peltier, W.A., 1990, An alternative astronomical calibration of the lower Pleistocene time scale based on ODP Site 677: Transactions of the Royal Society of Edinburgh Earth Science, v. 81, p. 251–261.

Shackleton, N.J., Crowhurst, S., Hagelberg, T., Pisias, N.G., and Schneider, D.A., 1995, A new late Neogene time scale: Application to Leg 138 sites, *in* Pisias, N.G., et al., Proceedings of the Ocean Drilling Program, Scientific Results: Texas A&M University, College Station, Texas, USA, v. 138, p. 73–101.

Shackleton, N.J., Crowhurst, S.J., Weedon, G.P., and Laskar, J., 1999, Astronomical calibration of Oligocene-Miocene time: Philosophical Transactions of the Royal Astronomical Society of London, series A, v. 357, p. 1907–1929.

Shackleton, N.J., Hall, M.A., Raffi, I., Tauxe, L., and Zachos, J., 2000, Astronomical calibration age for the Oligocene-Miocene boundary: Geology, v. 28, p. 447–450.

Shipboard Scientific Party, 1998, Site 1051, *in* Norris, R.D., et al., Proceedings of the Ocean Drilling Program, Initial Reports: Texas A&M University, College Station, Texas, USA, v. 171B, p. 171–239.

Wing, S.L., Bao, H., and Koch, P.L., 2000, An early Eocene cool period? Evidence for continental cooling during the warmest parts of the Cenozoic, *in* Huber, B.T., et al., eds., Warm climates in Earth history: Cambridge, UK, Cambridge University Press, p. 197–237.

Zachos, J.C., Lohmann, K.C., Walker, J.C.G., and Wise, S.W., Jr., 1993, Abrupt climate change and transient climates during the Paleogene: A marine perspective: Journal of Geology, v. 101, p. 191–213.

Manuscript Accepted by the Society August 13, 2002

Printed in the USA

Index

C

I

O

Q

R

T

X

Y

Z